PRINCIPLES OF
CELLULAR AND MOLECULAR IMMUNOLOGY

PRINCIPLES OF CELLULAR AND MOLECULAR IMMUNOLOGY

■

JONATHAN M. AUSTYN and KATHRYN J. WOOD

Nuffield Department of Surgery, John Radcliffe Hospital, Oxford

Oxford New York Tokyo

OXFORD UNIVERSITY PRESS

1993

Oxford University Press, Walton Street, Oxford OX2 6DP

Oxford New York Toronto
Delhi Bombay Calcutta Madras Karachi
Kuala Lumpur Singapore Hong Kong Tokyo
Nairobi Dar es Salaam Cape Town
Melbourne Auckland Madrid

and associated companies in
Berlin Ibadan

Oxford is a trade mark of Oxford University Press

Published in the United States
by Oxford University Press Inc., New York

A catalogue record for this book is available from the British Library

Library of Congress Cataloging in Publication Data
Austyn, Jonathan M.
Principles of cellular and molecular immunology / Jonathan M. Austyn and Kathryn J. Wood.
Includes bibliographical references and index.
1. Cellular immunity. 2. Molecular immunology. I. Wood, Kathryn J. II. Title.
[DNLM: 1. Immunity. 2. Immunity, Cellular. 3. Molecular Biology. QW 504 A938p]
QR185.5.A87 1991 616.07'9—dc20 91—3127

ISBN 0-19-854297-6 (h/b)
ISBN 0-19-854195-3 (p/b)

Typeset by The Charlesworth Group, Huddersfield, UK
Printed in Hong Kong

There are instances where we have been unable to trace or contact the copyright holder for certain figures before our printing deadline. We apologize for this. If notified the publisher will be pleased to rectify any errors or omissions at the earliest opportunity.

PREFACE

The idea for this book originated from a series of lectures given by the authors in the Nuffield Department of Surgery at the University of Oxford. These were designed to introduce immunology to students, and to update the immunological knowledge of their colleagues. We started by dividing the field of cellular and molecular immunology into a manageable number of largely self-contained topics. By beginning with the basics without assuming a background knowledge of immunology and then progressing to a more sophisticated level we found that it was possible to satisfy the needs of most people attending these lectures. We have taken a similar approach with this book and hope that it will be of equal value to readers from a wide variety of backgrounds and with different requirements; it should be accessible to students, clinicians, research workers in immunology and related areas, and all who wish to gain insight into this fascinating and rapidly-evolving field.

It is our hope that the reader with little or no immunological background should be able to gain a sufficient overview from the Introduction to read each chapter in sequence, or just a single chapter covering a topic of interest, perhaps omitting the boxed text which contains more detailed information. Once the basics have been grasped, the chapters can be re-read to include the boxed text, and by following up the reviews cited in the reference lists at the end of each chapter the reader should acquire a detailed understanding of the field or area. It should thus be possible to build up an integrated picture of current immunology. Research workers in immunology should be able to read the main chapters and study the original research papers for experimental or other details, and gain access to related literature. Throughout, the figures are intended to reinforce the text with a visual impression as an aid to learning and for clarification or elaboration of the points covered. At the beginning of the main chapters there is a 'map' of some of the cells and molecules of the immune system, with the relevant areas highlighted, to orientate the reader while traversing this territory.

When we were approached by Oxford University Press, and agreed to write this book, we had no idea of the difficulty of the task that lay ahead of us. Over a period of some eight years this book has been completely rewritten four times, and, up to the last set of page proofs, it has been continuously updated. Keeping up with the remarkable pace of immunological progress in the different areas has been a real challenge. Despite this, the book is as factually correct and as up-to-date as we could make it at the time it went to press.

There is a great deal that is still not known about immunology. When we took our own first steps into this field, we were frustrated that many books then available seemed to take an overly didactic and dogmatic stance. We read uncritical statements in immunology texts that said or implied something was known with certainty, and then discovered complete contradictions in the literature. Therefore we have tried, as far as possible, to distinguish between what is known with some certainty, what is thought reasonable and likely, and what is merely the subject of speculation.

Because immunology is an experimental science, and concepts in immunology have been developed over a number of years, we felt that it was important to give the reader some idea of the historical background and the experimental approaches that have been taken. It is clear that the experiments still to be performed will be designed according to the experiments, observations, interpretations, hypotheses, and theories of the past. Thus we have endeavoured to provide the reader with a framework within which future discoveries in immunology can be placed in context.

There are many people whom we wish to thank for their help and assistance during the writing of this book. First and foremost we would like to express our gratitude to Professor P.J. Morris, Nuffield Professor of Surgery and the Head of Department, who originally suggested the idea for a course on cellular and molecular immunology, who made available to us the resources and facilities of the Department, and who has supported us throughout. Without him, this book would not have been written. We are also indebted to Dr C. Crawford and Dr. S.V. Hunt who read drafts of this book in their entirety, and made many valuable comments and suggestions. Without them, its structure would have been weaker. In addition, we express our thanks to Dr N. Barclay, Professor S. Gordon, Dr A. Hamblin, Dr G. MacPherson, Dr H. Pure, Dr R. Sim, Professor R. Steinman, Professor J. Unkeless, and Dr M. Yudkin, who read drafts of one or more individual chapters, or who provided us with invaluable information. Without them, there would have been some essentials missing. We would also like to thank the numerous members of our own department who have read chapters, provided information, and given us feedback during the eight years of preparation of this book. We would like to acknowledge the co-operation and helpful comments of those people who have provided material for the illustrations in this book. Finally, we acknowledge the artistic skill of Debra Woodward, who prepared several of the figures, John Grandidge who worked studiously on preparation of the galleys and page proofs, and all the staff of Oxford University Press who were involved in this project.

Oxford
January 1993

J.M.A.
K.J.W.

CONTENTS

For Friends

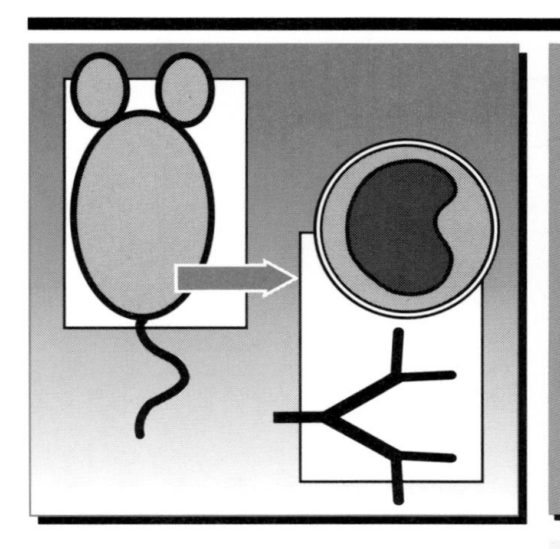

INTRODUCTION

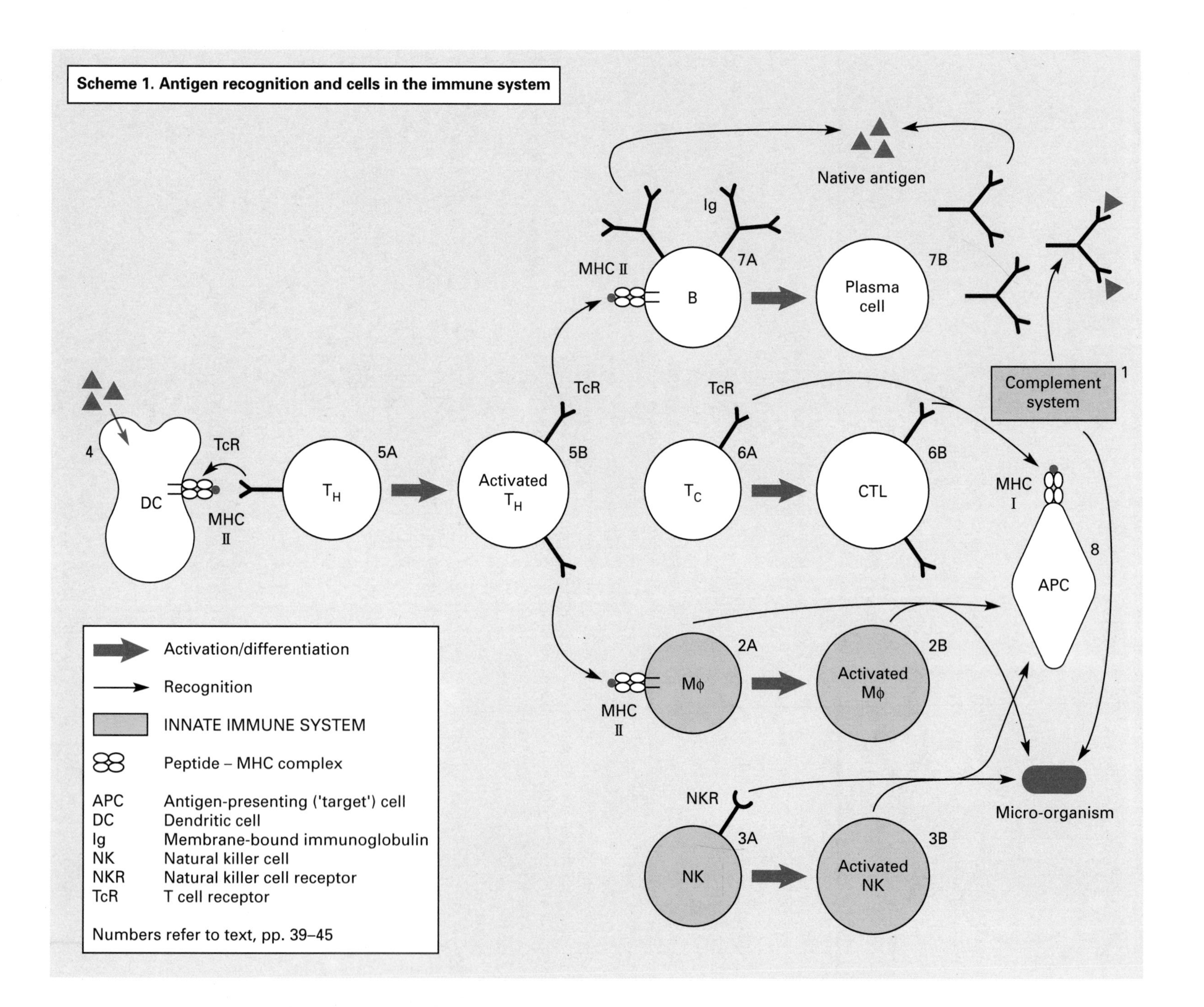
Scheme 1. Antigen recognition and cells in the immune system
Native antigen
Ig
MHC II
B
7A
Plasma cell
7B
Complement system
1
4
DC
TcR
MHC II
T_H
5A
Activated T_H
5B
TcR
TcR
T_C
6A
CTL
6B
MHC I
APC
8
MHC II
Mϕ
2A
Activated Mϕ
2B
NKR
NK
3A
Activated NK
3B
Micro-organism
Activation/differentiation
Recognition
INNATE IMMUNE SYSTEM
Peptide – MHC complex
APC Antigen-presenting ('target') cell
DC Dendritic cell
Ig Membrane-bound immunoglobulin
NK Natural killer cell
NKR Natural killer cell receptor
TcR T cell receptor
Numbers refer to text, pp. 39–45

1.1 General features of the immune system

1.1.1 Immunity

Immunology is the study of the immune system. The immune system is composed of a number of different cell types and a variety of molecules which are spread throughout the body. Normally the immune system defends us against infection and gives protection from cancerous cells.

Our bodies, and those of other animals, are subject to a continuous onslaught from the outside world by an immense variety of organisms. Many of these are pathogens which will cause disease if they enter the body. They range in size and complexity from the smallest viruses through bacteria, fungi, and protozoa, up to the largest parasites (Fig.1.1). A common feature of all these organisms, as far as the immune system is concerned, is that they are **foreign** to the body and so too are any of the individual's own cells that have become infected. More recently, organs transplanted from other individuals have joined the list of foreign things that can have access to our bodies. Normally, of course, the skin and other epithelia, such as those lining the gut, genital tract, and airways, provide a barrier to foreign agents from the outside world. If this barrier is breached, the immune system is activated (Fig.1.2) and the ensuing **immune response** generally leads to the destruction and removal of the foreign organism.

It was realized many centuries ago that when someone recovered from a disease caused by a particular pathogen, they often became **resistant** to the disease and could not subsequently be reinfected by the same organism: the person was said to have become **immune**, or to have acquired **immunity**. A classic example is measles: those who have had this disease know they will not get it again. It is as if the body remembers its previous battle with the organism, and is now prepared and lying in wait 'just in case' the organism is encountered again. This immunological **memory**, which can last for a lifetime, is one feature of so-called **adaptive immune responses** (Table 1.1).

However, recovery from one disease, such as measles, does not mean we are automatically protected from others like diphtheria or tetanus. Nor are we protected if another form of the organism arises to which we have not previously been exposed. Hence we can have repeated bouts of flu when new strains of the influenza virus appear. Such **specificity** is another aspect of adaptive immune responses. Nevertheless, most of us do eventually recover from flu, and many other diseases, and we can feel reasonably confident that when another strain of virus appears, our immune systems will be able to cope with it. (However, the epidemics of the 1920s when millions died from flu provide a dramatic example of the consequences of being unable to mount an adequate immune response.) The immune system can respond to an immense variety of different organisms, and an enormous **diversity** of immune responses is possible.

It is possible to exploit the immune system to protect us from a number of diseases, and we can therefore be spared the frequently disfiguring, disabling, and life-threatening consequences of diseases such as smallpox. In

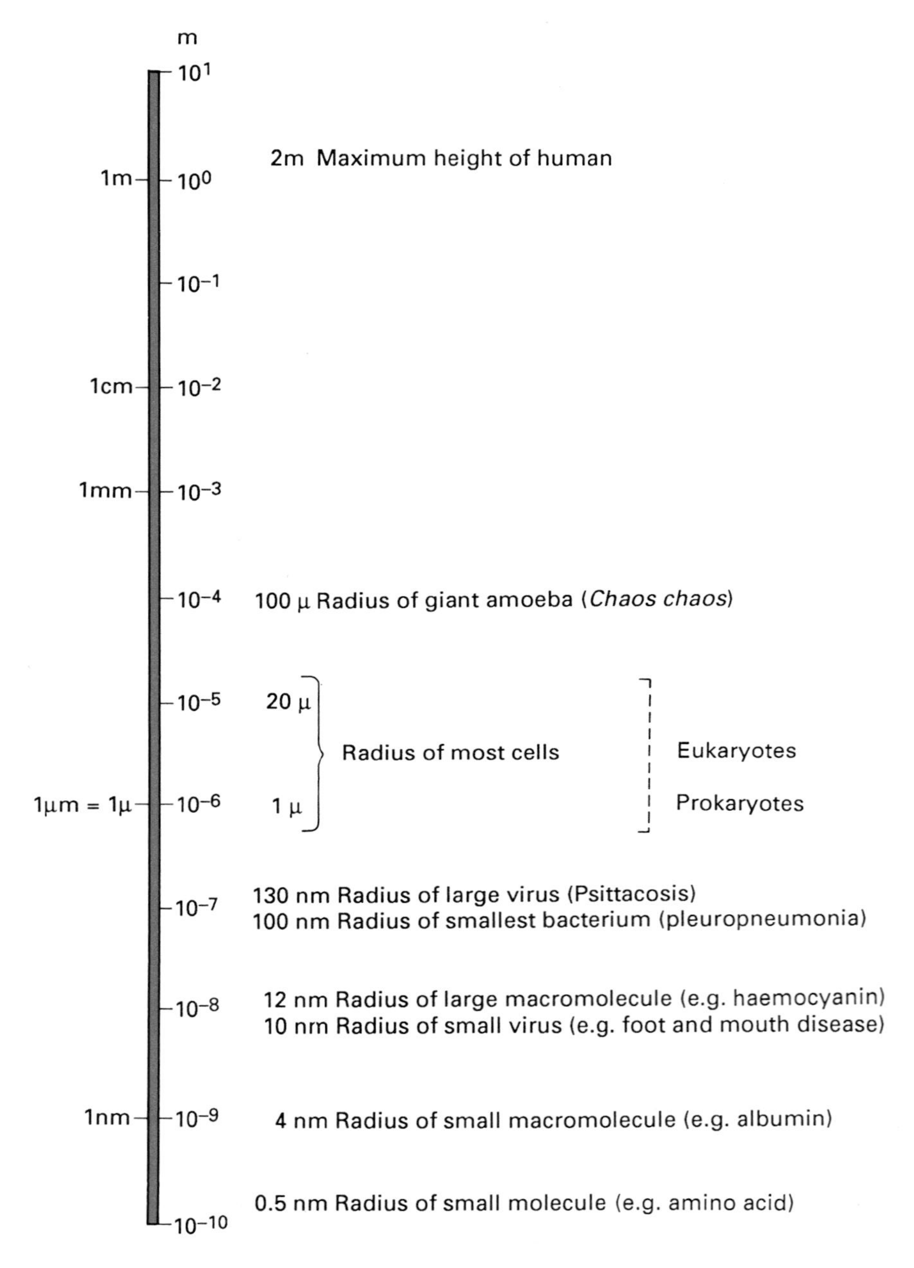

Fig.1.1 Sizes of various molecules, cells, and organisms

ancient India and China it was realized that people who recovered from one variant of this disease, now known to be caused by a less virulent form of the smallpox virus (the variola virus), were resistant to further infection; so snuff was prepared from dried crusts of smallpox pustules from infected individuals and taken by others in the hope of combating the disease. Even until the middle of the last century, the practice of variolation was carried out, with live organisms being taken from the active lesions of smallpox

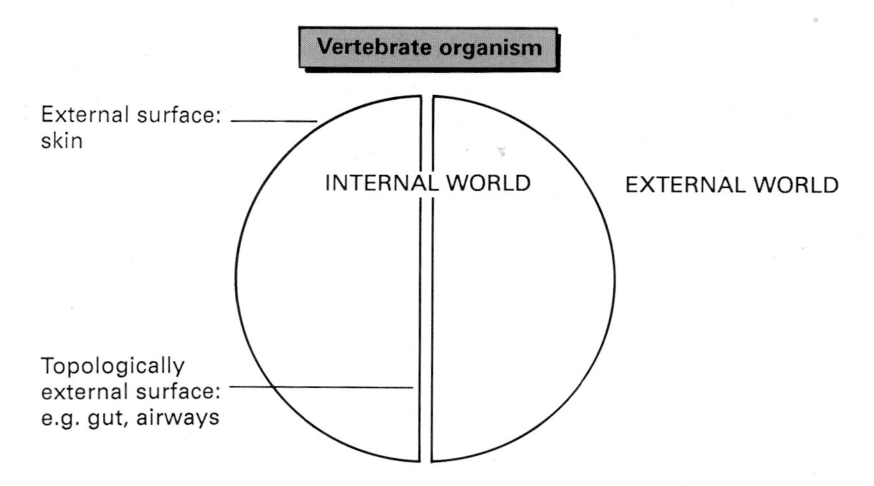

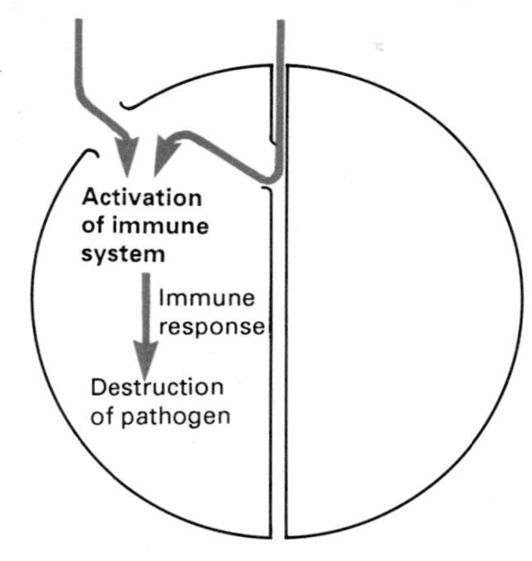

Fig.1.2 Activation of the immune system

Table 1.1 Some features of adaptive immune responses	
Memory	Recovery from infection by one pathogen frequently protects against subsequent infection by the same organism (= 'immunity')
Specificity	Recovery from infection by one pathogen does not usually provide protection against another (unless the organisms are closely related – e.g. cowpox and smallpox viruses)
Diversity	Responses can be made against a multitude of different organisms

patients and administered to others by pricking the skin. Needless to say, the risk of infection from this procedure generally outweighed the chances of protection.

Rather than expose the individual to live organisms, a safer and more successful approach, introduced in the latter part of the nineteenth century, was to expose the individual to an organism that had been inactivated or **attenuated** to reduce its potency. This, it was hoped, would produce a less virulent form of the disease, but the subsequent immune response, and the resultant immunological memory, would protect the individual against disease caused by the live, 'pathogenic' (= disease causing) organism. This was first used successfully to protect against anthrax.

An alternative approach was to administer a closely related but less pathogenic or a non-pathogenic organism to induce immunity. This was pioneered at the end of the eighteenth century by Jenner, who showed that exposure to the cowpox virus (vaccinia) afforded protection to the related smallpox virus (variola). These procedures became known as **vaccination**, and the agent used to induce the immunity became known as a **vaccine**. The immunity that results from the induction of clinical or subclinical infections is called **active immunity**.† Nowadays **vaccines** against a variety of pathogenic organisms are available, and the development of others is being actively pursued in many laboratories. **Immunization**, or the induction of immunity by vaccination, has become a common procedure which has led to the eradication of many diseases in the world today.

1.1.2 Innate and adaptive immunity

Some background

Immunity is mediated by a variety of different cells and molecules. This section gives a brief outline of the cell and its constituent parts. More detailed information can be obtained from any biochemistry text if required (see Further reading). Many molecules in the immune system are **proteins**, which are made up of chains of amino acids in a **polypeptide chain** (Fig.1.3). They can be soluble or cell surface molecules and can be modified either by the addition of carbohydrates, which are made up of multiple sugar units, to produce **glycoproteins** (Fig.1.4); or by the addition of lipid tails, made of fatty acids, to form **lipoproteins** (Fig.1.5).

Different parts of a cell carry out different functions (Fig.1.6). The **plasma membrane** which covers the cell surface is not only responsible for maintaining the cell's shape and integrity. Molecules are anchored in this membrane, and via these molecules an immune cell can 'receive information' about the outside world, and make an appropriate response. These surface components also mediate interactions with other cell types, allowing the cells to communicate with each other. This is critical for the co-ordination and integration of immune responses.

The **cytoplasm** (cytosol) of the cell contains all the synthetic machinery for making proteins and other molecules that are required for functions such as growth and reproduction, activities which are basic to all cell types. Other molecules are synthesized in the cytoplasm and are either stored awaiting their release into the environment or they are exported to other parts of the

† So-called **passive immunity** can result when preformed antibodies against an organism are administered to an individual; this form of immunity is relatively short-lived as the antibody is cleared from the circulation, and the cells that normally produce antibodies within that individual have not been stimulated.

cell. Some of these molecules are only found in certain cell types, depending on the cell's particular functions. For instance, each type of cell in the immune system contains, or can produce when required, a particular set of molecules that are needed to help destroy and remove invading organisms.

Within the cytoplasm lies the **nucleus**. The nucleus contains the **chromosomes** which carry **genes**, the basic substance of which is, of course, DNA, a nucleic acid. It also contains the machinery for duplicating chromosomes and making copies of the genes when the cell divides, DNA **replication**, and for switching the genes on and off to allow selective expression of particular genes in different cells. Every **somatic** (non-reproductive) cell in the body has an identical set of genes. However, only certain combinations of these genes are switched on or 'expressed' in any given cell, depending on what type of cell it is and what it needs to do. Some genes are active in virtually all cells of the body, especially those concerned with what used to be termed 'housekeeping' functions, while other genes are active only in particular cell types.

It really goes without saying that DNA carries the 'genetic message'. This is simply a code made up of four units, called **bases**, which are arranged in a particular order for each gene (Fig.1.7). These bases are linked to deoxyribose units and the bases in DNA are adenine, thymine, guanine, and cytosine. Their precise order is known as the **DNA sequence** of the gene. By rough analogy, each sentence in this book is composed of 26 (rather than four) units, the letters of the alphabet plus spaces, and the message differs depending on how these units are arranged.

When a gene is active in a cell, the DNA is 'copied' into a related nucleic acid called RNA (also composed of just four bases, but these now contain ribose rather than deoxyribose and the base uracil instead of thymine), in a process called **transcription** (Fig.1.8). After some changes have been made to the newly-synthesized RNA molecule (**RNA processing**) it is transported from the nucleus into the cytoplasm. Here it is 'read' by components of the protein synthetic machinery called **ribosomes**, which are often attached to a system of membranes called the **endoplasmic reticulum** (Fig.1.6). The ribosomes 'decode' the message by inserting one amino acid for each group of three bases in the RNA to form a polypeptide chain of increasing length and eventually the finished protein. This process of decoding or translating the RNA message is, not surprisingly, called **translation**, and in this way the original information in the DNA of the gene can be decoded and turned into proteins. This is what is meant by the often-used phrase 'genes code for (or encode) protein products'.

Some proteins are destined to be inserted into the plasma membrane, others are stored in the cytoplasm until they are released from the cell as **secretory proteins**. Before they reach their final destination in the cell, many proteins are modified to more complex molecules such as glycoproteins. Much of this modification and sorting occurs in a specialized part of the cell called the **Golgi apparatus** (Fig.1.6).

Finally, we should mention that many genes are organized differently in higher organisms (that is in animals, which are collectively termed eukaryotes) than in bacteria (prokaryotes). In the latter, genes are continuous stretches of DNA coding directly for proteins. In eukaryotes, however, genes are commonly divided into a number of segments or **exons**, and each exon

Proteins are composed of **amino acids**. Amino acids contain an amino group, a carboxyl group, and a particular side-chain (R) characteristic for each amino acid attached to a central (or α) carbon atom:

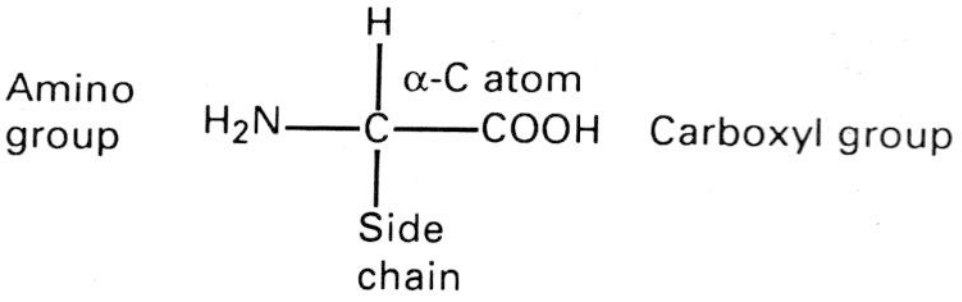

$$\underset{\text{Amino group}}{H_2N}-\overset{\overset{\displaystyle H}{|}\ \alpha\text{-C atom}}{\underset{\underset{\text{Side chain}}{|}}{C}}-\underset{\text{Carboxyl group}}{COOH}$$

In proteins, amino acids are joined by peptide linkages to produce a polypeptide chain:

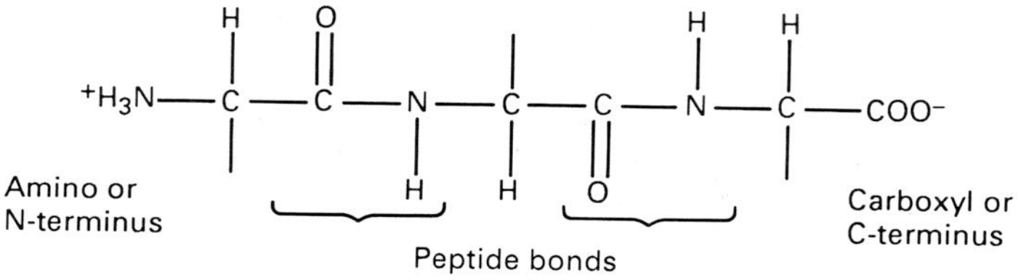

$$^{+}H_3N-\overset{H}{\underset{|}{\overset{|}{C}}}-\overset{O}{\overset{\|}{C}}-\underset{H}{\underset{|}{N}}-\underset{H}{\underset{|}{\overset{|}{C}}}-\underset{O}{\underset{\|}{C}}-\overset{H}{\overset{|}{N}}-\overset{H}{\underset{|}{\overset{|}{C}}}-COO^{-}$$

Amino or N-terminus — Peptide bonds — Carboxyl or C-terminus

There are 20 amino acids, which can be grouped according to whether their side chains are acidic (negatively-charged), basic (positively-charged), uncharged polar, or non-polar. They are given single letter or three letter abbreviations:

Abbreviations

Single letter code	Three letter code	Amino acid	Nature of side chain, R	Single letter code	Three letter code	Amino acid	Nature of side chain, R
A	Ala	Alanine	Non-polar	M	Met	Methionine	Non-polar
C	Cys	Cysteine	Non-polar	N	Asn	Asparagine	Uncharged polar
D	Asp	Aspartic acid	Acidic	P	Pro	Proline	Non-polar
E	Glu	Glutamic acid	Acidic	Q	Gln	Glutamine	Uncharged polar
F	Phe	Phenylalanine	Non-polar	R	Arg	Arginine	Basic
G	Gly	Glycine	Non-polar	S	Ser	Serine	Uncharged polar
H	His	Histidine	Basic	T	Thr	Threonine	Uncharged polar
I	Ile	Isoleucine	Non-polar	V	Val	Valine	Non-polar
K	Lys	Lysine	Basic	W	Trp	Trptophan	Non-polar
L	Leu	Leucine	Non-polar	Y	Tyr	Tyrosine	Uncharged polar

Thus, an example of part of a protein (the N-terminus of the enzyme lysozyme) can be written: KVFGRCELAAA

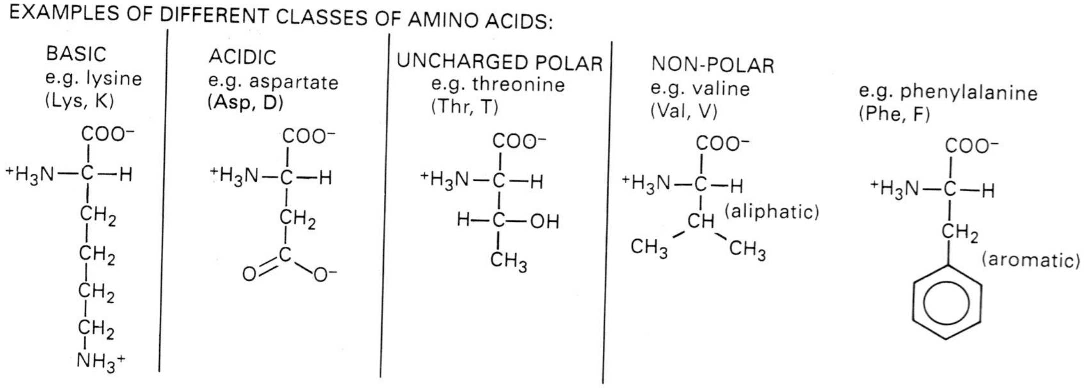

EXAMPLES OF DIFFERENT CLASSES OF AMINO ACIDS:

BASIC e.g. lysine (Lys, K): COO^{-}, $^{+}H_3N-C-H$, CH_2, CH_2, CH_2, CH_2, NH_3^{+}

ACIDIC e.g. aspartate (Asp, D): COO^{-}, $^{+}H_3N-C-H$, CH_2, $C(=O)O^{-}$

UNCHARGED POLAR e.g. threonine (Thr, T): COO^{-}, $^{+}H_3N-C-H$, $H-C-OH$, CH_3

NON-POLAR e.g. valine (Val, V): COO^{-}, $^{+}H_3N-C-H$, CH (aliphatic), CH_3 CH_3

e.g. phenylalanine (Phe, F): COO^{-}, $^{+}H_3N-C-H$, CH_2 (aromatic), benzene ring

Fig.1.3 Amino acids and proteins

codes for one part of the final protein (Fig.1.9). The exons are separated from each other by intervening sequences of DNA or **introns**, which do not code for any part of the final protein molecule. Such 'interrupted genes' are very common. Relatively few genes have no exons at all, while most contain at least one, and sometimes even 20 or more, exons. When a gene is

Amino acids can exist in mirror-image forms, called the L-isomer and D-isomer. All naturally occurring proteins are composed exclusively of L-amino acids.

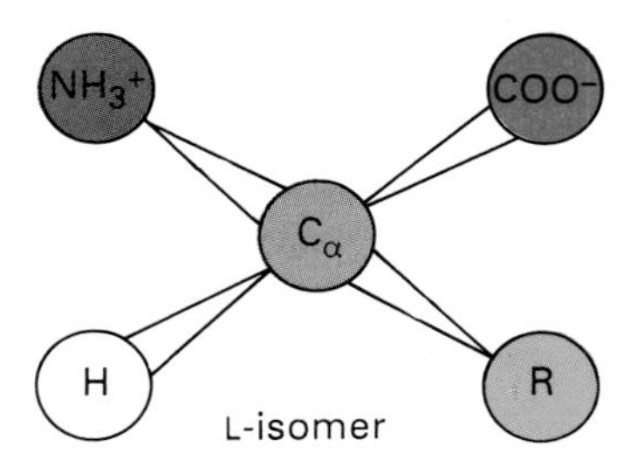

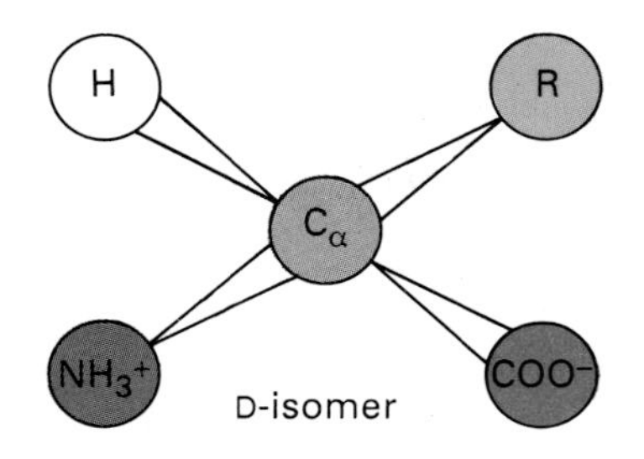

There are four recognized levels of **protein structure**.

PRIMARY STRUCTURE:
the amino acid sequence and location of disulfide (S–S) bonds, if present;
e.g. part of the 1° sequence of strand B of the $\alpha 3$ domain of an MHC class I molecule [C = cysteine: can form part of an S–S bond]

SECONDARY STRUCTURE:
the spatial arrangement of amino acids close to each other in the linear sequence such as α helices and β-pleated sheets;
e.g. folding of the polypeptide chain in the $\alpha 3$ domain of an MHC class I molecule

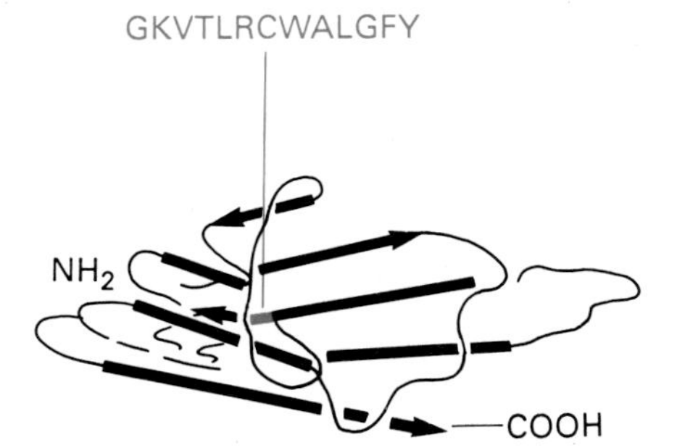

TERTIARY STRUCTURE:
the spatial arrangement of amino acids far apart from each other in the linear sequence, such as domains within the same polypeptide chain;
e.g. the $\alpha 1$, $\alpha 2$, and $\alpha 3$ domains of an MHC class I molecule

QUARTERNARY STRUCTURE:
the spatial relationship between different polypeptide chains (subunits) in a multi-subunit molecule; e.g. the association of the α chain of an MHC class I molecule with β_2-microglobulin (β_2m)

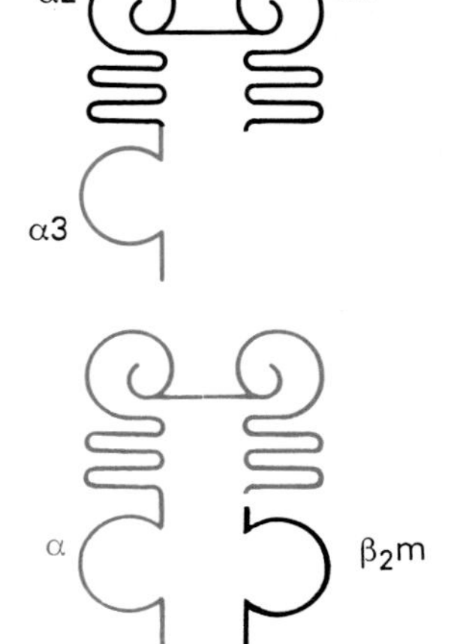

transcribed, the entire DNA sequence is usually copied into RNA, including the sequences for both exons and introns. The RNA copies of the introns are removed during RNA processing, and the sequences that correspond to the exons are joined contiguously before they are translated.

Two forms of immunity

The majority of cells in the immune system were formerly classified as white blood cells or **leukocytes**. These cells are initially produced in the bone marrow as less mature forms or **precursors**. Some of these precursors

The simplest carbohydrates, or sugars, are **monosaccharides**. These are aldehydes [$\mathrm{C}\begin{smallmatrix}\diagup \mathrm{H}\\ \diagdown \mathrm{O}\end{smallmatrix}$] or ketones [$>\mathrm{C{=}O}$] that have two or more hydroxyl [–OH] groups.

They can be categorized according to the number of carbon atoms they contain: for example trioses (3-carbon), pentoses (5-carbon), and hexoses (6-carbon). Monosaccharides containing 5 or more carbons can form ring structures.

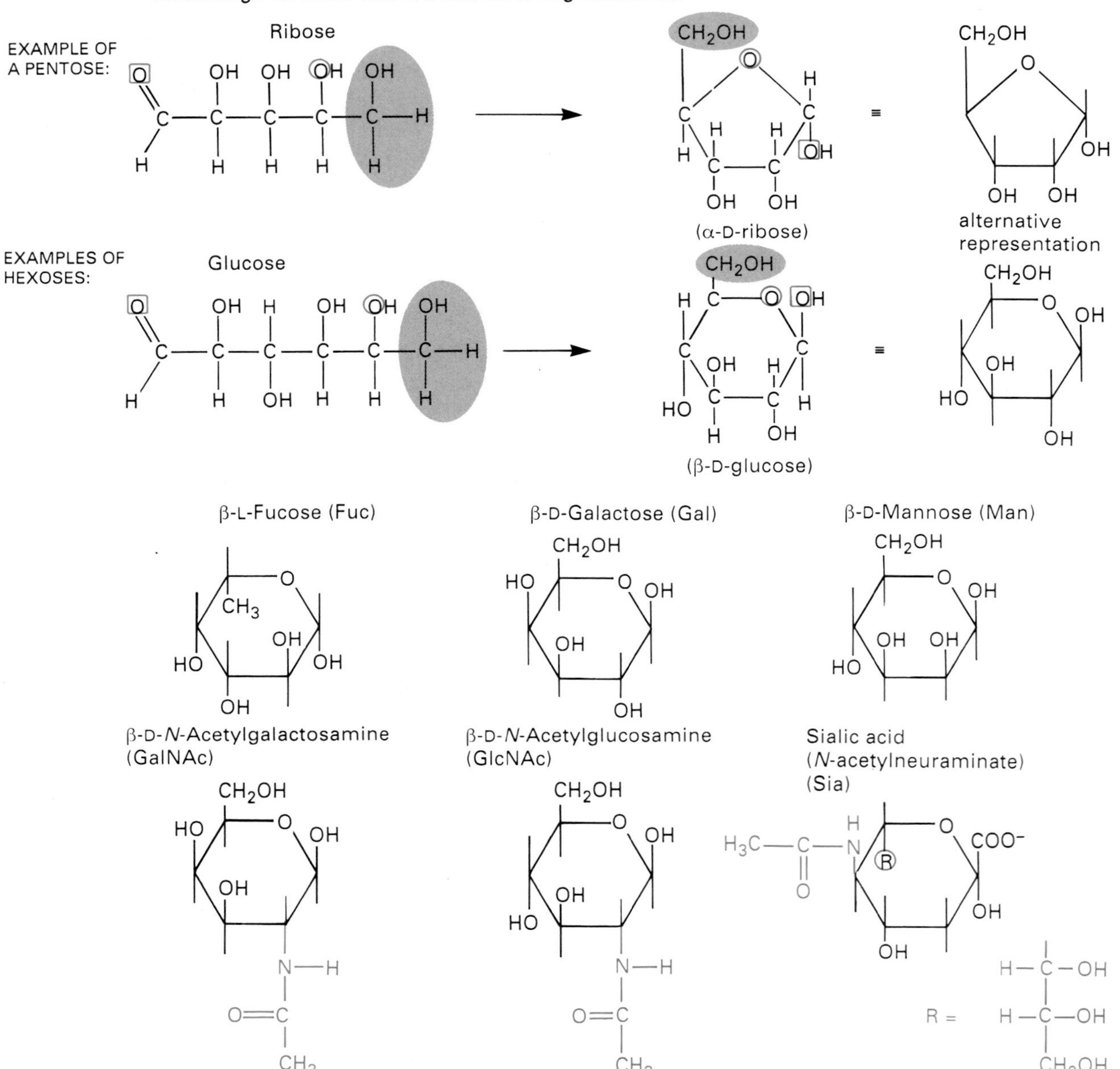

Fig.1.4 Sugars and glycoproteins

develop into more mature cells in the bone marrow, but many are carried in the blood to other areas of the body where they can develop further. The bone marrow is 'blood-forming', or haemopoietic, so bone marrow-derived cells can also be termed **haemopoietic cells**, and this term may now be preferable to the older term 'leukocytes'.

Disaccharides, **trisaccharides**, etc. consist of 2, 3, etc. sugars joined together by **glycosidic linkages**. Short chains of sugars are called **oligosaccharides**, longer chains **polysaccharides**, and these can form branched molecules. **Complex oligosaccharides** are groups of different types of sugar molecules joined together.

Glycoproteins are composed of oligosaccharides linked to polypeptide chains. They can be 'N-linked', in which case the carbohydrate is typically linked to asparagine (Asn; N), or 'O-linked' and joined to a serine (Ser; S) or threonine (Thr; T) residue. The hexoses shown above (apart from glucose) are commonly found in glycoproteins.

EXAMPLE OF N-LINKAGE (through Asn)

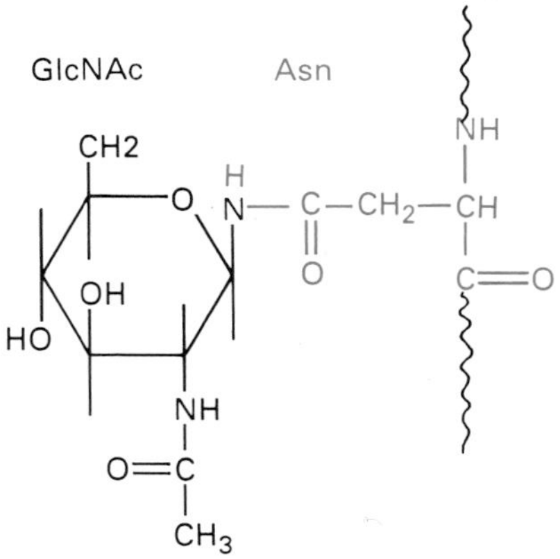
GlcNAc
Asn
CH2
NH
H
O
N
C
CH2
CH
OH
HO
O
C=O
NH
O=C
CH3

EXAMPLE OF O-LINKAGE (through Ser)

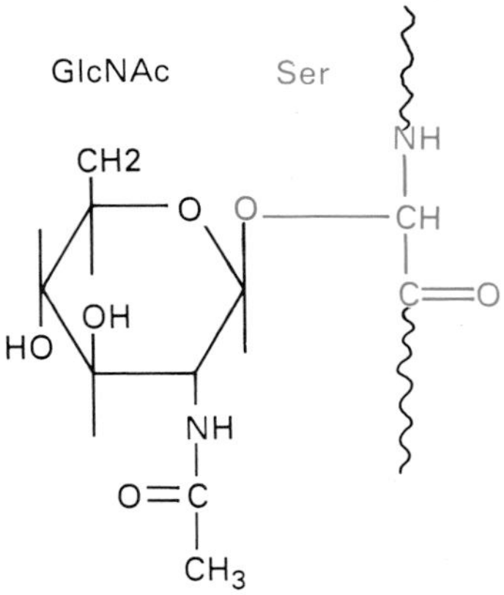
GlcNAc
Ser
CH2
NH
O
O
CH
OH
HO
C=O
NH
O=C
CH3

EXAMPLES OF GLYCOPROTEINS:

N-linked **high mannose type**

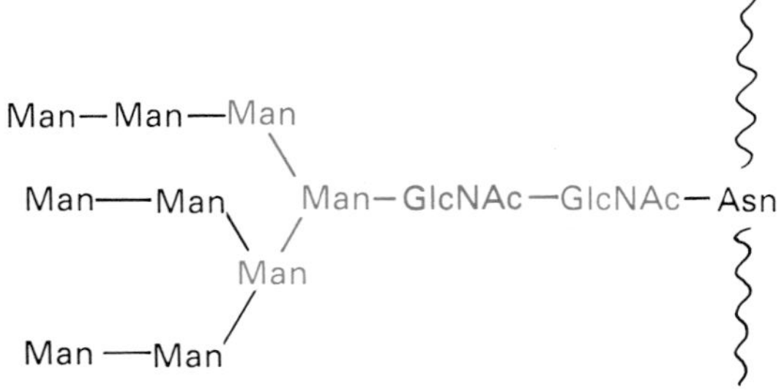
Man—Man—Man
Man—Man
Man–GlcNAc—GlcNAc—Asn
Man
Man—Man

N-linked **complex type**

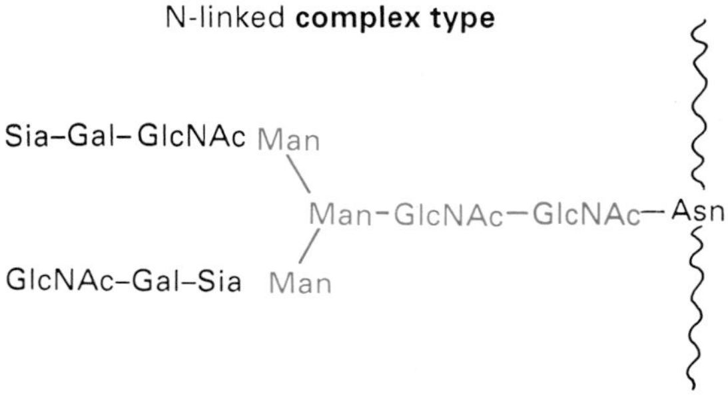
Sia-Gal-GlcNAc
Man
Man-GlcNAc-GlcNAc-Asn
GlcNAc-Gal-Sia
Man

The cells and molecules of the immune system as a whole are responsible for two different, but interrelated, forms of immunity. In fact, they can be roughly divided into two subdivisions of the immune system, one that is responsible for **innate** immunity, and the other that is responsible for **adaptive** immunity (Fig.1.10). The innate system is fully functional before infectious agents enter the body. In some cases the activity of this arm alone can be sufficient to destroy an invading organism. In contrast, the adaptive arm is activated only after a pathogen has entered the body, particularly if it has evaded the innate system and escaped destruction.

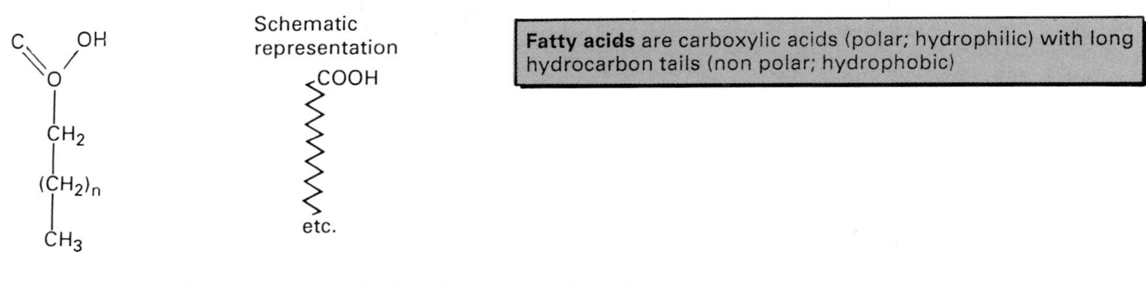

Schematic representation
COOH
etc.
Fatty acids are carboxylic acids (polar; hydrophilic) with long hydrocarbon tails (non polar; hydrophobic)

Saturated fatty acids contain no double bonds in their hydrocarbon tails, e.g. stearic acid

$CH_3(CH_2)_{16}COOH$ COOH

Unsaturated fatty acids contain one or more double bonds in their hydrocarbon tails, e.g. oleic acid

$CH_3(CH_2)_7CH{=}CH(CH_2)_7COOH$ COOH

Polyunsaturated fatty acids contain multiple double bonds

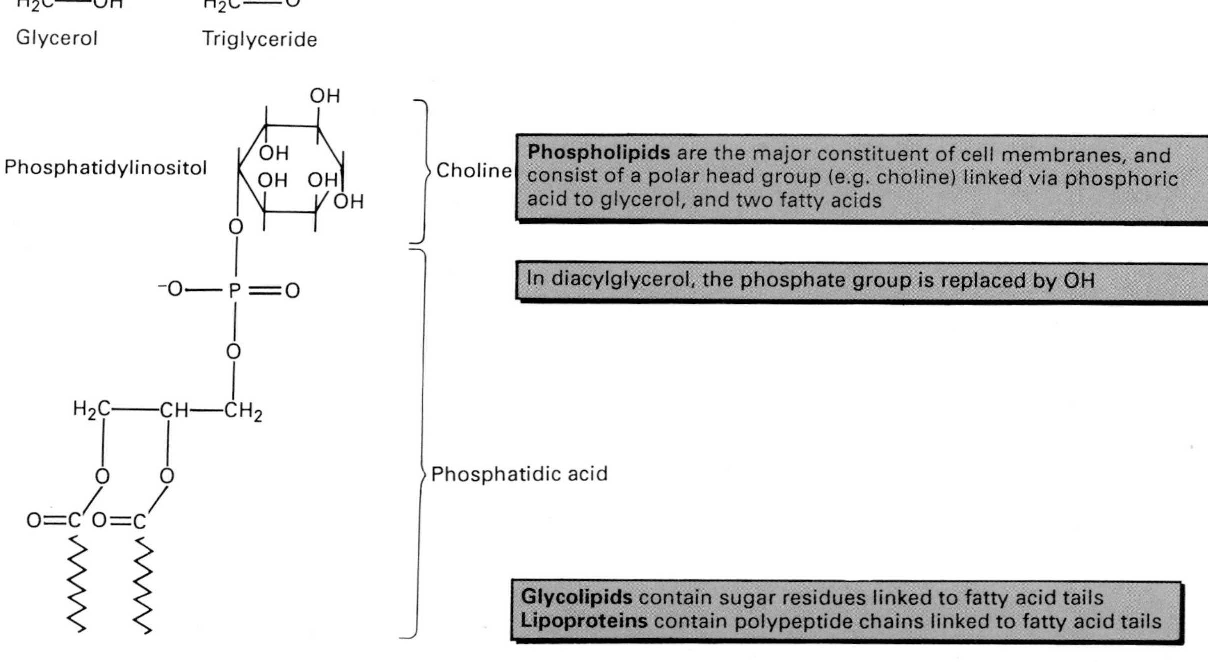

Triglycerides are fatty acids linked to glycerol
Glycerol
Triglyceride
Phosphatidylinositol
Choline
Phospholipids are the major constituent of cell membranes, and consist of a polar head group (e.g. choline) linked via phosphoric acid to glycerol, and two fatty acids
In diacylglycerol, the phosphate group is replaced by OH
Phosphatidic acid
Glycolipids contain sugar residues linked to fatty acid tails
Lipoproteins contain polypeptide chains linked to fatty acid tails

Fig.1.5 Fatty acids and phospholipids

The phenomenon of **immunological memory**, which was introduced above, is one important feature of the adaptive immune system. Once an adaptive immune response has occurred, the individual acquires the capacity to 'recognize' and destroy the organism very rapidly the next time it is encountered, and the person is now said to be immune to it. Not only can the body 'remember' having seen that particular organism, but the response to it on subsequent occasions is much more vigorous and effective than the first time. As a consequence of the two arms of the immune system, we are frequently unaware that a particular virus or bacterium may have entered our bodies, because it is so efficiently dealt with that we have no clinical symptoms; this is referred to as a 'subclinical infection'.

Cells and molecules of the innate immune system. The innate immune system includes a group of leukocytes called **phagocytes**, and another group called **natural killer cells** (Table 1.2). It also includes proteins like the **complement** components which are said to be 'soluble molecules' because they are dissolved in the body fluids rather than being primarily associated with, or stored in, cells.

One of the main functions of the phagocytes is to surround and engulf particles by the process of **phagocytosis**. In this way micro-organisms can be taken up and digested within the phagocyte, and complete organisms can often be destroyed in this manner. Equally, however, some phagocytes are important in producing soluble molecules that mediate **inflammatory responses** which frequently occur where there is infection, leading to the classic symptoms of pain, heat, redness, and swelling.

Phagocytes are classified into two main types depending on the shape of their nucleus (Fig.1.11). **Mononuclear phagocytes** have a simple-shaped nucleus and they include **macrophages** and their precursors, such as **monocytes** in the blood. (The word 'macrophage' comes from the Greek for 'big eaters', and these cells certainly are the biggest 'eaters' of all.) On the other hand, **polymorphonuclear leukocytes**, or **granulocytes**, have segmented nuclei. These are subdivided into **neutrophils**, **basophils**, and **eosinophils** depending on how they stain with particular dyes: Romanowsky stains contain a mixture of eosin, which stains eosinophilic structures, and azure or methylene blue, which stain basophilic structures; structures that stain with neither component, are said to be 'neutrophilic'. Granulocytes are also phagocytic and some, such as eosinophils, seem especially important in responses to certain types of infection. **Natural killer cells** belong to a different lineage (Fig.1.31) and these cells can destroy certain virally-infected, embryonic, or tumour cells. In addition to these cells, there are **megakaryocytes** producing **platelets** which are involved in blood clotting and inflammation.

Cells and molecules of the adaptive immune system. The predominant cell type in the adaptive immune system is the **lymphocyte**. Lymphocytes are responsible for the immunological memory, specificity and diversity of immune responses (Tables 1.1 and 1.2).

Historically, immunologists have distinguished between two types of immune responses. **Humoral responses** are mediated in part by **antibodies** in the serum. In contrast, **cell-mediated immune responses** result from the actions of different types of cells. This distinction reflects the existence of

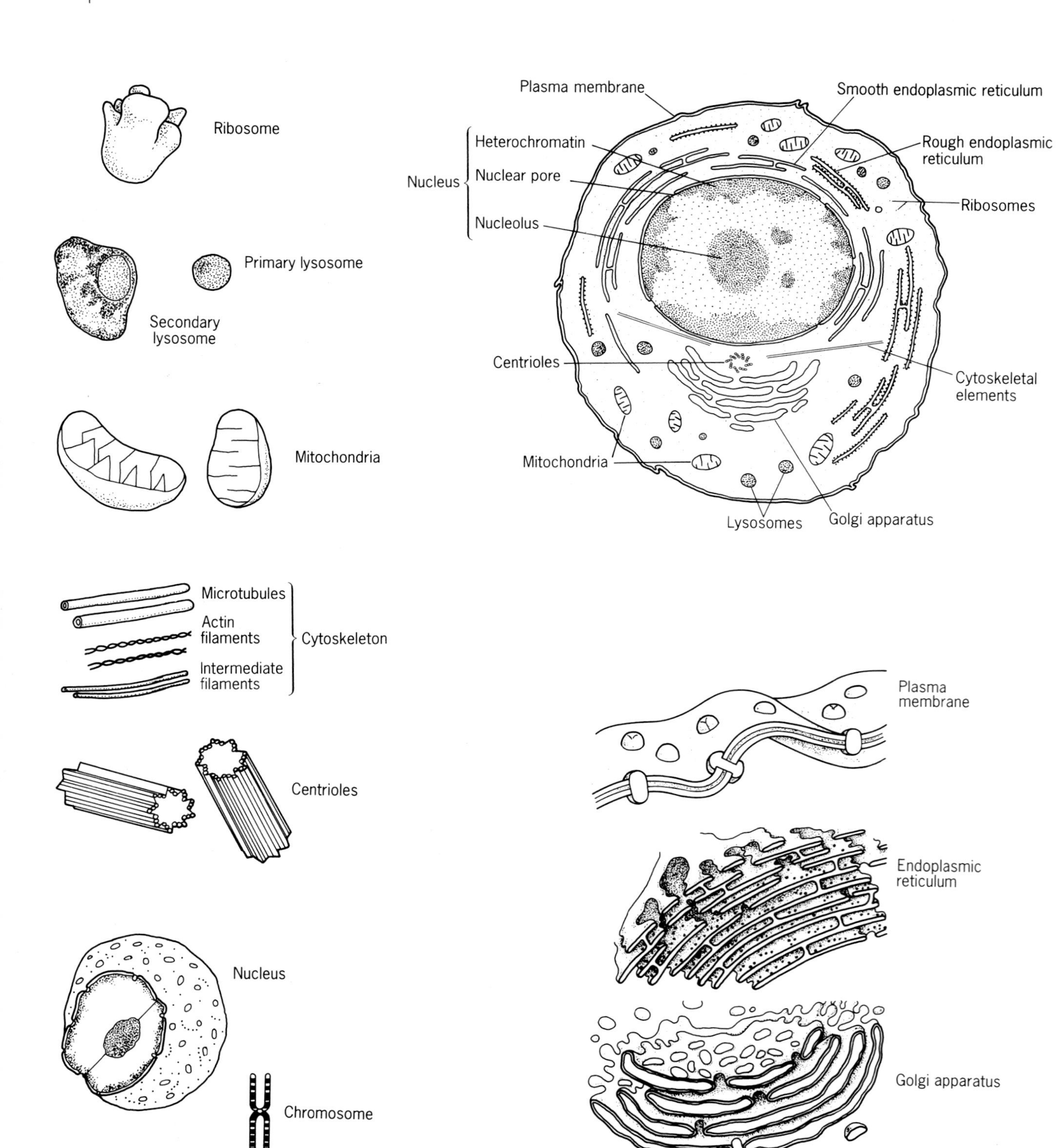
Ribosome
Primary lysosome
Secondary lysosome
Mitochondria
Microtubules
Actin filaments
Intermediate filaments
Cytoskeleton
Centrioles
Nucleus
Chromosome
Plasma membrane
Smooth endoplasmic reticulum
Nucleus
Heterochromatin
Nuclear pore
Nucleolus
Rough endoplasmic reticulum
Ribosomes
Centrioles
Cytoskeletal elements
Mitochondria
Lysosomes
Golgi apparatus
Plasma membrane
Endoplasmic reticulum
Golgi apparatus

Fig.1.6 Schematic view of a cross-section through an animal cell to show various cell organelles

Plasma membrane
This is composed of a continuous phospholipid bilayer in which molecules such as glycoproteins are embedded. It encloses the cytoplasm and cell organelles.

Endoplasmic reticulum (ER)
A system of membranes extending throughout the cell, in continuity with the outer membrane of the nucleus. It can occur both as smooth ER, and as rough ER when ribosomes are attached. The endoplasmic reticulum is involved in the biosynthesis and transport of lipids and glycoproteins. In the latter case the polypeptide chains of proteins, which are synthesized by the ribosomes of the rough ER, cross the membrane and enter the lumen for addition of carbohydrates, some of which are further modified in the Golgi apparatus.

Golgi apparatus
A system of flattened stacks of membranes which is involved in modifying and sorting macromolecules such as glycoproteins, which are then transported to other cellular compartments in vesicles (50 nm or larger).

Ribosomes
These can occur free, or in groups called polyribosomes, within the cytoplasm, or can be attached to the endoplasmic reticulum (in rough ER). Ribosomes are composed of two subunits and are responsible for protein synthesis, messenger RNA becoming associated with them and being translated into the polypeptide chain.

Lysosomes
These are membrane-bounded vesicles containing hydrolytic enzymes. When they fuse with other vesicles containing internalized material they form endolysosomes in which digestion of these contents occurs and ultimately secondary lysosomes.

Mitochondria
Produce energy for the cell in the form of adenosine triphosphate (ATP).

Cytoskeleton
Arrays of protein filaments that often extend from the area close to the nucleus containing the centrioles (called the microtubule organizing centre — MTOC) through the cytoplasm to the plasma membrane. They are involved in determining cell shape and motility, and internalization of molecules from the surface of the cell. They are of three main types: microtubules, actin filaments and intermediate filaments.

Centrioles
The pair of centrioles, each of which contains nine triplet tubules, are orientated at right angles to each other. During cell division (mitosis) the nuclear membrane dissolves, the chromosomes condense, and one of each pair of chromosomes becomes attached to a centriole via microtubules. The centrioles then move to opposite poles of the cell taking the chromosomes with them, and forming the mitotic spindle.

Nucleus
This is bounded by the nuclear membrane, which is composed of an inner and outer phospholipid bilayer, these becoming continuous at nuclear pores where the nuclear contents can communicate with the cytoplasm. The nucleus contains the chromosomal DNA which is associated with proteins called histones and packaged into chromatin fibres. It also contains one or more nucleoli which are involved in synthesis of ribosomes.

Chromatin
In interphase cells, between cell divisions or mitoses, the chromatin is packaged into differently condensed forms. The most highly condensed form, which is called heterochromatin, is transcriptionally inactive. The less condensed form is called euchromatin, and most RNA synthesis occurs at the border between these two types of chromatin.

Chromosomes
During cell division or mitosis the chromatin decondenses into pairs of chromosomes which have characteristic banding patterns after staining with certain dyes.

The **nucleic acids**, DNA and RNA, are composed of **nucleotides**, joined by **phosphodiester linkages.** Each nucleotide consists of a base, a sugar, and a phosphate group (a base plus sugar alone is called a **nucleoside**)

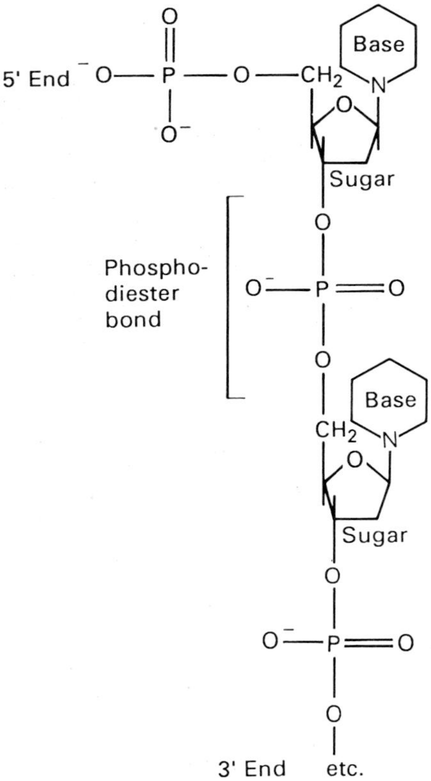

5' End
Base
Sugar
Phospho-diester bond
Base
Sugar
3' End etc.

The **bases** are N-containing ring compounds, that can be classified as **purines** or **pyrimidines**

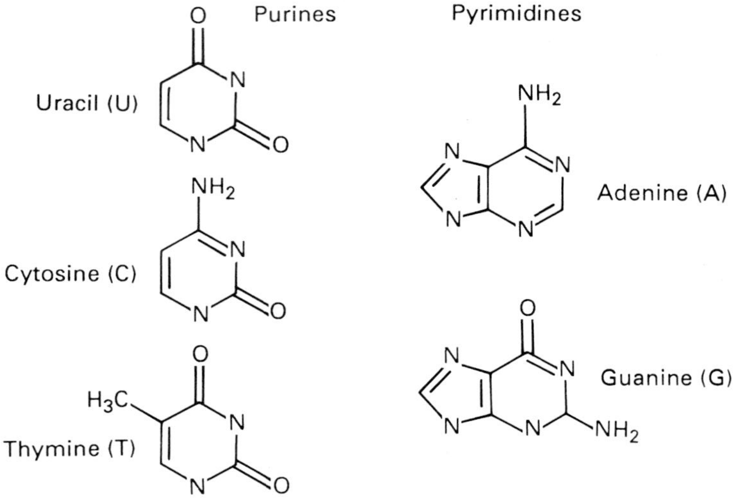

Purines Pyrimidines

Uracil (U)
Cytosine (C)
Thymine (T)
Adenine (A)
Guanine (G)

In the DNA double helix, A pairs with T, and G with C; in RNA, A pairs with U. These are termed **base pairs**

In **DNA** the sugar is β-D-deoxyribose (hence DNA is a deoxynbonucleic acid) and in **RNA** the sugar is β-D-ribose (thus RNA is a ribonucleic acid)

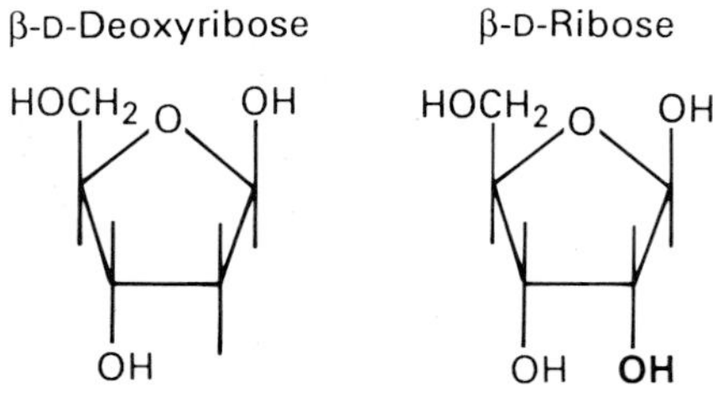

β-D-Deoxyribose β-D-Ribose

Three nucleotides (called a **codon**) encode one amino acid, or a stop codon (a signal to end the polypeptide chain) and the correspondence between these triplets and amino acids or stop codons is known as the **genetic code**

1st position (5' end)	2nd position U	C	A	G	3rd position (3' end)
U	Phe Phe Leu Leu	Ser Ser Ser Ser	Tyr Tyr STOP STOP	Cys Cys STOP Trp	U C A G
C	Leu Leu Leu Leu	Pro Pro Pro Pro	His His Gin Gin	Arg Arg Arg Arg	U C A G
A	Ile Ile Ile Met	Thr Thr hr Thr	Asn Asn Lys Lys	Ser Ser Arg Arg	U C A G
G	Val Val Val Val	Ala Ala Ala Ala	Asp Asp Glu Glu	Gly Gly Gly Gly	U C A G

Fig.1.7 Nucleic acids

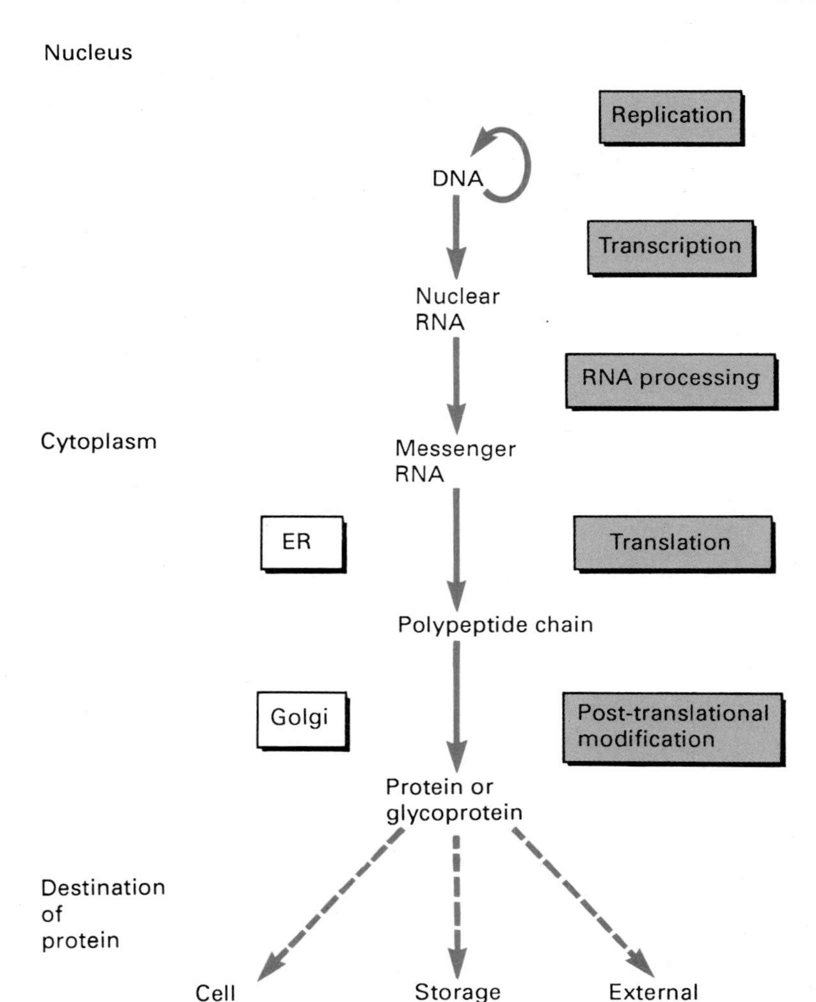

Fig.1.8 DNA to RNA to protein

two types of lymphocyte, **B lymphocytes** (B cells for short) and **T lymphocytes** (T cells) (Fig.1.12). B cells can develop into **plasma cells** which make antibodies, and these molecules attach to foreign organisms to facilitate their destruction. T cells do not make antibodies but they are responsible for cell-mediated responses.

T lymphocytes are divided into distinct subsets which have **regulatory** or **effector** functions (Fig.1.12). The former subsets can be said to control immune responses, while the latter are involved in bringing about or 'effecting' the removal of foreign organisms. Any cell in the immune system that effects the removal of something foreign can be termed an **effector cell**. Most of these cells are relatively inert or inefficient effector cells until they receive a particular signal or stimulus causing them to be activated or to develop further. **Cytotoxic T cells** are effector cells that can kill infected cells

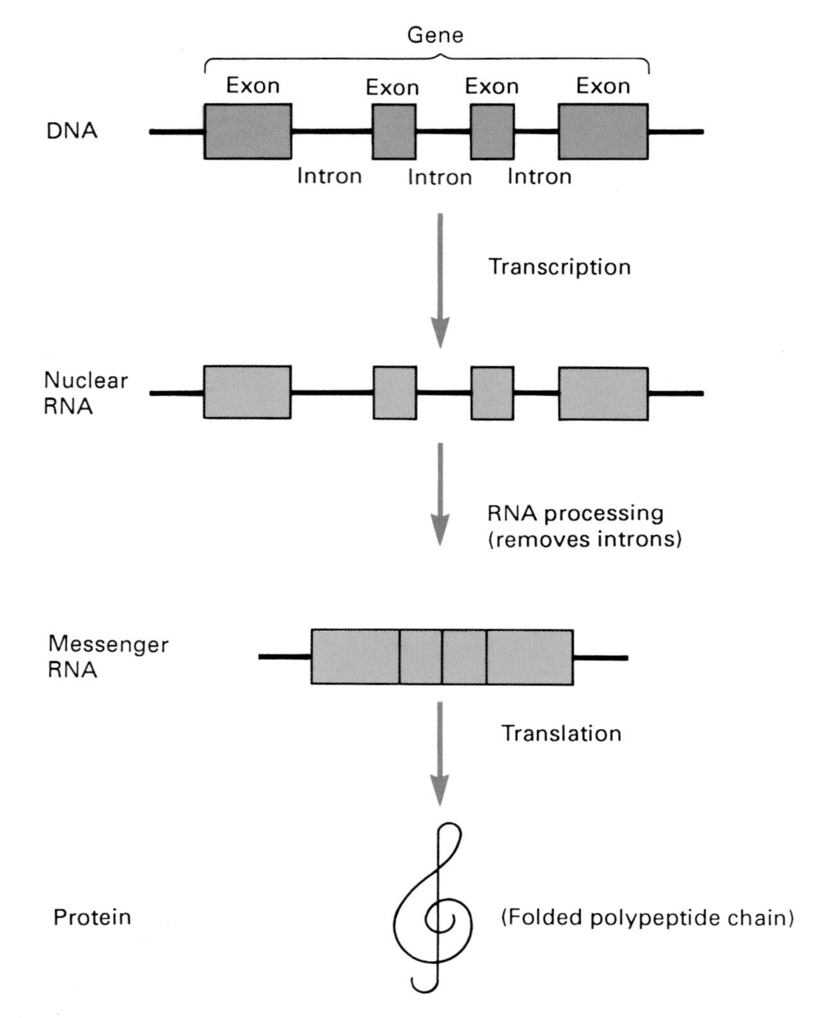

Fig.1.9 Transcription and translation

of the body, as well as other **target** cells like some tumour cells. **Helper T cells**, as their name implies, are regulatory cells that cause other cells of the immune system to become better effector cells. For example, helper T cells 'help' cytotoxic T cell precursors to develop into killer cells, they also help B cells make antibodies, and increase the effector functions of other cells like macrophages. (Note that precursor cytotoxic T cells and B cells have no effector activity until they receive 'help'.) A third, rather mysterious subset of T lymphocytes, **suppressor T cells**, is thought to inhibit immune responses, but it is not clear whether this is a distinct subset.

The innate and adaptive immune systems are linked. The cells and molecules of the innate and adaptive immune systems are closely linked. For example, antibodies produced by B cells of the adaptive system can bind to

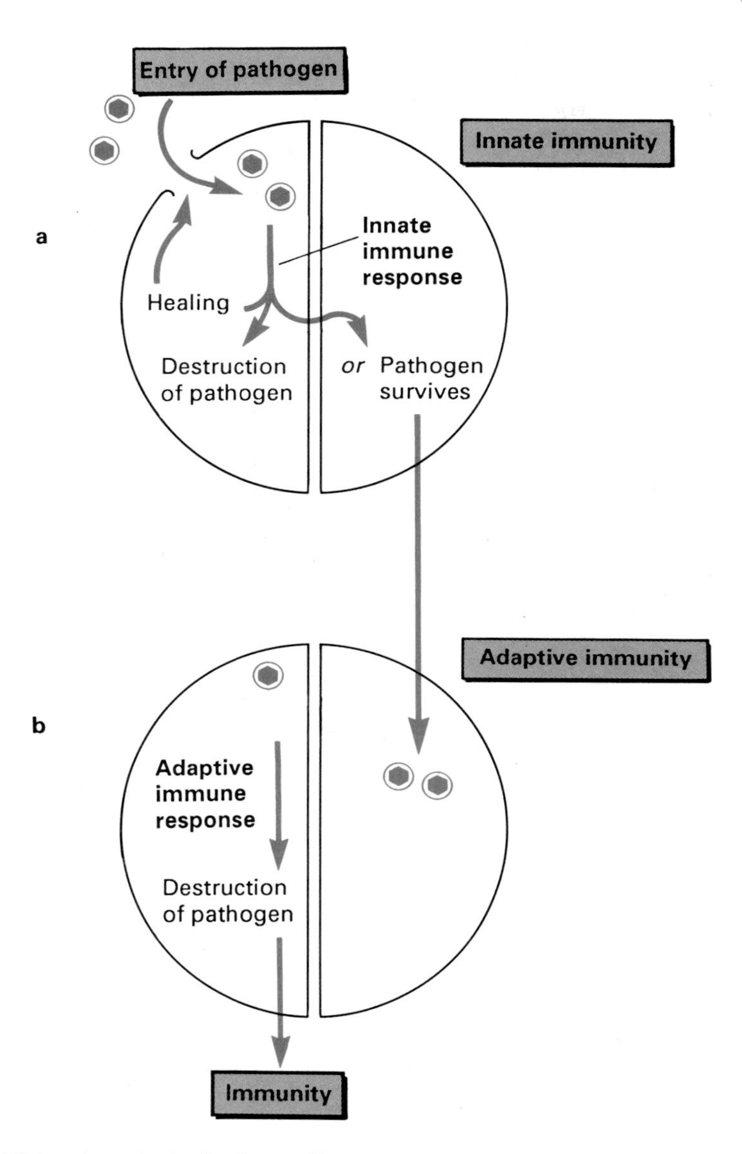

Fig.1.10 Innate and adaptive immunity

Table 1.2 Examples of cells and molecules of the immune system

	Cells	Molecules
Innate immunity	Phagocytes, natural killer cells	Complement
Adaptive immunity	Lymphocytes	Antibodies

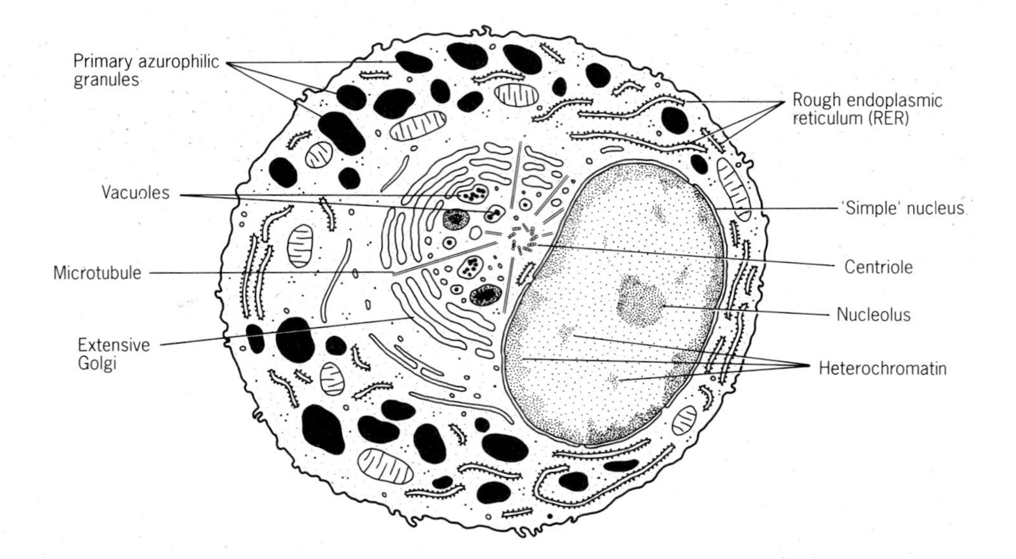

Promyelocytes

These are precursors to all the granulocytes — the neutrophils, basophils, and eosinophils. A generalized promyelocyte is illustrated although it is thought possible to discriminate between the respective precursors on ultrastructural and cytochemical criteria. Promyelocytes develop from myeloblasts in the bone marrow, and may be up to 18 μm in diameter.

The nucleus is not segmented and contains a rim of dense heterochromatin, as well as clumps within the nucleoplasm. The nucleolus is often not very clearly defined, being composed of a coarse intertwining strand containing dense granules (the neucleolonema) surrounding a spherical and finely granular body (the pars amorpha) (not shown).

The cytoplasm contains a large Golgi apparatus. Microtubules extend from the nearby centrioles, and in this area a number of vacuoles are present that often contain a core of dense material. Rough ER, free ribosomes and polyribosomes, and mitochondria are clearly visible. It is believed the earlier neutrophil promyelocytes have a basophilic cytoplasm with a few azurophilic granules, whereas later stages (such as shown) do not have a basophilic cytoplasm but contain more granules. These 'primary' azurophilic granules are lysosomes with a diameter of about 800 nm containing myeloperoxidase, elastase, lysozyme, cathepsin G, and many acid hydrolases.

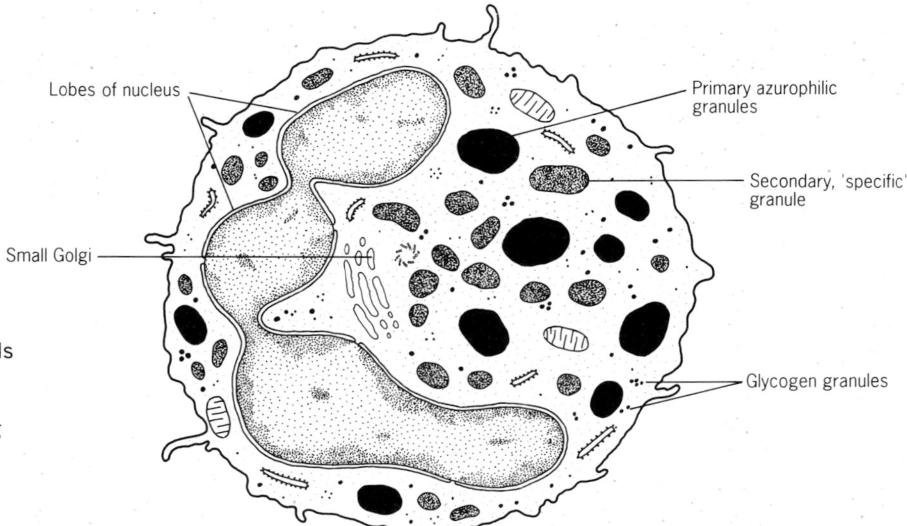

Fig.1.11 Schematic view of different cells of the innate and adaptive arms of the immune system (by transmission electron microscopy. The accompanying descriptions are more detailed than strictly necessary for this introduction, but may be referred to when these cells are mentioned elsewhere in this book)

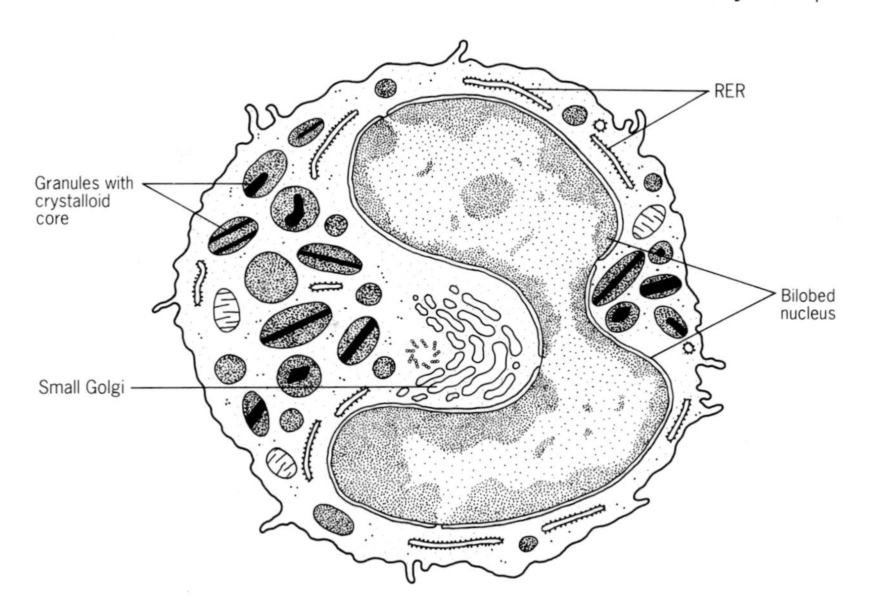

Eosinophils
Eosinophils are particularly prevalent in parasitic infections. They are capable of phagocytosis and can kill parasites and a variety of micro-organisms. Eosinophils are about 12 μm in diameter, and their basic granules stain a bright orange-pink with eosin, hence their name ('eosin-liking').

The nucleus is typically bi-lobed with a rim of chromatin and a fairly large nucleolus. Within the cytoplasm there are many vesicles, as well as ribosomes, mitochondria, small profiles of rough ER, and glycogen granules. The Golgi apparatus is not very extensive and centrioles and microtubules may be seen. A prominent feature of these cells is the presence of characteristic granules, which may be spherical, oval, or ellipsoid in shape. Many contain a core of extremely dense crystalloid material, composed of a basic protein(s) within a matrix of medium density which contains hydrolytic enzymes. These granules have a high content of peroxidase but lack lysozyme, unlike those of neutrophils. There are also smaller granules containing acid phosphatase and aryl sulfatase. When degranulation occurs, the latter enzyme can inactivate a mast cell-derived bronchoconstrictor (slow-reacting substance A), and other molecules that are released can neutralize histamine and possibly inhibit mast cell degranulation, so eosinophils may abrogate immediate-type hypersensitivity.

Neutrophils (left)
Neutrophils (polymorphonuclear leukocytes — PMN) are phagocytic cells that are recruited to sites of inflammation. They are 7–9 μm in diameter, 10–12 μm in blood smears, and their name comes from the fact that their cytoplasm and granules do not stain with Romanowsky dyes (i.e. their contents are at a relatively neutral pH).

A prominent feature of these cells is their segmented nucleus, which may be divided into three to five lobes. There is a peripheral rim of heterochromatin which is interrupted by channels of euchromatin, and nucleoli are not well-delineated.

The cytoplasm contains a very small Golgi apparatus, the centrioles may not be visible, there is little rough ER, and few mitochondria and ribosomes. The granules of mature neutrophils are heterogeneous. About one-third are 'primary' granules, as found in promyelocytes (600–800 nm) while the remainder are smaller 'secondary' or (neutrophil-) 'specific' granules (300–500 nm) that contain lactoferrin as well as lysozyme which can digest certain bacterial cell walls. When degranulation occurs, for example at sites of acute inflammation, their contents can be delivered to the cell's environment, thereby aiding destruction of pathogens and contributing to the inflammatory response. The granules can also fuse with phagocytic vacuoles, leading to degradation of internalized material. There are also numerous, very small granules (25 nm) of glycogen, a primary source of energy for the cell.

Very small Golgi

Granules with crystalloid and concentric lamellar structures

Short profile of RER

Glycogen granules

Basophils

Basophils are a trace cell type in peripheral blood, about 10 μm diameter, with acidic (or 'base-liking') granules that stain a deep violet-blue with Romanowsky dyes. They have structural and functional similarities to mast cells.

The nucleus is segmented, unlike that of mast cells in serous cavities which is round or oval, and the chromatin organization is similar to that of neutrophils and eosinophils. A nucleolus is only rarely seen in these cells.

The Golgi apparatus is not well-developed, and there are a few short profiles of ER, small numbers of free ribosomes, some mitochondria, and some glycogen granules. Aggregates of very small particles can be found free within the cytoplasm or incompletely surrounded by numerous membranes.

The granules of basophils differ in different species, and perhaps after different fixation procedures. Sometimes they contain crysalloid material (but not crystals) and can be filled partially or completely with particles of uniform size. Those of human basophils contain carbohydrates that are reactive with periodic acid–Schiff reagent, peroxidase, heparin, and histamine (mast cells contain the enzymes alkaline phosphatase and acid phosphatase, which are not present in basophils). Degranulation occurs in response to antibody–antigen reactions, releasing heparin, which is an anticoagulant, and histamine which causes vasodilation and increased vascular permeability, features of inflammatory reactions. It would appear that the primary function of basophils is thus secretion, for the granule contents are discharged into the environment rather than to phagocytic vacuoles within the cell (whereas the latter commonly occurs in neutrophils, eosinophils, and monocyte/macrophages); phagocytosis is not very obvious in basophils although it can occur.

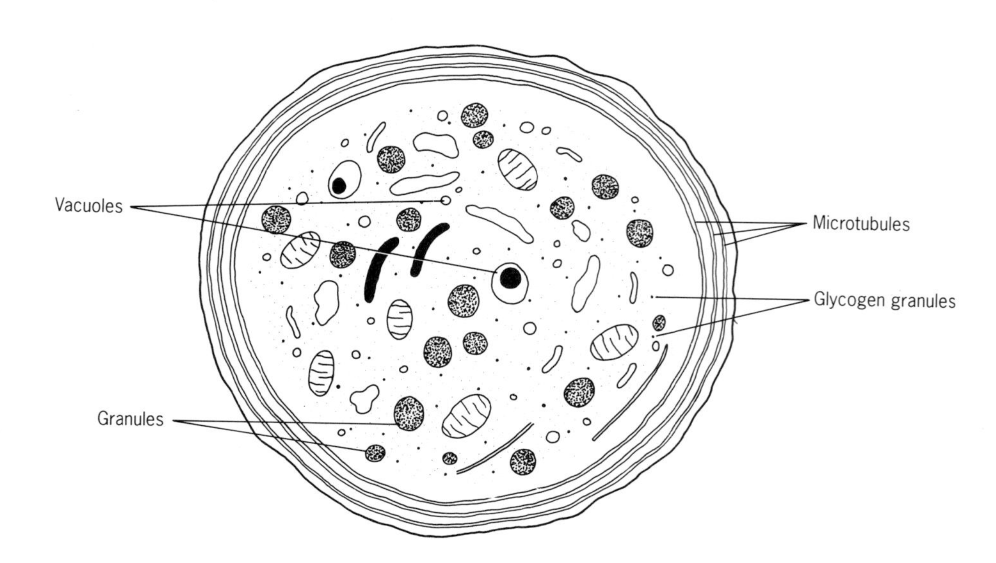

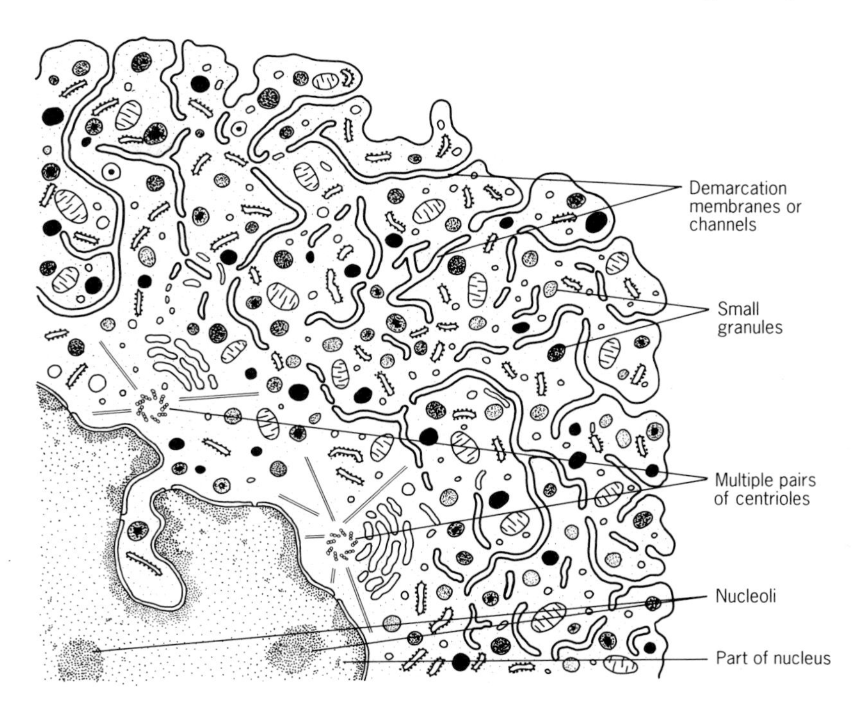

Megakaryocytes

Megakaryocytes are giant cells, with a diameter of 20–150 μm in the bone marrow, and they are the source of circulating platelets. The nucleus is extremely lobulated, with peripheral as well as random dense clumps of heterochromatin, and numerous nucleoli are present.

A remarkable feature of these cells is that repeated rounds of DNA synthesis occur with no division of cytoplasm so that the cells become polyploid, containing up to 64 copies of chromosomal DNA. After DNA synthesis has stopped, cytoplasmic maturation occurs. In the perinuclear region there are many ribosomes, rough ER profiles, multiple groups of vesicles, and tubules forming part of the Golgi apparatus, and several paired centrioles and microtubules. Perhaps the main characteristic of the intermediate cytoplasm, in which many small granules are found, is the presence of 'demarcation membranes' which bound narrow channels defining the boundaries of future platelets. In cells that are not actively producing platelets, in the peripheral zone of cytoplasm adjacent to the plasma membrane is a clear 'marginal zone' that is almost devoid of organelles but which contains many actin filaments (not shown). In megakaryocytes producing platelets, platelets can be seen 'budding' from the main body of the cell (not shown).

Platelets (left)

Platelets are portions of megakaryocyte cytoplasm, lacking nuclei, that exist as small bioconcave discs about 3 μm in diameter. They have a complex structure that includes microfilaments, microtubules, a dense tubular system, an open canalicular system, vesicles, glycogen granules, and small mitochondria, but very few ribosomes. Dense membrane-bounded granules are also present that contain hydrolytic enzymes and may be lysosomes.

Platelets have a number of important functions, being involved in hemostasis, due to their ability to adhere to cell surfaces and to each other, coagulation and inflammation, the latter being mediated by the histamine and/or serotonin they release.

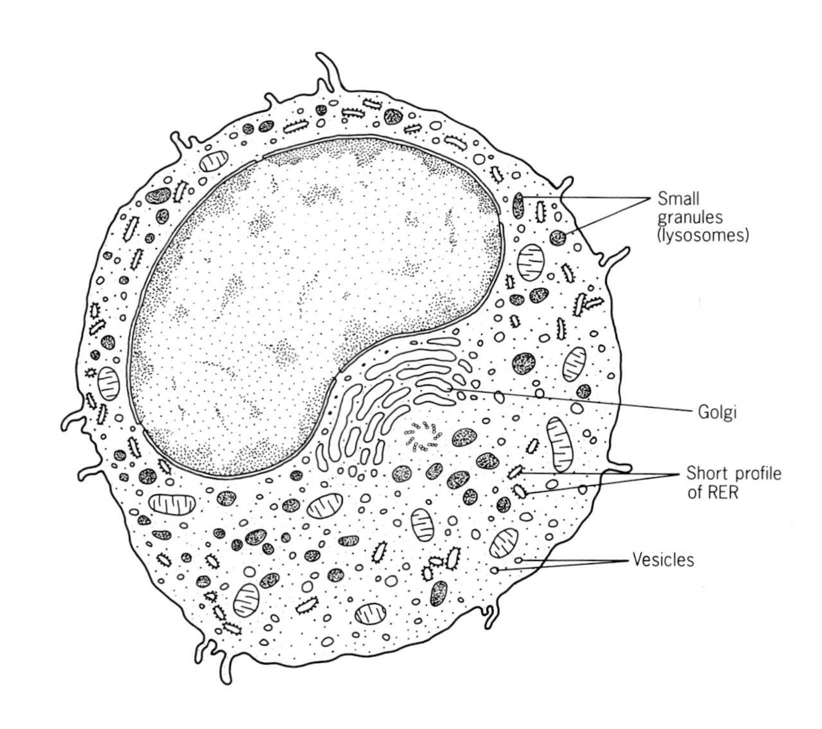

Monocytes

These are circulating precursors to the tissue macrophage that are 9–12 μm in diameter (16–22 μm in blood smears). They have an eccentric round or kidney-shaped nucleus with moderately condensed heterochromatin around the rim and dispersed throughout the nucleoplasm, but nucleoli are not often seen. A 'rosette' of variably shaped, azurophilic granules and centrioles are present near the Golgi at the 'hof' (indentation) of the nucleus. A number of vesicles, short profiles of rough ER, and mitochondria are scattered throughout the cytoplasm.

Macrophages (see Chapter 9)

Macrophages in tissues are involved in defense against pathogens and in inflammatory reactions, and they can adopt different morphologies and functions depending on precisely where they are situated in the body. The nucleus can be round, kidney-shaped, or horseshoe-shaped, with a nucleolus. Many small vesicles and dense granules are present near the Golgi apparatus, together with the centrioles. Dispersed throughout the cytoplasm are mitochondria, profiles of rough ER, microfilaments, lysosomes, lipid droplets, and endocytic vesicles containing soluble molecules and particles of different sizes that have been internalized from the cell's environment.

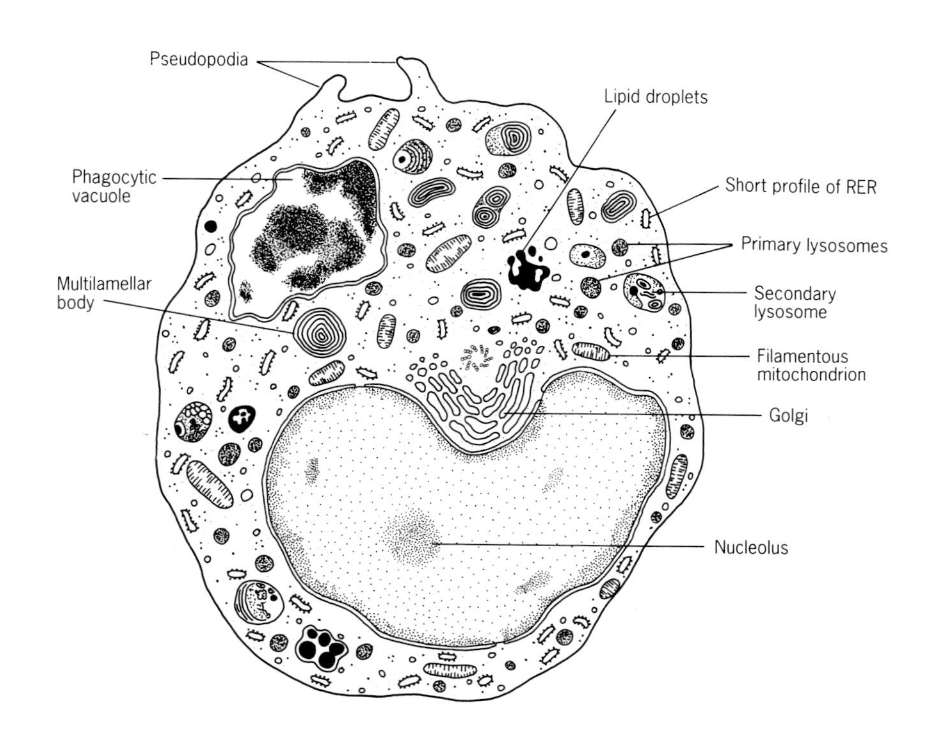

Lymphocytes (discussed throughout this book)
For many years lymphocytes attracted relatively little attention, perhaps in part because they are arguably the least inspiring cells from a morphological point of view. Small lymphocytes are spherical cells of 8–12 μm diameter. The bulk of the cell is occupied by the nucleus which is nearly spherical, the heterochromatin being distributed thoughout the nucleoplasm, and contains a nucleolus. Close to the small 'hof' or nuclear indentation are the centrioles and a small Golgi apparatus. The relatively small rim of cytoplasm contains many free ribosomes, a little ER, a few scattered mitochondria, and occasional granules (0.2 μm diameter). Small lymphocytes are 'resting' cells, awaiting contact with antigen and activation signals that trigger their development into large lymphoblasts and/or (in the case of B lymphocytes) plasma cells.

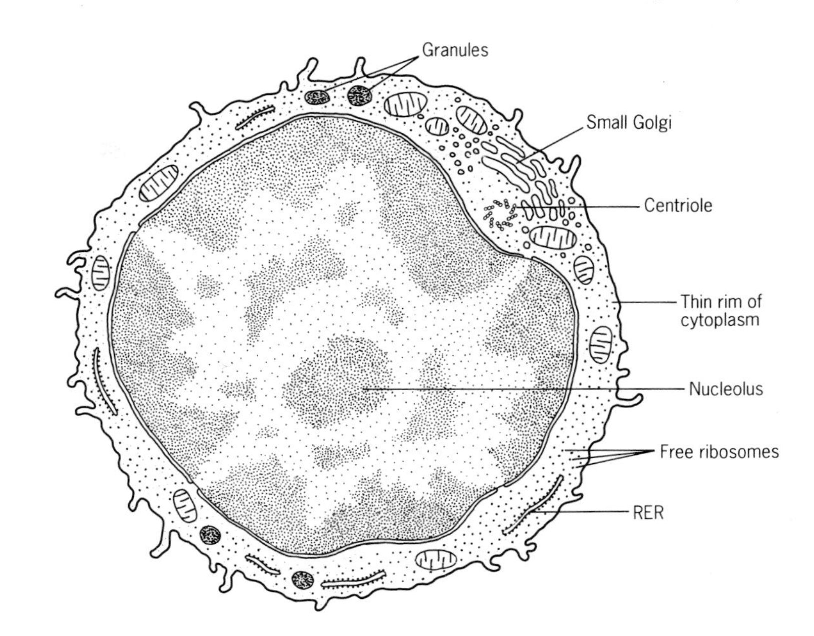

Plasma cells (see Chapter 7)
Plasma cells are large, with a diameter of 12–16 μm, and develop from resting B lymphocytes; they are responsible for antibody secretion. They occur primarily in secondary lymphoid tissues and are only rarely seen in peripheral blood.

The nucleus exhibits a characteristic radial or cartwheel distribution of heterochromatin, often with a large mass in the centre of the nucleoplasm, together with a nucleolus.

The most notable feature of the cytoplasm, which stains a royal blue with Romanowsky dyes, is the abundance of rough ER that is required for synthesis of antibody molecules which are secreted by the cell at rates of up to 2000 molecules per second. Free ribosomes, mitochondria, and occasional granules are seen, together with a well-developed Golgi apparatus and a pair of centrioles.

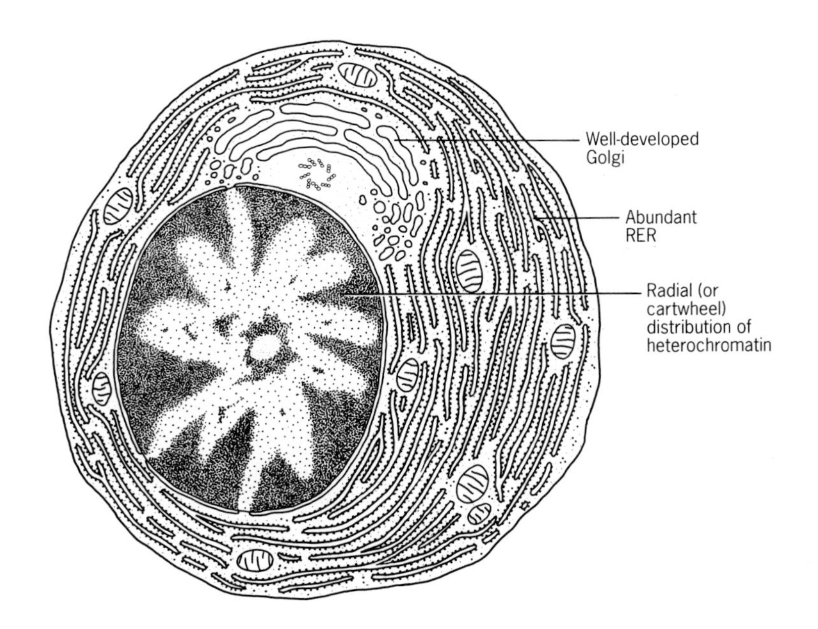

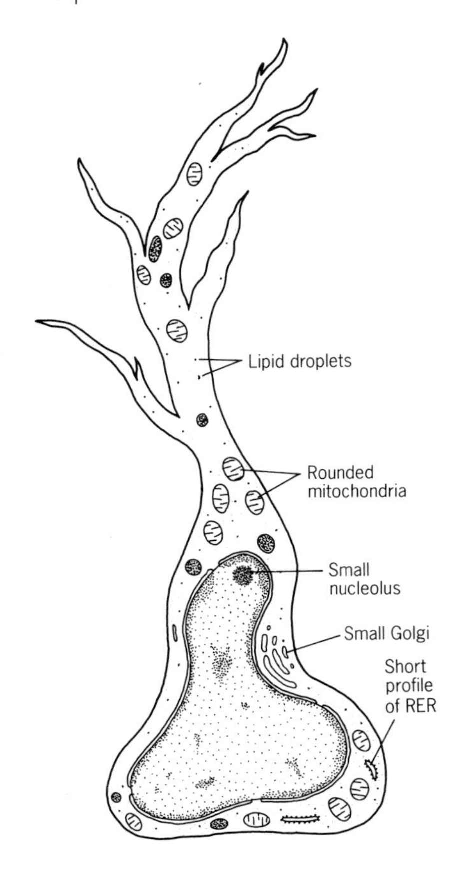

Dendritic cells (see Chapter 3)
These cells are present in lymphoid tissues, cells of this lineage are also present in most non-lymphoid tissues such as Langerhans cells in skin epidermis. Dendritic cells deliver an activation signal(s) to resting T cells that triggers their development into lymphoblasts. They can be isolated from tissues such as spleen, although small numbers are present in peripheral blood. When adherent they frequently exhibit long processes (such as illustrated).

The nucleus can be very irregular in shape with a peripheral rim of heterochromatin and a small nucleolus. The cytoplasm is clear, lacking large numbers of ribosomes, glycogen granules, and microfilaments. However, it contains relatively large numbers of rounded microchondria, a few short profiles of rough ER, lipid droplets, a pair of centrioles, and Golgi apparatus. Some dendritic cells also contain vacuoles with multilamellar and multivesicular structures (see Chapter 3; Fig. 3.35) and the Langerhans cell contains a characteristic organelle called the Birbeck granule (see Chapter 3; Fig. 3.36).

Cellular composition of normal human peripheral blood

Cell type	Number of cells per litre of blood
Total white cells	**4–10** × **10**^{9}
Neutrophils	**2–7** × **10**^{9}
Eosinophils	**0.4–4** × **10**^{9}
Basophils	**0.1–1** × **10**^{9}
Monocytes	**0.2–0.8** × **10**^{9}
Small lymphocytes	**1.5–3.5** × **10**^{9}
Platelets	**15–400** × **10**^{9}

the surface of phagocytes in the innate system to increase or potentiate their function, and the functions of lymphocytes can in turn be regulated by molecules produced by phagocytes. Furthermore, complement components interact with antibodies, and with cells from both branches of the immune system. Nevertheless, one essential difference remains (Table 1.3). The innate immune system is designed to mount an immediate response when challenged by something foreign and its activity remains at the same level during subsequent challenges. The adaptive immune system only becomes activated later, but once primed it can produce a more rapid and vigorous memory response when challenged by the same organism in the future. The response of the adaptive immune system to an initial encounter with something foreign is called the **primary response**; the subsequent more efficient memory responses that ensue in repeated exposure to the same foreign agent are called **secondary responses**.

Activation of cells of the immune system. Normally, lymphocytes are in an inactive or 'resting' state, unless an immune response is occurring. When an immune response is triggered, for example during a virus infection, lymphocytes that can respond to the virus become **activated** to carry out their specific functions (Fig.1.13). After they have been activated, the lymphocytes divide or 'proliferate' and their progeny or descendants give rise to a **clone** of cells in a process termed **clonal expansion**. These changes

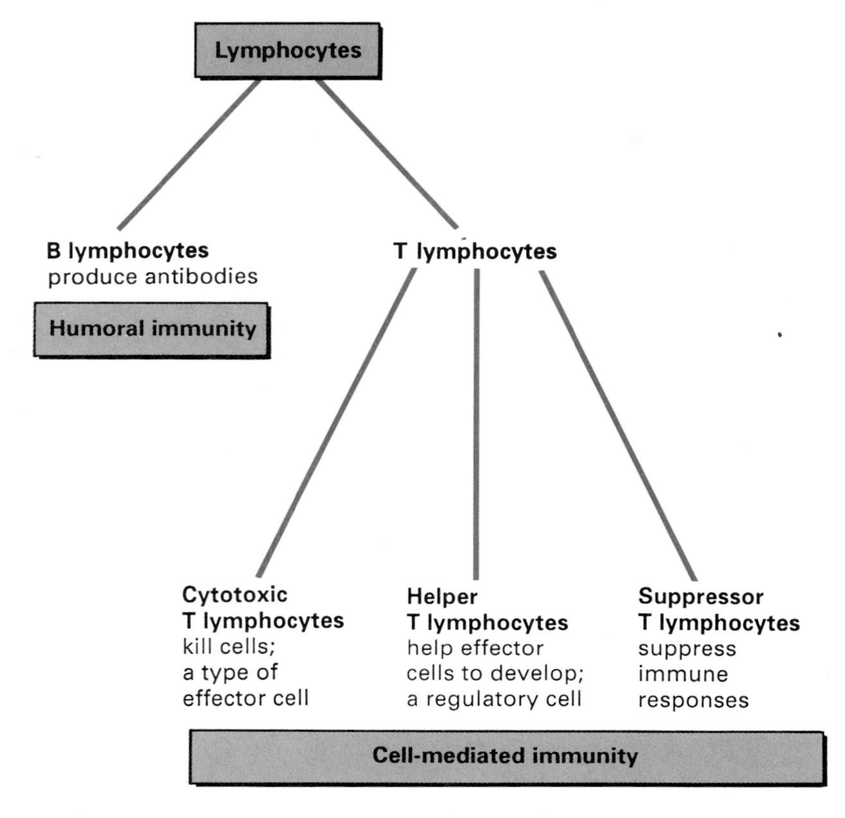

Fig.1.12 Lymphocytes can be divided into different subsets

correspond to movement through certain stages of the **cell cycle**, known as G_0, G_1, S, G_2, and M (Fig.1.14). Resting lymphocytes are usually said to be in the G_0 phase. When they are activated, they enter G_1 and become much larger cells called **blast cells** or **lymphoblasts**. They may then enter S phase of the cell cycle, when DNA is replicated, and subsequently progress via G_2 into M or mitosis when cell division actually occurs. As a result of activation, lymphocytes become able to carry out their specialized functions. Activated helper T cells are able to 'help' the generation of the effector phase of the response. B cells develop into plasma cells and make soluble antibodies

Table 1.3 Comparison of innate and adaptive immune responses

	Ability to mount a response:		
	Before pathogen is present	**When pathogen is first encountered**	**If pathogen is subsequently encountered**
Innate immunity	+	+	+
Adaptive immunity	−	+ (Primary response)	+++ (Secondary response)

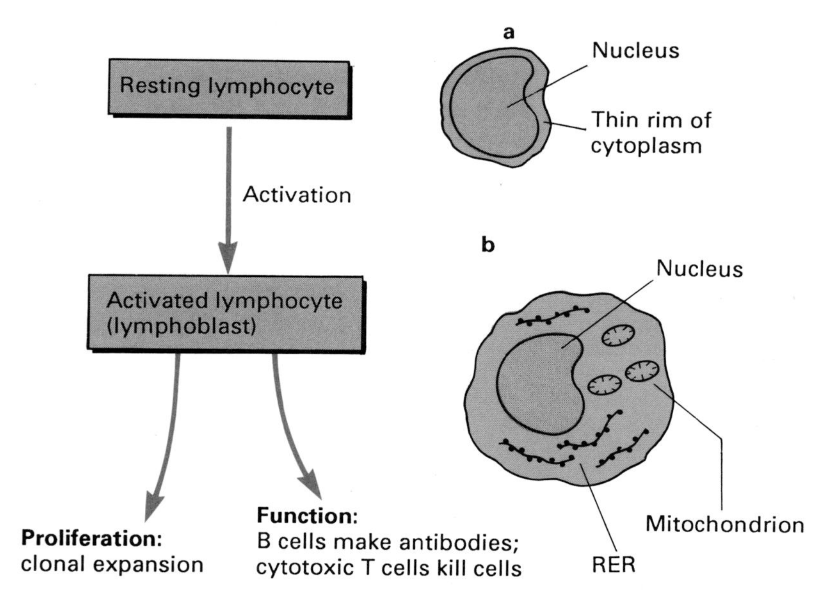

Fig.1.13 The features of lymphocyte activation

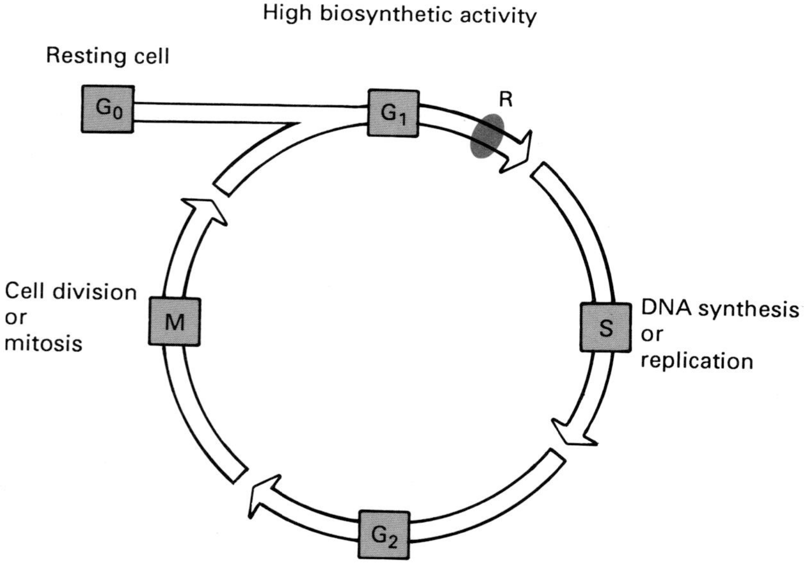

High biosynthetic activity

Resting cell

G_0

G_1

R

Cell division or mitosis

M

S

DNA synthesis or replication

G_2

$G_1 + S + G_2$ = 'Interphase'
R: An example of a restriction point in the cell cycle; once cells have reached this 'point of no return' they progress to a different point of, or through, the cell cycle irrespective of external conditions

Note: stages are not usually of equal length as shown.
For a more detailed and representative example see Fig. 8.20.

Fig.1.14 The cell cycle

against the virus, for example, and cytotoxic T cells develop full effector function and kill virally-infected cells; in their resting state, B cells and cytotoxic T cell precursors have no effector function.

Other cells of the immune system can also be activated during immune responses. Macrophages, for example, can be relatively quiescent cells, but if an immune response occurs they can develop into 'activated' macrophages with the ability to kill cells and micro-organisms.

Once the immune response has ceased, for example because the virus has been successfully eliminated and virally-infected cells have been killed, the lymphocytes revert once more to an inactive state. However, some of them form part of an expanded population of 'memory' cells which are specific for the particular virus. Memory cells are able to mount a vigorous and effective secondary response should the same virus be encountered in the future (Table 1.3). In summary, there are three states in lymphocyte activation: resting, activated, and memory cells (which morphologically resemble resting T cells).

1.1.3 Receptors, antigens, and lymphocyte antigen receptors

Immune recognition and receptors

Adaptive immune responses are **specific** (Table 1.1). Someone vaccinated against the smallpox virus (variola) becomes immune to this disease but not to polio, which is caused by the poliovirus, and vice versa.† The degree of specificity of immune responses is remarkable. Until now we have only considered immune responses to particular intact organisms, but the immune system is able to discriminate even between optical isomers of a very small molecule like an amino acid (i.e. between its D and L forms; Fig.1.3). The immense **diversity** of immune responses that can be generated is equally remarkable. The immune system can 'recognize' an extraordinarily large number of different molecules and it even responds to newly-created, man-made substances and newly-evolved organisms (e.g. new strains of the influenza virus) that have never existed before.

How is it that adaptive immune responses are so specific, yet so diverse? To answer this question we can start by thinking about it another way and asking how **immune 'recognition'** comes about. This central issue is addressed, in a straightforward manner, in the next few pages. We have used the term 'recognition' for convenience, although at first sight it seems rather anthropomorphic, and because it is very common for immunologists to talk about the cells and molecules of the immune system as 'recognizing' other cells and molecules in the body. Several other anthropomorphic terms are also used by immunologists, for instance we talk about immunological 'memory' and how the immune system is able to 'see' something as foreign.

What we actually mean by 'recognition' is that cells and molecules **interact** with or **bind** to each other in a more-or-less specific way (Fig.1.15). To take an example, if one cell attaches to another, we might say the first cell was able to 'recognize' the second. If, however, it was unable to associate

† Obviously if two organisms are related and share common features, a response made against one organism may 'cross-react' with the other, as in the case of vaccination with vaccinia, the cowpox virus, to combat smallpox.

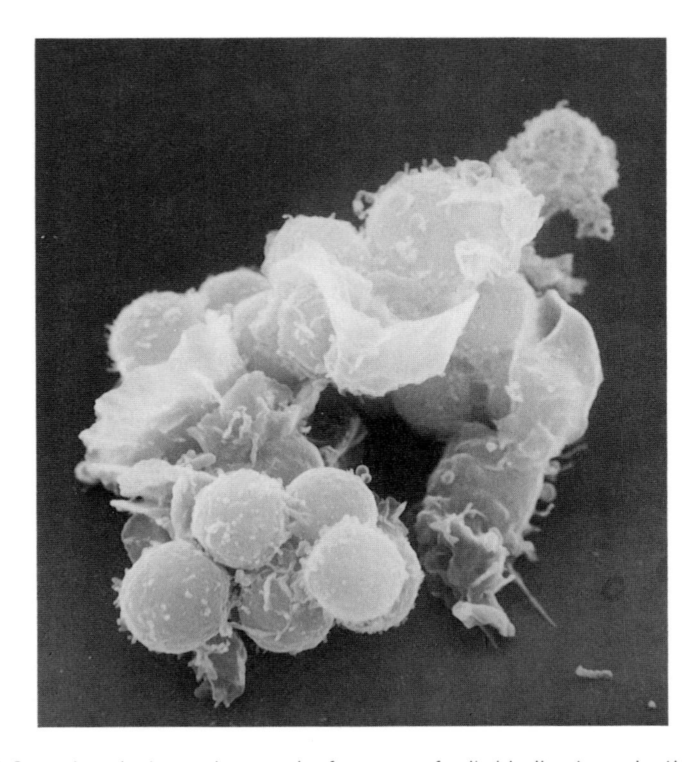

Fig.1.15 Scanning electron micrograph of a group of cells binding to each other. The photograph shows a cluster of lymphocytes, the smaller spherical cells, attached to three or more dendritic cells, with flaps or veils of cytoplasm, one of which is seen more completely lying on the right-hand side of the cluster

with a third cell, we could say that recognition by the first cell was specific for the second but not for the third cell. Equally, we could have taken the interactions between molecules as an entirely analogous example.

A fundamental principle in the immune system is that cellular responses normally follow recognition events (Fig.1.16); some other systems in the body also work in a similar manner. Indeed, it is precisely when these recognition events occur that signals are delivered to the inside of the cell to trigger the response. The cell might, for example, then start to make soluble antibodies if it is a B cell, or to produce the molecules needed to kill its specific target if it is a cytotoxic T cell. Cells therefore have the means of interacting with or 'recognizing' other cells but how, in molecular terms, does this occur?

Cells recognize things through molecules inserted in their plasma membranes (Fig.1.16). These molecules, in general, extend from the cell surface and interact with other molecules in the cell's environment; the latter can be either cell-attached or soluble molecules. This specific interaction often leads to a signal being transmitted into the cell via another part of the same molecule or by a different, but associated, membrane component.

Molecules involved in recognition functions are called **receptors**. In simple terms, each receptor can be thought of as having a complementary 'shape' or conformation to the molecule being recognized, which is termed the

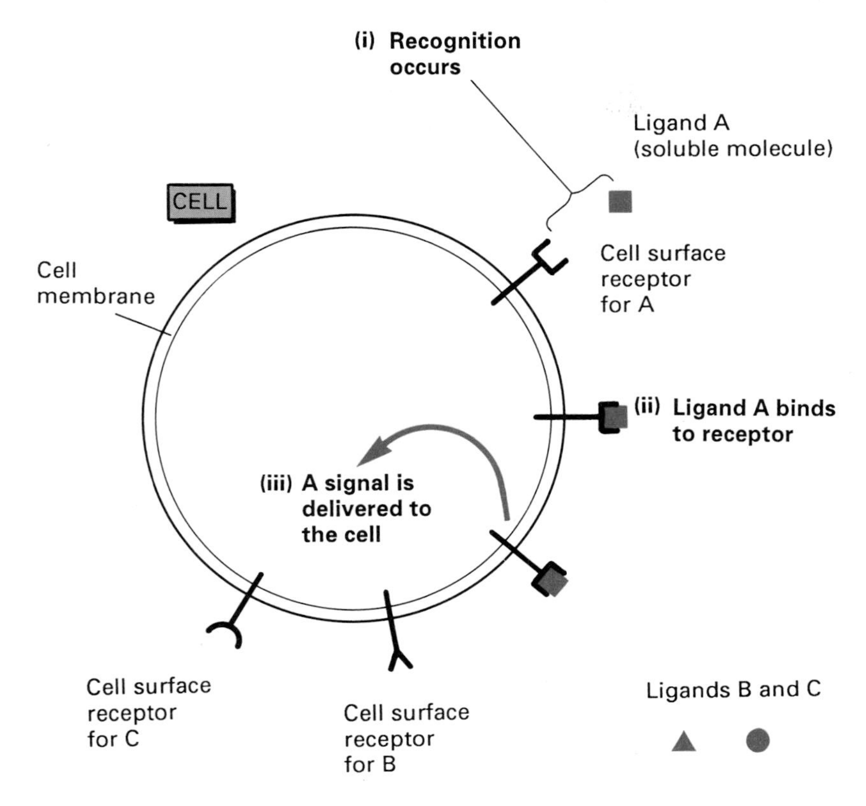

Fig.1.16 Molecular recognition

ligand. We say that a given receptor is 'specific' for its ligand. When recognition occurs, the ligand becomes bound to the receptor. The analogy of fitting a key into a lock is often used for a receptor–ligand interaction: usually one receptor interacts with one particular ligand, just as one key normally fits one particular lock.

The size and shape of the ligand that can be accommodated by the receptor depends of course on the size, charge and shape of that part of the receptor molecule that binds the ligand. A very small protein, for example, might be composed of a hundred or so amino acids, but receptors in the immune system can accommodate just a few amino acids (see Fig.1.1). This means that only part of a whole molecule or ligand is recognized by any given receptor, and when we talk about a receptor in the immune system being able to 'recognize a certain organism', what we really mean is that the receptor can bind to a very small part of the whole organism.

Antigens and lymphocyte antigen receptors

The immense diversity and specificity of immune responses implies there must be a very large number of different and specific receptors. We now know that the receptors that recognize foreign organisms, for example, are present on the surface of lymphocytes inserted into their plasma membranes.

Of course, not all receptors on immune cells can recognize things that are foreign to the individual. Probably every cell in the body has a number of

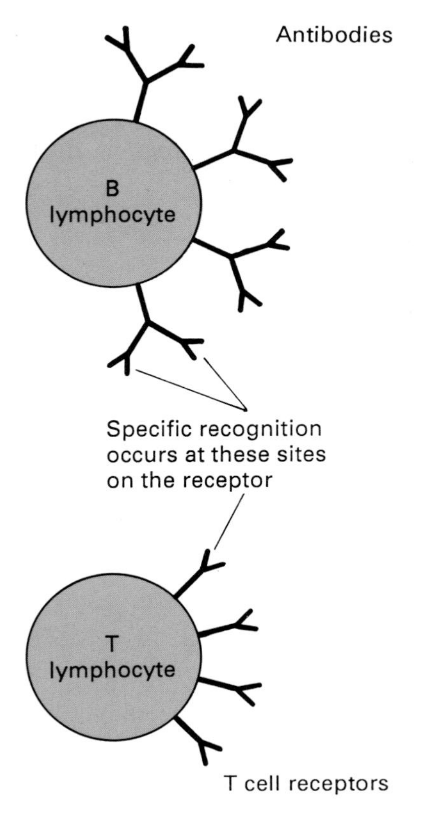

Fig.1.17 Receptors for foreign antigens on lymphocytes

receptors specific for particular molecules produced by other cells in the body. So, for example, there are receptors for growth factors that are needed for cellular proliferation; other receptors that allow the cell to adhere to other cell types and different substrates; and so on. However, as far as we know, only lymphocytes have membrane receptors that can recognize foreign molecules derived from outside of the body, and it is these that concern us now. It is now known that the membrane receptors on the surface of B lymphocytes that recognize foreign things are **antibodies**. Antibodies can also be made in a soluble form and secreted when B cells are activated and develop into plasma cells. Antibodies can also be called **immunoglobulins** (Ig). T cells also have membrane receptors for foreign things which are somewhat similar to those of B cells. The receptors on the surface of T lymphocytes are called **T cell receptors**, but, unlike antibodies, T cell receptors are not secreted. Each type of lymphocyte (B or T cell) thus has a characteristic type of receptor on its membrane (Fig.1.17).

Until now, we have talked about immune responses against particular organisms or molecules, such as viruses or optical isomers of amino acids. This becomes cumbersome and inconvenient, especially if we do not know precisely what a given receptor is specific for, or if we want to talk in general about things that are 'recognized as foreign' by the immune system. Fortunately, immunologists have a general term to cover anything and everything that is recognized in this way: **antigens.** Foreign antigens are recognized by specific receptors on lymphocytes, which are therefore called **antigen receptors.** Thus the antigen receptors of B cells are antibodies (also referred to as membrane immunoglobulins; mlg), while those of T cells are T cell receptors (TcR) (Fig.1.17).

Historically, an antigen was defined as any molecule that stimulated the production of antibodies. However, the definition is now more general. Roughly speaking, *the term antigen includes anything that can be recognized by a lymphocyte antigen receptor*. This definition in general excludes molecules that are normally produced within the body, such as the growth factors or adherence molecules which were mentioned above. A more accurate, if rather circular, definition of an antigen is any molecule that can be recognized by an antibody or by a T cell receptor.

As we said earlier, every lymphocyte receptor recognizes only a small part of the antigen which, in its entirety, could be a very large molecule. The different parts of an antigen that can be recognized by antigen receptors are called **antigenic determinants** or epitopes (Fig.1.18). Each and every lymphocyte receptor recognizes one particular antigenic determinant.

Immune responses can be directed to just a few antigenic determinants on an antigen. Other responses made against the same molecule, at a different time or in a different individual or species, could well involve recognition of a completely different set of antigenic determinants. It is also possible that some individuals might not be able to make an immune response to a particular antigen, and a special class of antigens to which this applies are simple molecules with a few, repetitive determinants. The molecule is said to be **immunogenic** if the individual responds to it, and it is referred to as an **immunogen**, but it is said to be **non-immunogenic** if no response is made against it (Fig.1.19). This means that any given antigen may or may not be an immunogen.

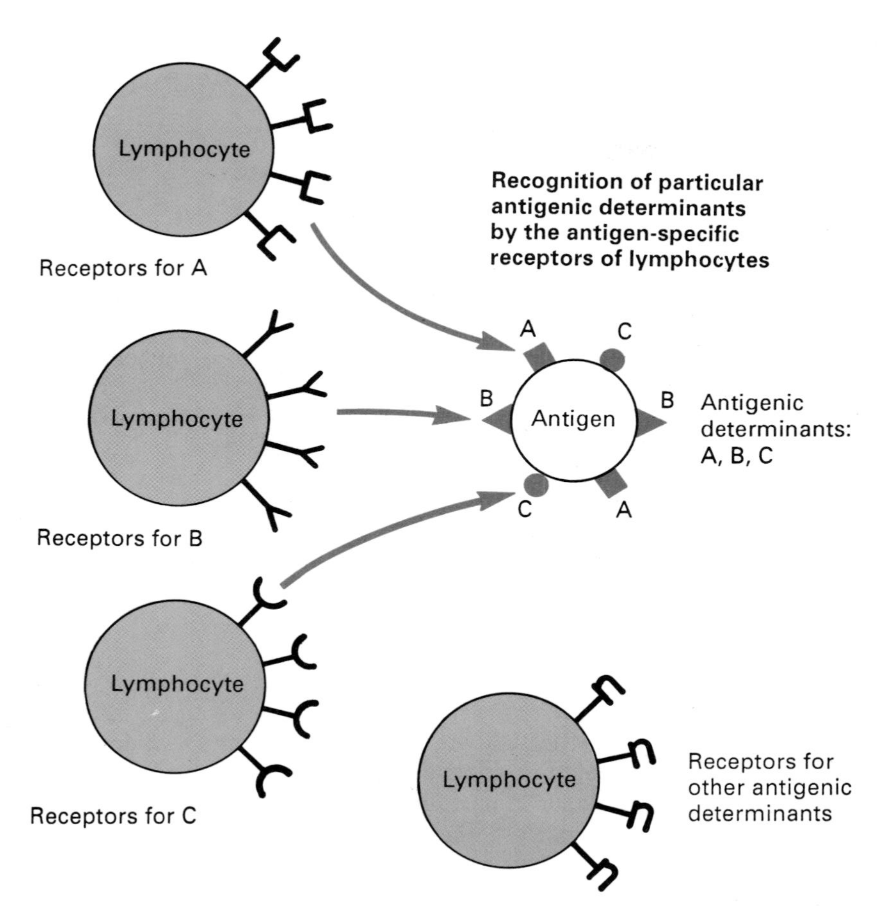

Fig.1.18 Antigen receptors and antigenic determinants

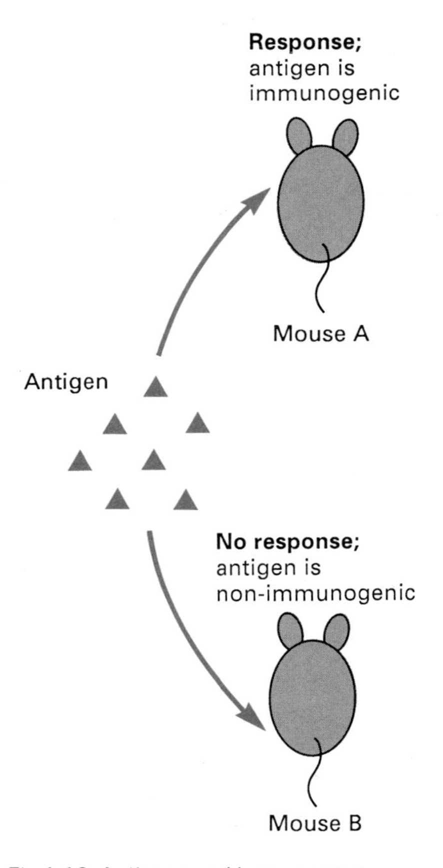

Fig.1.19 Antigens and immunogens

There is a major difference between the types of antigen that can be recognized by B cells and T cells (Fig.1.20). Antibodies appear to have evolved to recognize 'free' or soluble antigens, such as foreign proteins on viruses and bacteria, or foreign molecules dissolved in the body fluids. Such proteins in an unmodified form are referred to as **native antigens**. In contrast, T cell receptors recognize antigenic determinants on the surface of cells of the body, such as viral antigens (i.e. proteins made by the virus) on the surface of virally-infected cells. This allows B cells to monitor what might be called the 'fluid phase' of the body, and T cells to monitor the surface of cells in the body, for foreign antigens (Table 1.4).

In actual fact, T cells recognize degraded forms of foreign antigens, as peptides, bound to a special type of molecule on the surface of cells, the **MHC molecules**. This form of antigen is referred to as **processed antigen**. There are thus three types of molecule within the immune system that can bind antigens, and each has distinct properties (Table 1.5). MHC molecules bind processed antigens as peptides, T cell receptors apparently bind peptide–MHC complexes, and antibodies can bind native antigens. While any antibody or T cell receptor is, in general, highly specific for a particular antigen or pep-

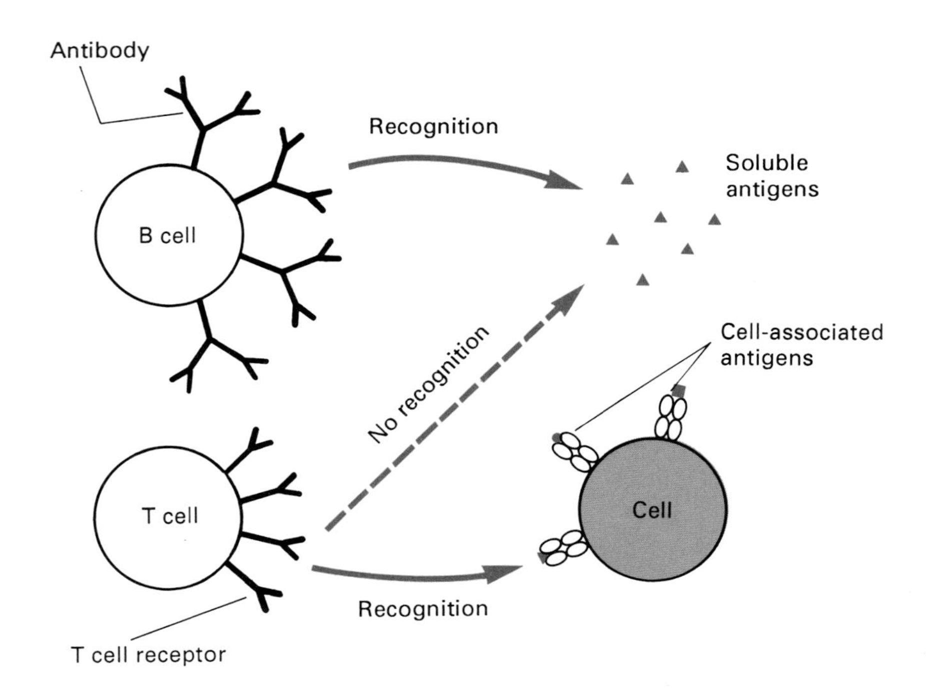

Fig.1.20 Antibodies can recognize 'soluble' antigens; T cell receptors recognize 'cell-associated' antigens

tide–MHC complex, each MHC molecule can bind a range of peptides. Moreover, the enormous diversity of antibodies and T cell receptors within any individual results in whole, or in part, from complex rearrangements of the genes encoding them. In contrast, there are relatively few MHC molecules within each individual, and their genes are not rearranged, (although they show considerable polymorphism within the species, meaning that multiple different forms or alleles exist in the population as a whole). There are two classes of MHC molecules, class I and II; both bind peptides but they have different functions as discussed below.

Table 1.4 Antigen recognition by B and T lymphocytes

	Recognition of antigens in the body:		
	In 'fluid phase'	**On cell surfaces**	**Form recognized**
B cells *(via membrane-bound antibodies)*	Yes	No*	Native; soluble
T cells *(via membrane-bound T cell receptors)*	No	Yes	Processed; peptides bound to MHC molecules

* Note that B cells can recognize antigens on the surface of *foreign* cells (e.g. micro-organisms)

Table 1.5 Antigen-binding molecules

	Different molecules/ individual	Genes rearranged	Form of antigen bound	Specificity of binding
Antibodies	c. 10^{12}	Yes	Native	Often very high
T cell receptors	c. 10^{12}	Yes	Peptide–MHC	Very high
MHC molecules	c. 10–20	No	Peptide	Broader

Clonal selection of lymphocyte antigen receptors

How is it possible for the immune system to generate such an enormous number of different antigen receptors? One early idea was the receptor somehow folded itself into a new conformation, or shape, around the antigen. In other words, the antigen was thought to 'instruct' the receptor to adopt a particular (complementary) shape and this idea, logically, was known as the **instruction hypothesis**. This hypothesis turns out to be incorrect. We now know that *antigen receptors are preformed* and are present in the immune system even before the antigen is encountered. In other words, there is an immense number of 'ready-made' antigen receptors in the body, and this contrasts greatly with the idea of 'made-to-measure' receptors predicted by the instruction hypothesis. What happens in an immune response is that clones of lymphocytes with receptors that 'fit best' to the antigen are selected from the enormous number, or **repertoire**, of cells with different receptors. This idea was put forward in the **clonal selection theory**, which is a cornerstone of immunology. The process is somewhat analogous to a customer choosing the best-fitting, ready-made article in a clothes store.†

Do lymphocytes have a variety of different antigen receptors on their surfaces? If so, this would mean any lymphocyte could recognize many different antigens. Or does each lymphocyte carry or 'express' just one type of receptor specific for one antigenic determinant? This would mean that any particular lymphocyte would recognize only one antigen. The latter turns out to be the case.

The antigen receptors of each lymphocyte are specific for just *one* antigenic determinant, even though there are hundreds of thousands of copies of the same receptor on every cell. Lymphocyte antigen receptors are thus said to be **clonally distributed** (Fig.1.21). The term came about because when a lymphocyte divides to form a clone of cells, each of the progeny should have identical receptors.‡

The fact that receptors are clonally distributed is important for the efficient working of the immune system as a whole. If each lymphocyte could recog-

† It is important to note that receptors can also be produced *during* immune responses, at least in antibody responses. Nevertheless, the clonal selection still applies. It is as if a new set of articles was delivered to a clothing store, on the off-chance that one of them would fit the customer better than what was originally in stock.

‡ In fact the term is retained even though we now know that some B cells in a clone may come to express receptors with a different specificity to that of the original cell.

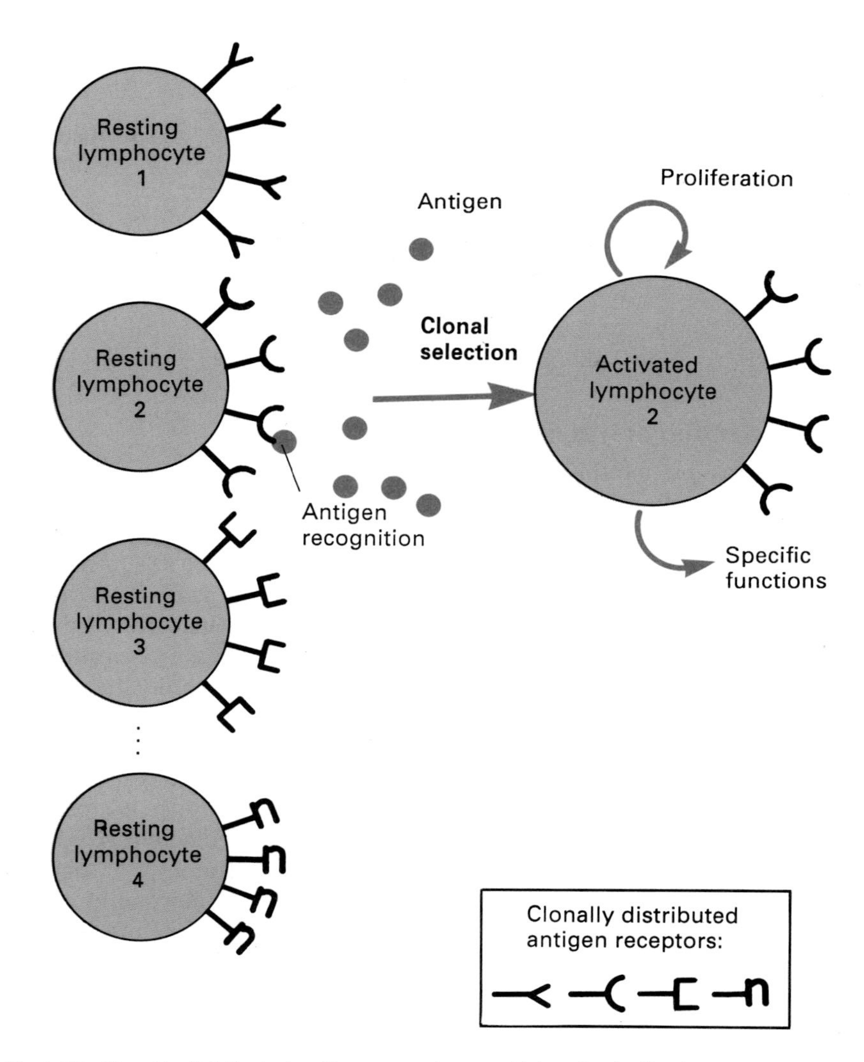

Fig.1.21 Clonally distributed antigen receptors and 'clonal selection'

nize many antigens, the response to a particular virus, for example, would involve many or all of the lymphocytes in the body, and this would clearly be an extravagant and wasteful process. If only a limited number of lymphocytes usually responds to a given antigen in any particular immune response, this is a much more efficient use of resources.

A corollary to the idea of clonally distributed antigen receptors (above) is that there must be an enormous number of lymphocytes present in the immune system to generate the immense diversity of responses. In fact, the total number of antigen receptors actually available within the immune system is probably of the order of at least hundreds or thousands of millions, and it has been estimated that several orders of magnitude more could potentially be produced. There are some 2×10^{12} (two million million) lymphocytes spread throughout the human body, although some are more concentrated in certain areas, the lymphoid tissues. Put together, the total mass of lymphocytes in the body is roughly similar to that of the brain or liver. Since immune responses are so diverse, and they can be generated against a remarkably large number of different antigenic determinants, the immune system must contain an equally immense variety or **diversity** of antigen receptors—the **immune repertoire**.

Most specific immune responses involve relatively few of the enormous number of clones of lymphocytes that exist, and are therefore said to be **oligoclonal** (oligo = few) in nature. **Polyclonal** (poly = many) responses, involving a far greater number of clones, as well as **monoclonal** (mono = single) responses due to a single clone of lymphocytes, are less common but they do occur in some experimental and pathological situations.

1.1.4 Self and non-self discrimination

The immune system recognizes and eliminates things foreign to the body. Because this is primarily a destructive function, it is important that a distinction is made between what is normally present in the body and what should not be there, respectively described as **'self'** and **'non-self'** (foreign) components. If there were no capacity for **self versus non-self discrimination**, immune responses could be directed against constituents of the individual's own body and ultimately lead to 'self-destruction'.

There are a number of ways that self–non-self discrimination might come about. First, clones of lymphocytes with receptors that react with self components could be destroyed. Since those receptors are thus 'deleted' from the repertoire, this mechanism is termed **clonal deletion**. Alternatively, the function of those lymphocytes that are self-reactive could be inhibited, or **suppressed**, so the individual would not be able to make a response even if they were present. Suppression implies an active process—it is also possible that self-reactive lymphocytes could be present in a 'switched-off' mode, and this is referred to as **clonal anergy**. It is also conceivable there could be genetic constraints so that the immune system would not be able to produce lymphocytes with receptors for some self components, e.g. one might not have the genes for them, but this idea is largely speculative. An individual does not normally make immune responses to self components, the phenomenon of **tolerance**. From what we currently know about the receptors and responses of lymphocytes, it seems likely that self tolerance results primarily from clonal deletion and clonal anergy.

The concept of what is 'self' might seem obvious. Normally, this refers to the individual's own body. However, in immunology, a better definition of 'self' is *the particular environment in which the immune system developed* (Fig.1.22). To understand why, it is important to realize there are circumstances, both clinically and experimentally, when the immune system can develop in a different environment. An example is someone who has had a bone marrow transplant, perhaps because they have a condition like aplastic anaemia. As we said earlier (p.9), all cells of the immune system develop from the bone marrow, but the marrow of patients with aplastic anaemia is defective and they cannot produce these cells. However, they can often be cured if they receive a transplant of bone marrow from a normal individual (the donor). The cells which develop and reconstitute the recipient's immune system are, of course, genetically identical with the donor, but the recipient

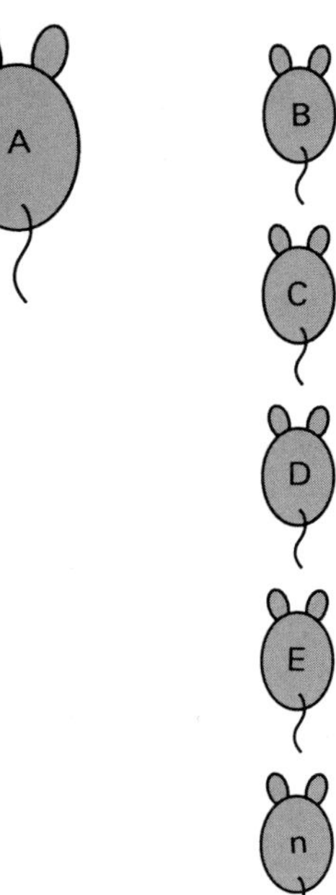
To mouse A,
its own body
is **self**
A
B
C
D
E
n
To mouse A,
all other
members of
the species
(B, C ... n)
are **non-self**

Fig.1.22 Self and non-self

Table 1.6 Responses to self and non-self (foreign) components

	Immune response generated in:	
	'Normal situations'	**Pathological situations**
Self-components	No (tolerance)	Yes (autoimmunity)
Non-self components	Yes (antigen elimination; graft rejection)	No (disease and/or death due to a pathogen that evades the immune response)

Syngeneic

Strain C Strain C

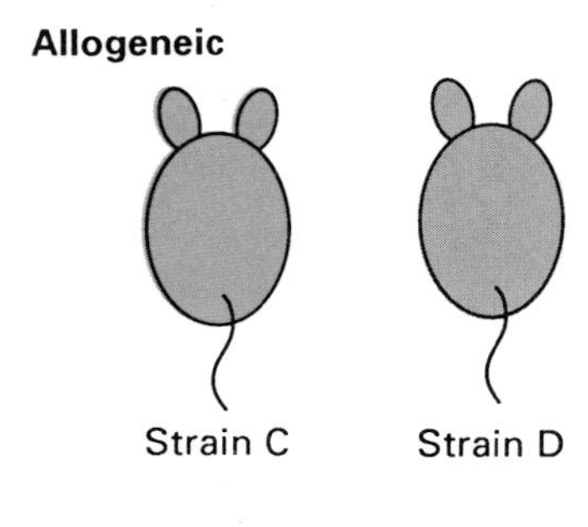

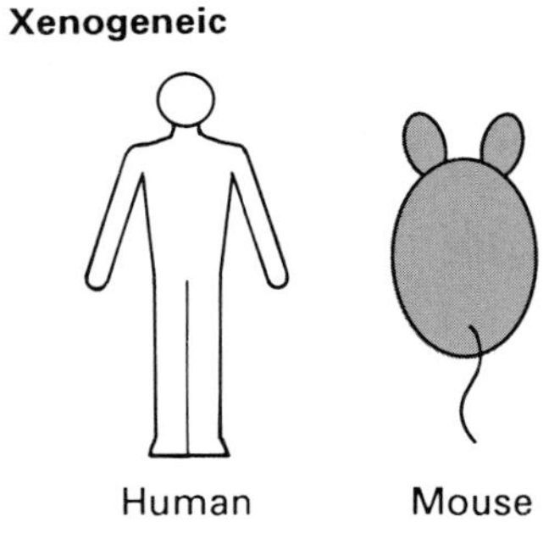

Fig.1.23 Genetic relationships

in whom they develop may be genetically different. As we shall see in detail elsewhere (Section 5.3), the immune cells which develop now recognize their new environment (the recipient's body) as 'self'. In the case of T cells, they are tolerant of the new 'self' MHC molecules (with or without peptides derived from the cells expressing them) but they *can* recognize peptides derived from foreign antigens bound to these self MHC molecules—the phenomenon of **self-restricted antigen recognition** (Table 1.7).

While the ability to discriminate self from non-self is vital for normal functioning of the immune system, there are occasions when this breaks down and self-components are attacked. The result of this 'breakdown of self-tolerance' is **autoimmunity** (Table 1.6). Not surprisingly this may be associated with severe clinical symptoms, and sometimes autoimmune diseases are fatal. In other cases the immune system reacts to non-self (foreign) antigens, but the response is somehow inappropriate or misdirected. This condition is called **hypersensitivity**, or **allergy**. One of the best-known examples is hay fever which is triggered by pollens, the symptoms of which are known to most people. As far as we know, it makes little biological sense for the immune system to mount such a vigorous response against an apparently innocuous agent like pollen, and in this respect the response seems inappropriate. Alternatively, an immune response to a particular infectious agent may result in damage to neighbouring tissues of the body and this is another cause of hypersensitivity reactions.

A final consideration is that a significant proportion of T cells in any normal animal can recognize non-self MHC molecules (and/or the bound peptides) from any genetically-different member of the species—the phenomenon of **alloreactivity**† (Table 1.7). Alloreactivity is thought to be responsible for the **rejection** of tissues transplanted between genetically dissimilar individuals of the same species.

† This term means 'reactive with allo . . .'. It derives from one of the terms used in immunology to describe the genetic relationship between different individuals: syngeneic, allogeneic (hence alloreactive), and xenogeneic. Genetically-identical individuals are said to be syngeneic, e.g. identical twins or members of an inbred strain of mice; genetically-different individuals of the same species are said to be allogeneic, e.g. most human beings; individuals of different species are said to be xenogeneic, e.g. mice in relation to humans (Fig.1.23).

Table 1.7 Features of T cell recognition and responses

MHC type: and definition	Phenomena: in absence of antigen	in presence of antigen
Self MHC: MHC molecules of the individual	Tolerance	Self-restricted (MHC-associative) recognition
Non-self MHC: MHC molecules of other members of the species	Alloreactivity	Allorestricted (MHC-associative) recognition

1.2 An overview of immune responses (Scheme 1)

1.2.1 A map of cellular and molecular interactions during immune responses

The aim of this section is to introduce some of the key players in the immune system very briefly in the context of immune responses in general. An additional function is to provide a simple scheme that can be used as a 'framework' for understanding the immune response (Scheme 1) and modified versions of this appear at the beginning of each of the following chapters. This cannot of course be really comprehensive, and other schemes could equally well have been presented, but it is hoped this framework will provide a map for orientation and a basis for thinking about cellular and molecular immunology. Here we consider, in a highly simplified form, what happens when an antigen, such as a pathogenic micro-organism, is present in the body. Each aspect of the immune response that results is discussed in relation to the framework provided by Scheme 1 (see p. 2). The area of the scheme under discussion is indicated in the text, e.g. **1**.

1.2.2 Innate responses (shaded elements of Scheme 1)

Pathogenic organisms such as viruses and bacteria can enter the body when tissue damage occurs. The body's initial response to injury and trauma is manifest as **inflammation**, a function of innate immunity. One consequence of the inflammatory response is the entry of soluble molecules like **complement** components (**1**), and the recruitment of cells such as **phagocytes** (**2**), from the blood into the site of injury.

Some pathogens like certain yeasts and bacteria can be phagocytosed by cells such as macrophages and can then be destroyed intracellularly (Fig.1.24). In addition, complement can be activated directly on the surface of certain organisms, via the so-called alternative pathway, causing pores to be produced in the cell wall or outer membrane and resulting in death of the organism by osmotic lysis. Phagocytosis can be promoted by activated

Uptake by phagocytes and intracellular destruction

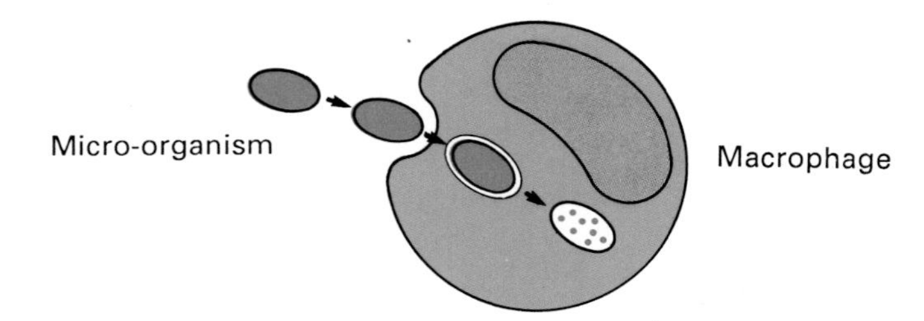

Activation of complement via the alternative pathway and lysis of the micro-organism

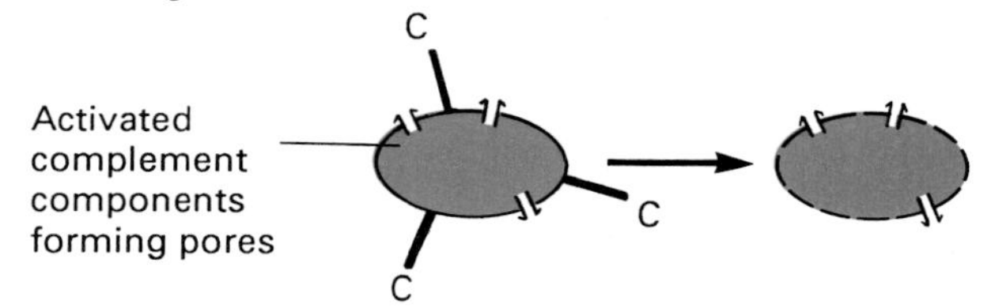

Activation of complement via the alternative pathway, opsonization, and uptake by phagocytes

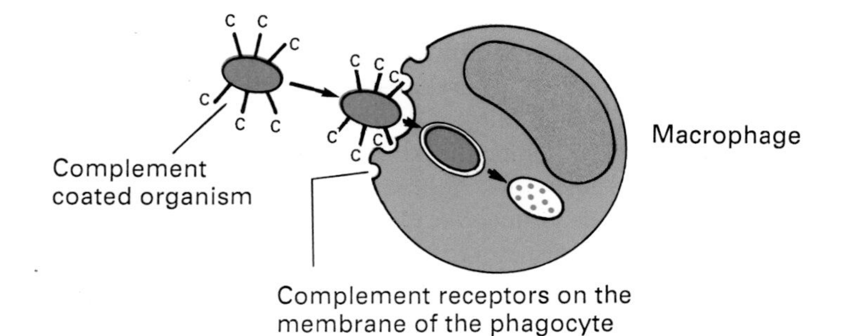

Killing of virus-infected cells by natural killer (NK) cells

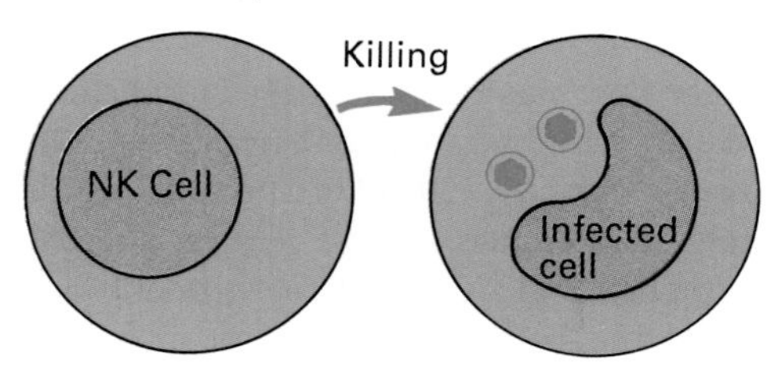

Fig.1.24 Mechanisms for elimination of pathogens during innate immune responses

complement components when they are bound to the surface of an organism, providing an additional mechanism for uptake, called **opsonization**. This process can result also in the elimination of the organism in case it does not trigger the alternative pathway of the complement cascade, or if the organism has evolved mechanisms to avoid internalization by phagocytic cells directly. Should any cells of the body become infected, for example at the point of entry of a virus, they may be killed by **natural killer cells** (**3**), thus limiting the spread of infection (Fig.1.24).

It seems likely that certain micro-organisms, or various antigenic portions of them, can also be taken up at sites of inflammation and infection (e.g. skin) by a specialized group of cells belonging to the lineage of **dendritic cells** (**4**). These cells then enter the lymphatics or bloodstream and travel to lymphoid tissues, the lymph nodes or spleen respectively (see Section 1.3), where they can initiate the adaptive immune response.

1.2.3 Adaptive responses (non-shaded elements of Scheme 1)

Until a micro-organism has been destroyed completely, or if it escapes attack by the innate arm of the immune system, the antigen may be present in the body in two basic forms. First, antigen may be present in body fluids either as free particles, like intact virions, or in soluble form, like bacterial cell wall products and toxins. Second, it may be present inside cells, either because the micro-organism has been internalized by cells such as phagocytes, which may or may not have been able to destroy it, or in the case of a virus for example, because it has successfully infected cells of the body. The problem posed for the immune system is how to recognize free antigens as well as cells that contain foreign antigens within them. The solution, apparently, has been the evolution of the two different types of lymphocyte—T cells and B cells (Table 1.4) (**5**, **6**, and **7**).

Because T cells cannot, as it were, 'look inside' cells of the body to see if antigen is present, essentially every cell needs to be able to display foreign antigens, when present, on the plasma membrane where they can be recognized by T cells. In this sense the cell is said to **present** or show antigen to T cells, and at this stage the cell is referred to as an **antigen-presenting cell** (**2A**, **4**, **7**, and **8**).

Inside antigen-presenting cells, protein antigens are degraded to peptides by mechanisms termed **antigen processing**. Processed antigen, in the form of one of a number of small peptides, then becomes bound to a particular type of molecule within the cell, an **MHC** molecule: this can then be transported to the cell surface as a **peptide–MHC complex**. A characteristic and unique feature of **antigen recognition** by **T cell receptors** is that they specifically recognize peptide–MHC complexes, and hence cell-associated antigens. This is known as **MHC-restricted antigen recognition** and it contrasts with recognition by antibodies on B cells which is not restricted to MHC-associated antigens (Fig.1.25). Other molecules on the surface of the T cell also play important roles in antigen recognition, cellular activation, and subsequent responses. A molecular complex called **CD3** is associated with the T cell receptor. Another molecule called **CD4** is present on many helper T cells (**5A**, **5B**) and **CD8** is expressed by many cytotoxic T cells (**6A**, **6B**) respectively. These molecules are sometimes called 'phenotypic markers'

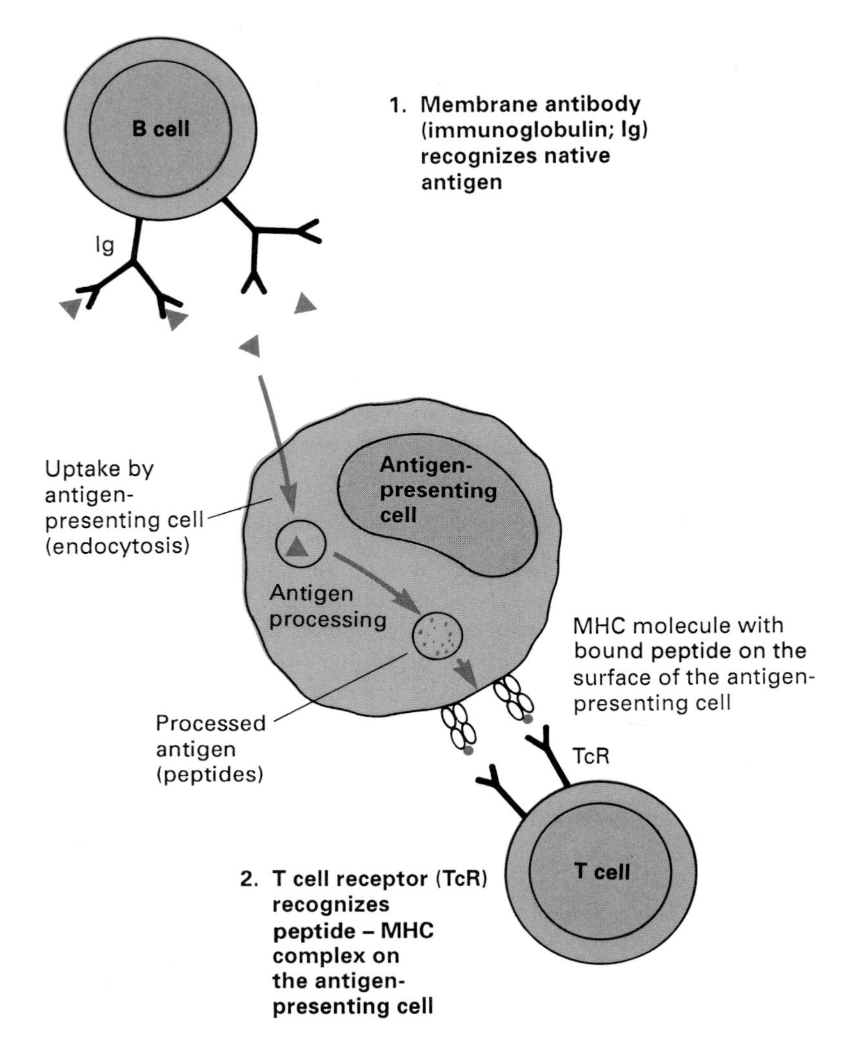

Fig.1.25 B cells can recognize native (unmodified) antigens; T cells recognize peptide–MHC complexes on antigen presenting cells

or 'markers' of T cells, because possession of the molecule can be used to define the cell type, e.g. **CD3** is a marker of T cells.

Helper T cells are considered to be central for initiation of much of the adaptive immune response. In general, this subset of T cells (**5A**) *initially* recognizes antigenic peptides associated with one of the two types of MHC molecule, **MHC class II**, on the surface of a specialized subset of antigen-presenting cells, **dendritic cells** (**4**). These cells apparently acquire antigens at a precursor stage before migrating to lymphoid tissues (e.g. lymph nodes and spleen, Section 1.3) where they interact with the antigen-specific helper T cells. Following recognition of peptide–MHC complexes on the dendritic cell, signals are delivered to the helper T cell which cause it to develop into an **activated helper T cell** (**5B**). These T cells then become able to 'help' immune responses by secreting soluble molecules called **cytokines** which act on other cells of the immune system. These include **B cells** (**7A**), **precursor cytotoxic T cells** (**6A**), **NK cells** (**3A**), and **macrophages** (**2A**), all of which are effector cells (p. 17) capable of participating in the elimination of pathogenic micro-organisms.

In some cases the activated helper T cell (**5B**) needs to interact directly with the MHC class II-expressing antigen-presenting cell (**7A, 2A**), and the cells come into very tight and physical contact with each other. (Before it is activated, the helper T cell (**5A**) seems unable to form such associations.) This interaction can have a number of consequences.

1. **B lymphocytes** (**7A**) can become antigen-presenting cells in part because they can internalize the antigens that are recognized and bound via antibody molecules on their cell surface. The internalized antigen is then degraded and the resulting peptides presented in association with MHC class II molecules at the cell surface. In response to cytokines produced by activated helper T cells (**5B**), B cells (**7A**) become activated and may develop into **plasma cells** (**7B**). These cells secrete antibodies that can be of different types (or classes), some of which can bind to antigens and opsonize them so they can be taken up by phagocytes (**2A**) via membrane receptors called **Fc receptors** which are specific for certain classes of antibody. Other classes of antibody, when bound to antigens, can activate complement (**1**) by the so-called classical pathway and bring about the lysis of micro-organisms, or their opsonization and internalization via **complement receptors** on phagocytes (**2A**). Thus, the production of antibodies during the adaptive immune response provides an additional means of eliminating antigens that may have escaped the innate response (Section 1.2.2). Even if a pathogen such as a virus is not directly eliminated in this manner, binding of antibodies can prevent its entry into cells of the body and thereby inhibit viral replication.

2. Macrophages (**2A**) become activated (**2B**) and acquire a dramatically enhanced capacity for destroying the intracellular micro-organisms they may have phagocytosed, either directly or via Fc, complement, or other membrane receptors.

3. Certain other antigen-presenting cells that can express MHC class II molecules may develop into, or become better, effector cells, with a greater capacity to destroy intracellular pathogens.

   In addition to the effects noted above, which can result from direct cellular interactions between activated helper T cells and MHC class II-positive antigen-presenting cells, the cytokines produced can also act on other cell types that may be unable to interact directly with helper cells. This can have a number of other consequences (4 and 5 below).

4. Precursor **cytotoxic T cells** (**6A**) that may have encountered peptides bound to the other type of MHC molecule, **MHC class I**, on the surface of cells containing pathogens (such as virally-infected cells, e.g. **8**) develop into **cytotoxic T lymphocytes** (**6B**). When the mature cytotoxic T cell recognizes antigen–MHC class I complexes, it can secrete another group of soluble molecules called **cytolysins** on to the antigen-presenting cell, in this case termed a **target cell** (**8**). Cytolysins bring about the destruc-

tion of the target cell, which may not have the capacity to deal with intracellular pathogens. In the case of a virally infected cell, this obviously prevents the ability of the virus to replicate and reduces its capacity to infect other cells.

5. Natural killer cells (**3A**) can be activated (**3B**), and their ability to kill certain types of target cell, perhaps including tumour cells, as well as some micro-organisms, is enhanced.

In general, antigens taken up from the extracellular environment by cells such as macrophages (e.g. by phagocytosis) or B cells (via their surface antibody molecules), are called **exogenous** antigens. Peptides produced from exogenous antigens tend to associate with MHC class II molecules (Fig.1.26). On the other hand, antigens that are produced within the cell (e.g. viral antigens) are called **endogenous** antigens, and their peptides tend to associate with MHC class I molecules. Because, generally, helper T cells recognize peptide–MHC class II complexes, and cytotoxic T cells recognize peptide–MHC class I complexes (Fig.1.27) it is possible for the immune system to discriminate between cells that need to be helped in their effector functions (**2A**, **7A**), or be killed (e.g. **8**). A summary of some mechanisms used for elimination of pathogens is shown in Fig.1.28.

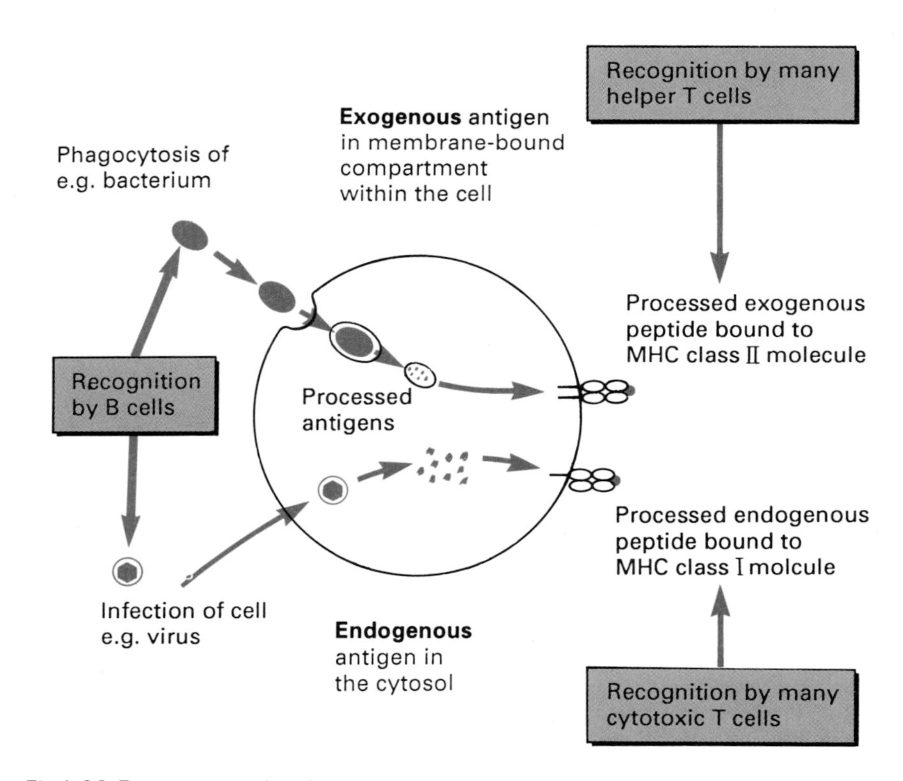

Fig.1.26 Exogenous and endogenous antigens generally associate with different classes of MHC molecule and be recognized by different types of T cell

Lymphocyte activation occurs in the so-called **afferent phase** of the adaptive immune response, and this necessarily precedes the **efferent** or **effector phase** during which the pathogenic micro-organism, or other antigen, is eliminated. The immune response then either wanes of its own accord or is switched off by suppressor mechanisms that may include the production of inhibitory cytokines and/or the action of cells such as suppressor cells whose function is poorly understood. However, memory cells will have been generated in the course of the primary immune response and these can mount a secondary response if the antigen is encountered at some time in the future.

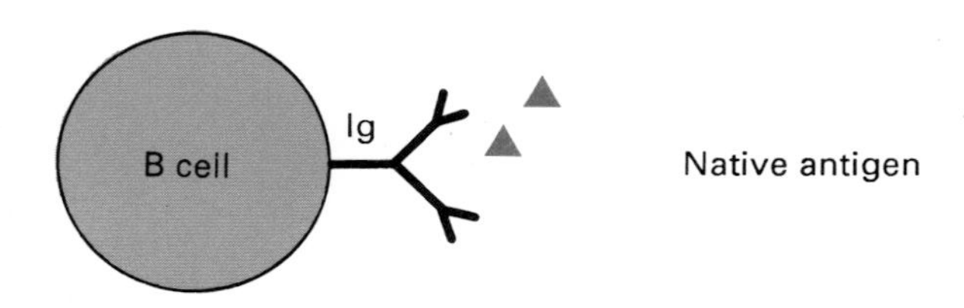

B cells tend to recognize free, native antigens

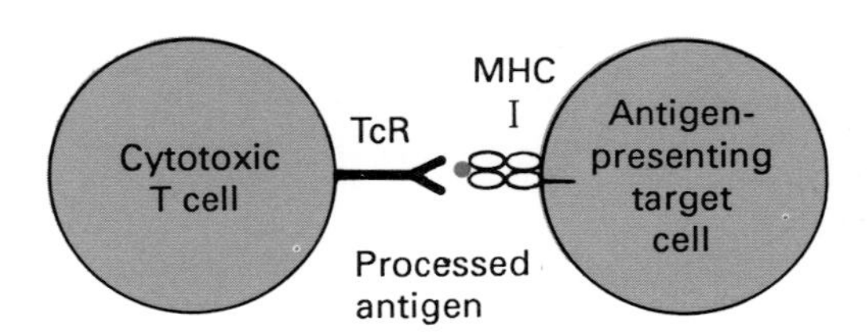

Cytotoxic T cells tend to recognize peptides bound to MHC class I molecules which are expressed on most cells of the body

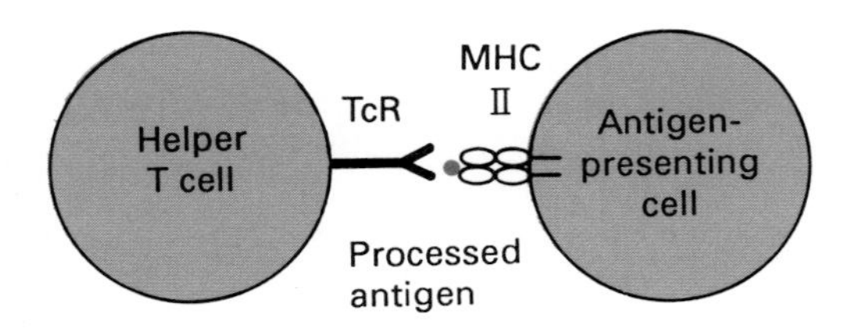

Helper T cells tend to recognize peptides bound to MHC class II molecules which are expressed mainly on cells that can develop effector functions

Fig.1.27 Antigen recognition by B cells, cytotoxic T cells, and helper T cells

a
Binding of antibody, opsonization, and uptake by phagocytes

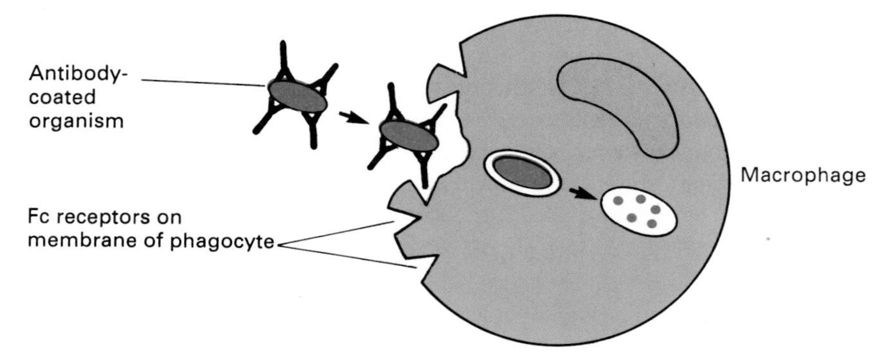

Binding of antibody, activation of complement via the classical pathway, and lysis of the organism

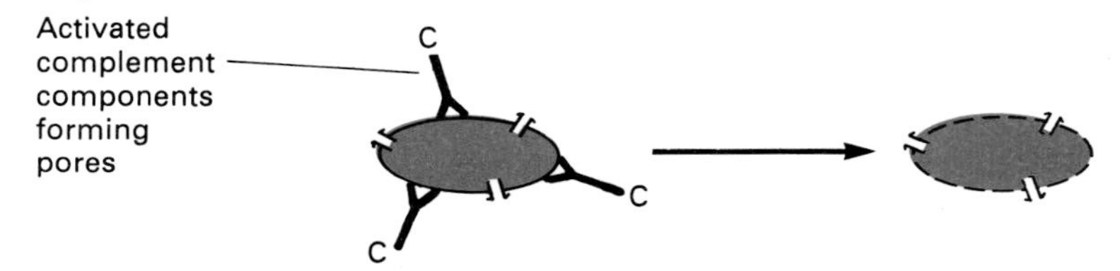

Binding of antibody, activation of complement, opsonization, and uptake by phagocytes

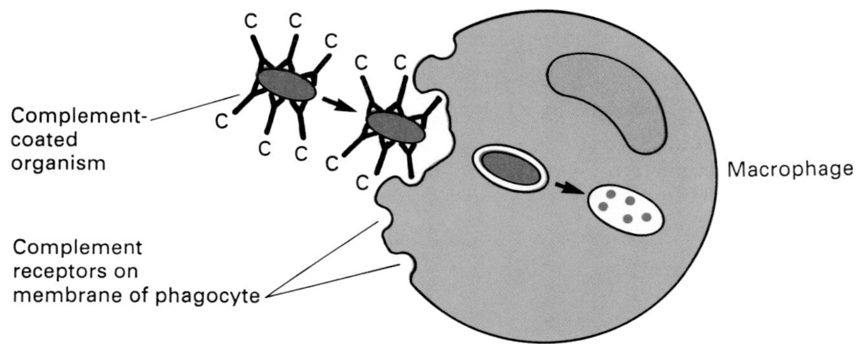

b
Cells containing antigens, such as virus-infected cells, can be recognized by cytotoxic T cells, which produce cytolysins that mediate their killer function

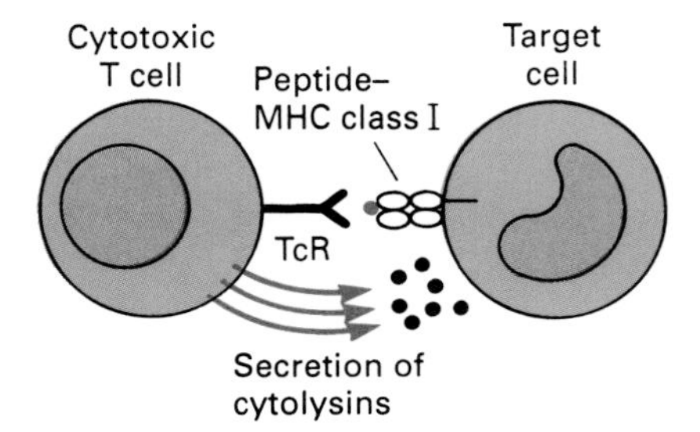

Cells containing antigens that can develop effector functions, such as macrophages, can be recognized by helper T cells, which produce helper cytokines that act on the antigen-presenting cell to induce or increase its effector capacity

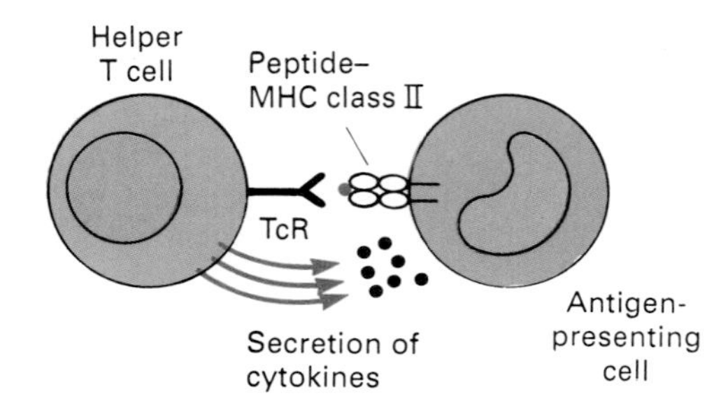

Fig.1.28 Some mechanisms for elimination of pathogens during adaptive immune responses

1.3 Lymphoid tissues

1.3.1 Bone marrow origin of cells of the immune system

All cells of the immune system are haemopoietic cells (the older terminology leukocytes) which originate from just one cell type in the bone marrow of adult animals during the process of **haemopoiesis** (=blood formation). This cell is called the **haemopoietic stem cell** and because it has the potential to produce *all* these cell types it is sometimes referred to as a **totipotential** (toti=all) stem cell. Since it gives rise to many different cell types, of different **lineages**, it is also referred to as a pluripotent, or sometimes multipotential, stem cell.

Each stem cell undergoes several rounds of division and thereby multiplies. However, during a particular division, one of the two daughter cells may become irreversibly changed so that it will eventually give rise to a different cell type (or to a more limited number of cell types). During this process of **differentiation** particular combinations of genes become switched on or off as the cell divides, and one of the daughter cells becomes **committed** to a particular lineage. The other daughter cell is another stem cell. This is an unusual mode of cell division because it is **asymmetric** (Fig.1.29). Stem cells are very special, because normally when a cell divides the progeny are the same. Hence, a loose definition of a stem cell is a cell that can give rise to one daughter cell, that begins differentiation into another cell type, and another that produces more stem cells. Stem cells therefore have the capacity for **self-renewal**.

What we have said so far applies to the adult. Before this stage, in many species, stem cells are successively found in the blood islands of the yolk sac, in the embryonic liver and fetal spleen, and finally in neonatal bone marrow

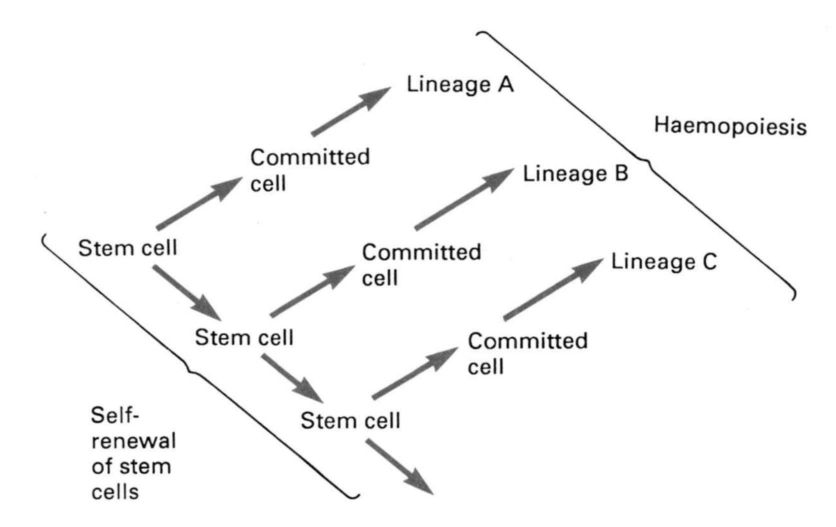

Fig.1.29 Asymmetric cell division of stem cells and haemopoiesis

(Fig.1.30). So the formation of blood cells frequently begins in the liver and then shifts sometimes to the spleen and then to the bone marrow. However, there are species differences. For example, in addition to bone marrow the spleen remains a **haemopoietic organ** in adult mice but not in humans.

Other stem cells in adult bone marrow also have the capacity for self-renewal, but give rise to a more restricted number of cell types. Nevertheless these are all derived from the earliest, pluripotent stem cell. For example, there is a more restricted stem cell committed to producing mononuclear and polymorphonuclear phagocytes, that is, cells of the **myeloid** lineage, and there may be another that produces the lymphocytes of the **lymphoid** lineage. A different lineage of **erythroid** cells includes the red blood cells (erythrocytes) and their precursors. A summary of haemopoietic stem cell differentiation is shown in Fig.1.31.

After a period of development in the bone marrow, the more fully differentiated cells enter the bloodstream and become distributed throughout the body to various tissues, where some of them may undergo further differentiation or other changes which are loosely referred to as **maturation** (Fig.1.32). For example, the macrophage differentiates from earlier stages in the bone marrow that include the monoblast and promonocyte, and it leaves the marrow as a blood monocyte. When the monocyte leaves the blood and enters the tissues, it becomes a mature macrophage, the characteristics of which depend on where it is sited: if it goes to the lung it becomes an alveolar macrophage, whereas if it goes to the liver it becomes a Kupffer cell, or in the brain a glial cell.

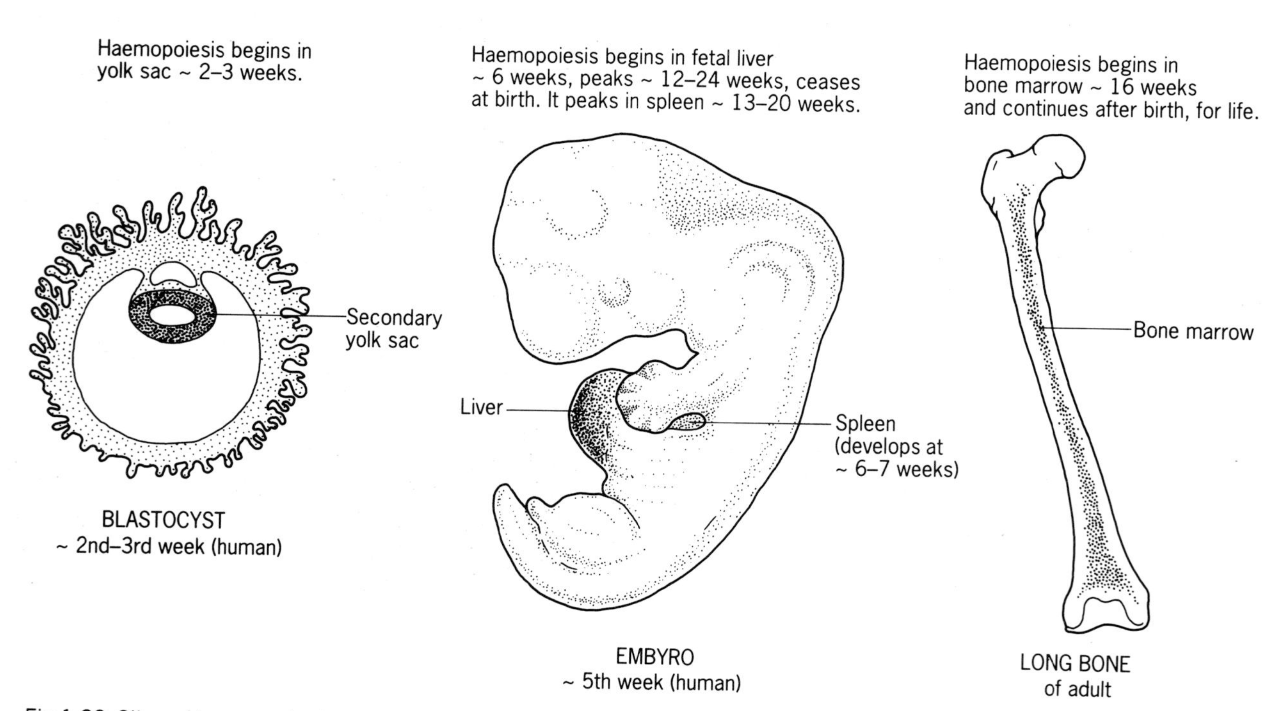
Haemopoiesis begins in yolk sac ~ 2–3 weeks.
Secondary yolk sac
BLASTOCYST
~ 2nd–3rd week (human)
Haemopoiesis begins in fetal liver ~ 6 weeks, peaks ~ 12–24 weeks, ceases at birth. It peaks in spleen ~ 13–20 weeks.
Liver
Spleen (develops at ~ 6–7 weeks)
EMBYRO
~ 5th week (human)
Haemopoiesis begins in bone marrow ~ 16 weeks and continues after birth, for life.
Bone marrow
LONG BONE
of adult

Fig.1.30 Sites of haemopoiesis during development

In adult mammals, B lymphocytes develop in the bone marrow before they enter the blood as mature cells (Fig.1.32). T lymphocytes are different, since T cell precursors leave the marrow and travel via the blood to a specialized organ, the **thymus**, where they differentiate further. It is believed the thymus is initially colonized by one or a few of these cells which then proliferate extensively to give rise to mature T cells, development being partly under the control of the **thymic hormones** which are produced within this organ. Once mature, T cells leave the thymus and enter the circulation or the **periphery**.

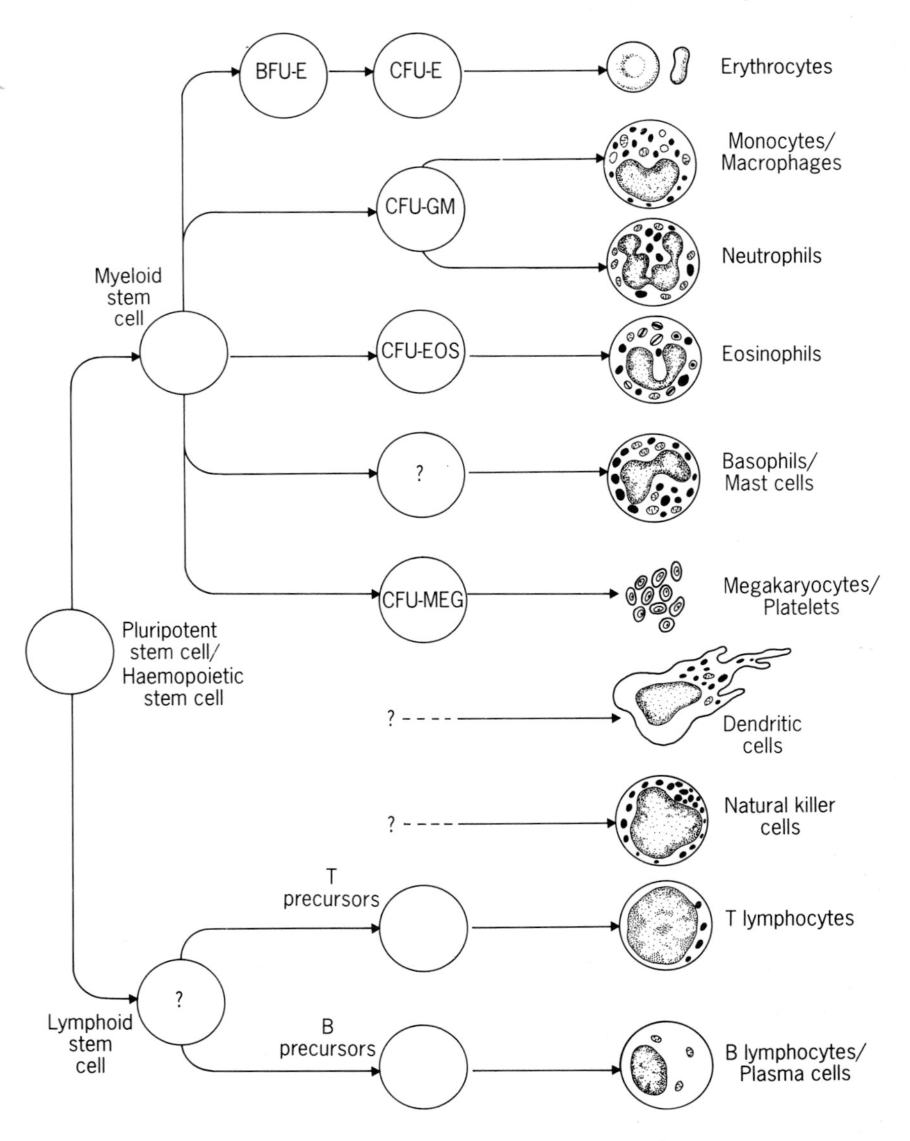

Fig.1.31 Bone marrow origin of cells of the immune system. BFU-E, burst-forming unit, erythroid; CFU-E, colony-forming unit, erythroid; CFU-GM, colony-forming unit, granulocyte-macrophage; CFU-EOS, colony-forming unit, eosinophil; CFU-MEG, colony-forming unit, megakaryocyte-platelets

T cells do not develop in the absence of a functional thymus. This situation is found in humans in certain clinical conditions, and in 'nude' mice which are frequently used experimentally. In these cases, not surprisingly, T cell responses are virtually non-existent. B cells are present but their responses are also impaired, mainly because T cells are often required to 'help' B cells make responses. *T* cells were so named precisely because their development is dependent on the *t*hymus.

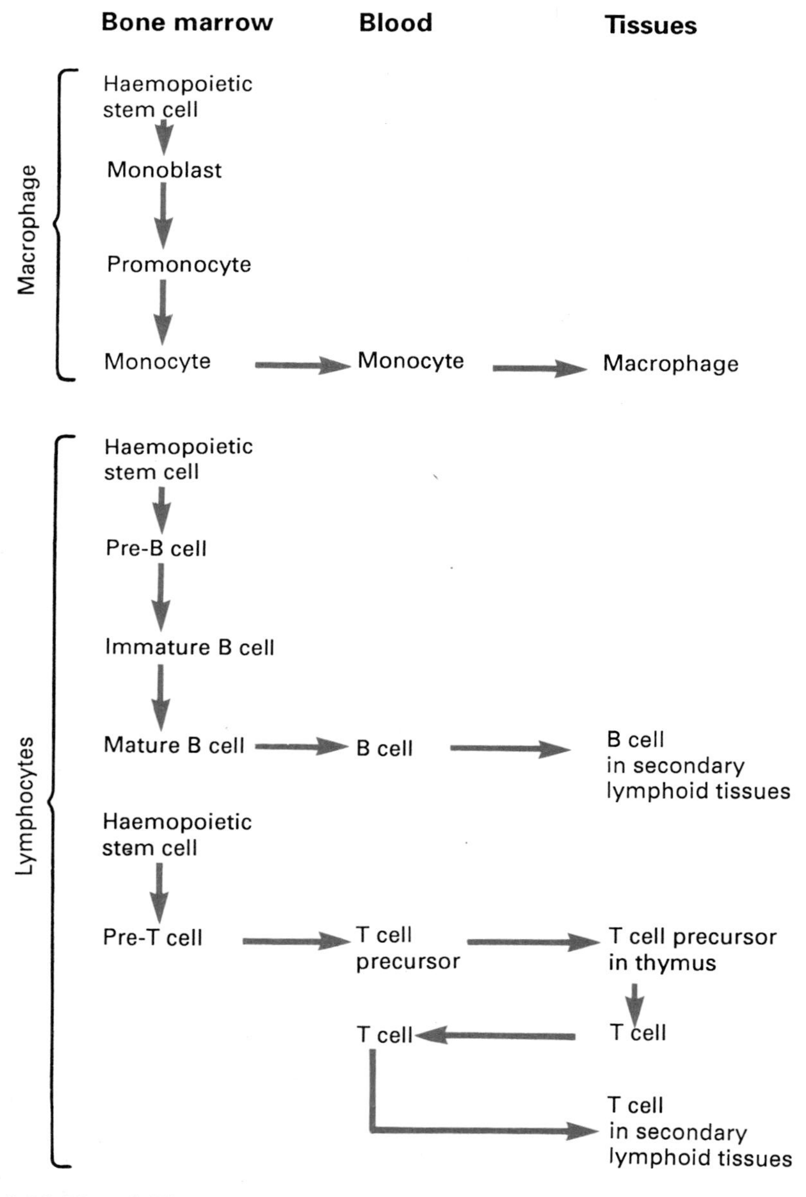

Fig.1.32 Sites of differentiation and maturation of macrophages and lymphocytes (adult mammals)

In birds, B lymphoctyes develop in a specialized organ called the *b*ursa of Fabricius, hence the term *B* cells. Like the thymus, this tissue is colonized by precursors which develop within its unique environment. Depending on their precise localization in this organ, some precursors differentiate into myeloid and erythroid cells, while others develop into B cells. If the bursa is surgically removed in embryonic life, B cells and antibodies are not produced. There is no known equivalent to this organ in mammals, where the bone marrow appears to have assumed its role, but the term B cell has been retained.

1.3.2 Primary and secondary lymphoid tissues

The bone marrow and thymus, the bursa of Fabricius in birds and, unusually, Peyer's patches in sheep, are called **primary** (or **central**) **lymphoid tissues** because mature lymphocytes are actually produced within these organs (Fig.1.33). The mature T and B cells then enter the blood and travel to other sites called the **secondary** (or **peripheral**) **lymphoid tissues**. These include the **spleen** and **lymph nodes**, as well as specialized **mucosal-associated lymphoid tissue** (abbreviated MALT) which consists of **Peyer's patches** in the gut (also termed gut-associated lymphoid tissue or GALT) and Waldeyer's ring, which includes the tonsils and adenoids.

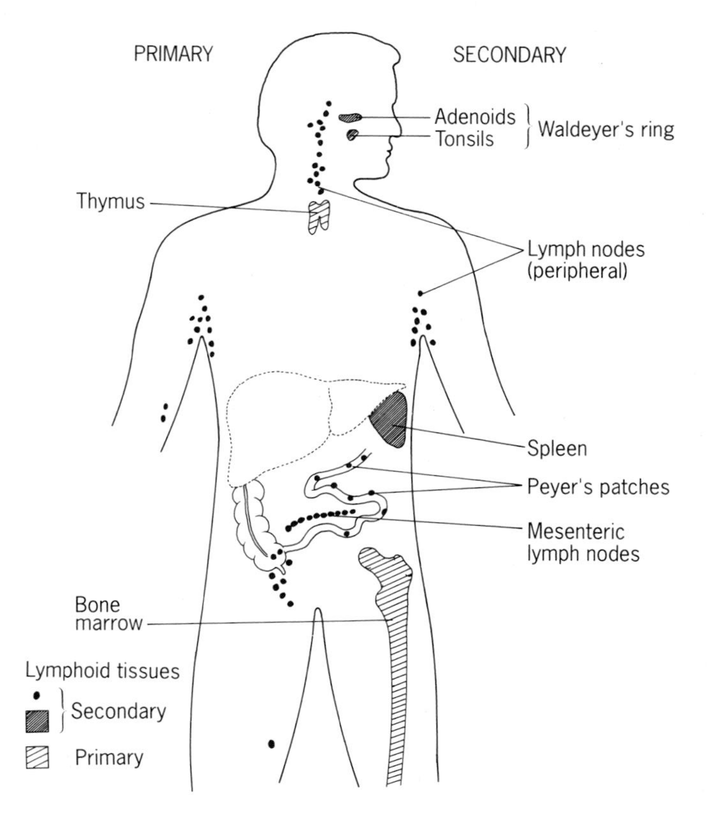

Fig.1.33 Primary and secondary lymphoid tissues

There is often no absolute distinction between primary and secondary tissues. For instance, in mammals the bone marrow is a secondary as well as a primary lymphoid tissue, since it contains some mature lymphocytes, while in rodents the spleen is a primary as well as secondary lymphoid tissue because it is a site of haemopoiesis.

Lymphoid tissues have beautifully organized structures which are presumably critical for their function. In primary tissues, like thymus, this seems likely to be required so that the developing T cells (thymocytes) can receive the appropriate signals for differentiation into mature T cells. The structures of secondary tissues are thought to promote efficient contact of mature B cells and T cells with antigens and antigen-presenting cells for the generation of lymphocyte responses. We understand relatively little about these processes *in situ*, that is, as they occur in intact tissues and live animals. Many of the current ideas in immunology have come from experiments carried out with isolated cells *in vitro*, that is, in culture after the tissues have been removed from the animal and dissociated. This is important to remember if one tries to extrapolate from results obtained *in vitro* to an understanding of the whole animal.

1.3.3 Lymphocyte recirculation

There are two circulatory systems in the body, the blood and the lymph (Fig.1.34). Blood, as we know, circulates throughout the body in vessels as large as the aorta and venae cavae, down to the size of the smallest capillaries, and in so doing it perfuses all the tissues. Fluid leaves the blood vessels and enters the tissues as extracellular fluid. Some of this fluid drains from the tissues and is collected into vessels called **lymphatics**, which connect with each other and become progressively larger. The fluid in the lymphatics is called **lymph**. **Lymph nodes** are found at the points where the lymphatics join each other, and they frequently occur in chains which form a network throughout the body (Fig.1.34). Often several vessels, called **afferent lymphatics**, bring the lymph to a particular node, but usually only one vessel, called the **efferent lymphatic**, leaves the node. Eventually the lymph finds its way into the **thoracic duct** which joins the left subclavian vein (just below the level of the neck) and this is where the lymph enters the blood. The afferent lymph is important for transporting antigens, in one form or another, from the tissues to the nodes where immune responses are generated. These antigens may be transported within, or on the surface of, specialized cells that play a critical role in triggering immune responses, **veiled cells**. These cells are found primarily in afferent lymph, and belong to the family of dendritic cells.

The blood transports several different cell types with various functions. For instance, erythrocytes circulate and transport oxygen in the blood until they are eventually removed by the liver and spleen at the end of their life. Polymorphonuclear leukocytes are transported to sites of inflammation. These cells are rapidly produced in the bone marrow and released into the blood in a mature form in response to inflammatory stimuli. Monocytes are also carried in the blood to inflammatory sites as well as to normal tissues, where they mature into tissue macrophages as mentioned earlier (Fig.1.32).

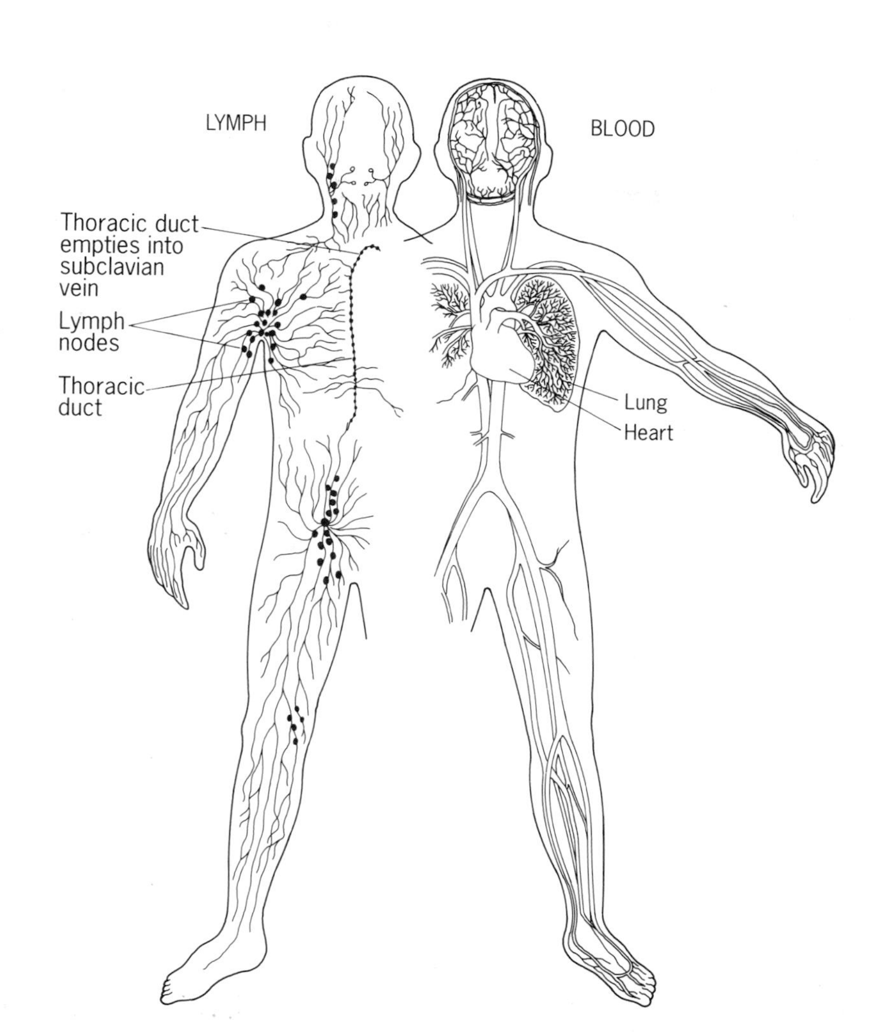

Fig.1.34 Blood and lymph

A notable and unique feature of lymphocytes is that they can cross from the blood directly into the lymphatic circulation and, of course, they find their way back into the blood via the thoracic duct. The traffic of lymphocytes from blood to lymph, and thence back to the blood is referred to as **lymphocyte recirculation** (Fig.1.35). There are specialized sites in the body where lymphocytes leave the blood and enter the lymph during the process of recirculation. These sites are called **high endothelial venules** (abbreviated to **HEV**) which are found especially in lymph nodes and Peyer's patches. The HEV are blood vessels in which the endothelial cells that line them have a different structure to those lining blood vessels in other areas, although endothelium in other sites of the body can assume a similar structure in response to inflammation. Even though this is a major site of lymphocyte traffic from blood to lymph, the presence of HEV and even lymph nodes is not essential for lymphocyte recirculation. It occurs in sheep via lymph nodes which do not have HEV, and it also occurs in birds and fish which do not

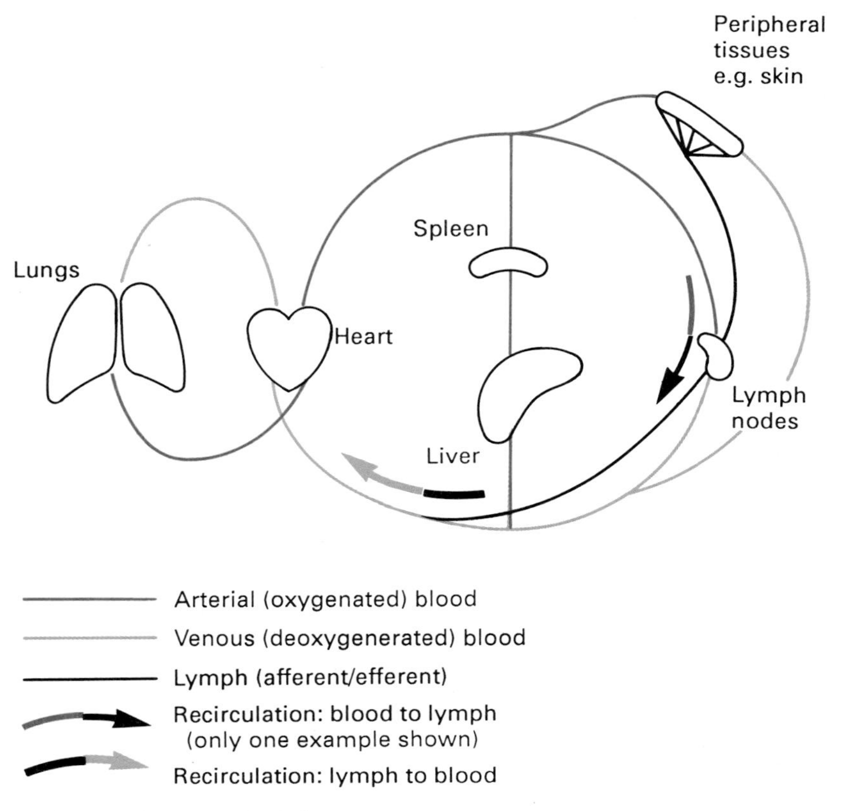

Fig.1.35 Lymphocyte recirculation

possess lymph nodes. Thus, the functional significance of the HEV structure is uncertain.

Lymphocytes, and other leukocytes, have special membrane molecules called **homing receptors** which allow them to bind to particular HEV and other endothelia. As a result, lymphocytes adhere to the walls of vessels in a process called **margination** and then pass between the cells of the blood vessels, in a process called **diapedesis**, into the surrounding tissues, from whence they return via the lymph back to the blood (Fig.1.36).

The movement, or **traffic**, of cells through the body to certain areas is called **migration**. T cells have a tendency to **migrate** to particular areas. If one isolates T cells from the spleen, when they are injected back into the blood a large number of the cells return to this organ rather than to other sites. Likewise, T cells isolated from peripheral lymph nodes (i.e. those nodes draining sites other than the gut) preferentially go back to their site of origin. The directed migration of cells to specific tissues, or regions within those tissues, is called **homing**. B cells, as another example, tend to **home** to the spleen and Peyer's patches.

There are important differences in the ability of lymphocytes to migrate and recirculate depending on whether they are resting, activated, or memory cells (Fig.1.13). What we have just discussed pertains especially to resting cells. However, lymphocyte migration is arrested when antigen enters a

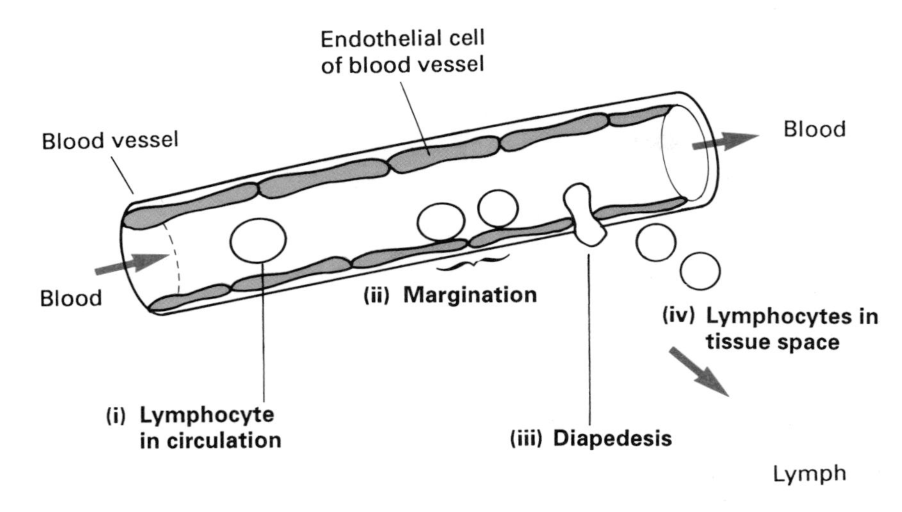

Fig.1.36 Traffic of lymphocytes from blood to lymph

lymph node. The specific lymphocytes are probably retained in this way to maximize the chance they will come into contact with antigen-presenting cells bearing the appropriate antigens, and be activated. Activated lymphocytes may temporarily lose their ability to recirculate. Memory cells, however, are long-lived cells that probably have an increased ability to recirculate from blood to lymph.

A special feature of lymph nodes is that they have both a blood and a lymph input. A large number of lymphoctyes enter nodes from the blood. **Veiled cells** which are thought to carry antigens enter the node via the afferent lymph. Particulate materials that enter the node via the afferent lymph are cleared by phagocytes that line their points of entry.

1.3.4 The structures of some lymphoid tissues

Thymus—a primary lymphoid tissue (Fig.1.37)

The thymus is where T lymphocytes develop from their precursors. Why there should be a specialized organ for T cell development is not known but there seems little doubt that within the thymus the repertoire of receptors carried by mature T cells is selected, and thus how they recognize antigens is determined (see Sections 1.1.4 and 5.3).

The thymus is a bilobed structure in the chest cavity (Fig.1.37). One surface is attached to the superior surface of the heart, another to the chest wall. It is relatively large in young animals but it decreases in size proportionately with age. The thymus is divided into many **lobules** by connective tissue which forms **trabeculae**, and each lobule contains an outer **cortex** and an inner **medulla**. The lobules are filled with **thymocytes**, which are predominantly immature T cells at various stages of development. Most of the immature T cells are found in the cortex, while the more mature or fully developed T cells are localized in the medulla. Hence, one view of thymocyte maturation holds that it progresses from the cortex inward to the medulla, although this is not universally accepted.

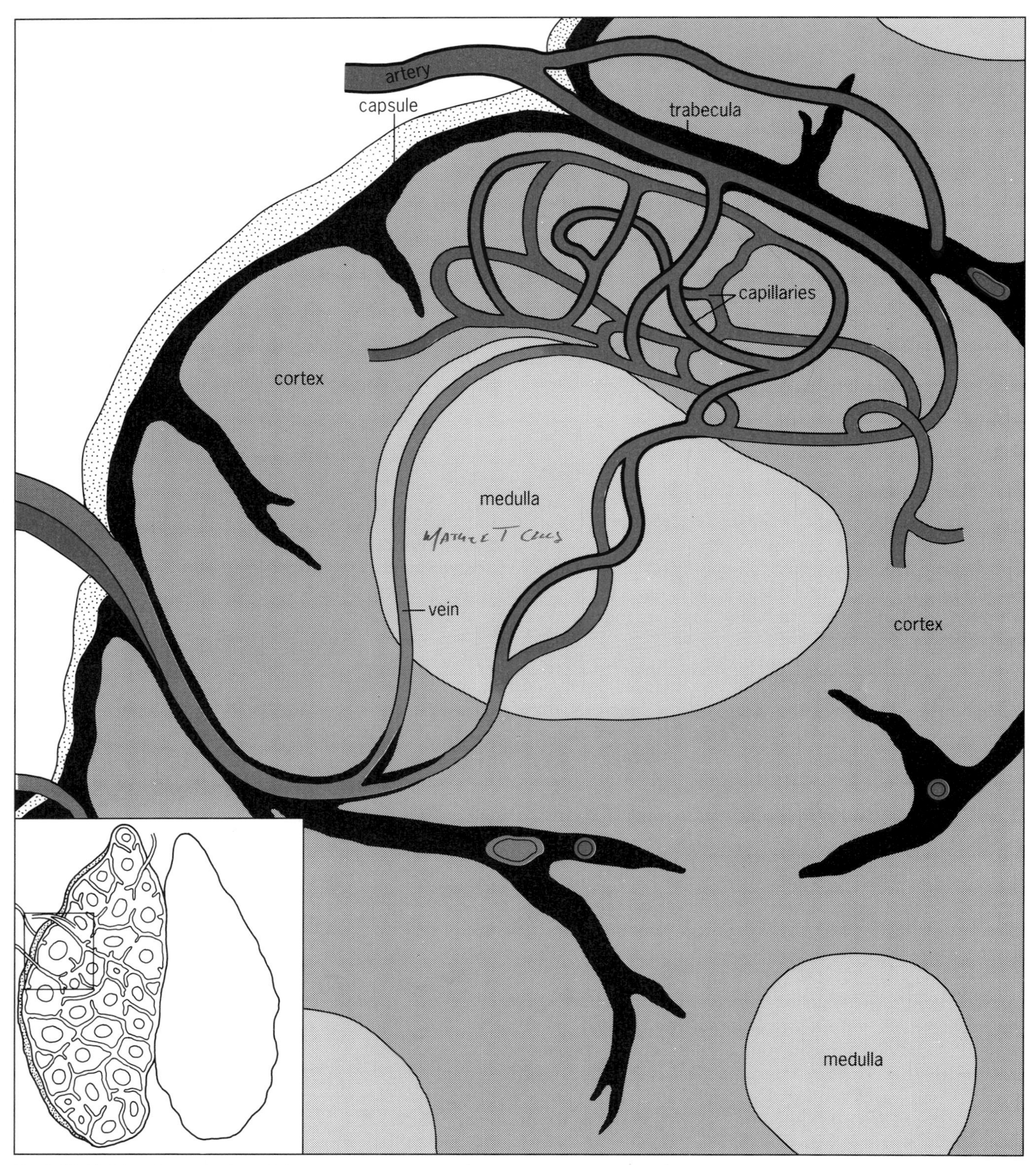

Fig.1.37 Highly schematic view of **thymus** structure (e.g. of mouse or human). Arteries enter the thymus and run within the trabeculae, which are continuous with the capsule and divide the thymus into lobules. The arteries then leave the trabeculae and enter the cortex to form a complex network of capillaries. The endothelium of these vessels is modified to constitute a blood–thymus barrier that is impermeable to the passage of most soluble molecules from the plasma; the cortical thymocytes, which are relatively immature, thus develop in a secluded environment. Blood then leaves the cortex and is carried in veins mostly within the corticomedullary junction, between the cortex and the medulla, which contains the more mature thymocytes, before leaving the thymus

Throughout each lobule, **epithelial cells**, which form an extensive network, are closely intimated with developing T cells. One type of epithelial cell may be present throughout both the cortex and the medulla or is restricted to the cortex, while two other types of epithelial cell are present only in medulla; the relationship between these cell types is unclear. Epithelial cells, unlike lymphocytes, are not produced by the bone marrow. Within the cortex and medulla, other cell types are tightly associated with the developing lymphocytes. In the cortex these cells include **thymic nurse cells** whose function is unknown, but thymocytes appear to be contained *within* these nurse cells and may require them for their development, although there is still some dispute about this. In the medulla, and at the cortico-medullary junction where most blood vessels are situated, there are **interdigitating cells** which have a bone marrow origin and belong to the family of dendritic cells. So-called **Hassal's corpuscles** are also located within the medulla, and they appear to be agglomerations of epithelial cells.

Spleen—a secondary lymphoid tissue (Fig.1.38)

The spleen is surrounded by a **capsule** and contains densely-packed areas of mature lymphocytes, with roughly equal numbers of T cells and B cells interspersed with other blood cells. It contains trabeculae, although the spleen is not divided into discrete lobules like the thymus (above). It is organized into **red** and **white pulp**: red pulp appears reddish in colour because it contains so many erythrocytes, while the white pulp is devoid of red cells but contains most of the lymphocytes.

Blood enters the spleen via the splenic artery which divides into more and more branches to become arterioles. Around each arteriole there is a **periarteriolar sheath** (abbreviated PALS) where most of the lymphocytes are found, that is, the PALS is largely the same as the white pulp. If a section is taken across the PALS, the **central arteriole** is mainly surrounded by T cells, while the majority of B cells are found further away from the arteriole beyond the T cells. The B cells are often organized into discrete regions called primary and secondary **follicles**. Primary follicles contain unstimulated B cells, while secondary follicles are sites of B cell activation and proliferation and contain a central **germinal centre**. The division of white pulp into areas of T and B lymphocytes is not absolute, since some B cells are present in predominantly T areas, and vice versa. Most probably, T cells and B cells interact with each other at these sites during immune responses.

Beyond both the T and B cell areas, and surrounding them, is the **marginal zone** which is situated between the white and the red pulp. This is where lymphocytes carried in the blood leave the arterioles (which have become smaller and smaller and have eventually traversed the PALS); the lymphocytes can then migrate into the T and B areas. The blood then either percolates through the red pulp (in an 'open' circulation) or it is carried in connecting vessels (a 'closed' circulation) into **venous sinuses** of the red pulp. Whichever is the case, blood eventually leaves the spleen via the splenic vein.

Lymphocytes in the white pulp are tightly associated with different and distinct cell types. **Interdigitating cells** (which are dendritic cells) form a network in T cell areas, while B cells can be similarly associated with **follicular dendritic cells** (which, despite their name, are unrelated to other types

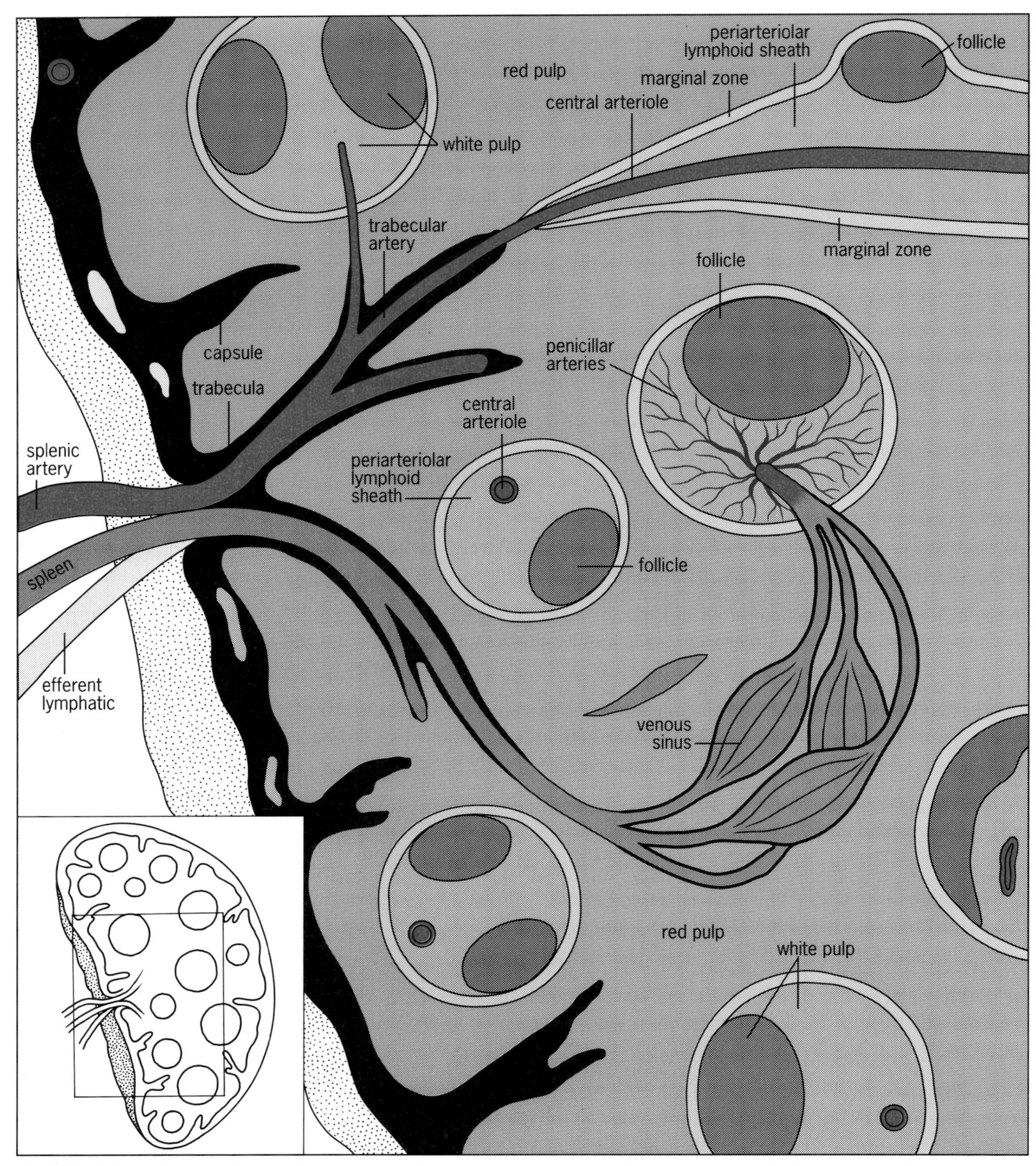

Fig.1.38 Highly schematic view of **spleen** structure (e.g. of mouse or human). The splenic artery divides into several branches, of which one is shown, which enter the spleen and run within the trabeculae which are continuous with the capsule. These trabecular arteries branch and leave the trabeculae to become central arterioles, which are surrounded by lymphocytes (mostly T cells) to form the periarteriolar lymphoid sheath (PALS). Within the PALS are spherical follicles, consisting predominantly of B cells. Together, the PALS and follicles comprise the white pulp, which is separated from the red pulp by the marginal zone. Within the PALS, the central arterioles give rise to the penicillar arteries which extend to the marginal zone. Venous blood ultimately enters the venous sinuses, in which the vein is surrounded by cells to form a structure resembling the staves of a barrel (indicated in 3D·view), and leaves the spleen. Lymph collects into efferent lymphatics; the spleen has no afferent lymphatics

of dendritic cell!). The function of each of these types of non-lymphoid (i.e. non-lymphocyte) cells is uncertain. Antigens apparently in their native conformation (the form of antigen recognized by B cells) persist on follicular dendritic cells in the spleen for long periods of time, and these cells may be involved in the generation of memory B cells. It seems likely that antigens are present on interdigitating cells as peptides bound to MHC molecules (the form recognized by T cells) and that these cells may be involved in initiation of T cell responses (Fig.1.25).

The marginal zone contains a specialized type of macrophage that may be required for some B cell responses. These **marginal zone macrophages** also retain certain antigens that stimulate B cells. The marginal zone also contains a distinct subset of dendritic cells. Towards the white pulp is found another type of macrophage called the **marginal metallophil**. In the red pulp, macrophages are responsible for clearing dead or damaged red blood cells from the circulation and they contain haemosiderin as a consequence of the large amount of iron which is digested this way. The red pulp also contains lymphoblasts.

Lymph nodes—secondary lymphoid tissues (Fig.1.39)

We have already mentioned lymph nodes in relation to lymphocyte recirculation (Section 1.3.3). Lymph nodes, like other secondary lymphoid tissues, contain a mixture of lymphocytes some of which are in the process of recirculating through them while others are organized into more specific structures (analogous to the white pulp of spleen). Peripheral lymph nodes, as opposed to mesenteric lymph nodes which drain the gut, generally contain more T cells than B cells (Fig.1.33).

The two routes by which cells enter lymph nodes are via the blood and lymphatics (Fig.1.34). This of course is different from the spleen, which lacks a lymphatic supply, and another difference between these tissues is that nodes do not have red pulp. Like spleen, however, lymph nodes are divided by trabeculae, and T cells and B cells are again compartmentalized into distinct areas, although perhaps not as discretely as in spleen.

Immediately beneath the capsule of the node is a space called the **subcapsular sinus**. Beneath this is the outermost layer, the **cortex**. The cortex contains B cells that tend to be localized in discrete primary follicles and secondary follicles with germinal centres. Germinal centres are found especially in antigen-stimulated lymph nodes and may be surrounded by a 'mantle zone' which is oriented towards the capsule. The function of germinal centres is still open to much speculation, but they are thought to be important for generating memory B cells rather than for triggering primary B cell responses.

T cells are situated predominantly in the **paracortex** of the node, which includes the area between B cell follicles (thus termed the **interfollicular regions**) as well as in the layer beneath the follicles. High endothelial venules (see p. 53) are found in the paracortex, which is the area where lymphocytes enter the body of the node from the blood. Once in the node, most T cells home to the T areas, while B cells go to B areas. As in the spleen, T and B lymphocytes in their respective areas are associated with interdigitating cells and follicular dendritic cells, respectively. Beneath the cortex and para-

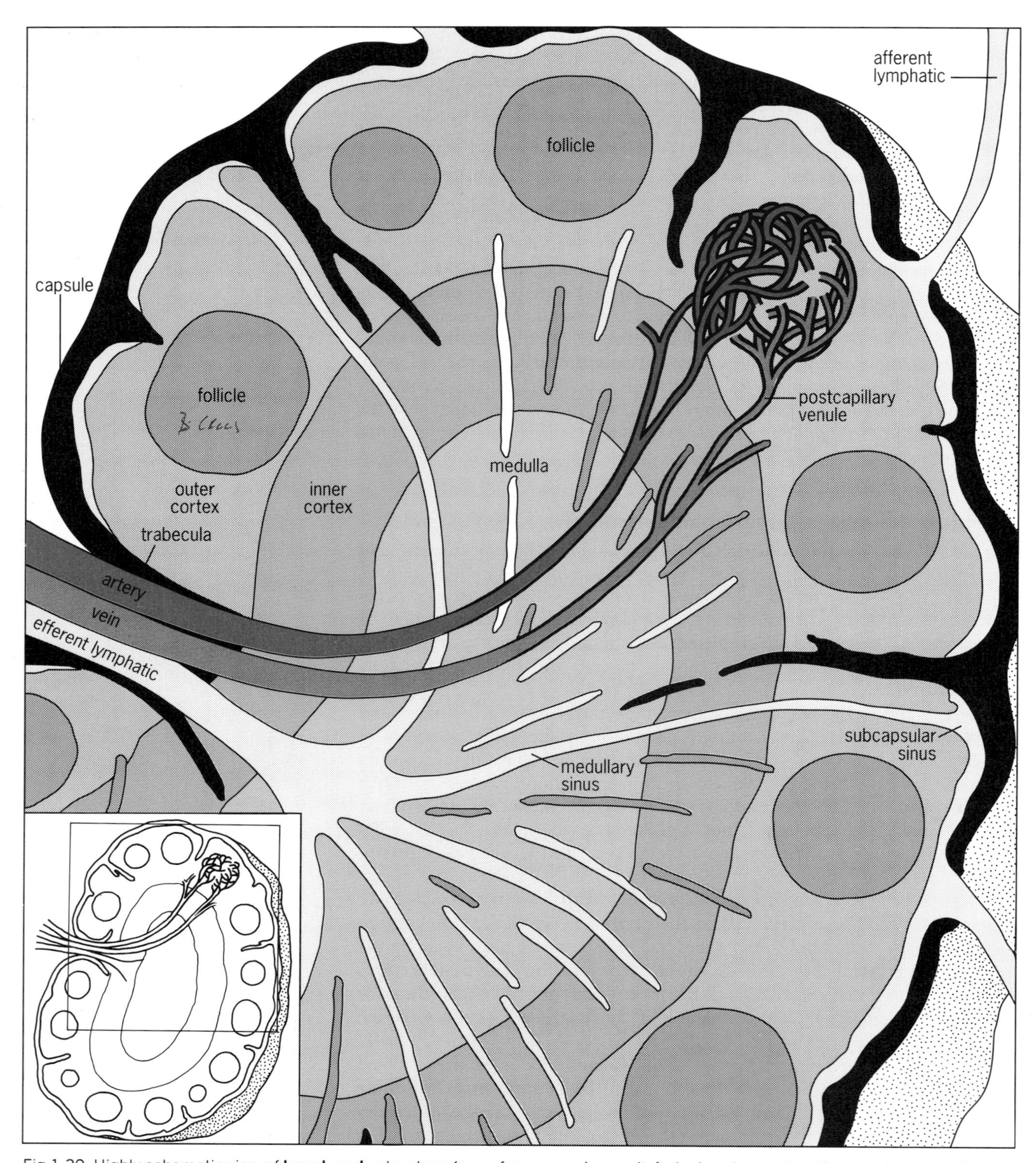

Fig.1.39 Highly schematic view of **lymph node** structure (e.g. of mouse or human). A single artery enters the lymph node at the hilus, divides, and the branches run within trabeculae which are continuous with the capsule. After branching, the vessels pass through the medulla within the medullary cords, and into the inner and outer cortex, the latter of which includes the interfollicular regions between the follicles; the inner and outer cortex are in fact continuous. Most T cells are present in the cortex whereas the follicles contain predominantly B cells. The vessels branch into capillaries around the follicles (an example is indicated in 3D-view), and return as postcapillary venules which are lined with high endothelium as they pass through the inner cortex. These vessels then unite into a vein which leaves the node at the hilum. Lymph enters the node through several afferent lymphatics which empty into the subcapsular sinus, percolates through the medullary sinuses and the lymphoid tissue, and ultimately leaves the node in one or more efferent lymphatics

cortex is the **medulla**. The medulla contains some T and B cells, but most of the B cells in this region are antibody-secreting **plasma cells** which are arranged along the **medullary cords**, together with macrophages.

Lymph enters the node via afferent lymphatics which drain into the subcapsular sinus, another region where macrophages and other non-lymphoid cells are found. The afferent lymph that drains from the skin and other epithelia contains 'veiled cells' (see p. 52) and these cells migrate to the T cell areas of the node and probably become interdigitating cells. Many veiled cells in lymph draining the skin are probably derived from **Langerhans cells** of the epidermis that may have acquired antigens.

After passage through the node, the lymph leaves via the efferent lymphatic which passes through the **hilum**. The efferent lymph contains lymphocytes but is almost depleted of veiled cells. The arterial blood also enters the node at the hilum, and the vessels extend up to the level of the subcapsular sinus and then return the venous blood back through the same point to the circulation.

Further reading

Cell and molecular biology texts

Alberts, B., Bray, M., Lewis, J., Raff, M., Roberts, K., and Watson, J.D. (1989). *Molecular biology of the cell* (2nd edn). Garland, New York.

Darnell, J., Lodish, H., and Baltimore, D. (1990). *Molecular cell biology* (2nd edn). Scientific American Books, USA.

Mathews, C.K. and van Holde, K.E. (1990). Biochemistry. The Benjamin/Cummings Publishing Company, Redwood City, California.

Singer, M. and Berg. P. (1991). *Genes and genomes*. Blackwell, Oxford.

Watson, J.D., Gilman, M., Witkowski, J., and Zoller, M. (1992). *Recombinant DNA*. (2nd edn). Scientific American Books, W.H. Freeman and Co., New York

Weir, D.M., Herzenberg, L.A., Blackwell, C., and Herzenberg L.A. (1986). *Handbook of experimental immunology* (4th edn). Blackwell, Oxford.

Immunology texts

Benjamini, E. and Leskowitz, S. (1991). *Immunology: a short course*. Wiley, USA.

Golub, E.S. (1982). *Immunology and synthesis*. Sinamer Associates, Massachusetts.

Klein, J. (1990). *Immunology*, Blackwell, Oxford.

More advanced

Paul, W.E. (1989). *Fundamental immunology* (2nd edn). Raven, New York.

Sell, S. (1987). *Immunology, immunopathology and immunity*. Elsevier, New York.

Clinical and applied immunology texts

Lachmann, P.J. and Peters, D.K. (1982). *Clinical aspects of immunology* (4th edn). Blackwell Scientific Publications, Oxford.

Weatherall, D.J. and Ledingham, J.G.G. (1991). *Oxford textbook of medicine* (2nd edn). Oxford University Press, Oxford.

Series

In Focus (ed. D. Male). IRL, Oxford.

Review journals

Advances in Immunology
Annual Review of Immunology
Immunological Reviews (Munksgaard, Copenhagen)
Immunology Today

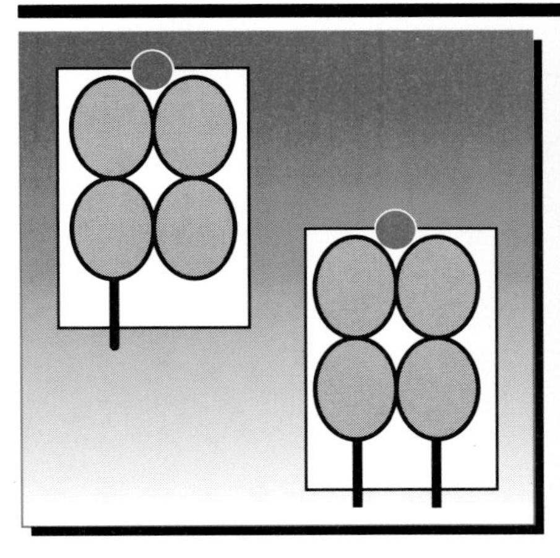

2

THE MAJOR HISTOCOMPATIBILITY COMPLEX

Scheme 2. MHC molecules

A schematic diagram of the peptide-binding groove of an MHC molecule, viewed from above

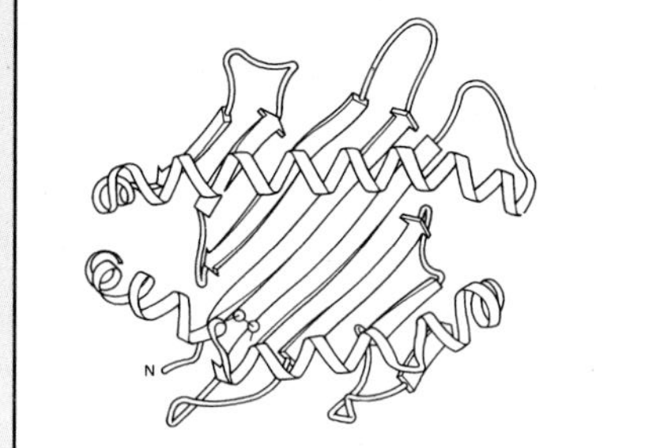

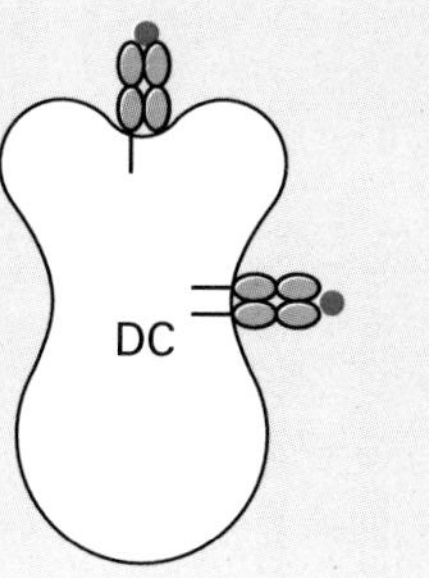

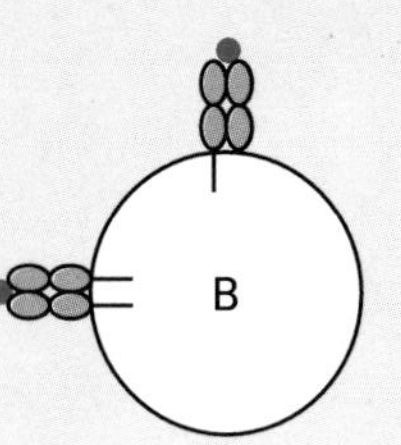
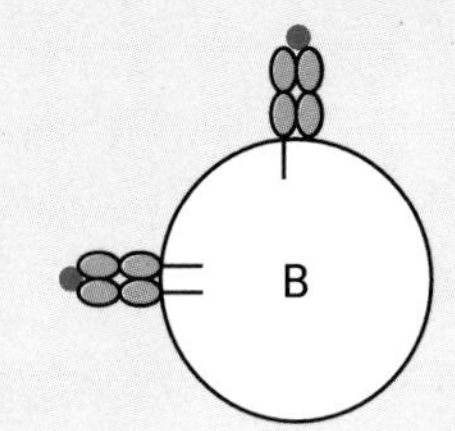
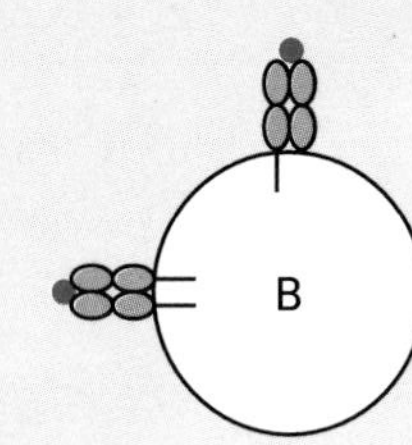
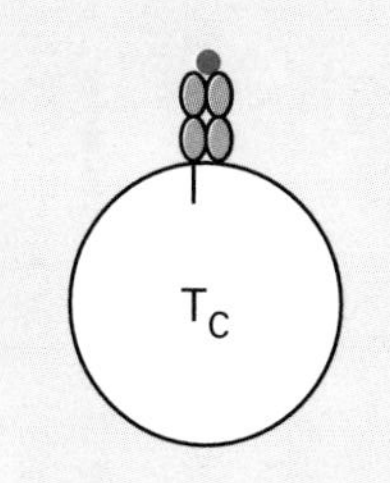
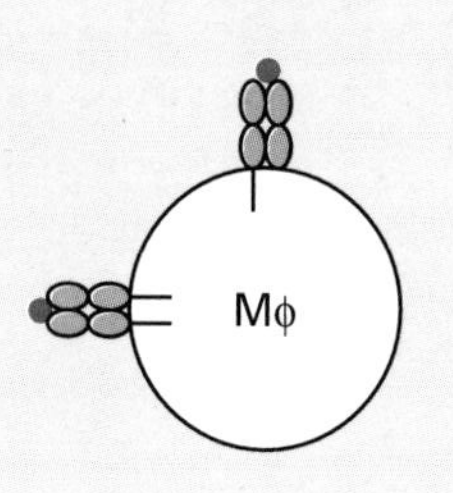
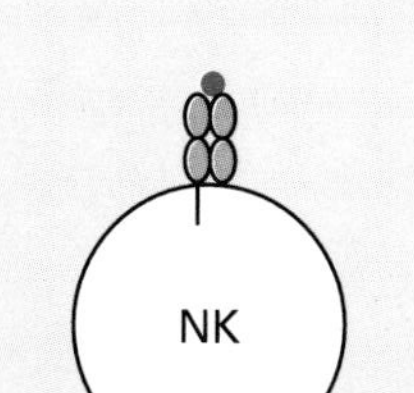

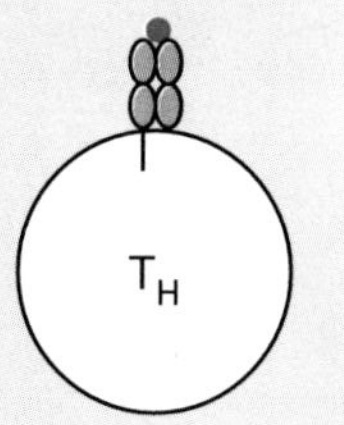
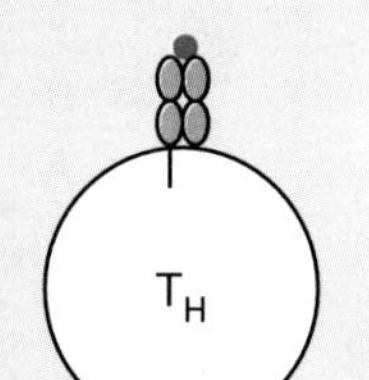
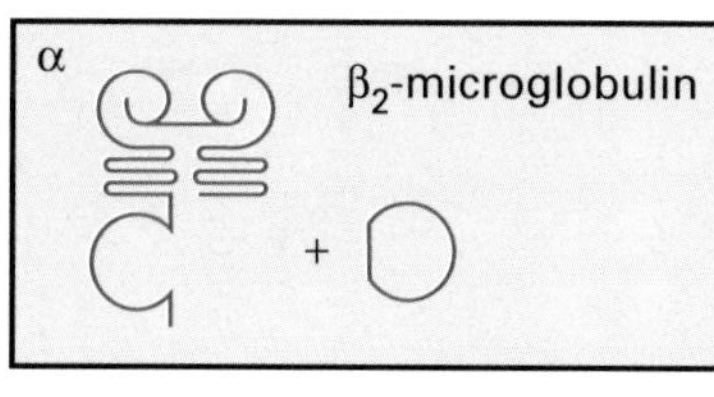
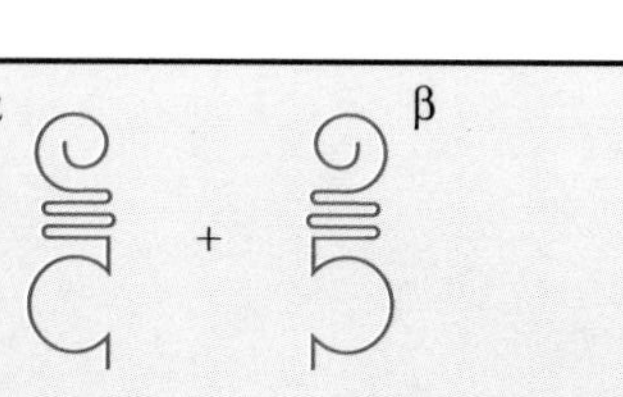

2.1 Introduction

One cannot study immunology for very long without encountering the **major histocompatibility complex, MHC** for short. The MHC contains a set of genes located together on one chromosome as a 'complex'. MHC genes code for several series or families of polymorphic glycoproteins, including two families of molecules that are expressed at the cell surface, the class I and class II molecules. These specialized membrane proteins act as guidance systems that allow T cells to recognize antigen (Figs.1.25 and 1.27).

Historical perspective. The term MHC derives from studies designed to investigate the fate of tissues (grafts) transplanted between individuals, As a result of these experiments the MHC was recognized as an important ('major') set of genes ('complex') responsible for controlling whether grafts are accepted between individuals whose tissues are genetically similar ('histocompatible') or rejected by individuals who are not ('incompatible'). These early transplantation studies were facilitated by the development of inbred strains of mice. Unlike outbred populations such as humans, every individual from the same inbred strain is genetically identical. Thus it became possible to transplant tissue between animals of the same inbred strain without any risk of rejection. The tissues of all mice of one strain are said to be **histocompatible** with respect to each other, but **(histo)incompatible** with those of any other inbred strain.

As the rejection of incompatible tissue was investigated more extensively, it became clear that the MHC was not the only influence on tissue compatibility. A large number of genetic loci were identified on different chromosomes that also played a role in graft rejection. These were called **minor histocompatibility genes** and the molecules they encode, minor histocompatibility antigens. In general, grafts are more rapidly rejected between individuals that are incompatible for the MHC. Incompatibility for minor histocompatibility antigens is cumulative, and multiple minor histocompatibility antigen differences can also result in a vigorous rejection response.

It is now recognized that MHC molecules play a fundamental role in physiological immune responses by enabling T cells to recognize antigens. MHC molecules bind or associate with antigen in the form of peptide fragments, and it is this peptide–MHC complex that is recognized by T cells. It is evident that MHC molecules both participate in and control physiological immune responses, such as those involved in the recognition and elimination of virus-infected cells. Graft rejection is simply a non-physiological manifestation of the normal immune response that occurs in every individual when they encounter antigen (see discussion of alloreactivity, Section 5.1.2).

As the MHC has such an important role to play in adaptive immune responses, it is perhaps not surprising that this gene complex has been identified in all of the vertebrate species that have been examined. Interestingly, some invertebrates, such as colonial tunicates, as well as some of the flowering plants, also appear able to recognize genetically dissimilar members of their species (Burnet 1971). Although as yet a gene complex similar to the MHC has not been identified in invertebrates, these observations suggest that MHC molecules, which play such an important role in the immune responses of higher mammals, may have a very distant evolutionary origin.

This chapter will focus on the structure of MHC molecules and the organization of the genes within the complex in mouse and human.

2.2 General features of the MHC

The MHC codes for three families of glycoproteins known as **class I, class II**, and **class III MHC molecules**. The members of two of these families, the class I and class II molecules, are also sometimes referred to as MHC antigens or **alloantigens** because they can be recognized by the immune system, for example during the rejection of tissue transplanted between MHC incompatible individuals. The class I and class II MHC molecules are expressed mainly as membrane glycoproteins at the cell surface, whereas the products of class III genes are usually soluble molecules. Class III molecules include some of the components of the complement system, one of the major effector mechanisms of the humoral immune response (see Section 9.4); soluble effector molecules such as tumour necrosis factors, TNFα and TNFβ (see Section 9.5.3); the enzyme 21-hydroxylase; and the heat shock proteins HSP70 (Fig.2.22).

One of the important features of MHC molecules is their **polymorphism**. That is, within each class of molecules and even at one locus, a large number of variants (polymorphic forms or alleles) exists in the population as a whole. Thus if a large number of individuals from the population was examined, many genes for each type of product would be found, each coding for a separate MHC allele or variant. However, it is important to remember that each individual only has a very small set of different MHC genes and expresses a maximum of two alleles for each locus.

Here we will be concerned exclusively with the class I and class II molecules and their genes. The products of these two families have different though related structures, and very different cellular distribution patterns.

A summary of the important features of the MHC is given in Table 2.1.

2.3 Nomenclature

The nomenclature used to describe the MHC is, to the uninitiated, extremely complex. However, much of this complexity has arisen simply because different nomenclature systems were developed to describe the MHC in different species, rather than because of any major underlying differences between the structure and function of the molecules encoded.

Table 2.1 MHC — major histocompatibility complex

1. The MHC is a complex of closely-linked genes
2. It is present in the genome of all vertebrates
3. The MHC codes for 3 main families of molecules:

   Class I molecules — cell surface glycoproteins that can be expressed by virtually all somatic, nucleated cells

   Class II molecules — cell surface glycoproteins expressed by certain cells involved in the initiation and effector phase of the immune response

   Class III molecules — some components of the complement system and other soluble molecules
4. MHC molecules are **polymorphic** — many variant forms of each type of molecule are found in the population
5. MHC molecules can associate with or bind peptides
6. MHC molecules allow peptides derived from foreign antigens to be recognized by T cells

The MHC of different species. The terminology used to describe the MHC in a number of species is given in Table 2.2.

The majority of nomenclature systems are modelled on that designated for *human* — **HLA**, which is short for human leukocyte antigens or, strictly, human leukocyte associated antigen A. Thus, the MHC in *rabbit* is designated **RLA** and in *guinea pig*, **GPLA**.

That used for the MHC in the *mouse*, **H-2**, is retained for historical reasons as it was the original term used by Gorer in 1936 when he first described this antigen system (Gorer 1936). Gorer identified several histocompatibility genes in the mouse and H-2 was simply the second he defined. The designation **RT1**, which replaces the old nomenclature, AgB, is used to describe the *rat* MHC.

A simplified linear map showing an outline of the organization of the MHC in three different species is given in Fig.2.1.

MHC regions. The MHC can be divided into a series of **regions**. A region is simply part of a chromosome, or a chromosome segment, that can be defined within the gene complex. For example, the H-2I region in the mouse and the HLA-D region in humans both encode the class II MHC molecules (Fig.2.1).

Table 2.2 Nomenclature for the MHC in different species

Species	MHC designation	
Mouse	H-2	Histocompatibility System 2
Rat	RT1	Formerly AgB
Human	HLA	Human Leukocyte associated Antigens
Rabbit	RLA	Rabbit Leukocyte Antigens
Guinea pig	GPLA	Guinea pig Leukocyte Antigens

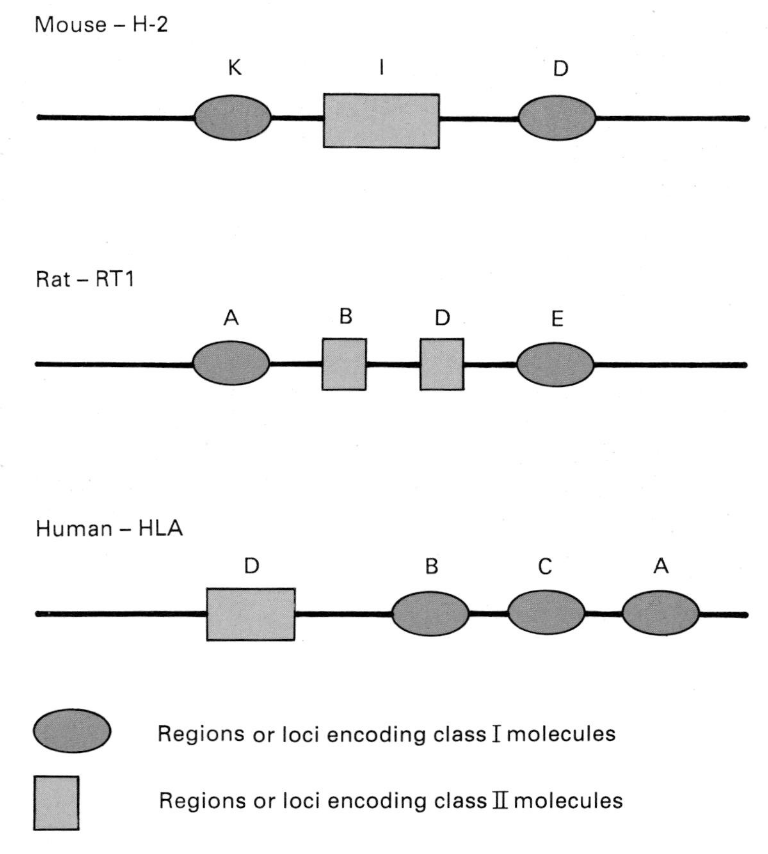

Fig.2.1 Simplified linear maps to show the organization of the major histocompatibility complex in three different species

MHC loci. Each region can be sub-divided into one or more **locus** or **loci.** A locus is a single gene, coding for a single product, which can be genetically separated from all neighbouring genes. (This term is sometimes used rather loosely to describe the MHC.) In all species each MHC locus is designated by an upper case letter or letters (Fig.2.2). For example, in certain strains of mouse several loci have been identified in the H-2D region, each of which encodes a functional class I α or heavy chain gene; these are designated the D, L, and R loci. In humans the class I loci are designated HLA-A, B, and C. The class II region, HLA-D, contains groups or families of loci, including HLA-DR, DQ, and DP. Each of these groups contains at least two functional genes (Table 2.3 and Fig.2.22).

(To clarify the difference between regions and loci further, it should be noted that, in humans, HLA-D is a region that encodes multiple loci including HLA-DRA and DRB1; on the other hand, the HLA-B locus is not classified as a region because it only contains a single gene: Fig.2.1.)

MHC alleles. Within the population numerous **alleles** exist for each MHC locus. In other words, different individual members of the population may each have a variant form of the product of a particular locus. MHC alleles can be distinguished by functional tests, either serologically (in other words

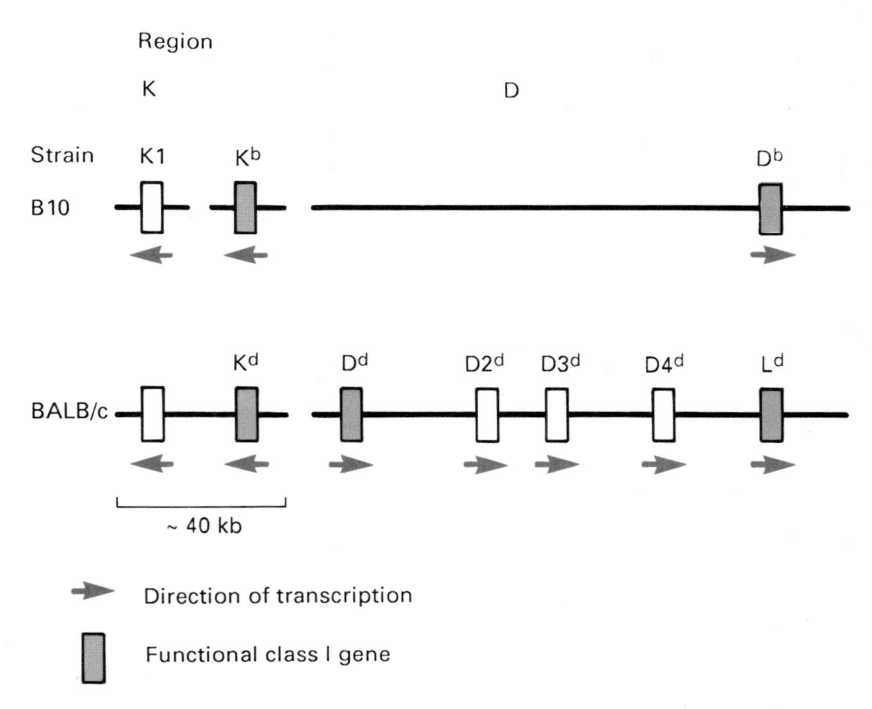

Fig.2.2 In certain strains of mice, several genes have been identified within the H-2D region. B10=C57BL/10

using antibodies specific for each particular gene product) or by cellular analyses *in vitro*, such as in mixed leukocyte cultures (see Section 5.1.2 and Appendix). The variety of approaches used to define MHC polymorphisms has contributed to the complexity of the nomenclature. For example, the existence of regions or loci originally identified by classical genetic techniques or serology in some cases may not be confirmed by molecular genetics or biochemistry; however, the nomenclature system adopted must take both pieces of data into account. This is particularly pertinent in the context of the I-J locus in the mouse (see Section 2.8).

In inbred strains of mice the allele for each MHC or H-2 locus is designated by the same lower case letter, written as a superscript following the locus designation. For example, all the alleles within the H-2 complex of the C57BL/10 strain of mice are designated b. So the alleles for the H-2K and D loci of the C57BL/10 mouse are designated H-2K^b and H-2D^b (read as 'K of b', 'D of b'; Fig.2.2).

In humans, each locus specificity is given a number *rather* than a letter. They are written in the form: locus designation, followed by the number, for example HLA-A2, HLA-B7, and HLA-DR4. Alleles that are only defined in a limited way, e.g. by only a small number of alloantisera, are prefixed by a 'w', e.g. HLA-DRw11, to indicate they are still being characterized.

As more HLA genes have been cloned and characterized the nomenclature system has been extended to allow each gene that is associated with a particular specificity to be identified. For class I alleles, each gene is identified as follows: e.g. A*1101, A*1102, B*2701, etc.; locus designation/*/specifi-

Table 2.3 Names for the genes in the HLA region

	Name	Molecular characteristics
Class I	HLA-A	Class 1 α chain
	HLA-B	Class 1 α chain
	HLA-C	Class 1 α chain
	HLA-E	Class 1 — related α chain
	HLA-F	Class 1 — related α chain
	HLA-G	Class 1 — related α chain
	HLA-H	Class 1 — related α chain
Class II	HLA-DRA	DRα chain
	HLA-DRB1	DRβ1 chain-determining specificities DR1, DR3, DR4, DR5, etc.
	HLA-DRB2	pseudogene; DRβ-related
	HLA-DRB3	DRβ3 chain-determining specificities DRw52
	HLA-DRB4	DRβ4 chain determining the specificity DRw53
	HLA-DRB5	DRβ chain found in the DR2 haplotype
	HLA-DQA1	DQα chain (expressed)
	HLA-DQB1	DQβ chain (expressed)
	HLA-DQA2	DQα-chain-related sequence (not known to be expressed)
	HLA-DQB2	DQβ-chain-related sequence (not known to be expressed)
	HLA-DOB	DOβ chain
	HLA-DNA	DNα chain
	HLA-DPA1	DPα chain (expressed)
	HLA-DPB1	DPβ chain (expressed)
	HLA-DPA2	pseudogene; DPα-chain-related
	HLA-DPB2	pseudogene; DPβ-chain-related

city/gene number. Thus, two genes for the specificity A11 have been identified in the population. The second gene A*1102 differs from A*1101 by a single nucleotide substitution at codon 19. The cell from which the new gene was isolated is of Indonesian origin. The nomenclature for class II alleles is similar but takes into account that in the DQ and DP subregions there are two α chains genes, DQA1, DQA2; DPA1 and DPA2—only DQA1 and DPA1 are functional, and in the DR, DQ, and DP subregions more than one β chain gene exists (Fig.2.23). Thus for α chains, each gene associated with a particular specificity is identified as follows: e.g. DQA1*0401, DPA1*0101, etc; and for β chains: e.g. DRB1*0301, DRB1*0302, DRB3*0101, DQB1*0201, DPB1*0401, etc.; that is locus designation and gene/*/specificity/gene number. For a complete listing of all the HLA specificities defined at the time of writing see Bodmer *et al.* (1990*a*,*b*, 1992).

MHC haplotypes. When all the alleles within the H-2 complex of an inbred strain of mouse are considered together, they constitute what is termed the **H-2 haplotype** of the mouse. A haplotype is a group of alleles that are inherited together on a single chromosome. So the haplotype of the C57BL/10 mouse is H-2^{b}, and that of the BALB/c mouse is H-2^{d}. (Note that when the letter is designating the haplotype of the mouse it is also superscripted.) Some of the common haplotypes of other strains of inbred mice are given in Table 2.4.

Table 2.4 H-2 haplotypes of some common strains of mice

Strains	Haplotypes	H-2 gene complex				
		K	A	E	S	D
C57BL/6; C57BL/10;	b	b	b	–	b	b
BALB/c; DBA/2	d	d	d	d	d	d
AKR; CBA; C3H;	k	k	k	k	k	k
SWR	q	q	q	q	q	q
SJL	s	s	s	–	s	s

Extended haplotypes. As the human population is outbred, haplotype designations cannot be made on the same scale as for inbred animals. However, since the genes coded within the MHC are tightly linked, they are frequently inherited together. Patterns of inheritance amongst certain groups of HLA antigens have been identified and these are called **extended haplotypes**. They occur when one allele is associated with another at a frequency greater than would be expected by chance, for example HLA-A1, B8, and DR3, are frequently inherited together.

MHC recombinants. Recombination events can occur within the MHC. Recombinant mice are those in which genetic recombination, or crossing over between chromosomes, has taken place within the H-2 complex. (Strictly, these are 'MHC recombinant' mice because, in general, recombination events can occur anywhere in the genome, not only within the H-2; however, these mice are selected for recombination events within the H-2.) For these strains, the alleles are designated according to the original inbred strain of mouse from which the recombinant strain was derived (Table 2.5).

Table 2.5 Recombinant H-2 haplotypes

Strain	Haplotype	Parental haplotypes	H-2 gene complex				
			K	A	E	S	D
A; B10.A	a	k/d	k	k	k	d	d
B10.HTG	g	d/b	d	d	d	d	b
B10.A(2R)	h2	a/b	k	k	k	d	b
B10.A(4R)	h4	a/b	k	k	b	b	b
B10.A(3R)	i3	b/a	b	b	k	d	d
B10.A(5R)	i5	b/a	b	b	k	d	d
B10.AKM	m	k/q	k	k	k	k	q

2.4 Cellular distribution of MHC molecules

Class I molecules can be expressed to a greater or lesser extent on virtually all somatic nucleated cells. It is important to note that the level of expression (i.e. the number of molecules on the cell surface) can vary enormously between different types of cells. For example, muscle cells and neurons have very low expression, whereas lymphocytes express high levels of class I MHC antigens. Furthermore, expression may be increased during an immune response (Fig.2.3). Increased expression of class I molecules presumably increases the chance of the cell presenting an antigenic peptide and hence the chance of it being recognized and killed by cytotoxic T cells (see Section 1.2.3), if it has been infected or is otherwise abnormal.

In contrast, class II MHC molecules are not expressed by all cells, but only by a limited number of cell types, for example dendritic cells, macrophages, and B cells (Fig.1.11). In general, class II molecules are expressed by cells involved in handling antigen taken up from the outside of the cell and/or initiating immune responses (see Scheme 1). In addition, expression of class II MHC molecules can be induced or increased by cytokines such as interferon γ (IFNγ). Increased or induced expression of class II MHC molecules is often found in the vicinity of an ongoing immune response, for example in tissue transplants undergoing rejection (Fig.2.3). The *phsysiological* significance of the induced expression of class II antigens is not clearly understood, but it seems likely that increased expression of class II molecules will increase the chance of the cell presenting an antigenic peptide to helper T cells thereby increasing the level of the immune response to the foreign antigen in general, and/or augmenting the response of the cell involved.

2.5 Structure of MHC molecules

2.5.1 Class I

General features

The products of the class I MHC genes (e.g. HLA-A, B, and C loci in human, H-2K locus and D region in the mouse) are membrane glycoproteins. Each class I molecule is a heterodimer composed of an **α** or **heavy chain** polypeptide (M_r 45 kDa) and **β_2-microglobulin** (M_r 12 kDa) which is noncovalently associated with the α chain. The class I α chain is polymorphic and encoded within the MHC, whereas polymorphism of β_2-microglobulin is limited (only one allele has been identified in human and seven in mouse). The β_2-microglobulin gene is not encoded by the MHC.

Our knowledge of the structure of MHC molecules has been evolving since 1981, when class I molecules were purified by affinity chromatography and subjected to amino acid sequence analysis (Coligan *et al.* 1981) (see

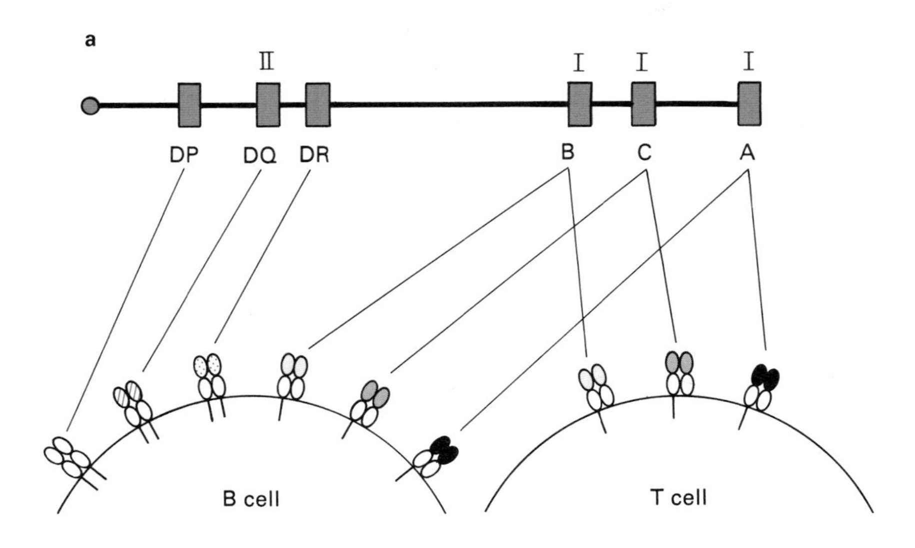

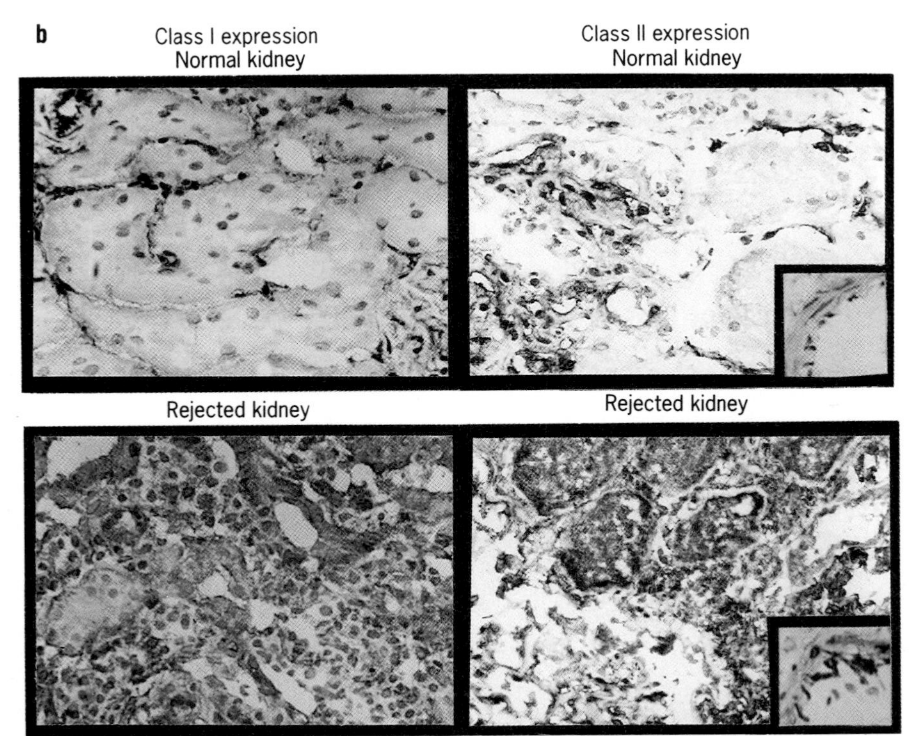

Expression of both class I and class II molecules is induced during rejection of tissue transplanted into an individual who is not identical with the kidney donor for major and minor histocompatibility antigens

Fig.2.3 Examples of tissue distributions of class I and class II MHC molecules

Appendix, p. 704). A schematic diagram of the structure of a class I antigen is shown in Fig.2.4.

The α chain is a transmembrane polypeptide chain that can be divided into five distinct structural regions, or domains. Three of these domains, α_1, α_2, and α_3, are exposed on the outside of the cell, and are known as the extracellular domains. The α_3 domain is found closest to the plasma membrane, and can be thought of as 'sitting' on the membrane; it is also referred to as the membrane proximal domain. β_2-microglobulin is associated with the extracellular portion of the α chain and 'sits' on the membrane next to the α_3 domain (Fig.2.4).

The fourth and fifth domains of the α chain comprise the transmembrane and cytoplasmic portions of the class I molecule. These regions show only a limited sequence relationship (low homology) with the other three domains. The transmembrane portion of all class I molecules contains a hydrophobic sequence of approximately 25 amino acids that spans the lipid bilayer. This region terminates in a stretch of five (hydrophilic) basic amino acids, including the residues arginine (R) and lysine (K), which are thought to be important for anchoring the polypeptide in the plasma membrane. A similar sequence of basic amino acid residues immediately following the transmembrane region of the polypeptide chain has been found in other transmembrane glycoproteins. The cytoplasmic domain is between 30 and 40 amino acids long, and about half of these are polar residues (Fig.1.3).

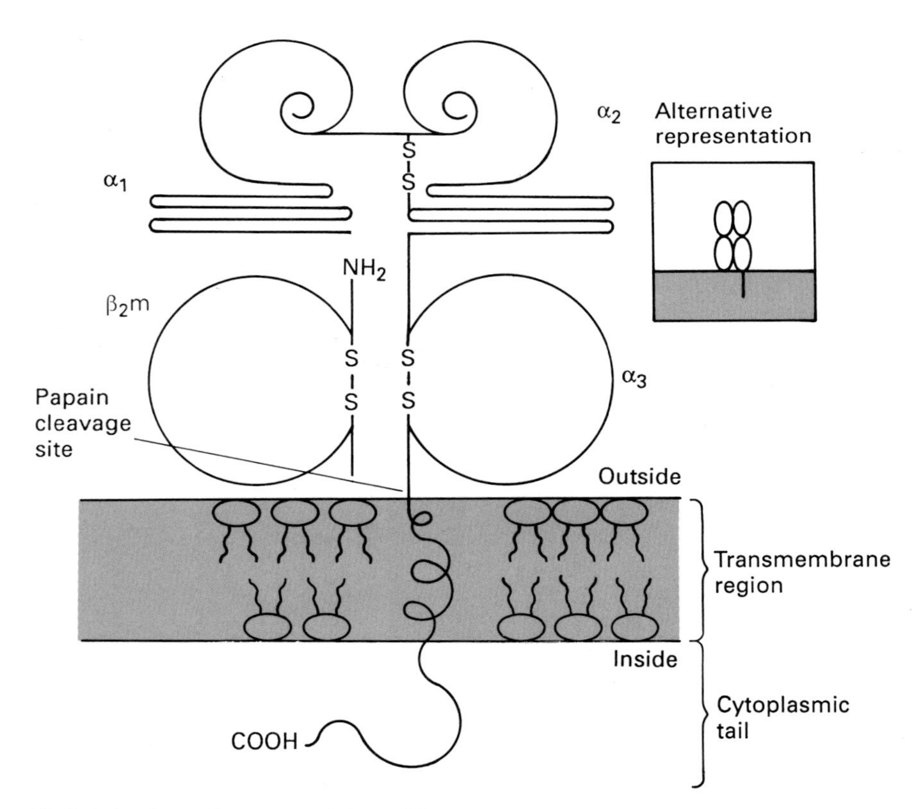

Fig.2.4 A schematic representation of the structure of a class I molecule

Class I molecules, being glycoproteins, are glycosylated (Fig.1.4). Carbohydrate moieties are associated with the polypeptide chain in the α_1 and α_2 domains in the mouse at amino acid residues 86, 176, and in some alleles 256, and with the α_1 domain in human at amino acid residue 86. In humans each carbohydrate moiety is composed of up to 15 sugar residues.

Although the structure of a protein can to some extent be predicted from its amino acid sequence, knowledge of the precise three-dimensional (3D) structure of the protein can be of enormous value in understanding its function. Such information can only be obtained by X-ray crystallography (For a discussion of this technique see van Holde 1985). The first 3D structure of a human class I MHC antigen, HLA-A2, was reported in 1987 (Bjorkman *et al.* 1987*a* and *b*) (for review see Townsend and McMichael 1987) the structures of two other class I alleles have also been determined, HLA-Aw68 (Garrett *et al.* 1989) and HLA-B27 (Madden *et al.* 1991).

Crystal structure

For X-ray crystallography of HLA-A2 a soluble fragment of the molecule was purified from a human lymphoblastoid B cell line, after digestion of the plasma membranes with the enzyme papain (Fig.2.4). Papain digestion released the extracellular portion of the molecule, a soluble fragment 271 amino acids long, composed of the α_1, α_2, and α_3 domains of the heavy chain and β_2-microglobulin. Crystals were prepared of this fragment of A2 and their structure determined by X-ray crystallography at 3.5 Å resolution. The primary structure or amino acid sequence of HLA-A2 was then 'fitted' or modelled to the electron density map. At this resolution it was possible to recognize the amino acid backbone of the polypetide chain, locate the disulphide bonds, orientate some of the side chains, and determine the α helix or β-pleated sheet character of the structure.

The structure of HLA-A2 is shown in Fig.2.5. It consists of two pairs of domains. The α_1 and α_2 domains have a similar tertiary structure and form one pair, while the α_3 domain and β_2-microglobulin form the other. Viewed from the outside of the cell, the α_1 and α_2 domains would comprise the top or exposed part of the molecule, while the α_3 domain and β_2-microglobulin would be located closest to the membrane. The structure of the α_3 domain and β_2-microglobulin both resemble that of an immunoglobulin constant domain (see Section 6.2.1, p. 339), composed of two anti-parallel β-pleated sheets, one of four β strands and the other of three (Fig.2.5). The structural similarity between these two domains was predicted from amino acid sequence data, as they show significant homology both to one another and immunoglobulin domains, but the X-ray data confirm that MHC class I molecules are members of the immunoglobulin superfamily (see Section 7.4). The α_1 and α_2 domains each consist of four β strands, spanned by a long α helix (Figs.2.5, 2.6). Structurally, these two domains are the mirror image of one another; thus the top half of the molecule consists of a platform of eight β strands topped by two α helices (Fig.2.6). These basic structural features of class I molecules have been confirmed by X-ray data obtained from crystals of another A locus allele, HLA-Aw68, and a B locus allele, HLA-B27.

A large groove or cleft in the structure was identified between the two α helices of the α_1 and α_2 domains, located on the top surface of the A2

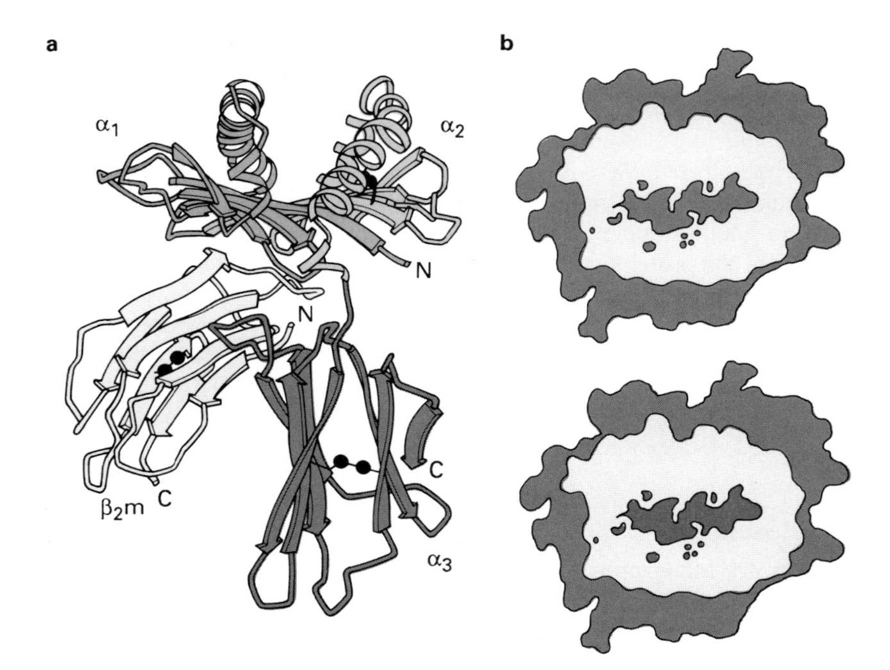

Fig.2.5 Structure of HLA-A2 (Adapted from Bjorkman *et al.* (1987*a*), with permission).
(a) Schematic representation.
(b) Top of the HLA-A2 molecule showing the peptide-binding groove in the absence or presence of a peptide (red)

molecule (Fig.2.6). This groove is most likely the binding site for antigen in the form of short peptides. The groove is 25 Å long and 10 Å wide. These dimensions fit very well with the idea that the form of antigen recognized by a T cell receptor is a peptide of 8–9 amino acids, generated from the intact molecule during antigen processing, associated with an MHC molecule (see Section 3.3). When the X-ray data from the A2 crystals were analysed, extra electron density of unknown origin was found in the groove formed by the

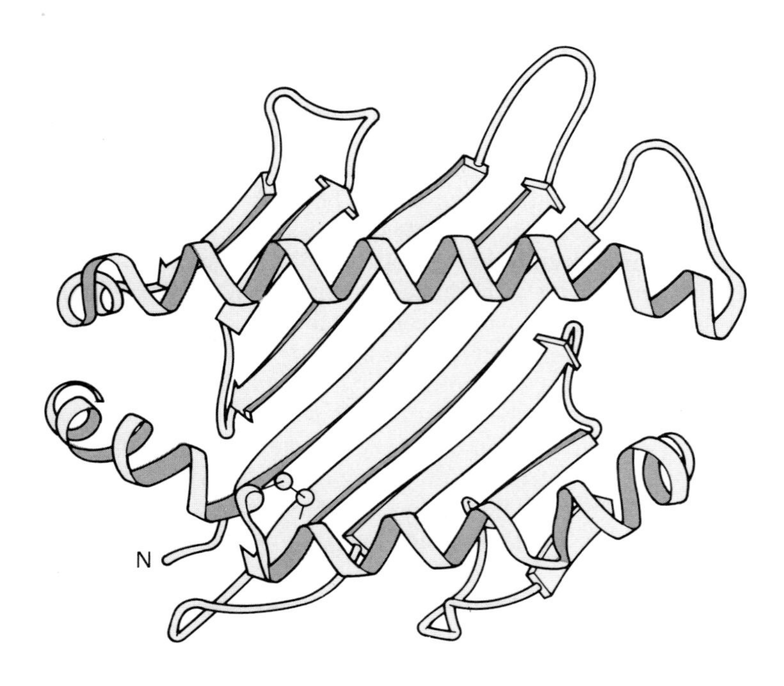

Fig.2.6 Diagrammatic representation of the peptide-binding groove of an MHC class I molecule, in the absence of a peptide, and viewed from above; compare with Fig.2.5. (Adapted from Bjorkman *et al.* (1987*a*), with permission)

α helices. Thus it appears that a peptide (or peptides, because the extra electron density could not be resolved into a single structure) was already occupying the supposed antigen binding cleft in this preparation of HLA-A2 molecules. Furthermore, extra electron density that was not part of the primary protein structure, was also found in the groove of the Aw68 and B27 molecules. These observations are very exciting and have important implications for the molecular basis of antigen recognition by T cells (see Sections 3.3 and 4.4). If the nature of the electron density occupying the groove could be defined, it would be the first direct demonstration of the interaction between a peptide and MHC molecule at the structural level. Analysis of the B27 crystals suggests that a more limited set of peptides is bound in the cleft of this class I allele than is found in the cleft of either HLA-A2 or Aw68. The peptides present in the cleft of B27 molecules are composed of nine amino acid residues and appear to be in an extended conformation (Madden *et al.* 1991; Jardetzky *et al.* 1991).

Alternative approaches to identify the electron density found in the HLA-A2, Aw68, and B27 crystals are being explored. One would be to take a peptide that is known to be recognized by cytotoxic T cells in the association with a particular allele, e.g. HLA-A2, and to try and generate a crystal containing the A2 molecules associated with the known peptide (Silver *et al.* 1991). Using this approach, because the amino acid sequence of the peptide is known, it should be possible to locate the peptide and resolve its structure from the electron density map. This information would provide further confirmation that the cleft formed by the α_1 and α_2 domains is the peptide binding site on class I MHC molecules, and would allow the nature of the interaction between the peptide and the MHC molecule to be investigated further. A second approach would be to elute naturally occurring self peptides from a

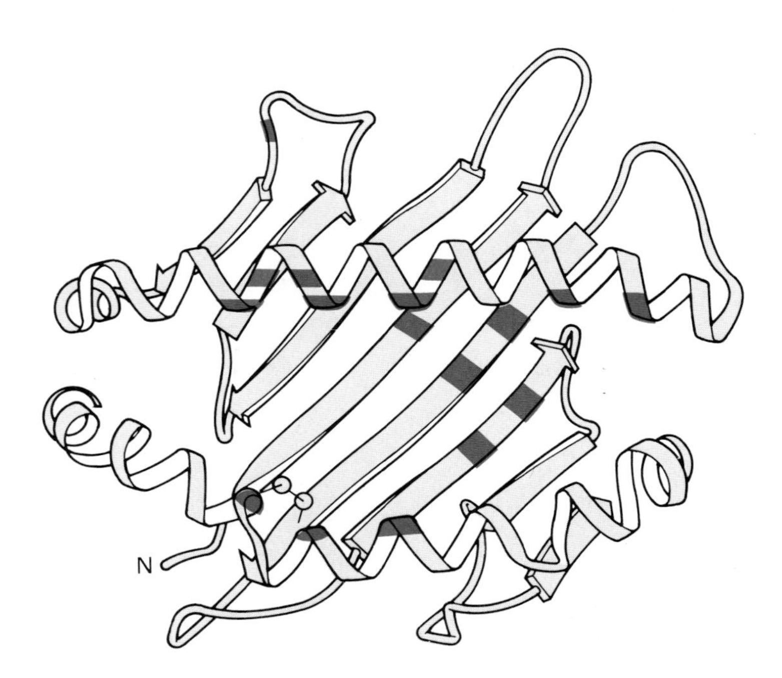

Fig.2.7 Location of polymorphic amino acid residues in the peptide binding groove of HLA-A2. (Adapted from Bjorkman *et al.* (1987*b*), with permission)

preparation of purified class I molecules and determine their amino acid sequences, and hence the peptide binding motif associated with that HLA allele. This approach has been applied successfully to analyse the self peptides bound to HLA-B27 (Jardetzky *et al.* 1991).

Evidence in support of the idea that the groove formed by the α_1 and α_2 domains binds antigen in the form of peptide is also provided by mapping the location of the polymorphic amino acid residues (i.e. those that vary between alleles) on the class I structure (Fig.2.7). In the A2 α chain, 17 amino acids are considered to be polymorphic, this information being obtained by comparing amino acid sequence data from different alleles; of these, 15 are located in the region of the groove, either in the α helices lining the sides or in the β sheets that form the floor. Thus the majority of the polymorphic amino acids would be able to contact the peptide present in the groove. These amino

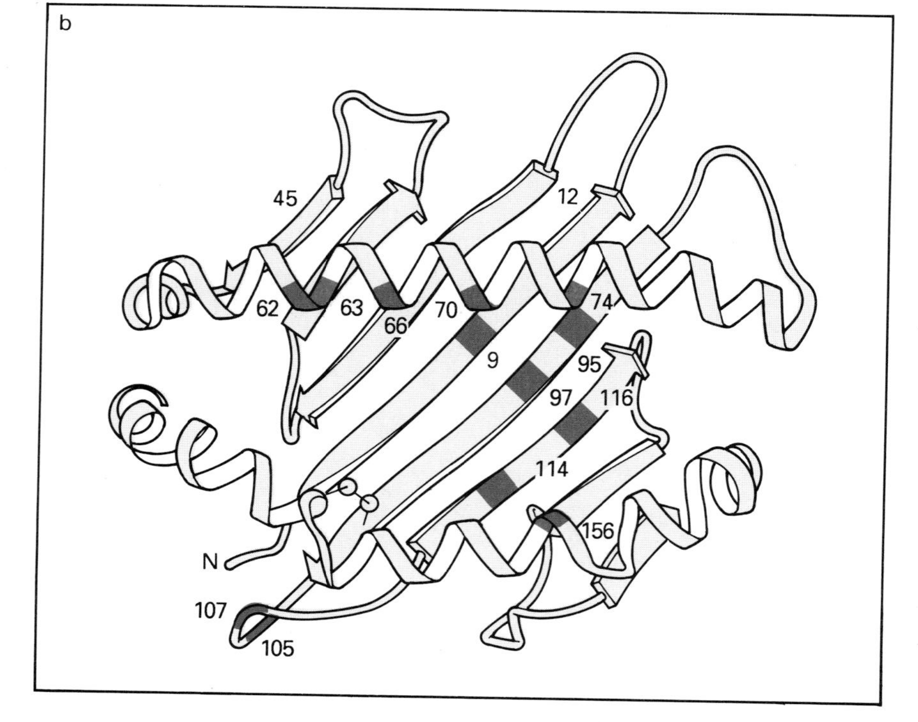

a

		▼		▲	▼	▼	▼	▼	▼	▼			▼	▼	▼	
		9	12	62	63	66	70	74	95	97	105	107	114	116	156	245
				+				−					+	−		
Aw68	(Aw68.1)	Tyr	Val	Arg	Asn	Asn	Gln	Asp	Ile	Met	Ser	Gly	Arg	Asp	Trp	Val
	Aw68.2	—	Met	—	—	—	—	—	—	Arg	Pro	—	His	Pyr	—	—
	AW69	—	—	—	—	—	—	—	Val	Arg	—	Trp	His	Tyr	Leu	Ala
A2	(A2.1)	Phe	—	Gly	Glu	Lys	His	His	Val	Arg	—	Trp	His	Tyr	Ala	Ala
					−	+	(+)	(+)		+			(+)			
								△		△				△		
		α_1							α_2							α_3

Fig.2.8 Amino acid substitution between HLA-A2 and Aw68.
(a) Comparison of the amino acid sequences for HLA-Aw68.1, Aw68.2, Aw69, and A2.1. For clarity, only the positions at which the sequences of Aw68.2, Aw69, and A2.1 are different from Aw68.1 are shown. The substituted amino acids which line the peptide binding groove are indicated (▼). The amino acid substitution at position 62, postulated to affect T cell receptor contacts, is indicated (▲) and amino acid residues affecting the 74 pockets (Fig.2.9) are shown (△). Charged amino acids are also shown (+ or −).
(b) Schematic diagram to show the α_1 and α_2 domains of the HLA-A locus molecules. Differences between HLA-A2.1 and Aw68.1 are indicated in solid red.
(Adapted with permission from *Nature*, Volumes 342 and 329, pp.617 and 508–9 respectively)

acids may therefore influence the nature of the peptides that can associate with each class I allele.

A comparison of the 3D structure of the α_1 and α_2 domains of A2 and Aw68 supports these conclusions. A2 and Aw68 are closely related molecules and as a result very similar in amino acid sequence. They differ from each other by only 13 amino acids, 6 of these differences being in the α_1 domain, 6 in α_2 and 1 in α_3. Of the 12 substitutions in the α_1 and α_2 domains, 10 are located in positions lining the floor and sides of the antigen binding groove (Fig.2.8). The differences in the chemistry, size, and polarity of the amino acids at these positions has a dramatic effect on the topography of the putative peptide binding sites (Fig.2.9).

If the cleft is responsible for binding peptides, the differences between A2 and Aw68 imply that these two HLA-A locus alleles will bind different sets of peptides. This idea is strongly supported by comparing the extra electron density present in the two crystal structures—they are quite different—and is consistent with the data obtained for HLA-B27. (For review see Rothbard and Gefter 1991).

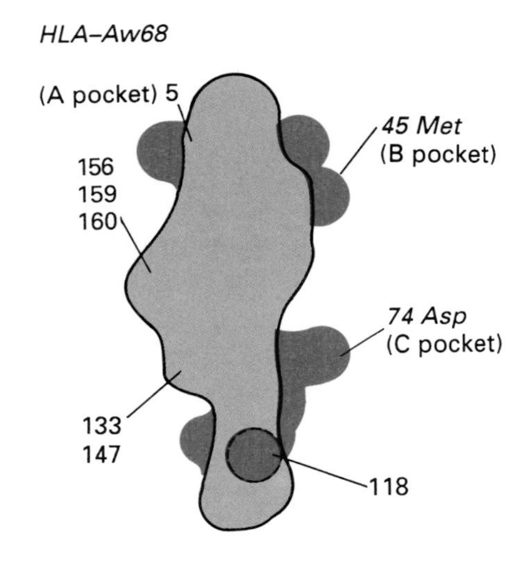

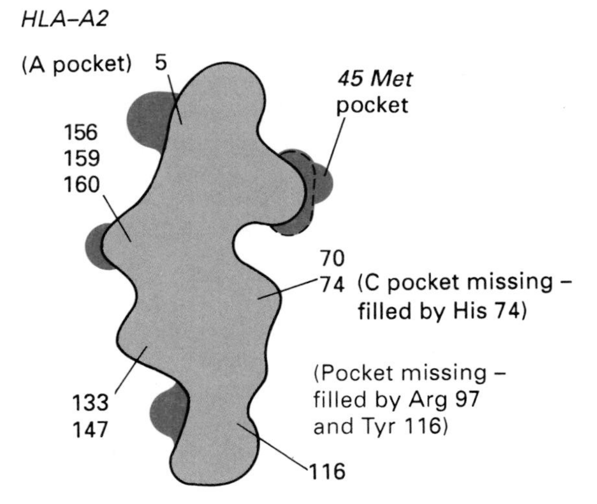

Fig.2.9 Schematic comparison of the antigen binding groove in HLA-Aw68 and A2 (adapted from Garrett *et al.* (1989), with permission). HLA-Aw68 shows prominent pockets in the peptide binding cleft near amino acid positions 74 and 116, and a ridge (not indicated) between amino acids 156 and 70. These pockets are absent in HLA-A2 because they are filled by amino acids His74, and Arg97 and Tyr116 respectively. The pocket located near amino acid residue 45 is present in both Aw68 and A2

While the majority of the polymorphic amino acid residues are located to the sides and floor of the antigen binding groove, the amino acid residues that are conserved between alleles and locus products are largely confined to other parts of the structure (Fig.2.10), but interestingly 10 of the amino acids that point into the groove are completely conserved in all human class I sequences. It may be that these residues are important for maintaining the structure or the conformation of the α_1 and α_2 domains of the class I molecule.

The determination of the structures of HLA-A2, Aw68, and B27 has given immunologists a model for the structure of other MHC molecules. Although many of the general structural features are likely to be shared by all class I molecules, it is important to remember that A2, Aw68, and B27 are only three out of more than 80 human class I alleles.

Serological determinants

As well as providing information about how MHC molecules might present antigens to T cells, structural analysis has also allowed the parts of the molecule that stimulate antibody production to be identified. Serology, or the analysis of MHC molecules using antibodies, is employed extensively to define the HLA alleles expressed by different individuals. This technique is known as **HLA typing** or **tissue-typing**.

Antibodies specific for different MHC alleles can be prepared by either deliberate or passive immunization between different members of the same species. For example, in animals, alloantisera and monoclonal antibodies are prepared by deliberately immunizing one animal with cells taken from an animal of a different strain (see Appendix, pp. 695–8). In the human, where

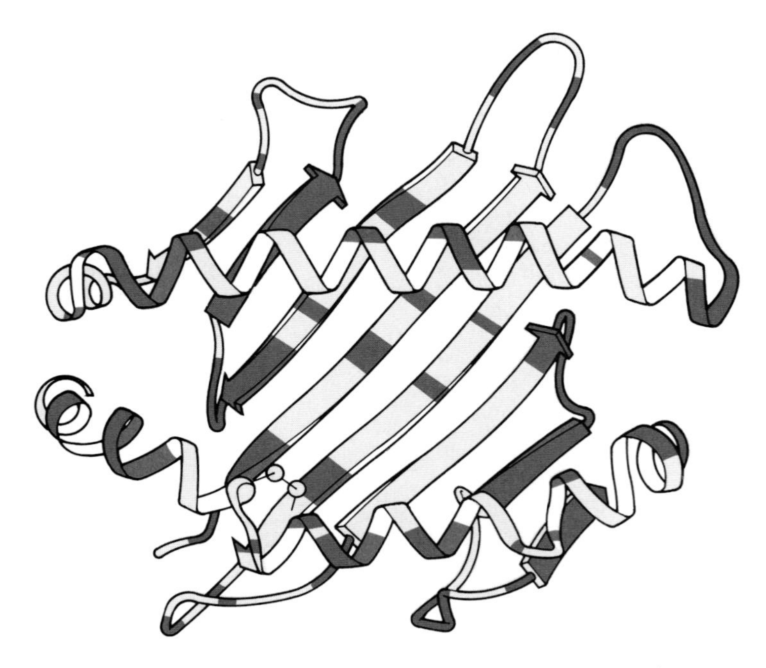

Fig.2.10 Location of conserved amino acids in the α_1 and α_2 domains of class I molecules. (Adapted from Bjorkman *et al.* (1987*a*), with permission)

this is not ethical, alloantisera for HLA typing are obtained from pregnant women who have made antibodies to the 'foreign', paternally-derived, HLA antigens expressed by the fetus or from blood-transfused individuals where the blood cells used for transfusion were of a different HLA type. Alloantibodies specific for a particular HLA allele can also be produced by the recipient during the rejection of a transplanted organ, such as a kidney, heart, or liver that is MHC incompatible. The binding of alloantibodies of known HLA specificity to cells of the individual being tissue typed is routinely detected by the addition of complement, which will lyse cells that have bound the antibody (see Appendix, p. 706). Not surprisingly, the epitopes of HLA-A2 that are recognized by antibodies are, in general, found on the exposed surfaces of the structure and not buried inside the molecule (Fig.2.11). This is consistent with the idea that antibodies interact with external conformational epitopes of proteins (see Section 1.1.3). Epitopes present on HLA class II molecules that are recognized by monoclonal antibodies have also seen mapped (Marsh and Bodmer 1989).

Epitopes that are unique to a particular MHC allele are known as *polymorphic* or **'private'** determinants, whereas those that are shared by more than one MHC allele are referred to as *monomorphic* or **'public'** determinants. This concept is illustrated schematically in Fig.2.12.

Polymorphism

MHC class I antigens are polymorphic because the amino acid sequence of the α chain differs between alleles. For example, A2 and A3 differ from each other by 15 amino acids, and A2 and Aw68 differ by 13 residues. As already noted, the majority of the variations between alleles are located in the α_1

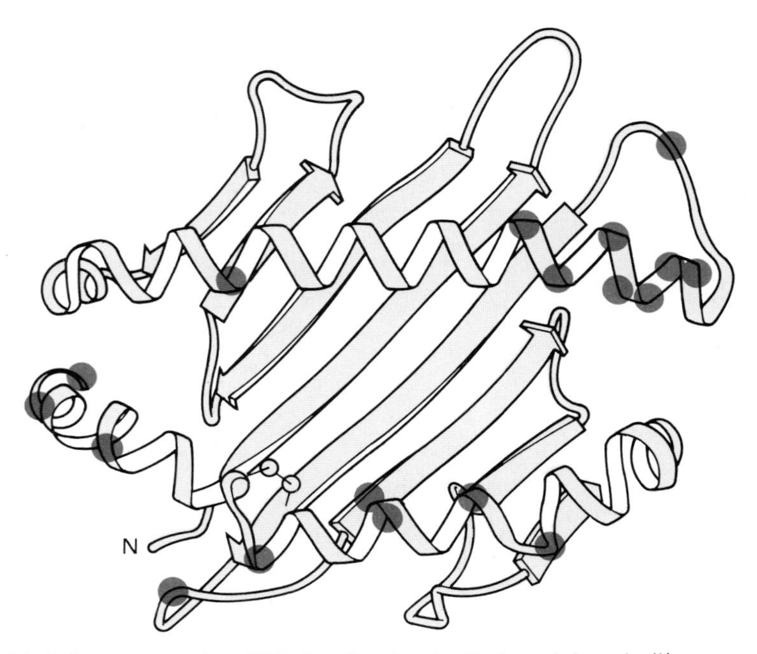

Fig.2.11 Epitopes present on HLA class I molecules that can interact with antibodies — serological epitopes. (Adapted from Bjorkman *et al.* (1987*a*), with permission)

and α_2 domains. The α_3 domain is much more highly conserved, which may reflect the importance of α_3 in its association with β_2-microglobulin and for interaction with the CD8 molecule on T cells (Salter *et al.* 1990) (see Section 4.2.4). The variant amino acids are clustered together at positions 62–83 in α_1, 105–116 in α_2 and 174–194 spanning the α_2 and α_3 domains. Clustering of the polymorphic residues is also seen in class II molecules (see Section 2.5.2).

Mutant class I molecules

In both mouse and human, class I glycoproteins have been identified that are very similar to a particular allele, but which contain additional muta-

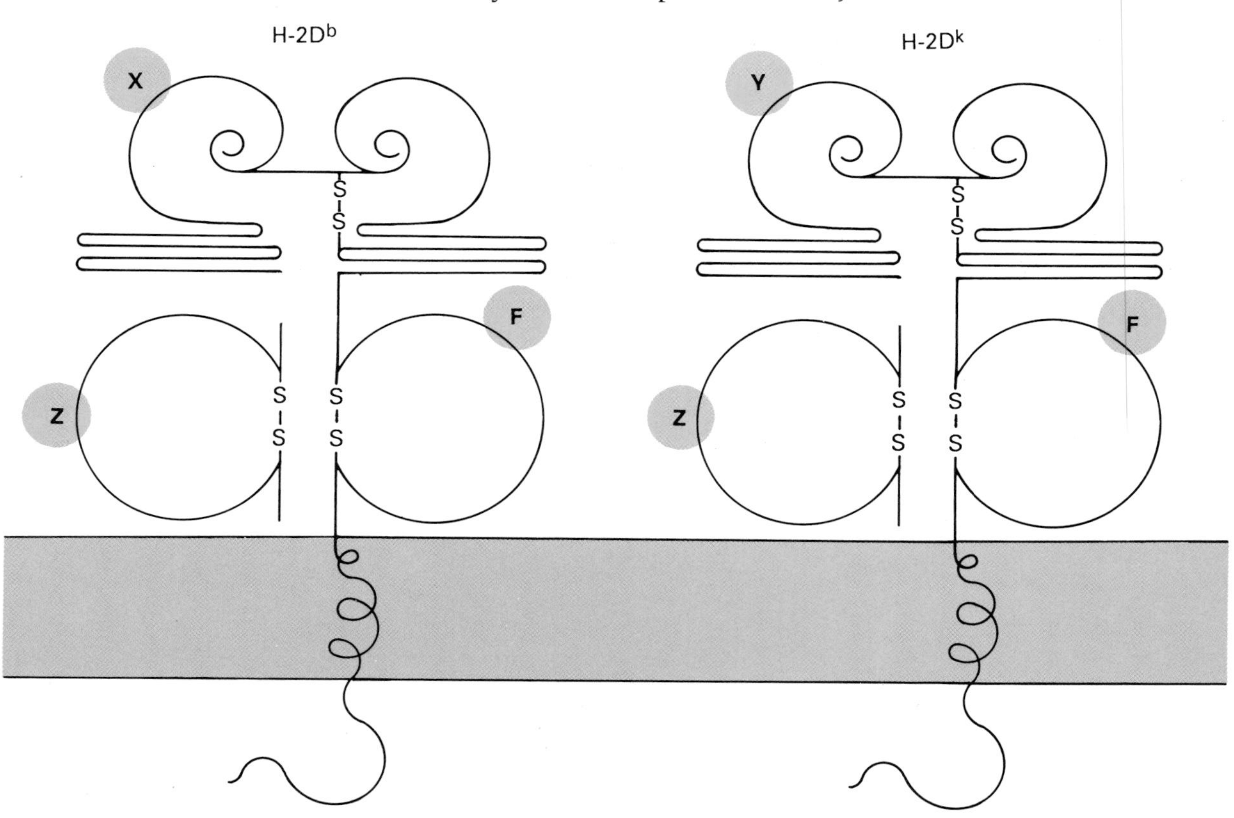

Determinant

X	Polymorphic	–	unique to D^b
Y	Polymorphic	–	unique to D^k
Z	Monomorphic	–	shared by all mouse class I molecules
F	Monomorphic	–	shared by all H-2D molecules

Fig.2.12 Diagram to illustrate the theoretical location of polymorphic and monomorphic serological determinants in class I molecules

tions. These are known as mutant MHC molecules. In the mouse some of the most carefully studied mutations are those that have occurred in H-2K^b molecules, where 17 spontaneous mutations have been detected (for review see Nathenson *et al.* 1986). The amino acid sequence changes found in some of these variant alleles, known as bm mutants, are shown in Table 2.6. Some of the mutants differ from K^b at only a single amino acid residue. A comparison of the nucleotide sequences of the normal and mutant molecules reveals that, in some cases, the mutation results from only a single nucleotide substitution, for example the bm5 mutation. Mutants of HLA-A2 exist in the human population (Table 2.6), and these have been very useful for exploring

Table 2.6 Mutant class I molecules

(a) Location of amino acid changes in mutant H-2K^b molecules

Haplotype	Locus product affected	Position in the polypeptide chain	Amino acid charges	Domain affected
bm1	K	152	E–A	
		155	R–Y	α_2
		156	L–Y	
bm3	K	77	D–X	α_1
		89	K–A	
bm5	K	116	Y–F	α_2
bm6	K	116	Y–F	α_2
		121	C–R	
bm8	K	22	Y–F	α_1
		23	M–I	
		24	E–S	
bm10	K	165	V–M	α_2
		173	K–X	

(b) Location of amino acid changes in some of the HLA-A2 mutants

Cell line	Allele affected	Position in the polypeptide chain	Amino acid changes	Domain affected
DK1	A2	149	A–T	α_2
		152	V–E	
		156	L–X	
DR1	A2	43	Q–R	α_2
M7	A2	43	Q–R	α_1

A–Alanine
C–Cysteine
D–Aspartic acid
E–Glutamic acid
F–Phenylalanine
I–Isoleucine
K–Lysine
L–Leucine
M–Methionine
R–Arginine
S–Serine
V–Valine
Y–Tyrosine
X–Unknown amino acid replacement

the potential interaction of the T cell receptor with the MHC–antigen complex now that the structure of HLA-A2 has been determined.

Mutant class I molecules have also been used extensively to investigate the interaction between peptides and MHC molecules, and the interaction of these complexes with T cell receptors (see Section 3.3.3).

2.5.2 Class II

General features

Class I and class II MHC molecules have many common structural features. In both cases the molecules are transmembrane glycoproteins, each with four extracellular domains arranged in pairs. Class II molecules are also composed of two polypeptides, the α and β chains, but in this case both chains span the plasma membrane and both chains are encoded by genes located in the MHC (see Section 2.8; Figs.2.21 and 2.22). At present no data for the 3D structure of class II molecules are available but they are likely to appear soon; however, by molecular modelling using the framework structure established for HLA-A2 and amino acid sequences for class II polypeptides, the structure of class II MHC molecules can be predicted (Fig.2.13) (Brown *et al.* 1988). Each polypeptide can be divided into four regions, two extracellular domains (α_1 and α_2; β_1 and β_2) each containing 90–100 amino

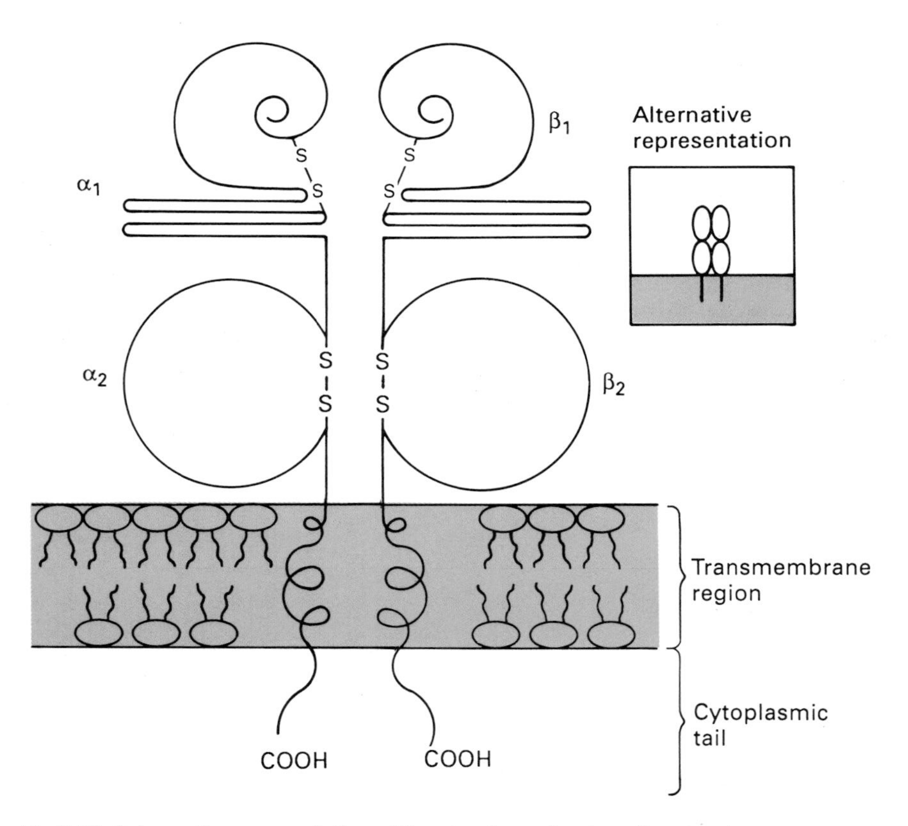

Fig.2.13 Schematic representation of the structure of a class II molecule

acids, a transmembrane region of 20–25 residues and a cytoplasmic tail of 3–15 residues in α chains and 8–20 in β chains. The structural model shows that the α_1 domain is paired with β_1 and the α_2 and β_2 domains are also closely associated and sit on the membrane. The α_2 and β_2 membrane proximal domains are predicted to adopt the immunoglobulin fold structure, as has been shown for the α_3 domain and β_2 microglobulin in class I molecules (Fig.2.5). The α_1 and β_1 domains are predicted to form an antigen binding platform by folding in a similar way to the α_1 and α_2 domains of class I molecules, each domain folding into an α helix and four β strands that forms half of the antigen binding cleft.

Both the α and β chains are glycosylated. The carbohydrate side chains are complex, involving galactose, fucose, and glucosamine units (Fig.1.4). The α and β chains differ in size (M_r 31–34 kDa and 26–29 kDa respectively) largely as a result of the fact that the α chains have two carbohydrate substitutions, one on the α_1 and the other in the α_2 domain, while β chains contain only one carbohydrate unit in the β_1 domain. Heterogeneity in the composition of the carbohydrate units may account for the differences observed in the molecular weights of class II molecules in different species and individuals, and the extent and type of glycosylation may vary between cell types. The function of the carbohydrate units is not known. Glycosylation is not required for the association of the α and β chains, nor for their expression on the cell surface. It has been suggested that the carbohydrate moieties may play an important role in cellular interactions, although this has not been determined at present.

Polymorphism

Class II, like class I, molecules are polymorphic. In general, the class II β chains exhibit a much greater degree of polymorphism than α chains. For example, in human more than 18 variants or polymorphic forms of HLA-DR β chains have been identified, whereas no variants of the DR α chain have been found. This is also true for H-2 IE molecules in the mouse. In contrast, both the α and β chains are polymorphic in HLA-DQ and H-2 IA molecules.

The polymorphic determinants are mainly located in the N-terminal domain of each chain, the α_1 domain in DQα chains and the β_1 domain of the DP, DQ, and DQβ chains. These regions of sequence variability are located in defined positions in the first domain of each chain (so-called 'allelic hypervariable regions'). For example, three allelic hypervariable regions have been identified in the DRβ_1 domain, the first including amino acids 9–13, the second 26–32, and the third around amino acid 70. Sequence data from some haplotypes is shown in Fig.2.14 (Todd *et al.* 1987; Marsh and Bodmer 1990*a*,*b*). When all of the sequence data are taken together, it can be seen that the different human class II alleles are generated by mixing and matching allelic hypervariable regions within the first domain. Thus for example, the first and the third allelic hypervariable regions of the DRB1 genes for DR5 and DR6 are identical, but the second region is different between the two alleles. This idea is represented diagrammatically in Fig.2.15. The model for the structure of class II antigens predicts that the majority of the polymorphic amino acids are located in and around the putative antigen binding groove (Fig.2.16).

```
                             10                     20                          30                       40                     50
DRB1*0101  G D T R P R F L|W Q   L K F|E C N F F N G   T E R V R|L L E R C   I Y N Q E E S V|R F   D S D V G E Y R A V
DRB1*1501  - - - - - - - -|Q -   D - Y|- - - - - - -   - - - - -|F - H - D   - - - - - - D L|- -   - - - - - - - - - -
DRB3*1501  - - - - - - - -|- -   P - R|- - - - - - -   - - - - -|F - D - Y   F - - - - - - -|- -   - - - - - - F - - -
DRB1*1502  - - - - - - - -|Q -   D - Y|- - - - - - -   - - - - -|F - N - G   - - - - - - N -|- -   - - - - - - - - - -
DRB3*1502  - - - - - - - -|- -   P - R|- - - - - - -   - - - - -|F - D - Y   F - - - - - - -|- -   - - - - - - F - - -
DRB1*1601  - - - - - C - -|Q -   D - Y|- - - - - - -   - - - - -|F - H - G   - - - - - - N -|- -   - - - - - - - - - -
DRB3*1601  - - - - - - - -|- -   P - R|- - - - - - -   - - - - -|F - D - Y   F - - - - - - -|- -   - - - - - - - - - -
DRB1*0301  - - - - - - - -|E Y   S T S|- - - - - - -   - - - - -|Y - D - Y   F H - - - - N -|- -   - - - - - - F - - -
DRB3*0101  - - - - - - - -|E L   R - S|- - - - - - -   - - - - -|Y - D - Y   F H - - - - F L|- -   - - - - - - - - - -
DRB3*0201  - - - - - - - -|E L   - - S|- - - - - - -   - - - - -|F - - - N   F H - - - - Y A|- -   - - - - - - - - - -
DRB1*0401  - - - - - - - -|E -   V - N|- - - - - - -   - - - - -|F - D - Y   F - H - - - Y -|- -   - - - - - - - - - -

                             60                    70                   80
DRB1*0101  T E L G R P D A E Y   W N S Q K D|L L E Q   R R R A V D T|Y C R   H N Y G V G E S F T   V Q R R
DRB1*1501  - - - - - - - - - -   - - - - - -|F - - D   - - - - - - -|- - -   - - - - - - - - - -   - - - -
DRB3*1501  - - - - - - - - - -   - - - - - -|I - - -   A - - - - - -|- - -   - - - - - V - - - -   - - - -
DRB1*1502  - - - - - - - - - -   - - - - - -|F - - D   - - - - - - -|- - -   - - - - - - - - - -   - - - -
DRB3*1502  - - - - - - - - - -   - - - - - -|I - - -   A - - - - - -|- - -   - - - - - - -
DRB1*1601  - - - - - - - - - -   - - - - - -|I - - -   A - - - - - -|- - -   - - - - A V - - - -   - - - -
DRB3*1601  - - - - - - - - - -   - - - - - -|F - - D   - - - - - - -|- - -   - - - - - - - - - -   - - - -
DRB1*0301  - - - - - - - - - -   - - - - - -|- - - -   K - G R - - N|- - -   - - - - - V - - - -   - - - -
DRB3*0101  - - - - - - V - - S   - - - - - -|- - - -   K - G R - - N|- - -   - - - - - - - - - -   - - - -
DRB3*0201  R - - - - - - - - -   - - - - - -|- - - -   R - G Q - - N|- - -   - - - - - V - - - -   - - - -
DRB1*0401  - - - - - - - - - -   - - - - - -|- - - -   K - - - - - -|- - -   - - - - - - - - - -   - - - -
```

Fig.2.14 Sequence data of the β_1 domain of HLA-DRB genes DRB1 and DRB3 to show the existence of three allelic hypervariable regions in the polypeptide chain (from Todd *et al.* (1987), with permission). The single letter amino acid code is used (see Fig. 1.3). Numbers above indicate amino acid position; a dash indicates identity with DRB1*0101 sequence

Polymorphism in class II molecules can be detected in several ways. At the protein level they can be detected using alloantisera or monoclonal antibodies as tissue typing reagents or by two-dimensional gel electrophoresis. At the DNA level they can be detected by examining restriction fragment length polymorphisms (RFLP) (for review see Bidwell 1987) or by using allele-specific oligonucleotides to detect DNA amplified using the polymerase chain reaction (PCR) (for review see Bell 1989). Each of these techniques is described in the Appendix.

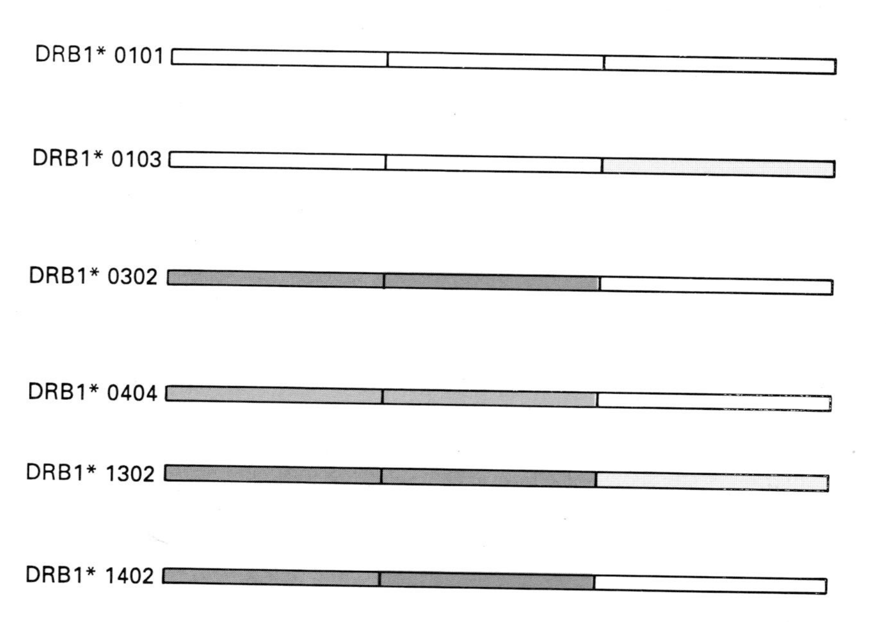

Fig.2.15 Diagrammatic representation of the mixing and matching of allelic hypervariable regions in the first domain of the HLA-DRB chain to generate different HLA-DR alleles

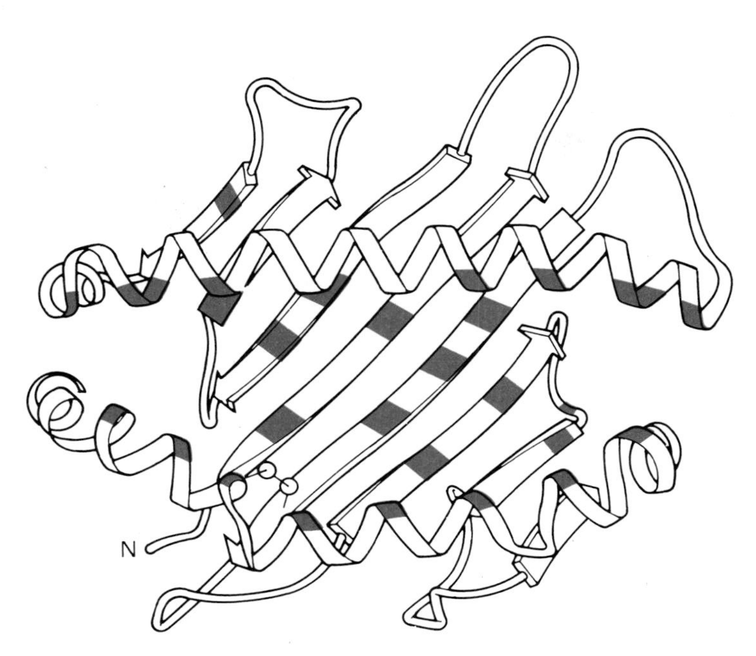

Fig.2.16 Predicted location of the polymorphic amino acid residues of MHC class II molecules (adapted from Brown *et al.* (1988), with permission) mapped onto the outline structure of a class I molecule

Interaction with superantigens

The term superantigen is used to describe certain molecules that can stimulate T lymphocytes by novel mechanisms (see Section 5.1). Superantigens are distinguished from (i) conventional antigens because they activate a high frequency of responding T cells, between 5 and 25%, and (ii) polyclonal T cell mitogens because they bind to a specific site in the variable portion of the T cell receptor β chain (Herman *et al.* 1991). Superantigens can be divided into two main categories at present, (i) enterotoxins produced by certain bacteria (Marrack and Kappler 1990) and (ii) Mls determinants that arise from the integration of retroviral sequences derived from mouse mammary tumour viruses (MMTV) into the mouse genome (see Section 5.2).

Superantigens bind to MHC class II molecules. The enterotoxins derived from *Staphylococcus aureus* do not bind to all class II molecules with equal affinity. For example, staphylococcal enterotoxin B (SEB) binds well to HLA-DR and less well to HLA-DQ alleles but not at all to HLA-DP alleles. In the mouse the toxins bind well to H-2 IE (the HLA-DR equivalent, Table 2.8), but less well to H-2 IA. In addition there is some evidence for weak haplotype specificity amongst the enterotoxins in mice. For example, toxins bound to H-2 IA^k stimulate T cells less well than toxins bound to H-2IA^d or IA^b. The superantigens do not bind to class II molecules in the peptide binding groove, but to the side of the molecule in association with the β chain (Fig.2.17).

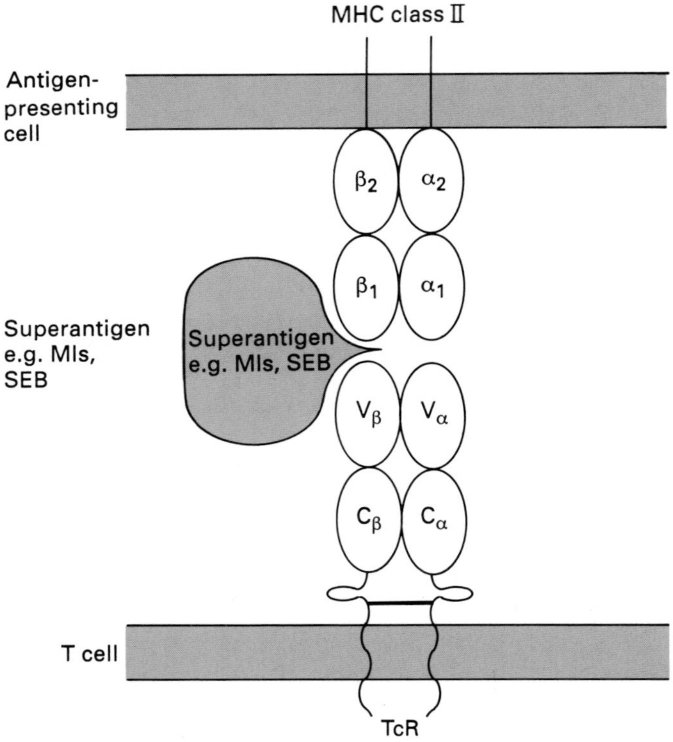

Fig.2.17 Model for the interaction of superantigens with class II molecules

2.6 Biosynthesis and surface expression

2.6.1 Class I molecules

β_2-microglobulin and the role of peptides

β_2-microglobulin plays an important role in the expression of class I molecules. Cells that have deletions or mutations in the β_2-microglobulin gene do not express class I molecules at the cell surface, although α chains are found in the cytoplasm. If the β_2-microglobulin gene is transfected (see Appendix) into these class I deficient cells, expression of class I molecules at the cell surface is returned to normal. Thus β_2-microglobulin is thought to play an important role in the assembly and transport of class I MHC molecules within the cell.

Several mutant cell lines that are deficient or express only low levels of class I at the cell surface have been described (see pp. 148, 149). Interestingly, these cells have normal levels of mRNA for the α chain and β_2-microglobulin. If peptides are incubated with these cells, class I is expressed at the cell surface (Townsend *et al.* 1989, 1990). Thus it would appear that association of antigenic peptides with the peptide binding groove is required for the correct folding of the α chain, its association with β_2-microglobulin, and transport of class I molecules to the cell surface (for review see Parham 1990*a*). If the intracellular antigen processing machinery of the cell is defective and it cannot produce peptides or deliver them to the endoplasmic reticulum where α chains and β_2-microglobulin are synthesized, expression of class I molecules at the cell surface is defective. Genes that may encode peptide transporter proteins have been mapped to the MHC (Fig.2.22 and Section 3.2.2). It is interesting to speculate as to the role of these molecules in the production of peptides required for the correct assembly of class I molecules (for review see Parham 1990*b* and 1992; Monaco 1992).

A mutant strain of mouse with defective β_2-microglobulin genes has been developed using the technique of homologous recombination (see Appendix) (Zijlstra *et al.* 1990). These mice do not express class I molecules at the cell surface, although α chains are found in the cytoplasm. Furthermore, as a result of inactivation of the β_2-microglobulin gene, the $CD8^+$ T cell population is virtually absent from the peripheral lymphoid tissue, preventing these mice from making normal cytotoxic T cell responses. Thus expression of class I molecules is critical for the development of the $CD8^+$ T cell repertoire (see Section 5.2.3).

Adenovirus

Cells transformed with human adenovirus 12 have greatly reduced expression of class I at the cell surface. The level of class I mRNA production is also decreased, although the rate of initiation of transcription remains the same as in uninfected cells. The adenovirus E3/19K gene product binds to class I molecules and retains them in the endoplasmic reticulum. A number of other viruses can also decrease membrane expression of MHC class I molecules on infected cells, and presumably decrease their capacity to present viral peptides to cytotoxic T cells.

2.6.2 Class II molecules

The invariant chain

The α and β chains of class II molecules are synthesized independently in the endoplasmic reticulum. During biosynthesis the α and β chains become associated with a third type of polypeptide chain, usually called the **invariant chain**, and designated Ii (it can also be referred to as the gamma chain, p31 or p33). The invariant chain is not encoded within the MHC of either mouse or human. In mouse the gene maps to chromosome 18 and in human to chromosome 5. It is a basic transmembrane glycoprotein, and is synthesized in large amounts by cells that express class II molecules. Interestingly, the invariant chain is anchored in the membrane by a stretch of hydrophobic residues at the N-terminal end of the polypeptide, such that at the cell surface the C-terminal portion is extracellular; in other words, the orientation of the chain in the membrane is opposite to that of the majority of glycoproteins. In normal cells, the expression of the α, β, and invariant chains is coordinately regulated. In human, recent evidence has suggested that four distinct forms of the invariant chain exist. Molecular studies have shown that the three species are produced from a single mRNA. Two of the products, which differ by 16 amino acids, arise because of different initiation codons, both of which are in phase and can be used to initiate translation, and the other two arise as a result of alternative splicing.

The invariant chain associates with the α and β chains of the class II molecules shortly after translation in the endoplasmic reticulum. This association appears to be very important for the intracellular transport of class II heterodimers and the invariant chain contains signals for the correct targeting of the $\alpha\beta$ complex to particular cellular compartments (see Section 3.3.2). The invariant chain remains associated with the $\alpha\beta$ complex during its transport through the Golgi apparatus and is thought to dissociate from the class II $\alpha\beta$ dimer in an acidic, protease-containing compartment. It is probable that immunogenic peptides bind to class II molecules in the endosomal compartment only after the invariant chain has dissociated. Thus the invariant chain prevents the early association of peptides with class II molecules as they are transported from the endoplasmic reticulum after biosynthesis (Roche and Cresswell 1990). By preventing the binding of peptides until class II molecules have entered the endosomal compartment, the invariant chain could be important for maintaining the differences between the antigen presentation pathways for class I and class II antigens (see Section 3.2; Teyton *et al.* 1990; Neefjes and Ploegh 1992).

In some cases the invariant chain may not be absolutely necessary for cell surface expression of class II molecules. For example, mouse L cells or 3T3 cells, the cells that are commonly used for gene transfection experiments (see Appendix), do not synthesize α, β, or invariant chains. When the α and β chain genes were transfected into these cells, class II molecules were expressed at the cell surface. However, when the gene for the invariant chain was transfected together with the α and β genes, cell surface expression of the class II molecules was *not* increased. Nevertheless, experiments using *Xenopus* oocytes did not support these findings. In this system, in the absence of the invariant chain, fewer class II molecules were expressed at the cell surface and abnormalities in the glycosylation of the α and β chains occurred. The reasons for these conflicting observations are not yet clear.

Association of α and β chains

Association of the α and β chains appears to be a prerequisite for membrane expression of class II molecules. This has been shown using strains of mice that are unable to produce one of the class II chains (Table 2.7). Mice of most haplotypes can express both IA and IE class II molecules. In contrast, mice of the b or s haplotypes express IA but do not express IE molecules (Table 2.4).

Table 2.7 Strains of mice with abnormal expression of class II MHC molecules

H-2 haplotype	Defective gene
b	Eα
s	Eα
f	Eα+Eβ
q	Eα+Eβ

This is because they do not produce α chains, due to a deletion in the Eα gene. Although these mice have a defective Eα gene they have functional Eβ genes, and IEβ chains can be detected in the cytoplasm of cells isolated from H-2^b and H-2^s mice.

The expression defect in the Eα gene can be complemented in F1 hybrids (in other words when H-2^b mice are crossed with mice from a strain that does express H-2 IE, such as BALB/c) or in transgenic mice (see Appendix) where a functional IEα gene is introduced into an H-2^b mouse embryo. In both of these cases cell surface expression of IE molecules is returned to normal. This implies that the introduction of a functional α chain gene allows the α chains to associate with the endogenously-produced IEβ chains and a functional H-2 IE molecule can therefore be expressed.

Two other strains of mice, the f and q haplotypes, also fail to express H-2 IE class II molecules. In these haplotypes, cells express neither IEα nor

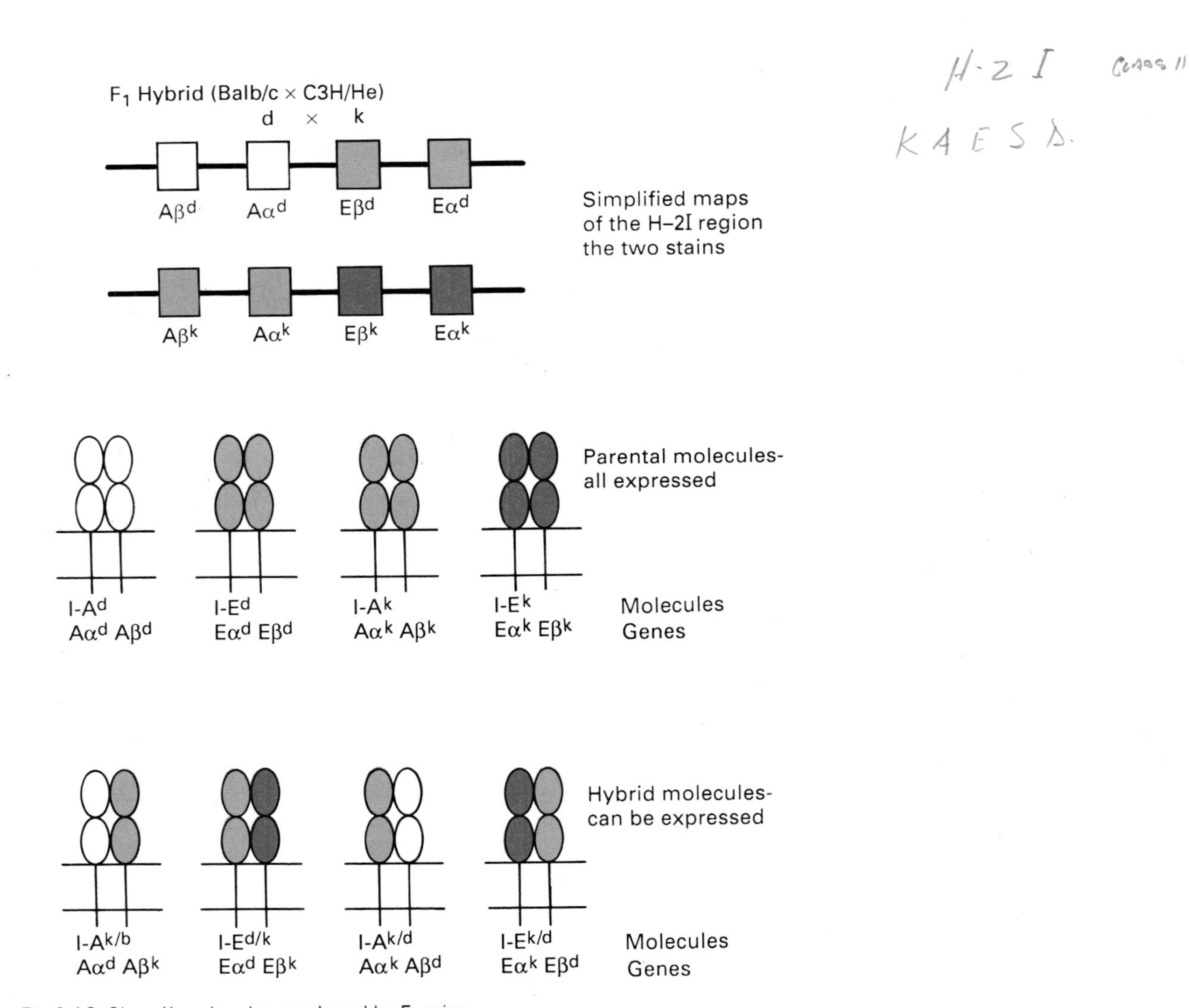

Fig.2.18 Class II molecules produced by F_1 mice

IEβ chains due to defects in *both* of these genes; the molecular basis for these defects awaits characterization.

Wild mice have also been examined for defects in class II expression. It has been estimated that approximately 20% of wild mice do not express cell-surface IE molecules. The lack of IE expression in these mice appears to correlate with the presence of **t chromosomes**. These are abnormal regions of chromosome 17 that affect gene expression and development in approximately 40% of wild mice.

The number of class II MHC molecules expressed at the cell surface can vary between individuals. Heterozygous individuals express more molecules than homozygotes because, in addition to the expected maternal and paternal α and β combinations, hybrids of the maternal and paternal α and β chains are also found. Thus F1 mice, for example, can in fact produce a total of eight class II heterodimers, all of which can be detected at the cell surface (Fig.2.18). These hybrid molecules are functional, have been shown to participate in antigen recognition by helper T cells, and presumably increase the range of peptides that can be presented to T cells in F1 animals (see Section 3.3).

It has commonly been assumed that class II molecules are always expressed on plasma membranes as heterodimers generated from the products of one group of $\alpha\beta$ loci. Thus in the mouse, the class II products expressed would be the IA molecule, formed by the association of the Aα and Aβ chains, and the IE product, Eα–Eβ. (In other words the class II molecules are formed from isotype matched $\alpha\beta$ pairs.) Indeed, hybrid molecules, such as Aα–Eβ or Eα–Aβ have not been detected on normal cells, nor have homodimeric molecules such as Aα–Aα or other hybrids such as Aα–Eβ. Similar isotype-matched pairing of the class II α and β chains was also predicted in man. Recently, the possibility that mismatched pairs of α and β chains, for example Aα–Eβ or DRα-DQβ, can exist at the cell surface has been raised.

Using cloned α and β chain genes from the mouse, transfection experiments (see Appendix) have established that mismatched pairs of α and β chains from different loci can in fact assemble and be expressed at the cell surface. In these experiments, only the mismatched α and β chain genes were introduced into L cells (which do not normally express class II molecules). Thus the normal, matched partners for the products of the transfected genes were not available for pairing, and the only option was for the mismatched α and β chains to pair with one another. In normal cells the matched chain is also available in the cytoplasm of the cell, so the matched $\alpha\beta$ pair may form preferentially. Obviously, it is not possible to rule out the existence of such hybrid molecules on normal cells completely, as they may be present on the membrane but at levels that are too low to be detected (for review see Lechler 1988). In human there is some evidence that mixed DR/DQ molecules do exist on lymphoblastoid B cell lines (Lotteau *et al.* 1987).

2.7 Structure and regulation of expression of MHC genes

2.7.1 Class I genes

Structure

All of the murine and human class I genes analysed have a similar structure (Fig.2.19), consisting of eight exons spanning 4000–5000 bases (4–5 kb). The first exon is short and contains the leader sequence encoding the signal peptide. The function of the signal peptide is to guide the nascent polypeptide chain through the endoplasmic reticulum during biosynthesis. It is cleaved from the mature protein.

Exons 2, 3, and 4 are all similar in size and code for the three extracellular domains, $\alpha1$, $\alpha2$, and $\alpha3$ (Figs.2.19 and 2.5). Each of these exons codes for an entire domain of the class I heavy chain and the exon-intron boundaries in the genomic DNA correlate precisely with those of the protein domains. This relationship between the exon/intron structure of the genomic DNA and the domain structure of the protein is also found in the genes coding for lymphocyte antigen receptors, the immunoglobulins and T cell receptors, and in the genes that code for some of the other members of the immunoglobulin gene superfamily (see Section 7.4). Exon 5 encodes the hydrophobic transmembrane region, and exons 6, 7, and 8 the cytoplasmic tail and 3′ untranslated region (Fig.2.4).

The gene for β_2-microglobulin is coded outside the MHC on chromosome 15 in human and 2 in mouse. It is present as a single copy. The structure of the β_2-microglobulin gene is shown in Fig.2.19. It consists of four exons, but most of the coding region is found in the second exon.

Regulation of expression—control of transcription

Several DNA sequence elements that control gene transcription have been identified upstream from class I MHC genes. These are known as ***cis*-acting regulatory elements**. Three sequences are present in the promoter region of the gene that is located 5′ to the signal sequence: (i) the cap site; (ii) the TATAAA box; and (iii) the CAAT sequence. Mouse class I α genes have the expected CAAT and TATAAA sequences 50 and 25 bp upstream of the 5′ cap site (e.g. Kimura *et al.* 1986). The majority of human class I α genes have a variant TATAAA sequence, T**C**TAAA. In some HLA-C genes the CAAT sequence has been replaced by **CGG**T and the TATAAA box by T**C**T**G**AA. The presence of these variant promoters may lead to lower levels of transcription and therefore decreased expression at the cell surface.

In the mouse additional sequences that control transcription have been found 5′ to some class I genes. For the H-2K^b gene, several sequences that have homology with those known to regulate other eukaryotic genes have been identified. The activity of these sequences resembles those of **tissue specific promoters** that determine whether a gene is expressed in a particular tissue. In the 5′ region of the H-2L^d gene a **negative regulatory element** has been found that is under developmental control. This element is highly conserved in class I genes and thus, depending on the state of development, this DNA element may act to turn class I transcription on or off.

In addition to *cis*-acting control elements, ***trans*-acting regulatory factors** encoded by separate genes that are distinct from the class I genes are also thought to control class I gene expression. For example, some human T lymphoblastoid cell lines express very low levels of class I antigens, whereas their B cell counterparts express high levels. If low and high expressing cell lines are fused together, there is increased expression of class I MHC antigens by the T cell line. This finding is interpreted as indicating that soluble factor(s) capable of activating class I expression and encoded by the *trans*-acting regulatory sequences is/are present in the B cell line.

2.7.2 Class II genes

Structure

The genomic organization of the genes encoding the α and β chains of class II molecules is, like that for class I, striking. Each structural domain of the polypeptide chain is encoded by a separate exon in genomic DNA (Fig.2.20). Exon 1 codes for the hydrophobic signal or leader peptide of approximately 30 amino acids; this is important during biosynthesis, but is removed from the mature polypeptide chain. The second and third exons code for the extracellular domains of the α and β chains: the N-terminal domains, α_1 or β_1, and the membrane proximal domains, α_2 or β_2, respectively. The next exon codes for the hydrophobic transmembrane region. In α chains, this exon also contains the sequence of nucleotides that codes for the cytoplasmic tail. In the β chain genes, however, the amino acids which comprise the cytoplasmic tail are encoded by a separate exon (see Fig.2.20). A final exon codes for the 3′ untranslated region of the α or β chains.

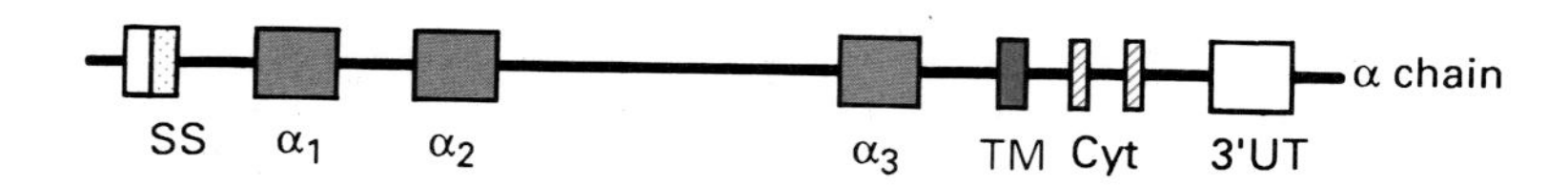

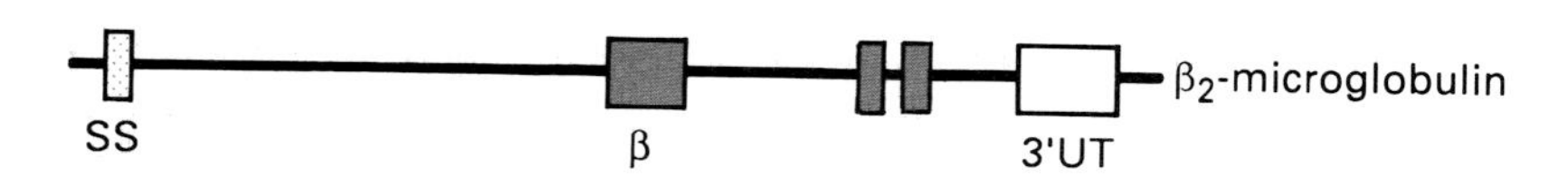

Exons encoding:

SS = signal sequence
TM = transmembrane region
Cyt = cytoplasmic tail
3'UT = 3' untranslated region

Fig.2.19 Exon–intron organization of the genes encoding the α chain of class I MHC molecules and β_2-microglobulin

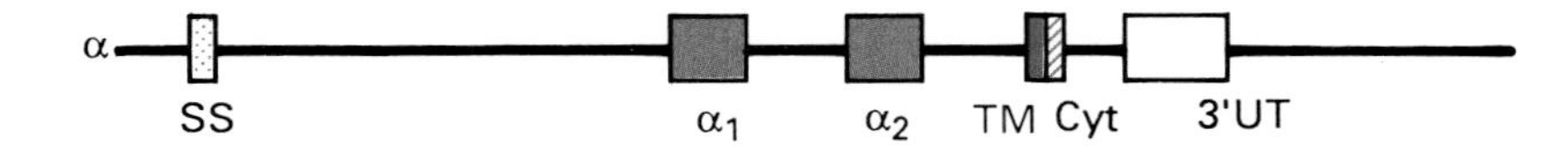

Exons encoding:

SS = signal sequence
TM = transmembrane region
Cyt = cytoplasmic tail
3'UT = 3' untranslated region

Fig.2.20 Exon–intron organization of the genes encoding the α and β chains of class II molecules

Regulation of expression

Alterations in chromatin structure have been reported to correlate with class II gene expression. DNAase hypersensitive sites associated with the looser chromatin structure have been identified after IFNγ treatment. These map to either the first intron or a site upstream of the promoter (Fig.2.21). DNA can be modified by methylation of particular nucleotide sequences, CpG sites, and the pattern of DNA methylation appears to be involved in controlling gene expression. In some situations demethylation of class II genes, for example DRA, appears to correlate with increased expression. It is important to note that this correlation is not absolute, as demethylation of at least one CpG site in the DRA gene *blocks* the expression of the gene (this is unusual, because demethylation more commonly results in increased gene expression).

All α and β genes are regulated in the same way, and this has allowed deductions to be made about the significance of various regulatory elements that are present (some of these are shown in Fig.2.21). Two highly conserved sequences, known as the X (14 bp) and Y (14 bp) boxes, spaced by a specific 18–19 bp sequence in β genes and a specific 19–20 bp sequence in α genes, are located between base pairs 115 and 100 and 80 and 70, on the 5′ side of the transcription initiation site of the α or β genes respectively. Interestingly, these elements are also found upstream from the invariant chain gene. It has been suggested that both the X and Y boxes are positive regulatory sequences that are part of a common mechanism for the regulation of expression of the α, β, and invariant chain genes.

An IFNγ regulatory sequence is also located upstream from the class II genes, e.g. DRA, in the same region as another IFNγ regulatory sequence known as the W box. The nucleotide sequence of the W box corresponds to part of the IFN consensus sequence (Freidman and Stark 1985). It is found to a greater extent in β genes than either α or invariant chain genes.

In addition, cell type specific enhancer elements have been identified *within* the HLA-DQA, DQB, and DRA genes and upstream from H-2 IEβ gene, suggesting that expression of class II antigens can be specifically regulated, correlating with the more limited tissue distribution of class II molecules (see Section 2.4).

Trans-acting regulatory factors also play a role in the control of class II expression. Evidence for this comes from various sources, for example patients with severe combined immunodeficiency syndrome (SCID) do not express HLA-DR, DQ, or DP molecules. Family studies have shown that the genetic defect is not linked to the MHC but is due to the action of a *trans*-acting factor. Interestingly, the level of invariant chain expression is normal in the SCID patients. Several *trans*-acting factors have been identified. For example, two nuclear proteins produced by B cells interact specifically with the X box, and a nuclear protein common to B cells, T cells, and a human carcinoma cell line can bind to the W box; the Y box binds a nuclear factor thought to be present in all cell types (Fig.2.21).

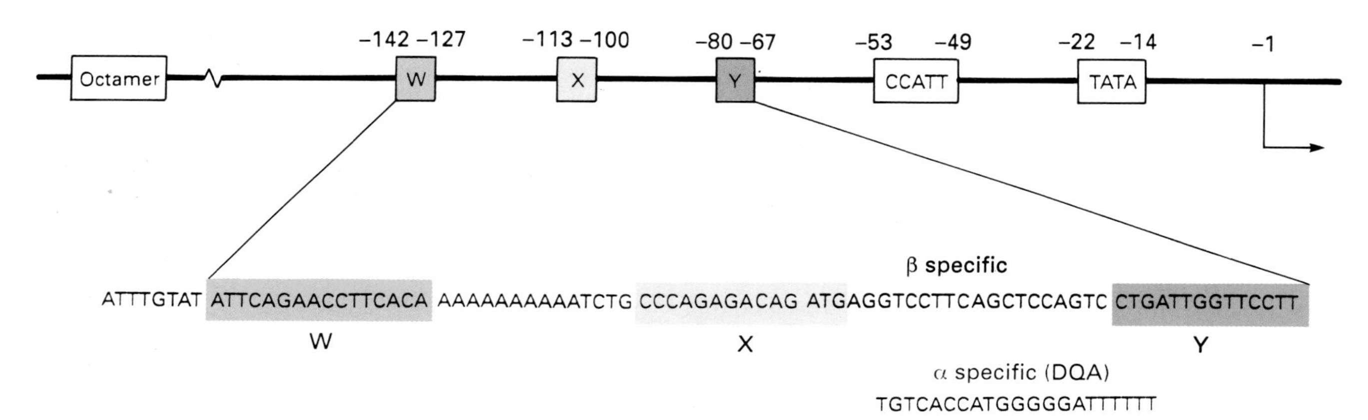

Fig.2.21 Regulatory sequences associated with class 11A and B genes. Other *cis*-acting elements, including J and S have been identified for some class II genes

2.8 Genomic organization of the MHC

Detailed maps of the organization of the mouse and human major histocompatibility complexes are given in Figs.2.22 and 2.23, respectively. Each complex comprises multiple loci (sometimes within one region) encoding class I, class II, and class III products (Table 2.8). In mouse the H-2 maps to chromosome 17, and in human the HLA region is located on the short arm of chromosome 6. The HLA complex is between two and three times larger than H-2 which covers approximately 3000 kb.

The organization of the genes within each of these complexes has been determined using a variety of techniques, including those of classical genetic analysis, and more recently molecular genetic approaches have been adopted

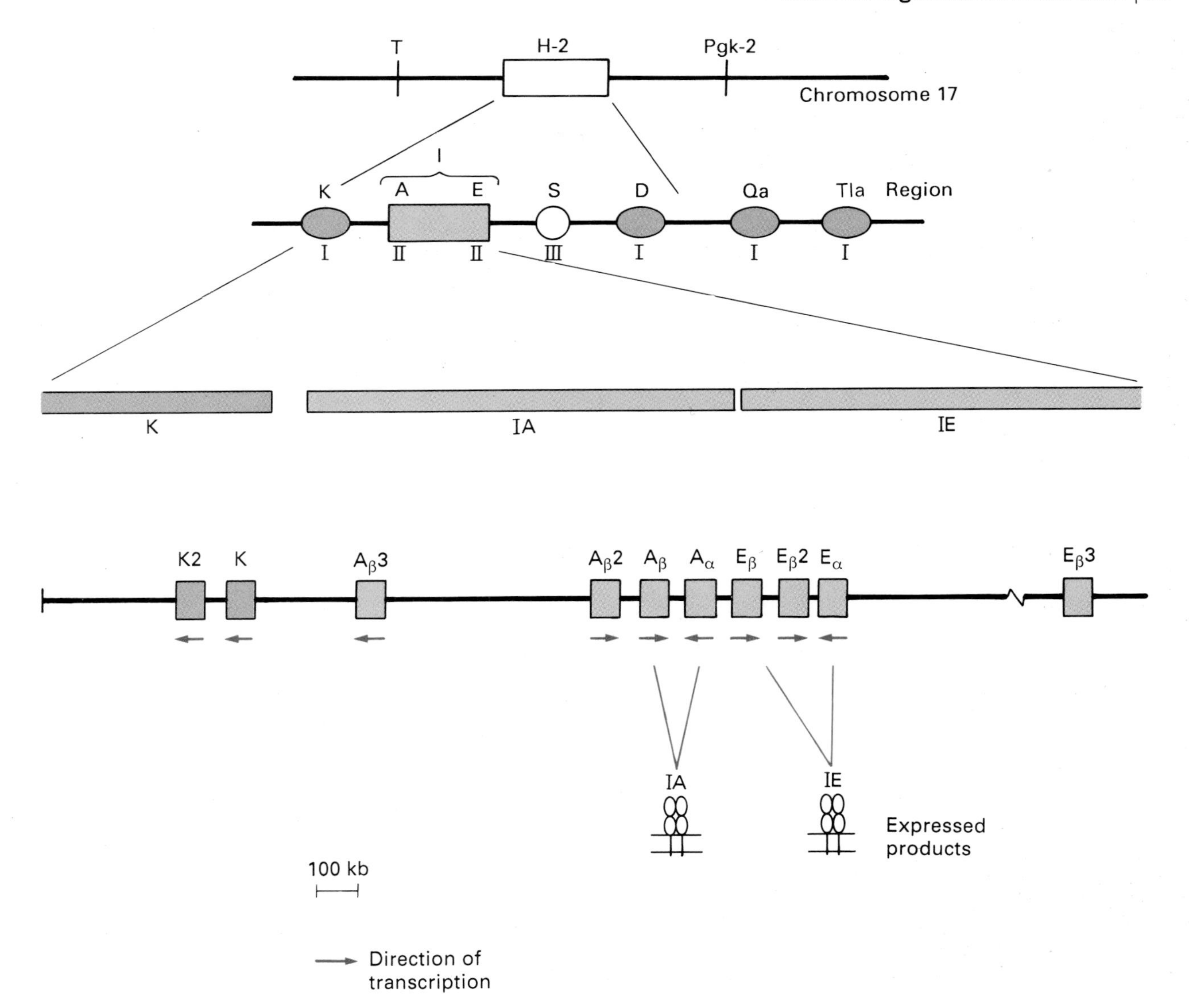

Fig.2.22 Genomic organization of the H-2 complex

with great success. The general organization of the complexes is different in mouse and human (Fig.2.1). In human the class II loci are located between the centromere and the class I loci, whereas in the mouse the class II loci are positioned between the class I loci. The organization of the human complex seems to be the more common pattern because the arrangement pattern of the H-2 has only been found in one other species, the rat.

2.8.1 Class I loci

The class I loci in mouse and human form a multigene family. There are many more class I genes than had been predicted from the number of locus

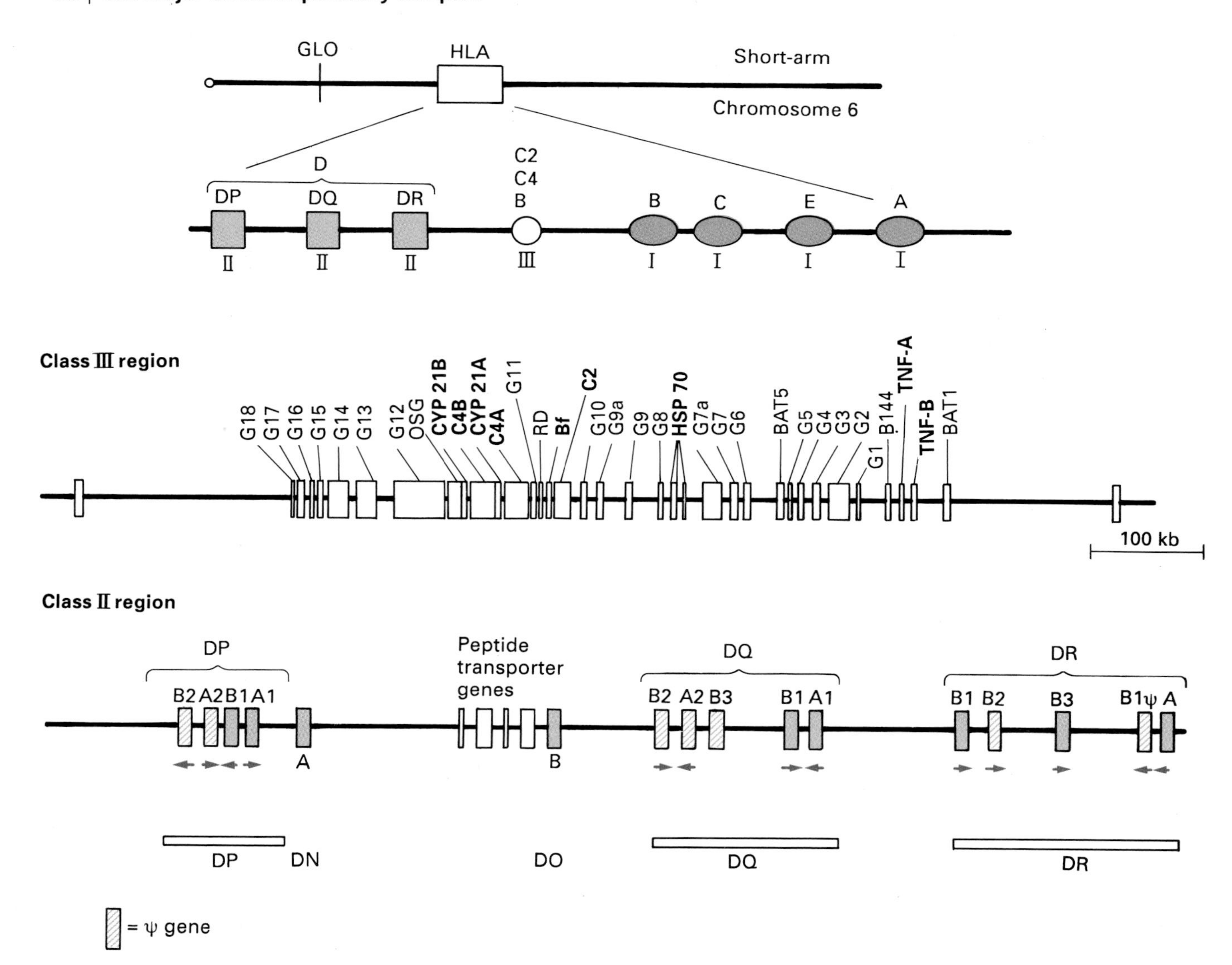

Fig.2.23 Genomic organization of the HLA complex. From Trowsdale *et al.* (1991), with permission

products identified serologically. In human, 17 class I genes have been mapped to the short arm of chromosome 6, and in the mouse 33 genes have been detected in BALB/c H-2^d mice and 26 in C57BL/10 H-2^b mice.

The evolution of class I MHC genes has been reviewed by Klein and Figueroa (1986) and Lawlor *et al.* (1990). They are presumed to have arisen from a single ancestral gene.

Class I loci in the mouse

Four loci or regions code for class I molecules, H-2K, D/L, Qa 2–3, and Tla.

H-2K and D loci. The K and D regions of H-2 encode the heavy or α chains of class I molecules (Fig.2.22). Two genes are present in the K region in BALB/c and B10 mice. The two genes are arranged in the reverse orientation to one another and are separated by about 20 kb. Gene transfection studies have been used with great success to investigate which genes code for antigens previously characterized by serology. The genes have been introduced or transfected into cells, usually mouse L cells, by DNA-mediated gene transfer (see Appendix p. 716), and expression of the antigens has been detected at the cell surface with monoclonal or alloantibodies (for review see Malissen 1986). Such studies have shown that one of the genes in the H-2K region is expressed as the H-2K molecule recognized serologically, while the other gene designated H-2K1 is located further along chromosome 17 closer to the I region. The H-2K1 gene has a functional promoter and may be expressed in some but not all inbred strains of mice (Fig.2.2).

The other class I region, H-2D, is **telomeric** with respect to H-2K, that is, this region is further along the chromosome from the centromere (Fig.2.22). The number of class I genes in this region varies between strains. For example, B10 H-2^b mice only have a single D locus gene whereas mice of the H-2^d and H-2^q haplotypes have at least three genes in the H-2D region, indeed five have been identified in BALB/c H-2^d mice.

In BALB/c mice two genes encode the serologically defined D^d and L^d antigens. These two genes are separated from each other by 170 kb of DNA, D^d lying closest to the I region and L^d being located within 60 kb of the Qa region. The three additional class I genes $D2^d$, $D3^d$ and $D4^d$ are found in the DNA between the D^d and L^d genes. It is not yet known whether the additional three genes are functional (Fig.2.2).

As the molecular analysis of the H-2 from different inbred strains of mice continues, it may become apparent that expression of different numbers of genes from each region of the H-2 is a common phenomenon. The cloning and analysis of class I genes in the BALB/c strain of mice has been reviewed by Hood *et al.* (1983).

Qa and Tla loci. Additional class I loci have been identified on the telomeric side of the H-2 complex. These loci are contained within the Qa and Tla regions (Fig.2.24).

When DNA probes for the H-2K and D locus class I genes were used to analyse mouse DNA, by the technique of Southern blotting (see Appendix, p. 713), many more genes than just those coding for the H-2K and D region

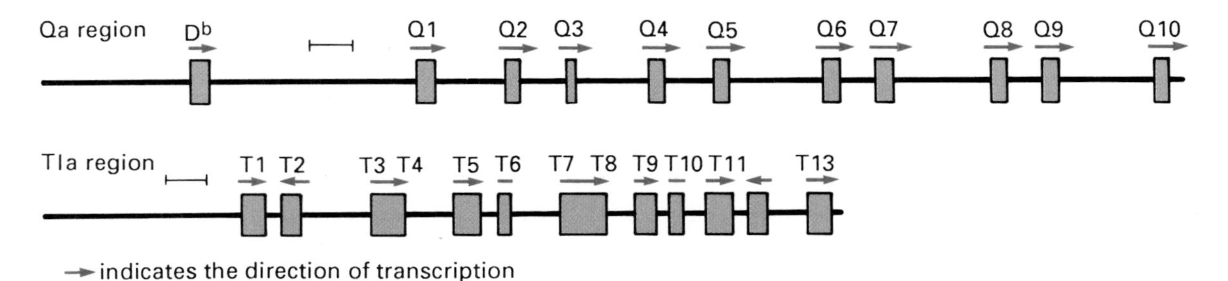

Fig.2.24 Loci encoded in the Qa and Tla regions in C57BL/10 mice.
NB The organization of the Qa and Tla genes in the BALB/c and C57BL/10 genomes is not identical

products were detected. In other words the DNA for, say, the H-2K gene, could hybridize or bind to not only the H-2K gene itself, but also to related or homologous sequences elsewhere in the H-2 complex. The nucleotide sequences of these extra class I genes are highly homologous to those that code for the 'classical' class I MHC molecules. Remarkably, the majority of the class I genes identified at the DNA level are actually encoded in the Qa and Tla regions of chromosome 17, not within H-2 complex itself in the K or D regions. For example, between 23 and 26 class I loci have been identified in the Qa and Tla regions in the C57BL/10 and BALB/c genomes (compared to merely 2–5 in the H-2K and D regions), and of these eight are located in the Qa region in BALB/c mice and ten in B10 mice (Fig. 2.24). In the Tla region 15 class I genes are present in B10 mice and 18 in BALB/c, but not all of these genes are expressed (For reviews see Mellor 1986 and Stroynowski 1990).

Protein products have only been identified for between six and nine of the 35 or so genes characterized in these two regions and not all of these are expressed at the cell surface. Some are apparently only expressed in the cytoplasm, eg. the soluble Q10 molecule, while others, Q4, Q6, 7, 8, 9, and 10 are expressed at the cell surface (Robinson 1987). Some of the Qa region class I molecules have a lower molecular weight than the 'classical' class I products. This is because they are linked to the membrane by a phosphatidyl inositol anchor (Fig.1.5), an alternative to the transmembrane region of K and D locus products.

The discrepancy between the (relatively small) number of class I polypeptides identified as cell surface molecules and the (relatively large) number of class I gene sequences identified in these two regions may be explained in different ways.

Firstly, it is possible that all of the genes in these two regions are actually expressed, but that their products cannot be detected by the reagents that are currently available. For instance, if some of the class I products of the Qa, Tla region genes were not sufficiently polymorphic, in other words the number of variant forms that exist in different strains of inbred mice is limited, then it would be difficult to generate serological reagents by alloimmunizations that could subsequently be used to detect them (see Appendix). If no alloantibodies are available, the products of the Qa and Tla region genes will remain undetected. This may explain why only four products were detected after each of the ten genes from the Qa region of B10 mice were transfected into L cells. All of the products identified reacted with anti-Qa-2 sera, thus the products of the other six genes may have been undetectable because of the lack of specific sera.

A second, equally plausible, possibility is that the 'extra' class I genes identified are pseudogenes. If this were the case the genes would not be expressed either as cell surface or as soluble molecules. This possibility should be clarified by sequencing these class I-related genes, since pseudogenes have characteristic features that prevent their transcription and/or subsequent translation into protein. Another approach to determining if these genes are pseudogenes is to prepare cDNA libraries from different tissues, liver, kidney, and so on, and to examine them for transcribed class I products. If clones are identified the genes they originate from can be mapped.

The properties of most of the Qa and Tla class I-related molecules are not the same as those of the 'classical' H-2K and H-2D class I molecules. One difference is that unlike H-2K and D antigens, some of the Qa, Tla genes are only expressed on cells of the haemopoietic lineage. Thus, the expression of some of the Qa, Tla genes is tissue specific, in other words the gene is transcribed and translated in some, but not all tissues in the body. As another example, expression of one of the Q10 genes has been shown to be liver specific. Nevertheless, there is now at least one example of Qa and Tla gene products that have a more widespread distribution body. For example, mRNA for the Q4 gene can be detected in a large number of tissues, including liver, kidney, and brain.

Another difference between Qa and Tla antigens and conventional class I molecules is that they are less polymorphic. For example if the sequences of the Q10 gene from two strains of mice, B10 and SWR, are compared, they show >90% homology. These differences suggest that the Qa, Tla class I molecules may have a different function to that of other class I glycoproteins, but as yet this is unknown. While there are differences, there are also some similarities between these different class I molecules—the most obvious is that β_2-microglobulin is associated with these class I-like α chains when they are expressed at the cell surface.

Class I loci in the human

Three HLA class I loci have been detected serologically, HLA-A, B, and C (Fig.2.23; Table 2.3). HLA-A and B were the first two HLA class I loci to be characterized. Extensive serological polymorphism has been detected in the products of both of these loci; 24 A locus and 52 B locus products have been defined so far. The HLA-C locus was detected subsequently and is located between A and B. It is less polymorphic; currently 11 C locus products have been identified. The A and B loci are separated by 1 centiMorgan (cM) or 10^6 bases. Seventeen class I genes have been identified in human in total: three code for the HLA-A, B, and C products and of the other 14, four code for intact class I molecules HLA-E, F, G, and H (Table 2.3). The HLA-E gene is located between the C and A loci and mRNA from this gene has been detected in several different cell types. The other, functional genes are located downstream from HLA-A and are known as non-classical class I genes; they may turn out to be similar to genes in the Qa or Tla regions in the mouse.

2.8.2 Class II loci

In the mouse H-2 complex, the class II molecules are encoded within the I region, and in the HLA complex by the D region (Fig.2.1). In both species the class II region of the MHC contains several loci, three in man and two in mouse. The evolution of class II MHC genes has been reviewed by Figueroa and Klein (1986).

Class II MHC molecules are also known as Ia antigens. The term Ia comes about because the mouse class II antigens map into the H-2 I region. The I region had been defined previously as the region into which the so-called Ir (*i*mmune *r*esponse) genes mapped (see Section 5.4.1). When class II molecules were identified serologically and were shown to map to the same region, they became known as *Ir*-*a*ssociated antigens. This term was then extended to other species: for example the class II molecules of man are also sometimes referred to as human Ia antigens.

Class II loci in the mouse

A map of the genes within the I region of the H-2 complex is shown in Fig.2.22. The diagrams show the genes for which products have been serologically and biochemically identified: these are located within the IA and IE **subregions**. The genetics and expression of murine class II molecules have been reviewed by Mengle-Gaw and McDevitt (1985).

Earlier genetic maps identified five subregions within the H-2 I region, not only IA and IE but also IB, IC, and IJ. Protein products encoded by the IB and IC subregions could not be demonstrated biochemically, and it is now thought likely that these two subdivisions of the I region were artefacts. However, the IJ subregion presents a paradox that has still to be resolved, and possible reasons for this will be discussed below.

IA and IE subregions. The IA and IE subregions were first defined by classical genetic mapping. They contain the genes that code for the four polypeptide chains required to produce IA and IE class II molecules. Two genes, an α gene and a β gene, are required for each class II product. The α and β chain genes of IA molecules are encoded within the genetically defined IA subregion (Aα and Aβ genes, Fig.2.22). Part of the β chain gene of IE is also encoded within the IA subregion. However, the remainder of the Eβ chain and the complete Eα chain are encoded within the IE subregion (Eα and Eβ genes, Fig.2.22). In addition to the functional IA and IE α and β genes, both subregions also contain extra α and β genes (Fig.2.22). These extra genes are not expressed and their function, if any, is not yet understood.

Recombination events within the H-2 complex frequently occur in the H-2 IE subregion (Table 2.5). **Hotspots for recombination** have been identified within or closely associated with the Eβ gene, the Aβ3 gene, and between Aβ3 and Aβ2 and Eβ2 and Eα.

Genes other than α and β chain genes also map to this region of the H-2 complex. The products of these genes are required for antigen processing and two of the four genes identified so far, TAP1 and TAP2, are thought to be involved in the transport of peptides into the endoplasmic reticulum (Monaco *et al.* 1990; Monaco 1992). The other two genes encode subunits of the low-molecular-mass polypeptide (LMP) complex, which is closely related to the proteasome (multicatalytic proteinase complex), an intracellular protein complex that has multiple proteolytic activities (Brown *et al.* 1991) (see Section 3.3.2).

IJ. The IJ subregion of the H-2 complex was originally defined serologically using antisera prepared by immunizations between two strains of inbred recombinant mice, called B10.A (3R) and B10.A (5R) (Table 2.5). In these strains a recombination event occurred between the IA and IE subregions within the H-2I region; this region/locus was called IJ. Originally this was thought to be the only recombination event to have occurred in the whole mouse genome. The antisera prepared using this immunization protocol reacted with a small number of T cells but not B cells (B cells express conventional IA and IE class II molecules). In particular, the IJ determinants seemed to be expressed by suppressor T cells and their factors (see Section 11.2).

Alloantisera and monoclonal antibodies raised between the two IJ disparate strains have been used in attempts to characterize the products of the IJ locus and to study their expression on suppressor T cells. The molecule recognized by these reagents was reported to have a molecular weight of M_r 25 kDa, which is lower than would be expected if it were a conventional class II MHC molecule, but a complete characterization is still awaited.

A major problem with IJ arose when the sequence of the entire mouse I region, and beyond, was determined. The classical IA and IE α and β chain genes were found, but the IJ subregion appeared to map to a section of DNA that was only 2 kb in length. This section of DNA was contained entirely within the IE β chain gene itself. To eliminate the possibility of artefacts in the cloning procedures, the analysis has now been performed in at least ten different recombinant strains of mice, and identical results have been obtained for each. It is clear that the molecular and serological data are extremely difficult to interpret, and they have resulted in much speculation as to the true identity and location of the IJ gene, since no discrete gene for the putative IJ molecule has been identified in the H-2 I region (for review see Murphy 1987).

While the IJ paradox remains unresolved several hypotheses have been proposed to try and rationalize the serological and molecular data. Some of these are as follows:

1. The 1–2 kb of DNA into which the IJ molecule maps is not large enough to encode the 25 kDa polypeptide chain of the IJ molecule apparently identified biochemically. It is, however, possible that the IJ molecule is transcribed from a section of DNA located entirely within the H-2 IEβ chain gene or on the opposite strand of DNA. Even so, probes that span the IEβ gene have failed to hybridize with mRNA from any of the IJ$^+$ cells examined. In addition, no DNA rearrangements have been observed within or near the Eβ gene.

2. The 1–2 kb region of DNA encodes only part of IJ molecule, the remainder being encoded elsewhere in the genome.

3. The mapping of IJ to the H-2 may be anomalous. IJ may only appear to map the I region of the H-2 complex, because a gene that regulates its expression maps to the same region. There is some evidence that non-H-2 genes can influence IJ determinants.

4. As a result of the controversy over the nature of IJ, the recombinant mouse strains B.10.A(3R) and B10.A(5R) have been re-examined. It was found that recombination events had also occurred in other parts of the genome and were not confined to the H-2 I region in these strains, so genes outside the MHC could influence IJ. Furthermore, the recombination event between the 3R and 5R strains occurred in a 1 kb segment of DNA located in the intron between the β_1 and β_2 exon of the Eβ gene. Thus no polymorphic gene or coding element maps to the IJ subregion in the 3R and 5R strains.

5. It has also been proposed that IJ is a determinant on the T cell antigen receptor and that the selection of this receptor is controlled by the H-2 I region. These receptors could either recognize self class II, or other T cell receptors specific for other HLA class II molecules. This could result in IJ mapping to the H-2I region.

Class II loci in the human

The D region of HLA has been the subject of much investigation and three subregions, HLA-DP, DQ, and DR, have been defined (Fig.2.23).

On the basis of serological and biochemical analyses, together with functional studies using T cell clones and lines, there are four distinct class II products or antigens in human. Two of these, HLA-DR and related antigens, are homologous to IE molecules in the mouse; one of them, HLA-DQ, is homologous to the IA molecules; a fourth, HLA-DP, shares homology with the Aβ3 pseudogene (Table 2.8). The genetic organization of the loci encoding these molecules is shown in Fig.2.23.

When the HLA-D and H-2I regions are compard, the HLA-D region is found to be much larger (about 1100 versus 250 kb) and more complex. Molecular analysis of HLA-D suggested that it contains four or five different types of α chain gene, and at least three different types of β chain gene. The class II region of the HLA complex has been reviewed by Kaufman *et al.* (1984) and Kappes and Strominger (1988).

DR subregion. Within the DR family of genes there is only one α gene (DRA). The product of the DRA gene associates with the product of each of the functional β genes (DRB1, DRB3, DRB4 and DRB5); DRB2 is a pseudogene (Table 2.3). The number of DRB genes present in the genome varies depending on the DR haplotype of the cell. For example, four DRB genes have been identified in cells of the DR4 haplotype, but only three are present in most other haplotypes, although in DR8 only two DRB genes have been found (Fig.2.23).

At least 18 HLA-DR specificites can be detected serologically. These result from antigenic determinants present on the polypeptide chain encoded by the DRB1 gene. The DRB3 and DRB4 genes encode the supertypic specificities

Table 2.8 Loci encoding homologous MHC molecules in the H-2 and HLA complexes

	Mouse: H-2	Human: HLA
Class I loci	K, D, L, (R)	A, B, C
	Qa–1	
	Qa–2	
	Tla	
Class II loci	I–A	DQ
	I–E	DR
	Aβ3ψ	DP
Class III loci	Ss	C4
		C2
	Slp	B

HLA-DRw52 and HLA-DRw53 respectively. These are referred to as 'supertypic' specificities by serologists because they are shared by a group of alleles. DRw52 is expressed in association with DR3, w11, w12, w13 and w14, while DRw52 is present on cells expressing DR4, DR7 and DR9. Cells with these DR specificities will express both the DRB1 gene and either the DRB3 or DRB4 gene. The products of both of these DRB genes will associate with the DRA gene product. The DRB5 gene is expressed in association with DRw15 and 16.

DQ and DP subregions and other loci. The DQ and DP subregions each have two α chain genes and two β genes (DQA1, DQB1, DQB2 and DQB2; DPA1, DPB1, DPA2 and DPB2). So far, only one of the two pairs of genes, one in each region, has been found to produce a functional class II product (DQA1 and DQB1 and DPA1 and DPB1). Nine HLA-DQ specificities can be detected serologically, but as yet no alloantisera are available to detect HLA-DP alleles. HLA-DP specificities are characterized using monoclonal antibodies, cellular assays (primed lymphocyte typing), and now most commonly using the polymerase chain reaction (PCR—see Appendix, p. 715) with allele specific oligonucleotide probes.

An additional α chain gene, DNA (originally known as DZ), and a β chain gene, DOB, have also been identified (Table 2.3). The DNA gene is located close to the DP subregion and the DOB gene close to the DQ subregion, thus the two genes are well separated from one another in the complex (Fig.2.23). Although there is evidence that these genes are transcribed and expressed in low amounts, the function of these genes or indeed whether they can associate as an $\alpha\beta$ pair at the cell surface, and thus form a further HLA class II product, is not clear.

A secreted form of the DQβ chain that arises as a result of alternative RNA splicing has been reported, as have the existence of short polypeptides that are structurally distinct from class II antigens resulting from alternative splicing of the DNA gene.

Two putative peptide transporter genes (TAP1 and TAP2) that are related to the ABC (ATP-binding casette) superfamily (see p. 150) have been mapped to the class II region between the DNA and DOB genes (Trowsdale *et al.* 1990; Spies *et al.* 1990). These genes may encode two halves of a heterodimer that functions as a peptide transporter. One of the genes, TAP1, has been shown to correct a defect in class I expression in one mutant cell line but not in another (Spies *et al.* 1990), suggesting that more than one gene is involved. A gene with sequence homology to proteasome components has also been mapped to this region, and lies between the two putative transporter genes (Glynne *et al.* 1991) (see Section 3.3.2).

2.9 Mechanisms for the generation of polymorphisms

The mechanisms by which the polymorphism of MHC molecules is generated and maintained are undoubtedly important, but as yet little evidence is available to explain this phenomenon. Within a population, multiple point mutations and the shuffling of sequences of nucleotides between members of the same gene family has been observed for both class I and class II genes. The data suggest that **point mutations**, **recombination**, and **gene conversion** (Fig.2.25), are the mechanisms likely to be important in the generation of polymorphism in MHC molecules on an evolutionary scale.

2.9.1 Generation of class I polymorphisms

Analysis of the protein and gene structures of naturally occurring class I mutants (see Section 2.5.1) has provided information about the mechanisms for the generation of class I polymorphism (for review see Nathenson *et al.* 1986). At the protein level the majority of the class I mutants contain multiple amino acid substitutions which occur in clusters. Interestingly, identical amino acid substitutions have also been found in similar positions

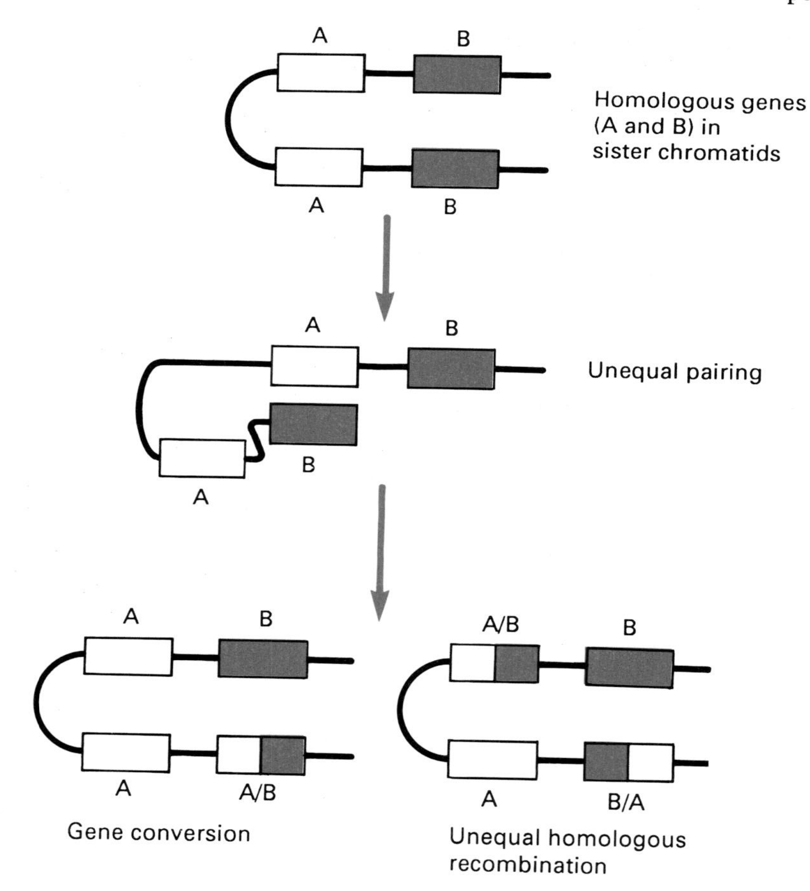
A
B
Homologous genes (A and B) in sister chromatids
A
B
A
B
Unequal pairing
B
A
A
B
A
A/B
Gene conversion
A/B
B
A
B/A
Unequal homologous recombination
Recombination between mispaired genes leads to expansion and contraction of the gene family. Hybrid genes (A/B and B/A) are generated.
Sequence divergence between alleles results from correction of a mispaired gene (acceptor gene, B) by other (donor gene, A).

Fig.2.25 Mechanisms that contribute to increasing the polymorphism of MHC molecules

in other class I molecules. An hypothesis was formulated to account for these findings, involving non-reciprocal genetic recombination, often referred to as **gene conversion**.

Recombination is the *reciprocal* exchange of information between two genes. In other words, it involves the transfer of information in both directions, so that the genes involved both lose and gain a block of nucleotides. This is represented in Fig.2.25. In contrast, gene conversion or copy substitution is a *non-reciprocal* genetic process. In this case information is transferred in one direction only, from the donor to the acceptor gene. During the process of gene conversion, a particular segment of a given gene, known as the acceptor gene, becomes identical to the corresponding portion of a related, but non-identical gene, the donor gene. The donor gene is usually located in a different part of the genome to the acceptor gene (Fig.2.25). After gene conversion the particular nucleotide sequence involved is present in both the donor and the acceptor gene. The molecular mechanism by which the nucleotide sequences are transferred from the donor to the acceptor gene is unknown. It may be mediated through homologous flanking regions in the two genes. However, other possibilities do exist, including the transfer of nucleotides by a *trans*-acting mediator.

Supporters of the gene conversion model as the mechanism responsible for the generation of polymorphism in MHC genes suggest that short segments of DNA are transferred between different MHC genes of the same class, either between alleles of the same locus or between different loci. If this process was repeated frequently during evolution, an extremely polymorphic family of molecules would be generated. If only short stretches of nucleotides were involved in this process, this would correlate with observations that the polymorphic amino acid residues present in MHC molecules are located in clusters, mainly in exons 2 and 3 of class I heavy chain genes and exon 2 of the class II β chain genes (Figs.2.19 and 2.20). Numerous examples are recorded in the literature where the donor and acceptor genes implicated in the gene conversion event have been identified. For example, for class I MHC molecules a donor gene Q10, mapping to the Qa2.3 locus (Fig.2.24), has been identified in the event that resulted in the generation of the K^b mutant, bm 1 (Mellor *et al.* 1983). Alleles of certain HLA class I molecules, e.g. HLA-B27 and HLA-A3, have features similar to those of the H-2K^b mutants and may also have arisen through gene conversion events. Nevertheless whether gene conversion is the most important mechanism for the generation of MHC polymorphism remains a matter for debate.

In human the HLA-A2/A28 family of alleles (Table 2.9) has been studied

Table 2.9 Members of the HLA-A2/A28 family

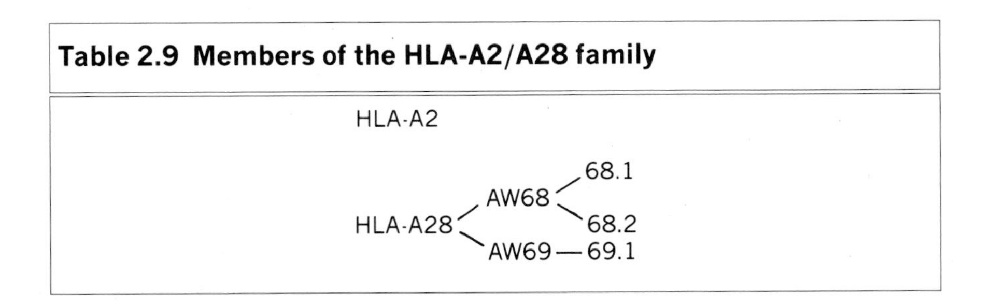

in detail, as these related molecules (which includes HLA-Aw68; see Section 2.5) seem to have diverged from one another quite recently in evolutionary terms (Fig.2.26). The following mechanisms are thought to contribute to the generation of polymorphism in this group: (i) single nucleotide substitutions; (ii) reciprocal recombination; and (iii) gene conversion (Fig.2.25) (Holmes *et al.* 1987).

2.9.2 Generation of class II polymorphisms

Two mechanisms have been proposed to account for the generation of polymorphism in class II genes: (i) multiple point mutations and (ii) gene conversion (see Section 2.9.1). In class II molecules the polymorphic residues are clustered in three groups in the first domain of each polypeptide chain, the allelic hypervariable regions (Fig.2.14). The pattern of substitution implicates gene conversion as the mechanism responsible for generating these polymorphisms (Erlich and Gyllensten 1991). For example, there is evidence that the bm 12 mutation was generated by a gene conversion event involving the transfer of 30 nucleotides from the exon encoding the first domain of the IE^b β chain to the corresponding region of the IA^b gene (Mengle-Gaw *et al.* 1984). However, the number of class II genes identified within the MHC is far fewer than the number of class I genes identified, and may be too low to serve as a reservoir of donor genes for the gene conversion process if it were the only mechanism in operation. Thus, point mutations in class II genes have also been suggested to play an important role in generating new polymorphisms.

2.9.3 Selection pressure

If the immune system fails to recognize a lethal pathogen because peptides from the pathogen cannot associate with the particular MHC molecules of that individual, the individual will not survive the infection. On a population basis it is likely that some individuals will possess an MHC allele that can associate with the antigenic peptides in question; otherwise the species would be eliminated. It is therefore supposed that there must be selection pressure for increased polymorphism to allow the recognition of any new antigen by the immune system in some members of the species.

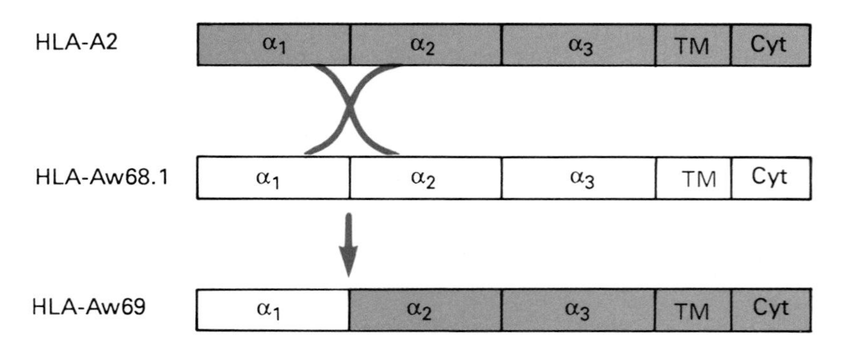

Fig.2.26 Intrallelic recombination between HLA-A2 and Aw68.1 gives rise to HLA-Aw69

If selection pressure was operating, one would predict that the mutation rates in the different exons of the class I and class II genes would be different. Thus the ratio of the number of replacement substitutions, compared with silent substitutions, would be higher in the exons coding for the α_1 and α_2 domains of the class I heavy chain and α_1 and β_1 domains of class II molecules, since these generate the peptide binding groove (Fig.2.5). For class I molecules the ratio for the α_1 and α_2 domains is higher than expected, suggesting that there is a positive selection pressure for variation in these regions. However, for class II molecules this is not the case, because the calculated ratio for the α_2 and β_2 domains is much lower than expected. This would suggest that selection pressure acts primarily to maintain the immunoglobulin domain structure of the second domains, whereas substitutions in the α_1 and β_1 domains can occur at random to produce new polymorphisms.

An examination of MHC polymorphism in other species can also provide interesting information about selection pressure. Some species, for example Syrian hamsters and cheetahs, lack class I polymorphism. In the case of the Syrian hamsters their ability to survive in the absence of polymorphism may be explained by their solitary life-style in a restricted environment and their reduced exposure to pathogens, particularly viruses. Cheetahs, however, have suffered a reduction in class I polymorphisms because of the reduction in the number of breeding pairs due to over-hunting. As a result, the cheetah as a species is more susceptible to viral infections. These observations illustrate the importance of selection pressure in generating new MHC polymorphisms during evolution. It will also be interesting to monitor the progress of mice that have been rendered class I (Zijlstra *et al.* 1990) or class II deficient (Grusby *et al.* 1991) by using homologous recombination (see Appendix, p. 720) to disrupt the β_2-microglobulin and E_α and A_β genes respectively, when they are exposed to infection.

2.10 Disease associations

For certain diseases, an increased frequency of particular HLA alleles has been noted in affected individuals in both population and family studies (for reviews see Solheim *et al.* 1982; Tiwari and Terasaki 1985). The strongest association is found in patients with narcolepsy (a disease characterized by sudden periods of sleep during normal waking hours) where all have the specificity HLA-DR2 (HLA-DRB1*02 to be precise!). Ankylosing spondylitis is an inflammatory disease affecting the joints of the spine and pelvis; it is more common in males than females. Approximately 96% of Causasians with the disease carry the class I allele HLA-B27, whereas HLA-B27 is present in only 7–9% of healthy individuals. (For a discussion of how B27 may be involved in disease pathogenesis see Benjamin and Parham 1990). Many of the autoimmune diseases show an increased frequency with particular class II alleles and some of these are shown in Table 2.10. Insulin

Table 2.10 Disease susceptibility and HLA-DR associations

Condition	DR antigen	Relative risk
Celiac disease	3	10.8
Graves' disease	3	3.7
Insulin dependent diabetes mellitus	3	3.3
	4	6.4
	2	0.2
Rheumatoid arthritis	4	4.2
Systenic lupus erythematosus	3	5.8

dependent diabetes mellitus, type I, (IDDM, type I) will be discussed in more detail here (for review see Todd 1990).

Type I IDDM results from the selective destruction of islets of Langerhans in the pancreas, the cells that are responsible for producing insulin. It has been proposed that the development of this disease involves an autommune response to an antigen from the islet cells and that it is likely to be T cell mediated. Genetic studies have shown that the inheritance of IDDM is polygenic, in other words it involves more than one gene, but it appears that the HLA-D region contributes significantly. When individuals with IDDM are tissue-typed, 95% of patients possess either HLA-DR3 or DR4 alleles compared to 45–54% in the normal population. At least one gene that is thought to confer susceptibility to IDDM has been mapped to the MHC, and is most tightly linked to the HLA-D region. In an attempt to identify the gene involved, Todd and his colleagues have sequenced the four major expressed polymorphic gene products, DRB1 and DRB3, DQA and DQB genes (Fig.2.23) isolated from IDDM patients, and compared the sequence data obtained with those from control, normal individuals of the same DR type. This analysis showed that there were no class II sequences

DR	DQ	IDDM association	Amino Acid Position																													
			3	9	13	14	26	28	30	37	38	45	46	47	52	53	55	56	57	66	67	70	71	74	75	77	84	85	86	87	89	90
4	W8	POSITIVE	S	Y	G	M	L	T	Y	Y	A	G	V	Y	P	L	P	P	A	E	V	R	T	E	L	T	Q	L	E	L	T	T
3	W2	POSITIVE	–	–	–	–	–	S	S	I	V	–	E	F	L	–	L	–	–	D	L	–	K	A	V	R	–	–	–	–	–	–
1	W5	POSITIVE	–	–	–	L	G	–	H	–	V	–	–	–	–	Q	R	–	V	–	–	G	A	S	V	R	E	V	A	Y	G	I
16	W5	POSITIVE	–	–	–	L	G	–	H	–	V	–	–	–	–	Q	R	–	S	–	–	G	A	S	V	R	E	V	A	Y	G	I
4	W7	NEUTRAL/	–	–	A	–	Y	–	–	–	–	E	–	–	–	–	–	–	D	–	–	–	–	–	–	–	–	–	–	–	–	–
5	W7	NEGATIVE	–	–	A	–	Y	–	–	–	–	E	–	–	–	–	–	–	D	–	–	–	–	–	–	–	–	–	–	–	–	–
2	W6	NEGATIVE	–	F	–	–	–	–	–	–	–	–	–	–	–	Q	R	–	D	–	–	G	–	–	–	–	E	V	A	F	G	I
NOD	AβNOD	POSITIVE	–	H	–	E	–	–	–	–	L	–	E	–	E	–	R	A	S	Y	*	–	–	–	–	–	E	E	T	E	P	–
B10	Aβ^b	NEGATIVE	–	Y	–	E	Y	–	–	–	V	–	E	H	E	–	R	–	D	E	I	–	–	–	–	–	E	G	P	E	H	–

Fig.2.27 Comparison of amino acid sequences in MHC alleles from diabetic and normal individuals and NOD mice. A comparison of the polymorphic residues of some of the DQβ alleles associated with DR haplotypes that have been studied in IDDM and control populations. The strength of association is to some extent dependent on the amino acid at position 57, but also on other amino acids in the DQβ and α chains. The asterisk in the NOD mouse Aβ sequence denotes a deleted amino acid. A dash indicates identity with the DQW8 sequence. The single letter amino acid code is used (see Fig.1.3)

unique to the IDDM patients; however, in the HLA-BQB1 gene there was an amino acid substitution at a single position in the polypeptide chain that correlated with the disease. This was located at position 57 in the β_1 domain of the DQB1 chain, and disease susceptibility was dependent on the identity of the amino acid at that position. Individuals with an aspartic acid residue at position 57 were found to be *resistant* to the disease, whereas individuals with a neutral amino acid, such as serine, alanine, or valine, rather than an acidic residue at that position were *susceptible* (Fig.2.27).

Interestingly, two animal models of IDDM exist, the NOD mouse and the BB rat. When the homologue of the HLA-DQB gene in these two strains was sequenced, a neutral amino acid residue, serine, was identified at position 57. In other mouse and rat strains that are not susceptible to IDDM, aspartic acid has been found at this position. Thus, it has been suggested that the amino acid at position 57 determines a critical function of the HLA-DQ molecule in IDDM. It is not clear whether the aspartic acid at position 57 in normal individuals is sufficient on its own to confer decreased susceptibility to IDDM: it may be that other amino acid residues contribute to the conformation of the DQ molecule and prevent the presentation of islet cell antigens to the immune system.

The story is obviously complex, but this type of approach should provide clear information about the role of MHC molecules in disease.

Further reading

Class I

Arnold, B. and Hämmerling, G.J. (1991). MHC Class I transgenic mice. *Annual Review of Immunology*, **9**, 297–322.

Bjorkman, P.J. and Parham, P. (1990). Structure, function, and diversity of Class I major histocompatibility molecules. *Annual Review of Biochemistry*, **59**, 253–88.

Hood, L., Steinmetz, M., and Malissen, B. (1983). Genes of the major histocompatibility complex of the mouse. *Annual Review of Immunology*, **1**, 529–68.

Mellor, A. (1986). The class I MHC gene family in mice. *Immunology Today*, **7**, 19–24.

Nathenson, S.G., Geliester, J., Pfaffenbach, G.M., and Zeff, R.A. (1986). Murine major histocompatibility complex class I mutants: molecular analysis and structure-function implications. *Annual Review of Immunology*, **4**, 471–502.

Parham, P. (1990*a*). Transporters of delight. *Nature*, **348**, 674–5.

Parham, P. (1990*b*). Peptide feeding and cellular cookery. *Nature*, **348**, 793–5.

Parham, P. (1992). Flying the first class flag. *Nature*, **357**, 193–4.

Rothbard, J.B. and Gefter, M.L. (1991) Interactions between immunogenic peptides and MHC. *Annual Review of Immunology*, **9**, 527–65.

Stroynowski, I. (1990). Molecules related to class-1 Major Histocompatibility Complex antigens. *Annual Review of Immunology*, **8**, 501–30.

Townsend, A. and McMichael, A. (1987). News and views. *Nature*, **329**, 482–3.

van Holde, K.E. (1985). *Physical biochemistry*, 2nd edn, Chapter 11. Prentice Hall, New Jersey.

Class II

Glimcher, L.A. and Kara, C.J. (1992). Sequences and factors: a guide to MHC class-II transcription. *Annual Review of Immunology*, **10**, 13–49.

Hammerling, G.J. and Moreno, J. (1990). The function of the invariant chain in antigen presentation by MHC Class II molecules. *Immunology Today*, **11**, 337–40.

Herman, A., Kappler, J.W., Marrack, P., and Pullen, A.M. (1991) Superantigens: mechanisms of T cell stimulation and role in immune responses. *Annual Review of Immunology*, **9**, 745–72.

Kappes, D. and Strominger, J.L. (1988). Human class II major histocompatibility complex genes and proteins. *Annual Review of Biochemistry*, **57**, 991-1028.

Kaufman, J.F., Auffray, C., Karman, A.J., Shackelford, D.A., and Strominger, J. (1984). The class II molecules of the human and murine major histocompatibility complex. *Cell*, **36**, 1–13.

Lechler, R.I. (1988). MHC class II molecular structure—permitted pairs. *Immunology Today*, **9**, 76–8.

Marrack, P. and Kappler, J. (1990). The Staphylococcal enterotoxins and their relatives. *Science*, **248**, 705–11.

Mengle-Gaw, L. and McDevitt, H.O. (1985). Genetics and expression of murine Ia antigens. *Annual Review of Immunology*, **3**, 367–96.

Murphy, D.B. (1987). The I-J puzzle. *Annual Review of Immunology*, **5**, 405–27.

Evolution of MHC genes

Erlich, H.A. and Gyllensten, U.B. (1991). Shared epitopes among HLA class II alleles: gene conversion, common ancestry and balancing selection. *Immunology Today*, **12**, 411–14.

Figueroa, F. and Klein, J. (1986). The evolution of class II MHC genes. *Immunology Today*, **7**, 78–81.

Klein, J. and Figueroa, F. (1986). The evolution of class I MHC genes. *Immunology Today*, **7**, 41–4.

Lawlor, D.A., Zimmer, J., Ennis, P.D., and Parham, P. (1990). Evolution of class I MHC genes and proteins; from natural selection to thymic selection. *Annual Review of Immunology*, **8**, 23–63.

Detection of MHC genes

Bell, J. (1989). The polymerase chain reaction. *Immunology Today*, **10**, 351–5.

Benjamin, R. and Parham, P. (1990). Guilt by association: HLA-B27 and ankylosing spondylitis. *Immunology Today*, **11**, 137–42.

Bidwell, J. (1987). DNA-RFLP analysis and genotyping of HLA-DR and DQ antigens. *Immunology Today*, **8**, 18–23.

Malissen, B. (1986). Transfer and expression of MHC genes. *Immunology Today*, **7**, 106–12.

Disease associations

Nepom, G.T. and Erlich, H. (1991) MHC class II molecules and autoimmunity. *Annual Review of Immunology*, **9**, 493–525.

Solheim, B.G., Ryder, L.P., and Svejgaard, A. (1982). HLA and Disease Associations. In *HLA typing: methodology and clinical aspects* (ed. S. Ferrone and B.G. Solheim), pp 167–75.

Tiwari, J.L. and Terasaki, P.I. (ed.) (1985). *HLA and disease associations*. Springer-Verlag, New York.

Todd, J. (1990). Genetic control of autoimmunity in type 1 diabetes. *Immunology Today*, **4**, 122–9.

Literature cited

Arnold, B. and Hämmerling, G.J. (1991). MHC Class I transgenic mice. *Annual Review of Immunology*, **9**, 297–322.

Bell, J. (1989). The polymerase chain reaction. *Immunology Today*, **10**, 351–5.

Benjamin, R. and Parham, P. (1990). Guilt by association: HLA-B27 and ankylosing spondylitis. *Immunology Today*, **11**, 137–42.

Bidwell, J. (1987). DNA-RFLP analysis and genotyping of HLA-DR and DQ antigens. *Immunology Today*, **9**, 18–23.

Bjorkman, P.J. and Parham, P. (1990). Structure, function, and diversity of Class I major histocompatibility molecules. *Annual Review of Biochemistry*, **59**, 253–88.

Bjorkman, P.J., Saper, M.A., Samraoui, B., Bennett, W.S., Strominger, J.L., and Wiley, D.C. (1987*a*). Structure of the human class I histocompatibility antigen, HLA-A2. *Nature*, **329**, 506–12.

Bjorkman, P.J., Saper, M.A., Samaroui, B., Bennett, W.S., Strominger, J.L., and Wiley, D.C. (1987*b*). The foreign antigen binding site and T cell recognition regions of class I histocompatibility antigens. *Nature*, **329**, 512–18.

Bodmer, J.G., Marsh, S.G.E., and Albert, E. (1990*a*). Nomenclature for factors of the HLA system 1989. *Immunology Today*, **11**, 3–10.

Bodmer, J.G., Marsh, S.G.E., Albert, E.D., Bodmer, W.F., Dupont, B., Erlich, H.A., Mach, B., Mayer, W.R., Parham, P., Sasazuki, T., Schreuder, G.M.T., Strominger, J.L., Svejgaard, A. and Terasaki, P.I. (1990*b*). Nomenclature for factors of the HLA system 1990. *Tissue Antigens*, **35**, 1–8.

Bodmer, J.G., Marsh, S.G.E., Albert, E.D., Bodmer, W.F., Dupont, B., Erlich, H.A., Mach, B., Mayr, W.R., Parham, P., Sasazuki, T., Schreuder, G.M. Th., Strominger, J.L.,

Svejgaard, A., and Terasaki, P.I. (1992). Nomenclature for factors of the HLA system 1991. *Tissue antigens*, **39**, 161–73.

Brown, J.H., Jardetsky, T., Saper, M.A., Samraoui, B., Bjorkman, P.J., and Wiley, D.C. (1988). A hypothetical model of the foreign antigen binding site of class II histocompatibility molecules. *Nature*, **332**, 845–50.

Brown, M.G., Driscoll, J., and Monaco, J.J. (1991). Structural and serological similarity of MHC-linked LMP and proteasome (multicatalytic proteinase) complexes *Nature*, **353**, 355–7.

Burnet, F.M. (1971). Self recognition in colonial marine forms and flowering plants in relation to their evolution and immunity. *Nature*, **232**, 230–5.

Coligan, J.E., Kindt, T.J., Uehara, H., Martinko, J., and Natherson, S.G. (1981). Primary structure of a murine transplantation antigen. *Nature*, **291**, 35–9.

Erlich, H.A. and Gyllensten, U.B. (1991). Shared epitopes among HLA class II alleles: gene conversion, common ancestry and balancing selection. *Immunology Today*, **12**, 411–14.

Freidman, R.L. and Stark, G.R. (1985). α-Interferon induced transcription of HLA and metallothionein genes containing homologous upstream sequences, *Nature*, **314**, 637–9.

Figueroa, F. and Klein, J. (1986). The evolution of class II MHC genes. *Immunology Today*, **7**, 78–81.

Garrett, T.P.J., Saper, M.A., Bjorkman, P.J., Strominger, J.L., and Wiley, D.C. (1989). Specificity pockets for the side chains of peptide antigens in HLA-Aw68. *Nature*, **342**, 692–6.

Glynne, R., Powis, S.H., Beck, S., Kelly, A., Kerr, L.-A., and Trowsdale, J. (1991). A proteasome-related gene between two ABC transporter loci in the class II region of the human MHC. *Nature*, **353**, 357–60.

Gorer, P.A. (1936). The detection of antigenic differences in mouse erythrocytes by the employment of immune sera *Br. J. Exp. Pathol.*, **17**, 42–50.

Grusby, M.J., Johnson, R.S., Papaioannou, V.E., and Glimcher, L.H. (1991). Depletion of $CD4^+$ T cells in major histocompatibility complex class II-deficient mice. *Science*, **253**, 1417–20.

Hämmerling, G.J. and Moreno, J. (1990). The function of the invariant chain in antigen presentation by MHC Class II molecules. *Immunology Today*, **11**, 337–340.

Herman, A., Kappler, J.W., Marrack, P., and Pullen, A.M. (1991). Superantigens: mechanisms of T cell stimulation and role in immune responses. *Annual Review of Immunology*, **9**, 745–72.

Holmes, N., Ennis, P., Wai, A.M., Denney, D.W., and Parham, P. (1987). Multiple genetic mechanisms have contributed to the generation of the HLA-A2/A28 family of class I MHC molecules. *Journal of Immunology*, **139**, 936–41.

Hood, L., Steinmetz, M., and Malissen, B. (1983). Genes of the major histocompatibility complex of the mouse. *Annual Review of Immunology*, **1**, 529–68.

Jardetzky, T.S., Lane, W.S., Robinson, R.A., Madden, D.R., and Wiley, D.C. (1991). Identification of self peptides bound to purified HLA-B27. *Nature*, **353**, 326–9.

Kappes, D. and Strominger, J.L. (1988). Human class II major histocompatibility complex genes and proteins. *Annual Review of Biochemistry*, **57**, 991–1028.

Kaufman, J.F., Auffray, C., Karman, A.J., Shackelford, D.A., and Strominger, J. (1984). The class II molecules of the human and murine major histocompatibility complex. *Cell*, **36**, 1–13.

Kimura, A., Israel, A., LeBail, O., and Kourilsky, P. (1986). Detailed analysis of the mouse H-$2K^b$ promoter:enhancer-like sequences and their role in the regulation of class I gene expression. *Cell*, **44**, 261–72.

Klein, J. and Figueroa, F. (1986). The evolution of class I MHC genes. *Immunology Today*, **7**, 41–4.

Lawlor, D.A., Zimmer, J., Ennis, P.D., and Parham, P. (1990). Evolution of class I MHC genes and proteins; from natural selection to thymic selection. *Annual Review of Immunology*, **8**, 23–63.

Lechler, R.I. (1988). MHC class II molecular structure—permitted pairs. *Immunology Today*, **9**, 76–8.

Lotteau, Y., Teyton, L., Burroughs, D., and Charron, D. (1987). A novel HLA class II molecule (DRα–DQβ) created by mismatched isotype pairings. *Nature*, **329**, 339–41.

Madden, D.R., Gorga, J.C., Strominger, J.L., and Wiley, D.C. (1991). The structure of HLA-B27 reveals nonamer self peptides bound in extended conformation. *Nature*, **353**, 321–5.

Malissen, B. (1986). Transfer and expression of MHC genes. *Immunology Today*, **7**, 106–12.

Marrack, P. and Kappler, J. (1990). The staphylococcal enterotoxins and their relatives. *Science*, **248**, 705–11.

Marsh, S.G. and Bodmer, J.G. (1989). HLA-DR and -DQ epitopes and monoclonal antibody specificity. *Immunology Today*, **10**, 305–12.

Marsh, S.G. and Bodmer, J.G. (1990). HLA-DRB nucleotide sequences, 1990., *Immunogenetics*, **31**, 141–4.

Mellor, A. (1986). The class I MHC gene family in mice. *Immunology Today*, **7**, 19–24.

Mellor, A.L., Weiss, E.H., Ramachondran, K., and Flavell, R.A. (1983). A potential donor gene for the bm1 gene conversion event in the C57BL mouse. *Nature*, **306**, 792–5.

Mengle-Gaw, L. and McDevitt, H.O. (1985). Genetics and expression of murine Ia antigens. *Annual Review of Immunology*, **3**, 367–96.

Mengle-Gaw, L., Conner, S., McDevitt, H.O., and Fathman, C.G. (1984). Gene conversion between murine class II MHC loci: functional and molecular-evidence from the bm12 mutant. *Journal of Experimental Medicine*, **1600**, 1184–94.

Monaco, J.J. (1992). A molecular model of MHC class-I-restricted antigen processing. *Immunology Today*, **13**, 173–8.

Monaco, J.J., Cho, S., and Attaya, M. (1990). Transport protein genes in the murine MHC: Possible implications for antigen processing. *Science*, **250**, 1723–6.

Murphy, D.B. (1987). The I-J puzzle. *Annual Review of Immunology*, **5**, 405–27.

Nathenson, S.G., Geliester, J., Pfaffenbach, G.M., and Zeff, R.A. (1986). Murine major histocompatibility complex class I mutants: molecular analysis and structure-function implications. *Annual Review of Immunology*, **4**, 471–502.

Neefjes, J.J. and Ploegh, H.L. (1992). Intracellular transport of MHC class II molecules. *Immunology Today*, **13**, 179–84.

Nepom, G.T. and Erlich, H. (1991). MHC class II molecules and autoimmunity. *Annual Review of Immunology*, **9**, 493–525.

Parham, P. (1990*a*). Transporters of delight. *Nature*, **348**, 674–5.

Parham, P. (1990*b*). Peptide feeding and cellular cookery. *Nature*, **348**, 793–5.

Parham, P. (1992). Flying the first class flag. *Nature*, **357**, 193–4.

Robinson, P.J. (1987). Structure and expression of polypeptides encoded in the mouse Qa region. *Immunol. Research*, **6**, 46–56.

Roche, P.A. and Cresswell, P. (1990). Invariant chain association with HLA-DR molecules inhibits immunogenic peptide binding. *Nature*, **345**, 615–18.

Rothbard, J.B. and Gefter, M.L. (1991). Interactions between immunogenic peptides and MHC. *Annual Review of Immunology*, **9**, 527–65.

Salter, R.O., Benjamin, R.J., Wesley, P.K., Buxton, S.E., Garrett, T.P., Clayberger, C., Krensky, A.M., Norment, A.M., Littman, D.R., and Parham, P. (1990). A binding site for the T cell co-receptor CD8 on the α_3 domain of HLA-A2. *Nature*, **345**, 41–6.

Silver, M.L., Parker, K.C. and Wiley, D.C. (1991). Reconstitution by MHC-restricted peptides of HLA-A2 heavy chain with β_2-microglobulin, in vitro. *Nature*, **350**, 619–22.

Solheim, B.G., Ryder, L.P., and Svejgaard, A. (1982). HLA and Disease Associations. In *HLA typing: methodology and clinical aspects* (ed. S. Ferrone and B.G. Solheim), pp 167–75.

Spies, T. and DeMars, R. (1991). Restored expression of major histocompatibility class I molecules by gene transfer of a putative peptide transporter. *Nature*, **351**, 323–4.

Spies, T., Bresnahan, M., Bahram, S., Arnold, D., Blanck, G., Mellins, E., Pious, D. and DeMars, R. (1990). A gene in the human Major Histocompatibility Complex class II region controlling the class I antigen presentation pathway. *Nature*, **348**, 744–7.

Stroynowski, I. (1990). Molecules related to class-1 Major Histocompatibility Complex antigens. *Annual Review of Immunology*, **8**, 501–30.

Teyton, L., O'Sullivan, D., Dickson, P.W., Lotteau, V., Sette, A., Fink, P., and Peterson P. (1990). Invariant chain distinguishes between the exogenous and endogenous antigen presentation pathways. *Nature*, **348**, 39–44.

Tiwari, J.L. and Terasaki, P.I. (ed.) (1985). *HLA and disease associations*. Springer-Verlag, New York.

Todd, J. (1990). Genetic control of autoimmunity in type 1 diabetes. *Immunology Today*, **4**, 122–9.

Todd, J., Bell, J.I., and McDevitt, H.O. (1987). HLA-DQβ gene contributes to susceptibility and resistance to insulin-dependent diabetes mellitus. *Nature*, **329**, 599–604.

Townsend, A. and McMichael, A. (1987). News and views. *Nature*, **329**, 482–3.

Townsend, A.R.M., Ohlen, C., Bastin, J., Ljunggren, H-G., Foster, L., and Karre, K. (1989). Association of class I major histocompatibility heavy and light chains induced by viral peptides. *Nature*, **340**, 443–8.

Townsend A. R.M., Elliot T., Cerundolo V., Foster L., Barber B., and Tse A. (1990). Assembly of MHC class I molecules analysed in vitro. *Cell*, **62**, 285–295.

Trowsdale J., Hanson I., Mockridge I., Beck S., Townsend A. and Kelly A. (1990). Sequences encoded in the class II region of the MHC related to the 'ABC' superfamily of transporters. *Nature*, **348**, 741–3.

Trowsdale, J., Ragoussis, J., and Campbell, R.D. (1991). *Immunology Today*, **12**, 443-6.

Van Holde, K.E. (1985). *Physical biochemistry*, 2nd edn, Chapter 11. Prentice Hall, New Jersey.

Zijlstra, M., Bix, M., Simister, N.E., Loring, J.M., Raulet, D.H., and Jaenisch, R. (1990). β_2 microglobulin deficient mice lack CD4-8 + cytolytic T cells. *Nature*, **334**, 742–6.

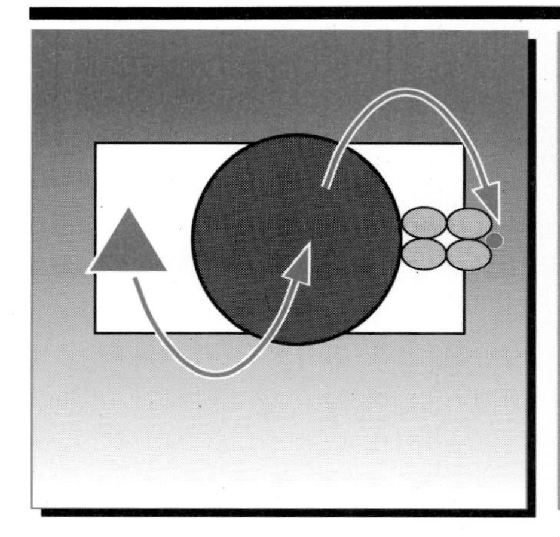

3

ANTIGEN-PRESENTING CELLS

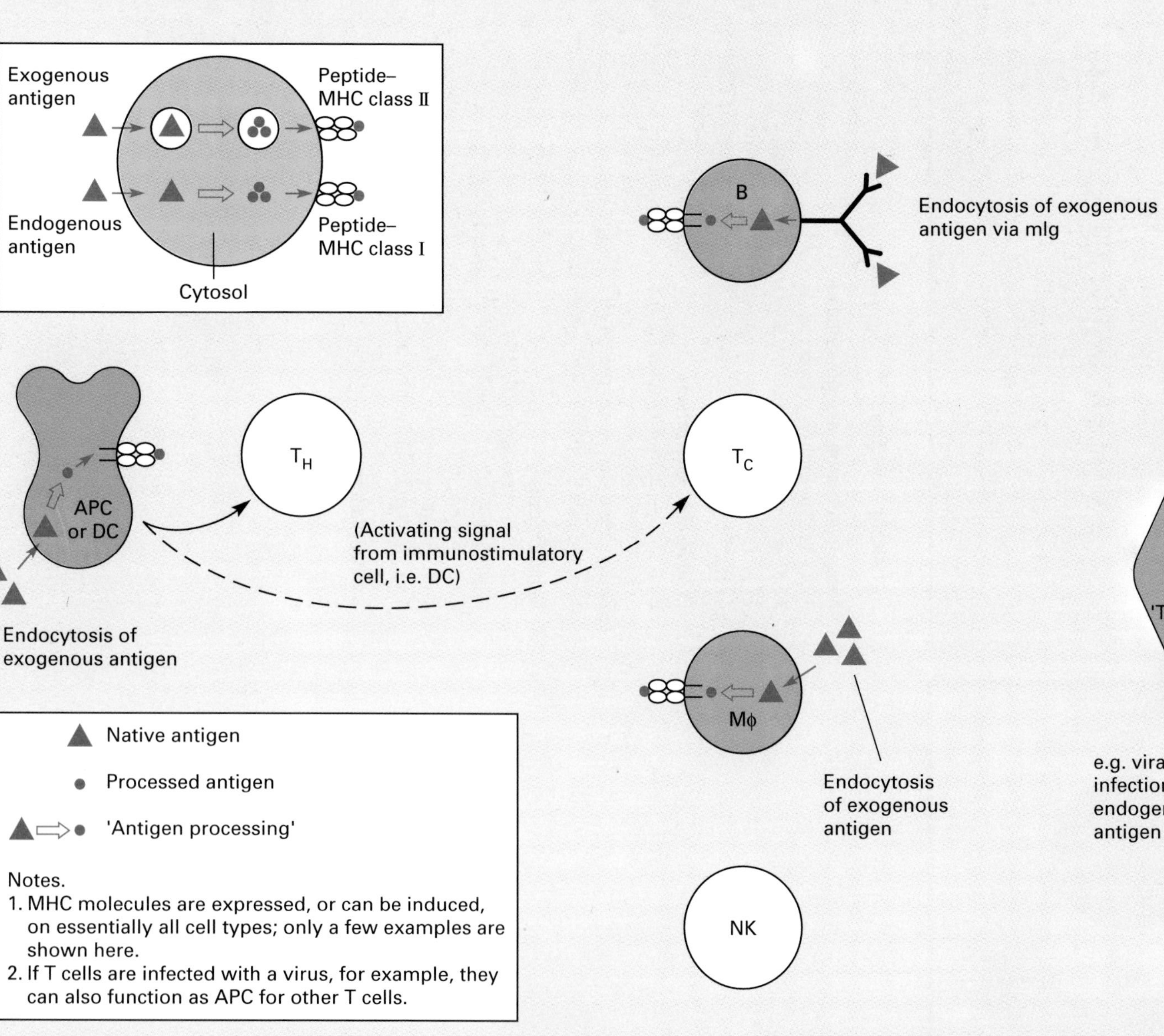
Scheme 3. Antigen-presenting cells
Exogenous antigen
Endogenous antigen
Peptide–MHC class II
Peptide–MHC class I
Cytosol
B
Endocytosis of exogenous antigen via mIg
APC or DC
T_H
T_C
(Activating signal from immunostimulatory cell, i.e. DC)
Endocytosis of exogenous antigen
Mϕ
Endocytosis of exogenous antigen
'Target' APC
e.g. viral infection; virus = endogenous antigen
NK
Native antigen
Processed antigen
'Antigen processing'
Notes.
1. MHC molecules are expressed, or can be induced, on essentially all cell types; only a few examples are shown here.
2. If T cells are infected with a virus, for example, they can also function as APC for other T cells.

3.1 Historical perspective

One early observation made by researchers who were studying *in vitro* immune responses was that antibodies were produced by suspensions of spleen cells from immunized animals if they were cultured in the presence of the appropriate antigen. It was then noted that similar responses were obtained if the spleen cells were incubated for a short time with the antigen and washed to remove the excess before culture. Such cells were called **antigen-pulsed cells**. From this it was evident that only an extremely small amount of antigen remained associated with the pulsed cells, and yet the magnitude of the response stimulated by these cells was similar to that obtained when the antigen was present in the culture continuously. It appeared that the antigen had been somehow processed by the cells into a more immunogenic form.†

In the mid 1960s it became apparent that two different populations of lymphocytes were required for antibody production *in vivo*, and that these acted synergistically. Claman *et al.* (1967) showed that irradiated mice could make antibodies against sheep erythrocytes if they were reconstituted with bone marrow cells together with thymus cells, but not when they were reconstituted with either population alone (Fig.3.1). This idea was then developed by Mitchell and Miller (1968), who showed that thymocytes or cells from the thoracic duct could restore immune responses if they were given to thymectomized recipients, but that antibodies were actually produced by another cell population derived from the bone marrow inoculum (Fig.3.2). These and other studies ultimately led to the identification of T cells and B cells as distinct subpopulations of lymphocytes, and the concept that T cells were needed to help B lymphocytes develop into antibody-secreting cells.

At about this time, Mosier (1967) was studying antibody responses in culture. He, too, demonstrated synergism between two distinct populations of lymphocytes but found that a different cell type was also required, and he proposed a 'three-cell interaction model' for these responses. He observed that purified lymphocytes in culture were unable to make antibodies to sheep erythrocytes unless an additional population of adherent cells was present (Table 3.1). These cells, which he termed **accessory cells**, consisted mainly of macrophages although we now know this was not a homogeneous population. They could stimulate lymphocyte responses after they were antigen-pulsed, whereas the same treatment of lymphocytes was ineffective. A similar requirement for adherent cells was also shown for T cell proliferative responses by Hersch and Harris (1968).

It was envisaged that **macrophages**, which avidly take up and destroy an immense variety of particles and molecules, would phagocytose antigens

† An initial explanation put forward for this phenomenon was that the antigen became complexed to RNA within the cell to produce a 'super-antigen'. This was based on the finding that antigens extracted from antigen-pulsed cells were sometimes associated with RNA. It was even considered that this RNA might be transferred from these cells to B lymphocytes as 'immune RNA' which ultimately coded for specific antibodies. We now know that the RNA was a contaminant and that this process does not occur.

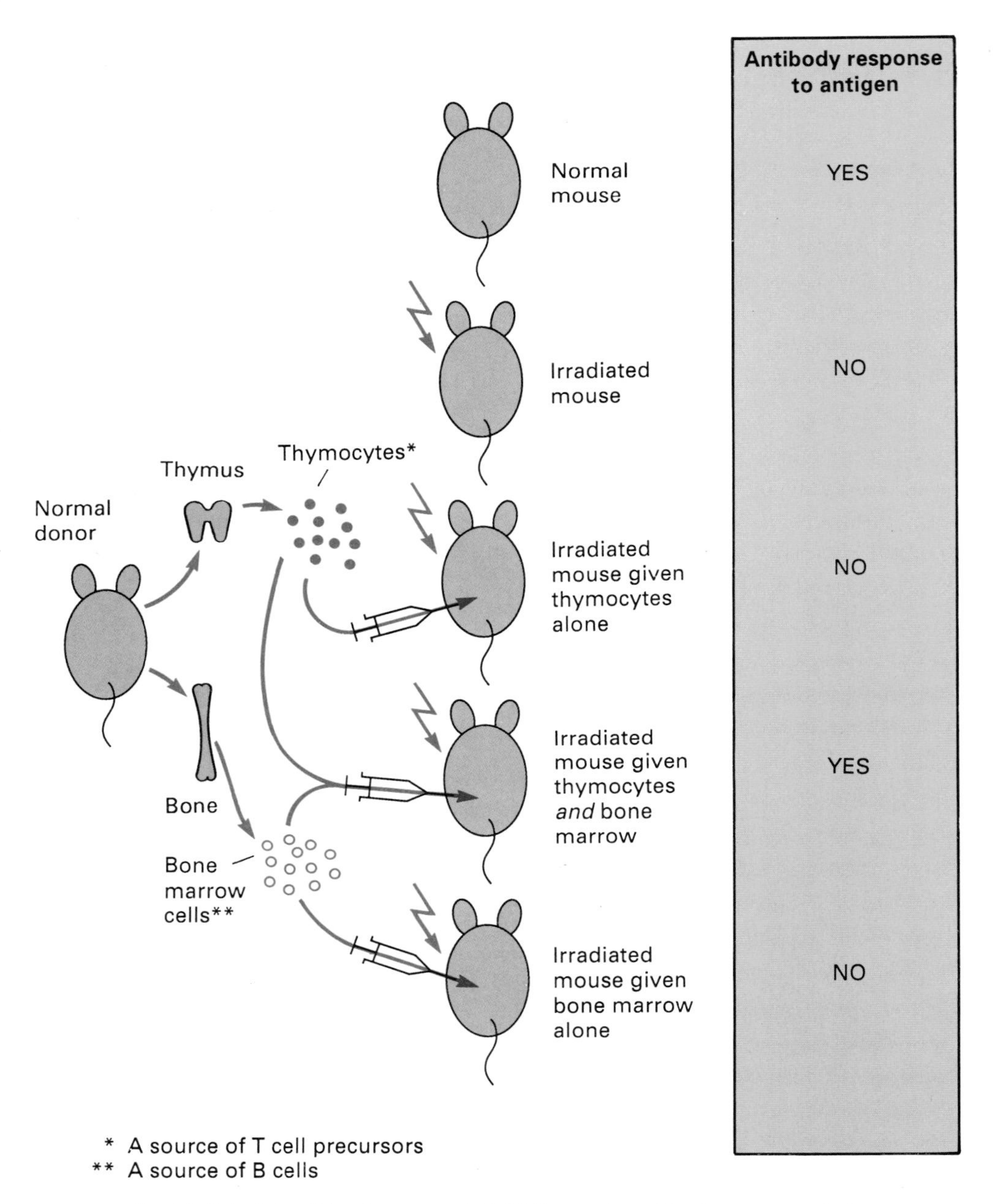

Fig.3.1 Principle of the Claman *et al.* (1967) experiment to show synergy between different lymphocyte populations during antibody responses *in vivo*

like sheep erythrocytes, digest them, and then 'present' immunogenic molecules to the lymphocytes. Hence these accessory cells also came to be known as **antigen-presenting cells**. The idea that macrophages were important in immune responses was reinforced in the 1970s when the MHC restriction of T cells was recognized, since macrophages were found to express MHC class II molecules that are required for antigen recognition by helper

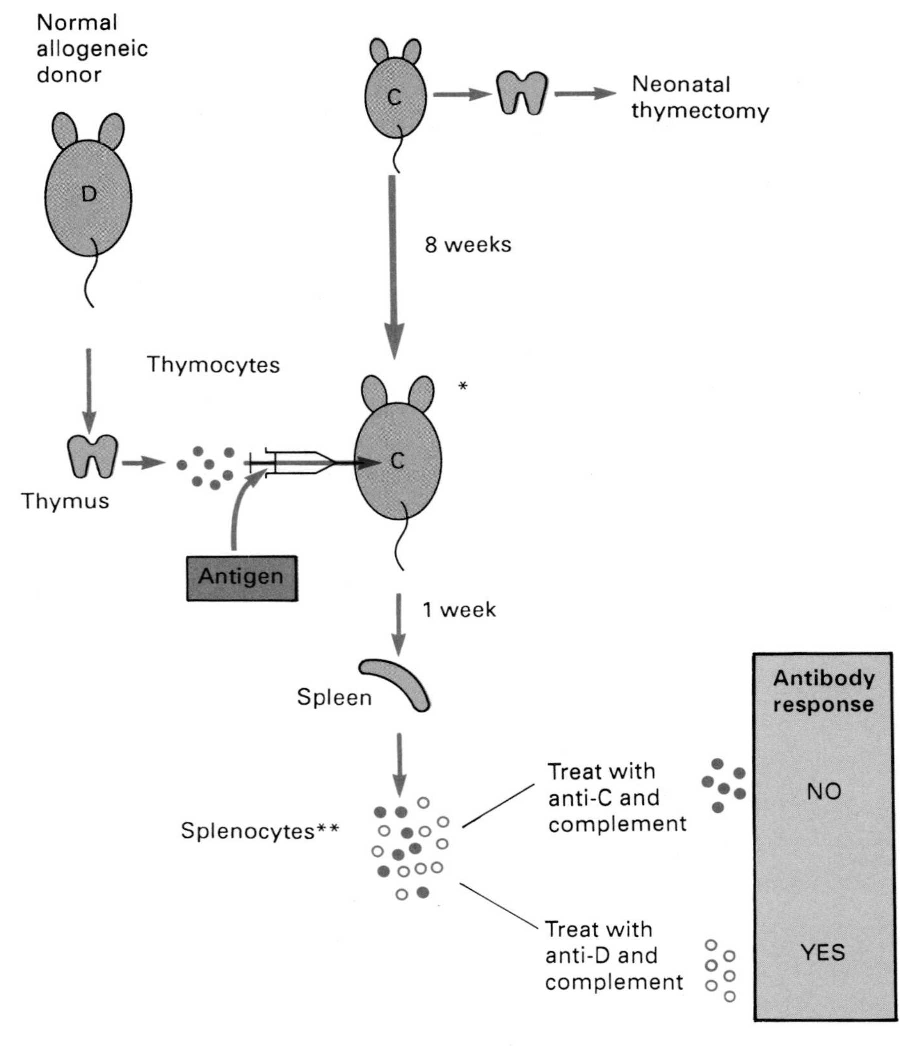

Fig.3.2 Principle of the Mitchell and Miller (1968) experiment to show that thymus-derived cells are needed for antibody responses, but do not actually produce antibodies themselves

T cells (Fig.1.27). Soon afterwards, it was found that macrophages could also produce a soluble molecule, or **cytokine**, called 'lymphocyte activating factor' now known as interleukin-1 (IL-1), which can in certain instances trigger lymphocyte proliferation, in part by stimulating the secretion of another cytokine, interleukin-2 (IL-2), from the activated T cells (Smith *et al.* 1980).

Put together, these and other observations led to the following scheme for the role of accessory cells in immune responses (Fig.3.3). Antigens are taken

Table 3.1 Principle of Mosier's experiments showing that adherent accessory cells are required for antibody responses in culture

Cells/antigen	Antibody response
Lymphocytes* + Ag	No
Lymphocytes + adherent cells** + Ag	Yes
Adherent cells + Ag	No
Lymphocytes + adherent cells	No
Ag-pulsed lymphocytes + adherent cells	No
Lymphocytes + Ag-pulsed adherent cells	Yes

* A mixture of B and T lymphocytes
** A mixture of macrophages and dendritic cells
Ag = antigen

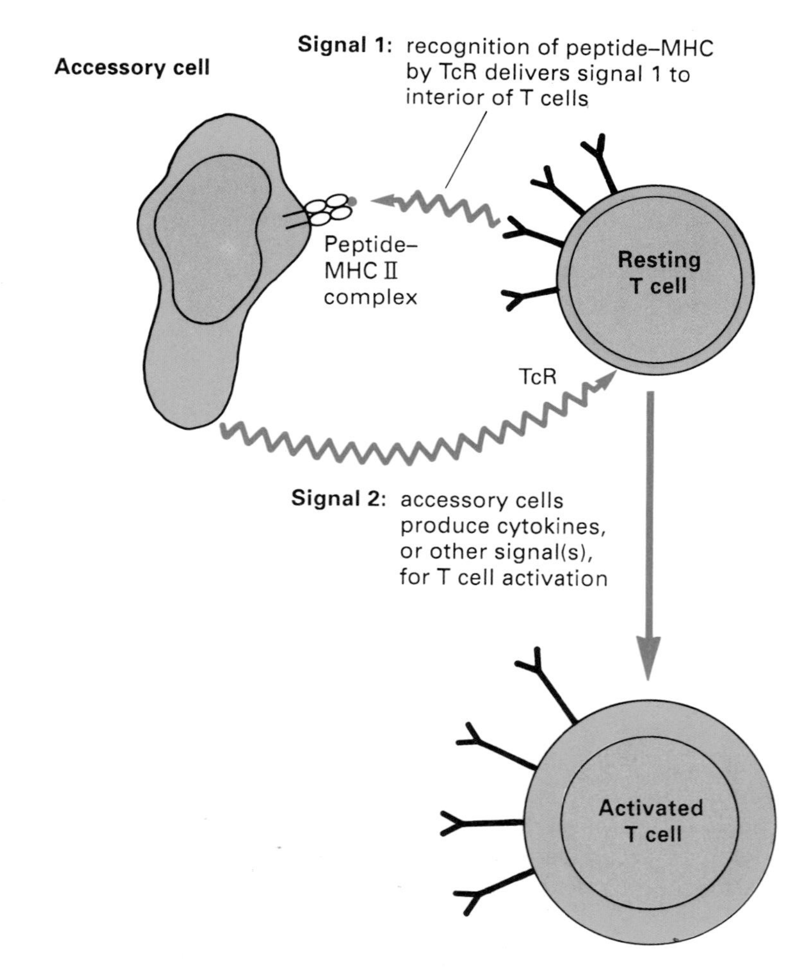

Fig.3.3 A 'two-signal' hypothesis for T cell activation

up by antigen-presenting cells, particularly by macrophages, and they are then somehow altered, or processed, from their native form and presented in association with MHC class II molecules to specific helper T cells. This provides a **first signal** for T cell activation. The antigen-presenting cells then elaborate one or more soluble molecules that provide the **second signal** also required for T cell activation. Once helper T cells are activated, they can help B cells make antibodies and cytotoxic T cells to develop from their precursors, thus triggering the effector stage of immune responses.

While the scheme just outlined, or a close variant of it, is accepted by many immunologists today, it fails to take account of important developments that were taking place in the 1970s and early 1980s in a different direction. These began with the discovery of **dendritic cells** (Steinman and Cohn 1973) as a trace cell type contained within many adherent macrophage populations from lymphoid tissues (Fig.3.4). When dendritic cells were separated from macrophages, and each population was tested for its ability to act as accessory cells for several primary immune responses *in vitro*, it emerged that it was the dendritic cells that actually stimulated these responses, while the macrophages had little if any of this activity. This was found especially when responses of initially resting T cells were examined, as opposed to responses of T cells that had already been activated, such as those isolated from recently immunized animals.

It now seems that dendritic cells are required to deliver a particular activation signal(s) to *resting* T cells that cannot be supplied by most other cell types, as well as being able to present the relevant antigens. Once the T cells have been activated in this way, they can respond to any other cell type expressing the same antigen–MHC complex. The former process can be termed **immunostimulation** and dendritic cells are therefore called **immunostimulatory cells**. However, most cells in the body (including

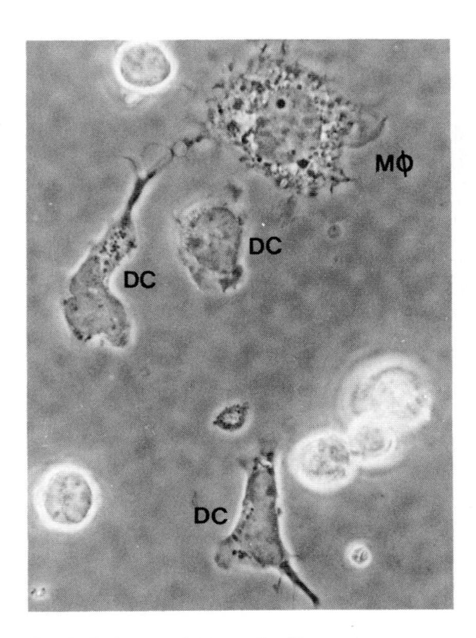

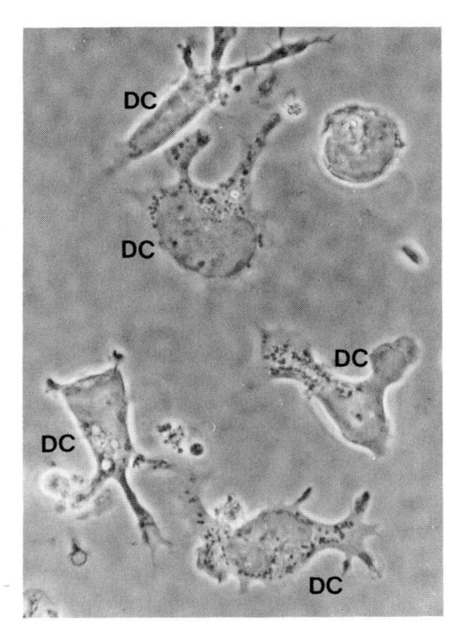

Fig.3.4 A population of adherent accessory cells from mouse spleen showing dendritic cells (DC) and a macrophage (Mφ)

dendritic cells) may be able to present antigens to *activated* T cells and thus act as antigen-presenting cells (Fig.3.5). The term 'accessory cell' can be used for any cell that is required together with antigen to stimulate a lymphocyte response, and it thus encompasses both antigen-presenting cells in general as well as specialized immunostimulatory cells.

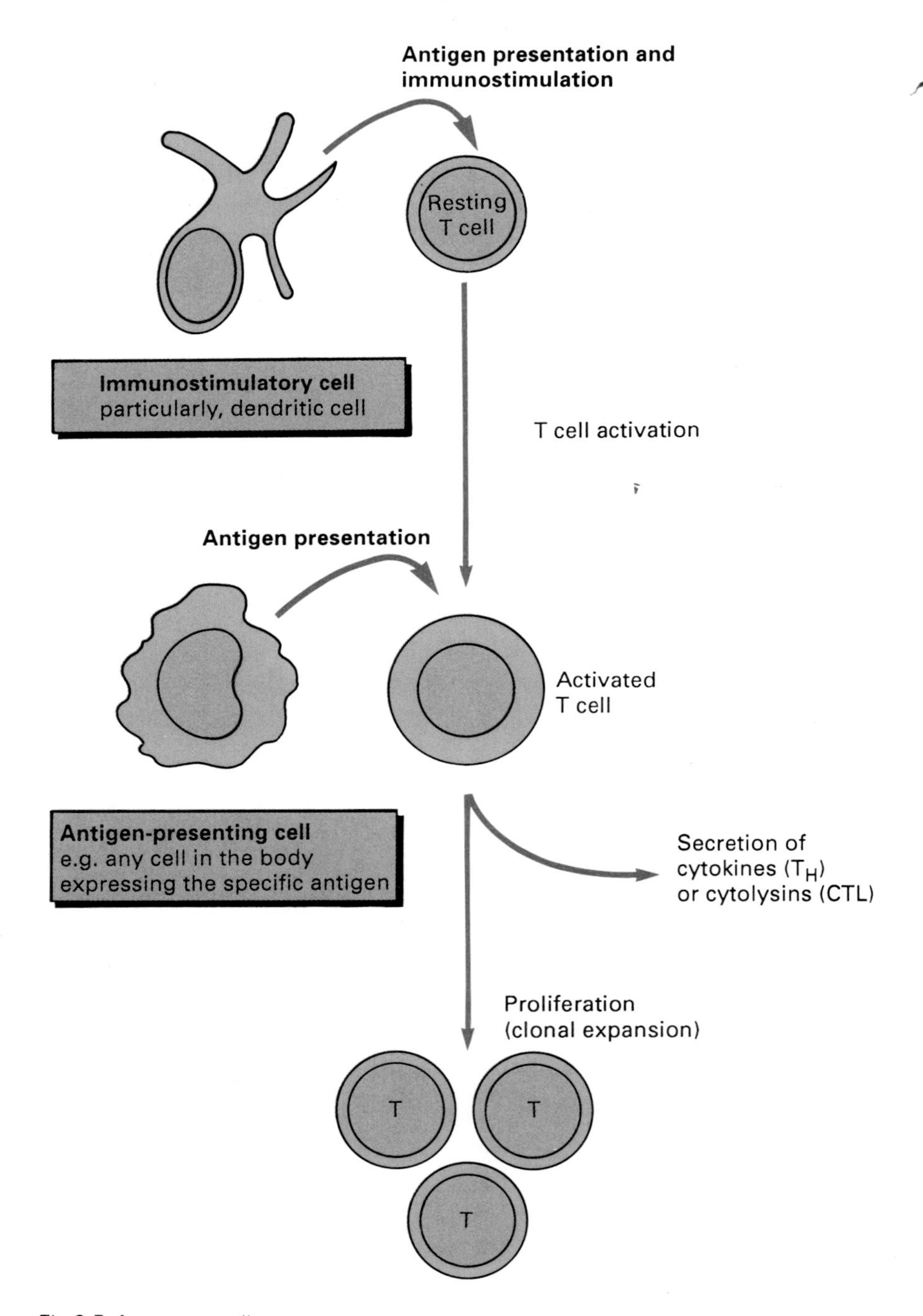

Fig.3.5 Accessory cells

Old and new definitions of antigen presenting cells. Antigen presenting cells were orginally thought of as cells that could process foreign antigens, express MHC class II molecules and activate helper T cells, thereby initiating the effector phase of immune responses. Cells expressing MHC class I molecules only, that could be targets for cytotoxic T cells, were not included in this category, especially as it was formerly believed that cytotoxic T cells recognized membrane-associated 'native' rather than processed antigens. There is now no good reason for maintaining this restrictive definition of an antigen presenting cell, first, because it is known that cytotoxic T cells recognize peptide–MHC complexes, as do helper T cells, and secondly, because any cell in the body probably has the capacity to process and present foreign antigens to activated helper and/or cytotoxic T cells. The term is therefore better used in a more general sense. *In other words, an antigen presenting cell is any cell that expresses peptide–MHC complexes that can be recognized by specific T cells.*

3.2 Antigen processing and presentation

3.2.1 T cell recognition of processed antigens

Before it was realized that lymphocytes could be subdivided into T cells and B cells, immune responses had been categorized as 'cellular' and 'humoral' responses. Gell and Benacerraf (1959) obtained evidence that different forms of antigen stimulated these two types of response (Fig.3.6).

Guinea pigs were immunized against a protein that was either in its **native** (intact) conformation, or after it had been **denatured**. It was found that the antibody response (humoral immunity) was specific for the particular form of antigen used for immunization. In other words, antibodies produced against a native protein did *not* react with the denatured molecule and vice versa. On the other hand T cell responses (cellular immunity) could be produced to *both* the native and denatured forms of antigen regardless of which had been used for immunization. This was demonstrated in delayed-type hypersensitivity reactions, now known to be mediated by T cells, by measuring a local response, the swelling that occurred at the site of antigen injection. These results can now be explained in that T cells mediate classical cellular responses and B cells mediate humoral responses.

Antibodies, the antigen receptors of B cells, frequently recognize antigenic determinants or epitopes that are directly accessible in the native three-dimensional structure of an antigen. However, T cell receptors recognize epitopes that may be exposed when the molecule is unfolded or degraded (Fig.3.7). These were respectively termed **conformational** and **sequential epitopes**. When an animal was immunized with antigen in its native conformation, antibodies were produced against this form, whereas T cells recognized the antigen only after it had been taken up by accessory cells and unfolded, degraded, or otherwise

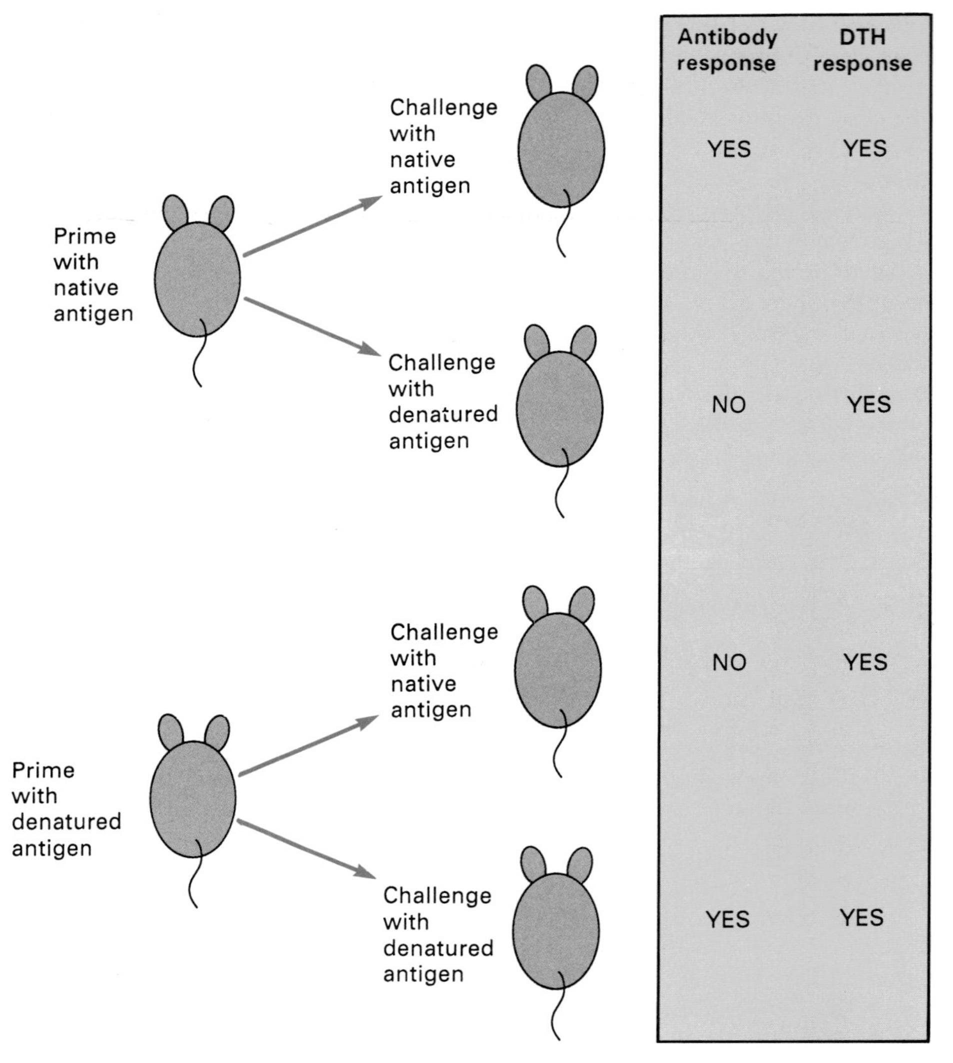

Fig.3.6 Principle of the Gell and Benacerraf experiment to show that B lymphocytes and T lymphocytes respond to different forms of antigen

'processed'. Thus, to a large extent, the same form of antigen (peptide–MHC) was ultimately presented to the T cells whether the animal was originally immunized with native or denatured antigen. These *in vivo* observations were subsequently confirmed in studies using isolated T cells *in vitro*.

T cells obtained from animals immunized with one form of antigen were found to respond in culture to native, denatured, or chemically modified forms in the presence of accessory cells. In contrast, antibodies generally showed little or no cross-reactivity between these forms. Moreover, the very same clone of T cells proliferated when either the native or denatured forms of antigen were added to the cultures, provided that accessory cells were present. These findings, for proliferative responses, also applied to helper responses: T cells from animals primed against a native antigen were able to help antibody and/or DTH responses to modified forms of the same antigen.

It was believed that chemical modification, denaturation, or enzyme digestion of the antigen in experimental situations mimicked an intracellular process that was required to expose peptides that could be buried within the three-dimensional structure of the native molecule. This molecular 'processing' was carried out *in vivo* by the antigen-presenting cell. Ultimately, it was realized that what was recognized by a helper T cell was a peptide bound

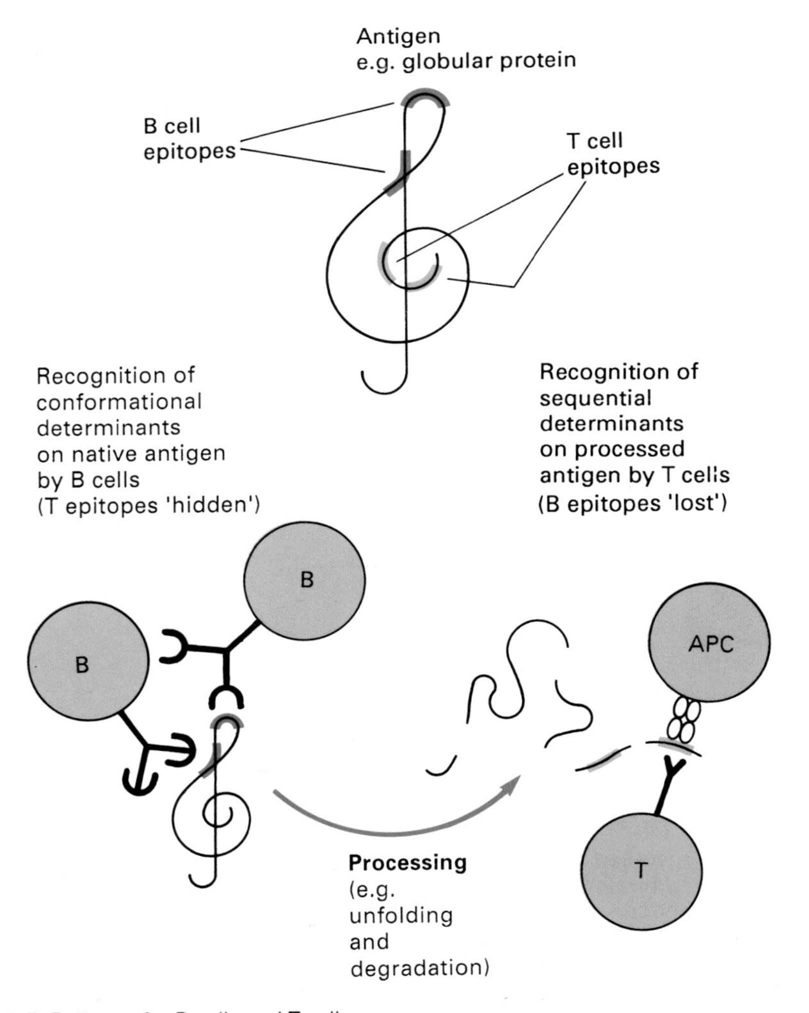

Fig.3.7 Epitopes for B cells and T cells

to an MHC molecule on an accessory cell. Of course, cytotoxic T cells also recognize peptide–MHC complexes, but the cell they recognize is more commonly referred to as a target cell.

It now seems probable that any cell in the body can process foreign antigens into peptides, some of which become bound to MHC molecules and these complexes can then be recognized at the cell surface by T cells. In fact it is likely that many self molecules (e.g. the cell's normal proteins) are handled in the same way (Fig.3.8). If so, most MHC molecules at the cell surface are normally occupied by self peptides and in some cases these have been eluted from MHC molecules and characterized (Jardetzky *et al.* 1991).

Foreign antigens are conventionally categorized as **exogenous** or **endogenous antigens** depending on whether they are derived from outside or synthesized within the cell (Fig.3.9). An example of an exogenous antigen would be a protein from a bacterium that is internalized by a cell such as a phagocyte. An example of an endogenous antigen is a viral protein that is synthesized within a cell after it has been virally infected.

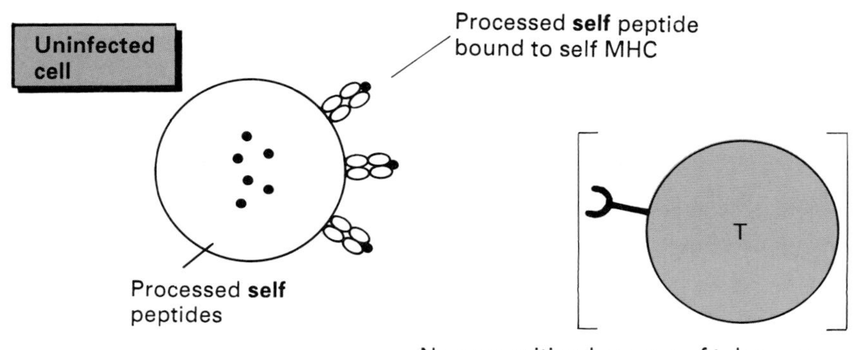

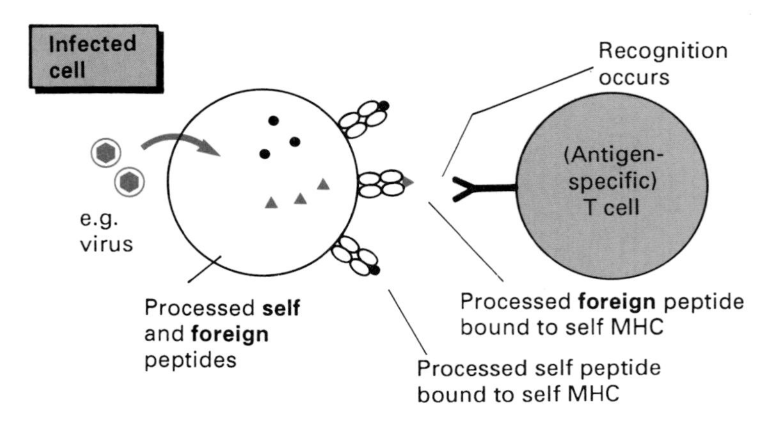

Fig.3.8 MHC molecules may bind a representative sample of self peptides as well as non-self (foreign) peptides, when present, for perusal by T cells

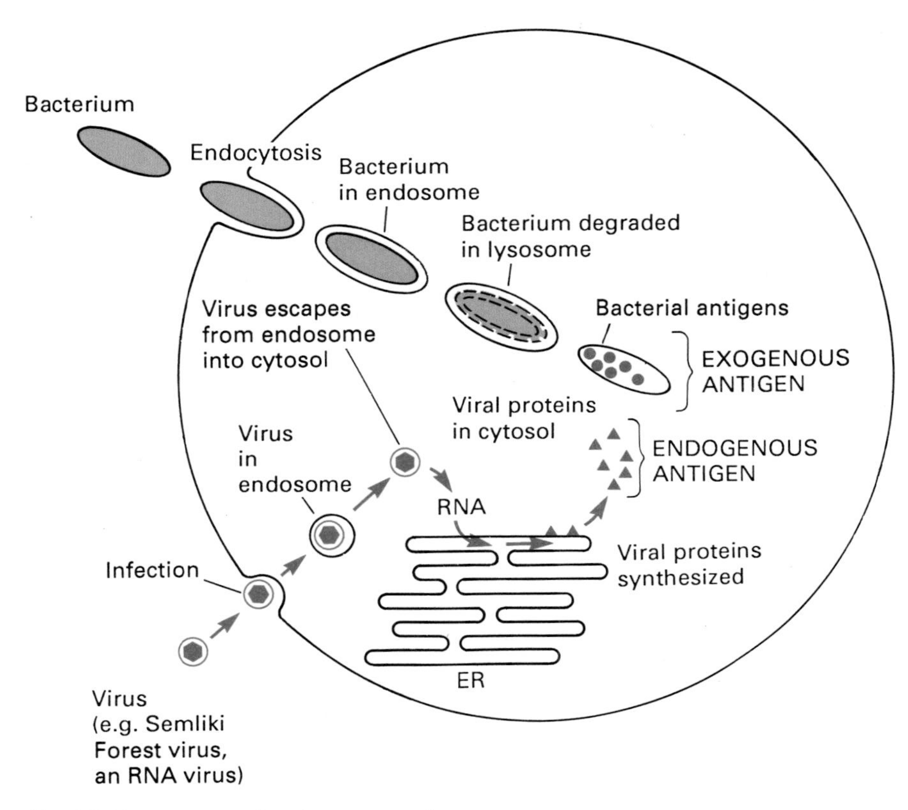

Fig.3.9 Exogenous and endogenous antigens

But what exactly do we mean by the term 'antigen processing'? *Quite simply, this is the mechanism by which the native (intact) form of an antigen is degraded to a set of peptides, some of which can bind to an MHC molecule.* These peptide–MHC complexes may then be presented to T cells (Fig.3.10). It is important to note that not all peptides will be able to bind to a given MHC molecule, and even if a certain peptide–MHC complex is formed, there may not be a T cell expressing a receptor specific for this complex in the immune system of that individual. Presumably the production of a variety of different peptides from a foreign antigen increases the chances that they can associate with the MHC alleles of that individual and be recognized by the T cells in the repertoire.

In this section some pathways by which antigen may be processed will be considered, focusing particularly on the uptake and presentation of *exogenous* antigens by macrophages and B cells, and the processing of *endogenous* antigens by fibroblasts and other cells in the body.

3.2.2 Presentation of exogenous antigens by macrophages

We start by briefly considering endocytosis in macrophages as one example of how exogenous antigens can be taken up by cells.

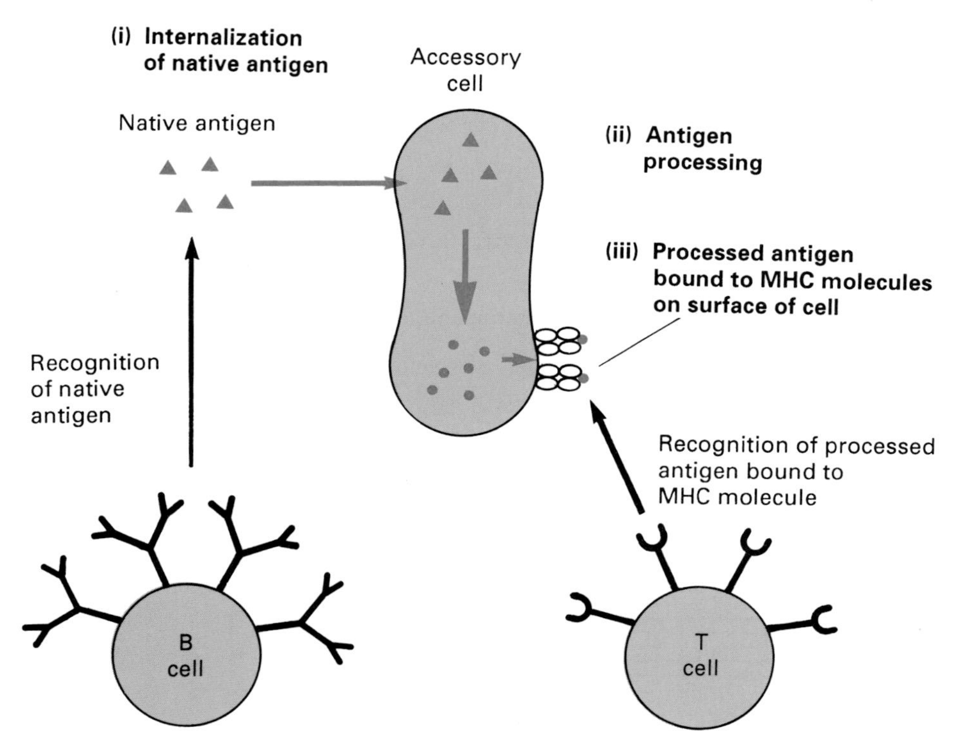

Fig.3.10 An outline of antigen processing

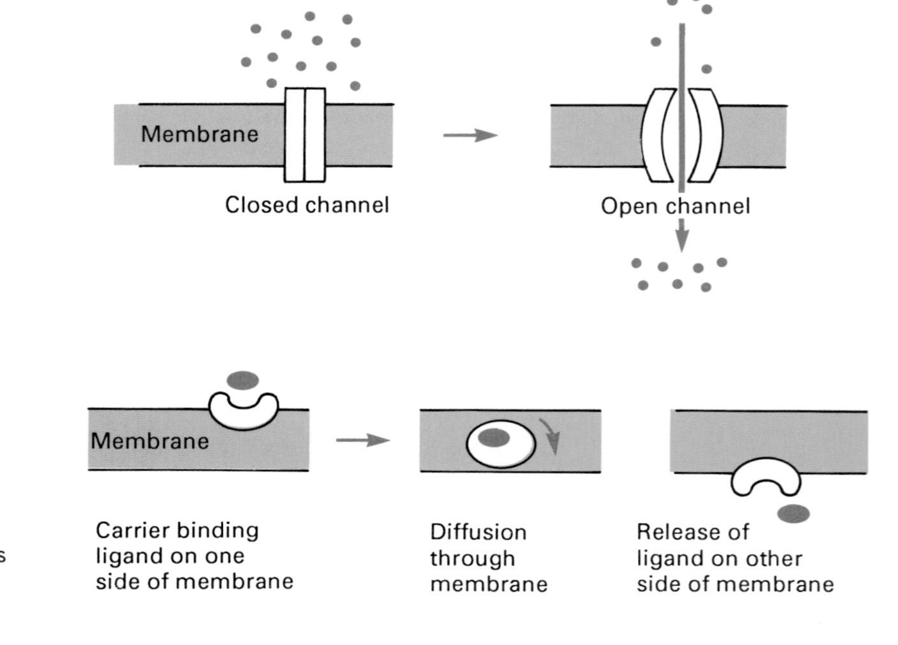

Fig.3.11 Transport through membranes via channels or by carrier molecules

Endocytosis and membrane recycling in macrophages

Carrier proteins and channels. The plasma membrane presents a barrier to the entry of most molecules into the cytoplasm of the cell (Fig.3.11). To allow small molecules and ions to be transported into and out of the cell a number of specific 'carrier proteins' and 'channels' are present in the cell membrane. For instance, calcium ions may enter, and potassium ions leave, T cells through specific ion channels that are opened when the cells become activated.

Vesicular transport. Cells are divided into specialized compartments by membranes, and membrane-bound **vesicles** are thought to carry molecules through the cell from one compartment to another (Fig.3.12). (More recent evidence, however, suggests that some vesicular compartments may be more

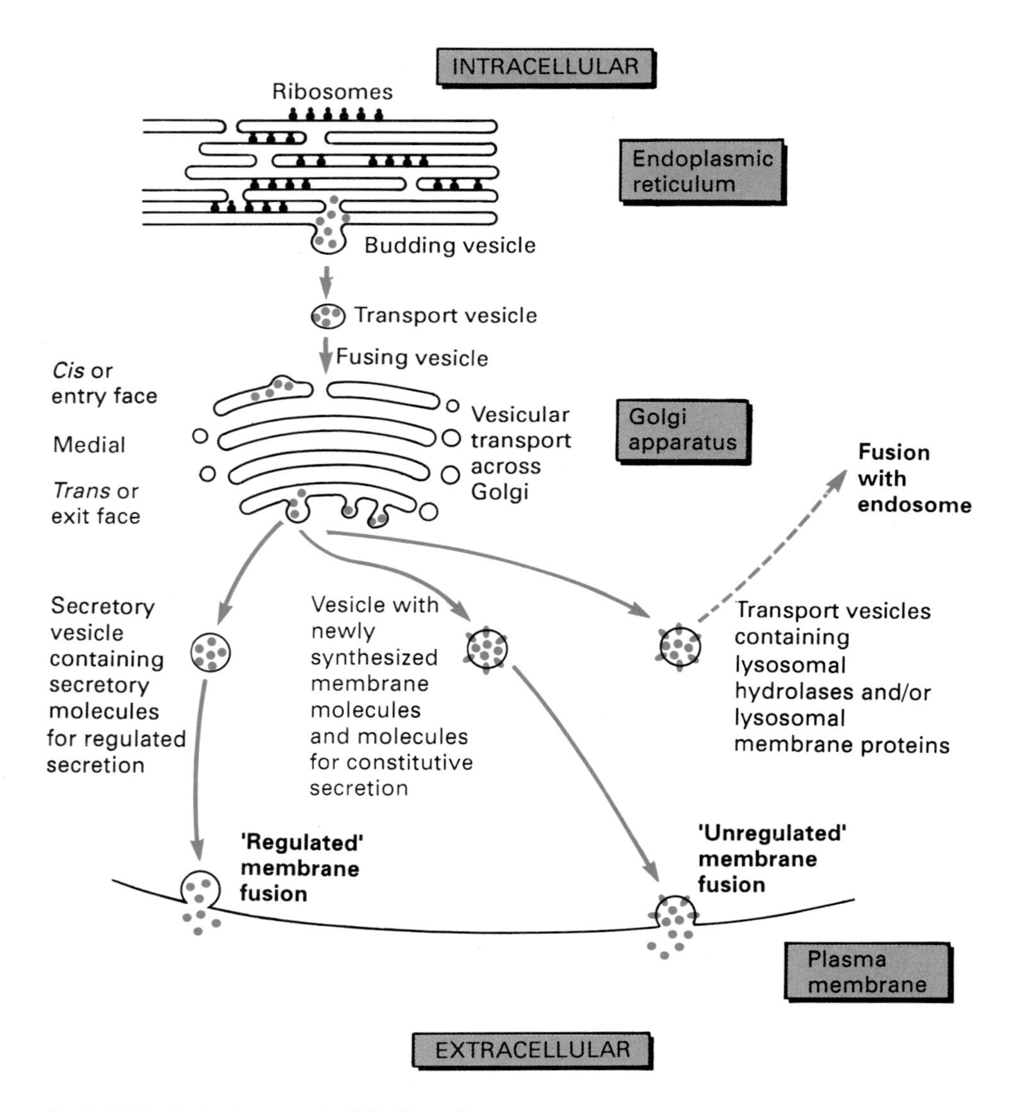

Fig.3.12 Vesicular transport within the cell

continuous than was originally thought; in some cells they appear to form a highly-connected tubular-vesicular network). When the vesicles reach their destination, the membranes surrounding them fuse with target membranes and their contents are delivered into the new compartment. For example, during the biosynthesis of glycoproteins like antibodies, protein molecules are transported in vesicles from the endoplasmic reticulum (ER) where they are synthesized, to the Golgi apparatus (Fig.1.6). Here modifications are made to the newly-synthesized product, particularly to the carbohydrate chains attached to the protein core. These glycoproteins are then transported away from the Golgi in vesicles and, depending on their fate, the molecules they contain may: (i) be stored in secretory granules in the cytoplasm awaiting a signal for secretion; (ii) become inserted into the plasma membrane (e.g. membrane-bound antibodies); or (iii) may be transported through the cell in other membrane-bound vesicles that were formally called primary lysosomes (e.g. certain degradative enzymes).

Exocytosis and endocytosis. During **exocytosis**, cytoplasmic vesicles transport their contents to the plasma membrane, fuse with it, and discharge their contents to the exterior of the cell (Fig.3.13). It is usual to discriminate

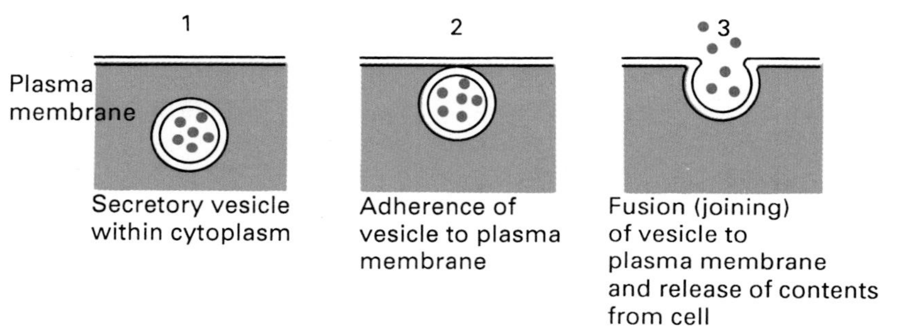

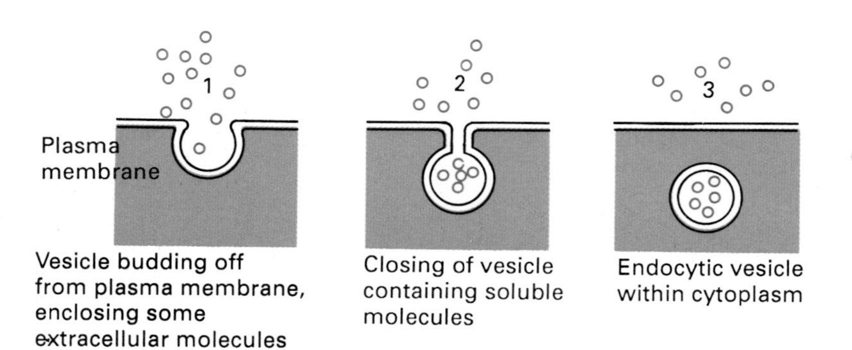

Fig.3.13 Exocytosis and endocytosis

between **constitutive** and **regulated pathways** of exocytosis. Molecules that are continously secreted from a cell follow the former route, while molecules that are stored within the cell awaiting a signal for secretion follow the latter route. The reverse process, referring to the uptake rather than the release of molecules, is called **endocytosis**. Particles and macromolecules become surrounded by portions of the plasma membrane which invaginate and pinch off to form vesicles containing the ingested material. Endocytosis can be divided into **phagocytosis** ('cell eating') and **pinocytosis** ('cell drinking') although now this term often refers only to the latter.

Phagocytosis occurs in specialized cell types such as macrophages and polymorphonuclear leukocytes which are therefore called phagocytes (Fig.1.11). During phagocytosis large particles like micro-organisms and cellular debris are taken into the cell within membrane-bounded phagocytic vacuoles or **phagosomes**. This uptake may be mediated by specific receptors on the plasma membrane of the cell. **Receptor-mediated phagocytosis** is a localized response of the membrane that proceeds by a 'membrane-zippering' mechanism involving sequential contact between the receptors and their ligands, as the membrane spreads over and engulfs the particle (Fig.3.14). During Fc receptor-mediated phagocytosis (see below), talin, a 225-kDa cytosolic protein becomes associated with actin filaments of the cytoskeleton in the region of the forming phagosome, and these components are presumably involved in the phagocytic process.

In contrast to phagocytosis, pinocytosis is a property of most or all cell types and it is constitutive, meaning that it does not need to be induced by the molecules to be internalized. This is in contrast to phagocytosis, which is only triggered once the particle has been encountered. During pinocytosis, pinocytic vacuoles or **pinosomes** are formed which contain samples of the extracellular fluid together with its dissolved molecules. Soluble molecules can also become attached to the membrane during **absorptive pinocytosis**, or they may be bound to specific receptors and internalized by **receptor-mediated pinocytosis**.

Receptor-mediated endocytosis often occurs at specialized regions of the membrane. These are called **coated pits** because, by electron microscopy, they consist of indentations in the plasma membrane that are coated with a thick material on their cytoplasmic face. (Fig.3.15). This material turns out to be a coat of several proteins, one of which is clathrin, which forms a basket-like molecular structure. Some membrane receptors are always associated with coated pits, but others perhaps only become localized in these regions after they have bound their ligands. The receptor–ligand complexes are then endocytosed in **coated vesicles** that are formed as the membrane in the region of the coated pits becomes invaginated. Endocytosis at coated pits appears to provide a specialized pathway for the uptake of certain molecules, since these molecules enter the cell faster than the general extracellular fluid enters during fluid phase pinocytosis.

Once inside the cytoplasm, the coat is lost from coated vesicles and they fuse with others to form larger vesicles called **endosomes** (Fig.3.16). Initially (e.g. 2–10 minutes after internalization) the endocytosed contents are found in endosomes close to the plasma membrane. These are called peripheral endosomes and they have a pH of 6–6.5 and are receptive to fusion with other vesicles entering the cell. Later (e.g. 10–30 minutes after internalization) they are present deeper in the cell within perinuclear or 'internal' endosomes, comprising a 'pre-lysosomal compartment', which has a pH of 5.5 and is unable to fuse with newly-formed vesicles. Peripheral and internal endosomes are also called early and late endosomes, respectively. Transport vesicles containing hydrolytic enzymes and/or other proteins, that were formally called primary **lysosomes,** may then fuse with the endosome to form an **endolysosome** and ultimately a terminal lysosome (originally termed a secondary lysosome). In macrophages, hydrolytic enzymes in this

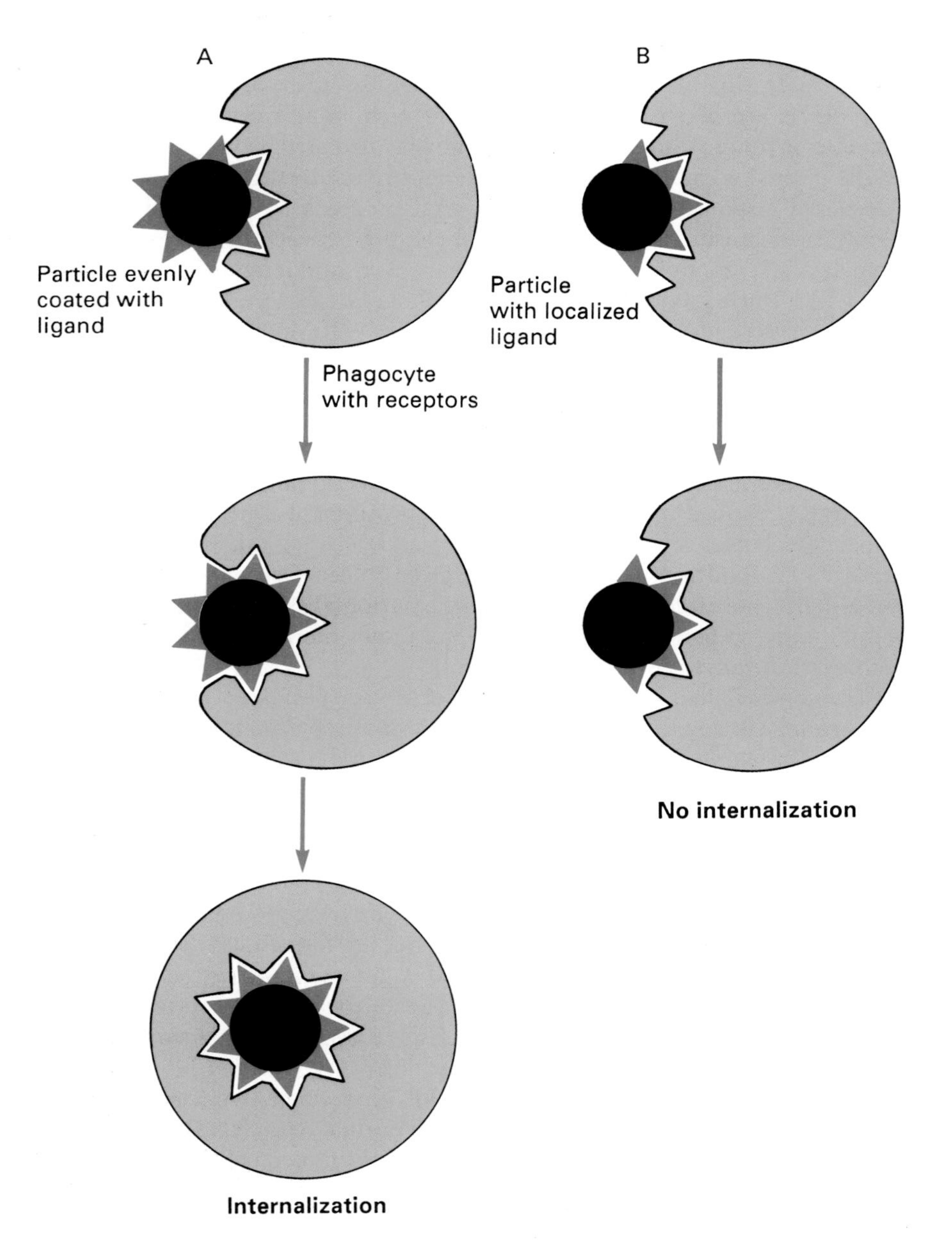

Fig.3.14 Evidence for a 'membrane-zippering' mechanism of phagocytosis

highly acidic environment (pH 4.5–5) degrade the endocytosed contents to amino acids, dipeptides, sugars, and nucleotides. These are then transported to the cytoplasm, or they may be transported in vesicles and exocytosed from the cell.

In summary, endocytosed contents are transferred sequentially from coated vesicles to early endosomes, late endosomes, and endolysosomes and/or terminal lysosomes.

Membrane recycling. During endocytosis, large volumes of extracellular fluid are taken into the cell together with large areas of plasma membrane. Macrophages, for example, constitutively internalize about 25% of their total volume every hour and the equivalent of their entire surface area every half an hour, yet the cell's overall volume and membrane area remain relatively constant. This is achieved by **membrane recycling**. As fast as membrane is internalized, there is a continuous shuttle of membrane vesicles back to the plasma membrane in compensation, either directly, or indirectly via the Golgi apparatus.

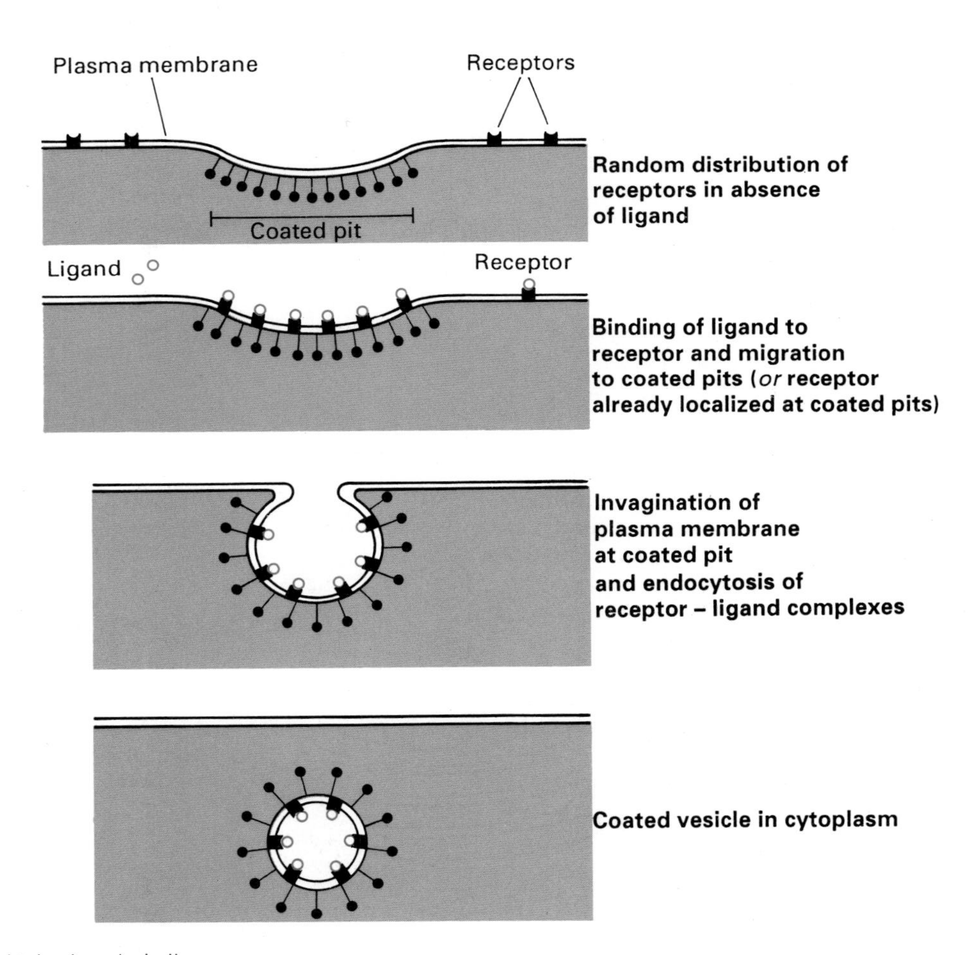

Fig.3.15 Endocytosis at coated pits

Endocytosis in macrophages. Macrophages have a variety of surface receptors that enable them to internalize particles by receptor-mediated endocytosis (Fig.3.16). The receptors involved in this process include: **Fc receptors** which bind particles and antigens complexed to certain classes of antibody; **complement receptors** which bind complement-coated

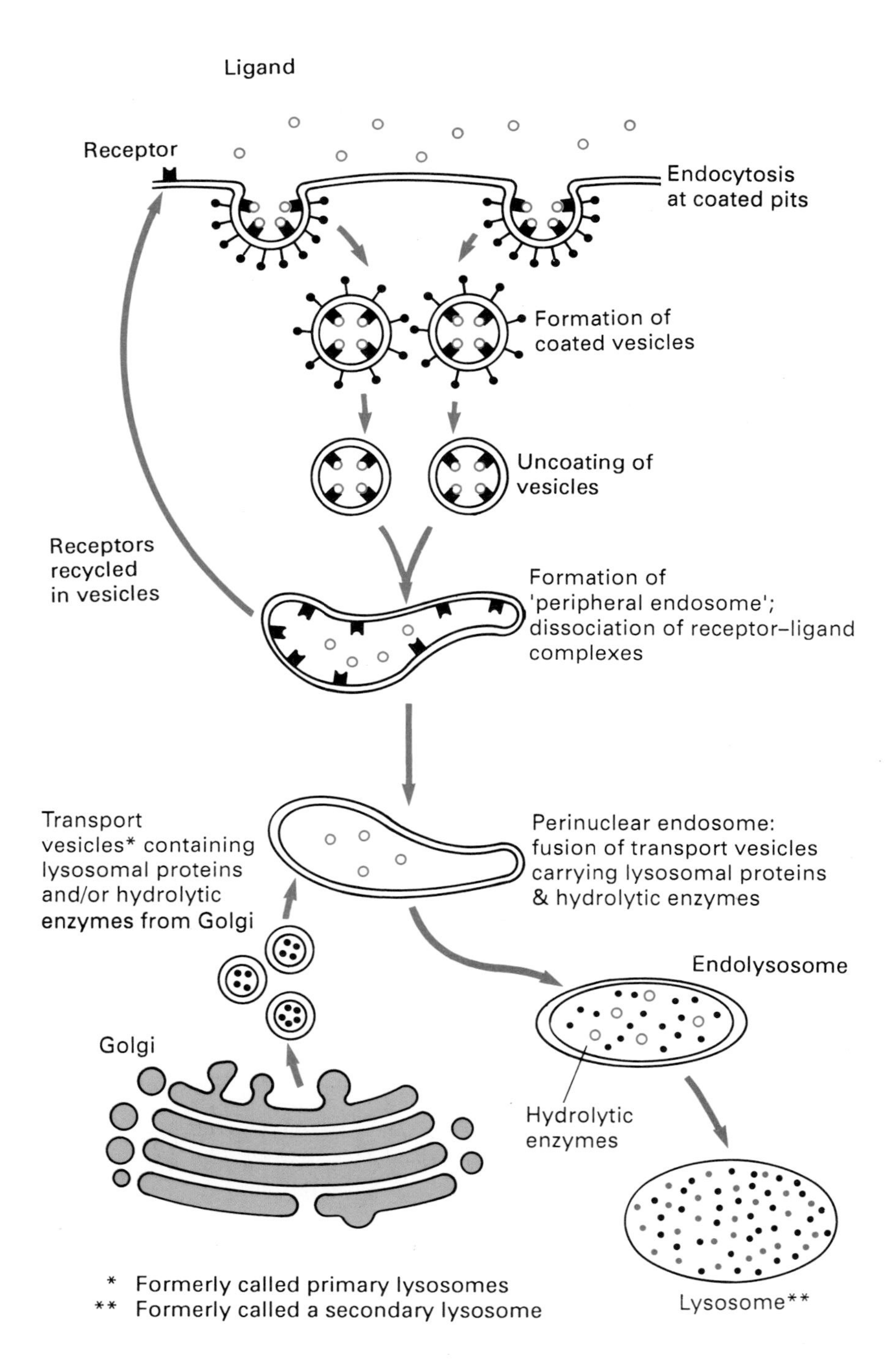

Fig.3.16 Endocytosis at coated pits and delivery to lysosomes (see Figs.3.15 and 3.17)

particles; and sugar-specific carbohydrate recognition systems, such as the **mannosyl-fucosyl receptor** through which some glycoproteins and particles like yeasts can be endocytosed. (Macrophage Fc and complement receptors are discussed in Section 10.4.5.) Moreover, macrophages can engulf a wide variety of particles and cellular debris by several less well-defined mechanisms. Much of the phagocytosed material is degraded within terminal lysosomes, and most of the contents of pinosomes end up in the same place. Some of the digested products enter the cytoplasm, but a large proportion of the degraded material is exocytosed and released from the cell as small fragments including peptides. Surface binding of ligands does not necessarily lead to receptor-mediated endocytosis, and other sequelae can also result, such as enzyme secretion.

Receptors are internalized during receptor-mediated endocytosis together with their ligands. Some of these receptors are degraded within lysosomes, an example being the Fc receptor. This provides a means by which surface expression of some membrane receptors can be reduced or 'down-regulated'. Other receptors, such as the mannosyl-fucosyl receptor, are degraded but rapidly replaced from intracellular pools. In other cases, the receptors can escape destruction altogether and may be recycled (from early endosomes) back to the surface of the cell so they can be reutilized, whereas their ligands dissociate and become degraded within lysosomes (Fig.3.17). Sorting of receptors and ligands occurs within the late endosomal compartment, and the existence of a specialized 'compartment for uncoupling of receptors and ligands' or **CURL endosome** has also been postulated. Indigestible material can persist with macrophages for long periods as 'residual bodies', and there is evidence to suggest that some endocytosed molecules in one form or another can remain cell-associated for several days.

Antigen presentation by macrophages

What is known of antigen processing in macrophages, as measured by their capacity to modify antigens to a form that can be recognized by T cells? Some of the earliest experiments to address this question directly were performed by Unanue (see Further reading) and colleagues, who studied the ability of macrophages to present antigens from the bacterium *Listeria monocytogenes* to T cells isolated from Listeria-immunized mice. In culture the macrophages bound, phagocytosed, and degraded the bacteria, and they were subsequently able to stimulate T cell responses.

Evidence for a 'lag phase'. It was found that macrophages were unable to present bacterial antigens to T cells if tested soon after they had bound or phagocytosed the bacteria (Fig.3.18). However, they became efficient presenting cells after incubation at 37°C for about 0.5–1 hour. These results indicated that an antigen-dependent and temperature-dependent **lag phase** was required for processing the bacterial antigens to a form that could be recognized by the T cells. Antigen processing only occurred in metabolically active cells, since macrophages that were killed by glutaraldehyde soon after binding the bacteria were unable to present antigens. However, they could still present the antigens if they were fixed after the lag phase, once processing had taken place.

Sensitivity to lysosomotropic agents. Ziegler and Unanue (1982) then studied the effect of treating macrophages with **lysosomotropic amines** on their ability to present antigens to T cells (Fig.3.19). These chemicals, such as ammonium chloride and chloroquine, are weak bases that can cross the

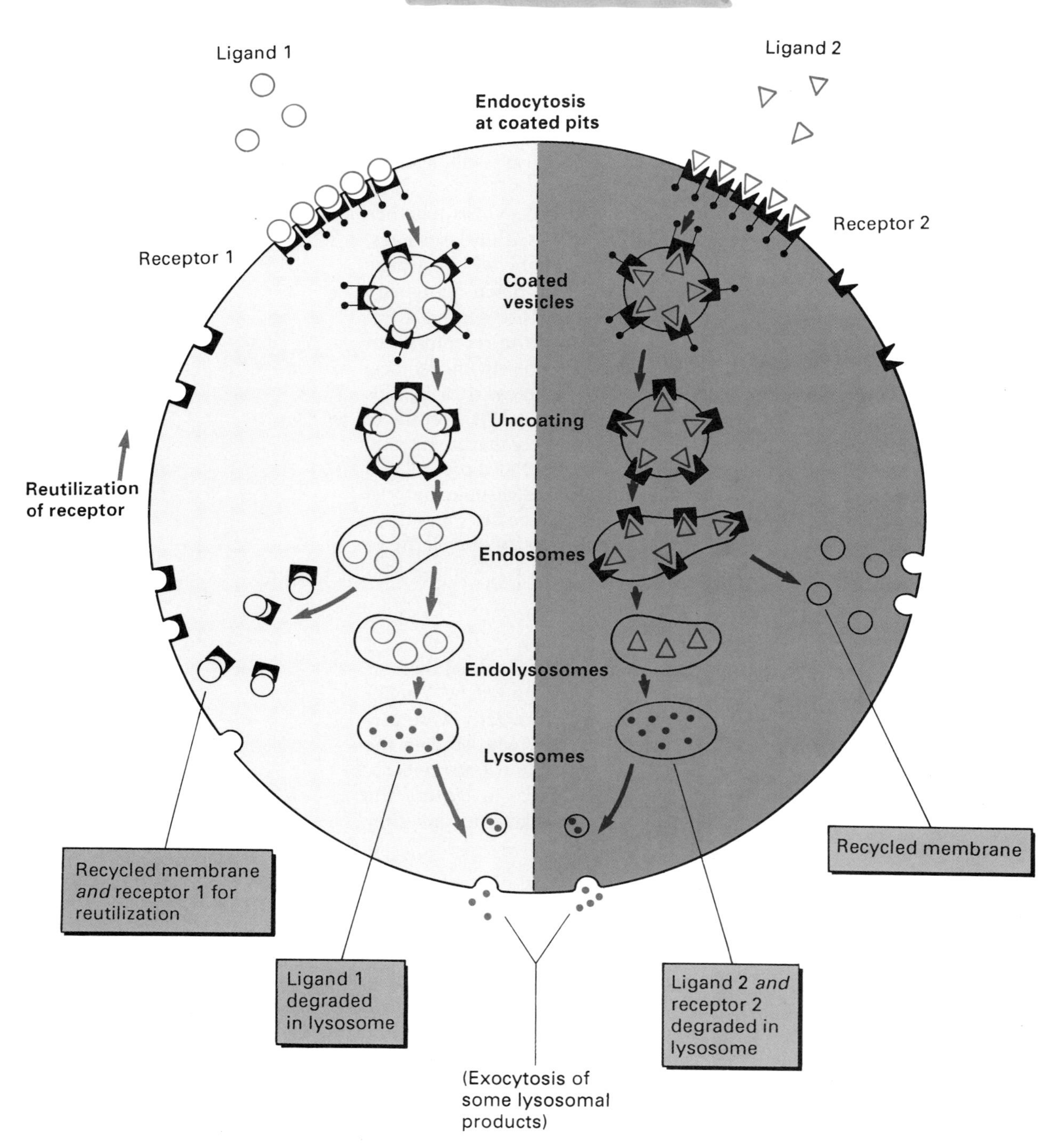

Fig.3.17 Recycling of membrane and of some receptors

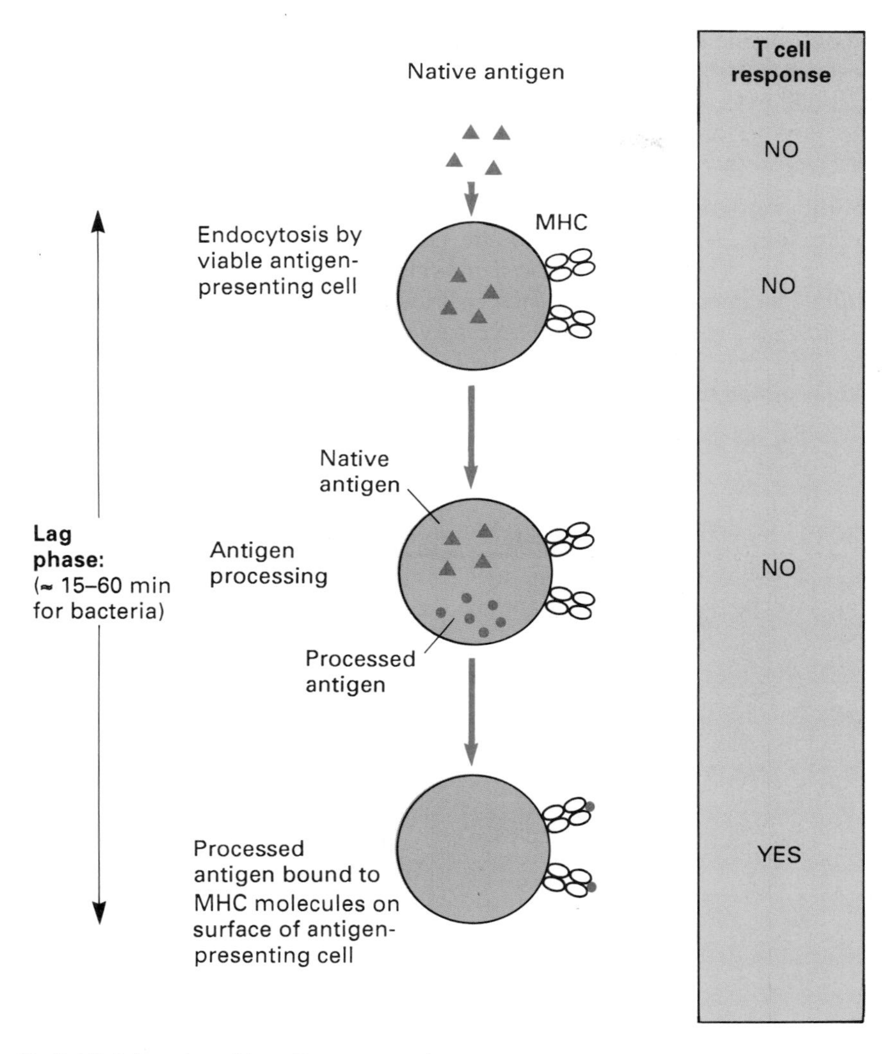

Fig.3.18 A 'lag phase' in antigen processing

plasma membrane and accumulate in acidic vesicles. Much of the degradation of endocytosed particles occurs in what were formerly called secondary lysosomes which provide an acid environment essential for acid hydrolase function (Fig.3.16). The lysosomotropic amines disrupt lysosomal function by increasing the pH within the lysosomes, as do chemicals like the Na^{+}-H^{+} ionophore monensin, and they therefore inhibit the degradation of antigens. The lysosomotropic amines had no effect on the uptake of bacteria by macrophages, but degradation of the endocytosed bacteria was inhibited as predicted. At the same time, it was found that macrophages treated with these agents were unable to present bacterial antigens to the T cells. In contrast, presentation was relatively unaffected if phagocytosis was allowed to proceed and the cells were incubated for an hour or so before treatment. These data were taken as evidence that antigens are taken up into an acid compartment of the macrophage for processing. One problem is that lysosomotropic amines are not specific in that some may inhibit phagosome–lysosome fusion, receptor–ligand dissociation, receptor recycling, and other cellular processes. For example, chloroquine also inhibits dissociation of the invariant chain from the $\alpha\beta$ dimer of MHC class II molecules (Section 2.6.2), and this may alter their cell surface expression.

Soluble antigens. Subsequent studies by other investigators extended the observations described above to the presentation of soluble antigens like lysozyme and ovalbumin to T cell hybridomas. As an assay, these investigators examined the ability of the hybridomas to secrete the growth factor IL-2 in response to antigen and accessory cells. They found that presentation of soluble proteins was inhibited when the macrophages were fixed or treated with lysosomotropic agents, as had been found in the studies with whole bacteria.

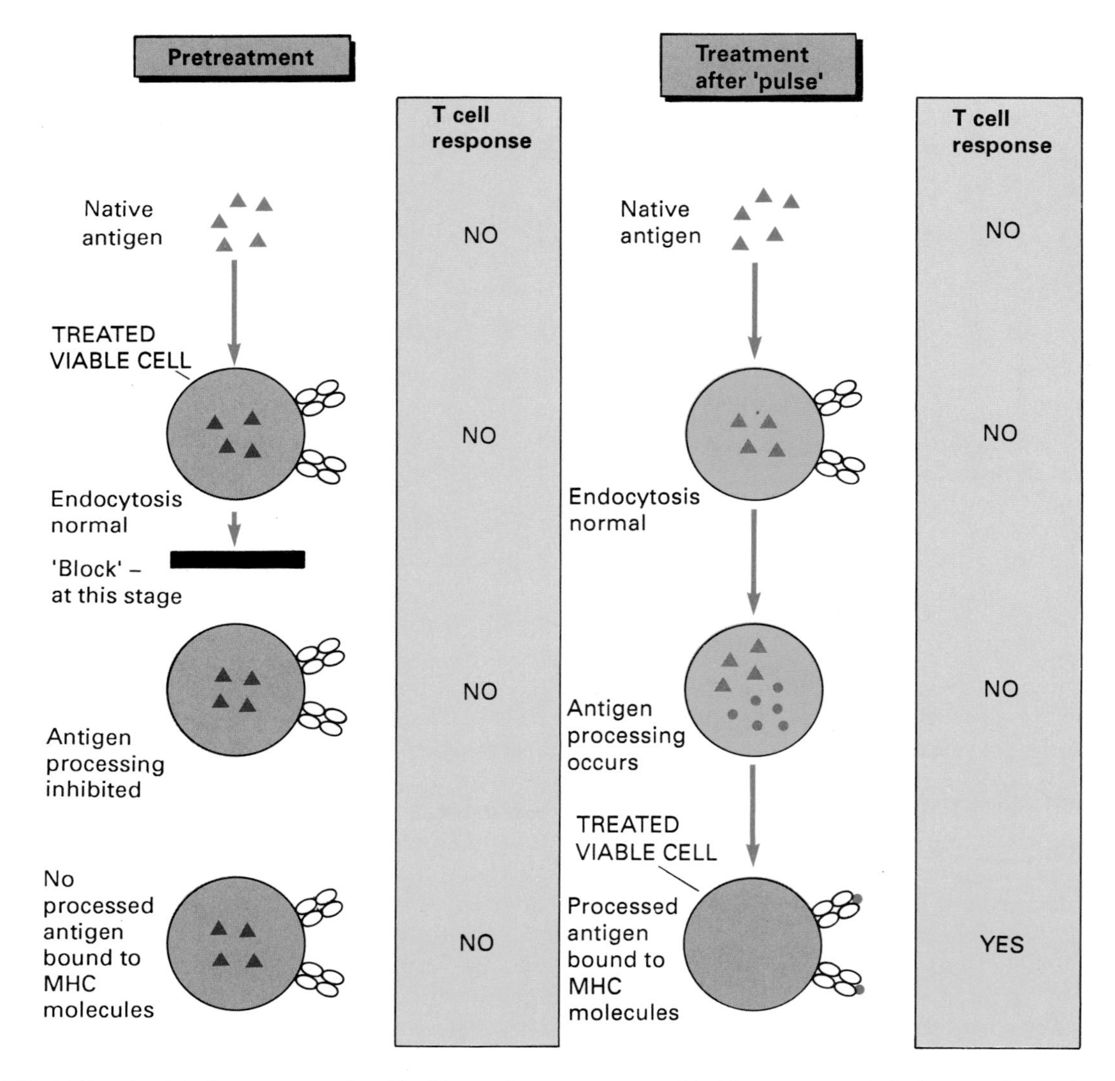

Fig.3.19 Effect of treatment of accessory cells with chloroquine, ammonium chloride, or monensin before or after antigen pulsing

Denatured antigens or fragments may not need cellular processing. A further advance came from observations of Shimonkevitz and colleagues (1983), later to be confirmed by others. Using a system in which ovalbumin was presented to T cell hybridomas, they found that fixed macrophages were unable to present native ovalbumin, as expected, but they could present the denatured molecule or enzyme digests of it (Fig.3.20). This indicated that one important event in antigen processing may be unfolding or cleavage of the native molecule. Moreover, it showed that the very same T cell clones that responded to native antigen in the presence of viable macrophages could recognize a modified or processed form in the presence of fixed cells. Presumably, the events occurring in live cells had been reproduced experimentally by prior denaturation or proteolytic cleavage of the antigen.

A model for antigen processing by macrophages. Based on these studies, it has been proposed that exogenous antigens are endocytosed and processed intracellularly within an acidic compartment which is lysosomal or perhaps prelysosomal in nature. Antigen processing results in some degree of unfolding or cleavage of the native molecule, and in most cases almost certainly the production of peptides (Fig.3.21). Some of these peptides become bound to MHC molecules and are transported to the cell surface where they can be recognized by specific T cells.

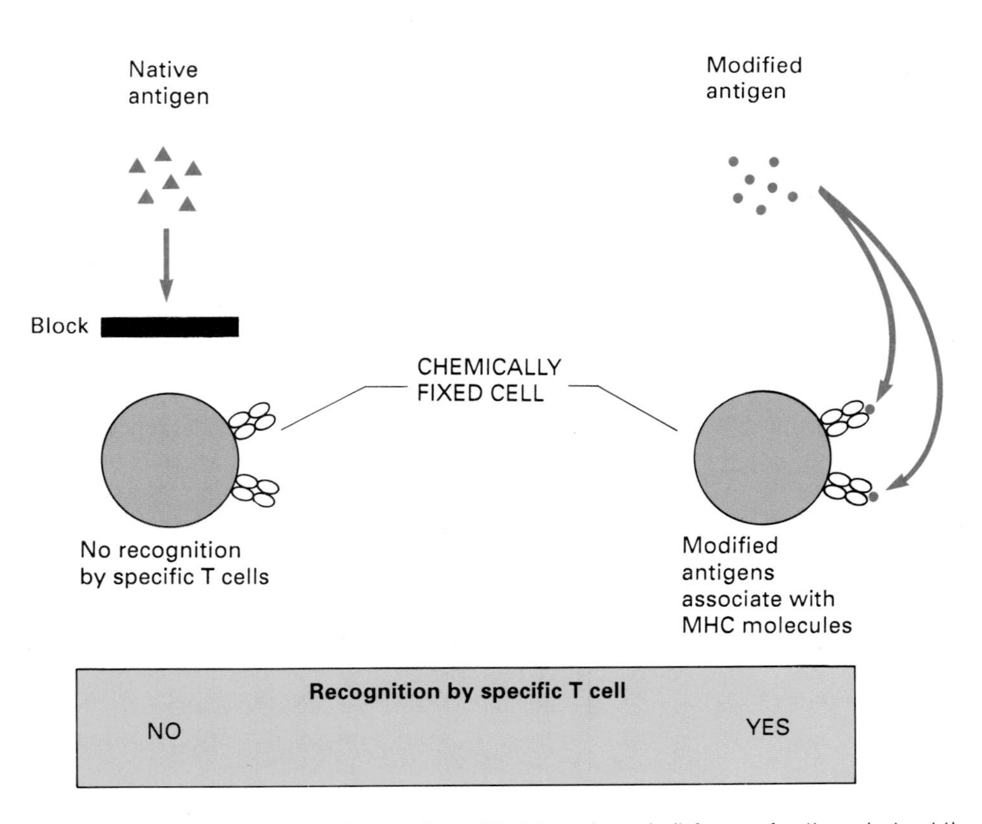

Fig.3.20 Glutaraldehyde-fixed accessory cells can 'present' modified (e.g. degraded) forms of antigen, but not the native antigen

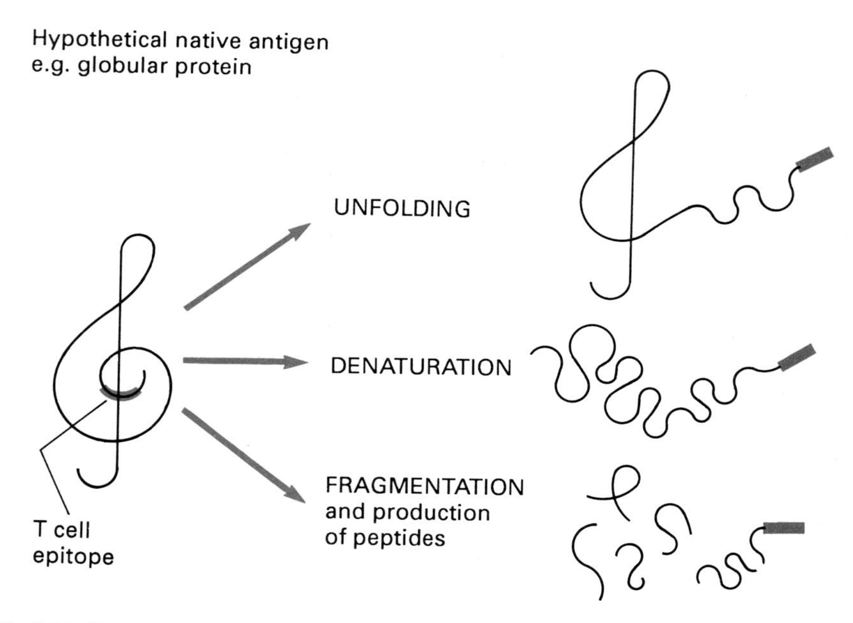

Fig.3.21 Presumed conformational changes during antigen processing

3.2.3 Processing and presentation of exogenous antigens by B lymphocytes

B cell tumours can process and present exogenous antigens to T cells, and normal B cells, particularly large B blasts, can also internalize and process antigens. Although B cells are non-phagocytic, there are two or more routes by which they can endoctyose antigens (Fig.3.22). One is *via non-specific pinocytosis* and another is *via their specific antigen receptors* (membrane-bound antibodies; immunoglobulins—Ig). Regarding non-specific pinocytosis, Chesnut and Grey (1982), and others, have shown that B cell lymphomas can present antigens to T cell hybridomas, irrespective of the B cells' antigen specificity.

However, they found that B cells handled antigens very differently to macrophages. For example, macrophages degraded ovalbumin and keyhole-limpet haemocyanin (KLH) to peptides and this could be inhibited by chloroquine. In contrast, there was little degradation of these antigens in B cells and the small amount that did occur was not chloroquine sensitive. Despite this, no significant differences were found in the ability of B cells and macrophages to stimulate T cell hybridomas, even though macrophages internalize and digest larger amounts of KLH than B cells. This could suggest that only a relatively small number of peptide—MHC complexes are required to stimulate responses of specific T cells.

What of uptake via membrane antibodies? It has been proposed that the function of membrane-bound antibodies on B cells is to bind antigen, which then is internalized and processed within the B cell (Fig.3.23). *The expression of antibodies on the membrane of a B cell may thus be viewed as a means of*

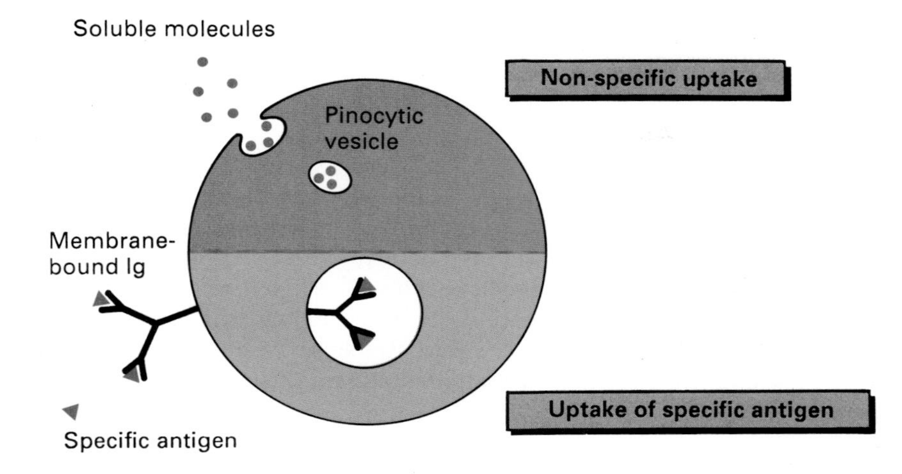

Fig.3.22 Two pathways for endocytosis of antigens in B lymphocytes

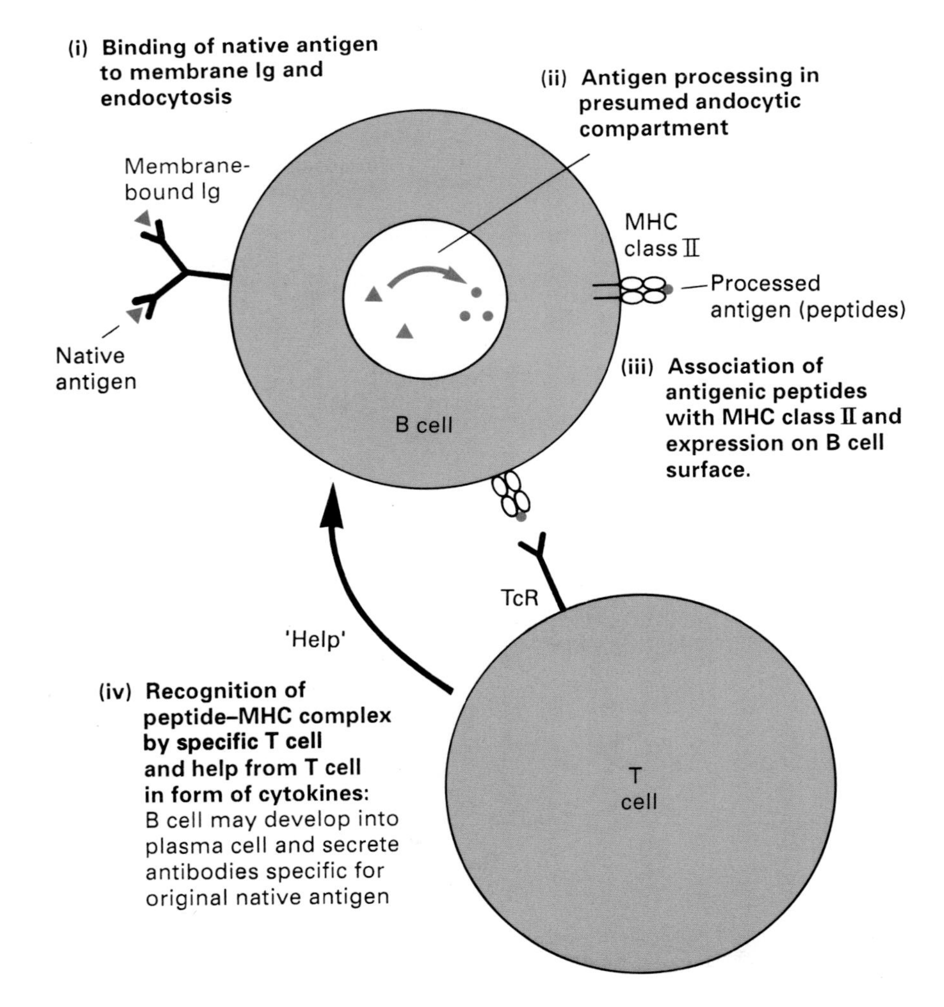

Fig.3.23 Antigen processing and presentation by B cells

capturing and concentrating antigen within the cell. The processed antigen is subsequently re-expressed on the surface of the B cell as peptides, bound to MHC molecules, in a form that can be recognized by the T cells.

The experiments that supported this model were carried out initially by Lanzavecchia (1985), who used B cell clones specific for the antigen tetanus toxoid, and T cell clones specific for the same antigen in association with MHC class II molecules of the B cells. When the B cells were pulsed with antigen they could stimulate the T cells. However, these responses were inhibited if antibodies against B cell Ig were added during the pulse period, presumably because antigen binding was blocked, but not when they were added afterwards. Responses were also inhibited when the cells were treated with chloroquine or fixed during the pulse period but not at a later stage.

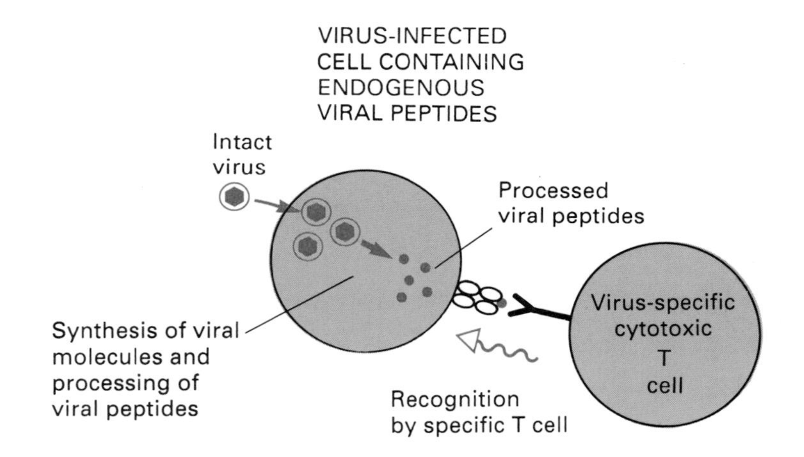

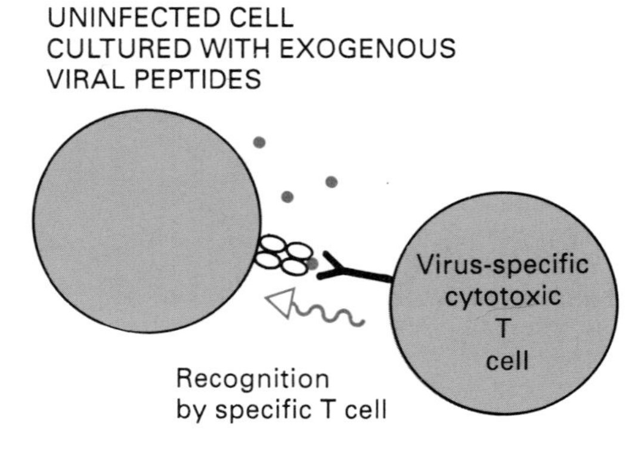

Fig.3.24 Cytotoxic T cells can recognize virus-infected cells as well as uninfected cells that have been cultured with preformed viral peptides

3.2.4 Presentation of endogenous antigens by fibroblasts and other cells

Most studies mentioned so far examined the ability of helper T cells to proliferate or to secrete IL-2 when they recognized processed antigens in the form of peptide–MHC complexes. It was originally believed, however, that cytotoxic T cells, unlike helper T cells, recognized intact (i.e. native) antigens inserted into cell membranes. Part of the reason for this view was because, historically, cytotoxic responses were examined using virally-infected or chemically-modified target cells and it was thought that during these responses the cytotoxic T cells recognized intact viral proteins in the cell membrane, or chemically modified integral membrane proteins, either of which had somehow become associated with membrane-bound MHC molecules. Townsend and colleagues (1986) challenged these ideas when they showed that cytotoxic T cells could kill fibroblasts infected with virus, as well as uninfected fibroblasts that had been incubated with pre-formed viral peptides (Fig.3.24). Thus it became clear that cytotoxic T cells did in fact recognize peptide–MHC complexes.

The precise pathway(s) for processing of endogenous antigens within the cytoplasm has not been fully defined (Fig.3.9). Generally, the evidence suggests that such processing often does *not* require an acidic compartment, unlike that needed for processing of many exogenous antigens (although this is not excluded; see below). One route may involve **ubiquitin-dependent proteolysis** within the cytoplasm (Fig.3.25). This pathway is normally involved in the regulation of turnover of cellular proteins, including those that may be incorrectly folded after biosynthesis or which have been routed to an incorrect cellular compartment.

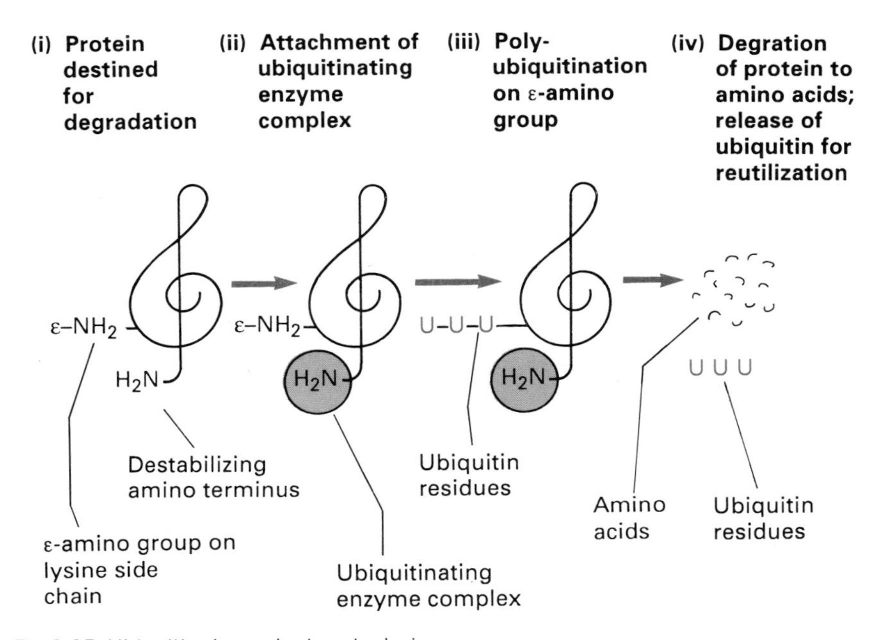

Fig.3.25 Ubiquitin-dependent proteolysis

Cytosolic proteins to be degraded by this pathway are modified by the attachment of multiple copies of the 8.5 kDa polypeptide ubiquitin, which allows them to be recognized by a specific protease. It has been shown that the half-life of ubiquitin-modified cytosolic proteins sometimes depends on the particular amino acid at their N-terminal end (Table 3.2). For example, if this residue is arginine the half-life may be less than 2 minutes, whereas if it is methionine or serine the half-life of the protein may be greater than 20 hours. In one study, by making a construct of a viral protein attached to ubiquitin, it was shown that the resultant rapid degradation could overcome defective presentation to cytotoxic T cells. However, it is not clear to what extent the ubiquitin pathway is involved in the processing of endogenous antigens in general.

Table 3.2 Protein degradation via the ubiquitin pathway—the N-end rule

Residue in ubiquitin–X–protein	Half-life of X–protein
Met-Ser-Ala-Thr-Val-Gly	>20 h
Ile-Glu	~ 30 min
Tyr-Gln	~ 10 min
Phe-Leu-Asp-Lys	~ 3 min
Arg	<2 min

A number of other routes for processing of endogenous antigens have also been proposed. One of these may involve **proteasomes**, large multimolecular enzyme complexes that have been visualized within the cytoplasm by electron microscopy, some components of which may be encoded within the MHC (Goldberg and Rock 1992; see Further reading).

In addition to non-lysosomal (e.g. ubiquitin-dependent) proteolysis, it is also possible that antigens may be translocated directly into lysosomes or into a prelysosomal compartment. This is known to involve 70 kDa heat shock proteins (Section 6.3.2; p. 380) which recognize consensus sequences on the protein to be degraded. Conceivably this could occur in any 'stressed' cell which expresses heat shock proteins, such as a virally-infected cell, and is likely to be a chloroquine-sensitive pathway. Alternatively, degradation could occur within the endoplasmic reticulum, for example in the case of newly-synthesized viral antigens, or within autophagocytic vesicles that can be formed from ER membranes and which ultimately contain lysosomal proteases.

3.3 Peptide–MHC interactions

3.3.1 Presentation of exogenous and endogenous antigens by different MHC molecules

It is notable that cytotoxic and helper T cells tend to recognize peptides bound to MHC class I and class II molecules, respectively, although this is by no means an absolute rule (Fig.1.27). Cytotoxic T cells may have evolved particularly to recognize and kill cells such as virally-infected cells, and because any cell in the body could in principle be infected by a virus, virtually all cells can express MHC class I molecules. On the other hand, helper T cells seem to have evolved to recognize cells that can develop into better effector cells in response to the cytokines that helper T cells secrete. Thus, MHC class II molecules are expressed or induced primarily on cells that can develop into potent effector cells, with the possible exception of mature dendritic cells (Section 3.4) which have no known effector function.

One further idea is that peptides derived from endogenous antigens (e.g. viral peptides) tend to associate with MHC class I molecules, whereas peptides from exogenous antigens (e.g. endocytosed molecules) tend to bind to MHC class II molecules (Fig.3.26) Therefore the pathways for processing of the two types of antigen and their association with the respective class of MHC molecule for subsequent presentation may, but need not, be separate.

Data in support of discrete pathways was obtained by Braciale and colleagues (1987), who examined the presentation of influenza virus antigens to class I-restricted and the less common subset of class II-restricted cytotoxic T cells. They found that viable antigen-presenting cells infected with the virus could present viral antigens to both types of T cell. However, presentation to class II-restricted cells could be inhibited by chloroquine, while presentation to class I-restricted cells could not, suggesting the existence of different antigen processing pathways. Moreover, fixed accessory cells could present *exogenous* viral peptides to class II-restricted cells but not to class I-restricted cells, while accessory cells infected with a recombinant vaccinia virus containing an influenza gene could present the *endogenous* antigen to class I-restricted but not class II-restricted cytotoxic T cells.

Even though the results of this study are quite clear, other studies seem to indicate that the pathways are not always discrete (e.g. Weiss and Bogen 1991). For example, some *endogenous* antigens such as an influenza matrix protein have been shown to have access to the *class II pathway*. This can occur by two routes, one of which is chloroquine sensitive, the other being insensitive, and these may correspond to different pathways for processing of endogenous antigens discussed in Section 3.2.4. It is possible to introduce soluble proteins such as ovalbumin into the class I pathway provided the antigen enters the cytoplasm, for example either because it is synthesized there (e.g. after transfection of the cell with a cDNA for the protein), or because the intact soluble molecule is experimentally introduced into the cytosol.

3.3.2 Intracellular sites of peptide–MHC associations

A current model for intracellular peptide–MHC association proposes that this can occur in different parts of a cell depending on whether MHC class I or class II molecules are involved (Fig.3.27). *Endogenously-produced* peptides can bind to class I molecules within the *endoplasmic reticulum* during or soon after synthesis of the MHC molecule (how peptides gain access to the lumen of the ER from the cytoplasm is discussed below). On the other hand, *exogenous peptides* can bind to class II molecules at a later stage, probably within an *endosomal compartment*, while these MHC molecules are being transported to the cell surface along their so-called export pathway.

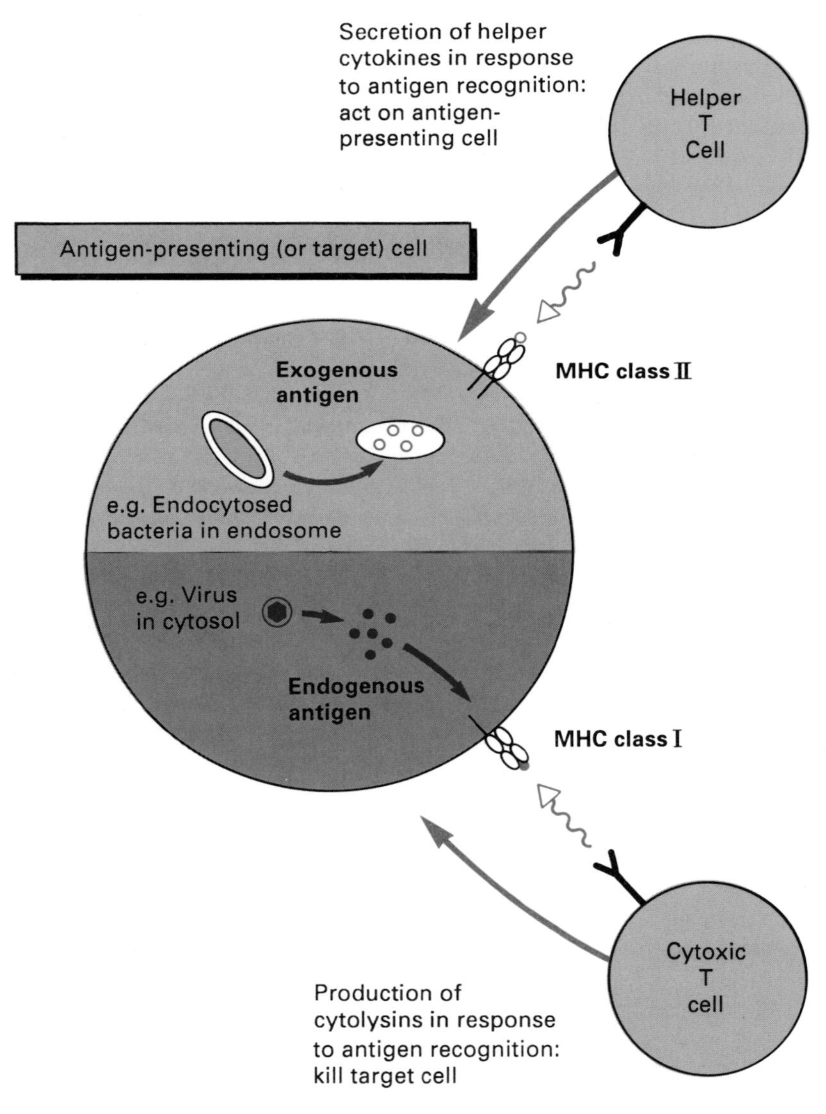

Fig.3.26 Exogenous and endogenous antigens tend to associate with different classes of MHC molecules

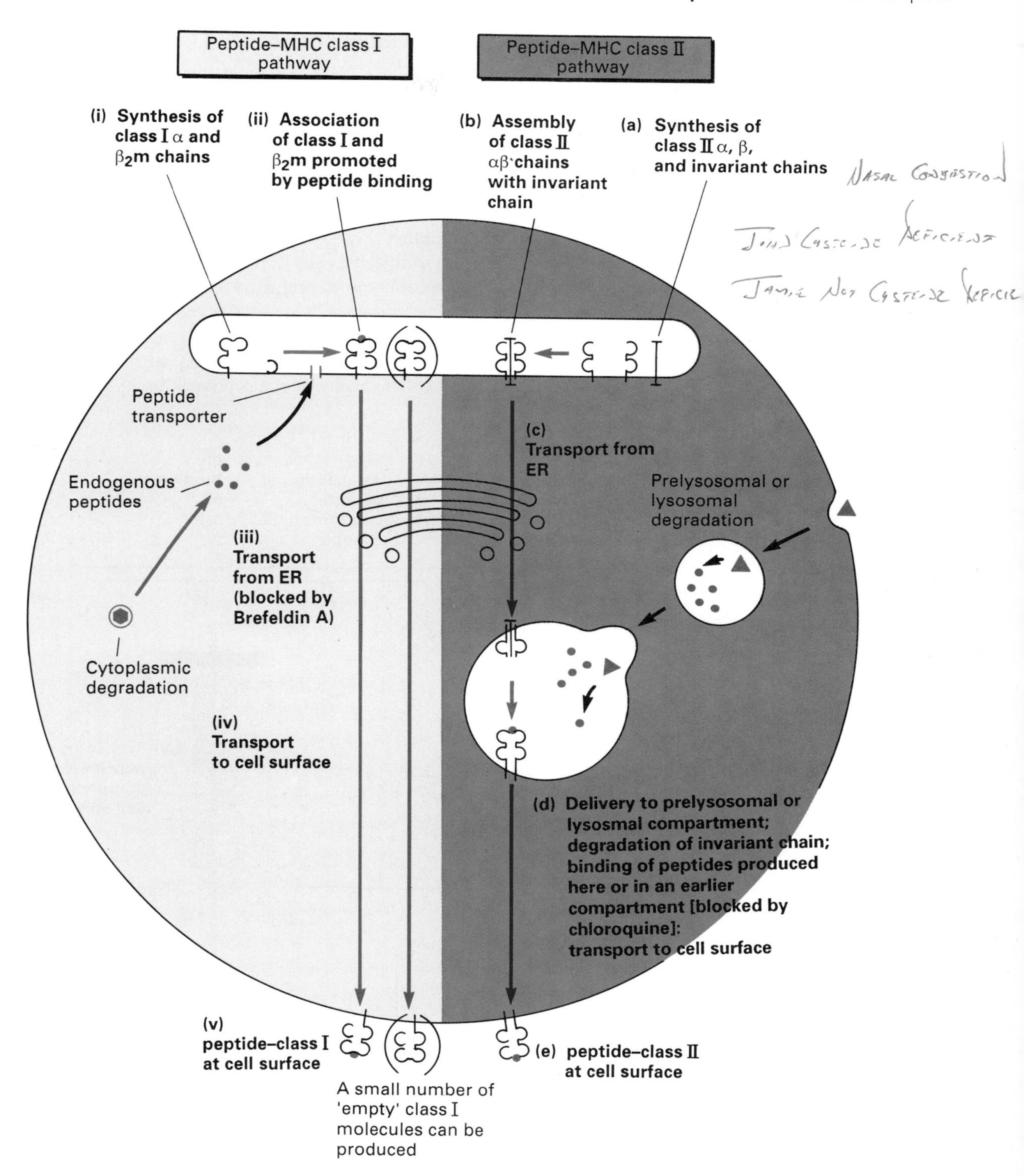

Fig.3.27 Peptide binding pathways for MHC class I and class II molecules

Peptide–MHC class I interactions

Evidence that peptides bind to class I molecules at an early stage in MHC biosynthesis has come from studies with **Brefeldin A**. This compound inhibits the transport of molecules from the ER to the Golgi apparatus and specifically inhibits exocytosis of membrane and secretory proteins. Brefeldin A was found to inhibit class I-restricted T cell recognition of virally-infected cells (Nuchtern *et al.* 1989). One interpretation of this is that endogenous peptide–MHC complexes were formed within the ER, but their transport to the surface was inhibited (Fig.3.28). However, Brefeldin A had no effect on T cell recognition of uninfected cells that were incubated with synthetic peptides which, by a second route, might have become associated with MHC molecules either after endocytosis or at the cell surface (see discussion of 'empty' MHC molecules, below).

Some intriguing results have been obtained with a mutant cell line, **RMA-S**, which suggest the presence of a peptide is *essential* for correct folding

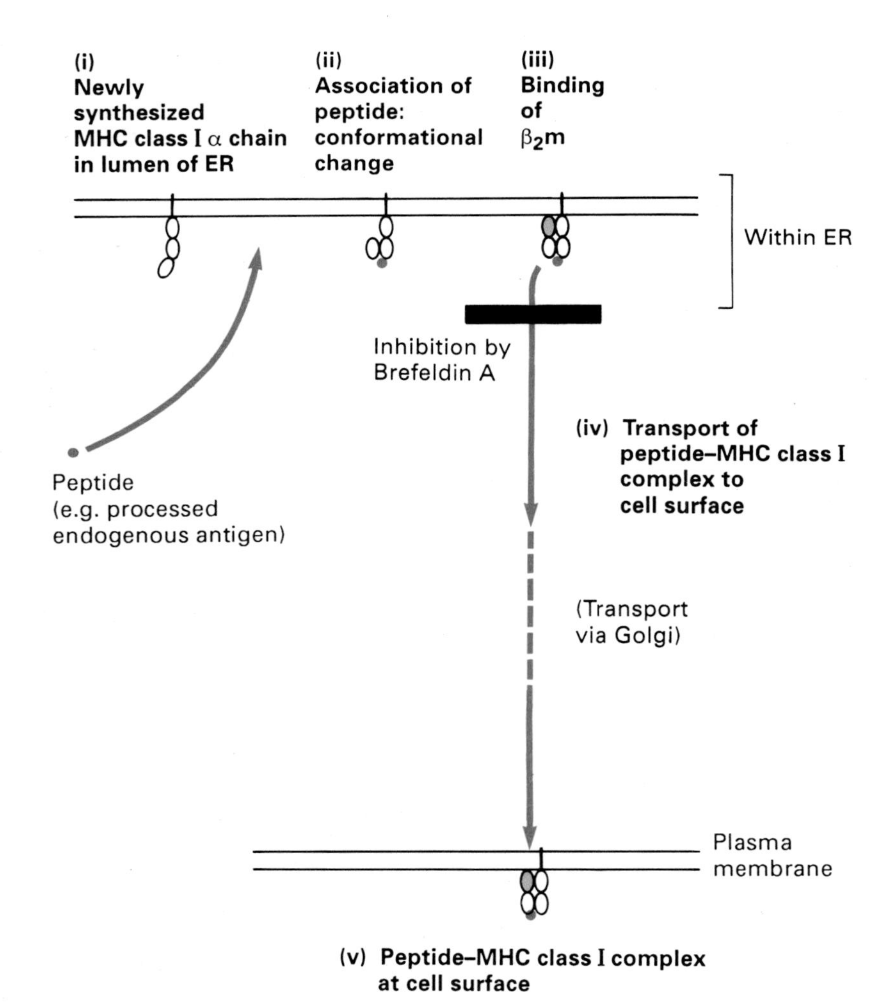

Fig.3.28 Peptide–MHC class I interactions

of the MHC molecule and for subsequent transport of the complex with β_2-microglobulin to the cell surface (Townsend *et al.* 1989).

The cell line in question appeared to have a defect in its ability to process viral antigens and/or transport of peptides from the cytoplasm into the lumen of the ER (but see p. 150). It expressed only low levels of MHC class I molecules at the cell surface and could not be recognized and killed by antigen-specific cytotoxic T cells after viral infection. However, culturing the cell in the presence of preformed viral peptides that could bind to these MHC molecules resulted in dramatically increased expression at the cell surface, and the cell could then be killed by specific cytotoxic T cells. It was suggested from these and other experiments that binding of the peptide facilitated correct folding of the MHC class I α chain, its association with β_2-microglobulin, and transport of the complex from the ER to the cell surface.

The fact that peptide may be required for correct folding of an MHC molecule is reminiscent of earlier 'instructive' theories for folding of antibodies around an antigen, although we now know that the latter does not occur. A particular conformation is apparently required before the newly-synthesized MHC class I α chain can associate with β_2-microglobulin and be transported to the cell surface (Fig.3.28). It may be that within the ER the MHC class I molecule interacts with one peptide after another until it finds one that can bind and permit it to adopt the necessary conformation. There may be a time limit on this process, for it is known that a class I α chain has to associate with β_2-microglobulin within about 30 minutes of its synthesis or it will adopt a conformation that can no longer permit the association. Thus, it is possible to view the bound peptide as a necessary subunit of a trimolecular complex formed between the MHC class I α chain, β_2-microglobulin, and the peptide itself.

'Empty' MHC molecules. Further studies with the RMA-S mutant cell line (above) revealed that class I molecules *lacking* bound peptides could be transported from the ER to the cell surface at 26°C, and to a more limited extent at 37°C; these molecules were termed 'empty' MHC molecules (Ljunggren *et al.* 1990). It was shown that empty class I molecules could bind peptides at the surface of intact cells, and that the efficiency of binding to these MHC molecules was greater than for peptide-occupied molecules. Binding of peptides to empty class I molecules resulted in a conformational change and an apparent increase in the density of class I molecules at the cell surface. Furthermore, empty class I molecules in detergent lysates were stabilized in the presence of β_2-microglobulin and peptides, and binding of peptides was shown to alter the conformation of the MHC molecule such that it had an increased affinity for β_2-microglobulin. Importantly, empty class I molecules were also shown to be expressed on the surface of the *normal* parent cell line, RMA, from which the RMA-S mutant was derived. This suggests that the concentration of peptides in the ER may be limiting in normal cells so that some empty class I molecules may always be produced. However, the more typical binding of peptides within the ER probably results in a significantly greater expression of peptide-bound class I molecules, compared to empty molecules, on the plasma membrane of normal cells.

Peptide transporters. How do peptides gain access to the lumen of the ER from the cytoplasm, for binding to newly-synthesized MHC class I molecules? Diffusion of molecules from the cytoplasm into compartments such as the ER is, of course, prevented by intracellular membranes. To accomplish delivery of a wide variety of different molecules into various membrane-bounded compartments there is a superfamily of **ATP-dependent transport proteins** or pumps. Two loci have been identified within the MHC region of mice, rats, and humans (Fig.2.22) that are very homologous to genes encoding these ATP-driven transport proteins. These genes have been designated TAP1 and TAP2 in humans (formerly RING4, Y3, or PSF1; and RING11, Y1, or PSF2) and in other species, equivalent to HAM1 and HAM2 in the mouse, and mtp1 and mtp2 in the rat. Like MHC molecules themselves, these genes are polymorphic. Evidence that the TAP1 product might comprise part of a peptide transporter came from transfection experiments.

The TAP1 gene was transfected into a human lymphoblastoid mutant cell line, LCL 721.134 (Spies and DeMars 1991). This cell line has a phenotype similar to the RMA-S mutant, described above, due to a deletion of one entire MHC haplotype and a non-functional TAP1 gene in the other. Transfection of this mutant cell line with the TAP1 gene resulted in the expression of normal levels of MHC class I molecules (encoded by the one intact, but TAP1-defective, MHC haplotype) on the cell surface. This suggests that the TAP1 product is, or contributes to, a peptide transporter molecule. However, when another mutant cell line, LCL 721.174, containing deletions of both TAP1 genes, was transfected with TAP1, there was *no* increased expression of class I molecules on the plasma membrane. This suggests that the TAP1 product is necessary but not sufficient to transport peptides into the ER.

It seems most likely that another gene in the vicinity of TAP1, probably the second homologous and closely-linked TAP2 gene is also required to constitute a functional pump. In support of this, transfection of the TAP2 gene into the RMA-S cell line restored expression of MHC class I levels, and recognition by cytotoxic T cells, to normal levels.

Each product of the MHC-encoded putative transporter genes is predicted to consist of a single ATP-binding domain and a single hydrophobic transmembrane domain (which crosses the membrane seven times). However, functional pumps encoded by the superfamily of peptide transporters are composed of two ATP-binding and two transmembrane domains, which can be constructed from one to four associated polypeptide chains. Since, in the case of TAP1, one chain is not sufficient, it seems most probable that this molecule may normally be associated with the product of the other linked gene (and/or other molecules) to constitute a functional pump. Experiments in which both genes are transfected into the same mutant cell will undoubtedly yield a clear picture in the near future. It should be noted, however, that at the time of writing there is no direct evidence that these MHC-encoded proteins are involved in transport of peptides (see also p. 159).

Peptide–MHC class II interactions

In contrast to endogenous peptide–class I interactions, exogenously-derived peptides can interact with class II molecules at a much later stage in their biosynthesis (Fig.3.29). Some evidence that this occurs after these MHC molecules have been transported from the ER and through the Golgi was obtained by allowing a human B cell line to endocytose a complex of neuraminidase and iron-saturated transferrin, via the transferrin receptor (Cresswell 1985). It was found that MHC class II molecules on the cell were desialylated, indicating that the neuraminidase had access to the class II molecules *after* they had been processed in the Golgi, because this is where sialic acid residues are added to newly-synthesized glycoproteins. Electron microscopic as well

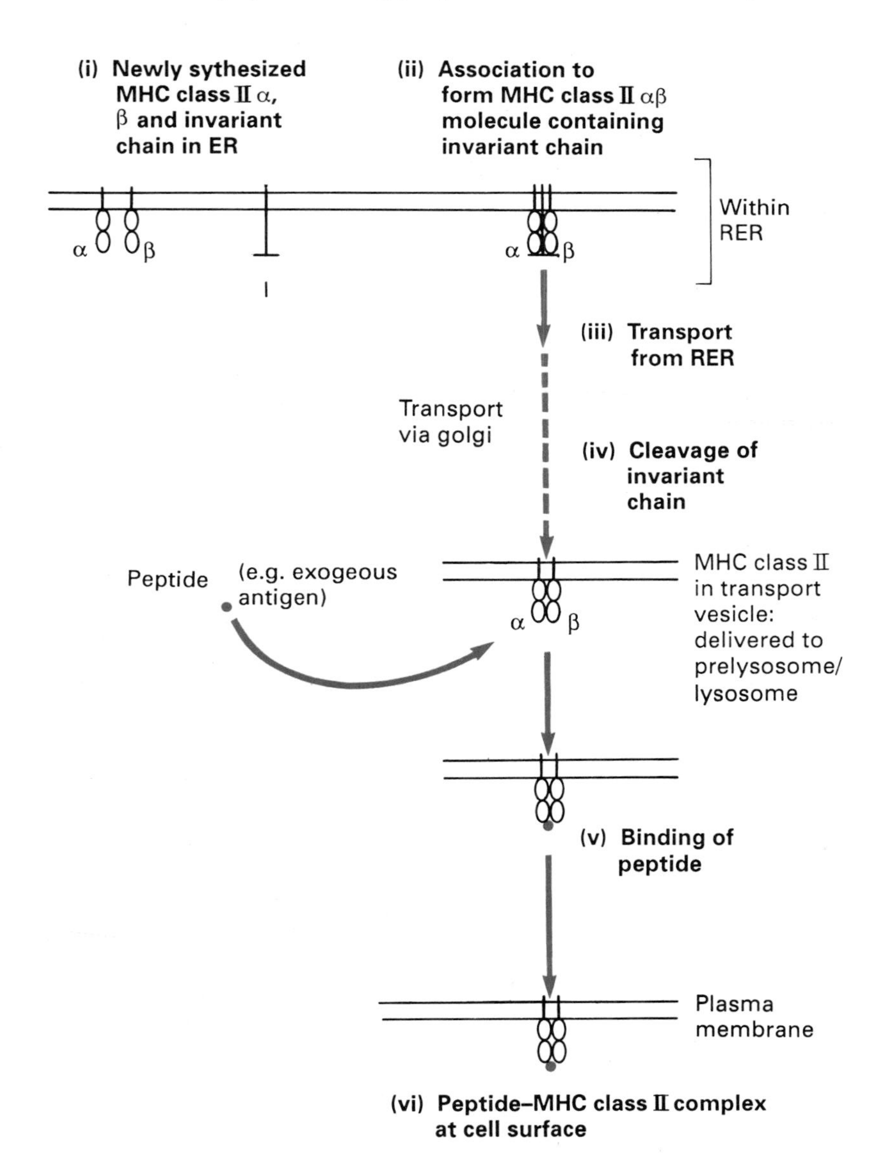

Fig.3.29 Peptide–MHC class II interactions

as biochemical evidence has also been obtained in favour of the view that endocytosed molecules encounter MHC class II molecules in their export pathway.

In one study of a B cell line, binding of antigen to membrane Ig was mimicked by cross-linking with an anti-Ig reagent. It was found that the anti-Ig was rapidly internalized by the cell, and the data were consistent with co-localization of the complexes to *early endosomes* containing MHC class II molecules, invariant chains, and the proteolytic enzymes cathepsins B and D. However, for reasons that are not yet clear, another study failed to demonstrate the presence of class II molecules in early endosomes (Neefjes *et al.* 1990). Instead the newly-synthesized class II molecules were routed to compartments resembling *lysosomes*, since they contained the lysosomal enzyme β-hexosaminidase and typical lysosome-associated proteins such as lamp-1. (An additional finding was the demonstration of vesicles close to the Golgi rich in class II molecules but lacking class I molecules; presumably the export pathways of these molecules diverge before this stage.) Of course, as yet, neither these or other studies have shown precisely where peptides associate with class II molecules.

As discussed in Section 2.6.2, newly-synthesized MHC class II molecules are associated with a third chain, the invariant chain. This becomes associated with the α and β chains of class II molecules within the ER, and it remains associated while they are transported through the Golgi. At a subsequent stage, the invariant chain is proteolytically cleaved by proteases such as cathepsins before the class II molecules are expressed on the plasma membrane (although there is evidence that a small number of unassociated invariant chains can also reach the cell surface).

The function of the invariant chain is now reasonably well understood. Studies with monoclonal antibodies have demonstrated that MHC class II molecules adopt a different conformation after cleavage of the invariant chain. It has also been shown that association of the invariant chain with class II $\alpha\beta$ molecules in the ER prevents them from binding peptides prematurely (Roche and Cresswell 1990). This imposes some degree of segregation on the peptide-binding pathways for MHC class II and class I molecules. In addition, the invariant chain contains sequences that act as signals to direct the transport of class II molecules into particular cellular compartments, including an endosomal compartment where they meet endocytosed molecules (Bakke and Dobberstein 1990).

This is consistent with the experimental finding that presence of the invariant chain is, in some cases, essential for the ability of a cell to present peptide–class II complexes (Stockinger *et al.* 1989). Thus, fibroblasts, which do not normally express MHC class II molecules, were found to be able to present antigens and stimulate antigen-specific class II-restricted T cell clones only if they were transfected with the genes for both the mouse IE α and β chains together with the invariant chain gene.

This discussion has focused on the fact that peptides can encounter MHC class II molecules in their export pathway, but peptides may also bind to class II molecules at the cell surface or during recycling of MHC molecules from the plasma membrane. Certainly, experiments to be described in the next section (Section 3.3.3) indicate that peptides can bind to intact, isolated class II molecules in the absence of the invariant chain, although how 'physiological' this route may be is uncertain. Moreover, competition between peptides for binding to the same MHC molecule can occur. For example, presentation of a particular peptide–MHC complex to a specific T cell hybridoma can be inhibited in the presence of a peptide that can bind to the same MHC molecule but which is not recognized by the T cell. Evidence that peptides can bind directly to MHC molecules is considered in the next section.

NB

3.3.3 Binding of peptides to MHC molecules

So far, it has not been possible to visualize immunogenic peptides directly on the surface of antigen presenting cells or target cells; rather, their presence has been inferred from their ability to stimulate T cell responses. There are likely to be several reasons for this (Fig.3.30). First, one would not expect to detect a T cell epitope using an antiserum or a monoclonal antibody because the epitopes recognized by T cells and B cells are different: an antibody raised against the native protein is unlikely to react with a 'processed' form of the antigen (Section 3.2.1). Second, even if it was possible to make a monoclonal antibody against the peptide that was recognized by the T cell, it seems unlikely that the antibody would have access to the peptide once it was bound to an MHC molecule, and in any case the conformations of the free and bound peptide could be different. A third reason relates to precisely how many molecules of peptide are bound on the surface of the cell. Fluorochrome-labelled or radiolabelled peptides generally cannot be detected on the surface of antigen-presenting cells even under conditions where T cell recognition occurs. Thus, it would seem that relatively few peptide–MHC complexes may be required to stimulate a T cell response; precisely how many molecules are required is not known, but the figure of a few 100–1000 has been suggested by some investigators.

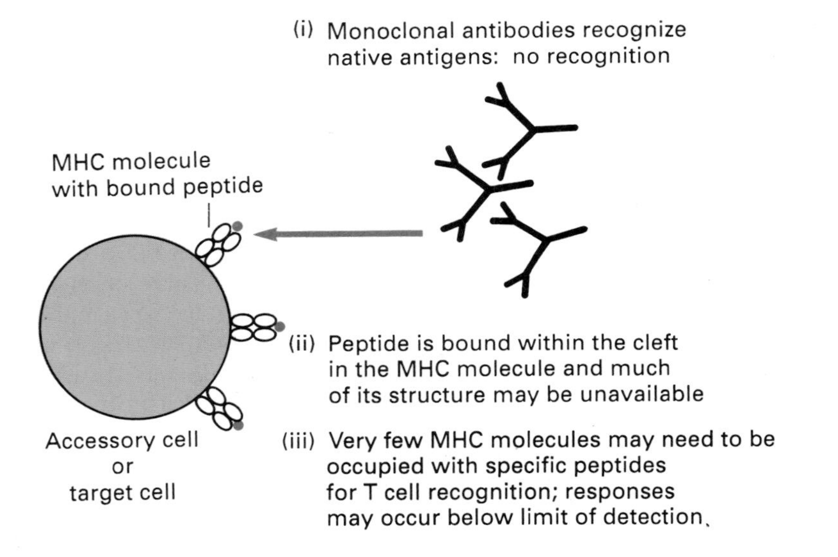
(i) Monoclonal antibodies recognize native antigens: no recognition
MHC molecule with bound peptide
(ii) Peptide is bound within the cleft in the MHC molecule and much of its structure may be unavailable
Accessory cell or target cell
(iii) Very few MHC molecules may need to be occupied with specific peptides for T cell recognition; responses may occur below limit of detection.

Fig.3.30 Reasons why surface antigens recognized by T cell receptors cannot be visualized at the cell surface

Interaction of antigen with MHC on accessory cells

What, then, is the evidence that MHC molecules can bind immunogenic peptides? Some of the first observations relating to this came from peptide competition experiments in which it was shown that some peptides can compete with others for binding to the same MHC molecule.

> Werdelin (1982) obtained T cells from a guinea pig immunized against the antigen DNP-PLL (dinitrophenylated poly-L-lysine). These cells responded in culture to this antigen, in the presence of the appropriate accessory cells, and they did not respond to another antigen, a polymer of glutamic acid and lysine, GL. However, a T cell response was not elicited if the accessory cells were pulsed with GL before they were used to present DNP-PLL, so it appeared that GL competed with DNP-PLL for binding to MHC on the accessory cells. (Presumably, GL could bind to these MHC molecules, but T cells capable of recognizing this complex were not present in the repertoire). In other experiments, Rock and Benacerraf (1983) studied T cell hybridomas that secreted IL-2 in response to the antigen GAT, a polymer of glutamic acid, alanine, and tyrosine, in the presence of accessory cells expressing H-2 I-A^{d} or I-A^{b}. Responses of I-A^{d}-restricted hybridomas to GAT were found to be blocked in the presence of another antigen, GT, a polymer of glutamic acid and tyrosine, whereas responses of I-A^{b}-restricted cells were not. These results implied that GT could compete with GAT for binding to I-A^{d} but not to I-A^{b} molecules.

Peptide competition only occurred when the cells were exposed to one antigen before, or at the time of, adding the other. Therefore, competition between peptides may occur while they are associating with an MHC class II molecule, but once the peptide–MHC complex has formed it is relatively stable. Other studies have extended these findings and shown that several unrelated peptides can compete for binding to a single site or at least a small number of overlapping sites on the same MHC molecule, presumably all within the peptide-binding groove.

Direct binding of antigen to MHC molecules

As discussed in Section 2.5.1, when MHC class I molecules were examined by X-ray crystallography, electron density of unknown origin was detected in the antigen-binding groove, and this was interpreted as a bound peptide(s) that had co-crystallized with the MHC molecules. Direct evidence for antigen binding to MHC molecules has been obtained in studies, amongst others, of the association of peptides with MHC molecules in solution, reconstituted into artificial planar lipid membranes, or in cell membranes (Fig.3.31). We shall take these in this order (affinity chromatography and elution experiments have also been performed; see p. 126 and below). In the vast majority of cases the early data concerned binding of peptides to MHC class II molecules, although direct binding of peptides to class I molecules has subsequently been demonstrated (Chen and Parham 1989; Bouillot *et al.* 1989).

Binding to isolated MHC molecules. Possibly the first direct evidence that a defined peptide can bind specifically to an MHC molecule was obtained by Babbitt and colleagues (1985). They used equilibrium dialysis to study the interaction between a fluoresceinated peptide (hen egg lysozyme—HEL; residues 46–61) and purified MHC class II molecules in a detergent solution.

This peptide bound to H-2 I-A^k molecules but not to I-A^d molecules, correlating with the fact that it is immunodominant (i.e. one of the most immunogenic) in strains of mice expressing I-A^k but not I-A^d. These findings were extended by Buus and colleagues (1986) who found that another peptide (ovalbumin 323–339), which could be recognized by T cells in association with I-A^d, could bind to these molecules but not to I-A^k or I-E^k.

However, the *specificity* of binding of peptides to isolated MHC molecules is not always clearcut. In addition, peptide binding to isolated class II molecules is often a weak interaction with a dissociation constant of about 2×10^6 M^{-1}; this value should be compared to the dissociation constant for antigen-antibody binding which is around 10^9 M^{-1}.

The results of antigen competition experiments, outlined towards the beginning of this section (3.3.3) suggested that different peptides should be able to compete for binding to isolated MHC molecules. Indeed, peptides that inhibited the response of T cells to the HEL 41–61 peptide in association with H-2 IA^k molecules were found to inhibit binding of this peptide to that MHC molecule, and peptides that were unable to inhibit T cell recognition did not inhibit peptide binding.

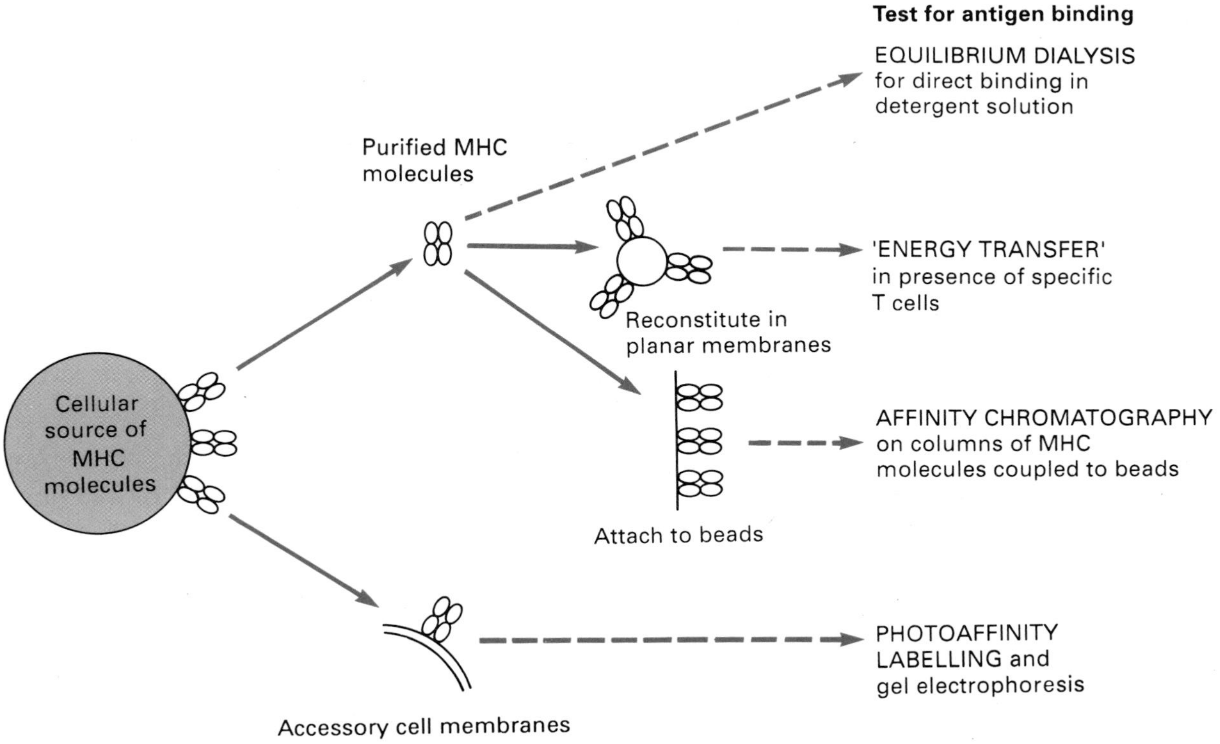

Fig.3.31 Some approaches to investigation of binding of peptides to MHC molecules

Binding to MHC in planar membranes. Watts and colleagues (1986) examined the interaction between peptides and class II molecules reconstituted in artificial membranes. Antigen-specific T cells responded to their antigenic peptide (ovalbumin 323–339) in the presence of phospholipid vesicles containing MHC (I-A^d) molecules. In the absence of serum proteins it was possible to show, using a technique that measured 'energy transfer' between the peptide and the MHC class II molecule, that the two molecules were in close proximity to each other, within about 4 nm (for scale, see Fig.1.1). However, in the presence of serum proteins, energy transfer only occurred when T cells were added, suggesting that undefined serum components might have competed for binding of the specific peptide, but that the interaction with the specific peptide was stabilized in the presence of the specific T cell receptor, although it is not clear whether this actually occurs.

Binding to MHC in accessory cell membranes. It was found that membranes isolated from accessory cells could also stimulate T cell hybridomas after they had been incubated with specific peptide. To examine which membrane components of intact cells bound immunogenic peptides, photoaffinity labelling was used by Phillips and colleagues (1986).

Beef insulin was radio-iodinated, labelled with a photoreactive probe, and added to mouse accessory cells. After exposure to light, to cross-link the insulin (peptides?) to adjacent membrane molecules, the cells were lysed and their membranes were solubilized. The radiolabel was found to be associated with the α and β chains of class II MHC molecules when analysed by gel electrophoresis. However, there appeared to be no haplotype specificity in this interaction in that the insulin became linked to class II molecules from strains that responded to this antigen as well as to those that did not. This is in apparent contrast to the results of Babbitt and Buus (see above). A likely explanation is that 'non-responder' animals are unable to respond to insulin bound to their MHC molecules because they do not have T cells specific for these complexes (see Section 5.4.3).

Kinetics of peptide–MHC interactions

How can different peptides compete for binding to an MHC molecule, but remain stably associated once they have been bound? Even though peptide competition can occur, cells pulsed with peptides can be targets for specific cytotoxic T cells for at least 72 hours, showing that peptide–MHC associations are often long-lived.

The dissociation constant for peptide–MHC interactions is of the order of 10^{-5}–10^{-6} M^{-1}. The dissociation constant is the ratio of the rate at which the molecules associate (the on-rate) compared to the rate at which they dissociate (the off-rate). It has been estimated that the overall dissociation constant for peptide–MHC interactions derives from a slow on-rate coupled with a very slow off-rate. This means simply that the peptides bind slowly but dissociate very slowly. The association rate is four or five orders of magnitude lower than that for an antibody–antigen interaction, while the dissociation rate is much slower than this, such that the half-life of a bound peptide (the time taken for half the peptide molecules to dissociate) is 30 hours or so. As noted in Section 3.3.2, peptides can associate with MHC class I molecules during their biosynthesis, which would account for the slow on-rate particularly if the class I molecule needs to 'find' an appropriate peptide and adopt a different conformation after peptide binding (Fig.3.28). Similar conformational changes also occur after the invariant chain is removed from MHC class II molecules, and this may also be associated with a 'tighter' interaction with the bound peptide. Thus, once bound, some peptides may remain associated for the life-time of the MHC molecule itself, explaining the very slow off-rate. An alternative or additional possibility is that different peptides bind to MHC molecules with a wide range of affinities. Peptides that are bound weakly may be more readily displaced by peptides that bind with a higher affinity, accounting for the results of competition experiments.

The size of T cell epitopes

Synthetic peptides. T cells can respond in culture to antigenic peptides between 8 and 25 amino acid residues in length. Molecules of approximately this length could be accommodated within the peptide binding groove of an MHC molecule, as visualized by X-ray crystallography (Section 2.5.1), depending on their conformation since the length of an α helix of about 20 amino acids is similar to that of a β sheet of 8 residues.

One approach to define the size of a T cell epitope has been to examine the response of a T cell clone of known antigenic specificity to defined peptides in the presence of accessory cells (Fig.3.32). If a series of peptides is prepared by proteolytic cleavage of the protein, the fragments to which the clone responds can be sequenced and the position of an epitope can be determined. A 'nested set' of peptides of different lengths can then be synthesized to examine the effect of changing the size of the molecule on the cellular response. A set of peptides can also be synthesized to encompass this epitope, but to contain modifications of sequence in order to determine which particular residues are most important.

Generally, a minimal peptide length is required for T cell recognition (above). A peptide that includes two or three residues that are known to be critical for T cell recognition may not actually trigger the response unless it is lengthened by the addition of one or two more amino acids. These additional residues may be needed, for example, for the peptide to establish a particular conformation and bind to the MHC molecule.

Sometimes making a change in the peptide at a single amino acid position can completely abolish a T cell response; this residue is therefore assumed to be in a critical region of the T cell epitope (Fig.3.33). The same is also true for MHC molecules. For example, a number of mutant strains of mice exist whose MHC molecules differ from each other in sequence by only one or a few residues (see bm mutants, Section 2.5.1). Accessory cells from one strain are often unable to stimulate peptide-specific T cells from the other strain. Until we knew the structure of an MHC molecule, it was not possible to decide whether these alterations resulted in the peptide no longer being able to bind to the MHC molecule, or whether they were required for direct interactions with the T cell receptor. We can now determine where these amino acids are situated within the peptide binding groove of the MHC molecule. Those residues at the bottom of groove are most likely required for binding of the peptide, while those at the top of the groove, or which point

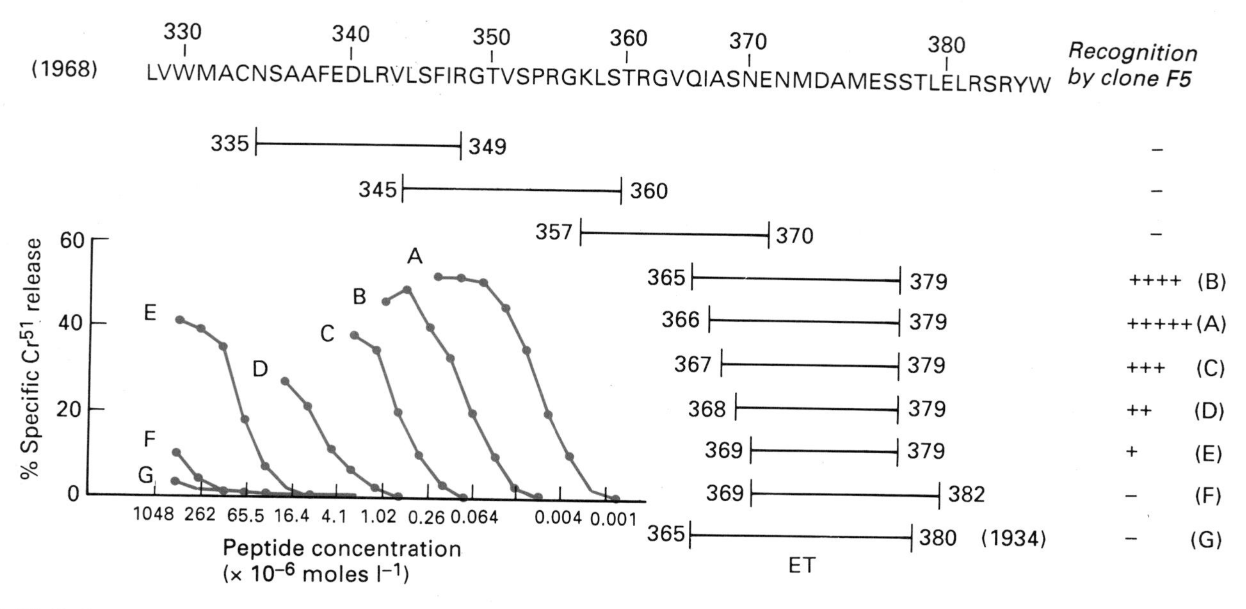

Fig.3.32 Definition of the epitope recognized by an influenza nucleoprotein (NP)-specific cytotoxic T cell clone from an H-2b mouse. The epitope of the NP molecule (from a 1968 strain of the influenza virus) recognized by the H-2D^b-restricted T cell clone A5, was initially mapped to the region shown (residues 328–386). This was accomplished by studying a series of deletion mutants of the NP molecule (not shown) transfected into a mouse fibroblast cell line that also expressed the H-2D^b gene. The A5 clone killed target cells expressing the whole NP molecule, residues 1-498, or residues 1-386. However, they could not recognize cells expressing residues 1-327, thus mapping the epitope between residues 327 and 386, and this was confirmed by their ability to kill cells expressing residues 328-498. A series of short synthetic peptides was then synthesized and the ability of A5 cells to kill H-2b target cells (and to proliferate) in the presence of these molecules was tested. An example of the data from cytotoxicity assays is shown, and the results are summarized. Note that the letters A–G in the summary correspond to the relevant titrations in the example. The maximum response was obtained with peptide 366-379, and diminished as the peptide was progressively shortened in the direction shown. There was no response to the NP 365-380 peptide from a 1934 strain of the virus which contains two amino acid differences (positions 372 and 373) compared to the 1968 strain. (Based on Townsend *et al.* (1986))

away from it, are more likely to be needed for recognition by the T cell receptor (for example, see Section 5.2.2; Figs.5.7, 5.8). Changes in any of these sites can abolish the T cell response, and most of these residues are polymorphic (Fig. 2.7). Ultimately it should be possible to determine what actually happens by studying crystals of defined peptides bound to MHC molecules, ideally in a ternary complex with the T cell receptor.

Naturally processed peptides. Naturally processed peptides isolated from normal or virus-infected cells and analysed by high-performance liquid chromatography are surprisingly homogeneous in length e.g. Jardetzky *et al.* 1991. Generally, those that stimulate cytotoxic T cell responses are composed of only eight or nine amino acids, although they are present within the often longer sequences defined using synthetic peptides. Peptides isolated from purified MHC class I molecules by acid extraction are also the same length (Falk *et al.* 1991), and in addition, empty class I molecules from the RMA-S mutant (Section 3.3.2) preferentially bind synthetic peptides of nine amino acids from a heterogeneous mixture (Schumacher *et al.* 1991). Whatever the explanation, it is clear there are mechanisms in normal and infected cells

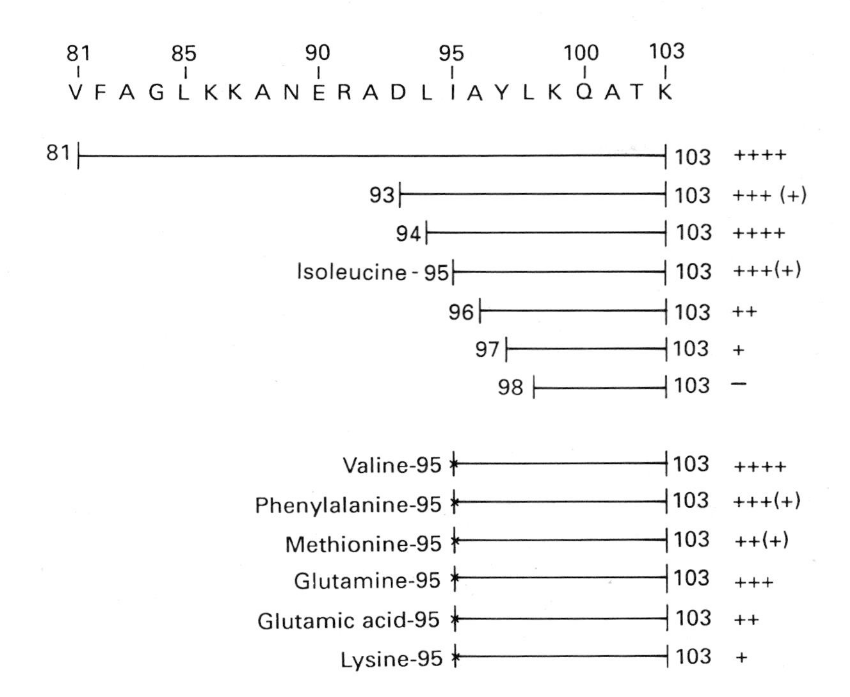

Fig.3.33 Definition of the epitope of moth cytochrome c recognized by a mouse helper T cell clone. The T cell clone A.E7 that was originally defined as being 'specific' for pigeon cytochrome c and IE^k was found also to recognize moth cytochrome c, and the epitope was mapped to the region shown. The size of the epitope necessary for T cell recognition (determined by the T cell proliferative response in the presence of antigen and H-2k antigen-presenting cells) was then examined by using synthetic peptides of different lengths. A summary is shown. (Note: the number of + signs should be considered on a logarithmic rather than linear scale). A response was elicited by peptide 97–103, the shortest molecule to stimulate the clone, but the response was dramatically increased when the peptide was lengthened by two amino acids (95–103) or more. The effect of changing the residue at position 95 was then examined. There was little effect on the response if this residue was valine or phenylalanine, both of which are uncharged amino acids, whereas recognition was markedly reduced if this was glutamic acid or lysine, both charged amino acids. Recognition of the moth cytochrome c peptides can occur in the absence of residue 95 (see 97–103 and 96–103), so it is thought that alterations in this amino acid change the affinity of the peptide for the MHC molecule and/or T cell receptor. (Based on Schwartz, R.H., Fox, B.S., Fraga, E., Chen, C., and Singh, B. (1985). *J. Immunol.*, **135**, 2598–608)

that determine the precise length of peptides that are bound to MHC molecules and presented to T cells.

One possibility is that the size of peptides entering the ER is governed by the specificity of the putative peptide transporter molecules (Section 3.3.2). There is a precedent for this in that members of this family of 'ABC' (ATP-binding cassette) transporters include a bacterial oligonucleotide permease which may be selective for peptides 3–5 amino acids long, and a yeast transporter protein specific for a defined peptide sequence (note that consensus sequences have been defined in certain T cell epitopes; see below).

Although it has been assumed that structural features of the MHC molecules *alone* select which peptides are bound (Section 2.5.1; e.g. Fig.2.9) it now has been shown that polymorphisms in the MHC-encoded putative peptide-transporter genes (see p. 150) can *also* determine the nature of peptides bound by class I molecules. The first direct evidence for this came from studies in which it was demonstrated that a sequence polymorphism in the rat TAP2 gene correlated with a change in the spectrum of the peptides that could be eluted from MHC class I molecules (Powis *et al.* 1992). It has been speculated that polymorphisms in these transporter genes could be very important in human disease, particularly autoimmunity.

NB

The conformation of peptide in the MHC binding site

As a general rule, epitopes recognized by antibodies tend to have a relatively high degree of atomic mobility and hydrophilicity. This means that to some extent they can adopt different configurations, and that they contain charged or polar residues. In marked contrast, T cell epitopes can often be buried

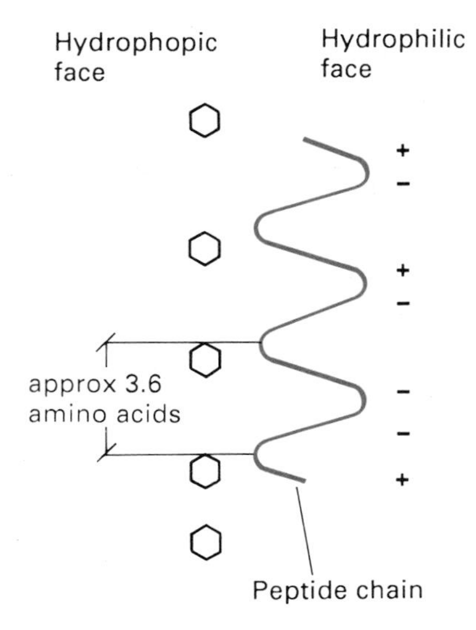

Fig.3.34 An amphipathic α helix

within the three-dimensional structure of a protein, and as such they are inherently more hydrophobic in nature. Antigen processing is required to expose these peptides, which may then bind to an MHC molecule, and some processing events such as cleavage of the protein may also allow these normally constrained regions to become more flexible and adopt a suitable conformation for binding to a particular MHC allele.

It would be immensely useful to be able to predict which sequences in proteins are T cell epitopes, for example it would provide a short-cut to preparing immunogenic peptides for vaccination. DeLisi and Berzofsky (1985) predicted from theoretical considerations that T cell epitopes might form **amphipathic α helices**. Amphipathic structures are those in which the hydrophilic residues are aligned on one side or face of the molecule with the hydrophobic residues on the other (Fig.3.34). An α helix is, of course, one of the main forms of secondary structural elements of a protein (Fig.1.3), in which the polypeptide chain makes an average of one complete turn for every 3.6 residues; it is said to have a periodicity of 3.6. (Other forms are β-pleated sheets, with a periodicity of 2.0, and random coils).

DeLisi and Berzofsky determined the periodicity of hydrophobic residues in the primary amino acid sequences of known T cell epitopes and were able to model most of these as amphipathic α helices. These epitopes do not necessarily have this configuration within the native molecule, but potentially they can adopt this structure when they are, for example, excised from the intact globular protein. They proposed that hydrophilic residues of the helix might be exposed to the aqueous solvent, while the newly-exposed hydrophobic residues could be stablized within the groove of the MHC molecule.

Taking a rather different tack, Rothbard and Taylor (1988) used an empirical approach to determine a consensus sequence for T cell epitopes. When primary sequences of known T cell epitopes were examined, an extraordinarily large number (27/28) contained a particular four-residue sequence or **motif**. This motif was initially proposed to contain two hydrophobic residues flanked on the N-terminal side by a charged residue or glycine, and on the C-terminal side by a polar residue or glycine. Using this model it was possible to predict where previously undetermined T cell epitopes were located in a number of proteins. An uncannily high number of these have turned out to be recognized by T cells in practice. Other models have also been proposed. However, none of these models is mutually exclusive and the same binding site might be able to accommodate any or all of the proposed peptide configurations.

In contrast to the predictive algorithms that have been used to define T cell epitopes, based on the sequence of the antigen alone (above), an apparently more accurate *experimental* technique has been developed which additionally takes into consideration polymorphisms of the MHC molecule (Gammon *et al.* 1991). In this approach peptides are synthesized, on pin heads, as a series that progresses in single amino acid steps (e.g. 1–10, 2–11, 3–12, etc.), and this has revealed the presence of **core sequences** common to all the immunogenic peptides in a continuous sequence.

Is processing always required?

Antigen processing is required to produce peptides that can bind to the MHC molecule and be recognized by the T cell receptor. The portion of the antigen that interacts with the MHC molecule has been termed an **agretope**, while the part that interacts with the T cell receptor is called an **epitope** (p.322). Thus antigen processing may be needed to generate sufficient conformational freedom to allow formation of an agretope and an epitope in the same peptide. According to this idea, molecules that have areas with a reasonable degree of conformational freedom may not need to be processed. In fact, human fibrinogen, a large molecule with a molecular weight of about 340 kDa, can be presented without processing, since fixed or chloroquine-treated accessory cells can present the *native* molecule to some T cells. The antigenic determinant was found to be located close to the C-terminal of one of the fibrinogen chains. This area is very hydrophilic and is predicted not to have a defined secondary structure in the native molecule (i.e. it is neither an α helix nor a β sheet) and thus it may already have sufficient conformational freedom.

There are other more-or-less well substantiated examples of proteins that do not appear to require processing before they can be recognized by some T cells, although other T cells might be able to recognize processed (unfolded or fragmented) forms of the same molecule. In addition, Walden *et al.* (1985) found that apparently native antigens can be presented to T cells by liposomes containing purified MHC molecules. Even so, the latter experiments have been criticized because it is very difficult to prove that every single molecule is actually in its native form.

Finally, we should perhaps consider whether antigen processing, always occurs *intracellularly*. There seems little doubt that antibodies can recognize the native conformation of antigens, so that some intact antigen must be seen by the immune system as a whole.† Adjuvants are often used for immunization in experimental animal models. While their action is not well defined, some, such as complete Freunds' adjuvant, induce an inflammatory response. This rapidly recruits polymorphonuclear leukocytes (especially neutrophils) to the site of injection, later to be followed by an influx of monocytes (Fig.1.11). These inflammatory cells secrete a vast number of proteases and other molecules that one might expect to alter the structure of many native proteins. It seems almost inevitable that some antigen would be degraded *extracellularly* to some extent in these sites. Thus by the time a T cell response is induced some of the antigen may already be in a 'processed' form, so that T cell responses *in vivo* may inevitably be strongly biased towards recognition of processed antigens.

3.4 Dendritic cells

As discussed at the beginning of this chapter (Section 3.1) accessory cells include a subset of specialized antigen-presenting cells, called immunostimulatory cells, that are required to present antigens and deliver activation

† Indeed, native antigens can be detected on the surface of follicular dendritic cells in B cell areas of lymphoid tissues—see Table 3.4—for considerable lengths of time after the immune response has apparently ceased (Section 8.3.1).

signals to resting T cells and thus *initiate* the immune response. The most potent immunostimulatory cell known is the dendritic cell.

Dendritic cells were first isolated from mouse lymphoid tissues in the course of studies on the function of accessory cells in immune responses in culture. Adherent spleen cells, which were used as accessory cells in many cases, contain a large proportion of macrophages, but it was noted that they also contained smaller numbers of a different cell type (Fig.3.4). These irregularly-shaped cells continuously extended and contracted dendritic (tree-like) processes in culture, in contrast to the relatively sessile and more rounded macrophages. Because of their lymphoid origin and their shape these cells were called **lymphoid dendritic cells**. By definition, this term was applied to the cells isolated from lymphoid tissues and studied in culture, rather than to any particular cell type *in situ* (e.g. in sections of frozen tissue). Over the next few years, methods were developed to purify and culture dendritic cells, and their properties were studied in detail. It was found they could be readily distinguished from other cell types in a number of ways, and it became apparent that they represented a new cell lineage. It was then found that dendritic cells were unusually potent accessory cells for immune responses in culture and *in vivo*. In this section we review some of the immunobiology of dendritic cells and discuss the evidence that they not only present antigens to T cells but also have an essential role as immunostimulatory cells in triggering the adaptive immune response.

3.4.1 Properties of dendritic cells

Lymphoid dendritic cells can be isolated from various lymphoid tissues including spleen, lymph nodes, tonsil, Peyer's patches, and thymus (see Sections 1.3.2 and 1.3.4) from different species such as mice, rats, and humans. In many cases these cells initially adhere to tissue culture surfaces but detach after a few hours in culture, in contrast to macrophages which are much more strongly and persistently adherent.

One notable feature of lymphoid dendritic cells in culture is their inability to phagocytose particles (Table 3.3). For example, if latex beads are added to cultures of dendritic cells and macrophages, only the latter cells are found to internalize the particles. In general, dendritic cells isolated from lymphoid tissues are considered to be weakly pinocytic and they have very few if any lysosomes; since they also lack a number of intracellular enzymes that are normally associated with phagocytic cells, they can be distinguished from macrophages cytochemically. By electron microscopy, dendritic cells have a number of characteristic features including very little rough ER, which would seem to indicate they are not particularly active secretory cells, and an irregularly-shaped nucleus (Figs.1.11 and 3.35).

The phenotypic markers of dendritic cells, i.e. molecules expressed on the cell surface, provide further evidence that they belong to a unique cell lineage. We know they are produced in the bone marrow and as leukocytes they express the CD45 leukocyte common antigen (Section 4.2.5). However, lymphoid dendritic cells do not have cytochemically demonstrable Fc receptors, for example, which are found on myeloid cells (mononuclear and polymorphonuclear leukocytes), and they do not react with several monoclonal antibodies against macrophages. Dendritic cells do not express T cell

Table 3.3 Some comparisons between lymphoid dendritic cells and macrophages

Feature	Dendritic cells	Macrophages
Cytology		
Processes	Actively form dendrites, pseudopodia and veils (lamellipodiae)	More sessile, form 'ruffles'
Nucleus	Oval or irregular, pulsatile within cytoplasm	Kidney-shaped, more stationary
Mitochondria	Rounded	More filamentous
Endoplasmic reticulum	Some smooth ER	More rough ER
Endosomes, lysosomes	Few	Many
Cytochemistry		
Non-specific esterase	Weak or negative	Positive
Membrane ATPase	Negative	Positive
Alkaline phosphatase	Negative	Positive
Adherence		
In culture	Variably weak or non-adherent	Strongly adherent
Endocytosis		
Phagocytosis	Non-phagocytic	Actively phagocytic
Pinocytosis	Relatively weak	Very active
Membrane markers (e.g. mouse)		
Lineage-restricted	33D1, NLDC145, N418	F4/80
Fc receptors	Undetectable	Present
Complement receptor (CD11b/18)	Weak	Present
Leukocyte common antigen (CD45)	Present	Present
MHC class I	Present	Present
MHC class II	Present (constitutive)	Present (inducible)

markers such as CD3 and T cell receptors, and they do not express membrane immunoglobulins characteristic of B lymphocytes nor do they label with a number of antibodies specific for B cells.

Another characteristic of dendritic cells is that they express high levels of MHC class II molecules, but expression is constitutive, contrasting with macrophages where expression is inducible. Dendritic cells also express MHC class I molecules and they have low levels of type 3 complement receptors (specific for iC3b; Section 10.4.5.). Unfortunately there are at present relatively few specific markers for dendritic cells in the mouse and not all populations are labelled, so that dendritic cells *in situ* are often defined as

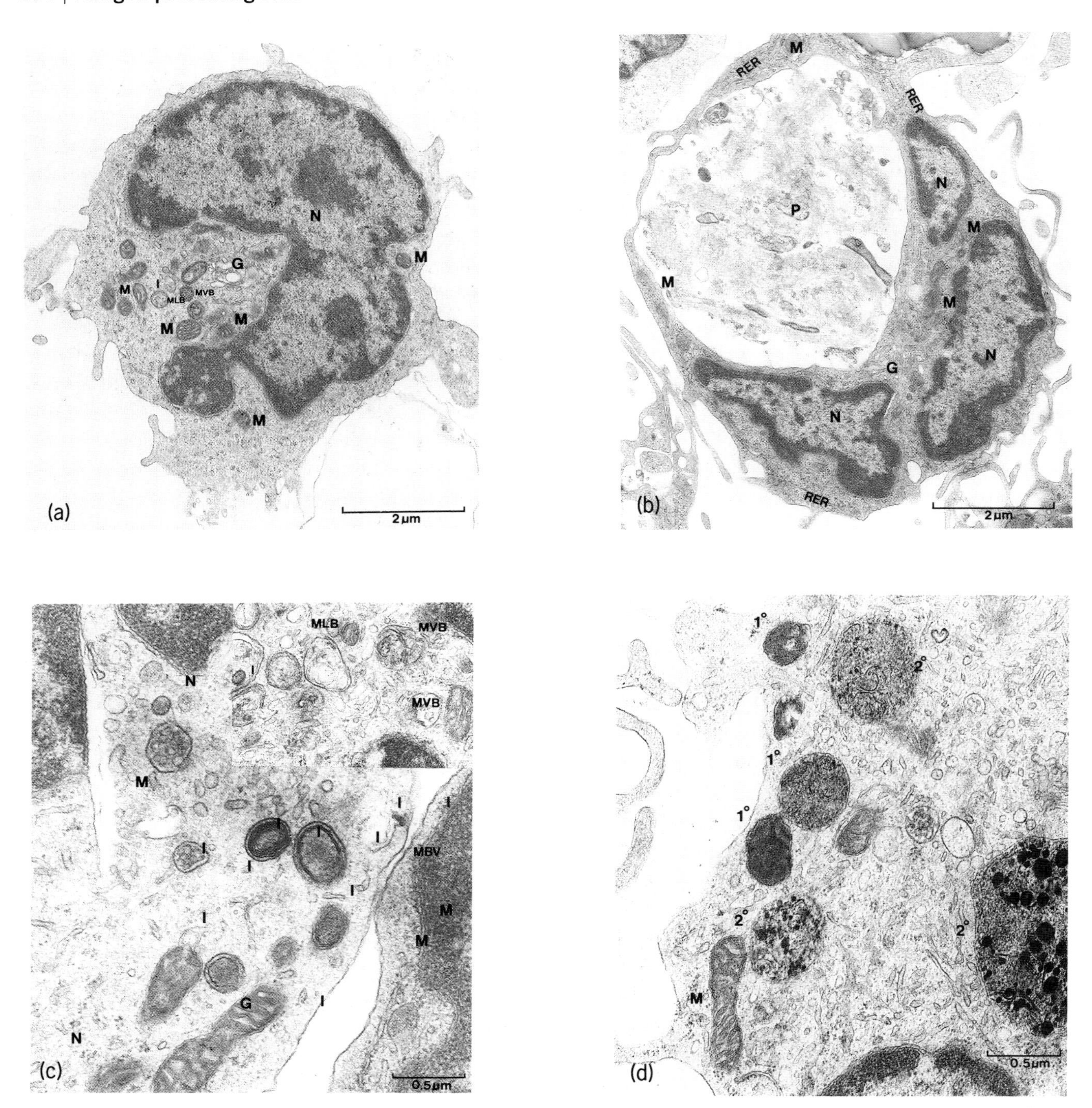

Fig.3.35 Transmission electron micrographs of splenic dendritic cells (a,c) and macrophages (b,d). Within the cytoplasm of the dendritic cell (a) can be seen the nucleus (N), Golgi region (G) and several round mitochondria (m), as well as multilamellar bodies (MLB), multivesicular bodies (MVB) and other inclusions (I) which are more evident in the higher-power view of another DC (c). The macrophage (b) contains a large membrane-bounded phagosome (P) and appears to be in the process of engulfing inorganic material from the preparation: note the rough endoplasmic reticulum (RER) which is much sparser in DCs. Within the cytoplasm of another macrophage (d) can be seen a spindly mitochondrion and presumptive primary 1° and secondary 2° lysosomes. From Austyn (1989) (see Further reading)

MHC class II-positive, CD45-positive cells with a dendritic morphology, characteristics that are not definitive.

3.4.2 Distribution of dendritic cells and related cells

Lymphoid tissues contain reasonably discrete T and B cell areas (Section 1.3.4) and different types of accessory cell are present in each of these: interdigitating cells in T areas and follicular dendritic cells in B areas. The cells that most closely resemble isolated lymphoid dendritic cells in phenotype were first thought to be the interdigitating cells of T areas of spleen (in the white pulp) and lymph node (in the paracortex), as well as in the medulla of thymus where most mature T cells seem to be located (Table 3.4) Nevertheless another population of dendritic cells has also been identified in the marginal zone of the spleen (Metlay *et al.* 1990), and it is possible these represent the majority of splenic dendritic cells that are conventionally isolated *in vitro*, rather than the interdigitating cells. It is likely that follicular dendritic cells of B areas are not related to lymphoid dendritic cells.

It is now clear that dendritic cells are not confined to lymphoid tissues, and that related cells are present in non-lymphoid tissues throughout the body. In this respect the family of dendritic cells constitutes a distinct lineage that is as widespread and diverse as the mononuclear phagocyte system (monocytes and macrophages; Section 10.4).

The most well-studied dendritic cells of non-lymphoid tissues are **Langerhans cells** of skin, which are thought of as immature cells, and similar stages may also be present in other non-lymphoid sites of the body (Fig.3.36). When they are isolated and cultured, Langerhans cells develop into cells that resemble dendritic cells in almost every respect (Schuler and Steinman 1985) (Table 3.5). What is particularly interesting about this maturation process is that Langerhans cells initially have a number of features in common with macrophages, including expression of Fc receptors, a number of cytochemical markers (e.g. lysosomal enzymes), and some macrophage

Table 3.4 Types of dendritic cells and their distribution

Site	Cell	Localization
*Lymphoid tissues**	Interdigitating cell	T areas of secondary lymphoid tissues; medulla of thymus
	Marginal zone DC	Splenic marginal zone
Non-lymphoid tissues	Langerhans cell	Epidermis of skin
	Non-lymphoid (interstitial) dendritic cell	Other non-lymphoid tissues, e.g. heart, kidney (but not the bulk of brain)
Body fluids	Veiled cell	Afferent lymph
	Blood dendritic cell	Peripheral blood

*Within B areas of lymphoid tissues is a different type of cell called a follicular dendritic cell which, despite its name, may not belong to this lineage

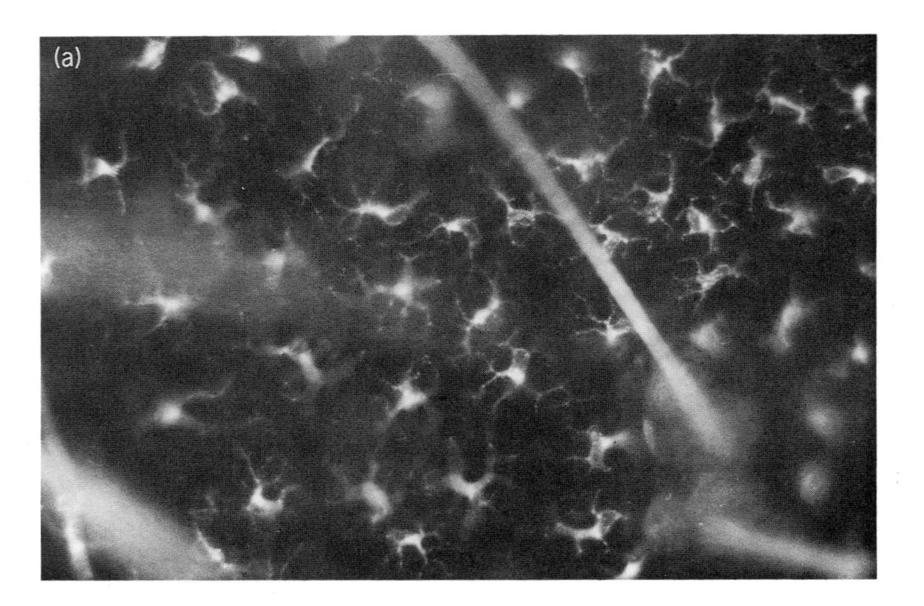

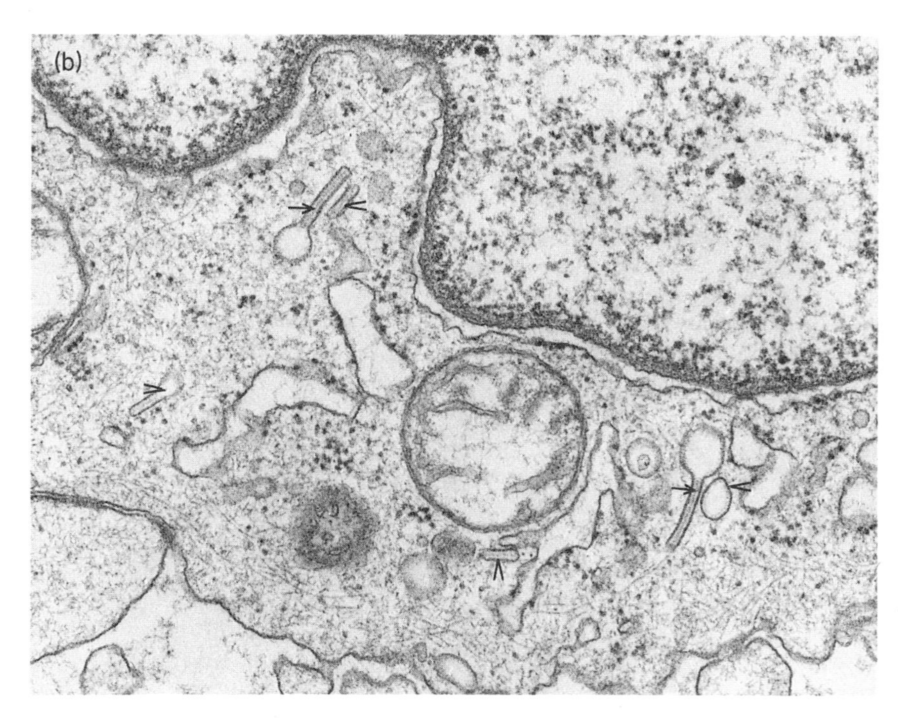

Fig.3.36 Langerhans cells in epidermis of skin. (a) Langerhans cells visualized within an epidermal sheet by immunofluorescence with anti-MHC class II monoclonal antibody. (b) Transmission electron micrograph of part of the cytoplasm and nucleus of a Langerhans cell illustrating cytoplasmic Birbeck granules which have been completely (arrows) or partially (arrowheads) sectioned. Photographs courtesy of (a) C. Larsen and R. Steinman and (b) N. Romani and G. Schuler

Table 3.5 Features and functions of freshly-isolated and cultured Langerhans cells (LC) compared to macrophages (Mφ) and lymphoid dendritic cells (DC) from mouse

	Mφ	Fresh LC	Cultured LC	DC
Features				
Birbeck granules	−	+	−	−
Non-specific esterase	+	+	−	−
Membrane ATPase	+	+	−	−
Antigen F4/80	+	+	−	−
FcγRII receptor (CDw32)	+	+	−	−
CR3 receptor (CD11b/18)	+	+	+	+
LCA (CD45)	+	+	+	+
MHC class I	+	+	+	+
MHC class II	+/−	+	+	+
Antigen 33D1	−	−	−	+/−
Functions				
Immunostimulation	−	−	+	+
Antigen processing	+	+	−	−
Endocytosis	hi	hi	lo	lo

markers like F4/80 in the mouse. These traits are lost when Langerhans cells mature into dendritic cells, and the cells never develop into macrophages indicating that they are committed to the dendritic cell lineage.

It is believed that Langerhans cells *in vivo* begin to mature into dendritic cells in response to inflammatory events, such as occurs after a contact sensitizing agent (Section 10.5.4) has been applied to the skin. This maturation process is induced at least in part by a soluble molecule called granulocyte–macrophage colony-stimulating factor (GM-CSF) which can be produced by other cells of the skin such as keratinocytes. This, and another cytokine, TNFα (Section 9.5.3) also sustains the viability of Langerhans cells in culture. The maturing dendritic cells, which may be carrying antigens acquired from the skin (e.g. a contact sensitizing agent, or a pathogen if injury and infection has occurred) then enter afferent lymphatics and travel to the draining lymph nodes where they become interdigitating cells in the T areas (Fossum 1989). The cells in afferent lymph are closely related to dendritic cells, although they are more heterogeneous and their phenotype can be considered as transitional between that of Langerhans cells and dendritic cells (Pugh *et al.* 1983). These cells in lymph were originally called **veiled cells** because they possess cell processes resembling large flaps or veils.

Cells that are indistinguishable from lymphoid dendritic cells have also been isolated from human peripheral blood. There is evidence that at least some of these cells are also on their way to a lymphoid tissue, but in this case to the spleen, where again they localize in T cell areas. It seems quite likely that many dendritic cells in the blood originate from non-lymphoid tissues. Cells resembling dendritic cells (or their immature stages) have been found in all non-lymphoid organs examined except for a large portion of the brain. These cells are MHC class II-positive, CD45-positive, and non-phagocytic with a dendritic morphology, and when isolated from the heart or kidney they can acquire the immunostimulatory function characteristic of dendritic cells.

Evidence that dendritic cells in non-lymphoid tissues can travel via the blood to the spleen, which lacks a lymphatic supply, has come from experiments in which allogeneic hearts were transplanted into mice, and allogeneic dendritic cells were later found in the recipients' spleens associated with $CD4^+$ T cells (Larsen *et al.* 1990). There is, therefore, considerable traffic of dendritic cells from the skin and other non-lymphoid organs into lymphoid tissues. There appear to be two parallel migratory routes for dendritic cells from non-lymphoid tissues: one via lymph into lymph nodes, and another via blood into the spleen (Fig.3.37). These migrating cells are thought to carry antigens in one form or another, and are likely to be required to initiate immune responses within the lymphoid tissues (see below). At present it is not clear in what form dendritic cells enter non-lymphoid tissues, but MHC class II-negative blood precursors have been identified which develop into dendritic cells in the presence of GM-CSF. MHC class II-negative precursors have also been found in the thymus, and these cells can be induced to mature in culture in the presence of the cytokine IL-1.

3.4.3 The role of dendritic cells in initiating immune responses

The first indication that dendritic cells play an important role in stimulating immune responses came from studies of the mixed leukocyte reaction (MLR) in mice (see alloreactivity—Sections 1.1.4 and 5.2.2). It was found that

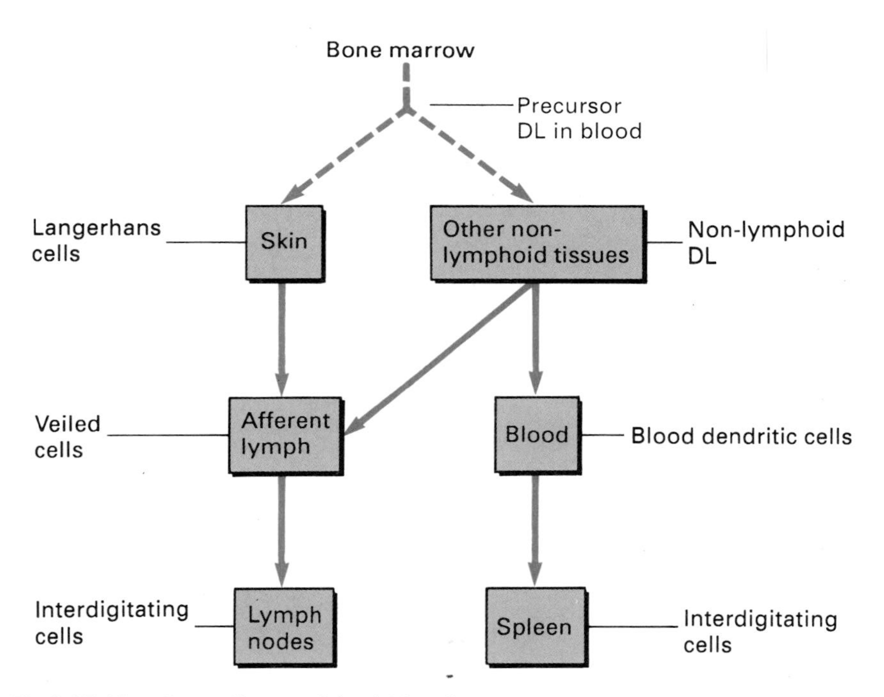

Fig.3.37 Migration pathways of dendritic cells

T cells proliferated when they were cultured with enriched populations of allogeneic splenic dendritic cells but they did not respond to enriched populations of splenic macrophages. Subsequent studies comparing the ability of dendritic cells and macrophages to stimulate a variety of other immune responses in culture have led to the same conclusion, irrespective of whether rodent or human cells were used. For example, dendritic cells are potent accessory cells for the development of cytotoxic T cells, antibody formation by B cells, and some polyclonal (mitogenic) responses like oxidative mitogenesis (Table 3.6). In contrast, macrophages are weak or inactive in these systems and are sometimes suppressive.

These results were confirmed and extended when the first cytotoxic monoclonal antibody specific for dendritic cells, 33D1, was used to deplete the cells from heterogeneous populations (Steinman *et al.* 1983). It was already known that treatment of accessory cells with anti-MHC class II antibodies inhibited their stimulatory function, showing that an MHC class II-positive cell was involved. Treatment of unfractionated spleen cell suspensions (which contain $<1\%$ dendritic cells) or spleen adherent cells (which contain up to 30% dendritic cells; Fig.3.4) with 33D1 and complement ablated the accessory function of these populations to essentially the same degree as anti-MHC class II antibodies and complement. Thus, even though the treated spleen cells contained some 50% class II-positive B cells, and the majority of treated spleen adherent cells were class II-positive macrophages, they were unable to stimulate responses such as the MLR and oxidative mitogenesis. This indicated that dendritic cells were the major stimulatory cell type within these populations.

The converse type of experiment to the above has been done with human

Table 3.6 Stimulation of immune responses by lymphoid dendritic cells *in vitro*

Response	Conditions
Allogeneic MLR	Alloreactive T cells proliferate when cultured with allogeneic DC
Syngeneic MLR	Class II-restricted T cells proliferate when cultured with syngeneic DC (autoreactivity or response to antigens in culture?)
Development of CTLs	Alloreactive CTL develop in the allogeneic MLR; antigen specific CTL develop when cultured with hapten-modified syngeneic DC
T-dependent Ig responses	Antibody-secreting cells develop when B cells are cultured with hapten/carrier complexes, T helper cells and DC
Mitogen responses	Polyclonal proliferation occurs when oxidized T cells are cultured with DC (oxidative mitogenesis), or when T cells are cultured with Con A or PHA and DC
Memory responses	Memory T cells proliferate when they are restimulated with specific antigen and DC

cells using a monoclonal antibody against the CD14 antigen expressed by human monocytes and macrophages (Van Voorhis *et al.* 1983). Treatment of peripheral blood cells with this antibody and complement to kill essentially all the monocytes, but to spare dendritic cells, had little or no effect on their ability to stimulate immune responses. Similar results were also obtained when cell populations from human spleens and tonsils were treated in the same way. Again, these studies emphasized the potency of dendritic cells in stimulating T cells and the inefficacy of B cells and monocytes for these responses.

There are two main conclusions that can be drawn from these early studies. *First, dendritic cells seem to be the principle inducers of immune responses* in culture (and other experiments suggest this is the same *in vivo*), at least in comparison to splenic macrophages and small resting B cells. *Second, expression of MHC class II molecules is necessary but not sufficient to trigger responses of resting T cells; rather, it appears these molecules must be expressed on dendritic cells.*

3.4.4 Antigen-presenting cell requirements for resting and activated T cells

How can those conclusions (Section 3.4.3) be reconciled with a large body of data suggesting that macrophages and other cells can stimulate immune responses in culture?

One simple explanation is that in many early studies of accessory cell function, the macrophages, or other accessory cell populations were not homogeneous and they contained small numbers of dendritic cells which were actually responsible for the stimulation observed. This may explain some discrepancies, but does not account for all of them. In a large number of other studies, pure populations of B cell tumours or fibroblasts transfected with MHC class II genes have been used as antigen-presenting cells. They were found to present antigens very well to a wide variety of different T cell clones and hybridomas, as well as to sensitized T cells isolated from recently immunized animals. In addition, MHC class II molecules in liposomes and planar membranes could be used to present some antigens to sensitized T cells and T cell clones. An additional point is that the magnitude of antigen-dependent proliferative responses using different antigen-presenting cells was found to be directly proportional to the number of MHC molecules expressed by the cell, as well as the concentration of antigen, in apparent contrast to the earlier conclusion (see the end of Section 3.4.3) that mere expression of MHC molecules is not sufficient to trigger an immune response.

Some of the contradictions mentioned above have now been resolved with the understanding that *resting T cells are stimulated primarily by dendritic cells, whereas activated T cells can be stimulated by a wider spectrum of antigen-presenting cells.* Resting T cells may be defined as those obtained from an animal that has not been deliberately immunized, whereas activated T cells are those stimulated in culture by dendritic cells or obtained from recently-primed lymphoid tissues (e.g. from lymph nodes draining the site of antigen administration). Perhaps T cell clones and hybridomas should also be included within this definition of activated cells: many T cell clones generally respond directly to IL-2, unlike resting T cells, and T cell hybridomas have an activated T cell as one of the fusion partners and can proliferate spontaneously.

An increasing amount of data is compatible with this idea. For example, Inaba and Steinman (1984*a*) compared the ability of dendritic cells and other accessory cell populations to stimulate the proliferation of resting T cells in the allogeneic MLR, compared to T cell blasts that had been activated in similar cultures (Fig.3.38). Resting T cells responded only to dendritic cells, but the T cell blasts could respond to the other accessory cells tested, macrophages and B cell blasts, as well as dendritic cells. When the T cell blasts were cultured without antigen they reverted to resting 'memory' cells. These cells, like virgin T cells, were once again primarily activated by dendritic cells (Table 3.7). These observations strongly suggest that dendritic cells are required for activation of resting and memory T cells, while activated T cells can respond to perhaps any antigen presenting cell provided that it expresses the relevant peptide–MHC complexes. Thus, one possibility is that most antigen-presenting cells merely present antigens to activated T cells, whereas dendritic cells provide the additional signals required for T cell activation.

This may not, however, be the whole story. Many studies comparing the accessory function of dendritic cells and other leukocytes have used splenic accessory cells, particularly splenic macrophages and resting B lymphocytes. It is conceivable that B blasts and cells from other sites can activate resting T cells to some extent (e.g. see Fig.3.38), but further experimentation is needed to clarify this issue.

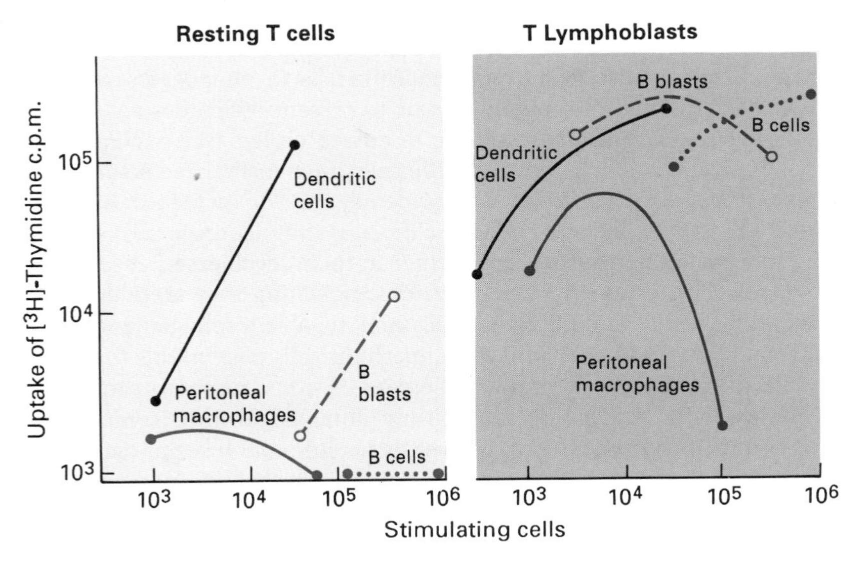

Fig.3.38 Populations of dendritic cells, peritoneal macrophages, B cell blasts and B cells were tested for their ability to stimulate proliferation ([^{3}H]thymidine incorporation) of resting T cells in the MLR (left) and alloreactive CD8- lymphoblasts from the MLR (right). The macrophages had been cultured in IFNγ to induce high levels of MHC class II expression. Note macrophages do not stimulate resting T cells, but are accessory cells for lymphoblasts, although they are inhibitory at high doses. Based on Inaba and Steinman (1984*a*) by copyright permission of the Rockefeller University Press

Table 3.7 Responses of resting, activated and memory T cells to different types of accessory cell (based on data of Inaba and Steinman 1984*a*, and Schuler and Steinman 1985)

Accessory cell	Response of:		
	resting T	activated T	memory T
Dendritic cells	++	++	++
Langerhans cells (fresh)	–	++	–
Macrophages	–	++	–
Resting B cells	–	++	–
B cell blasts	+	++	?

One further important point to consider is that immature dendritic cells, such as Langerhans cells, are unlike mature lymphoid dendritic cells, in that they are essentially unable to stimulate responses of resting T cells (Table 3.5, bottom). Thus, freshly isolated Langerhans cells have little stimulatory activity in the MLR, but this increases during culture when they develop into dendritic cells. Therefore the maturation of dendritic cells may be an important step in regulating when and where immune responses are generated.

3.4.5 Mechanism of dendritic cell stimulation

Different functions of immature and mature dendritic cells

It is not yet entirely clear how dendritic cells stimulate primary T cell responses. The inability of mature dendritic cells to phagocytose particles and their lack of lysosomes would appear to present difficulties for schemes of antigen processing similar to those discussed earlier (see Sections 3.2.2 and 3.2.3). The possibility that dendritic cells have membrane ectoenzymes for antigen processing has been considered, but no evidence has been obtained one way or the other. The likelihood is that dendritic cells can take up and process antigens at an earlier stage in their development (see below). Some veiled cells in lymph have inclusions indicating they were endocytic at some stage, while interdigitating cells may be able to phagocytose under certain conditions *in vivo*. Moreover, Langerhans cells may be able to endocytose molecules via their Fc and complement or other receptors; they have been shown to be capable of internalizing antibody–antigen complexes as well as certain viruses, bacteria, and yeasts *in vitro*, and it seems likely these antigens are subsequently processed by as yet largely undefined pathways. The processed antigens are then presumably bound to MHC molecules (i.e. as peptides) and expressed at the cell surface as the cells mature and lose their endocytic ability, but acquire their new immunostimulatory function (Table 3.5, bottom).

A pertinent observation is that freshly-isolated Langerhans cells can process and present intact proteins to T cell clones but dendritic cells cannot (see below), whereas both cell types can present exogenously-added specific peptides. Therefore, a reasonable hypothesis is that *dendritic cell precursors can endocytose and process antigens but cannot initiate T cell responses, whereas the mature cell is unable to process additional antigens but has the ability to initiate T cell responses to those peptide–MHC complexes it now expresses* (Fig.3.39). (This is further discussed below.) In general, immature dendritic cells may predominate in non-lymphoid organs, while mature dendritic cells are more commonly found in lymphoid tissues. It seems likely that the localization of members of the dendritic cell family at different stages of maturation in different sites, and their migration pathways, ensure that immune responses are primarily generated within lymphoid tissues. This may be one way of increasing the efficiency of the immune system, for it should not be forgotten that in humans, for example, there are perhaps 10^{12} T cells, and just a few of these have to locate their specific peptide–MHC complexes on a limited number of dendritic cells before an immune response can be initiated. It is clearly advantageous to ensure that T cells and dendritic cells with immunostimulatory function localize within the same tissue, e.g.

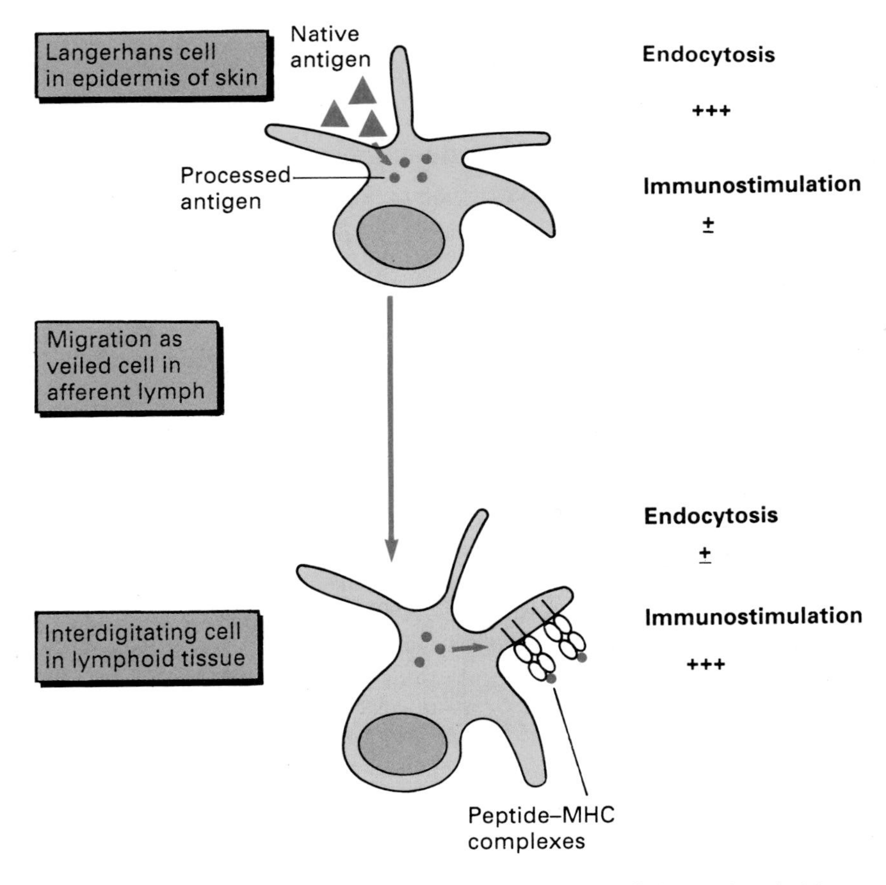

Fig.3.39 Possible functional differences between dendritic cells in non-lymphoid tissues, e.g. skin, and lymphoid tissues

the spleen or lymph nodes, in order to optimize the chances that they can contact each other and initiate the adaptive immune response.

Immunostimulation by mature dendritic cells

Role of IL-2 in immune responses. When T cells in culture are activated by dendritic cells they progressively release the lymphocyte growth factor IL-2 into the culture medium. At the same time the T cells become responsive to IL-2, presumably because they now express high affinity IL-2 receptors. When IL-2 binds to its receptors, this stimulates T cell proliferation resulting in clonal expansion of the specific T cells. It also causes the T cells to release other cytokines that can, for example, help B cells to make antibodies in the presence of their specific antigen, while others stimulate cytotoxic T cell precursors to develop into CTL. Therefore, by virtue of their ability to control the secretion of IL-2, as well as their responsiveness to it, dendritic cells most likely play a pivotal role in the initiation of immune responses. IL-2 and its receptor are discussed in Section 4.3.4.

Lack of IL-1 production by dendritic cells. One scheme for T cell activation suggests that accessory cells elaborate a soluble molecule that provides a second signal for lymphocyte activation (Fig.3.3). This molecule has been proposed to be IL-1. However, it is now known that dendritic cells do not secrete IL-1, nor do they contain messenger RNA for this molecule, and anti-IL-1 antibodies and IL-1 receptor inhibitors (Section 9.5.1) do not inhibit *in vitro* responses stimulated by dendritic cells. Thus this scheme does not explain the stimulatory activity of dendritic cells. It is of course possible that dendritic cells can produce another cytokine that activates T cells, but there is no evidence for this as yet. An alternative is that they have a unique membrane component that can act as a receptor or ligand for a complementary T cell molecule, or that a particular combination of membrane interactions is important for T cell activation.

Dendritic cell–T cell clustering. During stimulation of immune responses, dendritic cells physically associate with the responding lymphocytes in cell clusters. For example, they cluster with allogeneic T cells during an MLR, and with T cells and B cells during T-dependent antibody responses. This clustering event has been shown to be essential for subsequent T cell activation. Dendritic cells can express a number of adhesion molecules including LFA-1 (CD11a/CD18), ICAM-1 (CD54), LFA-3 (CD58), and B7/BB1 (Sections 4.2.6 and 4.2.8), which may be required for clustering with, and signalling to, the T cell.

There is evidence that activated T cells (T lymphoblasts) can cluster with a variety of accessory cells, including dendritic cells, macrophages, and B cell blasts. However, clustering with accessory cells other than dendritic cells can only occur in the presence of the specific antigen to which the T cells were activated. In contrast, dendritic cells seem able to cluster with T cell blasts in an *antigen-independent* manner, even though the T blasts are only stimulated to make a response if they subsequently recognize their specific antigen on the dendritic cell (Table 3.8).

These findings have been extrapolated to the situation that might occur with resting T cells. It is possible that dendritic cells can cluster with any

Table 3.8 Antigen-dependent and -independent clustering of activated T cells with different types of accessory cell (based on data of Inaba and Steinman 1984*b*)

Accessory cell	Antigen-dependent clustering	Antigen-independent clustering
Dendritic cells	++	++
Langerhans cells (fresh)	++	−
Macrophages	++	−
Resting B cells	++	−

resting T cell in an antigen-independent manner. Conceivably, it is only after this initial clustering event that T cell receptors can recognize their specific peptide–MHC complexes on the dendritic cell and the T cells are activated.† To date, the only soluble molecule known to influence the association of dendritic cells with T cells is IL-1, which promotes clustering by acting on the dendritic cell and thereby increases the *magnitude* of such T cell responses.

A further demonstration of the importance of dendritic cell-T cell clustering has come from studies on Langerhans cells. Freshly-isolated Langerhans cells have only weak immunostimulatory activity and they cluster with T cells very poorly. When they mature in culture, their immunostimulatory activity increases, even to surpass that of splenic dendritic cells, concomitant with an increased ability to cluster with T cells. (Tables 3.5 (bottom) and 3.8).

Endocytosis, MHC class II and invariant chain expression, and acidic organelles in dendritic cells

Freshly-isolated Langerhans cells can endocytose native antigens such as myoglobin and conalbumin, and can present immunogenic peptides derived from these antigens to primed T cells and T cell hybridomas, although they lack immunostimulatory activity and are unable to initiate responses of resting T cells (see above). The ability of fresh Langerhans cells to present these foreign antigens, presumably as peptide–MHC complexes, can be inhibited in the presence of chloroquine (Section 3.2.2) or cycloheximide (which, however, has the converse effect on antigen presentation by B cell blasts). Cultured Langerhans cells, which resemble mature dendritic cells (see above) lack the capacity to present these native antigens (Romani *et al.* 1989). However, both fresh and cultured Langerhans cells can pinocytose equivalent amounts of rhodamine-conjugated ovalbumin, used as an endocytic tracer, so uptake *per se* does not account for their differences in antigen presenting capacity.

It has been shown that the biosynthesic rate of MHC class II α and β chains and invariant chain is very high in fresh Langerhans cells (Puré *et al.* 1990). However, synthesis of these molecules is virtually undetectable after a day of culture, despite the fact that immunogenic peptides, derived from endocytosis of native antigens by fresh Langerhans cells, can be retained for at least two

† There may be parallels here with one scheme for cytotoxic T cell–target cell interactions, during which the T cell initially adheres to the target cell in an antigen-independent manner before specific recognition occurs and the target is killed (see Section 10.3.2).

days of culture. This suggests that the high rate of synthesis of class II molecules and invariant chains is associated with the capacity of Langerhans cells to present immunogenic peptides derived from foreign antigens, and that these peptide–MHC complexes may remain stably-associated as the cell matures into an immunostimulatory dendritic cell (see Section 3.3.2 for a discussion of the role of the invariant chain in antigen presentation).

The distribution of acidic compartments (Section 3.2.2) has also been studied in fresh and cultured Langerhans cells (Stössel *et al.* 1990). These compartments were visualized by electron microscopy using the so-called DAMP technique. DAMP is a DNP-coupled derivative of a weak base that readily diffuses into viable cells, is concentrated in acidic compartments at a density roughly proportional to the pH, becomes covalently linked to proteins, and can be visualized using an appropriate combination of anti-DNP and gold-conjugated antibodies.

Using the DAMP technique, a number of acidic organelles were readily detected in fresh Langerhans cells. These included, in order of increasing pH, early endosomes, late endosomes, and phagolysosomes and lysosomes (Section 3.2.2), structures which are associated with sequential processing of endocytosed antigens, cleavage of the invariant chain from newly-synthesized class II molecules, and their association with immunogenic peptides. Tubular structures in continuity with endosomal vesicles that may correspond to CURL endosomes (Section 3.2.2) were also labelled, as were the Langerhans cell-specific Birbeck granules. However, the majority of these acidic organelles were never encountered in cultured Langerhans cells, particularly the early endosomes; the number of Birbeck granules was also markedly reduced in human, or absent in mouse, cultured Langerhans cells. The fact that disappearance of these organelles paralleled the loss of antigen processing capacity on culture, suggests they may be critical for antigen processing, and subsequent presentation to activated T cells, by fresh Langerhans cells.

Clearly, these observations indicate that the ability of dendritic cells to present antigens and initiate primary T cell responses is associated with endocytosis and processing of native antigens at an immature stage. However, the mechanism of immunostimulation by mature dendritic cells remains something of a mystery at the time of writing.

Further reading

Antigen processing and related topics in cell biology

Braciale, T.J. and Braciale, V.L. (1991). Antigen presentation: structural themes and functional variations. *Imunology Today*, **12**, 124–9.

Brodsky, F.M. and Guagliardi, L. (1991). The cell biology of antigen processing and presentation. *Annual Review of Immunology*, **9**, 707–44.

Burgess, T.L. and Kelly, R.B. (1987). Constitutive and regulated secretion of proteins. *Annual Review of Cell Biology*, **3**, 243–93.

Dice, J.F. (1990). Peptide sequences that target cytosolic proteins for lysosomal proteolysis. *Trends in Biochemical Science*, **15**, 305–9.

Ellis, R.J. (1990). The molecular chaperone concept. *Seminars in Cell Biology* **1**, 1–9.

Fine, R.E. (1989). Vesicles without clathrin: intermediates in bulk flow exocytosis. *Cell*, **58**, 609–10.

Goldberg, A.L. and Rock, K.L. (1992). Proteolysis, proteasomes and antigen presentation. *Nature*, **357**, 375–9.

Gruenberg, J. and Howell, K.E. (1989). Membrane traffic in endocytosis: insights from cell-free assays. *Annual Review of Cell Biology*, **5**, 453–81.

Kornfeld, S. and Mellman, I. (1989). The biogenesis of lysosomes. *Annual Review of Cell Biology*, **5**, 483–525.

Monaco, J.J. (1992). A molecular model of MHC class-I-restricted antigen processing. *Immunology Today*, **13**, 173–9.

Neefjes, J.J. and Ploegh, H.L. (1992). Intracellular transport of MHC class II molecules. *Immunology Today*, **13**, 179–84.

Pelham, H.R. (1989). Control of protein exit from the endoplasmic reticulum. *Annual Review of Cell Biology*, **5**, 1–23.

Pernis, B., Silverstein, S.C., and Vogel, H.J. (ed.) (1988). *Processing and presentation of antigens.* Academic Press, San Diego, California.

Rechsteiner, M. (1987). Ubiquitin-mediated pathways for intracellular proteolysis. *Annual Review of Cell Biology*, **3**, 1–30.

Rothbard, J.B. and Gefter, M.L. (1991). Interactions between immunogenic peptides and MHC proteins. *Annual Review of Immunology*, **9**, 527–66.

Schwartz, A.L. (1990). Cell biology of intracellular protein trafficking. *Annual Review of Immunology*, **8**, 195–229.

Townsend, A. (1992). A new presentation pathway? *Nature*, **356**, 386–7.

Antigen-presenting and immunostimulatory cells

Austyn, J.M. (1989). *Antigen-presenting cells.* In Focus Series (ed. D. Male). IRL Press, Oxford.

Lanzavecchia, A. (1990). Receptor-mediated antigen uptake and its effect on antigen presentation to class II-restricted T lymphocytes. *Annual Review of Immunology*, **8**, 773–93.

Metlay, J.P., Pure, E., and Steinman, R.M. (1989). Control of the immune response at the level of antigen-presenting cells: a comparison of the function of dendritic cells and B lymphocytes. *Advances in Immunology*, **47**, 45–116.

Pierce, S.K., Morris, J.F., Grusby, M.J., Kaumaya, P., von Buskirk, A., Srinivasan, M., Crump, B., and Smolenski, L.A. (1988). Antigen-presenting function of B lymphocytes. *Immunological Reviews*, **106**, 149–80.

Sprent. J. and Schaefer, M. (1989). Antigen presenting cells for unprimed T cells. *Immunological Reviews*, **10**, 17–23,

Steinman, R.M. (1991). The dendritic cell system and its role in immunogenicity. *Annual Review of Immunology*, **9**, 271–96.

Townsend, A. and Bodmer, H. (1989) Antigen recognition by class-I-restricted T lymphocytes. *Annual Review of Immunology*, **7**, 601–24.

Unanue, E.R. (1984). Antigen-presenting function of the macrophage. *Annual Review of Immunology*, **2**, 395–428.

Literature cited

Babbit, B.P., Allen, P.M., Matsueda, G., Haber, E., and Unanue, E.R. (1985). The binding of immunogenic peptides to Ia histocompatibility molecules. *Nature*, **317**, 359–61.

Bakke, O. and Dobberstein, B. (1990). MHC class II-associated invariant chain contains a sorting signal for endocytic compartments. *Cell*, **63**, 707–16.

Bouillot, M., Choppin, J., Cornille, F., Martinon, F., Paps, T., Gomard, E., Fournie-Zaluski, M.C., and Levy, J.P. (1989). Physical association between MHC class I molecules and immunogenic peptides. *Nature*, **339**, 473–5.

Braciale, T.J., Morrison, L.A., Sweetser, M.T., Sambrook, T., Gething, M.J., and Braciale, V.L. (1987). Antigen presentation pathways to class I and class II MHC-restricted T lymphocytes. *Immunological, Reviews*, **98**, 95–114.

Buus, S., Sette, A., Colon, S.M., Jenis, D.M., and Grey, H.M. (1986). Isolation and characterization of antigen-Ia complexes in T cell recognition. *Cell*, **47**, 1071–7.

Chen, B.P. and Parham, P. (1989). Direct binding of influenza peptides to class I HLA molecules. *Nature*, **337**, 743–5.

Chesnut, R.W. and Grey, H.M. (1982). Antigen presentation by B cells and its significance in T-B interactions. *Advances in Immunology*, **39**, 51–94.

Claman, H.N., Chaperon, E.A., and Triplett, R.F. (1967). Immunocompetence of transferred thymus-marrow cell combinations. Journal of Immunology, **97**, 828–32.

Cresswell, P. (1985). Intracellular class II HLA antigens are accessible to transferrin-neuraminidase conjugates internalized by receptor-mediated endocytosis. *Proceedings of the National Academy of Sciences USA*, **82**, 8188–92.

DeLisi, C. and Berzofsky, J. (1985). T-cell antigenic sites tend to be amphipathic structures. *Proceedings of the National Academy of Sciences USA*, **82**, 7048–52.

Falk, K., Rötzschke, O., Stevanovic, S., Jung, G., and Rammensee, H.G. (1991). Allele-specific motifs revealed by sequencing of self-peptides eluted from MHC molecules. *Nature*, **351**, 290–6.

Fossum, S. (1989). Lymph-borne dendritic cells do not recirculate, but enter the lymph node paracortex to become interdigitating cells. *Scandinavian Journal of Immunology*, **27**, 97–105.

Gammon, G., Geysen, H.M., Apple, R.J., Pickett, E., Palmer, M., Ametani, A., and Sercarz, E.E. (1991). T cell determinant structures: cores and determinant envelopes in three mouse major histocompatibility complex haplotypes. *Journal of Experimental Medicine*, **173**, 609–17.

Gell, P.G. and Benacerraf, B. (1959). Studies on hypersentivity. Delayed hypersensitivity to denatured proteins in guinea pigs. *Immunology*, **2**, 64–70.

Grey, H.M., Colon, S.M., and Chesnut, R.W. (1982). Requirements for the processing of antigen by antigen-presenting B cells. II. Biochemical comparison of the fate of antigen in B cell tumors and macrophages. *Journal of Immunology*, **129**, 2388–9.

Hersh, E.M. and Harris, J.E. (1968). Macrophage–lymphocyte interaction in the antigen-induced blastogenic response of human peripheral blood leukocytes. *Journal of Immunology*, **100**, 1184–94.

Inaba, K. and Steinman, R.M. (1984*a*). Resting and sensitized T lymphocytes exhibit distinct stimulatory (antigen-presenting cell) requirements for growth and lymphokine release. *Journal of Experimental Medicine*, **160**, 1717–35.

Inaba, K. and Steinman, R.M. (1984*b*). Resting and sensitized T lymphocytes exhibit distinct stimulatory requirements for growth and lymphokine release. *Journal of Experimental Medicine*, **160**, 1711.

Jardetzky, T.S., Lane, W.S., Robinson, R.A., Madden, D.R. and Wiley, D.C. (1991). Identification of self peptides bound to purified HLA-B27. *Nature*, **353**, 326–9.

Lanzavecchia, A. (1985). Antigen-specific interaction between T and B cells. *Nature*, **314**, 537–9.

Larsen, C.P., Morris, P.J., and Austyn, J.M. (1990). Migration of dendritic leukocytes from cardiac allografts into host spleens: a novel pathway for initiation of rejection. *Journal of Experimental Medicine*, **171**, 307–14.

Ljunggren, H.G., Stam, N.J., Ohlen, C., Neefjes, J.J., Hoglund, P., Heemels, M.T., Bastin, J., Schumacher, T.N.M., Townsend, A., Karre, K., and Ploegh, H. (1990). Empty MHC class I molecules come out in the cold. *Nature*, **346**, 476–80.

Metlay, J.P., Witmer-Pack, M.D., Agger, R., Crowley, M.T., Lawless, D. and Steinman, R.M. (1990). The distinct leukocyte integrins of mouse spleen dendritic cells as identified with new hamster monoclonal antibodies. *Journal of Experimental Medicine*, **171**, 1753–71.

Mitchell, G.F. and Miller, J.F. (1968). Cell to cell interaction in the immune response. II The source of hemolysin-forming cells in irradiated mice given bone marrow and thymus or thoracic duct lymphocytes. *Journal of Experimental Medicine*, **128**, 821–37.

Mosier, D.E. (1967). A requirement for two cell types for antibody formation *in vitro*. *Science*, **158**, 1573–5.

Neefjes, J.J., Stollorz, V., Peters, P.J., Geuze, H.J., and Ploegh, H.L. (1990). The biosynthetic pathway of MHC class II but not class I molecules intersects the endocytic route. *Cell*, **61**, 171–83.

Nuchtern, J.G., Bonifacino, J.S., Biddison, W.E., and Klausner, R.D. (1989). Brefeldin A implicates egress from endoplasmic reticulum in class I restricted antigen presentation. *Nature*, **339**, 223–6.

Phillips, M.L., Yip, C.C., Shevach, E.M., and Delovitch, T.L. (1986). Photoaffinity labeling demonstrates binding between Ia molecules and nominal antigen on antigen-presenting cells., *Proceedings of the National Academy of Sciences USA*, **83**, 5634–8.

Powis, S.J., Deverson, E.V., Coadwell, W.J., Ciruela, A., Huskisson, N.S., Smith, H., Butcher, G.W., and Howard, J.C. (1992). Effect of polymorphism of an MHC-linked transporter on the peptides assembled in a class I molecule. *Nature*, **357**, 211–15.

Pugh, C.W., MacPherson, G.G., and Steer, H.W. (1983). Characterization of nonlymphoid cells derived from rat peripheral lymph. *Journal of Experimental Medicine*, **157**, 1758–79.

Puré, E., Inaba, K., Crowley, M.T., Tardelli, L., Witmer-Pack, M.D., Ruberti, G., Fathman, G., and Steinman, R.M. (1990). Antigen processing by epidermal Langehans cells correlates with the level of biosynthesis of major histocompatibility complex class II molecules and expression of invariant chain. *Journal of Experimental Medicine*, **172**, 1459–69.

Robertson, M. (1991). Antigen processing. Proteosomes in the pathway. *Nature*, **353**, 300–1.

Roche, P.A. and Cresswell, P. (1990). Invariant chain association with HLA-DR molecules inhibits immunogenic peptide binding. *Nature*, **345**, 615–618.

Rock, K.L. and Benacerraf, B. (1983). Inhibition of antigen-specific T lymphocyte activation by structurally related Ir-gene controlled polymers. Evidence of specific competition for accessory cell antigen presentation. *Journal of Experimental Medicine*, **157**, 1618–34.

Romani, N., Koide, S., Crowley, M., Witmer-Pack, M., Livingstone, A.M., Fathman, C.G., Inaba, K., and Steinman, R.M. (1989). Presentation of exogenous protein antigens by dendritic cells to T cell clones. Intact protein is presented best by immature, epidermal Langerhans cells. *Journal of Experimental Medicine*, **169**, 1169–78.

Rothbard, J.B. and Taylor, W.R. (1988). A sequence pattern common to T cell epitopes. *EMBO Journal*, **7**, 93–100.

Rotzschke, O., Falk, K., Deres, K., Schild, H., Norda, M., Metzger, J., Jung, G., and Armmensee, H.G. (1990). Isolation and analysis of naturally processed viral peptides as recognized by cytotoxic T cells. *Nature*, **348**: 252-4.

Schuler, G. and Steinman, R.M. (1985). Murine epidermal Langerhans cells mature into potent immunostimulatory dendritic cells *in vitro*. *Journal of Experimental Medicine*, **161**, 526–46.

Schumacher, T.N., De Bruijn, M.L., Vernie, L.N., Kast, W.M., Melief, C.J., Neefjes, J.J., and Ploegh, H.L. (1991). Peptide selection by MHC class I molecules. *Nature*, **350**, 703–6.

Shimonkevitz, R., Kappler, J.W., Marrack, P. and Grey, H.M. (1983). Antigen recognition by H-2 restricted T cells. I. Cell-free antigen processing. *Journal of Experimental Medicine*, **158**, 303–16.

Smith, K.A., Lachman, L.B., Oppenheim, J.J., and Favata, M.F. (1980). The functional relationship of the interleukins. *Journal of Experimental Medicine*, **151**, 1551–6.

Spies, T. and De Mars, R. (1991). Restored expression of major histocompatibility class I molecules by gene transfer of a putative peptide transporter. *Nature*, **351**, 323–4.

Steinman, R.M. and Cohn, Z.A. (1973). Identification of a novel cell type in peripheral lymphoid organs of mice. I. Morphology, quantitation, tissue distribution. *Journal of Experimental Medicine*, **137**, 1142–62.

Steinman, R.M., Gutchinov, B., Witmer, M.D., and Nussenzweig, M.C. (1983). Dendritic cells are the principal stimulators of the primary mixed leukocyte reaction in mice. *Journal of Experimental Medicine*, **157**, 613–27.

Stockinger, B., Pessara, U., Lin, R.H., Habicht, J., Grez, M., and Koch, N. (1989). A role of Ia-associated invariant chains in antigen processing and presentation. *Cell*, **56**, 683–9.

Stössel, H., Koch, F., Kampgen, E., Stoger, P., Lenz, A., Heufler, C., Romani, N., and Schuler, G. (1990) Disappearance of certain acidic organelles (endosomes and Langerhans cell granules) accompanies loss of antigen processing capacity upon culture of epidermal Langerhans cells. *Journal of Experimental Medicine*, **172**, 1471–82.

Townsend, A.R., Rothbard, J., Gotch, F.M., Bahadur, G., Wraith, D. and McMichael, A.J. (1986). The epitopes of influenza nucleoprotein recognized by cytotoxic T lymphocytes can be defined with short synthetic peptides. *Cell*, **44**, 959–68.

Townsend, A., Ohlen, C., Bastin, J., Ljunggren, H.G., Foster, L., and Karre, K. (1989). Association of class I major histocompatibility heavy and light chains induced by viral peptides. *Nature*, **340**, 443–8.

Van Voorhis, W.C., Valinsky, J., Hoffman, E., Luban, J., Hair, L.S. and Steinman, R.M. (1983). Relative efficacy of human monocytes and dendritic cells as accessory cells for T cell replication. *Journal of Experimental Medicine*, **158**, 174–91.

Walden, P., Nagy, Z.A., and Klein, J. (1985). Induction of regulatory T-lymphocyte responses by liposomes carrying histocompatibility complex molecules and foreign antigen. *Nature*, **315**, 327–9.

Watts, T.H., Gaub, H.E., and McConnell, H.M. (1986). T-cell-mediated association of peptide antigens and major histocompatibility complex protein detected by energy transfer in an evanescent wave-field. *Nature*, **320**, 179–81.

Weiss, S. and Bogen, B. (1991) MHC class II-restricted presentation of intracellular antigen. *Cell*, **64**, 767–76.

Werdelin, O. (1982). Chemically related antigens compete for presentation by accessory cells to T cells. *Journal of Immunology*, **129**, 1883–91.

Ziegler, H.K. and Unanue, E.R. (1982). Decrease in macrophage associated antigen catabolism caused by ammonia and chloroquine is associated with inhibition of antigen presentation to T cells. *Proceedings of the National Academy of Sciences USA*, **79**, 175–8.

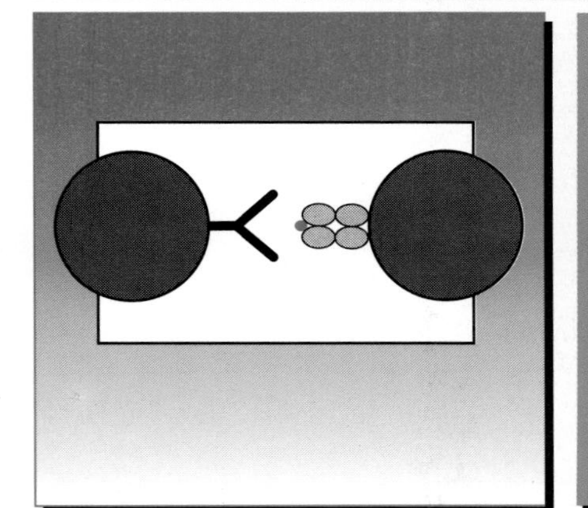

4

THE MOLECULAR BASIS OF T CELL RESPONSES

Scheme 4. T cell membrane molecules and cytokines

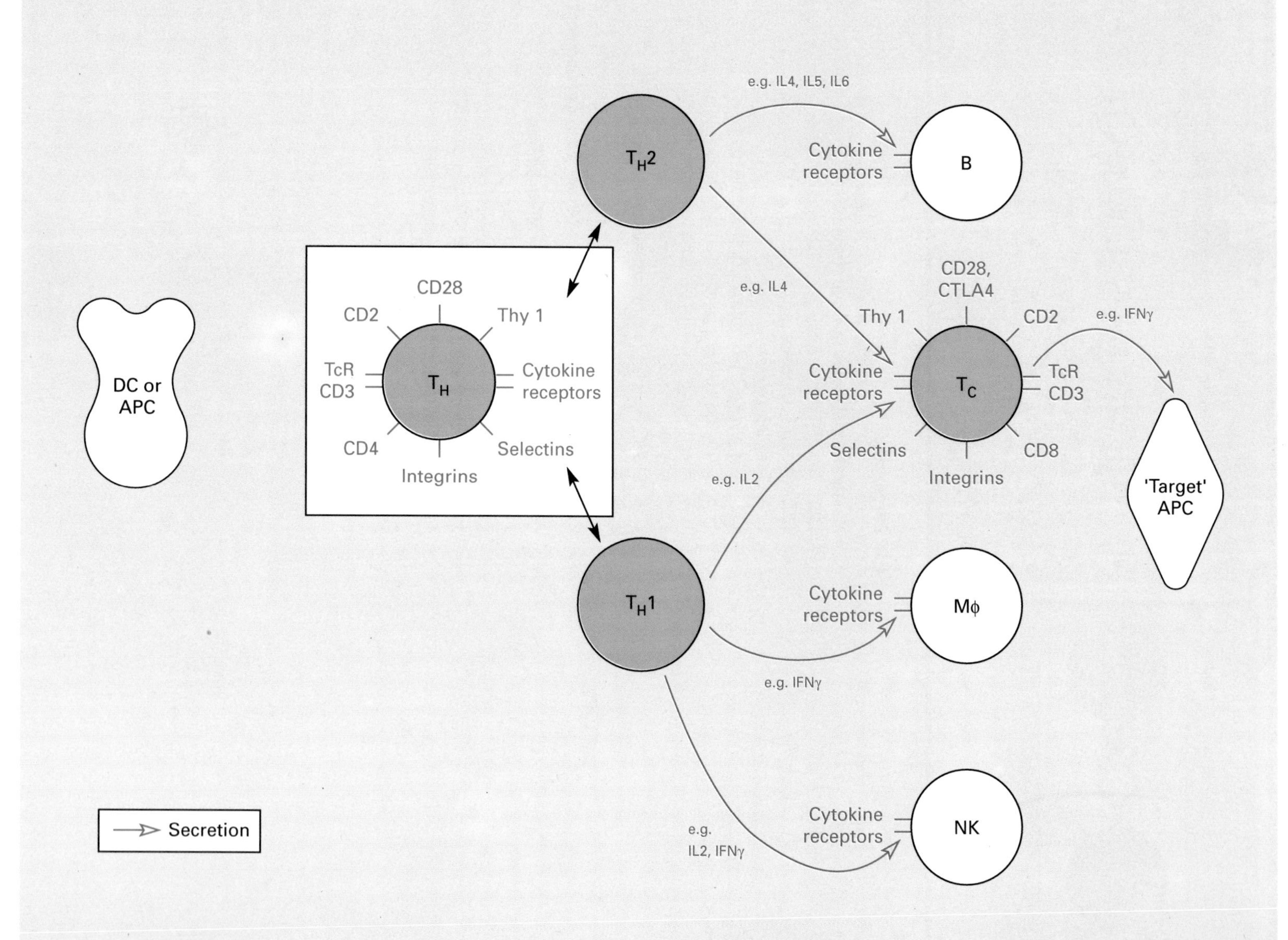

4.1 Introduction

T cell receptors recognize processed antigens in the form of peptides bound to MHC molecules (Sections 1.1.3 and 3.2.1) in contrast to antibodies which recognize native antigens. In general, T cell receptors and antibodies are highly specific for their antigenic determinants (Table 1.5). A third type of antigen recognition is seen in the case of MHC molecules which can bind a variety of different peptides at a single binding site; it has been estimated in one case that the antigen-binding groove of MHC molecules can potentially accomodate at least 10^7 different peptides, although only one binds at a time. The molecular basis for these different forms of antigen recognition is discussed in Chapter 2 for MHC molecules, and in Chapter 6 for antibodies and T cell receptors.

It is now known that MHC-restricted antigen recognition by T cells is mediated by a heterodimeric molecule called the $\alpha\beta$ T cell receptor; this is briefly considered in Section 4.2.1. However, not all T cells have this structure. A second type of heterodimer is expressed on other T cells, called the $\gamma\delta$ T cell receptor (Fig.4.1). This molecule also appears to mediate antigen recognition, but not in the 'classical' MHC-restricted fashion of $\alpha\beta$ T cell receptors. What we presently know about recognition by this molecule is discussed briefly in Section 4.2.2. T cell antigen receptors obviously play a

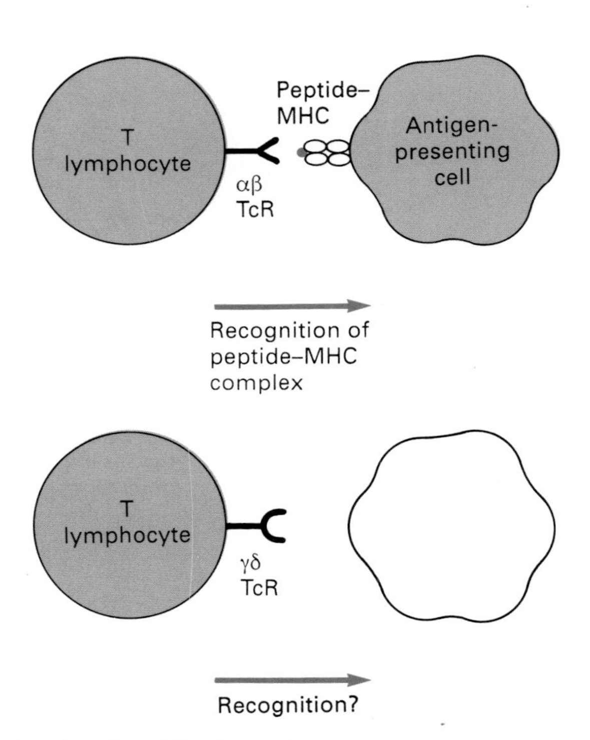

Fig.4.1 T cells bearing $\alpha\beta$ or $\gamma\delta$ T cell receptors

central role in most T cell responses, in that specific recognition of antigen is a *sine qua non* for physiological immune responses, but a large number of other membrane molecules have important roles in T cell function. These molecules fall into two types: (i) those that interact with other membrane molecules, generally on a different cell, and (ii) receptors for soluble molecules or cytokines that may be secreted by the same or a different cell (Fig.4.2). These molecules are discussed in this chapter. Table 4.1 is a partial summary of them.

It will be immediately obvious from Table 4.1 that these molecules are often designated by a **CD number**, CD standing for 'cluster of differentiation'. By definition the CD nomenclature (Fig.4.3) was originally used for a group (cluster) of antibodies directed against the same antigen, e.g. 'CD2 antibodies' but now also referred to as 'anti-CD2 antibodies'; the latter terminology will be used throughout this book for clarity. It is additionally used to describe the molecule that is recognized, e.g. 'the CD2 molecule'.

The CD nomenclature system can be used to define **homologous molecules** in different species (Fig.4.3). Homologous molecules are those whose amino acid sequence and therefore structure, and often function, has been conserved through evolution. For example, in mice, rats, and humans, CD3 is associated with the T cell receptor of all T cells, while CD4 and CD8 are expressed on reciprocal subsets of T cells. Each of these molecules has a similar structure as well as gene organization in different species, and seems

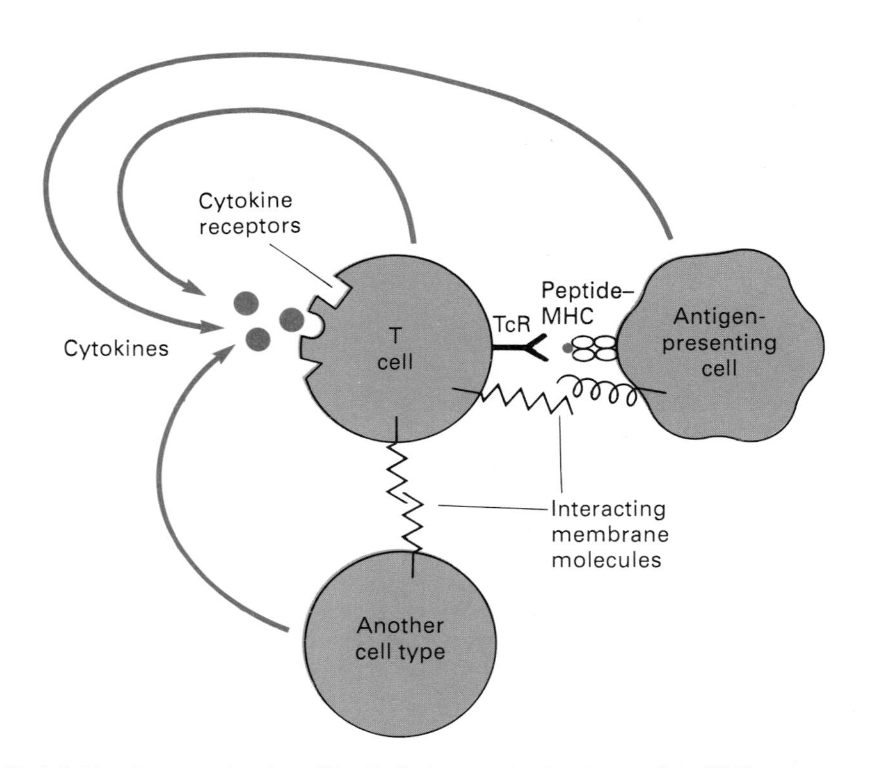

Fig.4.2 Membrane molecules of T cells that recognize foreign peptide–MHC complexes, interact with membrane molecules on other cells, or are cytokine receptors

Table 4.1 Summary of some membrane molecules on T cells involved in antigen recognition, interactions with other membrane molecules, and T cell responses

Molecule	Function	Synonyms
TcR	$\alpha\beta$: specific recognition of peptide–MHC complex	
TcR	$\gamma\delta$: recognition functions	
CD2	Interacts with CD58 (LFA3); mediates T cell activation	T11 (human)
CD3	Signal transduction from TcR via $p59^{fyn}$	T3 (human)
CD4	Binds MHC class II; mediates cell adhesion (indirectly?) and signal transduction via $p56^{lck}$	T4 (human) L3T4 (mouse)
CD5	T cell activation	T1 (human) Ly1 (mouse)
CD8	Binds MHC class I; mediates cell adhesion (indirectly?) and signal transduction via $p56^{lck}$	T8 (human) Ly2 (mouse)
CD11a/18	Binds CD54 (ICAM-1), ICAM-2, others?; mediates cell adhesion	LFA-1
CD28	Binds B7/BB1; regulates cytokine gene expression	9.3 (human)
CD45	A protein tyrosine phosphatase involved in the regulation of T cell activation	Leukocyte Common Antigen T200 (mouse)
Thy 1	T cell activation; cell adhesion?	

to play virtually the same role during antigen recognition and T cell responses. On the other hand some species differences do exist: CD4, for example, is also expressed by macrophages in rats and humans but not in mice, so some divergence of expression and perhaps function has occurred.

Much of what we initially learnt about the function of T cell molecules came from studies with specific monoclonal antibodies. When a monoclonal antibody binds to a particular molecule it can have several effects on cell functions (Fig.4.4). For example, it can block the binding of a natural ligand and thus inhibit cell function, such antibodies often being referred to as blocking antibodies; it may mimic binding of a ligand and deliver stimulatory or inhibitory signals to the cell; or it may have no effect.

Another consequence of antibody binding is that the molecule may be **modulated** (i.e. its level of expression is altered) or lost from the surface of the cell. Bivalent or multivalent antibodies such as IgG or IgM, respectively, can **cross-link** membrane molecules which may then aggregate into patches on the cell surface (**patching**) and form a 'cap' over one pole of the cell (**capping**), after which they may be shed from the cell surface or endocytosed (see Fig.4.5, and Section 3.2.2 for a discussion of endocytosis). These processes can be visualized by ultraviolet or electron microscopy if the antibody is tagged with a fluorochrome or particle such as ferritin, respectively. It will

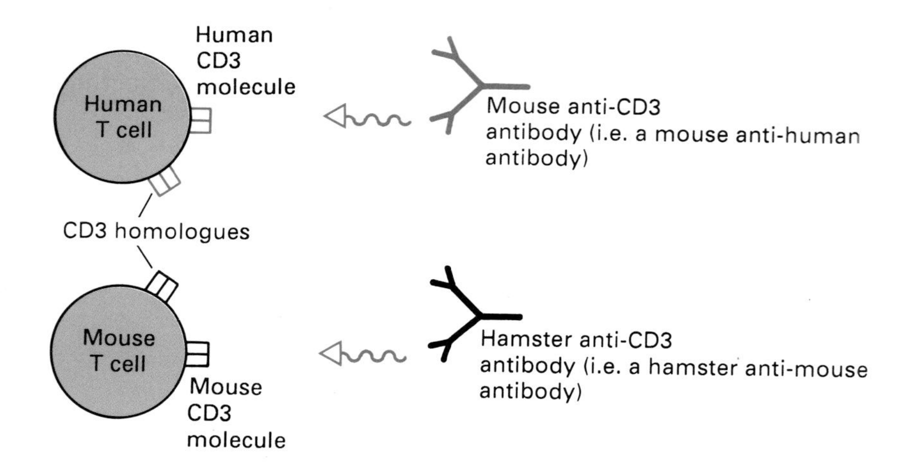

Originally the term 'anti-CD3' referred to an antibody prepared against a CD3 antibody, e.g. an anti-idiotype:

CD3 antigen — CD3 antibody — Anti-CD3 antibody

Fig.4.3 CD nomenclature (e.g. CD3)

be obvious that the loss of a molecule from the cell surface in this way can have profound effects on its associated cellular functions.

By examining the effect of specific monoclonal antibodies on cellular functions it has often been possible to deduce the function of membrane molecules. In some cases the function of the molecule was then confirmed after the gene was cloned and transfected into cells. A summary of some of the effects of monoclonal antibodies specific for molecules discussed in this section is in Table 4.2.

Table 4.2 Summary of effects on T cell responses of monoclonal antibodies against some T cell membrane molecules

TcR	Inhibition	Stimulation
CD2	Inhibition	Stimulation
CD3	Inhibition	Stimulation
CD4	Inhibition	
CD5		Stimulation
CD8	Inhibition	
CD11b/18	Inhibition	
CD28		Stimulation
CD44		Stimulation

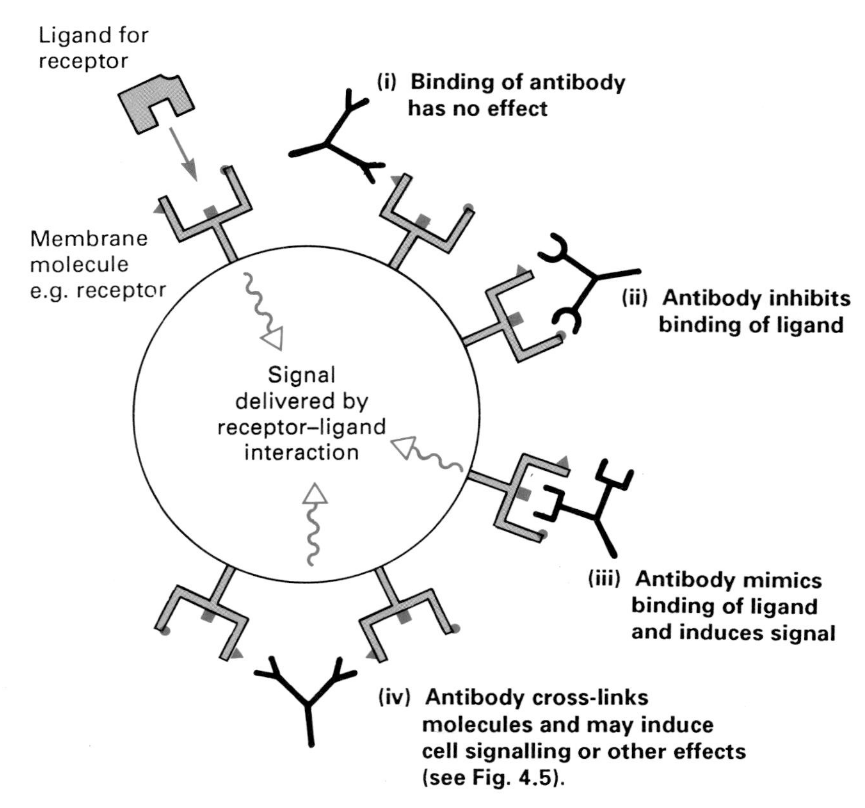

Fig.4.4 Some effects of monoclonal antibody binding to a cell membrane molecule

Overview. Many molecules expressed by T cells have been defined, and reports providing additional insights into their function, and of newly-identified molecules, appear continuously in the current literature. Clearly, it is not possible to cover all of these here. We have chosen instead to focus on about a dozen molecules expressed by T cells to give some idea of their different functions, and many of the principles that emerge will apply to other molecules as well.

We start by providing an overview of T cell antigen receptors (TcR; detailed in Chapter 6) and the CD3 complex which is associated with these molecules and is involved in delivering signals to the T cell during antigen recognition (Sections 4.2.1 to 4.2.3). The CD4 and CD8 molecules are closely involved in T cell recognition and subsequent responses by virtue of their capacity to bind to MHC molecules (Section 4.2.4). Next we consider the CD45 molecule which is expressed by all leukocytes, but which in T cells is also involved in the signalling pathways sometimes leading to T cell activation, and perhaps the function of memory T cells as well (Section 4.2.5).

Many molecules discussed in this chapter are members of the Ig gene

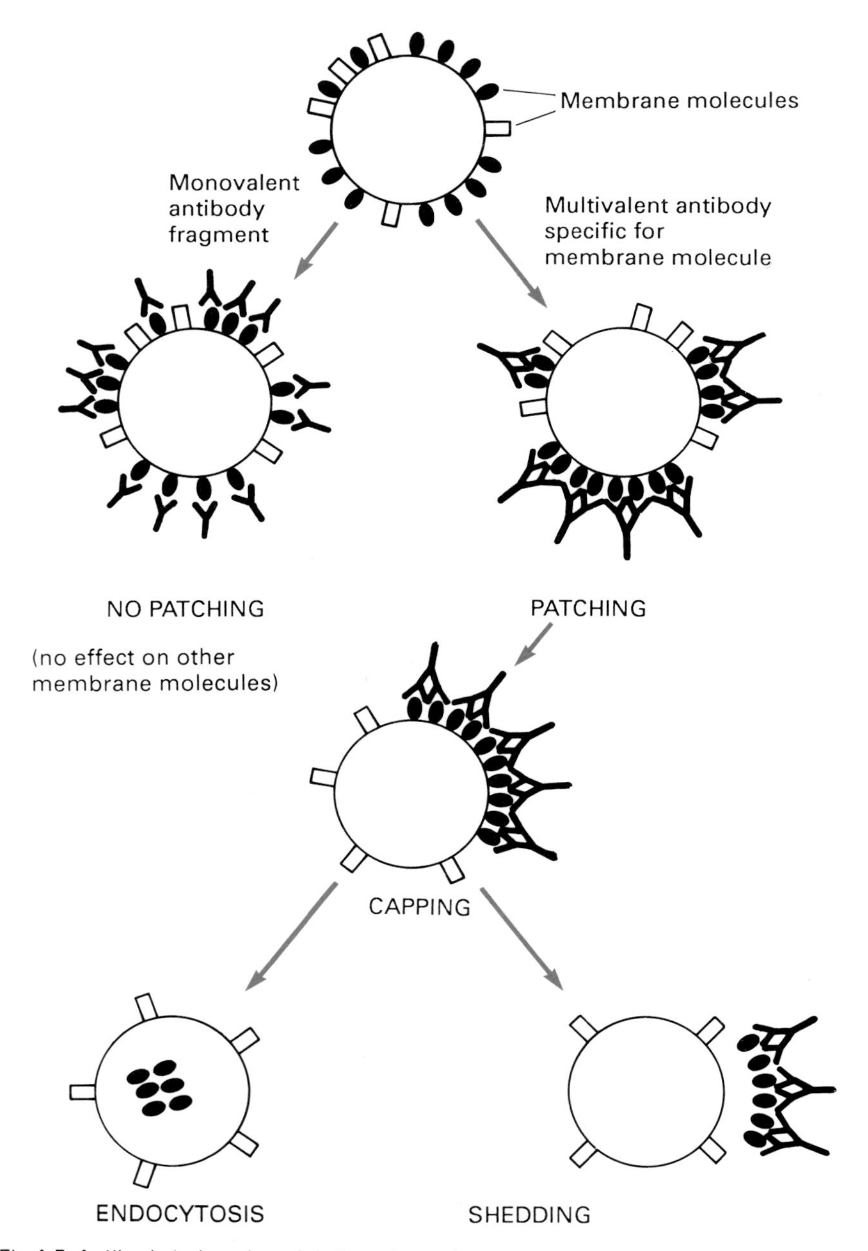

Fig.4.5 Antibody-induced modulation of membrane molecules

superfamily (Section 7.4), and two other families of molecules, the selectins and integrins, are discussed in Section 4.2.6. Some of these molecules are required for migration of leukocytes into tissues, and some are responsible for adhesion of T cells to antigen-presenting cells before antigen recognition can occur. We next consider two molecules, CD2 and CD28, whose expression is more restricted to T cells, and which are involved in signalling functions. However, these are mediated by different signalling cascades, that of CD2 (Section 4.2.7) being coupled to the TcR–CD3 complex, but CD28 (Section 4.2.8) having an alternative pathway that results in altered expression of genes for soluble molecules, the cytokines. Following an introduction to cytokines, the receptor for one of them, the IL-2 receptor, is considered in Section 4.4.3. This is an example of an **activation antigen** of T cells, being expressed in a high-affinity form by activated but not resting T cells. In addition, one of the first molecules to be defined on T cells, but whose function is still obscure, Thy-1, is discussed in Section 4.2.9.

For further information on these molecules, and those expressed by other leukocytes, the reader should follow the current literature, but may also wish to consult the Proceedings of the most recent International Workshop and Conference on Human Differentiation Antigens, entitled *Leukocyte typing*, which in 1989, for example, reported the data gathered by 525 laboratories on many of the 1100 monoclonal antibodies submitted to the workshop (Knapp *et al.* 1992—see Further reading).

4.2 Membrane molecules in T cell responses

4.2.1 The $\alpha\beta$ T cell receptor (Table 4.3)

The genes for the α and β chains of T cell receptors have been cloned, and their structure is discussed in detail in Chapter 7. Direct evidence that the $\alpha\beta$ heterodimer is necessary and sufficient for antigen–MHC recognition has been obtained from transfection experiments (Fig.4.6).

Table 4.3 T cell receptor: $\alpha\beta$

Distribution	T cells (non $\gamma\delta$)
Structure	Heterodimer (α plus β); disulfide-bonded
Approx. M_r	*Human:* $\alpha+\beta$ 90 kDa α 45–60 kDa (inc 6 N-CHO*) β 40–50 kDa (inc 2 N-CHO) *Mouse:* $\alpha+\beta$ 90 kDa α 45–55 kDa (inc 4 N-CHO) β 40–55 kDa (inc 2 N-CHO)
Homology to Ig	Yes: 2 extracellular Ig-like domains
Chromosome localization of genes**	*Human* α 14q11 β 7q32-35 *Mouse* α 14 C-D β 6B
Comments	Associated with p59^{fyn} protein tyrosine kinase

* N-CHO = N-linked oligosaccharide
** For comparison, localization of Ig genes:
in human—H 14, κ 2, λ 22
in mouse—H 16, κ 6, λ 12
(note linkages, underlined)

T cell
clone 1
TcR1
α1
β1
ag X +
MHC C

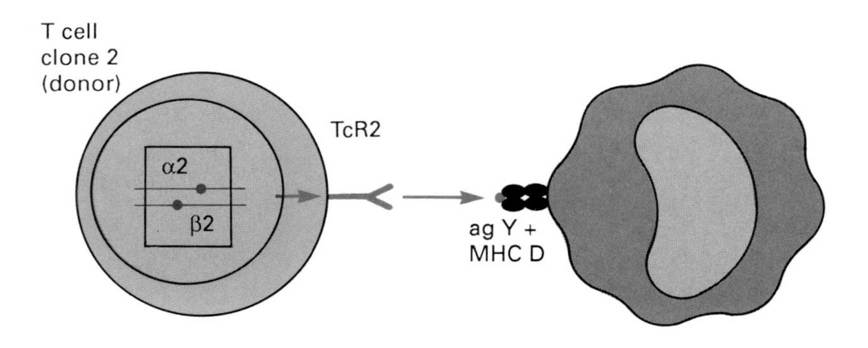

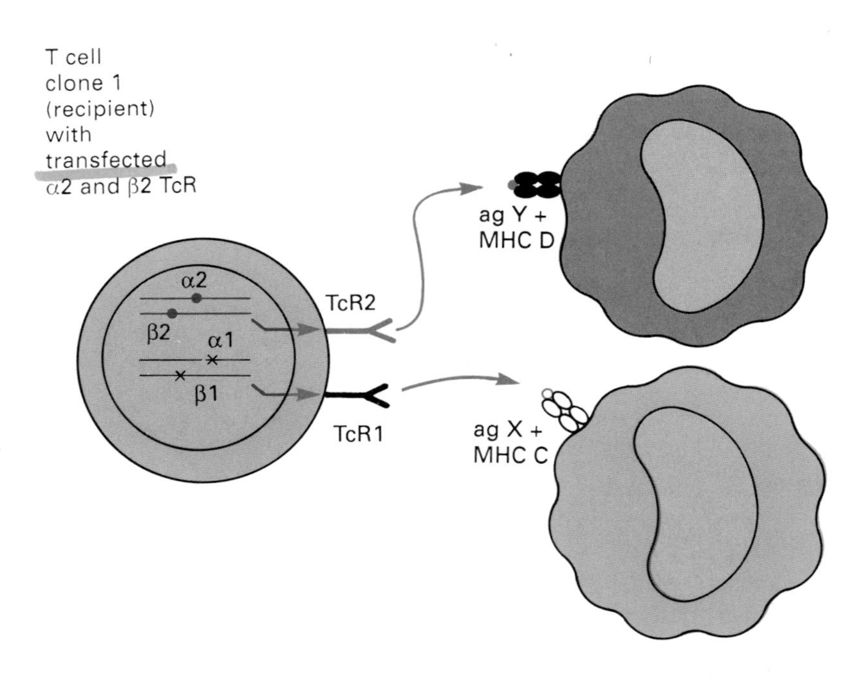

Fig.4.6 Transfection of TcR α and β chain genes into a T cell clone endows the recipient cell with the ability to recognize the peptide–MHC combination recognized by the donor (principle of Dembic *et al.* and Saito *et al.*)

For example, when the α and β chain genes from a cytotoxic T cell clone that was specific for a hapten plus H-2D^d were transfected into a T cell clone of a different specificity, they endowed the recipient cell with the ability to recognize the new hapten–MHC combination (Fig.4.6) (Dembic *et al.* 1986). Similarly, transfection of the α and β chain genes from a mouse helper cell clone into a human T cell enabled the transfectant to respond to the same peptide–MHC class II complex as the original mouse clone (Saito *et al.* 1987).

Before these genes were cloned, monoclonal antibodies had been produced that were specific for T cell receptors on particular T cell clones. These antibodies were originally called **clonotypic antibodies** because they defined structures that were 'typical' of a particular T cell clone, this distribution being expected for a molecule that could mediate antigen-specific recognition (Fig.4.7) (Meuer *et al.* 1983). Clonotypic antibodies were used for the early characterization of T cell receptors. Using them, it was possible to immunoprecipitate a heterodimer from the respective T cell clone (see Appendix), and comparison of the molecules from different clones revealed the presence of variable and constant regions within each polypeptide chain, the detailed structure of which is discussed in Section 6.3.1 (Fig.4.8). These clonotypic antibodies had profound effects on T cell antigen responses. On one hand, the response of the relevant T cell clone to antigen and accessory cells could be inhibited by some antibodies. On the other, some of these antibodies could stimulate T cell responses in the absence of antigen when they were coupled to an immobile support such as Sepharose beads. Presumably in the latter case the antibodies cross-linked the T cell receptors, thus mimicking antigen recognition, and this was sufficient to induce a response.

Experiments with clonotypic antibodies also provided evidence that the $\alpha\beta$ heterodimer was responsible for antigen–MHC recognition. For example, in

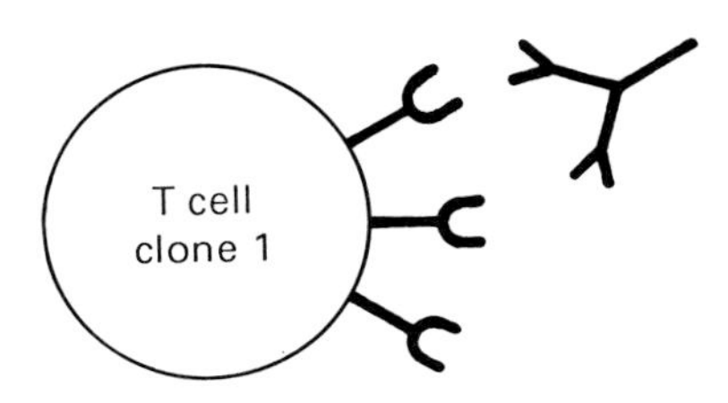

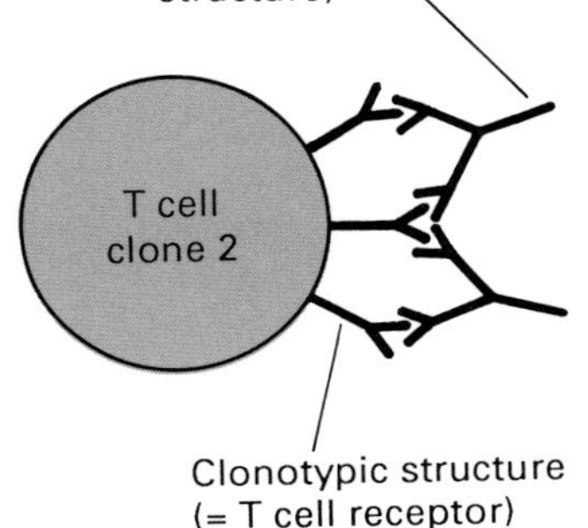

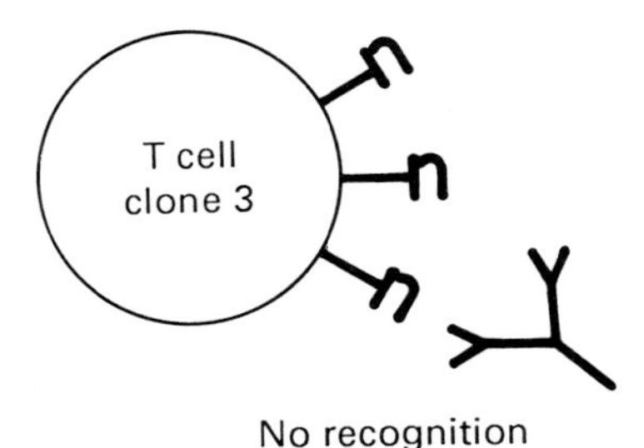

Fig.4.7 Clonotypic structures (T cell receptors) identified by monoclonal antibodies

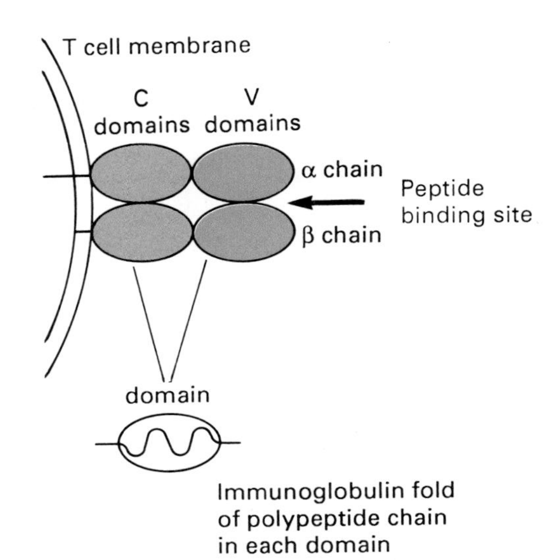

Fig.4.8 Outline structure of α and β chains of the T cell receptor

one study an antibody against a particular T cell clone with a certain antigen–MHC specificity was used to screen a large number of independently-derived T cell clones. The few clones that reacted with the antibody, and only those cells, were found to recognize the same antigen–MHC combination as the original clone (Yague *et al.* 1985). Moreover, they were all found to have the same **'fine-specificity'**, in that both they and the original clone showed the same pattern of alloreactivity (Sections 1.1.4 and 5.2.2), recognizing certain alleles of allogeneic MHC molecules but not others. Therefore these experiments also demonstrated that the $\alpha\beta$ heterodimer mediates self-restricted antigen recognition as well as alloreactivity, phenomena which are thus due to **cross-reactive recognition** by the same T cell receptor (see Section 5.2.2).

Experiments with clonotypic antibodies, and later with $\alpha\beta$ TcR transfectants, resolved a long-standing controversy as to how T cells were able to recognize an antigen and MHC molecule at the same time. Although some models initially proposed were based on the idea that one chain of the TcR might recognize an antigen while the other recognized an MHC molecule, there was earlier evidence that, even if this hypothesis was correct, recognition by the two molecules could not be independently reassorted. For example, when a T cell specific for an antigen X plus MHC C† (X+C) was fused to another specific for antigen Y plus MHC D (Y+D), the resultant hybridoma recognized only these combinations and was unable to respond to X+D or Y+C (Fig.4.9) (Kappler *et al.* 1981). Similar findings were made when liposomes containing T cell receptors were fused to a clone of different specificity. Furthermore, some T cell clones have been derived that are both self-restricted and allorestricted‡ in that they can recognize X+C as well as Y+D; these, too, did *not* respond to the other combinations (X+D, Y+C).

Of course, what in retrospect these experiments really showed was that neither the α nor the β chains of the T cell receptor can *independently* recognize either the antigen alone or the MHC molecule alone. It is possible that a cell expressing two different heterodimers, say $\alpha\beta$ and $\alpha'\beta'$ (specific for X+C and Y+D), can also express $\alpha\beta'$ and $\alpha'\beta$ dimers. However, because the precise antigen specificity is (at least in most cases) determined by the particular combination of the two chains, the $\alpha\beta'$ dimer for example may recognize a completely different antigen–MHC combination, such as Z+E.

4.2.2 The $\gamma\delta$ T cell receptor (Table 4.4)

The first T cell receptor gene to be cloned encoded the β chain (see Chapter 7 for references). In the ensuing search for an α chain gene, which was

† MHC alleles will be designated as C, D, E..., and antigenic peptides as ..., X, Y, Z; this obviates problems otherwise encountered in phrases such as 'A B cell of strain C...' etc.

‡ The term allorestriction refers to recognition of a foreign peptide bound to a non-self MHC molecule, e.g. a T cell from strain C that recognizes Y + D. It contrasts with the term alloreactivity that was originally used to describe recognition of a non-self MHC molecule alone, e.g. E. Detailed discussions of these concepts are in Sections 5.2.4 and 5.2.2 respectively.

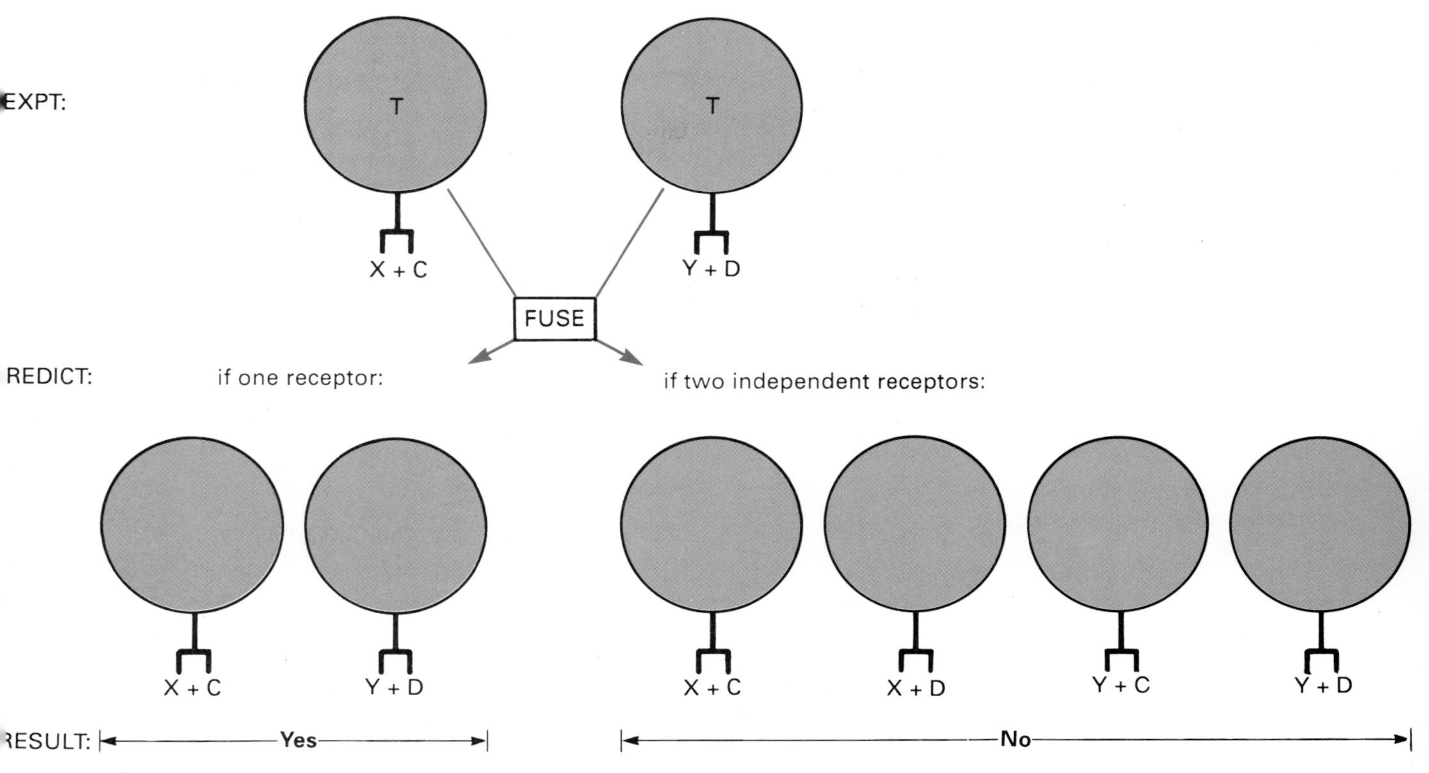

Fig.4.9 Some evidence that the T cell receptor behaves as a single receptor for the combination of antigen plus MHC

Table 4.4 T cell receptor: $\gamma\delta$

Distribution	T cells (non $\alpha\beta$)		
Structure	Heterodimer (γ plus δ); disulfide-bonded (plus a non-disulfide-bonded form in mice)		
Approx. M_r	*Human:* $\gamma+\delta$ 90 kDa γ 45–60 kDa (inc 4 N-CHO*) δ 40–60 kDa (inc 2 N-CHO) *Mouse:* $\gamma+\delta$ 90 kDa γ 45–60 kDa (inc 3 N-CHO) δ 40–60 kDa (inc 2 N-CHO)		
Homology to Ig	Yes: 2 extracellular Ig-like domains		
Chromosome localization of genes**	*Human* *Mouse*	γ 7p14 γ 13A2-3	δ 14q11 δ 14AC+D

* N-CHO = N-linked oligosaccharide
** For comparison, localization of Ig genes:
in human—H <u>14</u>, κ 2, λ 22
in mouse—H <u>16</u>, κ 6, λ 12
(note linkage, underlined)

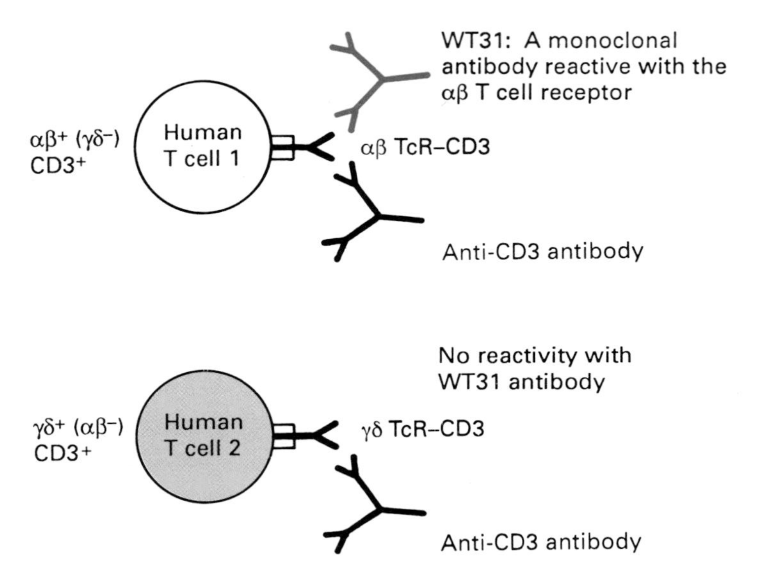

Fig.4.10 $\gamma\delta$ T cell receptors are expressed on $CD3^+$ T cells lacking $\alpha\beta$ T cell receptors

assumed to exist because the T cell receptor was by this time known to be a heterodimer of *two* glycoprotein molecules, a gene was found that came to be known as the gamma (γ) chain gene. This gene was homologous to the β gene, but it could not be the α chain gene because it encoded a molecule with no glycosylation sites. Needless to say, by the time the α chain gene was eventually cloned there was much confusion, because now there were three T cell receptor molecules whereas only two were expected, as well as some merriment because it was now possible to invent even more models for antigen–MHC recognition by T cells. However, it soon became apparent that, like the $\alpha\beta$ receptor, the γ chain was associated with another molecule, which is now termed the δ chain, to form a heterodimer.

By the time the γ chain gene was cloned, a monoclonal antibody, WT31, had been produced against the $\alpha\beta$ receptor on human T cells (Fig.4.10). The $\alpha\beta$ receptor was invariably associated on the T cell surface with CD3, a multi-chain complex to which a variety of antibodies had also been produced (Section 4.2.3). However, some T cells were found that expressed CD3 but which did not react with WT31, and it was therefore thought they might express a different T cell receptor molecule in association with CD3. T cells of this type were initially obtained from fetal umbilical cord blood and from the peripheral blood of patients with immunodeficiency diseases. These $\alpha\beta$ TcR-negative cells were treated with a cross-linking agent that had been shown previously to chemically link CD3 to the T cell receptors, thus permitting the T cell receptor chains to be immunoprecipitated by monoclonal antibodies prepared against a synthetic peptide corresponding to part of the predicted γ-chain sequence. This showed that the γ chain was in fact expressed on the cell surface together with CD3. Closer examination of $CD3^+$ $WT31^-$ T cells revealed that the γ chain was associated with another molecule which is now known as the delta (δ) chain. The gene for this chain was subsequently identified within the TcRα locus and because it and the γ chain genes are homologous to the α and β chain genes, as discussed in Section 6.3.1, and are expressed on T cells, it seems reasonable to assume the $\gamma\delta$ chains encode a T cell antigen receptor structure.

Though first detected in fetal blood and disease states, it is now clear that the $\gamma\delta$ receptor is expressed on mature T cells and thymocytes of normal adult animals. The $\alpha\beta$ and $\gamma\delta$ heterodimers are not expressed on the same T cell, and appear to define separate lineages, although $\gamma\delta$-bearing cells are the smaller subset. The vast majority of peripheral T cells that express the $\alpha\beta$ receptor ($\alpha\beta^+$ T cells) also express either CD4 or CD8 molecules (Section 4.2.4). In the thymus many of these developing $\alpha\beta^+$ T cells express both CD4 and CD8, and in this sense they are said to be 'double-positive'. In contrast, many $\gamma\delta$-bearing ($\gamma\delta^+$) peripheral T cells are 'double-negative' in that they express neither CD4 nor CD8, although there is a subpopulation, particularly in mouse gut epithelium, that expresses CD8. In mice, this molecule can be expressed as a homodimer or a heterodimer (Section 4.2.7).

The population of $\gamma\delta^+$ T cells expressing CD8 homodimers is unusual in that it does not require a thymus for development (e.g. it is present in nude mice) and it has been suggested that the gut epithelium may attract precursors of this lineage and promote their differentiation, perhaps playing a role similar to that of thymic epithelium (Section 5.3) (Guy-Grand *et al.* 1991).

In mice, T cells expressing $\alpha\beta$ and $\gamma\delta$ receptors have roughly reciprocal tissue distributions, but this is not the case in humans. In adult mice $\alpha\beta^+$ T cells are predominant in the blood and lymphoid tissues. In contrast, $\gamma\delta^+$ T cells are rare in peripheral lymphoid tissues but predominate in epithelial sites such as skin and gut (Fig.4.11). The distribution of $\gamma\delta^+$ T cells in

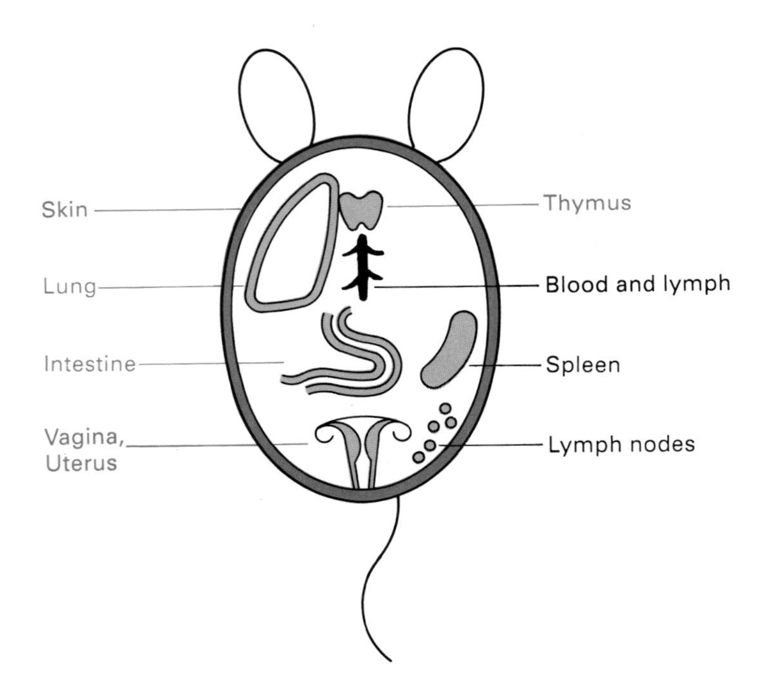

Fig.4.11 Predominant distributions of $\gamma\delta^+$ T cells (red) and $\alpha\beta^+$ T cells (black) in the adult mouse

humans, however, is more similar to that of the $\alpha\beta^+$ cells. For example, $\gamma\delta^+$ T cells are found together with $\alpha\beta^+$ T cells in lymphoid tissues, as revealed by staining with specific monoclonal antibodies, although $\gamma\delta^+$ T cells are by far the smaller population (Groh *et al.* 1989). Moreover, epidermal T cells in mice are $\gamma\delta^+$, whereas in humans these cells predominantly belong to the $\alpha\beta$ lineage.

So far, perhaps the major function described for T cells expressing $\gamma\delta$ receptors is cellular cytotoxicity, but unlike killing by 'conventional' $\alpha\beta^+$ cytotoxic T cells this function does not seem to be restricted by 'classical' MHC molecules since it is not inhibited by anti-MHC antibodies. In this respect $\gamma\delta^+$ T cells are similar to natural killer cells, which can kill a variety of different targets (Section 10.2.2), and they may be capable of antibody-dependent cellular cytotoxicity (Section 10.2.7).

However, much information on the function of $\gamma\delta$-bearing cells has come from studies of T cell clones or lines that were cultured in interleukin-2. Under these conditions $\alpha\beta^+$ T cells can also develop the ability to kill targets in an *MHC-unrestricted* manner so it is not clear if the activity of $\gamma\delta^+$ T cells is the same *in vivo*. For example, a $\gamma\delta^+$ T cell line was derived and shown to be initially alloreactive, but after culture in interleukin-2 it developed non-specific lytic activity and was able to kill target cells even of its own MHC type (see also Section 10.2.5).

It is conceivable that the $\gamma\delta$ receptor recognizes antigens bound to 'non-classical' MHC molecules, e.g. Qa and Tla in the mouse (Fig.4.12) (Vidovic *et al.* 1989; He *et al.* 1991). It is also possible that a molecule called CDl, which is expressed at a certain stage of T cell development in the thymus (Section 4.3) and by Langerhans cells in the skin, might be able to act as a

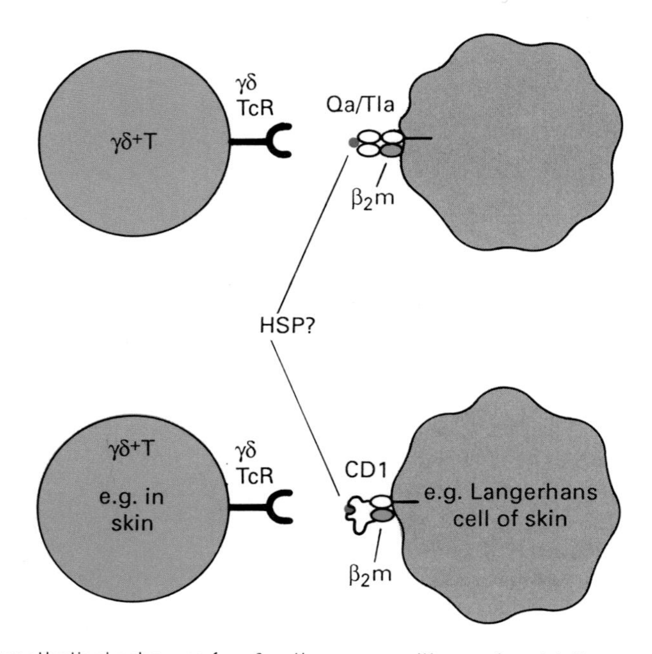

Fig.4.12 Hypothetical schemes for $\gamma\delta$ antigen recognition and restriction

restricting element (i.e. function like an MHC molecule) and another target molecule for $\gamma\delta^+$ T cells called TCT.1 is encoded in the same genetic region (Del Porto *et al.* 1991). The possibility that the $\gamma\delta$ receptor might recognize heat shock proteins is discussed in Section 6.3.2 (Born *et al.* 1990). Undoubtedly, a better understanding of the function of the $\gamma\delta$ receptor will come from transfection experiments similar to those already performed with the α and β chain genes or from the production of $\gamma\delta$ transgenic mice (see Appendix).

4.2.3 The CD3 complex (Table 4.5)

Relation of CD3 to the T cell receptor

CD3 was first discovered on human T cells, where it was termed T3. CD3 is in fact a complex of several closely-associated molecules. These components are believed to 'transduce' a signal from the T cell receptor when it recognizes the appropriate peptide–MHC complex into the interior of the cell, thus contributing to T cell activation. Antibodies against CD3 have been found to inhibit or stimulate T cell responses in a similar manner to the clonotypic antibodies produced against the $\alpha\beta$ T cell receptor (Section 4.2.1). Thus CD3 antibodies can block or induce T cell proliferation and IL-2 secretion. How-

Table 4.5 CD3

Distribution	T cells
Structure	Multichain complex of γ, δ and ε chains, also associated with ω during biosynthesis and with the signal transduction complex of ζ, η (the latter = P21 in mouse) at the cell surface
Approx. M_r	*Human:* γ 25–28 kDa (inc 2 N-CHO*) δ 20 kDa (inc 2 N-CHO) ε 20 kDa (ζ 16 kDa, ω 28 kDa) *Mouse:* γ 21 kDa (inc 1 N-CHO) δ 28 kDa (inc 3 N-CHO) ε 25 kDa (inc O-CHO?**) ζ 16 kDa (ω 28 kDa, P21 21 kDa)
Homology to Ig	Yes; γ, δ, ε have 1 extracellular Ig-like domain
Chromosome localization of genes	*Human* γ, δ, ε 11q23 *Mouse* γ, δ, ε 9
Comment	ζ and η not considered as part of CD3 complex *per se*

* N-CHO = N-linked oligosaccharide
** O-CHO = O-linked oligosaccharide

ever, while clonotypic antibodies act only on specific T cell clones, CD3 antibodies affect the function of all T cells.

CD3 proved to be an invariant structure on all T cells that was tightly associated with the T cell antigen receptor; this complex is referred to as the TcR-CD3 complex. Thus CD3 and the $\alpha\beta$ receptor could be co-modulated from the surface of T cells, antibodies against one component were able to immunoprecipitate the other, and CD3 and the $\alpha\beta$ chains could be chemically cross-linked, as could CD3 and the $\gamma\delta$ T cell receptor chains. It was also found that some cell lines that had lost the ability to express the $\alpha\beta$ heterodimer on their surface no longer expressed membrane CD3.

This was extensively studied in mutants derived from a human T cell leukaemia cell line called Jurkat which normally expresses both the $\alpha\beta$ receptor and CD3. Some mutant lines were unable to transcribe the β chain of the T cell receptor, failed to express CD3 or the T cell receptor on their surface, and were therefore unable to secrete IL-2 in response to antibodies against these molecules. However, when the Jurkat mutants were transfected with a functional β chain gene, a functional $\alpha\beta$ receptor was expressed on the cell surface in association with CD3, and the cells regained their ability to respond to the antibodies (Ohashi *et al.* 1985). This shows that surface expression of one molecule depends on the other.

It may seem paradoxical that a particular antibody against the CD3 complex can both stimulate and inhibit T cell responses (see above). Although we do not have precise molecular explanations for this, *inhibition* of T cell function generally seems to occur when T cells are treated with *soluble* anti-CD3 antibodies which probably induces capping of CD3 molecules (Fig.4.5), causing them to be shed from the surface or to be internalized by the cell. Since the T cell receptor is co-modulated with CD3, it is not surprising that the T cell is now unable to respond to its specific peptide–MHC complex. In contrast, *stimulation* frequently occurs when T cells are cultured with anti-CD3 antibodies that have been *immobilized* on a solid support, for example if they are coupled to Sepharose beads, or bound to Fc receptors on macrophages. This may induce cross-linking of TcR-CD3 complexes (see below) and the delivery of intracellular signals.

Structure of CD3

CD3 comprises three non-covalently linked polypeptide chains on the T cell surface, designated γ, δ (not to be confused with the TcR γ and δ chains), and ε (Table 4.4) (Minami *et al.* 1987). These molecules are structurally related, and their genes are closely-linked and considered to be members of the Ig gene superfamily (Fig.7.4). They are associated with two other molecules, zeta (ζ) and eta (η), which can exist both as disulfide-linked $\zeta\zeta$ homodimers and as $\zeta\eta$ (as well as $\zeta\gamma$) heterodimers; approximately 90% of ζ chains are homodimeric, the remainder forming heterodimers. These molecules are involved in delivery of signals to the T cell, during antigen recognition via the T cell receptor, which are required for T cell activation (Irving and Weiss 1991). Within the T cell cytoplasm, an additional chain, omega (ω), can be associated with different combinations of the CD3 γ, δ, and ε chains and the TcR α and β chains, during assembly of the TcR–CD3 complex.

The ζ and η polypeptides are produced by differential transcription of the same gene (Jin *et al.* 1990). Although they were originally thought to be components of the CD3 complex, these molecules are now considered to form independent subunits on the basis that they are structurally unrelated to the three CD3 chains (above), have a separate chromosomal location, and the ζ chain can also associate with other signal-transducing receptors such as $Fc_{\gamma}RIII$ receptors (Section 10.4.5). The ζ and η polypeptides contain a conserved amino acid motif within their cytoplasmic domains that is also present in the γ chain of $Fc_{\gamma}RIII$ and $Fc_{\varepsilon}RI$ (Section 10.4.5), and the MB-1 (IgM) chain associated with IgM on B cells (Section 8.4.2.). This may be a binding site for cytosolic proteins involved in the generation of signals after receptor cross-linking (see below and Section 4.2.5).

The stoichiometry of the TcR–CD3 complex has been investigated in transgenic mice expressing human ε chain genes (de la Hera *et al.* 1991). This has

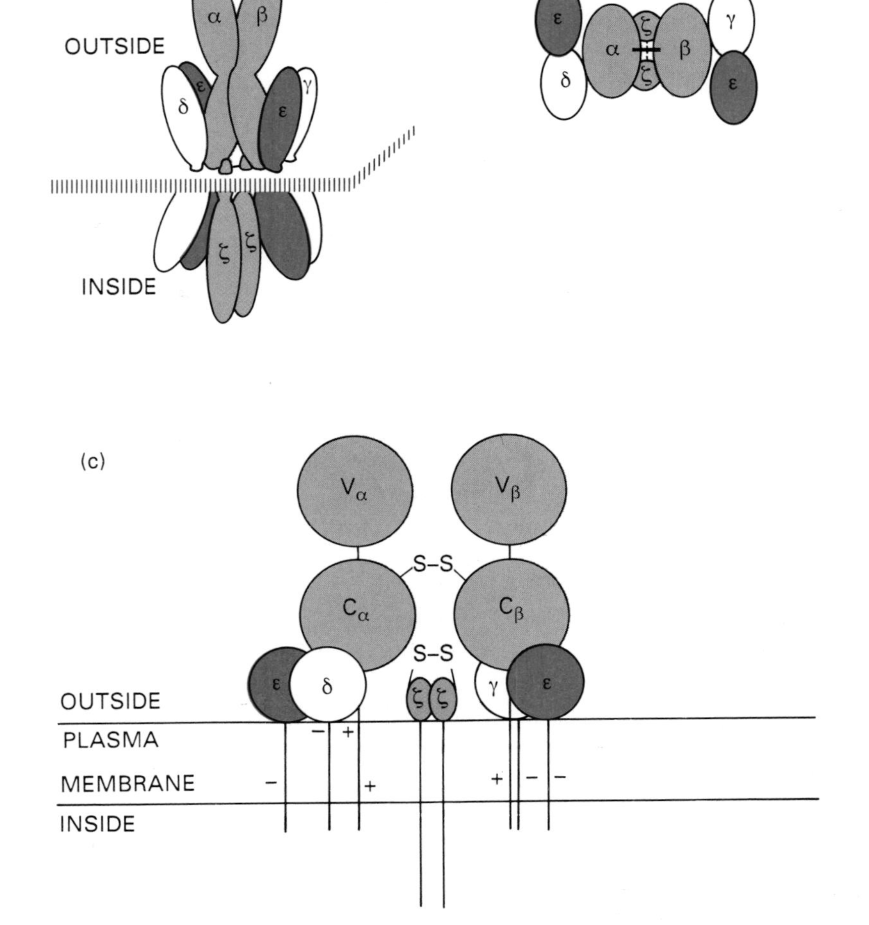

Fig.4.13 Proposed minimal structure of the $\alpha\beta$ TcR–CD3 complex. Stoichiometry and associations of TcR and CD3 subunits in (a) cross-section through the membrane and (b) plan view from outside the T cell. From de la Hera *et al.* (1991) by copyright permission of the Rockefeller University Press. (c) Alternative representation, indicating Ig-like domains (grey circles; see also Fig.6.1), to show interactions between subunits through extracellular disulfide bonding (S–S) and hypothetical ion bridges (⊕ ⊖) within the plane of the membrane. The eta (η) chain, which can form heterodimers with the zeta (ξ) chain, is not illustrated.

led to a minimal model for the complex in which the α and β T cell receptor molecules are associated respectively with a $\delta\varepsilon$ and a $\gamma\varepsilon$ heterodimer (Fig.4.13). This is of interest because most anti-CD3 antibodies produced to date are specific for the ε chain, suggesting that binding of these antibodies can potentially cross-link two ε chains within the same TcR–CD3 complex and/or lead to the formation of strings of TcR–CD3 complexes on the cell membrane, and mediate intracellular signalling. In addition, functional TcR-CD3 complexes contain ζ homodimers and/or heterodimers.

Association of CD3 with the antigen receptor

The CD3 γ, δ, and ε molecules each contain a single negatively-charged amino acid in their otherwise hydrophobic transmembrane regions, glutamic acid in the γ and aspartic acid in the δ and ε chains (Fig.4.13). It seems likely that these residues permit them to interact with the T cell receptor within the plane of membrane, since all four T cell receptor chains (α, β, γ, δ) contain a positively-charged lysine residue in the corresponding position, and the TcR α and δ chains also contain an arginine residue. Ion pairs formed between CD3 and the T cell receptor may thus facilitate a signal to be transduced from the TcR via CD3 and/or the $\zeta\zeta$ and $\zeta\eta$ subunits during peptide–MHC recognition. The cytoplasmic portions of the CD3 γ, δ, and ε chains are considerably larger than those of the T cell receptor chains, and are likely to interact with cytoskeletal or other cytoplasmic components.

Intracellular events during T cell activation

When T cells are activated, a complex series of intracellular biochemical events is initiated, which also occur during B cell activation as well as in responses of other cell types; these events are outlined here.

The phosphatidylinositol pathway. When lymphocytes are activated there is an increased turnover of lipids in the cell membrane (Fig.4.14). It is thought that ligation of antigen receptors results in activation of a GTP-binding protein (or G-protein) that in turn activates the enzyme phospholipase c. This results in the conversion of phosphatidylinositol (4,5)-biphosphate in the membrane to inositol (1,4,5)-trisphosphate (IP_3) and diacylglyerol (DAG) both of which are critical for the events that follow. First, IP_3 diffuses from the membrane into the endoplasmic reticulum, and as a result calcium ions (Ca^{2+}) are released, leading to an increase in the cytoplasmic concentration of these ions. Second, free Ca^{2+} ions together with DAG are thought to activate the enzyme protein kinase C (PKC) which moves from the cytoplasm and associates with the plasma membrane. Here PKC is likely to phosphorylate a number of different membrane molecules (see below), including ion transport systems. The consequent changes in ionic concentrations contribute, in part, to the subsequent changes in gene expression that are ultimately manifested as lymphocyte activation. T cell activation can sometimes be induced by chemicals that bypass signalling via membrane molecules such as the TcR-CD3 complex, and (i) stimulate an increase in concentration of Ca^{2+} in the cytoplasm and (ii) activate PKC. These, respectively, include calcium ionophores (e.g. A23187 and ionomycin) and phorbol esters (e.g. phorbol myristate acetate, PMA). It is important to note that an increase in Ca^{2+} concentration alone is *not* sufficient to lead to a T cell response.

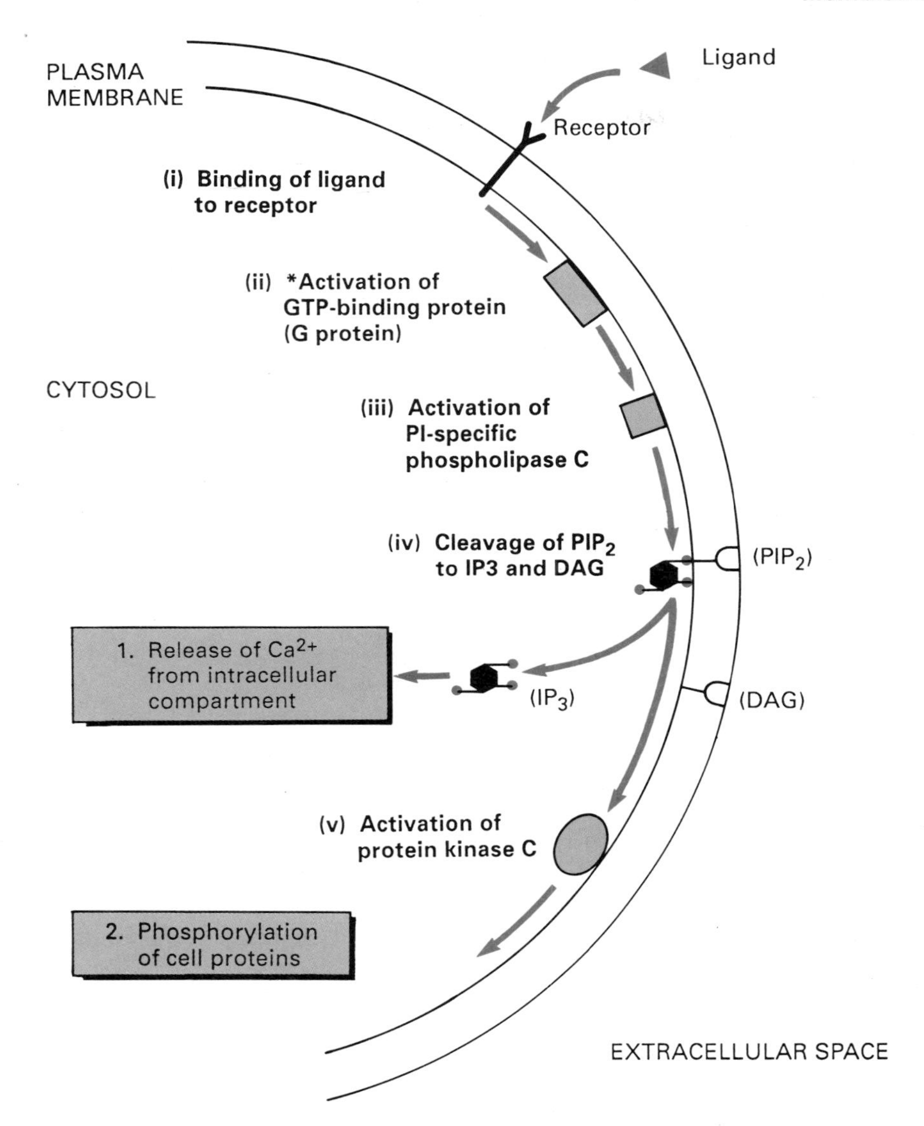

*Whether or not a G protein is involved in T cell activation is presently controversial

Fig.4.14 Outline of the phosphatidylinositol pathway

Phosphorylation of membrane molecules. Phosphorylation of membrane molecules plays an important role in the regulation of cellular responses and is seen during T cell activation. Phosphorylation of the cytoplasmic tails of these molecules occurs on serine or tyrosine residues, and is mediated by enzymes known as **protein kinases**. Some membrane molecules have an intrinsic tyrosine protein kinase activity. In addition, mammalian cells contain families of genes encoding a variety of protein kinases, some of which are expressed in a cell-type specific manner. These molecules can become associated with the intracellular portions of membrane molecules, resulting in their phosphorylation, and protein kinases themselves can be regulated in a similar manner, sometimes by autophosphorylation.

One family of tyrosine protein kinases is called Src. In mammalian cells the genes encoding these molecules are called **proto-oncogenes** because they are also used as transforming (i.e. tumour-inducing) proteins by certain oncogenic retroviruses where they are known as **oncogenes**. During evolution these viruses acquired the genes from their host cells and integrated them into their own genomes. The Src family is so called because these genes are related to the transforming gene of the Rous sarcoma virus which is designated v-src; its cellular counterpart is called c-src. Src-related tyrosine protein kinases expressed by T cells include **lck** and **fyn**, the former being associated with CD4 and CD8 molecules (Section 4.2.4) and the latter with the T cell receptor; and those expressed in B cells include **lyn**, which is asociated with membrane Ig, and **blk** (section 8.4.2). These molecules are expressed particularly by lymphocytes, lck and lyn being cell-specific.

Enzymes causing dephosphorylation, the removal of phosphate groups, are known as **phosphatases**. Less is known about these molecules than the kinases, but one particular membrane molecule, CD45, has intrinsic protein tyrosine phosphatase activity. Regulation of T cell responses by phosphorylation and dephosphorylation, in relation to the TcR-CD3 complex, CD4, CD8, and CD45 is discussed in Section 4.2.5.

Calcium ions and CD3

Binding of monoclonal antibodies to CD3 on T cells brings about a rapid increase in the cytoplasmic concentration of free Ca^{2+} ions. This can also be induced by antibodies against $\alpha\beta$ T cell receptors, perhaps mimicking their recognition of specific peptide–MHC complexes and triggering signal transduction via CD3. The initial rise in Ca^{2+} concentration seems to be due to mobilization from intracellular stores, although these and other ions later enter the cell through specialized membrane channels (Section 3.2.2). At one time it was thought that the ε chain of CD3 might form part of an ion channel because it contains a large hydrophobic transmembrane domain, but no proof of this has been obtained. The influx of Ca^{2+} is accompanied by an efflux of K^+, resulting in hyperpolarization (a reversal of charge) of the cell membrane which in turn may regulate other ionic fluxes (Kuno and Gardner 1987).

Phosphorylation of CD3

When T cells are activated, some components of the CD3 complex are phosphorylated on their intracellular portions by protein kinases. When mouse or human T cells recognize their specific peptide–MHC complexes, or are stimulated with lectins or combinations of calcium and PMA, the γ chain becomes phosphorylated on one or two serine residues; phosphorylation of the mouse ε chain can also occur. It has been shown that phosphorylated γ chains are endocytosed by the cell, whereas non-phosphorylated chains are not. Thus γ chain phosphorylation may control endocytosis and re-cycling of the entire T cell receptor–CD3 complex (Krangel 1987) (see Section 3.2.2 for a discussion of endocytosis and recyling).

4.2.4 CD4 and CD8: interaction with MHC molecules (Tables 4.6 and 4.7)

T cell subsets

T cells were first divided functionally into helper and cytotoxic subsets. Some T cells were found to be required for antibody responses of B cells in culture and for the development of delayed-type hypersensitivity *in vivo*, while other T cells could kill targets such as virus-infected cells (see pages 259 and 291). Attempts were made to produce antisera against T cell membrane molecules that could be used as markers for each subset, partly in the hope that one might be able to predict the function of a T cell from its phenotype, but also to separate T cells with different functions.

Some of the first antisera against mouse T cells defined membrane molecules that were called Lyt-1, Lyt-2, and Lyt-3, later abbreviated to Ly1, Ly2, and Ly3. These appeared to be expressed on reciprocal subsets of peripheral T cells. It was initially thought that Ly1 was expressed on T cells with helper activity while Ly2 and Ly3 were both expressed by cytotoxic and suppressor T cells. When T cells were treated with the appropriate antisera and complement to kill one subset of T cells, the respective functions were diminished. It also proved possible to inhibit cytotoxic T cells simply by adding antisera against the Ly2 molecule as a source of 'blocking antibodies'.

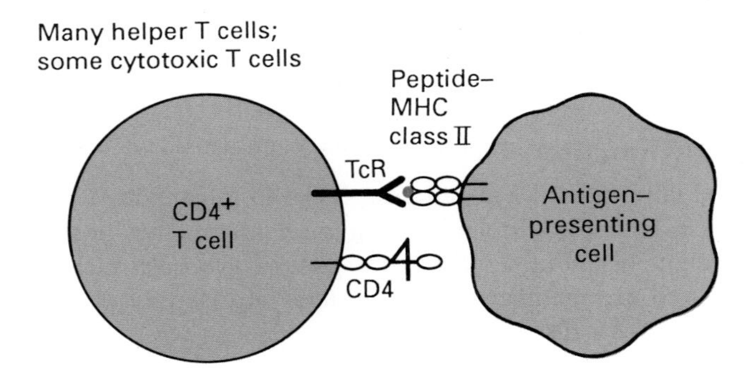

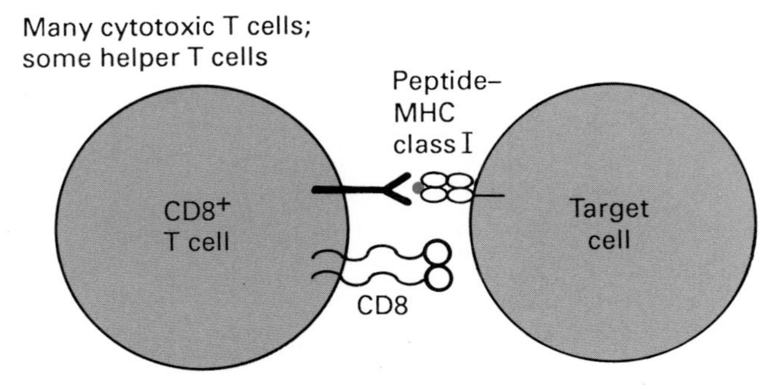

Fig.4.15 Expression of CD4 and CD8 on T cells correlates closely with MHC class II and class I restriction, rather than function

Therefore the phenotype of mouse helper T cells was initially defined as $Ly1^+$, $Ly2^-$ 3^- and that of cytotoxic cells as $Ly1^-$, $Ly2^+$ 3^+. In fact, Ly1 has now been shown to be present on *all* mouse T cells, albeit only at low levels on cells that were originally thought to be negative. Ly1 is homologous to the human T1 molecule and this molecule, which is now called CD5, might be involved in helper cell function because antibodies against human CD5 can enhance B cell responses in the presence of T cells.

Later, monoclonal antibodies were produced against human T cell subsets. Some antibodies defined a molecule called T8 which was found to be homologous to mouse Ly2 (Table 4.1). Expression of T8, like Ly2, seemed to correlate broadly with cytotoxic function, although suppressor cells also appeared to express this molecule. The nomenclature CD8 was assigned to these homologous molecules. A reciprocal subset of human T cells with helper function was found to express a molecule called T4, but this was not homologous to mouse Ly1. Human T4 is now known to be homologous to the mouse L3T4 molecule and these homologues have been designated CD4.

CD4 and CD8 are virtually non-polymorphic in structure and are expressed on reciprocal subsets of peripheral T cells. For some time it was thought that CD4 was a marker for helper T cells while CD8 was a marker for the cytotoxic and suppressor cell subsets. However, this turns out to be only a rough approximation, because expression of these molecules actually correlates more closely with the MHC restriction of T cells rather than with their function. With very few exceptions, T cells that recognize peptides bound to MHC class II molecules express CD4, while class I-restricted T cells express CD8 (Fig.4.15). As yet no membrane-associated phenotypic marker that correlates precisely with T cell function has been found.

Structure and function of CD4 (Table 4.6)

CD4 was originally called T4 in humans, L3T4 in mice, and W3/25 in rats. In these species the CD4 molecule is a monomeric glycoprotein with a molecular weight of about 55 kDa. CD4 genes have been cloned from all three species and are members of the Ig gene superfamily (Section 7.4.2). The structure of CD4 is discussed below.

Table 4.6 CD4

Distribution	T cells; macrophages and some dendritic cells (in rat and human, not in mouse)
Structure	Single chain
Approx. M_r	*Human:* 55 kDa (inc 2 N-CHO*) *Mouse:* 55 kDa (inc 2 N-CHO)
Homology to Ig	Yes: 4 extracellular Ig-like domains
Chromosome localization of genes	*Human* 12 *Mouse* 6
Comment	Associated with $p56^{lck}$ protein tyrosine kinase

* N-CHO = N-linked oligosaccharide

CD4 is expressed by mature T cells and thymocytes. In the human and rat it is also expressed by monocytes, macrophages, and some dendritic cells, but this does not seem to be so in the mouse for reasons that are not understood. Our ideas about the function of CD4 derive mostly from studies of T lymphocytes, and we have little understanding of its role on the macrophage or other cell types.

Monoclonal antibodies against the CD4 molecule inhibit the functions of MHC class II-restricted T cells, such as proliferation in mixed leukocyte reactions (MLR; see Appendix) and their ability to provide help for B cell responses; they also inhibit killing by the small subset of class II-restricted cytotoxic T cells. However, anti-CD4 antibodies do not inhibit the function of MHC class I-restricted T cells exhibiting either helper or cytotoxic functions. Because of this association and the fact that CD4 is a non-polymorphic molecule, it has been proposed that CD4 interacts with a non-polymorphic region of MHC class II molecules (see below). There is also evidence that CD4 can transmit a signal to the interior of the T cell that affects cellular activation (Fig.4.16).

Interaction with MHC class II molecules. The interaction between CD4 and MHC class II molecules can increase the overall avidity of a T cell for an antigen-presenting cell. This means that if the affinity (strength of binding) of a T cell receptor for its specific peptide–MHC complex is relatively weak, binding of CD4 to the MHC molecule can augment the interaction.

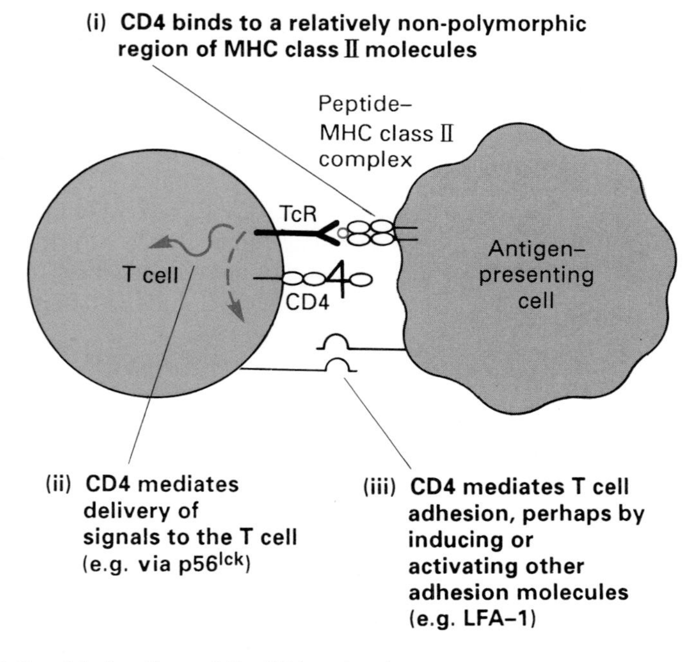
(i) CD4 binds to a relatively non-polymorphic region of MHC class II molecules
Peptide–MHC class II complex
TcR
T cell
CD4
Antigen-presenting cell
(ii) CD4 mediates delivery of signals to the T cell (e.g. via p56lck)
(iii) CD4 mediates T cell adhesion, perhaps by inducing or activating other adhesion molecules (e.g. LFA-1)

Fig.4.16 Possible functions of the CD4 molecule

In one set of experiments, the capacity of T cells to respond to different concentrations of their specific antigen in the presence of MHC class II-bearing accessory cells was taken as a measure of the affinity of the antigen receptor. T cell clones that responded to low doses of antigen, and which were thus thought to have a *high affinity* receptor, were *poorly inhibited* by a CD4 antibody, while clones with a *lower affinity* receptor were *more readily inhibited*. In other studies, a mouse T cell hybridoma was examined that recognized an MHC class I molecule but which, unusually, expressed the CD4 molecule. The hybridoma was more easily inhibited by anti-CD4 antibodies when the relevant class I molecules were expressed on cells that *also* expressed MHC class II molecules, implying the CD4 molecule interacted with the latter.

Direct evidence that CD4 interacts with MHC class II molecules first came from transfection experiments.

For example, a mouse hybridoma that normally recognized a class I molecule, although it expressed neither CD4 nor CD8, was transfected with cDNA for *human* CD4. Then a mouse target cell expressing the appropriate H-2 class I molecule was transfected with *human* MHC class II genes (HLA-DRA and B; see Section 2.3). The hybridoma responded weakly to the target cells if neither or only one of the genes was transfected, but the response was greatly increased when both genes were used. This suggested that human CD4 interacted with human MHC class II molecules on the target cell and that this interaction facilitated the response of the hybridoma (Gay *et al.* 1987). In other experiments a mouse hybridoma was infected with a retrovirus containing the human CD4 gene, and the cellular response to a human class II molecule was found to be increased, especially when the antigen concentration was suboptimal.

The interaction between CD4 and MHC class II molecules can mediate cell adhesion. This does not mean that cellular adhesion results directly from the association between CD4 and MHC class II, but that other adhesion molecules are induced or activated through this interaction.

In one study, a human CD4 molecule (in an SV40 recombinant vector) was introduced into a fibroblast cell line. Human B cells which express HLA-DR were found to bind to monolayers of these infected cells but not to uninfected cells, and adhesion was specifically inhibited by antibodies against CD4 or HLA-DR (Doyle and Strominger 1987).

Signalling to the T cell. Anti-CD4 antibodies can inhibit T cell responses. This may be due both to blocking of the interaction between CD4 and MHC class II molecules, perhaps reducing the avidity of T cell recognition (see above), and/or interruption of signalling to the T cell through the CD4 molecule which is normally required for T cell activation.

The cytoplasmic tail of CD4 does not possess tyrosine kinase activity (Section 4.2.3). However it is associated with a src-like tyrosine protein kinase, $p56^{lck}$, which is expressed at high levels only in T cells and is involved in the events of signal transduction during T cell activation (Glaichenhaus *et al.* 1991).

Ligation of CD4 results in activation of the phosphatidylinositol pathway (Section 4.2.3), mobilization of Ca^{2+} from intracellular stores, and activation of several protein kinases including $p56^{lck}$ and protein kinase C. However, the precise sequence of events and the intracellular signals involved are not clear (see also Section 4.2.5). It has been shown that when it is bound to CD4, $p56^{lck}$ is autophosphorylated, and in many T cells it dissociates from CD4 when protein kinase C is activated.

Structure of CD4 (Fig.4.17). The sequence of CD4 predicts a molecule composed of four Ig-related extracellular domains, a single transmembrane domain, and a short cytoplasmic tail. While the predicted structure of the first, membrane-distal, domain (V1) closely resembles that of an Ig V domain (Fig.7.23), the others (V2, V3, V4) are less similar. In particular, V2 has an unusual disulfide bond that links adjacent strands rather than the β-pleated sheets; V3 lacks a disulfide bond, the cysteine residues being replaced by hydrophobic amino acids that could point inwards and thus stabilize the Ig-like fold; and V4 is more similar to an Ig C domain. N-glycosylation sites are present in both the V3 and V4 domains.

To investigate the structure of CD4 by X-ray crystallography, crystals of the four extracellular domains have been prepared, but found to diffract poorly. However, the structure of crystals of a genetically-engineered V1V2 fragment has been determined at high resolution (Wang *et al.* 1990).

Analysis of these crystals has shown that the V1 domain closely resembles an Ig V domain, but it is unusual in that the region corresponding to the CDR-2 loop is longer and it contains a core of hydrophobic residues. These create a prominent ridge about 2.5 nm long on one face of the molecule. Moreover, the last β strand of the V1 domain is unusually long and forms the first strand of the V2 domain, thus providing a rigid spine to this structure. The V2 domain is truncated and yet, despite the unusual intra-chain disulfide bond (above), it maintains the overall structure of an Ig fold.

It seems likely that the intact CD4 molecule resembles a flail with two rigid V1V2 and V3V4 rods linked by a flexible region. The length of the V1V2 fragment is about 7 nm while that of the intact molecule is about 12.5 nm. Given that the length of a T cell receptor is about 6–7 nm, and that of an MHC molecule 7 nm, it seems likely that CD4 on the T cell membrane stretches past the T cell receptor and would be able to interact with both the membrane-distal and the membrane-proximal domains of the class II molecule. It has been shown that CD4 can bind to a region of class II β_2 domains that is structurally analogous to the CD8-binding loop of class I α_3 domains (see below) (Konig *et al.* 1992; Cammarota *et al.* 1992).

Although, as yet, other regions of the class II molecule for binding of CD4 have not been defined, the idea that CD4 interacts with both membrane-distal (α_1 or β_1) and membrane-proximal (α_2 or β_2) MHC class II domains is consistent with some experimental observations. It has been presumed that, because of its non-polymorphic nature, CD4 interacts with the conserved α_2 or β_2 domains of class II molecules, as now seems likely. However, anti-CD4 antibodies were found to block recognition of cells

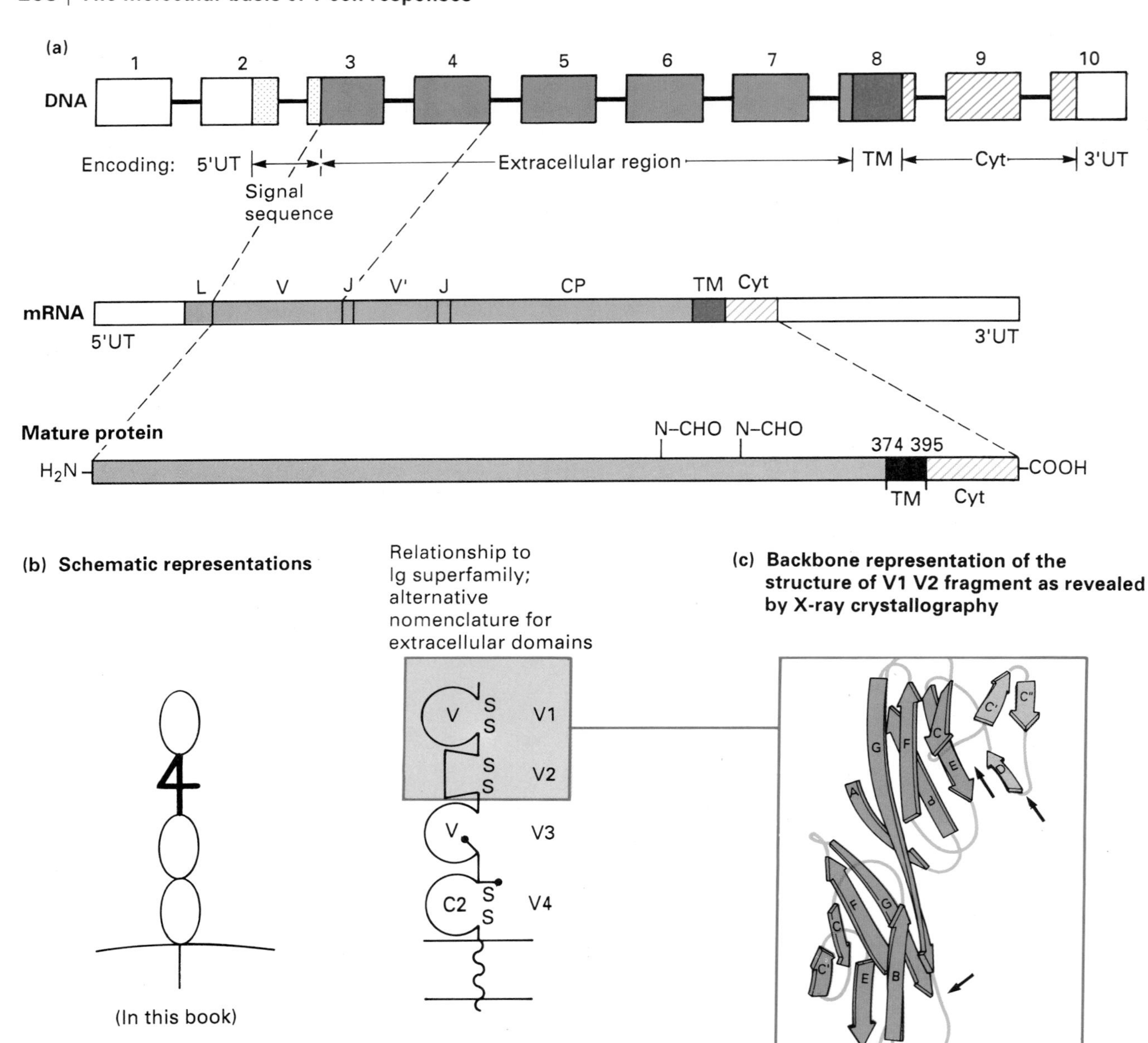

Fig.4.17 **CD4**. (a) The mouse CD4 gene, mRNA transcript and human polypeptide. The gene spans about 33 kilobases of DNA and the diagram does not show the size of the introns to scale; note the presence of an intron in the 5′ untranslated region as well as an intron dividing the sequence encoding the N-terminal V-like domain into two exons — both unusual, though not unique, features. Portions of mRNA encoding the untranslated regions (UT) and different protein domains are indicated: L–leader sequence; V and V′–sequences of homology to Ig V regions; J and J′–sequences homolgous to Ig J segments; CP–the connecting peptide (thought to resemble two Ig-like domains); TM–transmembrane region; Cyt–cytoplasmic tail. In the mature protein there are two potential N-glycosylation sites (N-CHO). (b) Schematic representation of CD4 indicating the relationship and nomenclature of Ig-like domains: V1 is the most homologous to an Ig V region, V2 is truncated relative to an Ig V region, V3 lacks an intrachain disulfide bond, and V4 is homologous to an Ig C2 region. (c) Representation of the backbone structure of a V1V2 fragment as revealed by X-ray crystallography. The prominent ridge in the V1 domain, which also contains the principal residues involved in HIV gp120 binding, is indicated (grey), as is the location of residues mutations in which are known to alter binding of CD4 to MHC class II molecules (arrows) [(c) is adapted from *Nature*, **348**, 413 (1990), with permission]

transfected with recombinant (exon-shuffled) mouse MHC molecules containing a polymorphic class II β_1 domain, but linked to the α_2 and α_3 domains of a class I molecule with which CD4 *does not* interact (Golding 1985). This experiment suggests that CD4 can, additionally, interact with elements within the β_1 domain that are perhaps conserved between MHC class II molecules.

CD4 as the HIV receptor. Infection of T cells by the human immunodeficiency virus, HIV, the causative agent of AIDS, apparently requires expression of the CD4 molecule which thus appears to be an HIV receptor. This is supported by several lines of evidence. First, CD4-negative cell lines can only bind and internalize HIV when they are transfected with CD4. Second, some antibodies against HIV precipitate the 55 kDa CD4 molecule from infected cells, and, conversely, some anti-CD4 antibodies precipitate the gp120 envelope glycoprotein of HIV which mediates cell attachment. Furthermore, binding of the virus can be inhibited by some antibodies against CD4 and by the soluble purified CD4 molecule itself. This explains the reported tropism of HIV for helper T cells and might account for viral entry to Langerhans cells as well as macrophages, both of which express CD4 in humans; this may be important for the spread of infection to the human brain and central nervous system. However, it is possible this is not the whole story and the virus may be able to enter cells by other routes. The binding site on CD4 for the HIV gp120 molecule has been located (Ryu *et al.* 1990).

Mutant recombinants of the CD4 molecule, and epitopes recognized by anti-CD4 antibodies that are known to inhibit binding of gp120, were mapped on to the structure determined by X-ray crystallography (above). Two binding sites were identified, both within the V1 domain. Most of the critical residues lie in the region corresponding to the Ig-like CDR2 loop, but which forms a ridge in CD4; this is thought to interact with a complementary groove on gp120. The second site is located in a cluster of negative-charged residues. These sites are obvious targets for the design of inhibitors of gp120 binding to prevent HIV infection.

Structure and function of CD8 (Table 4.7)

The CD8 molecule was first defined in the mouse by alloantisera against polymorphic forms of the Ly2 antigen (designated Ly2.1, Ly 2.2), and subsequently by monoclonal antibodies specific for Ly2 and its homologues in other species, T8 in humans and OX8 in rats. CD8 is expressed on the reciprocal subset to mature $CD4^+$ T cells and by some natural killer cells. Anti-CD8 antibodies can inhibit killing by class I-restricted cytotoxic T cells, but not by the small subset of class II-restricted cytotoxic cells, and they inhibit helper functions of the less frequent population of class I-restricted helper T cells. Therefore expression of CD8 correlates more closely with class I MHC restriction than with T cell function (Fig.4.15).

The CD8 gene has been cloned from human, rat, and mouse and, like CD4, it is a member of the Ig gene superfamily (Section 7.4.2). Its predicted structure is discussed below, and structural data for CD8 from X-ray crystallography are now becoming available (Leahy *et at.* 1992).

Table 4.7 CD8

Distribution	T cells; some NK cells
Structure	Homodimers and heterodimers; the mouse α chain is differentially transcribed to form α and α' products differing in their cytoplasmic portions; the human α chain can also be differentially transcribed, but the α and α' products respectively do and do not contain the transmembrane region; the predicted human CD8β chain exists in two forms (β and β') differing in their cytoplasmic tails, through alternative splicing resulting in a frame shift. (mouse: α=Ly2; β=Ly3)
Approx. M_r	*Human:* α 34 kDa (inc 1 O-CHO** site, but may not be used) *Mouse:* α 38 kDa (inc 3 N-CHO*) α' 34 kDa (inc 3 N-CHO) β' 30 kDa (inc 1 N-CHO)
Homology to Ig	Yes
Chromosome localization of genes	*Human* 2 *Mouse* 6
Comment	Associated with p56^{lck} protein tyrosine kinase

* N-CHO = N-linked oligosaccharide
** O-CHO = O-linked oligosaccharide

Interaction with MHC class I molecules. *CD8 has been shown to bind to the membrane-proximal α_3 domain of MHC class I molecules* (see below). This interaction triggers signalling through CD8 (see below), and it would appear that this also facilitates the cellular response if the T cell receptor has a low affinity for its specific peptide–MHC complex, analogous to the role of CD4–class II interactions (see above).

Evidence for the latter has been obtained from transfection experiments (Dembic *et al.* 1987). For example, when T cell receptor genes were transfected into a CD8$^-$ cytotoxic T cell hybridoma it was found that the cells could lyse targets provided they expressed antigen at a high density, but co-transfection of CD8 permitted them to lyse targets expressing a much lower density of antigen. Moreover, when T cell receptor genes from a CD8$^+$ cytotoxic T cell were transfected into a CD4$^+$ cell it was found that the class I specificity of the original cell could only be transferred if the

CD8 gene was co-transfected. It is possible that CD8-class I interactions mediate cellular adhesion, because anti-CD8 antibodies can prevent cytotoxic T cells binding to their targets, and this is one way in which they may inhibit T cell functions. However, CD8 also has signalling functions that may be disrupted through binding of anti-CD8 antibodies.

Signalling to the T cell. Like CD4, the cytoplasmic tail of CD8 is associated with the src-like tyrosoine protein kinase $p56^{lck}$ (Section 4.2.3). This strongly suggests that direct signalling through CD8 occurs.

However, the properties of $p56^{lck}$ when bound to CD8 differ from when it is bound to CD4. Thus, in some T cells, binding to CD8 appears to be associated with less vigorous autophosphorylation of the kinase, and in many T cells it does *not* dissociate when protein kinase C is activated (compare with CD4, above; see also Section 4.2.5).

Structure (Fig.4.18). The sequence of CD8 predicts a structure containing only one extracellular Ig-like domain linked to the membrane by a sequence of 40 amino acids, a single transmembrane region, and a cytoplasmic tail that is shorter than that of CD4 (27 compared to 40 amino acids respectively).

Unlike CD4, the CD8 molecule can form homodimers, and perhaps other multimers, as well as heterodimers with molecules encoded by separate genes. Moreover, CD8 can be differentially transcribed to produce CD8α and CD8α' chains, which may be expressed at different levels in mature T cells and thymocytes. In addition, CD8 can be produced as a soluble molecule by T cells, but its function is currently unknown. Although not yet shown in every case, it seems likely that these observations pertain to mouse, rat, and human CD8.

The mouse CD8 (Ly2) molecule forms a heterodimer with the mouse Ly3 molecule; these molecules are now termed CD8α and CD8β respectively. The 35–38 kDa α and 30–34 kDa α' products of Ly2 can associate with the 28–30 kDa Ly3 β chain in the form $\alpha\beta$ and $\alpha'\beta$. Ly3 is not necessary for expression of Ly2 because when the Ly2 gene was transfected into fibroblasts, multimers of α and α' chains were expressed at the cell surface. The rat 32 kDa CD8 molecule, first defined by an antibody called MRC OX8, forms a heterodimer with the homologue of mouse Ly3, which however is a larger molecule, 37 kDa. In the human, the 34 kDa human CD8α molecule on mature T cells can form heterodimers with the CD8β chain, and can also be expressed in association with the 45 kDa CD1 molecule and β_2-microglobulin on thymocytes.

The binding site for CD8α on the MHC class I molecule has been mapped (Salter *et al.* 1990). Three clusters of residues exclusively within the α_3 domain of the class I molecule were defined as a minimal binding site for CD8. Two of these are topologically associated in the intact molecule and contain highly conserved residues that form an exposed loop. The third cluster of amino acids is spatially separated from these residues but on the same face of the α_3 domain. It has been speculated that the CD8β molecule might additionally interact with β_2-microglobulin.

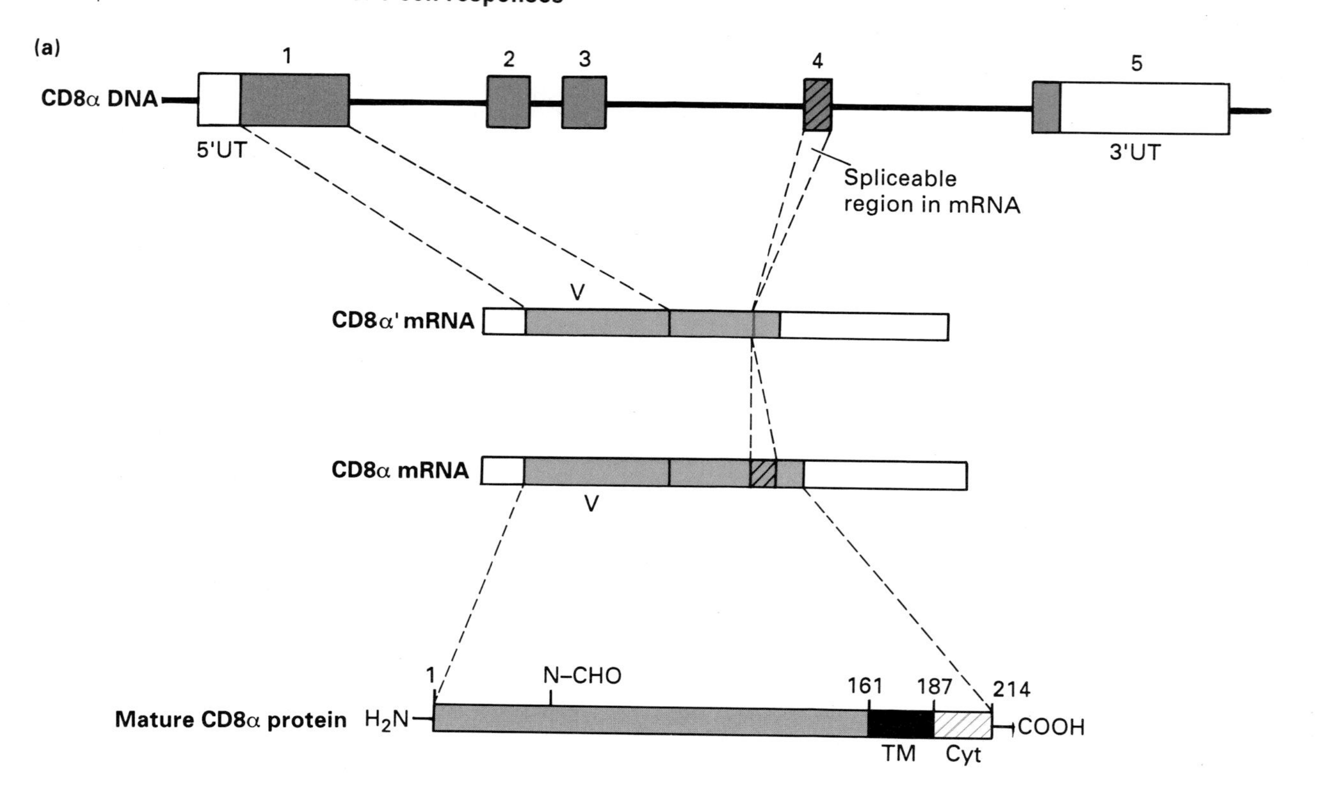

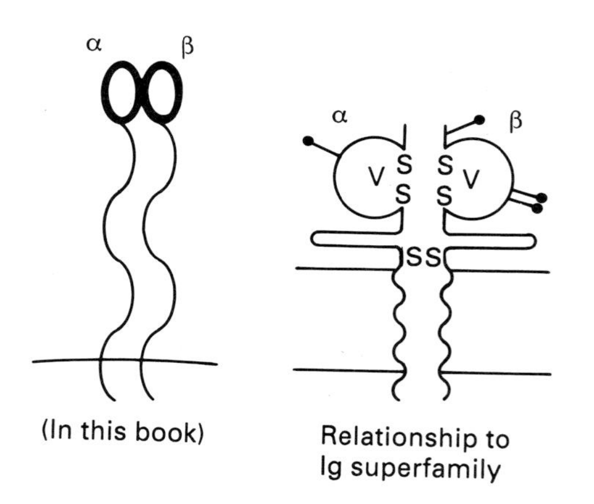

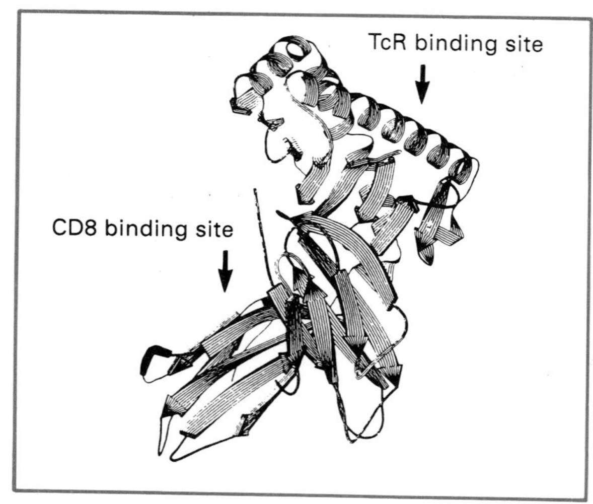

Fig.4.18 **CD8**. (a) The mouse CD8 gene, α and α′ mRNA transcripts, and human polypeptides. The diagram of the gene does not show the lengths of introns and exons to scale; note exon 4 encodes a spliceable region in the mRNA, the two transcripts producing the α and α′ forms of mouse CD8 which differ in the cytoplasmic region. The exon and portion of the mRNA encoding the V-like domain is indicated. The mature polypeptide contains one potential N-glycosylation site (N-CHO). (b) Schematic representation of CD8 indicating the single Ig-like domains of the α and β chains. (c) Sites of MHC class I molecule that interact with CD8, and with the T cell receptor for comparison. A representation of the backbone structure of MHC class I is shown. The CD8α binding site is located in the membrane-proximal α_3 domain whereas the T cell receptor interacts with residues of the α_1 and α_2 domains which form the peptide-binding groove [(c) is reproduced from *Nature*, **345**, 44 (1990), with permission]

For these mapping studies, a series of cells transfected with point-mutated human HLA-A2 genes was constructed. Binding of CD8 to these mutant MHC molecules was then assessed by studying the capacity of fibroblasts transfected with the human CD8α gene to bind to these cells, their lysis by CD8$^+$ cytotoxic T cell clones, and inhibition by anti-CD8 antibodies.

To span the approximately 9 nm distance necessary for CD8 to interact with the membrane-proximal α_3 domains of class I molecules, the sequence within CD8 that links the single Ig-like domain to the membrane would have to be in a fairly extended confirmation, and detailed models for CD8–class I interactions can now be based on crystallographic data (Parham 1992; see Further reading).

4.2.5 CD45, the leukocyte common antigen: regulation of signal transduction (Table 4.8)

Protein Tyrosine Phosphatase

CD45 is a molecule expressed by all leukocytes, giving rise to its name, the leukocyte common antigen (LCA). However, different types of leukocyte express different forms of CD45, and a number of studies indicate that this molecule regulates intracellular signalling cascades and plays an important role in T cell responses.

Structure

The CD45 gene has been cloned from humans, rats, and mice. It spans 130 kilobases of DNA and contains 33 exons. Exons 1, 2, and 3 encode the 5′ untranslated region, the signal sequence of 23 amino acids, and the first 7 residues of the mature protein. A most important point is that *exons 4, 5, and 6, can be differentially transcribed* to produce RNA encoding 220, 205, 200, 190, and 180 kDa products, or **isoforms** (Fig.4.19). The 220 kDa polypeptide contains the sequence encoded by all three exons, the 205 kDa product contains the sequence encoded by exons 4 and 5 only, while the 180 kDa isoform lacks all three. Although eight different transcripts can be produced, some of the glycosylated molecules are similar in molecular weight, accounting for five main forms being detected at the cell surface in most studies. These products are differentially expressed by different cell types (see below).

Table 4.8 CD45

Distribution	All leukocytes, but different isoforms are expressed by different cells
Structure	Single chain; three exons can be differentially spliced to generate eight mRNAs encoding different isoforms
Approx. M_r	170–240 kDa
Homology to Ig	No
Chromosome localization of genes	Human Mouse 1
Comments	Possesses protein tyrosine phosphatase activity

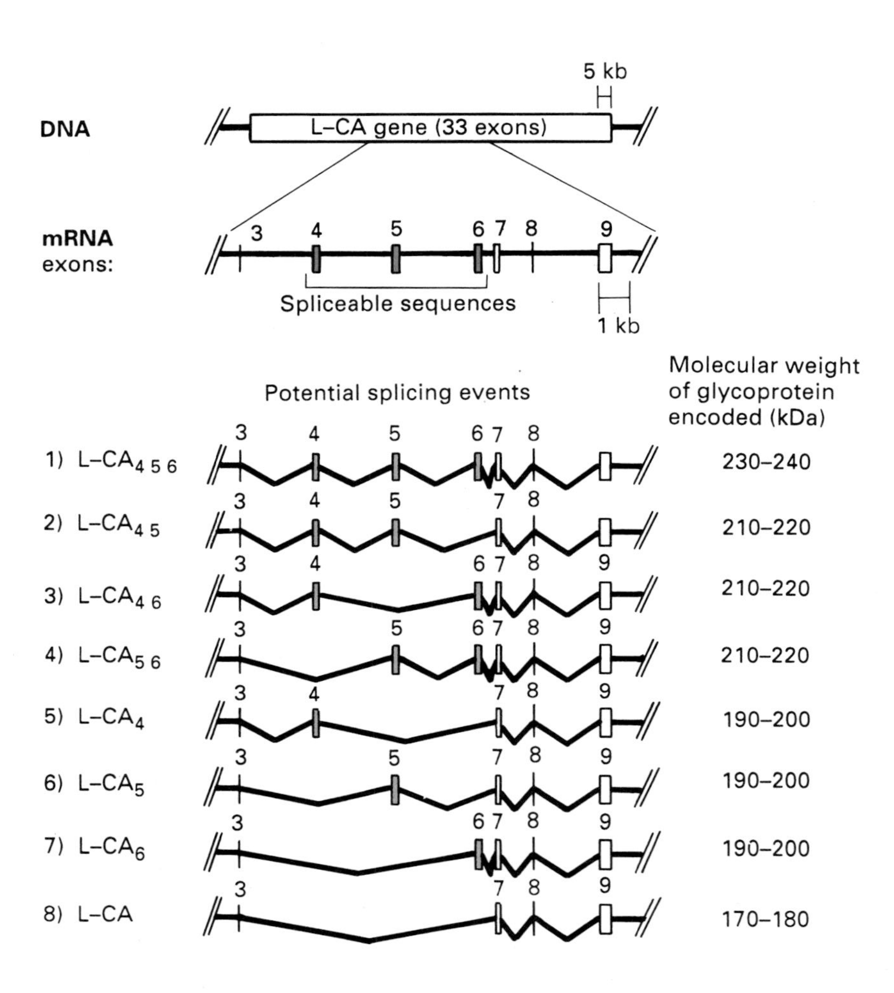

Polypeptide

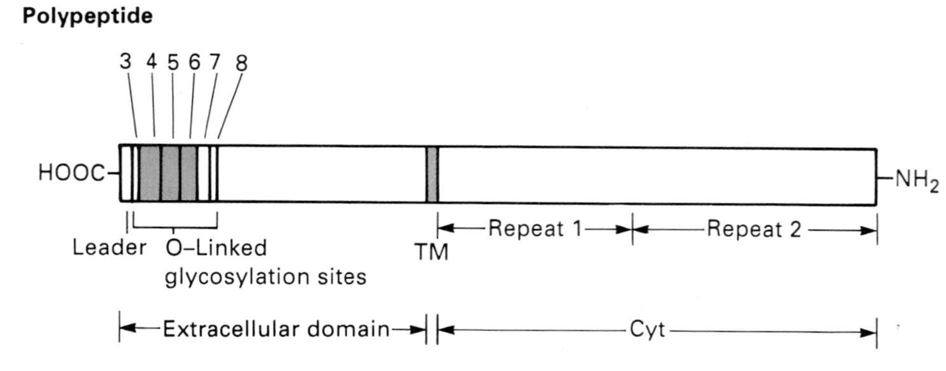

Fig.4.19 **CD45.** The gene, mRNA transcripts, and polypeptide isoforms. The mouse CD4 gene spans approximately 130 kilobases of DNA and comprises 33 exons: an amplified representation of exons 3–9 is shown. Exons 3, 4, and 5 encode mRNA sequences that can be differentially transcribed, potentially to produce 8 different mRNA species; the predicted molecular weights of the glycoproteins encoded by these is indicated. Note the orientation of the molecule, the N-terminal being cytoplasmic. (Adapted from Thomas and Lefrancois (1988); see Further reading.)

The sequence of CD45 predicts a molecule consisting of a heavily glycosylated, cysteine-rich extracellular region of 400–550 amino acids, a single transmembrane portion of 22 residues, and an unusually large intracytoplasmic portion of 700 amino acids composed of two highly homologous globular domains (Fig.4.19). Many of the glycosylation sites are present within the extracellular sequence encoded by exons 4, 5, and 6, suggesting that cell-specific expression of differently glycosylated products may be of functional importance; glycosylation of the mature molecule also explains the existence of CD45 molecules with molecular weights above 220 kDa. The molecule is not related to Ig, and its N-terminus is extracellular. Electron microscopy of the purified molecule reveals that CD45 resembles a comma and in the intact cell it would resemble a tadpole with its tail extending outside of the cell; in addition, the complete molecule containing sequences encoded by exons 4, 5, and 6 has a short flail at the end.

CD45 nomenclature. Monoclonal antibodies have been produced which recognize different isoforms of CD45. Strictly speaking, the peptide sequences corresponding to the transcripts for exons 4, 5, and 6 are designated CD45RA, CD46RB, and CD45RC respectively, and the whole molecule lacking all three is CD45RO. However, the nomenclature for anti-CD45 antibodies currently in use does *not* adhere to this scheme (perhaps not surprisingly because the same exon may be differently transcribed to encode different epitopes).

The so-called CD45 group of antibodies recognize all isoforms of this molecule, while the more cell-restricted CD45RA antibodies precipitate 220 and 205 kDa products; CD45RB the 220, 205 and 190 kDa products; and CD45RO the 180 kDa product. (Originally, anti-CD45 antibodies were categorized as the CD45 group, which recognized CD45 on all leukocytes, this nomenclature being retained, whereas the group of antibodies recognizing CD45 in a cell-specific manner was designated CD45R, R standing for 'restricted').

Cellular expression of CD45 isoforms

The cellular distribution of CD45 isoforms is complex. For example, B cells express the highest molecular weight form (220 kDa), immature thymocytes are thought to express only the lowest molecular weight form (180 kDa, CR45RO) but at least three different isoforms can be detected on mature thymocytes, and different populations of mature T cells also express different forms (see below). Some anti-CD45 antibodies recognize epitopes on carbohydrate residues expressed in a cell-restricted manner.

For example, one such antibody called CT1 inhibits cytolysis by CTL; this determinant is also developmentally regulated during thymocyte differentiation and is expressed on a subpopulation of $CD8^+$ T cells in the gut epithelium. Another antibody called NK-9 inhibits cytolysis by NK cells, and the epitope is also expressed by some lymphocytes.

CD45 expression on mature T cells. The expression of different CD45 isoforms by mature T cells has been correlated with apparently different cellular functions, although this is by no means absolute. $CD4^+$ T cells can be divided into two populations according to their expression of CD45RA or CD45RO. It has been suggested that resting (naïve) T cells express CD45RA (220 and 205 kDa) whereas memory T cells express CD45RO (180 kDa), these forms of human CD45 being recognized by antibodies such as 2H4 and UCHL1 respectively. These CD45 isoforms also appear to be differentially expressed by two subsets of mature $CD4^+$ T cells that have been designated T_H1 and T_H2 (Sections 4.4.2, 8.4.5, and 11.3); the relationship of these subsets to naïve and memory cells is controversial. However, the expression of CD45RO by T cells is not a marker, *per se*, of 'memory' cells, because this molecule is also expressed on thymocytes and other types of leukocytes.

Memory T cells, produced as a consequence of a primary immune response, are not only present in increased numbers within the periphery, due to clonal expansion, but they also have a lower threshold for activation. A number of other molecules involved in T cell functions are expressed at higher levels by memory compared to resting T cells, including CD2, CD11a/CD18 (LFA-1), and CD28 (Sections 4.2.7, 4.2.6, and 4.2.8), all of which may contribute to the capacity of memory cells to respond more readily to lower concentrations of antigen. In addition, the pattern of cytokine gene expression in neonatal, presumably naïve, and adult mouse T cells (containing memory cells) has been shown to be different (Ehlers and Smith 1991). Using the polymerase chain reaction, a sensitive technique for detecting mRNA transcripts (see Appendix), it has been demonstrated that in response to immobilized anti-CD3 antibodies (Section 4.2.3) both neonatal and adult T cells transcribe the genes for IL-2, but that adult T cells additionally transcribe those for IL-3, IL-4, IL-5, IL-6, IFNγ, and GM-CSF. It has been suggested that this provides a more reasonable criterion for distinguishing between resting and memory T cells. Nevertheless, it seems likely that different CD45 isoforms have different signalling functions on resting, activated, and memory T cells, as well as during thymocyte differentiation.

Signalling functions of CD45

An important insight into the potential signalling functions of CD45 came from the discovery that this molecule possesses intrinsic protein tyrosine phosphatase (PTPase) activity (Section 4.2.3) (Tonks *et al.* 1988).

The cytoplasmic portion of CD45 contains sequences that are homologous to known PTPases; these sites are also present in protein kinases (Section 4.2.3). Confirmation that CD45 has PTPase activity came from the finding that the purified molecule has this enzymic activity *in vitro*. The cytoplasmic tail of CD45 also contains several sites that can be phosphorylated (e.g. by protein kinase C). These observations indicate that CD45 has the potential both to be phosphorylated by protein kinases and to dephosphorylate these and other molecules, important events in the signalling cascades that regulate cellular responses.

Signalling in mature T cells. CD45 has been implicated in the regulation of signalling through the $\alpha\beta$ TcR–CD3 complex (Section 4.2.3), CD4, and CD8 (Section 4.2.4) which results in T cell activation (Ledbetter *et al.* 1988).

These molecules are associated with protein kinases, the $\alpha\beta$ T cell receptor with fyn and CD4 and CD8 with lck. Although the precise signalling cascades mediated by these molecules are not worked out, clues as to possible functions of CD45 come from a number of observations which, in outline, include the following.

1. Normally CD45 is immobile within the plane of the membrane, possibly because of its association with the cytoskeletal element **fodrin**. However, its mobility is altered by agents such as phorbol esters, which lead to phosphorylation of CD45. This may regulate its function, and CD45 may then be able to associate with other membrane molecule.
2. Ligation of the $\alpha\beta$ T cell receptor results in phosphorylation of the CD3 ζ chain. This is mediated by activation of fyn, as well as of $p56^{lck}$ associated with CD4 or CD8 which come into close proximity with the TcR–CD3 complex during T cell activation. Cross-linking of CD45 to CD4 *enhances* CD4-mediated signalling, and correlates with dephosphorylation of $p56^{lck}$ and perhaps also of the ζ chain.
3. Binding of monoclonal antibodies specific for CD2, CD3, and CD28 on resting T cells can result in an increase in Ca^{2+} concentration and sometimes proliferation. Cross-linking of CD45 to these molecules *inhibits* these events. Anti-CD45 antibodies also inhibit the increase in cytoplasmic Ca^{2+} triggered in resting B cells through Ig or CD19, as well as activation through CD40 which is not associated with a rise in intracellular Ca^{2+} (Section 8.4.2).

It is possible that the different isoforms of CD45 expressed by resting, activated, and memory T cells can bind to different surface molecules and regulate their intermolecular associations, behaviour (e.g. recycling), or function. For example, CD45RA on resting T cells may facilitate its association with the TcR-CD3 complex, CD45RO on memory cells with CD4, and CD45 on activated cells with the IL-2 receptor (since anti-CD45 antibodies alter signalling through the IL-2 pathway). However this is speculative, and ligands for different forms of CD45 have yet to be identified.

Signalling during thymocyte development. Different stages of thymocyte differentiation are associated with the expression of different isoforms of CD45. The least mature thymocytes express CD45RO but not TcR, CD4, and CD8, i.e. they are 'double-negative' for the latter markers (Section 4.3). Expression of these and other molecules is differentially regulated during the subsequent stages of development into mature T cells. Different isoforms of CD45 expressed during this process may be involved in various signalling cascades that result in different cellular events.

Although little is known of the function of CD45 during thymocyte development, it is clear that while ligation of TcR on mature T cells often results in proliferation and cytokine secretion, many double-positive cells undergo programmed cell death through apoptosis instead (Section 10.3.5). The

$\alpha\beta$ TcR is identical on mature and double-positive thymocytes, but this molecule appears to be uncoupled from CD3 in the latter case, perhaps due to the extent of phosphorylation of the ζ chain (Nakayama *et al.* 1989). This, and differential signalling through CD4, CD8, and perhaps the CD45 isoforms, may determine whether a thymocyte bearing a TcR with a particular specificity will be selected for development into a mature T cell (positive selection), or deleted or rendered anergic (negative selection; for example, see Section 5.3.5).

4.2.6 Integrins and selectins: cell adhesion and activation molecules (Table 4.9)

Different subpopulations of T cells migrate preferentially through and into particular tissues. This phenomenon of **homing**, a property of all leukocytes,

Table 4.9 Simplified overview of families of adhesion molecules

Family	Examples	General distribution	Type of ligand	Outline structure
Selectins				
L	MEL-14	leukocyte	CHO	
E	ELAM-1	endothelium	CHO	
P	GMP-140 (CD62)	endothelium	CHO	
Integrins				
$\beta1$	VLA1-6 (CD49a–f/CD29)	broad	matrix proteins/ other	
$\beta2$	LFA-1 (CD11a/CD18)	leukocyte	various?/ Ig superfamily	S-S α β
	CR3; p150, 95 (CD11b; c/CD18)	leukocyte	various	
$\beta3$	vitronectin R (CD51/CD61)	endothelium leukocyte	matrix protein	
Ig superfamily				
	ICAM-1	endothelium leukocyte	LFA-1 (=$\beta2$ integrin)/ other	
	VCAM-1	endothelium	VLA-4 (=$\beta1$ integrin)/ other	
	ATHERO-ELAM	endothelium		
	PECAM	endothelium		
Cartilage link protein-related				
	CD44	leukocyte	matrix protein	

o short consensus repeats resembling complement-binding proteins
● epidernal growth growth factor-like domain
● lectin (sugar-binding) domain
Ω Ig-like domain

is essential for the capacity of the immune system to protect tissues against invading pathogens, in other words for **host defence**.

Resting T cells, generated in the thymus, generally become localized within secondary lymphoid tissues (Figs.1.32, and e.g. 1.38, 1.39). This, and their compartmentalization to particular areas within these tissues, facilitates their interaction with accessory cells bearing peptide–MHC complexes for which they are specific. This may result in T cell activation (see immunostimulation, Section 3.4.3), and the initiation of the effector phase of the adaptive immune response. *Activated T cells* are also present within lymphoid tissues but show an increased propensity for migration into the blood and thence to sites of inflammation and infection. In contrast, *memory T cells*, produced within secondary lymphoid tissues, constantly *recirculate* throughout the blood and lymph (see lymphocyte recirculation, Section 1.3.3 and Fig.1.35). Of course these are broad generalizations because, for example in the mouse, $\gamma\delta$ T cells tend to segregate at epithelial sites, but these general principles are probably correct.

For a T cell to enter a non-lymphoid tissue, it must first adhere to the endothelium of a blood vessel at that site, and cross the vessel wall by diapedesis (Fig.1.36). Within the tissue, the T cell then has to interact with antigen-presenting cells as well as elements of the extracellular matrix (see below), and receive the requisite signals for it to produce the appropriate response (e.g. help or cytotoxicity); many of these events, of course, also apply to other leukocytes such as monocytes and granulocytes (e.g. neutrophils). These processes are controlled by receptor-ligand interactions mediated by specialized molecules on the T cell membrane and complementary molecules on endothelial cells, antigen-presenting cells, and the extracellular matrix.

Endothelial cell molecules that act as adhesion molecules for leukocytes have been termed **vascular addressins**. The expression of some of these molecules can be induced on endothelial cells close to sites of inflammation, and is sometimes biased to particular tissues. Thus different populations of leukocytes can be recruited specifically into certain areas of the body depending on their differential expression of molecules complementary to vascular addressins. Chemotaxis (directed cell movement) of these cells into the tissues is further controlled by the **chemotactic molecules** produced at sites of inflammation (Section 9.2.2); movement of the cells along a chemotactic gradient in a concentration-dependent manner is termed chemokinesis.

Families of adhesion molecules

Vascular addressins, as well as the molecules expressed by leukocytes that mediate their adhesion to endothelial cells and their interaction with other cell types and the extracellular matrix of tissues, can be grouped into structurally-related families (Table 4.9).

Some vascular addressins are constitutively expressed on endothelial cells, others are only induced after inflammation and blood clotting (Sections 9.2 and 9.3). Two major vascular addressins for T cells are members of the **immunoglobulin gene superfamily**, ICAM-1 (CD54) and VCAM-1 (see below), which respectively contain five and seven extracellular Ig-like domains. The VCAM-1 molecule can be alternatively spliced to produce a molecule with six Ig-like domains, ATHERO-ELAM, which is localized to atherosclerotic lesions in rabbits. Another member of this family, PECAM-1, containing six Ig-like domains, is an addressin for monocytes and neutrophils.

The **integrins** are a large family of heterodimeric glycoproteins which can be subdivided according to the particular β subunit they possess, which is shared by all members of the group (see Table 4.10). On this basis, this family can be subdivided into the $\beta1$, $\beta2$, and $\beta3$ integrins. The $\beta2$ integrins are expressed particularly by leukocytes, giving rise to their alternative name, the **leukocyte integrins**, whereas the others, in general, are more widely distributed. These molecules, like members of the Ig gene superfamily (Section 7.4), are particularly involved in protein–protein interactions.

Another family of glycoproteins, the **selectins**, also known as **leukocyte endothelial-cell cell-adhesion molecules** (LEC-CAMS), is involved in recognition of carbohydrate structures. One interesting fact is that many more carbohydrate than protein structures can be created, by virtue of the ability of monosaccharides to form branched and linear arrays (e.g. Fig.1.4). This makes carbohydrates particularly well-suited for specific recognition functions, and although in the past immunologists have tended to focus primarily on protein–protein interactions, and considerably less is currently known about carbohydrate recognition, this situation is beginning to change. A third major vascular addressin for some T cells, ELAM-1 (see below), belongs to this family.

In the remainder of this section we will discuss, in turn, the selectins, the $\beta2$ leukocyte integrins, and finally the $\beta1$ and $\beta3$ integrins, with particular emphasis on T cell functions.

The selectins

One particular selectin, designated **L-selectin** but originally termed MEL-14, is expressed by *T cells* (Lasky *et al.* 1989). It functions as a **homing receptor** and is involved in lymphocyte recirculation, by selectively facilitating the adhesion of T cells to high endothelial venules of peripheral lymph nodes for example (Section 1.3.3); it is also involved in neutrophil emigration to inflammatory sites. This 90 kDa molecule was also variously known as gp90^{mel}, TQ1, Leu8, LAM-1, LEC-CAM-1, and LECAM-1, the latter three designations being derived from variations of the term LEC-CAMS (see above) but with the L also standing for 'lymphocyte'.

Two selectins expressed by *endothelial cells* are **E-selectin** or ELAM-1 and **P-selectin** or CD62, the latter also being known as GMP-140 or PADGEM. ELAM-1 is synthesized by endothelial cells in response to inflammatory agents, particularly in skin. It mediates the adhesion of *memory* T cells, as well as NK cells, neutrophils, and monocytes at these sites. In contrast, CD62 (Johnston *et al.* 1989) is a pre-formed molecule, stored within granules of both endothelial cells and platelets, which is mobilized to the cell surface in response to thrombogenic agents produced during clotting reactions at sites of injury. It mediates the adhesion of neutrophils and monocytes, but probably *not* T lymphocytes, to these cells.

The selectins are transmembrane glycoproteins with a highly characteristic structure (Table 4.9). The extracellular portion contains an N-terminal sugar-binding or **lectin** domain, followed by an epidermal growth factor-like domain, and then a variable number of short consensus repeats (60 amino acids) that are homologous to those in proteins that regulate complement activation (Fig.9.13) (Jutila *et al.* 1992). There are nine of these domains resembling complement-binding proteins in CD62, six in ELAM-1, and two in MEL-14. These three molecules are encoded within a gene cluster on human chromosome 1, and like all selectins they have carbohydrate ligands.

ELAM-1, for example, binds to a family of polysaccharide derivatives containing sialic acid and fucose residues linked to polylactosamine (e.g. Fig.1.4). Some of these ligands are closely related to structures found at the non-reducing termini of glycolipids and O-linked and N-linked sugars of glycoproteins (Figs.1.4 and 1.5), designated Lewis x (CD15) and its derivative sialylated Lewis x, as well as CD65; it seems likely that a set of related, but different, structures is also recognized by CD62. One endothelial ligand for ELAM-1 (i.e. L-selectin) has been identified as a mucin-like molecule, that is structurally unrelated to the Ig superfamily members, integrins, or selectins (Lasky *et al.* 1992).

Less is currently known about the ligand(s) on endothelial cells for the T cell selectin MEL-14, but this likewise appears to be a fucosylated and/or mannosylated polysaccharide(s). T cells and neutrophils express the same MEL-14 molecule, and the affinity of binding to one carbohydrate-based ligand, PPME, is dramatically increased after activation of these cells. Inhibitors of these adhesion-promoting molecules might prove useful as anti-inflammatory and anti-thrombogenic drugs.

The β2 leukocyte integrins (Table 4.10)

The three heterodimeric molecules identified in this family share the 95 kDa β2 integrin subunit, also designated CD18; the α chains are designated CD11a (180 kDa), CD11b (165 kDa), and CD11c (150 kDa). Although the α chains are distinct, all three contain a sequence close to the N-terminus that is homologous to repeats in other molecules including complement factor B and von Willebrand factor (Section 9.3.1), and common epitopes have also been defined by monoclonal antibodies.

Table 4.10 CD11/CD18: A family of heterodimeric molecules sharing a common β chain; β2 (leukocyte) integrins

Molecule	Expression on		$M_r\alpha$	$M_r\beta$
	lymphoid cells	myeloid cells		
CD11a/CD18 LFA1	Yes	Yes	170–177 kDa	95 kDa (CD18)
CD11b/CD18 CR3	Few	Yes	165 kDa	95 kDa
CD11c/CD18 p150, 95	?	Yes	150 kDa	95 kDa

The intact CD11a/CD18 molecule is the lymphocyte function-associated antigen-1, LFA-1; CD11b/CD18 is the complement receptor type 3, CR3; and CD11c/CD18 is called p150,95. Here, for clarity, these alternative names will be used instead of the CD nomenclature. LFA-1 is expressed by lymphocytes (including T cells), myeloid cells (monocytes, macrophages, and granulocytes), and a variety of other cell types. In contrast, CR3 and p150,95 are *not* expressed by T cells (except for a small subset of CD4$^+$ cells expressing CR3) but are included here for completeness.

Leukocyte adhesion deficiency, LAD. The importance of the $\beta 2$ integrins for functioning of the immune system in host defence is illustrated by an immunodeficiency disease of humans, LAD. This inherited condition is due to a defect in the gene for CD18 which results in defective expression of all three molecules, LFA-1, CR3, and p150,95, at the cell surface (Anderson *et al.* 1987). Levels of expression may be much reduced, to 5–10% of normal (the moderate phenotype), or essentially absent, <0.3% of normal (the severe phenotype). This is associated with a number of clinical symptoms, and LAD patients are particularly susceptible to recurrent bacterial infections that may be fatal.

LFA-1 (CD11a/CD18) (Table 4.11). The genes for CD11a and CD18 have been cloned (Larson *et al.* 1989; Kishimoto 1987). LFA-1 expression is confined exclusively to certain hemopoietic cells. This molecule is expressed by T and B lymphocytes, NK cells, granulocytes, and some dendritic cells and macrophages, but *not* by precursors to myeloid and erythroid cells. It was called the lymphocyte function-associated antigen-1 because monoclonal antibodies specific for the α chain of this molecule were found to inhibit the function of helper and cytotoxic T lymphocytes; of course anti-CD18 antibodies can have similar effects. For example, anti-LFA-1 antibodies can inhibit T cell proliferation in response to specific antigens and mitogens, T-dependent B cell responses (although T-independent responses, Section 8.3.2, are unaffected), and cytotoxicity by CTL as well as NK cells (Section 10.3.2). This contrasts to anti-CD4 and anti-CD8 antibodies which predominantly inhibit helper or cytotoxic responses respectively (Section 4.2.4).

NB

LFA-1 mediates the adhesion of T cells to other cell types, including antigen-presenting cells. Although it is constitutively expressed by T cells (Dustin and Spinger 1989), LFA-1 may be *inactive* until the TcR–CD3 complex is ligated by recognition of specific peptide–MHC complexes, perhaps in association with other T cell activation signals (e.g. Section 3.4.3).

Table 4.11 CD11b/CD18: LFA-1

Distribution	T cells, B cells, NK cells, some Mφ, granulocytes
Structure	Heterodimer ($\alpha+\beta$): α=CD11b, β=CD18
Approx. M_r	α 170–177 kDa; β 95 kDa
Homology to Ig	No: a member of the integrin gene superfamily
Chromosome localization of genes	*Human* α 16 β 21 *Mouse*

Cross-linking of the TcR-CD3 complex by anti-CD3 antibodies induces transient activation of LFA-1 (by 20 min) followed by deactivation (within 2 h) on a small subset of peripheral T cells. Interestingly, a similar phenomenon is seen in macrophages after cross-linking of molecules such as CD14 (Section 10.4.1). A putative cation-binding site on CD11a (also present in CD11b and CD11c) has been implicated in the transition of LFA-1 from the 'off' to the 'on' state, but this may not be the only requirement for activation of its adhesion functions.

There are two or more ligands for LFA-1, those defined to date being **ICAM-1** (**CD54**) and **ICAM-2**, which are members of the Ig gene superfamily; ICAM-1 contains five extracellular Ig-like domains and ICAM-2 contains two. Both molecules are expressed by endothelial cells. ICAM-1 is one of the major vascular addressins for T cells (see above) and its expression can be induced on endothelial cells by cytokines such as IL1, TNFα, and IFNγ (Section 9.5). In addition, ICAM-1 is expressed weakly by resting peripheral lymphocytes, but is up-regulated after mitogen-stimulation of T cells or Epstein–Barr virus (EBV) transformation of B cells. Expression of ICAM-1 is thus important for adhesion of T cells to endothelial cells and LFA-1$^+$ antigen-presenting cells. It also turns out that ICAM-1 is the receptor for the major group of rhinoviruses, responsible for about one-half of all common colds (Staunton *et al.* 1990).

ICAM-1 was first defined by a monoclonal antibody raised in mice by immunization with LFA-1$^-$ cells from LAD patients (see above). The existence of this ligand was predicted from the capacity of LAD cells to adhere to normal LFA-1$^+$ T cells (**heterotypic adhesion**). This antibody precipitated the heavily-glycosylated 90–114 kDa ICAM-1 molecule (which contains a 55 kDa polypeptide). Subsequently the anti-ICAM-1 antibody was used in a cloning strategy to isolate the ICAM-2 gene (Staunton *et al.* 1989). The existence of the ICAM-2 ligand was predicted because it was found that the adhesion of cells of one particular cell line to each other (**homotypic adhesion**) could be inhibited by anti-LFA-1 but not by anti-ICAM-1 antibodies. A cDNA library was prepared from mRNA isolated from endothelial cells, and expressed in COS cells (see Appendix). These cells were then incubated in petri dishes coated with LFA-1, in the presence of the inhibitory anti-ICAM-1 antibody. The cells that attached to the dishes thus expressed cDNA for an LFA-1 ligand distinct from ICAM-1, and this strategy allowed the ICAM-2 gene to be cloned.

CR3 (CD11b/CD18) and p150,95 (CD11c/CD18). As noted above these molecules are expressed by myeloid cells, but *not* by the majority of T cells. However, like LFA-1, they play an important role in adhesion reactions and in emigration of these cells to extravascular sites.

CR3, as its name implies (see above), can bind complement components, particularly iC3b, and it mediates phagocytosis of complement-coated particles by macrophages and granulocytes (Section 10.4.5). However, this molecule has a wide variety of other ligands. These include a ligand on endothelial cells, fibrinogen of the extracellular matrix (see below), the filamentous hemagglutinin of *Bordetella pertussis* (the bacterium which causes whooping cough), and gp63 of *Leishmania donovani* (the protozoan which causes leishmaniasis). In addition to these protein ligands, CR3 probably has distinct binding sites for carbohydrate ligands such as bacterial lipopolysaccharide (endotoxin) and the lipophosphoglycan of Leishmania. Like LFA-1, CR3 can exist in active and inactive states, and 'on/off' cycles of this molecule can be stimulated by chemoattractants such as C5a (p. 552) produced at inflammatory sites, as well as by a soluble form of ELAM-1 (see above). Presumably, activation of this molecule facilitates the attachment of myeloid cells to endothelial cells and their homing to inflamed tissues.

The diversity and distribution of ligands for CR3 suggests that this molecule is likely to play an important role in emigration of neutrophils and macrophages to extravascular sites, and in host defence. This has been confirmed from *in vivo* experiments (reviewed in Rosen 1990). Pretreatment of mice with an inhibitory anti-CR3 antibody, 5C6, was found to inhibit the migration of intravenously-injected macrophages into the peritoneal cavity; the recruitment of neutrophils to this site in response to an inflammatory stimulus; and delayed-type hypersensitivity responses (Section 10.5.4) to sheep erythrocytes. However, pretreatment dramatically increased the susceptibility of mice to infection by the bacterium *Listeria monocytogens*: whereas normal mice recovered after injection of several thousand bacteria, mice pre-treated with 5C6 died when injected with as few as ten organisms. Remarkably, 5C6 treatment also reversed the diabetes that can be induced in mice after injection of T cells from the NOD strain of mouse, which is pre-disposed to type I diabetes. This was

Table 4.12 The integrin family of cell adhesion receptors*

Sub-family	Receptor/subunits	Other names	R-G-D recognition	Distribution	Ligand/function
β1 integrins	VLA-1 ($\alpha^1\beta_1$)	CD49a?	No?	Widespread	Adhesion to collagen, laminin
	VLA-2 ($\alpha^2\beta_1$)	CD49b Ia/IIa ECMRI	No?	Widespread	Adhesion to collagen, laminin
	VLA-3 ($\alpha^3\beta_1$)	CD49c? ECMRII	Yes?	Widespread	Adhesion to fibronectin, laminin (collagen)
	VLA-4 ($\alpha^4\beta_1$)	CD49d LPAM-2	No	Lymphocytes, monocytes and other	Cell–cell adhesion; Peyer's patch HEV; adhesion to fibronectin
	VLA-4_{alt} ($\alpha^4\beta_p$)†	LPAM-1	No?	Mouse leukemic cell lines, lymph node	Peyer's patch HEV
	VLA-5 ($\alpha^5\beta_1$)	CD49e? Ic/IIa FNR	Yes	Widespread	Adhesion to fibronectin
	VLA-6 ($\alpha^6\beta_1$)	CD49f Ic/IIa	No?	Widespread	Adhesion to laminin

	VLA-6_{alt} ($\alpha^6\beta_4$)†	—	No?	Epithelial	?
β2 integrins	LFA-1 ($\alpha^L\beta_2$)	CD11a	No	Lymphoid, myeloid	Leukocyte adhesion; binds ICAM-1, ICAM-2
	MAC-1 ($\alpha^M\beta_2$)	CD11b MAC-1 MO1	Yes?	Myeloid	Leukocyte adhesion; binds C3bi, leishmania, factor X, fibrinogen, LPS, others
	p150, 95 ($\alpha^X\beta_2$)	CD11c CR-4	No	Myeloid	Leukocyte adhesion, (C3bi)?
β3 integrins	IIB/IIIA ($\alpha^{IIB}\beta_3$)	CD41a	Yes	Platelets	Adhesion to fibronectin, fibrinogen, Von Willebrand factor, vibronectin
	VNR ($\alpha^V\beta_3$)	CD51	Yes	Endothelial, tumour, and other cells	Adhesion to vibronectin, Von Willebrand factor, fibrinogen, others
	VNR_{alt} ($\alpha^V\beta_5$)†	$\alpha^V\beta_1$	Yes?	Carcinomas, others	fibronectin, vibronectin

* Adapted from Hemler *et al.* 1990 — see Further reading.
† Examples of the usage of alternative β subunits that can bind to the same α subunit.

Table 4.12 (continued) Molecular weights (kDa) of integrin subunits under reducing and nonreducing conditions

Subunit	Reduced	Nonreduced
α^1	210	200
α^2	165	160
α^3	130/25	150
α^4	150	140
α^5	135/25	155
α^6	120/30	140
α^L	180	170
α^M	170	165
α^X	150	145
α^V	125/24	150
α^{IIB}	120/25	145
β_1	130	110
β_2	95	90
β_3	105	90
β_4	220	210
	(180)	(165)
	(145)	(125)
β_5	110	100
β_p	?	100

associated with inhibition of macrophage and T cell infiltration of the insulin-producing islets of Langerhans in the pancreas of these animals. Clearly, to label CD11b/CD18 simply as a 'complement receptor' does not do justice to its actual function.

At present, less is known about the functions of **p150,95**, but it seems likely that at least some of these will be similar to those of CR3. The cellular distribution of p150,95 is somewhat different to that of CR3. One intriguing observation is that, in mice, expression of p150,95 appears to be restricted primarily to dendritic cells (Section 3.4.1), and it may be the principal $\beta2$ integrin they express. An antibody specific for this molecule has been used to identify a subset of dendritic cells within the marginal zone of spleen that is phenotypically distinct from the interdigitating cells of T areas; whether these subsets have different functions is not yet known (Section 3.4.2).

The $\beta1$ and $\beta3$ integrins

The $\beta1$ integrins include the VLA glycoproteins. These molecules, which share the 130 kDa β_1 subunit (CD29) but have distinct α chains (CD49), were named after two 'very late antigens', VLA-1 and VLA-2, expressed by T cells after some time in culture. VLA-1 and VLA-2, as well as a $\beta3$ integrin, the vitronectin receptor (CD51/CD61), are expressed by activated but not resting T cells, and can thus be classified as **activation antigens**. In contrast, four other VLA proteins, VLA-3 to VLA-6, appear to be expressed constitutively by resting T cells; VLA-4 to VLA-6 are also expressed by monocytes. While expression of most $\beta2$ integrins is widespread, VLA-4 may be relatively restricted to these two cell types. The $\beta1$ and $\beta3$ integrins are receptors for components of the extracellular matrix.

The extracellular matrix. The extracellular matrix holds tissues together but is more than just an inert scaffolding. It greatly influences the development, behaviour, and polarity (cell directionality) of the cells it contains. The matrix is essentially a hydrated gel containing fibres composed of different proteins that are secreted by cells such as fibroblasts. The three major proteins of the extracellular matrix are collagen, fibronectin and elastin. **Collagen** is composed of triplets of different types of polypeptide chains, about 1000 amino acids long; **fibronectin** is a disulfide-bonded dimer of 2500 amino acid polypeptides, which is also deposited by macrophages at inflammatory sites; and **elastin** is a protein of 830 amino acids. Basal laminae, a specialized form of matrix, contain another major protein called **laminin** which is composed of disulfide-bonded dimers. The matrix also contains other molecules such as **vitronectin**.

Recognition by $\beta1$ and $\beta3$ integrins. Once T cells, and other leukocytes, have entered extravascular sites, recognition of the extracellular matrix is mediated by $\beta1$ and $\beta3$ integrins. These molecules are specific for one or more of the matrix components, although there can be species differences in the patterns of recognition by homologous molecules. Some integrins recognize amino acid sequences containing the **RGD motif** (arginine–glycine–aspartic acid, sometimes with serine, RGDS) within their ligands, for example VLA-5 recognizes this motif in fibronectin. However, other VLA molecules recognize different sequences, for example VLA-4 can recognize an alternatively-spliced form of fibronectin lacking the RGD motif, so the same matrix component can have distinct cell-binding domains. Moreover, some integrins have additional specificities for non-matrix components; thus VLA-4 is also a receptor for VCAM-1, a member of the Ig gene superfamily (see above).

Costimulatory function of β1 and β3 integrins for T cell activation. It has become clear that β2 integrins constitutively expressed by T cells can synergize with other T cell molecules such as CD2, CD4, CD8, and LFA-1 (Sections 4.2.7, 4.2.4., and above), and sometimes with each other, to induce proliferation of T cell subsets; this phenomenon is termed **costimulation**. This also applies to VLA-3, VLA-4, VLA-5, and VLA-6, as well as the vitronectin receptor, a β3 integrin (for example, Nojima *et al.* 1990; Roberts *et al.* 1991). These **costimulatory functions** for T cell activation may be a general feature of extracellular matrix proteins. Thus signals delivered through integrins to T cells adhering to the extracellular matrix, can facilitate the T cell response to antigen-presenting cells that may be present within, or recruited to, these extravascular sites during inflammation and infection.

4.2.7 CD2: an alternative pathway for T cell activation (Table 4.13)

Human T lymphocytes were first identified by the seemingly strange property of being able to adhere to sheep erythrocytes (SRBC) and form 'rosettes'. Most human T cells can rosette with chemically-treated or neuraminidase-treated SRBC (neuraminidase is an enzyme which cleaves terminal sialic acid residues from glycoproteins—see Fig.1.4), while thymocytes and T cell blasts can rosette with untreated SRBC. Monoclonal antibodies were then produced that inhibited rosetting, and which defined a molecule which was first termed the 'SRBC receptor', later called T11 and now designated CD2 (Table 4.1). It was presumed that the natural ligand for CD2 was mimicked by a molecule on SRBC, and both have now been identified. The physiological ligand for CD2 is now known to be LFA-3, which has been designated CD58,

Table 4.13 CD2

Distribution	T cells, NK cells, some Mφ (rat)
Structure	Single chain
Approx. M_r	*Human:* 54 kDa (inc 3 N-CHO) *Mouse:* 60 kDa (inc 5 N-CHO)
Homology to Ig	Yes: 2 extracellular Ig-like domains
Chromosome localization of genes	*Human* *Mouse* 3
Comment	Signal transduction is connected to the TcR-CD3 cascade

while the related molecule on SRBC is called the T11 Target Structure, T11TS. The genes for both CD2 and LFA-3 (this terminology being used here for clarity) have been cloned, and found to be members of the Ig gene superfamily. Their structures are discussed below. The expression of CD2 is restricted to T cells, thymocytes and NK cells in the human, although B cells in the mouse and splenic macrophages in the rat also express this molecule.

Structure and function of CD2

Anti-CD2 antibodies can have both inhibitory and stimulatory effects on T cell functions. For example, some antibodies *inhibit* proliferation of helper T cells, secretion of IL-2 and expression of the high affinity IL-2 receptor (Section 4.4.3), as well as adhesion of cytotoxic T cells to target cells and their cytolysis (Section 10.3). Because of these inhibitory effects, the CD2 molecule was originally termed lymphocyte-function associated antigen-2 (LFA-2; compare with LFA-1, Section 4.2.6) by some investigators.

In contrast, combinations of other anti-CD2 antibodies were found to *stimulate* T cell responses; for example, they could induce proliferation of helper T cells, cytolysis by cytotoxic T cells, and expression of IL-2 receptors by thymocytes *in vitro*. Soluble anti-CD2 antibodies are capable of stimulating these responses, unlike anti-CD3 antibodies which first need to be immobilized on synthetic beads or to interact with macrophage Fc receptors (Section 4.2.3). The inhibitory and stimulatory effects of anti-CD2 antibodies indicate that *CD2 serves a dual role in mediating both T cell adhesion (e.g. to antigen-presenting cells) and activation.*

Because of the T cell activating properties of stimulatory anti-CD2 antibodies, the CD2 molecule is said to provide an **alternative pathway** for T cell activation, meaning an alternative to that triggered through the TcR–CD3 complex (Meuer *et al.* 1984). Even so, these pathways are interconnected in mature T cells (see below). Other molecules have also been implicated in CD2-mediated responses of T cells. These include CD28 (Section 4.2.8), CD45RA (Section 4.2.5) and CD44, a homing receptor (Section 4.2.6) also known as Hermes or Pgp-1; antibodies specific for these molecules can augment T cell activation via CD2. However, the physiological *ligand* for CD2 is LFA-3.

Interaction between CD2 and LFA-3 (CD58)

The LFA-3 molecule was first defined by monoclonal antibodies that inhibited T cell functions by binding to the surface of the target cells rather than to the T cells and its name was due to these functional effects (compare with LFA-1, Section 4.2.6). It is now thought that the interaction between CD2 and LFA–3 provides a mechanism for early cell adhesion, in addition to that mediated by the interaction between LFA-1 and ICAM–1 (Section 4.2.6). These two pathways are different in that the latter requires Mg^{2+} ions and is temperature-dependent, whereas the interaction between CD2 and LFA-3 is independent of both. These initial adhesion events between T cells and antigen-presenting cells or target cells precede antigen recognition via the TcR.

Evidence for an interaction between CD2 and LFA-3 was obtained from experiments with the purified molecules (Dustin *et al.* 1987). Purified CD2 molecules induced human erythrocytes to aggregate, and this was specifically inhibited by anti-LFA-3 antibodies (which bind to erythrocytes). Furthermore, LFA-3 in planar membranes mediated attachment of T cells via the CD2 molecule. It was also found that binding of radio-iodinated purified CD2 molecules to an LFA–3^{+} cell line could be inhibited by antibodies specific for LFA-3 (or, of course, for CD2). In the reciprocal experiment, binding of purified LFA-3 to the same cells coated with purified CD2 was inhibited by antibodies to CD2.

LFA-3 is expressed on essentially all hemopoietic cells, as well as connective tissue and epithelial cells. This 55–70 kDa molecule contains two Ig-like extracellular domains which can be joined to the membrane either via a conventional transmembrane portion or via a glycosyl-phosphatidylinositol (GPI) linkage (Fig.1.5); these forms contain polypeptide chains of molecular weight 29 kDa and 25 kDa respectively, and presumably mediate different functions. Other members of the Ig gene superfamily also have these alternate forms, including the $Fc_{\gamma}RIII$ receptor of human neutrophils (Section 10.4.5) and N-CAM, which is expressed in the nervous system but not the immune system; Thy-1 (Section 4.2.7) also has a GPI anchor. GPI-linked molecules can be released from the cell surface by a specific phospholipase c enzyme, but the significance of this is not yet clear.

Signalling in T cells and NK cells. In mature T cells, signalling through CD2 apparently requires the presence of TcR at the cell surface: mutant T cell lines lacking TcR cannot be stimulated by anti-CD2 antibodies, but CD2 triggering is restored if they are transfected with the relevant TcR gene. Moreover, the signalling cascades mediated via CD2 and the TcR-CD3 complex (Section 4.2.3) are connected (Bierer *et al.* 1988). This is evident from the effects of antibodies specific for one component on signalling through the other (see above) and from other observations. For example, stimulation of mature T cells by anti-CD2 antibodies results in phosphorylation of the CD3 γ and ζ chains associated with the TcR–CD3 complex, and an increase in intracellular Ca^{2+} concentration and the induction of IL-2 gene expression.

Despite this association between CD2 and the TcR–CD3 complex in mature T cells, anti-CD2 antibodies can induce cytotoxic activity in NK cells which express neither TcR nor CD3 (Section 10.2.4). Furthermore, thymocytes at an early stage of development also fail to express these genes and yet a rise in intracellular Ca^{2+} concentration can be stimulated via CD2; this pathway has been implicated in the differentiation of CD3^{-} thymocytes towards the $\gamma\delta^{+}$ TcR lineage. Clearly, signalling through CD2 can occur in the absence of the TcR–CD3 complex, and in these cases CD2 is presumably coupled to other signal transduction elements. In NK cells, one candidate is a molecule associated with the CD16 $Fc_{\gamma}RIII$ receptor expressed by these cells, called the $Fc_{\gamma}RIII\gamma$ chain (Section 10.4.5) (Arulanandam *et al.* 1991).

Mast cells express three different Fc receptors, $FC_{\gamma}RII$, CD16, and $Fc_{\varepsilon}RI$; the γ chain of CD16 is also associated with the latter receptor (Section 10.4.5). This molecule is structurally homologous to the CD3 ζ and η chains and is a member of the same gene family. However, mast cells do not express these CD3-associated molecules nor CD2. When mast cells were transfected with the CD2 gene, a combination of stimulatory anti-CD2 antibodies (see below) was found to induce a rise in intracellular

Ca^{2+}, histamine release and the production of IL-6 by the transfectants. Moreover, cross-linking of the CD2 and $Fc_{\gamma}R$ receptors triggered IL-6 production. It has therefore been suggested that the $Fc_{\gamma}RIII\gamma$ chain, or a related structure, is involved in signal transduction via CD2 in $CD3^{-}$ NK cells and thymocytes, and perhaps also in B cells of the mouse, which express this molecule.

Structure and functional regions of CD2 (Fig.4.20). CD2 on T cells is a monomeric glycoprotein. Genes for CD2 have been cloned from human, rat, and mouse (for example, Sewell *et al.* 1987). The sequence of human CD2 predicts an extracellular region of 185 amino acids forming two Ig-like domains, a transmembrane region of 25 amino acids, and a large cytoplasmic portion of 117 amino acids. The large cytoplasmic tail of CD2 is probably involved in signalling functions and/or interactions with cytoskeletal elements. Some functions of different regions of the extracellular and cytoplasmic domains of CD2 have been defined.

1. The extracellular domain. Three functional regions of CD2 can be defined from the effects of anti-CD2 antibodies on its adhesion and T cell activating properties (see above). These antibodies can be grouped according to their function and are designated anti-$T11_1$, anti-$T11_2$, and anti-$T11_3$ antibodies (these and their corresponding determinants sometimes also being described as CD2-1, CD2-2, and CD2-3 respectively). Anti-$T11_1$ antibodies inhibit rosetting (adhesion) between T cells and SRBC. Anti-$T11_2$ antibodies bind to a separate site and do not inhibit rosetting, but in combination with anti-$T11_3$ antibodies, which bind to a different determinant, induce helper T cell proliferation and the secretion of cytokines, and activate cytotoxic T cells and NK cells. (It is important to note, however, that the cytotoxic activity of $\gamma\delta^{+}$ T cells can be induced by anti-CD2 antibodies specific for only one epitope.)

Three functional regions of CD2 have also been mapped by constructing a series of COS cells transfected with mutant CD2 genes (Fig.4.20) (Peterson and Seed 1987). These cell lines were tested for their ability to be recognized by a panel of anti-CD2 antibodies and their capacity to bind to cells expressing LFA-3, to determine whether the mutation was of structural significance. Anti-$T11_1$ antibodies bound to a sequence containing residues 86-101; anti-$T11_2$ antibodies to a sequence closer to the N-terminus containing residues 42–57; and anti-$T11_3$ antibodies to two residues in the membrane-proximal domain, a tyrosine at position 104 and a glutamate at position 141. LFA-3 is thought to bind at least in part at, or close to, the sequence containing residues 42–57 (i.e. the anti-$T11_2$ binding site). It is of interest that if this region is modelled according to an Ig κ chain, to which it has greatest homology, it is the equivalent of CDR-2 and CDR-3 loops (Fig.6.16) which interact directly with antigen. (As an aside, note that the region equivalent to the CDR-2 loop in CD4 creates a prominent ridge in that molecule; Section 4.2.4.)

2. The cytoplasmic domain. Signal transduction through CD2 is dependent on its cytoplasmic tail, as shown by studies of deletion mutants lacking this region.

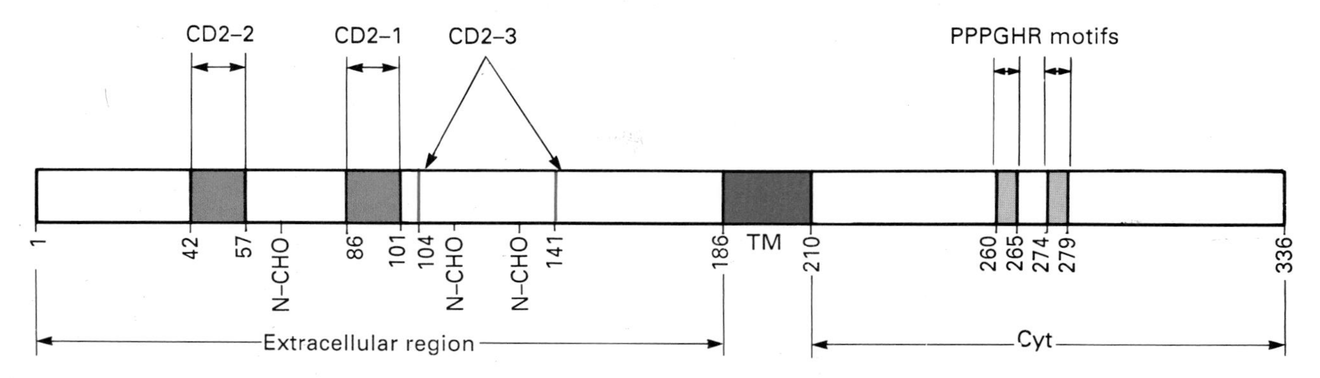

Fig.4.20 Functional determinants of the human CD2 molecule (see text)

Experiments with site-directed CD2 mutants defined a segment between residues 253 and 287 that is required for the CD2-mediated increase in intracellular Ca^{2+} and induction of IL-2 gene expression in T cells, noted above. This region contains two repeats of a particular sequence motif, PPPGHR, (positions 260–265 and 274–279), and the first motif is required for these responses (Chang *et al.* 1990). Its precise function, however, is not yet known.

4.2.8 CD28: regulation of cytokine gene expression (Table 4.14)

CD28, also known as T44 or Tp44, is expressed by mature T cells, thymocytes, and plasma cells. This 44 kDa glycoprotein contains a single Ig-like extracellular domain, and is expressed as a disulfide-linked homodimer at the cell surface. Its ligand is the B7/BB1 molecule which contains two extracellular Ig-like domains. To date, expression of the latter molecule has only been detected on activated and malignant B cells, and dendritic cells.

Table 4.14 CD28

Distribution	T cells, thymocytes, plasma cells
Structure	Disulfide-linked homodimer
Approx. M_r	44 kDa (monomer)
Homology to Ig	Yes: 1 Ig-like extracellular domain
Chromosome localization of genes	Human Mouse
Comments	Ligand is B7/BB1; signal transduction is independent of TCR-CD3 cascade

CD28 expression on thymocytes and mature T cells. CD28 is expressed at low levels on immature CD3$^-$ double-positive thymocytes, but at higher levels on the more mature CD3$^+$ single-positive cells. Approximately 80% of human peripheral blood T cells also express CD28, and this has been correlated with various functional capacities. About 95% of CD4$^+$ T cells express CD28 to a varying extent, and those with the lowest levels can secrete a particular spectrum of cytokines (see below) after appropriate stimulation. It has been suggested that the small subpopulation of CD28$^-$ CD4$^+$ T cells may not be selected in the thymus (Section 5.3.1). About 50% of CD8$^+$ T cells express CD28 and these cells can develop cytotoxic activity, whereas the CD28$^-$ subset can suppress Ig synthesis by B cells, thereby apparently discriminating between at least some cytotoxic and suppressor T cells (Section 11.2). A curious observation is that mature T cells lacking CD28 express the CD11b/CD18 molecule (Section 4.2.6).

Effects of anti-CD28 antibodies on T cell responses. Anti-CD28 antibodies exhibit both costimulatory and inhibitory functions in T cell responses (Moretta *et al.* 1985). However, unlike CD2 which is connected to the TcR-CD3 signalling cascade in mature T cells (Section 4.2.7), CD28 uses distinct signalling pathways. Moreover, stimulation of CD28 increases cytokine production at both transcriptional (DNA) and post-transcriptional (mRNA) levels.

The effects of anti-CD28 antibodies on T cells include the following.

(a) On their own, they do not stimulate T cells. However, if their valency is increased by aggregation there is an increase in cytoplasmic Ca^{2+}, activation of the phosphatidylinositol pathway (Section 4.2.3) and increased expression of the IL2 receptor p55 chain (Section 4.4.3) in T cells. This pathway appears to be unlinked to the TcR-CD3 signalling cascade (compare with CD2; Section 4.2.7).

(b) *Augmentation* of T cell and thymocyte proliferation triggered by suboptimal concentrations of mitogens, anti-CD2 or immobilized anti-CD3 antibodies.

(c) *Inhibition* of antigen-specific responses (i.e. via TcR) of T cell clones and proliferation in the MLR, depending on the extent of cross-linking: monovalent Fab fragments inhibit but anti-CD28 $F(ab')_2$ fragments stimulate.

(d) Optimal concentrations of anti-CD3 antibodies for T cell proliferation lead to very little cytokine production *in vitro*. The addition of anti-CD28 antibodies greatly increases accumulation of cytokines, particularly IL-2, in the culture medium. CD28$^+$ T cells can produce IL-2, IFNγ, TNFα, TNFβ, GM-CSF and IL-3, resembling the cytokines produced by the putative subset of T_H1 cells in the mouse (Section 4.4.2).

(e) At least for IL-2, anti-CD28 antibodies augment cytokine production by initially inhibiting degradation of the mRNA (i.e. stabilizing the message) and later by enhancing transcription. In T cells stimulated with these antibodies, a protein complex is induced which binds to a site in the 5′ flanking region of the IL-2 gene, and increases IL-2 enhancer activity fivefold (Fraser *et al.* 1991). This site is upstream of the transcription start site, and is similar to conserved regions in several other cytokine genes.

The B7/BB1 ligand for CD28. The B7/BB1 or B7 molecule is an activation antigen of B cells, being expressed by activated but not resting B cells. This molecule can be induced on small resting B cells in the human by cross-linking HLA-DR (MHC class II) molecules. The importance of the B7-CD28 interaction for T cell activation and cytokine production has been demonstrated in studies with fusion proteins of B7 or CD28 linked to human Ig $C_\gamma 1$ domains to produce soluble molecules (Linsley *et al.* 1991*a*). B7/BB1 is also a ligand for CTLA-4, expressed by many cytotoxic T cells (Linsley *et al.* 1991*b*).

The distribution of the B7/BB1 antigen, and the inhibitory effects of antibodies specific for this molecule or for CD28, is intriguing. Dendritic cells and perhaps B cell blasts are the only cell types so far shown capable of initiating primary T cell responses (immunostimulation; Section 3.4.3). If expression of B7/BB1 is in fact restricted to these cell types, an essential 'second signal' for T cell activation (Fig.3.3) could be delivered through the interaction of CD28 with B7/BB1. In the absence of this signal, antigen recognition may result in an unresponsive state or anergy of T cells (Harding *et al.* 1992); this could be one mechanism for inducing T cell tolerance (Section 5.2.5).

4.2.9 Thy-1 (Table 4.15)

The Thy-1 molecule was first discovered on *mouse* thymocytes and mature T cells, and was originally called the theta (θ) antigen. This molecule, which may dimerize, is a 25 kDa glycoprotein with a single extracellular Ig-like domain anchored to the membrane via a GPI linkage. Its function is currently unknown (but see below) and matters are complicated by species differences in expression on various cells and tissues. For example, it is important to note that Thy-1 is *not* expressed by mature T cells in the rat and human, although it is expressed on thymocytes. In addition, Thy-1 is expressed on brain and kidney in all three species; on fibroblasts and hemopoietic stem cells in the mouse; and on immature B lymphocytes in the human.

In trying to understand the function of this molecule, one strategy has been to construct transgenic mice containing recombinant Thy-1 genes, but although these have provided some insights into the tissue-specific control of expression of Thy-1, as yet they have not revealed its function. Nevertheless, a role for mouse Thy-1 in T cell proliferation and thymocyte adhesion has been suggested by some *in vitro* studies.

Table 4.15 Thy-1

Distribution	Mature T cells in mouse (but not in human or rat); thymocytes; brain and kidney; others
Structure	Single chain
Approx. M_r	19–25 kDa
Homology to Ig	Yes: 1 Ig-like extracellular domain
Chromosome localization of genes	Human Mouse 9
Comments	Not a transmembrane molecule: GPI-linked

Regarding the former, a few antibodies specific for Thy-1 can induce proliferation of T cell hybridomas in the absence of accessory cells (Pont *et al.* 1985). However, other anti-Thy-1 antibodies do not stimulate mitogenesis even after cross-linking by anti-Ig antibodies, a procedure that can trigger T cell responses in the case of some monoclonal antibodies specific for other T cell molecules.

Some members of the Ig gene superfamily have important roles in molecular and cell adhesion (e.g. LFA-1 and ICAM-1, CD2 and LFA-3 in the immune system, V-CAM on endothelial cells, and N-CAM on neural tissue; see elsewhere in this chapter and Table 7.4). This has led to the speculation that Thy-1 may also have adhesive functions. Since Thy-1 appears to be expressed ubiquitously on thymocytes from different species, this led some investigators to examine a possible role for Thy-1 in adhesion of thymocytes (and some lymphoma cell lines) to a thymic epithelial cell line (He *et al.* 1991). It was found that purified, soluble Thy-1 molecules as well as Fab fragments of an anti-Thy-1 antibody significantly, though not completely, inhibited binding to the epithelial cells; this adhesion pathway was Ca^{2+}-independent and occurred at 4°C. While these experiments suggest that Thy-1 can have adhesion functions, its ligand(s) remain to be identified.

4.3 Expression of membrane molecules during T cell development

Before the T cell repertoire develops (Chapter 5) it is thought that a few stem cells initially colonize the embryonic thymus. These give rise to at least four major subpopulations of thymocytes in the adult mouse which can be defined according to the expression of a combination of different markers. One complication is that these markers are not expressed in an all-or-nothing manner, but often to a greater or lesser degree on different cells, and there may be overlap between the different populations. The markers studied most thoroughly include: CD4, CD8, CD3, and T cell receptors (Sections 4.2.1 to 4.2.4), the latter especially at the gene level; growth factor receptors, especially the IL-2 receptor (Section 4.4.3); homing receptors, MEL-14 in particular (Section 4.2.6); and some molecules with poorly understood functions such as Thy-1 (Section 4.2.9) and CD5 (Section 4.2.4). One approach has been to use immunocytochemistry (see Appendix) to localize expression of these molecules within the thymus.

A small subpopulation (*c.* 2–3%) of thymocytes express neither CD4 nor CD8 and are localized in the thymus cortex (Fig.4.21). Some of these **double-negative cells** express IL-2 receptors and are believed to be intrathymic stem cells which give rise to all thymocyte populations. However, by far the majority of adult thymocytes (*c.* 85%) express both CD4 and CD8 and are not responsive to antigen. Many of these **double-positive cells** are small *cortical* cells that may not develop further, dying (or being killed) within the thymus. Some double-positive cells, however, are blast cells within the *subcapsular region* which may be intermediates between the most immature double-negatives and the mature **single-positive T cells**. The precise relationship between these different populations is not completely clear but ultimately some cells with the mature single-positive $CD4^+CD8^-$ *or* $CD4^-CD8^+$ phenotypes are produced, the predominant T cells in the *medulla.* It is these antigen-responsive cells that are thought to be transported to the periphery, although this has not been demonstrated directly. According to one view, therefore, the most immature cells are located in the thymic cortex while the most mature cells are confined to the medulla.

The idea that T cell development proceeds from the cortex to the medulla has been challenged in studies of the expression of the MEL-14 antigen, a homing receptor expressed on peripheral T cells (Section 4.2.6). Cells that express high levels of MEL-14 in the thymus could be viewed as being the most mature T cells and might be expected to predominate in the medulla where, according to expression of CD4 and CD8, the most mature cells are found. However, by immunocytochemistry, $MEL\text{-}14^+$ cells have a cortical location. Although this is currently a controversial area, an alternative view of T cell development holds that T cells mature in *both* the cortex and the

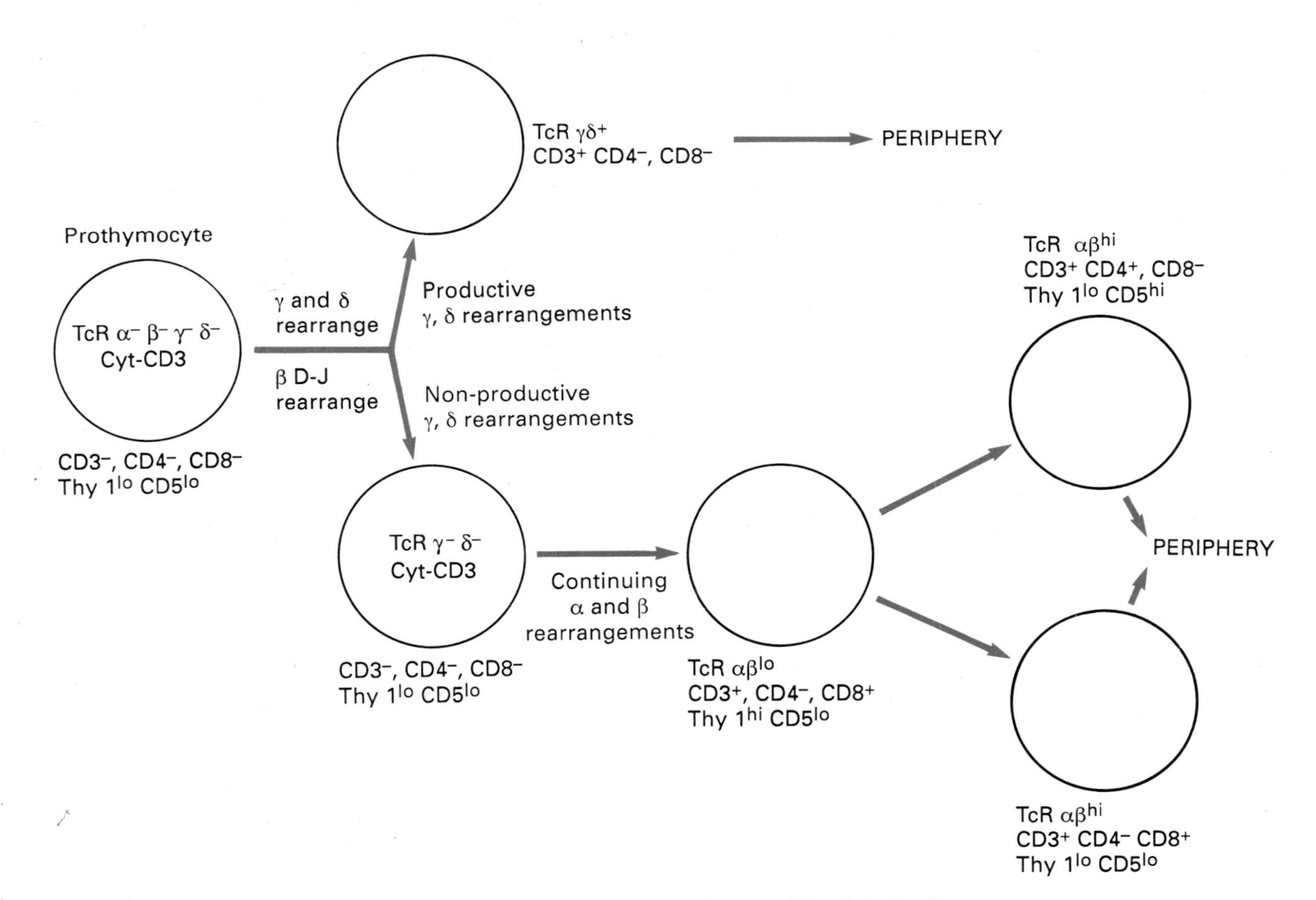

Fig.4.21 A highly schematic view of the differentiation pathways of $\alpha\beta^+$ and $\gamma\delta^+$ T cells in the thymus

medulla, and perhaps leave the thymus via the cortico-medullary junction where the major blood vessels are found. Clearly, much remains to be learnt about this differentiation process.

An alternative approach to studying subpopulations in the thymus *in situ*, has been to examine the phenotype of isolated cells *in vitro*, together with T cell receptor gene arrangements occurring within them, and to try to correlate this with their function. By examining the expression of different T cell molecules, three stages in thymic maturation were initially defined (although these have since been subdivided) (Fig.4.22). Cells at these stages comprise about 10% (stage I), 65% (stage II), and 30% (stage III) of the total thymic population, respectively.

One of the earliest markers to be expressed on thymocytes is CD2 (stage I) and this molecule is retained throughout thymic development (Section 4.2.7). Because CD2 is expressed on double-negative cells which do not express CD3 and have unrearranged T cell receptor genes, it has been suggested that this molecule could be involved in driving thymocyte development at this stage (e.g. proliferation).

It has been shown that some combinations of anti-CD2 antibodies can induce IL-2R expression in $CD3^-$ thymocytes, thus implicating a function for CD2 in an IL-2-dependent pathway of development. However, T cells have been isolated that do not express CD2 but which have T cell receptor-CD3 complexes and are fully functional, and on many thymocytes the expression of one chain of the IL-2R (Section 4.3.4) precedes CD2 expression. Moreover, single-positive T cells expressing the T cell receptor–CD3 complex can develop in neonatal mice or cultured embryonic thymus lobes that have been treated with anti-CD2 antibodies to inhibit CD2 function, so the situation is far from clear.

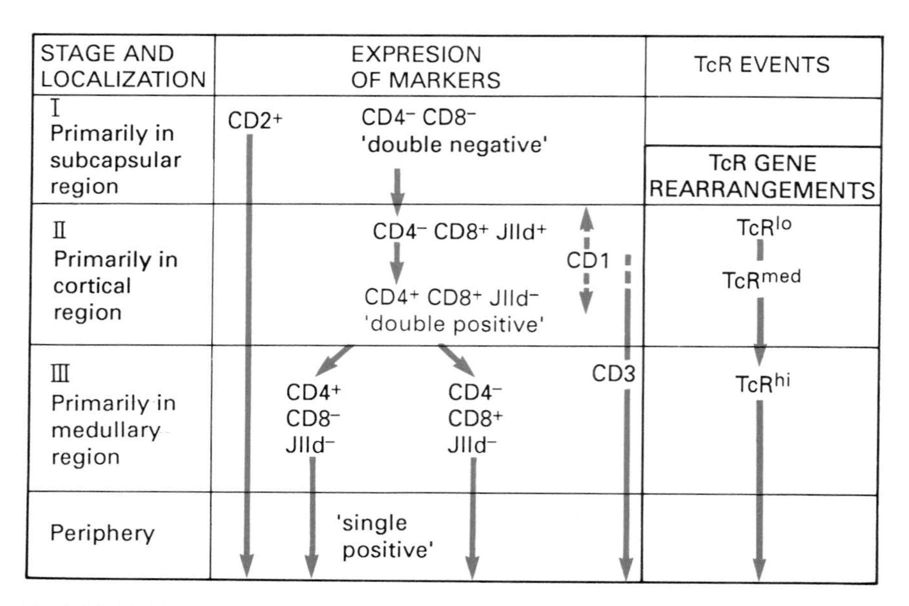

Fig.4.22 Highly simplified scheme of expression of T cell molecules during development in the thymus

During stage II, CD4, and CD8 are coexpressed on the same cells. In the human, the CD1 (T6) molecule is expressed only at this stage in association with both β_2-microglobulin and CD8. It is at stage II that the T cell receptor genes start to be rearranged (see Chapter 7).

In the mouse embryo, the first rearrangements of γ genes are detectable at d14, reaching a maximum at d15 and then declining to very low levels at birth (around d21). β chain rearrangements are also detectable at d14, but these reach a maximum at d17, perhaps because extra joining events are required to produce a functional variable region in the β chain but not in the γ chain (Section 7.3.2).

It is not until stage III that CD3 is expressed, and presumably functional T cell antigen receptor heterodimers can then be expressed on thymocyte membranes. At about this time these cells become single-positive and express either CD4 or CD8$^+$, and only the stage III thymocytes are antigen responsive. Development of the T cell repertoire, and cellular interactions within the thymus, are considered in detail in Chapter 5.

4.4 Cytokines and their receptors in T cell responses

4.4.1 Cytokines

Cells of the immune system can 'communicate' with each other during immune responses via soluble molecules or **cytokines** that are secreted by one cell and which bind to specific membrane receptors on either the same or a different cell (Fig.4.23). Cytokines can be thought of as the hormones of the immune system. They can mediate long-range communication between cells that may be widely distributed throughout the body, as well as short-range communication between cells in close contact with each other. In general, many cytokines act in a **paracrine** fashion in that they are produced by one cell and act on another, but in some cases they can act in an **autocrine** manner, acting on the very same cell from which they were derived.

When immune cytokines were first studied, a distinction was made between those that were thought to be produced by lymphocytes and which were therefore called **lymphokines**, and those produced by monocytes or macrophages, thus called **monokines**. Further work then revealed that many cytokines formally designated as lymphokines or monokines could also be synthesized by cells of other lineages. In most cases, then, these terms are no longer applicable in this restrictive sense and they are now generally avoided. An additional complication became evident when it was found that some activities formally ascribed to different molecules on the basis of quite diverse biological assays could in fact be mediated by the same cytokine.

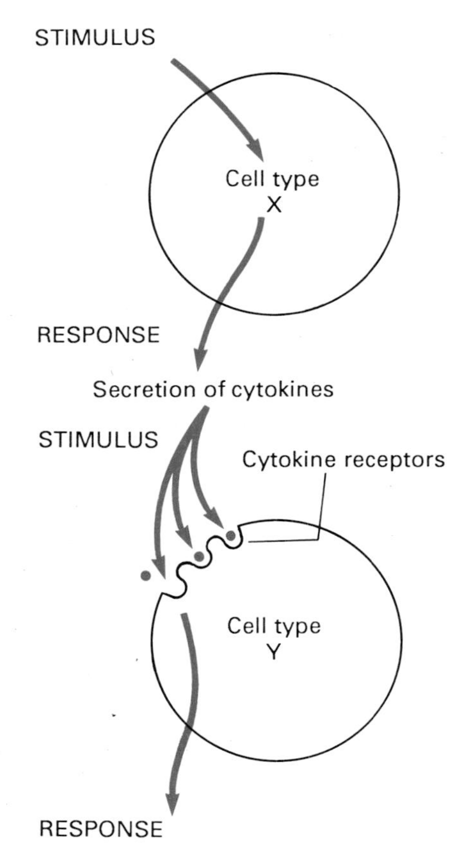

Fig.4.23 Cytokines can 'communicate' between leukocytes and other cell types

A new nomenclature system was then adopted in an attempt to rationalize the data: the term **interleukin** was proposed for any molecule that 'communicated between leukocytes' (*inter* + *leuk*ocytes). It was suggested that as the various cytokines were characterized, and preferably as the molecules were gene cloned, they should be numbered sequentially. In this way there is now an increasingly large group of cytokines designated IL-1, IL-2, IL-3, and so on. However, this nomenclature has not been applied universally, so that other cytokines are still known by their original names, such as the interferons (IFN) and tumour necrosis factors (TNF), even though their genes have been cloned.

Further characterization of various cytokines has revealed that even the original definition of an interleukin is no longer tenable. It is now clear that many of these molecules can also be produced by, and act on, cells other than leukocytes. For example, some interleukins are produced by epithelial cells, and others can stimulate cells in non-lymphoid organs such as the brain and liver, and have a multitude of different effects throughout the body. Indeed, it could be argued that one of the most important advances

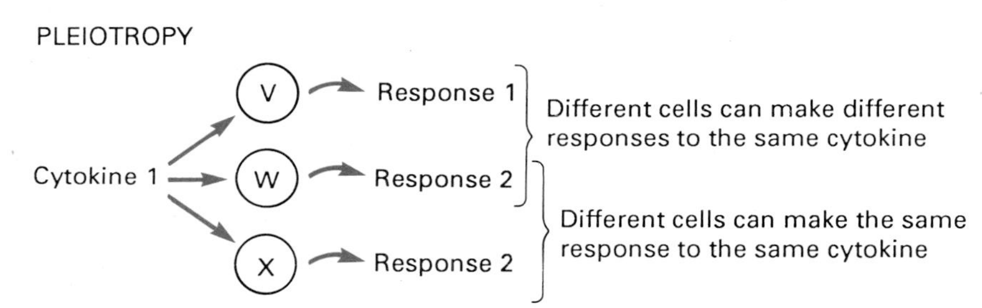

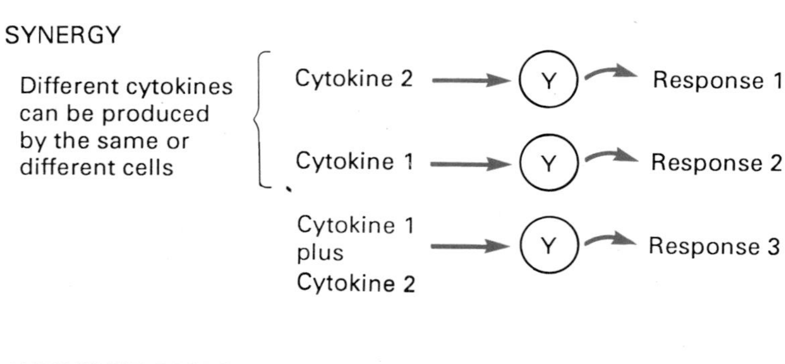

Different cell types represented V, W, X, Y, and Z.

Fig.4.24 Pleiotropic, synergistic, and autocrine effects of cytokines

in this area was the realization of the multiple and diverse, or **pleiotropic** effects, of most cytokines (Fig.4.24). That the same cytokine can have multiple effects on different cells in a variety of tissues underlines the importance of these molecules in regulating and co-ordinating cellular responses throughout the body, and not just within the immune system.

It is becoming possible to dissect the function of various cytokines as they are purified, and as they and their receptors are gene cloned. This has, in some cases, clarified the role these molecules play in cellular responses. Sections 4.4.3 and 4.4.4 consider some interleukins that are involved in T cell responses, although because they have pleiotropic functions this is an artificial distinction. Most attention will be given to IL-2 and its receptor (Section 4.4.3). Then we briefly outline the effects of IL-4, IL-6, and IL-7 on T cells (Section 4.4.4). Further details of IL-4 and IL-6 (and the receptors for these cytokines) are given in Chapter 8, particularly in relation to B cell responses. IL-10 is considered in Chapter 11. Before this however, we need to discuss the finding that different subsets of T cells can secrete different cytokines (Section 4.4.2).

4.4.2 Different T cells can produce different cytokines

In general, it is possible to distinguish between naïve, resting and memory T cells on the basis of their responsiveness to different types of accessory cell (Section 3.1) and different concentrations of antigen, as well as their phenotype although definitive markers have yet to be identified. Thus, naïve and memory T cells need to recognize specific peptide–MHC complexes on immunostimulatory cells in order to generate a response, whereas activated T cells may respond to any antigen-presenting cell expressing the relevant antigen (Section 3.4.3); memory T cells can be distinguished from naïve T cells because of their capacity to respond more efficiently to lower doses of antigen, and by their differential expression of new or elevated levels of certain molecules at the cell surface (e.g. Section 4.2.5); and activated T cells express activation antigens, such as the high affinity IL-2 receptor (Section 4.4.3), that are not present on naïve, or expressed at only low levels on memory, T cells. In addition, these different populations exhibit different recirculation patterns *in vivo* (Mackay *et al.* 1990).

It may also be possible to discriminate between different T cell populations according to their patterns of cytokine synthesis. For example, it has been suggested that stimulation of neonatal ('naïve') T cells results primarily in IL-2 gene transcription, whereas adult ('memory') T cells transcribe the genes for a complete spectrum of cytokines (Section 4.2.5). This situation may, however, be complicated by the relative immaturity of the neonatal immune system in mice compared, for example, to that of humans. An additional complication results from the different functions of helper and cytotoxic T cells, and the lack of definitive markers for these subpopulations. While, in general, these cells respectively secrete cytokines and cytolysins after stimulation (Section 3.3.1), there is no doubt that $CD8^+$ T cells also produce cytokines. In fact, it seems likely that the current division of T cells into distinct helper and cytotoxic subpopulations is an oversimplification, because it is possible to induce both helper and cytolytic functions in the same T cell clone (Eljaafari *et al.* 1990). Despite these reservations, T cells can undoubtedly produce different cytokines, and the pattern of cytokine synthesis determines the nature of the ensuing immune response (Table 4.16; see Section 8.4.5 and Fig.8.36). The pattern of cytokine synthesis by helper T cells can be determined by their microenvironments within lymphoid tissues.

CD4 subsets in mouse: T_H1 and T_H2 cells

It has been suggested that subsets of $CD4^+$ T cells produce distinctly different cytokines. Some mouse T cell clones have been categorized according to the particular pattern of cytokines secreted after stimulation, for example by mitogens (Mosmann *et al.* 1986). So-called T_H1 cells secrete IL-2 in particular, and tend also to produce IFNγ, TNFα, TNFβ, and GM-CSF. In contrast, T_H2 cells produce IL-4, IL-5, IL-6 (as well as IL-10; Section 11.3); both T_H1 and T_H2 cells can produce IL-3. In addition, some monoclonal antibodies appear to discriminate between these putative T cell subsets. About 30% of mouse peripheral T cells express a marker called 3G11 and appear to correspond to T_H1 cells, whereas about 10% of T cells express another marker called 6C10 and correspond to T_H2 cells (Hayakawa and Hardy 1988). However, there is suggestive evidence that these populations may not be so precisely demarcated; about 40% of peripheral T cells express both 3G11 and 6C10, and produce a mixture of cytokines associated with both T cell subsets.

Table 4.16 Some features of $CD4^+$ T_Hl and T_H2 cells

	T_H1	T_H2
Phenotype	High molecular weight form of CD45 (CD45RA); e.g. $OX22^+$ in rat	Low molecular weight form of CD45 (CD45RO); e.g. $OX22^-$ in rat
	Other markers: $3G11^+$ $6C10^-$ (mouse)	Other markers: $3G11^-$ $6C10^+$ (mouse)
Cytokines	IL-2, IL-3, IFNγ, TNFβ (and tend to produce TNFα and GM-CSF)	IL-3, IL-4, IL-5, IL-6, IL-10
APC requirements	Require 'non-B' APC	Respond to B and 'non-B' APC, and IL-1
Functions	Mediate GvH in rats	
	Participate in primary T-dependent antibody responses	Mainly responsible for secondary antibody responses
	Contain suppressor–inducer cells (i.e. induce suppressor function in $CD8^+$ T cells)	Possess helper–inducer activity
	Respond well to alloantigens (e.g. MLR) and mitogens	Include memory cells responding to recall antigens

CD4 Subsets in rat and human

$CD4^+$ T cells with different functions appear to express different isoforms of CD45 (Section 4.2.5). Although T_H1 and T_H2 subsets have not been well-defined in rat and human, there is a rough correlation between these phenotypically different subsets and T_H1 and T_H2 cells in the mouse. Subsets approximating to mouse T_H1 cells express CD45RA, defined by monoclonal antibodies such as 2H4 in human and MRC OX22 in rat, whereas putative T_H2 cells express CD45RO, defined in the human by the UCHL1 antibody.

In general, $CD45RA^+$ T cells produce IL-2, participate in primary T-dependent antibody responses and the allogeneic MLR, induce graft-versus-host disease in rats (Section 5.3.2), but respond poorly to recall antigens. In contrast, $CD45RO^+$ T cells respond well to recall antigens, generating good proliferative responses and providing help for secondary T-dependent antibody responses. These observations suggest that naïve and memory T cells express the $CD45RA^+$ and $CD45RO^+$ phenotype respectively. Consistent with this, T cells from human umbilical cord blood express the high molecular weight forms of CD45 (see Section 4.2.5) whereas T cells expressing the low molecular weight form are only present in the peripheral blood of adults, presumably after prior exposure to antigen. There is, in addition, some evidence in the rat that $OX22^+$ T cells are the precursors to $OX22^-$ T cells (Powrie and Mason 1989), although conflicting evidence has been obtained from experiments in which the purified populations were transferred to nude mice: each population appeared to be capable of giving rise to both $OX22^+$ and $OX22^-$ T cells in these animals.

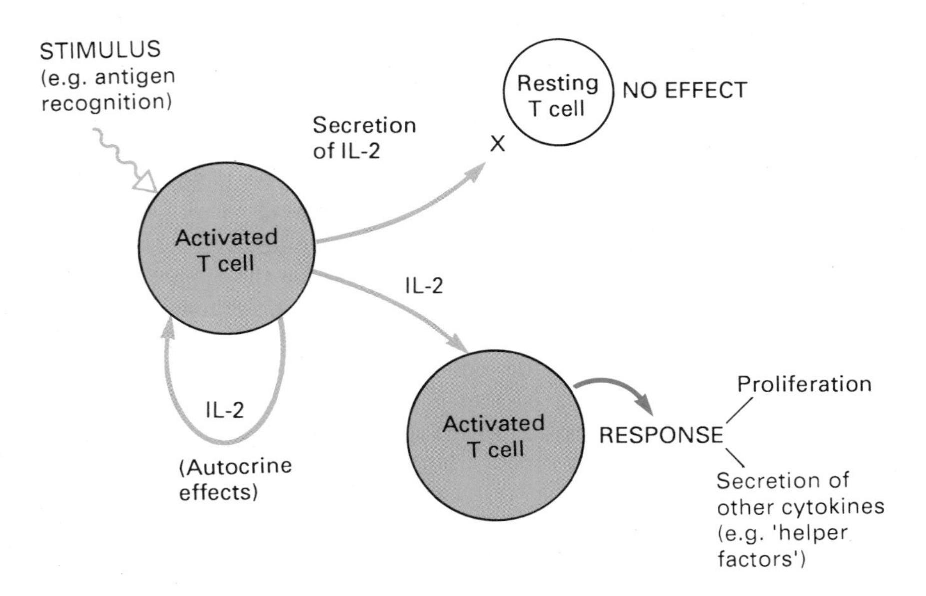

Fig.4.25 IL-2 is produced by, and acts on, activated T cells

4.4.3 Interleukin-2 and its receptor

Interleukin-2

In 1976 it was found that human T cells could be grown continuously in culture if cell-free supernatants were added from leukocytes stimulated with the mitogen phytohemagglutinin, PHA (Fig.4.25). The discovery of a soluble molecule that stimulated the proliferation of activated T cells made it possible to produce T cell lines and clones starting with normal, antigen-stimulated T cells, a finding that has been crucial to the development of cellular immunology. The molecule stimulating the growth of antigen-sensitized T cells was initially called T cell growth factor (TCGF) and was later designated interleukin-2, IL-2. This cytokine (Section 4.4.1) could be assayed by measuring the proliferative response of IL-2-dependent cell lines, the magnitude of which is proportional to the concentration of IL-2. However, this assay is not completely specific for IL-2 because other cytokines such as IL-4 can also stimulate T cell proliferation.

Mouse IL-2 has an apparent molecular weight of 23 kDa, whereas human IL-2 is smaller, 14–15 kDa depending on the particular source. There is evidence for microheterogeneity of IL-2, resulting from variable amounts of glycosylation and sialylation of the mature polypeptide (Fig.4.26). Gene cloning of human and mouse IL-2 (Taniguchi *et al.* 1983) revealed that there were no sites for N-glycosylation but that O-glycosylation could occur (Fig.1.4), so that post-translational modifications would explain the small variations in size that had been observed. An unusual feature of mouse IL-2 is the occurrence of 12 consecutive glutamine residues (amino acids 139–152) in the polypeptide chain; a similar sequence is also present in rat IL-2.

IL-2 was shown to be a T cell product because, for instance, T cell mitogens such as PHA and Con A stimulated its secretion by heterogeneous populations of leukocytes but when these populations were treated with anti-T cell antibodies and complement to deplete the T cells, it was not produced. Moreover, IL-2 is only produced by activated T cells and T cell lines, clones, and hybridomas (although it may also be secreted by large granular lymphocytes and/or natural killer cells; see Section 10.2.6). While all these T cell populations proliferate in response to IL-2, it has no effect on resting cells, at least in rodents. In humans, the situation is not quite as clear, since there are reports that IL-2 can induce T cell proliferation in the apparent absence of any other stimulus, and it has been suggested that there may be a species difference in its action; an alternative explanation is that it acts on activated T cells that may be present in the individual because of infections.

For some time it was believed that IL-2 was produced by helper T cells and that it acted on cytotoxic T cells. Subsequent work has supported the idea that helper T cells can produce large amounts of IL-2, but it is now known that some cytotoxic cells can also secrete this cytokine, and it seems likely that IL-2 acts on both helper and cytotoxic subsets in a paracrine and an autocrine manner. In addition to stimulating the proliferation of activated T cells, IL-2 induces the secretion of other cytokines that mediate the helper functions of T cells.

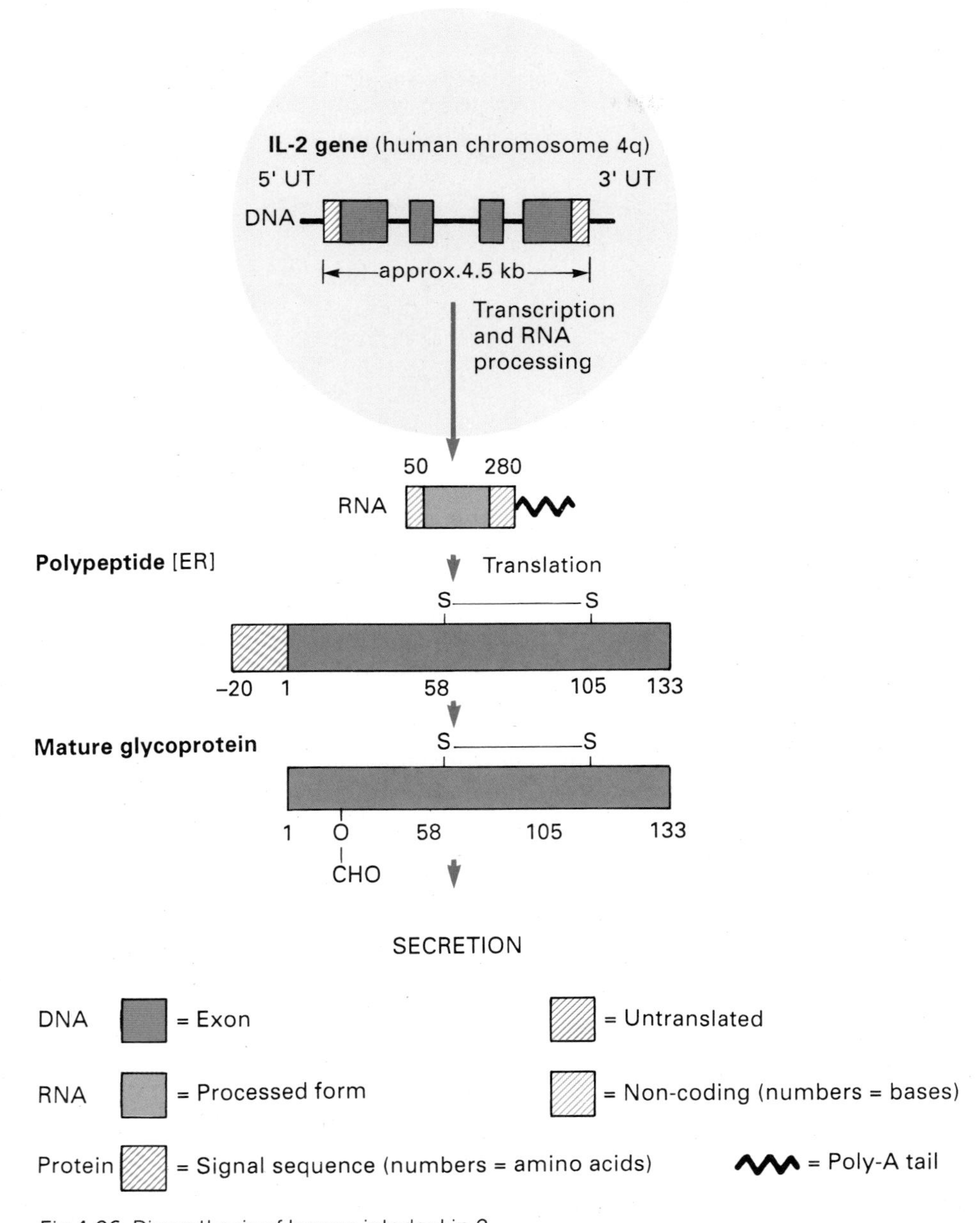

Fig.4.26 Biosynthesis of human interleukin-2

The interleukin-2 receptor (Table 4.17)

The interleukin-2 receptor (IL-2R) is a heterodimer consisting of an α chain of molecular weight 55 kDa (also called p55, or Tac, and now designated CD25) and a β chain of molecular weight 75 kDa (also called p75) (Fig.4.27).

Table 4.17 IL-2 receptor

Distribution	T cells; B cells; some macrophages (e.g. rat) and dendritic leukocytes (e.g. rat, mouse)
Structure	Heterodimer $=\alpha+\beta$ (linked non-covalently) $\alpha=$ p55 (CD25; Tac) $\beta=$ p75
Approx. M_r	α 55 kDa β 75 kDa
Homology to Ig	No
Chromosomal localization of genes	*Human* IL-2Rα (p55): 10p15–p14 IL-2Rβ (p75): 22(q12–13)

In heterodimeric form, the $\alpha\beta$ molecule can bind IL-2 with a high affinity ($K_d = 10\text{–}50 \times 10^{-12}$ M), whereas the affinity of the α chain alone is low ($K_d = 10 \times 10^{-9}$ M) and that of the β chain alone is intermediate ($K_d = 0.5\text{–}1 \times 10^{-9}$ M).

It is now thought that the $\alpha\beta$ high affinity receptor is exclusively responsible for mediating the proliferative response of T cells in the presence of physiological concentrations of IL-2. However with high concentrations of IL-2, the β chain, which is constitutively expressed on resting T cells in the absence of the α chain, can mediate internalization of IL-2 and transduction of signals leading to altered gene expression by the T cell. The α chain appears to be necessary simply to convert the intermediate affinity receptor into a high affinity receptor.

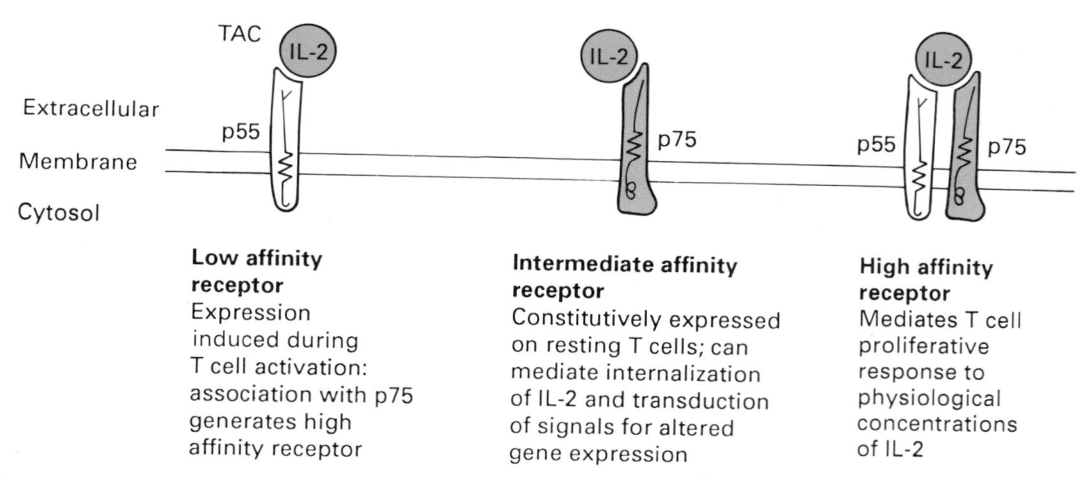

Fig.4.27 The IL-2 receptor is a heterodimer

The α chain of the IL2 receptor: the Tac antigen, CD25. The first chain of the IL2 receptor to be identified was the α chain.† When monoclonal antibodies were prepared against activated human T cells, one of them bound to an antigen that was originally called Tac (Uchiyama *et al.* 1981). This antibody against CD25 inhibited T cell proliferation in response to IL-2 as well as the binding of radiolabelled IL-2 to the cells, and it precipitated a molecule of 55 kDa from activated human T cells and T cell lines.

Subsequent biochemical studies showed that the primary transcript of the Tac antigen has a molecular weight of about 34.5 kDa which is reduced to 33 kDa after cleavage of the leader sequence. This molecule is then N-glycosylated to two intermediate forms (35 kDa and 37 kDa), and O-glycosylation produces the mature product (55 kDa); there is evidence that the molecule is phosphorylated and sulphated as well. All of these forms, except the primary transcript, are able to bind IL-2.

The conclusions from biochemical studies were confirmed when the CD25 gene was cloned. An anti-Tac antibody was used to purify sufficient protein from human T cell lines to allow sequencing of the N-terminal amino acids of the molecule. Oligonucleotide probes were then prepared and used to screen cDNA libraries constructed from activated T cells. Using this strategy, cDNA clones for CD25 were successfully isolated. These clones were used to show there is only one gene for CD25 and that this is localized on human chromosome 10, unlinked to the IL-2 gene on chromosome 4 (Leonard *et al.* 1985).

The CD25 gene can be transcribed to produce mRNAs of different length by alternative splicing, through the use of different polyadenylation signals, and a region within the mRNA can also be spliced out, after which the molecule that is produced is unable to bind IL-2, but the biological significance of this has yet to be defined (Fig.4.28). The genomic organization of the CD25 gene has been established in some detail (Fig.4.26).

The β chain of the IL-2 receptor, p75. When quantitative studies were carried out, a discrepancy became apparent between the binding of anti-Tac antibodies and the amount of radiolabelled IL-2 that bound to activated T cells: there appeared to be more binding sites for the Tac antibody than IL-2-binding sites. Subsequent studies showed that activated T cells have a mixture of low and high affinity IL-2 receptors, and that while anti-Tac binds to both forms of receptor, only the latter was detectable in IL-2 binding studies.

When cDNA for human CD25 was transfected into mouse T cell lines, both high and low affinity receptors for human IL-2 were expressed. However, only low affinity receptors were produced when cDNA for CD25 was transfected into non-T cells. It therefore became apparent that T cells contained a second chain that associated with the CD25 molecule to turn it into a high affinity, biologically-functional IL-2 receptor. This molecule, the β

† The α chain was so-called because it was found before the β chain. In terms of biochemical nomenclature, however, these designations should be reversed because the α chain is smaller than the β chain. Some investigators retain the original nomenclature, while others now use the latter, leading to some confusion in the literature. At the time of writing there are also reports of other molecules being associated with the IL-2 receptor αβ heterodimer, but the significance of this is not yet clear.

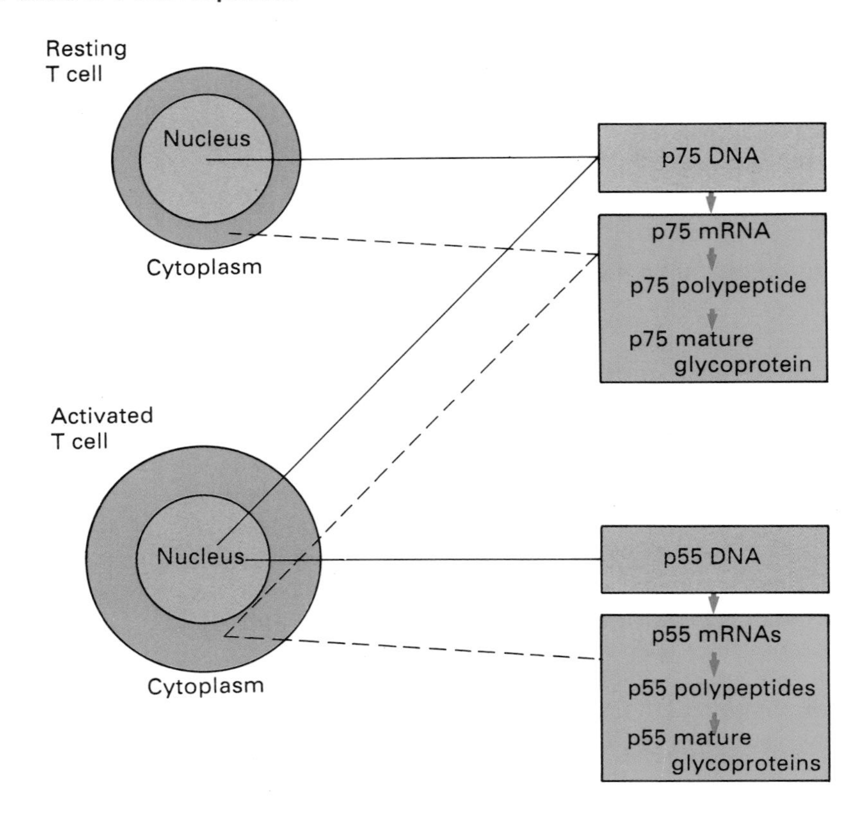

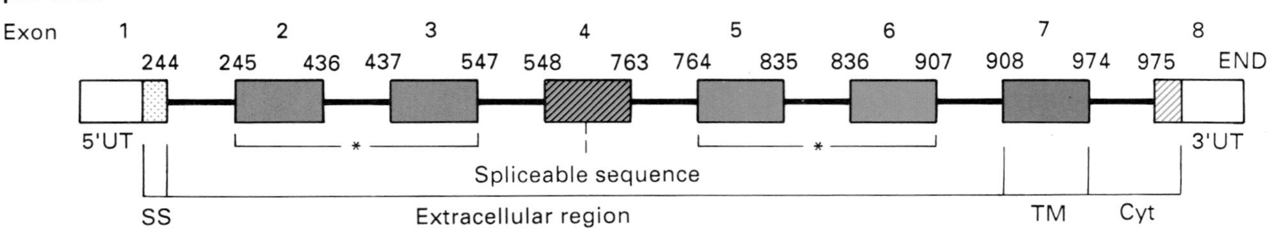

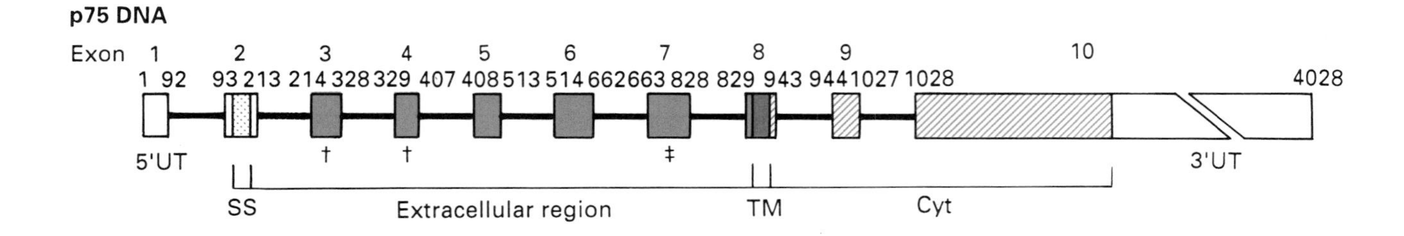

Fig.4.28 Biosynthesis of the IL-2 receptor

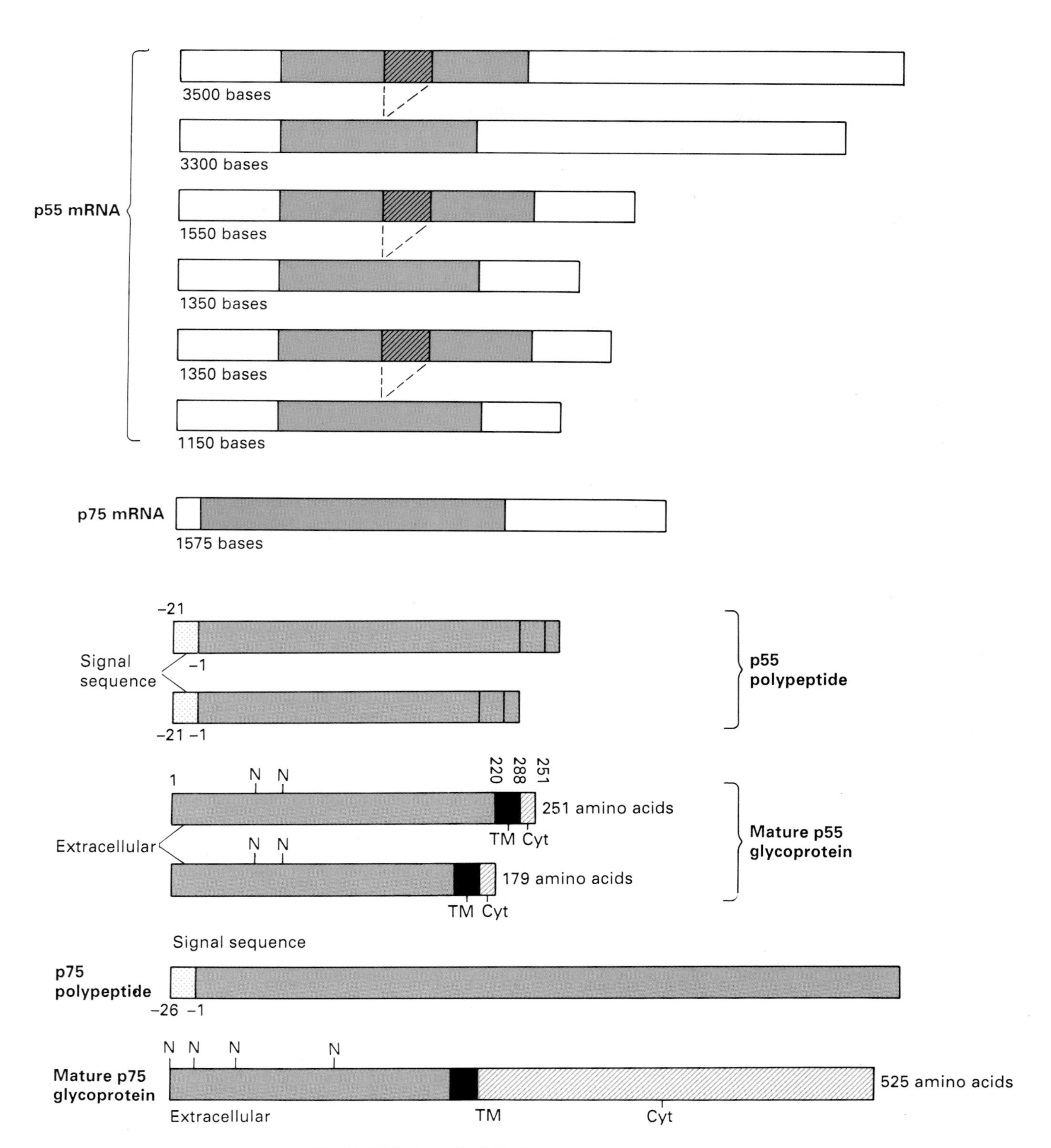

p55 mRNA
3500 bases
3300 bases
1550 bases
1350 bases
1350 bases
1150 bases
p75 mRNA
1575 bases
−21
Signal
sequence
−1
−21 −1
p55
polypeptide
1
N N
220
288
251
251 amino acids
Extracellular
TM Cyt
N N
179 amino acids
TM Cyt
Mature p55
glycoprotein
Signal sequence
p75
polypeptide
−26 −1
N N N N
Mature p75
glycoprotein
525 amino acids
Extracellular
TM
Cyt
N = N-CHO sites; O-CHO sites not shown

chain, was characterized using cell lines that bound IL-2 but which did not express the Tac antigen. When radiolabelled IL-2 was cross-linked to the binding proteins on these cells, it became linked to a component of about 75 kDa. Subsequent studies revealed the details of the affinity and biological activity of the β chain that were outlined at the beginning of this section (Fig.4.27).

The IL-2R β chain has been gene cloned, and found to have significant homology to other members of a distinct cytokine receptor gene superfamily which includes the receptors for IL-3, IL-4, IL-6 (gp130 chain), IL-7, erythropoietin, and GM-CSF (Bazan 1989). The IL-2 receptor may not be unique in having two chains that interact to make a high affinity receptor. There is evidence that the receptors for at least one other interleukin (IL-3) and nerve growth factor have two chains, and the IL-1 receptor may also exist in high and low affinity forms (Section 9.5.1).

Regulation of IL-2 responsiveness

Much has been learnt about how IL-2 receptors and responses to IL-2 are regulated in T cells.

1. The continued expression of IL-2 receptors by activated T cells depends on antigen and accessory cells being present. When antigen is removed, the high affinity receptors disappear from the cell surface and proliferation ceases. In fact, despite the original description of IL-2 as a factor which allows continuous culture of T cells, many T cell lines grown in IL-2 alone eventually become unresponsive to IL-2 and regain responsiveness only when they are restimulated with specific antigen and accessory cells.

2. IL-2 regulates the expression of its own receptors, meaning that if only a few IL-2 receptors are expressed on the T cell, binding of IL-2 to these can induce the expression of more high affinity receptors on the cell. However, when IL-2 binds to high affinity receptors it is internalized and degraded within the T cell. Concomitantly, the high affinity receptors are removed from the cell surface and are replaced by low-affinity IL-2 receptors which do not mediate proliferation.

3. There is yet another level of control of IL-2 responsiveness in that $CD4^+$ T cells become refractory to stimulation by IL-2, in marked contrast to $CD8^+$ T cells. In other words, both subsets can express a similar number of receptors of similar affinity, but the $CD4^+$ cells stop responding to IL-2 before the $CD8^+$ cells, irrespective of how many IL-2 receptors they express. Since these $CD4^+$ cells are an important source of IL-2, this, together with the removal of IL-2 via high affinity receptors, may contribute to control of the responses of $CD8^+$ T cells.

4. A soluble form of the IL-2 receptor has been described in the mouse and human which reacts with anti-IL-2 receptor antibodies and binds IL-2. Its function is not known.

4.4.4 Effects of other cytokines on T lymphocytes

For several years it was thought that IL-1 played an essential role in T cell responses by acting as a 'second signal' for activation following ligation of

Table 4.18 Summary of some effects of cytokines on T lymphocytes

	Mature T cells	Thymocytes
IL-1	Potentiates IL-2 release from activated T cells	Costimulatory: stimulates proliferation in presence of suboptimal concentration of mitogen, e.g. PHA
IL-2	Costimulatory: stimulates proliferation of resting T cells, e.g. in presence of phorbol ester PMA Drives proliferation of activated T cells Induces secretion of other cytokines, e.g. IL-4, from activated T cells Induces NK-like (LAK) activity of cytotoxic T cells	Costimulatory: stimulates proliferation in presence of suboptimal concentration of mitogen, e.g. PHA
IL-4	Maintains viability of resting T cells Costimulatory: stimulates proliferation of resting T cells in presence of e.g. phorbol ester PMA, mitogens, anti-CD3 Increases IL-2 mRNA and IL-2R expression in mitogen-activated cells Induces cytotoxic activity in primary and memory T cells stimulated by alloantigen (more potent than IL-2) but not LAK activity	
IL-6	Costimulatory: induces proliferation of monocyte-depleted T cells in presence of mitogen PHA Increases CTL activity of T cells stimulated by alloantigen **in presence of IL-2**	Costimulatory: stimulates proliferation in presence of suboptimal concentration of mitogen, e.g. PHA Increases CTL activity stimulated by alloantigens **in presence of IL-2**
IL-7	Costimulatory: induces proliferation of T cells in presence of phorbol ester PMA Induces proliferation of T cells activated *in vivo* (but not *in vitro*) by alloantigen or anti-CD3 Induces cytotoxic activity in human peripheral blood lymphocytes	Stimulates proliferation of isolated fetal thymocytes in organ culture, and growth of early thymocytes Stimulates proliferation of isolated fetal thymocytes **in presence of IL-2**
IL-10	Produced by T_H2 cells and inhibits cytokine secretion by T_H1 cells	

the T cell receptor–CD3 complex (see Section 3.1; Fig.3.3). However, there is little evidence for a role of IL-1 in T cell activation from the resting state. Instead, IL-1 may be able to potentiate T cell responses, perhaps by inducing release of IL-2 from activated helper T cells, and also apparently by a direct effect on the immunostimulatory function of dendritic cells (Section 3.4.5). IL-1 and its receptor are considered in Section 9.5.1. In addition to IL-1 and IL-2, there are now at least four other cytokines with well-defined effects on thymocytes and/or mature T cells: IL-4, IL-6, IL-7, and IL-10 (Table 4.18).

IL-4 is a potent helper factor for the generation of cytotoxic T lymphocytes (CTL) in the MLR, in both primary and memory responses to alloantigens (Widmer and Grabstein 1987). In this respect it is considerably more potent than IL-2. Unlike IL-2, IL-4 does not induce lymphokine-activated killer (LAK) activity (Section 10.2.5). Indeed IL-4 can inhibit the IL-2-induced generation of LAK cells when both cytokines are present simultaneously, although conflicting reports have appeared concerning the capacity of IL-4 to inhibit the development of IL-2-induced antigen-specific CTL. In addition, IL-4 maintains the viability of small, dense, apparently resting T cells, and it acts as a costimulatory factor, stimulating the proliferation of T cells in the presence of phorbal myristate acetate (PMA) and of T cells activated by the mitogens PHA, Con A, and anti-CD3 antibodies. Generally similar effects have been noted in human and mouse.

IL-6 has a wide spectrum of effects on a number of cell types, rather similar to IL-1 (Section 9.5.2). For thymocytes and T cells, on one hand it is a costimulatory factor (Lotz *et al.* 1988), inducing proliferation in the presence of suboptimal doses of PHA, and on the other it increases the cytotoxic activity of T cells stimulated by alloantigens but only in the presence of IL-2.

IL-7 was first defined as a 25 kDa molecule that stimulates the proliferation of surface immunoglobulin-negative pre-B cells (Section 8.2.1) from bone marrow cultures, but which does not stimulate their development into surface immunoglobulin-positive B cells nor the proliferation of mature B cells. It was then found to have a costimulatory effect on T cells, driving proliferation in the presence of PMA (which also induces responsiveness to IL-2 and IL-4; see above) and inducing cytotoxic activity in peripheral blood T cells. IL-7 stimulates the growth of thymocytes with an 'early' phenotype in fetal thymus organ culture (Watson *et al.* 1989), and it can stimulate proliferation of isolated fetal thymocytes in the presence of IL-2.

These activities are summarized in Table 4.18; IL-10 is discussed in Section 11.3.

Further reading

General

Barclay, A.N., Beyers, A.D., Birkeland, M.L., Brown, M.H., Davis, S.J., Somoza, C., and Williams, A.F. (1992). *Leucocyte antigen factsbook.* Academic Press, London.

Doherty, P.C., Allan, W., Eichelberger, M., and Carding, S.R. (1992). Roles of $\alpha\beta$ and $\gamma\delta$ T cell subsets in viral immunity. *Annual Review of Immunology*, **10**, 123–51.

Knapp, W., Dorken, B., Gilks, W.R., Rieber, E.P., Schmidt, R.E., Stein, H., and von dem Borne, A.E. (ed.) (1992). *Leukocyte typing IV. White cell differentiation antigens.* Oxford University Press, Oxford.

Owen, M.J., and Lamb, J.R. (1988). *Immune recognition.* In Focus Series (ed. D. Male). IRL Press, Oxford.

Terhorst, C., Alarcon, B., de Vries, J., and Spits, H. (1988). T lymphocyte recognition and activation. In *Molecular Immunology* (ed. B.D. Hames and D.M. Glover). IRL Press, Oxford.

Membrane molecules of T cells

See Chapters 6 and 7 for T cell receptor

CD3

Clevers, H., Alarcon, T., Wileman, T., and Terhorst, C. (1988). The T cell receptor/CD3 complex: a dynamic protein ensemble. *Annual Review of Immunology*, **6**, 629–62.

CD4, CD8

Bierer, B.E., Sleckman, B.P., Ratnofsky, S.E., and Burakoff, S.J. (1989). The biologic roles of CD2, CD4, and CD8 in T-cell activation. *Annual Review of Immunology*, **7**, 579–99.

Littman, D.R. (1987). The structure of the CD4 and CD8 genes. *Annual Review of Immunology*, **5**, 561–84.

Parham, P. (1992). The box and the rod. *Nature*, **357**, 538–9.

Parnes, J.R. (1989). Molecular biology and function of CD4 and CD8. *Advances in Immunology*, **44**, 265–311.

Robey, E. and Axel, R. (1990). CD4: collaborator in immune recognition and HIV infection. *Cell*, **60**, 697–700.

Rudd, C.E. (1990). CD4, CD8 and the TCR-CD3 complex: a novel class of protein-tyrosine kinase receptor. *Immunology Today*, **11**, 400–6.

Travers, P. (1990). One hand clapping (News). *Nature*, **348**, 393–4.

CD45

Clark, E.A. and Ledbetter, J.A. (1989). Leukocyte cell surface enzymology: CD45 (LCA, T200) is a protein tyrosine phosphorylase. *Immunology Today*, **10**, 225–8.

Janeway, C.A. Jr. (1992). The T cell receptor as a multicomponent signalling machine: CD4/CD8 coreceptors and CD45 in T cell activation. *Annual Review of Immunology*, **10**, 645–74.

Thomas, M.L. (1989). The leukocyte common antigen family. *Annual Review of Immunology*, **7**, 339–69.

Thomas, M.L. and Lefrancois, L. (1988). Differential expression of the leucocyte-common antigen family. *Immunology Today*, **9**, 320–6.

CD11/CD18, other integrins, selections, and lymphocyte homing receptors

Dustin, M.L. and Springer, T.A. (1991). Role of lymphocyte adhesion receptors in transient interactions and cell locomotion. *Annual Review of Immunology*, **9**, 27–66.

Haynes, B.F., Telen, M.J., Hale, L.P., and Denning, S.M. (1989). CD44—a molecule involved in leukocyte adherence and T-cell activation. *Immunology Today*, **10**, 423–8. [Erratum: *Immunology Today* (1990) **11**, 80.]

Hemler, M.E. (1990). VLA proteins in the integrin family: structures, functions, and their role on leukocytes. *Annual Review of Immunology*, **8**, 365–400.

Mackay, C.R. (1991). Skin-seeking memory T cells (News). *Nature*, **349**, 737–8.

Pardi, R., Inverardi, L., and Bender, J.R. (1992). Regulatory mechanisms in leukocyte adhesion: flexible receptors for sophisticated travelers. *Immunology Today*, **13**, 224–30.

Picker, L.J. and Butcher, E.C. (1992). Physiological and molecular mechanisms of lymphocyte homing. *Annual Review of Immunology*, **10**, 561–91.

Ruoslahti, E. and Pierschbacher, M.D. (1987). New perspectives in cell adhesion: RGD and integrins. *Science*, **238**, 491–7.

Shimizu, Y., van Seventer, G.A., Horgan, K.J., and Shaw, S. (1990). Roles of adhesion molecules in T-cell recognition: fundamental similarities between four integrins on resting human T cells (LFA-1, VLA-4, VLA-5, VLA-6) in expression, binding, and costimulation. *Immunological Reviews*, **114**, 109–143.

Springer, T.A. (1990). Adhesion receptors of the immune system. *Nature*, **346**, 425–34.

Springer, T.A. and Lasky, L.A. (1991). Sticky sugars for selectins (News). *Nature*, **349**, 196–7.

Stoolman, L.M. (1989). Adhesion molecules controlling lymphocyte migration. *Cell*, **56**, 907–10.

Other

June, C.H., Ledbetter, J.A., Linsley, P.S., and Thompson, C.B. (1990). Role of the CD28 receptor in T-cell activation. *Immunology Today*, **11**, 211–16.

Makgoba, M.W., Sanders, M.E., and Shaw, S. (1989). The CD2-LFA-3 and LFA-1-ICAM pathways: relevance to T-cell recognition. *Immunology Today*, **10**, 417–22.

Moingeon, P., Chang, H.C., Sayre, P.H., Clayton, L.K., Alcover, A., Gardner, P., and Reinherz, E.L. (1989). The structural biology of CD2. *Immunological Review*, **111**, 111–44.

Reif, A.E. and Schlesinger, M. (ed.) (1989). *Cell surface antigen Thy-1*. Marcel Dekker, New York.

T cell activation and intracellular events

Alexander, D.R. and Cantrell, D.A. (1989). Kinases and phosphatases in T-cell activation. *Immunology Today*, **10**, 200–5.

Altman, A., Coggeshall, K.M., and Mustelin, T. (1990). Molecular events mediating T cell activation. *Advances in Immunology*, **48**, 227–360.

Gardner, P. (1989). Calcium and T cell activation. *Cell*, **59**, 15–20.

Klausner, R.D. and Samelson, L.E. (1991). T cell antigen receptor activation pathways: the tyrosine kinase connection. *Cell*, **64**, 875–8.

Rao, A. (1991). Signalling mechanisms in T cells. *Critical Reviews of Immunology*, **10**, 521–59.

Robinson, P.J. (1991). Phosphatidylinositol anchors and T-cell activation. *Immunology Today*, **12**, 35–41.

Ullman, K.S., Northrop, J.P., Verweij, C.L., and Crabtree, G.R. (1990). Transmission of signals from the T lymphocyte antigen receptor to the genes responsible for cell proliferation and immune function: the missing link. *Annual Review of Immunology*, **8**, 421–52.

Ullrich, A. and Schlessinger, J. (1990). Signal transduction by receptors with tyrosine kinase activity. *Cell*, **61**, 203–12.

T cell development (see also Chapter 5)

Carding, S.R., Hayday, A.C., and Bottomly, K. (1992). Cytokines in T-cell development. *Immunology Today*, **12**, 239–45.

Clevers, H.C. and Owen, M.J. (1991). Towards a molecular understanding of T-cell differentiation. *Immunology Today*, **12**, 86–92.

Finkel, T.H., Kubo, R.T., and Cambier, J.C. (1991). T-cell development and transmembrane signalling: changing biological responses through an unchanging receptor. *Immunology Today*, **12**, 79–85.

Nikolic-Zugic, J. (1991). Phenotypic and functional stages in the intrathymic development of *alpha-beta* T cells. *Immunology Today*, **12**, 65–70.

Cytokines and receptors (see also Chapters 8, 9, and 11)

Arai, K., Lee, F., Miyajima, A., Miyatake, S., Arai, N., and Yokota, T. (1990). Cytokines: coordinators of immune and inflammatory responses. *Annual Review of Biochemistry*, **59**, 783–836.

Balkwill, F.R. and Burke, F. (1989). The cytokine network. *Immunology Today*, **10**, 299–304.

Hamblin, A.S. (1988). *Lymphokines*. In Focus Series (ed. D. Male). IRL Press, Oxford.

Kudlow, J.E., MacLennon, D.H., Bernstein, A., and Grotlieb, A.I. (ed.) (1988). *Biology of growth factors: molecular biology, oncogenes, signal transduction and clinical applications*. Plenum Press, New York.

Miyajima, A., Kitamura, T., Harada, N., Yokota, T., and Arai, K.i. (1992). Cytokine receptors and signal transduction. *Annual Review of Immunology*, **10**, 295–331.

Smith, K.A. (1989). The interleukin 2 receptor. *Annual Review of Cell Biology*, **5**, 397–425.

Waldmann, T.A. (1989). The multi-subunit interleukin-2 receptor. *Annual Review of Biochemistry*, **58**, 875–911.

T cell subsets (see also Chapters 8 and 11)

Bottomly, K. (1988). A functional dichotomy of CD4+ T lymphocytes. *Immunology Today*, **9**, 268–74.

Moller, G. (ed.) (1991). T-helper cell subpopulations. *Immunological Reviews*, Vol.123. Munksgaard, Copenhagen.

Powrie, F. and Mason, D. (1988). Phenotypic and functional heterogeneity of CD4+ T cells. *Immunology Today*, **9**, 274–7.

Sprent, J. and Webb, S.R. (1987). Function and specificity of T cell subsets in the mouse. *Advances in Immunology*, **41**, 39–133.

Vitetta, E.S., Berton, M.T., Burger, C., Kepron, M., Lee, W.T., and Xin, X.M. (1991). Memory B cells and T cells. *Annual Review of Immunology*, **9**, 193–218.

Literature cited

Anderson, D.C. and Springer, T.A. (1987). Leukocyte adhesion deficiency: an inherited defect in the Mac-1, LFA-1 and p150,95 glycoproteins. *Annual Review of Medicine*, **38**, 175–94.

Arulanandam, A.R., Koyasu, S., and Reinherz, E.L. (1991). T cell receptor-independent CD2 signal transduction in FcR+ cells. *Journal of Experimental Medicine*, **173**, 859–68.

Bazan, J.F. (1989). A novel family of growth factor receptors: a common binding domain in the growth hormone, prolactin, erythropoietin and IL-6 receptors, and the p75 IL-2 receptor beta-chain. *Biochemical and Biophysical Research Communications*, **164**, 788–95.

Bierer, B.E., Peterson, A., Gorga, J.C., Herrmann, S.H., and Burakoff, S.J. (1988). Synergistic T cell activation via the physiological ligands for CD2 and the T cell receptor. *Journal of Experimental Medicine*, **168**, 1145–56.

Born, W., Hall, L., Dallas, A., Boymel, J., Shinnick, T., Young, D., Brennan, P., and O'Brien, R. (1990). Recognition of a peptide antigen by heat shock-reactive gamma delta T lymphocytes. *Science*, **249**, 67–9.

Cammarota, G., Schierle, A., Takacs, B., Doran, D.M., Knorr, R., Bannwarth, W., Guardiola, J., and Sinigaglia, F. (1992). Identification of a CD4 binding site on the β_2 domain of HLA-DR molecules. *Nature*, **356**, 799–801.

Chang, H.C., Moingeon, P., Pedersen, R., Lucich, J., Stebbins, C., and Reinherz, E.L. (1990). Involvement of the PPPGHR motif in T cell activation via CD2. *Journal of Experimental Medicine*, **172**, 351–5.

de la Hera, A., Muller, U., Olsson, C., Isaaz, S., and Tunnacliffe, A. (1991). Structure of the T cell antigen receptor (TCR): two CD3 epsilon subunits in a functional TCR/CD3 complex. *Journal of Experimental Medicine*, **173**, 7–17.

Del Porto, P., Mami-Chouaib, F., Bruneau, J.M., Jitsukawa, S., Dumas, J., Harnois, M., and Hercend, T. (1991). TCT.1, a target molecule for gamma delta T cells is encoded by an immunoglobulin superfamily gene (Blast-1) located in the CD1 region of human chromosome 1. *Journal of Experimental Medicine*, **173**, 1339–44.

Dembic, Z., Haas, W., Weiss, S., McCubrey, J., Kiefer, H., von Boehmer, H., and Steinmetz, M. (1986). Transfer of specificity by murine alpha and beta T cell receptor genes. *Nature*, **320**, 232–8.

Dembic, Z., Haas, W., Zamoyska, R., Parnes, J., Steinmetz, M., and von Boehmer, H. (1987). Transfection of the CD8 gene enhances T-cell recognition. *Nature*, **326**, 510–11.

Doyle, C. and Strominger, J.L. (1987). Interaction between CD4 and class II MHC molecules mediates cell adhesion. *Nature*, **330**, 256–9.

Dustin, M.L., Sanders, M.E., Shaw, S., and Springer, T.A. (1987). Purified lymphocyte function-associated antigen-3 binds to CD2 and mediates T lymphocyte adhesion. *Journal of Experimental Medicine*, **165**, 677–92.

Dustin, M.L. and Springer, T.A. (1989). T-cell receptor cross-linking transiently stimulates adhesiveness through LFA-1. *Nature*, **341**, 619–24.

Ehlers, S. and Smith, K.A. (1991). Differentiation of T cell lymphokine gene expression: the *in vitro* acquisition of T cell memory. *Journal of Experimental Medicine*, **173**, 25–36.

Eljaafari, A., Vaquero, C., Teillaud, J.L., Bismuth, G., Hivroz, C., Dorval, I., Bernard, A., and Sterkers, G. (1990). Helper or cytolytic functions can be selectively induced in bifunctional T cell clones. *Journal of Experimental Medicine*, **172**, 213–18.

Fraser, J.D., Irving, B.A., Crabtree, G.R., and Weiss, A. (1991). Regulation of interleukin-2 gene enhancer activity by the T cell accessory molecule CD28. *Science*, **252**, 313–6.

Gay, D., Maddon, P., Sekaly, R., Talle, M.A., Godfrey, M., Long, E., Goldstein, G., Chess, L., Axel, R., Kappler, J., and Marrack, P. (1987). Functional interaction between human T-cell protein CD4 and the major histocompatibility complex HLA-DR antigen. *Nature*, **328**, 626–9.

Glaichenhaus, N., Shastri, N., Littman, D.R., and Turner, J.M. (1991). Requirement for association of $p56^{lck}$ with CD4 in antigen-specific signal transduction in T cells. *Cell*, **64**, 511–20.

Golding, H., McCluskey, J., Munitz, T.I., Germain, R.N., Margulies, D.H., and Singer, A. (1985). T-cell recognition of a chimaeric class II/class I MHC molecule and the role of L3T4. *Nature*, **317**, 425–7.

Groh, V., Porcelli, S., Fabbi, M., Lanier, L.L., Picker, L.J., Anderson, T., Warnke, R.A., Bhan, A.K., Strominger, J.L., and Brenner, M.B. (1989). Human lymphocytes bearing T cell receptor gamma delta are phenotypically diverse and evenly distributed throughout the lymphoid system. *Journal of Experimental Medicine*, **169**, 1277–94.

Guy-Grand, D., Cerf-Bensussan, N., Malissen, B., Malassis-Seris, M., Briottet, C., and Vassali, P. (1991). Two gut intraepithelial $CD8^+$ lymphocyte populations with different T cell receptors: a role for the gut epithelium in T cell differentiation. *Journal of Experimental Medicine*, **173**, 471–82.

Harding, F.A., McArthur, J.G., Gross, J.A., Raulet, D.H., and Allison, J.P. (1992). CD28-mediated signalling co-stimulates murine T cells and prevents induction of anergy in T-cell clones. *Nature*, **356**, 607–9.

Hayakawa, K. and Hardy, R.R. (1988). Murine $CD4^+$ T cell subsets defined. *Journal of Experimental Medicine*, **168**, 1825–38.

He, H.T., Naquet, P., Caillol, D., and Pierres, M. (1991). Thy-1 supports adhesion of mouse thymocytes to thymic epithelial cells through a Ca^{2+}-independent mechanism.

Journal of Experimental Medicine, **173**, 515–18.

Irving, B.A. and Weiss, A. (1991). The cytoplasmic domain of the T cell receptor zeta chain is sufficient to couple to receptor-associated signal transduction pathways. *Cell*, **64**, 891–901.

Ito, K., Van Kaer, L., Bonneville, M., Hsu, S., Murphy, D.B., and Tonegawa, S. (1990). Recognition of the product of a novel MHC TL region gene (27b) by a mouse gamma delta T cell receptor. *Cell*, **62**, 549–61.

Jin, Y.J., Clayton, L.K., Howard, F.D., Koyasu, S., Sieh, M., Steinbrich, R., Tarr, G.E., and Reinherz, E.L. (1990). Molecular cloning of the CD3 eta subunit identifies a CD3 zeta-related product in thymus-derived cells. *Proceedings of the National Academy of Sciences USA*, **87**, 3319–23.

Johnston, G.I., Cook, R.G., and McEver, R.P. (1989). Cloning of GMP-140, a granule membrane protein of platelets and endothelium: sequence similarity to proteins involved in cell adhesion and inflammation. *Cell*, **56**, 1033–44.

Jutila, M.A., Watts, G., Walchek, B., and Kansas, G.S. (1992). Characterization of a functionally important and evolutionarily well-conserved epitope mapped to the short consensus repeats of E-selectin and L-selectin. *Journal of Experimental Medicine*, **175**, 1565–73.

Kappler, J.W., Skidmore, B., White, J., and Marrack, P. (1981). Antigen-inducible, H-2-restricted, interleukin-2-producing T cell hybridomas. Lack of independent antigen and H-2 recognition. *Journal of Experimental Medicine*, **153**, 1198–214.

Kishimoto, T.K., O'Connor, K., Lee, A., Roberts, T.M., and Springer, T.A. (1987). Cloning of the beta subunit of the leukocyte adhesion proteins: homology to an extracellular matrix receptor defines a novel supergene family. *Cell*, **48**, 681–90.

Konig, R., Huang, L.Y., and Germain, R.N. (1992). MHC class II interaction with CD4 mediated by a region analogous to the MHC class I binding site for CD8. *Nature*, **356**, 796–8.

Krangel, M.S. (1987). Endocytosis and recycling of the T3-T cell receptor complex. The role of T3 phosphorylation. *Journal of Experimental Medicine*, **165**, 1141–59.

Kuno, M. and Gardner, P. (1987). Ion channels activated by inositol 1,4,5-trisphosphate in plasma membrane of human T-lymphocytes. *Nature*, **326**, 301–4.

Larson, R.S., Corbi, A.L., Berman, L., and Springer, T.A. (1989). Primary structure of the lymphocyte function associated antigen-1 alpha subunit: an integrin with an embedded domain defining a protein superfamily. *Journal of Cell Biology*, **108**, 703–12.

Lasky, L.A., Singer, M.S., Yednock, T.A., Dowbenko, D., Fennie, C., Rodriguez, H., Nguyen, T., Stachel, S., and Rosen, S.D. (1989). Cloning of a lymphocyte homing receptor reveals a lectin domain. *Cell*, **56**, 1045–55.

Lasky, L.A., Singer, M.S., Dowbenko, D., Imai, Y., Henzel, W.J., Grimley, C., Fennie, C., Gillett, N., Watson, S.R., and Rosen, S.D. (1992). An endothelial ligand for L-selectin is a novel mucin-like molecule. *Cell*, **69**, 927–38.

Leahy, R., Axel, R., and Henderickson, W.A. (1992). Crystal structure of a soluble form of the human T cell coreceptor CD8 at 2.6 Å resolution.

Ledbetter, J.A., Tonks, N.K., Fischer, E.H., and Clark, E.A. (1988). CD45 regulates signal transduction and lymphocyte activation by specific association with receptor molecules on T or B cells. *Proceedings of the National Academy of Sciences USA*, **85**, 8628–32.

Leonard, W.J., Depper, J.M., Kanehisa, M., Kronke, M., Peffer, N.J., Svetlik, P.B., Sullivan, M., and Green, W.C. (1985). Structure of the human interleukin-2 gene. *Science*, **230**, 633–9.

Linsley, P.S., Brady, W., Grosmaire, L., Aruffo, A., Damle, N.K., and Ledbetter, J.A. (1991*a*). Binding of the B cell activation antigen B7 to CD28 costimulates T cell proliferation and interleukin 2 mRNA accumulation. *Journal of Experimental Medicine*, **173**, 721–30.

Linsley, P.S., Brady, W., Urnes, M., Grosmaire, L., Damle, N.K., and Ledbetter, J.A. (1991*b*). CTLA-4 is a second receptor for the B cell activation antigen B7. *Journal of Experimental Medicine*, **173**, 721–30.

Lotz, M., Jirik, F., Kabouridis, P., Tsoukas, C., Hirano, T., Kishimoto, T., and Carson, D.A. (1988). B cell stimulating factor 2/interleukin-6 is a costimulant for human thymocyte and T lymphocytes. *Journal of Experimental Medicine*, **167**, 1253–8.

Mackay, C.R., Marston, W.L., and Dudler, L. (1990). Naïve and memory T cells show distinct pathways of lymphocyte recirculation. *Journal of Experimental Medicine*, **171**, 801–17.

Meuer, S.C., Fitzgerald, K.A., Hussey, R.E., Hodgdon, J.C., Schlossman, S.F., and Reinherz, E.L. (1983). Clonotypic structures involved in antigen-specific human T cell function. Relationship to the T3 molecular complex. *Journal of Experimental Medicine*, **157**, 705–19.

Meuer, S.C., Hussey, R.E., Fabbi, M., Fox, D., Acuto, O., Fitzgerald, K.A., Hodgdon, J.C., Protentis, J.P., Schlossman, S.F., and Reinherz, E.L. (1984). An alternative pathway of T cell activation: a functional role for the 50kd T11 sheep erythrocyte receptor protein. *Cell*, **36**, 897–906.

Minami, Y., Weissman, A.M., Samelson, L.E., and Klausner, R.D. (1987). Building a multichain receptor: synthesis, degradation, and assembly of the T-cell antigen receptor. *Proceedings of the National Academy of Sciences USA*, **84**, 2688–92.

Moretta, A., Pontaleo, G., Lopez-Botet, M., and Moretta, L. (1985). Involvement of T44 molecules in an antigen-independent pathway of T cell activation. Analysis of the correlation to the T cell antigen receptor complex. *Journal of Experimental Medicine*, **162**, 823–38.

Mosmann, T.R., Cherwinski, H., Bond, M.W., Giedlin, M.A.,

and Coffman, R.L. (1986). Two types of murine helper T cell clone. I. Definition according to profiles of lymphokine activities and secreted proteins. *Journal of Immunology*, **136**, 2348–57.

Nakayama, T., Singer, A., Hsi, E.D., and Samelson, L.E. (1989). Intrathymic signalling in immature CD4$^+$ CD8$^+$ thymocytes results in tyrosine phosphorylation of the T-cell receptor zeta chain. *Nature*, **341**, 651–4.

Nojima, Y., Humphries, M.J., Mould, A.P., Komoriya, A., Yamada, K.M., Schlossman, S.F., and Morimoto, C. (1990). VLA-4 mediates CD3-dependent CD4$^+$ T cell activation via the CS1 alternatively spliced domain of fibronectin. *Journal of Experimental Medicine*, **172**, 1185–92.

Ohashi, P.S., Mac, T., Van den Elsen, P., Yanagi, Y., Yoshikai, Y., Calman, A.F., Terhorst, C., Stobo, J.D., and Weiss, A. (1985). Reconstitution of an active surface T3/T-cell antigen receptor by DNA transfer. *Nature*, **316**, 606–9.

Peterson, A. and Seed, B. (1987). Monoclonal antibody and ligand binding sites of the T cell erythrocyte receptor (CD2). *Nature*, **329**, 842–6.

Pont, S., Regnier-Vigouroux, A., Naquet, P., Blanc, D., Pierres, A., Marchetto, S., and Pierres, M. (1985). Analysis of the Thy-1 pathway of T cell hybridoma activation using 17 rat monoclonal antibodies reactive with distinctive Thy-1 epitopes. *European Journal of Immunology*, **15**, 1222–8.

Powrie, F. and Mason, D. (1989). The MRC OX-22$^-$ CD4$^+$ T cells that help B cells in secondary immune responses derive from naïve precursors with the MRC OX-22$^+$ CD4$^+$ phenotype. *Journal of Experimental Medicine*, **169**, 653–62.

Roberts, K., Yokoyama, W.M., Kehn, P.J., and Shevach, E.M. (1991). The vitronectin receptor serves as an accessory molecule for the activation of a subset of gamma delta T cells. *Journal of Experimental Medicine*, **173**, 231–40.

Rosen, H. (1990). Role of CR3 in induced myelomonocytic recruitment: insights from *in vivo* monoclonal antibody studies in the mouse. *Journal of Leukocyte Biology*, **48**, 465–9.

Ryu, S.E., Kwong, P.D., Truneh, A., Porter, T.G., Arthos, J., Rosenberg, M., Dai, X.P., Xuong, N.H., Axel, R., Sweet, R.W., and Hendrickson, W.A. (1990). Crystal structure of an HIV-binding recombinant fragment of human CD4. *Nature*, **348**, 419–26.

Saito, T., Weiss, A., Miller, J., Norcross, M.A., and Germain, R.N. (1987). Specific antigen-Ia activation of transfected human T cells expressing murine Ti alpha beta-human T3 receptor complexes. *Nature*, **325**, 125–30.

Salter, R.D., Benjamin, R.J., Wesley, P.K., Buxton, S.E., Garrett, T.P., Clayberger, C., Krensky, A.M., Norment, A.M., Littman, D.R., and Parham, P. (1990). A binding site for the T-cell co-receptor CD8 on the alpha 3 domain of HLA-A2. *Nature*, **345**, 41–6.

Sewell, W.A., Brown, M.H., Owen, M.J., Fink, P.J., Kozak, C.A., and Crumpton, M.J. (1987). The murine homologue of the T lymphocyte CD2 antigen: molecular cloning, chromosome assignment and cell surface expression. *European Journal of Immunology*, **17**, 1015–20.

Staunton, D.E., Dustin, M.L., and Springer, T.A. (1989). Functional cloning of ICAM-2, a cell adhesion ligand for LFA-1 homologous to ICAM-1. *Nature*, **339**, 61–4.

Staunton, D.E., Dustin, M.L., Erickson, H.P., and Springer, T.E. (1990). The arrangement of the immunoglobulin-like domains of ICAM-1 and the binding sites for LFA-1 and rhinovirus. *Cell*, **61**, 243–54. [Erratum: *Cell*, **61**, 1157.]

Taniguchi, T., Matsui, H., Fujita, T., Takaoka, C., Kashima, N., Yoshimoto, R., and Hamuro, J. (1983). Structure and expression of a cloned cDNA for human interleukin-2. *Nature*, **302**, 305–10.

Tonks, N.K., Charbonneau, H., Diltz, C.D., Fischer, E.H., and Walsh, K.A. (1988). Demonstration that the leukocyte common antigen CD45 is a protein tyrosine phosphatase. *Biochemistry*, **27**, 8695–701.

Uchiyama, T., Broder, S., and Waldmann, T.A. (1981). A monoclonal antibody (anti-Tac) reactive with activated and functionally mature human T cells. *Journal of Immunology*, **126**, 1293–7.

Vidovic, D., Roglic, M., McKune, K., Guerder, S., MacKay, C., and Dembic, Z. (1989). Qa-1 restricted recognition of foreign antigen by a gamma delta T cell hybridoma. *Nature*, **340**, 646–50.

Wang, J., Yan, Y., Garrett, T.P., Liu, J.H., Rodgers, D.W., Garlick, R.L., Tarr, G.E., Husain, Y., Reinherz, E.L., and Harrison, S.C. (1990). Atomic structure of a fragment of human CD4 containing two immunoglobulin-like domains. *Nature*, **348**, 411–18.

Watson, J.D., Morrissey, P.J., Namen, A.E., Conlon, P.J., and Widmer, M.B. (1989). Effect of IL-7 on the growth of fetal thymocytes in culture. *Journal of Immunology*, **143**, 1215–22.

Widmer, M.B., and Grabstein, K.H. (1987). Regulation of cytolytic T-lymphocyte generation by B-cell stimulatory factor 1. *Nature*, **326**, 795–8.

Yague, J., White, J., Coleclough, C., Kappler, J., Palmer, E., and Marrack, P. (1985). The T cell receptor: the alpha and beta chains define idiotype, and antigen and MHC specificity. *Cell*, **42**, 81–7.

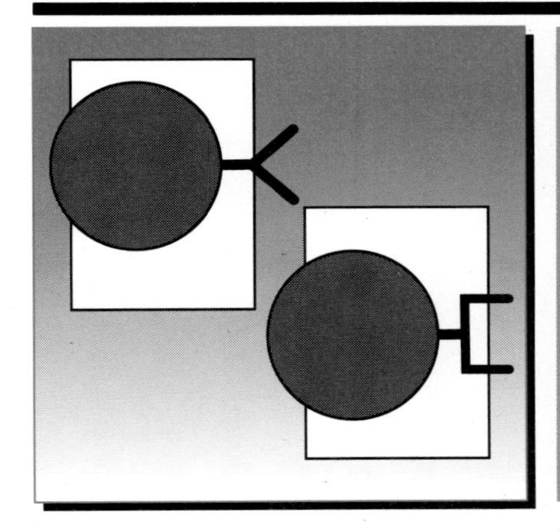

5

THE T CELL REPERTOIRE

Scheme 5. The T cell repertoire

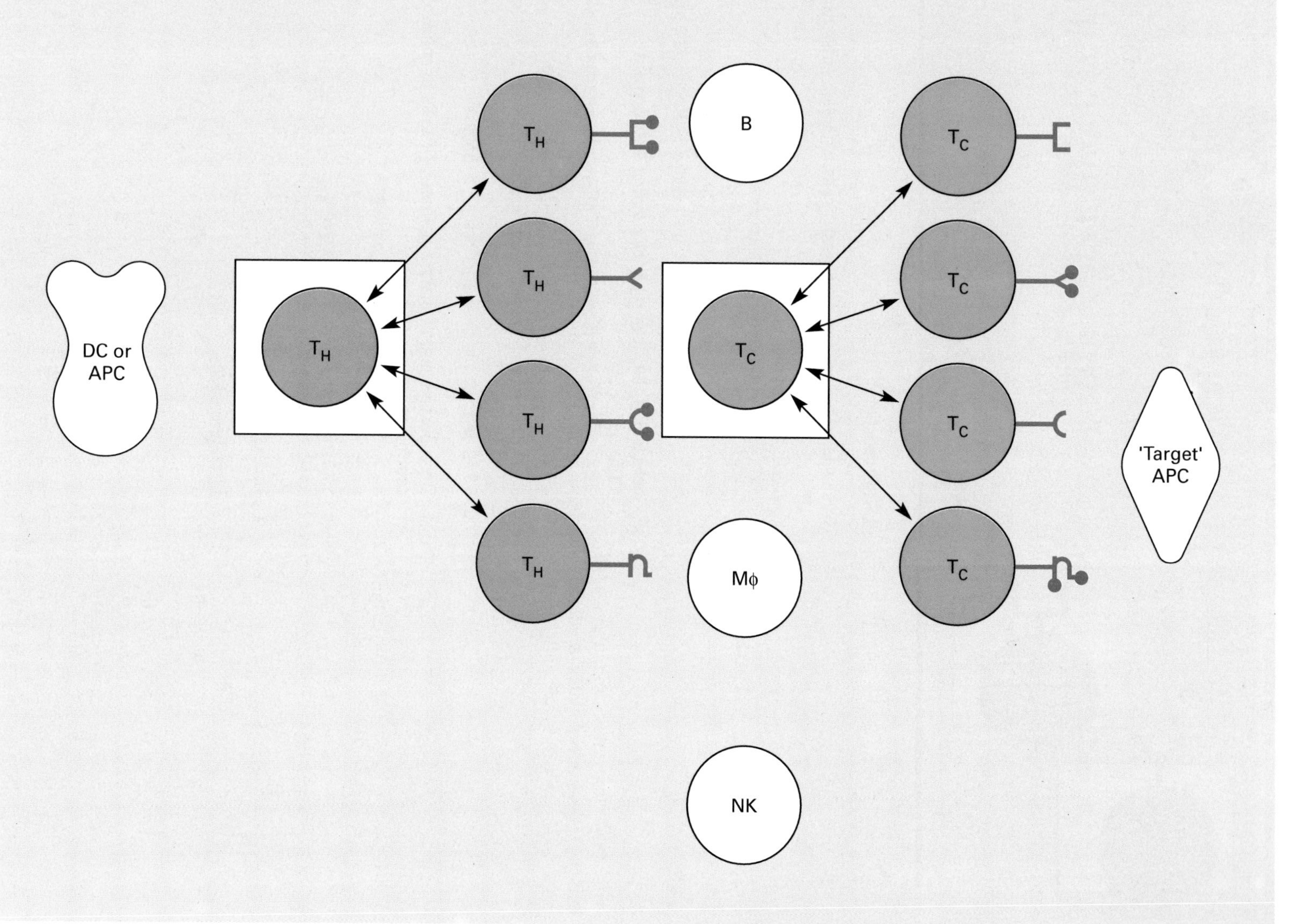

5.1 Introduction

This chapter considers properties of the T cell repertoire, and how the repertoire is generated. The term **T cell repertoire** refers to the total population of mature and developing T cells with different antigen (peptide–MHC) specificities in the immune system. For a general introduction to this area, see Sections 1.1.3 and 1.1.4.

We first consider features of antigen recognition by mature T cells in the periphery, that is, in all tissues other than the thymus (Section 5.2), and outline the events that occur during development. Then we turn to the thymus and discuss in detail how the precise set of antigen specificities is selected during T cell development (Section 5.3); this section could be omitted by readers who do not require a detailed understanding of the experimental techniques used. Finally, mechanisms of genetically-controlled unresponsiveness of mature peripheral T cells to certain antigens are addressed (Section 5.4).

5.2 Responses of mature peripheral T cells to foreign antigens

5.2.1 Discovery of MHC restriction

The first real understanding that T cells recognize foreign antigens associated with MHC molecules came about in the early 1970s. The experiments that led to the concept of 'MHC-associative recognition of antigens' by T cells, or **MHC restriction**, are usually attributed to Zinkernagel and Doherty (1974) and, independently, to Shearer and colleagues (1974). These investigators were studying the ability of cytotoxic T cells to kill virus-infected target cells and chemically-modified target cells, respectively.

What Zinkernagel and Doherty found is shown schematically in Fig. 5.1. They infected mice of a certain strain, we shall call it C (see footnote, p. 192), with the lymphocytic choriomeningitis virus (LCMV). They then isolated cytotoxic T cells from the spleens of these mice and tested them in cytotoxicity assays for their ability to kill various ^{51}Cr-labelled target cells. The targets used were macrophages from the same or a different strain of mouse (i.e. C or D) that were either infected with virus or uninfected. As expected, the cytotoxic T cells killed strain C target cells infected with LCMV, but they did not kill uninfected targets. The T cells were shown to be antigen-specific, since C strain cells infected with a different virus were not lysed. But the surprise, at that time, was that the T cells were unable to kill D strain cells infected with LCMV. This result would not have been obtained if the T cells were able to recognize the virus alone. Shearer obtained similar results using hapten-modified (trinitrophenyl-coupled) target cells, and these results were soon confirmed by other investigators studying cytotoxic responses against minor histocompatibility antigens (Section 2.1).

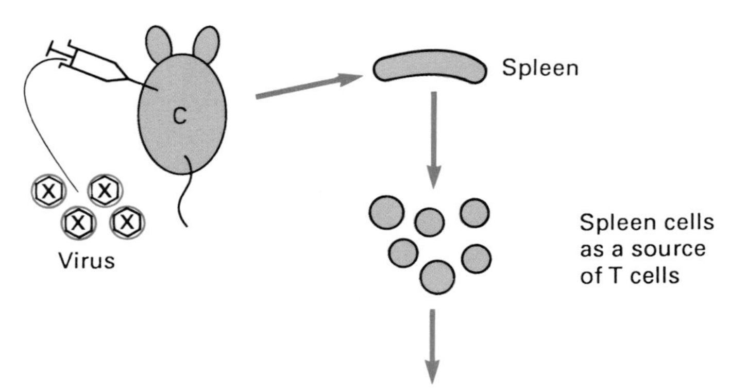

Test for killing of virus-infected target cells:

Prediction:	Virus X Strain C	Uninfected Strain C	Virus X Strain D	Virus Y Strain C
If recognition of virus only	Yes	No	Yes	No
If recognition of strain only	Yes	Yes	No	Yes
If recognition of virus plus strain component	Yes	No	No	No
Observed result	Yes	No	No	No

Fig.5.1 Principle of Zinkernagel and Doherty's experiments (1974) to show cytotoxic T cells recognized antigen plus a 'strain-specific component' (MHC)

The results can be generalized by saying that cytotoxic T cells obtained from strain C immunized against antigen X could kill only target cells of that combination, but not X plus a different strain, D, nor antigen Y plus C, or Y plus D. These investigators reasoned that if the T cells had been able to recognize the antigen alone, as might be expected for B cells, they should have killed target cells *of any strain* that contained or expressed that antigen. But in fact the T cells were specific for both the antigen *and* a 'strain-specific component'. By genetic mapping, using defined strains of inbred mice (see Appendix), this component was mapped to the MHC region of the genome, and particularly to the MHC class I loci (Fig.2.1) which had been identified some years previously (Table 5.1).

The early mapping studies did not formally prove that MHC molecules were involved in T cell antigen recognition, because the inbred strains of

Table 5.1 Principle of genetic mapping experiments to show MHC class I-restricted recognition of antigens by cytotoxic T cells

Strain of target cells	MHC genotype of virus-infected target cells				Killing by immune T cells from strain C
	H2K class I	H2I class II	H2Ss class III	H2D class I	
C	**C**	C	C	**C**	**Yes**
D	D	D	D	D	No
Recombinant 1*	D	C	C	D	No
Recombinant 2*	**C**	D	D	**C**	**Yes**

*Congenic strains, identical at all genetic loci except within the MHC region

mice that were used also differed in other regions of their genomes. Proof that the MHC was involved came from studies of T cell responses in mutant strains of mice (Section 2.5.1; p. 82) that differed simply by substitutions of one to three amino acids in one of their MHC class I genes.

For example, the bm1 and bm6 strains are identical except for a small change in sequence in the H-2K^b molecule (Table 2.6). It was found that if a mouse from one strain (e.g. bm1) is infected with virus, its cytotoxic T cells cannot lyse infected target cells from the other strain (bm6) because of the differences in the H-2K^b molecule. These experiments clearly showed that class I molecules are required for T cell recognition. In addition, cell lines transfected with genomic clones containing the appropriate class I genes (see Appendix) have been used as target cells in assays of killing by cytotoxic T cells (Forman *et al.* 1983). These cells, which contain a single transfected class I molecule (e.g. H-2L^d), can indeed, after infection by a virus, be lysed by cytotoxic T cells isolated from virus-infected H-2^d mice. Furthermore, killing can be inhibited by monoclonal antibodies directed specifically to the class I molecule in question, H-2L^d.

Such experiments led to our present understanding that *antigen recognition by many CTL is MHC class I-restricted.*

In contrast to CTL-target cell recognition, it was found that *the interaction between helper T cells and antigen presenting cells is predominantly MHC class II-restricted.* This was initially addressed by Rosenthal and Shevach (1973) in studies of antigen-specific proliferative responses of T cells (primarily helper cells) from guinea pigs of different strains (Fig.5.2).

Two strains were examined, strain 2 and strain 13 which differed at the MHC region (Fig.5.2). They primed each strain with antigen and obtained sensitized T cells, which were then cultured with antigen-pulsed macrophages as antigen-presenting cells (see Section 3.1). It was found that the sensitized T cells from strain 2 guinea pigs only proliferated in response to specific antigen and antigen-presenting cells from strain 2, but not from strain 13, and vice versa.

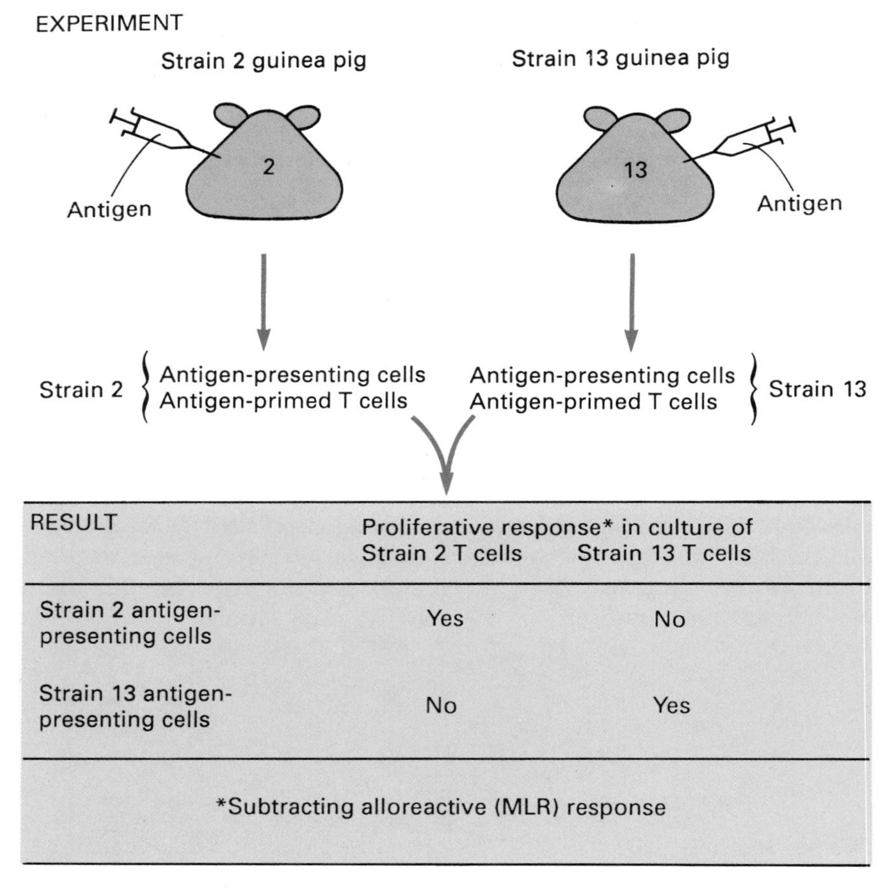

RESULT	Proliferative response* in culture of Strain 2 T cells	Strain 13 T cells
Strain 2 antigen-presenting cells	Yes	No
Strain 13 antigen-presenting cells	No	Yes

*Subtracting alloreactive (MLR) response

Fig.5.2 Principle of Rosenthal and Shevach's experiments (1973) to show MHC restriction of T cell proliferative responses to antigen-presenting cells

This type of response was then mapped to the MHC class II region in the mouse (Fig.2.1), where many more genetically-defined strains were available.

The interaction between helper T cells and B lymphocytes in antibody responses was likewise shown to be MHC class II-restricted. Some of the earliest experiments that led to this understanding were by Katz and colleagues (1974), who examined hapten-carrier responses in mice (Fig.5.3).

They isolated sensitized T cells from a strain C animal that had been primed to a carrier protein X. They also isolated B cells from another animal of the same strain that had been primed to a hapten H coupled to carrier Y. In the presence of the hapten–carrier complex, H-X, the sensitized T cells could help primed B cells from the same strain (C) make antibodies against the hapten when the T cells and B cells were transferred to a (C × D) F1 animal that had been irradiated to destroy its own lymphocytes, which thus acted as an '*in vivo* test tube' (for further details see Section 8.3.1 on 'carrier priming'). However, if the hapten-primed B cells were from a different strain (D), the carrier-primed T cells from strain C were unable to provide help for the antibody response. It was subsequently found that the animals from which the T cells and B cells were obtained had to be genetically identical at the MHC class II region in order for the cells to be able to 'co-operate' with each other in the antibody response, thus mapping the response to the class II region.

5.2.2 Alloreactivity

A large proportion of T cells from any individual can recognize and respond to the MHC molecules of any other genetically different member of the same species. This reactivity against allogeneic MHC molecules is called

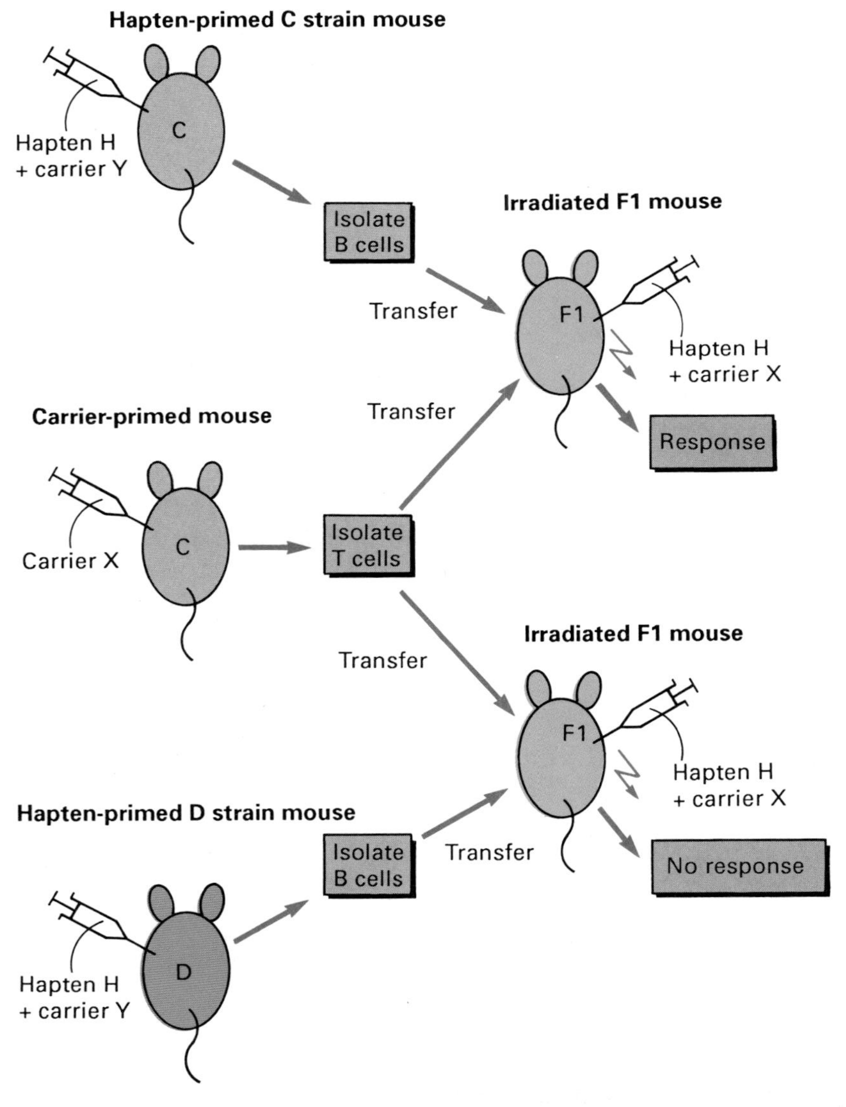

Fig.5.3 Principle of Katz *et al.* experiment (1973) to show MHC restriction of T cells for B cell antibody responses

alloreactivity and the T cells that can respond in this way are known as **alloreactive T cells**. This phenomenon is responsible for graft rejection (Section 2.1) and graft-versus-host reactions *in vivo*. In the latter case, if T cells from one individual are injected into a genetically different individual, the alloreactive T cells recognize the recipient's MHC molecules, become activated, and mount a vigorous and sometimes fatal attack against their host (see Section 5.3.2 for further details). It can also be demonstrated in the allogeneic mixed leukocyte reaction (MLR) *in vitro*, in which T cells cultured with allogeneic cells proliferate in response to the foreign MHC molecules (see Appendix).

It has been estimated that perhaps from 1 to 10% of the T cell repertoire is reactive against any complete MHC disparity. This compares with frequencies of one in many thousands or even millions for many antigen-specific responses. Since there are at least a hundred different MHC alleles in species such as mice and humans (Section 2.3; p. 68), it is evident that the vast majority of T cells in the repertoire are likely to have alloreactive specificities. *It is now clear that in physiological responses most if not all of these T cells can respond to foreign peptides bound to the individual's own MHC molecules.* The latter form of recognition is called self-restriction and is discussed in Section 5.2.3 (Fig. 5.4).

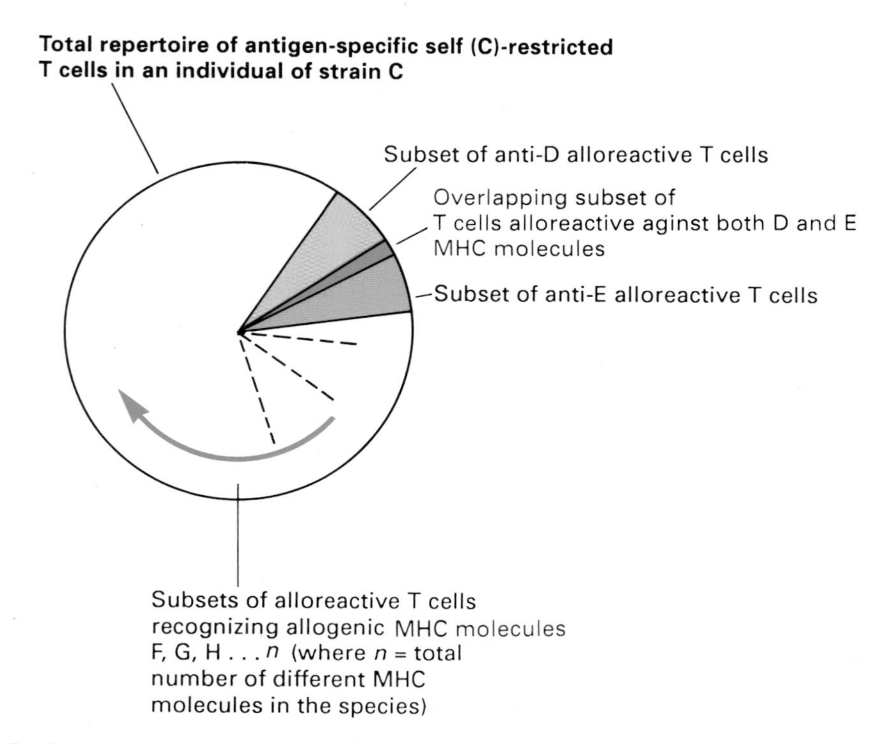

Fig. 5.4 Alloreactive T cells specific for a given MHC disparity normally recognize different foreign antigens bound to self-MHC molecules during physiological (antigen-specific) immune responses

The subset of alloreactive T cells that reacts with a particular allogeneic MHC molecule has been shown to overlap with T cells specific for antigen plus self MHC molecules (Fig.5.4) from experiments in which alloreactive T cells were positively and negatively selected (by *in vivo* filtration—see Section 5.2.4). While the positively-selected cells were enriched for alloreactive cells, and the negatively-selected population was correspondingly depleted of reactivity, there was no difference in the content of antigen-specific helper T cells between the populations. Thus the alloreactive population contains a representative sample of helper T cells, and essentially similar conclusions have been reached regarding cytotoxic T cells.

The simplest explanation for alloreactivity is that it results from the cross-reaction of T cells, normally specific for antigen and self MHC, with the allogeneic MHC molecules. This concept is shown schematically in Fig.5.5 and is encapsulated in the pseudo-mathematical jargon that 'allo = self + X' (where allo and self refer to the MHC type and X is an antigen). There are now many examples of T cell clones that can use the same T cell receptor molecule to recognize (i) an antigen in the context of the MHC molecules of the individual from which they were derived and (ii) an allogeneic MHC molecule (Ashwell *et al.* 1986; Braciale *et al.* 1981).

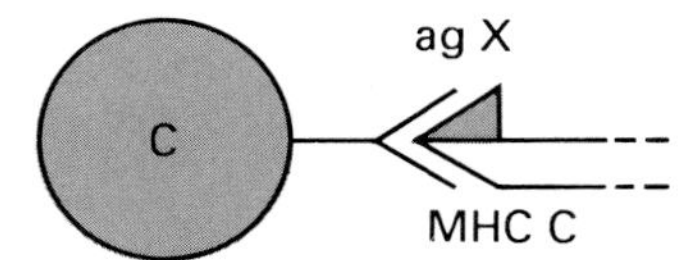

The fact that the allogeneic MLR can be inhibited by antibodies against the foreign MHC demonstrates that these molecules are indeed the targets of alloreactive T cells. In addition, it has been shown that monoclonal antibodies specific for the T cell receptor of a particular alloreactive T cell clone can inhibit the clone's response to antigen plus self MHC as well as its alloreactivity, indicating that the same T cell receptor mediates both types of recognition (Kay and Janeway 1984). Even more convincingly, transfection of the α and β chain genes of a T cell receptor (Section 4.2.1) is necessary and sufficient to endow the recipient cells with both the antigen–MHC specificity and alloreactivity of the cell of origin.

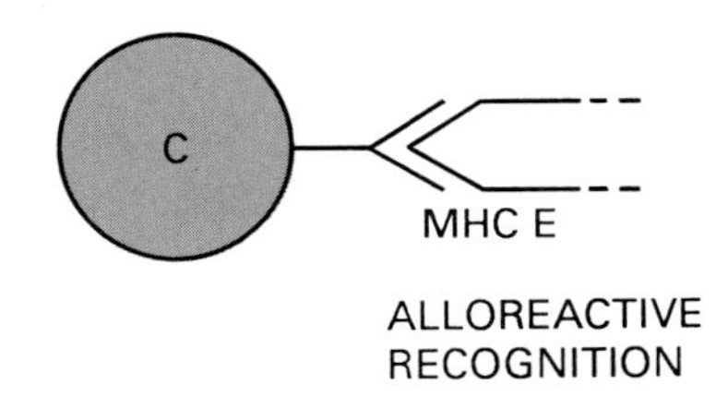

Originally it was believed that alloreactive T cells recognized determinants *on the foreign MHC molecule itself*, possibly animo acid sequences that were not present in the MHC molecules of the individual from which the T cells were obtained. However, following the elucidation of the structure of MHC molecules (Section 2.5) and other work (Section 3.2.1), it has become apparent that the majority of MHC molecules expressed on the surface of a cell are likely to be occupied by peptides (Fig.3.8). Thus allogeneic MHC molecules display a variety of different peptides within their peptide binding grooves, many of which will be derived from that individual's self proteins, and at least some of which may not be present in the individual from which the alloreactive T cells were obtained. Thus an alternative explanation for the phenomenon of alloreactivity is that alloreactive T cells actually recognize foreign peptide–MHC complexes rather than (or as well as) determinants of the allogeneic MHC molecule itself (Fig.5.6). Evidence has now been obtained that supports this idea, although the extent to which this form of recognition occurs relative to that of the MHC molecule alone has not been completely clarified.

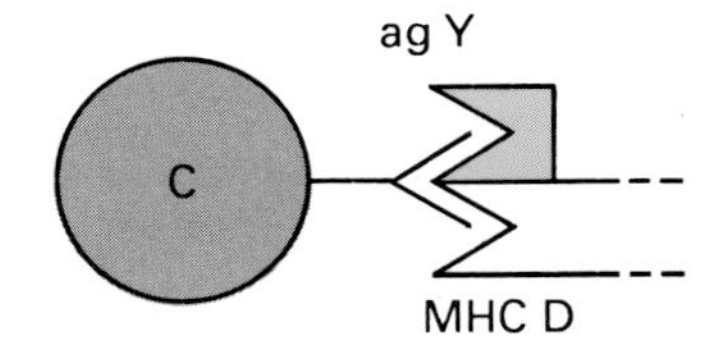

Fig.5.5 Very stylized diagram showing how a particular T cell from strain C might recognize different combinations of antigen (ag) and/or MHC

To investigate the molecular basis of alloreactivity one major approach has been to examine the reactivity of alloreactive T cell clones or hybridomas to different MHC molecules, compare their amino acid sequences, and then to determine the position of the polymorphic residues that differ between MHC molecules that are recognized and those that are not. Their response to cells transfected with chimeric ('exon-shuffled') or single-site mutated MHC molecules (Pierres *et al.* 1989; Buerstedde *et al.* 1989), as well as to cells from mutant strains of mice (e.g. bm mutants) has also been examined

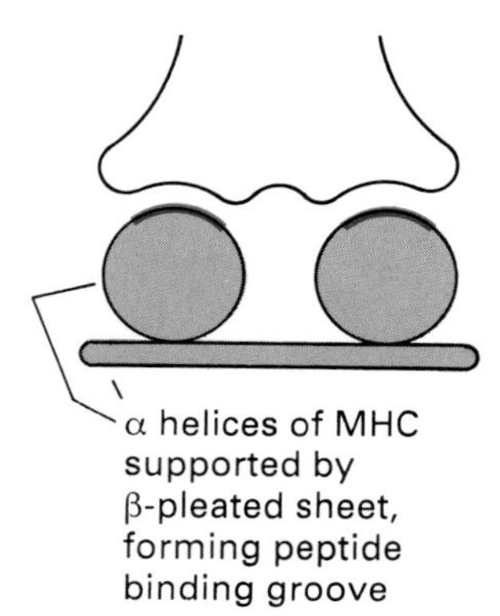

ALLOREACTIVITY (according to original definition): recognition of determinants (red) on foreign MHC α helices (grey) with no contribution from bound peptide even if present

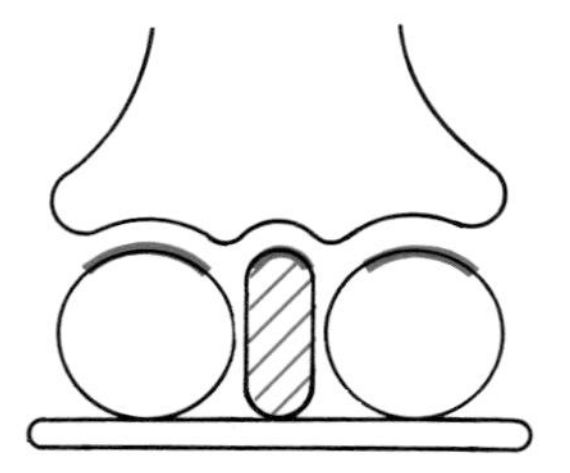

SELF-RESTRICTED ANTIGEN RECOGNITION: recognition of determinants (red) of self MHC α helices (open) and foreign peptide (hatched).

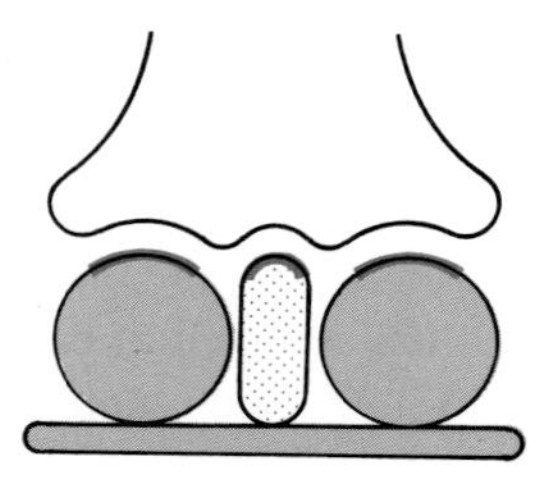

ALLORESTRICTED ANTIGEN RECOGNITION: recognition of determinants (red) of non-self-MHC α helices (red tint) and foreign peptide (dots).

Fig.5.6 Models for different forms of T cell recognition according to current ideas for TcR-peptide–MHC interactions

(Bill *et al.* 1989). In this way, potential determinants for the alloreactive T cells can be identified and mapped onto the 3D structure (Section 2.5, Figs.2.5, 2.16). If the polymorphisms are present at the top of the α helices of the peptide-binding groove, or on the side pointing away from the cleft, they could potentially form contacts with the T cell receptor itself. If they are situated on the side of the α helices pointing into the groove, or on the β-pleated sheets forming the bottom of the cleft they would be unavailable to interact with the T cell receptor, but instead could determine which particular peptide was bound to the MHC molecule (Fig.5.7).

In general, such analyses have shown that the polymorphic residues determining whether or not the MHC molecule can be recognized by alloreactive T cells are commonly located *within* the peptide binding groove rather than outside of this region (Figs.2.7, 2.16). Very frequently, these residues are positioned on the floor of the cleft or on the side of the α helices forming the sides of the cleft and pointing into it. It seems highly unlikely that most of these polymorphic residues could interact directly with the T cell receptor, but they are likely to be involved in interactions with the bound peptide. Therefore, it seems probable that alloreactive recognition can result from interactions of the T cell receptor with residues of the bound peptide, the nature of which is determined by the precise sequence of the amino acids forming the cleft, perhaps with additional contacts also occurring between the T cell receptor and the MHC molecule.

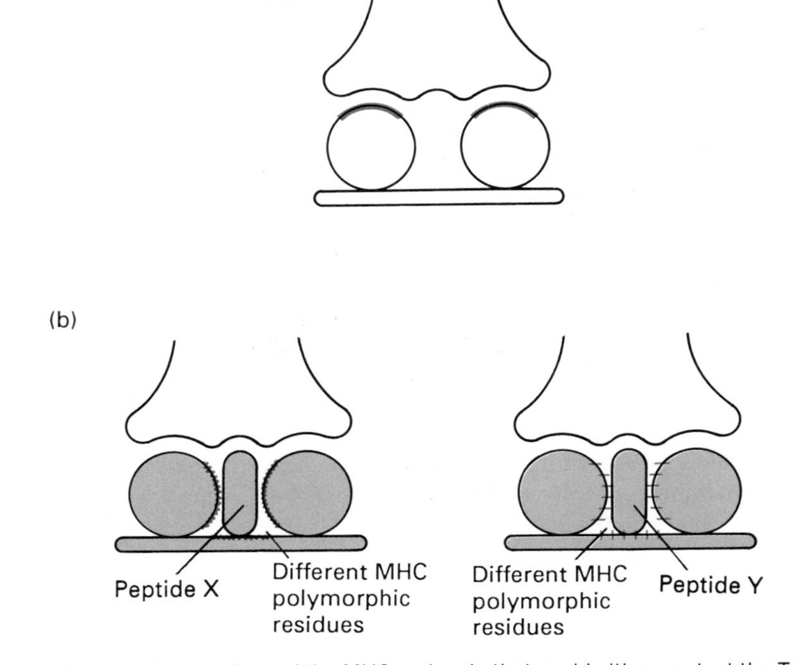

Fig.5.7 Polymorphic residues of the MHC molecule that could either contact the T cell receptor directly (a) or determine which peptide can bind to the MHC molecule

One particular study that examined responses of a panel of T cell clones, however, gave a different insight into which residues of the MHC molecule were involved in alloreactivity and self-restricted antigen recognition by the same T cell (Lombardi *et al.* 1986; Fig.5.8). The clones were derived from a DR4Dw4/DRw13Dw19 (here abbreviated DR4/DR13) heterozygous individual and were alloreactive against the DR1Dw1 (here abbreviated DR1) molecule (Section 2.3). It was found that one of these alloreactive clones also responded to the antigen *Candida albicans* in the context of DR4, and thus exhibited both self-restricted antigen recognition and alloreactivity. The polymorphisms of these DR molecules were then examined. There were no differences in the DRα chain because this is non-polymorphic in the human. One particular residue in the DRβ chain was different between DR1 and DR4, but this residue is predicted from modeling of the structure of MHC class II molecules (Fig.2.16) to point into the groove; all other differences were on the floor of the groove. Therefore, it seems likely that each MHC molecule would be occupied by a different set of peptides, but that the residues of the MHC molecules that interacted with the T cell receptor were likely to be the same. In this sense the alloreactive response to DR1 would result from recognition of a peptide (bound by DR1 but not DR4) that was seen in the context of a region of the foreign MHC molecule that actually resembled *self* (there being no contact residues different).

In summary, alloreactivity probably results from recognition of a variety of different peptides bound to allogeneic MHC molecules, and the density of these is likely to be significantly higher than that of foreign peptides from a given antigen bound to self MHC molecules. The unusual magnitude of the response is due to the wide variety of different peptides that can bind to a particular MHC allele, and the high frequency T cells that are normally specific for peptides bound to self MHC molecules but which can cross-react with these foreign peptide-allogeneic MHC complexes. In some studies, peptides involved in alloreactive recognition have been purified and characterized (Rotzschke *et al.* 1991).

Recognition of an antigen in the context of an allogeneic MHC molecule is actually called **allorestriction** (Fig.5.5). This contrasts with the original concept of alloreactivity which referred to recognition of the allogeneic MHC molecule *itself*. Therefore, alloreactivity and allorestriction could be the same phenomenon. An important proviso, however, derives from the results of the last experiment described above (see box), in which at least one T cell clone appeared to recognize a foreign peptide plus a region of the MHC molecule resembling self. Such a T cell would be regarded as 'self-restricted'. Only further experimentation will be able to determine whether this is a general phenomenon. Self-restriction and allorestriction are discussed further in Sections 5.2.3 and 5.2.4.

5.2.3 Self-restriction and mechanisms of positive selection

It is commonly believed that T cells from any individual are biased towards recognition of foreign peptides bound to the particular MHC molecules

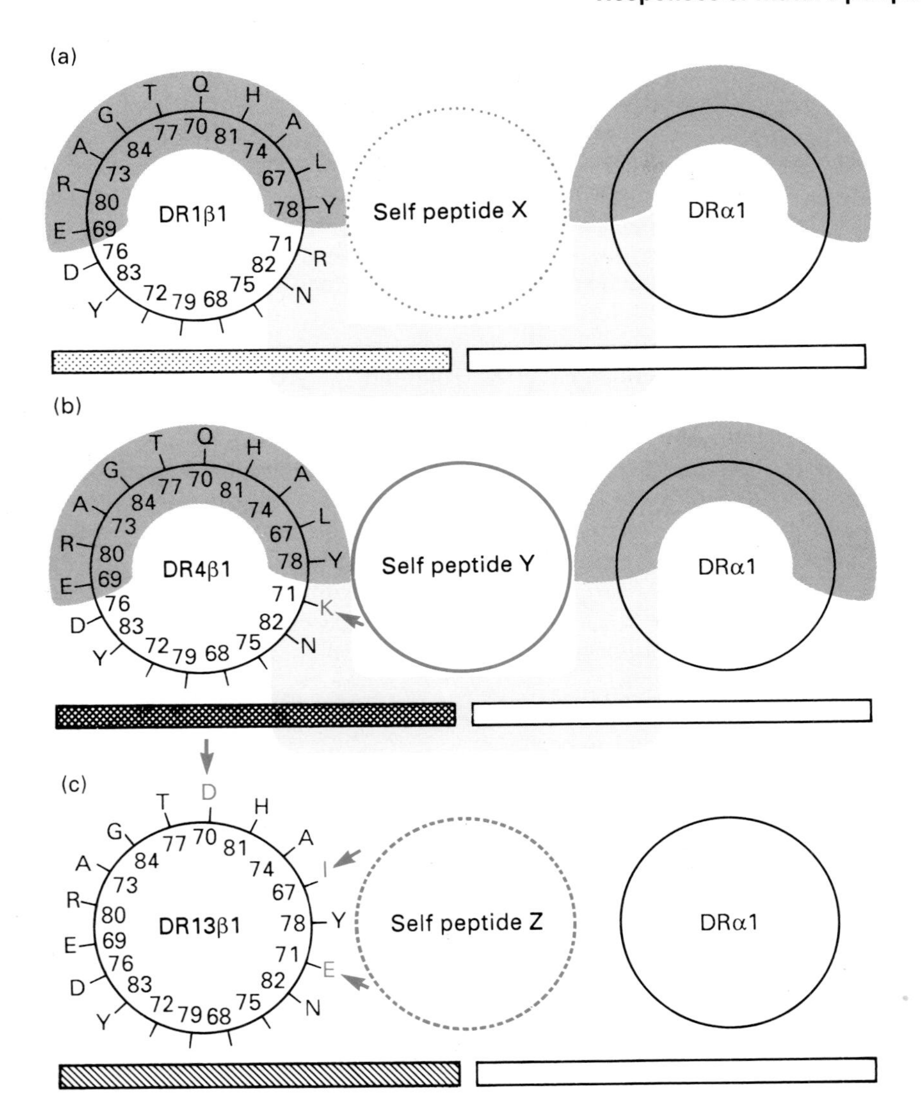

Fig.5.8 Schematic representation of a cross-section through the α helices and the floor of anti-parallel strands that comprise the amino-terminal domains of DR1Dw1 (DR1; a), DR4Dw4 (DR4; b), and DRw13Dw19 (DR13; c). Putative 'self' peptides are shown lying between the two α helices. The orientation of individual amino acid residues on the β_1 domain α helices from positions 67 to 83 is shown. Residues that are conserved between these DR types are shown as black letters and those that differ as red symbols. The sequences of the anti-parallel strands comprising the floor of the domains include multiple differences in the exposed regions as indicated by the use of different patterns in the rectangular boxes. The surface of the molecules that is predicted to interact with the T cell receptor is highlighted with dark shading, and the surface predicted to interact with bound peptides with lighter shading. The amino acid differences are indicated by red arrows. From Lombardi *et al.* (1989), with permission. See text for details of this study; although not discussed, DRw13 is shown here (c) for comparison

expressed in that individual, rather than MHC molecules of genetically-different members of the species. *This tendency towards recognition of foreign peptides bound to self MHC molecules is termed self-restriction.* Self-restriction of the T cell repertoire is due to **positive selection**, in contrast to negative selection which results in self tolerance as discussed in Section 5.2.5 (Table 5.2). During positive selection, developing T cells with the potential to recognize foreign peptides bound to self MHC molecules are preferentially selected in favour of those that recognize foreign peptides bound to allogeneic MHC molecules.

Why have mechanisms for positive selection evolved? As yet there are no clear answers, but one hypothetical possibility is as follows. T cell receptors are generated by gene rearrangement (Section 7.3) and it is quite possible

Table 5.2 Consequences of results of positive and negative selection

Phenomenon	Consequence
Positive selection	**Self-restriction**: preferential recognition of foreign peptides bound to *self* MHC molecules by T cells
Negative selection	**Self-tolerance**: failure of T cells to recognize self peptides bound to self MHC molecules (and/or self MHC alone)

that some receptors could have the potential to recognize peptides alone (compare with antibodies; Section 7.2) rather than peptide–MHC complexes; the latter is of course essential for the special capacity of T cells to monitor *cell surfaces* for the presence of foreign antigens (compare with B cells). Therefore, unless T cell receptor genes encode molecules that are already biased towards MHC-restricted antigen recognition, positive selection may be required to select a repertoire of MHC-restricted T cell receptors. In this case, the apparent self-restriction of the repertoire may arise more by accident than by design, since T cell receptors are selected exclusively by self MHC molecules, and do not encounter other, polymorphic, non-self MHC molecules during development.

It is known that developing thymocytes need to interact with MHC molecules for positive selection to occur.

Different subsets of T cells in (C × D) F1 mice are restricted to recognizing foreign peptides bound to either C *or* D strain MHC molecules (Fig.5.10). The development of the subset of C-restricted T cells in an F1 mouse can be inhibited by administering antibodies specific for C strain MHC molecules to the neonate. One interpretation is that the antibodies bind to thymic MHC molecules and inhibit recognition by thymocytes that would otherwise be selected by these molecules; these thymocytes are then eliminated by programmed cell death (apoptosis; Section 10.3.5). Moreover, experimental mice that do not express MHC class I molecules, because of a deficiency of β_2-microglobulin, lack $CD8^+$ T cells (Ziljstra *et al.* 1990). Likewise, CD4 and CD8 molecules are involved in positive selection, because this process is inhibited in mice treated with anti-CD4 or anti-CD8 antibodies, or in which the molecules are not expressed because of genetic manipulations (e.g. in 'gene knock-out' experiments).

In addition, it is presumed that the MHC molecules responsible for positive selection contain self peptides that may or may not be involved in this process. Inevitably this has to occur in the absence of the foreign peptide for which the T cells will be specific, because the repertoire of T cell receptors is generated before exposure to the antigen. The mechanisms for positive selection are currently unclear, but some hypotheses are presented below and in Section 5.2.6.

At first sight, the concept of a *self*-restricted T cell repertoire seems to be incompatible with what we discussed earlier (Section 5.2.2): alloreactive T cells recognize peptides bound to *allogeneic* MHC molecules (i.e. allorestricted antigen recognition). Later we shall describe other experimental situ-

Table 5.3 'Levels' of self restriction

Intrinsic bias in the repertoire	Bias in immune response
Develops during ontogeny	Imposed during response
Due to positive selection of maturing thymocytes	Due to selection of mature T cells
Occurs primarily in the thymus	Occurs in the periphery (i.e. secondary lymphoid tissues)
Controlled by MHC molecules expressed on thymic epithelial cells	Controlled by MHC molecules expressed by antigen-presenting cells in the periphery
Occurs in the absence of foreign antigen	Occurs only in the presence of the specific antigen

ations in which allorestricted T cells can be clearly demonstrated (Section 5.2.4). Nevertheless, the current view is that the T cell repertoire is *intrinsically biased* or **skewed** towards self-restricted antigen recognition. In addition, a further bias towards self-restriction is imposed *in the course of* physiological immune responses as a consequence of the so-called **'priming environment'** (see later in this section).

There are therefore two different levels on which one may consider self-restriction (Table 5.3). The first, or '*intrinsic bias in the repertoire*', is generated during T cell ontogeny and is largely controlled by the MHC molecules expressed in the thymus (Section 5.3.3). The second comes about *during immune responses* and is determined by the MHC molecules expressed by antigen-presenting cells in the periphery (Fig.5.9) While the first is due to selection of a particular population of *developing* T cells, in the *absence* of nominal† antigen (see above), the second is due to selection and activation of a subset of antigen-specific *mature* T cells and requires the *presence* of antigen (Fig.5.9). We shall briefly consider each level in turn.

† In the case of a T cell specific for a peptide derived from an antigen X plus an MHC molecule, the antigen X can be referred to as the 'nominal' antigen. This term encompasses the idea that the antigen itself is not recognized by the T cell receptor (rather, it is a peptide derived from the antigen bound to MHC) as well as the possibility that the T cell might have another specificity that has yet to be defined (i.e. a different peptide bound to a different MHC molecule).

Self-restriction of the repertoire

The concept of an intrinsically self-restricted T cell repertoire comes in part from studies with thymus-reconstituted nude mice and bone marrow chimeras, discussed in section 5.3.2. The general finding is that (C × D) F1 T cells become restricted to C-strain MHC molecules, but *not* to D, if they develop in a C-strain thymus.

Peripheral T cells from a heterozygotic (C × D) F1 animal comprise different populations, some of which are restricted to recognizing foreign peptides bound to one parental MHC type, C, while other T cells are restricted to D

(Fig. 5.10). If an animal of one parental strain (e.g. a C × C homozygote) is irradiated to destroy its existing T cells and given a transplant of bone marrow from a (C × D) F1 animal, the T cells that develop, which are genetically of F1 origin, preferentially recognize foreign peptides bound to MHC molecules of strain C but not strain D. The same result can be obtained if (C × D) F1 nude mice are transplanted with a C strain thymus (see Section 5.3.3), showing that this is where positive selection of developing thymocytes can occur. In the control experiment, if an F1 thymus is transplanted into an F1 nude mouse, the restriction of the T cells is found to be similar to those of a normal F1 mouse containing its own thymus, i.e. a '**euthymic**' animal.

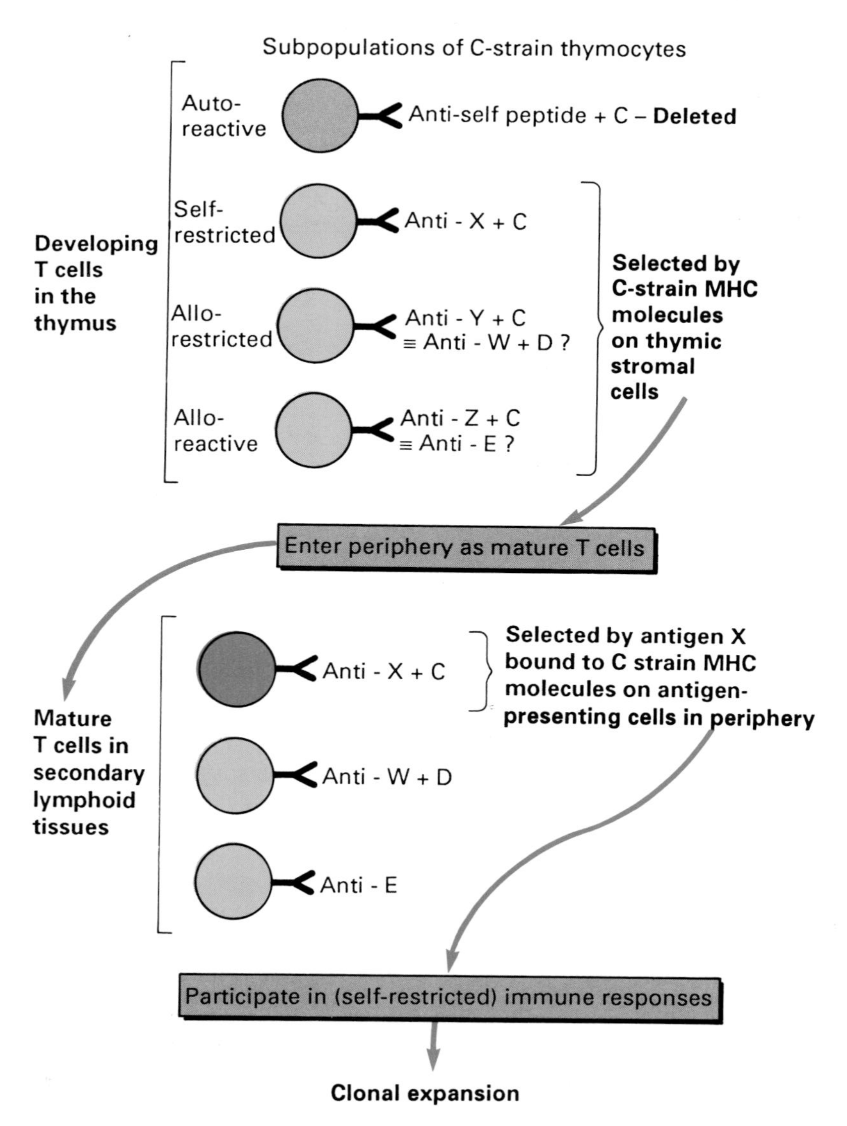

Fig. 5.9 Levels of self-restriction

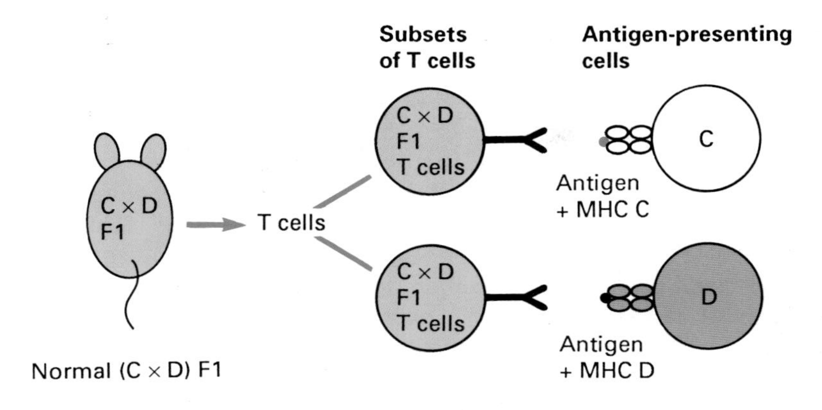

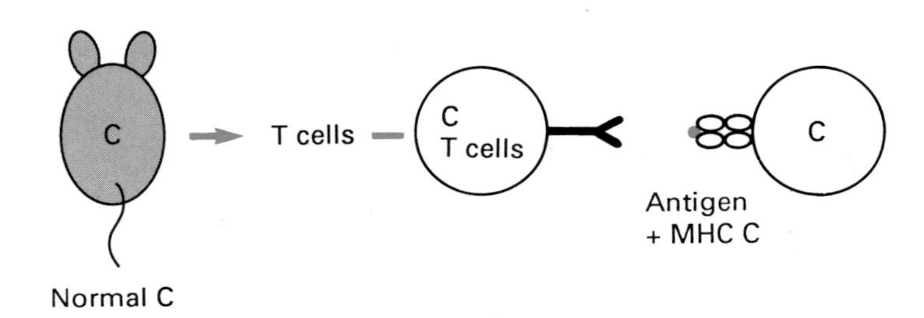

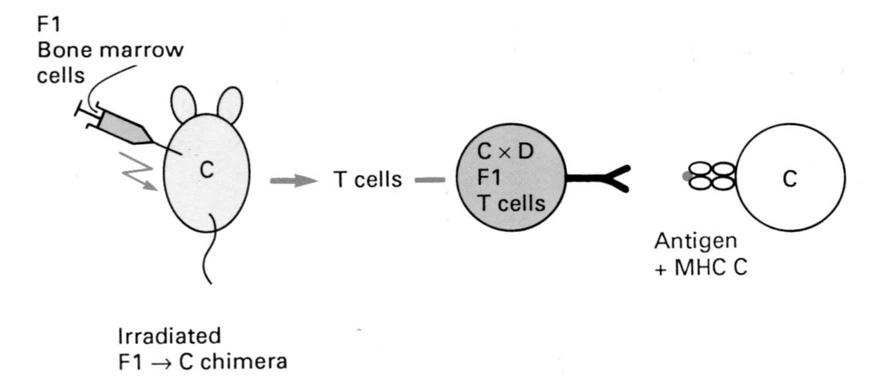

Fig.5.10 F1 T cells that develop in a C strain environment become restricted to C strain MHC molecules

More recently, the results of experiments with transgenic mice (see Appendix) have also been taken to support the idea of an intrinsically self-restricted repertoire.

For example, transgenic mice have been constructed that contain the genes for the α and β chains of a T cell receptor specific for the male-specific, minor histocompatibility antigen, H-Y, in the context of H-2D^b

molecules (Kisielow *et al.* 1988). As discussed in the section on self-tolerance (5.2.5), it was found that $CD8^+$ T cells expressing this transgene were clonally deleted in male H-2^b mice, because they were autoreactive. However, the finding of relevance to this section was that in females there was an *increase* in the proportion of $CD8^+$ transgene-expressing thymocytes in H-2^b mice but *not* in other genotypes tested. Apparently the MHC molecules in the thymus of H-2^b mice (which of course included H-$2D^b$) were able to select the transgene-expressing T cells, but other MHC molecules could not. (Note, in addition, that positive selection occurred in the absence of the antigen for which the transgenic T cell receptor was specific, because H-Y is not expressed in female mice.

As discussed later (Section 5.3.5), it is thought that positive selection is due to interactions between developing T cells and thymic epithelial cells. It has been suggested that the MHC molecules expressed by thymic epithelial cells contain a variety of peptides that are produced within this particular cell type, many of which might be different from those produced by other cell types (Fig.5.11), and there is a limited amount of experimental evidence to support this view. T cells specific for these epithelial cell peptides bound to self MHC molecules might be selected and expanded, and then enter the periphery. Because these thymic epithelial cell peptides are not encountered once the T cells have left the thymus, it has been argued there is little or no

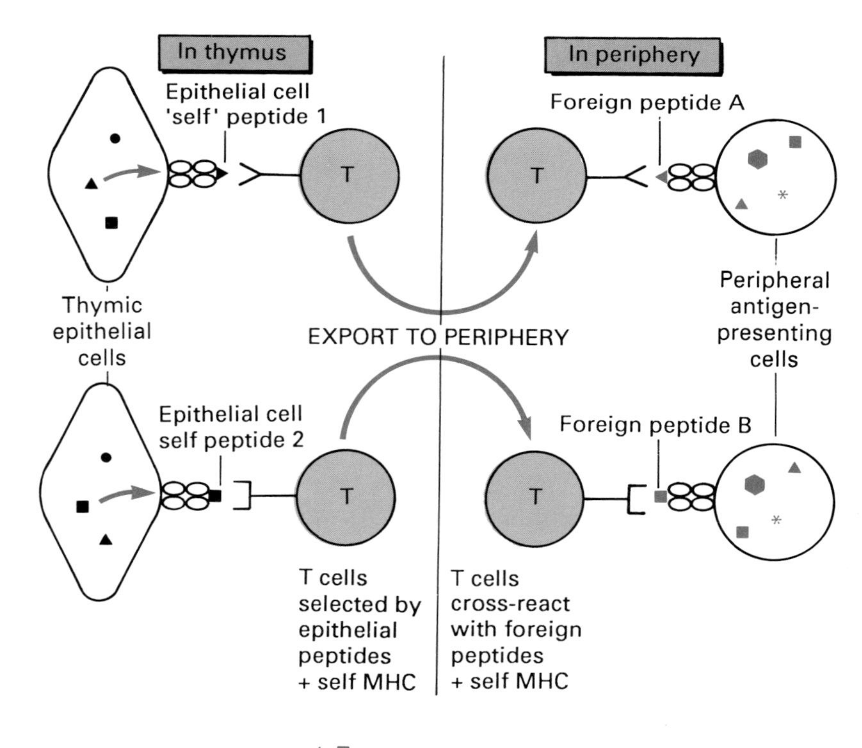

Epithelial cell-derived peptides

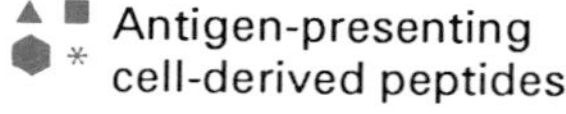

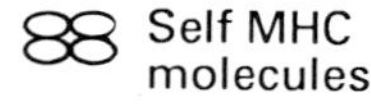

Fig.5.11 One hypothesis for positive selection of T cells (see text)

risk of autoreactivity. Self-restriction of the repertoire could thus be due to cross-reactive recognition by T cell receptors, actually selected by thymic epithelial cell peptide–self MHC complexes, with foreign peptide–self MHC complexes on antigen presenting cells in the periphery. The cell types responsible for positive (and negative) selection in the thymus are considered further in Section 5.3.5.

Self-restriction as a consequence of the priming environment

In early experiments demonstrating the MHC restriction of T cells (Section 5.2.1) it was observed that cytotoxic T cells from a virally-infected C strain animal could kill virally-infected target cells from strain C but not strain D. It would, however, be a mistake to take these results as evidence for an intrinsically self-restricted T cell repertoire. Let us suppose that C strain animals do in fact contain allorestricted T cells that can potentially recognize viral peptides in the context of D strain MHC molecules (Fig.5.6). When C strain mice are immunized, their antigen-specific T cells recognize antigen that is presented by antigen-presenting cells, but these cells express only C strain MHC molecules. *These animals do not contain antigen-presenting cells expressing D strain MHC, and thus D-restricted cells can not be activated* (Fig.5.12). The immune response is therefore inevitably biased towards self-restricted antigen recognition because only C-restricted T cells are activated and clonally expanded in a C strain 'priming environment'. The former observations, then, have no bearing on whether or not the repertoire itself is intrinsically self-restricted, but demonstrate that antigen-specific T cell *responses* are inevitably self-restricted.

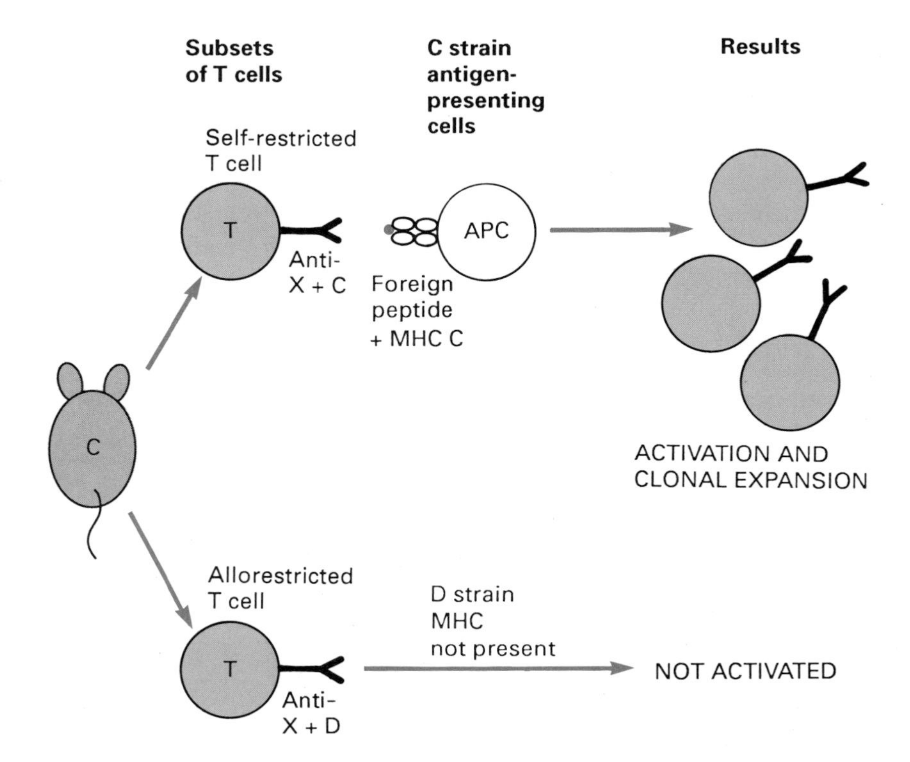

Fig.5.12 Self-restriction as a consequence of the priming environment

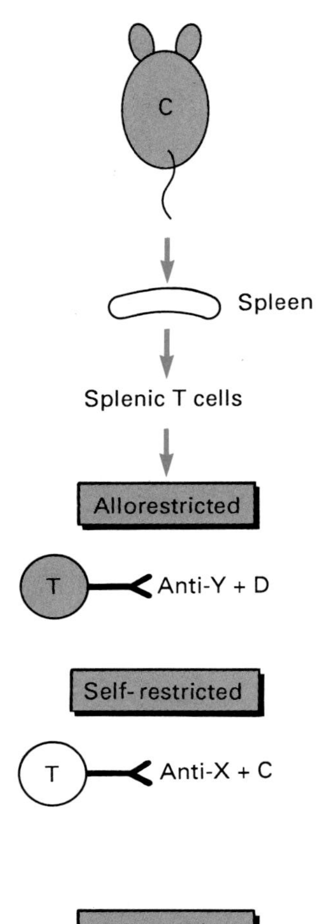

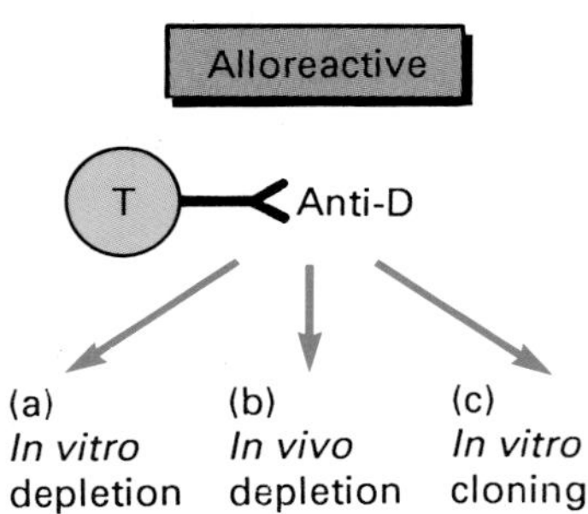

Fig.5.13 Procedures to deplete alloreactive T cells and demonstrate allorestricted T cells

5.2.4 Allorestriction

Does the T cell repertoire contain allorestricted T cells? The answer to this would appear to be 'yes' and thus one can question whether the repertoire is truly self-restricted. In the section on alloreactivity (5.2.2) we saw that this phenomenon is at least in part due to recognition of peptides bound to allogeneic MHC molecules (Fig.5.6). Moreover, we previously cited cases of T cell clones that recognized foreign antigens in a self-restricted manner and which were also alloreactive, for example T cells from a C strain animal that recognized antigen X plus MHC C, and also allogeneic MHC molecules from strain E. It has been possible to isolate T cell clones that can also recognize a different foreign antigen Y, in the context of another MHC, D. In other words they appear to have three specificities: X + C, Y + D and E (+ peptide Z?). Almost certainly this is due to cross-reactive recognition by the same T cell receptor because the precise epitopes recognized all have the same configuration, even though other regions of the particular peptide–MHC complexes involved could be very different from each other (Fig.5.5).

Allorestricted cells have also been detected in other experimental situations. Normally it is difficult to detect allorestricted cells because the response of T cells from strain C against a defined antigen, Y, in the presence of antigen-presenting cells from strain D, will be masked by the vigorous anti-D alloreactive response (the allogeneic MLR). If these alloreactive cells are first depleted, it is possible to test whether the remaining T cells can recognize a different antigen Y + D. Several approaches have been used to remove alloreactive T cells, and have often revealed the presence of allorestricted cells in the repertoire (Fig.5.13).

Alloreactive cells can be depleted *in vitro*. One approach is to culture T cells from strain C with antigen presenting cells from strain D, and then to kill the proliferating alloreactive cells by adding bromodeoxyuridine and exposing the cells to light (Fig.5.14). Bromodeoxyuridine is incorporated into newly-synthesized DNA and, after activation by light, the DNA becomes cross-linked, preventing the cells from replicating and ultimately killing them. Alternatively, the cells can be cultured with highly-radioactive nucleotides which become incorporated into the DNA and kill the cell. Another approach is to incubate the T cells with allogeneic cell monolayers, to which the alloreactive cells bind, and then to recover the unbound, negatively-selected cells. The negatively-selected C strain cells can then be primed and tested for their ability to respond to antigen Y in the presence of D strain cells. Using these procedures, it has been possible to demonstrate T cell proliferative responses to haptens and simple polymeric antigens in the presence of allogeneic accessory cells. Some of the most extensive experiments were carried out by Ishii and colleagues (1982), who tested over 30 different strain combinations and found in every instance that T cells could respond to antigens plus allogeneic antigen-presenting cells. In some but not all cases cytotoxic T cells that can lyse virally-infected or chemically-modified allogeneic cells have also been found.

Limiting dilution analysis *in vitro*. The idea of this technique is to dilute the cells until there is statistically only one responding cell per culture well. This cell is then grown in the presence of preformed 'helper factors', typically culture supernatants from stimulated T cells containing IL-2. Using this approach, it has been possible to demonstrate a surprisingly high frequency of allo-restricted cytotoxic T cells that can recognize the trinitrophenyl hapten, TNP, or viral peptides in association with defined MHC mutant molecules (Reimann *et al.* 1985).

Alloreactive cells can be depleted *in vivo*. Sprent (1978) used a system of '*in vivo* filtration' in which mature C strain T cells were injected into irradiated allogeneic (D) or semi-allogeneic (C × D) recipients. The specific alloreactive cells were retained in the spleens for about 2 days and began to clonally expand, whereas the 'negatively-selected' cells recirculated through the animal and could be obtained from the thoracic duct lymph at this time, i.e. after 2 days (Fig.5.15). These T cells were then primed in secondary recipients, of strain D or strain C × D, that were irradiated to kill their own lymphocytes, and infected with virus to provide a source of antigen presenting cells with the relevant antigen. Using these techniques, some investigators have obtained good cytotoxic responses to virally-infected or TNP-modified allogeneic cells, in some but not all strain combinations. (As an aside, we should note that the same system of *in vivo* filtration can be used to obtain positively-selected cells. If one waits longer than 2 days, large numbers of specifically-sensitized T cells are released from the lymphoid tissues, such as the spleen, and enter the central lymph. These can be obtained from the thoracic duct if it is cannulated at about 4–5 days).

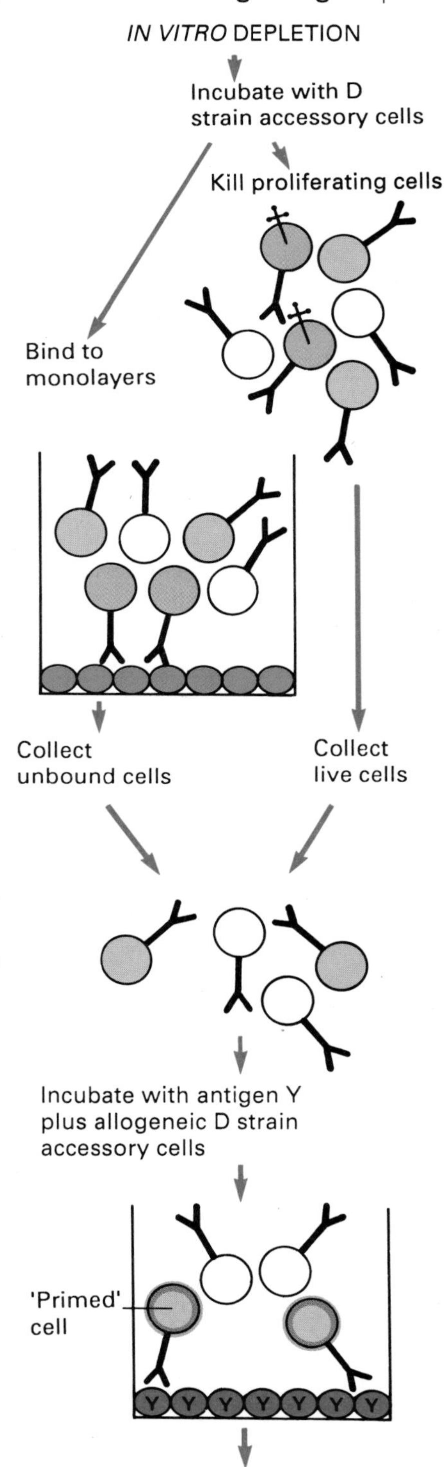
IN VITRO DEPLETION
Incubate with D strain accessory cells
Kill proliferating cells
Bind to monolayers
Collect unbound cells
Collect live cells
Incubate with antigen Y plus allogeneic D strain accessory cells
'Primed' cell
Y
TEST RESPONSE

Fig.5.14 *In vitro* depletion of alloreactive cells to reveal allorestricted T cells

Explanations for allorestriction

Several explanations have been put forward to account for the presence of allorestricted T cells in the repertoire. First, these cells may be 'truly' allorestricted and unable to cross-react with another antigen in the context of self MHC. This would mean that a portion of the T cell repertoire could never be used physiologically, and while this might appear wasteful one might argue that these cells could just be a by-product of the normal diversification process that generates a large number of antigen receptor specificities (see Chapter 7). Second, it was proposed that allorestricted cells might actually be generated during the experimental procedures, perhaps originating from cells with a different specificity. While there are precedents for this in B cells, in that some of their clonal progeny can acquire new specificities through somatic mutation (see p. 449—'affinity maturation'), there is no evidence for somatic mutation occurring in genes for T cell receptors, and this seems unlikely. Probably the most favoured explanation for the existence of allorestriction is according to the concept discussed earlier (Section 5.2.2) with reference to alloreactivity, i.e. cross-reactive recognition (also called 'aberrant recognition') by T cells that under physiological circumstance are specific for peptide–self MHC complexes (Figs.5.4, 5.5, 5.6).

Perhaps it is not surprising that the repertoire does not appear to be completely self-restricted. Even if it was generated as such (i.e. the primary specificity of every T cell is in fact for a foreign peptide bound to a self MHC molecule) it seems almost inevitable that at least some of these cells would

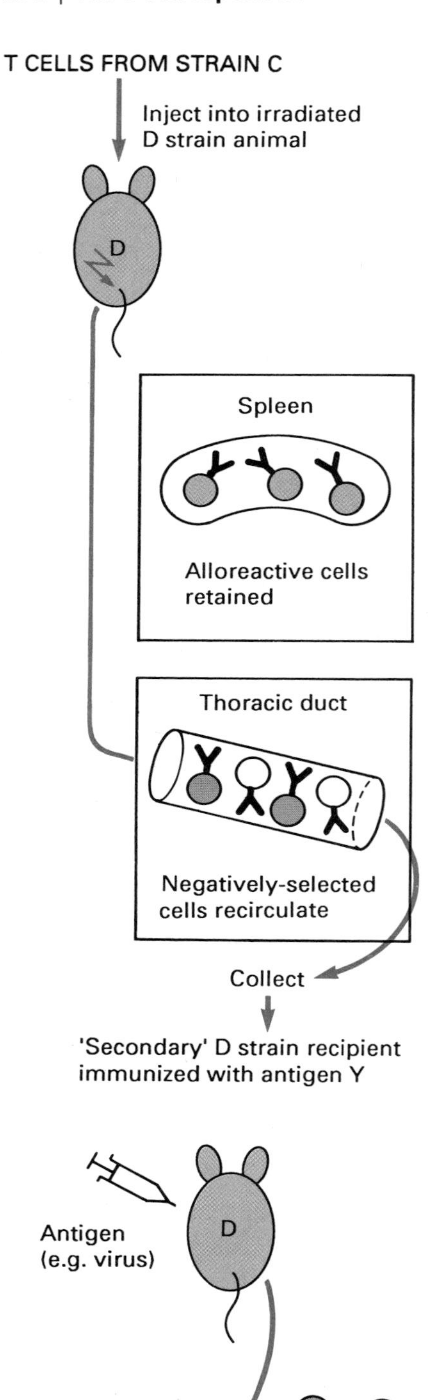

Fig.5.15 *In vivo* depletion of alloreactive cells to reveal allorestricted T cells

be able to recognize combinations of different peptides bound to allogeneic MHC molecules, because of the cross-reactivity or degeneracy of T cell recognition. Thus it turns out that even the demonstration of allorestricted T cells within the repertoire is not sufficient to answer the question as to whether the repertoire is really self-restricted. It becomes a question of which came first—the chicken or the egg? Is the repertoire generated for self-restricted antigen recognition as such, and other specificities are simply due to cross-reaction, or is it primarily alloreactive (possibly because of some intrinsic bias in the germ-line towards recognition of foreign MHC) and these alloreactive T cells can then 'cross-react' with peptide–self MHC complexes? As yet there is no clear answer to this question.

5.2.5 Self-tolerance and mechanisms of negative selection

The majority of MHC molecules expressed by normal cells are occupied by self peptides (Fig.3.8). Given that the immune system is primarily destructive in nature, and that many immune responses depend on T cells as regulatory or effector cells (Section 1.2.3), it is important that the T cell repertoire is able to distinguish between self and non-self antigens, and that it is unresponsive to self components. *The normal state of unresponsiveness of an individual's T cells to self peptides bound to self MHC molecules is called* ***self-tolerance***. The term tolerance can also be used to describe the unresponsiveness of B cells towards certain self components (Section 8.6) and, in a more restricted sense, to describe a state of unresponsiveness that can result from previous exposure to an antigen (Section 11.2.2). In this section, however, we are concerned only with self-tolerance T cells.

In a series of classical studies it was shown, using a veritable menagerie of different species, that exposure to foreign tissues early in life was sufficient to induce a state of tolerance. Some of the earliest work was by Owen (1945) on dizygotic, or non-identical twin, cattle. These animals share a common blood supply, referred to as **parabiosis**, and they are therefore exposed to each other's cells *in utero* while their immune systems are developing. It was found that, as adults, they were unable to reject tissues that were transplanted from one to the other and thus a state of tolerance had been established. Similar observations were made by Hasek and Hraba (1955), who used a model of experimental parabiosis in chickens. Then Billingham, Brent, and Medawar (1956) demonstrated that it was possible to induce tolerance in mice by injecting them soon after birth with lymphoid cells from another strain; this form of induced tolerance is referred to as **neonatal tolerance**. In another series of experiments Tripplett (1962) removed pituitary glands from frog embryos and transplanted them back into the same individuals when the animals were mature adults. The glands were rejected, showing that tolerance to self components can be established at an early period of development. A number of other studies have shown that both T cells and B cells can be tolerized, and that it is sometimes possible to induce unresponsiveness to antigens even in adult animals (see Section 11.2).

Mechanisms and sites of tolerance induction: overview

In principle, tolerance to self components could result from three or more mechanisms (Table 5.4). First, potentially autoreactive cells could be elimin-

Table 5.4 Overview of mechanisms of tolerance

Mechanism*	Predominant site of induction	Predominant stage of T development affected	Predominant type of antigen involved
Clonal deletion	Thymus	Immature	Self
Clonal anergy (or, functional inactivation)	Periphery	Mature	Self
Suppression	Periphery	Mature	Non-self

*Does not include genetically-determined unresponsiveness to certain antigens; e.g. Ir gene phenomena — see Section 5.4.

ated from the repertoire, **clonal deletion**. Second, these cells could be present in the periphery but rendered unresponsive, **clonal anergy or functional inactivation**. Third, the function of these cells could be inhibited by other cells, **suppression**. There is reasonably clear evidence that the first two mechanisms are responsible for tolerance to self components. At the time of writing, the general consensus is that *clonal deletion occurs primarily in the thymus, during T cell development, while functional inactivation of mature T cells occurs primarily in the periphery*. However, some evidence is emerging that clonal deletion can also occur in the periphery and there may, in addition, be non-deletional mechanisms of tolerance induction (e.g. functional inactivation) in the thymus, so this view might change in the future. It is generally difficult to demonstrate suppressive mechanisms of self-tolerance, although this can certainly occur for foreign antigens in the periphery; this is discussed in Chapter 11 and will not be further considered here. Additional mechanisms of self-tolerance have also been proposed (Liu and Janeway 1990).

In this section, we first make some general comments about the experimental approaches that have been used to demonstrate clonal deletion and clonal anergy of T cells. Then the evidence for each of these mechanisms of self-tolerance, and the sites where they occur, will be considered in turn. Finally, we will address the issue of whether or not the T cell repertoire is tolerant to all self components. The precise cell types responsible for negative seletion (and for positive selection, Section 5.2.3) are discussed in Section 5.3.5.

Experimental approaches

An apparent revolution in our understanding of self-tolerance came about with the development of monoclonal antibodies specific for a variety of T cell receptor V_β regions, and the realization that expression of a given V_β region could sometimes bias T cell recognition towards a particular 'self' antigen–MHC combination (e.g. Table 5.5). It thus became possible to trace the development of T cells, from the thymus to the periphery, in animals that naturally expressed these 'self' antigens, and to construct transgenic animals with T cells bearing receptors expressing particular V_β regions and follow the fate

Table 5.5 Examples of reactivity of different TcR V_β regions to superantigens*

V_β region	Superantigen reactivity	MHC molecule required
Mouse:		
$V_\beta 17a$	Undefined component	IE
$V_\beta 6$, $V_\beta 8.1$, $V_\beta 9$	Mls-1a (Mtv-7)	IE > IA
$V_\beta 3$	Mls-2a (Mtv-13)/Mls-3a (Mtv-6)	IE ≫ IA
$V_\beta 5.1$, $V_\beta 5.2$, $V_\beta 11$	Etc-1 (Mtv-9)	IE
$V_\beta 3$, $V_\beta 8$ family	Stapphylococcal enterotoxin B (SEB)	class II
Human:		
$V_\beta 3$, $V_\beta 12$, $V_\beta 14$	Stapphylococcal enterotoxin B (SEB)	class II
$V_\beta 2$	*Staph. aureus* toxic shock syndrome toxin-1 (TSST)	?

*See text. Further examples and details are in:
Janeway C. (1991). Mls: makes a little sense (News). *Nature*, **349**, 459–61
Herman A., Kappler J.W., Marrack P., Pullen A.M. (1991). Superantigens: mechanism of T-cell stimulation and role in immune responses. *Annual Review of Immunology*, **9**, 745–52

of these cells. Such analyses have revealed that *T cells bearing these autoreactive receptors are generated in the thymus but they can be deleted before final maturation and/or inactivated in the periphery*. These mechanisms maintain unresponsiveness to the 'self' antigen–MHC complexes in question.

In many studies, the majority of antigens examined are not defined self protein antigens, but so-called **superantigens** (Fig.2.17; Table 5.5). These are not 'typical' antigens in that they do not localize within the peptide-binding groove of an MHC molecule, and they stimulate polyclonal proliferation of T cells; this group includes bacterial enterotoxins and the Mls retroviral superantigens (see below). The nature of other antigens remains unknown at the time of writing, such as an undefined component (which may or may not turn out to be a superantigen) associated with an MHC molecule like H-2 IE, or the male-specific antigen H-Y. Clearly, there must be reservations as to the relevance of these experiments to the situation pertaining with typical self peptides. However, there are as yet no V_β associations with defined, conventional, self peptide antigens, and at present it is difficult to trace the induction of tolerance to these molecules.

Evidence for clonal deletion

Studies of V_β expression in normal and neonatally-tolerant mice. The first experiments along these lines were reported in 1987. Kappler and Marrack noted that a high proportion of T cell clones and hybridomas that recognized allogeneic H-2 IE molecules expressed the $V_\beta 17a$ element, as detected by specific monoclonal antibodies. Subsequently it became apparent that some component expressed particularly by B cells, that is unidentified at the time of writing, was associated with these MHC molecules. It was then found that T cells expressing these $V_\beta 17a^+$ autoreactive receptors were not present in the periphery of IE^+ strains of mice, although they were present in IE^- strains (Section 2.6.2), clear evidence that clonal deletion had occurred.

Further analysis then revealed that immature thymocytes expressing $V_\beta 17a^+$ receptors *were* present in the thymus of IE^+ mice, but these receptors were not detectable on mature thymocytes (Kappler *et al.* 1987). This has been a consistent observation in many other studies (e.g. see Mls, below), and suggests that clonal deletion occurs at an immature stage of thymocyte development before final development into mature single-positive $CD4^+$ or $CD8^+$ T cells, perhaps at the double-positive stage (Section 4.3). The fact that *in vivo* treatment of mice with anti-CD4 antibodies alone inhibits deletion of T cells from *both* the CD4 and CD8 single-positive subsets provides support for deletion of a common (double-positive) precursor.

At about the same time these experiments were described, there were reports that T cells expressing other V_β regions were deleted in strains of mice expressing particular alleles of **Mls** (see below) in association with MHC class II molecules, particularly IE (MacDonald *et al.* 1988). For example, peripheral T cells and mature thymocytes expressing $V_\beta 6$ and $V_\beta 8.1$ were deleted in strains expressing the Mls-1a locus product, and $V_\beta 3$ in Mls-2a (Table 5.5). A completely different method of detection, quantitative RNA hybridization, was also used to demonstrate deletion of $V_\beta 9^+$ T cells in Mls-1a mice. Both subsets of single-positive mature T cells were deleted in response to Mls, and deletion of both subsets was prevented by treatment with anti-CD4 antibodies, again suggesting that deletion occurs at the double-positive stage of thymocyte differentiation (see above). (However, it is now known that both $CD4^+$ and $CD8^+$ T cells can respond to Mls *in vitro* and *in vivo*, even though an association of Mls with MHC class I molecules has yet to be shown).

Both Mls and the component associated with IE that is reactive with $V_\beta 17a$ are endogenous 'self' antigens (i.e. intrinsic to the cell). Clonal deletion has also been demonstrated using exogenous superantigens. For example, **Staphylococcal enterotoxin B** (SEB) stimulates proliferation of $V_\beta 3^+$ and $V_\beta 8^+$ T cells *in vitro* (Table 5.5). When SEB is administered to neonatal mice, T cells expressing these receptors are clonally deleted (White *et al.* 1989). In addition, treatment of Mls^b mice with cells from Mls^a strains induces tolerance to Mls^a and, in neonatal mice, this is likewise associated with deletion of the relevant (e.g. $V_\beta 6^+$) T cell clones.

Mls and related products are encoded by retroviruses. Mls, the 'minor lymphocyte-stimulating locus' of the mouse is unusual because it is not linked to the MHC and yet it can stimulate vigorous, polyclonal proliferation of Mls-disparate T cells *in vitro*. Mls differences do not trigger graft rejection, and as yet Mls has not been defined in other species.

Mls was originally thought to be a single gene with four alleles designated a, b, c, and d; Mls^a appeared to be the most stimulatory allele, while

Mls^b was non-stimulatory, the others being weaker. However, it is now known that Mls products are encoded by at least two independent and unlinked loci, Mls-1 and Mls-2, corresponding to Mls^a and Mls^c respectively, and Mls^d is due to coexpression of both determinants; an Mls-3 locus has also been defined. The Mls nomenclature can be rather confusing because each of the Mls-1 and Mls-2 loci has a stimulatory allele, a, and a non-stimulatory allele, b (Table 5.6). Thus Mls^a can also be designated as Mls-1a, and Mls^c as Mls-2a.

Stimulation by Mls requires an MHC class II molecule. The Mls product is thought to bind to the lateral face of the T cell receptor V_β region, rather than the area encoded by the D, J, and N insertional sequences that interact with peptide–MHC complexes (Sections 6.3.1 and 7.3.3). It may also bind to the side of the MHC class II molecule rather than within the peptide-binding groove, thus bypassing the normal requirements for T cell receptor–peptide–MHC interactions.

For many years the nature of Mls determinants was completely unknown. There is now good evidence that *Mls is encoded by endogenous retroviruses belonging to the mammary tumour virus (Mtv) family*. The murine Mtv (MMTV) viruses are rampant in wild mice, and are transmitted vertically from mother to offspring in the milk. *In vivo*, the initial infection may stimulate extensive proliferation of T cells, perhaps allowing them to be infected. These cells may then act as reservoirs for the virus which cannot infect mammary epithelial cells until lactation begins, when MMTV is transmitted to the offspring. Activated B cells and dendritic cells appear to be particularly good stimulators of Mls responses *in vitro* so presumably they also harbour these viruses. Alternatively, other cell types may be activated by the cytokines produced by proliferating T cells and become infected. In addition, production of infectious provirus particles may be activated by irradiation, permitting infection of other cell types and the apparent transfer of Mls between cells.

The original evidence that Mls is encoded by Mtv can be summarized as follows. (i) Clonal deletion of T cells bearing V_β regions associated with recognition of Mls does *not* occur in mice cured of Mtv infection by foster nursing. (ii) Mls, detected by stimulation of T cells *in vitro* or by clonal deletion of T cells bearing particular V_β regions *in vivo*, is not genetically separable from Mtv loci. (iii) Cell lines that stimulate Mls responses contain Mtv transcripts, whereas otherwise genetically-identical cell lines that do not stimulate, do not possess them.

The Mls product appears to be encoded by an open reading frame in the Mtv 3′ long terminal repeat (Fig.5.16). For example, if the MMTV 3′ long terminal repeat, also designated *orf*, is transfected into B cell lymphoma cells, they acquire the expected Mls stimulatory activity (Choi *et al.* 1991). Furthermore, transgenic mice containing an intact MMTV virus (Mtv-2), or *orf* alone, have been constructed, and in both cases clonal deletion of T cells bearing the appropriate V_β region (in this case $V_\beta 14$) is observed (Acha-Orbea *et al.* 1991). The *orf* sequence is highly conserved between different members of the Mtv family, but discrete differences can be detected. This sequence could encode a variety of polypeptides from 19–37 kDa in molecular weight, but it is not yet clear whether peptides can also be produced that are capable of binding to the peptide-binding groove of MHC molecules and being presented to T cells in a more typical manner.

In addition to these observations on the association between Mls and Mtv (*orf*), B cell tumours containing a different retrovirus, MuLV (a defective murine leukemia virus), stimulate $V_\beta 5^+$ T cells, and this response can be inhibited by antibodies specific for the gag p30 product of this virus. Therefore, Mls-type responses may be restricted not only to Mtv-encoded products but could be a more general phenomenon, although so far the majority of known associations are with Mtv.

Studies of V_β expression in T cell receptor transgenic mice. Transgenic mice have been constructed containing the genes for T cell receptors that express particular V_β regions or which are specific for particular antigen–MHC combinations. Thus the fate of these cells in animals expressing the relevant antigens can be followed using antibodies specific for V_β regions, or for idiotypic determinants expressed by these receptors (Section 11.1.3). In general, transgene-expressing T cells are deleted from the periphery, and the subset of mature thymocytes, in the presence of the relevant antigens. However, in contrast to the results of experiments described above, deletion of double-positive immature thymocytes has *also* sometimes been observed. As yet it is not entirely clear whether this is a consequence of the early expression of the transgene in the thymus of these animals, or whether it is because most transgenic mice studied to date contain T cell receptors specific for class I-restricted antigens.

Table 5.6 Mls nomenclature*

'Old' nomenclature	'New' nomenclature		Typical strains
	Mls-1	Mls-2	
Mls^a	a	b	AKR/J
Mls^b	b	b	C57BL/6
Mls^c	b	a	C3H/He BALB/c
Mls^d	a	a	CBA/J DBA/2

Explanation: The a alleles of Mls-1 and Mls-2 are stimulatory, the b alleles are non-stimulatory. Thus Mls^a will stimulate T cells from Mls^b and Mls^c, and Mls^c can stimulate Mls^a and Mls^b. However, by far the strongest proliferative responses are stimulated by Mls^a.

*Based on Janeway *et al.* (1988) *Immunology Today*, **9**, 125.

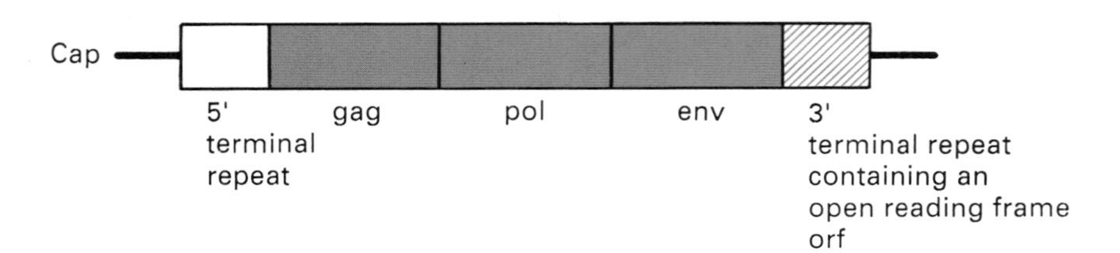

Fig.5.16 Outine structure of the mouse mammary tumour virus (Mtv) family of retroviruses. Mtv is a typical retrovirus in that it contains gag, pol, and env genes in its RNA genome: gag encode a polyprotein that is subsequently cleaved to produce the capsid proteins; pol encodes the reverse transcriptase and an enzyme necessary for integration into the host cell genome; and env encodes the envelope glycoprotein. However, Mtv is unusual in that while, like other retroviruses, it contains repeated sequences (long terminal repeats — LTR) at the 5′ and 3′ ends, Mtv also contains an open reading frame in the 3′ LTR. This has been shown to encode an Mls product (see text)

In some cases, transgene-expressing T cells do appear in the periphery but they are unresponsive to the relevant antigen (see clonal anergy; below). This can be associated with low levels of expression of CD8 or an apparently double-negative phenotype, so modulation of this molecule may contribute to the unresponsive state. This was observed, for example, in male H-2D^b mice that expressed transgenic T cell receptors specific for the male-specific H-Y antigen and H-2D^b (for the result of the experiment in female mice see Section 5.2.3), and in H-2L^d mice expressing an allo-reactive (anti-H-2L^d) T cell receptor, respectively. Deletion of T cells expressing these receptors did not occur in mice of other MHC genotypes.

Transgenic mice containing T cell receptors specific for viral antigens have also been studied. For example, one T cell receptor was specific for LCMV plus H-2D^b, and another was bispecific for both LCMV and Mls^a. Deletion of both double-positive and mature thymocytes was seen in infected mice of the relevant genotype, although in the latter case double-positive cells were *not* deleted in uninfected Mls^a mice. It may be that the timing of deletion varies according to the antigen, and that thymocytes are sensitive to negative selection at different stages of development. However, these studies provide additional evidence for clonal deletion as a mechanism of tolerance.

The site of clonal deletion. The studies outlined above indicate that clonal deletion can occur within the *thymus*.

Evidence to support this view has come from experiments with radiation-induced bone marrow chimeras and thymus-reconstituted nude mice, described in Sections 5.3.2 and 5.3.3. In addition, clonal deletion is aborted in thymectomized animals (as noted in the next section), and moreover, clonal deletion of $V_\beta 3^+$ T cells occurs in euthymic Mls-2a mice (see above) but not in their athymic, nude counterparts (which do, however, contain T cell precursors). While these experiments show that clonal deletion, at least for the antigens studied, does almost certainly occur in the thymus, there is additional evidence that *mature* T cells may also be deleted *extrathymically*. Transfer of cells expressing Mls^a to adult Mls^b recipients that had been thymectomized led to an initial expansion in mature peripheral $V_\beta6^+$ T cells, followed by their disappearance (Webb *et al.* 1990).

Evidence for clonal anergy or functional inactivation

Functional inactivation of apparently mature T cells provides an additional mechanism for self-tolerance.

Significant deletion of $V_\beta6^+$ T cells does *not* occur in adult Mls^b mice tolerized to Mls^a by administration of cells from these strains (in contrast to the neonatal situation; above), but the mature $V_\beta6^+$ T cells present in the periphery are *unresponsive* to Mls stimulation (Blackman *et al.* 1990). Likewise, administration of SEB to adult mice induces tolerance to this antigen, but the number of $V_\beta8^+$ T cells in the periphery of these animals is reduced only by about 50% (again in contrast to the neonatal situation) and these cells are also unresponsive to this superantigen and various other stimuli. Furthermore, in another set of experiments, $CD4^+$ T cells bearing a single transgenic $V_\beta8.1^+$ T cell receptor chain were found to be expressed in the periphery of Mls^a mice, but they were anergic; it is not entirely clear why these T cells were not deleted, but this may have been due to the relatively low affinity of this particular V_β chain for Mls.

These experiments provide strong evidence for functional inactivation of mature peripheral T cells, as do others noted below, and in some cases the defect in responsiveness has been linked to the IL-2 pathway in these cells. For example, in some cases the T cells expressed IL-2 receptors after stimulation, but the production of IL-2 was impaired. It has also been possible to induce tolerance across Mls and minor histocompatibility antigen barriers in adult mice using a regime of bone marrow transplantation and treatment with anti-CD4 and anti-CD8 antibodies, despite the continued presence of mature $V_\beta6^+$ T cells in the periphery.

Site of induction of clonal anergy. Functional inactivation of mature T cells can occur *extrathymically*.

Adult-thymectomized mice expressing IE and/or Mls do contain peripheral T cells expressing the relevant V_β regions that are normally deleted in the presence of these antigens (Jones *et al.* 1990). However, these T cells are unresponsive to stimulation by the appropriate V_β antibodies or stimulator cells. Therefore self-tolerance is maintained in the absence of a thymus by functional inactivation of potentially autoreactive T cell clones. Clonal anergy has also been demonstrated in transgenic mice that express H-2 IE only on certain cells of the pancreas but not other tissues, including the thymus, reinforcing the idea of a peripheral mechanism (Burkly *et al.* 1989). It seems unlikely that these results were due to soluble class II molecules being released from the pancreas and gaining access to the thymus because other studies have shown that administration of genetically-engineered soluble MHC class I molecules does not induce tolerance.

In other experiments, radiation-induced bone marrow chimeras (Section 5.3.2) have been grafted with two neonatal thymi expressing different minor histocompatibility antigens, so that T cells developing from one thymus were not exposed to antigens of the other until *after* they had matured and emigrated from the thymus (Zamoyska *et al.* 1989). Nevertheless, the T cells originating from one thymus were unresponsive to antigens of the other, thus demonstrating a mechanism for functional inactivation of mature T cells. However, some studies with bone marrow chimeras have additionally indicated the existence of a non-deletional mechanism for tolerance induction within the *thymus* (Ramsdell *et al.* 1989). Clearly, the relative importance of clonal deletion versus clonal anergy, in the thymus and periphery, to the tolerant state awaits further investigation.

Self peptide—MHC complexes are expressed in the periphery of tolerant mice

In addition to the mechanisms of tolerance already discussed (clonal deletion and functional inactivation), unresponsiveness to some self antigens could result from the failure of antigen-presenting cells in the periphery to express the relevant self peptide–MHC complexes. In principle, for example, some self proteins might not be processed by normal cells, the peptides might not have access to self MHC molecules, or the processing pathways might be able to discriminate between self and non-self proteins. At present, however, these possibilities seem unlikely. Moreover, there is clear evidence for expression of self peptide–MHC complexes by cells in the periphery of tolerant mice.

It has been shown that self haemoglobin peptide–MHC class II complexes are generated *in vivo*, and that these complexes can be expressed by cells in the periphery (Lorenz and Allen 1988). Additional evidence has come from studies on the presentation of the complement C5 component (Lin and Stockinger 1989). C5-deficient strains of mice have a genetic defect which results in the absence of this complement component from the serum, and they are not tolerant to C5. Thus when they are immunized, these mice can generate $CD4^+$ T cells specific for C5 presented by MHC class II molecules, as well as C5-specific antibodies. However, C5-sufficient strains, which express normal levels of C5 in the serum, are tolerant to this molecule and are unable to generate C5-specific responses when they are immunized. Nevertheless, MHC class II-positive cells isolated from these animals *can* stimulate C5-specific T cell clones and hybridomas (in the absence of exogenously-added antigen). However, it is not yet clear, in these cases, whether tolerance is due to clonal deletion or functional inactivation.

T cells are only tolerant of physiologically-produced self peptides

In normal cells, self peptides may be generated from self proteins by the same pathways used for processing of foreign antigens (Section 3.2). It seems likely that only some of these peptides would be capable of binding to self MHC molecules and being presented to T cells during the induction of tolerance.

There is good evidence that some peptides derived from self proteins by non-physiological means can be recognized by self T cells.

CTL lines that recognize self peptide–MHC complexes have been produced by culturing spleen cells with syngeneic accessory cells and IL-2 in the presence of peptides derived from self proteins, such as a synthetic peptide from β_2-microglobulin, hydrolytic fragments of haemoglobin, and a tryptic digest of total liver proteins (Schild *et al.* 1990). The CTL lines were able to kill syngeneic cells in the presence of these self peptides but not in their absence, indicating that they were not normally produced by physiological processing.

Another study (Benichou *et al.* 1990) has demonstrated that peptides corresponding to the N-termini of the α and β chains of a particular mouse MHC class II molecule could bind to the same, intact, class II molecule (as shown by peptide competition experiments; Section 3.3.3). When they were used to immunize syngeneic adult mice, these peptides elicited strong proliferative responses, but they induced tolerance after injection into syngeneic neonatal mice. This indicates that T cells were susceptible to tolerance induction by these self peptides, but presumably they are not produced physiologically. Of interest, another synthetic peptide corresponding to one particular region of an MHC class I molecule was also able to bind to these class II molecules, but it did not elicit responses in adult mice. This suggests that this particular peptide *is* physiologically produced in this strain, and thus it can induce tolerance.

MHC molecules and some self proteins are expressed in the thymus, and other molecules have access to this tissue, for example via the blood, so it is perhaps not surprising that tolerance can be induced to peptides derived from them. However, many proteins are sequestered in other tissues and may not have access to the thymus. How is tolerance induced to peptides from these molecules? It is possible this comes about through peripheral mechanisms such as clonal anergy (see above), perhaps explaining the necessity for having tolerance-inducing mechanisms in both thymic and extrathymic sites.

A further idea, which has yet to be tested experimentally, is that thymic tolerance could be particularly important to ensure that peripheral T cells are rendered unresponsive to the self peptides normally expressed by dendritic cells (DC) and B cells. This is because both cell types (the latter perhaps to a more limited extent, and only after activation) have the capacity to initiate immune responses against the peptide–MHC complexes they express (immunostimulation: Section 3.4.3), and in the periphery it is important the immune responses are only generated to the foreign antigens they may acquire.

These ideas have important implications for autoimmunity. For example, it is conceivable that autoimmunity could develop (i) if a peptide that was normally sequestered from DC gained access to those cells; (ii) if DC in the periphery started to produce an aberrant self protein that was also presented by other cell types, for example after viral infection; or (iii) if a viral peptide presented by DC resembled a self component on other cell types. However, these ideas have yet to be tested experimentally.

The syngeneic MLR. One situation in which T cells appear to react against self components occurs when they are cultured with dendritic cells either from the same (autologous) or a genetically-identical (syngeneic) individual. This results in a proliferative response which is called the autologous or syngeneic MLR respectively (Weksler *et al.* 1981). Typically, this response is slower to develop, and of a smaller magnitude, than the more vigorous allogeneic MLR (see Appendix). Nevertheless, the syngeneic MLR is a typical MHC class II-restricted T cell response in that it exhibits both specificity and memory, and can be inhibited by anti-class II antibodies.

It has been suggested that the syngeneic MLR is an antigen-specific response to peptides derived from the culture medium or endogenous viruses. However, this is difficult to reconcile with several observations, including the fact that T cells from fully allogeneic bone marrow chimeras, which are immunoincompetent and unable to mount antigen responses (Section 5.3.2), can nevertheless respond in a syngeneic MLR. Another possibility is that the responding T cells are in fact specific for self peptide–MHC complexes expressed by dendritic cells. This is in keeping with other observations pointing to the existence of autoreactive T cells that can be detected early in immune responses and from which autoreactive clones can be generated. Possibly, also, the T cells responding in the syngeneic MLR *in vitro* are removed from regulatory influences that operate *in vivo*.

5.2.6 The sequence of positive and negative selection in the thymus

Both positive and negative selection can occur within the thymus and we will confine our discussion to this organ (but note that negative selection, as well as positive selection of some cytotoxic T cells, can also occur extra-thymically; Sections 5.2.5 and 5.3.4). Both positive and negative selection of thymocytes depends on interactions between their T cell receptors and self MHC molecules expressed by thymic stromal cells (Section 5.3.5); the former process occurs in the absence of foreign peptides (Section 5.2.3) while the latter presumably requires recognition of bound self peptides (Section 5.2.5). In general, positive selection is thought to occur before negative selection, although this has not been proven unequivocally, and the precise sequence of events might also depend on the antigen in question.

During positive selection, it is believed that interactions with self MHC molecules rescue the thymocytes from programmed cell death, and promote their maturation towards the more mature ($CD4^+$ or $CD8^+$) single-positive phenotype. During negative seletion, autoreactive T cells are thought to be killed by apoptosis (Section 10.4.5). One model for how this might occur is as follows (Table 5.7). Thymocytes bearing T cell receptors with a wide range of affinities are selected during interactions with stromal cells such as epithelial cells (Section 5.3.5; see also Fig. 5.11). Subsequently, those thymocytes bearing receptors with a high affinity for self peptide–MHC complexes are negatively selected by interactions with different stromal cells such as dendritic cells (Section 5.3.5). The surviving cells, with a low affinity for self peptide–MHC complexes, but perhaps a high affinity for foreign peptides bound to self MHC molecules, then enter the periphery to constitute the

Table 5.7 Some models for positive and negative selection*

Model	Positive selection	Negative selection
Quantitative differences	A range of T cells, with TcR of low to high affinities for self MHC, is selected	T cells with high affinity TcR for self peptide–MHC complexes are deleted
Qualitative differences	Signals delivered, e.g. by cortical epithelium, promote survival and further development	Different signals, e.g. from medullary epithelium, lead to cell death
Developmentally programmed response	Signals delivered to thymocytes at one stage of development result in a particular intracellular response and rescue from death	The same signals delivered at another stage result in a different response, and induce programmed cell death

*Note that these models are not necessarily mutually exclusive

mature T cell pool. Of course, participation of CD4 or CD8, and adhesion molecules, could also contribute to the avidity of these interactions.

Another model proposes that developing thymocytes receive qualitatively different signals, produced by different types of stromal cells, perhaps resulting in positive selection in the cortex and negative selection in the medulla. A third model is based on the idea that the maturing thymocytes undergo a developmentally programmed response depending on their level of expression of T cell receptors and/or other molecules and on how these molecules are linked to different signalling pathways. This could determine whether they are positively selected or deleted (Table 5.7). In support of this concept, for example, two biochemically distinct populations of immature thymocytes have been identified, depending on whether or not their T cell receptors are functionally coupled; only the former cells appear susceptible to clonal deletion (Section 4.2.5).

The thymic stromal cells responsible for positive and negative selection are considered in Section 5.3.5.

Support for the idea that positive selection precedes negative selection has come from studies on the development of one particular thymocyte subpopulation in the presence of different selection elements (Guidos *et al.* 1990; see below). The earliest intrathymic progenitors do not express T cell receptors, CD4 or CD8 (Section 4.3). Subsequently, these cells develop into double-positive ($CD4^+$ $CD8^+$) thymocytes and acquire low levels of the T cell receptor (TcR^{lo}). These cells comprise two populations: small cells (the majority) and blast cells. Only the blast cells contain precursors that can develop into T cells; the small ('non-mature') thymocytes die intrathymically. Before developing into mature single-positive T cells with high levels of T cell receptors (TcR^{hi}), the double-positive cells pass through a transitional stage in which they express intermediate levels of T cell receptors (TcR^{med}) and lower levels of either CD8 or CD4. These TcR^{med} $CD4^+$ $CD8^{lo}$ and TcR^{med} $CD4^{lo}$ $CD8^+$ thymocytes are committed to the mature $CD4^+$ and $CD8^+$ single-positive lineages, respectively.

Purified $CD4^+$ $CD8^+$ thymocyte blasts expressing low levels of T cell receptors (TcR^{lo}) were injected into thymi expressing different selecting elements (e.g. H-2 IE for $V_\beta 17a^+$, or Mls^a for $V_\beta 6^+$ T cells; Section 5.2.5). After allowing their development *in vivo*, the donor T cells were isolated (on the basis of a Thy-1 marker different from that of the recipient), and their levels of expression of T cell receptors and the appropriate V_β regions were analysed. There were three main conclusions. First, based on the frequency of TcR^{med} cells expressing particular V_β regions, compared to the frequency in the mature TcR^{hi} population, the data indicated that maturation of TcR^{lo} $CD4^+$ $CD8^+$ to the TcR^{med} transitional stage occurred as a result of positive selection. Second, maturation from the TcR^{med} to the TcR^{hi} stage was accompanied by clonal deletion of T cells expressing the appropriate V_β regions, i.e. negative selection. Third, positive and negative selection occur sequentially during thymocyte maturation. It seems likely that different intracellular signalling events could be mediated by the different levels of expression of TcR (coupled or uncoupled to CD3), CD4, and CD8 on the developing thymocytes (Sections 4.2.3, 4.2.5). A model is shown in Fig.5.17.

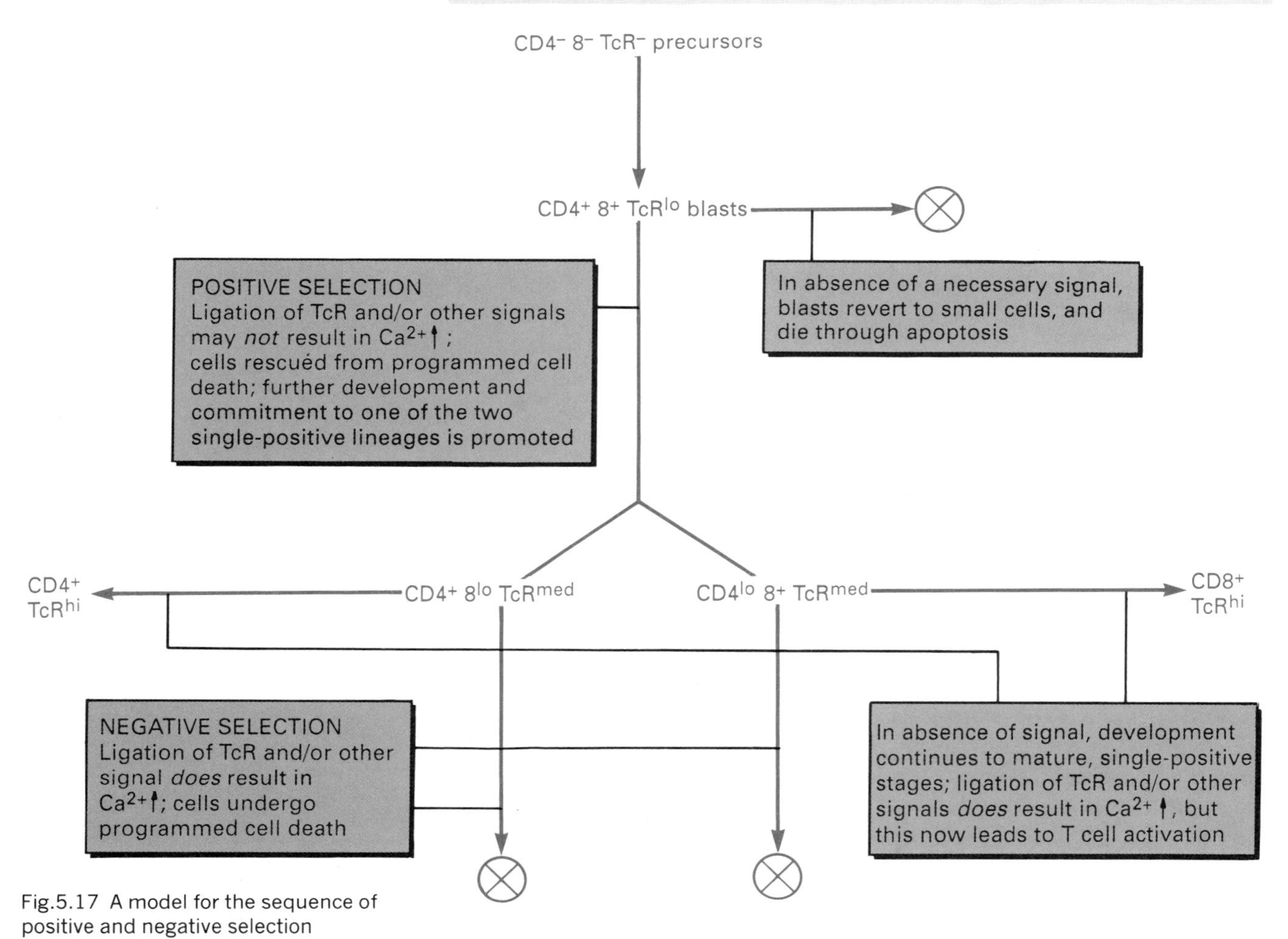

Fig.5.17 A model for the sequence of positive and negative selection

5.3 Selection of the T cell repertoire during development

5.3.1 Introduction

After the MHC restriction of T cells was understood, studies were carried out using radiation-induced bone marrow chimeras and thymus-reconstituted nude mice that revealed an important feature of the T cell repertoire that was originally referred to as its 'plasticity'. These experiments showed that *the self-restriction and tolerance of T cells is not genetically predetermined, but rather is a function of the MHC molecules expressed in environment in which they develop*; this idea has already been encountered in Section 5.2. This phenomenon was originally termed **T cell education** or, because it was shown to develop largely in the thymus, **thymic education**. These terms are somewhat inappropriate because they could imply that the specificity of *mature* T cells can be changed, and it is known that this does not occur. Instead, thymic education is due to selection from the population of *developing* T cells of a subset of cells that then matures and enters the periphery to constitute the T cell repertoire of that individual, so a better term is perhaps **selection of the repertoire** (Fig.5.18). This term encompasses both phenomena of positive and negative selection which generate the apparently self-restricted and self-tolerant repertoire (Table 5.2).

The particular MHC alleles that are recognized, together with bound peptides, during conventional antigen-specific T cell responses can be referred to as the **restriction specificity** of those T cells (cf. self-restriction and allorestriction, Sections 5.2.3 and 5.2.4). As we shall see, it turns out that the restriction specificity of *helper T lymphocytes* is determined by the MHC molecules encountered during their development in the *thymus*. However, that of *cytotoxic T cells* appears to be determined, at least in part, *outside of the thymus* although precisely where is not known. We should, however, point out that while many experiments are consistent with these conclusions, other data are not. This could be due to differences in the experimental systems that have been used, or a biological complexity yet to be understood. In what follows we will emphasize the more consistent findings so that the 'exceptions' may be more readily comprehended.

The experiments that were carried out with bone marrow chimeras and thymus-reconstituted nude mice which support the ideas just presented are discussed in some detail in this section, particularly as these experimental systems, especially the former, are now being used to complement various molecular and cellular studies investigating the T cell repertoire (Section 5.2). **Some readers who do not require a detailed knowledge of these systems may wish to proceed directly to Section 5.4.**

5.3.2 The restriction specificity of T cells is acquired: studies with bone marrow chimeras

Production of radiation-induced bone marrow chimeras

The main evidence that the restriction specificity of T cells is 'acquired' rather than 'inherent' comes from experiments with T cells from radiation-induced

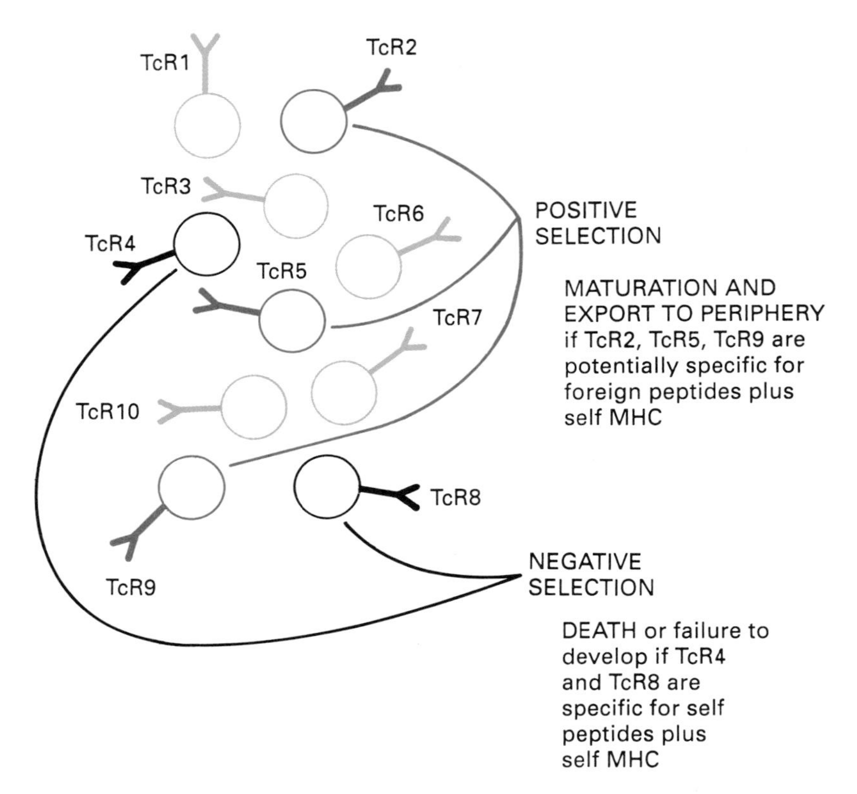

Fig.5.18 One view of 'thymic education' or selection of the T cell repertoire

bone marrow chimeras. If an animal is exposed to a sufficiently high dose of radiation, its existing lymphocytes, peripheral antigen presenting cells, and their precursors are killed. However, other haemopoietic cells (leukocytes) and cells such as thymic epithelial cells, which are not bone marrow-derived, escape destruction. Part of the reason why some cells are more radiosensitive than others is because they die or 'turn-over' at a relatively high rate in normal animals and are replaced by precursor cells that are killed in irradiated animals.

Irradiated animals become susceptible to infections that can be fatal because after a short time they no longer possess lymphocytes or peripheral antigen-presenting cells (e.g. dendritic cells, Section 3.4), which of course are required to mount immune responses against pathogens. However, if they receive a transplant of bone marrow from a normal individual these cells develop from the graft and repopulate or **reconstitute** the recipient's immune system. In this case the recipient can once again generate immune responses and resist infections and, providing the original dose of radiation was not high enough to damage other tissues such as the gut, which is quite radiosensitive, the animal will continue to thrive.

Bone marrow transplantation is used clinically to treat patients born with a defective immune system, or others treated, for example, with radiation as therapy for certain forms of cancer. The individual becomes a **chimera**, so-called after a mythological creature that was described by Homer as a monster with a lion's head, a goat's body, and a serpent's tail. This term is used in immunology because the individual contains cells from two or more genetically distinct individuals. A bone marrow chimera consists of leukocytes from one individual, the **donor** coexisting with tissues of another, the **host** or recipient. If the recipient was irradiated to destroy its own cells, the chimera is called a **radiation-induced bone marrow chimera**. In this way, lymphocytes from an animal of one strain can develop within an individual of a different strain, and the effect of experimentally changing the environment on the MHC restriction of developing T cells may be studied.

Since we shall be discussing the thymus, it is perhaps worthwhile at this point to discuss what happens in this organ in a radiation-induced bone marrow chimera (Fig.5.19). Radiation, at the doses commonly used to eliminate lymphocytes and peripheral accessory cells, *kills* thymocytes but normally *spares* the thymic epithelial cells together with some haemopoietic cells within the thymus that appear to turn over more slowly than in the periphery. The latter certainly include macrophages and perhaps, at relatively early times, thymic dendritic cells. However, these radioresistant, thymic haemopoietic and/or non-haemopoietic cells are sometimes called **thymic restricting cells,** because the precise cell type that determines the restriction specificity of developing T cells is not clearly defined (but see Section 5.3.5).

After an allogeneic bone marrow transplant the thymus becomes repopulated with allogeneic thymocytes (plus the mature T cells that develop from them), and allogeneic macrophages and dendritic cells also enter this organ. This creates a potential complication for interpreting the results of experiments with radiation-induced bone marrow chimeras (Table 5.8). The thymus contains epithelial cells of host type only, but at certain times it may contain macrophages and dendritic cells expressing two different types of MHC molecule (donor and host). Clearly, the cellular composition of the thymus will vary depending on the time when the chimeras are analysed (Table 5.8). Moreover, the thymocytes and T cells themselves express MHC molecules of donor type. These considerations become important in attempts to define the precise cell types involved in positive and negative selection (see Section 5.3.5).

Graft-versus-host disease

Early attempts to make chimeric mice often met with little success because the animals frequently succumbed to a wasting disease and died. This was not just because they were more susceptible to infections after irradiation, but because of an immune response known as the **graft-versus-host (GVH) reaction** (Fig.5.20). This phenomenon can occur in allogeneic situations in which the recipient for some reason is unable to reject the transplant. For example, a 'classical' situation is when bone marrow from a mouse of strain C is injected into a genetically-tolerant but semi-allogeneic C × D F1 mouse. GVH also occurs in irradiated animals that received bone marrow from a fully allogeneic donor (e.g. C into D): normal animals reject allogeneic marrow transplants, but irradiated mice are unable to reject the foreign tissue because they are immunosuppressed as a result of radiation treatment.

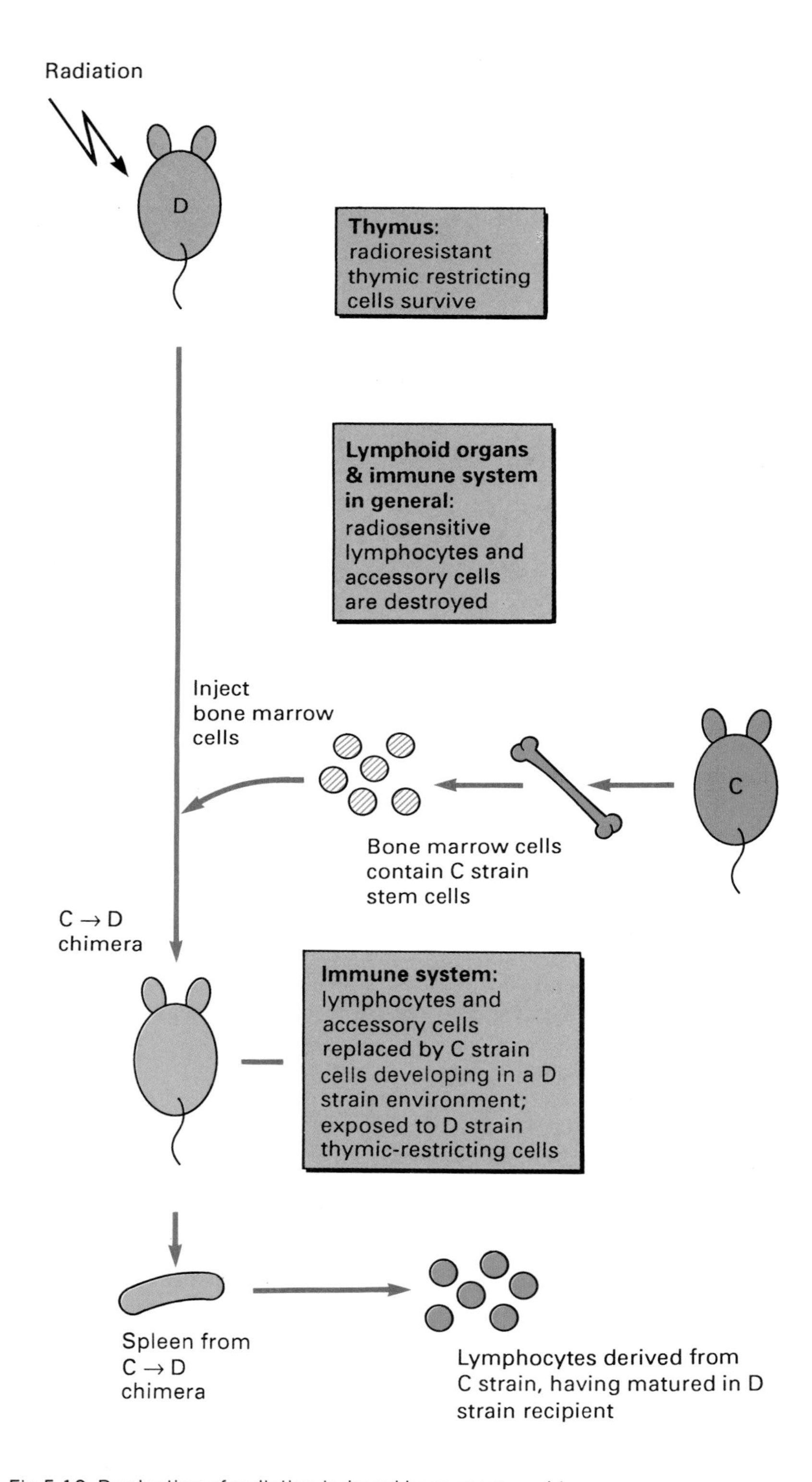
Radiation
D
Thymus: radioresistant thymic restricting cells survive
Lymphoid organs & immune system in general: radiosensitive lymphocytes and accessory cells are destroyed
Inject bone marrow cells
C
Bone marrow cells contain C strain stem cells
C → D chimera
Immune system: lymphocytes and accessory cells replaced by C strain cells developing in a D strain environment; exposed to D strain thymic-restricting cells
Spleen from C → D chimera
Lymphocytes derived from C strain, having matured in D strain recipient

Fig.5.19 Production of radiation-induced bone marrow chimeras

Table 5.8 MHC molecules expressed by different cell types in the thymus of a radiation-induced bone marrow chimera at different times after reconstitution

Cell types	MHC molecules expressed		
	Before irradiation	**Soon after irradiation**	**At later times**
Epithelial cells (= radioresistant, non-haemopoietic)	Host	Host	Host
Macrophages, thymic dendritic cells (= relatively radio-resistant, haemopoietic)	Host	Host	Donor
Thymocytes, T cells (= radiosensitive, haemopoietic)	Host	Not present	Donor

GVH occurs because the recipient is attacked by mature T cells, present in the bone marrow inoculum, that developed in the donor's thymus but which were carried back to the marrow in the blood. GVH can also be induced by injecting other sources of mature T cells such as spleen or lymph node cells. When donor T cells are transplanted into an allogeneic recipient, they become activated in the new environment as a consequence of their alloreactivity (Section 5.2.2), and they mount a vigorous and often fatal attack on host tissues. (In practice, for some rather complex reasons, GVH is in fact maximally induced if the recipient is treated with a low dose of radiation). GVH is also a problem in clinical bone marrow transplantation, manifesting as a variety of more-or-less severe and sometimes fatal symptoms which include skin lesions, damage to the gut, and enlargement of the spleen.

It was found experimentally that GVH could be overcome simply by removing mature T cells from the bone marrow inoculum before transplantation. For example, T cells can be killed by treating the marrow with monoclonal antibodies against markers of mature T cells, and complement, and this approach is now used clinically. Alternatively, fetal liver can be used to reconstitute irradiated animals. This tissue contains haemopoietic stem cells but not mature T cells, so these animals are not prone to GVH. Depletion of mature T cells enabled radiation-induced bone marrow chimeras to be constructed successfully, and the restriction specificity and tolerance of their T cells could then be examined.

Convention

For convenience we shall use a particular convention to describe various types of chimera. The donor strain will be put to the left side of an arrow with the recipient strain to the right. The cells, treatments, or animals used will be put in brackets after the strain; for example, bone marrow = bm; irradiation = rad; thymus = thy; nu = athymic nude; and so on. Thus, a chimera formed by transplanting C strain bone marrow into an irradiated D strain mouse can be represented as: C (bm)→D (rad). For simplicity this will sometimes be abbreviated C→D if the type of chimera is obvious.

Assays of restriction

Several assays can be used to determine the restriction specificity of mature T cells (Table 5.9). In general these measure the ability of the T cells of inter-

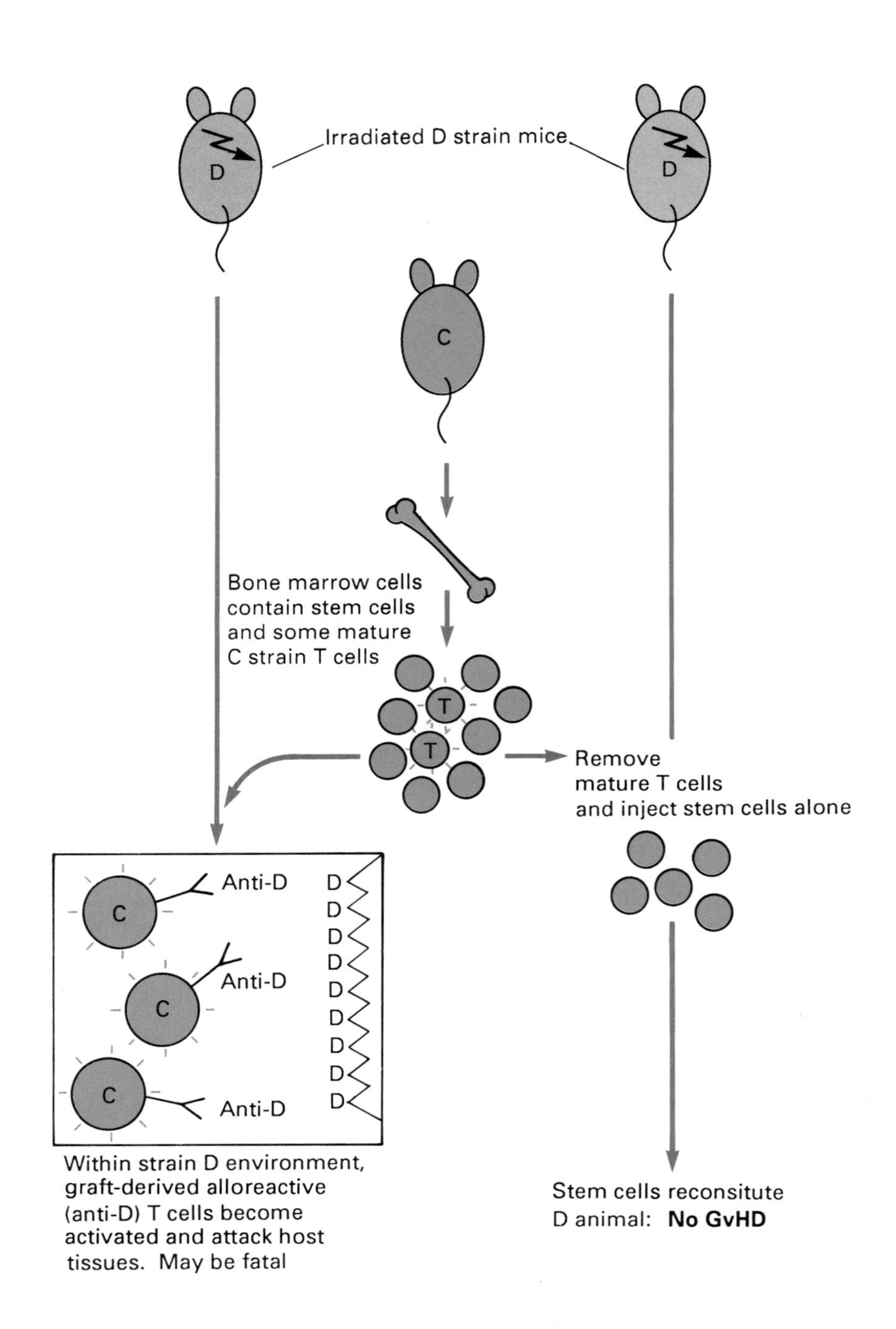

Fig.5.20 Graft-versus-host disease (GVHD)

est to recognize cells of different strains and to kill them, in the case of cytotoxic T cells, or to initiate a number of immune responses in the case of helper T cells (see Appendix).

Cytotoxicity assays. Cytotoxic T cells from lymphoid organs of animals immunized against a virus or chemically-modified cells can be assayed for their ability to kill target cells in culture (see Appendix). This assay was used to determine the restriction specificity of cytotoxic T cells for MHC class I molecules on target cells, as discussed in Section 5.2 (e.g. see Fig. 5.1).

Antibody responses to hapten–carriers. Helper T cells can be obtained from lymphoid organs of carrier-primed animals, and transferred to irradiated secondary recipients together with hapten-primed B cells from another individual (see Section 8.3.1). The ability of these recipients to make an antibody response when they are immunized with the hapten–carrier conjugate can be used as an assay of the helper T cells' restriction to MHC class II molecules on B cells *in vivo* (e.g. see Fig. 5.3). Alternatively this can be assayed *in vitro*, by culturing T cells from carrier-primed animals with hapten-primed B cells, and determining the plaque-forming cell (PFC) response which is a measure of the frequency of antibody-secreting cells.

Delayed-type hypersensitivity (DTH) responses. DTH responses are inflammatory responses (see Section 10.5.4) mediated by helper T cells (formerly called T_{DTH} or T_D cells). The helper cells respond to antigen and MHC class II molecules on accessory cells by releasing cytokines that recruit and activate inflammatory cells such as monocytes and macrophages to the site where antigen is localized. The response is manifested as tissue swelling and local cellular proliferation (in part due to proliferation of the T cells). To assay the restriction of helper cells in DTH responses, T cells from immunized animals can be transferred to naïve animals which are then challenged with the same antigen used to sensitize the transferred T cells. The response is often measured as swelling at the site of injection and compared to a control site that does not receive antigen, e.g. one ear compared to the other, or by the uptake of a radioactive nucleotide at these sites as a measure of cellular proliferation.

Ir gene responses. (See Section 5.4). When inbred strains of mice were immunized with certain antigens, especially small molecules containing simple repetitive determinants, it was found that T cells from some strains could make a response to the antigen but those from other strains could not. These strains were defined as 'responders' (R) and 'non-responders' (NR), respectively. In general, F1 animals with responder and non-responder strain parents (R × NR) have a responder phenotype. Most of these responses were mapped to the MHC class II region, and the antigens were said to be 'under Ir gene control'; explanations for this phenomenon are given in Section 5.4. The ability of T cells to proliferate in culture in response to such an antigen and antigen-presenting cells is an assay of their restriction to the MHC class II molecules expressed by the latter, and defines their 'Ir phenotype'.

Table 5.9 Assays of T cell restriction specificity

T cell population	Assay	Measure
Cytotoxic T cells	Cytotoxicity assay	Release of radioisotope, e.g. chromium-51, from radiolabelled target cells, such as a fibroblast cell line
Helper T cells	T-dependent antibody response	Amount of antibody produced *in vivo*, or frequency of 'plaque-forming cells' *in vitro*
	Delayed-type hypersensitivity	Tissue swelling or incorporation of a radionucleotide (e.g. tritiated thymidine) at the site of antigen administration in a sensitized animal
	Ir gene response	Proliferative response, frequently determined by uptake of radionucleotide (e.g. tritiated thymidine) by T cells from an animal injected with an antigen under Ir gene control, e.g. synthetic copolymers or some small proteins

The restriction repertoire of T cells in F1 individuals

T cells from normal F1 (C × D) animals recognize antigens in the context of MHC molecules from either parental strain: some F1 T cells are restricted to antigen plus MHC C, others to antigen plus MHC D (Fig.5.10 top). It is now clear that this is because the T cells developed in an environment containing both types of MHC molecule, rather than because they are of F1 genotype.

One approach used to demonstrate these different populations of T cells in F1 mice was **cold-target competition**. If one sets up a cytotoxicity assay (see Appendix) using cytotoxic T cells and ^{51}Cr-labelled target cells, and then titrates in unlabelled or 'cold' target cells, the chromium release is progressively inhibited because the cold targets compete with labelled cells for binding to the killer cells. It was found that bulk populations of cytotoxic T cells from virally-infected (C × D) F1 mice could lyse virus-infected target cells from both parental strains. However, killing of C strain targets was inhibited by adding unlabelled infected cells from strain C, but not from strain D, and vice versa. Thus an F1 animal has discrete populations of cytotoxic T cells, each of which recognizes antigen bound to just one of the parental strain (C or D) MHC class I molecules.

Similar analyses have also shown that in a homozygous parental strain mouse (e.g. C), some T cells are restricted to one MHC class I molecule (e.g. H-2D), while others are restricted to a different class I molecule (e.g. H-2K). Obviously in an F1 mouse (e.g. H-2^a × H-2^b) there are several subpopulations of T cells, each of which restricted to just one parental strain molecule (e.g. H-2D^a, H-2K^a, H-2D^b, or H-2K^b). Subpopulations of class I-restricted and class II-restricted T cells in F1 mice are shown in Fig.5.21.

The presence of different populations of F1 T cells has also been demonstrated in antigen-specific proliferative responses. For example, antigen-primed F1 T cells that were restimulated with antigen in culture in the presence of antigen-presenting cells from one parental strain (e.g. C) proliferated strongly when they were rechallenged with antigen and strain C cells but poorly with strain D cells, and vice versa (Paul *et al.* 1977). This type of experiment showed that F1 T cells contained different populations that could be separately stimulated.

The presence of discrete T cell populations in F1 animals has also been confirmed in negative selection experiments *in vitro* and *in vivo*, respectively, using bromodeoxyuridine treatment and exposure to light, or filtration of cells through irradiated recipients to deplete the C-restricted subpopulation of T cells but to leave the D-restricted population intact (see Section 5.2.4, Figs. 5.14, 5.15). In addition, T cells cloned from F1 animals are generally either C-restricted or D-restricted in their antigen recognition.

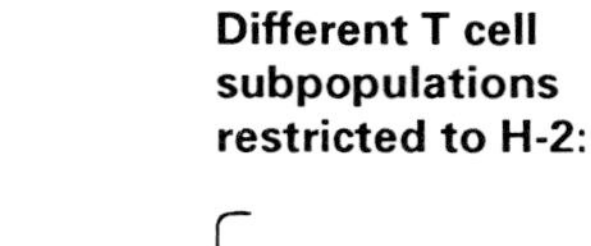

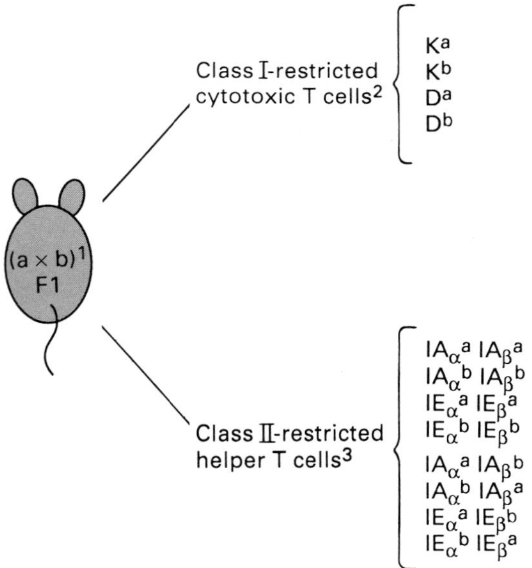

Notes:
1. a and b are fully-incompatible MHC types.
2. Additional similar populations also exist in mouse strains expressing H-2L.
3. Some mouse strains do not express IE and thus lack the correspondingly restricted cells.

Fig.5.21 Subpopulations of self-restricted T cells in F1 heterozygotic mice

The repertoire of F1 T cells maturing in a parental-strain environment is contracted

If (C × D) F1 T cells develop from bone marrow that was transplanted to an animal of one parental strain, say C, they can be shown to recognize antigens preferentially in association with strain C MHC molecules, and they are no longer restricted to D (Fig.5.10, Table 5.10) (Singer *et al.* 1981). Thus, the restriction repertoire of T cells from these chimeras seems to be 'contracted' relative to that of normal F1 animals, since they are more-or-less exclusively C-restricted.

This means, for example, that if the chimera is immunized with a virus its cytotoxic T cells kill virally-infected target cells from strain C, but they do not kill targets from strain D. (Note that this is not an artefact due to the priming environment, see Section 5.2.3, because antigen-presenting cells expressing both C and D strain MHC molecules are present in the chimeras.) Furthermore, the helper T cells from these chimeras can help B lymphocytes from strain C but not strain D to make antibody responses, and they transfer DTH only to C strain secondary recipients. In addition, the chimeric T cells acquire the Ir phenotype of the recipient (C). This means that even if normal (R × NR) F1 T cells can proliferate in response to an Ir antigen (Table 5.9 and Section 5.4) and F1 accessory cells, T cells from an F1→C chimera fail to respond under the same circumstances if the recipient (C) is a non-responder.

This type of chimera demonstrates that the restriction repertoire that T cells acquire (in this case C) is independent of their own genotype (in this case F1). However, these T cells are tolerant to MHC molecules from both parental strains (C,D) in the absence of antigen, but alloreactive to 'third-

Table 5.10 Summary of self restriction and tolerance of helper T cells from radiation-induced bone marrow chimeras*

	Restriction of (class II-restricted) helper T cells	Strain of peripheral antigen-presenting cells	Observed restriction specificity		Tolerance to MHC types
			When tested in the animal, *in vivo*	In the presence of APC from C and D, *in vitro*	
Chimera					
F1→C	C	(C × D) F1	C	C	C and D
C→F1	C and D	C	C	C and D	C and D
C→D	D	C	(No response)	D	C and D
For comparison:					
Normal F1	C and D	(C × D)F1	C and D	C and D	C and D
Normal C	C	C	C	C	C **not** D

*C = (C × C) homozygote; F1 = (C × D) heterozygote

party' MHC molecules (e.g. from strain E). This can be shown in the MLR *in vitro* when accessory cells from the various parental and third-party strains are used as stimulators. We shall see that the mechanisms leading to acquisition of the restriction repertoire are distinct from those resulting in the development of tolerance.

The repertoire of parental-strain T cells maturing in an F1 environment is expanded

T cells from normal animals of strain C are in general biased towards recognition of foreign peptides bound to C strain MHC molecules (Section 5.2.3). However, if these cells develop from C strain bone marrow that was transplanted to an F1 host (in a C→F1 chimera) they become restricted to both C and D strain MHC (Table 5.10). They therefore have an 'expanded' repertoire relative to T cells from normal C animals. While we now know this is true, when these chimeras were first studied their T cells only seemed to be restricted to C strain MHC molecules.

To explain the discrepancy between these results and what we now know to be the case, let us consider in more detail the helper cells and accessory cells in these chimeras. Assume, for the moment, they actually do contain both C-restricted and D-restricted T cells. To be activated, the D-restricted helper cells need to recognize antigen plus D strain MHC molecules on accessory cells. But all the pre-existing host (F1) accessory cells were killed by irradiation, and those that developed from the (C) bone marrow are of strain C origin. Since these chimeras have no D strain accessory cells, it is not possible to activate D-restricted T cells within these animals (see Section 5.2.3, and Fig.5.12). Note that this problem does not arise in the case of an F1→C chimera.

The solution to this problem is to prime the C→F1 chimeric T cells to antigen in an environment that also contains D strain accessory cells. If this is done, helper T cells are indeed found to be restricted to both parental strains.

For example, T cells from the chimeras can be transferred to normal F1 recipients *before* they are primed. When these secondary recipients (which express C and D MHC molecules) are immunized, both C- and D-restricted helper cells are activated, and the chimeric T cells can provide help for B cells from both strains in antibody responses. (However, when this was done with F1→C T cells, only C-restricted cells were detected showing that the repertoire in these animals really was 'contracted', as discussed above).

D-restricted cytotoxic T cells from C→F1 chimeras can also be primed if they are transferred to a suitable environment.

The way this is commonly done, in the case of viral responses, is to transfer the mature T cells to a *freshly*-irradiated, virus-infected F1 animal. Unlike a long-term radiation-induced bone marrow chimera, *acutely*-irradiated animals contain sufficient numbers of antigen-bearing F1 accessory cells that have not yet died to allow antigen-specific T cells to be activated. If chimeric T cells are primed in these secondary recipients and subsequently tested in cytotoxicity assays, they are found to kill virally-infected targets from both strain C and strain D.

We should stress that this approach involves the transfer of *mature* T cells to the secondary recipient: unlike developing T cells, the restriction specificity of mature T cells cannot be altered in the new environment.

T cells from C→F1 chimeras are like normal F1 T cells in that they contain discrete populations of T cells restricted to each parental strain. This can, for example, be shown by cold target inhibition experiments (see above). These T cells are tolerant to the MHC molecules of both parental strains (unlike T cells from normal C animals which are alloreactive to D strain MHC molecules), but alloreactive to E.

Parental (C) strain T cells maturing in a fully allogeneic environment (D) become restricted to D, but the chimera is immunoincompetent

One might think the 'cleanest' system to show that T cells can acquire a different restriction specificity during development would be in a fully allogeneic chimera, such as C (bm)→D (rad). These turned out to be very difficult animals to produce, partly because of GvH disease. Then, to the chagrin of some immunologists, when they were eventually obtained they were found to be immunoincompetent and unable to make cytotoxic T cell or antibody responses, being thus even more susceptible to infections and difficult to maintain. The reason they are immunoincompetent should not be hard to guess from our discussion of the population of D-restricted T cells in a C→F1 chimera.

It turns out that fully allogeneic chimeras do contain (C strain) T cells that are restricted to host-type (D) MHC molecules, but they lack the requisite D strain accessory cells to respond (Table 5.10).

In a study by Singer and colleagues (1981) it was shown that unfractionated spleen cells from fully allogeneic chimeras were unable to make antibodies to the hapten–carrier TNP-KLH in culture. However, normal responses were obtained if spleen cells from the reciprocal chimeras C→D and D→C were cultured together, so that accessory cells from both strains were present. In addition, T cells from C→D chimeras could respond to antigens *in vitro* if they were cocultured with D strain antigen-presenting cells. Thus there was no intrinsic defect in the ability of the chimeric T cells to respond to antigens (Fig. 5.22). Furthermore, 'double-donor chimeras' (for example C (bm) + D (bm)→D (rad)) were constructed and spleen cells from these animals could make normal antibody responses.

At this point, it might be thought that one should be able to transfer C→D chimeric T cells to an F1 animal for priming. However, the situation is complicated by the fact that although these T cells are tolerant of both C strain and D strain MHC molecules, they are not tolerant of unique F1 MHC class II determinants (Kruisbeek *et al.* 1982) which are formed when one MHC class II chain (e.g. α from C) associates with another (e.g. β from D) (Fig. 5.21). Hence, when they are transferred to F1 animals, what are loosely-termed 'allogeneic effects' occur which make it difficult to examine the restriction of the developing T cells. However, it did prove possible to transfer chimeric (C→D) T cells to virus-infected, acutely-irradiated animals of strain D and show that D-restricted cytotoxic T cells were present.

What else do experiments with bone marrow chimeras tell us?

To summarize our discussion so far, the following statements can be made.

1. The restriction repertoire of T cells is determined by the environment in which they mature (precisely where is considered below).
2. The genotype of a T cell is irrelevant to its restriction specificity: this indicates that no matter which T cell receptor genes an individual possesses, the molecules can be diversified sufficiently and selected in any environment to produce an apparently self-restricted repertoire.

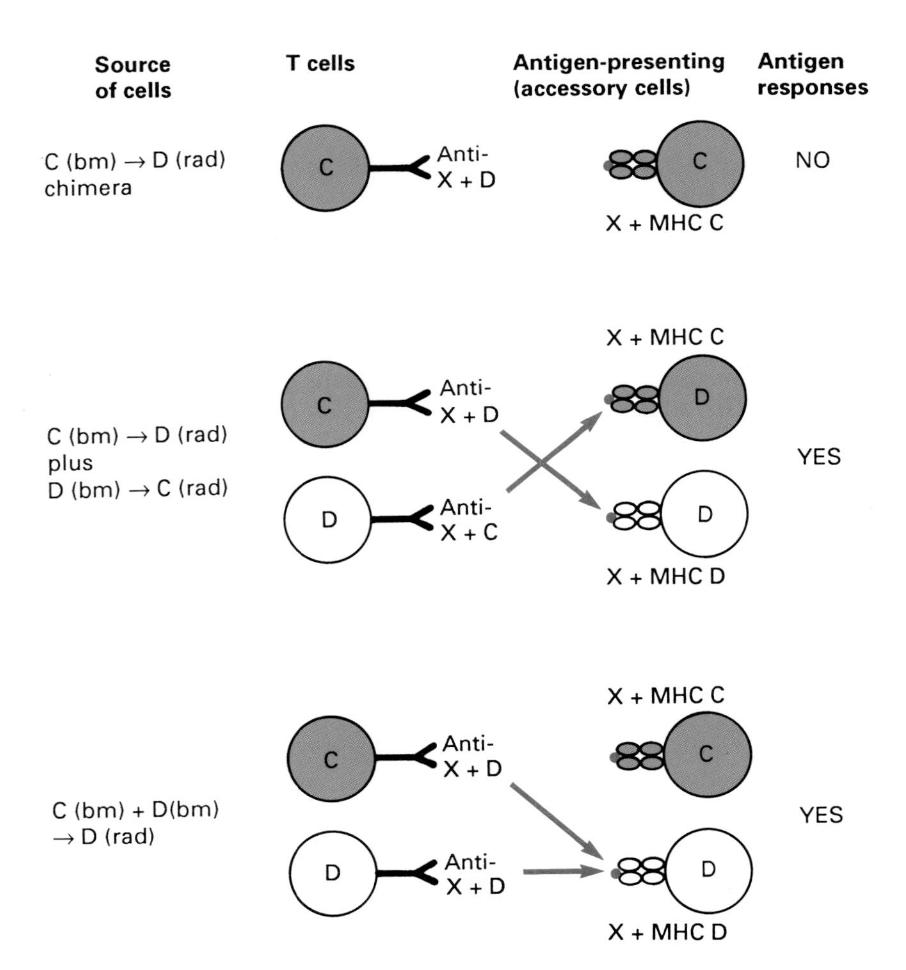

Fig.5.22 T cells from fully allogeneic radiation-induced bone marrow chimeras can respond to antigens in the presence of host strain accessory cells

3. T cells become tolerant to both their own MHC molecules and those of the environment in which they develop. What is notable is that merely being tolerant to a particular set of MHC molecules is not sufficient to make the T cells restricted to these molecules for antigen recognition, and induction of tolerance and self-restriction thus result from different processes (negative and positive selection, respectively).

This has also been confirmed in double-donor chimeras of the type C (bm) + D(bm)→D (rad). These animals contain two discrete populations of T cells, genetically of strain C or strain D, which have matured together (Fig. 5.22 (bottom)). They are tolerant of each other, but both sets are restricted to the recipient's MHC (D). This contrasts to an F1→D chimera, where all T cells are genetically of F1 origin.

5.3.3 The restriction repertoire of helper T cells is determined in the thymus

Studies with ATXBM mice

The experiments with bone marrow chimeras showed that helper T cells preferentially recognize antigens in association with the MHC molecules that are encountered at some stage during their development. Since T cells depend on the thymus for maturation, it was not unreasonable to consider this as the place where the restriction specificity of these cells might be determined.

One way to examine this question is to make a bone marrow chimera, and then to replace its thymus with one from a different strain (Fig. 5.23). In practice, adult mice are irradiated to kill their pre-existing lymphocytes, and reconstituted with bone marrow from the same strain as a source of precursors; remember, that in these experiments the developing T cells have to be of the same strain as the other tissues so that one examines only the effect of introducing a different thymus (Table 5.11). At some stage its thymus is removed (adult thymectomy:AT) to produce an **adult-thymectomized, X-irradiated bone marrow-reconstituted (ATXBM)** animal. A thymus is then obtained from a normal animal, irradiated to kill pre-existing thymocytes, and transplanted into the ATXBM recipient to produce a **thymus-reconstituted ATXBM mouse**.

Some studies have looked at the restriction of helper cells from F1 ATXBM animals transplanted with a C strain thymus, i.e. C (thy) + F1 (bm)→F1 (ATX) chimeras. The F1 T cells that developed were found to be restricted to the thymic (C) MHC molecules but were not restricted to D. These studies are consistent with the idea that the MHC molecules encountered in the thymus, rather than the environment in general, impart the restriction on helper T cells. In some cases, however, there was only a *preferential* rather than *absolute* restriction of the T cells to these molecules.

Studies with nude mice

Another way to examine the role of the thymus is to transplant a thymus into a naturally occurring athymic mutant, the nude mouse (Fig. 5.24). This animal does not have a functional thymus (it possesses only thymic rudiment

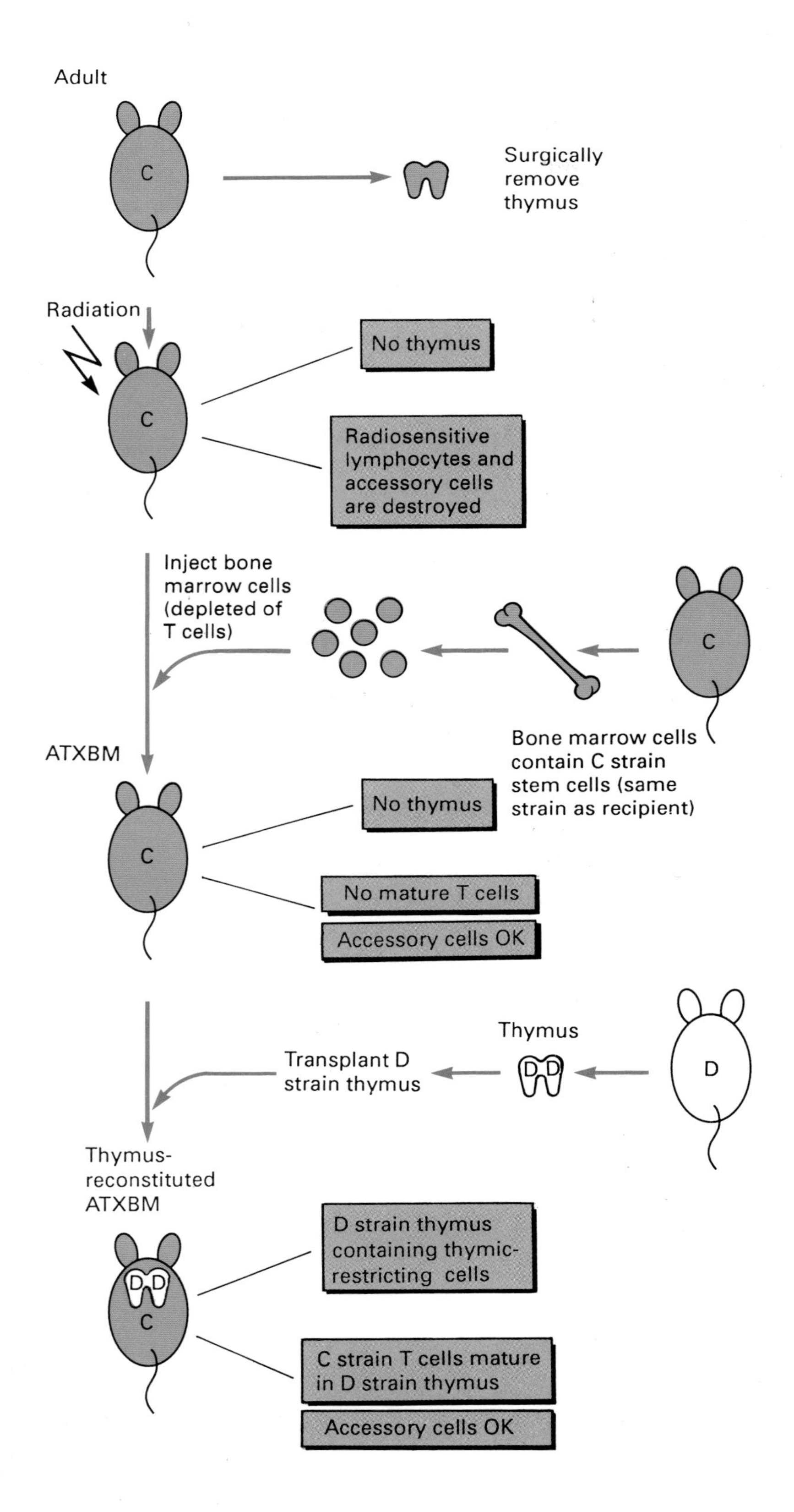

Fig.5.23 Production of thymus-reconstituted ATXBM animals

that is devoid of thymocytes) and consequently it lacks mature T cells so there should be no necessity for irradiation. However, the nude mouse does have apparently normal stem cells in the bone marrow which can give rise to T cells when a thymus is transplanted (Fig.1.31, 1.32). Thus, one might think that this would be a 'clean' system in which to study the effect of transplanting a normal thymus from a different strain on the restriction of developing T cells (Table 5.11). But there are, as always, complications. For instance, some T cell precursors in nude mice can develop into mature T cells in older mice (Ikehara *et al.* 1984) for reasons that are not understood, and there are other complications when the restriction of cytotoxic cells is considered (see Section 5.3.4). These difficulties may account in part for the contradictory results reported for experiments with thymus-reconstituted nude mice.

F1 nude mice transplanted with a parental thymus. The mature T cells that developed in F1 nude mice reconstituted with a C strain thymus were found to be restricted to the MHC type of the thymus (Singer *et al.* 1982). These results are compatible with those obtained using the 'equivalent' bone marrow chimera F1 (bm)→C (rad) (Section 5.3.2), and with the ATXBM data discussed above. They suggest that the repertoire of F1 T cells maturing in a parental thymus becomes 'contracted' relative to normal F1 T cells, consistent with the idea that the thymus determines the restriction of helper T cells.

Parental strain nude mice with a fully allogeneic thymus. Compatible data to the above have also been obtained in some cases where nude mice were reconstituted with a completely allogeneic thymus.

Table 5.11 Comparison between experimental systems*

Experimental animal	Combination	Radiation	MHC type of		
			Thymic-restricting cells	General environment (e.g. non-lymphoid cells)	T cells and peripheral accessory cells
Radiation-induced bone marrow chimera	C(bm)→D(rad)	Yes	D	D	C
Thymus-reconstituted ATXBM animal	D(thy)+F1(bm) →F1(ATX)	Yes	D	C	C
Thymus-reconstituted nude animal	D(thy)→C(nu)	No	D	C	C

*C = (C × C) homozygote; F1 = (C × D) heterozygote

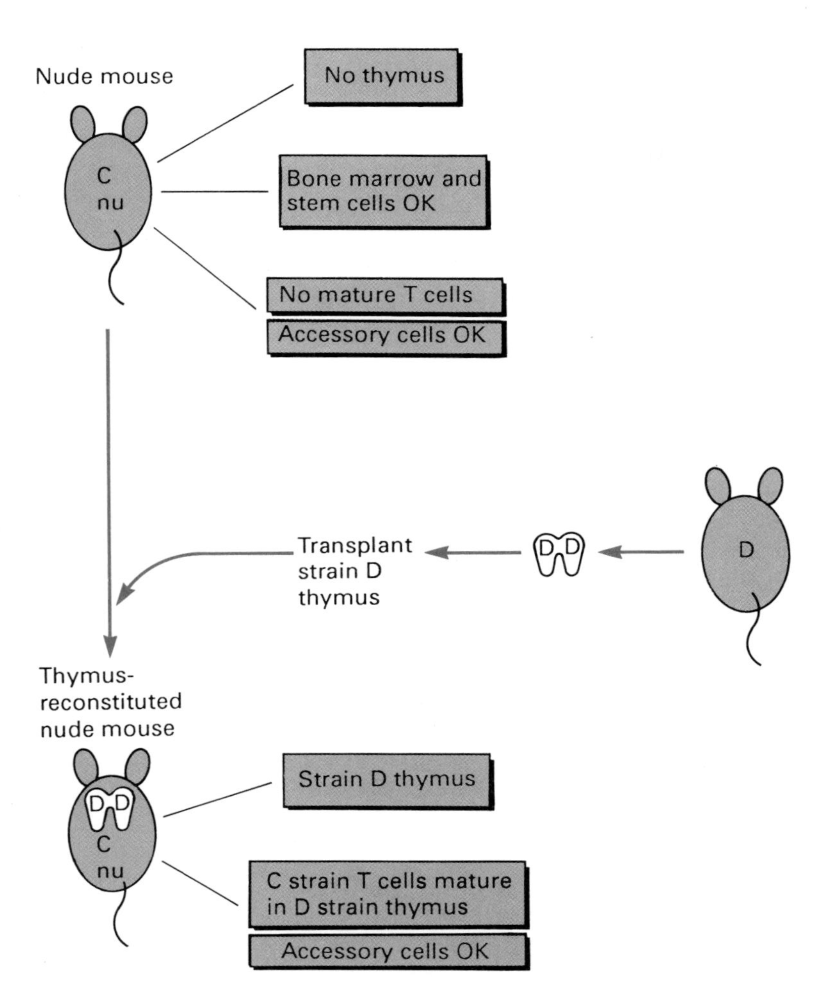

Fig.5.24 Production of thymus-reconstituted nude mice

For example, the ability of T cells from C (thy)→D (nu) animals to help antibody responses in culture has been examined. The T cells were found, in some studies, to be restricted to the MHC of the thymus, provided that C strain accessory cells were added (see discussion of fully allogeneic bone marrow chimeras—Section 5.3.2) and there was no evidence for restriction of helper cells to host-type MHC. In contrast, other investigators have found exactly the opposite: helper cells for antibody responses from these animals appeared to be restricted to host-type MHC. These contradictory results have yet to be fully explained but they may be due to the different approaches that were used: some investigators used *in vitro* assays, whereas others adoptively transferred the cells for priming *in vivo*.

5.3.4 The restriction repertoire of cytotoxic T cells can be determined outside of the thymus

The experiments discussed so far in this section (5.3) confirm a unique role for the thymus in helper T cell development but this may not apply to cytotoxic T cells. Many cytotoxic T cells seem able to undergo at least part of their development in the absence of a thymus, and at least some of these cells appear to acquire their ability to recognize antigens in association with self MHC molecules *extrathymically*.

Studies with bone marrow chimeras

In order to establish general principles, we have glossed over some discrepancies that are apparent in studies on bone marrow chimeras (Section 5.3.2), particularly in relation to the restriction of cytotoxic T cells (Table 5.12). For example, in some studies, the repertoire of cytotoxic T cell precursors isolated from the *thymus* of F1→C chimeras is seen to be 'contracted', whereas the *spleens* of the same animals contain cells restricted to *both* parental MHC types (Kruisbeek *et al.* 1983). Moreover, the repertoire of the converse type of chimera (C→F1) is often not fully 'expanded' and frequently the restriction of cytotoxic cells is preferentially to C (i.e. the *donor strain*) even after adoptive transfer for priming in secondary F1 recipients. Furthermore, cytotoxic T cells restricted to *both* MHC types can be obtained from C→D chimeras (rather than D alone). Therefore, it may be that the bone marrow and/or spleen contains some cytotoxic T cell precursors that are already restricted to the donor type and further positive selection of these cells does not occur in the thymus (Fig.5.25).

Several studies do point to a profound difference in the effects of the thymic environment on the restriction of class II-restricted or helper cells, compared to class I-restricted or cytotoxic T cells.

A good example is the study of Bradley and colleagues (1982) who showed, using bone marrow chimeras made from MHC recombinant mice, that the restriction repertoire of class II-restricted cells was strictly dictated by the host, but there was only a relative *preference* of class I-restricted cells for the host type MHC. As long as the class II requirement was fulfilled and 'help' could be provided, it was possible to generate cytotoxic T cells against donor-strain MHC molecules.

Studies with ATXBM mice

ATXBM mice (Section 5.3.3) that do *not* receive a thymus transplant (and which are constructed using marrow depleted of mature T cells) contain precursors that can develop into cytotoxic T cells, suggesting that some of these cells can develop to a certain point in the absence of a thymus. In fact both Thy-1-positive and CD8-positive cells have been found not only in older ATXBM mice without a thymus, but also in nude mice. Mature cytotoxic T cells were also found to develop from spleen cells obtained from ATXBM animals when they were stimulated with antigen in culture provided that a source of 'helper' cytokines was also added (Duprez *et al.* 1982). These cells were found to be restricted to self MHC molecules, just like those from

Table 5.12 Comparison of self restriction of helper T cells and cytotoxic T cells in different experimental systems*

System	Restriction of helper T cells	Restriction of cytotoxic T cells
Radiation-induced bone marrow chimeras		
(C × D)F1(bm)→C(rad)	C	Generally C
C(bm)→(C × D)F1(rad)	C and D	Often preferentially to C
C(bm)→D(rad)	D	C and D
Thymus-transplanted nude mice		
C(thy)→F1(nu)	C	Generally C
F1(thy)→C(nu)	C and D	Often preferentially to C

*C = (C × C) homozygote; F1 = (C × D) heterozygote

non-thymectomized mice. These findings again suggest that cytotoxic T cells can undergo at least part of their development in the absence of a thymus, and that at least some cytotoxic cells can acquire their restriction repertoire outside of the thymus. This results in only *preferential* restriction of cytotoxic T cells to the thymus MHC type in C (thy)→F1 (ATXBM) mice, for example.

Studies with nude mice

Some studies on the restriction of cytotoxic cells from thymus-reconstituted nude mice are consistent with findings made for helper cells (Section 5.3.3). For example, the restriction repertoire of thymic (but not splenic) precursor cytotoxic T cells from F1 nude mice transplanted with a parental (C) thymus is found to be 'contracted'. But other results are at variance (Table 5.12). This is particularly so in the case of cytotoxic T cells that develop in a parental strain (C) nude mouse transplanted with an F1 or fully allogeneic (D) thymus. It has been found that the T cell repertoire of these animals is not completely 'expanded' to include cytotoxic T cells restricted to D strain MHC molecules. (Lake *et al.* 1980). It will be recalled that this is just what happened in the 'equivalent' bone marrow chimeras (above). This implies that at least some cytotoxic T cells may be unable to expand their repertoire to acquire a new, additional, MHC restriction pattern. In general, other studies with nude mice also support the idea that the restriction of cytotoxic cells can be determined, at least in part, by an *extrathymic* pathway.

For example, spleen and bone marrow cells have been transferred from F1 nude mice (*without* a thymus transplant) into irradiated but euthymic (i.e. thymus-bearing) parental (C) strain animals. After 3 months the recipients were immunized and their T cells were cultured in the presence of added lymphokines. The helper cells were found to be restricted to thymus (C) MHC as expected, and had the Ir phenotype of the recipient. However, the cytotoxic cells were restricted to both C and D strain MHC. In other studies, the spleens of thymus-reconstituted nude mice have been found to contain cytotoxic T cells restricted to the host (nude) type no matter what strain of thymus was transplanted. In addition, clear evidence for distinct pathways determining the restriction of helper and cytotoxic T cells has been obtained (Kast *et al.* 1984). The bm1 and bm12 strains of mice have mutant MHC class I and class II molecules, respectively, which differ from the 'wild-type' MHCs in a few amino acids. As a consequence, helper T cells from the bm12 mouse are unable to respond to the male-specific H-Y antigen, and cytotoxic cells from the bm1 mouse cannot respond to Sendai virus; the wild-type responds to both. When a wild-type nude mouse was transplanted with a bm12 thymus, the helper T cells that developed were unable to respond to H-Y; i.e. they became restricted to the *thymus* MHC molecules. However, cytotoxic T cells from a wild-type nude mouse transplanted with a bm1 thymus were still able to respond to H-Y, i.e. they remained restricted to *host* MHC molecules. See also Speiser *et al.* 1992.

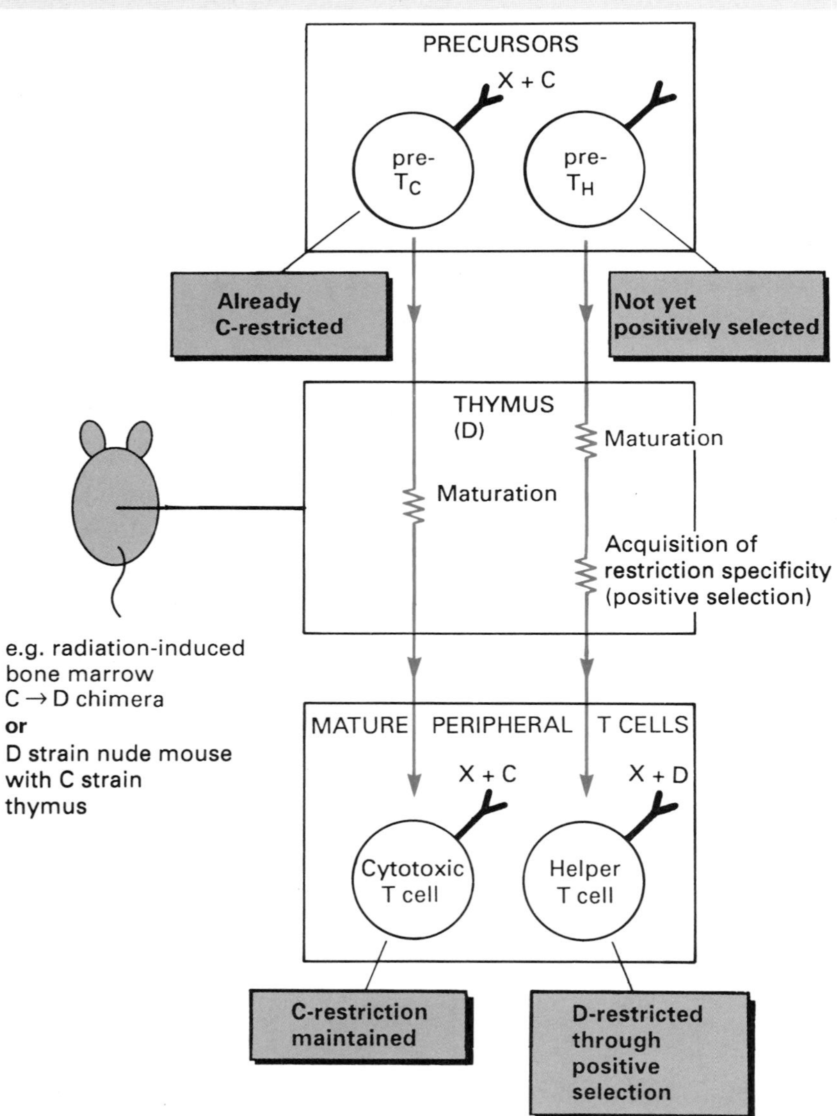

Fig.5.25 The restriction specificity of helper T cells is determined by the thymus: that of cytotoxic T cells can be determined extrathymically

The thymus determines the restriction of helper T cells, not just that of class II-restricted cells

Some cytotoxic T cells are class II-restricted. An important question is whether the thymus determines the restriction of class II-restricted T cells in particular or of helper cells in general. It has been shown that the less frequent CD8$^+$ (i.e. class I-restricted) subset of helper T cells is restricted to the thymic MHC type, but the uncommon CD4$^+$ (i.e. class II-restricted) subset of cytotoxic T cells is not (Golding *et al.* 1985). *This suggests that the thymus may determine which MHC molecules are used for antigen recognition by cells with helper function, irrespective of their CD4 and CD8 phenotype and MHC class restriction.*

5.3.5 The cells responsible for positive and negative selection in the thymus

It is clear from the data discussed in Sections 5.3.3 and 5.3.4 that positive selection (resulting in self-restriction; Section 5.2.3) of helper T cells occurs in the thymus, whereas positive selection and apparently also development of at least some cytotoxic T cells can occur extrathymically. Moreover, it is clear that negative selection (resulting in self-tolerance; Section 5.2.5) can occur both within the thymus and extrathymically (i.e. peripherally) and that both clonal deletion and clonal anergy are mechanisms contributing towards the state of self-tolerance.

Very little is currently known about the precise cell types within the periphery responsible for positive and negative selection (except perhaps for a possible role of gut epithelium in the development of some $\gamma\delta^+$ T cells; Section 4.2.2). In contrast, there is a large body of data relating to the cell types within the thymus that are responsible for these selection events, and this is the subject of this section. This proves to be a controversial area, the main debate centring around the relative roles of thymic epithelial cells and bone marrow-derived haemopoietic cells, which are collectively termed **thymic stromal cells**. The various non-lymphoid stromal cells have different distributions within the thymus (Fig.1.37).

Thymic stromal cells. In the thymic *cortex*, where the majority of immature thymocytes are localized, there is one predominant type of **epithelial cell** which is stellate in shape, and these cells form a ramifying network throughout this region. Thymic **nurse cells** of epithelial origin, which apparently contain thymocytes within their cytoplasm (!) are also present. Within the *medulla*, which mainly contains the more mature thymocytes, a morphologically and phenotypically different type of epithelial cell is present that is more voluminous and oval in shape, and which contains many secretory organelles. A third type of epithelial cell containing large vacuoles has also been identified, at lower frequency, in this area, together with **Hassal's corpuscles**, which are agglomerations of concentrically-arranged epithelial cells. One type of haemopoietic cell, the thymic **interdigitating cell** (a member of the dendritic cell family; Section 3.4.2) is localized to the medulla and the cortico-medullary junction. **Macrophages** are present throughout the thymus, as are some **B cells** but the latter are unusual in that many of these cells express CD5 (Section 8.5).

One view of thymocyte development holds that thymocytes mature as they progress from the cortex into the medulla, as suggested by the relative distribution of immature and mature cells in these areas (Section 4.3). In some studies, thymocytes in association with various stromal cells ('rosettes') have been isolated. It has been suggested from these studies that, as thymocytes mature, they sequentially migrate from the subcapsular region, enter the cortex and become associated with macrophages, then interact with cortical epithelial cells, and are eventually found in association with interdigitating cells in the medulla. However, this linear sequence of cell interactions has not been proven unequivocally, and its possible relationship to the order of positive and negative selection events is obscure.

The present consensus is that thymic epithelial cells are primarily responsible for inducing self-restriction of T cells (positive selection), whereas haemopoietic cells can induce self-tolerance (negative selection) (Table 5.13). The evidence for and against these ideas is reviewed in the following sections. It will be seen that there are many conflicts in the data, and generally the reasons for this are not clear. However, this could, for example, be due to the different experimental systems employed, the peculiarity of some of the antigens investigated (e.g. Mls; Section 5.2.5), differing susceptibilities of thymocytes of different lineages to different selection pressures at various points in their development (Section 5.2.6), and/or requirements for the integrated functions of more than one stromal cell type for selection events. Nevertheless, we shall try to provide a framework for understanding the conflicting lines of evidence.

Stromal elements for positive selection

A role for thymic epithelium. There is a considerable body of evidence that thymic epithelial cells are responsible for positive selection. For example, as discussed in Section 5.3.2, the repertoire of T cells in F1 (bm)→C (rad) chimeras (p. 295) is skewed towards the MHC type of the radiation-resistant thymic epithelium, and comparable results have been obtained for C (thy)→F1 (nu) mice. Both types of experiment implicate epithelial cells rather than haemopoietic cells in this process.

One persuasive line of evidence that epithelial cells are responsible for positive selection has come from experiments in which thymus lobes depleted of haemopoietic cells were transplanted to ATXBM mice (Fig. 5.23) and their T cells became restricted to MHC of the epithelium.

Table 5.13 Overview of cells implicated in positive and negative selection in the thymus

Cell type	Positive selection	Negative selection
Epithelial cells	Yes	?
Haemopoietic cells:	?	Yes
Dendritic cells		Anergy?
$CD5^+$ B cells		Deletion?
Macrophages		No
T cells		Deletion?

Thymus lobes from embryonic mice at 14 days of gestation were first cultured in 2′-deoxyguanosine (dGuo) which kills thymocytes, dendritic cells, and B cells, but spares the epithelial cells (Fig.5.26) (Lo and Sprent 1986). Lobes treated in this way have been shown to be competent to support the development of thymocytes into (CD4, CD8) double-positive and single-positive T cells. When dGuo-treated lobes from strain C were transplanted into (CxD) F1 ATXBM recipients, the mature T cells that developed were found to be restricted to C strain MHC molecules, even though the lobes were recolonized by F1 haemopoietic cells. Evidence that cortical, rather than medullary, epithelial cells are involved in positive selection has come from studies of transgenic mice bearing H-2 IE on the cortical epithelium (Bill *et al.* 1989). In these animals, $V_{\beta}6^{+}$ T cells (normally associated with Mls^{a}-reactivity, Section 5.2.5) were selected into the $CD4^{+}8^{-}$ mature T cell subset, whereas this did not occur in animals expressing IE within the medulla. Nevertheless, it is not clear how 'clean' these systems are; for example, some macrophages survive DGuo treatment and a small population of IE^{+} medullary interdigitating cells has been detected in some transgenic mice otherwise thought to express this molecule solely on the cortical epithelium.

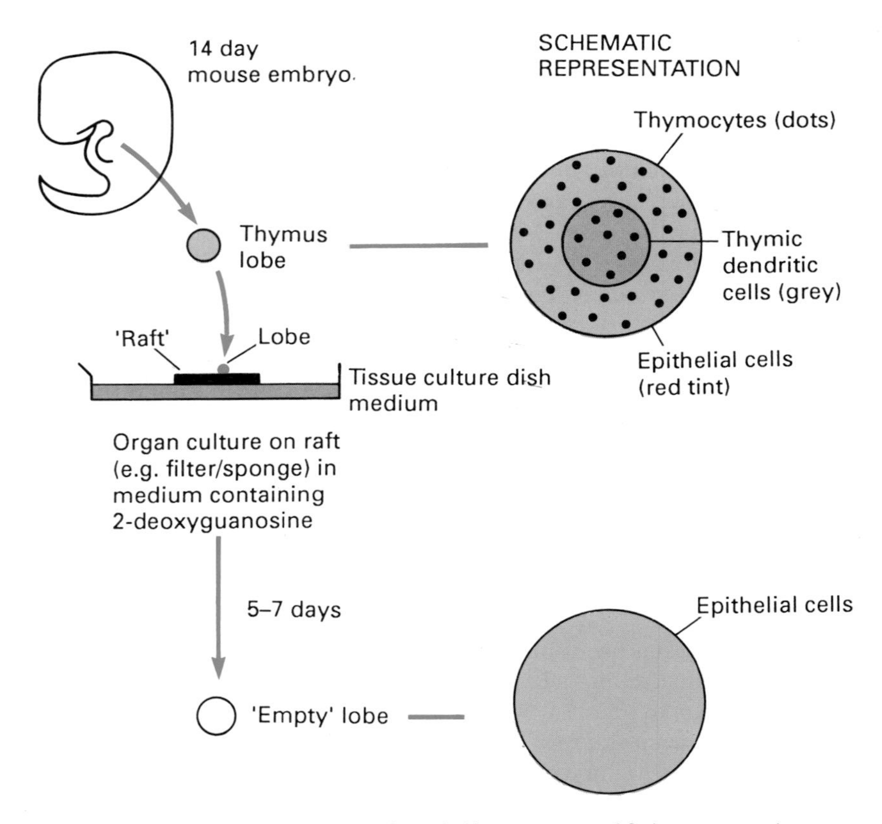

Fig.5.26 Embryonic thymus organ culture in the presence of 2-deoxyguanosine

A role for haemopoietic cells? A long-standing debate, concerning a possible role for haemopoietic cells in positive selection, originates from an earlier study by Longo and Schwartz (1980).

These investigators first showed that the restriction (Ir phenotype) of T cells in C (bm)–F1 (rad) chimeras was normally skewed towards C type MHC, consistent with other studies (Section 5.3.2). However, 3 months after reconstitution, they depleted these animals of pre-existing (C-restricted) T cells and found that the newly-produced cells were now restricted to the bone marrow F1 MHC type, correlating with repopulation of the thymus by F1 haemopoietic cells at this time (Table 5.8).

Since then, there has been a series of claims and counter-claims for and against their observations and, while the more recent studies appear to be *against* a role for haemopoietic cells in positive selection, the original findings remain largely unexplained.

Stromal elements for negative selection

A role for haemopoietic cells. By tracing the fate of T cells expressing particular V_β regions, investigators have attempted to unravel the contribution of haemopoietic cells and epithelial cells to one mechanism of tolerance induction, clonal deletion (Section 5.2.5). Such studies indicate that haemopoietic cells, rather than thymic epithelial cells, can induce clonal deletion (Marrack *et al.* 1988).

It has been shown that chimeras constructed with H-2 IE^+ bone marrow transplanted into irradiated IE^- recipients delete T cells expressing the appropriate $V_\beta 17a$ element. Likewise, if a thymus from an Mls-1a recipient is transplanted into an Mls-1b nude mouse, deletion of $V_\beta 6^+$ T cells is observed, whereas this is not seen after transplantation of a dGuo-treated thymus. That deletion occurs *before* entry of thymocytes to the medulla is suggested by the observation that $V_\beta 6^+$ T cells are present throughout the thymus of Mls-1b mice, whereas they are excluded from the medulla of the Mls-1a strain.

Of course, studies of V_β expression alone cannot address alternative models for tolerance induction such as clonal anergy. For this, one needs to examine tolerance in a functional manner, such as MLR or Mls responses in culture, and graft-versus-host responses (Section 5.3.2) or graft rejection *in vivo*. This has revealed a complexity of responses which is only beginning to be worked out (Speiser *et al.* 1991). However, it seems clear from most studies that haemopoietic cells can induce functional inactivation as well as clonal deletion of autoreactive T cells.

For example, it has been shown that clonal deletion of $V_\beta 6^+$ or $V_\beta 17a^+$ T cells does *not* occur in radiation-induced bone marrow chimeras that express Mls-1a or IE on host, but not haemopietic, tissues; this is consistent with results from the reciprocal type of chimera noted above. Despite the failure of these chimeras to delete potentially autoreactive T cells, which were present in substantial numbers in the periphery of these animals, these T cells may or may not be unresponsive when tested in Mls responses *in vitro*, depending on the H-2 haplotype of the bone marrow inoculum.

Other studies have demonstrated different forms of **split tolerance** in various experimental systems.

For example, if haemopoietic cells are responsible for tolerance induction, T cells from C (bm)–F1 (rad) should not be tolerant to the other parental MHC type (D). In practice, in some studies, significant anti-host responses can be demonstrated in the MLR, but the cytotoxic T cells are completely tolerant in terms of CTL activity. Another form of split tolerance has been found after transplantation of dGuo-treated thymus lobes to allogeneic mice. One might predict that, in the absence of donor haemopoietic cells, tolerance should not occur. In fact, T cell reactivity to donor class I and class II MHC molecules can indeed be detected in the MLR, and yet the thymus and other transplanted tissues of donor strain are not rejected. What is going on in these systems is not yet understood.

Different functions of different haemopoietic cell types. The capacity of purified populations of haemopoietic cells to induce tolerance has also been examined.

In one study, embryonic thymus lobes were cultured with allogeneic dendritic cells to allow their migration into the thymus, and then transplanted to SCID mice (Matzinger and Guerder 1989). The cytotoxic T cells that developed were found to be unresponsive to the allogeneic MHC molecules, implying they had been tolerized by the dendritic cells. However, in another set of experiments different populations of purified Mls^a cells were injected into the thymus of neonatal Mls^b mice, and the T cells that developed *in vivo* were examined both for clonal deletion of $V_\beta 6^+$ T cells and tolerance in a graft-versus-host response (Inaba *et al.* 1991). The results of this study indicate that different cell types can induce tolerance by different mechanisms. Dendritic cells induced tolerance but *not* clonal deletion of $V_\beta 6^+$ T cells, implying they were responsible for clonal anergy. Thymic $CD5^+$ B cells (Section 8.5), but not ($CD5^-$) splenic B cells, induced tolerance associated with deletion of $V_\beta 6^+$ T cells, implying they were responsible for clonal deletion. (It is also interesting in this respect that treatment of Mls^a mice with anti-μ antibodies to inhibit B cell development was shown, in other studies, to prevent deletion of $V_\beta 6^+$ T cells.) However, macrophages purified from the thymus induced neither deletion of $V_\beta 6^+$ T cells nor tolerance. Thymocytes and T cells have also been the subject of scrutiny. For example, Mls^a $CD8^+$ T cells transfused into Mls^b neonatal mice were found to induce a long-lasting state of tolerance associated with clonal deletion (Webb and Sprent 1990).

A role for thymic epithelial cells? There is general agreement, from studies of thymus grafts depleted of haemopoietic cells, that epithelial cells are *incapable* of tolerizing $CD8^+$ T cells. For other T cells, some studies of bone marrow chimeras treated with sublethal doses of radiation before transplantation, indicate that clonal deletion can be induced by a radiation-sensitive (presumably haemopoietic) thymic component, whereas clonal anergy can be induced by a radio-resistant (presumably epithelial) component. However, other studies of bone marrow chimeras preconditioned with supralethal doses of irradiation have led to the conclusion that both clonal deletion as well as a temporary form of clonal anergy can be induced by a radioresistant component, also assumed to be epithelial cells (Gao *et al.* 1990).

Although we have concentrated throughout this section on studies in mice, tolerance has of course been investigated in other species (Section 5.2.5). Although possible species differences add a further complication to an already complex area, we shall conclude by noting one set of experiments in which quail limbs were transplanted to neonatal chickens (Ohki *et al.* 1987). Normally these transplants are rejected within two weeks. However, when a quail limb was transplanted to a neonatal chicken bearing a quail thymus, that was obtained at a stage *before* colonization by haemopoietic cells, the limb was not rejected. This indicates that, at least in this situation, the thymic epithelium may play an important role in tolerance induction.

5.4 Unresponsiveness of mature peripheral T cells to antigens under Ir gene control

In Section 5.2 we discussed how T cells can recognize and respond to foreign antigens. In this section we consider how they are unable to respond to certain types of antigens that are said to be 'under Ir gene control'. We will confine ourselves here to **genetically-controlled** mechanisms of unresponsiveness, and will not consider suppressive mechanisms resulting in T cell unresponsiveness which are discussed in Chapter 11. We will focus on studies in mice, but note that similar findings have been made in other species, including humans.

5.4.1 MHC-linked immune response genes

Antibody responses

Early studies showed that the antibody responses of mice to antigens such as heterologous erythrocytes (i.e. erythrocytes from a different species) or bacterial antigens varied according to the MHC type. Sometimes the quantity and types (e.g. IgG or IgM) of antibodies produced were quite different in different strains. The pattern of responses was however complex, because these antigens contain many different antigenic determinants, and the genetic basis for the control of antibody responses was thus difficult to analyse. This was because the overall response that was measured was the sum of the different responses to each antigenic determinant on the molecule.

McDevitt and Sela (1967) and others then turned to the analysis of antibody responses against more simple molecules, and initially they examined a number of **synthetic polypeptides**. By examining the antibody responses to these molecules it was found that different mouse strains could be classified as **responders** (R) or **non-responders** (NR) depending on whether or not they produced a response.

These synthetic molecules consisted, for example, of tyrosine (T) or histidine (H) residues linked to a core of multiple glutamic acid (G), alanine (A), and lysine (L) residues in a repetitive structure (Fig. 5.27) and were respectively called (T,G)-AL and (H,G)-AL.† The (G)-AL core alone was non-immunogenic in mice, in that it did not induce the production of antibodies, but responses were elicited when the tyrosine or histidine residues were linked to this core. The immune responses to such molecules depended on the strain of mouse. For example, H-2^b mice responded well to (T,G)-AL but poorly to (H,G)-AL while for H-2^k mice the reverse was true. H-2^b mice were thus called *responders* to (T,G)-AL but *non-responders* to (H,G)-AL, while H-2^k mice had the converse phenotype. Other examples of synthetic polymers are GAT and GT. It was also possible to classify mice as R and NR strains when responses to naturally occurring, relatively small polypeptides, like ovalbumin, lysozyme, and insulin were examined, and similar a phenomenon was observed.

However this classification sometimes proved to be an oversimplification, for intermediate responses to some antigens could also be obtained (Table 5.14). Moreover, in most instances, this applied mainly to secondary (IgG) responses against an antigen, since R and NR animals often made identical primary (IgM) responses. This implies that the immune response defect in non-responder animals was due to the inability of their T cells to be activated to provide the requisite help for these antibody responses (see Section 8.3.1), or to generate memory T cells. Furthermore, the differences

† Note that this terminology was adopted before the single-letter amino acid code was introduced (Fig. 1.3), and it has since been retained.

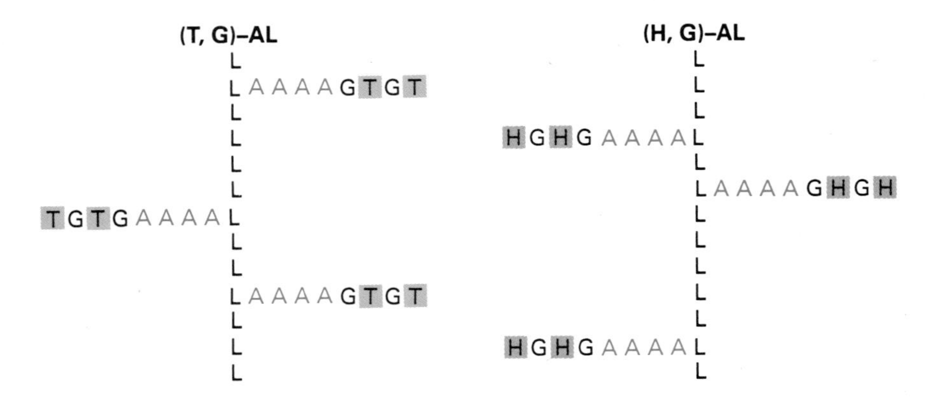

Fig. 5.27 The structures of the synthetic random copolymers (T,G)-AL and (H,G)-AL

Table 5.14 Examples of responses of different strains of various antigens 'under Ir gene control'

	C57BL (H-2^b)	BALB/C (H-2^d)	C3H/He (H-2^k)
(T,G)-AL	H	H/L	L
(H,G)-AL	L	M/L	H
GAT	H	H	H
GT	L	L	L
OVA	H	H/M	L
HEL	L	L?	H

H = high response; M = moderate response; L = low response; H/L, M/L, H/M = variable responses depending on experimental system; OVA = ovalbumin; HEL = hen egg lysozyme

between R and NR strains were often evident only at certain concentrations of antigen, and more similar responses could be elicited by very low or very high concentrations. Despite these reservations, however, there are often striking strain-dependent differences in antibody responses in R and NR strains.

T cell responses

The ability of an antigen to induce a response in some strains of mouse (R) but not in others (NR) is also seen if one measures a T cell response, rather than antibody formation (above). If T cells are isolated from an animal that one has tried to immunize with one of the antigens in question, these T cells may or may not proliferate when they are subsequently cultured with the antigen and accessory cells, and this also allows the animal to be classified as an R or NR strain.

Ir genes

The genes that control responses to synthetic polypeptides with repetitive determinants and to other, naturally-occurring molecules, were originally called **immune response (Ir) genes**; the antigens were therefore said to be 'under Ir gene control'. These genes were then mapped to the MHC class II region, by classical techniques in which responses were studied in mouse strains congenic at the H-2 complex. In fact it was precisely because these Ir genes mapped into the MHC class II (roman 2!) region of the mouse that it also became known as the I (capital i!) region. In most of this section we confine ourselves to a consideration of these MHC-linked Ir genes, but we shall see that a number of non-MHC-linked genes can also control immune responses.

Ia antigens

When mice of one strain were immunized with spleen cells from an MHC-mismatched strain of mouse, it was found that the antisera defined polymorphic determinants of particular cell surface molecules. Some of these molecules were also genetically mapped to the MHC class II region, where Ir genes were located, and were therefore originally termed immune-response-region (*I*)-*a*ssociated or **Ia molecules**. Because Ia molecules were defined by conventional antisera, and subsequently by monoclonal antibodies, they were also called **Ia antigens**, even in species other than the mouse. These molecules are of course the MHC class II molecules, discussed in Chapter 2, so these terms are synonymous.

MHC-linked Ir genes encode Ia antigens

The fact that the serologically-defined Ia antigens mapped into the same genetic region as the functionally-defined Ir genes suggested that Ia antigens might be the products of Ir genes. We now know this to be true, and the evidence for it is briefly reviewed here.

1. The most direct evidence has come from gene cloning and DNA sequencing of the MHC in the mouse. These studies revealed that the MHC class II region contains four functional genes which encode the chains of IA or IE molecules (Chapter 2). (But note the controversy over I-J which maps to the same region but for which no gene has been found; Section 2.8.2).

However, many other lines of evidence had pointed to the same conclusion before the techniques of molecular biology became available to immunologists as summarized in the following (points 2–5, below).

2. Ir genes and Ia antigens were not separable by genetic recombination.

3. Monoclonal antibodies produced against Ia antigens were found to inhibit Ir gene responses in culture.

4. Gene complementation studies. The α and β chains of the mouse IA and IE antigens are encoded by separate loci, and expression of the complete (IA or IE) molecule is therefore said to be 'under two-gene control'. Some strains of mice such as C57BL are unable to produce IEα chains and consequently the IEβ chain remains in the cytoplasm and intact IE molecules cannot be expressed at the cell surface. This deficiency can be corrected, and membrane-bound IE molecules expressed, if the appropriate IEα chain gene is made available, for example in an F1 hybrid or transgenic mouse. This effect is known as **gene complementation**. It turns out that C57BL mice, for example, are unable to respond to some antigens under Ir gene control, but they can respond when gene complementation occurs. Thus expression of an Ia molecule correlates with the ability to make an Ir response. This can be explained by the fact that some antigens under Ir gene control are recognized by helper T cells only when the antigen is bound IE molecules on antigen-presenting cells; since C57BL mice do not express these molecules their T cells cannot respond. Other examples of two-gene control of Ir gene responses are known.

5. The bm12 mouse mutant has a single amino acid change in one of its MHC Class II molecules, compared to the parent (wild-type) strain, and these mice have different Ir phenotypes. Hence a single alteration in an Ia molecule can result in a change in Ir phenotype. Ir genes also control the capacity to respond to protein antigens that may differ by just one amino

acid. Because of their apparently extreme specificity, at one time it was believed that Ir genes encoded T cell receptors although we now know this is not true

5.4.2 Non-MHC-linked Ir genes

Although we have been talking exclusively about MHC-linked Ir genes, many other immune response genes map outside of the MHC. In principle, of course, any gene that controls some stage of the immune response may become evident as an Ir gene. However, the term 'Ir gene' if used on its own usually refers to an MHC-linked Ir gene.

The resistance of certain strains of mice to infection by organisms such as mycobacteria, *Leishmania*, and *Salmonella* is controlled in part by a non-MHC gene(s) on chromosome 1 which may be involved in macrophage function. Even so, in these and other situations, MHC-linked Ir genes do contribute to their recovery from infection. At least 30 non-MHC-linked genes have been identified, and perhaps the best known are those of **Biozzi mice**, which were bred for their ability to make high or low antibody responses to certain antigens.

5.4.3 Function of MHC-linked Ir genes

Hypotheses

There are basically two hypotheses to explain the phenomenon of Ir gene responses, although these are not mutually exclusive (Fig.5.28)

(a) **The determinant selection hypothesis**. In general, helper T cells recognize peptides bound to MHC class II molecules. Determinant selection proposes that certain peptides are *unable* to associate with MHC class II molecules. Therefore, even if T cells were present that could potentially recognize this combination, these T cells are unable to respond.

Perhaps the simplest case has been discussed above, for IE, where there is a defect in expression of an MHC molecule. However, determinant selection applies more generally to instances in which MHC molecules are normally expressed but are unable to associate with the antigen. This could be controlled, for example, by polymorphisms in both the MHC molecules and the putative peptide transporters (see p. 159).

(b) **The clonal deletion, or 'hole-in-the-repertoire', hypothesis**. This hypothesis proposes that the peptides in question *can* bind to MHC molecules, but that T cells that could recognize this combination are not present in the repertoire. No particular mechanism is necessarily implied by the different names for this hypothesis, and they are used simply to refer to the fact that these T cells are functionally absent (this might result, for example, from clonal deletion or clonal anergy—see Section 5.2.5).

This could be due to a variety of reasons that include the following.

1. The conformation of the antigen under Ir gene control and the MHC molecule resembles that of one of the individual's own self peptides bound to a self MHC molecule, and T cells that recognize the latter are

deleted during induction of self tolerance (negative selection; Section 5.2.5).

2. The mechanism(s) that determines the apparent self-restriction of the T cell repertoire does not generate the requisite T cell clones (positive selection, Section 5.2.3).
3. T cells specific for the particular combination of the antigen under Ir gene control and an MHC molecule are present in the repertoire but are suppressed for some reason (see Chapter 11).
4. In addition, specific T cell receptors may not be formed because a particular T cell receptor gene (e.g. V, D, or J segment) is not present in the genome, or because the requisite exon rearrangements have not occurred (note, however, that this would *not* be controlled by Ir genes as such).

Evidence for determinant selection

The central idea in the determinant selection hypothesis is that a particular antigen (peptide) is or is not able to bind to an MHC molecule.

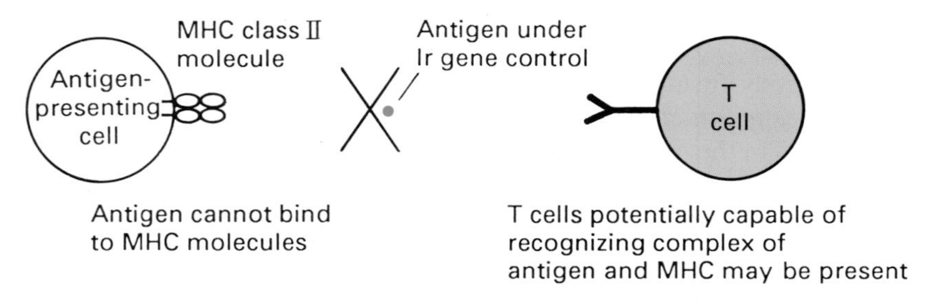
DETERMINANT SELECTION
MHC class II molecule
Antigen under Ir gene control
Antigen-presenting cell
T cell
Antigen cannot bind to MHC molecules
T cells potentially capable of recognizing complex of antigen and MHC may be present

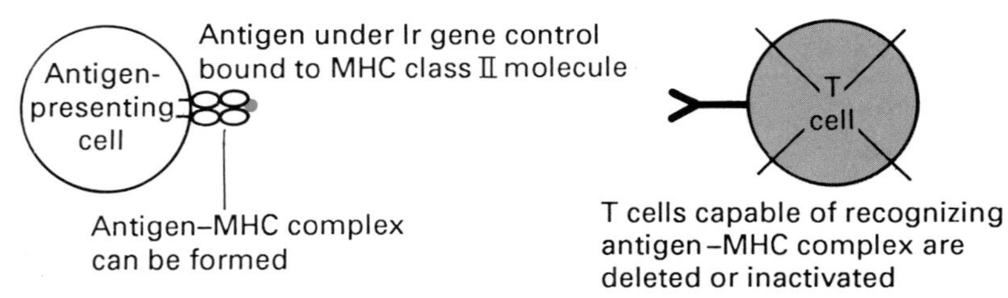
CLONAL DELETION
Antigen under Ir gene control bound to MHC class II molecule
Antigen-presenting cell
T cell
Antigen–MHC complex can be formed
T cells capable of recognizing antigen–MHC complex are deleted or inactivated

Fig.5.28 Two hypotheses for Ir gene control of antigen responses

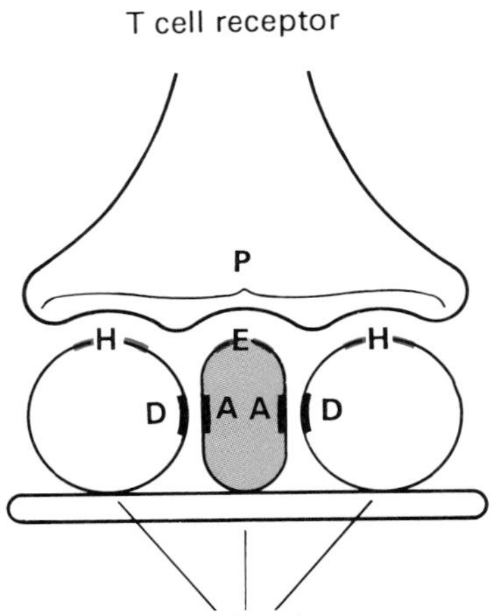

A Agretope
D Desetope
H Histotope
E Epitope
P Paratope (the antigen-binding site of the T cell receptor)

Fig.5.29 Terminology for sites of interaction between antigens, MHC molecules, and T cell receptors

Before we outline the evidence for this, we should note terms that have been used to describe the various parts of antigens, MHC molecules, and T cell receptors that interact with each other (Fig.5.29). We have avoided most of these where possible in this book, but they are found in the literature.

epitope—that portion of an antigen (peptide) that binds to (or is recognized by) the T cell receptor. This term is also used for the part of an antigen (i.e. the antigenic determinant) that binds to the combining site of an antibody.

histotope—that portion of an MHC molecule that binds to (or is recognized by) the T cell receptor. The term comes from the phrase *histo*compatibility molecule epi*tope*.

desetope—that portion of an MHC molecule that binds to (or interacts with) the antigen. The term comes from the phrase *de*terminant *se*lection site, plus '-tope'.

agretope—that portion of the antigen that binds to (or interacts with) the MHC molecule. This term comes from the phrase antigen (*ag*) *r*estriction *e*lement interaction site, plus '-tope'.

Evidence for determinant selection is based on attempts to show whether or not the antigen can physically associate with an MHC molecule.

1. Direct evidence that peptides can bind to MHC molecules (in detergent-solubilized form or reconstituted in synthetic lipid membranes, for instance) is discussed in Section 3.3.3. Some of these experiments also indicate that, in some cases, antigens under Ir gene control might be able to bind to MHC molecules from the responder strain but not from the non-responder strain.

2. Evidence that antigens can compete for binding to an apparently single site on an MHC molecule was discussed in Chapters 2 and 3. This has been shown directly (using purified MHC molecules as noted in 1), or inferred from studies in which the response of T cells to antigen and accessory cells was assayed in the presence of competing peptides (Section 3.3.3).

3. Results of studies that examined the **fine-specificity** of T cell clones have been taken to support the idea of determinant selection.

The principle of these experiments is to examine responses of T cell clones from a particular strain (say C) to antigens like pigeon cytochrome c (cyt c) and related molecules, like moth cyt c, in the presence of accessory cells from the same or different strains (C or D). In other words, cross-reactive recognition by the clone is examined. For example, it was found that certain clones responded better to moth cyt c than the original immunogen (pigeon cyt c) in the presence of accessory cells from strain C (Solinger *et al.* 1979). This type of response, which is greater to a different antigen than the original immunogen, is called a **heteroclitic response**. However, the same clone responded better to pigeon cyt c than to moth cyt c when D strain accessory cells were used. This was consistent with determinant selection, since the same T cell clone responded differently to the same antigen in the presence of different MHC molecules (on accessory cells), which thus appeared to control the response.

4. The magnitude of the response of a T cell clone has been shown to be a product of the antigen concentration and the density of MHC molecules on accessory cells. This is consistent with the idea that an interaction between the two governs the T cell response.

According to the determinant selection hypothesis as originally proposed, the defect that leads to unresponsiveness to certain antigens must be *at the level of the accessory cell*, since these are the cells which express the relevant MHC class II molecules. It is important to note, however, that in principle this hypothesis applies to the association of *any* antigen with *any* MHC molecule and since cytotoxic T cells also recognize peptides just like helper cells (but usually bound to MHC class I molecules) the idea of determinant selection can be applied more generally. Often the expression **antigen presentation by MHC molecules** is used to convey a similar idea.

Evidence for clonal deletion

The clonal deletion hypothesis provides an alternative explanation for unresponsiveness to antigens under Ir gene control. As we have seen earlier in Section 5.2.3, the (helper) T cell repertoire appears to be selected to recognize peptides bound to particular MHC (class II) molecules, according to the (self) MHC alleles that are encountered in the thymus. In one form the argument for clonal deletion runs something like this: Ir genes encode MHC class II molecules, so they are expressed in the thymus (as well as on peripheral antigen-presenting cells); thymic expression of class II molecules determines the precise T cell repertoire; thus the Ir defect may be *at the level of the T lymphocyte*, rather than (or as well as) the peripheral antigen-presenting cell. Several lines of evidence are consistent with the clonal deletion hypothesis.

One of the earliest studies that tried to answer whether Ir genes operated at the level of the accessory cell or the T cell was by Shevach and Rosenthal (1973). It was known that responder strain T cells proliferate when they are cultured with the antigen and their own (responder strain) accessory cells, but T cells from a non-responder do not respond to antigen in the presence of non-responder accessory cells. The ideal way of testing where the Ir defect lies would be to ask whether responder strain T cells can respond to antigen and non-responder strain accessory cells (and thus non-responder strain MHC molecules), or vice versa. But since these cells of necessity differ at the MHC class II region, antigen-specific responses would be masked by the allogeneic MLR. In the original experiments, therefore, T cells were obtained from F1 animals (R × NR guinea pigs) which were genetically tolerant to MHC molecules from both parental (R and NR) strains. It was found that these F1 T cells only responded to the antigen in the presence of responder strain accessory cells but not with non-responder accessory cells. It was argued that, since the T cells were the same, the defect must be at the level of the accessory cell.

While the foregoing experiment has often been quoted in support of determinant selection, the results can also be explained by clonal deletion: if the non-responder phenotype resulted from deletion of T cells that were capable of recognizing that antigen in association with non-responder MHC molecules, this would also happen in F1 animals, since MHC molecules of *both responder and non-responder* strains are codominantly expressed. Therefore this system does not in fact answer the question.

An alternative approach is to deplete alloreactive T cells (Section 5.2.4), and then to ask directly whether the remaining T cells can recognize antigens in the presence of non-responder strain accessory cells. This is of course the same as looking for a certain type of allorestricted T cells and, as noted in Section 5.2.4, Ishii and colleagues (1981) as well as others did this very experiment. They first depleted alloreactive cells by treatment with bromodeoxyuridine and exposure to light, and then asked whether the residual T cells could respond to antigen plus accessory cells from the non-responder strain. They found, in every strain combination tested, that they could. Therefore it was argued there was nothing intrinsically wrong with the ability of antigen to associate with non-responder MHC molecules, and that Ir gene defects must operate at the level of the T cell.

A similar approach was then taken by Dos Reis and Shevach (1983), for example, who made similar observations using guinea pig cells. Strain 2 animals do not respond to the B chain of beef insulin while strain 13 animals (which differ in the MHC class II region) do. After depleting alloreactive cells, it was found that responder T cells from strain 13 animals could be primed to this antigen in the presence of (non-responder) strain 2 accessory cells. (In addition, it was found that strain 13 T cell lines also responded to guinea pig (i.e. self) insulin in the presence of strain 2 accessory cells. This is not 'autoreactivity' because strain 2 T cells do not normally encounter their own insulin bound to strain 13 MHC molecules or vice versa). These observations are consistent with the idea that in strain 2 animals, T cells that react to self-insulin plus self (strain 2) MHC molecules might be clonally deleted in the thymus during T cell ontogeny. Since this complex of self-insulin bound to a self MHC molecule happens to resemble that of the beef-insulin B chain bound to a self MHC molecule, the animal is unable to respond to the latter.

5.4.4 Summary

As noted in Section 5.4.3, the two hypotheses put forward to explain Ir gene responses are not mutually exclusive. To summarize, it is probably worthwhile considering events leading to Ir gene-controlled respones, in the thymus and periphery respectively.

Within the *thymus*, a particular set of self peptides is bound to self MHC molecules, and thymocytes and T cells reactive with these combinations are deleted or functionally inactivated during negative selection (Section 5.2.5). If one of these complexes happens to resemble that of a foreign peptide bound to self MHC on an antigen-presenting cell in the periphery, a T cell response cannot be generated. This would be apparent as an *Ir gene defect at the level of the T cell repertoire*, compatible with the clonal deletion hypothesis for Ir responses. However, this actually results from determinant selection within the thymus during negative selection; the same could also apply to T cells specific for certain foreign antigens that might not be generated during positive selection (Section 5.2.3).

In the *periphery*, the ability of a particular foreign peptide to bind to MHC molecules on antigen-presenting cells, controls whether or not a T cell response can be generated. This would be apparent as an *Ir gene defect at the level of the antigen-presenting cell*, compatible with the determinant selection hypothesis.

Ir gene responses are, in general, made manifest because of the relative simplicity of the structures of the antigens concerned. During immune responses to more complex antigens, some antigenic determinants may not elicit T cell responses for the reasons considered in this section (5.4), but the response is generated by other T cells in the repertoire that are specific for different antigenic determinants on the same molecule.

Further reading

Acha-Orbea, H. and Palmer, E. (1991). Mls—a retrovirus exploits the immune system. *Immunology Today*, **12**, 356–61.

Cerottini, J.C. and MacDonald, H.R. (1989). The cellular basis of T cell memory. *Annual Review of Immunology*, **7**, 77–89.

Fleischer, B. (1989). Bacterial toxins as probes for the T-cell antigen receptor. *Immunology Today*, **10**, 262–4.

Fowlkes, B.J. and Pardoll, D.M. (1989). Molecular and cellular events of T cell development. *Advances in Immunology*, **44**, 207–64.

Hanahan, D. (1990). Transgenic mouse models of self-tolerance and autoreactivity by the immune system. *Annual Review of Cell Biology*, **6**, 493–537.

Haynes, B.F. (1990). Human thymic epithelium and T cell development: current issues and future directions. *Thymus*, **16**, 143–57.

Haynes, B.F., Donning, S.M., Singer, K.W., and Kurtzberg, J. (1989). Ontogeny of T-cell precursors: a model for the initial stages of human T-cell development. *Immunology Today*, **10**, 87–91.

Hunig, T. (1983). T-cell function and specificity in athymic mice. *Immunology Today*, **4**, 84–7.

Ikuta, K., Uchida, N., Friedman, J., and Weissman, I.L. (1992). Lymphocyte development from stem cells. *Annual Review of Immunology*, **10**, 759–83.

Lechler, R.I., Lombardi, G., Batchelor, J.R., Reinsmoen, N., Bach, F.H. (1990). The molecular basis of allorectivity. *Immunology Today*, **10**, 83–8.

Lo, D., Burkly, L.C., Flavell, R.A., Palmiter, R.D., and Brinster, R.L. (1990). Antigen presentation in MHC class II transgenic mice: stimulation versus tolerization. *Immunological Reviews*, **117**, 121–34.

Matis, L. (1990). The molecular basis of T cell specificity. *Annual Review of Immunology*, **8**, 62–5.

Miller, J.A.P. and Morahan, G. (1992). Peripheral T cell tolerance. *Annual Review of Immunology*, **10**, 51–69.

Mueller, D.L., Jenkins, M.K., and Schwartz, R.H. (1989). Clonal expansion versus functional clonal inactivation: a costimulatory signalling pathway determines the outcome of T cell antigen receptor occupancy. *Annual Review of Immunology*, **7**, 445–80.

Nossal, G.J. (1983). Cellular mechanisms of immunologic tolerance. *Annual Review of Immunology*, **1**, 33–62.

Ritter, M.A. and Crispe, N. (1991). *The thymus*. In Focus Series (ed. D. Male). IRL Press, Oxford.

Schwartz, R.H. (1986). Immune response (Ir) genes of the murine major histocompatibility complex. *Advances in Immunology*, **38**, 31–201.

Schwartz, R.H. (1989). Acquisition of immunologic self-tolerance. *Cell*, **57**, 1073–81.

Scollay, R., Wilson, A., D'Amico, A., Kelly, K., Egerton, M., Pearse, M., Li, W., and Shortman, K. (1986). Developmental status and reconstitution potential of subpopulations of murine thymocytes. *Immunological Reviews*, **104**, 81–120.

Tomonari, K. (1990). Tolerance *in vivo* and *in vitro*. *Immunological Reviews*, **116**, 139–57.

van Ewijk, W. (1991). T-cell differentiation is influenced by thymic microenvironments. *Annual Review of Immunology*, **9**, 591–616.

von Boehmer, H. (1990). Developmental biology of T cells in T cell-receptor transgenic mice. *Annual Review of Immunology*, **8**, 531–56.

Literature cited

Acha-Orbea, H., Shakhov, A.N., Scarpellino, L., Kolb, E., Muller, V., Vessaz-Shaw, A., Fuchs, R., Blocklinger, K., Rollini, P., Billotte, J., Sarafidou, M., MacDonald, H.R., and Diggelman, H. (1991). Clonal deletion of V 14-bearing T cells in mice transgenic for mammary tumour virus. *Nature*, **350**, 207–11.

Ashwell, J.D., Chen., C., and Schwartz, R.H. (1986). High frequency and nonrandom distribution of alloreactivity in T cell clones selected for recognition of foreign antigen in association with self class II molecules. *Journal of Immunology*, **136**, 389–95.

Benichou, G., Takizawa, P.A., Ho, P.T., Killion, C.C., Olson, C.A., McMillan, M., and Sercarz, E.E. (1990). Immunogenicity and tolerogenicity of self-major histocompatibility complex peptides. *Journal of Experimental Medicine*, **172**, 1341–6.

Bevan, M.J. (1984). High determinant density may explain the phenomenon of alloreactivity. *Immunology Today*, **5**, 128–30.

Bill, J. and Palmer, E. (1989). Positive selection of $CD4^+$ T cells mediated by MHC class-II-bearing stromal cell in the thymic cortex. *Nature*, **341**, 649–51.

Bill, J., Yague, J., Appel, V.B., White, J., Horn, G., Erlich, H., and Palmer, E. (1989). Molecular genetic analysis of 178 I-A^{bm12}-reactive T cells. *Journal of Experimental Medicine*, **169**, 115–34.

Billingham, R.E., Brent, L., and Medawar, P. (1956). Quantitative studies on tissue transplantation immunity. III. Actively acquired tolerance. *Royal Society of London, Philosophical Transactions*, Series B, **239**, 357–414.

Blackman, M.A., Gerhard-Burgert, H., Woodland, D.L., Palmer, E., Kappler, J.W., and Marrack, P. (1990). A role for clonal inactivation in T cell tolerance to Mls-1^a. *Nature*, **345**, 540–2.

Braciale, T.J., Andrew, M.E., and Braciale, V.L. (1981). Simultaneous expression of H-2-restricted and alloreactive recognition by a cloned line of influenza vira-specific cytotoxic T lymphocytes. *Journal of Experimental Medicine*, **153**, 1371–6.

Bradley, S.M., Knight, A.M., and Singer, A. (1982). Cytotoxic T lymphocyte responses in allogeneic radiation bone marrow chimeras. The chimeric host strictly dictates the self-repertoire of Ia-restricted T cells but not H-2K/D-restricted T cells. *Journal of Experimental Medicine*, **156**, 1650–64.

Buerstedde, J.M., Nilson, A.E., Chase, C.G., Bell, M.P., Beck, B.N., Pease, L.R., and McKean, D.J. (1989). A polymorphic residue responsible for class II molecule recognition by alloreactive T cells. *Journal of Experimental Medicine*, **169**, 1645–54.

Burkly, L.C., Lo, D., Kanagawa, O., Brinster, R.L., and Flavell, R.A. (1989). T-cell tolerance by clonal anergy in transgenic mice with nonlymphoid expression of MHC class II I-E. *Nature*, **342**, 564–66.

Choi, Y., Kappler, J.W., and Marrack, P. (1991). A superantigen encoded in the open reading frame of the 3′ long terminal repeat of mouse mammary tumour virus. *Nature*, **350**, 203–7.

Dos Reis, G.A. and Shevach, E.M. (1983). Antigen presenting cells from nonresponder strain 2 guinea pigs are fully competent to present bovine insulin B chain to responder strain 13 T cells: evidence against a determinant selection model and in favour of a clonal deletion model of immune response gene function. *Journal of Experimental Medicine*, **157**, 1287–99.

Duprez, V., Hamilton, B., and Burakoff, S. (1982). Generation of CTLs in thymectomized, irradiated, bone marrow-reconstituted mice. *Journal of Experimental Medicine*, **156**, 844–59.

Forman, J., Goodenow, R.S., Hood, L., and Ciavarra, R. (1983). Use of DNA-mediated gene transfer to analyze the role of *H-2L^d^* in controlling the specificity of anti-vesicular stomatitis virus cytotoxic T cells. *Journal of Experimental Medicine*, **157**, 1261–72.

Gao, E.K., Lo, D., and Sprent, J. (1990). Strong T cell tolerance in parent→F1 bone marrow chimeras prepared with supralethal irradiation. Evidence for clonal deletion and anergy. *Journal of Experimental Medicine*, **171**, 1101–21.

Golding, H., Munitz, T.I., and Singer, A. (1985). Characterization of antigen-specific, Ia-restricted, L3T4+ cytolytic T lymphocytes and assessment of thymic influence on their self specificity. *Journal of Experimental Medicine*, **162**, 943–61.

Guidos, C.J., Danska, J.S., Fathman, C.G., and Weissman, I.L. (1990). T cell receptor-mediated negative selection of autoreactive T lymphocyte precursors occurs after commitment to the CD4 or CD8 lineages. *Journal of Experimental Medicine*, **172**, 835–45.

Hasek, M. and Hraba, T. (1955). Immunological effects of experimental embryonal parabiosis. *Nature*, **175**, 764–5.

Ikehara, S., Pahwa, R.N., Fernandes, G., Hansen, C.T., and Good, R.A. (1984). Functional T cells in athymic nude mice. *Proceedings of the National Academy of Sciences USA*, **81**, 886–8.

Inaba, M., Inaba, K., Hosono, M., Kumamoto, T., Ishida, T., Muramatsu, S., Masuda, T., and Ikehara, S. (1991). Distinct mechanisms of neonatal tolerance induced by dendritic cells and thymic B cells. *Journal of Experimental Medicine*, **173**, 549–59.

Ishii, N., Baxevanis, C.N., Nagy, Z.A., and Klein, J. (1981). Responder T cells depleted of alloreactive cells react to antigen presented on allogeneic macrophages from non-responder strains. *Journal of Experimental Medicine*, **154**, 978–82.

Jones, L.A., Chin, L.T., Merriam, G.R., Nelson, L.M., and Kruisbeek, A.M. (1990). Failure of clonal deletion in neonatally thymectomized mice: tolerance is preserved through clonal anergy. *Journal of Experimental Medicine*, **172**, 1277–85.

Kappler, J.W., Roehm, N., and Marrack, P. (1987). T cell tolerance by clonal elimination in the thymus. *Cell*, **49**, 273–80.

Kast, W.M., De Waal, L.P., and Melief, C.J. (1984). Thymus dictates major histocompatibility complex (MHC) specificity and immune response gene phenotype of class II MHC-restricted T cells but not of class I MHC-restricted T cells. *Journal of Experimental Medicine*, **160**, 1752–66.

Katz, D.H., Hamaoka, T., and Benacerraf, B. (1974). Cell interactions between histocompatible T and B lymphocytes. Failure of physiologic cooperative interactions between T and B lymphocytes from allogeneic donor strains in humoral response to hapten-protein conjugates. *Journal of Experimental Medicine*, **134**, 1405–18.

Kay, J. and Janeway, C.A. (1984). The Fab fragment of a directly activating monoclonal antibody that precipitates a disulfide-linked heterodimer from a helper T cell clone blocks activation by either allogeneic Ia or antigen and self-Ia. *Journal of Experimental Medicine*, **159**, 1397–412.

Kisielow, P., Bluthmann, H., Staerz, U.D., Steinmetz, M., and von Boehmer, H. (1988). Tolerance in T-cell-receptor transgenic mice involves deletion of nonmature $CD4^+8^+$ thymocytes. *Nature*, **333**, 742–6.

Kruisbeek, A.M., Hathcock, K.S., Hodes, R.J., and Singer, A. (1982). T cells from fully H-2 allogeneic (A→B) radiation bone marrow chimeras are functionally competent and host restricted but are alloreactive against hybrid Ia determinants expressed on (AxB) F1 cells. *Journal of Experimental Medicine*, **155**, 1864–9.

Kruisbeek, A.M., Sharrow, S.O., and Singer, A. (1983). Differences in the MHC-restricted self-recognition repertoire of intra-thymic and extra-thymic cytotoxic T lymphocyte precursors. *Journal of Experimental Medicine*, **130**, 1027–32.

Lake, J.P., Andrew, M.E., Pierce, C.W., and Braciale, T.J. (1980). Sendai virus-specific, H-2-restricted cytotoxic T lymphocyte responses of nude mice grafted with allogeneic or semi-allogeneic thymus glands. *Journal of Experimental Medicine*, **152**, 1805–10.

Lin, R.H. and Stockinger, B. (1989). T cell immunity or tolerance as a consequence of self antigen presentation. *European Journal of Immunology*, **19**, 105–10.

Liu, Y., and Janeway, C.A. Jr. (1990). Interferon gamma plays a critical role in induced cell death of effector T cells: a possible third mechanism of self-tolerance. *Journal of Experimental Medicine*, **172**, 1735–40.

Lo, D. and Sprent, J. (1986). Identity of cells that imprint H-2-restricted T-cell specificity in the thymus. *Nature*, **319**, 672–5.

Lombardi, G., Sidhu, S., Batchelor, J.R., and Lechler, R.I. (1989). Allorecognition of DR1 by T cells from a DR4/DRw13 responder mimics self-restricted recognition of endogenous peptides. *Proceedings of the National Academy of Sciences, USA*, **86**, 4190–4.

Longo, D.L. and Schwartz, R.H. (1980). T-cell specificity for H-2 and Ir gene phenotype correlates with the phenotype of thymic antigen-presenting cells. *Nature*, **287**, 44–6.

Lorenz, R.G. and Allen, P.M. (1988). Direct evidence for functional self-protein/Ia-molecule complexes *in vivo*. *Proceedings of the National Academy of Sciences USA*, **85**, 5220–3.

McDevitt, H.O. and Sela, M. (1967). Genetic control of the antibody response. II. Further analysis of the specificity of determinant-specific control, and genetic analysis of the response to (H,G)-A-L in CBA and C57 mice. *Journal of Experimental Medicine*, **126**, 969–78.

MacDonald, H.R., Schneider, R., Lees, R.K., Howe, R.C., Acha-Orbea, H., Festenstein, H., Zinkernagel, R.M., and Hengartner, H. (1988). T-cell receptor V use predicts reactivity and tolerance to Mls^a-encoded antigens. *Nature*, **332**, 40–5.

Marrack, P., Lo, D., Brinster, R., Palmiter, R., Burkly, L., Flavell, R.H., and Kappler, J. (1988). The effect of thymus environment on T cell development and tolerance. *Cell*, **53**, 627–34.

Matzinger, P. and Guerder, S. (1989). Does T cells tolerance require a dedicated antigen-presenting cell? *Nature*, **338**, 74–6.

Ohki, H., Martin, C., Corbel, C., Coltey, M., and Le Douarin, N.M. (1987). Tolerance induced by thymic epithelial grafts in birds. *Science*, **237**, 1032–5.

Owen, (1945). Immunogenetic consequences of vascular anastomoses between bovine twins. *Science*, **102**, 400–401.

Paul, W.E., Shevach, E.M., Pickeral, S., Thomas, D.W., and Rosenthal, A.S. (1977). Independent populations of primed F1 guinea pig lymphocytes respond to antigen-pulsed parental peritoneal exudate cells. *Journal of Experimental Medicine*, **145**, 618–30.

Pierres, M., Marchetto, S., Naquet, D., Landais, D., Peccoud, J., Benoist, C., and Mathis, D. (1989). I-A alpha polymorphic residues that determine alloreactive T cell recognition. *Journal of Experimental Medicine*, **169**, 1655–8.

Pircher, H., Bürki, K., Lang, R., Hengartner, H., Zinkernagel, R.M. (1989). Tolerance induction in double specific T-cell receptor transgenic mice varies with antigen. *Nature*, **342**, 559–61.

Ramsdell, F., Lantz, T., and Fowlkes, B.J. (1989). A nondeletional mechanism of thymic self tolerance. *Science*, **246**, 1038–41.

Reimann, J., Kabelitz, D., Heeg, K., and Wagner, H. (1985). Allorestricted cytotoxic T cells. Large numbers of allo-H-$2K^b$-restricted antihapten and antiviral cytotoxic T cell populations clonally develop in vitro from murine splenic precursor T cells. *Journal of Experimental Medicine*, **162**, 592–606.

Rosenthal, A.S. and Shevach, E.M. (1973). Function of macrophages in antigen recognition by guinea pig T lymphocytes. I. Requirement for histocompatible macrophages and lymphocytes. *Journal of Experimental Medicine*, **138**, 1194–212.

Rotzschke, O., Falk, K., Faath, S., and Rammensee, H.G. (1991). On the nature of peptides involved in T cell alloreactivity. *Journal of Experimental Medicine*, **174**, 1059—71.

Schild, H., Rotzschke, O., Kalbacher, H., and Rammensee, H.-G. (1990). Limit of T cell tolerance to self proteins by peptide presentation. *Science*, **247**, 1587–9.

Shearer, G.M., Rehn, T.G., and Garbarino, C.A. (1974). Cell-mediated lympholysis of trinitrophenyl-modified autologous lymphocytes. Effector cell specificity to modified cell surface components controlled by the H-2K and H-2D serological regions of the murine major histocompatibility complex. *Journal of Experimental Medicine*, **141**, 1348–64.

Shevach, E.M. and Rosenthal, A.S. (1973). Function of macrophages in antigen recognition by guinea pig T lymphocytes. II. Role of the macrophage in the regulation of genetic control of the immune response. *Journal of Experimental Medicine*, **138**, 1213–29.

Singer, A., Hathcock, K.S., and Hodes, R.J. (1981). Self recognition in allogeneic radiation bone marrow chimeras. A radiation-resistant host element dictates self specificity and Ir phenotype of T helper cells. *Journal of Experimental Medicine*, **153**, 1286–301.

Singer, A., Hathcock, K.S., and Hodes, R.J. (1982). Self recognition in allogeneic chimeras. Self recognition by T helper cells from thymus-engrafted nude mice is restricted to the thymic H-2 haplotype. *Journal of Experimental Medicine*, **155**, 339–44.

Solinger, A.S., Ultee, M.E., Margoliash, E., and Schwartz, R.H. (1979). T-lymphocyte proliferative response to cytochrome c. I. Demonstration of a T-cell heteroclitic proliferative response and identification of a topographic antigenic determinant on pigeon cytochrome c whose immune recognition requires two complementing major histocompatibility complex-linked immune response genes. *Journal of Experimental Medicine*, **150**, 830–48.

Speiser, D.E., Chvatchko, Y., Zinkernagel, R.M., and MacDonald, H.R. (1991). Distinct fates of self-specific T cells developing in irradiation bone marrow chimeras: clonal deletion, clonal anergy, or in vitro responsiveness to self-Mls-1^a controlled by hemopoietic cells in the thymus. *Journal of Experimental Medicine*, **172**, 1305–14.

Speiser, D.E., Stubi, U., and Zinkernagel, R.M. (1992). Extrathymic positive selection of $\alpha\beta$ T-cell precursors in nude mice. *Nature*, **355**, 170-2.

Sprent, J. (1978). Role of H-2 gene products in the function of T helper cells from normal and chimeric mice. *Immunological Reviews*, **42**, 108–37.

Triplett, E.L. (1962). On the mechanism of immunogenic self-recognition. *Journal of Immunology*, **89**, 505–10.

Webb, S.R. and Sprent, J. (1990). Induction of neonatal tolerance to Mls^a antigens by $CD8^+$ T cells. *Science*, **284**, 1643–6.

Webb, S., Morris, C., and Sprent, J. (1990). Extrathymic tolerance of mature T cells: clonal elimination as a consequence of immunity. *Cell*, **63**, 1249–56.

Weksler, M.E., Moody, C.E. Jr., and Kozak, R.W. (1981). The autologous mixed-lymphocyte reaction. *Advances in Immunology*, **31**, 271–312.

White, J., Herman, A., Pullen, A.M., Kubo, R., Kappler, J.W., and Marrack, P. (1989). The V_β-specific superantigen staphyloccocal enterotoxin B: stimulation of mature T cells and clonal deletion in neonatal mice. *Cell*, **56**, 27–35.

Zamoyska, R., Waldmann, H., and Matzinger, P. (1989). Peripheral tolerance mechanisms prevent the development of autoreactive T cells in chimeras grafted with two minor incompatible thymuses. *European Journal of Immunology*, **19**, 111–17.

Zijlstra, M., Bix, M., Simister, N.E., Loring, J.M., Raulet, D.H., and Jaenisch, R. (1990). β_2-microglobulin deficient mice lack $CD4^-8^+$ cytolytic T cells. *Nature*, **344**, 742–6.

Zinkernagel, R.M. and Doherty, P.C. (1974). Restriction of in vitro T cell-mediated cytotoxicity in lymphocytic choriomeningitis within a syngeneic or semiallogeneic system. *Nature*, **251**, 547–8.

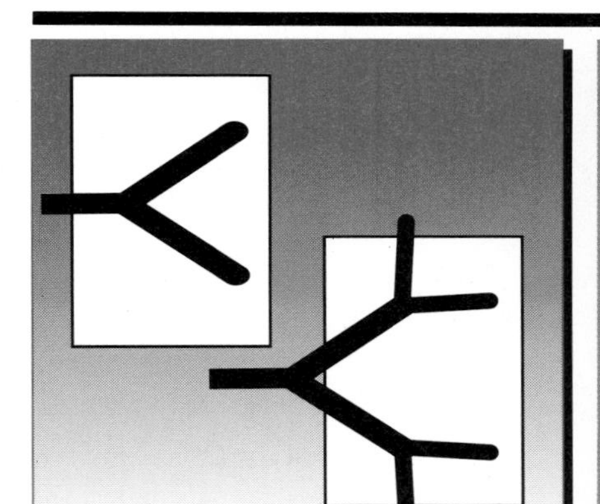

6

ANTIGEN RECEPTORS—STRUCTURE AND FUNCTION

Scheme 6. Antibodies and T cell receptors

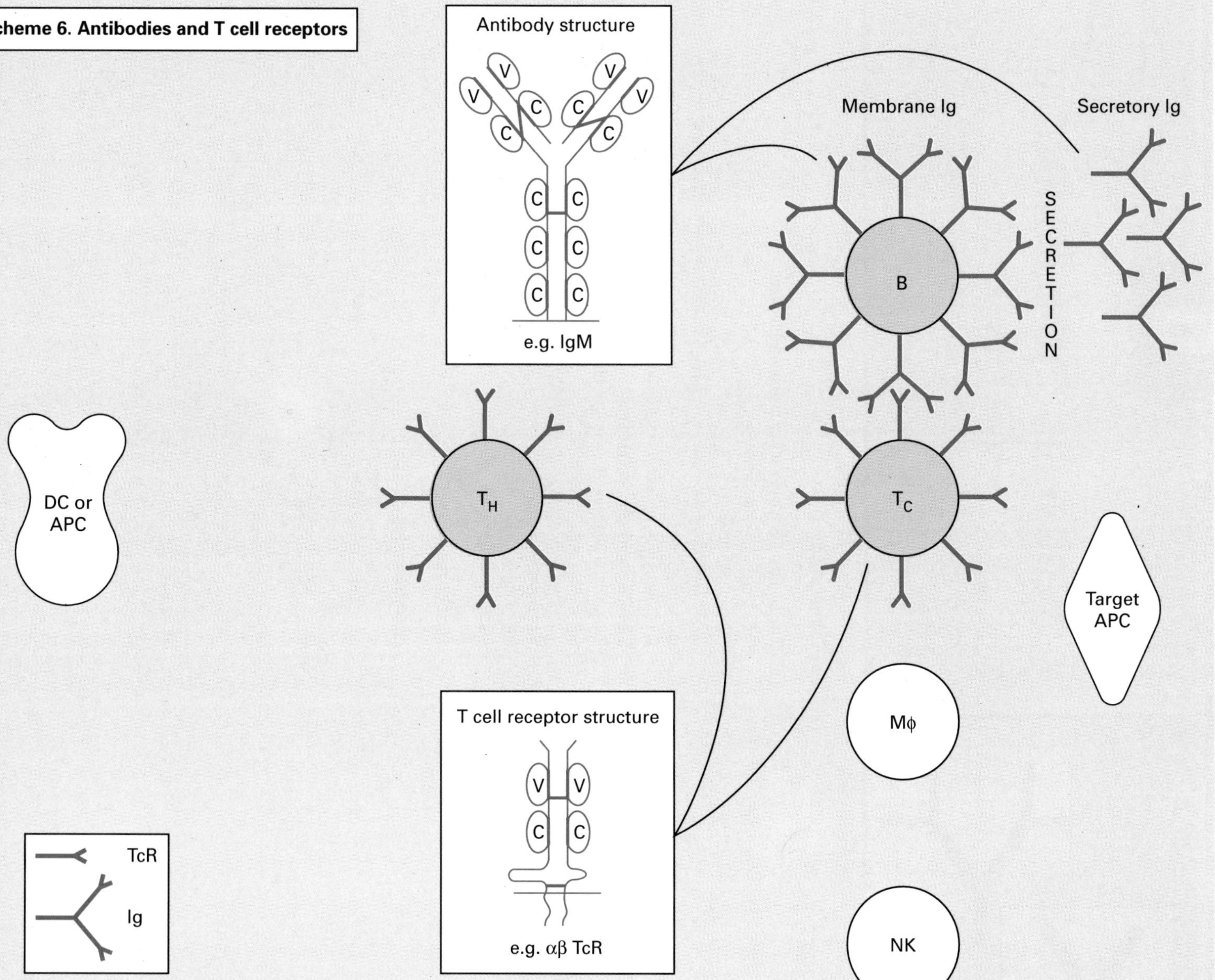

6.1 Introduction

One of the fundamental features of the immune system is its ability to recognize foreign antigens. Antigen recognition is mediated by receptors expressed by T and B lymphocytes and is highly specific.† Each clone of lymphocytes has a unique antigen receptor, and all the antigen receptors expressed by one clone are identical. The ability of the immune system to recognize and respond to an enormous number of different antigens results from the immense number of different clones of lymphocytes, each of which expresses a unique antigen receptor (Section 1.1.3).

B and T lymphocytes express different types of antigen receptor (Fig.6.1). Those of T lymphocytes are known as T cell receptors (TcR), while the antigen receptors expressed by B lymphocytes are known as antibodies or immunoglobulins (Ig for short). The two terms are used interchangeably by immunologists.

It was originally believed that the antigen receptors on all the progeny of a given lymphocyte were identical to those of the parent cell. We now know, at least for B lymphocytes, that when the cell divides, some of the progeny express antigen receptors that are structurally different from those of the parent cell; we shall return to this in Chapter 8.

†In this chapter we do not consider the third form of antigen recognition that takes place during an immune response, that of peptide binding to MHC molecules (see Sections 1.1, 3.3, and 5.1).

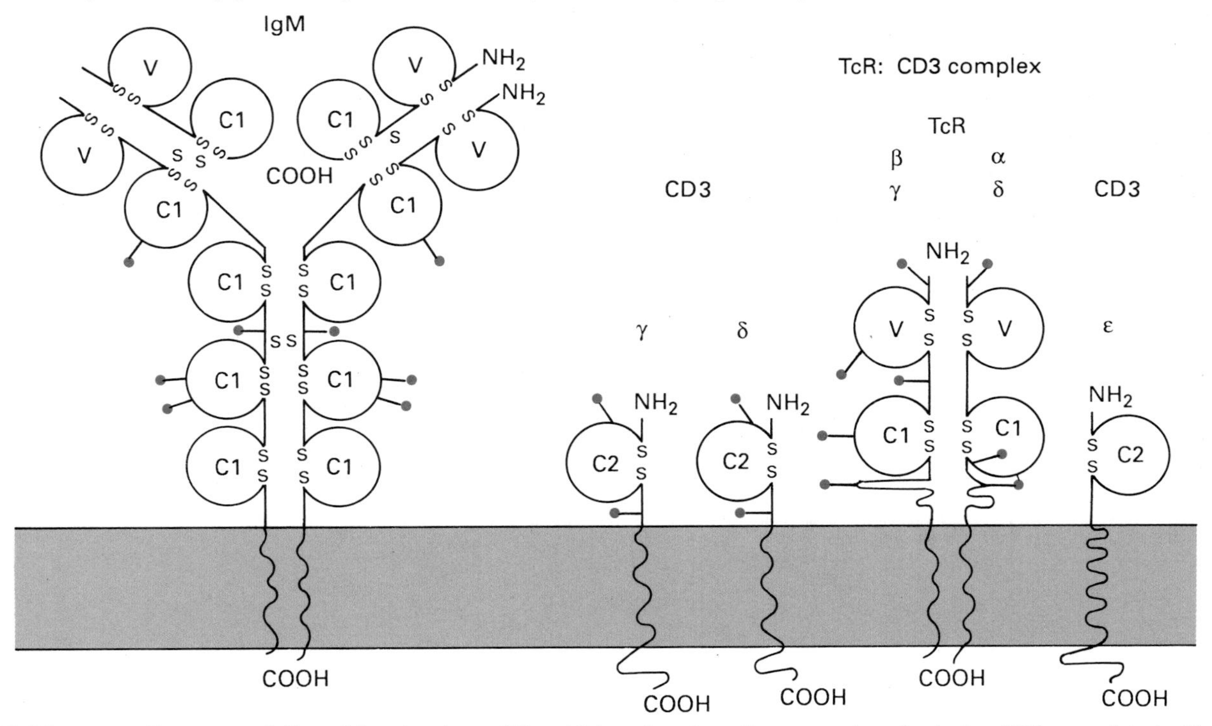

Fig.6.1 Diagrammatic representation of the structure of T and B lymphocyte antigen receptors (including CD3, associated with TcR). Membrane forms of both types of receptor are shown (TcR is either composed of α plus β, or γ plus δ chains). Antibodies can also exist as soluble molecules in the plasma and body fluids

The antigen receptors of T and B lymphocytes have certain properties in common. For example, they obviously both have the ability to recognize antigen and, as we shall see, they are structurally related. However, in other respects they differ quite markedly and these differences are briefly summarized here and in Table 6.1. (Unless otherwise stated, TcR refers to $\alpha\beta$ TcR.)

1. Antibodies are able to recognize soluble antigens as well as antigens on foreign cell surfaces. In contrast, T cell receptors are thought only to recognize antigens on the surface of other cells of the body (Fig.6.2).
2. Furthermore, T cell receptors recognize antigen *only* when it is bound to or associated with an MHC molecule (see Section 5.2.1). In contrast, recognition of antigen by antibodies does not usually depend on or involve an MHC molecule. Antibodies can recognize antigen directly (Fig.6.2).
3. Antibodies and T cell receptors recognize different parts of the antigen (called antigenic determinants or epitopes). Antibodies can recognize the intact antigen in its 'native' conformation and often recognize epitopes formed by the interaction of different parts of the polypeptide chain(s) on the outer, exposed surface of the molecule. These are known as conformational epitopes. On the other hand, T cell receptors only recognize antigen after it has been processed into peptides and become associated with an MHC molecule. T cell epitopes are sequential and often buried within the protein structure (Fig.6.3).
4. Antibodies exist in two forms, as membrane glycoproteins expressed by B cells (membrane Ig; mIg) and as soluble molecules circulating in the plasma or extracellular fluids (secreted Ig; sIg). T cell receptors are gener-

Table 6.1 A summary of general properties of T and B lymphocyte antigen receptors

	Antibody	T cell receptor
Predominant form of antigen recognized	Soluble	Cell surface
Is co-recognition of MHC required?	No	Yes
Location of the antigen receptor	B cell membrane (membrane Ig). Soluble antibody (secretory Ig)—in plasma and body fluids	T cell membrane
Conformation of the antigen	Native	Denatured
Antigen epitope recognized	Conformational—non-linear	Peptide fragments—linear
	Exposed	Buried
	~15 amino acids	9–12 amino acids

ally thought to function only as membrane molecules and soluble forms of T cell receptors have not been defined as yet (but see Section 11.2).

5. The antibodies on the surface of B cells are divalent, meaning that each molecule has two identical binding sites for antigen. T cell receptors are monovalent, each having a single binding site for the antigenic peptides + MHC (Fig.6.1).
6. T cell receptors are very tightly associated with other molecules present in the plasma membrane of the T lymphocyte, particularly CD3 which is important for T cell activation (see Section 4.2). The situation for B cell activation is not yet clearly established, although some molecules linked to membrane Ig have been described (see Section 8.4.2) (Venkitaraman *et al.* 1991).
7. One similarity between both types of antigen receptor is that each molecule can be divided into two parts: the so-called variable (V) and the constant (C) regions (Fig.6.1). The variable portion forms the antigen binding site, and in the case of secreted antibodies the constant region is responsible for what are termed the 'effector functions' of immunoglobulins, such as complement fixation (see Section 9.4). Whether the constant region of the T cell receptor is associated with any particular effector function is not known but as T cell receptors exist as membrane molecules this seem unlikely.

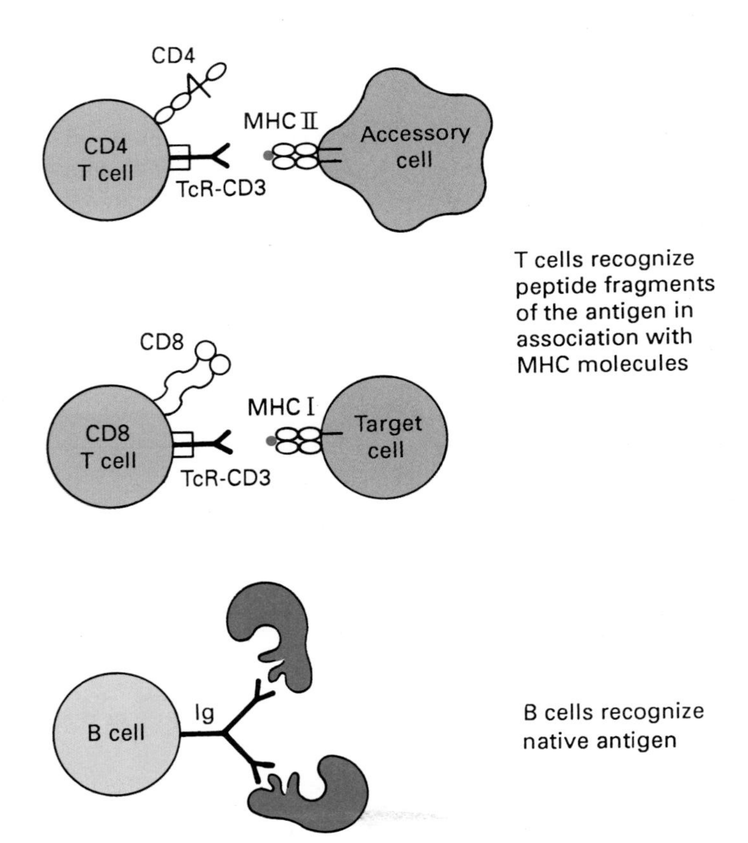

Fig.6.2 Requirements for antigen recognition by T cells and B cells

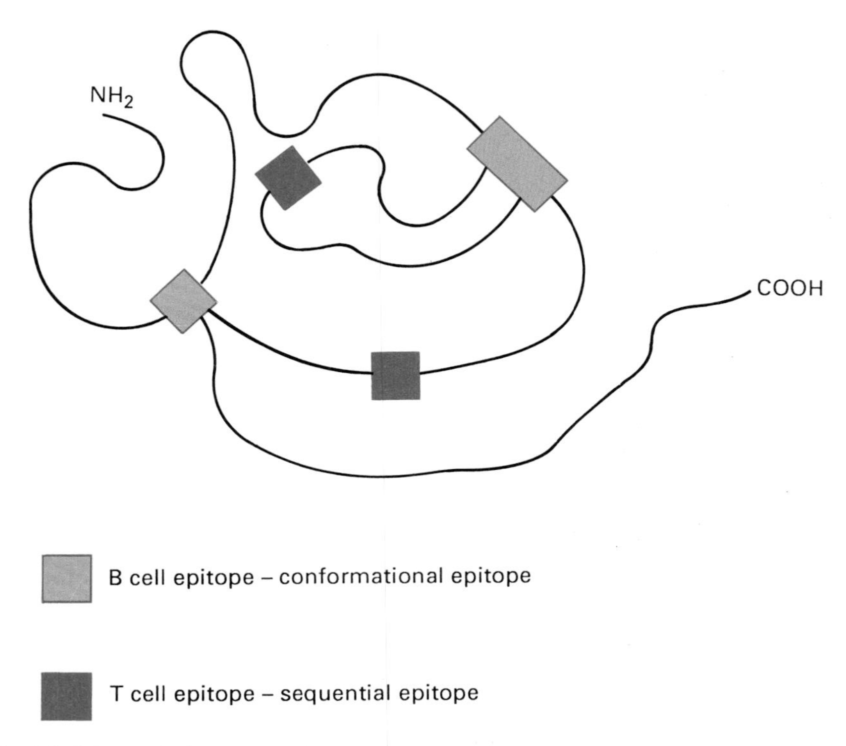

Fig.6.3 Schematic diagram to illustrate the location of T cell and B cell epitopes on a protein antigen (see also Figs.3.7 and 11.26)

In this chapter the structure of antigen receptors will be discussed, together with some aspects of their function, namely their ability to interact with antigen. The two types of receptors will be considered separately, in the order in which their structures were described. The organization of the genes coding for B and T cell antigen receptors is discussed in Chapter 7.

6.2 Antibodies—B lymphocyte antigen receptors

The main features of antibody structure will be outlined in this section. Data obtained from analysis of the structure of human antibodies will be discussed in most depth, as these are more complete, but interesting data obtained by analysing immunoglobulins from other species will also be mentioned.

6.2.1 Structure: an overview

All antibodies have the same basic structure—heavy and light chains

The basic structural unit of all antibodies is identical irrespective of their antigen specificity. Each unit comprises four polypeptide chains, of two different types (Fig.6.4). The larger of these is called the **heavy chain** for historical reasons (50–70 kDa) and the smaller, quite logically, the **light chain** (25 kDa) (if the structure of immunoglobulins were elucidated today, the two chains would probably be referred to as the α and β chains). Two *identical* heavy chains and two *identical* light chains are linked together to form a unit of structure—the immunoglobulin monomer. Each unit can be divided into two; each half is the mirror image of the other and comprises a heavy chain and a light chain held together by a disulfide bond as well as noncovalent forces. The two halves of the molecule are linked together by non-covalent interactions and one or more inter-heavy chain disulphide bond. In their soluble, secreted form some immunoglobulins can form polymers, where several monomer units (e.g. Fig.6.5) are joined together (Fig.6.6). For review see Nissonoff *et al.* (1975).

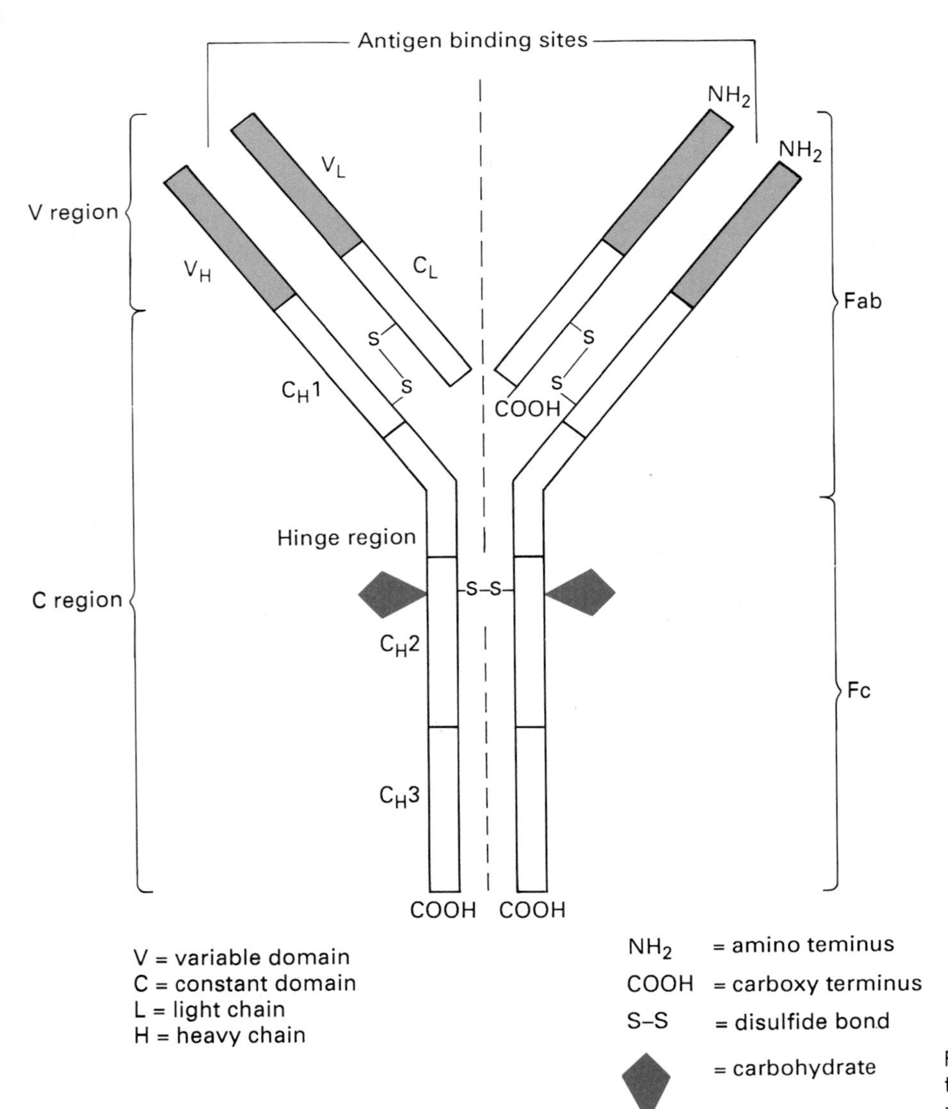

Fig.6.4 Schematic diagram to illustrate the structure of an antibody. An IgG molecule is shown

The structure of the immunoglobulin monomer can be divided into two parts with different functions

Studies investigating the susceptibility of antibodies to enzyme digestion revealed that the molecules could be divided into two different parts or **regions** (Porter 1959). The enzymes pepsin and papain were used for these studies and the pattern of fragmentation obtained with each enzyme is shown in Fig.6.7. The experiments demonstrated that:

1. One region of the immunoglobulin monomer is responsible for binding the antigen—the **Fab region** (**f**ragment with **a**ntigen **b**inding). Each monomer contains two identical Fab regions, and these can be separated from one another by digestion of the intact antibody molecule with papain.

The second region of the Ig monomer, produced by papain digestion, is called the **Fc region** (since the **f**ragment was readily **c**rystallized). This region mediates the effector functions of each immunoglobulin, such as binding complement components or Fc receptors on a variety of cell types (see Section 6.2.5).

2. If the enzyme pepsin is used, a fragment containing the two Fab regions **still** joined together is obtained. This is called the $\mathbf{F(ab')_2}$ **fragment**. The Fab regions remain joined together because pepsin cleaves the heavy chains on the other side of the inter-heavy chain disulfide to papain.

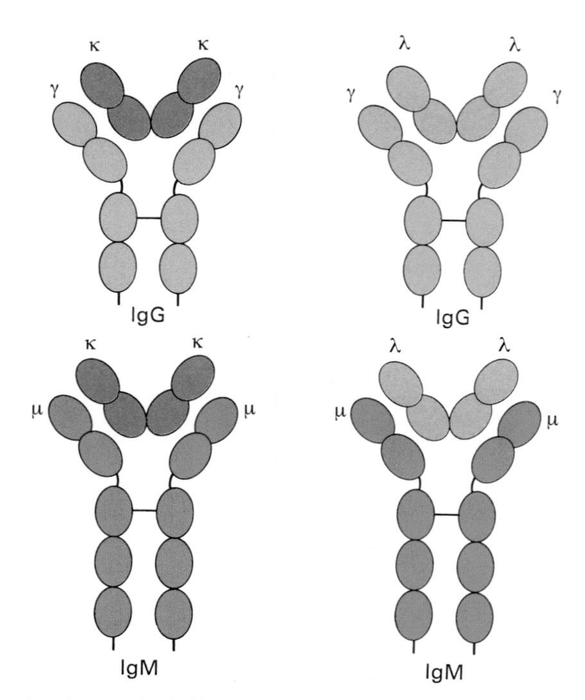

Fig.6.5 Basic structures of soluble monomeric IgG and IgM. The heavy chains of immunoglobulins (here, γ and μ respectively) can associate with two different types of light chain (κ, λ) to form the monomers

These early studies together with data obtained using electron microscopy led to the now classical model for the structure of the immunoglobulin monomer, where it is represented as a Y-shaped molecule with the two arms each representing a Fab region, and the stem the Fc region (Fig.6.4). They also demonstrated that each Fab region is composed of just part of one heavy chain linked to a complete light chain and that the interaction between the heavy chain and the light chain results in the formation of the antigen binding site at one end of the molecule. The Fc portion of the antibody contains the remainder of the two heavy chains.

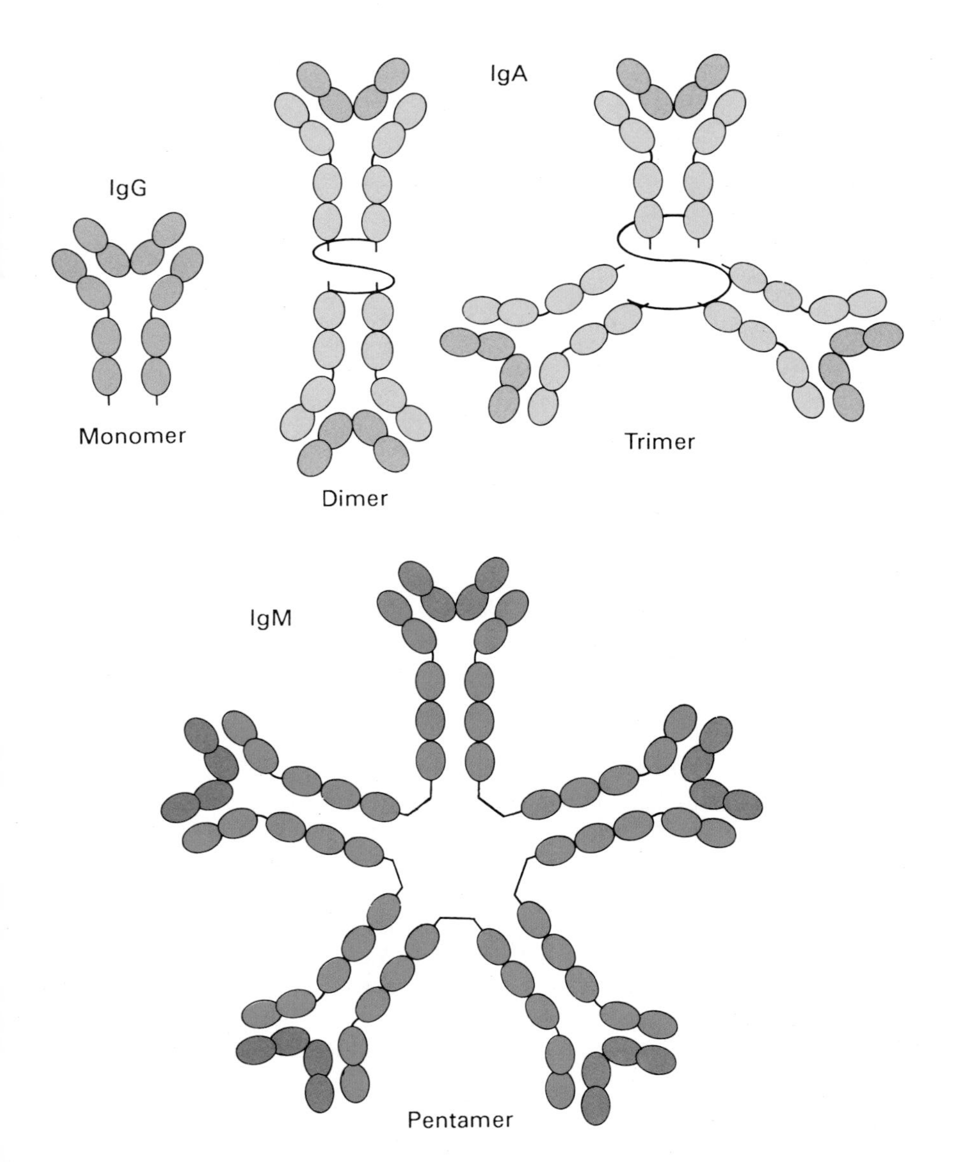

Fig.6.6 Soluble immunoglobulins can exist as monomers (IgG) or polymers–dimers, trimers (IgA) and pentamers (IgM)

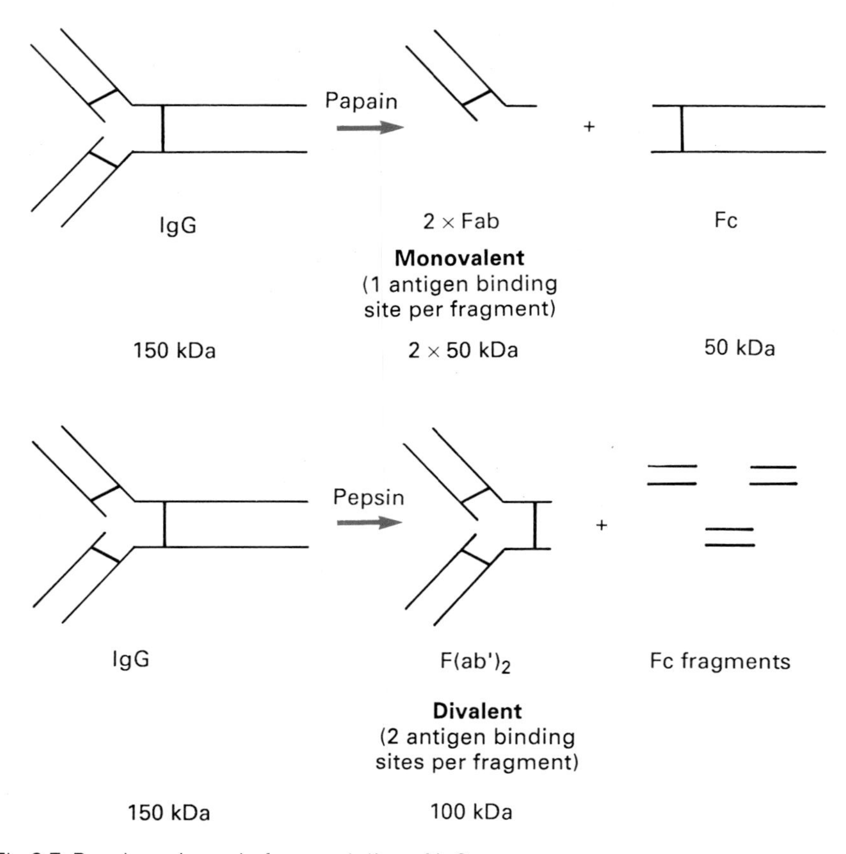

Fig.6.7 Papain and pepsin fragmentation of IgG

It became clear from these experiments, and later when the amino acid sequences of different **myeloma**† light chains were compared, that each immunoglobulin polypeptide could be divided into two distinct regions—**the variable** and the **constant regions**. The variable region is located at the N-terminal end of each chain and the constant region lies in the C-terminal portion (Fig.6.4).

A comparison of amino acid sequences revealed that the first 100 or so amino acids of each light chain were different from one another whereas the second 100 amino acids were more similar (Fig.6.8). The sequences could also be divided into two related groups, one for λ chains and the other for κ chains. Similar findings were also obtained when heavy chain amino acid sequences were compared. Thus a very large number of variable region sequences exist and these are associated with a much smaller number of constant region sequences. These data suggested that two pools of genetic information were required to synthesize immunoglobulin polypeptides and that an unusual genetic mechanism was involved to bring components from

†Myeloma—B cell tumour which gives rise to an overproduction of antibody often manifested by excretion of light chains in the urine, the Bence-Jones proteins (myeloma proteins). A source of large quantities of a homogeneous population of antibodies which proved invaluable for characterizing their structure.

Variable

	β-STRAND B ... C	E ... F
Ig Vλ	V T L T C R S S T G A V - T T S N Y A N W V Q Q K P - D H	- - N K A A L T I T G - A Q T E D E A I Y F C A L W Y
Ig V heavy	L S L T C T V S G S T F - - S N D Y Y T W V R Q P P - - G	- K N Q F S L R L S S - V T A A D T A V Y Y C A R N L
TcR Vα	T S L N C T F S D S A - - - - S Q Y F W W Y R Q H S - G K	- S L H F S L H I R D - S Q P S D S A L Y L C A V T L
TcR Vβ	V T L R C K P I S G - - - - H N S - L F W Y R Q T M M - -	N A S F S T L K I Q P - S E P R D S A V Y F C A S S F

Constant

Ig Cλ	A T L V C L I S D F Y P G A - - V T V A W K A D S S P - -	- N N K Y A A S S Y L S L T P - E Q W K S H R S Y S C Q V T H
Ig Cκ	A S V V C L L N N F Y P R E - - A K V Q W K V D N A L Q -	- D S T Y S L S S T L T L S K - A D Y E K H K V Y A C E V T H
Ig C heavy (I)	A A L G C L V K D Y F P E P - - V T V S W N S G A L T - -	- S G L Y S L S S V V T V - P - S S S L G T Q T Y I C N V N H
Ig C heavy (III)	V S L T C L V K G F Y P S D - - I A V E W E S N G Q P - -	- D G S F F L Y S K L T V D K - S R W Q Q G N V F S C S V M H
TcR Cβ	A T L V C L A T G F F P D H - - V E L S W W V N G K - - -	N D S R Y C L S S R L R V S A T F W Q N P R N H F R C Q V Q F
TcR Cγ	G T Y L C L L E K F F P D V - - I R V Y W K E K N G N - -	- K G T Y M K F S W L T V - P - E R A M G - K E H S C I V K H

Fig.6.8 Examples of variable and constant domain sequences from immunoglobulins and T cell receptor α and β chain genes. Amino acids are boxed where identity occurs (see Fig.1.3. for single-letter amino acid abbreviations)

each pool together. This is indeed the case and the mechanism for generating diversity in immunoglobulins will be discussed in Chapter 7.

Heavy and light chains are divided into different domains

Each intact polypeptide chain of the Ig monomer can be divided into a number of discrete structural units or **domains** as depicted in Figs.6.1 and 6.5. This domain structure was suggested because linear repeats of homologous units of amino acids were found within the polypeptide chain. In particular, the regular spacing of cysteine residues along the heavy and light chains, separated by about 60 amino acids, suggested that these might form intrachain disulphide bonds, resulting in loops in the polypeptide chain.

Each domain is roughly equal in size and contains approximately 110 amino acids. Light chains have two domains (Fig.6.5), while heavy chains have four or five domains, the number depending on the class of the antibody in question (Fig.6.5: for example, IgM has five domains whereas IgG has four).

The N-terminal domain of each polypeptide is called the **variable domain**, and is designated V_L for light chains and V_H for heavy chains (Fig.6.4). In the Ig monomer, the variable domain of one light chain is paired with the variable domain of a heavy chain to form the antigen binding site of the Fab region. This combination therefore represents the top half of one arm of the Y-shaped molecule (Fig.6.4).

All the other domains of the antibody chains are known as constant domains. Light chains have only one constant domain, designated C_L, attached to the V_L region. Heavy chains have three or four constant domains (depending on the class of antibody) attached to the V_H domain, designated C_H1, C_H2, C_H3 and C_H4 if present; Fig.6.4). In the Ig monomer the C_H1 domain of one heavy chain pairs with the C domain of one light chain and represents the bottom half of one arm of the Y-shaped molecule. The remaining constant region domains of the heavy chains interact with one another to form the Fc region (the stem of the Y).

Each domain has a similar 3D structure—the 'immunoglobulin fold'

The domains of the heavy and light chains have closely-related structures (Fig.6.9). If the amino acid sequences of constant domains from many different antibodies are compared, they show significant sequence similarities or homology. Each domain contains two highly-conserved cysteine residues which form the **intra-domain disulfide bond**—a feature of most immunoglobulin domains; amino acids at other positions in the domain are also conserved, such as a tryptophan (W) residue present in strand C located approximately 14 amino acids away from the first cysteine (C) residue (Fig.6.8). Moreover, the three-dimensional structure of each domain is very similar; the polypeptide chain forms what is called an **immunoglobulin fold** (Fig.6.9). The immunoglobulin fold is a very important structural feature that is found not only in immunoglobulins, but also in a large number of other molecules that are important for the function of the immune system or more generally for communication between cells. Molecules that share this structural feature are members of the immunoglobulin superfamily (see Section 7.4).

The structure of each immunoglobulin domain comprises two β-pleated sheets. These sheets interact with one another to form a globular structure, the core of which is hydrophobic. The two β-pleated sheets are held together by the intra-domain disulfide bond formed between the two conserved cysteine residues of the polypeptide chain referred to above (Fig.6.9).

In *constant* domains, one of the β-pleated sheets contains four anti-parallel strands while the other contains three. The *variable* domains are similar, but in this case both β-pleated sheets contain four anti-parallel strands. Joining each strand of a β-pleated sheet are loops of amino acids which take up a random structure. The amino acid sequence of the loops joining the β strands in the constant domains is less conserved than the sequences in the β strands themselves. In the variable domains, these loops contain the amino acid residues that form the antigen binding site. These are called **hypervariable** or **complementarity determining regions** of the variable domains (Fig.6.9). Very little α helix structure is found in the immunoglobulin molecule.

6.2.2 Classification

Immunoglobulins can be classified using antisera and antibodies

If an animal is immunized with an immunoglobulin from another species (xenoimmunization; see Appendix), the immune system will recognize it as just another foreign antigen. Alloantisera can also be prepared against certain immunoglobulins by immunizing other members of the same species (Fig.1.23; Appendix, p. 695). In this way it is possible to make antisera which can be used to characterize or classify the different forms of immunoglobulin present within each species. Using xenoantisera (e.g. rabbit anti-human Ig) one can define major species differences, while alloantisera (e.g. strain A anti-strain B Ig) can be used to detect polymorphisms that have arisen within the species. Monoclonal antibodies (see Appendix) have also been prepared against immunoglobulins. The molecular basis of the classification, outlined below, has been established by cloning the immunoglobulin genes themselves (see Section 7.2).

Class and subclass: heavy chains

Immunoglobulins can be divided into different classes and subclasses

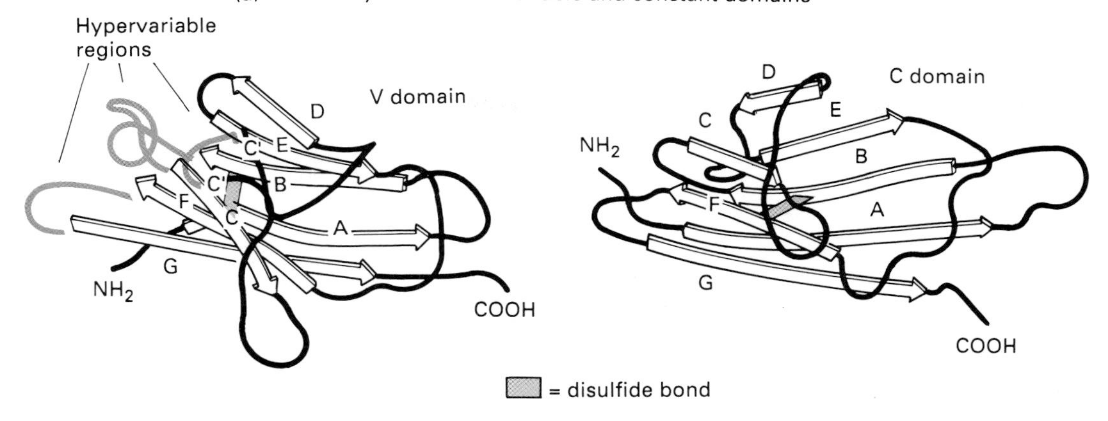

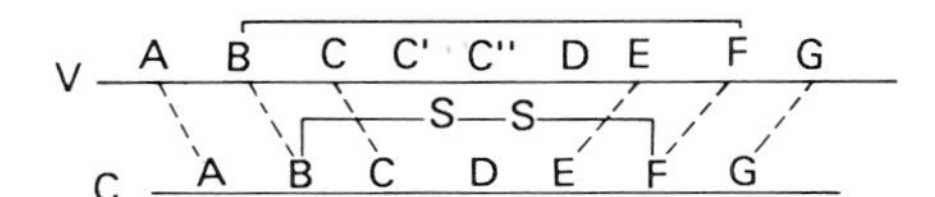

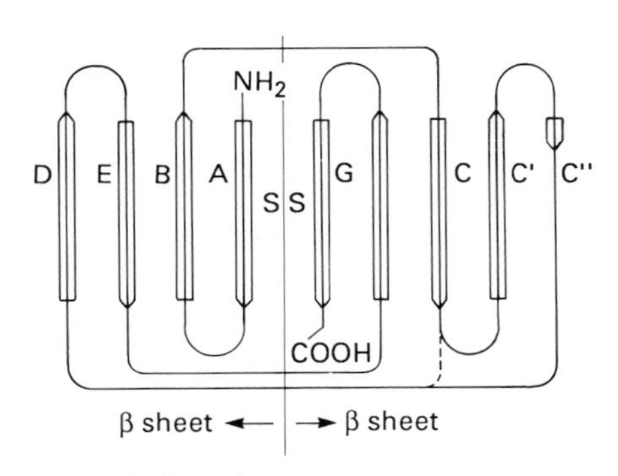

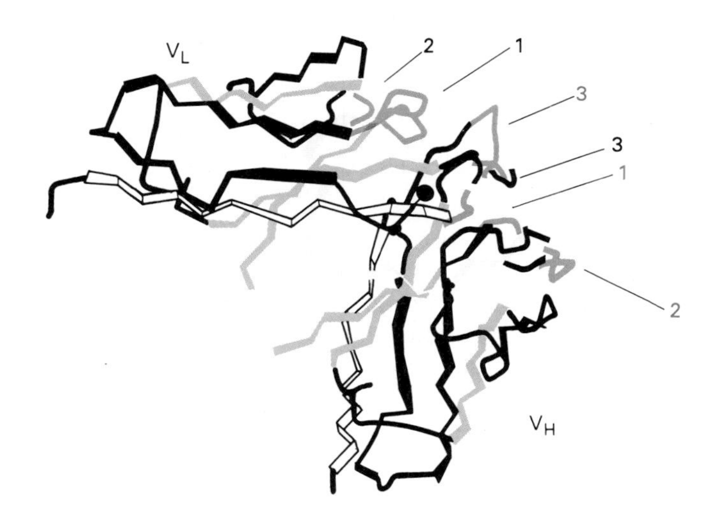

Fig.6.9 Polypeptide folding of immunoglobulin variable and constant domains. Parts (a)–(c) are adapted from Williams and Barclay (1988), *Annual Review of Immunology*, **6,** 381; © 1988 by Annual Reviews Inc. Part (d) is adapted from Winter and Milstein (1991), with permission

Table 6.2 Properties of human immunoglobulin heavy chains

Class of the heavy chain	Number of constant domains	Name of the immunoglobulin in which the heavy chain is present
μ	4	IgM
γ	3	IgG { IgG1, IgG2, IgG3, IgG4 }
α	3	IgA { IgA1, IgA2 }
ε	4	IgE
δ	3	IgD

depending on the antigenic determinants located on the heavy chain. Thus the class of an immunoglobulin can be determined solely on the basis of structural differences in the heavy chains.

There are five classes of heavy chain in mammals: μ, γ, α, ε, and δ. The class of the heavy chain determines which immunoglobulin is produced—μ=IgM; γ=IgG; α=IgA; ε=IgE; and δ=IgD (Table 6.2). Some classes are heterogeneous, and can be further divided into subclasses. Each subclass shares all of the structural determinants that define the immunoglobulin class, but in addition has extra determinants that result from small changes in the amino acid sequence of the heavy chain.

In human, for example, there are four subclasses of the γ heavy chain, called $\gamma1$, $\gamma2$, $\gamma3$, and $\gamma4$, and two subclasses of α chain, called $\alpha1$ and $\alpha2$. In mouse, γ heavy chains can also be divided into four subclasses, but in this species they are designated $\gamma1$, $\gamma2a$, $\gamma2b$, and $\gamma3$.

Type: light chains

Two **types** of light chain have been identified; these are called κ or λ. The type of light chain present in a particular antibody does *not* affect the class or subclass definition. For example, an IgG molecule by definition contains two identical γ heavy chains, but these may be linked to *either κ or λ* light chains. If it is necessary to discriminate between these, the type of light chain is stated in brackets thus: IgG(κ); IgG(λ) (Fig.6.5).

In human and mouse, λ light chains can be divided into subtypes on the basis of sequence similarities (for an explanation of the molecular basis for this see Section 7.2.1, Fig.7.3). Some of the properties of human light chains are given in Table 6.3.

Isotypes

The genes for every class and subclass of heavy chain, and for both types and all of the subtypes of light chains, are present in every normal individual of the species. Each of these distinct forms is known as an **isotypic variant**, or **isotype**, and is encoded by a separate gene that is present in every member of the species. The isotype of an immunoglobulin chain is synonymous with

Table 6.3 Properties of human immunoglobulin light chains

Types of light chain	K	λ
Molecular weight (M_r, kDa)	23	23
Total number of domains	2	2
Variable domains	1	1
Constant domains	1	1
Allotypic markers	Yes Km allotypes	No
Subtypes	No	Yes
Relative abundance	2	1

the class and subclass of heavy chain, and the type of light chain (Figs.6.10 and 6.11). Models of the structure of human immunoglobulins were reviewed by Pumphrey (1986*a* and *b*).

Allotypes

Within the human population there are allelic variants of some of the immunoglobulin polypeptide chains, just as for MHC molecules (see Sections 2.2 and 2.5). These allelic forms are called **allotypes** (Fig.6.10). Unlike isotypes, which are a feature of every individual, any individual has either one (homozygote) or two (heterozygote) of the allotypic forms of a particular immunoglobulin chain (Fig.6.11). Because a given allotype is expressed by only some members of the species, it can be defined by using alloantisera. (Of course, alloantisera cannot be used to define isotypic variants above, since these are present in all members of the species, and therefore are nonimmunogenic when transferred between members of the same species). In the human, allotypic variants have been defined for γ and $\alpha 2$ heavy chains, and for κ light chains, and these are known as the Gm, Am, and Km allotypes, respectively (e.g. Fig.6.10).

The molecular basis for the different allelic variants is known. They usually result from amino acid substitutions in the constant domains of the light or the heavy chains (Fig.6.10 and Tables 6.4, 6.5).

For example, the Km allotypes arise as a result of amino acid substitutions in the constant domain of the κ light chain at amino acid residues 153 and 191 (Table 6.4). These amino acid substitutions are conservative, in other words the substituted amino acids are of a similar size and charge to the original. The Km allotypes can be defined using alloantisera, suggesting that either the amino acids 153 and 191 are on the outer, exposed surface of the constant domain or that their substitution generates a new antigenic determinant. Their location to the outer surface of the constant domain has been confirmed by X-ray crystallography.

The molecular basis of the Gm allotypes of human IgG is also known and is shown in Table 6.5. These allotypic variants also result from amino acid substitutions in the constant region of the γ heavy chain.

Two allotypic forms of the $\alpha 2$ heavy chains have been identified in human, designated A2m(1) and A2m(2). The two allotypes differ by six amino acids in the constant domains.

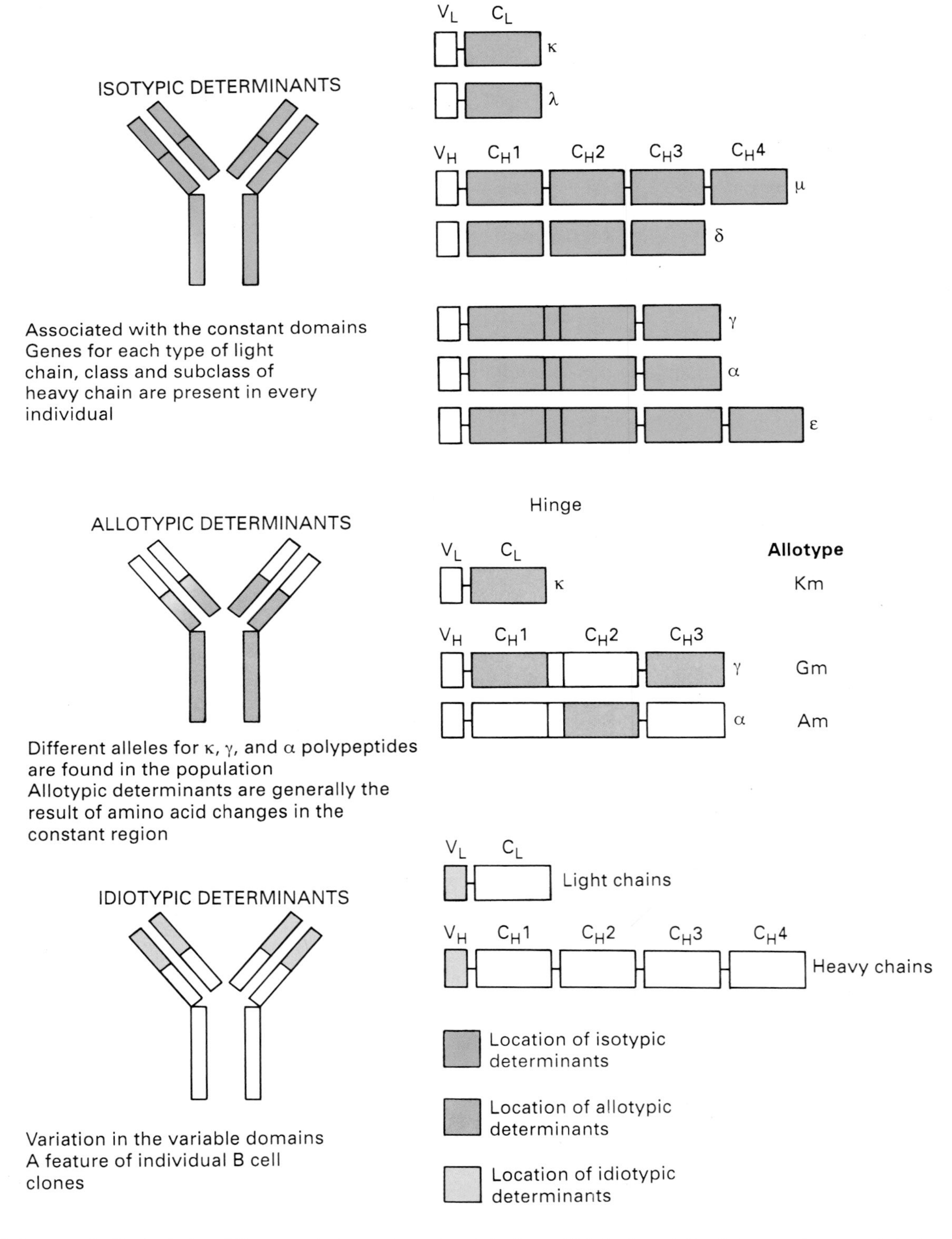

Fig.6.10 Diagram to illustrate the differences between isotypic, allotypic, and idiotypic variation in human immunoglobulins

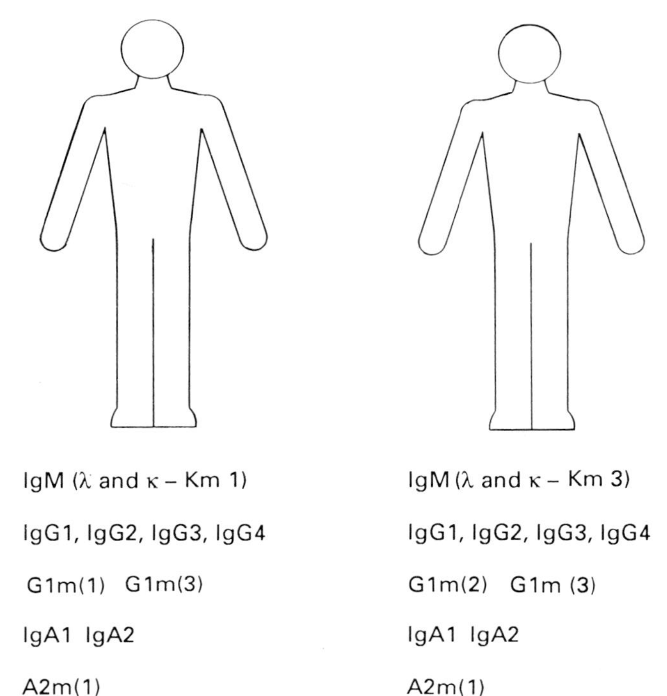

These two individuals are homozygous for Am allotypes, but heterozygous for Km and G1m allotypes. They are identical for the Am allotype A2m(1) but have different Gm and Km allotypes.

Fig.6.11 Isotypes and allotypes

Allotypes are inherited in a mendelian fashion. For example, the IgG1 molecules from one individual may carry the G1m(1) and G1m(3) allotypes (one from the maternal and one from the paternal genes) while IgG1 molecules from another may carry the G1m(2) and G1m(17) allotypes (see Fig.6.10 for a similar example). [It should be noted that any *given* Ig molecule from the first individual will carry either G1m(1) or G1m(3) but not both because of allelic exclusion—see Section 8.2.2]. In rabbits allotypic determinants are associated with the variable domains.

Table 6.4 Km allotypes of the κ light chain

Km allotype	Amino acid position	
	153	191
Kml	Val (V)	Leu (L)
Kml, 2	Ala (A)	Leu (L)
Km3	Ala (A)	Val (V)

Table 6.5 Amino acid changes associated with the Gm allotypes of human IgG

Locus	Chain and domain location	Allotype	Amino acid residues involved
G1m	γ1–C_H1	G1m (4)	Arg 214
		G1m (17)	Lys 214
	γ1—C_H3	G1m (1)	Asp 356, Leu 358
		G1m (−1)	Glu 356, Met 358
G2m	γ2—C_H2	G2m (23)	ND
G3m	γ3—C_H2	G3m (21)	Tyr 296
	γ3—C_H3	G3m (11)	Phe 436
G4m	γ4—C_H2	G4m (4a)	Leu 309
		G4m (4b)	Gap at posn. 309

Idiotypes

Idiotypes are also antigenic determinants present on immunoglobulins that are frequently associated with the antigen binding specificity of each molecule (Fig.6.10). These determinants arise as a result of particular amino acid sequences in the variable regions of the heavy and/or light chains and are usually characteristic of the antibodies made by a particular B cell clone. Each idiotype is usually restricted to one or a few antibodies within one individual. The 3D structure of an idiotope-anti-idiotope complex has been obtained (Bentley *et al.* 1990). This topic is discussed in Section 11.1.

6.2.3 Fine structure—hypervariable or complementarity determing regions

When the amino acid sequences of the variable domains of antibodies with different antigen specificities are compared, it is clear that the variable amino acids are neither randomly nor uniformly distributed throughout the variable domain, but are clustered in three or four particular, relatively small, areas known as the **hypervariable regions** or **complementarity determining regions** (CDRs) (Fig.6.12). The outline of the three-dimensional structure of immunoglobulin domains—the immunoglobulin fold—has already been described (Section 6.2.1). The CDRs are associated with the loops of polypeptide that join the strands of polypeptide chain forming the two β-pleated sheets of the immunoglobulin fold (Fig.6.9).

From X-ray crystallographic studies of antibody-antigen complexes, it is clear that the sites of contact between the antibody and antigen can involve amino acid residues in the CDRs and that these are located at the extremities of the Fab arms (see Fig.6.13(c) and Section 6.2.4). Furthermore, genetic and protein engineering techniques have recently been used to 'transplant' the three heavy chain CDRs from a mouse monoclonal antibody of known antigen specificity into the framework of a human antibody. After the CDR 'transplant' the human antibody acquired the same antigen binding specificity as the mouse monoclonal. This proved beyond doubt that the CDRs interact with the antigen (Jones *et al.* 1986).

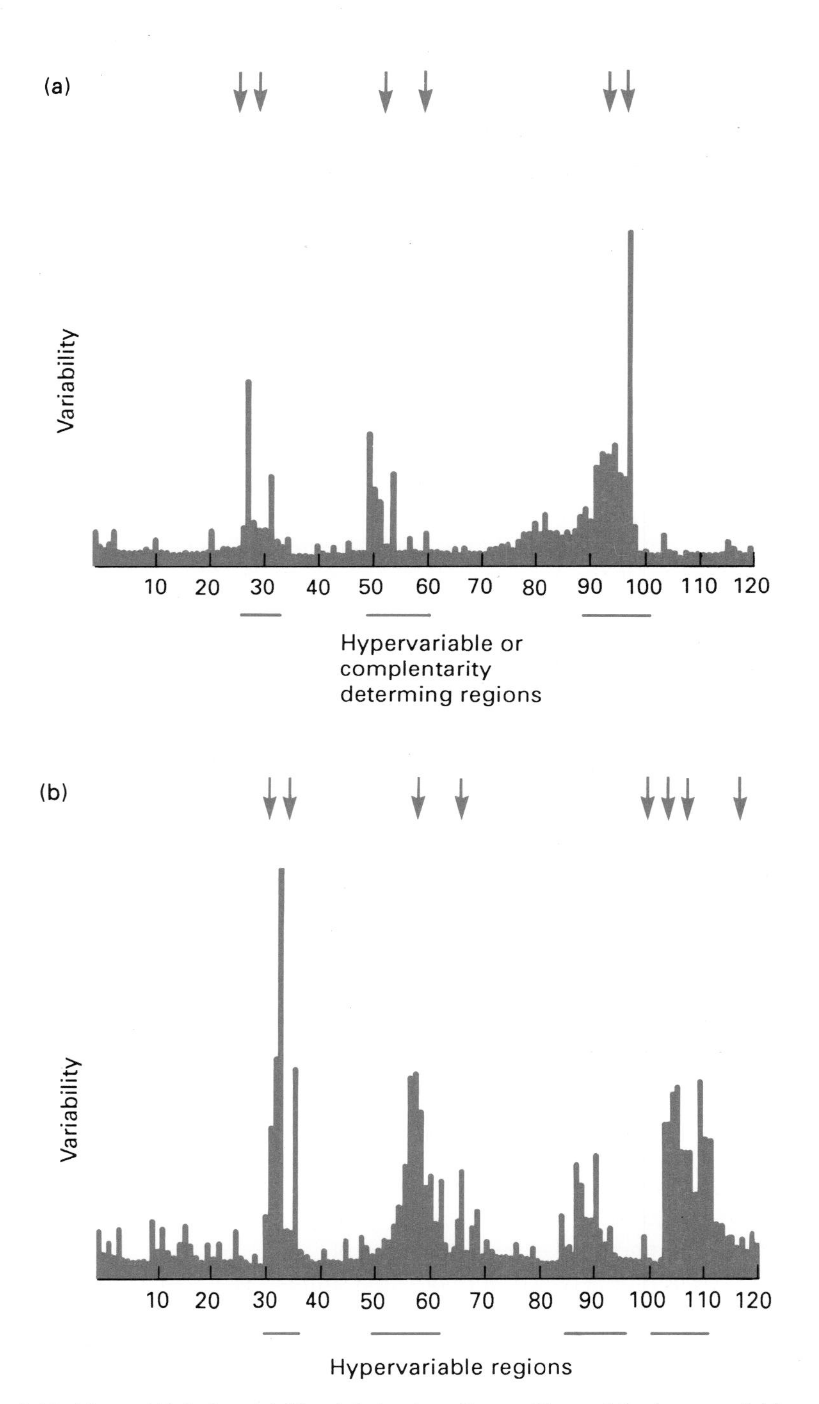

Fig.6.12 Wu and Kabat variability plots to show the positions of the hypervariable residues in immunoglobulin variable (V) domains of light chains (a) and heavy chains (b). Arrows indicate positions at which affinity labels have been localized (see Fig.6.14)

The concept of hypervariable regions was introduced by Wu and Kabat (1970), who compared large numbers of amino acid sequences from the variable domains of myeloma proteins (refer to footnote, p. 338). A parameter termed 'variability' was defined as the number of different amino acid residues found at any one position in the polypeptide chain, divided by the frequency of the most common amino acid at that position. If variability is plotted against amino acid position, a (Wu–Kabat) variability plot can be obtained (Fig.6.12). For light chains, amino acids 29–35, 51–65, and 93–103 comprise the CDRs (Fig.6.12(a)) and for heavy chains, CDRs include amino acids 31–37, 51–68, 84–91, and 101–110 (Fig.6.12(b)).

6.2.4 Functions

Membrane immunoglobulins bind antigen, leading to internalization, antigen processing and expression of antigenic peptides in association with the MHC molecules of the B cell. These peptide-MHC complexes can be recognized specifically by activated helper T cells, which can then provide 'help' for B cell responses by secreting cytokines (see Sections 4.3.1 and 8.4.3).

Secreted immunoglobulins have two main functions: (i) to bind antigen and (ii) to trigger effector mechanisms as a result of (i). These two functions are associated with the two regions of the molecule. The antigen binding site is formed by interaction of the variable domains of the light and heavy chains (Fig.6.4), and the constant regions are important for the effector functions.

Antigen binding

The binding of an antibody to its antigen results from a specific interaction between the **antigen binding site** (also known as the **antibody combining site**) and the antigenic determinant or epitope of the antigen that elicited the immune response (Fig.6.13).

The nature of this interaction has been the subject of intense investigation, (Absolom 1986) and has been shown to result from interactions between functional groups (e.g. amino acid or carbohydrate) on the surface of the antigen and particular amino acids in the CDRs of the heavy and light chains that form the antigen binding site (Fig.6.13(c)). No covalent bonds are formed or broken when an antibody binds antigen. The types of forces or bonds involved are all noncovalent, and include hydrophobic bonds, hydrogen bonds, dipole-induced dipole and electrostatic interactions. These forces all act over very short distances, in the order of a few nanometres (For scale refer to Fig.1.1), and this, together with the fact that multiple bonds are involved, can result in high affinity binding. The high affinity of antibody-antigen interaction is important because if the antibody forms a stable complex with the antigen, its elimination from the body will be more efficient. High affinity binding between antibody and antigen is essential for the triggering of many of the effector mechanisms of the humoral arm of the immune response.

Immunologists have asked a number of questions about the properties of the antigen binding site and the precise nature of the interactions between antibody and antigen, including (i) whether the size of the binding site is similar between antibodies of different specificities, (ii) whether the antigen

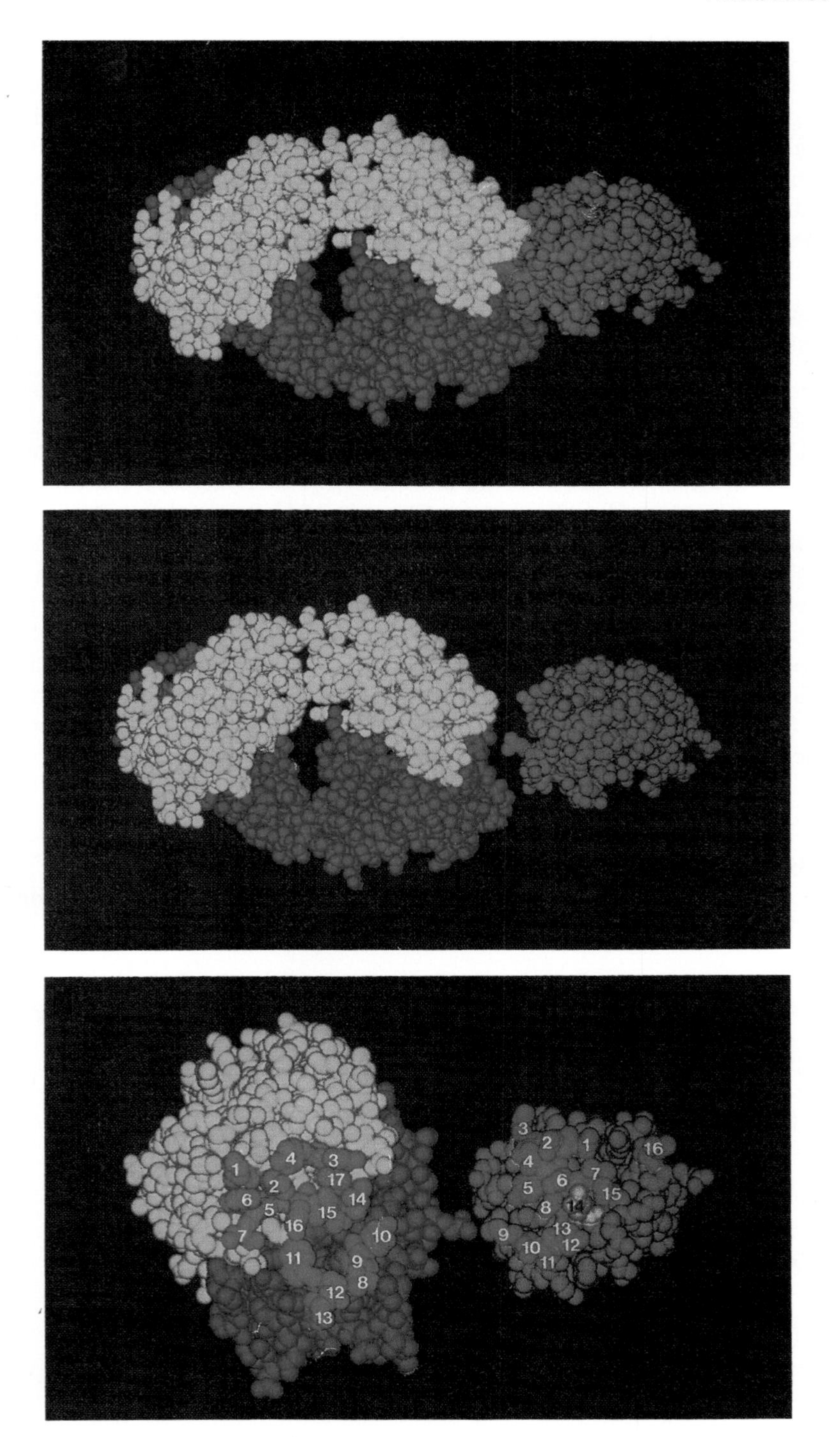

Fig.6.13 Antibody–antigen binding. (A) The crystallographic structure of the lysozyme-anti-lysozyme Fab complex; Gln121, an exposed residue, is shown in red, heavy chain in blue, light chain in yellow, and lysozyme in green. In (B) the Fab and lysozyme have been separated to show the glutamine residue. In (C) the contact residues in the Fab and lysozyme are numbered; Gln121 is shown in mauve. (From Amit *et al.* (1986), with permission; © 1986 by the AAAS)

binds to molecular clefts in the antigen binding site or through surface interactions, and (iii) if there are conformational changes in the antibody and/or antigen as a result of the interaction. The answers to these questions have been emerging over many years as a result of the application of a variety of techniques to this problem. In recent years X-ray crystallographic analysis of antibody–antigen complexes has refined the data considerably, and data from these studies will be considered in most detail here.

Specificity of antibody–antigen interaction. Haptens have been used to demonstrate antigen specificity.

For example, antibodies can be prepared to the hapten para-azobenzene arsonate (ABA) (Table 6.6), when it is coupled to a protein carrier, bovine IgG, through the para determinant. The specificity of the antibodies produced can then be tested using different ABA derivatives, *ortho-*, *meta-*, and *para*-methyl-ABA (Table 6.6). This approach allowed the structural features of the hapten required for antibody binding to be determined. Antibody binding of highest affinity was found with the para form of methyl-ABA, in other words the form of the hapten that mimics the immunizing antigen (Table 6.6). Substitution of the methyl group elsewhere in the benzene ring, at either the meta or ortho positions, resulted in much lower antibody binding affinity, presumably because substitution at these other positions interferes with the way the hapten interacts with the antigen binding site.

Table 6.6 Experimental data to illustrate the specificity of anti-hapten antibodies. Immunization was carried out using para substituted benzene arsonate

Test antigen	Structure	Binding of the anti-hapten antibody
Benzene arsonate	Ars	+
Ortho-methyl benzene arsonate	Ars, Me	−
Meta-methyl benzene arsonate	Ars, Me	+/−
Para-methyl benzene arsonate	Ars, Me	+ + +

[Data taken from Pressman *et al.* 1954]

The data demonstrate that the principal specificity of the antibodies synthesized in response to a particular antigen is generally for the form of the antigen or hapten used for the immunization. In some cases weak cross-reactions with related antigens or haptens can occur, but in general these bind with a lower affinity.

Amino acids in the CDRs interact with the antigen. Various techniques have been used to examine which particular residues are involved in antibody–antigen binding. The data obtained using the techniques of affinity labelling, X-ray crystallography, and sequence studies, will be discussed here. In the past, the interaction of antibodies or Fab fragments with small molecules, such as haptens, was investigated because the characterization of the interaction of antibodies with macromolecular antigens was much more difficult. The three-dimensional structure of the complexes formed between lysozyme and a Fab fragment of an anti-lysozyme monoclonal antibody (Fig.6.13) (Amit *et al.* 1986), and between the influenza virus neuraminidase glycoprotein and antibody (Colman *et al.* 1987*a* and *b*) have been determined. These have allowed a much more detailed analysis of the nature of the antibody–antigen interaction (see later in this section). More recently, the crystal structure of an antibody–peptide complex has also been solved (Stanfield *et al.* 1990).

Affinity labelling. Affinity labelling was one of the first techniques used to identify the amino acid residues present in the antibody combining site. The principle of this technique is illustrated in Fig.6.14 (see also Fig.6.12). A chemically-reactive hapten, or analogue of the hapten is prepared. Affinity labels can be prepared in various ways, but the most useful are photo-affinity reagents which remain inert until activated by UV light. To analyse the amino acid residues associated with the antigen binding site, the affinity labelled hapten is allowed to react with the antibody. Once binding occurs, the reactive hapten becomes covalently bound to the residue(s) in contact with it. The precise site of attachment of the hapten to the antibody can then be determined by separating the heavy and light chains, preparing peptides containing the labelled hapten by enzyme cleavage, and sequencing the amino acids in the peptides (Fig.6.14).

The results obtained using this approach showed that, for those antibodies studied, both the heavy and the light chains were involved in binding antigen, because amino acids from the variable domains of both the heavy and the light chains were found to be associated with the labelled hapten. Moreover, in some of these experiments the labelled amino acid residues were very close to, or within, the hypervariable regions or CDRs of the heavy and light chains (Fig.6.12). This provided evidence that these regions were likely to be part of the antigen combining site long before any structural data were available (for review see Givol 1974).

X-ray diffraction studies. The interaction between antibody and antigen can be studied in crystals by X-ray crystallography (for reviews see Poljack 1975; Amzel and Poljack 1979). Until recently the majority of information was obtained by examining Fab fragments, although a limited amount of information was also available for intact immunoglobulins. Using a human myeloma protein (see footnote, p. 338) called Kol, the structure of the antibody combining site determined for crystals of the Fab fragment was found to be very similar to that obtained later for the intact immunoglobulin. This suggested that X-ray crystallography of Fab fragments was a valid approach for investigating the structure of the whole molecule and for modelling the interaction between antibody and antigen. However, this approach is obviously not ideal, and much more information can be obtained by analysing

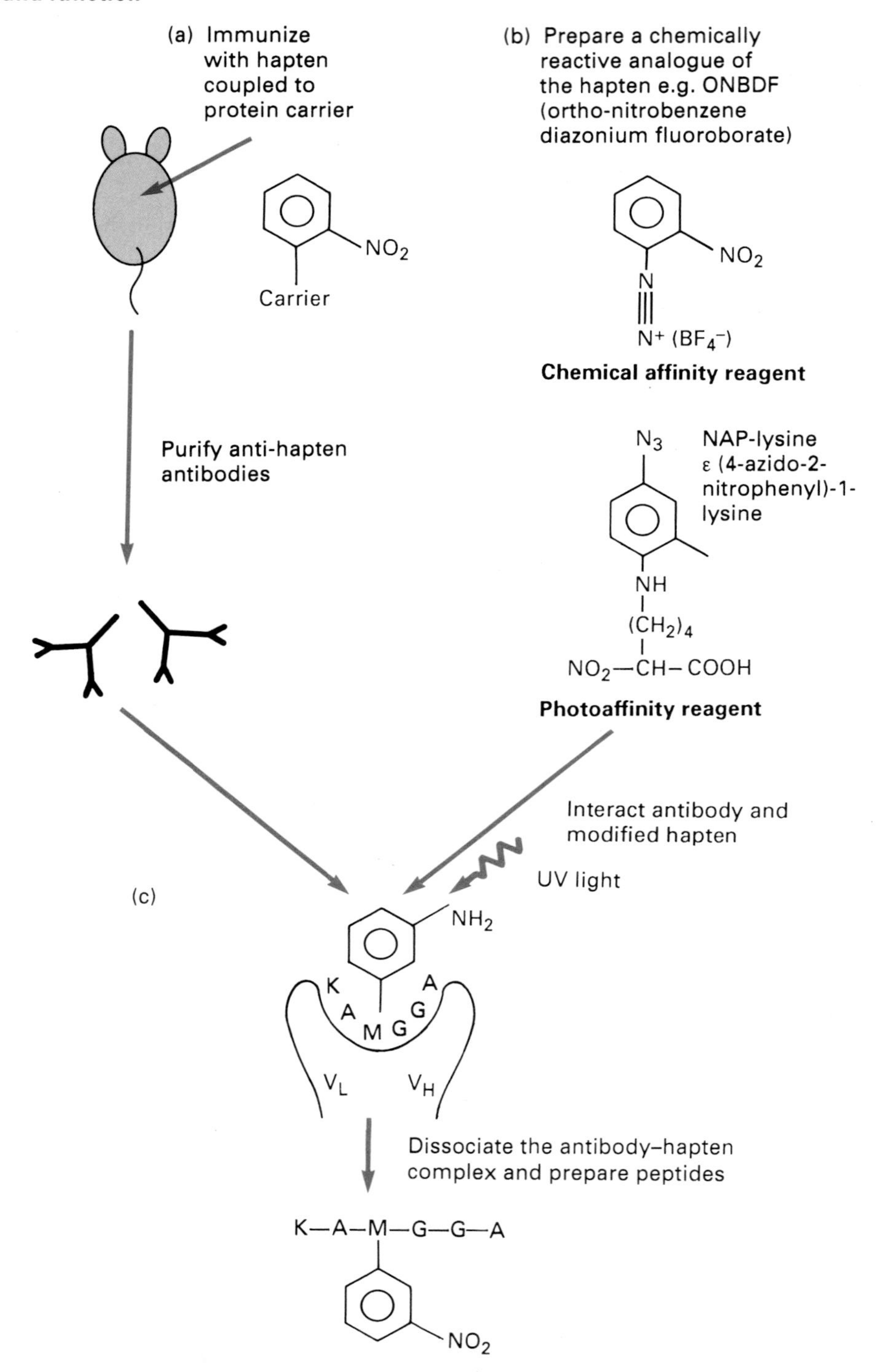

Fig.6.14 Affinity labels were originally used to identify amino acid residues involved in antibody–antigen interactions

crystals containing antibody–antigen complexes. Two examples where this has been achieved will be discussed briefly below, the lysozyme–anti-lysozyme and the influenza neuraminidase–anti-neuraminidase immune complexes (Figs.6.13 and 6.15).

Fab structure. The pairing of the V_L and V_H domains is determined by the amino acids present in both the framework regions and in the CDRs. Most of the amino acids buried at the interface of the two domains are conserved and this accounts for the very similar geometry observed between the various crystal structures examined so far. The small differences in the overall V_H–V_L structure found in different antibodies is mainly due to differences in their CDRs.

If the structure of a typical combining site is viewed from the direction of the antigen as it binds, the first and second CDRs (CDR1 and CDR2) of the heavy and light chain are spatially separate and the space between is filled with CDR3 from each chain (Fig.6.16). This view of the antigen binding site emphasizes the significance of CDR3 in determining the specificity of an immunoglobulin for its antigen.

A comparison of Fab structures in the presence or absence of antigen reveals that the Fab region of the antibody in some, *but not all*, immune complexes may undergo conformational changes when antigen binds. For example, analysis of the neuraminidase–anti-neuraminidase antibody complex (Fig.6.15) showed that there is a significant change in the structure when the neuraminidase binds to the antibody. The distance between the CDR2 of the heavy chain and CDR3 of the light chain is much larger, and the relative positions of the CDRs around the antigen binding surface are altered when the antigen is bound (Fig.6.15), compared to the Fab fragment in the absence of antigen. Data obtained using other techniques, for example

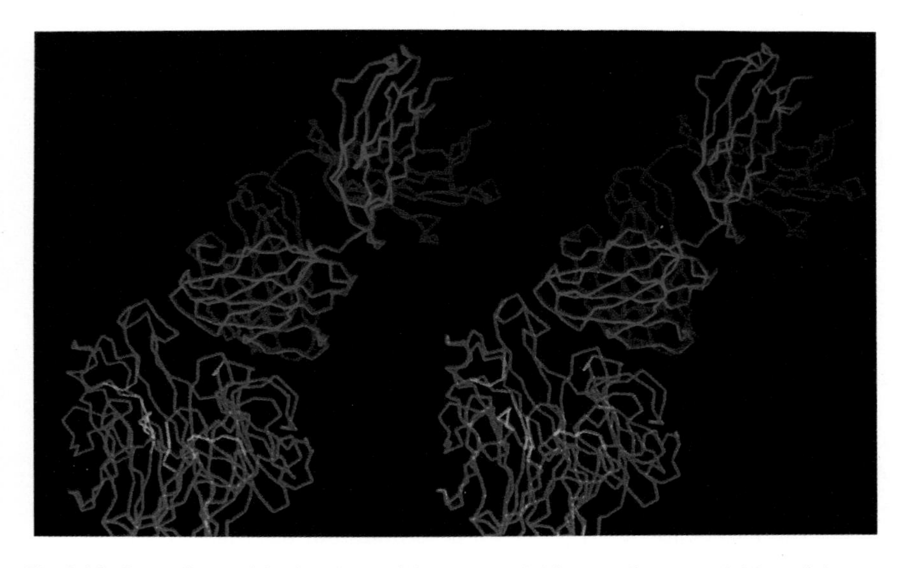

Fig.6.15 Crystallographic structure of the neuraminidase-anti-neuraminidase-Fab complex. Heavy chain is in purple, light chain in blue, and neuraminidase in green (active site in yellow). (From Colman *et al.* (1987*b*), with permission)

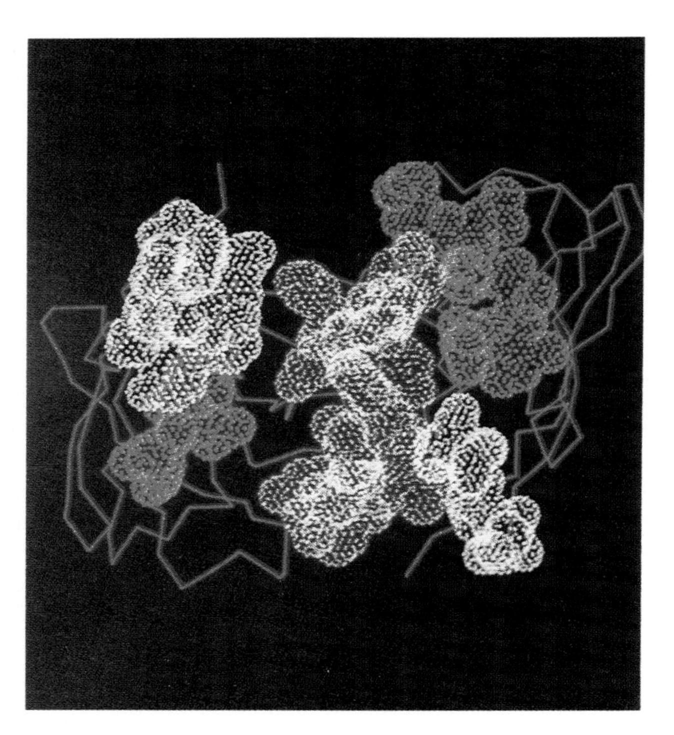

Fig.6.16 An immunoglobulin combining site viewed from the position of the antigen (CDR1, blue; CDR2, yellow; CDR3, pink). (From Davis and Bjorkman (1988), with permission)

circular dichroism, also support the idea that some antibodies can undergo a conformational change when they bind antigen. The biological significance of the changes in the V_H–V_L orientation that occur during the interaction is unknown, but it has been suggested they facilitate the formation of functional immune complexes. Interestingly in the neuraminidase–anti-neuraminidase complex the structure of the antigen is *also* distorted when it interacts with the antibody. The binding event has been described as a 'handshake', with both components involved adapting to each other to some extent.

In contrast, in the lysozyme–anti-lysozyme complex the arrangement of the V_H and V_L domains is not changed by the presence of antigen (Fig.6.13). This conclusion is supported by nuclear magnetic resonance data for other antibody–hapten interactions. Thus conformational changes *may* occur when antigen and antibody interact, but there are no absolute rules, and each situation will depend upon the properties of the antibody and the antigen involved.

In the lysozyme–anti-lysozyme complex the region of contact between the antibody and antigen is about 30 × 20 Å. The contact areas are complementary in shape and all six hypervariable regions of the antibody are involved, although the majority of the contacts with the antigen are by CDR3 of the heavy chain (Table 6.7). It is interesting to note that the third hypervariable loop, which significantly influences antigen binding in this complex, corresponds to the D gene segment which has a major influence on the generation of diversity of the heavy chains (see Section 7.2). Some of the structural features of the neuraminidase–Fab complex are similar to those described for the lysozyme–anti-lysozyme complex. The area of contact between neuraminidase and the antibody is equivalent, with a similar number of interactions involved, and all six of the CDRs are implicated.

Sequence and binding studies. The number of three-dimensional structures available for antigen–antibody complexes is very limited at present, so other methods are also used to investigate interactions occuring between an antibody and its antigen. For example, sequence comparisons of the variable regions of different antibodies that all interact with the same antigen can be informative.

This analysis was performed using heavy chains purified from 19 antibodies, all of which were specific for the hapten phosphocholine (Fig.8.10). Of these, ten had identical heavy chain variable domain amino acid sequences. The remainder differed by between one and eight amino acids. Two particular heavy chain residues, tyrosine-33 and arginine-52, were present in all molecules, and have been shown by other techniques to be in contact with the phosphocholine. Interestingly, when the light chain sequences of these 19 antibodies were compared, they were found to be

Table 6.7 Amino acid residues in contact in the lysozyme–anti-lysozyme complex

	Antibody residues	Lysozyme residues
Light chain		
CDR1	His 30	Leu 129
	Tyr 32	Leu 25, Gln 121, Ile 124
FR2	Tyr 49	Gly 22
CDR2	Tyr 50	Asp 18, Asn 19, Leu 25
CDR3	Phe 91	Gln 121
	Trp 92	Gln 121, Ile 124
	Ser 93	Gln 121
Heavy chain		
FR1	Thr 30	Lys 116, Gly 117
CDR1	Gly 31	Lys 116, Gly 117
	Tyr 32	Lys 116, Gly 117
CDR2	Trp 52	Gly 117, Thr 118, Asp 119
	Gly 53	Gly 117
	Asp 54	Gly 117
CDR3	Arg 99	Arg 21, Gly 22, Tyr 23
	Asp 100	Gly 22, Tyr 23, Ser 24, Asn 27
	Tyr 101	Thr 118, Asp 119, Val 120, Gln 121
	Arg 102	Asn 19, Gly 22

CDR = complementarity determing region
FR = framework region

markedly different. However, residues tyrosine-94 and leucine-96 were constant throughout, and are known to be in contact with the antigen. The data support the idea that for all of the antibodies examined, the molecular features of the antigen binding site for phosphocholine are very similar, even though for other parts of the variable domain they may be different (for review see Davies and Metzger 1983).

Antibodies specific for lysozyme and myoglobin have been examined in a similar way, by using variants of these well-characterized antigens from different species to determine which amino acids from the antigen are involved in binding.

For example, a monoclonal antibody specific for sperm whale myoglobin was prepared and the ability of myoglobins from different species to inhibit its binding to sperm whale myoglobin examined (Berzofsky *et al.* 1982). In this study only myoglobins which were similar to sperm whale myoglobin at amino acids 83, 144, and 145 inhibited binding (Fig.6.17), suggesting that these amino acids are involved in the antibody–antigen interaction.

Effector functions

The effector functions of immunoglobulins are those associated with inactivation and/or removal of foreign antigens from the body. They generally occur after the antibody has interacted with its specific antigen (see above).

When antibody binds to a membrane antigen, the cell membrane can become damaged, either as a result of complement activation (see Section 9.4) or antibody dependent cellular cytotoxicity (ADCC—see Section 10.2.7). The interaction of antibodies with soluble antigens can also provoke a vigorous response, which includes complement activation. In this case, tissue damage is usually limited because the immune complexes are solubilized.

When antibodies interact with soluble antigens a lattice is usually formed. A high degree of lattice formation leads to the generation of large immune complexes which eventually become insoluble and precipitate. In the laboratory this property is exploited for the characterization of new antigens, a technique known as immunoprecipitation (see Appendix). However, *in vivo* precipitation of immune complexes can have very harmful effects. Lesser degrees of lattice formation produce immune complexes which remain soluble. This usually occurs when excess antigen is present (it can also result under conditions of antibody excess), but can also arise when a low density of antigenic determinants is present on the antigen or when the affinity of the antibody is low.

The ability of immune complexes to cause disease, their immunopathogenic potential, is inversely related to the rate at which they are cleared from the circulation. The clearance rate depends upon a number of factors including: (i) their composition; (ii) their ability to interact with phagocytic cells; and (iii) their ability to activate complement. Soluble complexes remaining in the circulation are eventually deposited on capillary endothelium where tissue damage occurs as a result of the activation of the other plasma enzyme systems, including the coagulation pathway, fibrinolysis, and kinin formation (see Section 9.3). If the immune complexes are insoluble the tissue damage will usually be localized to the site where they are deposited.

Experimentally, the injection of a soluble antigen into an animal with high levels of circulating IgG induces localized cell death or necrosis. This is due to the formation of insoluble immune complexes and is known as the **Arthus reaction** (type III hypersensitivity; Section 10.5.3). The local tissue damage occurs because of the attraction of leukocytes, especially polymorphs (see Fig.1.11) to the site of injection. These engulf and digest the immune complexes, but in doing so contribute to the tissue damage.

The class and subclass of the immunoglobulin, characteristics dictated by the heavy chain constant regions, determine which effector mechanisms are triggered (Morgan and Weigle 1987). These include complement activation (see Section 9.4), Fc receptor binding (see Sections 10.2.7 and 10.4.5), elimination of allergens, and histamine release from mast cells. All of these mechanisms are involved in the elimination of foreign antigens. Denaturation of the antibody molecule nearly always results in loss of effector activity, implying that the quaternary structure of the molecule is extremely important.

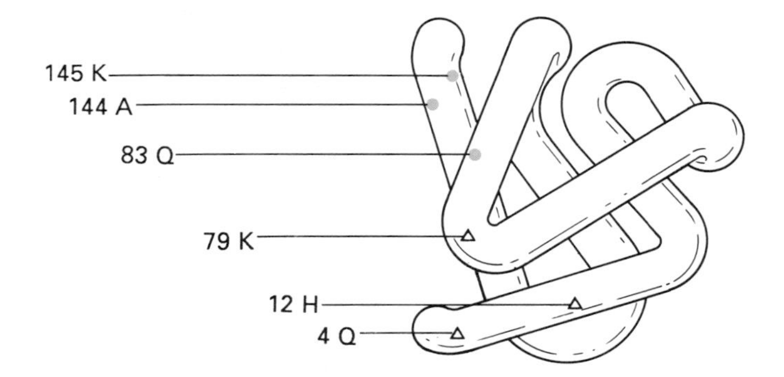

Myoglobin	Amino acid residue														Inhibition
	83	86	88	91	109	110	132	140	142	144	145	147	148	151	152	
Sperm whale	E	L	P	Q	E	A	N	K	I	A	K	K	E	Y	Q	Yes
Dwarf sperm whale							S									Yes
Goosebeaked whale					D		T							F	H	Yes
Killer whale	D													F	H	No
Sea lion	D						K	N				R		F		No
Human	I					C			M	S	N			F		No
Ox		V	H	E	D		S	N	A	E			V	F	H	No
Sheep		V	H	E	D		S	N	M		Q		V	F		No

Fig.6.17 Identification of conformational epitopes in myoglobin. Comparison of the amino acid sequences of myoglobin from different species with sperm whale myoglobin. Only the positions at which the sequences are different are shown for clarity. Myoglobins which differ at positions 83, 144 or 145 (●) did not inhibit binding of a monoclonal antibody specific for sperm whale myoglobin to its target antigen, implying that the monoclonal antibody recognized this epitope. (Adapted with permission from Owen and Lamb (1988) *Immune recognition*, IRL Press; based on data of Benjamin *et al.* (1984))

The constant region of the heavy chain, and consequently the class of the antibody, can change during the course of an immune response (although the antigen specificity of the antibody is unchanged) (Fig.6.18). This process is known as **class switching** (see Sections 7.2 and 8.2) and as a result, the effector functions performed by the antibody will also change (see Section 6.2.5). A summary of the effector functions performed by each class of immunoglobulin is listed in Table 6.8. The generation of class switch variants *in vitro* by manipulation of a single hybridoma clone has allowed the functional activity of different classes of antibody with identical antigen specificities to be compared (Kipps *et al.* 1985; e.g. Waldor *et al.* 1987).

Attempts have also been made to associate a particular effector mechanism with individual immunoglobulin domains. Because each domain interacts with others, most functions that are predominantly associated with one domain require the presence of other domains in the molecule for maximum activity. Experiments of this type are carried out by comparing the functional activity of fragments of immunoglobulins such as Fab, Fc, and Fv (V_HV_L) fragments, to that obtained using the whole molecule. The functions which are predominantly associated with one domain are summarized Table 6.9.

6.2.5 Relationship between structure and effector function

IgG

Structure. IgG is the predominant form of immunoglobulin in the plasma. The important structural features of human IgG are listed in Table 6.10. There are four subclasses of IgG in human (Fig.6.19) which differ from one another in the constant region of the γ heavy chains mainly in the **hinge region**. The hinge region joins the C_H1 and C_H2 domains of the γ heavy chain and determines the flexibility of the molecule.

In $\gamma 1$ heavy chains the hinge encompasses amino acids 216–231, but it can be much larger. For example, in $\gamma 3$ heavy chains the hinge region contains an extra 47 amino acids that appear to have been generated by a quadruplication of the gene segment encoding the $\gamma 1$ hinge region.

The inter-heavy chain disulfide bonds are located in the hinge region and the different subclasses of IgG have different numbers of disulfide bonds, two in IgG1 and IgG4, four in IgG2, and 11 in IgG3 (Fig.6.19). IgG1 differs from the other subclasses with respect to the position of the half-cysteine residue that forms the inter-heavy-light chain disulfide bond.

Functions. A summary of the functional properties of human IgG is given in Table 6.11.

IgG is the only class with the ability to cross the placenta, and therefore maternal humoral immunity can be transferred to the fetus to protect it during pregnancy.

When IgG is bound to its antigen, these complexes can activate the complement system. The different subclasses of IgG activate complement via different pathways and with different efficiencies. For example, in the human, IgG1, IgG2, and IgG3 can activate the classical pathway when they are present in an immune complex. The relative efficiency of activation by each subclass is IgG3 > IgG1 > IgG2. IgG4 activates the alternative pathway. The flexibility of IgG molecules seems to be important in determining the ability of Clq to bind to the $C_{\gamma}2$ domain and this is dictated by the hinge region. Indeed there is a correlation between the sequence of the hinge region and Clq binding in the IgGs of human, mouse, guinea pig, and rabbit (Table 6.12). Human IgG1 and rabbit IgG are both efficient activators of the classical pathway, and have six and five amino acids, respectively, in the upper hinge segment, whereas human IgG4 which does not bind Clq, has only three (Table 6.12).

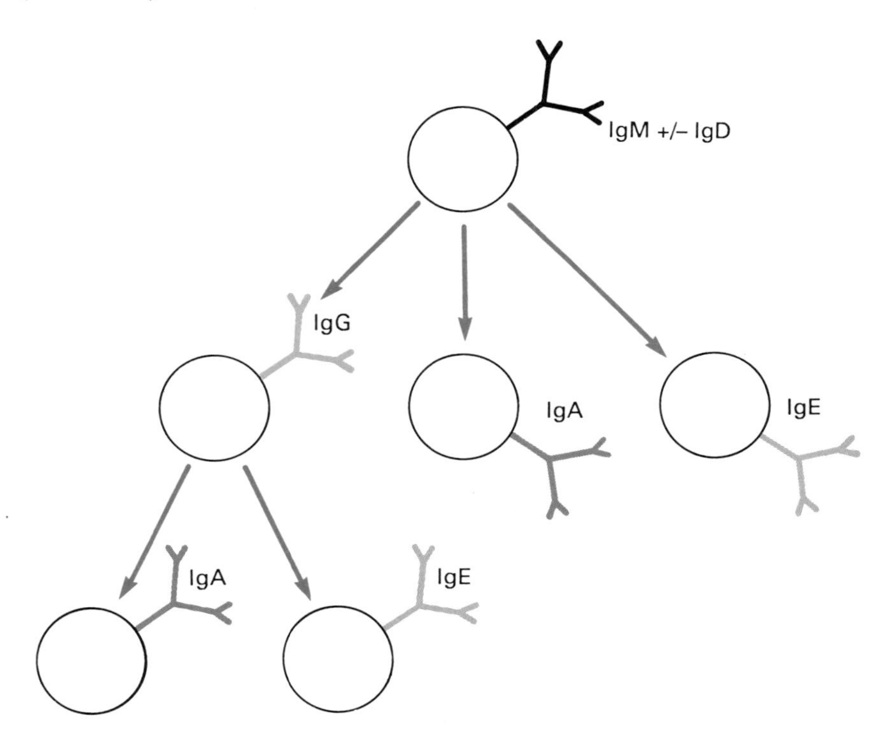

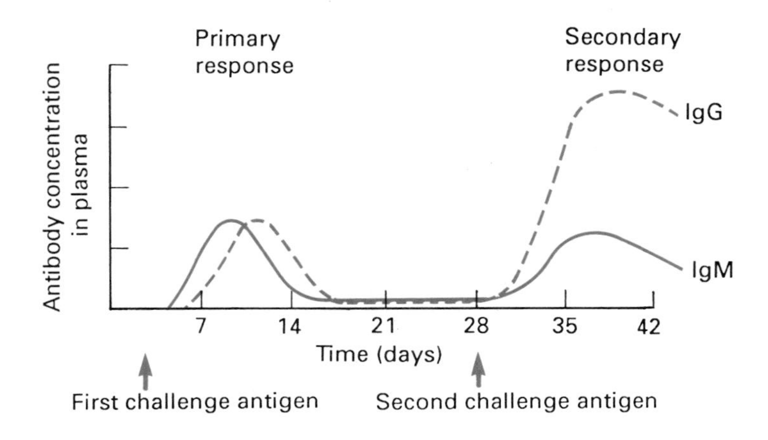

Fig.6.18 The class of the antibodies synthesized can change during B cell responses

Table 6.8 Effector functions of the different classes of antibody

Class	Function
IgM	Synthesized during the primary immune response. Efficient complement activation — classical pathway
IgG	Placental transfer to confer immunity to the fetus. Fc receptor binding Complement activation
IgA	Present in extravascular secretions
IgE	Interact with allergens giving complexes that bind to specific receptors on mast cells triggering degranulation, histamine release etc
IgD	Unknown

Table 6.9 The function of the domains of human IgG

Domain	Function
V_γ–V_L	Antigen binding
$C_\gamma 2$	Binding of Clq to initiate complement activation (see Section 9.4)
$C_\gamma 3$	Interaction with Fc receptors e.g. on macrophages and monocytes
$C_\gamma 1 + C_\gamma 3$	Interaction with protein A Interaction with Fc receptors on placental syncitiotrophoblasts, neutrophils, killer cells, and in certain species intestinal epithelial cells in neonatal animals

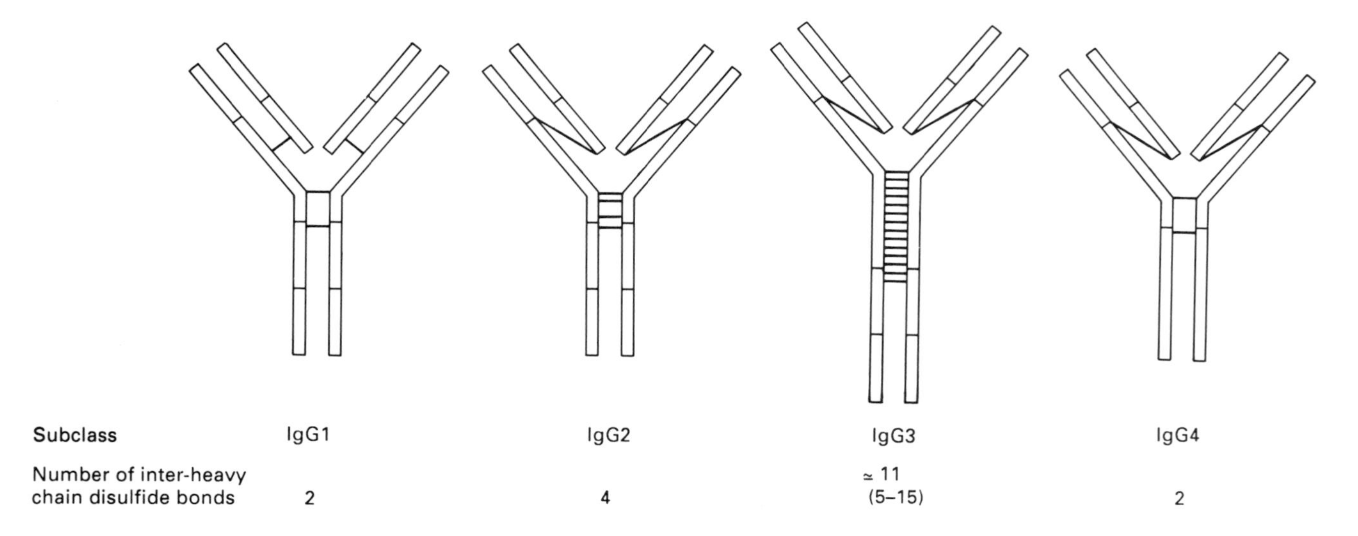

Fig.6.19 Diagram to illustrate the structure of the subclasses of human IgG, IgG1–4, and the relative positions of the interchain disulfide bonds

Table 6.10 Structural properties of human IgG

No. of polypeptide chains	4	2 identical heavy chains — γ 2 identical light chains — κ or λ
Molecular weight (M_r)	150 kDa	
No. of domains in each γ heavy chain	4	1 variable 3 constant (+ hinge region)
Carbohydrate content	3%	$C_\gamma 2$ domain residue 297
Serum concentration (mg/100 ml)	1250±300	
Allotypes	Yes	Gm
Subclasses or isotypes	4	IgG1, IgG2, IgG3, IgG4
% of each subclass		65% 24% 8% 3%
Heavy chains		$\gamma 1$ $\gamma 2$ $\gamma 3$ $\gamma 4$

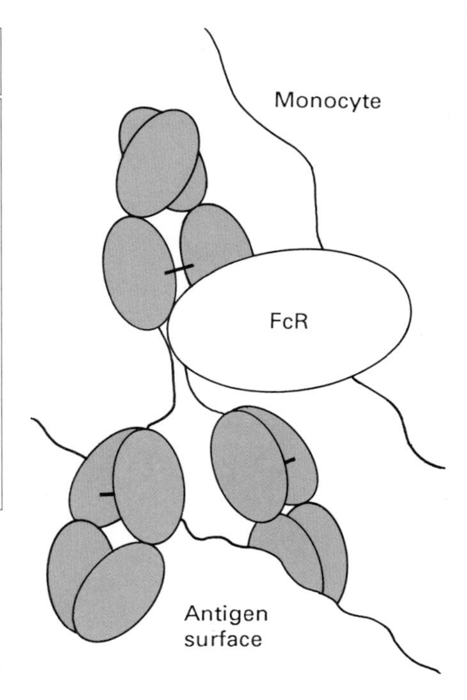

Fig.6.20 Model for the interaction of monocyte Fc receptor (FcR) with a human IgG1 bound to an antigen exposed on a cell surface. The FcR is shown interacting with part of the hinge region and $C_\gamma 2$ domain. From Burton (1986), with permission

IgG molecules also interact with Fc receptors (Section 10.4.5). Fc receptors, specific for different subclasses of IgG, are expressed by different cell types, including macrophages, monocytes, and B lymphocytes themselves (see Section 8.4.2) (Anderson and Looney 1987; Unkeless *et al.* 1988). In humans, IgG1 and IgG3 can bind to macrophage Fc receptors with greater avidity than IgG2 and IgG4 molecules (Table 6.11). In the mouse, macrophages have separate receptors for monomeric IgG2a (Fc_γRI), antigen-complexed IgG1 and IgG2b (Fc_γRII), and there is another receptor for IgG3 (Fc_γRIII). A model for the interaction of human IgG1 with a monocyte Fc receptor (Fc_γR1) is shown in Fig.6.20 (Burton 1986). The model predicts that the receptor binds to the hinge region of IgG1, on the C-terminal side of the inter-heavy chain disulfide bond, near to a bend in the $C_\gamma 2$ domain. This hypothesis is supported by data showing a loss in the affinity of binding of hinge-deleted IgG to Fc receptors. The cellular consequences of binding to each of these Fc receptors are different and are discussed in Section 10.4.5. Fc receptors are also found on placental trophoblasts, platelets, neutrophils, and killer cells, and in some cases on neonatal intestinal epithelial cells.

Table 6.11 Functional properties of human IgG1-4

	IgG1	IgG2	IgG3	IgG4
Classical pathway activation	+++	++	++++	+/−
Alternative pathway activation	+	+	+	+++
Fc receptor-binding; polymorphonuclear leukocytes	+	+	+	+
Fc receptor-binding; macrophages	+	−	+	−
Placental transfer	+	+	+	+
Susceptibility to proteolytic digestion	+	+	++++	+

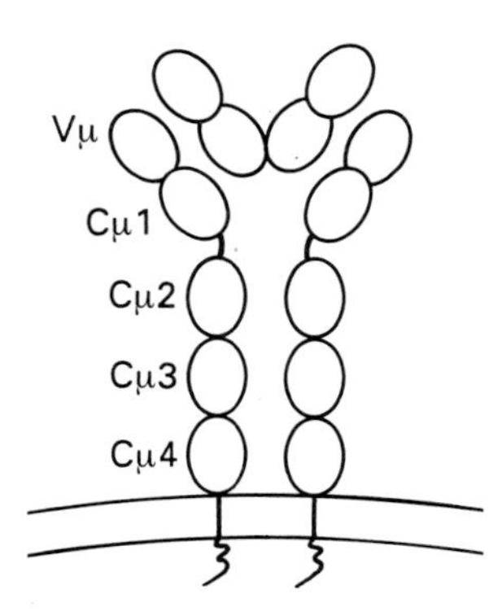

Fig.6.21 Schematic drawing of membrane-bound monomeric IgM

The protein A molecules of *Staphylococcus aureus* seem to mimic an Fc receptor and bind the C_H1 and C_H3 domains of IgG. This has probably evolved as a bacterial defence mechanism so that specific antibodies do not bind to the bacteria via their Fab portions but are bound via their Fc region and are thus rendered ineffective. This property of protein A is of enormous use experimentally, particularly for the purification of IgG.

IgM

Structure. A summary of the structural features of human IgM is given in Table 6.13. IgM, like other antibodies, can exist in two forms, either as a membrane-bound molecule on the surface of B lymphocytes or as a soluble molecule in the plasma (Fig.6.21). The membrane-bound and secreted forms are encoded by two different mRNAs, each of which has a different 3′ end (Rogers *et al.* 1980). Cell surface IgM is a monomer (two heavy and two light chains) and it contains a different amino acid sequence at the carboxyl terminus in its heavy chains to the secretory form. This additional sequence in the membrane-bound form comprises 41 amino acids and is encoded by a separate exon (see Section 7.2.1). Twenty-five of these 41 amino acid residues are hydrophobic and form the transmembrance portion of the molecule. These hydrophobic amino acid residues are then followed by three polar residues, which extend into the cytoplasm and anchor the IgM molecule in the plasma membrane. A similar arrangement of amino acids is found in other transmembrane glycoproteins, for example class I and class II MHC molecules, but they usually have longer cytoplasmic tails.

Most of the soluble IgM in the plasma is pentameric, that is, each molecule contains five monomeric IgM units (Figs.6.21, 6.6). A second type of polypep-

Table 6.12 Correlation of hinge region amino acid sequences and complement activation among IgG of different species

	216 Upper 224 Middle Lower 234	Complement fixation
Human IgG1	E P K S C D K T H T – C P P – – – – – – C P A P E L L G G P	+
Human IgG2	E R K – – – – – – – – C C V E C P P – – C P A P P V A G – P	–
Human IgG4	E S K Y G – – – – P P C P P – – – – – – C P A P E F L G G P	–
Mouse IgG1	V P R D C G – – – – – C K P C I – – – – C T V P – – E – V S	–
Mouse IgG2a	E P R G P T I K – – P C C P P K – – – – C P A P N L L G G P	+
Mouse IgG2b	E P S G P I S T I N P C P P C K E C H K C P A P N L E G G P	+
G pig IgG1	Q S W G H T – – – – – C P P C I P – – – C G A P Z L L G G P	–
G pig IgG2	E P I R T P Z B P B P C T C P K – – – – C P P P E N L G G P	+
Rabbit IgG	A P S T C S K P M – – C – – – – – – – – – P P P E L L G G P	+

↑ 1st Inter-H S-S ↑ last inter-H S-S

←Genetic hinge→

Alignment of sequences after Burton, D.R. (1985). *Mol Immunol.* **22**, 161–206

tide chain, called the J chain, is also present in pentameric IgM. In this form, each IgM monomer has a shorter C-terminal sequence of 18 amino acids and the penultimate cysteine residue of this segment seems to be essential for polymerization with the J chain into a pentameric configuration (for a review on the assembly of IgM see Davis and Shulman 1989). Electron microscopy of soluble IgM has revealed that the five IgM monomers are arranged in a circle, with the Fab arms of the molecules around the circumference (where they can bind antigen) and the Fc portions pointing into the centre. Pentameric IgM therefore potentially has ten antigen binding sites. However, only an average of five of these are usually used.

The J chain. The J chain is a polypeptide of low molecular weight (about 15 kDa) that is essential for the polymerization of IgM and, as we shall see below, also for IgA (for review see Koshland 1985). It is a glycoprotein composed of 129 amino acids that contains one carbohydrate unit and is synthesized by the plasma cells. The amino acid sequence of the J chain has been determined. Each molecule has eight cysteine residues that are involved in the formation of both inter-chain and intra-chain disulfide bonds. One of the cysteines forms a disulfide bond with the penultimate cysteine residue of one of the μ or α heavy chains, and this event probably triggers the polymerization of soluble IgM and IgA molecules. A membrane-bound sulfydryl oxidase enzyme, which is thought to catalyse this disulfide bond formation, has been found in the membranes of activated B lymphocytes, but not resting B lymphocytes. The gene which encodes the J chain has been cloned and it is not homologous to any of the immunoglobulin genes, nor is it encoded on any chromosome where they are encoded. It is located on chromosome 15.

Table 6.13 Structural properties of human IgM

Number of polypeptide chains per monomer	4	2 identical heavy chains — μ 2 identical light chains — κ or λ
Monomer molecular weight (M_r)	190 kDa	
No of domains in each μ heavy chain	5	1 variable, 4 constant } no hinge region
Carbohydrate content	10%	5 oligosaccharide units located 1-C_μ1, 3-C_μ3 + 1-tail piece of soluble IgM
Form of the molecule:		
Monomer — location	B cell membrane	
— valency	2	
Polymer — location	soluble molecule present in plasma	
— valency	10 — composed of 5 IgM molecules associated with the J chain	
Serum concentration (mg/100 ml)	125±50	
Allotypes	No	

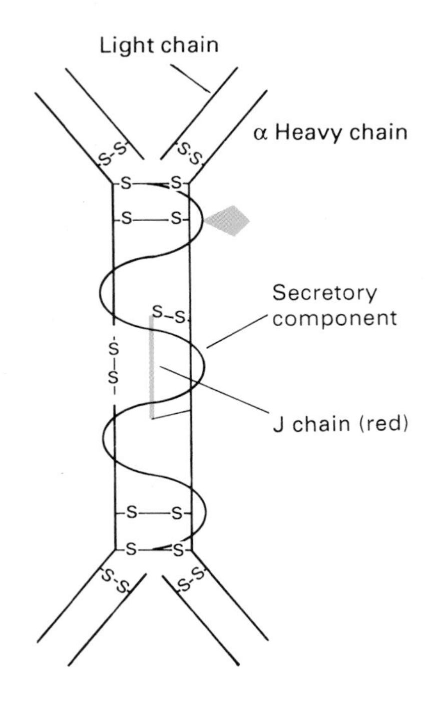

Fig.6.22 Schematic diagram to show components of an IgA dimer

Functions. IgM is one of the two predominant forms of immunoglobulins on the surface of B cells, the other being IgD. Cross-linking of IgM by polyvalent antigens results in B cell proliferation and differentiation (see Section 8.4.2). How membrane immunoglobulins, like IgM and IgD, which lack a cytoplasmic tail can trigger B cell activation is not entirely clear. However, membrane proteins that are noncovalently associated with IgM have been described, including those designated IgM-α and Ig-β; IgM-α is the product of the MB-1 gene that is known to have sequence homology with proteins of the T cell receptor-CD3 complex (see Section 4.2.3). These findings suggest that these membrane proteins may transduce signals from IgM (Section 8.4.2).

IgM is the first antibody produced during a primary immune response and the first class to be synthesized by neonatal animals. IgM–antigen complexes can activate the complement cascade. Activation of complement by IgM is about 15 times more efficient than activation by IgG. IgM is particularly effective in combatting bacterial infections during primary antibody responses (Fig.6.18).

IgA

Structure. The majority of IgA molecules in the serum are monomers, but polymers containing between two and five monomers of IgA have been identified in various body fluids (Figs.6.22 and 6.6). In polymeric forms of IgA, the α chains have an extra 18 amino acid residues at the C-terminus of the polypeptide which are also found in pentameric IgM molecules.

IgA is the main class of immunoglobulin found in secretions where it is present as a dimer associated with the J chain and the **secretory component** (Fig.6.22). Whereas the J chain is most probably synthesized by the B lymphocyte making the antibody (see above), the secretory component is synthesized by epithelial cells (Mestecky and McGhee 1987); the properties of the secretory component will be discussed below. The features of human IgA are listed in Table 6.14.

Table 6.14 Structural properties of human IgA

Number of polypeptide chains per monomer	4	2 identical heavy chains — α 2 identical light chains — κ or λ
Monomer molecular weight	160 kDa	
Number of domains per heavy chain	4	1 variable 3 constant (+ hinge region)
Carbohydrate content	7%	
Form of the molecule in plasma	Dimers and other multimers—associated with the J chain and secretory component	
Serum concentration (mg/100 ml)	210±50	
Subclasses or Isotypes	IgA1 IgA2	Predominant in serum Predominant in secretions
Allotypes	IgA2m (1) IgA2m (2)	

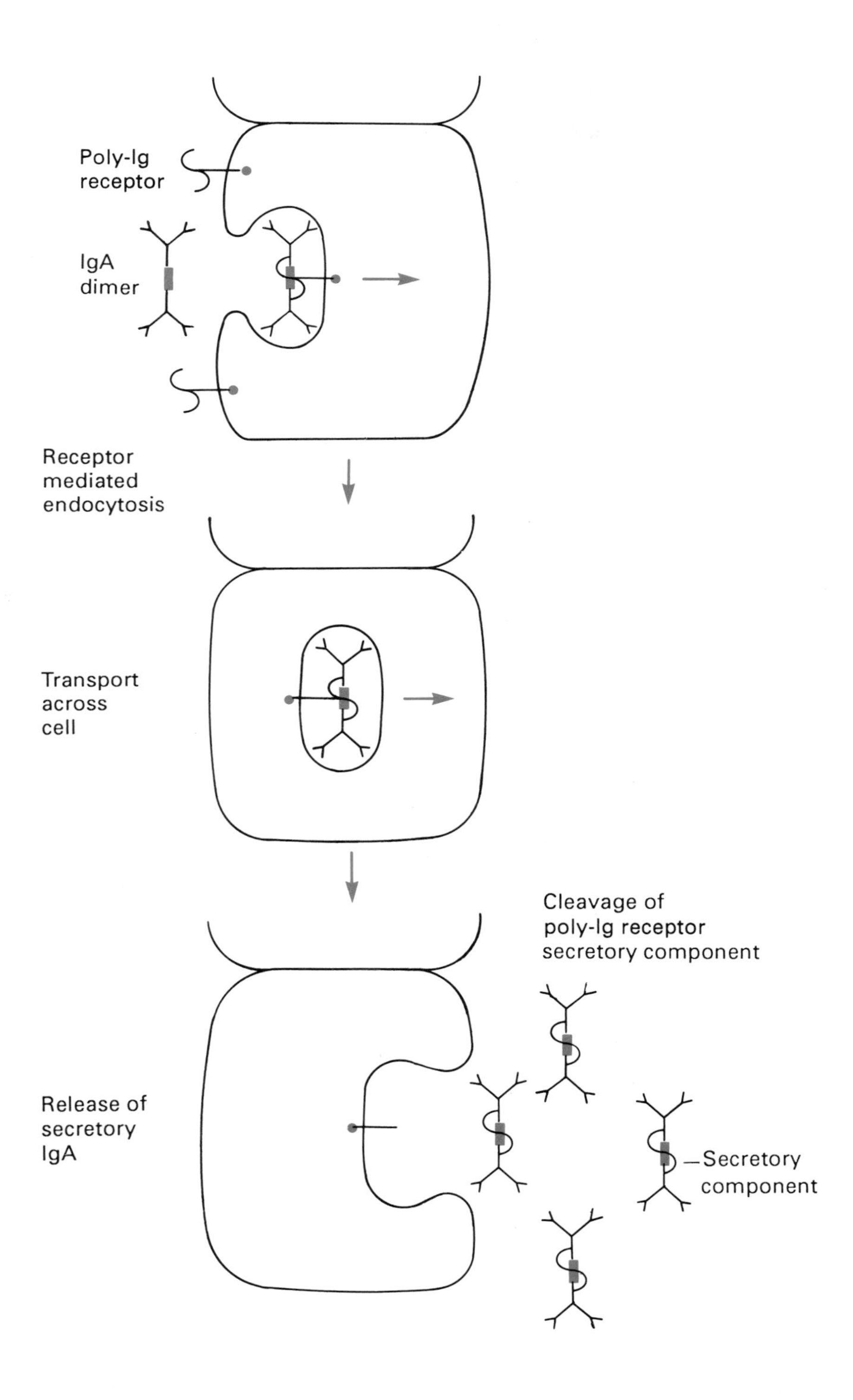

Fig.6.23 Schematic diagram to show the proposed mechanism of the secretion of IgA

The secretory component. The secretory component is a 70 kDa fragment of the poly-Ig receptor that is synthesized by epithelial cells and which is essential for the transport of IgA from the blood to the mucosa (e.g. of the gut).

The poly-Ig receptor acts as an Fc receptor for IgA dimers in the plasma. The IgA dimers become bound by their Fc portions to these membrane receptors on the plasma-side of epithelial cells which line the exocrine glands, and the complexes are endocytosed (see Section 3.2.2) (Fig.6.23). The internalized IgA–receptor complexes are then transported through the cytoplasm to the luminal-side of the cell where proteolysis of the poly-Ig receptor occurs. As a consequence, the IgA dimer is released into the digestive tract attached to the proteolytic fragment of the poly-Ig receptor, now called the secretory component. The other proteolytic fragment of the poly-Ig receptor is left on the cell surface.

The secretory component is highly glycosylated and lacks free sulfydryl groups, but it contains a number of intra-chain disulfide bonds. Depending on the species, the secretory component may either be disulfide-bonded to the IgA molecule or non-covently associated with it.

Functions. IgA is the predominant class of immunoglobulin in secretions such as saliva, tears, breast milk, and those of the digestive tract (Underwood and Schiff 1986).

IgA is produced predominantly by plasma cells, originating in Peyer's patches (see Section 1.3), which migrate out of the patch and became associated with the gut epithelial cells. IgA is fairly resistant to proteolytic digestion, which is an obvious advantage as it is present in the body fluids and secretions, many of which contain proteolytic enzymes.

Proteolytic enzymes secreted by bacteria in the digestive tract have been found to cleave IgA1 and IgA2 molecules in different ways. In saliva, molecules of the IgA2 isotype are more resistant to baterial proteases than IgA1; presumably bacteria evolved these proteases to effect the rapid removal of IgA1 antibodies from the saliva as one of their defence mechanisms. IgA1, however, is more resistant to attack from intestinal proteases than IgA2. These properties may represent beneficial environmental adaptations of the two molecules.

The presence of IgA in secretions is important functionally, as it allows antigenic material to be cleared from the organism before it can enter the bloodstream. Thus it provides local protection from potentially pathogenic micro-organisms. IgA is also important for transferring maternal immunity to newborn animals; it is secreted by the mother in colostrum which is fed to the neonate in the breast milk (compare with placental transfer of IgG to the fetus). IgA in human breast milk is contained in cells, and it has been suggested that these cells are transferred and attach to the intestinal mucosa where they can release IgA.

IgA–antigen complexes can activate the alternative pathway of complement.

IgE

Structure. IgE forms only a minor component of normal plasma immunoglobulin. A summary of the structural properties of IgE are given in

Table 6.15. The structure of IgE was worked out only after a myeloma-producing IgE was identified (see footnote, p. 338). In contrast to IgM the two inter-chain disulfide bonds of IgE are separately located between the C_H1 and C_H2 domains and between the C_H2 and C_H3 domains. Carbohydrate units are attached throughout the heavy chain. IgE does not have a hinge region and it is one of the least flexible of all the immunoglobulins.

Functions. IgE is the major class of antibody involved in 'arming' mast cells and basophils for their role in allergic reactions. IgE interacts through its Fc region with receptors on these cells (see Section 10.4.5). This interaction can be abrogated by reduction of the inter-heavy chain disulfide bonds or by heat denaturation of the IgE molecule. When IgE molecules that are attached to Fc receptors on mast cells ($Fc_\varepsilon RI$) (Metzger *et al.* 1986) and are cross-linked by antigen, signalling occurs (Eisenman and Bolen 1992) and the mast cell releases vasoactive amines, such as histamine and other pharmacologically active substances, which contribute to an inflammatory response which can be manifested as allergy (see Section 10.5.1). On monocyte/macrophages and eosinophils, the $Fc_\varepsilon RII$ (CD23) receptor mediates IgE dependent cytotoxicity against parasites (Conrad 1990). The $Fc_\varepsilon RII$ receptors can also promote phagocytosis of IgE-coated particles by monocytes.

IgE is important in the humoral responses to parasitic diseases, and is often found in high levels in the plasma of patients with parasitic infections. IgE does not cross the placenta, and immune complexes containing IgE do not activate complement.

IgD

Structure. The structural features of human IgD are listed in Table 6.16. It is present in very low concentrations in normal human plasma and is undetectable in mouse plasma. Like IgE it was only characterized when an IgD myeloma protein was identified. Human δ chains have three constant domains but mouse δ chains have only two of the three, C_H1 and C_H3; the C_H2 homologue has been deleted (Blattner and Tucker 1984). The hinge region of the δ chains is the largest of any heavy chain, comprising 64 amino acids.

Functions. The biological role of IgD is unknown. Because it is present at extremely low levels in the plasma, its functions are difficult to define. It is

Table 6.15 Structural properties of human IgE

Number of polypeptide chains	4	2 identical heavy chains — ε 2 identical light chains — κ or λ
Molecular weight	190 kDa	
Number of domains per heavy chain	5	1 variable, 4 constant; no hinge region
Carbohydrate content	13%	
Serum concentration (mg/100 ml)	0.03	
Allotypes	No	

Table 6.16 Structural properties of human IgD

Number of polypeptide chains	4	2 identical heavy chains — δ 2 identical light chains — κ or λ
Molecular weight	175 kDa	
Number of domains per heavy chain	4	1 variable 3 constant (+ hinge region)
Carbohydrate content	9%	$C_\delta2$ and $C_\delta3$ domains
Serum concentration (mg/100 ml)	4	
Allotypes	No	

present on the surface of a large proportion of B lymphocytes and may be involved in the activation and/or developmental state of the B cell (see Section 8.2.1). Higher levels of IgD are found in newborn animals, and it has been reported to cross the placenta.

It has been suggested that the long hinge region of IgD might be important in the cross-linking of IgD by antigen and that because this region is very susceptible to proteolysis, secretory IgD might be cleaved after binding antigen. The resulting complexes between the Fab region of IgD and the antigen may be involved in the generation of anti-idiotypic responses (see Section 11.1). IgD does not activate complement.

6.2.6 Biosynthesis and assembly

An investigation of the biosynthetic pathways for immunoglobulins was undertaken once the structure of these antigen receptors had been solved. Again, the availability of myeloma cells as a source of homogeneous immunoglobulin in large quantities was an enormous advantage. Heavy and light chain mRNAs are each transcribed as a single species, containing both the V and C region of each chain (see Section 7.2). Both chains insert directly into the rough endoplasmic reticulum (RER; Fig.1.6), via their hydrophobic leader sequences, located at the N-terminal end of the growing polypeptide chain (Fig.6.24). (The leader sequence is cleaved from the completed heavy and light chains expressed at the cell surface). After the completed heavy and light chains are released from the cisternal space of the RER, they assemble into the four-chain structure, through either H-L half molecules or H-H half dimers. Two proteins, α and β, are required to allow transport to the cell surface. The α chain is common to all five Ig classes, but the β chain is class specific (Venkitaraman *et al.* 1991). For each heavy chain isotype only one of these two pathways is used, for example μ chains assemble predominantly through H-L half molecules, while γ and α chains utilize the H-H route of assembly. Interestingly, in B cells synthesizing two heavy chain isotypes simultaneously, for example μ and δ, only the homologous heavy chains will assemble into the four chain molecule, i.e. μ-μ or δ-δ plus paired light chains. The carbohydrate moieties are transferred to the polypeptide chains from a lipid carrier, very soon after entry into the endoplasmic reticulum, and modification of the carbohydrate continues during the transport through the Golgi. Within the Golgi the mature immunoglobulins are packaged into vesicles for transport to the plasma membrane. Polypeptides containing a hydrophobic transmembrane region are inserted into the plasma membrane when the vesicle fuses with it. Heavy chains lacking the transmembrane region are transported as soluble molecules in the vesicles and are released into the extracellular fluid when the vesicles open up as they fuse with the plasma membrane during exocytosis (see Section 3.2.2 for a discussion of exocytosis). Experimental data suggest that the time from synthesis to secretion takes about 30 minutes, and that a mature plasma cell can produce about 10^5 antibody molecules per minute.

6.2.7 Level of expression of light chains in different species

Although both κ and λ light chains are present in all species, the percentage of each type of light chain that is expressed can vary considerably. For example, in the mouse the light chains are predominantly of the κ type, approximately 95% κ and only 5% λ, whereas in human the ratio of κ to λ is approximately 2:1. This correlates roughly with the number of genes present for each chain in different species. In the mouse there are a large number of variable region gene segments for κ light chains but relatively few for λ chains, while in human the number is more equal (see Section 7.2). The reason for these variations between species is unknown. The presence of a particular κ or a λ light chain (while contributing to antigen specificity) does not seem to endow the molecule with any special functions. This is unlike the heavy chain, where a change in the class of the heavy chain leads to dramatic changes in function of the Ig molecule as a whole (see Section 6.2.5).

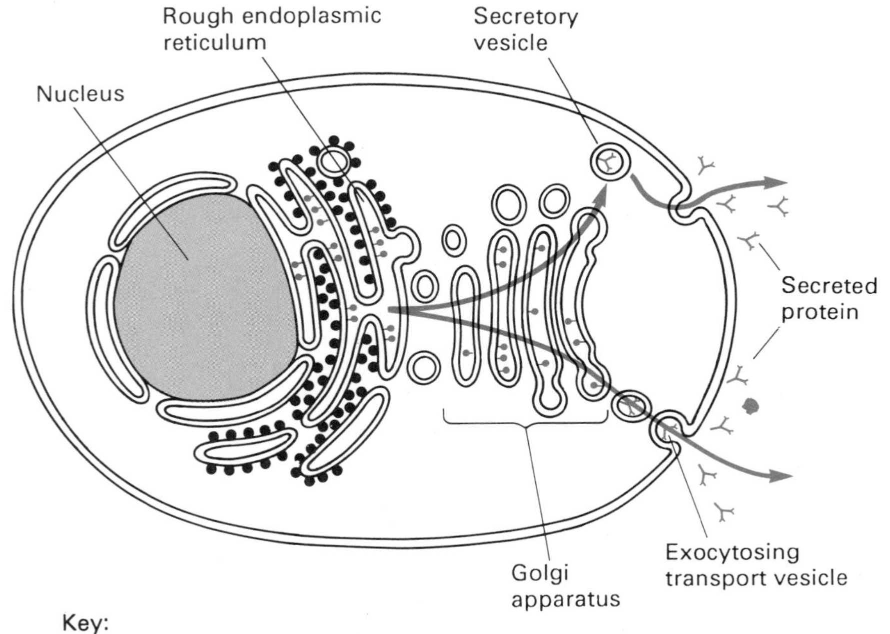

Fig.6.24 Biosynthesis and assembly of immunoglobulins

6.2.8 'Designer antibodies'

The exquisite antigen specificity of immunoglobulins gives them enormous potential as agents for targeting defined structures and for drug delivery. Work towards 'designer' antibodies which can be engineered to recognize a known target structure has been progressing for some time, but recently there have been several new advances which now make designer antibodies a real possibility (e.g. Fig.6.25).

1. Many V_H domains have been shown to bind antigen with high affinity *in the absence of the light chain.* This finding is unexpected, because in all the

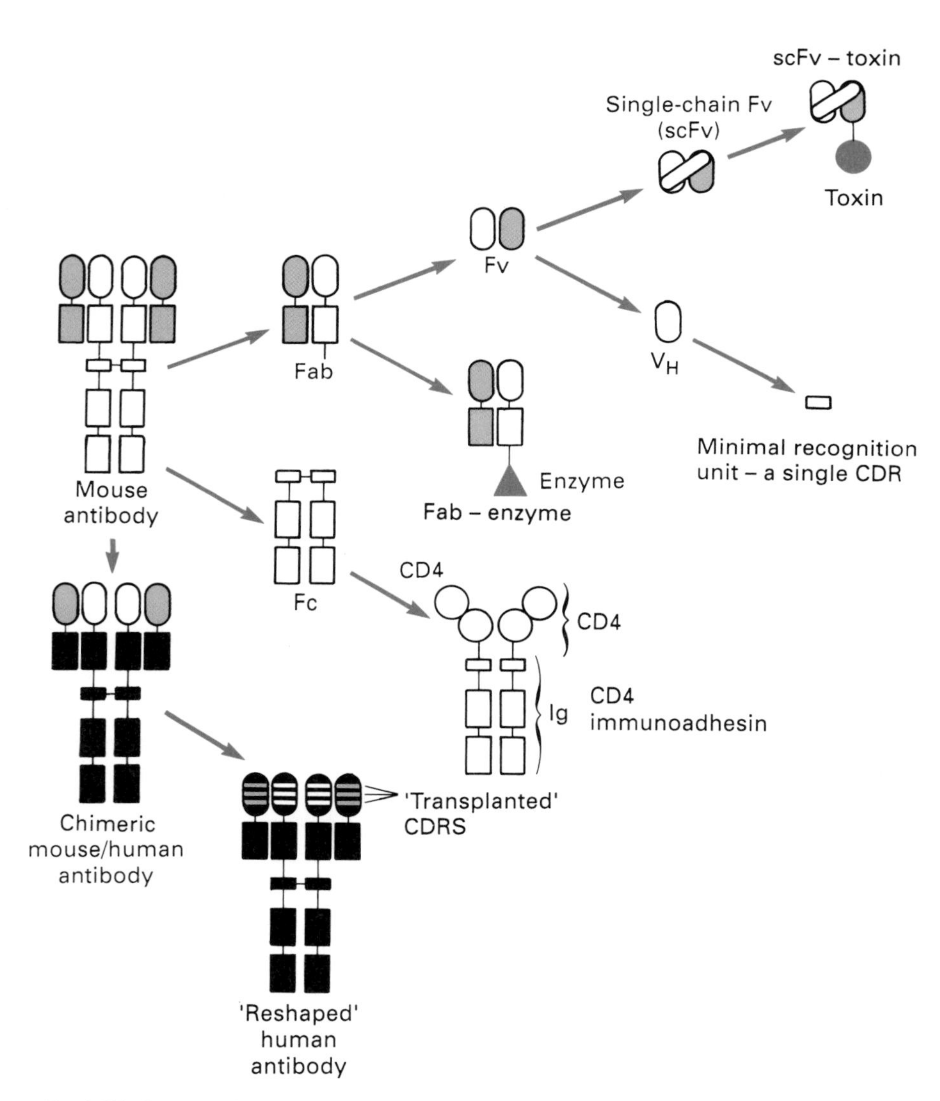

Fig.6.25 'Designer' antibodies and antibody fragments. (Adapted from Winter and Milstein 1991)

crystal structures of immune complexes examined to date the CDRs of both the heavy and the light chains are involved in antigen binding (Fig.6.16). Indeed, the crystal structures of the lysozyme–anti-lysozyme complex shows that the light chain contributes approximately 40% of the antigen binding surface (Fig.6.13). However, when the V_H domains from the anti-lysozyme immunoglobulin were prepared in isolation, the affinity of binding to lysozyme was only 10-fold less than that of the parent antibody, showing that the contribution made by the light chain in this case is relatively small (Ward *et al.* 1989). Isolated V_H domains with antigen binding activity have also been found in other systems.

V_H domains with high affinity antigen binding may not be useful in every situation but they may have advantages over monoclonal antibodies for some applications, for example clearance of toxins from body tissue, because they will not trigger effector mechanisms (see Section 6.2.5) that may give rise to undesirable side-effects. They may therefore prove extremely useful for constructing high affinity antibodies for use in human therapy.

2. A vector system has been discovered that allows the rapid production of antibody Fab fragments in bacteria (Huse *et al.* 1989). Using this system, it is possible to generate large numbers of monoclonal Fab fragments against a defined antigen in just 2 weeks. The soluble Fab fragments can then be screened and selected for antigen binding activity. V genes encoding Fab or Fv regions with antigen binding activity can also be expressed on the surface of bacteriophages, so-called phage display libraries (McCafferty *et al* 1990). In this way the cloned V genes can be selected directly with antigen by affinity chromatography (see Appendix), thereby revolutionizing the time taken to screen the Fv regions produced using this technology (Clackson *et al.* 1991).

3. Molecular modelling has shown that five of the six hypervariable regions adopt only a limited number of conformations (Chothia *et al.* 1990). Thus it should be possible to produce reasonably accurate models of the structure of an antibody combining site on the basis of amino acid sequence data alone. These findings should be enormously useful for modelling antibodies for use in therapy and may enable refinements to be made to the structure and therefore facilitate the development and production of antibodies with a higher binding affinity for the antigen.

4. The CDRs from a rodent monoclonal antibody of known antigen specificity can be 'transplanted' into a human framework (Fig.6.25). This technique offers a method for constructing human monoclonal antibodies of identical antigen specificity to well-characterized mouse monoclonal antibodies. Human monoclonal antibodies may be more appropriate therapeutic agents for clinical use, because they may eliminate many of the problems associated with the xenogeneic anti-mouse immunoglobulin response made by humans undergoing treatment with mouse monoclonal antibodies (For example, the side-effects associated with the use of the anti-CD3 monoclonal antibody, OKT3, as an immunosupressive agent to prevent the rejection of transplanted organs). For review see Winter and Milstein (1991).

6.3 T cell antigen receptors

T cell receptors are the molecules expressed by T lymphocytes that enable them to recognize antigen. The molecular and biochemical characterization of the T cell receptor eluded immunologists for many years. The original premise of scientists working in this area was that T cell receptors would have a similar structure to that described for immunoglobulins, because both molecules interact with antigen. Although we are now aware from the data published since late 1983 that this idea is essentially correct, unfortunately much of the early work failed to increase our understanding of the structure of the antigen receptor significantly, nor of antigen recognition by T lymphocytes. Fortunately, we now have a clear model for both the structure and, in general, the function of these molecules.

There are two types of T cell receptor. The $\alpha\beta$ receptor was the first to be identified. It is expressed by the majority of peripheral T cells and is the structure responsible for binding to the peptide–MHC complex (see Sections 4.2.1 and 5.2.1). A second form of TcR, the $\gamma\delta$ receptor, was identified later and found to be expressed on a minor population of peripheral T cells in mice and humans; estimates range between 1 and 10%. T cells bearing this TcR appear early in ontogeny and are thought to have a broader antigen specificity than T cells bearing the $\alpha\beta$ form of T cell receptor (see Section 4.2.2). An alternative nomenclature, TcR1 ($\gamma\delta$ receptor) and /TcR2 ($\alpha\beta$ receptor), is used by some laboratories to indicate the timescale of the gene rearrangements that take place during T cell differentiation (see Section 7.3). In this dicussion we will refer to the two forms of T cell receptor as the $\alpha\beta$ and $\gamma\delta$ receptors.

6.3.1 Structure

The $\alpha\beta$ T cell receptor

The breakthrough in our understanding of the structure of the $\alpha\beta$ T cell receptor molecule was the result of a number of different developments.

1. The cloning of T cells *in vitro* provided immunologists with a homogeneous population of T cells that could be used to characterize the structure of the TcR. Previous attempts to analyse the structure of the antigen receptor expressed by T cells were plagued by the same problems encountered during the characterization of immunoglobulin structure, that is, the heterogeneity of the molecules expressed by T cells in normal individuals. T cell clones can be thought of as equivalent to myelomas in this context (see footnote, p. 338).
2. The production of antisera and subsequently monoclonal antibodies to unique structures expressed by individual T cell clones, so-called clonotypic antibodies (see Section 4.2.1).
3. The observation that although one molecule present on the T cell surface was involved in antigen recognition, other cell surface molecules were

also involved in T cell activation, for example, in signalling to the cell that antigen recognition has occurred and in initiating the cellular response (see Section 4.2).

4. The isolation of polysome-associated, T cell specific mRNA. This was then cloned as cDNA and particular genes were shown to be rearranged in T cells *but not* in other cell types.

The $\alpha\beta$ T cell receptor was originally identified using clonotypic antibodies. These recognized a disulfide-bonded heterodimer (80–90 kDa) composed of an α and a β chain (43–49 kDa and 38–44 kDa, respectively, depending on the species) (Fig.6.1 and Table 6.17) (for review see Weiss 1990). Both chains are glycosylated and contain N-linked sugars. The genes encoding the α and β chains have both been cloned and further structural information has been obtained by sequencing (for review see Allison and Lanier 1987).

The α and β chains each have two extracellular domains, one variable and the other constant. Connecting peptides of about 20 amino acids containing cysteine residues join the constant domain to the transmembrane region. It is within this peptide that the interchain disulfide bond is formed. Each chain has a short cytoplasmic tail containing a positively charged lysine residue which may be important for interaction with CD3 polypeptides (see Section 4.2.3).

The variable domains of the α and β chains together form the antigen binding site of the T cell receptor. Sequence data suggest that the variable domains are folded into β-pleated sheet structures closely resembling immunoglobulin variable domains (Fig.6.9). Hypervariable or complementarity determining have been identified within variable domain sequences in at least three regions, equivalent to those of immunoglobulins (Fig.6.12), although the exact number remains controversial. No structural data for T cell receptor are available at present (although a number of laboratories are trying to engineer soluble TcR molecules for crystallization—Traunecker *et al.* 1989), but by analogy with immunoglobulins the CDRs of variable domains of the α and β chains are predicted to be in contact with the peptide-MHC complex (Section 6.3.2) (Davis and Bjorkman 1988).

The $\gamma\delta$ T cell receptor

The existence of an alternative set of presumptive T cell antigen receptors was first reported in 1986 (for review see Brenner *et al.* 1988). This second

Table 6.17 Characteristics of murine and human $\alpha\beta$ T cell receptors

	Murine	Human
No. of molecules per cell	25 000–30 000	25 000–30 000
No. of polypeptide chains	2	2
Disulfide bonded	Yes	Yes
Molecular weights		
α	40–50 kDa	43–49 kDa
β	40–50 kDa	38–44 kDa

type of T cell receptor was detected on a variety of cell types including cell lines from patients with immunodeficiency diseases, T cell clones established from fetal blood and thymocytes, and on a leukaemia T cell line, initially by four different research groups all working independently. All of these T cells were found to express the CD3 molecule, but not the $\alpha\beta$ T cell receptor described above. Instead, CD3 was found in association with another type of heterodimer, one chain of which was derived from the T cell receptor γ chain gene that had already been cloned (see Sections 4.2.2 and 7.3). Subsequently the second chain of this heterodimer, the δ chain, was characterized. It is now clear that $\gamma\delta$ TcRs are also expressed by some peripheral T cells and in fetal thymus. Many T cells expressing the $\gamma\delta$ receptor do not express CD4 or CD8 molecules that characterize mature functional $\alpha\beta$ T cells, although some do express CD8 at some sites in the body.

$\gamma\delta$ T cell receptors exist in several structural forms (Fig.6.26). The molecular basis for these variations is understood and will be discussed in Chapter 7.

Both the γ and δ chains contain one variable and one constant domain. The variable domains of the two chains interact with one another to form the antigen binding site. Both chains are inserted into the T cell membrane, and, as for the $\alpha\beta$ TcR, a lysine residue is present midway through the transmembrane region in each chain. This positively charged lysine residue may be important for interacting with conserved negatively charged residues (Asp and Glu) present in four of the CD3 polypeptides (see Section 4.2.3).

The $\gamma\delta$ receptor immunoprecipitated from human peripheral T cells using anti-γ chain antibodies is reported to be a disulfide-linked heterodimer (70–80 kDa). Under reducing conditions the two chains of the receptor are dissociated and have molecular weights of 55 kDa (γ) and 40 kDa (δ). Both chains are glycosylated. A second form of the γ chain can also be expressed by human peripheral T cells; it too is recognized by anti-γ chain antibodies, has the same isoelectric point as the 40 kDa chain, but a lower molecular weight, 36 kDa. It is also disulfide bonded to the δ chain (Fig.6.26). A third form of the $\gamma\delta$ receptor expressed by a cell line isolated from an immunodeficiency patient has also been described: this is a non-disulfide-linked heterodimer, composed of a 55–66 kDa γ chain and a 40 kDa δ chain (Fig.6.26). The human γ chain can therefore exist in several forms depending on the presence or absence of a cysteine residue needed to form a disulfide bond with the δ chain, the length of the polypeptide backbone, and the amount of glycosylation. On the surface of some $CD3^+$ T cells $\gamma\gamma$ receptors have also been identified.

6.3.2 Function

The function of T cell receptors is to enable T cells to recognize foreign antigens. The two forms of T cell receptor identified are both thought to recognize antigen, but most likely in different ways. Antigen recognition by mature peripheral T cells is discussed in detail in Section 5.2.

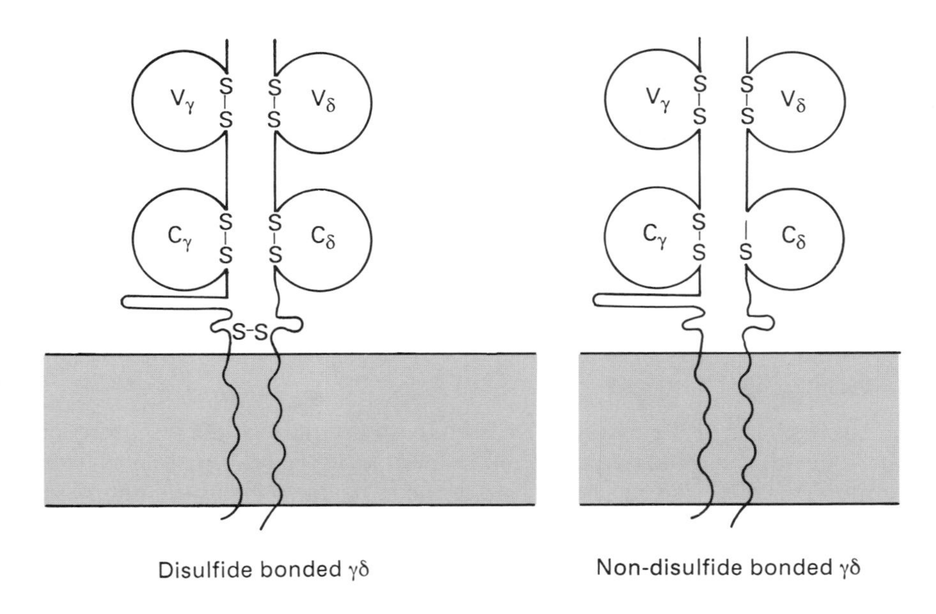

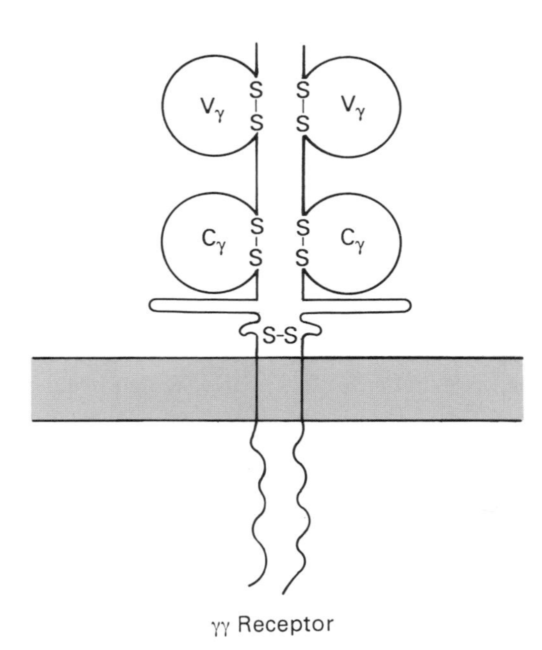

Fig.6.26 The predicted structures of $\gamma\delta$ receptors. (CD3 polypeptides are omitted for clarity)

The $\alpha\beta$ T cell receptor

The $\alpha\beta$ T cell receptor recognizes antigen in the form of peptide bound to an MHC molecule at the surface of an antigen presenting or target cell (see Section 3.2.1). Formal proof that the α and β chains are all that is required for antigen recognition has been obtained.

For example, the α and β chain genes from a cytotoxic T cell clone specific for H-2D^d and the hapten fluorescein were cloned and transferred to another cytotoxic T cell of a different specificity. After transfection, the second cytotoxic T cell acquired the ability to kill target cells expressing the fluorescein–H-2D^d complex, confirming that the α and β polypeptides were responsible for mediating antigen recognition (Dembic *et al.* 1986). Proof has also been obtained from experiments using transgenic mice (see Appendix) which have been constructed to express one $\alpha\beta$ T cell receptor on a large proportion of their T cells. In these T cell receptor transgenic mice the antigen specificity of the majority of the T cells is identical to that of the T cell clone from which the $\alpha\beta$ transgenes were isolated (e.g. Kisielow *et al.* 1988).

Although the $\alpha\beta$ T cell receptor has not been crystallized, the resolution of the three-dimensional structure of human MHC class I molecules (see Section 2.5) has allowed certain predictions to be made about the nature of the $\alpha\beta$ T cell receptor–peptide–MHC interaction (for review see Davis and Bjorkman 1988). The amino acids identified as important in V_H–V_L pairing of immunoglobulins are conserved and occur in identical positions in V_α and V_β. Thus it seems reasonable to assume that T cell receptor variable domains fold in a similar way to immunoglobulin variable domains. Hypervariable regions do exist in T cell receptor α and β polypeptides and presumably these are directly involved in contacting the peptide–MHC complex (Jorgensen *et al.* 1992). The three-dimensional structure of HLA class I molecules (see Section 2.5) shows that the peptide binding site is on the exposed, top surface of the molecule (Figs.2.5 and 2.6). The view of an MHC molecule as seen by an incoming T cell receptor is shown in Fig.6.27. Analysis of the peptide–MHC complex and the relative positions of CDR1 and CDR2 loops in immunoglobulin V_H and V_L domains shows that they are probably about the same distance apart in T cell receptors ($\sim$18.4 nm). The upper surface of the peptide–MHC complex might interact with a T cell receptor such that CDR1 and CDR2 are in contact with the two helices of the α_1 and α_2 domains of the MHC molecule (Hong *et al.* 1992). CDR3, which in immunoglobulins is more centrally placed (Fig.6.16), might then be able to contact the peptide bound in the groove of the MHC molecule. This model is represented in Fig.6.28. This is obviously only a model and may yet prove to be incorrect; but this the particular alignment is consistent with data from a number of studies.

Strenuous efforts are being made to produce a soluble form of T cell receptor through gene cloning techniques, so that crystals can be prepared for structural studies. So far progress has been relatively slow, but soluble molecules have now been isolated. One approach to the problem has involved fusing the genes for the extracellular domains of the T cell receptor α and β chains to the phospholipid tail of decay accelerating factor (DAF) of the complement system (see Section 9.4). The recombinant phosphatidylinositol-linked $\alpha\beta$ receptor can then be solubilized by the enzyme phospholipase. The possibility of engineering chimaeric proteins from both T cell receptors and antibodies has also been explored. The 3D structure of TcR is eagerly awaited.

In the mouse the presence of certain T cell receptor V_α and/or V_β polypeptides has been correlated with the recognition of particular antigens. For

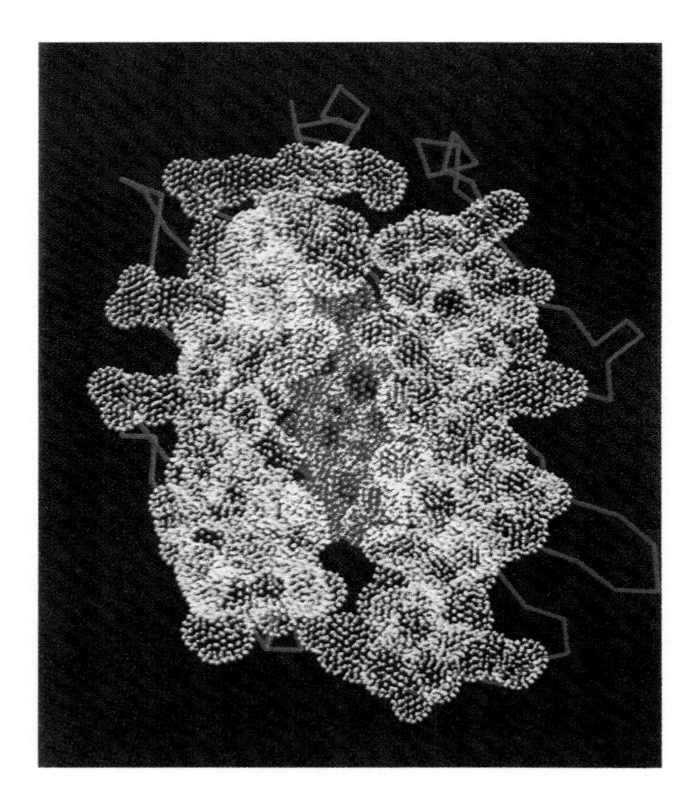

Fig.6.27 The view of the peptide–MHC complex seen by T cell receptor. Peptide is in pink, α helices in yellow, and β strands in blue. (From Davis and Bjorkman (1988), with permission)

example, two laboratories noted that T cells specific for a C-terminal peptide of cytochrome c plus H-2 IE preferentially used a single V_α segment to recognize the peptide–MHC complex. Other associations have also been documented, but in general these involve superantigens which may not be representative of conventional antigens (see Section 5.2). For a review of the molecular basis of T cell specificity see Matis (1990).

The $\gamma\delta$ T cell receptor

Expression of the $\gamma\delta$ T cell receptor precedes expression of the $\alpha\beta$ T cell receptor by 1 or 2 days during fetal ontogeny. In human, $\gamma\delta$ T cells are also present in the mature lymphocyte population in peripheral blood and lymphoid organs. In the mouse they predominate in epithelia of tissues not normally considered as lymphoid organs such as skin, intestine, and the lung. The distribution of $\gamma\delta$ T cells in the mouse is shown in Table 6.18; in humans the distribution of these T cells more closely parallels that of $\alpha\beta$ T cells. The antigen specificity and function of $\gamma\delta$ T cells remains largely unknown, although some progress is being made in both of these areas.

Hypervariable regions are identifiable in the variable domains of both the γ and the δ chain. Thus the association of the two variable domains is

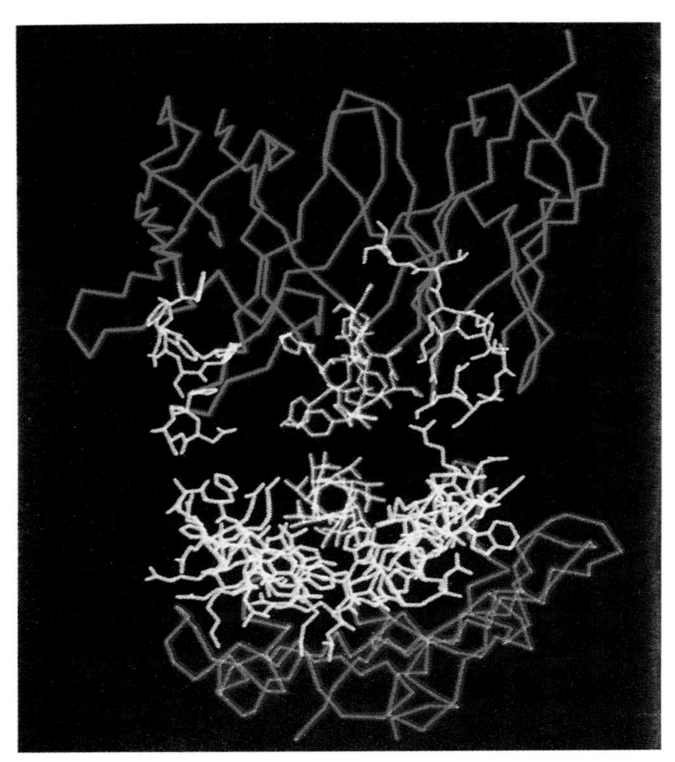

Fig.6.28 Model for the interaction of a T cell receptor (top: CDR1 and CDR2, yellow; CDR3, pink) with a peptide–MHC complex (bottom: α helices, yellow; peptide, pink). (From Davis and Bjorkman (1988), with permission). In this model the peptide bound to the MHC molecule is represented in an α helical configuration (pink), but note that peptides may actually be bound in an extended configuration (Madden *et al.* 1992)

thought to form an antigen binding site. Sequences corresponding to CDR3 have been implicated in influencing the antigen specificity of $\gamma\delta$ T cell receptors (Rellahan *et al.* 1991). Because many $\gamma\delta$ T cells do not usually express the T cell accessory molecules CD4 or CD8, it has been unclear whether they recognize antigen in association with an MHC molecule in the same way as $\alpha\beta$ T cells. It has been suggested that class I molecules encoded by the Tla region in mice (see Section 2.8) or other molecules associated with β_2-microglobulin, such as CD1 (Section 4.2.4), might be involved in antigen recognition by $\gamma\delta$ T cells. Indeed murine $\gamma\delta$ T cell clones have been isolated that are specific for an allelic MHC class I gene product encoded by the Tla region, and a human $\gamma\delta$ T cell clone that recognizes CD1 has been reported. Moreover, Tla molecules have been shown to be involved in the selection of $\gamma\delta$ T cells in the thymus.

For these experiments, rearranged $\gamma\delta$ TcR genes from a $\gamma\delta$ T cell clone of known specificity were used to construct transgenic mice. In mice that *did not* express the Tla determinant recognized by this $\gamma\delta$ TcR, transgenic $\gamma\delta$ T cells were found in the thymus and peripheral lymphoid organs. However, in mice that *did* express the Tla determinant transgenic $\gamma\delta$ T cells were not detectable (Dent *et al.* 1990). This transgenic experiment showed that self-reactive $\gamma\delta$ T cells can be eliminated in the thymus in a similar

Table 6.18 Distribution of $\gamma\delta$ T cells in the mouse

Organs	Presence of $\gamma\delta$ T cells	Contact with epithelial cells
Lymphoid organs		
Thymus	+	
Spleen	+	
Lymph-node	+	
Blood	+	
Digestive system		
Tongue	+	+
Oesophagus	+/−	+
Stomach	+	+
Small intestine	+	+
Large intestine	+	+
Liver	−	
Pancreas	−	
Reproductive system		
Ovary	+/−	
Uterus	+	+
Vagina	+	+
Testis	−	
Epididymis	+/−	−
Mammary gland during lactation	+	+
Urinary system		
Kidney	−	
Bladder	+	+
Others		
Skin	+	+
Brain	−	
Heart	−	

manner to that described for negative selection of $\alpha\beta$ self-reactive T cells (see Section 5.2.5). This and other observations suggest that at least some $\gamma\delta$ T cells undergo selection in the thymus and that self peptides may be actively involved in this process (Lafaille *et al.* 1990). However $\gamma\delta$ T cells have been shown to be present in β_2-microglobulin deficient mice which do not express class I molecules at the cell surface (Ziljstra *et al.* 1990). These data are obviously conflicting, and further investigation of the role of either Qa/Tla class I molecules or CD1 in antigen recognition by $\gamma\delta$ T cells is required.

There is now evidence that some $\gamma\delta$ T cells may not need a thymus to develop, and they may be positively selected by the gut epithelium, this perhaps playing an analogous role to the thymic epithelium (see Section 5.2.3). Many $\gamma\delta$ T cells in the mouse gut also express homodimeric ($\alpha\alpha$) CD8 molecules, and these differ from the $\gamma\delta$ cells that apparently develop in the the thymus which express the more common heterodimeric ($\alpha\beta$) CD8 molecule (see Section 4.2.4) (Guy-Grand *et al.* 1991). Moreover many of the $\alpha\alpha$ CD8 cells do not express Thy-1, otherwise thought to be a T cell marker in the mouse. The significance of the observations is, as yet, unclear.

Recently evidence has been accumulating to suggest that $\gamma\delta$ T cells participate in immune responses to mycobacteria and other infectious microorganisms. Many mouse $\gamma\delta$ T cells can be stimulated by a heat shock protein (HSP-65) from *Mycobacterium bovis*, and human $\gamma\delta$ T cells reactive with the same mycobacterial protein have been isolated from the joints of patients with rheumatoid arthritis and leprosy. One could speculate that peptides from heat shock proteins associate with Tla class I-like molecules, or the CD1 antigen, and that this complex is recognized by $\gamma\delta$ TcR.

Heat shock proteins (HSPs) are a highly conserved group of proteins found in both prokaryotic and eukaryotic organisms. The synthesis of HSPs is increased in response to many environmental stresses including temperature changes, inflammation, fever, and viral infections (for reviews see Kaufmann 1990, Young 1990, and Moller 1991). In bacteria, HSPs can leak from the cell wall following damage; these proteins have been identified as the major antigens responsible for triggering an immune response in several types of bacterial infection. In normal mice, 5% of the $\gamma\delta$ cells in the newborn thymus were found to be reactive with HSP-65. These observations suggest that individual $\gamma\delta$ populations might be specific for particular autologous HSPs and that the function of $\gamma\delta$ T cells might be to respond to certain stress signals. The reasons for, and consequences of, this reactivity are unknown. Cross-reactivity of $\gamma\delta$ T cells with heterologous HSPs might result in a breakdown of tolerance, and this might explain why $\gamma\delta$ T cells reactive with HSP-65 are found in patients with rheumatoid arthritis (for review see Born *et al.* 1990). There is much speculation as to the significance of $\gamma\delta$ T cells in autoimmune diseases.

Subpopulations of $\gamma\delta$ cells have been identified in both human and mouse based on the molecular structure of the $\gamma\delta$ TcR. In the mouse distinct TcR repertoire and homing properties of $\gamma\delta$ T cells have been described.

For example, the TcR expressed by $\gamma\delta$ T cells in skin are relatively homogenous, each consisting of one particular $V_\gamma5$ and one particular $V_\delta1$ chain. A second homogenous subset using the $V_\gamma6$-$J_\gamma1$ gene segments, distinct from that found in the skin, has been found in epithelia of the vagina, uterus, and tongue. This $\gamma\delta$ T cell subset, unlike that found in the skin, is selected late in the development of the fetal thymus. The reasons for the homing of particular subsets of $\gamma\delta$ T cells to different epithelia are not understood.

It has been suggested that the $\gamma\delta$ T cells might play a role in immune surveillance (for review see Janeway *et al.* 1988) and the localization of $\gamma\delta$ T cells to different epithelia in the body support this hypothesis (Table 6.18).

Several laboratories have isolated $\gamma\delta$ T cells that have cytotoxic activity. It has been suggested that these cells may account for some of the 'non-MHC restricted' cyotoxicity seen in human peripheral T cells. $\gamma\delta$ cells have been shown to express several mediators of cytoxicity (Section 10.3.4) including perforin and the serine esterases, SE1 and SE2 (Koizumi *et al.* 1991). This 'NK-like' activity (Section 10.2) may be significant if $\gamma\delta$ T cells are involved in immune surveillance.

Further reading

Immunoglobulins

Absolom, D.R. (1986). The nature of the antigen–antibody bond and the factors affecting its association and dissociation. *CRC Critical Reviews in Immunology*, **6**, 1–46.

Amzel, L.M. and Poljack, R.J. (1979). Three dimensional structure of immunoglobulins. *Annual Review of Biochemistry*, **48**, 961–98.

Burton, D.R. (1986). Is IgM dislocation a common feature of antibody function? *Immunology Today*, **7**, 165–7.

Colman, P.M., Air, G.M., Webster, R.G., Varghese, J.N., Baker, A.T., Lentz, M.R., Tulloch, P.A., and Laver, W.G. (1987*a*). How antibodies recognise virus proteins. *Immunology Today*, **18**, 323–6.

Conrad, D.H. (1990). Fc_{ε}RII/CD23: the low affinity receptor for IgE. *Annual Review of Immunology*, **8**, 623–46.

Davies, D.R. and Metzger, H. (1983). Structural basis of antibody function. *Annual Review of Immunology*, **11**, 87–117.

Davis, A.C. and Shulman, M.J. (1989). IgM molecular requirements for its assembly and function. *Immunology Today*, **10**, 118–28.

Givol, D. (1974). Affinity labelling and topology of the antibody combining site. *Essays in Biochemistry*, **10**, 73–103.

Koshland, M.E. (1985). The coming of age of the immunoglobulin J chain. *Annual Review of Immunology*, **3**, 425–53.

Mestecky, J. and McGhee, J.R. (1987). Immunoglobulin A (IgA): molecular and cellular interactions involved in IgA biosynthesis and immune response. *Advances in Immunology*, **40**, 153–245.

Metzger, H., Alcaraz, G., Hohman, R., Kinet, J-P., Pribluda, V., and Quarto, R. (1986). The receptor high affinity for immunoglobulin E. *Annual Review of Immunology*, **4**, 419–70.

Morgan, E.L. and Weigle, W.O. (1987). Biological activities residing in the Fc region of immunoglobulin. *Advances in Immunology*, **40**, 61–134.

Morrison, S.L. (1992). *In vitro* antibodies: strategies for production and application. *Annual Review of Immunology*, **10**, 239–65.

Nisonoff, A., Hooper, J.E., and Spring, S.B. (1975). *The antibody molecule*. Academic Press, New York.

Poljack, R.J. (1975). X-ray diffraction studies of immunoglobulins. *Advances in Immunology*, **21**, 1–33.

Pumphrey, R.S.H. (1986*a*). Computer models of the human immunoglobulins. 1. Shape and segmental flexibility. *Immunology Today*, **7**, 174–8.

Pumphrey, R.S.H. (1986*b*). Computer models of the human immunologlobulins. 2. Binding sites and molecular interactions. *Immunology Today*, **7**, 206–11.

Underwood, B.J. and Schiff, J.M. (1986). Immunoglobulin A: Strategic defence initiative at the mucosal surface. *Annual Review of Immunology*, **4**, 147–65.

Unkeless, J.C., Scilgliano, E., and Freedman, V. (1988). Structure and function of human and murine receptors for IgG. *Annual Review of Immunology*, **6**, 251–81.

Winter, G. and Milstein, C. (1991). Man-made antibodies. *Nature*, **349**, 293–9.

T cell receptor

Allison, J.P. and Lanier, L.L. (1987). Structure, function and serology of the T cell antigen receptor complex. *Annual Review of Immunology*, **5**, 503–40.

Born, W., Happ, M.P., Dallas, A., Reardon, C., Kubo, R., Shinnick, T., Brennan, P., and O'Brien, R. (1990). Recognition of heat shock proteins and gamma-delta cell function. *Immunology Today*, **11**, 40–3.

Brenner , M.B., Strominger, J.L., and Kragel, M.S. (1988). The $\gamma\delta$ T cell receptor. *Advances in Immunology*, **43**, 139–92.

Davis, M.M. and Bjorkman, P.J. (1988). T cell antigen receptor genes and T cell recognition. *Nature*, **334**, 395–402.

Janeway, C.A., Jones, B., and Hayday, A. (1988). Specificity and function of T cells bearing $\gamma\delta$ receptors. *Immunology Today*, **9**, 73–6.

Jorgensen, J.L., Reay, P.A., Ehrich, E.W., and Davis, M.M. (1992). Molecular components of T-cell recognition. *Annual Review of Immunology*, **10**, 835–73.

Kaufmann, S.H.E. (1990). Heat shock proteins and the immune response. *Immunology Today*, **11**, 129–36.

Matis L.A. (1990). The molecular basis of T cell specificity. *Annual Review of Immunology*, **8**, 65–82.

Möller, G. (ed.) (1991). Heat-shock proteins and the immune system. *Immunological Review* Vol. 121. Munksgaard, Copenhagen.

Raulet D.H. (1989). The structure function and molecular genetics of the $\gamma\delta$ T cell receptor. *Annual Review of Immunology*, **7**, 175–208

Weiss, A. (1990). Structure and function of the T cell antigen receptor. *Journal of Clinical Investigation*, **86**, 1015–22.

Yang, R.A. (1990). Stress proteins and immunology. *Annual Review of Immunology*, **8**, 401–420.

Literature cited

Absolom, D.R. (1986). The nature of the antigen–antibody bond and the factors affecting its association and dissociation. *CRC Critical Reviews in Immunology*, **6**, 1–46.

Allison, J.P. and Lanier, L.L. (1987). Structure, function and serology of the T cell antigen receptor complex. *Annual Review of Immunology*, **5**, 503–40.

Amit, A.G., Mariuzza, R.A., Phillips, S.E.V., and Poljack, R.J. (1986). Three-dimensional structure of an antigen-antibody complex at 2.8Å resolution. *Science*, **233**, 747–53.

Amzel, L.M. and Poljack, R.J. (1979). Three dimensional structure of immunoglobulins. *Annual Review of Biochemistry*, **48**, 961–98.

Anderson, C.L. and Looney, R.J. (1987). Human leukocyte IgGFc receptors. *Immunology Today*, **7**, 264–6.

Benjamin, D.C., Berzofsky, J.A., East, I.J., Gurd, F.R.N., Hannum, C., Leach, S.J., *et al.* (1984). The antigenic structure of proteins: a reappraisal. *Annual Review of Immunology*, **2**, 67–101.

Bentley, G.A., Boulot, G., Riottot, M.M., and Poljack, R.J. (1990). Three-dimensional structure of an idiotope-anti-idiotope complex. *Nature*, **348**, 254–7.

Berzofsky, J.A., Buckenmeyer, G.K., Hicks, G., Gurd, F.R.N., Feldman, R.J., and Minna, J. (1982). Topographic antigenic determinants recognised by monoclonal antibodies to sperm whale myoglobulin. *Journal of Biological Chemistry*, **257**, 3189–98.

Brenner , M.B., Strominger, J.L., and Kragel, M.S. (1988). The $\gamma\delta$ T cell receptor. *Advances in Immunology*, **43**, 139–92.

Blattner, F.R. and Tucker, P.W. (1984). The molecular biology of immunoglobulin D. *Nature*, **307**, 417–21.

Born, W., Happ, M.P., Dallas, A., Reardon, C., Kubo, R., Shinnick, T., Brennan, P., and O'Brien, R. (1990). Recognition of heat shock proteins and gamma-delta cell function. *Immunology Today*, **11**, 40–3.

Burton, D.R. (1986). Is IgM dislocation a common feature of antibody function? *Immunology Today*, **7**, 165–7.

Chothia, C., Lesk, A.M., Tramontaro, A., Levitt, M., Smith-Gill, S.J., Air, G. *et al.* (1990). Conformations of immunoglobulin hypervariable regions. *Nature*, **342**, 877–83.

Clackson, T., Hoogenboom, H.R., Griffiths, D., and Winter, G. (1991). Making antibody fragments using phage display libraries. *Nature*, **352**, 624–8.

Colman, P.M., Air, G.M., Webster, R.G., Varghese, J.N., Baker, A.T., Lentz, M.R., Tulloch, P.A., and Laver, W.G. (1987*a*). How antibodies recognise virus proteins. *Immunology Today*, **18**, 323–6.

Colman, P.M., Laver, W.G., Varghese, J.N., Baker, A.J., Tulloch, P.A., Air, G.M., and Webster, R.G., (1987*b*). Three-dimensional structure of a complex of antibody wth influenza virus neuraminidase. *Nature*, **326**, 358–63.

Conrad, D.H. (1990). $Fc_{\varepsilon}RII$/CD23: the low affinity receptor for IgE. *Annual Review of Immunology*, **8**, 623–46.

Davies, D.R. and Metzger, H. (1983). Structural basis of antibody function. *Annual Review of Immunology*, **11**, 87–117.

Davis, M.M. and Bjorkman, P.J. (1988). T cell antigen receptor genes and T cell recognition. *Nature*, **334**, 395–402.

Davis, A.C. and Shulman, M.J. (1989). IgM molecular requirements for its assembly and function. *Immunology Today*, **10**, 118–28.

Dembic, Z., Haas, W., Weiss, S., McCubrey, J., Kiefer, H., von Boehmer, H., and Steinmetz, M. (1988). Transfer of specificity by murine α and β T cell receptor genes. *Nature*, **320**, 232–8.

Dent, A.L., Matis, L.A., Hooshmand, F., Widacki, S.M., Bluestone, J.A., and Hedrick, S.M. (1990). Self-reactive $\gamma\delta$ T cells are eliminated in the thymus. *Nature*, **343**, 714–19.

Eisenman, E. and Boler, J.B. (1992). Engagement of the high affinity IgE receptor activates src protein-related tyrosine kinase. *Nature*, **355**, 78–80.

Givol, D. (1974). Affinity labelling and topology of the antibody combining site. *Essays in Biochemistry*, **10**, 73–103.

Guy-Grand, D., Gerf-Bensussan, N., Malissen, B., Malassis-Seris, M., Brottet, C., and Vassalli, P. (1991). Two gut intraepithelial $CD8^+$ lymphocyte populations with different T cell receptors: A role for the gut epithelium in T cell differentiation. *Journal of Experimental Medicine*, **173**, 471–81.

Hong, S-C., Chelouche, A., Lin, R-h, Shaywitz, D., Braunstein, N.S., Glimcher, L. and Janeway, C.A. (1992). An MHC interaction site maps to the amino terminal half of the T cell receptor α chain variable domain cell, **69**, 999–1009.

Huse, W.D., Sastry, L., Iverson, S.A., Kang, A.S., Alting-Mees, M., Burton, D.R., Benkovic, S.J., and Lerner, R.A. (1989). Generation of a large combinational library of the immunoglobulin repertoire in phage lambda. *Science*, **246**, 1275–81.

Janeway, C.A., Jones, B., and Hayday, A. (1988). Specificity and function of T cells bearing $\gamma\delta$ receptors. *Immunology Today*, **9**, 73–6.

Jones, P.T., Dear, P.H., Foote, J., Neuberger, M.S., and Winter, G. (1986). Replacing the complementarity determining regions in a human antibody with those of mouse. *Nature*, **321**, 522–5.

Jorgensen, J.L., Esser, U., Fazekas de St. Groth, B., Reay, P.A., and Davis, M.H. (1992). Mapping T cell receptor-peptide contacts by variant peptide immunisation of single chain TCR transgenics. *Nature*, **355**, 224–30.

Kaufmann, S.H.E. (1990). Heat shock proteins and the immune response. *Immunology Today*, **11**, 129–36.

Kisielow, P., Bluthmann, H., Staerz, U.D., Steinmetz, M., and von Boehmer, H. (1988). tolerance in T cell receptor transgenic mice involves deletion of nonmature CO4^{+}8^{+} thymocytes. *Nature*, **333**, 742–6.

Kipps, T.J., Parham, P., Punt, J., and Herzenberg, L.A. (1985). Importance of isotype in human ADCC directed by murine monoclonal antibodies. *Journal of Experimental Medicine*, **161**, 1–17.

Koizumi, H., Liu C-C., Zheng, L.M., Joug, S.V., Bayne, N.K. Holoshitz, J., and Young, J.D.E. (1991). Expression of perforin and serine esterases by human $\gamma\delta$ T cells. *Journal of Experimental Medicine*, **173**, 499–502.

Koshland, M.E. (1985). The coming of age of the immunoglobulin J chain. *Annual Review of Immunology*, **3**, 425–53.

Lafaille, J.J., Haas, W., Coufinho, A., and Tmegawa, S. (1990). Positive selection of $\gamma\delta$ T cells. *Immunology Today*, **11**, 75–8.

McCafferty, J., Griffiths, A.D., Winter, G., and Chiswell, D.J. (1990). Phage antibodies: filamentous phage displaying antibody variable domains. *Nature*, **348**, 552–4.

Madden, D.R., Gorga, J.C., Strominger, J.L., and Wiley, D.C. (1992). The three dimensional structure of HLA-B27 at 2.1 A resolution suggests a general mechanism for tight peptide binding to MHC. *Cell*, **70**, 1035–48.

Matis L.A. (1990). The molecular basis of T cell specificity. *Annual Review of Immunology*, **8**, 65–82.

Mestecky, J. and McGhee, J.R. (1987). Immunoglobulin A (IgA): molecular and cellular interactions involved in IgA biosynthesis and immune response. *Advances in Immunology*, **40**, 153–245.

Metzger, H., Alcaraz, G., Hohman, R., Kinet, J-P., Pribluda, V., and Quarto, R. (1986). The receptor high affinity for immunoglobulin E. *Annual Review of Immunology*, **4**, 419–70.

Möller, G. (ed.) (1991). Heat-shock proteins and the immune system. *Immunological Review* Vol. 121. Munksgaard, Copenhagen.

Morgan, E.L. and Weigle, W.O. (1987). Biological activities residing in the Fc region of immunoglobulin. *Advances in Immunology*, **40**, 61–134.

Nisonoff, A., Hooper, J.E., and Spring, S.B. (1975). *The antibody molecule*. Academic Press, New York.

Poljack, R.J. (1975). X-ray diffraction studies of immunoglobulins. *Advances in Immunology*, **21**, 1–33.

Porter, R.R. (1959). The hydrolysis of rabbit gamma-globulin and antibodies with crystalline papain. *Biochemical Journal*, **73**, 119–26.

Pumphrey, R.S.H. (1986*a*). Computer models of the human immunoglobulins. 1. Shape and segmental flexibility. *Immunology Today*, **7**, 174–8.

Pumphrey, R.S.H. (1986*b*). Computer models of the human immunologlobulins. 2. Binding sites and molecular interactions. *Immunology Today*, **7**, 206–11.

Rellahan, B.L., Bluestone, JA., Houlden, B.A., Cotterman, M.M., and Matis, L.A. (1991). Junctional sequences influence the specificity of $\gamma\delta$ T cell receptors. *Journal of Experimental Medicine*, **173**, 503–6.

Rogers, J., Early, P., Carter, C., Calame, K., Bond, M., Hood, L., and Wall, R. (1980). Two mRNAs with different 3′ ends encode membrane bound and secreted forms of immunoglobulin mu chain. *Cell*, **20**, 303–12.

Stanfield, R.L., Fieser, T.M., Lerner, R.A., and Wilson, I.A. (1990). Crystal structures of an antibody of peptide and its complex with peptide antigen at 2.8 Å. *Science*, **248**, 712–19.

Traunecker, A., Dolder, B., Oliveri, F., and Karjalarinen (1989). Solubilising the T cell receptor—problems in solution. *Immunology Today*, **10**, 29–32.

Unkeless, J.C., Scilgliano, E., and Freedman, V. (1988). Structure and function of human and murine receptors for IgG. *Annual Review of Immunology*, **6**, 251–81.

Venkitaraman, A.R., Williams, G.T. Dariavach, P., and Neuberger, M.S. (1991) The B cell antigen receptor of the five immunoglobulin classes. *Nature*, **352**, 777–81.

Waldor, M.K., Mitchell, D., Kipps, T.J., Herzenberg, L.A., and Steinman, L. (1987). Importance of IgG isotype in therapy of experimental allergic encephalomyelitis with monoclonal anti-CD4 antibody. *Journal of Immunology*, **139**, 3660–4.

Ward, E.S., Gussow, D., Griffiths, A.D., Jones, P.T., and Winter, G. (1989). Binding activities of single immunoglobulin variable domains secreted from *E. coli*. *Nature*, **341**, 544–50.

Weiss, A. (1990). Structure and function of the T cell antigen receptor. *Journal of Clinical Investigation*, **86**, 1015–22.

Winter, G. and Milstein, C. (1991). Man-made antibodies. *Nature*, **349**, 293–9.

Wu, T.T. and Kabat, E.A. (1970). An analysis of the sequences of the variable regions of Bence-Jones proteins and myeloma light chains and their implications for antibody complementarity. *Journal of Experimental Medicine*, **132**, 211—50.

Young, R.A. (1990). Stress proteins and immunology. *Annual Review of Immunology*, **8**, 401–20.

Ziljstra, M., Bix, M., Simister, N.E., Loring, J.M., Raulet, D.H., and Jaenisch, J. (1990). β_2-microglobulin deficient mice lack CD4^{+}8^{+} cytolytic T cells. *Nature*, **334**, 742–6.

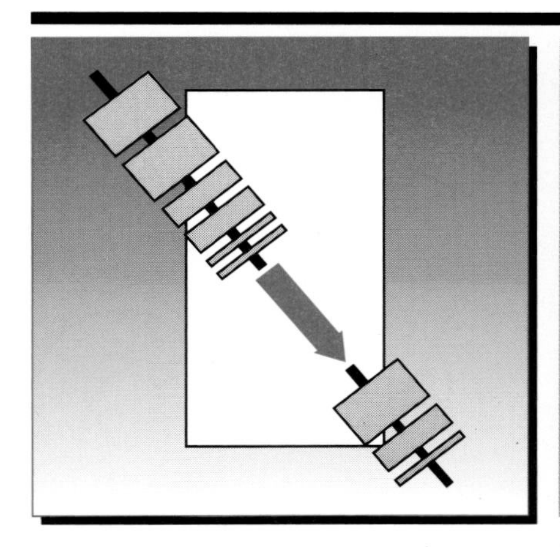

7

ANTIGEN RECEPTOR GENES AND THE IMMUNOGLOBULIN SUPERFAMILY

Scheme 7. Antigen receptor genes

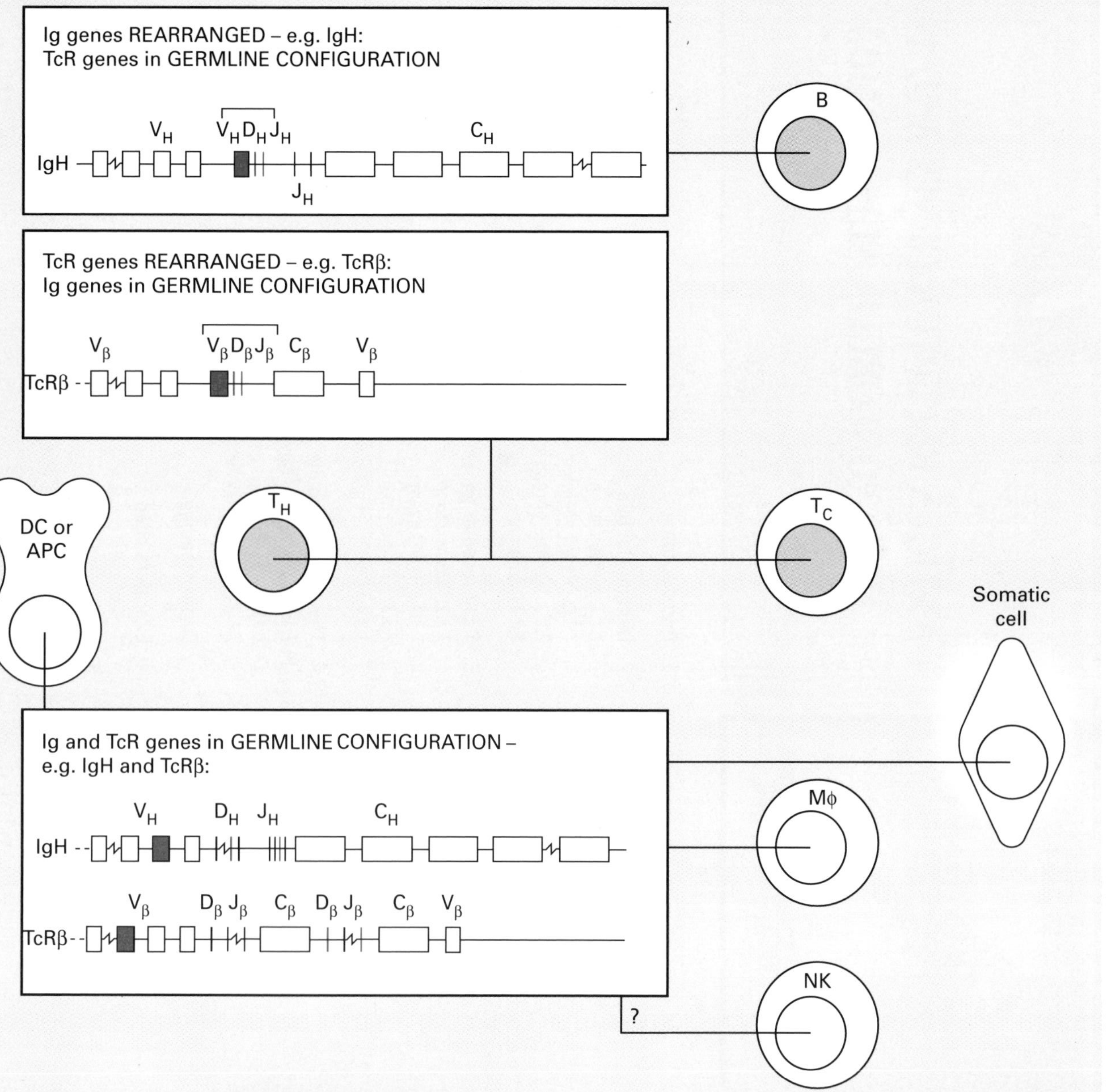

7.1 Introduction

One of the most important features of the immune system is its ability to mount a response against virtually any foreign antigen and even newly-evolved organisms. In previous chapters we have seen that antigen recognition is mediated by specific antigen receptors on lymphocytes–antibodies (immunoglobulins) and T cell receptors. There are many millions of B cells and T cells in the body and each clone of lymphocytes is able to recognize a particular antigenic determinant or epitope through its antigen receptor. The determinant recognized can be unique to that clone, and is probably different from those recognized by the antigen receptors of almost every other clone of lymphocytes in the same individual. These statements imply that the number of antigen receptors present in any individual must also be enormous. Yet in spite of this diversity, all the antigen receptors within each family, immunoglobulins, or T cell receptors, have a similar protein structure, as we have seen in Chapter 6 (Fig.7.1). Variability in the structure allows each receptor to interact specifically with a particular antigenic determinant. This variability is localized to one part of each molecule, the variable (V) region. Within the basic framework structure of the variable domain are hypervariable or complementarity determining regions (CDRs) which are directly involved in the interaction with antigen (see Section 6.2.3). The remainder of each antigen receptor molecule, the constant (C) region varies only between the different classes and types of immunoglobulin polypeptide chains and between the different forms of the T cell receptor. The genetic basis for these common structures and, at the same time, for the diversity of antigen receptors will be the focus of this chapter. It will be seen that diversity arises from a complex series of **DNA rearrangements** and other genetic changes that appear to be unique to the lymphocyte antigen receptor genes.

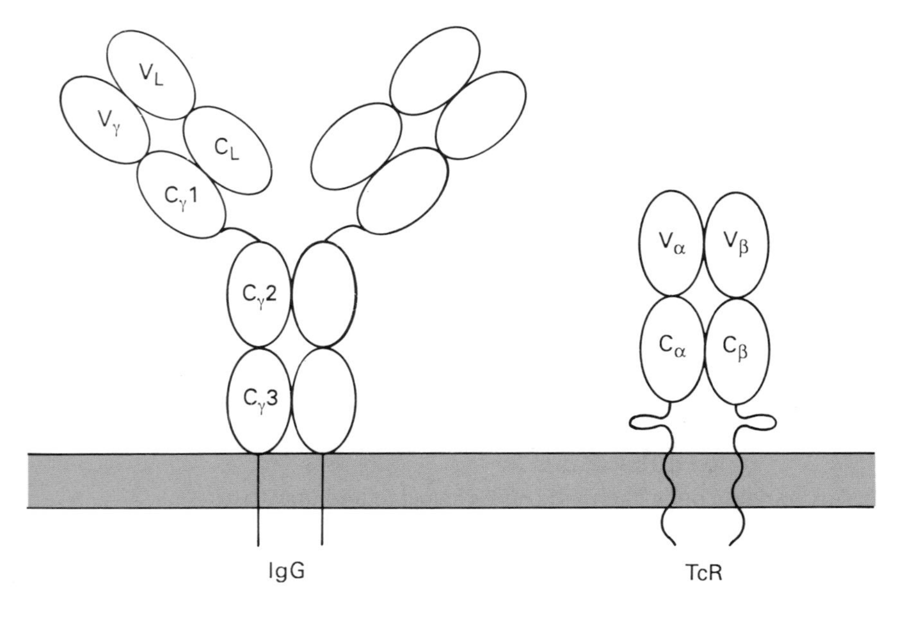

Fig.7.1 Basic structure of lymphocyte antigen receptors

7.2 Immunoglobulin genes

When the structure of immunoglobulins was first elucidated in the late 1950s and as amino acid sequence data accumulated (see Section 6.2), two basic problems became evident. How could variable and constant regions be present within the same polypeptide chain when the structural data available suggested that these two regions were selected from different pools of genetic information? Secondly, how did the enormous number of variable regions produced by any one individual come about? There was speculation as to how the genes for immunoglobulins might be arranged in the DNA, but it became clear that to provide definitive answers to these questions, experiments at the molecular level were required.

It was difficult, at first sight, to rationalize the features of the newly-discovered antibody structure with the ideas held at the time about the relationship between proteins and their genes. A long held view, known as the **germline theory**, envisaged that each polypeptide was encoded by a separate gene. Thus it was assumed that there was a separate gene for each immunoglobulin polypeptide produced by an individual, that each gene contained nucleotide sequences encoding a variable and a constant region, and that these were contiguous. Certainly this theory was consistent with the organization of genes in prokaryotes and there had been no reason to believe that these rules did not apply to the organization of eukaryotic genes as well. However, in the 1960s, this hypothesis did not seem compatible with the information that had been obtained about immunoglobulins.

In 1965, taking all the experimental and theoretical data into account, Dreyer and Bennett proposed that the information for each immunoglobulin polypeptide was in fact encoded by two separate pools of genes: one very large gene pool containing all the information required for the synthesis of the variable regions, and another much smaller pool containing the information for the constant regions, (Fig.7.2). This radically new concept suggested

Germline DNA

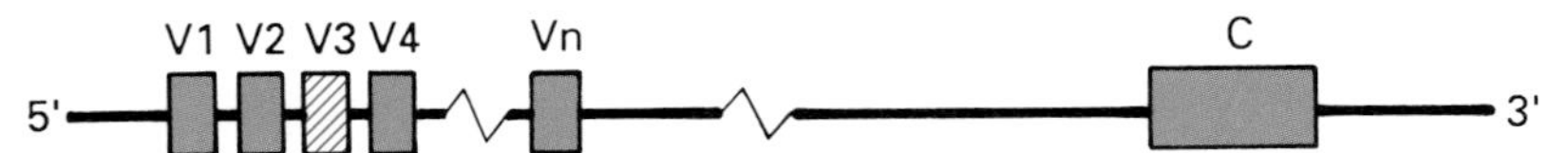

The heavy and light chain genes are encoded by two separate pools of information: one large pool containing the gene segments encoding the variable domain, and a much smaller pool encoding the constant domain.

B lymphocyte DNA

Any variable gene segment can be selected and the DNA rearranged so that it becomes joined to the constant region gene.

Fig.7.2 Dreyer and Bennett hypothesis for the organization of immunoglobulin genes

that one polypeptide chain could be encoded by two genes or, at least, by two 'gene segments'. By proposing that the products of different variable region genes could combine with the same constant region gene, it was possible to explain how immunoglobulins could recognize so many different antigenic determinants and yet have identical structures in other parts of the molecule, the constant regions. This hypothesis also implied that the variable and constant region genes had to be joined together before they could be transcribed and translated into protein.

Unfortunately, experimental data to support Dreyer and Bennett's hypothesis were not forthcoming until over 10 years later, when technical developments in molecular biology allowed the appropriate experiments to be performed. We now know that the principles of this hypothesis also apply to the genes encoding T cell receptors. As investigations into the organization of immunoglobulin genes developed it soon became clear that this idea alone could not completely explain the degree of diversity of lymphocyte antigen receptors found in the immune system, and molecular analyses have revealed that the situation is in many ways much more complex than was originally predicted by Dreyer and Bennett (1965).

7.2.1 Organization

The organization of mouse and human light chain genes is shown in Fig.7.3. The chromosomal location of these genes has been defined in both species: the κ chain genes are, respectively, on chromosomes 6 and 2 and the λ genes are located on chromosomes 16 and 22 (Table 7.1).

From experimental data obtained in the late 1970s we now know that in the embryo, i.e. the germline, the DNA coding for the variable region of an immunoglobulin light chain is physically separated on the chromosome from that which encodes the constant region, as predicted (Fig.7.2). For these experiments, DNA from mouse embryos (which was assumed to be representative of germline DNA) and adult myeloma cells (representative of fully differentiated B cells) was used for hybridization studies with purified λ light chain mRNA (Hozumi and Tonegawa 1976). The two types of DNA were digested with a restriction endonuclease, and the fragments of DNA generated were hybridized with the λ mRNA (Northern blots; see Appendix). The mRNA coding for the λ light chain hybridized to two separate fragments of the digested embryonic DNA, but to only a single fragment of the digested myeloma DNA (Fig.7.4). These data clearly suggested that the DNA sequences coding for the variable and constant region of the light chain were separated in the embryo, but had been *rearranged* and were found on the same fragment of DNA in mature B lymphocytes.

In the germline of undifferentiated B lymphocytes, and in fact in all cells of the body other than mature B cells, the gene segments coding for the variable and constant regions of immunoglobulin light chains are separated in the DNA. In developing B lymphocytes we shall see that some gene segments are brought closer together to form a functional immunoglobulin unit; this process is termed **DNA** or **gene rearrangement**. For reviews see Honjo *et al.* (1989). An example of this, for κ chain genes is shown in Fig.7.5.

Light chain genes

κ chains. On chromosome 6 in the mouse there is a single exon encoding the constant domain of κ chains (C_κ), together with a large number of **gene segments**† encoding the variable domains (Fig.7.3). The DNA encoding the variable domain is further divided into two different families of gene segments, **V_κ and J_κ**.

†The term gene segment is used to describe the exons V, D, and J encoding the variable domains of immunoglobulins and T cell receptors.

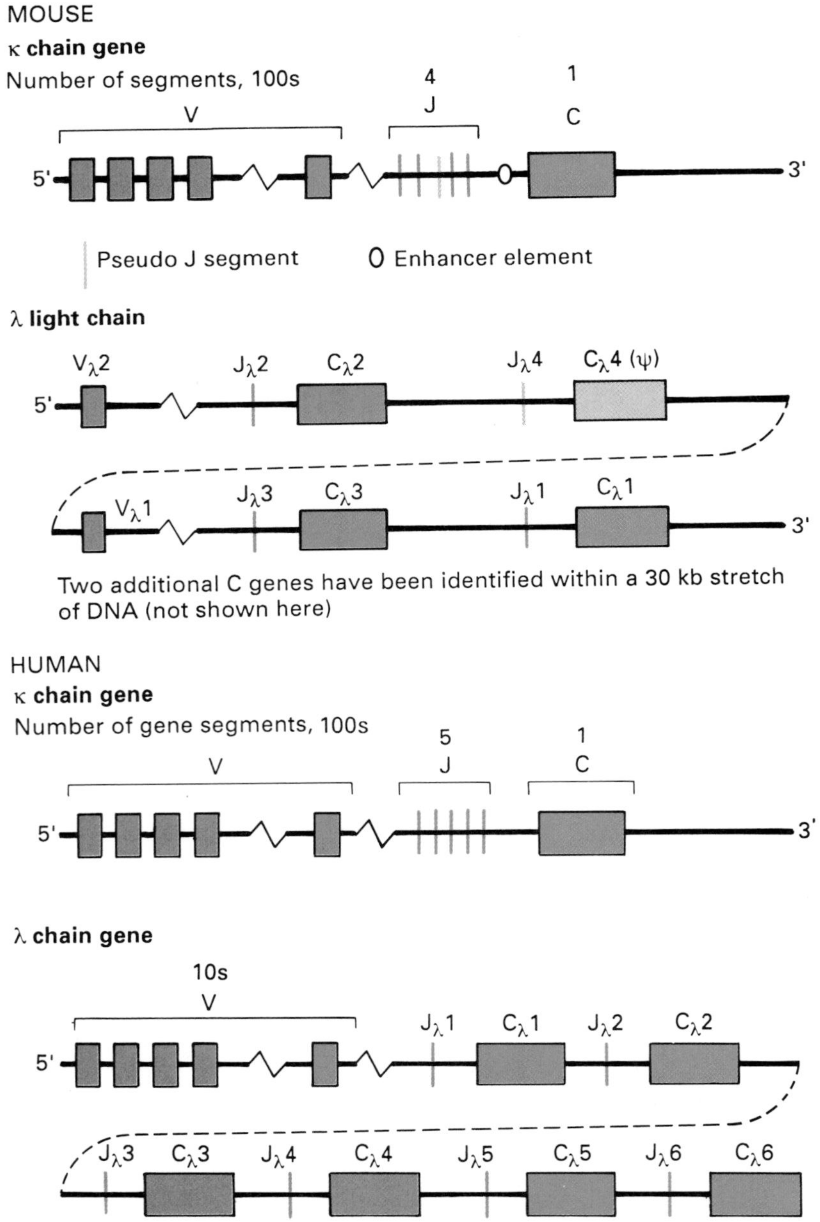

MOUSE

κ chain gene

Number of segments, 100s

V 4 J 1 C

5' 3'

Pseudo J segment

Enhancer element

λ light chain

$V_\lambda 2$ $J_\lambda 2$ $C_\lambda 2$ $J_\lambda 4$ $C_\lambda 4$ (ψ)

5'

$V_\lambda 1$ $J_\lambda 3$ $C_\lambda 3$ $J_\lambda 1$ $C_\lambda 1$

3'

Two additional C genes have been identified within a 30 kb stretch of DNA (not shown here)

HUMAN

κ chain gene

Number of gene segments, 100s

V 5 J 1 C

5' 3'

λ chain gene

10s V

$J_\lambda 1$ $C_\lambda 1$ $J_\lambda 2$ $C_\lambda 2$

5'

$J_\lambda 3$ $C_\lambda 3$ $J_\lambda 4$ $C_\lambda 4$ $J_\lambda 5$ $C_\lambda 5$ $J_\lambda 6$ $C_\lambda 6$

Fig.7.3 Organization of the mouse and human immunoglobulin light chain genes

Table 7.1 Chromosomal locations of immunoglobulin and T cell receptor genes

		Human	Mouse
Immunoglobulin	Heavy chain	14q32	12F1
	κ	2p12	6C2
	λ	22q11	16
T cell receptor	α	14q11	14C–D
	β	7q32	6B
	γ	7p15	13A2–3
	δ	14	14

The family of V_κ segments is by far the largest of the two, and encodes the majority of the 110 amino acids of the variable domain, from amino acids 1–95 or 96 (see Fig.6.12). The number of V_κ gene segments has been estimated by Southern blotting analysis to be in the order of 100, but a significant proportion of these are pseudogenes. A pseudogene is a defective gene that cannot be used to synthesize a functional protein product. Although the psuedogene segments cannot contribute to the generation of light chain

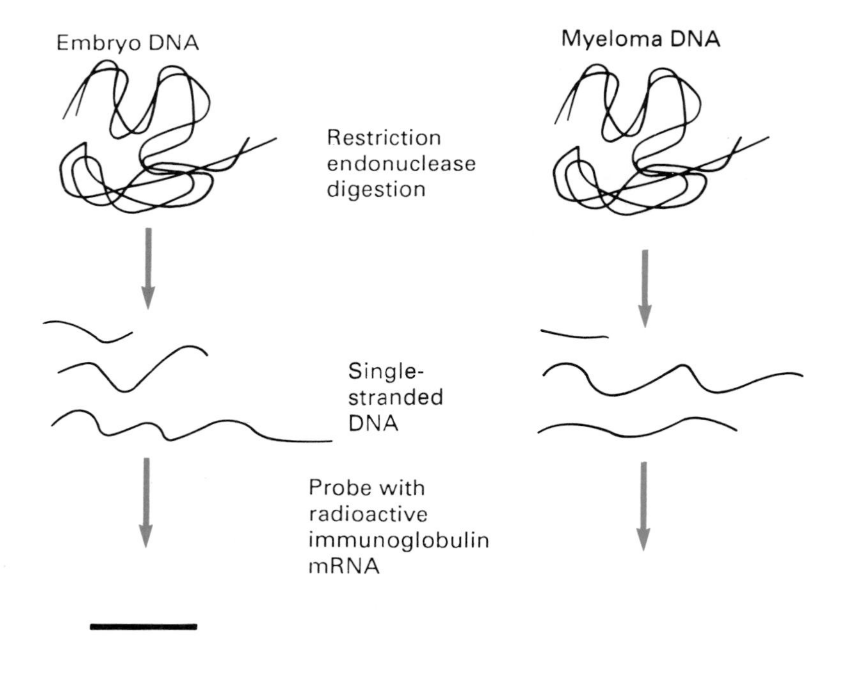

Fig.7.4 Experiment to show that V and C genes are arranged differently in embryonic (germ-line) and B cell DNA (Hozumi and Tonegawa 1976)

diversity directly, they may provide an extra pool of information that can be used to generate new V_K segments, either by gene conversion or unequal recombination. These two mechanism have also been implicated in the generation of new polymorphisms in MHC genes (see Section 2.9; Fig.2.25). The utilization of pseudogenes has also been shown to be a major contributor to the generation of diversity in chicken λ light chains. The remainder of the variable domain, from amino acids 95/96–110, is encoded by a small segment of DNA called the joining segment, J_κ. Five have been identified for murine κ chain genes; these are designated $J_{\kappa 1}$ to $J_{\kappa 5}$, but only four of these are functional; $J_{\kappa 3}$ is a pseudogene (Fig.7.3).

The large pool of V_κ segments is arranged one after another in one region of the chromosome, sometimes referred to as the V_K region, in what is termed a 'tandem array'. Each V_κ segment is preceded by another gene segment encoding the leader sequence, which is important for the biosynthesis of the light chain (see Section 6.2.6). The J_κ segments are located downstream, in a

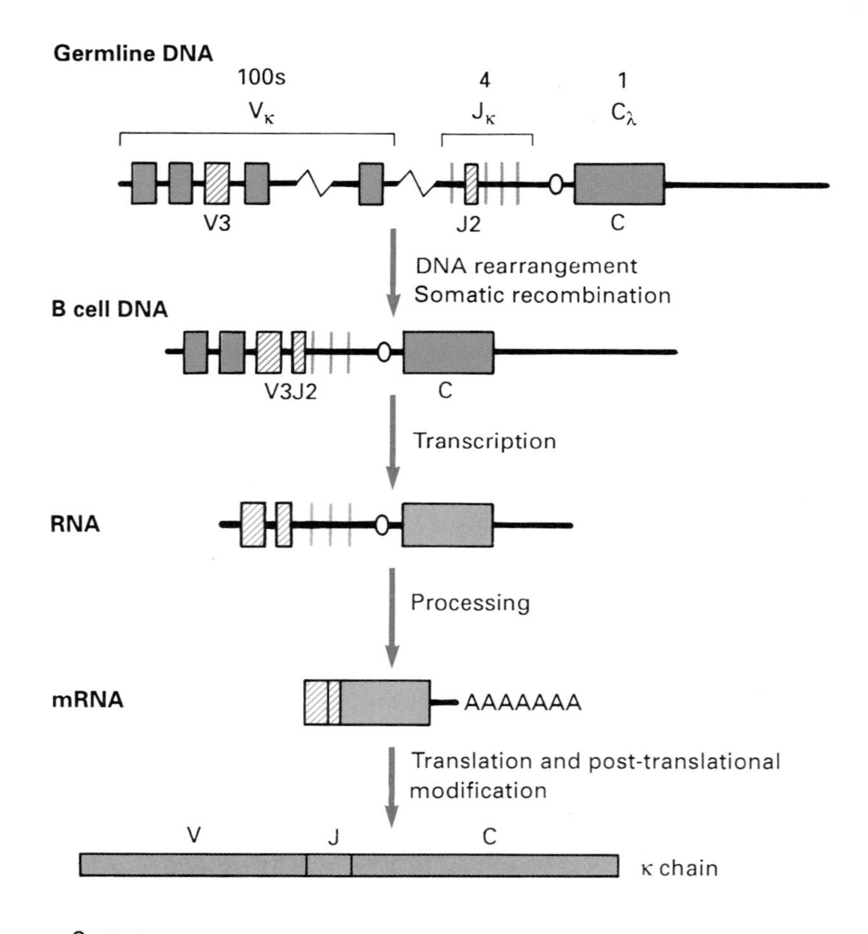

Fig.7.5 Diagram to illustrate the principles of immunoglobulin gene rearrangement and expression. The κ chain gene is used to illustrate this process

separate region on the same chromosome. They are also organized in a tandem array, and are separated from the V_κ segments by intervening DNA or intron sequences (Fig.1.9.) that are eliminated during rearrangement and processing of the κ light chain (Fig.7.3).

The organization of the human V_κ complex is similar to that of mouse.

λ chains. The DNA encoding the λ chains is also subdivided into three gene segments. As for κ genes, V and J segments are used to produce the variable domain and a separate exon encodes the constant domain, one for each of the subtypes of the λ light chains (Fig.7.3). In the mouse, there are only two V_λ gene segments compared to over 100 V_κ gene segments (see above, Fig.7.3). Thus the number of different λ light chains that can be produced by rearrangement of these gene segments is far fewer than the potential number of κ chains. This may not be a problem for the generation of diversity in the immunoglobulin repertoire in the mouse, as only about 5% of the light chains produced are of the λ type. A further difference in the organization of the mouse λ light chains is that the two V_λ and four J_λ gene segments are not grouped together in two separate regions of the chromosome; instead, each V_λ gene segment is associated with two J–C_λ segments (Fig.7.3). $V_\lambda 1$ joins preferentially to $J_\lambda 1 C_\lambda 1$ and $J_\lambda 3 C_\lambda 3$, and $V_\lambda 2$ with $J_\lambda 2 C_\lambda 2$. The $J_\lambda 4 C_\lambda 4$ pair is defective.

The organization of λ genes in human is mixture of those described so far (see above, Fig.7.3). Some 30 V_λ gene segments have been identified, together with at least four or five functional and four pseudo C_λ genes. Each C_λ gene is associated with its own J_λ segment.

The difference in the organization of the λ locus in the two species may reflect the greater use of this locus in human, where the ratio of antibodies containing each type of light chain is approximately 2 κ:1 λ. The chromosomal location of the light chain genes is given in Table 7.1.

Heavy chain genes

The key differences between the germline organization of the heavy and light chain genes are:

(1) the heavy chain variable region is encoded by **three** gene segments, the variable (V), diversity (D) and joining (J) segments (Fig.7.6); and

(2) the constant regions for all the heavy chain isotypes are encoded together as a cluster in a separate region of the chromosome downstream from the variable region gene segments (Fig.7.6).

The heavy chain genes are located on chromosome 12 in mouse and 14 in human (Table 7.1). The organization of the genes is similar but not completely identical in the two species (Fig.7.6).

Variable region genes. In the variable region gene cluster the members of each of the three families of gene segments, V_H, D_H, J_H, are ordered in tandem arrays sequentially along the chromosome (Fig.7.6). Together, these gene segments encode the variable domain of the heavy chain which comprises between 100 and 130 amino acids.

The V_H segments encode up to 98 of the amino acids at the N-terminal end of the variable domain, and include the complementarity determining regions, CDR1 and CDR2 (see Fig.6.12). Each V_H segment is split into two exons: the 5′ exon codes for all but four amino acids of the leader sequence, which is important for the insertion of the heavy chains into the membrane rough endoplasmic reticulum as they are being synthesized (see Section 6.2.6, Fig.6.24). The second V_H exon is separated from the first by a short intron of about 100 base pairs; this codes for the remaining four amino acids of the leader peptide as well as for the major portion of the V_H domain (for clarity the V_H segment is shown as a single exon in Fig.7.6 while a more detailed diagram is shown in Fig.7.7).

The total number of V_H segments present in the genome is difficult to quantitate. DNA sequence analysis and Southern blotting data have shown

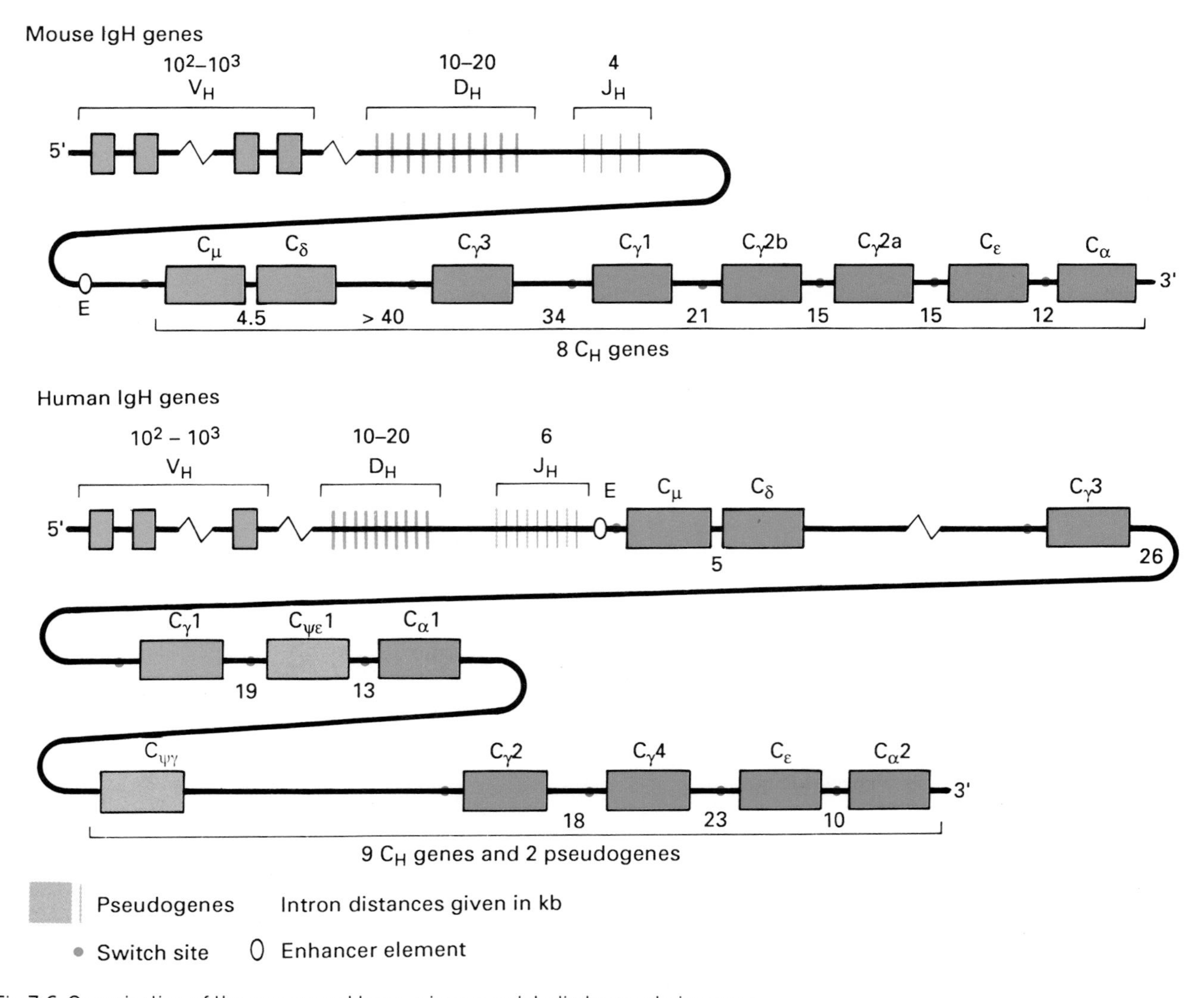

Fig.7.6 Organization of the mouse and human immunoglobulin heavy chain genes

that the V_H segments can be subdivided into seven or eight families, each containing between 4 and more than 100 members. All the members of each family show significant homology to one another and are clustered together, with an average spacing between family members of 10–20 kb. The number of pseudogenes within the V_H family is not known and cannot be determined by Southern blotting because the pseudogenes cross-hybridize with the V_H probes. Although the pseudogenes in the V_H pool cannot contribute directly to the generation of antibody diversity, they may do so in an indirect fashion, by providing a reservoir of information that can be utilized to generate new V_H gene segments (see above).

The D_H segments encode the third hypervariable region, or CDR3, of the heavy chain variable domain (Figs.6.12 and 6.16). They are located between the V_H and J_H clusters in genomic DNA and cover about 80 kb in the mouse (Fig.7.6). D_H segments are characterized by their variability in sequence and length, and encode between 1 and 15 amino acids. These two features have made determining the number of D_H segments extremely difficult, but current estimates suggest that there are between 10 and 20 in both the mouse and human genomes. D_H segments can also be grouped into families on the basis of similarities in their sequence, and in the mouse, families containing between one and ten members have been identified.

The cluster of J_H gene segments is located furthest downstream, in other words 3′ to the V_H and D_H clusters, and about 7 kb upstream from the constant region genes. Four J_H segments have been identified in the mouse and nine in the human genome, although three of these are pseudo J_H segments (Fig.7.6). The J_H segments encode the last 16–21 amino acids of the heavy chain variable domain.

To generate a functional gene for the heavy chain variable domain one member from each of these (V_H, D_H, and J_H) families has to be selected and joined together (Fig.7.7).

Constant region genes. The genes for each heavy chain isotype (see Section 6.2.2) are arranged sequentially along the chromosome downstream from the J_H segments, and span about 200 kb of DNA in the mouse (Fig.7.6). The organization of the C_H genes in the human is similar, although duplication of some of the genes has occurred and there are three pseudogenes (Fig.7.6). The genes encoding the μ and δ constant regions are located closest to the J_H segments. Downstream from C_δ are the genes for the constant regions of the various isotypes of γ chains, followed by those for the α and/or ε chains. The order of the constant region genes along the chromosome roughly reflects the order in which the different classes of immunoglobulin are synthesized during immune responses, for example IgM and IgD are expressed first (see Section 8.2.1).

Immunoglobulin heavy chain constant regions comprise between two and four domains depending on the class of the heavy chain, and in addition some also have hinge regions (see Section 6.2.6). Each structural domain is encoded by a separate exon (Fig.7.8). For example, μ chains have four constant domains but no hinge region, so there are four exons each encoding one of the domains; for mouse δ chains, there are only two constant domains, designated $C_\delta 1$ and $C_\delta 3$ (p. 368), each encoded by one exon. In other isotypes that contain a hinge region, the hinge can either be encoded

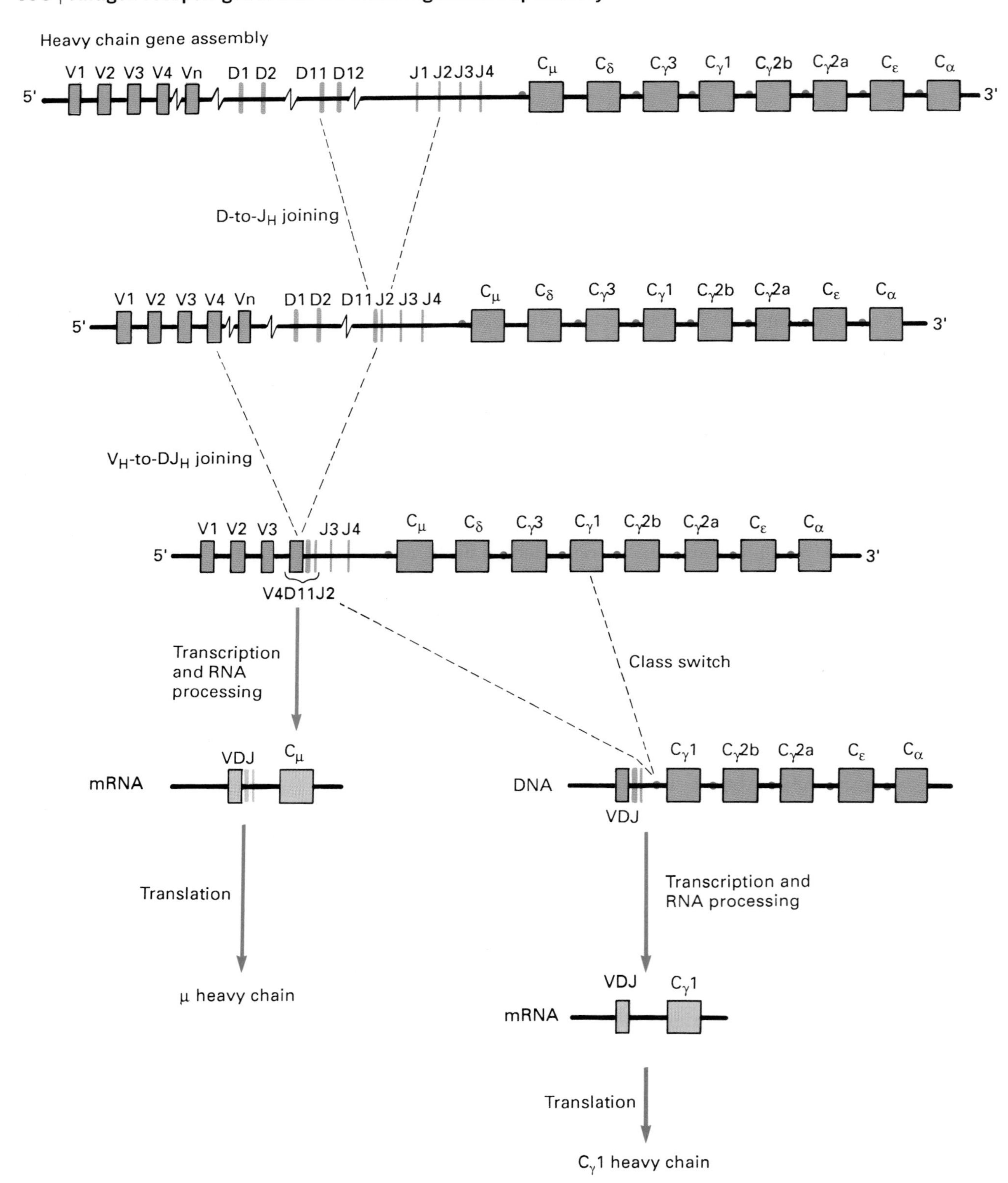

Fig.7.7 The events leading to the synthesis of μ and $\gamma 1$ heavy chains

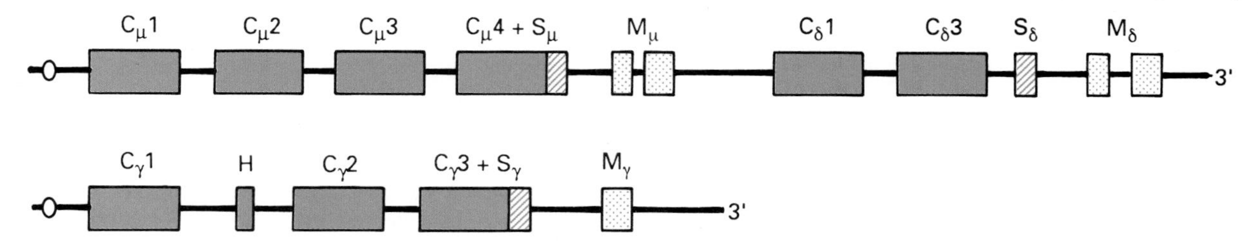

S: exon encoding secretory piece   H: exon encoding hinge region
M: exon(s) encoding transmembrane region and cytoplasmic tail

Fig.7.8 Exon–intron organization of the heavy chain constant regions $C_{\mu/\delta}$ and C_γ are shown as examples

as a separate small exon, between the first and second domains, e.g. for C_γ, or at the start of the second constant domain exon. Each of the exons coding for a C_H domain is separated in the DNA by introns that vary between 0.1 and 0.3 kb in length. Thus, for the constant region of immunoglobulin heavy chains there is a precise relationship between the exon–intron structure in the DNA and the domain structure of the protein (Fig.7.8).

Membrane and secretory pieces. In principle, each antibody can exist in two forms, as a membrane-bound molecule on the surface of the B cell (membrane Ig–mIg) and as a soluble, secreted molecule in the plasma or other body fluids. Which of the two forms an immunoglobulin will assume is determined by the amino acid sequence at the 3′ end of the heavy chain constant region. The molecular basis for this is evident from the organization of the constant region exons (Fig.7.8).

The heavy chains of membrane Igs have a slightly higher molecular weight than their secreted counterparts, as they contain additional amino acid residues which form the hydrophobic transmembrane portion and the cytoplasmic tail of the molecule. The exon or exons encoding the transmembrane region and the cytoplasmic tail are located downstream from the last C_H exon. These exons encode 26 hydrophobic amino acids that span the lipid bilayer and a hydrophilic cytoplasmic tail that varies in length depending on the class of the heavy chain; for example in μ chains it is extremely short and only comprises three amino acids, but γ chains have a 28-residue cytoplasmic tail. The information coding for the secreted forms of these molecules is contiguous with the final C_H exon although the C_δ gene is unusual, as the information for both the secreted and membrane forms of the δ chain is located in separate exons, downstream from the last constant region exon. Regulation of the production of membrane Ig versus secreted immunoglobulin is probably controlled at the RNA level and involves RNA processing (Fig.7.9). Indeed two different species of mRNA, one of 2.7 kb (membrane) and the other of 2.4 kb (secreted), can be detected in B cells producing IgM.

In all cases the transmembrane regions of membrane Ig heavy chains show a high degree of homology. They all adopt the conformation of an α helix as they span the lipid bilayer, and this together with the high level of homology in the different isotypes suggests that the interaction of this region of the immunoglobulin with other proteins in the membrane may be critical for transmembrane signalling and B cell activation (see Section 8.4.2).

7.2.2 Rearrangement

Variable domain rearrangements

Immunoglobulin genes remain in their germline configuration in all cells except developing and mature B lymphocytes. As both the heavy and light chain variable regions are encoded by multiple gene segments, a single gene segment from each pool has to be selected and the DNA rearranged to assemble a functional gene (Fig.7.7) (Schatz *et al.* 1992). During B lymphocyte differentiation, the heavy chain genes are assembled first. They are rearranged in two stages; the first involves the joining of a D_H and J_H gene segment together and the second the joining of a V_H segment to the D_H-J_H unit. One idea is that the rearrangement of the V_H-D_H-J_H segments on one chromosome is successful, rearrangement of the heavy chain gene variable gene segments on the other chromosome is suppressed, **allelic exclusion** (see Section 8.2.2), and the rearrangement of the light chain variable gene is triggered. In the mouse, the κ chain genes are rearranged in preference

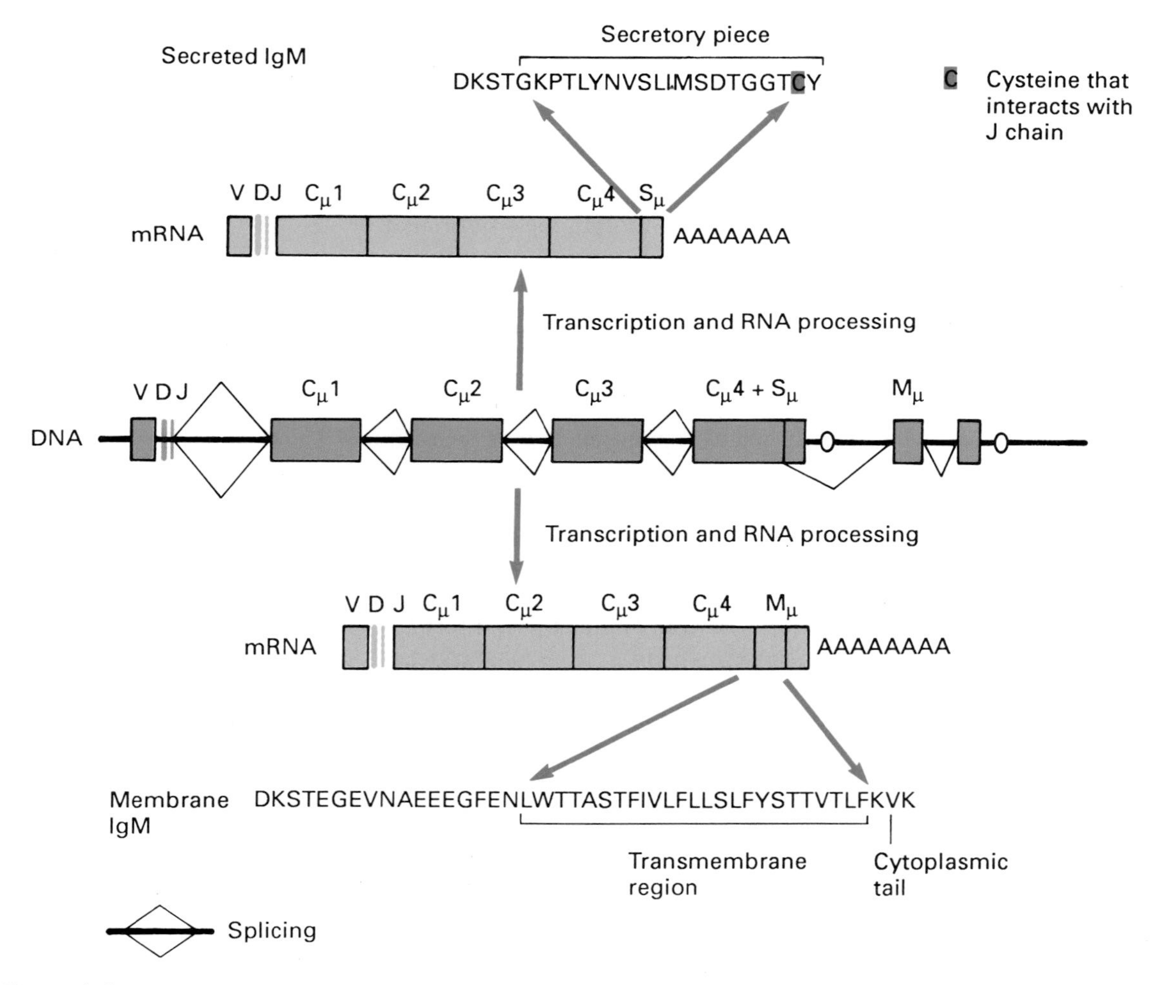

Fig.7.9 Heavy chain gene expression—generation of secreted and membrane forms of IgM

to the λ genes in most cases, **isotype exclusion** (see Section 8.2.2). If rearrangement of the heavy chain genes is unsuccessful, the pre-B cell will attempt to produce a successful or productive rearrangement on the second chromosome.

The 12–23 base pair rule

The precise mechanism for the joining of the V_H, D_H, and J_H or the V_L and J_L gene segments has not been completely elucidated. However, analysis of the nucleotide sequences flanking the gene segments has revealed that particular sequences are present downstream from each V_H and V_L gene segment, upstream from each J_H and J_L segment, and on either side of each D_H gene segment (Fig.7.10) (Tonegawa 1983). These flanking sequences are also referred to as **consensus sequences** and they are related to one another. Each consists of a palindromic **heptamer** (seven nucleotides) plus a **nonamer** (nine nucleotides) of defined sequence, separated from one another by a random sequence of a fixed number of nucleotides. Examples of the heptamer and nonamer consensus sequences flanking the light chain gene segments are shown in Fig.7.11; two different heptamer and nonamer sequences have been found for both κ and λ genes. The consensus sequences flanking either side of the gene segments to be joined are related to one another, as the nucleotide sequences are complementary to each other reading in the opposite direction. This means that they can potentially loop back on each other and form regions of double stranded DNA (Fig.7.11).

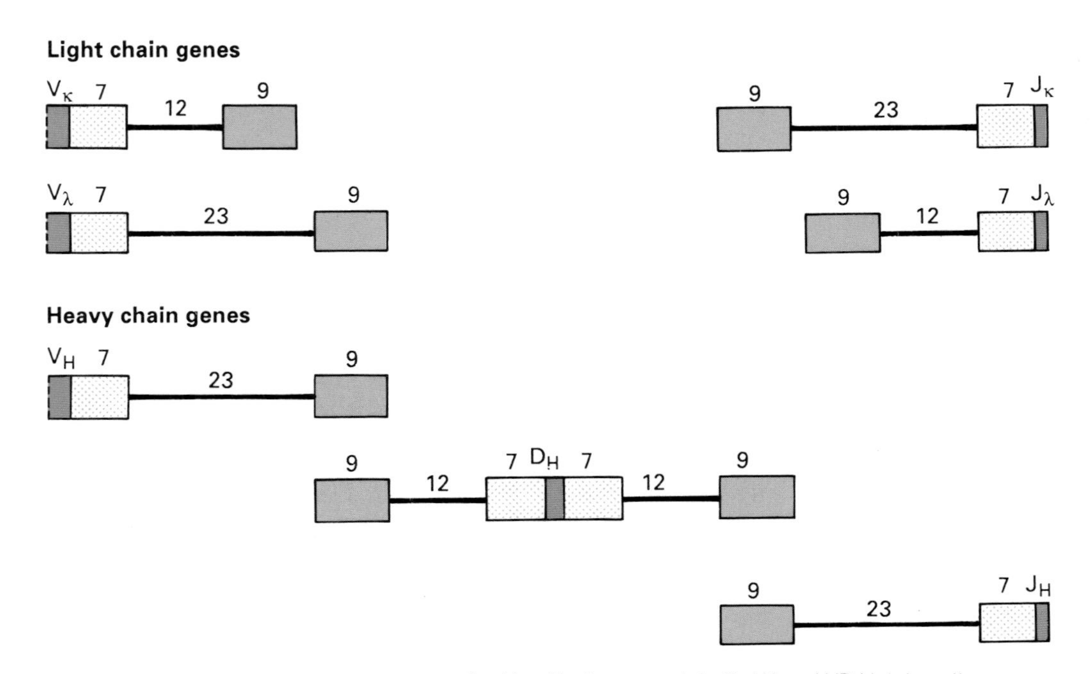

Fig.7.10 Heptamer and nonamer consensus sequences flanking the immunoglobulin VJ and VDJ joining sites

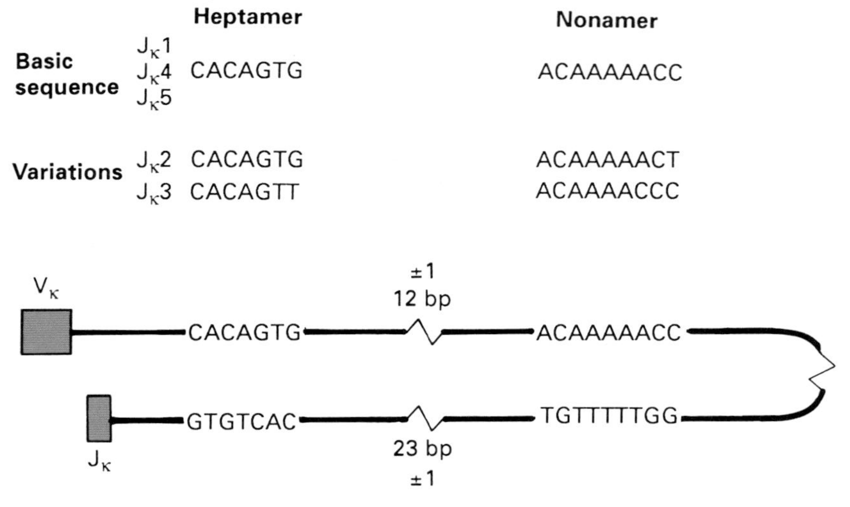

Fig.7.11 Examples of the heptamer and nonamer sequences flanking the κ light chain gene segments, and formation of double-stranded RNA

The other important feature of the consensus sequences is that they are separated from one another by a fixed number of nucleotides, 12 ± 1 or 23 ± 1. This number of nucleotides corresponds to either one or two turns of the DNA double helix. Thus the spacing and orientation of the consensus sequences around each of the gene segments to be joined is very precise (Fig.7.10) and appears to determine or control the recombination events. The phenomenon known as the **12–23 base pair rule** states that a gene segment with a 12 base pair (bp) spacer can **only** recombine or be joined to a segment flanked by a 23 bp spacer and vice versa. Thus in the case of the heavy chains, only D_H–J_H and V_H–D_HJ_H rearrangements are possible: abberant V_H–J_H and D_H–D_H rearrangements are not permitted according to the 12–23 base pair rule.

Several hypotheses have been proposed for the mechanism of joining the gene segments together and all suggest that the consensus sequences play an important role.

Deletion model. This mechanism involves only one of the chromatids in the cell and suggests that the information in between the two gene segments to be joined is *excised and deleted* as a result of the joining event (Kronenberg *et al.* 1986). For example, the consensus heptamer and nonamer sequences downstream of the selected V_L segment and upstream of the J_L segment come together and form a **stem and loop structure** (Fig. 7.12, panel 1). The stem brings the V and J gene segments selected for joining into close proximity, while all of the intervening V and J segments are present in the loop. It is proposed that this stem and loop structure is stabilized by specific DNA binding proteins, although these have yet to be characterized. The V and J segments are joined by the action of the recombinase enzyme that cuts and rejoins the DNA. This mechanism of gene rearrangement automatically leads to deletion of the DNA between the two segments being joined. The model is known as the **stem and loop** or **looping out and deletion model**

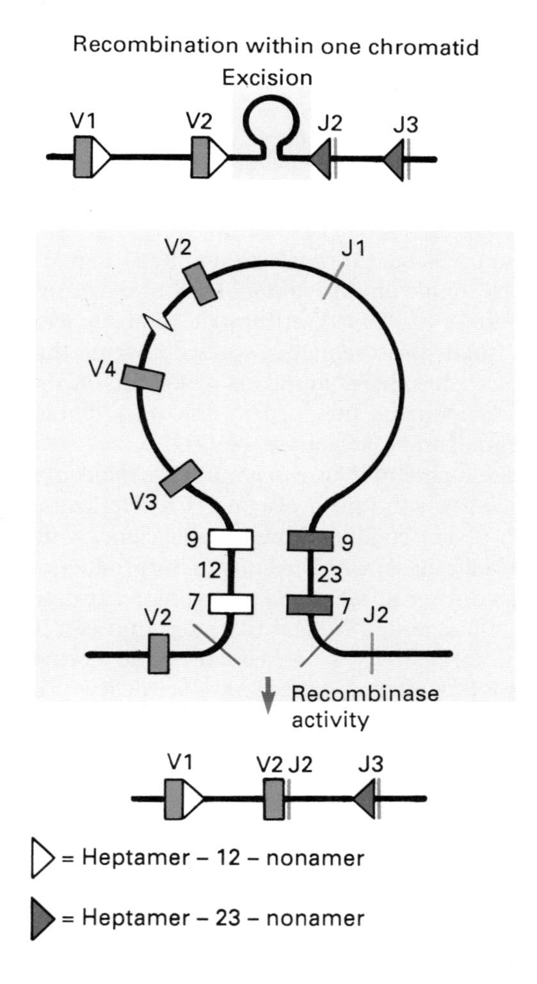

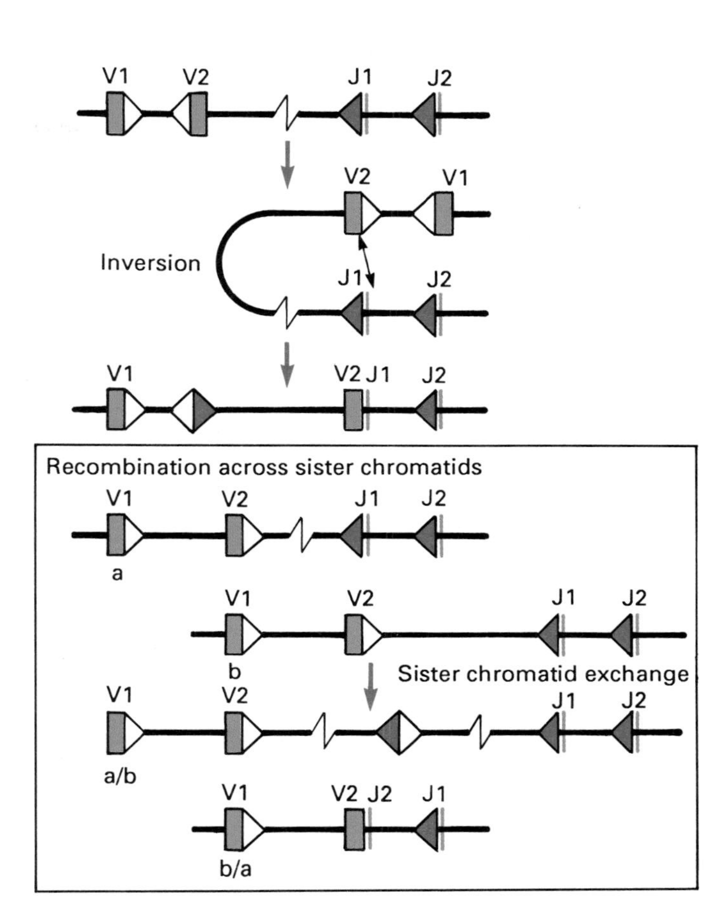

Fig.7.12 Mechanisms proposed for VJ and VDJ joining. VJ joining in a light chain gene is used as an example for clarity

(Fig.7.12). A variation on the theme of this model, the **inversion model**, is shown in panel 2. V–J, D–J, and V–DJ joining is usually thought to occur by this deletion mechanism and circular pieces of DNA, containing the excised gene segments, are generated as a consequence of the recombination event (Okazaki *et al.* 1987; Fujimoto and Yamagishi 1987).

Recombination across sister chromatids. This mechanism involves both of the chromatids present in the cell in the joining process. The two gene segments are joined as a result of an unequal crossing over or recombination event, whereby the DNA between the selected V and J gene segments is *not deleted* but transferred to the other chromatid at cell division. In this way there is no loss of information during the joining process. This mechanism is referred to as **sister chromatid exchange** (Fig.7.12, panel 3).

Several attempts have been made to isolate and characterize the recombinase enzyme and sequence-specific DNA binding proteins that are thought to be involved in these joining events. Proteins that bind the heptamer and nonamer sequences have been identified in nuclear extracts isolated from differentiating B cells, and named RAG-1 and RAG-2. When RAG-1 and

RAG-2 genes are cotransfected into fibroblasts the frequency of recombination is increased 1000-fold. The RAG-1 and -2 genes are conserved between all species in which V, D, J recombination takes place and the pattern of expression of these genes correlates precisely with recombinase activity. The precise roles of RAG-1 and RAG-2 are not clearly defined but they are thought to participate directly in the recombination reaction (Oettinger *et al.* 1990). A third gene, known as TdT is also involved. Cotransfection of RAG-1, RAG-2 and TdT seems to be sufficient to enable a non-lymphoid cell to rearrange its VDJ gene segment (Kallenbach *et al.* 1990). Although proteins with endonuclease activity have been isolated, no sequence-specific enzyme that recognizes the consensus sequences has been found as yet. Much work remains to be done before the components involved in the recombinase reaction are fully characterized (Zouali and MacLennan 1992).

An exciting development with great potential for elucidating the molecular events involved in the joining process was the discovery and characterization of a mutant strain of mouse with severe combined immunodeficiency—the **scid mouse**. These mice have a specific defect which results in the production of abnormal VJ and VDJ junctions during antigen receptor gene rearrangements and the inability to produce functional antigen receptors that can be expressed at the lymphocyte cell surface. As a consequence, the normal development of B cells and T cells is disrupted and these mice are deficient in normal lymphocyte responses. Further studies using these and other RAG-deficient mice may provide further information about the mechanisms involved in the joining process.

Heavy chain class switching

VDJ to C joining. The class of the immunoglobulin synthesized by a B cell can change during its development. This process is known as **class switching**. All B cells express IgM and/or IgD initially (both isotypes are often present on the same cell) as a result of their antigen-independent development (see Section 8.2.1). However, following antigenic stimulation, further DNA rearrangements can occur and the previously rearranged $V_HD_HJ_H$ segments can switch between different C_H genes. This enables the B cell to synthesize immunoglobulins with the same antigen specificity but with a different constant region which may mediate a different set of effector functions (see Section 6.2.5). The expression of each new heavy chain isotype, apart from δ, involves a DNA recombination mechanism that is distinct from that involved in VDJ and VJ rearrangement (for review see Shizumu and Honjo 1984).

Before antigenic stimulation, the rearranged VDJ block is transcribed into RNA together with the C_μ and/or C_δ genes, which results in the expression of IgM and/or IgD at the cell surface. The expression of IgD does not involve class switching; there is no switch region 5′ to the C_δ gene (see below). Instead the coexpression of IgM and IgD results from differential splicing of a long RNA transcript (Fig.7.13). In contrast, the expression of any of the other constant region genes does involve DNA rearrangement, (although intermediate stages in which the two isotypes involved in the switching event are expressed by processing of a long RNA transcript might also be involved).

Class switching appears to be mediated by a set of repeated sequences, **switch regions**, that are located 5′ to each set of constant region exons, except for C_δ (Fig.7.6) (Radbruch *et al.* 1986). The switch regions generally consist of short tandem repeat sequences, but these can vary in length (2–10 kb) and sequence for different heavy chain constant regions. The switch sequences for certain immunoglobulin constant region genes have been examined in detail.

For example, the μ switch sequence, S_μ, is located 1–2 kb upstream of the C_μ gene and contains tandem repeat sequences of the form $(GAGCT)_n(GGGGT)$. The repeat number of these sequences varies for the different constant region genes but is usually in the range 2–5, although it can be as high as 17, and the total number of nucleotides in each repeat varies from about 20 to 80.

The switch sequences are not flanked by any obvious consensus sequences (in contrast to the situation for the variable region gene segments) and

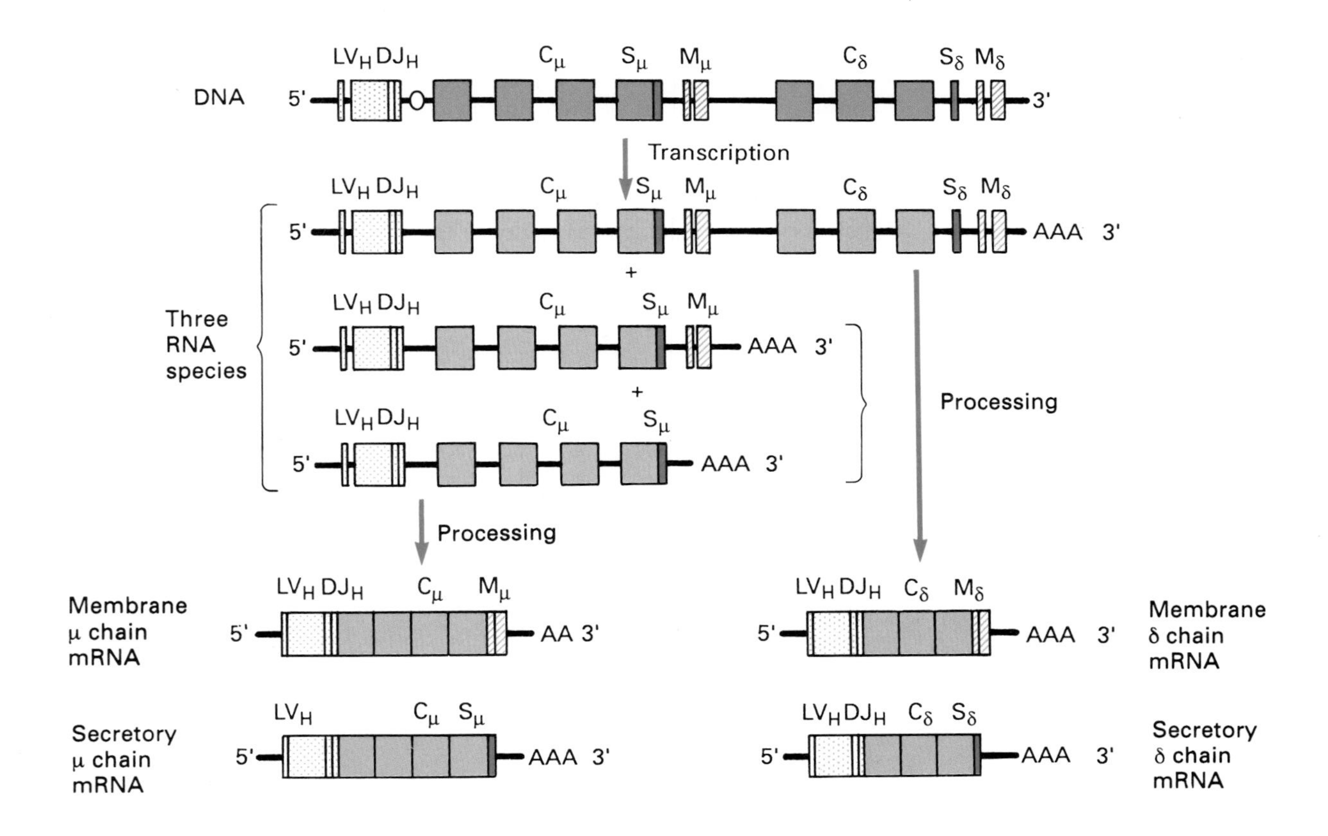

Fig.7.13 Expression of μ and δ chains

joining can occur at various points within the switch region. The identification of these sequences upstream from every C_H gene other than δ, led to the proposal that they are involved in the mechanism that mediates the class switch. It is known that when a switch event occurs these tandem repeats are deleted but as yet the precise mechanism of class switching remains unclear. One hypothesis suggests that these sequences may base pair with one another and promote class switching; however, the fact that switch sequences differ between isotypes make this less likely. Furthermore, no specific site within the switch sequences has been identified that might consistently be involved in the switch event. This implies that switching does not require the precise alignment of the rearranged variable gene segments (VDJ) with the new constant region. A second hypothesis proposes that the heavy chain switch may occur as a result of homologous recombination involving the switch regions. It has also been suggested that the switch sequences may be involved in binding class-specific switching proteins.

Several models for class switching have been proposed: some are similar to those described for VJ and VDJ joining (Fig. 7.12) and include the following.

Looping out and deletion models. The DNA containing the previously used constant region gene (e.g. C_μ) loops out during class switching, such that the rearranged VDJ sequences are brought close to the new constant region (Fig.7.14). Any intervening pieces of DNA (containing unused C_H genes between the original and the new C_H gene), together with the original C_μ gene, are then excised and deleted from the genome. Circular DNA containing the genes deleted during class switching has been detected in cytokine-treated B cells (Matsuoka *et al.* 1990; Iwasato *et al.* 1990)

The problem with this model is that it predicts that the C_H gene originally expressed by the B cell will be deleted as a result of the switch event. Although there is some experimental evidence to support this, other data show that deletion does not occur in every case. Indeed some B cells have been shown to be capable of switching back to producing IgM after previously synthesizing IgG. If the switch event had resulted in a deletion of the C_μ gene, re-expression would not be possible.

Sister chromatid exchange. An alternative explanation for the mechanism of class switching is sister chromated exchange, that is exchange of genes between the active and inactive chromatids. It is probably the more reasonable of the two hypotheses. The sister chromatid exchange model proposes that the C_μ gene in our example would not be deleted during the switch event. Instead, it would be transferred to the sister chromatid by a recombination at cell division. In this way the C_μ would be stored in the germline and would still be available for re-use at a later time (Fig.7.14).

Two further mechanisms involving either the splicing of a large RNA transcript or the *trans* splicing of two precursor RNAs have also been proposed to explain class switching. These models are not mutually exclusive and different mechanisms could operate at different stages of B cell development.

Generation of immunoglobulin diversity

A number of mechanisms contribute to the generation of diversity in immunoglobulins.

Combinatorial joining of heavy chain VDJ and light chain VJ segments. The variable domains of the heavy and light chain are encoded by three and two gene segments, respectively, (see Section 7.2.1). The presence of multiple V_H, V_L, D_H, J_H, and J_L gene segments in the germline contributes significantly to the number of heavy and light chain variable regions that can be produced by one individual (Table 7.2). Each B cell can potentially select and join together any gene segment from each of the pools available. Insertion of different D_H gene segments during the assembly of the heavy chain variable domain can dramatically alter the antigen specificity of the immunoglobulin produced, since the D_H region encodes CDR3. The combinatorial association of the different V, D, and J segments is not the only mechanism involved in the generation of diversity in the immunoglobulin

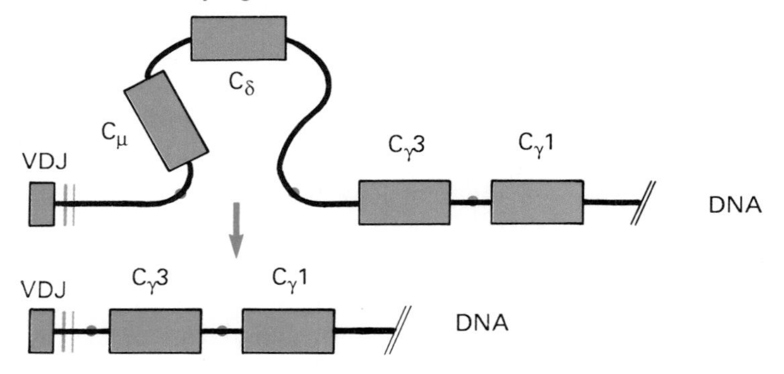

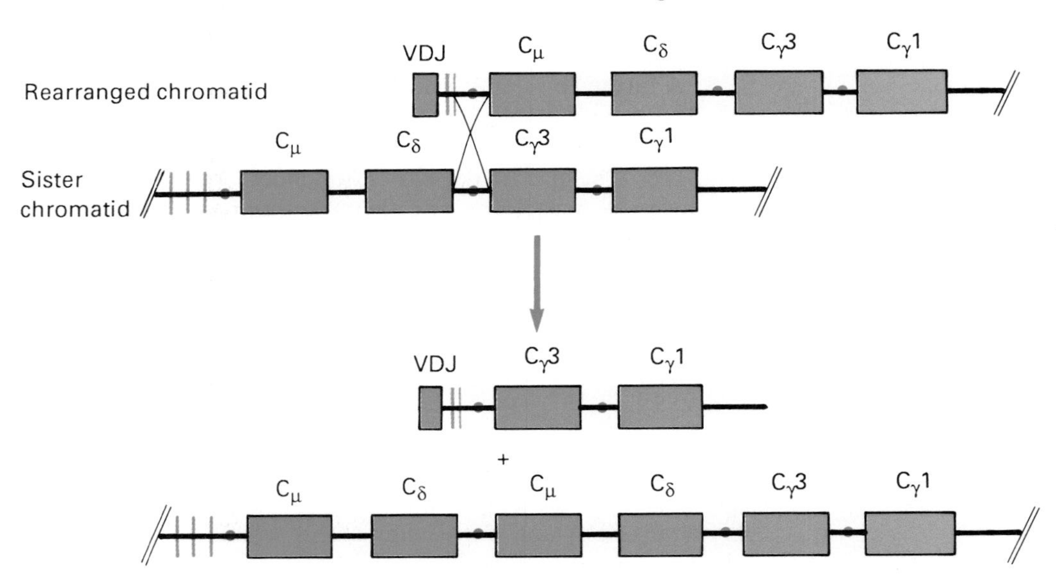

Fig.7.14 Models proposed for the mechanism of DNA rearrangement during class switching (switch sites, red dots)

Table 7.2 Estimate of the diversity of the immunoglobulin repertoire

	Heavy chain	κ chain
V gene segments	250–1000	250
D gene segments	10	0
J gene segments	4	4
Ds read in all frames	Rarely	–
N region addition	V–D, D–J	None
Variable region combinations	62 500–250 000	
Junctional combinations	10^{11}	

repertoire; other mechanisms can also significantly increase the diversity of the immunoglobulin repertoire.

Junctional diversity. Diversity is increased further by the imprecision of joining that occurs at the D_H–J_H, V_H–D_HJ_H, and V_L–J_L boundaries. Variability can be introduced in several ways by the joining process.

1. Joining can occur anywhere within the nucleotide codon at the position of the join. Joining of the same gene segments can therefore result in a different amino acid being introduced at the junction between the two segments (Fig.7.15).

2. Imprecise joining can, in some cases, lead to a frame-shift mutation which alters the reading frame of the DNA after the join (Fig.7.15). This occurs because the exact nucleotide used for joining can vary by as much as ten residues on either side of the junction. A deletion of nucleotides can occur during D_H–J_H or V_H–D_HJ_H joining; this can result in a complete change in the sequence of amino acids present in the variable domain. Of course in some cases such frame-shift mutations are nonproductive and introduce stop codons or termination signals into the nucleotide sequence.

3. A further mechanism has been found to contribute to the diversity of the heavy chains which involves the insertion of extra nucleotides, that are not part of the gene segments being joined, during the joining process (Fig.7.15). The insertion of extra nucleotides increases the size of the domain. The inserted nucleotides are called **N sequences** and the process is referred to as the **N region addition**. The addition of N sequences probably involves the enzyme terminal deoxynucleotidyl transferase and is independent of the template provided by the V, D, or J gene segments. The presence of these extra nucleotides in the heavy chain can have a dramatic effect on the antigen specificity of the immunoglobulin produced by the B cell. This process contributes to the diversity of the immunoglobulin repertoire but occurs only rarely.

Assuming there are some 250–1000 V_H segments, 10 D_H segments and 4 J_H segments in the genome, and taking into consideration the variability at the sites of D–J and V–DJ joining, this would readily allow a diversity in excess of 10 000 heavy chains ($250 \times 10 \times 4$ plus junctional diversity) (Table 7.2). In principle this could give rise to at least 10^{11} antibodies, as

1. Variation in the exact position of the VJ join can result in the insertion of different amino acids in the third hypervariable region

V	J	Amino acid 95	96	97
CCX CCC	TGG ACG	P	W	T
CCX CCC	TGG ACG	P	T	
CCX CCC	TGG ACG	P	P	T

2. Frame shift mutations can arise as a result of DJ or VDJ joining in the heavy chain

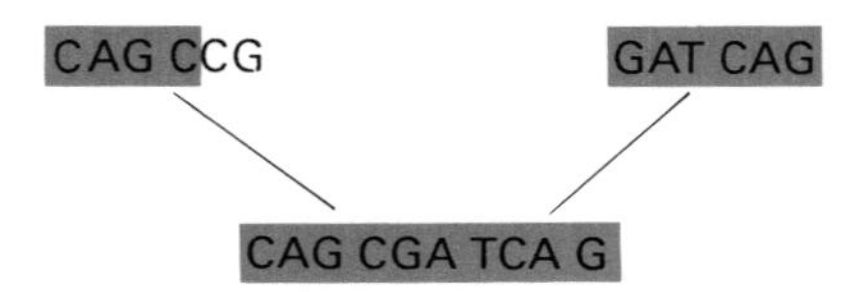

3. Insertion of extra nucleotides, N sequences, can occur during VDJ joining

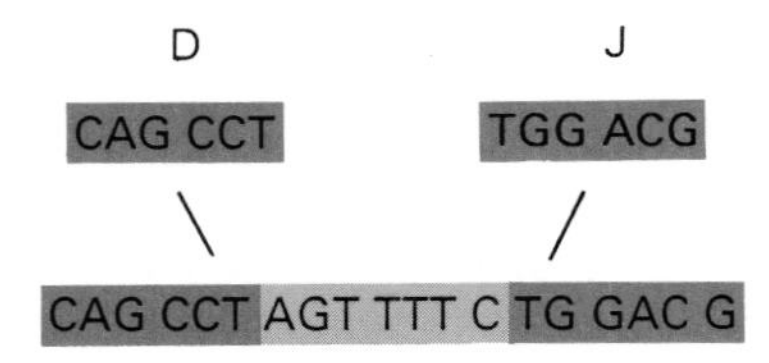

Fig.7.15 Generation of junctional diversity

theoretically any heavy chain can pair with any of the 10 000 or so light chains. However, the generation of antibody diversity does not stop here.

Somatic mutation. An additional mechanism, somatic mutation, also generates diversity in immunoglobulin heavy chains (Tonegawa 1983). Comparisons of the amino acid and nucleotide sequences of variable regions of different antibodies show that variants differing from the expected sequence by a single amino acid substitution exist. This substitution generally arises from a single nucleotide change that occurs at a late stage in B cell development, after antigenic stimulation.

Somatic mutation does contribute to the generation of diversity of immunoglobulins, although the mechanism by which it occurs is unknown. In the chicken and the rabbit, it appears that pseudo-V and -D gene segments may function as potential donor sequences for gene conversion. However, in other mammals there is little evidence in support of a significant role for this mechanism (Wysocki and Gefter 1989). Rearrangement of variable domain gene segments seems to be important for targeting the machinery for introducing somatic mutations. The overall frequency of mutations has been found to be similar for both productively and nonproductively rearranged genes. Any of the four nucleotides (A, T, C, and G) may be a target for somatic mutation, and both insertions and deletions of nucleotides have been documented. These mutations are not confined to the hypervariable regions that are directly involved in antigen binding (see Section 6.2.3), although mutations in these regions are detected more frequently in the expressed molecules. One reason for this could be that somatic mutation in other parts of the domain would result in a non-permissible alteration in the protein structure, resulting in abortion of the clone. Alternatively, there may be selection for production of only those immunoglobulins in which somatic mutation results in an increased affinity for antigen; indeed those B cells expressing receptors with a higher affinity for antigen are selectively expanded during the immune response, especially when the antigen concentration is limiting; this response is known as **affinity maturation** (see Section 8.3.1) (Nossal 1992). If this reasoning is correct, one would predict that immunoglobulins carrying mutations in the hypervariable regions would be detected more frequently.

Somatic mutation also occurs during secondary immune responses after class switching, and this correlates with the stage at which antibodies show an increased affinity for antigen. One could postulate that the DNA rearrangements leading to the expression of an IgM antibody by the primary B cell enable it to interact with the antigen with reasonable affinity. After antigenic stimulation, the B cell becomes activated and it may proliferate, isotype switching occurs, and somatic mutations could potentially arise during these events. The antibodies that are generated will have a range of affinities, and clones with mutations producing antibodies with an increased affinity for antigen will continue to proliferate in preference to clones producing low affinity antibodies. When all the antigen is cleared from the body these high affinity clones could revert to memory cells. Experimental evidence obtained by studying antibody responses to haptens supports these ideas.

For example, in the antibody response to oxazalone in the mouse, somatic mutations in the V_H–V_L pair occur as the response progresses and these changes can be correlated with an increase in antibody affinity for oxazolone (Berek and Milstein 1987).

In addition to replacement of single nucleotides, complete replacement of a functional V_H gene has been described for some $CD5^+$ pre-B cell lines and B cell lymphomas in the mouse (see Sections 8.3.1 and 8.5). The complete replacement of V_H sequences during maturation probably plays only a minor role in the generation of antibody diversity.

7.2.3 Control of rearrangements and tissue-specific expression of Ig genes

Control of gene rearrangements during development

The mechanisms underlying the control of the gene rearrangements that occur during antigen-independent B cell development are not understood. Control is definitely exerted, and examples of this are the observations that (V,D,J) gene rearrangements occur in an ordered manner and that allelic exclusion operates in B cells. These areas are discussed in more detail in Sections 8.2.2 and 8.4.3 and are only mentioned briefly here.

Allelic exclusion is the term used to describe the phenomenon whereby B cells express only one functional heavy chain and one functional light chain, resulting from the productive rearrangement of the genes on one of the two homologous chromosomes. The same phenomenon is found for the rearrangement and expression of T cell receptor genes. The other immunoglobulin genes either remain in their germline configuration or have aberrant, so-called non-productive rearrangements. Allelic exclusion is of critical importance for the immune response, as it maintains the monospecific recognition properties of each B cell clone for antigen—if the immunoglobulin genes on each chromosome were rearranged productively, each cell could express several antigen receptors each with a different specificity for antigen.

Gene rearrangements take place in an ordered sequence, as mentioned above (Fig.7.16). For example, for heavy chain variable region genes, D_H-J_H joining occurs first and this is followed by the joining of the V_H segment. Heavy chain rearrangements are thought to precede the rearrangement of the light chain genes. It would appear that B cells attempt to rearrange immunoglobulin genes until a functional rearrangement of both a heavy and light chain gene is produced. Once this has occurred, there is a block on further immunoglobulin gene rearrangements.

Tissue-specific control of expression

Immunoglobulin gene expression is very tightly controlled in a tissue specific manner, such that expression of immunoglobulin polypeptides occurs only in B cells, and only after the genes have been rearranged. Each V_H and V_L gene segment is preceeded by a transcriptional **promoter** that initiates transcription of the rearranged heavy chain gene. V gene segment promoters consist of an octanucleotide, $\mathrm{TATA}^{A}_{T}\mathrm{A}^{A}_{T}$, the TATA box, that lies between 90 and 160 base pairs 5′ from the site of initiation of transcription. The TATA box is recognized by the enzyme RNA polymerase II the enzyme that catalyses transcription of the immunoglobulin gene. A second stretch of 15 conserved nucleotides $\mathrm{TGCA}^{G}_{C}\mathrm{CTGTGNCCAG}$ is also found in this region, some 60 nucleotides further upstream. If the V gene promoters are deleted then transcription of immunoglobulin genes is prevented.

Nucleotide sequences, known as **enhancer sequences**, are critical for full function of the V gene promoter and for tissue-specific expression of immunoglobulin genes (Staudt and Lenardo 1991). These enhancer sequences are located in the intron between the J and C gene segments of both the heavy chain and κ light chain genes, but not in the λ genes in the mouse. They are usually present on the same strand of DNA as the genes they control, in other words they are *cis*-acting and not *trans*-acting, although they can be some distance away ($>$ 500 base pairs in some cases). More is becoming known of the mechanism by which these enhancers act; presumably the cell specificity results from the presence of tissue-specific enhancer proteins present only in B lymphocytes which act in concert with the enhancer sequence and the V region promoter (For review see Kadesch 1992). At least five sites within the κ chain enhancer region have been shown to bind nuclear factors. One binding site contains an octamer

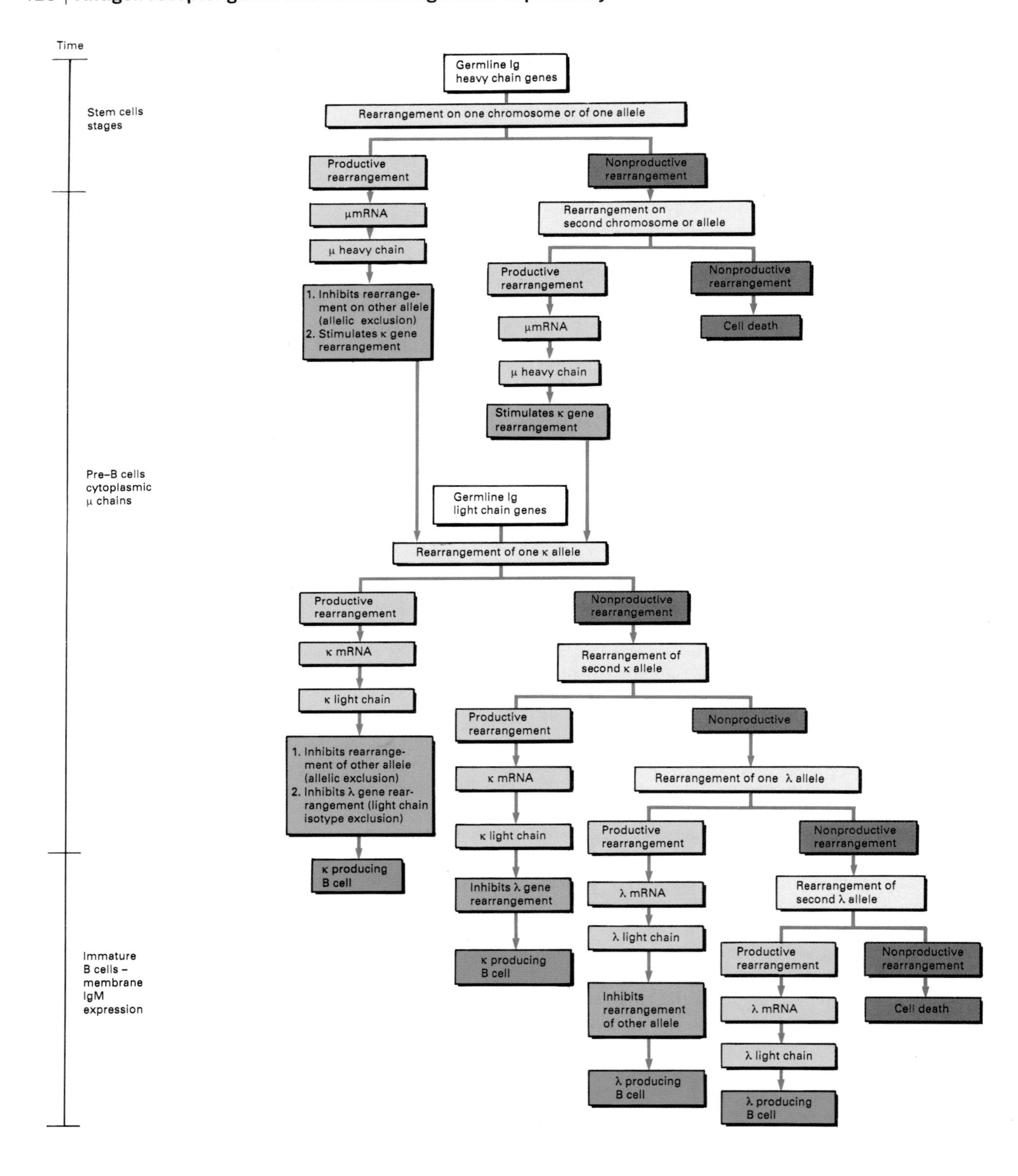

Fig.7.16 The order of immunoglobulin gene rearrangements during B cell development

consensus sequence and binds at least two factors, Oct-1 and Oct-2, one of which, Oct-2, is specific to lymphoid cells and appears to direct the tissue specificity of the enhancer element (Singh *et al.* 1986). In addition, transcription occurs only after rearrangement of the DNA in the B cell and presumably this is also linked to the tissue specificity of immunoglobulin expression.

We have seen that DNA rearrangements are of critical importance to the production of a transcriptionally active immunoglobulin gene. Once transcription is initiated, there are still intervening sequences present in the rearranged gene that have to be removed before translation. This final removal of extraneous nucleotide sequences is accomplished at the RNA level. It has been shown that nuclear RNA precursors of κ chain mRNA in the mouse undergo two rounds of RNA splicing, to join the leader peptide and V segments, followed by the V and C segments, before appearing as cytoplasmic RNA. A final RNA splicing step then removes the intron separating the V and C sequences, thus making the V and C region coding sequences contiguous.

7.2.4 Summary (Fig.7.17)

Light chains. The DNA encoding the variable region of light chains is divided into two families of gene segments, V_L and J_L. The V_L gene segment encodes the majority of the variable domains, 96 or so of the 110 amino acids. The remainder of the domain is encoded by the J_L segment. Pools of V_L and J_L segments are present for both κ and λ light chains, and each is arranged in a tandem array along the chromosome. The precise number of segments in each pool and their organization varies between different species. During B cell development a V_L and a J_L segment are selected from each pool and joined together. This joining process involves DNA rearrangement so that the V and the J segment are in close proximity both to one another and the gene coding for the light chain constant region.

Heavy chains. The DNA encoding the heavy chain variable region is divided into three gene segments, V_H, D_H, and J_H. One gene segment has to be selected from each pool and joined together to produce a functional heavy chain variable domain. D_H–J_H joining occurs first and this unit is then joined to the selected V_H segment. The genes encoding the heavy chain constant regions are also arranged in an ordered array along the chromosome—C_μ followed by C_δ, C_γ, and so on. The constant region gene for each isotype is coded by several exons, each exon encoding a separate functional domain. Once a functional V-D-J unit has been produced this can associate with different constant region genes, a phenomenon known as the class or isotype switch, resulting in a change in the effector function of the immunoglobulin produced without any alteration in the antigen binding specificity.

Generation of antibody diversity

1. Any V_H, D_H, J_H, or V_L, J_L gene segments from the pools present in the germ line can be selected and joined together to produce a functional heavy or light chain variable region.

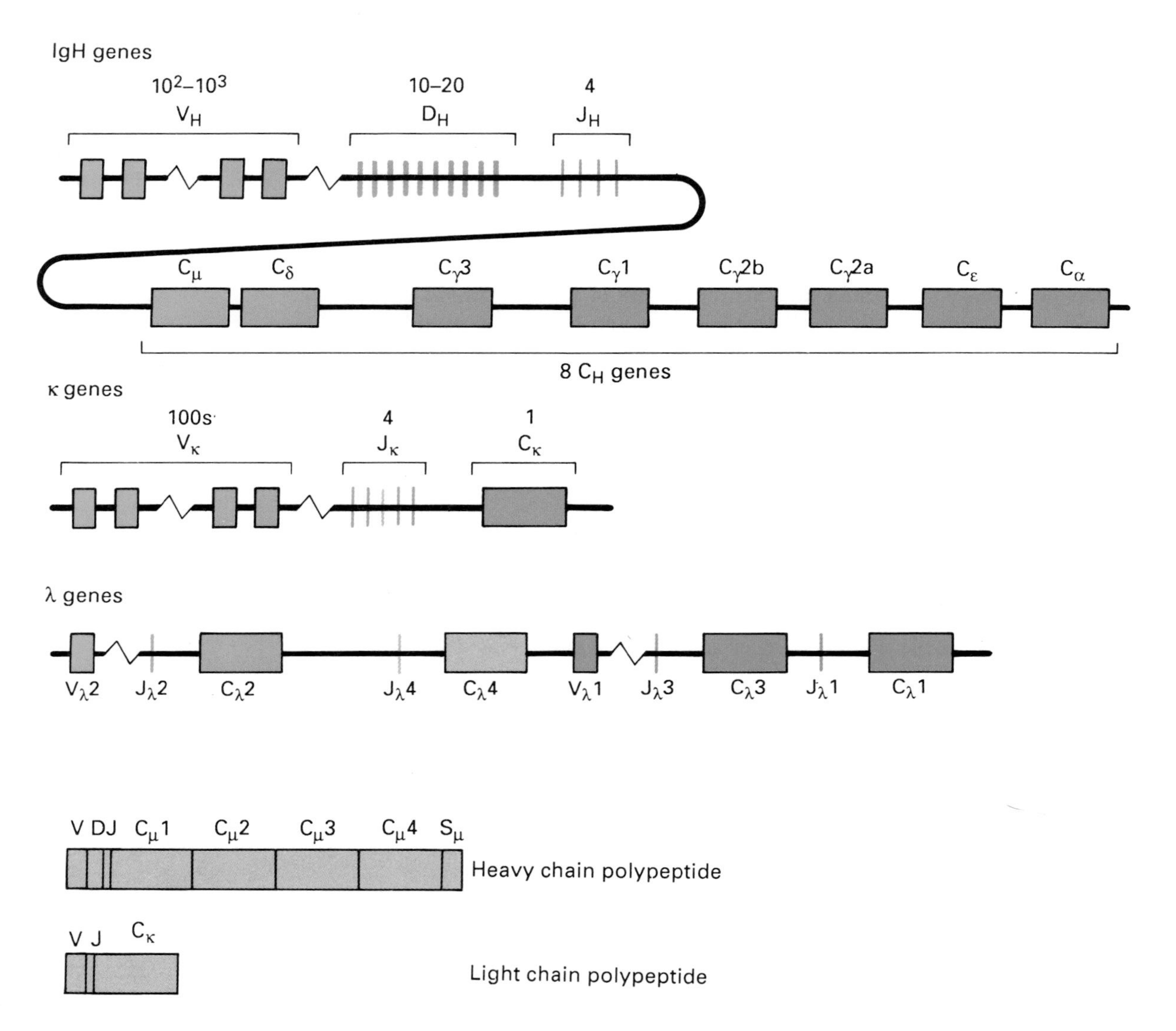

Fig.7.17 Immunoglobulin gene and polypeptide organization in the mouse — summary diagram

2. The joining mechanism is imprecise and N sequences can be added; thus joining the same V, D, and J or V and J segments can result in different variable regions, with different antigen binding specificities.
3. Somatic mutations can occur after DNA rearrangement.
4. Any light chain can associate with any heavy chain to form the antigen binding site.

7.3 T cell receptor genes

Two sets of T cell receptors have been described, the $\alpha\beta$ (also known as TcR2) and the $\gamma\delta$ (also known as TcR1) receptors. The structure of the $\alpha\beta$ and $\gamma\delta$ T cell receptors was described in Chapter 6, Section 6.3 and the $\alpha\beta$ receptor is shown again in Fig.7.1 (Saito *et al.* 1984*a*).

The repertoire of T cell receptors expressed by any individual can recognize an enormous number of antigenic peptides in association with MHC molecules. This requirement for extensive diversity of the antigen receptors of T cells suggested that similar genetic mechanisms to those described for immunoglobulins may operate in this system. In this section the organization of the genes that encode TcR and the rearrangements of these genes during T cell development will be discussed. The first successful attempt to isolate TcR genes was reported in 1984, so all of the information available has been obtained relatively recently.

Four different genes coding for T cell receptor polypeptides have been found, the α, β, γ, and δ genes. The genes for T cell receptors are arranged in a similar way to those coding for immunoglobulin light and heavy chains. In all cases, the coding information for each of the polypeptides is divided into two pools, one for the variable domains, which is further divided into multiple gene segments, and the other for the constant domains. T cell receptor genes also undergo rearrangement before they can be transcribed and translated. DNA rearrangements are therefore a common feature of the genes encoding the antigen receptors of the immune system.

7.3.1 Organization

The genes for the α, β, γ, δ chains have been cloned and their chromosomal locations identified (Table 7.1). The organization of each of these genes in the germ-line is discussed below.

Organization of the β chain genes

cDNA clones for the T cell receptor β gene were the first to be isolated (Hedrick *et al.* 1984; Yanagi *et al.* 1984). The genomic organization of the T cell receptor β genes is related to that described for the immunoglobulin heavy chain genes in that the β chain variable domain is encoded by three gene segments, V_β, D_β, and J_β (Fig.7.18, compare Fig.7.17 for example; see also Fig.7.19). As for the immunoglobulin heavy chain genes, a pool of V_β segments is arranged in a tandem array along the chromosome upstream from the D_β and J_β segments.

In the mouse all but one of the V_β gene segments map to about 330 kb of DNA located upstream (approximately 250 kb) from the remainder of the variable domain gene segments. The remaining V_β segment, $V_\beta 14$, is encoded in the opposite transcriptional orientation to all the others, downstream from the $C_\beta 2$ gene. Although this V_β segment is in an unusual position in relation to the others, it is used and is thought to form a functional V region by an inversion mechanism (see Section 7.2.2; Fig.7.12).

The repertoire of V_β gene segments is somewhat limited; there are 20 members of the cluster in the mouse and between 50 and 100 in the human.

The V_β gene segments can be divided into subfamilies, but each subfamily is small in size. In the mouse the V_β locus has been mapped (Fig.7.18). Each of the subfamilies has a single member except $V_{\beta 5}$ and $V_{\beta 8}$, each of which has three members. The human V_β gene segments can be divided into 15 subfamilies, most having one or two members and the largest having 6 members.

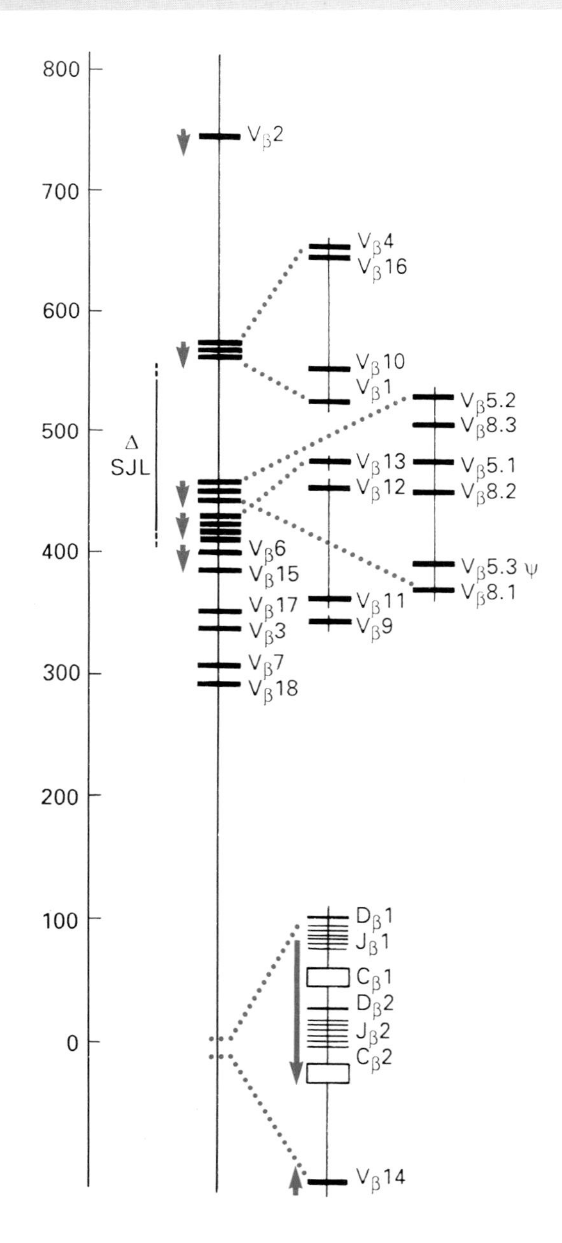

Fig.7.18 Organization of the T cell receptor β locus in mice. Coding regions are indicated by vertical bars. Map distances are given in kb. ΔSJL = deletion in SJL mice. → direction of transcription

The organization of the D_β and J_β segments and the C_β exons is rather *different* to that previously described for the immunoglobulin heavy chains (Fig.7.18, compare with Fig.7.6). In both mice and humans, there has been a duplication of the DNA containing one D_β, several J_β gene segments, and a C_β gene (Fig.7.18). Downstream from the V_β segments is a single D_β segment, $D_\beta 1$ separated from seven J_β segments, $J_\beta 1$–7, by approximately 600 base pairs in both mouse and human. Each J_β segment encodes 15–17 amino acids of the variable region, but only six of the seven are functional in both species. In close proximity to the last J_β segment, $J_{\beta 7}$, is the first of the C_β exons, $C_\beta 1$. The duplication of this section of the β chain gene results in the presence of a second set of D_β, J_β, and C_β segments 3′ to the first (Fig.7.18). Of the seven $J_\beta 2$ gene segments six are functional in mouse but all seven are functional in the human.

The two C_β genes are highly homologous, with only four amino acid differences between the two in mouse and six differences in human. Both genes have a similar exon–intron structure.

Each C_β gene is divided into four exons, the first coding for the immunoglobulin-like domain plus part of the peptide that connects the extracellular domain to the transmembrane region. The remainder of the connecting peptide is encoded by the second exon and part of the third, the remainder of which also encodes the transmembrane region. The fourth codes for the short cytoplasmic tail which contains five or six amino acids and the 3′ untranslated region.

The two C_β genes appear to be used equally, irrespective of the function of the T cell that expresses them, suggesting that the two genes produce domains that have a similar structure. No information is present in the $C\beta$ gene that would allow the synthesis of a secreted β chain, and consequently all naturally occurring TcR β chains are membrane bound.

Organization of the α chain genes

The variable domain of the α chain gene is composed of two segments, V and J (Sim *et al.* 1984; Saito *et al.* 1984*b*). The organization of the T cell receptor α genes in mice is shown in Fig.7.19; as for the other antigen receptor genes, each family of gene segments is located together on the chromosome in a tandem array (Kronenberg *et al.* 1984).

The number of V_α genes present in the germ line is difficult to estimate, but between 50 and 100 are thought to be present in the mouse and human. The V_α segments can be subdivided into at least ten families, each having between one and ten members.

The J_α gene segments stretch over a large distance in their genomic configuration, covering about 100 kb of DNA. The J repertoire is very large, comprising at least 50 gene segments, which is considerably larger than that of the other T cell receptor or immunoglobulin genes.

The C_α gene is also divided into exons which roughly correspond to the predicted protein domains.

The first and second exons code for the external part of the constant region, the first for the structural domain which is predicted to adopt the conformation of an immunoglobulin fold and the second for the connecting peptide that links the polypeptide chain to the plasma membrane. The third exon codes for the transmembrane region and short cytoplasmic tail, and the fourth for the 3′ untranslated region.

As for the C_β genes, the C_α genes also lack sequences that would produce a secreted form of the α chain.

Organization of the γ chain genes

Chronologically, the γ chain genes were the second set of T cell receptor cDNA clones to be isolated. At first they were thought to code for the α chain, but the polypeptides they produced did not correlate with data already available for the α chains expressed by mature T cells. Later this gene was shown to encode a third type of TcR polypeptide, the γ chain. In association with the δ chain, the γ chain forms a second type of T cell receptor expressed by immature thymocytes and some mature T cells.

The genomic organization of the γ chain genes has been studied extensively and is shown in Fig.7.20 (for review see Quatermous *et al.* 1985). As for the

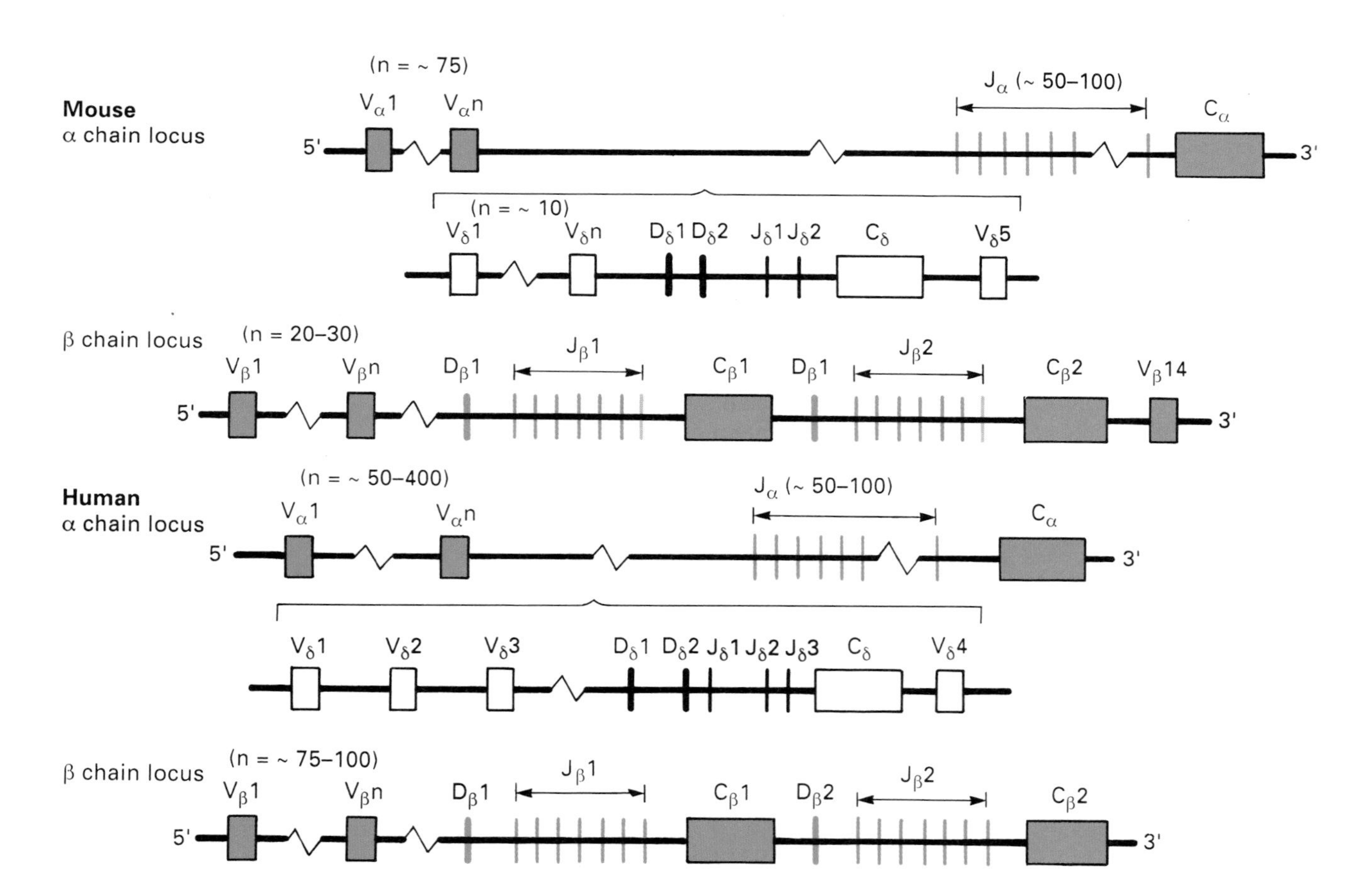

Fig.7.19 Organization of T cell receptor α and β chain loci in the mouse and human. (Leader sequences and pseudogenes are omitted for clarity)

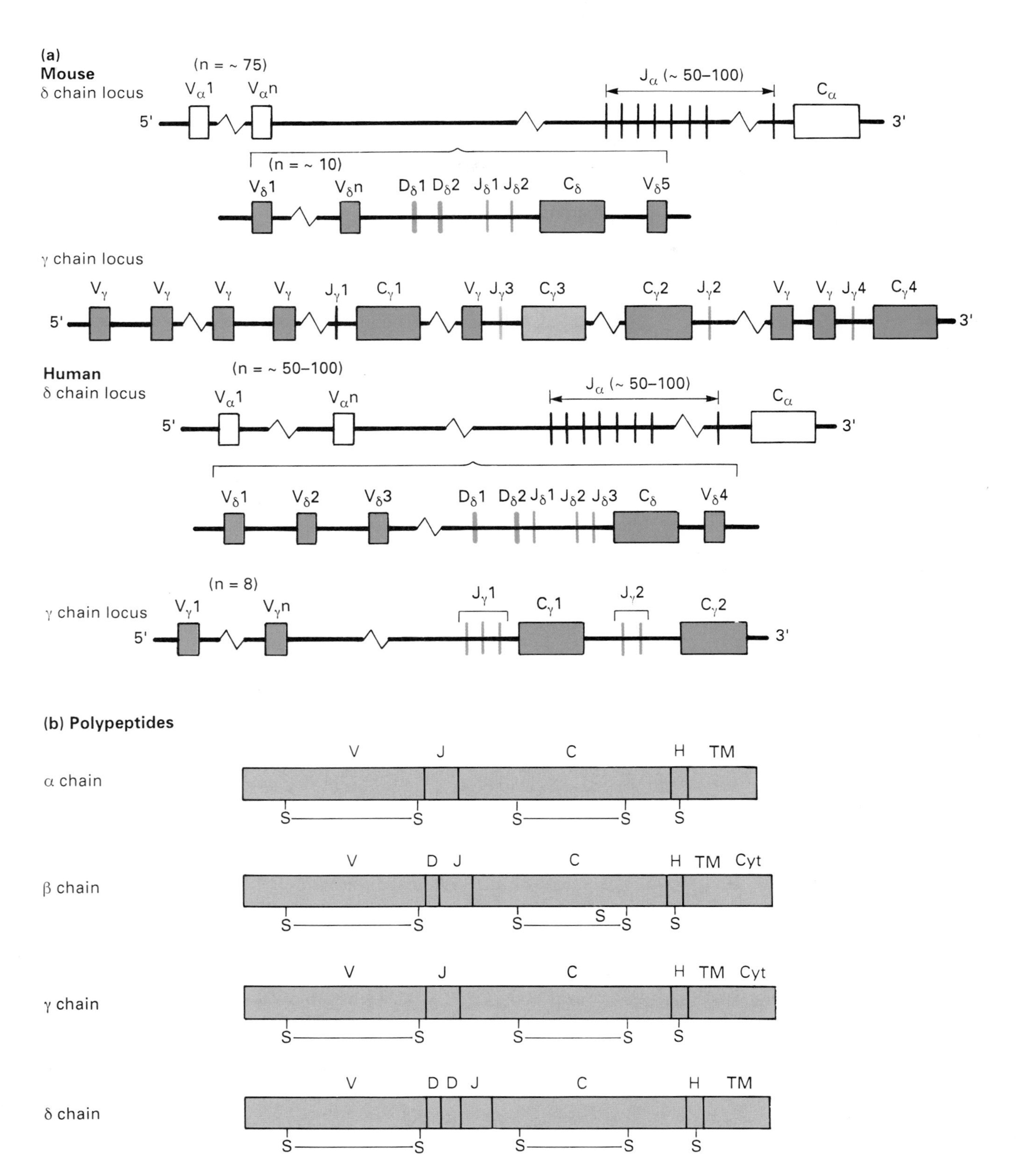

Fig.7.20(a) Organization of T cell receptor γ and δ chain loci in the mouse and humans. (Leader sequences are omitted for clarity).
(b) Comparison of structures of α, β, γ, and δ mature TcR polypeptides

α chain the variable domain is divided into two gene segments, V_γ and J_γ. D_γ segments have not been found in either the human or mouse genome. In the mouse a limited set of seven potentially functional V_γ segments is present, although it is likely that only six functional combinations are possible. Eight functional human V_γ genes have been found, and these can be divided into four subgroups (Fig.7.20).

The organization of the γ genes in the mouse is complex, and in several ways is similar to that described above for the immunoglobulin λ light chains genes.

Four C_γ genes have been found in mouse, the third of which is a pseudogene. There is also some doubt as to whether the fourth gene is expressed. The $C_\gamma 1$ and $C_\gamma 2$ genes both have cysteine residues in the connecting peptide which can be used to form disulfide bonds with the δ chain or with another γ chain at the cell surface (Fig.6.26). Each C_γ gene is associated with a single J gene segment and thus four J_γ–C_γ clusters have been identified in the mouse as shown in Fig.7.20. The V_γ segments appear to be associated with particular J_γ–C_γ clusters. There are seven potentially functional V_γ genes. Four are linked to $C_\gamma 1$ and two to $C_\gamma 4$. As the $C_\gamma 3$ gene is nonfunctional it seems likely that only six of these V_γ genes can be used to synthesize γ chains. The nomenclature used for γ gene segments at present is inconsistent between research groups. The nomenclature system used here is based on nucleotide sequence homology.

In the human the organization is slightly different: only two C_γ gene segments have been described and these are separated by about 20 kb of DNA. Three J_γ segments have been identified upstream of the $C_\gamma 1$ gene and two upstream of the $C_\gamma 2$ gene. All of these J gene segments appear to be used for the production of human γ chains. The human C_γ genes are divided into three exons, the first coding for the extracellular domain, the second for the connecting peptide, and the third for the remainder of the connecting peptide, the transmembrane region, the 12 amino acid cytoplasmic tail, and the 3′ untranslated region.

The human $C_\gamma 1$ gene contains a cysteine residue in the connecting peptide and can therefore form a disulfide bond with the δ chain or indeed another γ chain; the $C_\gamma 2$ gene does not contain this cysteine residue. This accounts for the two forms of the $\gamma\delta$ receptor that have been detected at the cell surface, one of which is disulfide bonded and the other that is not (Fig.7.21). The human $C_\gamma 2$ gene may contain between one and three 48-base pair repeats, each of which is highly homologous but not absolutely identical. These repeats code for an amino acid sequence containing N-glycosylation sites, and the presence or absence of these repeat sequences explains the differences in molecular weight values reported for the γ chain.

For example, assuming the 48 bp repeat codes for 16 amino acids and has a molecular weight of approximately 1.5 kDa, and that the N-linked carbohydrate that can be attached if this sequence is present has a molecular weight of 3.5 kDa, then the molecular weights of γ chains containing different numbers of these repeats would be approximately as follows:

0 repeats $M_r = 35$ kDa
1 repeat $M_r = 35 + 1.5 + 3.5 = 40$ kDa
2 repeats $M_r = 35 + [(1.5 + 3.5) \times 2] = 45$ kDa
3 repeats $M_r = 35 + [(1.5 + 3.5) \times 3] = 50$ kDa

These values generally agree with some of the molecular weights reported for γ chains isolated from T cells (see box, p. 375).

Organization of the δ chain genes

When the γ chain gene product was identified, much attention was focused on the partner polypeptide with which it formed a heterodimer on certain T cells. Investigation of the T cell receptor δ chain gene rearrangements that occur during thymic ontogeny revealed the presence of another T cell receptor gene, surprisingly coded *within* the δ gene locus, between the V and J clusters (Fig.7.18) (Chien *et al.* 1987). The organization of the δ chain gene is unusual in that the V_δ gene segments are located 3′ to the V_α segments. The remainder of the variable domain is encoded by a D_δ and a J_δ segment; two or three D_δ and J_δ segments are present in the mouse and human genomes. The location of the δ gene within the α locus is intriguing because the potential diversity of the δ gene products is enormous if the V_α segments could be used in addition to the V_δ segments to produce functional δ genes. However, in most cases the α and δ chains utilize distinct V gene segments.

7.3.2 Rearrangement

In many respects the mechanisms used by T cells to generate diversity in their repertoire of antigen receptors are similar to those used by B cells. A combination of the presence of multiple genes in the germline and a series of gene rearrangements allows sufficient diversity to be generated in the T cell repertoire. T cell receptor genes undergo rearrangement in T cells during

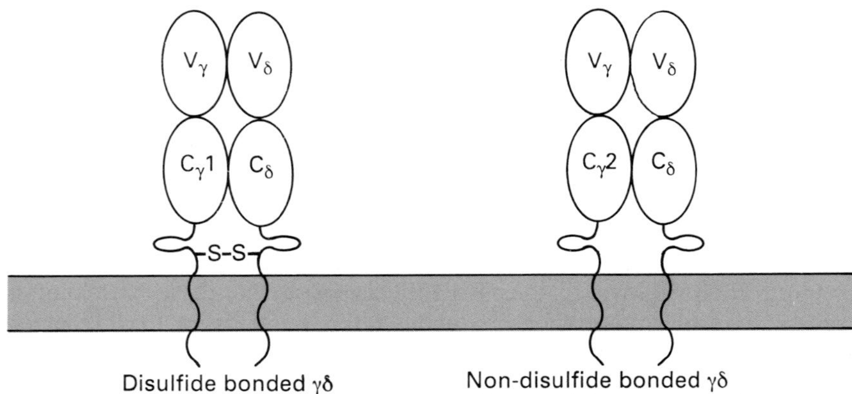

Fig.7.21 At least two forms of the human $\gamma\delta$ T cell receptor can be identified at the cell surface

the development and maturation of the cell (Leiden 1992). As for immunoglobulin genes, there are two phases in the rearrangement process. In the embryo and all non-T cells in the adult, for example liver, the various gene segments coding for T cell receptor polypeptides are separated from one another and form clusters of similar segments along the chromosome, whereas in the mature T cell the variable domain gene segments are rearranged so that they are much closer together and contiguous with the constant region genes. Firstly, the variable domain gene segments have to be joined together, and secondly, these then have to associate with the DNA coding for the constant domain of the T cell receptor polypeptide.

The mechanism of joining of TcR variable gene segments is most probably the same as that operating during immunoglobulin gene rearrangements (Yancopoulos *et al.* 1986). The heptamer and nonamer concensus sequences that flank both the heavy and light chain variable gene segments (see Section 7.2.2) also flank the T cell receptor variable region gene segments for all four T cell receptor genes, α, β, γ, and δ (Fig.7.22). Furthermore, these consensus sequences are also separated by either 12 or 23 nucleotides as was described for immunoglobulins and thus the '12–23' rule may also apply to the rearrangement of the T cell receptor variable region genes (see p. 399).

All T cell receptor V gene segments, irrespective of which gene they belong to, have the heptamer and nonamer sequences separated by 23 base pairs, or two turns of the DNA helix on their 3′ side (Fig.7.22). On the 5′ side of both of the D_β segments and the D_γ segment, the consensus sequences are in the reverse orientation and separated by 12 nucleotides, the single turn recognition signal; on the 3′ side they are separated by 23 nucleotides, the two-turn recognition signal. On the 5′ side of each of the J segments the sequences are separated from one another by a one-turn signal (Fig.7.22).

One important difference to note between the arrangement of the consensus sequences for the immunoglobulin heavy chain variable genes and the T cell receptor β and δ chain variable genes, is their spacing relative to one another flanking the D and the V and J segments (compare Fig.7.10 and 7.22). In the Ig heavy chain, inappropriate joining of V_H to J_H segments, that is without the insertion of a D_H segment, is *prevented* because the consensus sequences are incorrectly arranged assuming the 12/23 spacer rule is applied (Section 7.2.2). However, the arrangement of the consensus sequences for β and δ T cell receptor genes *does not preclude* these 'aberrant' rearrangements (Fig.7.22). Thus, if the proposed rules for V–D–J joining are correct, both V_β-D_β-J_β and V_β-J_β joining as well as V_β-D_β-D_β-J_β joining is permissible. Indeed all of these configurations have been found in T cells at the cDNA level, although the insertion of two D gene segments is relatively infrequent. This obviously has important implications as far as the T cell repertoire is concerned, because they could potentially increase the diversity of the T cell receptor β chains. However, the possibility also exists that these extra joining events may result in more non-productive rearrangements of T cell receptor genes.

For the β and δ genes, the D and J segments are selected and joined together first of all; these then join to the selected V segment, thus forming a functional V–D–J unit. The $D_\beta 1$ gene segment can rearrange with either of the J_β clusters, $J_\beta 1$ or $J_\beta 2$, apparently at random, providing a further mechanism for increasing the diversity of the β chains (see Fig.7.18). For the δ and γ genes that do not possess D gene segments, V–J rearrangements occur first. The order of T cell receptor gene rearrangements is very precisely controlled in the thymus, and is noted in Sections 4.3 (see box, p. 232) and 7.3.4.

7.3.3 Generation of T cell receptor diversity

In several respects the strategies employed for generating diversity in T cell receptors are similar to those used to generate the immunoglobulin repertoire. The potential diversity in the T cell receptor repertoire is estimated in Table 7.3, and arises through several mechanisms:

1. The presence of multiple gene segments encoding variable domains in the germline.
2. The possibility of combining any of the gene segments, one from each of the V, D, and J pools, to generate a functional gene.
3. For β and δ genes the possibility of multiple rearrangement combinations, for example V–D–J, V–J, V–D–D–J.

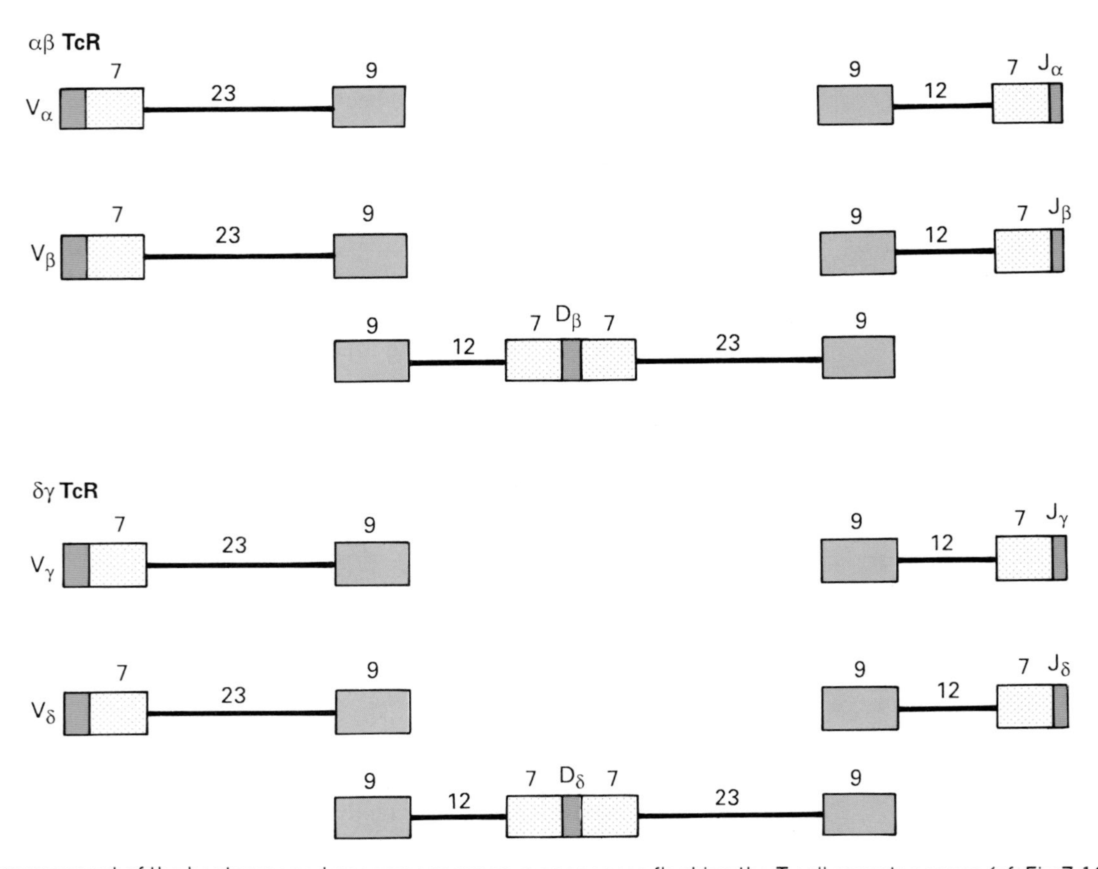

Fig.7.22 Arrangement of the heptamer and nonamer consensus sequences flanking the T cell receptor genes (cf. Fig.7.10)

Table 7.3 Estimate of the diversity of the T cell receptor repertoire

	TCR $\alpha\beta$		TCR $\gamma\delta$	
	α	β	γ	δ
V gene segments	100	25	7	10
D gene segments	0	2	0	2
J gene segments	50	12	2	2
Ds read in all frames	–	Often	–	Often
N region addition	V–J	V–D, D–J	V–J	V–D, D–D, D–J
V region combinations	2500		70	
Junctional combinations	10^{15}		10^{18}	

4. The imprecise nature of the joining process can generate increased variability:
   (i) joining of two segments (e.g. V–D, D–J or V–J) can occur anywhere within the gene at the joining position, thereby generating a new amino acid at the join;
   (ii) frameshift mutations may occur as a result of the joining process —these are relatively common in the D segments of TcR β and δ chain genes; and
   (iii) extra nucleotides can be inserted at the joining position, N-sequences —N-region diversity has been detected for all TcR genes.

   These latter mechanisms for the generation of diversity are particularly important for T cell receptor genes, which have a somewhat restricted germ line repertoire relative to the immunoglobulin genes. Many of the rearrangements occurring in T cells result in frame shift mutations, particularly in the β and δ chains. Furthermore, if two D_β segments are inserted into the β chain, then they appear to be utilized with equal frequency in all three reading frames. This is not found for immunoglobulin heavy chains, where the D_H segments are usually read in the same frame.

5. The formation of an antigen binding site through the association of the α and β, or γ and δ, chains in the T cell receptor protein, and the potential association of any α with any β chain, or of any γ with any δ chain. In addition, there is evidence to suggest that two γ chains may associate to form a receptor at the cell surface (Fig.6.26).

Somatic mutation has *not* yet been described as a mechanism for increasing the diversity of T cell receptors. Comparison of expressed variable region genes with their counterparts in the germ line has not revealed any point mutations. At present this conclusion may simply be due to a lack of information, as only relatively small numbers of T cell hybridomas and clones have been examined. Thus somatic mutation in TcR genes cannot be completely ruled out, but it may be that mutation events might be disadvantageous to the T cell repertoire where both antigen and MHC specificities are involved in antigen recognition.

7.3.4 Control of gene rearrangements

The order of rearrangement of the four TcR genes during T cell development has been investigated in the mouse. During T cell development the γ and δ chain genes are rearranged first. The transcription rate of these genes is high at day 14 of gestation and they are first detected at the cell surface by day 15. The kinetics of the α and β chain rearrangements during ontogeny are very similar but slightly delayed. The β chain genes rearrange before α and by day 14 of gestation D_β-J_β rearrangements are detectable. Complete V-D-Jβ rearrangements can be found from day 15. It is not yet clear whether the α and β chain genes only rearrange in the same cell if the γ chain gene rearrangements are non-productive, or whether this happens automatically in a proportion of T cells even if the γ–δ rearrangements have been productive (Fig.4.21). Obviously rearrangement of the α genes automatically disrupts and potentially eliminates the δ chain genes (Fig.7.18).

Lymphoid cell-specific transcription factors that modulate T cell receptor gene expression have been cloned. T cell receptor enhancer elements have also been described (Leiden 1992). However, much more information is required before a clear understanding of the molecular mechanisms involved is obtained.

7.4 The immunoglobulin superfamily

The immunoglobulin superfamily is a group of molecules that contain units or domains within their structure that are related to immunoglobulin variable or constant domains. A supergene family is defined as a set of genes that are related to one another by sequence, but not necessarily by function. The immunoglobulin superfamily includes the antigen receptors, immunoglobulins and T cell receptors, as well as several other lymphocyte cell surface molecules, for example CD4 and CD8 (Section 4.2.4), MHC class I and class II molecules, together with antigens found in the brain, such as Thy-1 (also present in mouse T cells) and NCAM. A listing of some of the members of the family is given in Table 7.4, and their structures are shown in Fig.7.24. The functions of the members of the immunoglobulin superfamily are diverse, but the majority play an important role in recognition at the cell surface. The genetic loci for members of the superfamily are not clustered together, but are widely spread throughout the genome, although several linkage groups have been identified. The number of molecules that belong to the superfamily is continually increasing as our knowledge of the structure of molecules expressed at cell surfaces expands as new genes are cloned and sequenced. The immunoglobulin superfamily is one of the important groups of molecules involved in cell surface recognition events, and as a result the regulation of the activity of cells in different tissues (Williams and Barclay 1988).

Table 7.4 Members of the immunoglobulin superfamily

Molecules and tissue expression	**Functions**	**Recognition within the superfamily**
Immunoglobulins:		
B lymphocytes only	B lymphocyte antigen receptors (membrane Ig) and in secreted form (secreted Ig)	No, antibodies recognize antigen without involvement of other molecules
T cell receptors:		
T lymphocytes and thymocytes	T lymphocyte antigen receptors; no known soluble forms	Yes, heterophilic; TcR binds MHC plus peptide but recognition does not involve the Ig-related MHC domains
CD3 γ, δ, ε chains:		
T lymphocytes and thymocytes	Part of the TcR complex; role in signal transduction	CD3 associates with TcR but no known recognition of other molecules
MHC antigens:		
Many cell types, induced by interferon	Present peptides from foreign antigen to the TcR; some soluble forms	Yes, heterophilic; TcR interacts with class I and class II MHC antigens
β_2-m associated antigens:		
Subsets of lymphoid cells	Functions not known ($\gamma\delta$ TcR recognition?)	No natural ligands known ($\gamma\delta$ TcR?)
T lymphocyte adhesion molecules:		
CD2, thymocytes and T cells (some macrophages in rat); LFA-3 (CD58), widespread expression; ICAM-1 (CD54), ICAM-2	CD2 of T cells interacts with LFA-3 on other cells in adhesion reactions. Anti-CD2 antibodies can trigger T cell activation (see Section 4.2.7)	Yes, heterophilic; CD2 binds LFA-3; ICAM-1 and ICAM-2 bind LFA-1, a member of the integrin family (see Section 4.2.6)
T subset markers:		
CD4 and CD8 on thymocytes and T cell subsets, CD4 on macrophages, CD8 on NK cells; CTLA4, CD28 on T cells	CD4 and CD8 control MHC class II- and class I-restricted antigen recognition by T cells. CTLA4, CD28 regulate cytokine genes	Perhaps heterophilic: CD4 and CD8 may bind the β_2 domain of class II and α_3 domain of class I MHC respectively. CTLA4 binds B7/BB1
Brain/lymphoid antigens:		
Thy-1, neurons, fibroblasts, various lymphoid cells; MRC OX-2, neurons, endothelium, various lymphoid cells	Anti-Thy-1 antibody triggers mouse T cell proliferation; MRC OX-2 function unknown	Possibly homophilic? No natural ligands known

Immunoglobulin receptors:		
Poly-Ig R, gut and liver epithelium. Several FcR on different cell types	Poly-Ig R transports multimeric IgA or IgM across epithelium. Different FcR bind various Igs for antigen uptake and signal transduction	Yes, heterophilic for both Poly-Ig R and other FcR which bind various domains of Ig
Neural-associated molecules:		
NCAM, neurons and glia, early embryo; MAG, peripheral and central myelin, some neurons; P_0 peripheral myelin, contactin, OBCAM	NCAM mediates adhesion of neural cells. MAG may function in myelination. P_0 constitutes 50% of peripheral myelin protein	Yes, homophilic for NCAM via Ig-related parts and perhaps for P_0. MAG not known
CEA:		
Epithelial cells and their tumors, early embryos	Tumour marker but function unknown	Natural ligand unknown
Growth factor and Cytokine receptors:		
PDGFR, widespread on mesenchymal cells; CSF-1R, monocyte lineage, 1L-1R, 1L-6R	Interact with growth factors or cytokines to trigger cell division or activation and other activities	No, PDGFR and CSF1R not known to react with molecules other than growth factors
Link protein:		
Basement membrane	Acts as a binding molecule between proteoglycan and hyaluronate	No
α_1B-glycoprotein:		
Found in serum	Function unknown	Natural ligands unknown
Invertebrate molecules:		
Fascilin 2, neuroglian, DLAR DPTP		
Muscle proteins:		
Twitchin		

7.4.1 Criteria for inclusion in the immunoglobulin superfamily

The main criteria for membership of the immunoglobulin superfamily are (i) that part of the molecule contains a sequence of the appropriate size for an immunoglobulin domain, 110 amino acids; (ii) that this shows significant similarity to either a variable or constant immunoglobulin domain sequence; and (iii) that the molecule is predicted to contain an immunoglobulin-like domain, in other words that the sequence shares the key structural features of the immunoglobulin fold. As discussed in Chapter 6, Section 6.2.1, the polypeptide chain of an immunoglobulin domain has a defined three-dimensional folding pattern (Fig.7.23). The strongest patterns of sequence similarity are found in the β strands B, C, E, and F. A statistical test must be used to evaluate the extent of the sequence similarities, and for these analyses the ALIGN program of Dayhoff *et al.* (1983) has proved very useful. On the basis of the ALIGN scores, sequences are assigned as related to either variable (the VSET) or constant (the C1SET and the C2SET) domains. The assignment of a sequence as belonging to the VSET does not indicate any sequence variability as is found in antigen receptor variable domains, but simply that it has a pattern of β strands similar to that in a variable domain (Fig.7.23).

Recently some molecules assigned as members of the family have not conformed to either a variable or constant domain pattern, but have appeared to be hybrids of the two. As more members of the family are identified, the criteria for inclusion may have to be broadened to account for these variants.

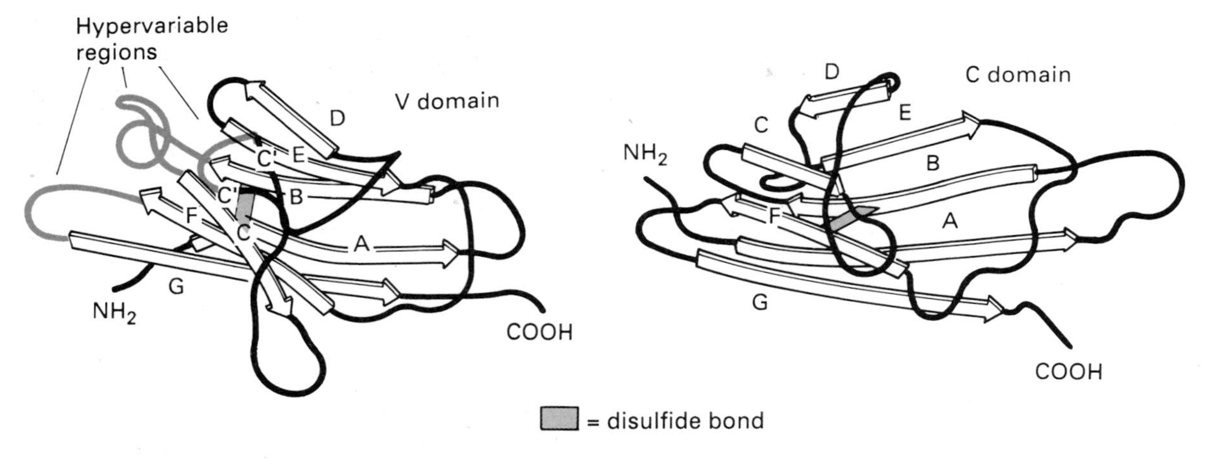

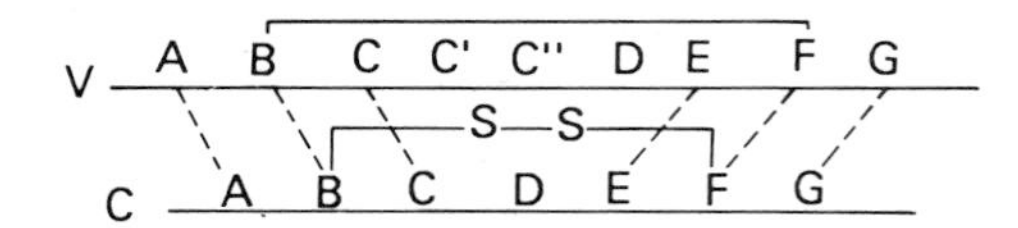

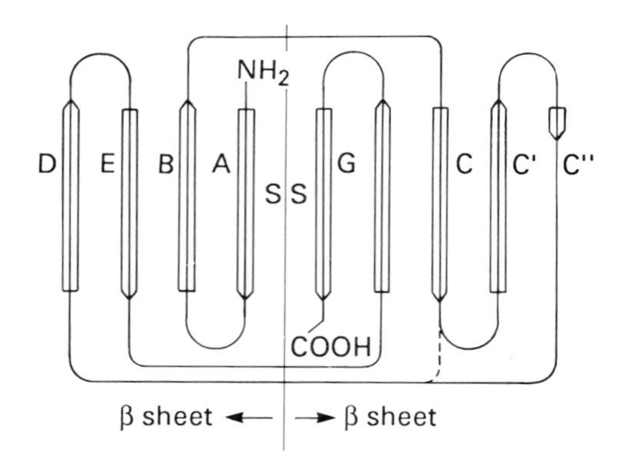

Fig.7.23 Structure of immunoglobulin variable and constant domains. Adapted from Williams and Barclay (1988), *Annual Review of Immunology*, **6,** 381; © 1988 by Annual Reviews Inc.

For example, the presence of an intra-domain disulfide bond that joins two of the strands of a β-pleated sheet (Fig.7.23) was once considered to be the distinguishing feature or hallmark of an immunoglobulin domain and was a requirement for assignment of a new molecule to the immunoglobulin superfamily. Recently a heavy chain variable domain that lacks the conserved disulphide bond has been found in a functional antibody. In this molecule a tyrosine residue is present instead of a cysteine in the F strand of the V_H domain (Fig.7.23). Thus structures can be related to immunoglobulin domains without containing the conserved disulfide bond. Members of the family that fit into this category are CD4 and CD2. In other cases the cysteine residues are replaced by hydrophobic amino acids that are also suitable for stabilizing the immunoglobulin fold. Members of the superfamily that fit into this group are LFA-3, platelet derived growth factor (PDGF), and the receptor for colony stimulating factor 1 (CSF-1) which lack cysteine residues in strands B and F.

The organization of the genes that code for members of the superfamily is interesting. In many cases the sequences related to immunoglobulin domains are encoded as separate exons in genomic DNA. This is true for the constant domains of immunoglobulins and T cell receptors (See Sections 7.2.1 and 7.3.1) and for 20 out of 30 domains described so far in non-antigen receptor genes. Thus in the genes coding for the majority of the members of the immunoglobulin superfamily, there is a precise exon/intron relationship that matches the domain structure of the protein. However, a number of examples have been described more recently, where introns are found between the sequences coding for the cysteine residues of the conserved disulfide bonds. These include CD4, poly-Ig receptor, Po glycoprotein, and NCAM. Nevertheless, overall, the genetic data strongly support the hypothesis that the members of the immunoglobulin gene family are related to one another, and can be used as another piece of evidence to support the assignment of molecules to the family.

7.4.2 Members of the immunoglobulin superfamily

Some of the members of the immunoglobulin superfamily, are listed in Table 7.4. Structural models for some of the molecules are shown in Fig.7.24 and are represented to show how the different parts of each molecule are related to either variable or constant domains. The members of the superfamily can be divided into the following categories:

1. Antigen receptors and other molecules that bind antigen. These include immunoglobulins and T cell receptors, both of which have been discussed in detail in this chapter and in Chapter 6. CD3 is also included in this group, as it is intimately associated with the TcR on the surface of T lymphocytes and the γ, δ, and ε chains all contain an immunoglobulin domain (see Section 4.2.3).

MHC molecules can also be thought of as antigen binding molecules, because they bind peptides to form complexes that can be recognized by T cells. The membrane proximal domains of class I α chains and the class II α and β chains are clearly related to immunoglobulin constant domains. In addition, β_2-microglobulin, the molecule associated with class I heavy chains at the cell surface, has been shown by X-ray crystallography

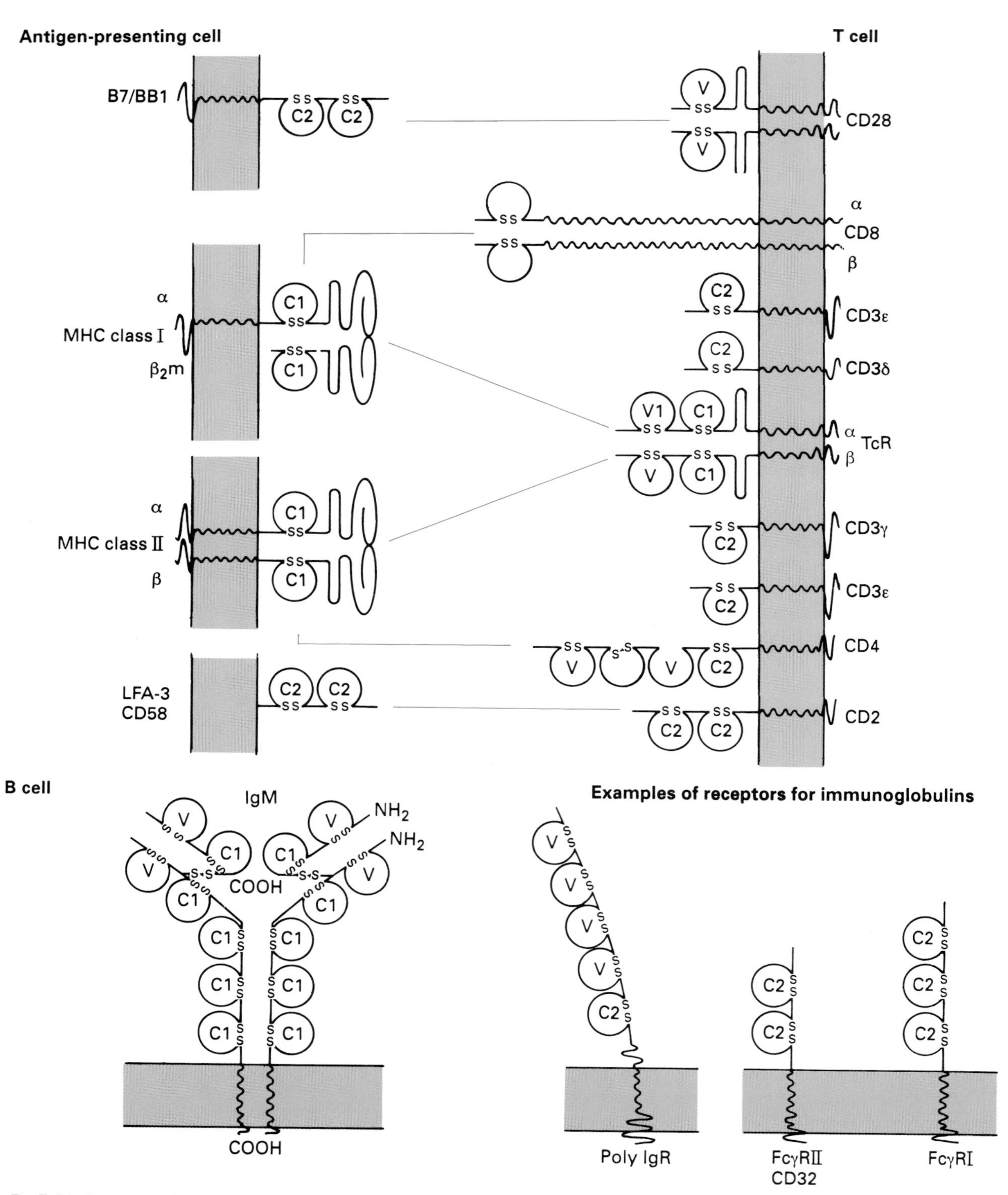

Fig.7.24 Some members of the immunoglobulin superfamily. Some of the molecules on this page are aligned to indicate the interactions that can occur between different members of the immunoglobulin superfamily. Adapted from Williams and Barclay (1988), *Annual Review of Immunology*, **6,** 381; © 1988 by Annual Reviews Inc.

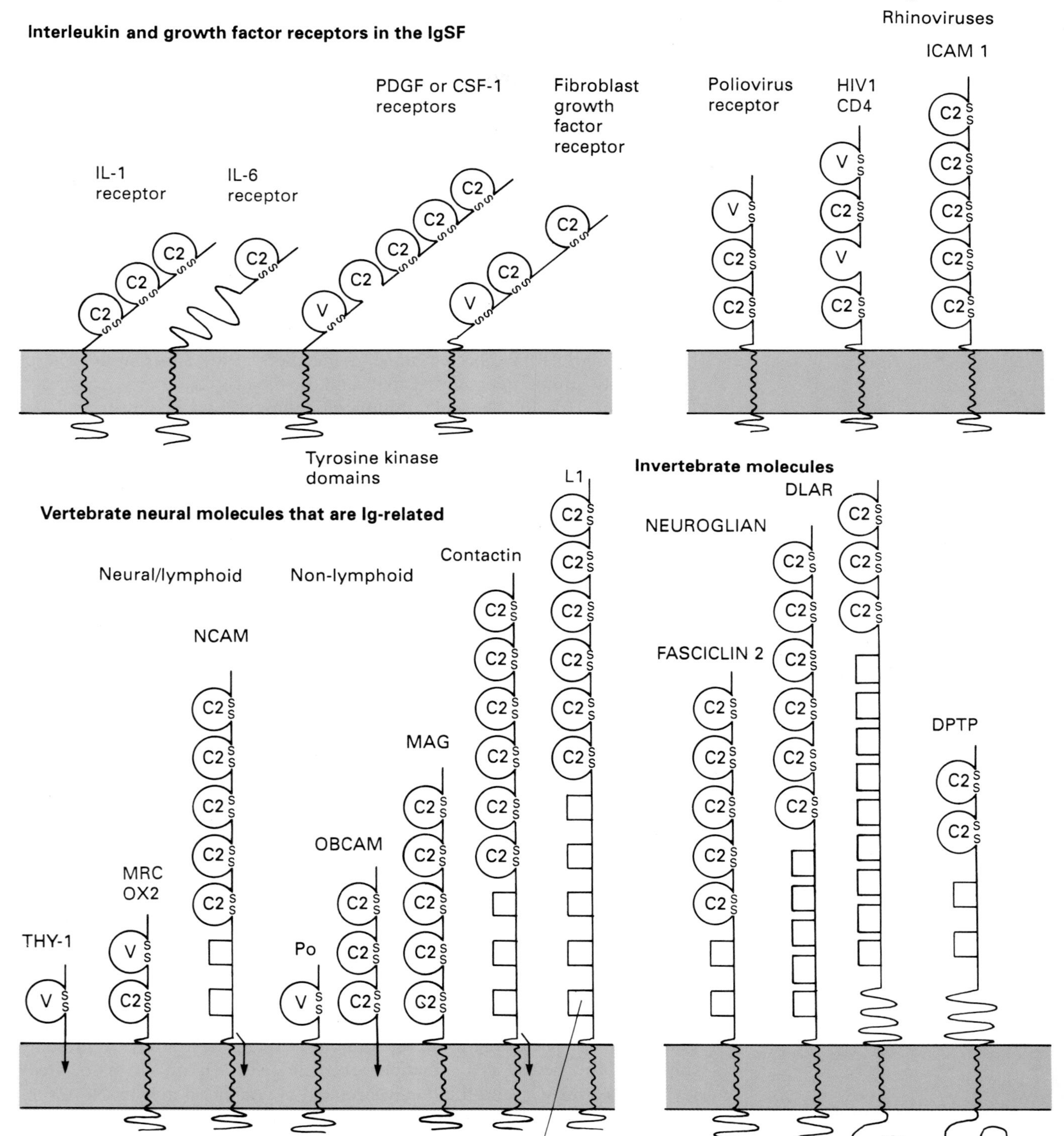

Interleukin and growth factor receptors in the IgSF
IL-1 receptor
IL-6 receptor
PDGF or CSF-1 receptors
Fibroblast growth factor receptor
Tyrosine kinase domains
Virus receptors that are Ig-related
Poliovirus receptor
HIV1 CD4
Rhinoviruses
ICAM 1
Vertebrate neural molecules that are Ig-related
Neural/lymphoid
Non-lymphoid
THY-1
MRC OX2
NCAM
Po
OBCAM
MAG
Contactin
L1
Fibronectin type III domain
Invertebrate molecules
FASCICLIN 2
NEUROGLIAN
DLAR
DPTP

to have a three-dimensional structure identical to that of the third immunoglobulin constant domain. The structure and function of these molecules are discussed in detail in Chapter 2.

The class I-related molecules in the mouse, Tla and Qa antigens, are associated with β_2-microglobulin. The functions of the Tla and Qa antigens are unknown, but several hypotheses proposing a role for these genes in the evolution of MHC molecules have been put forward. It has been suggested more recently that these molecules may be involved in the presentation of antigen to $\gamma\delta$ T cells. Human CD1 antigens are also associated with β_2-microglobulin and as such contain an immunoglobulin-related structure. The CD1 structure is not linked to the MHC, and thus forms a separate category of molecules that are associated with β_2-microglobulin. CD1 may also be involved in the presentation of peptides to $\gamma\delta$ T cells (see Section 6.3.2).

2. Receptors for immunoglobulin. Molecules that act as receptors for immunoglobulins have been defined as members of the superfamily. These are the poly-immunoglobulin receptor and the majority of other Fc receptors (see Section 10.4.5). The poly-immunoglobulin receptor is synthesized by epithelial cells and is involved in the transport of IgA to the extracellular fluids (see Section 6.2.5). Cellular FcR are involved in uptake of antigens and/or signed transduction.

3. T lymphocyte cell surface and adhesion molecules. The CD4 (Wang *et al.* 1990; Ryu *et al.* 1990), CD8, and CD2 (Driscoll *et al.* 1991) antigens (see Section 4.2) all contain immunoglobulin-related sequences. CD4 has a typical V-like domain, together with three other domains that are atypical but nevertheless contain immunoglobulin-related sequences. The CD8 antigen contains domains that are related to variable region sequences. CD2 is a T cell molecule that interacts with LFA-3 (CD58) expressed on other leukocytes, and as such both molecules can be considered as lymphocyte adhesion molecules. The two extracellular domains of CD2 exhibit immunoglobulin-related sequences. Other cell adhesion molecules, such as ICAM-1 (CD54) and ICAM-2, also belong to the superfamily (Section 4.2).

4. Brain and thymus antigens. This group includes Thy-1 and MRC OX2, whose function are not known at present. Thy-1 was the first antigen that could be used to separate murine T and B lymphocytes. It is expressed on different tissues in different species (Section 4.2.9). For example, in the mouse Thy-1 is present on a high percentage of thymocytes, on T cells and fibroblasts and is also expressed in the brain. However, in other species, although expression is conserved in the brain, it varies greatly on lymphoid cells. MRC OX2 is an antigen expressed by thymocytes, B cells, follicular dendritic cells, endothelium, neurons, and smooth muscle.

5. Neural cell adhesion molecules. Other molecules expressed in the brain have also been assigned to the immunoglobulin superfamily. At present these are the neural cell adhesion molecule (NCAM), myelin associated glycoprotein (MAG), and the Po glycoprotein of peripheral myelin. NCAM is the molecule responsible for the formation of aggregates by chick embryonic neural cells, and is therefore important for cellular interactions in neuronal tissue. The function of the MAG and Po glycoproteins are unknown. The Po glycoprotein constitutes approximately half of the protein in the multilayered membrane of the peripheral myelin sheath; it is not found in myelin of the central nervous system. MAG is expressed on peripheral and central myelin and some neurons and the molecule may play an important role in myelination.

6. Growth factor and lymphokine receptors. Immunoglobulin-related structures have been identified in two growth factor receptors, the receptor for platelet derived growth factor (PDGF) and the c-fms molecule that is the receptor for colony stimulating factor 1 (CSF-1). Five immunoglobulin-related domains can be identified in each of these receptors. The receptors for the cytokines IL-1 and IL-6 are also members of the superfamily (Sections 9.5.1, 9.5.2).

7. Virus receptors and viral cell surface molecules. Some members of the superfamily have been identified as the receptors for some important pathogenic viruses. For example CD4—HIV1; polio virus—polio virus receptor; and ICAM1—rhinovirus.

8. Tumour antigens. The tumour antigen 'carcinoembryonic antigen' (CEA) is also a member of the immunoglobulin superfamily, and contains both V- and C-like domains. It is expressed by epithelial cells and their tumours early in embryonic development. The function of the molecule is not known.

9. Extracellular molecules. So far, two molecules that are not thought to be cell surface structures have been identified as family members. These are the α_1 β-glycoprotein that is found in serum, but has no known function, and the basement membrane link protein that stabilizes the interaction between proteoglycan branch structures and a core of hyaluronic acid.

10. Muscle proteins. An invertebrate muscle protein called 'twitchin', which is thought to regulate the activity of myosin, has been found to contain 26 immunoglobulin domains, 32 fibronectin type 3 repeats, and a kinase domain!

7.4.3 Evolution and possible significance of the immunoglobulin superfamily

The evolutionary origin of this family is obviously a matter open to debate. An important question to answer is whether the family arose by divergent or convergent evolution. In other words, have the different members been produced as a result of divergence from a single primordial molecule by duplication of the original gene, or as a result of convergence to a favoured structure? If the latter hypothesis was correct, it is not clear why the production of a similar structural unit should result in conservation of amino acid sequence. If the immunoglobulin fold had arisen simply because it is an extremely stable structure it is likely that a range of amino acid sequences could have been used to produce it. Indeed there are molecules that resemble immunoglobulin domains at a structural level, but which share no sequence similarity at all, for example superoxide dismutase enzymes. These enzymes are not considered to be members of the family because they show no sequence similarity.

Sharing of sequence homology by the members of the superfamily would support the idea that they have been derived by gene duplication and subsequent divergence from a single domain. It has also been suggested that preceding the primordial domain, a half-domain structure may have existed. The half-domain is postulated to be like that of the β strands ABCC′ or GFED shown in Fig.7.23. Two half-domains are predicted to have associated to form a homodimer in the same way that the two β sheets of a V domain

(a) Heterophilic recognition from homophilic adhesion between cells

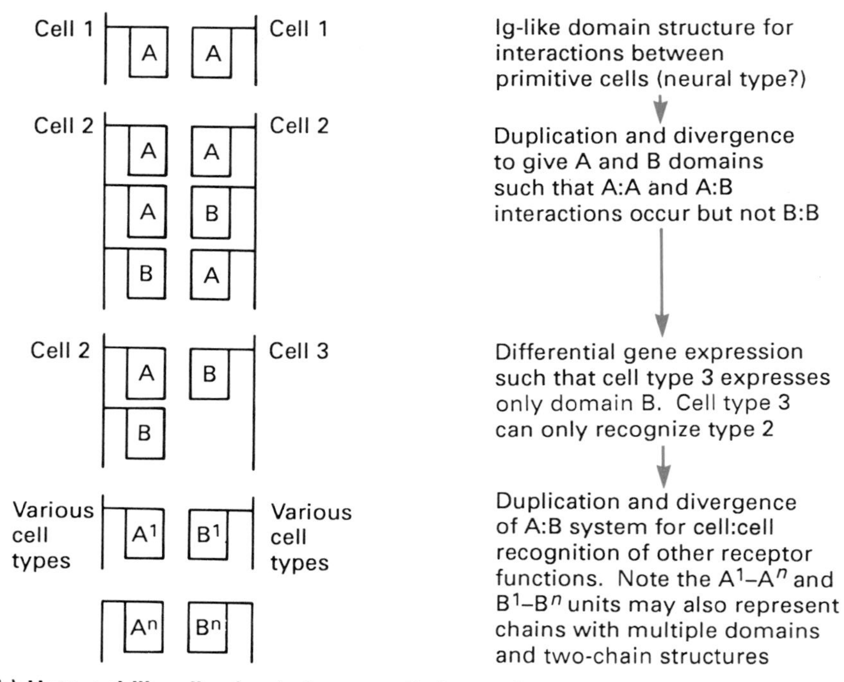
Cell 1
A
A
Cell 1
Ig-like domain structure for interactions between primitive cells (neural type?)
Cell 2
A
A
A
B
B
A
Cell 2
Duplication and divergence to give A and B domains such that A:A and A:B interactions occur but not B:B
Cell 2
A
B
B
Cell 3
Differential gene expression such that cell type 3 expresses only domain B. Cell type 3 can only recognize type 2
Various cell types
A1
B1
Various cell types
An
Bn
Duplication and divergence of A:B system for cell:cell recognition of other receptor functions. Note the A1–An and B1–Bn units may also represent chains with multiple domains and two-chain structures

(b) Heterophilic adhesion between cells from a homodimer on one cell

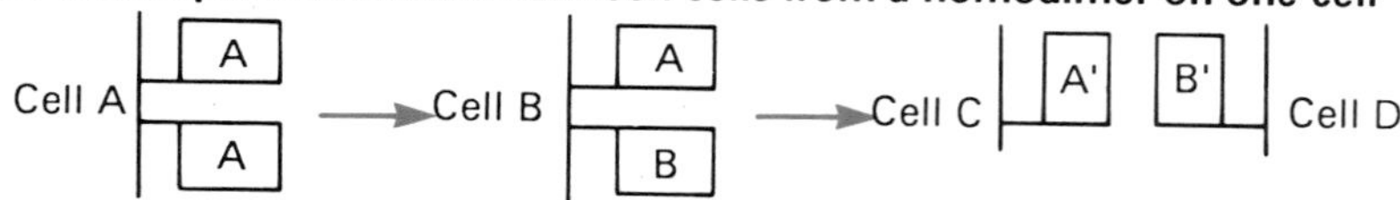
Cell A
A
A
Cell B
A
B
Cell C
A'
B'
Cell D

(c) Modification of heterophilic recognition to produce an immune system with similarities to the vertebrate T lymphocyte system

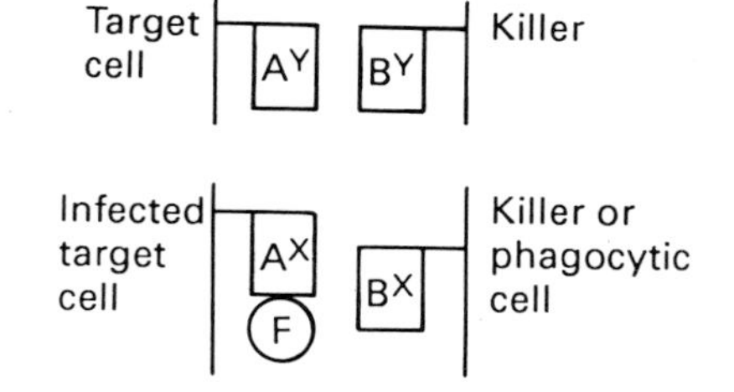
Target cell
AY
BY
Killer
Infected target cell
AX
F
BX
Killer or phagocytic cell

Postulated system of programmed cell death with specificity controlled by Ig-related domains

Specificity changed to incorporate a determinant of a common pathogen (F). Diversification of this system gives the Ig-related vertebrate immune system

Fig.7.25 Models for the evolution of the immunoglobulin superfamily (adapted from Williams and Barclay (1988), with permission)

associate. The 'half-domain hypothesis' is supported by the finding that the genes for certain members of the family have introns in the middle of the sequence encoding a particular domain as, for example, in CD4, Po, and the poly-Ig receptor. The existence of single-domain molecules, Thy 1 and Po, also support this hypothesis. Interestingly, Po is encoded by two exons and thus has an intron in the middle of the gene.

As the majority of family members are cell surface—associated molecules, it has been suggested that the sequence patterns have been conserved to protect the molecules from degradation by proteases in the extracellular environment. The stable domain structure has therefore been maintained, while changes have arisen in the loops and outer surfaces of the domains that are in contact with the external environment. In this way the biological properties of each family member could be modified, while maintaining the original structure of the domain unit. The structure of the domain may also allow members of the family to interact, e.g. CD2 and LFA-3 (Springer 1991).

If the hypothesis that these molecules have arisen as a result of divergence from a single precursor molecule is correct, a family tree of the evolutionary relationship between family members can be proposed. One such proposal is shown in Fig.7.25. Many factors have to be taken into consideration before a tree of this type can be put together, and as more sequences become available, modifications may have to be made to the original version.

The evolution of the superfamily with respect to its function is also unknown. It has been suggested that the primordial function involved a single domain interacting with itself, the two domains being present on opposed membranes (a homophilic interaction); indeed such a function has been suggested for the Po myelin protein. Interactions between dissimilar pairs of domains (heterophilic interactions) presumably evolved from these homophilic interactions, and thus highly specific recognition and interaction became possible between different cell types. All the available evidence points to the possibility that the first functions of members of the immunoglobulin superfamily were related to cell adhesion and the triggering of events at the cell surface to control the behaviour and interactions of cells in multicellular organisms. The identification of immunoglobulin-related molecules in neural tissues supports the idea that these molecules may be mediating functions of a more primitive type. Other hypotheses that are interesting in relation to the evolution of the immunoglobulin superfamily are reviewed by Williams and Barclay (1988).

Further reading

Immunoglobulin genes

Alt, F.W., Oltz, E.M., Young, F., Gorman, J., Taccioli, G., and Chen, J. (1992). VDJ recombination. *Immunology Today*, **13**, 306–14.

Blackwell, T.K. and Alt, F.W. (1989). In *Immunoglobulin genes in molecular immunology* (ed. B.D. Hames and D.M. Glover), pp. 1–60. IRL Press, Oxford.

Dunnick, W. and Stavnezer, J. (1990). Copy choice mechanisms of immunoglobulin heavy-chain switch combination. *Molecular and Cellular Biology*, **10**, 397–400.

Gottlieb, P.D. (1980). Immunoglobulin genes. *Molecular Immunology*, **17**, 1423–35.

Honjo, T., Alt., F.W., and Rabbitts, T.H. (1989). *Immunoglobulin genes*. Academic Press, London.

Kadesh, T. (1992). Helix-loop-helix proteins in the regulation of immunoglobulin gene transcription. *Immunology Today*, **13**, 31–6.

Lai, E., Wilson, R.K. and Hood, L.E. (1990). Physical maps of the mouse and human immunoglobulin-like loci. *Advances in Immunology*, **46**, 1–59.

Marcu, K.B. (1982). Immunoglobulin heavy chain constant region genes. *Cell*, **29**, 719–21.

Max, E.E. (1984). Immunoglobulins: molecular genetics. In *Fundamental Immunology* (ed. W.E. Paul), pp. 167–204.

Molgaard, H.V. (1980). Assembly of immunoglobulin heavy chain genes. *Nature*, **286**, 657–9.

Owen, M.J. and Lamb, J.R. (1988). *Immune recognition*. In Focus Series (ed. D. Male). IRL Press, Oxford.

Radbruch, A., Burger, C., Klein, S., and Muller, W. (1986). Control of immunoglobulin class switch recombination. *Immunological Reviews*, **89**, 69–83.

Schatz, D.G., Oettinger, M.A., and Schlissel, M.S. (1992). V(D)J recombination: molecular biology and regulation. *Annual Review of Immunology*, **10**, 359–83.

Shimizu, A. and Honjo, T. (1984). Immunoglobulin class switching. *Cell*, **36**, 801–3.

Standt, L.M., and Lenardo, M.J. (1991). Immunoglobulin gene transcription *Annual Review of Immunology*, **9**, 373–98.

Tonegawa, S. (1983). Somatic generation of antibody diversity. *Nature*, **302**, 575–81.

Wysocki, L.J. and Gefter, M.L. (1989). Gene conversion and the generation of antibody diversity. *Annual Review of Biochemistry*, **58**, 509–31.

T cell receptors

Davis, M.M. (1985). Molecular genetics of T cell receptor beta chain. *Annual Review of Immunology*, **3**, 537–60.

Davis, M.M. (1989). T cell antigen receptor genes. In *Molecular Immunology* (ed. B.D. Hames and D.M. Glover), pp. 61–80. IRL Press, Oxford.

Davis, M.M. and Bjorkman, P.J. (1988). T cell antigen receptor genes and antigen recognition. *Nature*, **234**, 395–402.

Dembic, Z., von Boehmer, H., and Steinmetz, M. (1986). The role of T cell receptor alpha and beta chain genes in MHC-restricted antigen recognition. *Immunology Today*, **7**, 308–11.

Hood, L., Kronenberg, M., and Hunkapiller, T. (1985). T cell antigen receptors and the immunoglobulin supergene family. *Cell*, **40**, 225–9.

Kronenberg, M., Sin, G., Hood, L.E., and Shastri, N. (1986). The molecular genetics of the T cell antigen receptor and T cell antigen recognition. *Annual Review of Immunology*, **4**, 529–91.

Leiden, J.M. (1992). Transcriptional regulation during T cell development: the αTCR gene as a molecular Model. *Immunology Today*, **13**, 22–30.

Mak, T.W. and Yanagi, Y. (1984). Genes encoding the human T cell antigen receptor. *Immunological Reviews*, **81**, 221–3.

Malissen, M., Trucy, J., Jouvin-Marche, E., Cazenave, P.A., Scollay, R., and Malissen, B. (1992). Regulation of TCR α and β gene allelic exclusion during T cell development. *Immunology Today*, **13**, 315–22.

Quatermous, T., Murre, C., Dialynas, D., Duby, A.D., Strominger, J.L., Waldman, T.A., and Seidman, J.G. (1985). Human T cell gamma chain genes: organisation, diversity and rearrangement. *Science*, **231**, 252–5.

Robertson, M. (1984). Receptor gene rearrangements and ontogeny of T lymphocytes. *Nature*, **311**, 305–6.

Immunoglobulin superfamily

Barclay, A.N., Johnson, P., McCaughan, G.W., and Williams, A.F. (1987). Immunoglobulin-related structures associated with vertebrate cell surfaces. In *T cell receptors* (ed. T.W. Mak), pp. 1.

Hood, L., Kronenberg, M., and Hunkapiller, T. (1985). T cell antigen receptors and the immunoglobulin supergene family. *Cell*, **40**, 225–9.

Williams, A.F. (1984). The immunoglobulin superfamily takes shape. *Nature*, **308**, 12–13.

Williams, A.F. and Barclay, A.N. (1988). The immunoglobulin superfamily—domains for cell surface recognition. *Annual Review of Immunology*, **6**, 381–405.

Literature cited

Immunoglobulins

Berek, C. and Milstein, C. (1987). Mutation drift and repertoire shift in maturation of the immune response. *Immunological Reviews*, **96**, 23–41.

Chien, Y., Becker, D.M., Lindsten, T., Okamura, M., Cohen, D.I., and Davis, M.M. (1984). A third type of murine T cell receptor gene. *Nature*, **312**, 31–5.

Chien, Y., Iwashima, M., Kaplan, K., Elliot, J., and Davis, M.M. (1987). A new T cell receptor gene located within the α locus and expressed early in T cell differentiation. *Nature*, **327**, 677–82.

Dayhoff, M.O., Barker, W.C., and Hunt, L.T. (1983) Establishing homologies in protein sequences. *Methods in Enzymology*, **91**, 524–45.

Dreyer, W.J. and Bennett, J.C. (1965). The molecular basis of antibody formation: a paradox. *Proceedings of the National Academy of Sciences*, **54**, 864–9

Driscoll, P.C., Cyster, J.G., Campbell, I.D., and Williams, A.F. (1991). Structure of domain 1 of rat T lymphocyte CD2 antigen. *Nature*, **353**, 762–5.

Fujimoto, S. and Yamagishi, H. (1987). Isolation of an excision product of T cell receptor α chain gene rearrangements. *Nature*, **327**, 242–3.

Hedrick, S.M., Cohen, D.I., Nielson, E., and Davis, M.M. (1984). The isolation of cDNA clones encoding T cell-specific membrane associated proteins. *Nature*, **308**, 149–53.

Hozumi, N., and Tonegawa, S. (1976). Evidence for somatic rearrangements of the immunoglobulin genes coding for variable and constant regions. *Proceedings of the National Academy of Sciences*, **73**, 3628–32.

Iwasato, T., Stimizu, A., Honjo, and Yamagishi, H. (1990). Circular DNA is excised by immunoglobulin class switch recombination. *Cell*, **62** 143–9.

Kadesch, T. (1992). Helix-loop-helix proteins in the regulation of immunoglobulin gene transcription. *Immunology Today*, **13**, 31–6.

Kallenbach, S., Goodhardt, M., and Rougeon, F. (1990). A rapid test for VDJ recombinase activity. *Nucleic Acid Research*, **18**, 6730–2.

Kronenberg, M., Sin, G., Hood, C.E., and Shastri, N. (1986). The molecular genetics of the T cell antigen receptor and T cell antigen recognition. *Annual Review of Immunology*, **4**, 529–91.

Leiden, J.M. (1992). Transcriptional regulation during T cell development: the α TCR gene as a molecular model. *Immunology Today*, **13**, 22–30.

Matsuoka, M., Yoshida, K., Maeda, T., Usada, S., and Sakano, H. (1990). Switch circular DNA formed in cytokine-treated mouse splenocytes: evidence for intramolecular DNA deletion in immunoglobulin class switching. *Cell*, **62** 135–42.

Nossal, G.J.V. (1992). The molecular and cellular basis of affinity maturation in the antibody response. *Cell*, **68**, 1–2.

Oettinger, M.A., Schatz, D.G., Gorka, C. and Baltimore, (1990). RAG-1 and RAG-2 adjacent genes that synergistically activate V(D)J recombination. *Science*, **248** 1517–23

Okazaki, K., Davis, D.D., and Sakano, H. (1987). T cell receptor β gene sequence in the circular DNA of thymocyte nuclei: direct evidence for intramolecular deletion in V-D-J joining. *Cell*, **49**, 477–85.

Quatermous, T., Murre, C., Dialynas, D., Duby, A.D., Strominger, J.L., Waldman, T.A., and Seidman, J.G. (1985). Human T cell gamma chain genes: organization, diversity and rearrangement. *Science*, **231**, 252–5.

Radbruch, A., Burger, C., Klein, S., and Muller, W. (1986). Control of immunoglobulin class switch recombination. *Immunological Reviews*, **89**, 69–83.

Ryn, S.E., Kwang, P.D., and Truneh, A. (1990). Crystal structure of an HIV-binding fragment of human CD4. *Nature*, **348**, 419–26.

Saito, H., Kranz, D.M., Takagaki, Y., Hayday, A.D., Eisen, H.N., and Tonegawa, S. (1984*a*). Complete primary sequence of a heterodimeric T cell receptor deduced from cDNA sequences. *Nature*, **309**, 757–62.

Saito, H., Kranz, D.M., Takagaki, U., Hayday, A.D., Eisen, H.N., and Tonegawa, S. (1984*b*). A third rearranged and expressed gene in a clone of cytotoxic T lymphocytes. *Nature*, **312**, 36–40.

Schatz, D.G., Oettinger, M.A., and Schlissel, M.S., (1992). V(D)J recombination: molecular biology and regulation., *Annual Review of Immunology*, **10**, 359–85.

Schimizu, A. and Honjo, T. (1984). Immunoglobulin class switching. *Cell*, **36**, 801–3.

Sim, G.K., Yagne, J., Nelson, J., Marrack, P., Palmer, E., Augustin, A., and Kappler, J. (1984). Primary structure of human T cell receptor α chain. *Nature*, **312**, 771–5.

Singh, H., Sen, R., Baltimore, D., and Sharp, P.A. (1986). A nuclear factor that binds to a conserved sequence motif in transcriptional control elements of immunoglobulin genes. *Nature*, **319**, 154–8.

Springer, T.A. (1991). A birth certificate for CD2. *Nature*, **353**, 704–5.

Tonegawa, S. (1983). Somatic generation of antibody diversity. *Nature*, **302**, 575–81.

Wang, J., Yan, U., and Garrett, T.P.J. (1990). Atomic structure of a fragment of human CD4 containing two immunoglobulin-like domains. *Nature*, **348**, 411–18.

Williams, A.F. and Barclay, A.N. (1988). The immunoglobulin superfamily-domains for cell surface recognition. *Annual Review of Immunology*, **6**, 381–405.

Wysocki, L.J. and Gefter, M.L. (1989). Gene conversion and the generation of antibody diversity. *Annual Review of Biochemistry*, **58**, 509–31.

Yanagi, U., Yoshikai, Y., Leggett, K., Clark, S.P., Aleksander, I., and Mark, T.W. (1984). A human T cell specific cDNA clone encodes a protein having extensive homology to immunoglobulin chains. *Nature*, **308**, 145–9.

Yancopoulos, G.D., Blackwell, T.K., Suh, H., Hood, L.E., and Alf, F.W. (1986). Introduced T cell receptor variable gene segments recombine in pre-B cells: evidence that B and T cells use a common recombinase. *Cell*, **44**, 251–9.

Zouali, M. and MacLennan, I.C.M. (1992). Molecular events in the development of the lymphocyte repertoire. *Immunology Today*, **13**, 41–3.

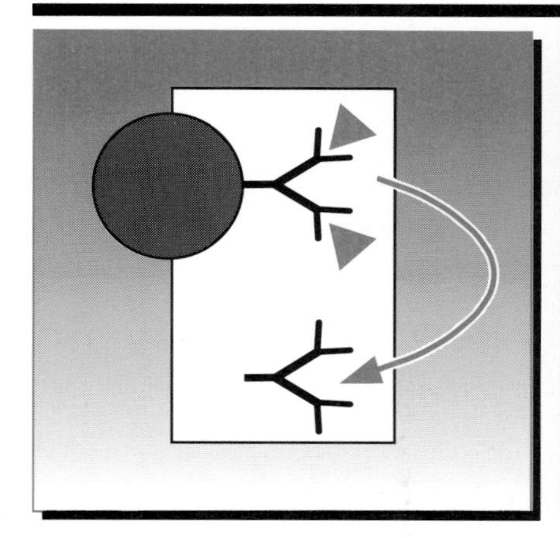

8

B CELLS AND ANTIBODY RESPONSES

Scheme 8. B cells and antibody responses

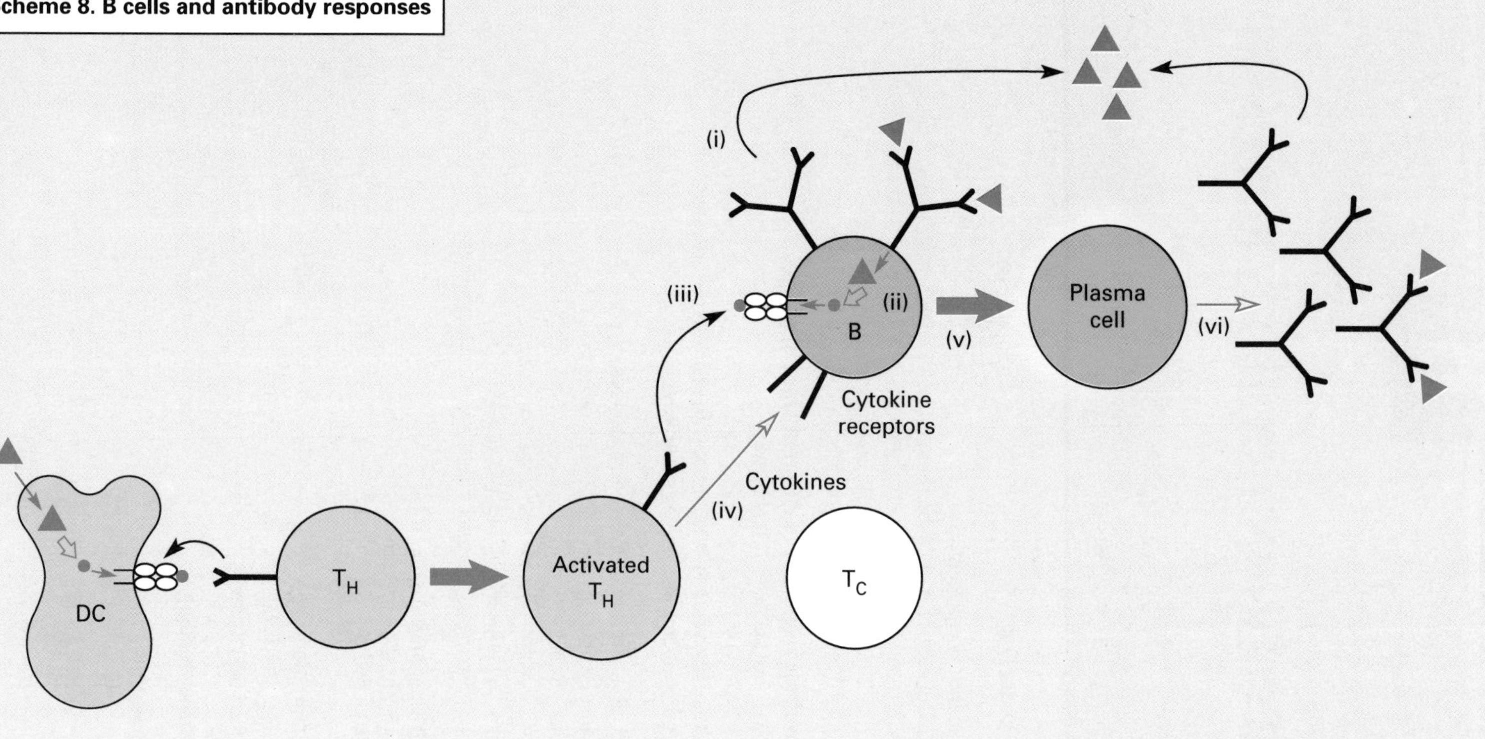

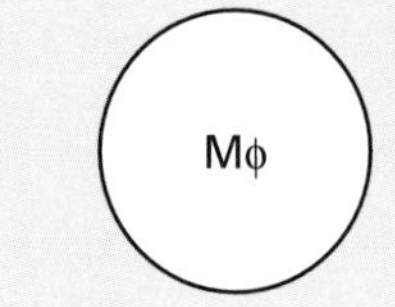

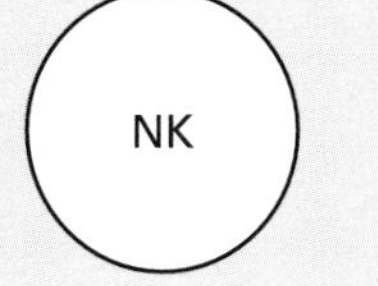

(i) Recognition of native antigen via mIg

(ii) Antigen internalization and processing

(iii) Presentation of peptide–MHC II complexes to activated T cells and/or

(iv) Binding of T cell-derived cytokines

(v) Blastogenesis (activation), proliferation (clonal expansion); class switching

(vi) Secretion of sIg

8.1 Introduction

The antigen receptors on B lymphocytes are membrane-bound antibody molecules (Section 1.1.3). The immense diversity of B cell antigen receptors, perhaps potentially of the order of 10^{10} different antibodies, comes about in part through rearrangement of germline genes (Section 7.2.2). The genes that encode these antibodies are rearranged in B cells, but not in other cell types where they remain in their germline configuration. Every B cell has about one or two hundred thousand antibody molecules on its surface, but every single one of these is specific for just one particular antigenic determinant (Fig.1.18). These membrane receptors are expressed on B cells before they encounter antigen, just as T cell receptors are preformed on T lymphocytes (Sections 1.1.3 and 7.3.2). They are also referred to as membrane-bound immunoglobulins (Ig).

B lymphocytes, like most other cells of the immune system, originate from hemopoietic stem cells (Figs.1.31, 1.32). One of the earliest cells that can be identified as belonging to the B cell lineage is the so-called 'pro-pre-B cell', which has already begun to rearrange some of its antibody genes (Fig.8.1). As the B cell develops from this stage into a mature, resting B lymphocyte, the *class* of antibody expressed by the cell can change, but these changes are *independent of antigen*. Once the *antigen specificity* has been determined it remains constant for that cell. When the mature B cell subsequently interacts with its specific antigen, the B cell becomes activated and it may proliferate (resulting in clonal expansion) and mature into an an antibody-secreting plasma cell. These *antigen-dependent stages* are also accompanied by changes in the class of antibody produced by the cell, and are dependent to a greater or lesser extent on 'help' from T cells and their products.

In this chapter, we first consider some of the antigen-independent events that occur as the B cell develops from its precursor forms into the mature resting B cell (Section 8.2). These stages are referred to as B cell development.† Then we discuss B cell responses to different types of antigen (Section 8.3). The nature of T cell help for antibody responses is discussed in Section 8.4, and here we attempt to dissect some of the stages that are involved in B cell responses (in particular activation, proliferation, and antibody secretion) as well as the different signals delivered to the B cell after binding T cell-derived soluble molecules or cytokines (Section 4.4.1). B cells, like T cells (Section 4.4.2), comprise different subsets, and one particular B cell subset ($CD5^+$ B cells) is described in Section 8.5. B cells, again like T cells (Section 5.2.5), are often tolerant to self molecules and other antigens, and consideration of this area concludes the chapter (Section 8.6)

† A variety of different terms are used in the literature for different stages in B cell development and B cell responses. We use the term 'development' in this sense because the expressions 'maturation' and 'differentiation', which might otherwise be suitable, are commonly used in reference to transition of a mature B cell into a plasma cell, a process that is also termed 'terminal differentiation'.

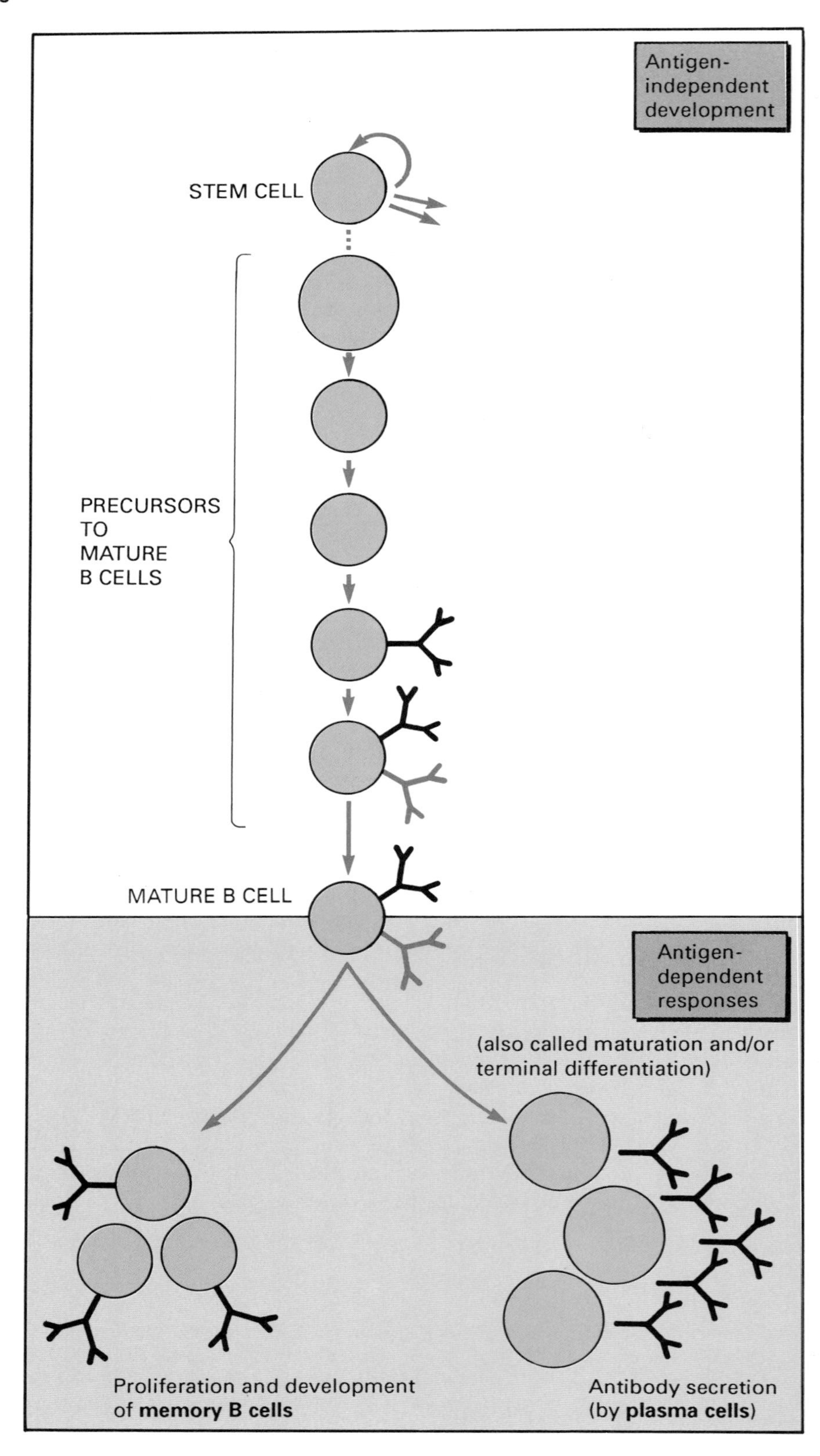

Fig.8.1 Antigen-independent development of B lymphocytes, and antigen-specific responses

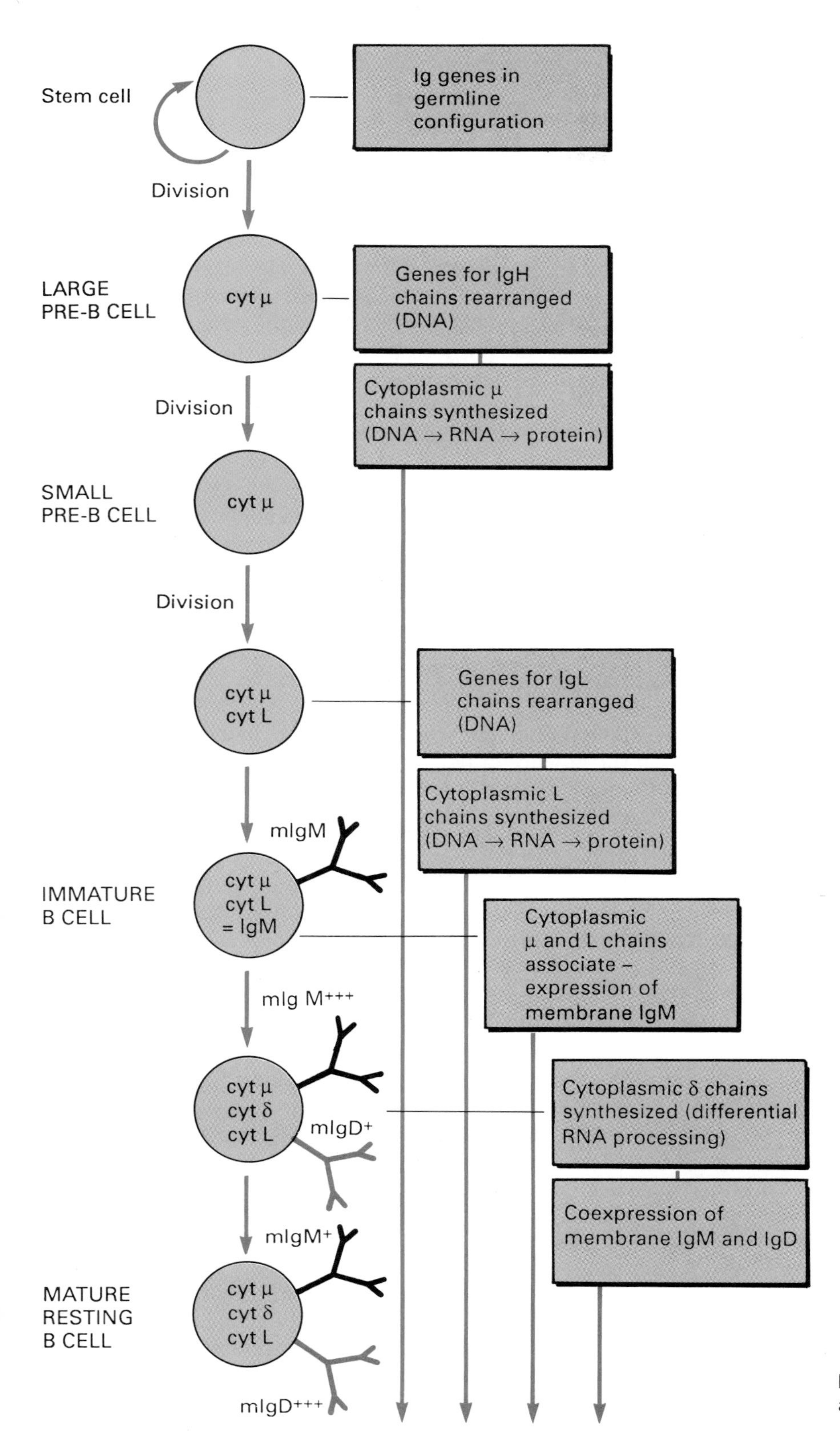

Fig.8.2 Expression of Ig genes and antibodies during B cell development

8.2 B cell development

8.2.1 Expression of antibody molecules during development of B cells

The immune system of the neonatal (new-born) mouse appears to be somewhat immature, in terms of its cellular functions and phenotypes, and it continues to develop for a few weeks after birth until the mature adult stage is reached. This is unlike the human, where the immune system is essentially fully mature at birth. Thus, the phenotypes of B cells in the adult mouse are rather different to those of the neonate, whereas these populations are more similar in human adults and neonates.

B cells, like other cells of the immune system, develop from haemopoietic stem cells which are normally found in the adult bone marrow (Fig.8.2). It is likely that these stem cells can undergo asymmetric cell division to produce one daughter cell of the original stem cell type, and another which matures along a certain lineage, such as lymphoid cells. However, it is unclear at what stage B lymphocytes diverge from T lymphocytes and other leukocytes.

One of the earliest stages in B cell development that can be defined is the **pre-pro-B cell**, which contains rearranged Ig μ heavy (μ_H) chains in its nucleus but has not yet started to produce the protein in the cytoplasm. This cell subsequently develops into a **pre-B cell**, defined as a cell that has μ chains in its cytoplasm, so that part of its future antigenic specificity is decided. It may also express very low levels of μ_H chains on its plasma membrane (i.e. in the absence of Ig light chains). The large pre-B cell then divides to give rise to the more mature, **small pre-B cells**.

Several polypeptides are associated with the μ_H chain within pre-B cells and at the cell surface of certain pre-B cell lines. Within the endoplasmic reticulum (ER), free Ig heavy chains (as well as other molecules) are associated with a molecule called BiP (Bole *et al.* 1986). When light chains are produced at the 'immature B cell' stage (see below) the associated HL chains no longer bind BiP and the complex is then transported to the Golgi, and thence to the cell surface. In the case of pre-B cells, the μ_H chains are associated with two other polypeptides, which permit transport of small amounts of μ_H chains to the cell surface in the absence of light chains. These molecules have been designated λ_5 and V_{preB}, and appear to be the products of pre-B-cell-specific genes termed omega (ω) and iota (ι) in the mouse. The 22 kDa λ_5 and 18 kDa V_{preB} products, respectively, have strong homology to the C and V regions of λ light chains and bind covalently and non-covalently to the μ_H chain, apparently to constitute a light chain structure and permit its expression at the cell surface (Tsubata and Reth 1990). The significance of this is not yet clear.

In the next stage of B cell development, cytoplasmic light chains, either κ or λ, are produced. The light chains associate with cytoplasmic μ chains to make intact IgM molecules which can now be expressed on the B cell surface. At this point the final antigenic specificity of the cell is determined, because the selected heavy chain variable region is now associated with the corresponding region of a light chain to form the antigen-binding site. Cells that express membrane-bound IgM are defined as **immature B cells** and these cells are exported from the bone marrow. Development of immature B cells, which express only membrane IgM, into **mature B cells** is accompanied by the appearance of IgD on their surface. Thus maturing B cells *co-express* IgM and IgD. Initially, only low levels of membrane IgD are expressed, relative to IgM, but these levels seem to be gradually reversed as the B cell develops. The most mature B cells are generally thought to be those cells with the highest levels of IgD but relatively low levels of IgM.

Virtually all mature B cells express two or more isotypes or classes of Ig on their surface but all of which, of course, have the same antigenic specificity. The majority of resting cells express IgM and IgD, but other classes (either IgG, IgA, or IgE) can *also* be expressed by apparently resting B cells. These different classes of antibody may have different or co-operative functions on the same cell. While it is generally accepted that acquisition of IgM and IgD is an *antigen-independent process* that occurs constitutively during B cell development, some investigators have argued that the other classes of antibody are only produced after antigen stimulation. Nevertheless, expression of membrane IgM and IgD by the B cell almost certainly precedes that of other classes.

8.2.2 Allelic exclusion of Ig genes

B cells are 'monospecific' in that any given B cell produces antibody molecules specific for just one particular antigenic determinant. This is essential for an immune system that operates according to the principles of clonal selection (see Section 1.1.3).

Some of the earliest evidence for monospecificity of B cells came from antigen-suicide experiments, in which highly radioactive antigens were prepared (Ada and Byrt 1969). When these antigens became bound to specific B cells these cells were killed, but the reactivity of the remaining B cells to unrelated antigens was found to be unaffected. This result would not have been obtained if each B cell could bind a variety of different antigens. More recent evidence comes from the fact that a specific monoclonal antibody is produced by a particular B cell hybridoma (see Appendix).

The monospecificity of B cells is due to the fact that each B cell expresses only *one* particular heavy chain and only *one* particular light chain, which of course associate to form the antigen receptor. However, every B cell in a heterozygous individual contains the genetic information for at least six different Ig chains, including the maternal and paternal alleles of H chains, κ chains and λ chains (Fig.8.3). If all of these were produced by a given B cell, eight different Ig molecules (each with a different antigen specificity) could potentially be expressed by the same cell. In fact, even more combinations are possible because there are, for example, four different classes of IgG and three different isotypes of λ L chains in every individual mouse (Section 6.2.2).

The process whereby *either* the maternal *or* the paternal allele of a H chain or L chain gene is selected for expression in a particular B cell is called

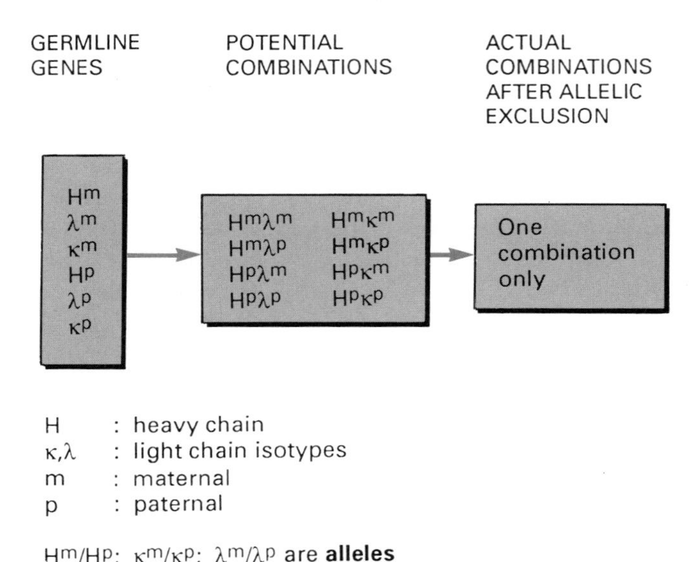

Fig.8.3 Allelic exclusion in B cells of a heterozygous individual

allelic exclusion. Light chain genes are also subject to **isotype exclusion** which determines whether a κ *or* a λ gene is expressed in that cell. Allelic exclusion of heavy chain genes occurs within pre-B cells, and this is followed by allelic and isotype exclusion of the light chain genes in immature B cells (Fig.8.2). It is important to note that allelic exclusion is a characteristic of lymphocyte antigen receptors in general because it also occurs for T cell receptor genes (Section 7.3.2). However, it does not pertain to MHC genes, for example, because *both* the maternal *and* paternal sets of these genes are codominantly expressed in all cells.

Mechanism of allelic exclusion. There are essentially two mechanisms by which allelic exclusion might operate.

1. **The stochastic model**. This suggests that the generation of the variable domains of heavy and light Ig chains (e.g. by DNA rearrangement; see Section 7.2.2) is intrinsically error prone and frequently results in the assembly of non-functional genes (e.g. by frame-shifts). Thus the probability of functional rearrangements could be so low that it is very unlikely two functional alleles would be produced in any given B cell.

2. **The regulated model** (Figs.8.4 and 7.16). This proposes that a functionally rearranged Ig gene, or the presence of the heavy or light chain which it encodes, *shuts off* further Ig gene rearrangements. Thus the B cell may 'attempt' to rearrange one allele of a heavy chain and if this is successful then rearrangement of the other allele is prevented. However, if this first attempt is abortive the B cell then 'tries' to rearrange the second allele. There is some evidence to support a *regulated* mechanism for allelic exclusion.

The first direct evidence in favour of the regulated model was obtained in studies of mouse κ genes (Ritchie *et al.* 1984). Transgenic mice were made by introducing an already-rearranged and functional κ chain transgene (see Appendix), hybridomas were prepared from spleens of these mice, and the monoclonal antibodies they produced were examined. It was found that hybridomas expressing the exogenously-introduced κ gene in their Ig molecules did *not* have rearranged endogenous κ chain genes, which appeared to have been 'turned off' in these cells.

Similar studies subsequently demonstrated that rearranged μ or δ chain genes could inhibit the rearrangement of the endogenous heavy chain genes. In addition, the presence of these transgenes *promoted* light κ chain rearrangement, so that both negative and positive regulation of antibody gene rearrangements can occur. However, only expression of the *membrane-bound* (rather than secretory) form of the μ chain induced allelic exclusion (Manz *et al.* 1988). Nevertheless, some conflicting evidence has been obtained, and it is also possible that various L chain alleles are sequentially rearranged, and that Ig rearrangements are inhibited when one of these products binds to a pre-existing H chain to form a stable, intact molecule (Harada and Yamagishi 1991).

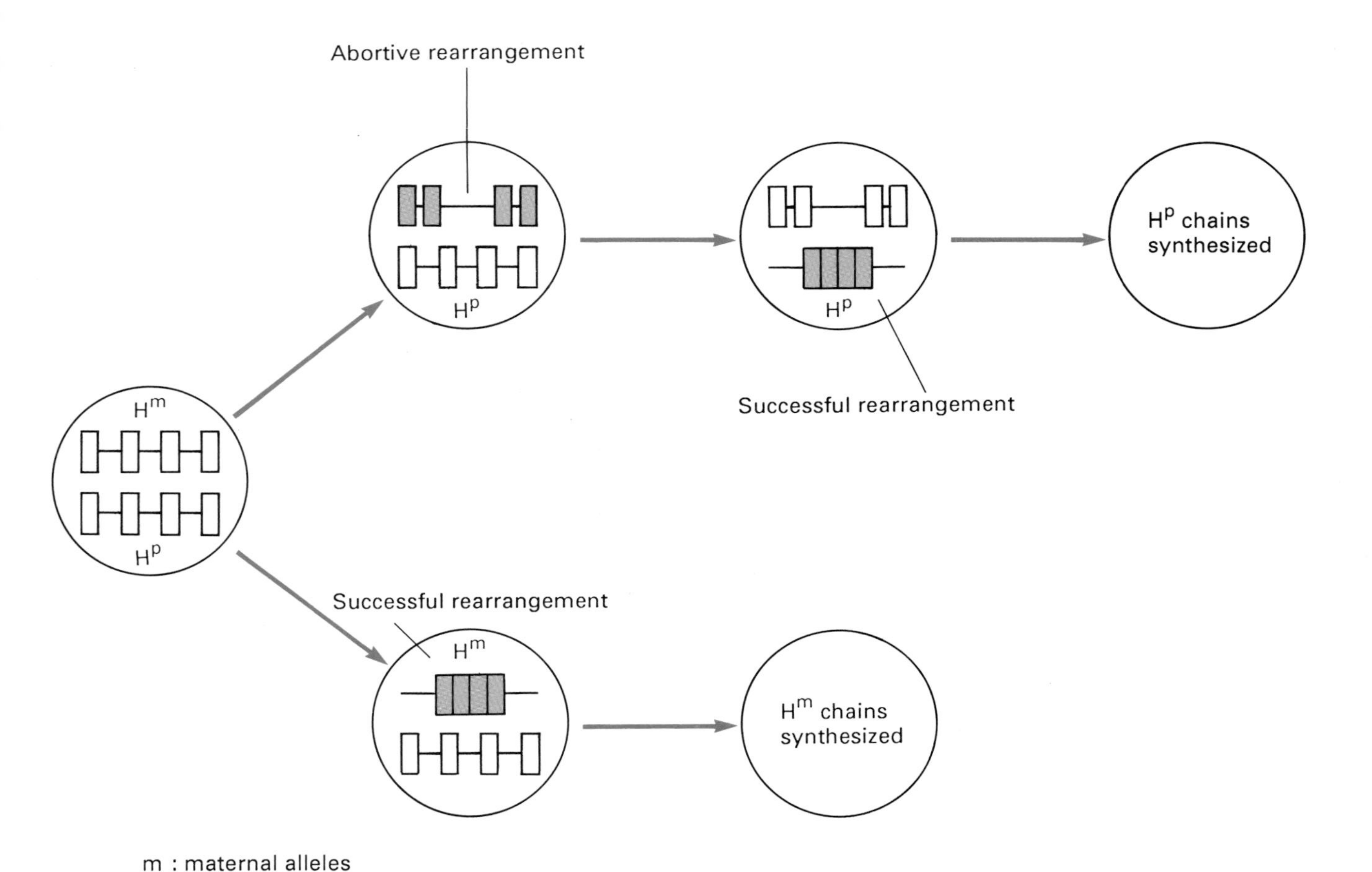

Fig.8.4 The regulated model for allelic exclusion

8.2.3 Other markers of B cell development

A number of other membrane molecules, in addition to Ig, are expressed at various times during B cell development (Fig.8.5). Perhaps the best marker for cells committed to the B cell lineage is the B-cell restricted form of CD45 (CD45RB, also called B220 or Ly5 in mice—see Section 4.2.5) that is first

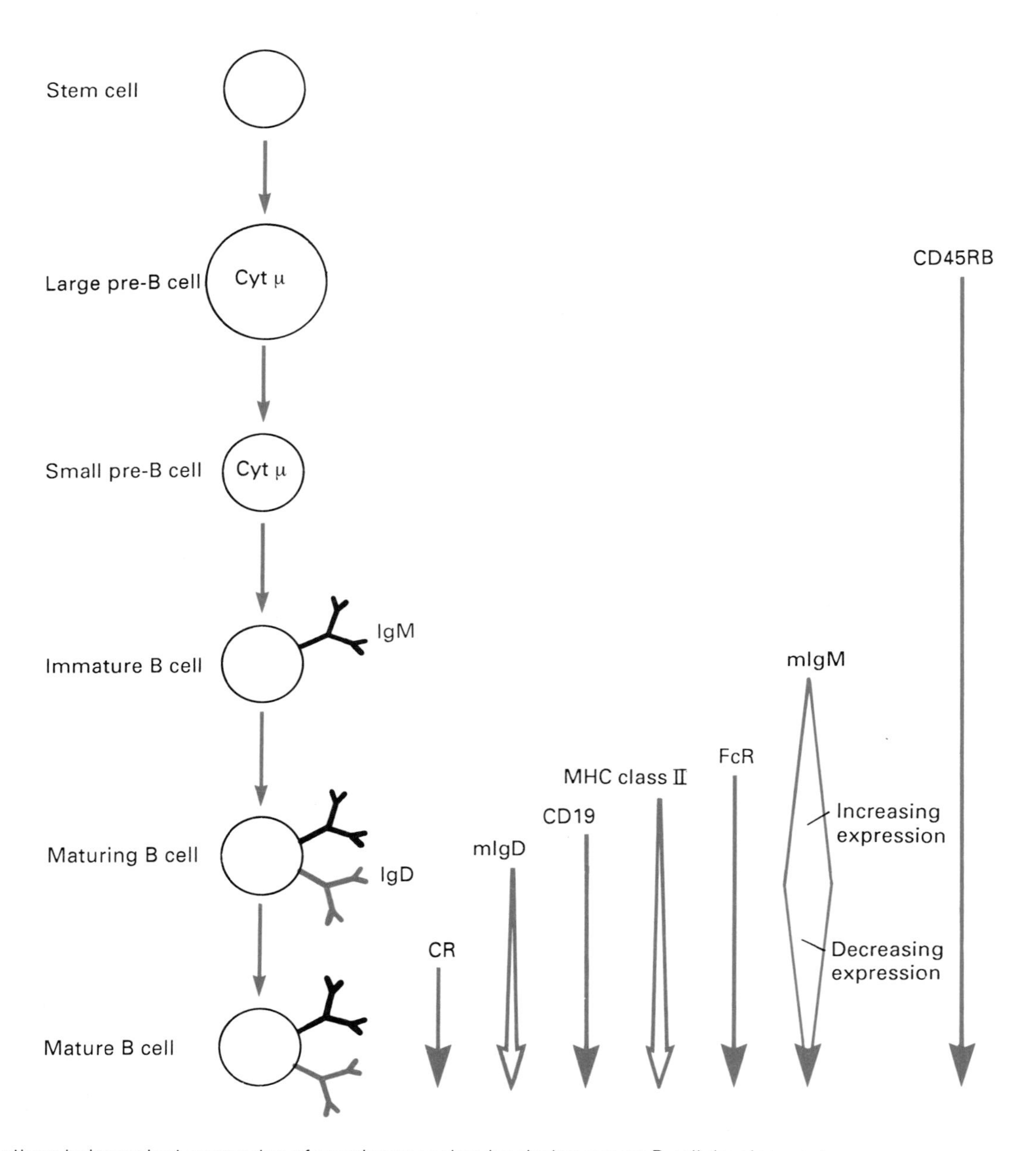

Fig.8.5 Antigen-independent expression of membrane molecules during mouse B cell development

expressed by the pro-pre-B cell and retained during subsequent development. In mice, following expression of membrane IgM by immature B cells, Fc receptors and then MHC class II molecules appear, and the density of the latter increases during development. In contrast, MHC class II molecules are expressed at an earlier stage in the human on pre-B cells. At this point a B cell specific marker, CD19, also begins to be expressed and this molecule is present at all subsequent stages of development and even after antigen stimulation of the B lymphocyte. This is followed by the acquisition of membrane IgD by mature B cells, and then complement receptors (CR1 :CD35 and CR2 :CD21—see Section 10.4.5) are expressed. All these stages appear to be antigen independent.

8.2.4 B cell lineages and subsets

B cells are phenotypically and functionally heterogeneous. This heterogeneity could reflect: (i) distinct B cell subsets, perhaps akin to the division of T cells into helper and cytotoxic subsets (Fig.1.12); (ii) different activation states, possibly analogous to the difference between T_H1 and T_H2 cells (e.g. Section 4.4.2); (iii) different developmental stages, such as is seen in the monocyte/macrophage lineage (Section 10.4). At the time of writing, the concensus is the following.

1. Two distinct lineages of B cells can be defined according to expression of CD5 (Lyl in the mouse; Section 4.2.4). $CD5^+$B cells are a relatively small, but significant, population in the adult which are discussed in detail in Section 8.5. *Most of what follows may (but need not!) pertain to the $CD5^-$ population of 'conventional' B cells.*

2. There is some evidence for two different subsets of (perhaps $CD5^-$) B cells, that were originally distinguished according to expression of an antigen, defined by conventional antisera, called Lyb5.† In the older literature these subsets are referred as $Lyb5^-$ and $Lyb5^+$ cells (Fig.8.6). However, monoclonal antibodies have yet to be produced against these presumptive subsets of B cells, and the Lyb5 molecule has not been characterized, so these markers should be viewed with extreme caution. Nevertheless, two major subsets of B cells have been defined in spleens of normal adult mice according to expression of other markers defined by monoclonal antibodies, such as those designated BLA-1 and BLA-2 (Hardy *et al.* 1984).

Despite the difficulty of defining subsets (other than $CD5^-$ and $CD5^+$ B cells) it does seem that two populations of (perhaps $CD5^-$) B cells exist in normal adult mice. These apparently differ according to when they appear in ontogeny, the types of antigen to which they can respond, and how they respond to regulatory signals, particularly those produced by helper T cells. Normal adult mice appear to have both subsets, whereas neonatal mice may have just one of these populations (originally designated $Lyb5^-$ B cells), which seem to require direct contact with T cells for many antigen responses and which are more readily tolerized by antigen alone. The other population

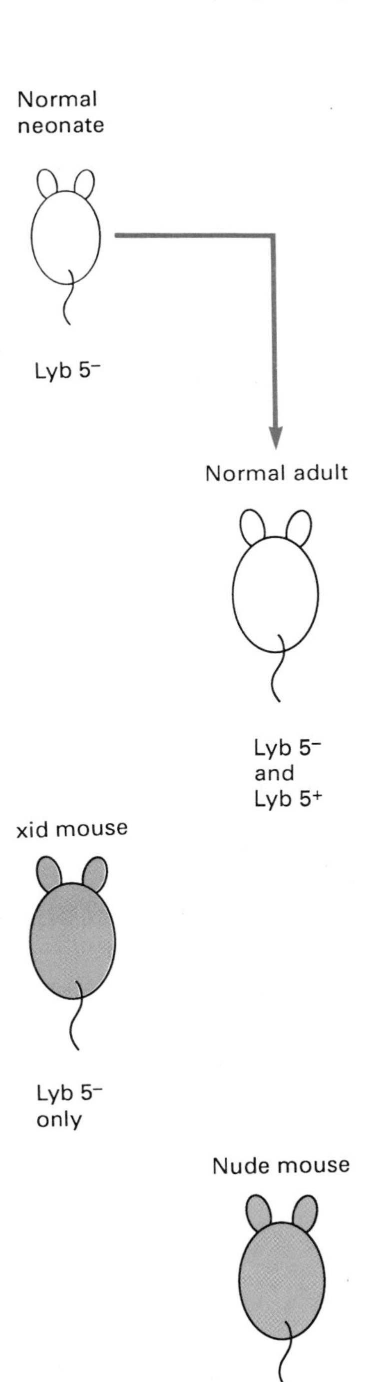

Fig.8.6 Possible subsets of B cells in mice (NB an important subset of $CD5^+$, i.e. Lyl^+, B cells is *not* shown)

† It is important not to confuse the three different markers Lyb5, CD5 (Lyl in the mouse), and Ly5 (CD45 or B220 in the mouse)!

(originally designated Lyb5$^+$ B cells) appears only later in ontogeny, over the first couple of months in normal neonates. It seems to respond more readily to T cell cytokines and is more difficult to tolerize (Table 8.1).

Much of the evidence for the existence of subsets of B cells other than CD5$^+$ B cells, has come from studies comparing the phenotype and function of B cells from normal adult and neonatal mice, and from animals with various genetic defects. Some strains, for instance CBA/N, have a particular X-linked immunodeficiency gene(s) called **xid** which results in their being unable to respond to certain types of antigen. This correlates with the apparent absence within these mice of the late-appearing (Lyb5$^+$) subset of B cells (Scher 1982). The situation is further complicated because CBA/N mice, which do not have Lyb5$^-$ B cells, also lack the CD5$^+$ subset, but this does not necessarily imply equivalence of these populations (i.e. two different subsets could be missing from these animals).

It is possible that nude mice, which are athymic and lack mature T cells, have a defect in the other (Lyb5$^-$) population, because mice with both nude and xid defects were found to be severely B cell deficient (Wortis *et al.* 1982). It has been suggested that the presumptive Lyb5$^-$ B cell subset, which appears to require close contact with T cells in antigen responses, may require the presence of T cells or even thymic hormones for normal development (Mond *et al.* 1982).

In summary, it is not clear at the time of writing how the apparently different subsets of Lyb5$^-$ and Lyb5$^+$ B cells relate to CD5$^+$ B cells. *We shall take the view, throughout this chapter, that CD5$^-$ and CD5$^+$ B cells are discrete and 'real' subsets.* Evidence to support this view is detailed in Section 8.5. *On the other hand, the Lyb5$^-$ and Lyb5$^+$ markers may define different stages of B cell maturation and/or activation.* This suggestion is consistent with the differential appearance of these cells during ontogeny, and also with reports that Lyb5$^-$ B cells of CBA/N mice can become Lyb5$^+$ cells after activation (see above). The reader should keep an eye on the literature for further clarification or resolution of these points.

Table 8.1 Comparison of presumptive populations of Lyb5$^-$ and Lyb5$^+$ B cells

	Lyb5$^-$	Lyb5$^+$
Appearance in ontogeny	early	late
T-dependency for responses and/or development	more	less
Optimal responses	need cognate interactions with T cells	respond to soluble molecules from T cells
Ability to be tolerized	readily	poorly

8.3 Antibody responses to different types of antigen

B cells can respond to antigen in two main ways: (i) they can proliferate (clonal expansion), and/or (ii) secrete specific antibodies, which aid removal of the antigen. In addition, (iii) memory B cells can be generated. In general, B cell responses to complex protein antigens require T cell help and these responses are therefore called **T-dependent (TD) responses** (Section 8.3.1). However, B cells can respond to other types of antigen in the absence of T cells and, quite logically, this type of response is termed **T-independent (TI)** (Section 8.3.2). Operationally, it is possible to distinguish between, and define, TI and TD antigens according to whether or not, respectively, they can stimulate responses in nude mice (Fig.8.7); TI antigens can be further classified as type 1 or type 2 (Section 8.3.2).

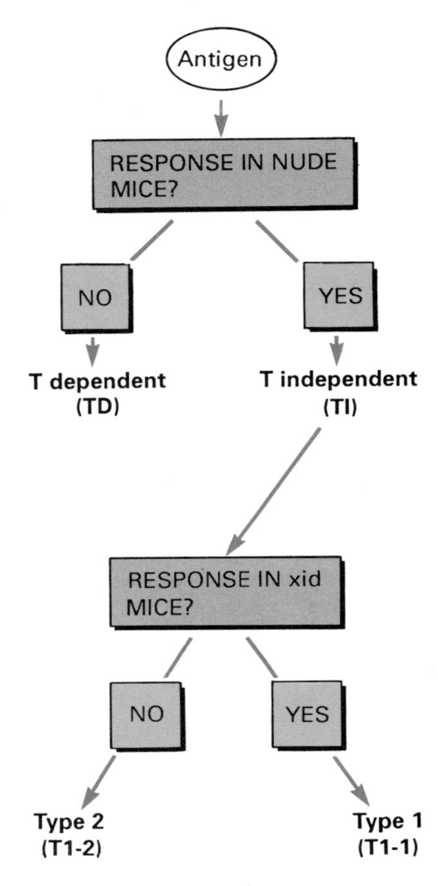

Fig.8.7 Conventional classification of antigens into T-dependent or T-independent types 1 and 2

8.3.1 T-dependent (TD) responses

Primary and secondary antibody responses

Primary responses. Animals can make primary antibody responses on their first exposure to a TD antigen such as a foreign protein (Fig.8.8). Before antibodies are produced there is a lag phase, partly due to the time required to activate specific T cells and B cells. In mice, the primary response normally peaks around 5–9 days after immunization and the main class of antibody produced is IgM. This class of antibody is able to fix complement, thus generating inflammatory mediators, and if the antigenic determinant is present on a micro-organism, complement sometimes kills the organism directly or stimulates its removal via complement receptors on phagocytes (Section 10.4.5). After a plateau of antibody production, the response declines with a variable time-course. As a consequence of the primary response, memory B lymphocytes are produced (see below).

Secondary responses. Animals make secondary responses when they encounter an antigen to which they have been primed (during a primary response). Secondary responses are characterized by a much shorter lag phase (e.g. 2–3 days), higher levels and increased persistence of antibody production, and there is a switch in the class of antibody produced, particularly to non-IgM classes such as IgG, which allows other effector arms of the immune system to be triggered (e.g. see Fig.1.28). These characteristics of secondary ('memory' or 'anamnestic') responses are due to the increased numbers of specific memory T and B lymphocytes that were produced during clonal expansion in the primary response, and to Ig class switching having occurred in the responding B cells, and because memory cells are more readily activated by antigen than resting cells. Although their exact nature is not understood, memory B cells can respond to relatively small doses of antigen and cannot be easily tolerized to it, perhaps because they express only low levels of, or no, membrane-bound IgD (see below).

Affinity maturation. Affinity maturation refers to the increased affinity of antibodies for their specific antigen as the antibody response continues,

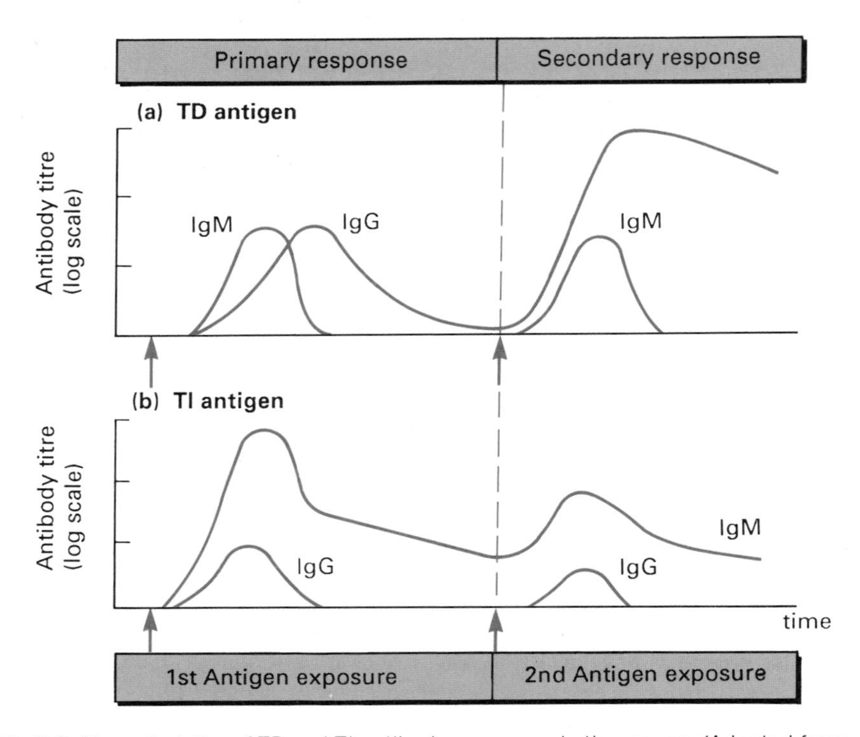

Fig.8.8 Characteristics of TD and TI antibody responses in the mouse. (Adapted from Klaus (1990) — see Further reading.) (a) Primary immunization with a TD antigen induces relatively short-lived IgM and IgG antibody responses. Secondary immunization induces a larger, higher affinity, predominantly IgG antibody response, resulting from re-stimulation of B memory cells. (b) TI antigens, in contrast, generally elicit long-lived primary IgM responses, poor IgG responses, and they do not induce classical memory

although this phenomenon is not seen in IgM responses. It has been suggested that B cell clones with a higher affinity are selected as the level of antigen falls, since the antibodies of highest affinity are generally seen after immunizing with low doses of antigen and their affinity increases with time after immunization. B cell clones with higher affinity are generated during primary and secondary immune responses, in large part due to somatic mutations that occur in V_H genes (Section 7.2.3) (Weiss and Rejewsky 1990). However, completely different sets of V genes can also be used in primary and secondary responses (Berek *et al.* 1985). Often the initial affinity of the antibodies produced in secondary responses is higher than that in primary responses as a result of selection of B cell clones using these new V-region genes.

Distinctions between resting and memory B cells. Resting and memory B cells differ according to the ease with which they respond to, or can be tolerized by, different concentrations of antigen (see above). They also differ in phenotype, and in their distribution in secondary lymphoid tissues and their recirculation patterns through the body (see below).

Resting (unstimulated) B cells are predominantly of the IgM^{lo} IgD^{hi} phenotype. In contrast, memory B cells are of the IgM^{hi} IgD^{lo} (or IgD^{-}) phenotype, and coexpress other classes of antibody (e.g. IgA or IgE) because class switching generally occurs after antigen stimulation. A variety of other phenotypic markers has also been used to discriminate between these populations. For example, resting B cells are thought to express low levels of complement receptor type 2 (CD21: Section 8.4.2) and Mel 14 (Section 4.2.6), but high levels of the 'heat-stable' antigen, J11d, which is also expressed by other haemopoietic cells including a subset of immature thymocytes, and CD44 (Camp *et al.* 1991). B cells participating in secondary responses have the reciprocal phenotype, $CD21^{hi}$, Mel 14^{hi}, $J11d^{lo}$, $CD44^{lo}$. An additional feature of memory B cells is that they express hypermutated or completely new V_H regions that are not found during the primary response (see above).

There are essentially two hypotheses concerning the origin of memory B cells. One proposes that when B cells are stimulated by antigen they first develop into lymphoblasts and/or antibody-forming plasma cells, and then revert to a resting state to become memory cells. The other proposes that memory cells and plasma cells develop from *distinct precursors*, one committed to memory B cell production and another to plasma B cells.

There is some experimental evidence to support the latter, two lineage, model. For example, when $J11d^{hi}$ and $J11d^{lo}$ resting B cells were adoptively transferred to SCID mice, it was found that these populations, respectively, generated predominantly primary and secondary antibody responses when the recipients were primed and boosted with antigen (Linton *et al.* 1989). The expression of different V_H genes in B cells participating in primary and secondary responses is also consistent with this model.

Anatomical localization of antibody responses (Fig.8.9). Primary antibody responses, the generation of memory B cells, and secondary antibody responses, *occur in different subcompartments* of secondary lymphoid tissues such as spleen and lymph nodes (Figs.1.38 and 1.39).

Some B cells are present in **T areas** (e.g. the PALS of spleen and the cortical interfollicular areas of lymph nodes), where the majority of T cells are localized together with interdigitating cells (IDC). However, most B cells are localized in **B areas**, with follicular dendritic cells (FDC) and some T cells. In the absence of antigen stimulation, most B cells are organized in the **primary follicles** of B areas, but after stimulation **germinal centres** develop in these areas. In addition, some B cells are present in the **marginal zones** of spleen which contain marginal zone (MZ) macrophages and a subset of dendritic cells, where lymphocytes enter from the bloodstream. Thus within each subcompartment, lymphocytes can be associated with different populations of accessory cells, IDC or FDC (Section 3.4.2), or MZ macrophages, which are thought to play different roles during the various stages of B cell responses.

Primary TD antibody responses are initiated within T cell areas, (as are TI-1 responses; Section 8.3.2). Here, it is likely that T cells recognize their specific peptide–MHC complexes on dendritic cells and become activated so they can provide the necessary help for B cell responses (Sections 3.2.3 and 8.4). Some B cells develop into antibody-forming plasma cells and remain within T cell

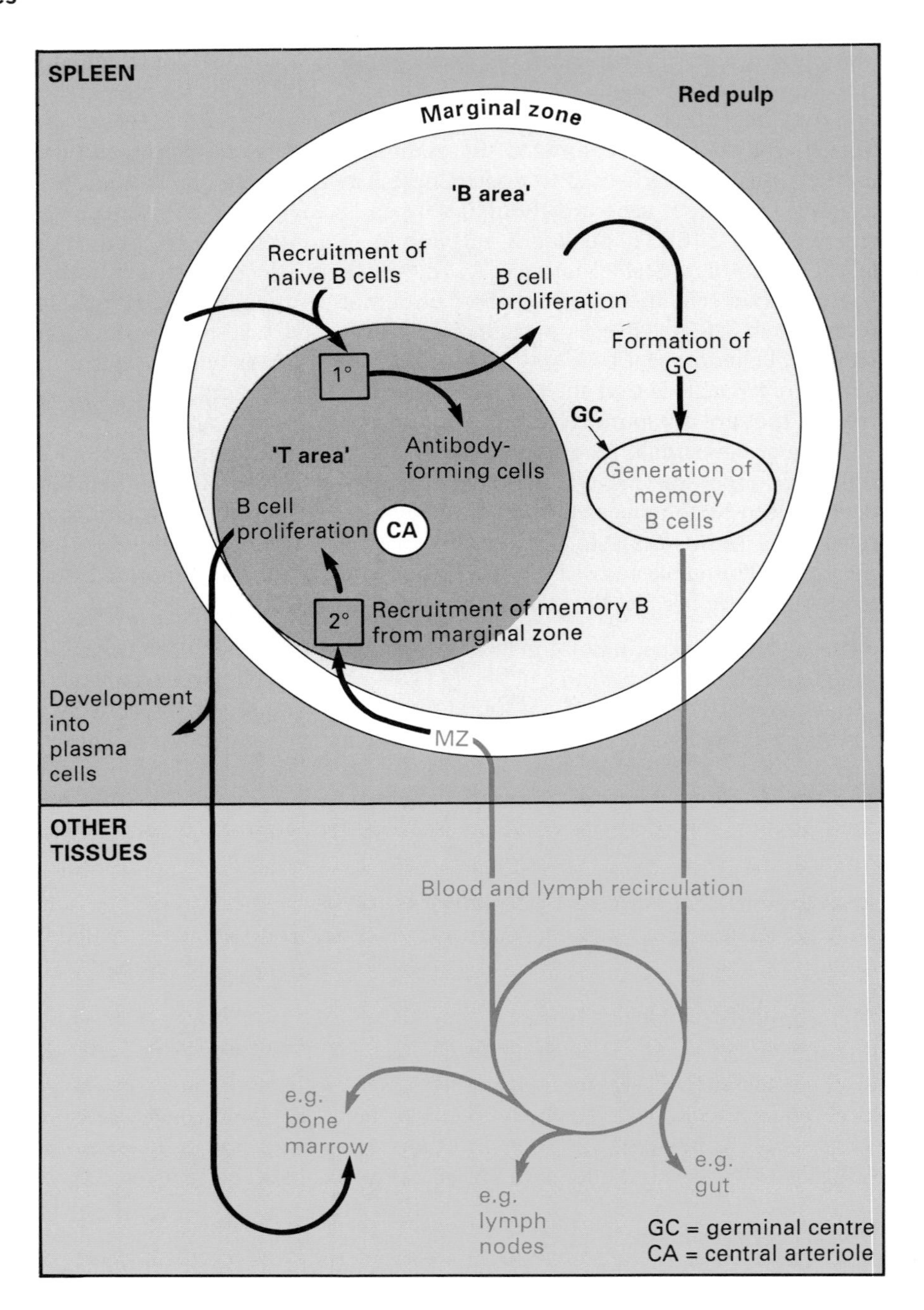

Fig.8.9 Anatomical compartments for primary (1°) B cell responses, the generation of memory B cells, and secondary (2°) responses

areas, while other B cells migrate into the follicles to undergo extensive proliferation, and apparently contribute to the formation of a germinal centre; formation of germinal centres is dependent on T cells, because they do not develop in athymic animals in response to TD antigens.

Memory B cells are generated primarily within germinal centres. High affinity B cell clones that have undergone somatic hypermutation, or which express new V_H gene segments, may be selected during interactions with FDC; only activated B cells can associate with these cells. An important feature of FDC is their capacity to retain native antigens on their cell surface for considerable periods of time, from weeks to months and even longer, a phenomenon that may be essential for the development of B cell memory (Gray and Skarvall 1988). Binding of antigen to FDC is dependent on its being complexed with antibody and complement, so these cells presumably only come into play after the primary response has been generated. FDC express three different types of complement receptor (types 1, 2, and 3; Section 10.4.5), as well as Fc receptors ($Fc_{\gamma}RI$; Section 10.4.5), which are required for binding of immune complexes (Schrieves *et al.* 1989). Localization of native antigens on these cells can be inhibited by treatment of mice with cobra venom factor, which inhibits C3 activation.

Once formed, memory B cells leave the germinal centres and recirculate through the lymph and blood as long-lived cells, in contrast to resting B cells which tend to remain localized in B areas of lymphoid tissues. Some memory cells migrate to particular lymphoid tissues, for example IgA^+ B cells are preferentially situated in gut-associated lymphoid tissues (Section 1.3.2) and IgG^+ B cells localize to peripheral lymph nodes, while some long-term B cells also become localized in the bone marrow. Other memory B cells become positioned in the marginal zone of the spleen (Lieu *et al.* 1988); a characteristic of these cells is that they do not express CD23, the $Fc_{\varepsilon}RII$ receptor (Section 10.4.5).

Memory B cells appear to be recruited specifically from the marginal zone during secondary antibody responses. Proliferation of these cells then occurs in the T areas closest to the marginal zone, where a subset of dendritic cells is located, in contrast to the follicular site for B cell proliferation during primary B cell responses (see above). These activated, proliferating lymphoblasts then migrate to the **red pulp** of the spleen and the bone marrow, where they develop into plasma cells.

The marginal zone also appears to be an important site for the generation of TI-2 antibody responses (Section 8.3.2) which in contrast to TD, and some TI-1, responses do *not* generate memory B cells (Lane *et al.* 1986). Generation of this type of response probably requires an interaction of the participating B cells with marginal zone macrophages, which have a specialized capacity to localize and retain certain types of polysaccharides (TI-2 antigens) (Humphrey and Grennan 1981). Analogous structures in lymph nodes are also involved in this type of response.

Hapten–carrier responses: a reinterpretation

The mechanism of T-dependent antibody responses was often difficult to study because they are directed against multiple, poorly-defined or unknown determinants on complex protein molecules. To try to dissect the respective roles of T cells and B cells in these responses, one approach was to modify a complex molecule with a defined determinant, and to study the antibody response against this determinant alone. Most frequently, haptens and carriers were used in these experiments.

A **hapten** is a molecule, or part of a more complex molecule, that *can bind to a lymphocyte receptor but which is unable to induce an immune response on its* *own*. In other words, haptens are **antigenic** but **not immunogenic**. Haptens are typically small determinants such as phosphocholine (PC), trinitrophenyl (TNP), and 4-hydroxy-5-iodo-3-nitrophenacetyl (mercifully shortened to NIP) groups (Fig.8.10). Although haptens cannot induce immune responses if they are administered to mice, for example, on their own, immune responses can be made against them when they are chemically coupled or 'conjugated' to immunogenic macromolecules as **carriers**. Examples of carriers are heterologous (i.e. a different species') proteins such as bovine serum albumin (BSA), ovalbumin (OVA), or chicken γ-globulin (CGG), and even cells like heterologous erythrocytes.

Phosphocholine (PC)

CH_3
$CH_3—N^+—CH_3$
CH_2
CH_2
O
$O=P—O^-$
OH

Trinitophenol (TNP)

O_2N NO_2
NO_2

4-Hydroxy-5-iodo-3-nitrophenacetyl (NIP)

O
$O—C—CH_3$
H NO_2
OH

Relative sizes of typical hapten and carrier

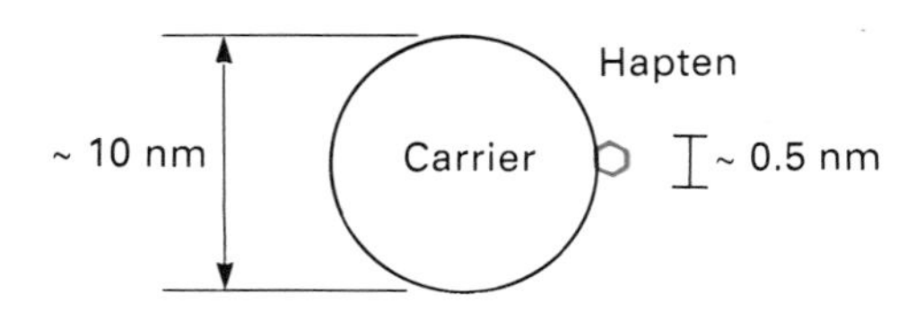

Fig.8.10 The structure of some haptens

Immune responses induced against haptens coupled to carriers are known as **hapten–carrier responses** (Fig.8.11). As we shall see, it was originally thought that B cells recognized the hapten while T cells recognized the carrier. We now know this is an oversimplification. It was presumed that these experimental responses mimicked those that were induced to unmodified proteins. For instance, substituted benzene rings in amino acids could represent haptens intrinsic to the protein, while determinants elsewhere in the molecule (now known to be small peptides) could be T cell epitopes.

Immunization against a hapten–carrier (i.e. priming) may induce a primary antibody response to the hapten (as well as to other determinants on the molecule, but one normally *only* measures the anti-hapten response). A second exposure to the same hapten–carrier conjugate (i.e. challenging) often results in a secondary response. We will indicate the conjugate of a hapten (H) coupled to one particular carrier (CI) as H-CI.

Mechanism of T-dependent B cell responses. The current model for how T cells help B cell responses is discussed in Section 3.2.3. In outline, the B cell recognizes a determinant on a native antigen via its mIg, and the antigen is then internalized and processed (Fig.3.23). Processed peptides become

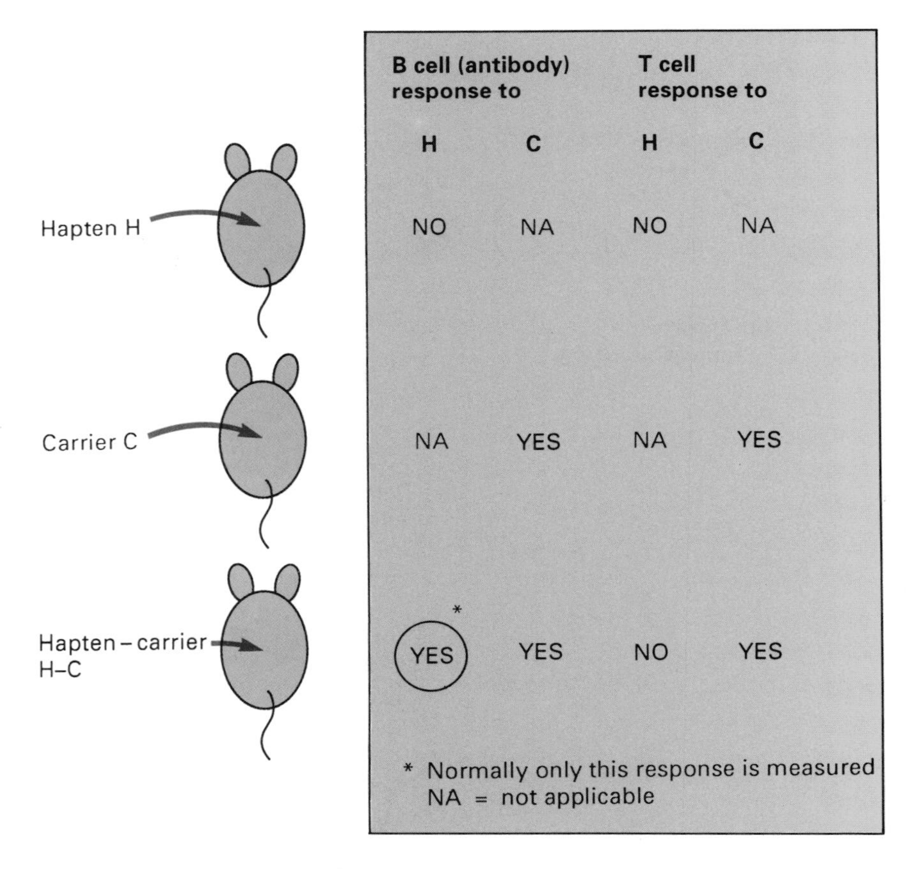

Fig.8.11 Responses of B and T cells to haptens, carriers, and hapten–carrier conjugates

bound to the B cell's MHC molecules and are presented to specific T cells which, in turn, provide the requisite 'help' to generate the antibody response. In the case of a hapten–carrier response, one can envisage that a hapten-specific B cell recognizes the complex via the hapten. Subsequent processing

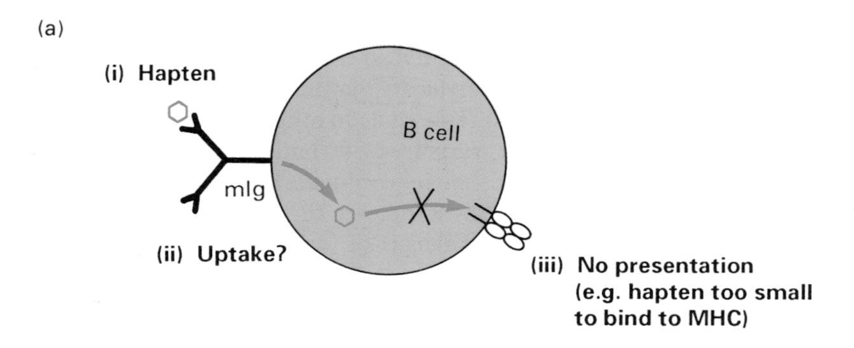

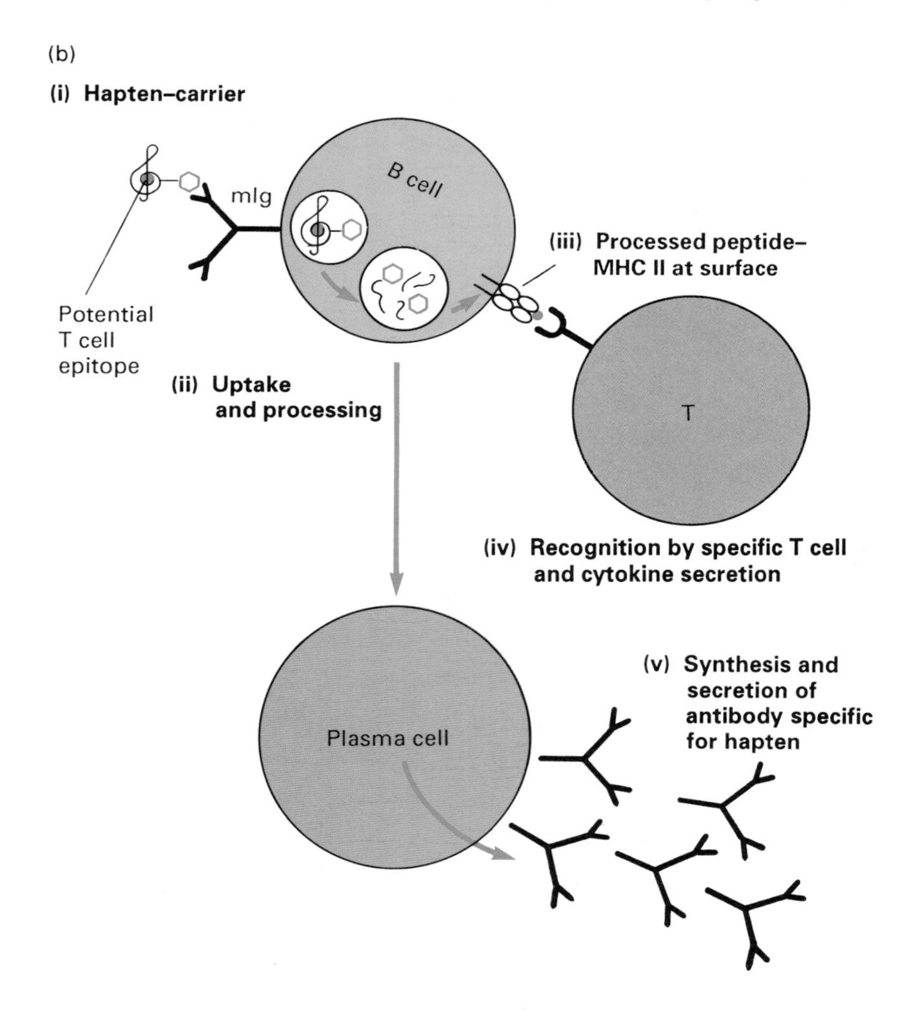

Fig.8.12 Mechanism of hapten–carrier responses

of the complex within the cell then generates peptides derived from the carrier that can bind to MHC molecules and be presented to T cells (Fig.8.12). Most probably, it is not possible to generate responses to a hapten alone because, even after its recognition and internalization by the B cell, the hapten cannot stably associate with an MHC molecule.

What is unclear about T-dependent B cell responses, however, is the precise nature of T cell help. As discussed in Section 8.4, it may be necessary for direct T cell–B cell contact to occur (a **cognate interaction**), presumably when the specific T cell recognizes its peptide–MHC complex on the B cell membrane. However, it is also possible that at a certain stage the B cell becomes responsive to cytokines that are produced when a T cell recognizes its peptide–MHC complex on another antigen presenting cell. In other words, B cells at different stages of activation may respond differentially to signals from T cells in immediate contact with the B cell and/or generated by T cells at a distance.

Whether the same B cell can respond to both types of signals is unclear. One idea was that $Lyb5^-$ B cells (Section 8.2.4) required direct cognate interactions with T cells, whereas $Lyb5^+$ B cells responded more readily to soluble molecules acting at a distance. Moreover, it was proposed that $Lyb5^+$ cells might be involved in responses of unprimed T cells and B cells (i.e. a primary response), whereas $Lyb5^-$ cells were involved in secondary responses with primed T cells and B cells. However, as noted in Section 8.2.4, these may represent either distinct subsets or different activation stages of the B cell, and it is not clear if the same cell can respond preferentially to soluble molecules at one stage of the response and to cognate interactions at another (e.g. in primary and secondary responses).

Further complications in dissecting T-dependent responses arise from the existence of different helper subsets or activation stages of the same T cell, typified by T_H1 and T_H2 helper cells, which may respond to different types of antigen presenting cells. As discussed in Section 4.4.2, T_H2 cells seem to provide more B cell help than T_H1 cells. They can, for example, secrete IL-4, which can act on resting B cells and may therefore be important in the initiation of the B cell response, although the same cytokine has a completely different spectrum of effects once the B cell has been activated (Section 8.4.4).

In the light of these uncertainties, it remains difficult to explain the findings of earlier studies of hapten–carrier responses according to current concepts (Fig.3.23). However, these findings are summarized below, partly because they are of historical interest and partly because they need to be explained by any future hypotheses; a brief attempt to reconcile these observations with our current thinking has also been made.

The carrier effect (Fig.8.13). It was found that when an animal is primed to H-CI (e.g. TNP-BSA) and then challenged with the same hapten coupled to an unrelated carrier, H-CII (e.g. TNP-OVA), a secondary anti-hapten response to TNP can *not* be generated. (Of course, a primary response can be generated to H-CII, but this is not measured.) This phenomenon was called the **carrier effect**, and it could also be demonstrated after adoptive transfer of primed cells to irradiated animals.

Mitchison (1971) for example, immunized mice with NIP-CGG, and after some months transferred their spleen cells to irradiated recipients of the same strain. When the latter were challenged with NIP-CGG, as little as 0.1 ng stimulated a secondary anti-NIP response, whereas up to 1 μg of NIP-BSA was required even to trigger a primary anti-NIP response in naive animals.

Presumably, immunization with H-CI sensitizes B cells, specific for H, that present processed CI (plus MHC), together with T cells specific for processed CI (plus MHC); these cells generate the primary response. However, after challenge with H-CII, the memory B cells can recognize H but then present processed CII (plus MHC), which of course cannot be recognized by memory T cells specific for processed CI (plus MHC). Thus a secondary anti-hapten response cannot be induced.

Phenomenon

Prime	Challenge	Secondary anti-hapten (H) response	
H	H	No	
H–CI	H	No	
H–CI	H–CI	Yes	The carrier effect
H–CI	H–CII	No	

Possible explanation

Priming with H–CI leads to production of memory helper T cells specific for processed CI (plus MHC)

e.g.

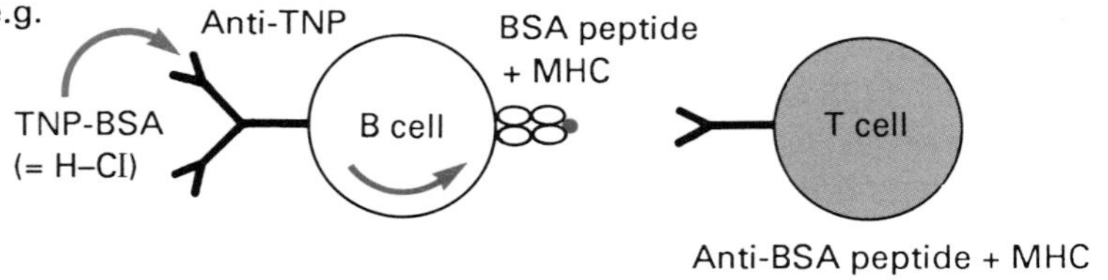

Anti-TNP

BSA peptide + MHC

TNP-BSA (= H–CI)

B cell

T cell

Anti-BSA peptide + MHC

Challenge with H–CII leads to presentation of processed CII by the B cells, which cannot be recognized by CI-primed memory helper T cells

e.g.

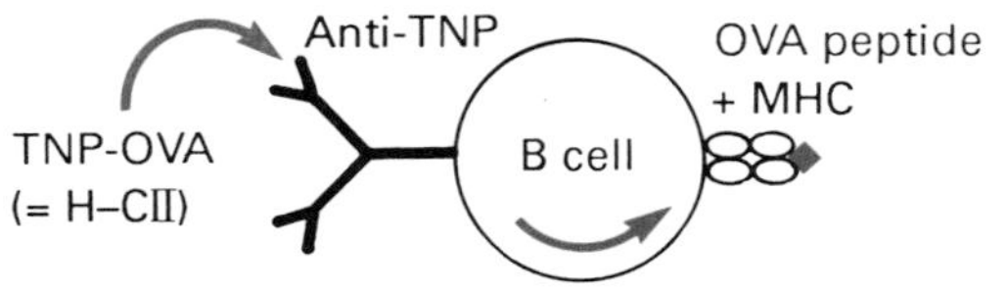

Anti-TNP

OVA peptide + MHC

TNP-OVA (= H–CII)

B cell

Fig.8.13 The carrier effect

Phenomenon

Prime	Challenge	Secondary anti-hapten response	
H-CI	H-CI	Yes	Carrier effect: see Fig. 8.13
H-CI	H-CII	No	Carrier priming overcomes carrier effect
H-CI + CII	H-CII	Yes	

Possible explanation

Priming with H–CI or CII generates different populations of memory helper T cells specific for processed CI or CII: the former provide help in the response to H–CI . . .

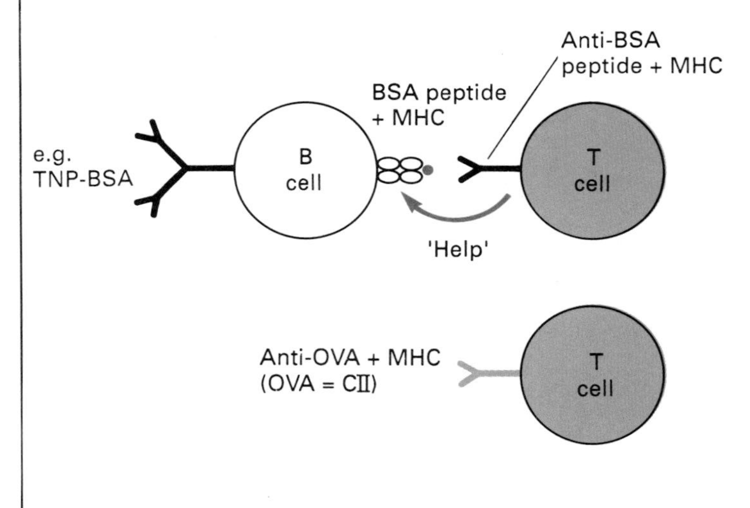

. . . the latter provide help for H–CII responses:

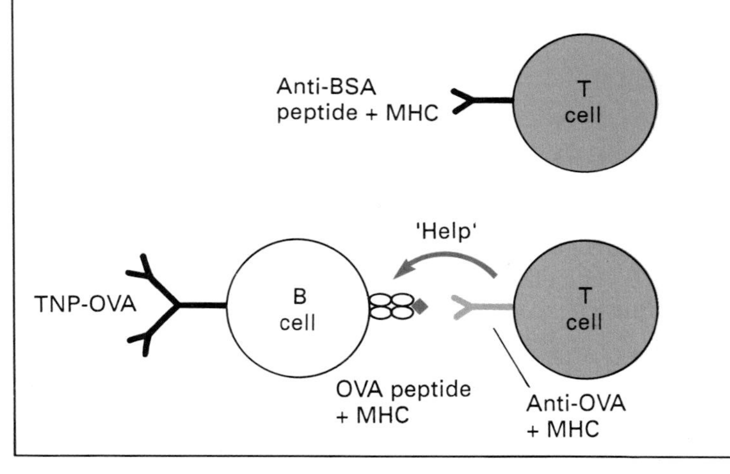

Fig.8.14 Carrier priming overcomes the carrier effect

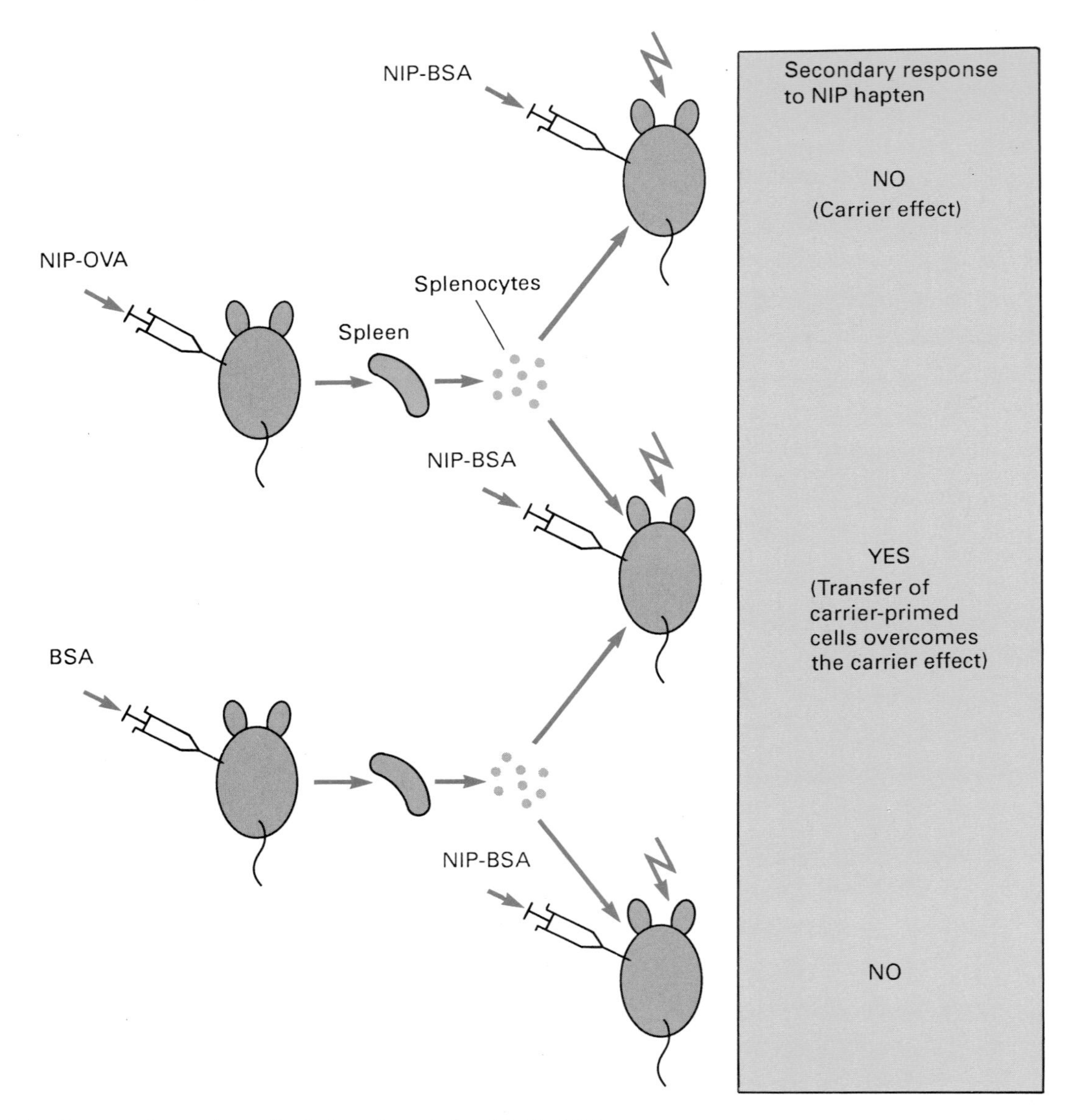

Fig.8.15 Adoptive transfer to show effect of carrier priming (cf. Fig.8.14)

Carrier priming (Fig.8.14). The carrier effect was found to be overcome if the animal was also primed to the second, unrelated carrier. In other words, if an animal is immunized with H-CI, it will make a secondary anti-hapten response when it is challenged with H-CII provided it was also primed against the second carrier CII. The effect of **carrier priming** can also be demonstrated in adoptive transfer experiments (Fig.8.15).

For example, if spleen cells from different animals that were separately primed to NIP-OVA and BSA are transferred together to irradiated recipients, the latter can make secondary anti-hapten responses to NIP-BSA (but not if the spleen cells from BSA-primed animals are omitted). Alternatively, spleen cells from an H-CI primed animal can be transferred into a recipient that was previously primed to CII, and this animal can now produce a secondary response to H-CII. Raff (1970) used the ability to adoptively transfer this effect to show that it was a consequence of transferring *T cells* that were primed to the carrier: treatment of CII-primed spleen cells with anti-Thyl and complement to kill the T cells before adoptive transfer prevented the recipients from responding to H-CII, whereas similar treatment of the H-CI-primed (B) cells had no effect (Fig.8.16).

It appears that carrier priming elicits a population of memory T cells specific for processed CII (plus MHC). After challenge with H-CII, these T cells can recognize processed CII (plus MHC) on the memory B cells that were primed by H-CI and which can also present H-CII because their mIg is specific

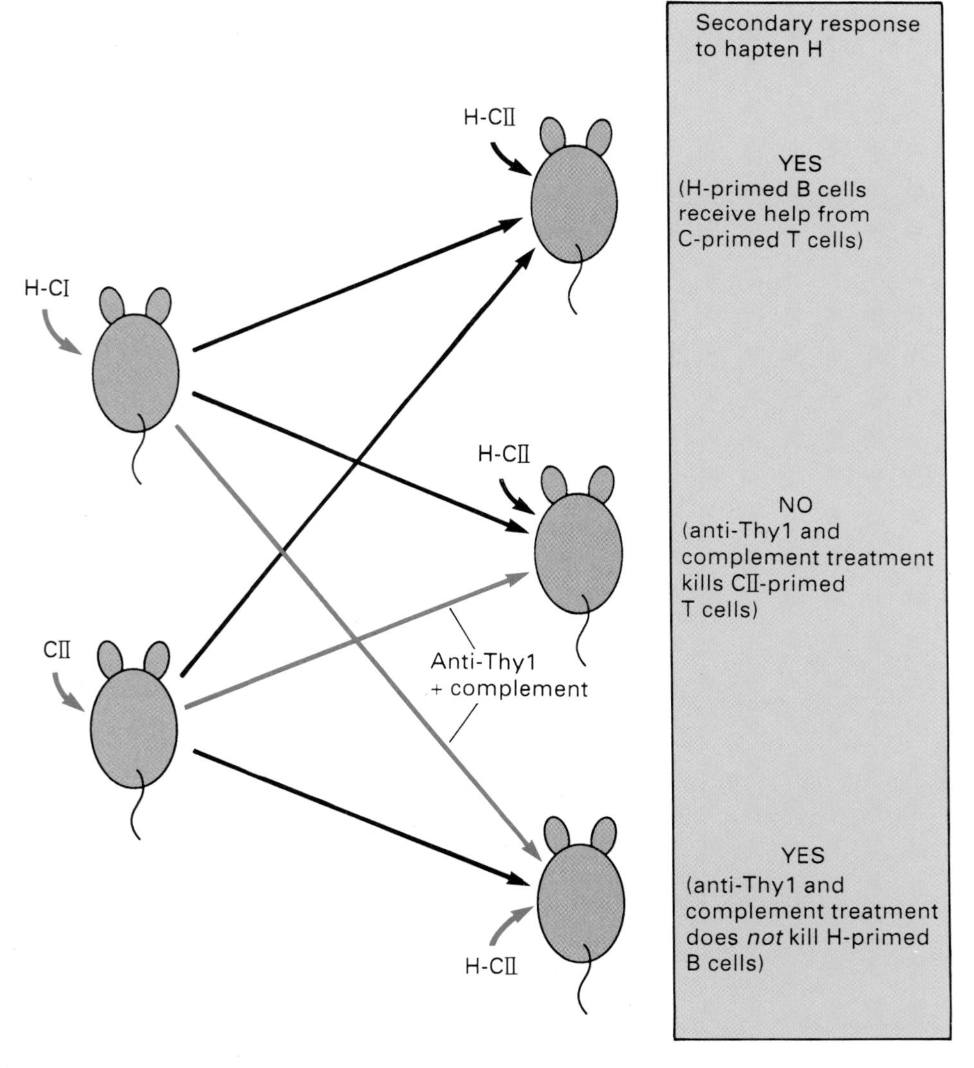

Fig.8.16 Principle of the Raff experiment to show that carrier-primed cells are T cells. Black arrows, splenocytes isolated and treated with complement alone as a control; red arrows, splenocytes treated with anti-Thy1 antibody and complement

for H, common to both molecules. Thus a secondary response can be generated.

Linked recognition (Fig.8.17). An additional observation was that primed B cells and T cells could only generate a secondary response if the hapten was *physically linked* to the carrier. These responses were therefore said to require **linked recognition** of the hapten–carrier.

For example, if H-CI-primed and CII-primed spleen cells were transferred to an irradiated recipient (see above), no response was obtained when the recipient was challenged separately with CII and the hapten on another carrier, H-CIII, rather than the H-CII conjugate.

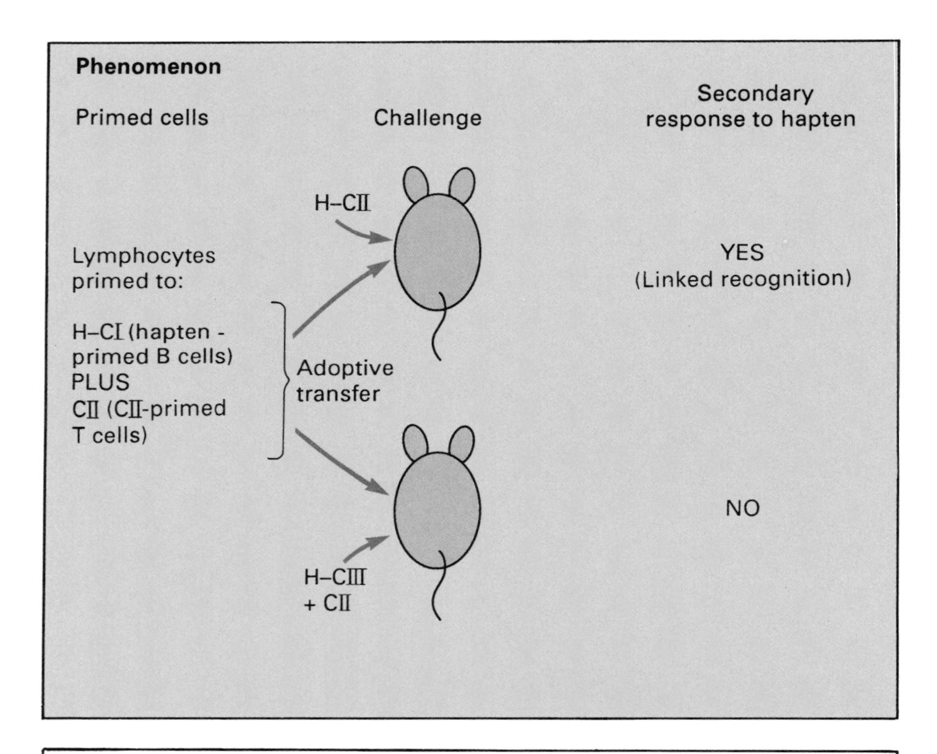

Possible explanation

Failure to respond to challenge with H-CIII may be because B cells present processed CIII peptides + MHC class II

Thus there is no recognition by memory helper T cells that were primed to processed CII (or CI) + MHC class II; no help; and no secondary response

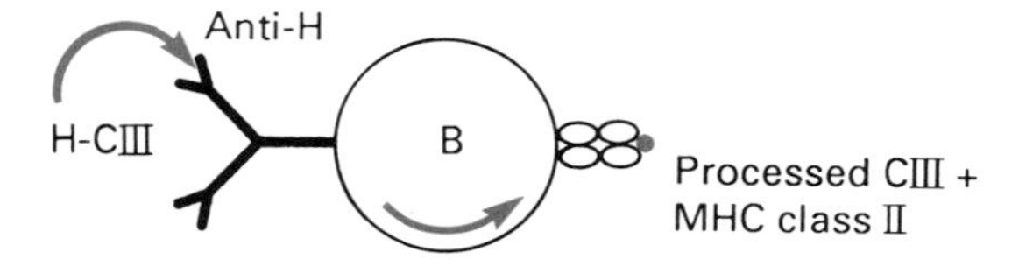

Fig.8.17 Linked recognition of hapten–carrier conjugates

It was originally thought that the hapten–carrier conjugate formed a 'bridge' between the responding B cells and T cells, thus allowing the respective lymphocytes to be brought into contact so that 'help' could be delivered by the specific T cell to the appropriate B cell. An alternative explanation, in keeping with current concepts, is that challenge with CII and H-CIII on separate molecules restimulates the population of memory T cells specific for

Phenomenon

Prime	Challenge	Secondary anti-hapten response	
H–CI	H–CI	Yes	Carrier effect: see Fig. 8.13
H–CI	H–CII	No	The allogenic effect (and 'unlinked recognition')
H–CI	H–CII plus ALLOANTIGEN	Yes	

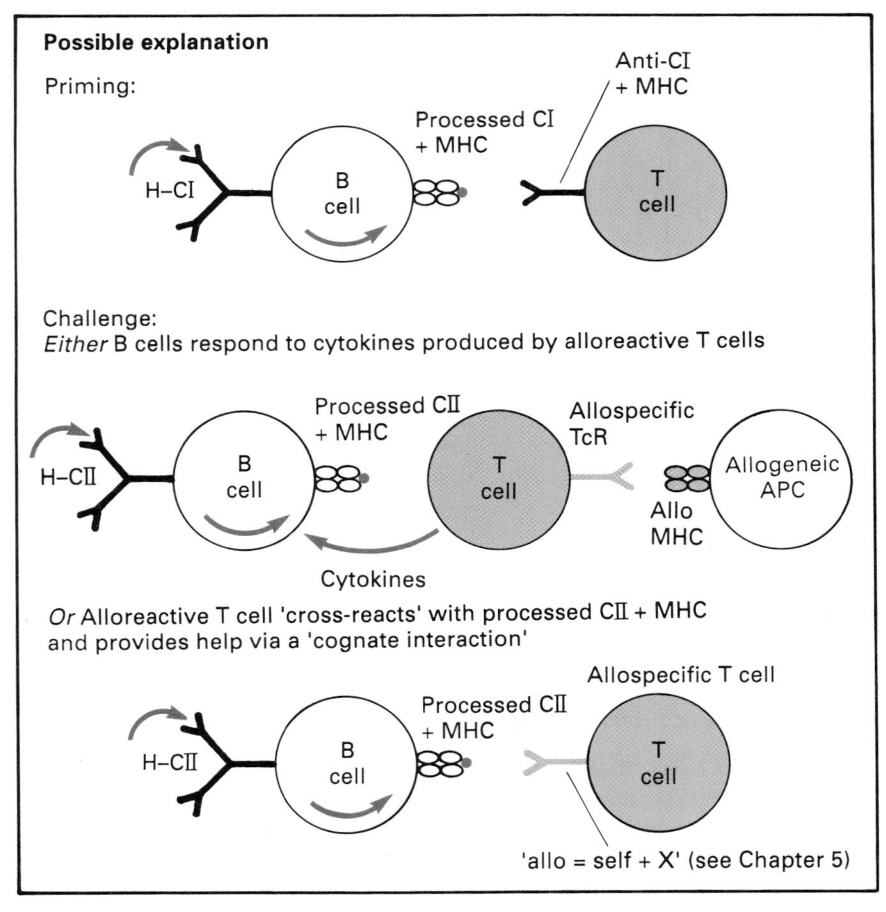

Fig.8.18 The allogeneic effect

processed CII (plus MHC), but these cannot, of course, recognize the hapten-specific memory B cells which now present processed CIII (derived from the H-CIII conjugate). Thus a secondary response cannot be generated.

The allogeneic effect (Fig.8.18). In some early experiments, allogeneic spleen cells were transferred to animals that had been primed to hapten-carrier conjugates. Under these circumstances it was found that animals primed to H-CI could produce secondary anti-hapten responses when they were challenged with H-CII *alone*. Thus, transfer of allogeneic cells overcame the carrier effect, and this phenomenon became known as the **allogeneic effect** (Hamaoka *et al.* 1973). In addition, hapten-carrier linkage was sometimes no longer required, in that it was possible to prime against H-CI and generate a response by challenging with CI and H-CIII.

There are at least two explanations for the allogeneic effect. First, it is possible that the *cytokines* produced when the allogeneic T cells mounted a graft-versus-host response (Section 5.3.2) could bypass the need for cognate T cell–B cell interactions. Support for this idea came from *in vitro* experiments in which it was found that MLR supernatants from cultures of allogeneic T cells and accessory cells (as a source of T cell-derived cytokines) could replace the function of T cells and stimulate antibody production by B cells. Second, it is possible that some of the alloreactive T cells that were activated after transfer could *cross-react* with 'antigen plus self MHC' (Fig.5.5) and provide help for hapten-specific B cells expressing processed CII or CIII (plus MHC) through cognate interactions.

Thus, there were two routes by which antibody responses could be generated.

1. Syngeneic B cells and T cells could collaborate and produce secondary antibody responses provided that the T cells were primed to the carrier, and the hapten and carrier were linked on the same molecule. This was originally termed **physiologic cooperation**, since it was thought to reflect the situation in normal animals.
2. Allogeneic B cells and T cells could also collaborate in antibody responses, but there was no requirement for T cells having been primed to the carrier, nor for hapten-carrier linkage. In the **allogeneic effect**, T cell recognition of allogeneic MHC molecules esssentially bypassed the need for helper cell recognition of carrier determinants.

These different routes may reflect the existence of different subsets of B cells, or B cells at different stages of activation, that can respond to different T cell signals during TD responses (e.g. Section 8.4.5).

8.3.2 T-independent (TI) responses

T-independent (TI) antigens can stimulate antibody responses in nude mice and therefore these responses seem not to require T cells. It is unclear, however, to what extent this type of response is truly independent of T cells, especially since nude mice do have T cell precursors and they may also have a B cell defect (Section 8.2.4). Thus, different investigators have variously suggested that TI antigens can only stimulate B cell responses if a few

Table 8.2 Examples of different types of antigen (*not* a rigid classification)

TD	TI-2	TI-1
Proteins most	*Polymerized flagellin* *Natural polysaccharides* pneumococcal SIII levan?	*Bacterial cell wall products* LPS PPD of BCG
Hapten–carriers most	*Synthetic polysaccharides* dextran ficoll	*Polyclonal B cell activators* several

'contaminating' T cells are present, or that they only stimulate previously activated B cells. What seems quite clear is that, even if a response to a TI antigen does occur in the complete absence of T cells, the quality and quantity of the response is markedly changed when T cells are added.

TI antigens can be subdivided into type 1 and type 2 antigens (Fig.8.7 and Table 8.2). **TI-2 antigens**, or simply **type 2 antigens**, can stimulate B cell responses in normal adult mice, but *not* in neonatal mice or those with an xid defect (Section 8.2.4). Typically, these molecules have repeating determinants that may cross-link B cell antigen receptors, thus providing signals for B cell activation. Examples of type 2 antigens include polymerized flagellin, natural polysaccharides such as pneumococcal polysaccharide SIII, and synthetic polysaccharides like dextran, levan, and ficoll (Table 8.2).

TI-2 antigens localize selectively on marginal zone macrophages *in vivo*, and accessory cells expressing MHC class II molecules may be required for type 2 antigen responses *in vitro*, although it is not yet clear whether these cells are necessary for cytokine production or act to 'immobilize' the antigen (Morrisey *et al.* 1981).

Neonatal mice become responsive to type 2 antigens at about two to three weeks of age, correlating with the appearance of a subset of (perhaps Lyb5^{+}) B cells that is absent from xid mice (Lindsten *et al.* 1979).

TI-1 antigens, or **type 1 antigens**, differ from type 2 antigens in that they can stimulate B cell responses in normal neonatal and xid mice, as well as normal adults. Type 1 antigens include bacterial cell wall products like lipopolysaccharide (LPS) and the purified protein derivative from the bacille Calmette-Guérin organism, BCG (Table 8.2). Many type 1 antigens can also stimulate polyclonal B cell responses (see below), particularly if they are used at high concentrations, whereas most type 2 antigens do not have this activity although there are exceptions. In addition, TI-1 antigens can sometimes induce memory cells, whereas TI-2 antigens do not.

A major difference between TD and TI responses (apart from their apparent dependency on T cells or otherwise) is that secondary responses are readily generated to TD antigens but it is unusual to elicit other than a primary response to most TI antigens. This implies that memory T cells cannot be generated during TI responses, probably because T cells recognize peptide–MHC complexes but many TI antigens are not proteins (although binding of carbohydrate residues to the peptide-binding groove of MHC molecules could conceivably occur).

In summary, it is possible to discriminate between type 1 and type 2 antigens depending on whether or not they elicit B cell (antibody) responses in neonates and xid mice. Both types of TI antigen stimulate responses in normal adults and nude mice (Fig.8.7).

8.3.3 Responses to mitogens and polyclonal B cell activators

Molecules that induce cellular proliferation are called **mitogens**. For example, the mitogen concanavalin A induces polyclonal proliferation of T cells. Other mitogens can stimulate both T cells and B cells in this way while some, such as LPS, predominantly stimulate B cell proliferation. Many B cell mitogens also trigger polyclonal antibody secretion. Agents that stimulate polyclonal responses of B cells are called **polyclonal B cell activators** (PBA), although sometimes the distinction between mitogens, which cause proliferation, and PBA, which stimulate proliferation and/or Ig secretion, is not clearly made. The extent to which these molecules act on B cells in the absence of T cells (and/or accessory cells) is controversial. In general the receptors for mitogens and polyclonal B cell activators are presumed to be *unrelated* to membrane-bound antibodies of the B cell, although anti-IgM, which does bind to membrane Ig on B cells, can be a mitogen at high doses. As a general rule, B cells from neonatal and xid mice respond poorly to mitogens, particularly to LPS, and are unresponsive to, or tolerized by, anti-IgM treatment.

8.3.4 Overlaps in classification of antigens

The classification of antigens as TD, TI type 1 or type 2, mitogens or PBAs is not absolute. There are several examples where a particular molecule can be put into one or another category depending on its concentration and other parameters such as the state of activation of the B cell. For instance, many (TI) type 1 antigens are also polyclonal B cell activators, as are a few type 2 antigens like polyI:polyC.

To take LPS as a further example, this molecule behaves as a type 1 antigen because at low (submitogenic) concentrations, *antibodies can be made against specific determinants on this molecule*. Moreover, if haptens are coupled to LPS, specific anti-hapten antibody responses can be produced. However at high concentrations LPS is a polyclonal activator, and it induces polyclonal mitogenesis and *the production of 'non-specific' antibodies* because it triggers many B cell clones to secrete Ig. Dextran sulfate exemplifies a molecule that can be type 1, type 2, or a polyclonal B cell activator depending on conditions.

The precise form of the antigen is also important. Thus, xid mice *can* respond to some molecules that would normally be classified as type 2 antigens if they are presented in a different form. For example, they cannot respond to *soluble* TNP-polyacrylamide or TNP-dextran, but they can respond to these molecules in an *insoluble* form. In addition, they cannot respond to pneumococcal polysaccharide SIII on its own, but they do respond if it is coupled to a TD carrier.

8.4 B cell activation and B cell responses

8.4.1 Different phases of B cell responses

Resting T lymphocytes are essentially inert until they have been activated by antigens or mitogens, when they develop into blast cells, proliferate, and

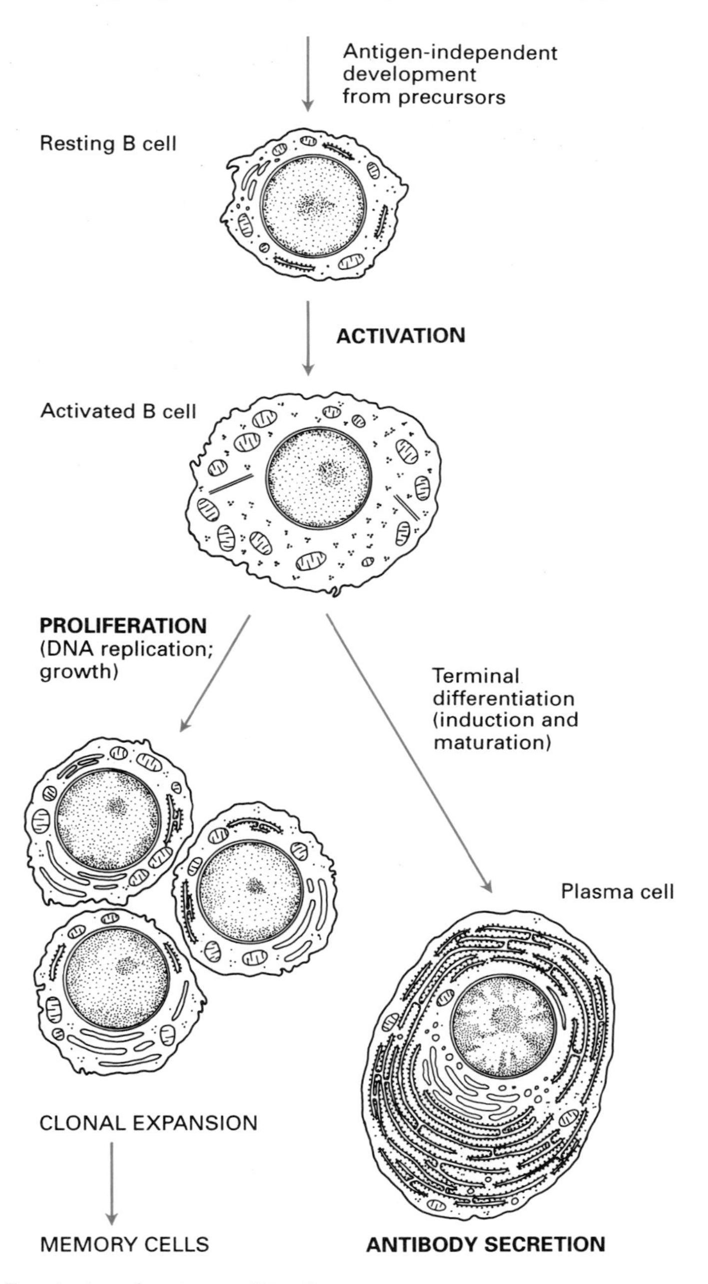

Fig.8.19 Terminology for stages of B cell responses

become able to carry out their specific functions, e.g. help B cell responses, activate macrophages, or kill target cells. Roughly speaking, the same principles apply to B lymphocytes. Resting B cells are also immunologically inert, but when they are activated by antigens or polyclonal B cell activators (Section 8.3.3), they can develop into blast cells, proliferate, and secrete antibodies which mediate their effector functions.

B cell responses therefore can be divided into several stages, which we will define as activation, proliferation, and antibody secretion (Fig.8.19). These *antigen-dependent events* are entirely unrelated to the *antigen-independent* **development** of mature resting B cells from their precursors (Section 8.2), and correspond to different stages of the cell cycle (Fig.8.20).

1. The term **activation** can be used to describe the earliest events *preceeding* B cell proliferation and/or Ig secretion. This corresponds to progression of the cell from G_0 into G_1 of the cell cycle, when small resting B cells develop into B lymphoblasts (**blastogenesis**).
2. **Proliferation** can be divided into DNA synthesis or **replication** (S phase), and mitosis (M phase) or **growth**, which may or may not include DNA synthesis depending on the assay. A cell undergoing several cycles of proliferation is sometimes called a 'cycling cell'.
3. **Antibody secretion** is often referred to as 'terminal differentiation' of the B cell, or sometimes as 'differentiation'. This can be divided into the **induction phase**, when class switching occurs (Sections 7.2.2 and 8.4.3) and a new class of membrane Ig is expressed by the cell, and **maturation** in which the soluble form of that Ig is secreted.

Resting B cells must be activated before they can proliferate and secrete Ig. However, B cells can proliferate without secreting antibodies, and proliferation and Ig secretion can be mutually exclusive, at least within the same cell cycle. Class switching seems to depend upon proliferation having

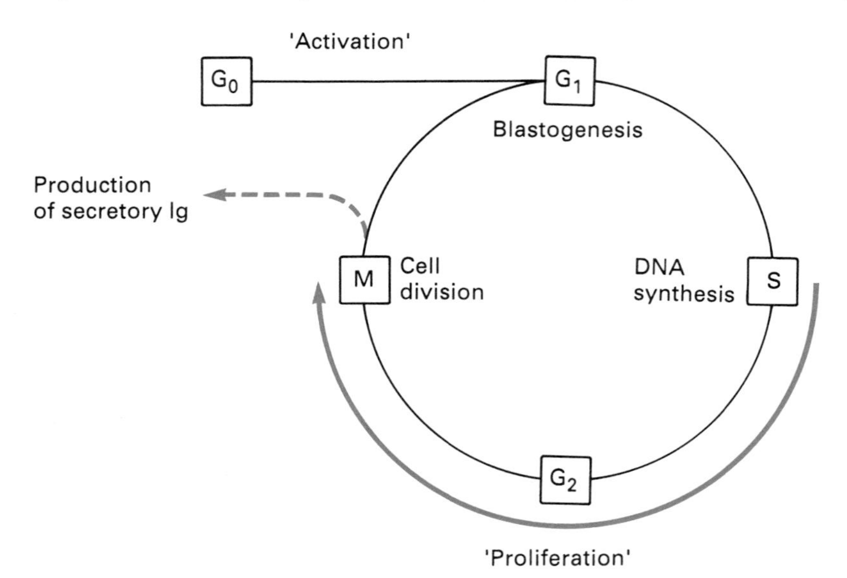

Fig.8.20 The cell cycle

occurred, but whether a B cell can secrete antibodies without first proliferating is controversial.

Cytokines and other stimuli. As a general rule, it is thought that when B cells are triggered by certain stimuli, they start to express membrane receptors for particular cytokines. These soluble molecules may be produced by T cells, accessory cells, and perhaps in some cases by B cells. When a particular cytokine binds to its specific receptor, the B cell is thought to progress to another point in the cell cycle (e.g. it may become a cycling cell) and then begins to express receptors for other cytokines. These receptors can mediate a different set of cellular responses (e.g. class switching may occur and the cell might begin to secrete Ig). In this way it is envisaged that there must be several **control points** for B cell responses. One difficulty has been to define these points precisely and to determine the precise effects of any particular cytokine on the B cell.

8.4.2 B cell activation

Stages in B cell activation

The G_0 and G_1 phases can be divided into several 'subcompartments' of the cell cycle (Fig.8.21). Truly resting B cells are sometimes said to be in the most senescent form of G_0 called **G_0Q**. When a resting B cell is activated from G_0 into the G_1 phase of the cell cycle, a number of changes occur. These include a dramatic increase in cell size (i.e. it becomes a lymphoblast), increased RNA synthesis, and the expression of a number of new molecules such as membrane receptors for cytokines that are required for subsequent proliferation and/or Ig secretion by the B cell.

Within a few hours of leaving G_0Q, the cell begins to express new antigens, such as CD23 (the Fc_εRII receptor) and the cell is said to be in the **G_0A**

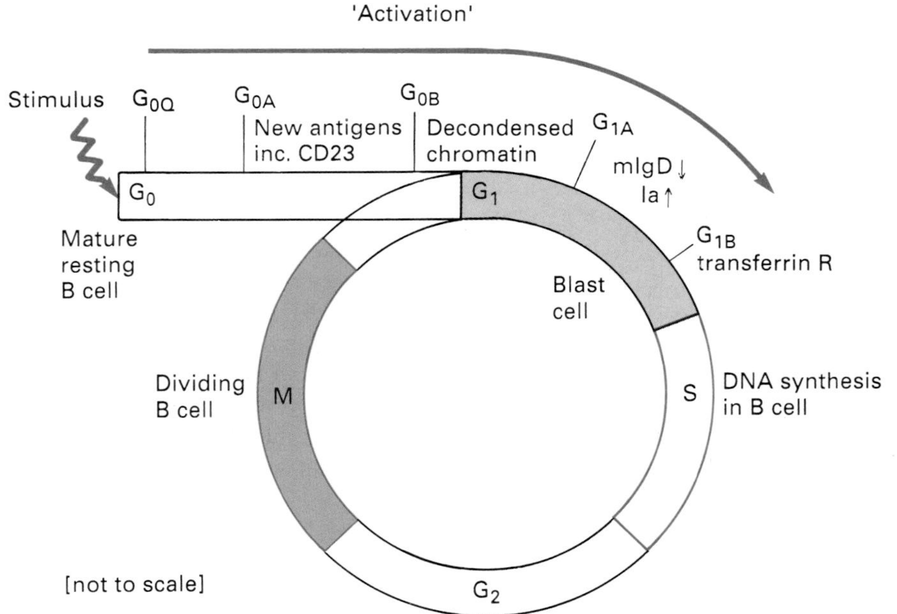

Fig.8.21 Subcompartments of G_0 and G_1 for B cells

phase. Sixteen hours or so later, in the **G_0B** phase, the chromatin in the nucleus decondenses, i.e. it becomes looser in structure, presumably to allow new gene transcription. Subsequent entry to G_1 is marked by the loss of membrane IgD and greatly increased expression of MHC class II molecules, and the cell is said to be in **G_1A** phase of the cycle. In later G_1, **G_1B**, the cells express transferrin receptors which mediate the uptake of transferrin-bound iron that is required in part for subsequent DNA synthesis. Some of these stages can also be defined by the expression other surface antigens identified by monoclonal antibodies.

It is important to note that different stimuli may cause the B cell to progress to different points of G_0 and G_1, and in many cases these points have not yet been well defined.

Activation of B cells via membrane-bound Ig molecules

B cells can be activated by a variety of routes. Under physiological circumstances, B cells are thought to become activated after they recognize specific antigen via their membrane-bound Ig molecules. Ligation of these delivers an activation signal to the B cell (see below) and blastogenesis is initiated. It is possible to activate B cells experimentally by this route by using anti-Ig or anti-idiotypic antibodies to cross-link membrane-bound Ig. Of course, anti-Ig will activate B cells polyclonally, while anti-idiotypes only activate specific clones of B cells (compare with T cell activation by anti-CD3 antibodies or clonotypic antibodies, respectively; Sections 4.2.3 and 4.2.1).

Anti-IgM is most commonly used to activate B cells experimentally. At low concentrations, soluble IgM causes the cells to become responsive to certain cytokines, while at higher concentrations some IgM antibodies alone can stimulate B cell proliferation. Thus, the degree to which the B cells are activated, and their progress through the cell cycle, depends on the concentration of anti-IgM as well as the particular determinants that are recognized, and it also makes a difference whether soluble or immobilized anti-IgM is used. B cells can also be activated via membrane-bound Ig by using the bacterium *Staphylococcus aureus* Cowan 1 strain (SAC), which can bind to the Fc and Fab portion of antibodies (see Section 6.2.1).

Mechanism of B cell activation via membrane-bound Ig. Intact antibodies against membrane-bound Ig on B cells, or their $F(ab')_2$ fragments, stimulate phosphatidylinositol turnover in the cell, and lead to a rapid increase in the concentration of intracellular free Ca^{2+} ions, and activation of protein kinase C (Wilson *et al.* 1987). These events are very similar to those following ligation of T cell receptors (Section 4.2.3; Fig.4.14). Depending on the particular specificity of the anti-Ig antibody, and whether or not it is immobilized on beads, both anti-μ and anti-δ chain antibodies can be effective (a soluble antibody having no effect may become stimulatory when it is coupled to beads). In contrast, monomeric Fab fragments of these antibodies do not stimulate these intracellular events, suggesting that some degree of receptor cross-linking is important.

Membrane-bound Ig molecules are non-covalently associated with at least three different polypeptide chains at the cell surface. In the case of IgM, these are the 32–34 kDa IgM-α (or MB-1), 37–39 kDa Ig-β, and 35 kDa

Ig-γ chains (Hombach *et al.* 1990). IgM-α has sequence homology to the CD3 γ, δ, and ε chains which are associated with the T cell receptor (Section 4.2.3), and forms a disulfide-bonded heterdimer with the Ig-β chain. IgM-α has one charged glutamic acid residue in its transmembrane domain, which may form a salt bridge with other components in the plane of the membrane, although unlike the T cell receptor, IgM lacks a corresponding charged residue in this region. It also has a cytoplasmic tail of 60 amino acids which contains tyrosines in a motif conserved between Ig molecules and the T cell receptor, and which could be a binding site for a G protein (see Biochemical events in T cell activation—Section 4.2.3). The α and β chains are apparently associated with all five classes of Ig, although the former may be differently glycosylated in each case (Venkitaraman *et al.* 1991).

The cytoplasmic portion of IgM is associated with a Src-like protein tyrosine kinase called Lyn, the product of the *lyn* gene, which is expressed preferentially in B cells but not T cells (Yamanishi *et al.* 1991). It exists in two forms, $p56^{lyn}$ and $p53^{lyn}$, and is related both to the $p56^{lck}$ tyrosine kinase expressed by T cells but not B cells, and to $p59^{fyn}$ which is associated with the T cell receptor (Section 4.2.3). During IgM signal transduction (e.g. after cross-linking by anti-IgM antibodies), there is evidence that the Lyn protein is autophosphorylated and then, in turn, it phosphorylates at least ten other proteins in the B cell. Antibodies specific for CD45, which has phosphotyrosine phosphatase activity (Section 4.2.5), can inhibit proliferation of B cells induced by anti-IgM and other stimuli, presumably by interfering with these phosphorylation events (Gruber *et al.* 1989). Another Src-like protein kinase gene called *blk* is also preferentially expressed by B cells, but involvement of this molecule in IgM signal transduction has yet to be shown.

Other routes for B cell activation

There are other ways of activating B cells, independently of their membrane Ig molecules (Table 8.3). Binding of polyclonal B cell activators such as LPS was discussed in Section 8.3.3. Cognate interactions with T cells can also result in B cell activation, and it seems likely that MHC class II molecules mediate delivery of a signal to the B cell when they are recognized, in association with a foreign peptide, by T cell receptors (Section 8.3.1; Fig.3.23). This is accompanied by a number of intracellular changes, associated with lymphocyte activation, which can also lead to proliferation of *primed* B cells (Lane *et al.* 1990). Resting B cells also express receptors for IL-4, and when this cytokine binds to the cell it initiates activation (Section 8.4.4). None of these stimuli triggers the intracellular events outlined above (See Mechanism of B cell activation via membrane-bound Ig), and how they act is something of a mystery at present (although LPS may be able to activate protein kinase C directly).

A number of other membrane molecules on B cells are also involved in signalling events and B cell activation (Table 8.3). It seems most likely that LFA-1, and its ligand ICAM-1 (Section 4.2.6), play an important role in interactions between B cells and other B cells (i.e. B cell–B cell, or *homotypic*, adhesions), and between B cells and different cell types such as T cells. In

Table 8.3 Some intermolecular interactions resulting in B cell activation*

B cell molecule	Ligand
Membrane Ig	Antigen (or experimental anti-Ig)
MHC class II	T cell receptor/CD4† on T cells
PBA 'receptors'	Polyclonal B cell activator (e.g. LPS)
Cytokine receptors	Cytokines (especially IL-4 for activation of resting B cells)
B7/BB1 (on activated B)†	CD28† and CTLA-4† on T cells
CD5† (on B subset)	CD72 on other B cells?
CD11b/CD18 (LFA-1)†	ICAM-1† on other cells
CD19-CD21 (CR2) complex	Complement (e.g. immune complexes)
CD21(CR2)-CD35(CR1) complex	Complement (e.g. immune complexes)
CD23 (Fc_{ε}RII)‡	IgE
CD58 (LFA-3)†	CD2† on T cells
CD72 (mouse Lyb2)	CD5† on T cells

*Further examples and details are in Clark and Lane (1991) — see Further reading
†See relevant sections of Chapter 4
‡See Section 10.4.5

addition, some of these interactions may be required for the capacity of B cell blasts to activate resting T cells (immunostimulation; Section 3.4.3). Activation of T cells by this route may involve binding of the B cell molecule B7/BB1 to CD28 or CTLA-4 on T cells (Section 4.2.8), as well as binding of CD72 (Lyb2) to its ligand CD5 (mouse Ly1; Section 4.2.4) on T cells (Van de Velde *et al.* 1991). Monoclonal antibodies specific for CD72 can augment antibody responses to TD antigens as well as B cell proliferation.

One molecule that appears to be coordinately regulated with LFA-1, and to have a role in B cell activation, is CD22; another important molecule in B cell activation is the complement receptor type 2, CR2 (CD21), which associates with CD19 in the plasma membrane to form a complex; CD22 and the CR2–CD19 complex are discussed below. Activation signals can also be delivered via the CD40 molecule (Rousset *et al.* 1991), a glycoprotein with homology to the nerve growth factor receptor and TNF receptors (Section 9.5.3). These molecules are reviewed in Clark and Lane (1991)—see Further reading.

CD22. CD22, also termed B lymphocyte cell adhesion molecule, BL-CAM, is a member of the immunoglobulin gene superfamily and is homologous to several homotypic cell adhesion molecules including myelin-associated glycoprotein, MAG, and carcinoembryonic antigen, CEA (Section 7.4). Its cytoplasmic tail is predicted to contain sites that can be phosphorylated, e.g. by protein kinase C, and these may be involved in the signalling functions of this molecule. Monoclonal antibodies against CD22 have been shown to facilitate entry of B cells into the cell cycle, and to increase intracytoplasmic Ca^{2+} levels stimulated by anti-IgM (see above). Other studies have shown that it is involved in homotypic adhesions of B cells (Wilson *et al.* 1991), and it may play a role in maintaining the architecture of B cell areas of lymphoid tissues.

The C21–CD19 complex. CR2 (CD21) is a 145 kDa receptor for iC3b and its C3dg fragment (Section 9.4.4); it is also the receptor for Epstein–Barr virus. Ligation of CR2 to surface-bound or cross-linked C3dg has been shown to prime B cells for proliferation in response to T cell-derived cytokines, phorbol myristate acetate, or anti-IgM antibodies in vitro. Moreover, cross-linking of CR2 (CD21) and membrane IgM results in a synergistic increase in intracellular Ca^{2+} levels in cultured B cells. Physiologically, this type of signalling could occur when the B cell recognizes immune complexes or antigens bound to complement components activated via the alternative pathway (Carter *et al.* 1988). In addition, monoclonal antibodies that can bind and modulate CR2 have been shown to *decrease* antibody responses *in vivo*. All these observations indicate that CR2 plays an important role in B cell activation.

CR2 can form a multimolecular complex with CD19, a B-cell-specific marker (Section 8.2.3) that is a member of the immunoglobulin gene superfamily, and at least three other components designated p130, p50 and p20 (and perhaps p14 in addition). It seems likely that the latter components are associated with CD19 as a pre-existing signal transduction complex in the membrane, which in turn can associate with CR2, the ligand-binding subunit of the complex (Matsumoto *et al.* 1991). After ligation of CR2, the CD19 molecule may transduce the signals leading to an increase in intracellular Ca^{2+} concentration. CR2 can also form a separate complex with complement receptor type 1 (CR1; CD35) on the cell surface and this may also mediate signal transduction to the B cell (Tuveson *et al.* 1991).

Regulation of B cell responses via Fc receptors

B lymphocytes express Fc receptors for IgG that can regulate B cell activation (Fig.8.22); these are called $Fc_{\gamma}RII$ receptors. In the mouse, they bind antigen-complexed IgG1, IgG2a, and IgG2b, and are similar to those on macrophages, except that the receptors on B cells contain an additional intracytoplasmic segment (Section 10.4.5). However, the Fc receptors on B cells do *not* mediate phagocytosis, antibody-dependent cellular cytotoxicity, or other cellular functions associated with macrophage Fc receptors. Binding of intact IgG to B cell $Fc_{\gamma}RII$ receptors causes a marked increase in MHC class II expression, suggesting that the cells are stimulated from G_0Q into G_0A of the cell cycle, but in the absence of other stimuli they do not progress into G_1 (Fig.8.21). When intact IgG binds to $Fc_{\gamma}RII$ receptors on B cells, it can also inhibit the delivery of activation signals by $F(ab')_2$ fragments of anti-Ig antibodies.

Fc receptor-mediated binding of IgG blocks phosphatidylinositol breakdown, although it has no effect on the increase in concentration of Ca^{2+} ions. 'Negative signalling' via the $Fc_{\gamma}RII$ receptor can be overridden by soluble molecules, especially IL-4 (Section 8.4.4), which by itself has no effect on phosphatidylinositol turnover, and the effect of IL-4 can itself be reversed by IFNγ (Fig.8.22) (O'Garra *et al.* 1987).

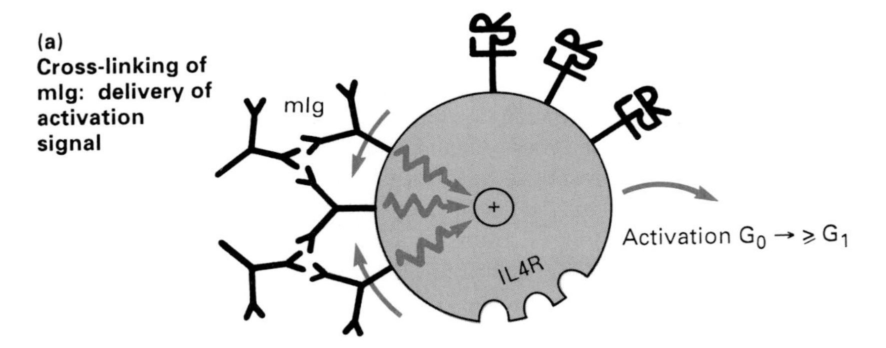

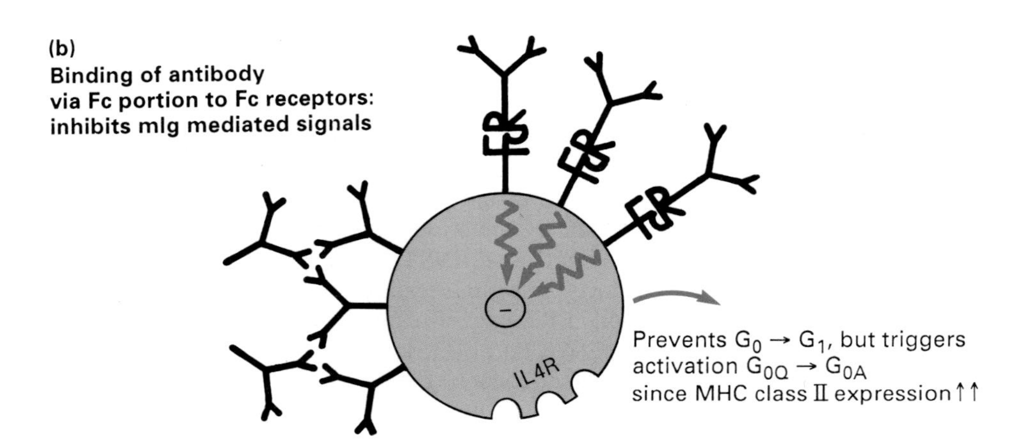

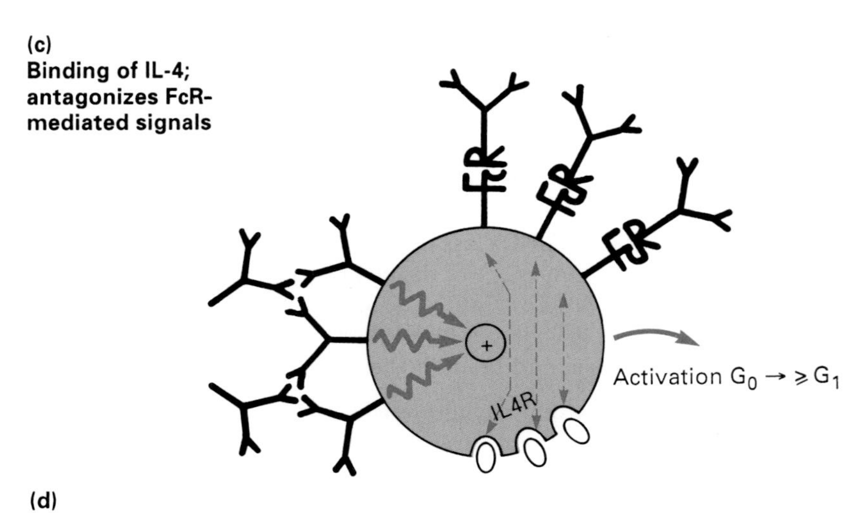

Fig.8.22 Regulation of B cell function by Ig and Fc receptors

An IgM receptor has also been identified on human pre-B and mature B cells, which is a phosphatidylinositol-linked 58 kDa membrane molecule specific for the CH3/CH4 domains of IgM (Ohno *et al.* 1990); it seems likely similar receptors are present on mouse B cells (and IgM receptors have also been reported on subpopulations of T cells). The $Fc_{\mu}R$ receptor may mediate delivery of a *positive* signal to the B cell, in contrast to negative signalling via the $Fc_{\gamma}R$ receptor, as evidenced by the enhancement of *in vivo* antibody responses induced by administration of IgM, but not IgG which leads to inhibition. Ligation of a third type of Fc receptor on B cells, $Fc_{\varepsilon}RII$ (CD23: Section 10.4.5) specific for IgE, may also *enhance* B cell responses (Cairns and Gordon 1990).

8.4.3 Control of B cell responses by cytokines

Assays

A variety of different assays were originally used in attempts to define the cytokines required for B cell responses. This undoubtedly led to confusion because often the same cytokine can stimulate different assays, although these were originally attributed to different activities (Section 4.4.1). The development of monoclonal antibodies specific for these molecules and cloning of their genes has led to a deeper, though as yet incomplete, understanding of their function in B cell responses.

In vitro assays that have been used to detect soluble molecules include the following.

1. **Co-stimulator assays.** B cells were treated with suboptimal concentrations of different agents such as low doses of anti-IgM or LPS that, alone, were not sufficient to induce a response. The ability of various cytokines to stimulate a response was then assessed, for example by measuring increased DNA synthesis, the expression of a new class of membrane Ig, or the total amount of Ig secreted by the B cells. For example, in the mouse, the cytokine originally designated BCGFI (now called IL-4) was found to stimulate proliferation of B cells activated with anti-IgM, whereas another cytokine designated BCGFII (now called IL-5) acted on cells triggered with dextran sulfate. Similar activities were defined in the human on the basis of co-stimulator assays triggered by anti-Ig, anti-idiotypic antibodies or polyclonal B cell activators.

2. **Tumour cell assays**. B cell tumours have been used as target cells in assays similar to those outlined in (a); these tumours are thought to represent 'frozen' stages of B cell development or activation.

3. **Preactivation assays**. Resting B cells were polyclonally stimulated with mitogens such as LPS and the effect of added cytokines on the response of the *activated* B blasts was assessed.

Using these assays, soluble molecules were at first defined according to whether they primarily stimulated B cell proliferation, Ig production, or both (Fig.8.23).

1. Molecules that stimulated B cell proliferation were called **B cell growth factors**, or BCGFs, but since DNA replication was often measured other investigators used the term **B cell replication factors**, BRFs. Different molecules with similar activities were distinguished by roman numerals (e.g. BCGFI and BCGFII).

2. Molecules that stimulated antibody production (hence terminal differentiation or maturation–see above) were called **B cell differentiation factors** or **B cell maturation factors**, BCDFs and BMFs, respectively. If a particular class of membrane-bound or secretory antibody was induced, the heavy chain was included in the name, thus $BCDF_{\mu}$ and $BCDF_{\gamma}$. The term **T cell replacing factor**, or TRF, has also been used, originally for a T cell cytokine(s) that induced T-depleted, antigen-primed B cells to secrete antibodies against the antigen (sheep erythrocytes), but later also for cytokines that induced Ig secretion in certain co-stimulator assays.

3. Molecules that stimulated both types of response were, quite logically, called **B cell growth and differentiation factors** or **B cell replication and maturation factors**, BCGDFs and BRMFs, respectively.

A little later, another nomenclature system came into existence. Soluble molecules that acted on B cells were termed **B cell stimulation factors**, or BSFs, and different molecules were then distinguished by numbers, thus BSF-1 and BSF-2.

There have been attempts to rationalize the designation of different cytokines by using the interleukin nomenclature, thus IL-4, IL-5, IL-6, these numbers being assigned as the molecules were gene cloned. It is now known there is considerable overlap between molecules defined by different assays (see above). Hence, to name just a few:

IL-1 = TRFII
IL-4 = BCGFI = $BCDF_{\gamma}$ = BSF-1
IL-5 = BCGFII = TRF
IL-6 = IFNβ2 (interferon β2) = BSF-2

These molecules are known to have pleiotropic functions (Section 4.4.1), and the original categorization of molecules into those that stimulated B cell proliferation versus antibody production has proven to be an oversimplification.

The effect of cytokines on antibody production by B cells

In the course of an immune response, a B cell may develop into a plasma cell and secrete Ig. This can be accompanied by a change in the class of antibody produced by the cell, and these processes are influenced or controlled by T cell cytokines. The most important points for the present purposes (detailed in Section 7.2) are as follows (Fig.8.24; compare with Fig.8.2).

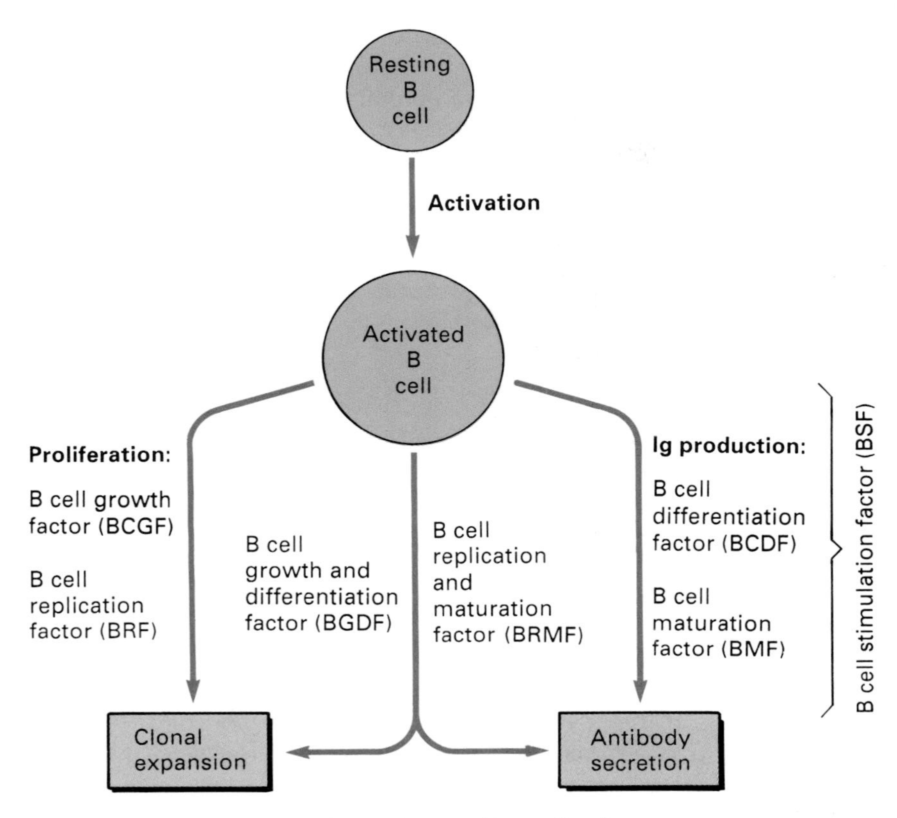

Fig.8.23 Some nomenclature for cytokines acting on B cells

1. Mature, resting B cells come to express membrane-bound IgM and IgD molecules through a series of *DNA rearrangements* that occur during their (antigen-independent) development from precursors.
2. When B cells are stimulated by antigens they often start to secrete IgM (and perhaps IgD). The conversion of a membrane-bound to secretory form of Ig occurs through *differential RNA processing.*
3. After antigenic stimulation, the B cell may undergo class switching and begin to express a different class of membrane-bound Ig (e.g. IgG). This involves further *DNA rearrangements* and the juxtaposition of the assembled VDJ region segments next to a new C region gene. The only exception to this is when the developing B cell begins to coexpress IgM and IgD, which is *regulated at the RNA level.*
4. The 'switched' B cell may secrete the new class of membrane Ig; this is again *regulated at the RNA level* [see (2) above].

Direction of class switching. Class switching generally proceeds in a direction that roughly corresponds to the 5′–3′ order of the Ig_HC genes (Fig.8.25). A major issue is to what extent class switching is a *spontaneous* process and to what extent it is *regulated* by T cells or their products (see below).

Different types of antigen and class switching (Fig.8.26). The predominant classes of antibody produced in response to many TI antigens are IgM and IgG3. This suggests that B cells of normal mice may be able to switch spontaneously to IgG3 production. (However, xid mice—Section 8.2.4—are unable to make IgG3 in response to TI as well as to TD antigens). Although TI antigens do not usually stimulate secondary responses, if T cells are added to B cells responding to TI antigens *in vitro* other Ig subclasses are produced at the expense of IgG3.

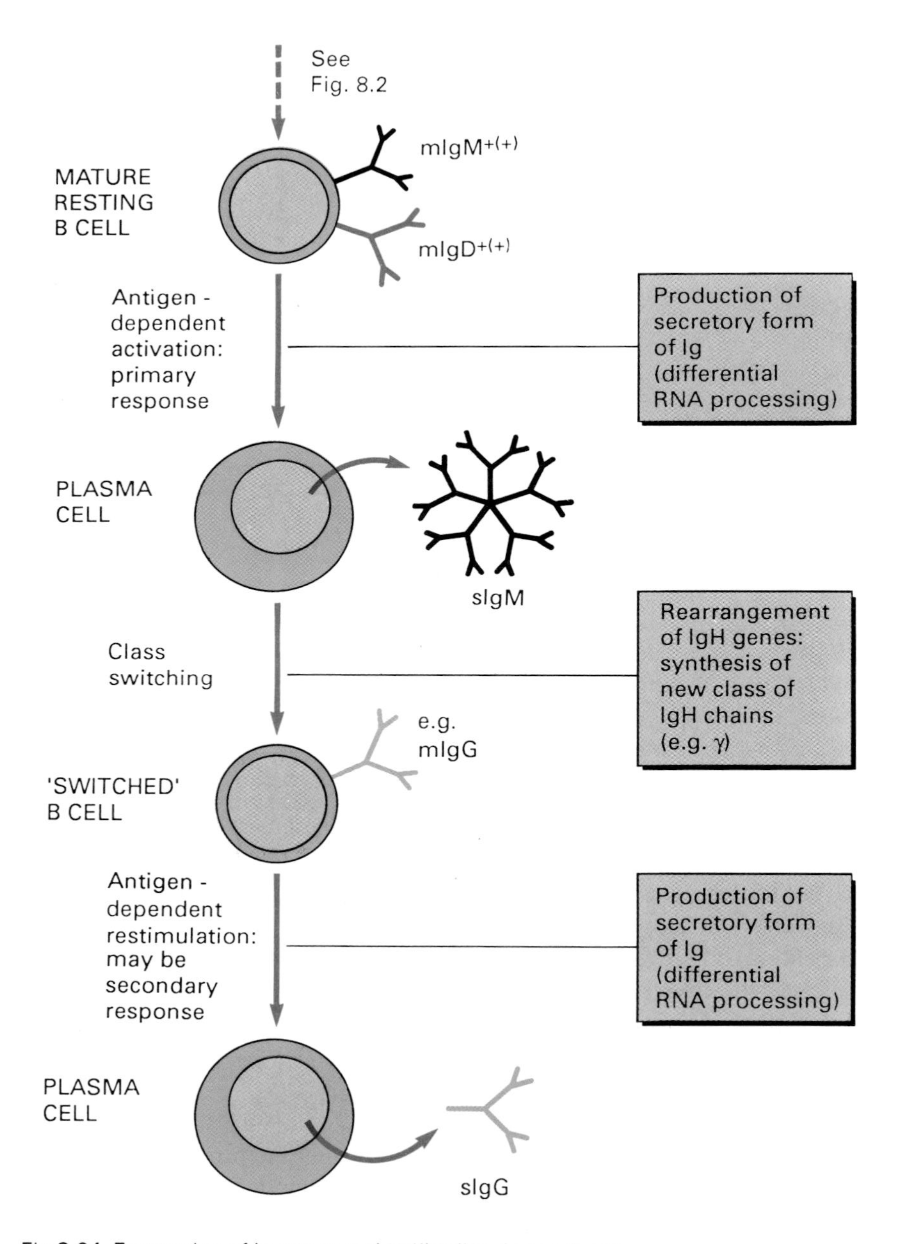

Fig.8.24 Expression of Ig genes and antibodies during B cell responses

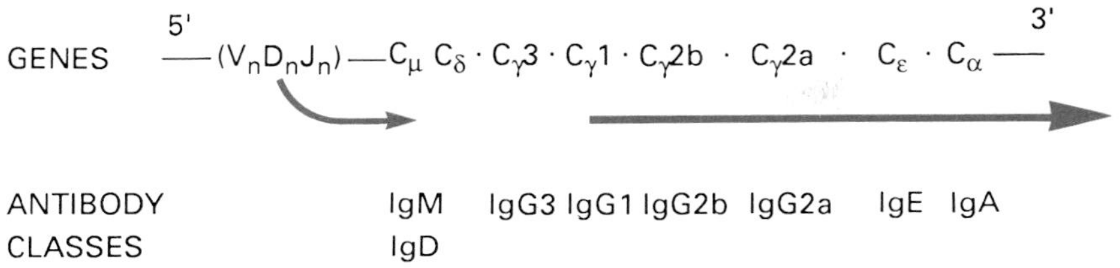

5' 3'

GENES —$(V_nD_nJ_n)$—C_μ $C_\delta \cdot C_\gamma3 \cdot C_\gamma1 \cdot C_\gamma2b \cdot C_\gamma2a \cdot C_\varepsilon \cdot C_\alpha$—

ANTIBODY CLASSES: IgM, IgD | IgG3 IgG1 IgG2b IgG2a IgE IgA

Fig.8.25 Isotype switching in mouse B cells generally occurs in the order shown, corresponding to the 5′ to 3′ gene sequence

The predominant antibody classes produced in TD responses are IgM, during primary responses, and *subclasses other than IgG3* in secondary responses. T cells seem to be very important for triggering the production of non-IgM classes, since nude or thymectomized mice can make IgM in response to TD antigens, but they have drastically impaired IgG responses. This TD defect can be more-or-less overcome in culture by adding T cells.

As a general rule, the production of non-IgM classes of antibody tends to follow the 5′–3′ order of the C genes (Fig.8.25). However, certain subclasses do seem to predominate in TD responses to particular antigens. Often, for example, soluble proteins tend to induce IgG1, viruses tend to induce IgG2a, and parasites tend to induce IgE production.

Role of cytokines. There are basically two extreme views as to the role of T cells and T cell-derived cytokines in class switching.

1. Switching occurs spontaneously in the 5′–3′ direction during successive rounds of proliferation, and T cells simply provide the factors that are required for proliferation.
2. Every switching event needs a specific T cell or its product (i.e. a cytokine), whether or not B cell proliferation is also required. Perhaps there could be distinct T cells for each switching event that occurs.

Intermediate possibilities must also be considered (Fig.8.27). One is that a T-cell causes the B cell to switch from expression of membrane-bound IgM

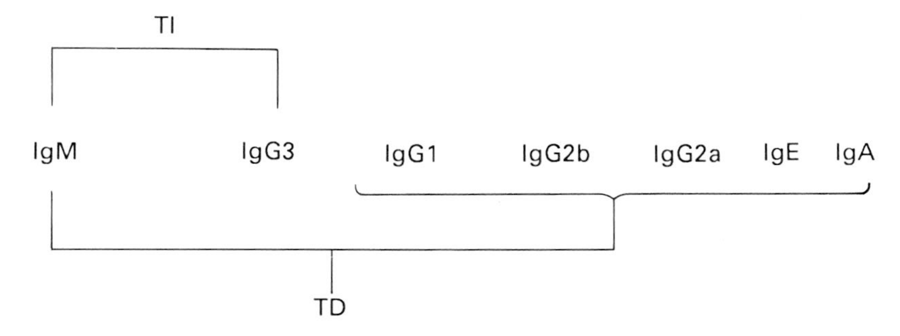

Fig.8.26 Subclasses of antibodies produced in response to different antigens

(or IgG3) to one of the other classes (e.g. IgG1) and switching then occurs spontaneously in a 5′–3′ direction (e.g. IgG1, IgG2b, IgG2a . . .). It is also possible that switching occurs up to a certain point (e.g. Ig2Ga in the example just given) and then stops until a different T cell induces another switch (e.g. from IgG2a to IgE).

Once the B cell expresses a particular class of membrane Ig, it seems to be committed to producing that antibody. However, it is not clearly established whether binding to antigen alone is sufficient to induce synthesis and production of the secretory form of the antibody, or whether T cells are required at this stage as well. For example, T cells could act non-specifically to cause secretion of *any class* of antibody that happens to be expressed on the membrane of an activated B cell. On the other hand, there could be T cells that specifically help B cells to secrete the *particular class* of membrane-bound antibody they express (e.g. one T cell may help IgG^+ B cells make secretory IgG; another may help IgA^+ B cells make secretory IgA; and so on).

These issues are not fully resolved, but a number of different activities of cloned T cells and/or their cytokines have been described (Fig.8.28).

1. Switching of IgM^+ B cells into cells that express membrane IgG, IgA, or IgE. For instance, IL-4 ($BCDF_\gamma$) can induce the expression of IgG1 by B cells (see Section 8.4.4). As another example, T cell clones derived from Peyer's patches (the main site of IgA^+ B cells) can induce IgM^+ B cells to become IgA^+ cells.

2. Secretion of IgM, IgG, IgA, or IgE from B cells that express the respective membrane-bound class of antibody. For example, $BCDF_\mu$ (not yet assigned an IL number) causes IgM^+ B cell tumours to secrete IgM, and IL-4 ($BCDF_\gamma$) can induce secretion of IgG1 (and in fact IgE) from B cells. Other T cell clones have been obtained from Peyer's patches that trigger secretion of IgA from 'post-switched' IgA^+ B cells (see above). Some of these clones were found to have Fc receptors for IgA and it has been suggested these receptors regulate the production of IgA by B cells.

3. Suppression of B cell production of particular classes of Ig. IgA and IgE suppressive factors have been reported. Some, but not all, of these molecules bind IgA or IgE and are produced by T cells that can express the respective Fc receptors. This suggests that Fc receptors might be involved in suppressive as well as stimulatory functions, conceivably depending on whether they are released by the T cell or expressed on its surface. However, other molecules, like IFNγ, also play an important role in suppressing the production of some subclasses of Ig but stimulating others (Section 8.4.4).

8.4.4 The function of defined cytokines

A number of cytokines that act on B cells have now been gene cloned and in every case they have turned out to have pleiotropic effects (Fig.4.24).

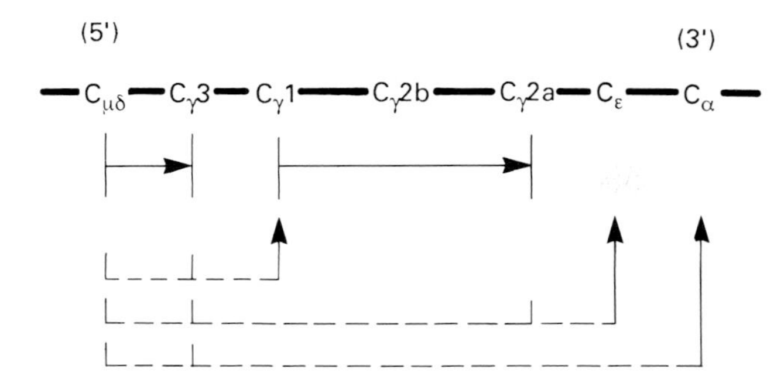

Fig.8.27 An hypothesis for both spontaneous (→) and specifically T cell-induced (↑) isotype switching

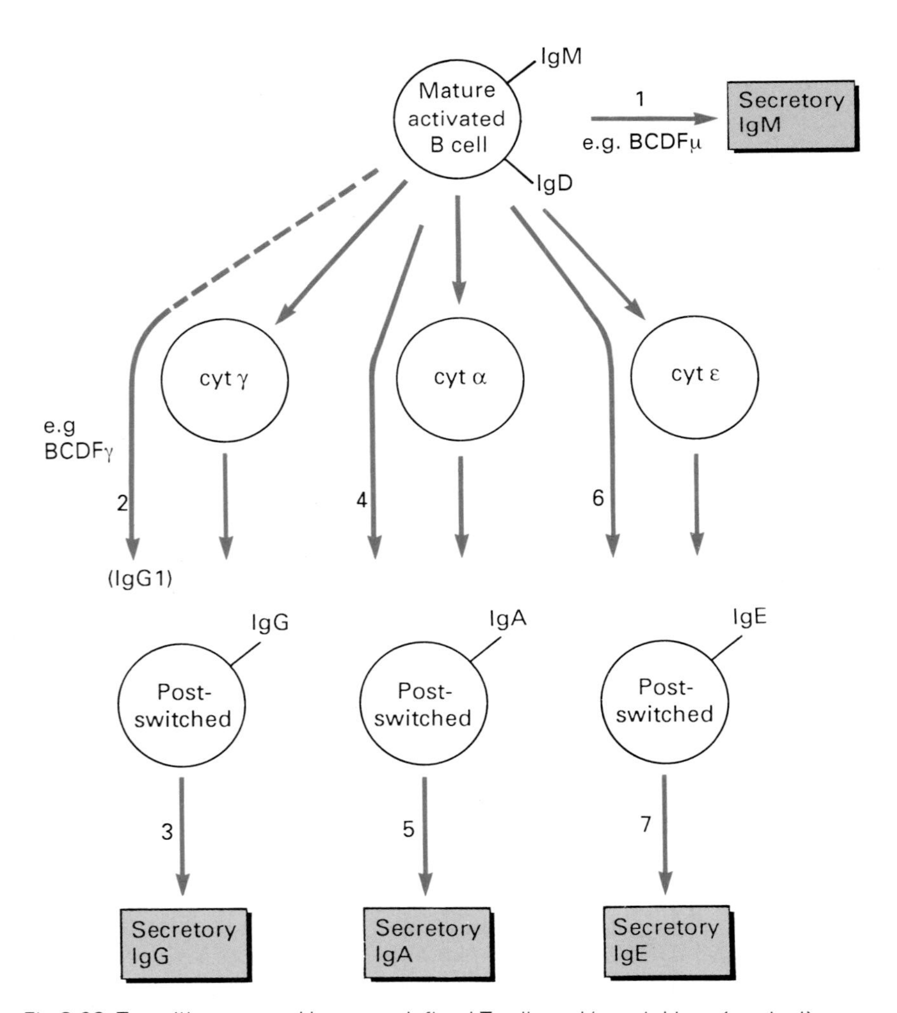

Fig.8.28 Transitions caused by some defined T cells and/or cytokines (see text)

Similar molecules have been obtained from both mouse and human but we focus mainly on activities in the former species.

Interleukin-4 (Fig.8.29)

IL-4 seems primarily to be a B cell activating factor that also strongly influences the class of antibody produced by B cells. It is produced by T cells, although there are reports that some B cell lines and mast cells may also be able to secrete it.

Effects on B cells (Fig.8.30). Cloning of IL-4 led to the discovery that this molecule has multiple effects on B cells (Noma *et al.* 1986; Lee *et al.* 1986).

1. IL-4 acts on *resting B cells* (G_0Q) and stimulates their entry into G_1 phase (Fig.8.21), and it is therefore a 'B cell activating factor'. It induces a greatly increased expression of MHC class II molecules on B cells and *primes* them to respond to anti-IgM. It also causes them to express the $Fc_{\varepsilon}RII$ receptor, CD23.
2. IL-4-treated B cells proliferate in the presence of anti-IgM (BCGFI, BSF-1 activity). However, xid B cells do not respond in this manner, even though they can express IL-4 receptors.
3. B cells stimulated with LPS make IgM and IgG3. If LPS-treated B cells are exposed to IL-4, IgM and IgG3 production is inhibited and they make IgG1 instead, as well as IgE ($BCDF_{\gamma}$ and $BCDF_{\varepsilon}$ activity; Snapper *et al.* 1988). Note that LPS-treated cells also produce IgG2a and IgG2b, and that production of these classes is likewise inhibited by IL-4.
4. The ability of IL-4 to overcome Fc receptor-mediated inhibition of B cell responses was noted in Section 8.4.2.

Effects on other cells. IL-4 acts on other cell types. Amongst its effects are the following.

1. It synergizes with IL-3 in causing proliferation of mast cells in connective tissues (i.e. mast cell growth factor-2, MCGF-II, activity). However IL-3 on its own leads to proliferation of mucosal mast cells and their differentiation from the bone marrow.
2. IL-4 is a growth factor for T cells (Section 4.4.5) and it induces cytotoxic activity (TCGFII activity).

Interleukin-5 (Fig.8.31)

IL-5 at present predominantly seems to be a growth factor for B cells, although it also influences Ig secretion. It is a T cell product that was originally isolated from some T cell clones and lines. The mouse and human IL-5 molecules cloned to date have similar functions but different structures (Kinashi *et al.*; Azuma *et al.* 1986), raising the possibility that IL-5 is a family of molecules.

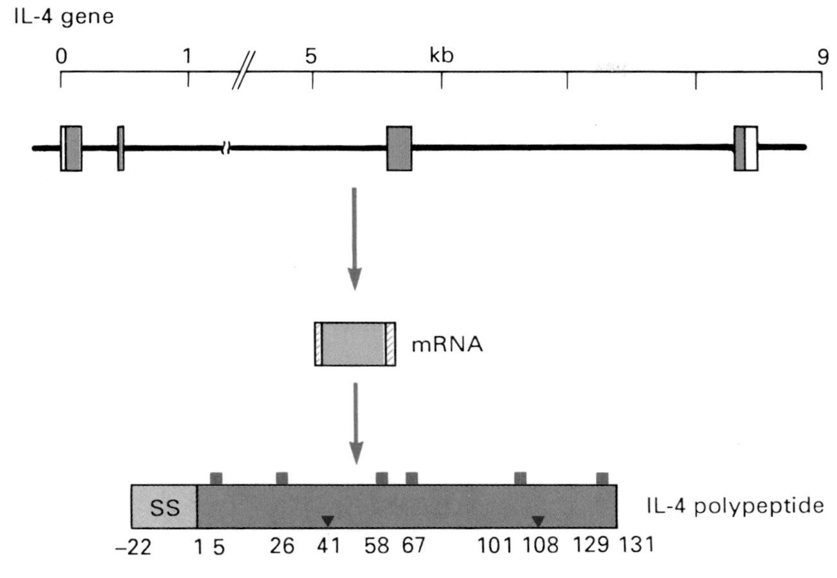

Fig.8.29 Human IL-4. The human IL-4 gene consists of four exons extending over 9 kb of DNA. The IL-4 polypeptide consists of 153 amino acid residues from which a signal sequence of 22 residues is cleaved. There are two N-glycosylation sites (▼) and six cysteine residues (■) (Redrawn from Hamblin, A.S. (1988) *Lymphokines*. In Focus Series (ed. D. Male). IRL Press, Oxford). Mouse IL-4 is 140 amino acids long (including a 20-residue signal sequence)

Effects on B cells (Fig.8.32). Mouse IL-5 acts primarily on *activated* B cells. However, an important difference between mouse and human IL-5 is that it stimulates B cells in the former species but not in the human (!).

1. IL-5 induces proliferation of B cells activated with dextran sulfate (BCGFII activity; Section 8.4.2). It also triggers proliferation of other activated B cells and B cell tumours.

2. It increases the secretion of antibodies by primed B cells in general (TRF activity). For example, IgM and IgG are produced by activated B cells and some B cell tumours in response to IL-5. However, it triggers IgA secretion from LPS-activated B cells ($BCDF_{\alpha}$ activity).

3. IL-5 induces expression of IL-2 receptors on B cells (Loughnan and Nossal 1989).

Effects on other cells. Other activities associated with IL-5 include the following.

1. In the mouse it induces growth and differentiation of eosinophils, and is thus known as eosinophil colony stimulating factor (Eo-CSF) and eosinophil differentiation factor (EDF).

2. IL-5 acts together with IL-2 to induce development of cytotoxic T cells from thymocytes (Section 4.4.5).

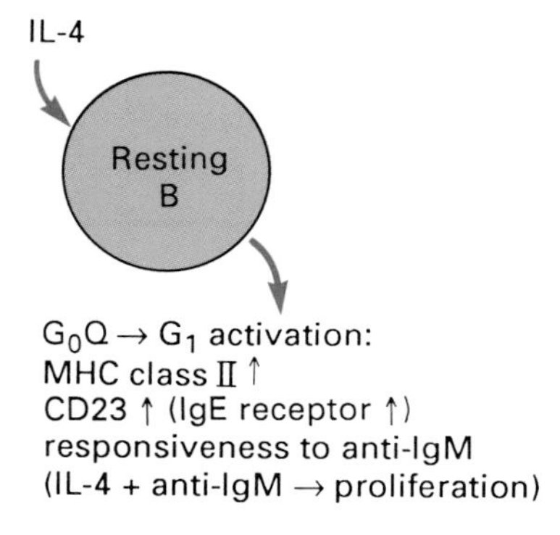

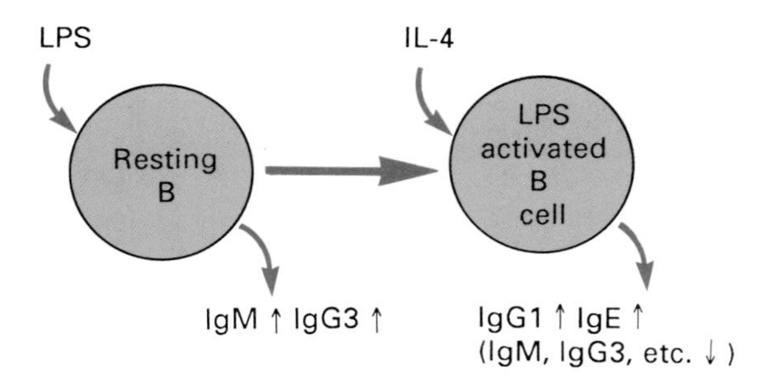

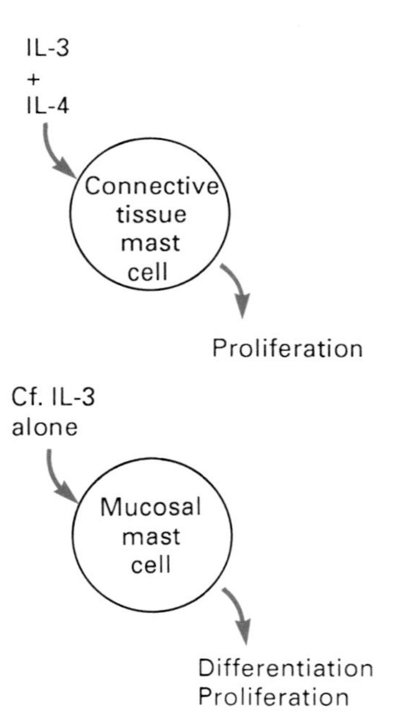

Fig.8.30 Effects of IL-4 on B cells and mast cells

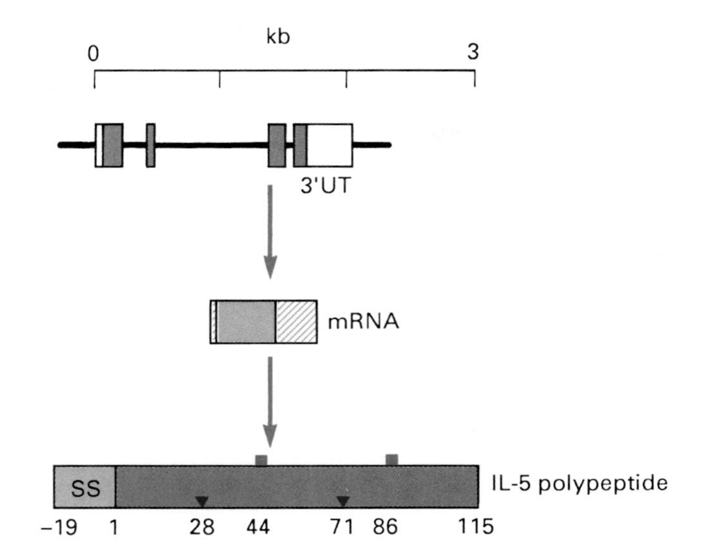

Fig.8.31 Human IL-5. The IL-5 gene consists of four exons which generate a polypeptide of 134 amino acid residues. A 19-residue signal sequence is present (SS) and there are two cysteine residues (■). There are N-glycosylation sites at residues 28 and 71 (▼) (Redrawn from Hamblin, A.S. (1988) *Lymphokines*. In Focus Series (ed. D. Male). IRL Press, Oxford). Mouse IL-5 is 133 amino acids long (including a 20-residue signal sequence)

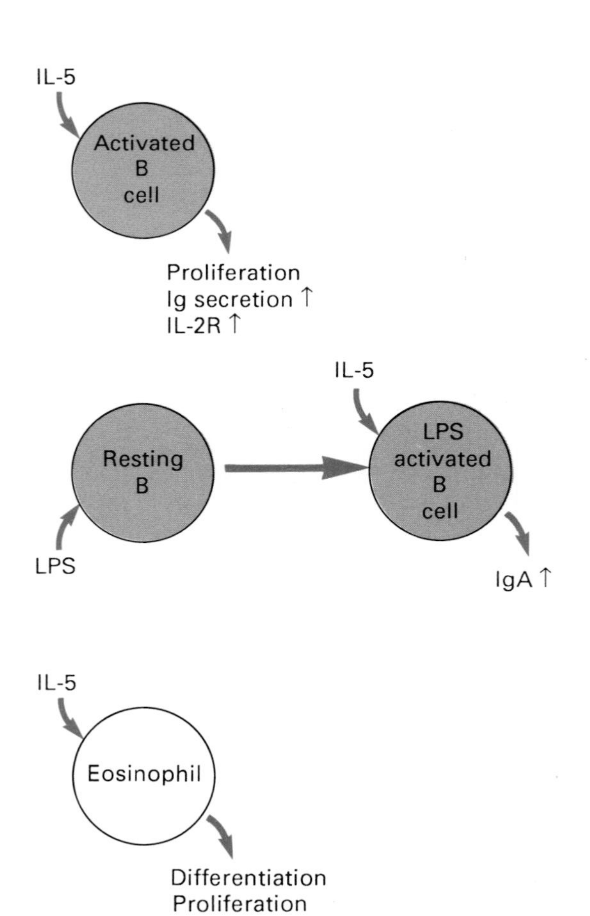

Fig.8.32 Effects of IL-5 on B cells and eosinophils

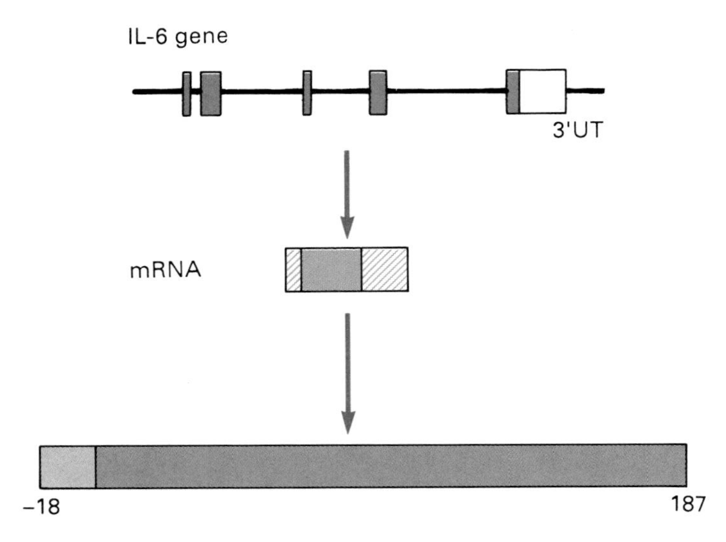

Fig.8.33 IL-6 gene. The human IL-6 gene consists of 5 exons. It has sequence homology, and a possible common evolutionary origin, to G-CSF: the gene for G-CSF gene also contains 5 exons and the location of cysteine residues is highly conserved in both molecules. The mouse IL-6 polypeptide contains 211 residues, including a 24 amino acid signal sequence

Interleukin-6 (Fig.8.33)

IL-6 currently seems especially important for inducing high rate Ig synthesis by activated B cells, and this can occur in the absence of any effect on B cell proliferation. It is produced by T cells, the P388D1 mouse macrophage line, and a number of human carcinoma cell lines, and the human gene has been cloned (Hirano 1986). (For a discussion of IL-6 and its role in inflammation, see Section 9.5.2).

Effects on B cells (Fig.8.34). IL-6 is said to act in the final 'maturation' or 'terminal differentiation' stage of B cell responses (Haraguchi *et al.* 1988).

1. It causes a generalized increase in Ig synthesis by activated B cells and lymphoblastoid cells, and promotes the secretion of IgM and IgG by B cells stimulated with pokeweed mitogen (BCDF/BSF-2 activity). IL-6 receptors are expressed on activated B cells but *not* on resting B cells (cf. T cells—see below).
2. IL-6 has no effect on the growth of normal activated B cells, but it is required for proliferation of plasmacytomas and hybridomas in culture.

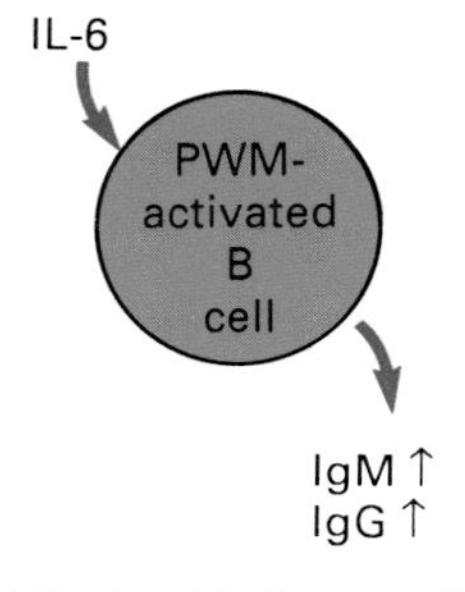

Fig.8.34 Effects of IL-6 on B cells

Effects on other cells. IL-6, apart from being known as BSF-2, has been shown to be identical to other activities that were called 26 kDa protein and IFNβ2 although, despite the latter name, it is greatly disputed whether IL-6 has any anti-viral activity. This molecule can be induced from a variety of cell types by different agents such as IL-1, with which it shares similar functions (Section 9.5.2), TNFα and poly(I):poly(C).

It is produced by human fibroblastoid cells and monocytes stimulated with LPS, as well as by T cells activated with mitogens or antigens. It has a variety of effects, including the following.

1. It acts on hepatocytes to induce acute phase proteins.
2. IL-6 acts on haemopoietic cells, apparently in G_0, after which they become dependent on IL-3.
3. It acts on nerve cells.
4. *Resting* T cells have been shown to express IL-6 receptors, and IL-6 induces IL-2 receptors on T cells (Section 4.4.5).

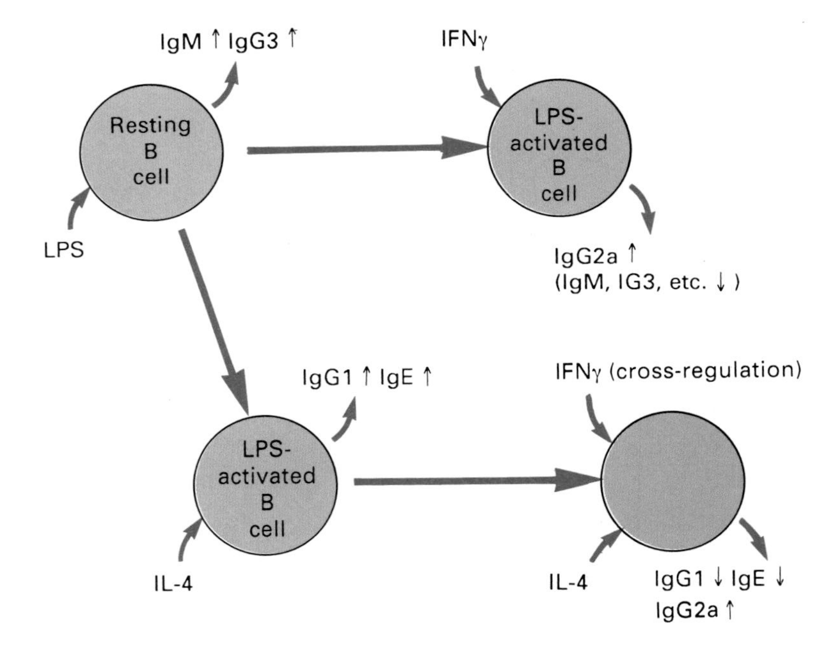

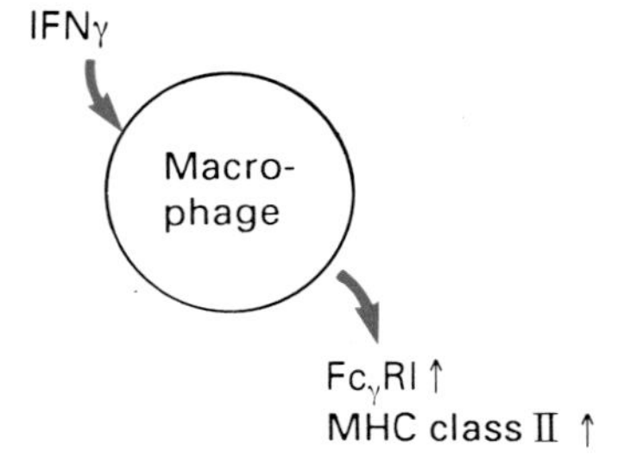

Fig.8.35 Effects of interferon-γ on B cells and macrophages

Effects of other cytokines on B cells

Interleukin-1. IL-1 can synergize with anti-Ig plus dextran sulphate, or anti-Ig plus IL-4, to trigger B cell proliferation. It apparently has no effect on Ig secretion. (For a discussion of IL-1 and its role in inflammation, see Section 9.5.1; for its role in T cell responses see Sections 3.4.5 and 4.4.4).

Interleukin-2. Activated B cells express IL-2 receptors, and recombinant IL-2 causes proliferation of mouse B cells activated with LPS *plus* anti-IgM, and of some activated human B cells. It can synergize with other molecules, such as IFNγ, to stimulate Ig synthesis by B cells. (For details of this cytokine, and its effects on T cells see Section 4.4.3).

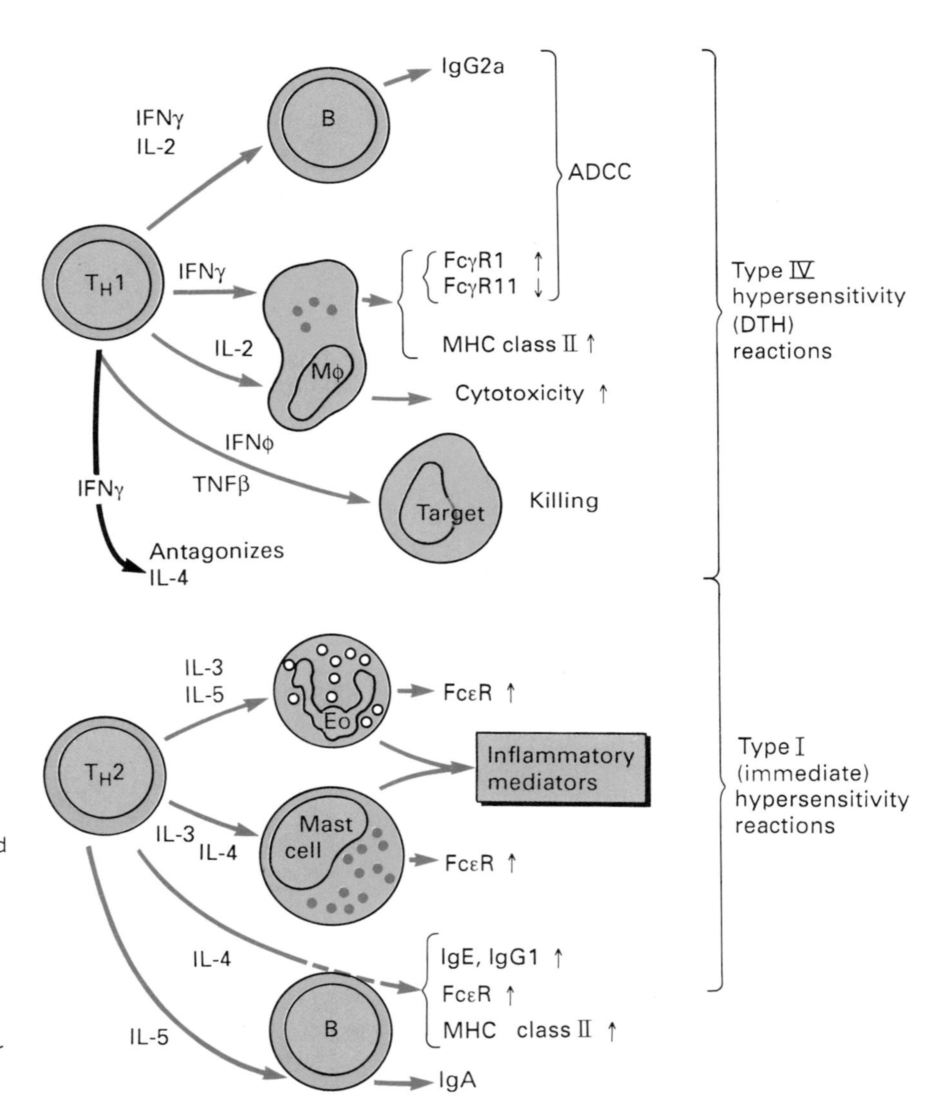

Fig.8.36 Predicted functions of T_H1 and T_H2 cells. These presumptive subsets of T cells produce different lymphokines in response to antigen-presenting cells, and may be important in different types of immune response, particularly type 4 (DTH) and type 1 (immediate) hypersensitivity reactions. Only some consequences of lymphokine secretion are indicated. In addition, lymphokines from the two subsets (e.g. interferon-γ and IL-4) may synergize, with still further effects. Eo, eosinophil; B, B cell.

Interferon-γ (Fig.8.35). General features of this cytokine are discussed in Section 9.5.4. One of the major effects of IFNγ on the B cell is to antagonize the various effects of IL-4.

For example, it inhibits B cell proliferation induced by anti-IgM and IL-4, IL-4-induced expression of $Fc_{\varepsilon}RII$ receptors, and IL-4-induced synthesis of IgG1 and IgE by LPS-stimulated B cells. At the same time, it shifts Ig production to IgG2a. This is important in the light of the effect of IFNγ on macrophages, where it up-regulates the $Fc_{\gamma}RI$ receptors, specific for monomeric IgG2a, but down-regulates $Fc_{\gamma}RII$ receptors, specific for IgG1 and IgG2b (Section 10.4.5). IFNγ increases expression of MHC class II molecules by macrophages and induces a number of phenotypic and functional changes recognized as macrophage activation (Section 10.4.4).

8.4.5 Two types of helper T cells and B cell responses

As discussed in Section 4.4.2, evidence has been obtained for the existence of distinct subsets of helper T cells that can secrete different combinations of cytokines, and these therefore regulate B cell secretion of different Ig isotypes (Stevens *et al.* 1988) (Fig.8.36). One subset, designated T_H1, seems to be the predominant source of IL-2, IFNγ, and TNFβ (Sections 4.4.3, 9.5.4, and 9.5.3 respectively), while the other, T_H2, subset is the main source of IL-4 and IL-5. Both make IL-3 and a variety of other cytokines to a greater or lesser extent.

The fact that T_H2 cells secrete IL-3, IL-4, and IL-5 suggests that they may be important in immediate hypersensitivity responses (Section 10.5.1). For example, IL-3 and IL-4 promote mast cell development and IL-5 stimulates eosinophil production. All these cells have IgE receptors, and these are induced on B cells by IL-4. Moreover, IL-4 and IL-5 would be expected to stimulate IgE production by B cells. Binding of IgE to its receptors on these cell types is important in this type of hypersensitivity.

The fact that T_H1 cells preferentially seem to make IL-2 and IFNγ, which antagonizes IL-4, suggests that these cells may play an important role in delayed-type hypersensitivity responses (Section 10.5.4). For example, IFNγ causes B cells to produce IgG2a and up-regulates the $Fc_{\gamma}RI$ receptor for this class of antibody on macrophages. This subclass can stimulate complement-mediated cytotoxicity, and antibody-dependent cellular cytotoxicity (ADCC) and phagocytosis by macrophages.

8.5 CD5-positive B cells

8.5.1 Distribution

In mice and humans there is a distinct population of B cells that expresses the CD5 antigen (Section 8.2.4). CD5 was originally thought of as a marker for T lymphocytes and was called Ly1 and Leu1 (T1) in mice and humans

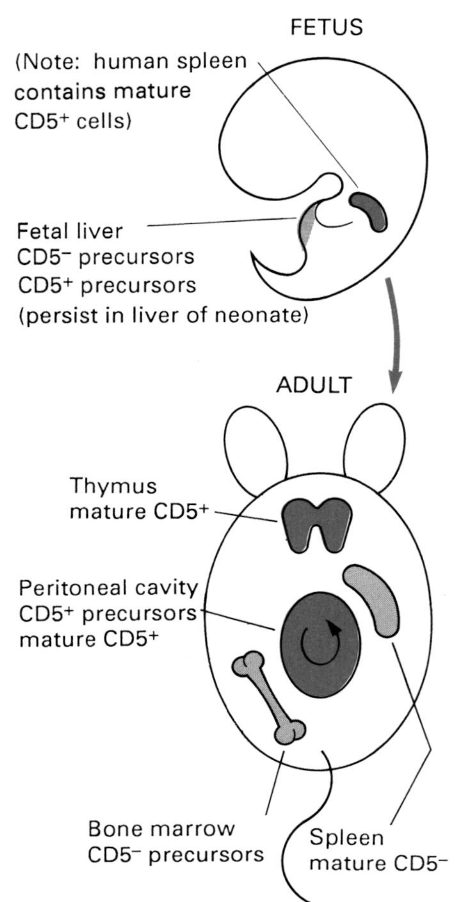

Fig.8.37 Location of CD5⁻ and CD5⁺ cells and their precursors in fetal and adult mice

(Section 4.2.4). These homologous molecules have 63% identity overall, increased to 90% in the C-terminal domain, and both have a conserved cysteine-rich N-terminus (Huang *et al.* 1987).

CD5 was first identified in the B cell lineage on malignant B cells, and was subsequently found to be expressed on a subpopulation of normal B cells. The CD5 antigen does not appear to be a cell cycle-related or activation marker of B cells and, although not yet absolutely certain, it does not seem to represent a developmental marker. In other words, there is no evidence as yet that $CD5^-$ B cells can develop into $CD5^+$ B cells (but with the possible exception that PMA can induce CD5 expression on $CD5^-$ B cells). Thus $CD5^+$ B cells are currently considered to be a distinct B cell lineage.

$CD5^+$ B cells comprise a population that appears early in ontogeny, and maintains itself independent of the bone marrow (Fig.8.37). These cells predominate in early B cell populations but are comparatively rare later, although CD5 is not expressed on the earliest B cell precursors that do not express membrane IgM or which express IgM in the absence of IgD (Fig.8.2). Thus in the human *fetus*, a majority of B cells in the spleen express CD5, but they comprise only 10–25% of total B cells in the *adult* spleen and blood circulation. In adult mice, only 1–2% of B cells in the spleen are $CD5^+$ and they are also rare in peripheral blood and usually undetectable in lymph nodes, Peyer's patches, and bone marrow (see Fig.1.33 for location of these tissues in humans). In marked contrast, $CD5^+$ cells are very frequent in the peritoneal cavity of adult mice where they comprise about 10–40% of the total B cell population. These cells express high levels of IgM but have lower levels of IgD. $CD5^+$ B cells are also present in the *thymus* (Section 5.3.5).

In *mice*, $CD5^+$ precursors originate in the *fetal liver* together with $CD5^-$ precursors (Fig.8.37). However, *$CD5^+$ precursors*, with the capacity for self-renewal, then move from the liver into the *peritoneal cavity* after birth. In contrast, *$CD5^-$ precursors* become localized in the *bone marrow*.

Some observations supporting these ideas are as follows. Haemopoietic stem cells are found in the fetal liver before they move to the bone marrow of adults. Although mouse and human fetal liver does not contain *mature* $CD5^+$ cells, it contains *precursors* to $CD5^-$ and $CD5^+$ B cells, since both populations develop in irradiated mice reconstituted with fetal liver (Hayakawa *et al.* 1985). However, only the $CD5^-$ B cell population is generated in irradiated mice reconstituted with *adult* bone marrow, suggesting this tissue does not contain $CD5^+$ precursors. In contrast, $CD5^+$ cells do develop if irradiated animals are reconstituted with *neonatal* mouse liver or adult peritoneal cells. Further evidence for distinct precursors of the $CD5^-$ and $CD5^+$ populations comes from experiments in which B cell development is inhibited at an early stage: treatment of mice with anti-IgM antibodies eliminates $CD5^+$ cells for life, whereas $CD5^-$ cells reappear as soon as the treatment is stopped.

A potentially important difference between mice and humans is that *$CD5^+$ precursors are present in human bone marrow* because $CD5^+$ B cells comprise the major population found soon after clinical bone marrow transplantation.

8.5.2 Function

Increased numbers of $CD5^+$ B cells are present in autoimmune strains of mice such as NZB (Hayakawa *et al.* 1983). In this strain there is abnormal B cell proliferation and splenomegaly, which is almost entirely due to $CD5^+$ cells; $CD5^+$ B cells also appear in the lymph nodes of these mice. These B cells seem to be responsible for most of the IgM autoantibodies that are produced in this strain that, for example, are specific for single-stranded DNA (ssDNA) and thymocytes. It is intriguing that $CD5^+$ B cells are not present in CBA/N xid mice (although this should not be taken as reason to equate them with $Lyb5^+$ cells; Section 8.2.4). NZB mice that have been made congenic for xid are also deficient in these cells, and there is a slight reduction in their numbers in nude mice. In humans, $CD5^+$ B cells are increased in patients with rheumatoid arthritis, and they produce predominantly IgM antibodies that bind with low affinity to the Fc portion of the IgG molecule, i.e. they have 'rheumatoid factor' activity (Hardy *et al.* 1984).

Because of these observations, there has been a tendency to consider the antibodies produced by $CD5^+$ B cells as **autoantibodies**. In contrast, $CD5^-$ B cells have been thought to produce non-IgM antibodies, particularly IgG, against exogenous antigens after immunization. It now appears that the antibodies produced by $CD5^+$ B cells are in fact '*polyreactive*', in that a single antibody may bind with a low affinity to a variety of different self antigens as well as to exogenous antigens; these antibodies thus resemble **natural antibodies**.

For example, monoclonal antibodies derived from $CD5^+$ B cells of healthy individuals that were selected for their ability to bind the Fc portion of IgG were found also to react with insulin, thyroglobulin, and ssDNA, as well as the exogenous antigen tetanus toxoid (Table 8.4). Moreover, while such antibodies are frequently IgM, similar IgG and IgA antibodies have been demonstrated. An important characteristic of these polyreactive antibodies in mice and humans, is the use of particular V gene segments in virtually unmutated configurations. Possibly these segments encode conserved sequences that generate a large binding site capable of accommodating a variety of different antigens. It has been argued that use of these unmutated gene segments might be required for the idiotypic cross-reactivity and 'connectivity' (Section 11.1.3) that seems essential for early development of the B cell repertoire.

Although $CD5^+$ B cells can produce polyreactive, low-affinity antibodies as described above, they can also produce *monoreactive*, high-affinity antibodies against defined cell antigens.

For example, some monoclonal antibodies derived from $CD5^+$ B cells from individuals with diabetes, Hashimoto's thyroiditis, and systemic lupus erythematosus were found to be specific for insulin, thyroglobulin, and ssDNA, respectively, which are considered to be autoantigens in these conditions (Table 8.4). These characteristics are more typical of autoantibodies.

Monoreactive antibodies produced by $CD5^+$ B cells might be generated by the accumulation of somatic mutations and subsequent clonal selection,

Table 8.4 Human CD5⁺ B cells can produce low-affinity polyreactive antibodies (resembling 'natural antibodies') and high-affinity monoreactive antibodies (more typical of autoantibodies): see text

Monoclonal antibody	Isotype	Source (individual)	Selecting antigen	Antigen binding activity (affinity) against				
				Fc IgG	Insulin	Thyroglobulin	ss DNA	Tetanus toxoid
Polyreactive								
A	IgM	normal	Fc IgG	low	low	–	'low'	low
B	IgA	normal	insulin	–	low	'low'	low	'low'
C	IgG1	normal	ss DNA	–	low	v. low	'low'	low
Monoreactive								
D	IgG1	diabetes	insulin	–	'low'	0	0	0
E	IgG1	Hashimoto's	thyroglobulin	–	0	high	0	0
F	IgG1	SLE	ss DNA	–	0	0	high	0
G*	IgG1	normal	tetanus toxoid	–	0	0	0	high

– = not determined; 0 = exceptionally low affinity; v. low, $K_d = 5 \times 10^{-3}$ M; low, $K_d = 1 \times 10^{-5}$–1×10^{-2} M; high, $K_d = < 10^{-9}$ M.
Hashimoto's = Hashimoto's thyroiditis; SLE = systemic lupus erythmatosus
Based on Casali and Notkins (1989)—see Further reading.
* Example of an antibody produced by CD5⁺ B cells specific for an exogeneous antigen

events that occur during affinity maturation of antibodies produced by CD5⁻ B cells in antigen-driven responses, although in these cells there is also a substantial contribution from antibodies using 'new' V segments in secondary responses (Section 8.3.1).

Polyreactive antibodies could provide a first line of defence against pathogens, before specific high-affinity monoreactive antibodies are produced later in the immune response.

8.6 B cell tolerance

In Section 5.2.5 we discussed T cell tolerance and mechanisms of negative selection including clonal deletion and clonal anergy. There is evidence that similar mechanisms can be responsible for B cell tolerance.

In the current chapter we have seen, e.g. in Section 8.5, that normal individuals do have B cells that can recognize self-components such as insulin and thyroglobulin, although in many cases their binding affinity is low; in Section 11.1 we also consider B cells that can react against self-idiotypes. In most cases, however, normal individuals do not have B cells specific for some abundant self-antigens like serum albumin; this applies particularly to B cells with high affinity receptors, for it is easier to induce tolerance of these cells than of those with low affinity receptors. In addition, it is easier to induce tolerance in immature B cells than mature B cells (see Fig.8.1 for a scheme of B cell development), although the latter is certainly possible, so that B cell tolerance may more commonly be induced in early development, although this is not entirely clear. Moreover, early studies demonstrated that B cell tolerance could be induced by different doses and forms of antigen (see, for example, Mitchison 1964), conceivably by different mechanisms.

Nevertheless, there are good examples where apparent B cell tolerance to an abundant self antigen, like the complement C5 component, is in fact due to tolerance at the *T cell* rather than B cell level (p. 286). Also, since the induction of B cell tolerance requires cross-linking of membrane-bound Ig molecules, the B cell repertoire could potentially contain B cells specific for *monomeric* self antigens, and clearly a state of T cell tolerance could be essential to avoid autoimmunity. Obviously if a B cell response to a particular antigen is T-dependent (see Section 8.3.1) tolerance to this antigen at the level of the T cell will result in the incapacity of autoreactive B cells to respond. It is important to note that smaller doses of soluble antigens can induce tolerance in T cells than in B cells. The contribution of T cell tolerance should always be borne in mind when considering unresponsiveness of B cells to certain antigens, although there are clear examples of tolerance induction in B cells which we now consider (Sections 8.6.1 and 8.6.2).

8.6.1 Clonal deletion

To consider a parellel with T cells, it is a well known fact that the majority of thymocytes generated within the thymus, perhaps 95% of the total, never enter the periphery. It is assumed that this is a result of the various selection pressures that are required to generate a self-tolerant, apparently self-restricted repertoire (Sections 5.2.5 and 5.2.3). It is perhaps less well known that, likewise, a similar proportion of B cells produced within the bone marrow does not enter the periphery.

It has been estimated that the total recirculating pool of B cells in mice contains about 2×10^8 cells. The bone marrow has the capacity to produce up to 5×10^7 primary B cells per day, but data suggest that only 5×10^6 virgin cells are recruited (Opstelten and Osmond 1983; reviewed in MacLennan and Gray 1986). Consequently, it would appear that 4.5×10^7 newly-produced B cells die each day, i.e about 90% of the total number produced. At least part of this could be due to clonal deletion of self-reactive B cells, although this has not yet been proven. Possibly the most persuasive evidence that clonal deletion of B cells can occur, came originally from studies with transgenic mice (see Appendix).

Nemazee and Burki (1989) constructed mice that were transgenic for the heavy and light chain genes of an IgM antibody reactive with H-2^k class I molecules. Some 20–50% of B cells in H-2^d transgenic mice expressed this transgenic mIg, as demonstrated by binding of an anti-idiotypic antibody specific for the receptor (Section 11.1.1), but they were not detectable in transgenic H-$2^d \times$ H-2^k mice. In addition, the number of peripheral B cells was reduced by 50% in the latter mice compared to the former, consistent with clonal deletion resulting in elimination of B cells.

Clonal deletion in these transgenic mice was probably due to the immature transgene-bearing B cells encountering their 'self antigen' *early* in development. In addition, the antigen (MHC) was expressed at a *high* concentration on most cells, and the transgenic receptor was probably of high affinity.

It is experimentally possible to delete peripheral B cells from neonatal mice by treatment with anti-μ chain antibodies. Some hints as to possible mechanisms of clonal deletion have come from studies with anti-μ *in vitro*. Such treatment of *immature* B cells in culture results in modulation of IgM from the surface of the cells, and they probably die. In contrast, similar treatment of *mature* B cells leads to patching and capping of IgM and its clearance from the surface, but membrane-bound IgM is then re-expressed and the cells survive. Other studies have shown that spontaneous proliferation of some immature B lymphomas can also be inhibited by treatment with anti-Ig, and that the inhibitory signal is delivered in G_1 of the cell cycle.

8.6.2 Clonal anergy

Evidence that clonal anergy may also contribute to B cell tolerance has come from other studies with transgenic mice.

Goodnow and colleagues (1989) produced mice that were transgenic for hen-egg lysozyme (HEL), which was secreted into the serum of these animals just like any other soluble protein. They also constructed mice that were transgenic for the heavy and light chain genes of an antibody specific for HEL. In the latter animals, about 90% of their B cells were reactive with HEL and expressed both IgM and IgD on their membranes. However, when the two types of transgenic animals were crossed, the double transgenics still contained many transgene-expressing B cells but they were found to be *unresponsive* to HEL, and these cells expressed much reduced levels of membrane-bound IgM, implicating this as contributing to the state of anergy. When mature B cells expressing the transgenic Ig were transferred to the HEL-transgenic mice, essentially the same phenomena were observed: clonal anergy and decreased levels of IgM on the cell surface.

Why did clonal deletion not occur, as was seen in the transgenic mice referred to earlier (Section 8.6.1)? The answer to this is not yet entirely clear but it has been suggested that in the HEL–anti-HEL transgenic mice (see box above), an intially *low* concentration or univalent form of the *soluble* antigen was encountered by *mature* B cells (compare with the *high* concentrations of *membrane-bound* MHC class I molecules encountered by *immature* B cells in the former case; Section 8.6.1). Nevertheless, there was some reduction in the numbers of transgenic B cells expressing anti-HEL Ig in the presence compared to the absence of antigen, which could suggest that some degree of clonal deletion had also occurred.

There is no doubt that clonal anergy can be induced in mature as well as immature B cells (Nossal and Pike 1980). For example, *deaggregated*, hapten-coupled human γ-globulin (HGG) is tolerogenic in mice (whereas *heat-aggregated* HGG is immunogenic) and this antigen can induce unresponsiveness of adult B cells, although higher concentrations are required than are needed to induce tolerance of immature B cells. A mechanism of tolerance induction of mature B cells may be required especially in the light of somatic hypermutation of their antigen receptors and the potential to produce auto-reactive and potentially deleterious B cell clones; this process, of course, does not occur during the generation of T cell receptors (Section 7.3.3).

Further reading

General

Bird, G. and Calvert, J.E. (ed.) (1988). *B lymphocytes in human disease*. Blackwell Scientific Publications, Oxford.

Klaus, G. (1990). *B lymphocytes*. In Focus Series (ed. D. Male). IRL Press, Oxford.

B cell development

Kincade, P.W. (1987). Experimental models for understanding B lymphocyte formation. *Advances in Immunology*, **41**, 181–267.

MacLennan, I.C., Oldfield, S., Liu, Y.J., and Lane, P.J. (1989). Regulation of B-cell populations. *Current Topics in Pathology*, **79**, 35–7.

Witte, O., Howard, M., and Klinman, N. (ed.) (1988). *B cell development*. UCLA Symposia on Molecular and Cellular Biology, vol.85. Alan R. Liss, New York.

B cell activation, antibody responses, and B cell memory

Clark, E.A. and Lane, P.J. (1991). Regulation of human B-cell activation and adhesion. *Annual Review of Immunology*, **9**, 97–128.

Colle, J.H., Le Moal, M.A., and Truffa-Bacchi, P. (1990). Immunological memory. *Critical Reviews in Immunology*, **10**, 259–88.

Esser, C. and Radbruch, A. (1990). Immunoglobulin class switching: molecular and cellular analyses. *Annual Review of Immunology*, **8**, 717–35.

Gray, D. and Sprent, J. (ed.) (1991). Immunological memory. *Current Topics in Microbiology and Immunology*, Vol.157. Springer-Verlag, Berlin.

Kishimoto, T. and Hirano, T. (1988). Molecular regulation of B lymphocyte response. *Annual Review of Immunology*, **6**, 485–512.

Liu, Y.J., Johnson, G.D., Gordon, J., and MacLennan, I.C.M. (1992). Germinal centres in T-cell-dependent antibody responses. *Immunology Today*, **13**, 17–21.

Mosier, D.E. and Subbarao, B. (1982). Thymus-independent antigens: complexity of B lymphocyte activation revealed. *Immunology Today*, **3**, 217.

Reth, M., Hombach, J., Wienands, J., Campbell, K.S., Chien, N., Justement, L.B., and Cambier, J.C. (1991). The B-cell antigen receptor complex. *Immunology Today*, **12**, 196–201.

Reth, M. (1992). Antigen receptors on B lymphocytes. *Annual Review of Immunology*, **10**, 97–121.

Sercarz, E. and Berzofsky, J. (ed.) (1988). *Immunogenicity of protein antigens: repertoire and regulation*. CRC Press, Boca Raton, Florida.

Szakal, A.K., Kosko, M.H., and Tew, J.G. (1989). Microanatomy of lymphoid tissue during humoral immune responses: structure function relationships. *Annual Review of Immunology*, **7**, 91–102.

Singer, A. and Hodes, R.J. (1983). Mechanisms of T cell-B cell interaction. *Annual Review of Immunology*, **1**, 211–41.

Vitetta, E.S., Fernandez-Botran, R., Myers, C.D., and Sanders, V.M. (1989). Cellular interactions in the humoral immune response. *Advances in Immunology*, **45**, 1–105.

Vitetta, E.S., Berton, M.T., Burger, C., Kepron, M., Lee, W.T., and Yin, X.M. (1991). Memory B cells and T cells. *Annual Review of Immunology*, **9**, 193–218.

Cytokines and B lymphocytes (see also Chapters 4, 9, and 11)

Callard, R.E. (ed.) (1990). *Cytokines and B lymphocytes*. Academic Press, Orlando, Florida.

Finkelman, F., Holmes, J., Katona, I.M., Urban, J.F., Bechmann, M.P., Park, L.S., Schooley, K.A., Coffman, R.L., Mosmann, T.R., and Paul, W.E. (1990). Lymphokine control of in vivo immunoglobulin isotype selection. *Annual Review of Immunology*, **8**, 303–33.

Gordon, J., Flores-Romo, L., Cairns, J.A., Millsum, M.J., Lane, P.J., Johnson, G.D., MacLennan, I.C. (1989). CD23: a multi-functional receptor/lymphokine? *Immunology Today*, **10**, 153–7.

Howard, H. and Paul, W.E. (1983). Regulation of B-cell growth and differentiation by soluble factors. *Annual Review of Immunology*, **1**, 307–33.

Ishizaka, K. (1988). IgE binding factors and regulation of the IgE antibody response. *Annual Review of Immunology*, **6**, 513–34.

Melchers, F. and Andersson, J. (1986). Factors controlling the B-cell cycle. *Annual Review of Immunology*, **4**, 13–36.

Metzger, H. (1988). Molecular aspects of receptors and binding factors for IgE. *Advances in Immunology*, **43**, 277–312.

Mossman, T.R. and Coffman, R.L. (1989). TH1 and TH2 cells: different patterns of lymphokine secretion lead to different functional properties. *Annual Review of Immunology*, **7**, 145–73.

Paul, W.E. (1989). Pleiotropy and redundancy: T cell-derived lymphokines in the immune response. *Cell*, **57**, 521–4.

Sideras, P., Noma, T., and Honjo, T. (1988). Structure and function of interleukin 4 and 5. *Immunological Reviews*, **102**, 189–212.

Snapper, C.M., Finkelman, F.D., and Paul, W.E. (1988). Regulation of IgG1 and IgE production by interleukin 4. *Immunological Reviews*, **102**, 51–75.

Van Snick, J. (1990). Interleukin-6: an overview. *Annual Review of Immunology*, **8**, 253–78.

B cell tolerance

Goodnow, C.C. (1992). Transgenic mice and analysis of B-cell tolerance. *Annual Review of Immunology*, **10**, 489–518.

Nossal, G.J. (1987). Bone marrow pre-B cells and the clonal anergy theory of immunologic tolerance. *International Review of Immunology*, **2**, 321–38.

$CD5^+$ B cells

Casali, P. and Notkins, A.L. (1989). CD5+ B lymphocytes, polyreactive antibodies and the human B cell repertoire. *Immunology Today*, **10**, 365–8.

Kipps, T.J. (1989). The CD5 B cell. *Advances in Immunology*, **47**, 117–85.

Literature cited

Ada, G.L. and Byrt, P. (1969). Specific inactivation of antigen-reactive cells with ^{125}I-labelled antigen. *Nature*, **222**, 1291–2.

Azuma, C., Tanabe, T., Konishi, M., Kinashi, T., Noma, T., Matsuda, F., Yaoita, Y., Takatsu, K., Hammarstrom, L., Smith, C.I., Severinson, E., and Honjo, T. (1986). Cloning of cDNA for human T-cell replacing factor (interleukin-5) and comparison with the murine homologue. *Nucleic Acids Research*, **14**, 9149–58.

Berek, C., Griffiths, G.M., and Milstein, C. (1985). Molecular events during maturation of the immune response to oxazolone. *Nature*, **316**, 412–18.

Bole, D.G., Hendershot, L.M., and Kearney, J.F. (1986). Post-translational association of immunoglobulin heavy chain binding protein with nascent heavy chains in nonsecreting and secreting hybridomas. *Journal of Cell Biology*, **102**, 1558–66.

Cairns, J.A. and Gordon, J. (1990). Intact, 45-kDa (membrane) form of CD23 is consistently mitogenic for normal and transformed B lymphoblasts. *European Journal of Immunology*, **20**, 539–43.

Camp, R.L., Kraus, T.A., Birkeland, M.L., and Puré, E. (1991). High levels of CD44 expression distinguish virgin from antigen-primed B cells. *Journal of Experimental Medicine*, **173**, 763–6.

Carter, R.H., Spycher, M.O., Ng, Y.C., Hoffman, R., and Fearon, D.T. (1988). Synergistic interaction between complement receptor type 2 and membrane IgM on B lymphocytes. *Journal of Immunology*, **141**, 457–63.

Goodnow, C.C., Crosbie, J., Jorgensen, H., Brink, R.A., and Basten, A. (1989). Induction of self-tolerance in mature peripheral B lymphocytes. *Nature*, **342**, 385–391.

Gray, D. and Skarvall, H. (1988). B cell memory is short-lived in the absence of antigen. *Nature*, **336**, 70–73.

Gruber, M.F., Bjorndahl, J.M., Nakamura, S., and Fu, S.M. (1989). Anti-CD45 inhibition of human B cell proliferation depends on the nature of activation signals and the state of B cell activation. A study with anti-IgM and anti-CDw40 antibodies. *Journal of Immunology*, **142**, 4144–52.

Hamaoka, T., Osborne, D.P., and Katz, D.H. (1973). Cell interactions between histoincompatible T and B lymphocytes. I. Allogeneic effect by irradiated host T cells on adoptively transferred histoincompatible B lymphocytes. *Journal of Experimental Medicine*, **137**, 1393–404.

Harada, K. and Yamagishi, H. (1991). Lack of feedback inhibition of V_κ gene rearrangement by productively rearranged alleles. *Journal of Experimental Medicine*, **173**, 409–415.

Hardy, R.R., Hayakawa, K., Parks, D.R., Herzenberg, L.A., and Herzenberg, L.A. (1984). Murine B cell differentiation lineages. *Journal of Experimental Medicine*, **159**, 1169–88.

Hayakawa, K., Hardy, R.R., Parks, D.R., and Herzenberg, L.A. (1983). The 'Ly-1 B' cell subpopulation in normal immunodefective, and autoimmune mice. *Journal of Experimental Medicine*, **157**, 202–18.

Hayakawa, K., Hardy, R.R., Herzenberg, L.A., and Herzenberg, L.A. (1985). Progenitors for Ly-1 B cells are distinct from progenitors for other B cells. *Journal of Experimental Medicine*, **161**, 1554–68.

Hirano, T., Yasukawa, K., Harada, H., Taga, T., Watanabe, Y., Matsuda, T., Kashiwamura, S., Nakajima, K., Koyama, K., Iwamatsu, A., Tsunasawa, S., Sakiyama, F., Matsui, H., Takahara, Y., Taniguchi, T., and Kishimoto, T. (1986). Complementary DNA for a novel human interleukin (BSF-2) that induces B lymphocytes to produce immunoglobulin. *Nature*, **324**, 73–6.

Hombach, J., Tsubata, T., Leclercq, L., Stappert, H., and Reth, M. (1990). Molecular components of the B-cell antigen receptor complex of the IgM class. *Nature*, **343**, 760–2.

Huang, H.J., Jones, N.H., Strominger, J.L., and Herzenberg, L.A. (1987). Molecular cloning of Ly-1, a membrane glycoprotein of mouse T lymphocytes and a subset of B cells: molecular homology to its human counterpart Leu-1/T1 (CD5). *Proceedings of the National Academy of Sciences USA*, **84**, 204–208.

Humphrey, J. and Grennan, D. (1981). Different macrophage populations distinguished by means of fluorescent polysaccharides. Recognition and properties of marginal zone macrophages. *European Journal of Immunology*, **11**, 212–28.

Kinashi, T., Harada, N., Severinson, E., Tanabe, T., Sideras, P., Konishi, M., Azuma, C., Tominaga, A., Bergstedt-Lindqvist, S., Takahashi, M., Matsuda, F., Yaoita, Y., Takatsu, K., and Honjo, T. (1986). Cloning of complementary DNA encoding T-cell replacing factor and identity with B-cell growth factor II. *Nature*, **324**, 70–3.

Lane, P.J., Gray, D., Oldfield, S., and MacLennan, I.C.M. (1986). Differences in the recruitment of virgin B cells into antibody responses to thymus-dependent and thymus-independent type 2 antigens. *European Journal of Immunology*, **16**, 1569–75.

Lane, P.J., McConnell, F.M., Schieven, G.L., Clark, E.A., and Ledbetter, J.A. (1990). The role of class II molecules in human B cell activation. Association with phosphatidyl inositol turnover, protein tyrosine phosphorylation, and proliferation. *Journal of Immunology*, **144**, 3684–92.

Lee, F., Yokota, T., Otsuka, T., Meyerson, P., Villaret, D., Coffman, R., Mosmann, T., Rennick, D., Roehm, N., Smith, C., Zlotnick, A., and Arai, K. (1986). Isolation and characterization of a mouse interleukin cDNA clone that expresses B-cell stimulatory factor 1 activities and T-cell- and mast-cell-stimulating activities. *Proceedings of the National Academy of Sciences USA*, **83**, 2061–5.

Lindsten, T. and Andersson, B. (1979). Ontogeny of B cells in CBA/N mice. Evidence for a stage of responsiveness to thymus-independent antigens during development. *Journal of Experimental Medicine*, **150**, 1285–92.

Linton, P.J., Decker, D.J., and Klinman, N.R. (1989). Primary antibody-forming cells and secondary B cells are generated from separate precursor cell subpopulations. *Cell*, **59**, 1049–59.

Liu, Y.J., Oldfield, S., and MacLennan, I.C. (1988). Memory B cells in T-cell dependent antibody responses colonize the splenic marginal zones. *European Journal of Immunology*, **18**, 355–62.

Loughnan, M.S. and Nossal, G.J. (1989). Interleukin 4 and 5 control expression of IL-2 receptor on murine B cells through independent induction of its two chains. *Nature*, **340**, 76–9.

Manz, J., Denis, K., Witte, O., Brinster, R., and Storb, U. (1988). Feedback inhibition of immunoglobulin gene rearrangement by membrane μ, but not by secreted μ heavy chains. *Journal of Experimental Medicine*, **168**, 1363–82. [Erratum *Journal of Experimental Medicine*, **169**, 2269.]

Matsumoto, A.K., Kopicky-Burd, J., Carter, R.H., Tuveson, D.A., Tedder, T.F., and Fearon, D.T. (1991). Intersection of the complement and immune systems: a signal transduction complex of the B lymphocyte-containing complement receptor type 2 and CD19. *Journal of Experimental Medicine*, **173**, 55–64.

Mitchison, N.A. (1964). Induction of immunological paralysis in two zones of dosage. *Proceedings of the Royal Society*, B **161**, 275–92.

Mitchison, N.A. (1971). The carrier effect in the secondary response to hapten-protein conjugates. I. Measurement of the effect with transferred cells and objections to the local environment hypothesis. *European Journal of Immunology*, **1**, 10–17.

Mond, J.J., Scher, I., Cossman, J., Kessler, S., Mongini, P.K.A., Hansen, C., Finkelman, F.D., and Paul, W.E. (1982). Role of the thymus in directing the development of a subset of B lymphocytes. *Journal of Experimental Medicine*, **155**, 924–36.

Morrissey, P.J., Boswell, H.S., Scher, I., and Singer, A. (1981). Role of accessory cells in B cell activation. IV. Ia+ accessory cells are required for the in vitro generation of thymic independent type 2 antibody responses to polysaccharide antigens. *Journal of Immunology*, **127**, 1345–7.

Muraguchi, A., Hirano, T., Tang, B., Matsuda, T., Horii, Y., Nakajima, K., and Kishimoto, T. (1988). The essential role of B cell stimulatory factor 2 (BSF-2/IL-6) for the terminal differentiation of B cells. *Journal of Experimental Medicine*, **167**, 332–4.

Nemazee, D.A. and Bürki, K. (1989). Clonal deletion of B lymphocytes in a transgenic mouse bearing anti-MHC class I antibody genes. *Nature*, **337**, 562–6.

Noma, Y., Sideras, T., Naito, T., Bergstedt-Lindqvist, S., Azuma, C., Severinson, E., Tanabe, T., Kinashi, T., Matsuda, F., Yaoita, Y., and Honjo, T. (1986). Cloning of cDNA encoding the murine IgG1 induction factor by a novel strategy using SP6 promoter. *Nature*, **319**, 640–6.

Nossal, G.L. and Pike, B.L. (1980). Clonal anergy: persistence in tolerant mice of antigen-binding B lymphocytes incapable of responding to antigen or mitogen. *Proceedings of the National Academy of Sciences USA*, **77**, 1602–6.

O'Garra, A., Rigley, K.P., Holman, M., McLaughlin, J., and Klaus, G.G.B. (1987). B-cell-stimulatory factor 1 reverses Fc receptor-mediated inhibition of B-lymphocyte activation. *Proceedings of the National Academy of Sciences USA*, **84**, 6254–8.

Ohno, T., Kubagawa, H., Sanders, S.K., and Cooper, M.D. (1990). Biochemical nature of an Fcμ receptor on human B-lineage cells. *Journal of Experimental Medicine*, **172**, 1165–76.

Opstelten, D. and Osmond, D.G. (1983). Pre-B cells in mouse bone marrow: immunofluorescence stathmokinetic studies of the proliferation of cytoplasmic μ-chain-bearing cells in normal mice. *Journal of Immunology*, **131**, 2635–40.

Raff, M.C. (1970). Role of thymus derived lymphocytes in the secondary humoral response in mice. *Nature*, **226**, 1257–8.

Ritchie, K.A., Brinster, R.L., and Storb, U. (1984). Allelic exclusion and control of endogenous immunoglobulin gene rearrangement in κ transgenic mice. *Nature*, **312**, 517–20.

Rousset, F., Garcia, E., and Banchereau, J. (1991). Cytokine-induced proliferation and immunoglobulin production of human B lymphocytes triggered through their CD40 antigen. *Journal of Experimental Medicine*, **173**, 705–10.

Schriever, F., Freedman, A.S., Freedman, G., Messner, E., Lee, G., Daley, J., and Nadler, L.M. (1989). Isolated follicular dendritic cells display a unique antigenic phenotype. *Journal of Experimental Medicine*, **169**, 2043–58.

Sher, I. (1982). CBA/N immune defective mice; evidence for the failure of a B cell subpopulation to be expressed. *Journal of Experimental Medicine*, **155**, 903–13.

Snapper, C.M. and Paul, W.E. (1987). Interferon-γ and B cell stimulatory factor-1 reciprocally regulate Ig isotype production. *Science*, **236**, 944–7.

Snapper, C.M., Finkelman, F.D., and Paul, W.E. (1988). Differential regulation of IgG1 and IgE synthesis by interleukin 4. *Journal of Experimental Medicine*, **167**, 183–96.

Stevens, T.L., Bossie, A., Sanders, V.M., Fernandez-Botran, R., Coffman, R.L., Mosmann, T.R., and Vitetta, E.S. (1988). Regulation of antibody isotype secretion by subsets of antigen-specific helper T cells. *Nature*, **334**, 255–8.

Tsubata, T. and Reth, M. (1990). The products of pre-B cell-specific genes (λ_5 and V_{preB}) and the immunoglobulin μ chain form a complex that is transported onto the cell surface. *Journal of Experimental Medicine*, **172**, 973–6.

Tuveson, D.A., Ahearn, J.M., Matsumoto, A.K., and Fearon, D.T. (1991). Molecular interactions of complement receptors on B lymphocytes: a CR1/CR2 complex distinct from the CR2/CD19 complex. *Journal of Experimental Medicine*, **173**, 1083–9.

Van de Velde, H., Van Hoegen, I., Luo, W., Parnes, J.R., and Thielmans, K. (1991). The B-cell surface protein CD72/Lyb2 is the ligand for CD5. *Nature*, **351**, 662–5.

Weiss, U. and Rajewsky, K. (1990). The repertoire of somatic antibody mutants accumulating in the memory compartment after primary immunization is restricted through affinity maturation and mirrors that expressed in secondary responses. *Journal of Experimental Medicine*, **172**, 1681–9.

Venkitaraman, A.R., Williams, G.T., Dariavach, P., and Neuberger, M.S. (1991). The B-cell antigen receptor of the five immunoglobulin classes. *Nature*, **352**, 777–81.

Wilson, H.A., Greenblatt, D., Poenie, M., Finkelman, F.D., and Tsien, R.Y. (1987). Crosslinkage of B lymphocyte surface immunoglobulin by anti-Ig or antigen induces prolonged oscillation of intracellular ionized calcium. *Journal of Experimental Medicine*, **166**, 601–6.

Wilson, G.L., Fox, C.H., Fauci, A.S., and Kehrl, J.H. (1991). cDNA cloning of the B cell membrane protein CD22: a mediator of B-B cell interactions. *Journal of Experimental Medicine*, **173**, 137–46.

Wortis, H.H., Burkly, L., Hughes, D., Roschelle, S., and Waneck, G. (1982). Lack of mature B cells in nude mice with X-linked immunodeficiency. *Journal of Experimental Medicine*, **155**, 903–13.

Yamanashi, Y., Kakiuchi, T., Mizuguchi, J., Yamamoto, T., and Toyoshima, K. (1991). Association of B cell antigen receptor with protein tyrosine kinase Lyn. *Science*, **251**, 192–4.

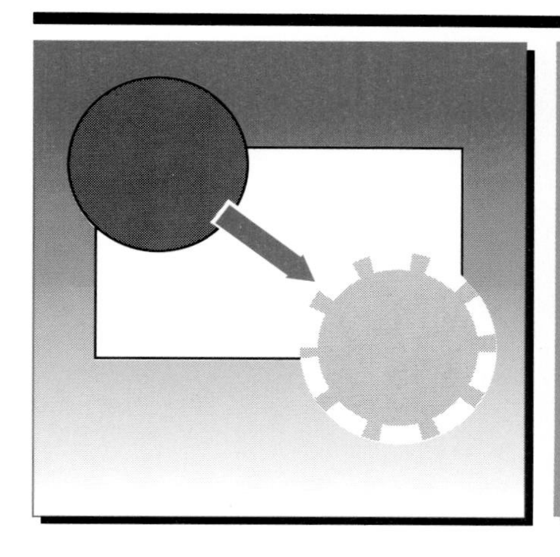

9

INFLAMMATORY MEDIATORS AND SOLUBLE EFFECTOR MECHANISMS

Scheme 9. Complement and inflammatory mediators

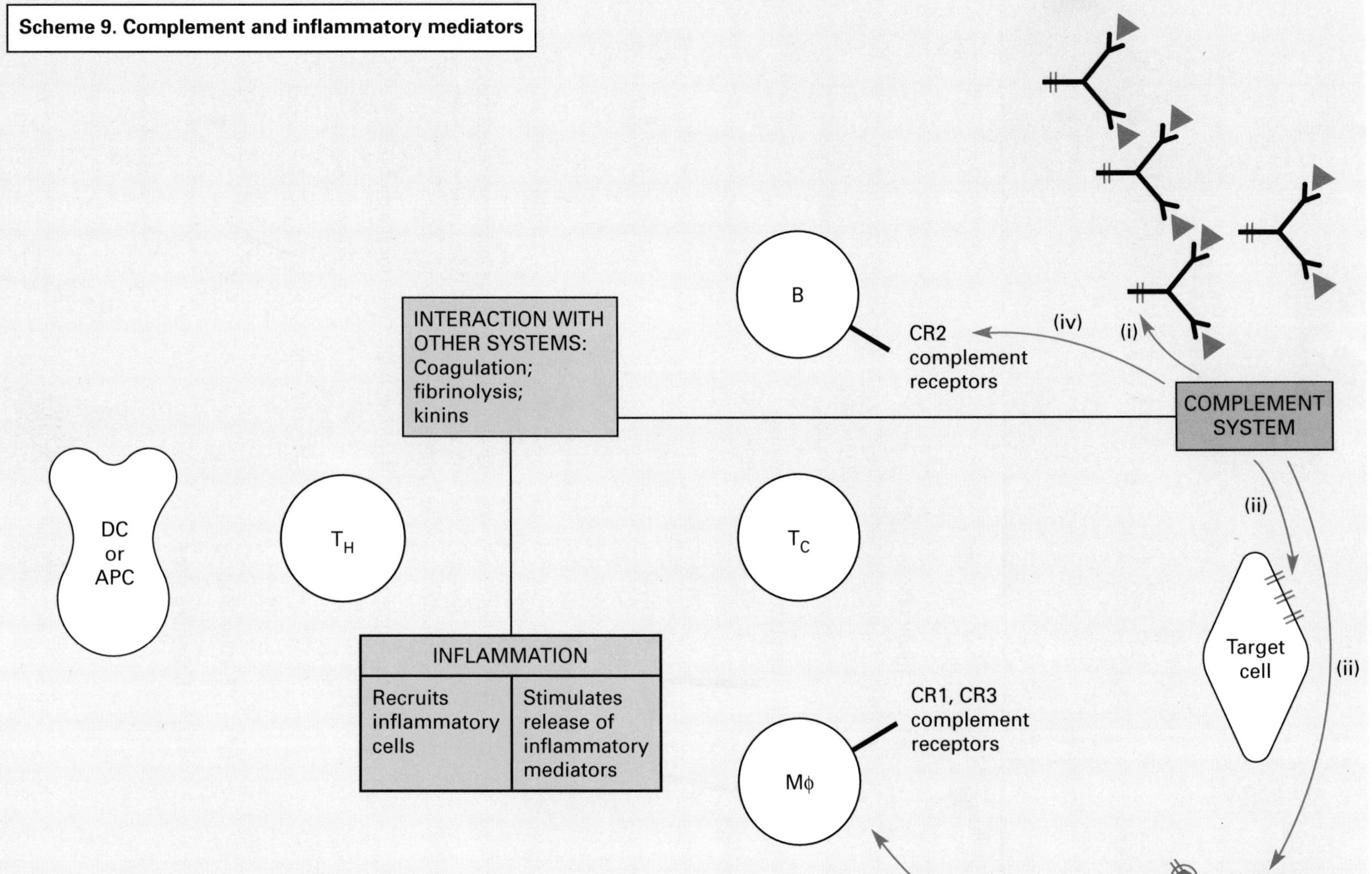

(i) Binding to antibody–antigen complexes and activation by direct pathway

(ii) Activation by direct pathway/indirect pathway in presence of Ig: pore formation

(iii) Opsonization and phagocytosis by e.g. Mϕ

(iv) Costimulatory signals for B cell activation

9.1 Introduction

An animal must defend itself against a multitude of different pathogens including viruses, bacteria, fungi, and protozoan and metazoan parasites as well as tumours (For scale see Fig.1.1). For this, a number of effector mechanisms capable of defending the body against such antigens have developed and these can be mediated by soluble molecules or by cells. Soluble mediators include those derived by activation of a number of different proteolytic cascade systems in the blood plasma, such as the **coagulation, kinin, fibrinolysis,** and **complement pathways**, together with other mediators of inflammation, such as **leukotrienes** and **prostaglandins** (products of arachidonic acid metabolism). Cellular effector mechanisms can be mediated by a variety of cell types including cytotoxic T cells (CTL), natural killer (NK) cells, and macrophages. Effector mechanisms either directly eliminate, or facilitate the elimination of, the foreign antigen from the body by mounting an attack either when the pathogen is present in the body fluids, i.e. extracellular, or after it has infected the host's cells. In many cases we are unsure of the relative contribution or importance of the different effector mechanisms in the fight against a particular infection, since many of the pathways are closely interrelated, both in terms of their activation and control mechanisms.

This chapter will focus on *soluble* molecules that contribute to the elimination of antigen from the body, and will include a brief discussion of those that trigger or mediate inflammatory responses (Section 9.2), an outline of the interaction of the different proteolytic cascade systems present in blood plasma (Section 9.3), as well as a more detailed discussion of the complement system (Section 9.4). Finally, some inflammatory cytokines will be described, these include IL-1, IL-6, TNFα, TNFβ, and the interferons (Section 9.5). *Cellular* effector mechanisms are discussed in Chapter 10.

9.2 Inflammation

9.2.1 The response to injury and infection

Inflammation is a manifestation of the body's response to tissue damage and infection. The mechanism for triggering the response the body makes to injury is extremely sensitive. Responses are made to tissue damage that might not normally be thought of as injury, for example when the skin is stroked quite firmly or if some pressure is applied to a tissue. In addition, the body has the capacity to respond to both minor injuries such as bruising, scratching, cuts, and abrasions, as well as to major injuries such as severe burns and amputation of limbs.

Depending on the severity of the tissue damage resulting from an injury, the integrity of the skin or internal surfaces may be breached and damage to the underlying connective tissues and muscle, as well as blood vessels, can occur. In this situation infection can, and frequently does result because the normal barrier to the entry of harmful organisms has been broken. It is obviously most important that the body can respond to injury by healing and repairing the damaged tissue, as well as by eliminating the infectious agents that may have entered the wound and their toxins. It is also important that the appropriate response to the tissue damage and infection can be made: it is no use bringing all of the body's defences into action to repair a minor scratch, just as one would not expect a single mechanism to be able to deal with the sudden loss of a limb or a major infection. The responses the body makes to tissue damage must be ordered and controlled. The body must be able to act quickly in some situations, for example to reduce or stop the loss of blood, whereas tissue repair and reconstruction can begin a little later. Therefore, a wide variety of interconnected cellular and humoral (soluble) mechanisms are activated when tissue damage and infection occur. *These mechanisms include four proteolytic cascade systems present in the blood plasma: the coagulation clotting, kinin, and fibrinolytic pathways, and complement.* The extent of these mechanisms, and the precise relationships between them, are only just beginning to be worked out, but we have a reasonable understanding of the basic principles underlying the activation of each system, and of some of the interactions that occur. Figure 9.1 summarizes some interactions between these four plasma enzyme systems during an inflammatory response.

In brief, when tissue injury occurs, the resulting exposure of collagen, vascular basement membrane, or certain negatively charged substances can result in activation of the **Hageman factor** (**factor XII**), as well as triggering the release of **histamine** from mast cells. Activation of the Hageman factor does not require contact with immune complexes and occurs *before* any immune response is mounted. Activated Hageman factor (factor XIIa) has strong procoagulant activity, i.e. it can activate the **clotting pathway** which results in activation of **factor X**, the generation of **fibrin**, clot formation and the prevention of further loss of blood. However, it is relatively inefficient with respect to kinin generation.

The **kinins** are important mediators of inflammatory responses. For kinin generation to proceed efficiently, activated Hageman factor activates prekallikrein via a series of prekallikrein activators, resulting in the production of **kallikrein**. In addition, activated Hageman factor activates factor XI, producing factor XIa of the clotting cascade and can also induce the release of small quantities of **plasmin,** the end product of the **fibrinolytic pathway**. Activated kallikrein, factor XIa, and plasmin all have the capacity to activate more Hageman factor and in this way an amplification loop is set in motion. The generation of kallikrein triggers kinin production, including the formation of **bradykinin**, which is responsible for inducing pain, increasing vascular permeability, and causing vasodilation. Kallikrein also activates the fibrinolytic pathway, leading to the removal of blood clots. The plasmin that is produced can then activate the **complement cascade** through the alternative pathway. All of these events can occur before an immune response has been mounted.

Once an immune response has been triggered, immunological events, including the production of antibodies and the formation of immune complexes, interact with the kinin, coagulation, and fibrinolytic pathways via

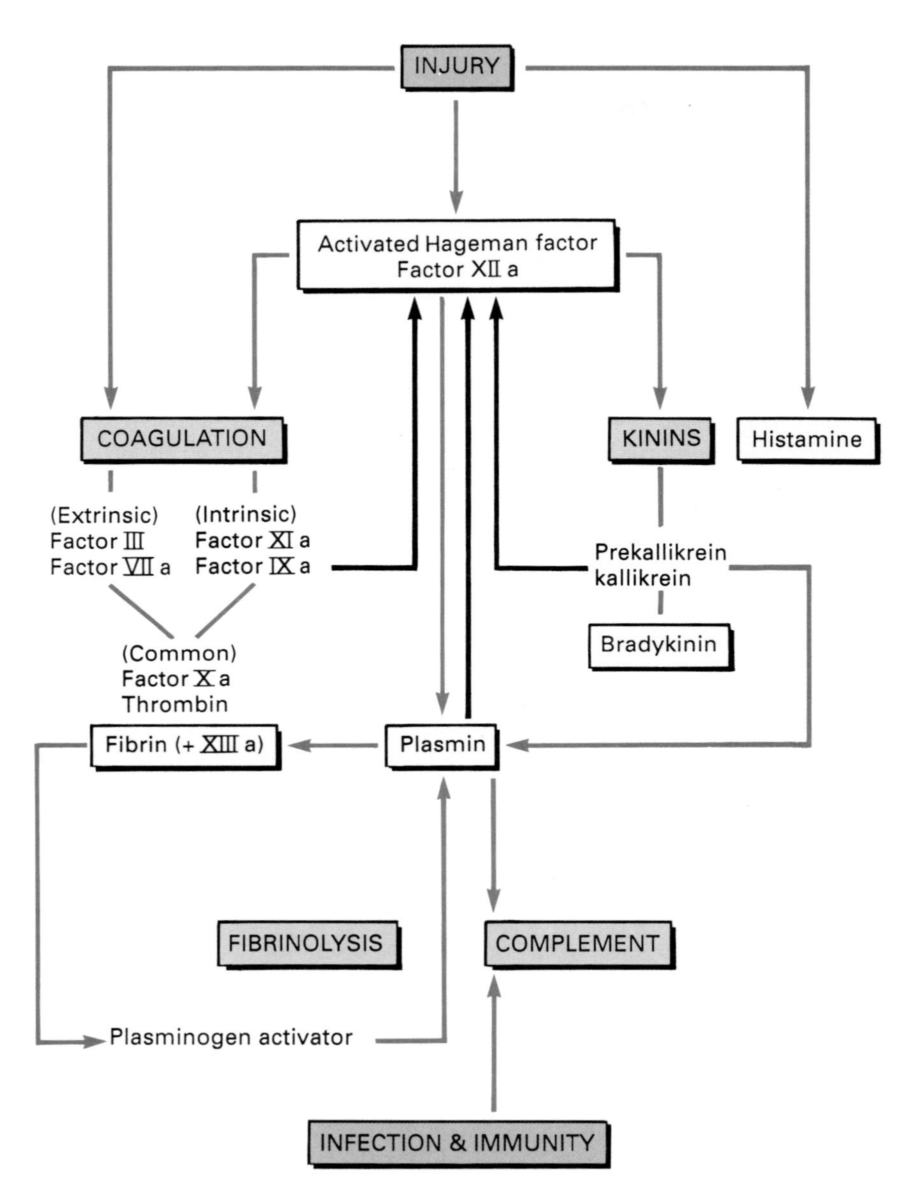

Fig.9.1 The interrelationship between the four plasma enzyme systems in inflammatory responses

activation of complement (Fig.9.1). Some products of complement activation have potent biological activity, including the ability to increase vascular permeability and to trigger mast cells to release histamine, resulting in vasodilation and the promotion of chemotaxis (cell movement) of inflammatory cells into the site of injury.

The **clotting system** is triggered very rapidly following injury, to prevent bleeding and further infection by forming clots in the damaged vessels (Fig.9.1). This can occur by two pathways: the **intrinsic pathway**, which is activated via the Hageman factor; and the **extrinsic pathway**, which is activated via tissue factor and factor VII (see Section 9.3.1). Clots are formed by the action of thrombin on fibrinogen and in part are composed of the protein fibrin. **Fibrin** acts as the beginnings of a scaffold on which tissues may subsequently be repaired and on which new capillaries can be constructed, a process known as **angiogenesis**. Although the rapid response of the coagulation pathway is essential, the extent of blood clotting must be limited so that it does not progress to undamaged vessels; in addition, the clots must ultimately be removed from the area of damage. This is controlled by **fibrinolysis** (fibrin breakdown) due to the enzyme **plasmin**. A number of different cell types are recruited into the area where damage has occurred, and these are responsible for removing the damaged tissues, inducing the formation of new tissue, and reconstructing the damaged cell matrix, including basement membranes and connective tissue. A new blood supply to the area is also established during the repair process.

If infection occurs as a consequence of the tissue damage, the innate and, later, the adaptive immune systems are triggered to destroy the infectious agent. For example, as part of the innate immune response, the complement cascade may be activated via the alternative pathway to destroy some micro-organisms. At the same time, cells such as polymorphonuclear leukocytes are recruited into the area, and this is followed by the recruitment of monocytes and macrophages, all of which can acquire potent anti-microbial activity (see Section 10.4). If the adaptive immune system is activated, B cells can make antibodies which aid the phagocytic cells to eliminate antigen and activate complement via the classical pathway, also leading to the destruction of the micro-organisms.

The most important point to note is that none of these mechanisms acts in isolation: products of the innate arm of the immune system act on cells of the adaptive immune system (see Section 1.1.2), while mechanisms that aid tissue repair can also help in overcoming infection.

9.2.2 The phases of inflammation

The body's response to injury and infection is manifest as **inflammation**. The main purpose of this immensely complex response seems to be to bring fluid, proteins, and cells from the blood into the damaged tissues. It should be remembered that the tissues are normally bathed in a watery fluid (extracellular lymph) that lacks most of the proteins and cells that are present in blood, since the majority of proteins are too large to cross the blood vessel endothelium. Thus there have to be mechanisms that allow cells and proteins to gain access to extravascular sites where and when they are needed if damage and infection has occurred. The main features of the inflammatory response are, therefore: **vasodilation**, i.e. widening of the blood vessels to increase the blood flow to the infected area; **increased vascular permeability**, which allows diffusible components to enter the site; and **cellular infiltration** by chemotaxis, or the directed movement of inflammatory cells through the walls of blood vessels into the site of injury. Of course, the degree to which these occur is normally proportional to the severity of the injury and the extent of infection

The classical signs of inflammation result, to a large degree, directly from these events. Four of the so-called five 'cardinal signs of inflammation' were recognized by the ancient Greeks and defined by Celsius. These are redness (rubor), swelling (tumour), heat (calor), and pain (dolor). Later the fifth sign, loss of function (functio laesa), was added. In general, it is now said that the redness and heat are caused mainly by the increased blood flow to the area; swelling and, in part, pain are caused by the increased permeability. The swelling and pain combined lead to loss of function. These are, of course, overgeneralizations, but help to chart the overt course of an inflammatory response.

Inflammation can be initiated by a variety of stimuli, including those we have discussed so far, namely trauma, necrosis (cell or tissue death), and infection. It is customary to divide the ensuing response into several phases; for the purposes of this discussion we shall consider the response to an external injury as an example. The earliest, gross event of an inflammatory response is temporary vasoconstriction, i.e. narrowing of blood vessels caused by contraction of smooth muscle in the vessel walls, which can be seen as blanching (whitening) of the skin. This is followed by several phases that occur over minutes, hours and days later, outlined below.

1. The **acute vascular response** follows within seconds of the tissue injury and lasts for some minutes. This results from vasodilation and increased capillary permeability due to alterations in the vascular endothelium, which leads to increased blood flow (hyperaemia) that causes redness (erythema) and the entry of fluid into the tissues (oedema). This phase of the inflammatory response can be demonstrated by scratching the skin with a finger-nail (the reader may wish to participate in this experiment). The 'wheal and flare reaction' that occurs is composed of (i) initial blanching of the skin due to vasoconstriction (ii) the subsequent rapid appearance of a thin red line when the capillaries dilate; (iii) a flush in the immediate area, generally within a minute, as the arterioles dilate; and (iv) a wheal, or swollen area that appears within a few minutes as fluid leaks from the capillaries.

2. If there has been sufficient damage to the tissues, or if infection has occurred, the **acute cellular response** takes place over the next few hours (hopefully the reader will not have acquired this degree of trauma!). The hallmark of this phase is the appearance of polymorphonuclear leukocytes, particularly neutrophils, in the tissues (see Fig.1.11). These cells first attach themselves to the endothelial cells within the blood vessels (margination) and they then cross into the surrounding tissue (diapedesis) (Fig.1.36). During this phase erythrocytes may also leak into the tissues and a haemorrhage can occur (e.g. a blood blister). If the vessel is damaged, fibrinogen and fibronectin are deposited at the site of injury, platelets aggregate and become activated, and the red cells stack together in what are called 'rouleau' to help stop bleeding and aid clot formation. The dead and dying cells contribute to pus formation.

3. If the damage is sufficiently severe, a **chronic cellular response** may follow over the next few days. A characteristic of this phase of inflammation

is the appearance of a mononuclear cell infiltrate composed of macrophages and lymphocytes. The macrophages are involved in microbial killing (as are the neutrophils), in clearing up cellular and tissue debris, and they also seem to be very important in remodelling the tissues (see Section 10.4).

4. Over the next few weeks, **resolution** may occur, meaning that the normal tissue architecture is restored. Blood clots are removed by fibrinolysis, and if it is not possible to return the tissue to its original form, *scarring* results from in-filling with fibroblasts, collagen, and new endothelial cells. Generally, by this time, any infection will have been overcome. However, if it has not been possible to destroy the infectious agents or to remove all of the products that have accumulated at the site completely, they are walled off from the surrounding tissue in *granulomatous tissue*. A **granuloma** is formed when macrophages and lymphocytes accumulate around material that has not been eliminated, together with epitheloid cells and giant cells (perhaps derived from macrophages) that appear later, to form a ball of cells.

Inflammation is often considered in terms of **acute inflammation** that includes all the events of the acute vascular and acute cellular response (1 and 2 above), and **chronic inflammation** that includes the events during the chronic cellular response and resolution or scarring (3 and 4).

In addition, a large number of more distant effects occur during inflammation. These include: the production of **acute phase proteins**, including complement components, by the liver; fever, caused by pyrogen acting on the hypothalamus in the brain; and systemic immunity, resulting in part from lymphocyte activation in peripheral lymphoid tissues.

9.2.3 Mediators of inflammation

Most of the events reviewed in the previous section are caused in large part by the action of various **inflammatory mediators** (Larsen and Henson, 1983). These are soluble, diffusible molecules that act locally at the site of tissue damage and infection, and at more distant sites, such as IL-1, IL-6, and TNF (see Section 9.5).

Bacterial products and toxins can themselves can act as **'exogenous' mediators** of inflammation. Notable amongst these is endotoxin, or lipopolysaccharide (LPS), which is one of the components of cell walls of Gram-negative bacteria. The immune system of higher organisms has probably evolved in a veritable sea of endotoxin, so it is perhaps not surprising that this substance evokes powerful responses. For example, endotoxin can trigger complement activation via the alternative pathway, resulting in the formation of the peptides C3a and C5a which cause vasodilation and increased vascular permeability. Endotoxin also activates the Hageman factor (factor XII), leading to activation of both the coagulation and fibrinolytic pathways as well as the kinin system. Endotoxins also elicit T cell proliferation, and have been described as 'superantigens' for T cells (see Section 5.2.5).

'Endogenous' mediators of inflammation are produced from within the (innate and adaptive) immune system itself, as well as other systems. For example, they can be derived from molecules that are normally present in the plasma in an inactive form, such as peptide fragments of some components of the complement, coagulation, and kinin systems. Mediators of inflammatory responses are also released at the site of injury by a number of cell types that either contain them as preformed molecules within storage granules, e.g. histamine, or which can rapidly switch on the machinery required to synthesize the mediators when they are required, for example to produce metabolites of arachidonic acid (see below).

Mononuclear phagocytes (monocytes and macrophages—see Section 10.4) are central to inflammation, as they produce many components which participate in or regulate the different plasma enzyme systems (Table 10.7), and hence the mediators of the inflammatory response. They are also actively phagocytic and are involved in microbial killing, as are polymorphonuclear neutrophils. While the latter can be thought of as short-lived 'kamikaze' cells that need to be continually replaced from the bone marrow, mononuclear phagocytes are long-lived and some can proliferate *in situ*. Other cells such as mast cells and basophils are much less phagocytic, but together with platelets, these cells are particularly important for secretion of vasoactive mediators. The function of these cell types is at least partially under the control of cytokines that are produced by lymphocytes and monocytes. All inflammatory cells have receptors for immunoglobulins and for complement components, and they possess specialized granules (Fig.1.11) containing an immense variety of products that are released perhaps by common mechanisms. Cytotoxic T cells and NK cells, in general, also possess granules which are important for their cytotoxic function (see Sections 10.2 and 10.3). In general, lymphocytes are involved much later in the **adaptive** response to infection, and the early events of inflammation are mediated in part by molecules produced by cells of the **innate** arm of the immune system.

Early phase mediators produced by mast cells and platelets

Mediators produced by **mast cells** and **platelets**, are especially important in **acute inflammation**. Some of these molecules contribute to vasodilation and vasopermeability, features of the acute vascular response (see Section 9.2.2), while others are responsible for recruiting other inflammatory cells to the area (Hanahan 1986; Braquet and Rola-Pleszczynski 1987).

Mast cells are easily recognized as their cytoplasm is full of membrane-bound granules containing preformed mediators (the word 'mast' is German for 'forced fattening'). Some of these are **vasoactive** and have very potent effects on smooth muscle cells. Mast cells are often situated next to arterioles and in submucosal sites, so that when injury triggers the release of the contents of the granules, the muscle wall of the arterioles relaxes causing the vessels to dilate, although the opposite effect (contraction) occurs in smooth muscle in other sites such as the airways and bronchi.

Platelets may contribute to inflammatory responses resulting as a consequence of tissue injury, through a variety of mechanisms including: (i) the release of vasoactive amines and other permeability factors; (ii) the release of lysosomal enzymes; (iii) the release of coagulation factors which lead to localized and generalized fibrin deposition; and (iv) the formation of platelet aggregates or thrombi which result in the blocking of vessels and capillaries.

The most important vasoactive mediators that are stored in mast cell granules, and are also present in platelets, are **histamine** in man, as well as **serotonin** or 5-hydroxytryptamine (5-HT) in rodents. Histamine release results in a series of responses consisting of: (i) primary, local dilation of small vessels; (ii) widespread arteriolar dilatation; (iii) local increased vascular permeability; and (iv) the contraction of nonvascular smooth muscle (Beer and Rocklin 1987). A number of different cells of the body have receptors for histamine. These can be of two types. The H_1 receptors mediate acute vascular effects together with smooth muscle constriction in the bronchi (histamine acts as a 'spasmogen'). In contrast, the H_2 receptors mediate a number of anti-inflammatory effects but cause the vasodilation. 5-HT is also capable of increasing vascular permeability, dilating capillaries and producing contraction of nonvascular smooth muscle.

Late phase mediators—prostaglandins and leukotrienes

The release of preformed histamine and serotonin and the cytokines, IL-1, IL-6, and TNF, is responsible for the very early events in inflammation, but vascular events also occur later (from about 6–12 hours after initiation). In part, the later vascular events are mediated by products of **arachidonic acid**. Arachidonic acid is derived from the metabolism of membrane phospholipids by the action of the enzymes phospholipase A_2 and C (Pruzanski and Vadas 1991). Arachidonic acid has a short half-life and the main products of its metabolism are **prostaglandins** and **leukotrienes**. These molecules are synthesized and released by macrophages when, for example, ligation of the Fc receptors (FcR) occurs or when both the mannosyl fucosyl receptors (MFR) and complement receptors (CR) are engaged (see Section 10.4.5). Arachidonic acid metabolites are not stored in significant quantities and all have to be newly synthesized before secretion.

Arachidonic acid can be metabolized by two major routes, the **cyclo-oxygenase** and **lipoxygenase pathways**. The cyclo-oxygenase pathway produces prostaglandins, prostacyclin, and thromboxanes (Figs.9.2, and e.g. 9.3); the lipoxygenase pathway produces leukotrienes, including slow-reacting substance (SRS, a mixture of two leukotrienes, LTC4 and LTD4), and hydro(pero)xyeicosatetranoate derivatives (H(P)ETEs) (Figs.9.4, and e.g. 9.5). The principal effects of these mediators are on blood vessel walls and muscles, but they can also act on other cell types as chemo-attractants and/or induce degranulation and the release of preformed mediators.

Prostaglandins. The prostaglandins (PG) are a family of lipid-soluble hormone-like molecules synthesized from arachidonic acid, a 20-carbon, unsaturated fatty acid (Fig.1.5). There are at least 16 different prostaglandins and these can be divided into nine different chemical classes, designated PGA to PGI; each contains a five-membered cyclopentane ring. The structure of PGE_2 is shown in Fig.9.3. Different prostaglandins are produced by different cell types in the body. For example, macrophages and monocytes are large producers of both PGE_2 and PGF_2, neutrophils produce moderate amounts of PGE_2, mast cells produce PGD_2, and lymphocytes may produce PGE_2 but this is controversial. It is important to note that, unlike histamine, prostaglandins do not exist free in tissues, but have to be synthesized and released in response to an appropriate stimulus. Prostaglandins are found in many human tissues, and are thought to act as important modulators and mediators of the inflammatory response by virtue of their potent biological activity.

The biological activity of the PGs is incredibly diverse, but as far as immunoregulation is concerned, PGE_2 has been most extensively studied (Phipps *et al.* 1991). It stimulates differentiation of immature thymocytes, B cells, and other haemopoietic precursors by the bone marrow. However, PGE_2 is inhibitory to mature cells and suppresses a wide variety of functions, including lymphocyte proliferation, CTL and NK cytotoxicity, leukocyte locomotion, and cell–cell interactions (including platelet aggregation). It also inhibits the release of other inflammatory mediators and interleukins. Nevertheless, prostaglandins can also promote inflammatory reactions, and PGE_2, for example, increases the chemotactic and vasoactive properties of several other mediators. Local injection of PGE_2 has been shown to produce the classic signs of an inflammatory response, as outlined above (Section 9.2.2), and it has also been shown to result in collagen biosynthesis, which may contribute to chronic inflammation. As a result of these complex activities, it seems most likely that prostaglandins can regulate the balance of inflammatory and non-inflammatory events.

It is notable that the inflammatory macrophage has an increased secretory activity for many substances, but it produces a much lower level of prostaglandins; this may facilitate the subsequent induction of immune responses in which the macrophage participates after activation (see Section 10.4).

Thromboxane A_2 (TXA_2) and prostacyclin (PGI_2). TXA_2 is produced by monocytes and macrophages, as well as by platelets. It causes platelets to aggregate and constricts blood vessels and airways. These effects are somewhat opposed by the action of prostacyclin (PGI_2) (Fig.9.2).

Leukotrienes. The enzyme 5′-lipoxygenase converts arachidonic acid into LTA4, which is the precursor to other leukotrienes including LTB4, LTC4, and LTD4. The mixture of LTC4 and LTD4, originally called slow reacting substance (SRS) of anaphylaxis, is produced by a wide variety of cells, including monocytes and macrophages. The mixture is spasmogenic and causes contraction of smooth muscle, for example in the bronchus, and it has effects on mucous secretion; all of these events occur in asthma for example (see Section 10.5). When LTC4 and LTD4 act individually rather than as a mixture, they can have opposing effects on blood vessels in skin, causing constriction and dilation, respectively. Both may also have distant effects on the nervous system.

LTB4 (Fig.9.5), as well as 5-hydroxyeicosatetranoate (5-HETE), causes the chemotaxis (directed locomotion) and/or chemokinesis (general cell movement) of a number of cell types including neutrophils. These mediators also induce degranulation of mast cells, which causes histamine to be released with its subsequent effects (histamine is discussed above).

Late-phase mediators of the inflammatory response are also produced by activation of the coagulation system (fibrin split products) and the kinin system (particularly bradykinin, that also acts on nerve endings to cause the sensation of pain), and from triggering of the complement cascade (anaphylatoxins). All of these pathways will be outlined or discussed in more detail later in this chapter. Some of these products are also **chemotactic factors**, that recruit cells and initiate the acute cellular phase.

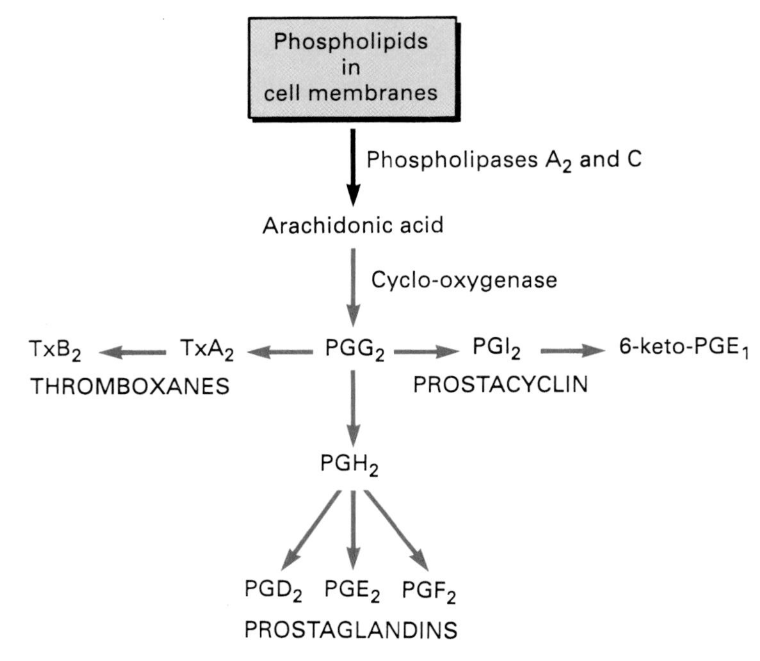

Fig.9.2 The cyclo-oxygenase pathway results in the generation of prostaglandins, prostacyclin, and thromboxane from arachidonic acid

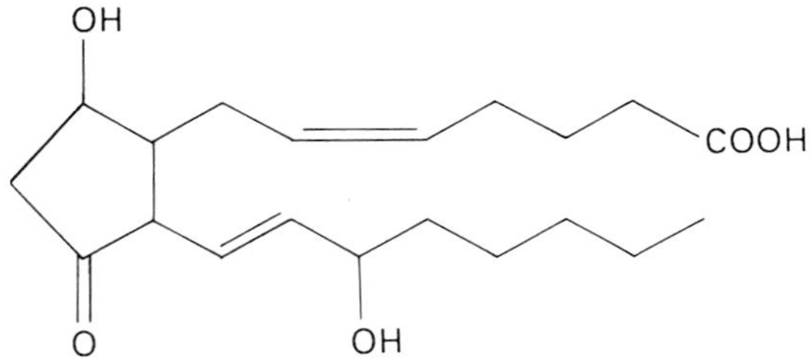

OH COOH O OH

Fig.9.3 An example of a prostaglandin. Structure of prostaglandin E_2 (PGE_2)

Mediators released by neutrophils

Neutrophils can be attracted to sites of injury and infection during the acute cellular phase of the inflammatory response by a number of different molecules, in addition to those from the systems we have just noted (above). For example, they can be attracted by degradation products of cell matrix proteins such as collagen and fibronectin, or by fibrin peptides. Molecules that are released by micro-organisms are also chemotactic for neutrophils, one of the most important classes being the f-Met-peptides which start with the concensus sequence f-Met-leu-phe.

Neutrophils have at least three different types of granules in their cytoplasm. Some of these contain powerful hydrolytic enzymes that are released when the cell phagocytoses bacteria, for instance. These enzymes are not only toxic to some micro-organisms but they also cause digestion of tissue in their vicinity and lead to **fibrinoid necrosis**. Neutrophils can also produce

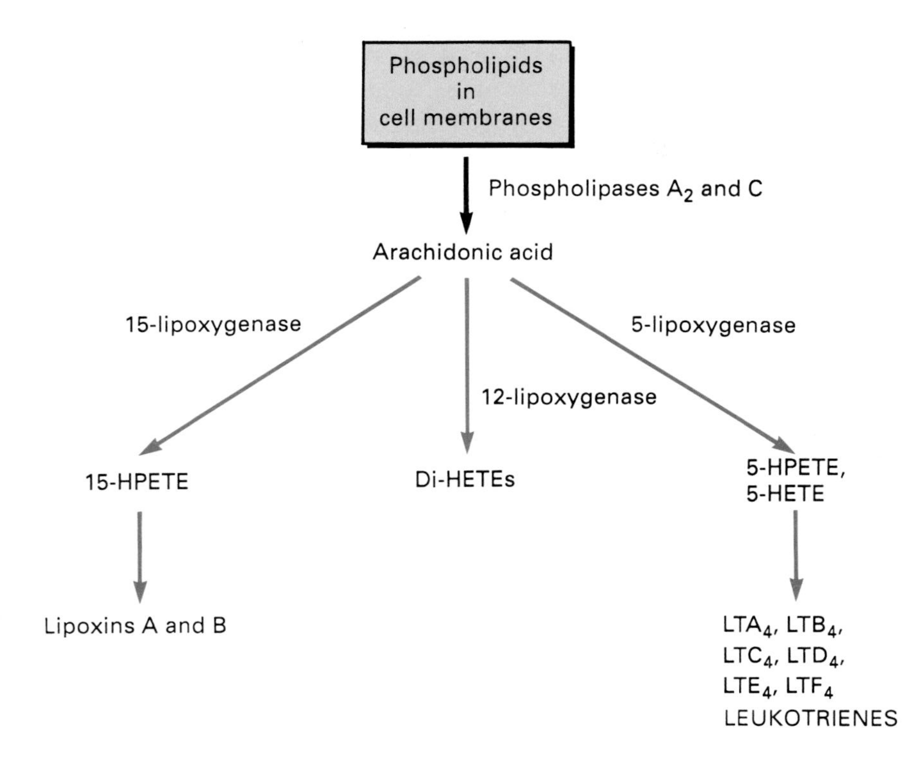

Fig.9.4 The lipoxygenase pathway results in the generation of hydroxyeicosatetranote (HETE) derivatives and leukotrienes (LT) (HPETE, hydroperoxyeicosatetranoate)

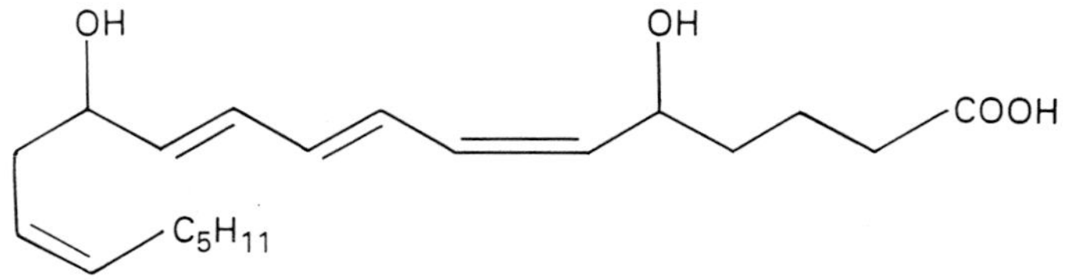

Fig.9.5 An example of a leukotriene. Structure of leukotriene B_4 (LTB_4)

'reactive oxygen intermediates' (see Section 10.4.4), that can be toxic to bacteria but also damage the neutrophils. The accumulation of dead and dying cells and micro-organisms, together with accumulated fluid and various proteins and other products in the area makes up what is seen as pus.

Involvement of the adaptive immune response

If an adaptive immune response has been induced, the humoral effector mechanisms (for example antibody and complement) can act in conjunction with neutrophils, macrophages, and other cells to bring about destruction of the infectious agent, since all of them possess antibody Fc receptors and complement receptors of one type or another (see Section 10.4.5).

Cells that may have been infected are attacked by cytotoxic T cells, and by other killer cells that were active before the T cells were activated. These cells, and macrophages, are discussed in Chapter 10.

9.3 The plasma enzyme systems

The four plasma enzyme systems discussed in this chapter are the coagulation (clotting), fibrinolytic, kinin, and complement systems. The first three are outlined in this section; complement is considered in Section 9.4

9.3.1 The coagulation mechanism

The blood clotting system or coagulation pathway, like the other plasma enzyme systems discussed in this chapter, is a **proteolytic cascade**. Each enzyme of the pathway is present in the plasma as a zymogen, in other words in an inactive form, which on activation undergoes proteolytic cleavage to release the active factor from the precursor molecule. The coagulation pathway functions as a series of positive and negative feedback loops which control the activation process. The ultimate goal of the pathway is to produce thrombin, which can then convert soluble fibrinogen into fibrin, which forms a clot (Fig.9.6). The generation of thrombin can be divided into three phases, the **intrinsic** and **extrinsic pathways** that provide alternative routes for the generation of factor X, and the final **common pathway** which results in thrombin formation. The sequence of events in the three parts of the clotting cascade are outlined below. The properties of some components of the clotting cascade are listed in Table 9.1.

The intrinsic pathway

The intrinsic system is an enzyme cascade that is activated when blood comes into contact with sub-endothelial connective tissues or with negatively charged surfaces that are exposed as a result of tissue damage. Quantitatively it is the most important of the two pathways, but is slower to cleave fibrin than the extrinsic pathway. The Hageman factor (factor XII; the two terms will be used interchangeably), factor XI, prekallikrein, and high molecular weight kininogen (HMWK) are involved in this pathway of activation. Thus this pathway provides a further example of the interrelationship between the various enzyme cascade systems in plasma. The events that occur during activation via the intrinsic system are shown in Fig.9.6. The first step is the binding of Hageman factor to a sub-endothelial surface exposed by an injury. A complex of prekallikrein and HMWK also interacts with the exposed surface in close proximity to the bound factor XII, which becomes activated.

The Hageman factor is a single chain protein, (80 kDa) that becomes activated on contact with certain negatively charged surfaces. The surface contact is important to the activation process, as it appears to induce a conformational change in the native Hageman factor, increasing the rate at which it is activated. During activation the Hageman factor is initially cleaved into two chains, (50 and 28 kDa), that remain linked by a disulfide bond. The light chain (28 kDa) contains the active site and the molecule is referred to as activated Hageman Factor (HFa) or factor XIIa. There is evidence that the Hageman factor can autoactivate, thus the pathway is self-amplifying once triggered (compare with the alternative pathway of

Table 9.1 Characteristics of some of the coagulation factors

Factor		Mol. wt (kDa)	Origin	Normal plasma concentration (μg/ml)	Function
I	Fibrinogen	340	liver	3000	Precursor of fibrin
II	Prothrombin	70	liver	100	Converts fibrinogen to fibrin: activates factors I, V, VIII, and XIII
V	Proaccelerin; labile factor	350	liver	15	Cofactor for the activation of factor II by factor Xa
VII	Proconvertin; stable factor	50	liver	0.3	Activates factors IX and X
VIII	Antihaemophilic factor	350	?	0.2	Cofactor for activation of factor X by factor IXa
IX	Christmas factor	55	liver	5	Activates factor X
X	Stuart–Prower factor	55	liver	10	Activates factor II
XI	Plasma thrombospondin antecedent	160	liver	5	Activates factors IX and XII, and prekallikrein
XII	Hageman factor	75	liver	30	Activates factor XI, prekallikrein, and plasminogen
XIII	Fibrin stabilizing factor	330	liver; platelets	20	Cross-links fibrin
Prekallikrein		85	liver	20	Precursor of kallikrein; activates factor XII, prekallikrein; cleaves HMWK
High molecular weight kininogen (HMWK)		150	liver	20	Cofactor for activation of factors XI, XII, and prekallikrein

complement, Section 9.4.5). In addition, the surface itself may create a favourable environment by increasing the local concentration of the three molecules required for activation and thus accelerating the reaction.

Activated Hageman factor in turn activates prekallikrein. The kallikrein produced can then also cleave factor XII, and a further amplification mechanism is triggered. The activated factor XII remains in close contact with the activating surface, such that it can activate factor XI, the next step in the intrinsic pathway which, to proceed efficiently, requires Ca^{2+}. Also involved at this stage is HMWK, which binds to factor XI and facilitates the activation process by bringing the two factors (XIIa and XI) into close proximity; in the absence of HMWK, activation proceeds very slowly. Activated factors XIa, XIIa, and kallikrein are all serine proteases, like many of the enzymes of the complement system. They are all present in plasma as inactive molecules, zymogens, and are converted to their active forms by limited proteolysis. The order or sequence of events is controlled because these enzymes are highly substrate specific and will only act on the next molecule in the sequence, thus factor XIa acts on factor IX and so on.

Eventually the intrinsic pathway activates factor X, a process that can also be brought about by the extrinsic pathway. Factor X is the first molecule of the common pathway (see below) and is activated by a complex of molecules

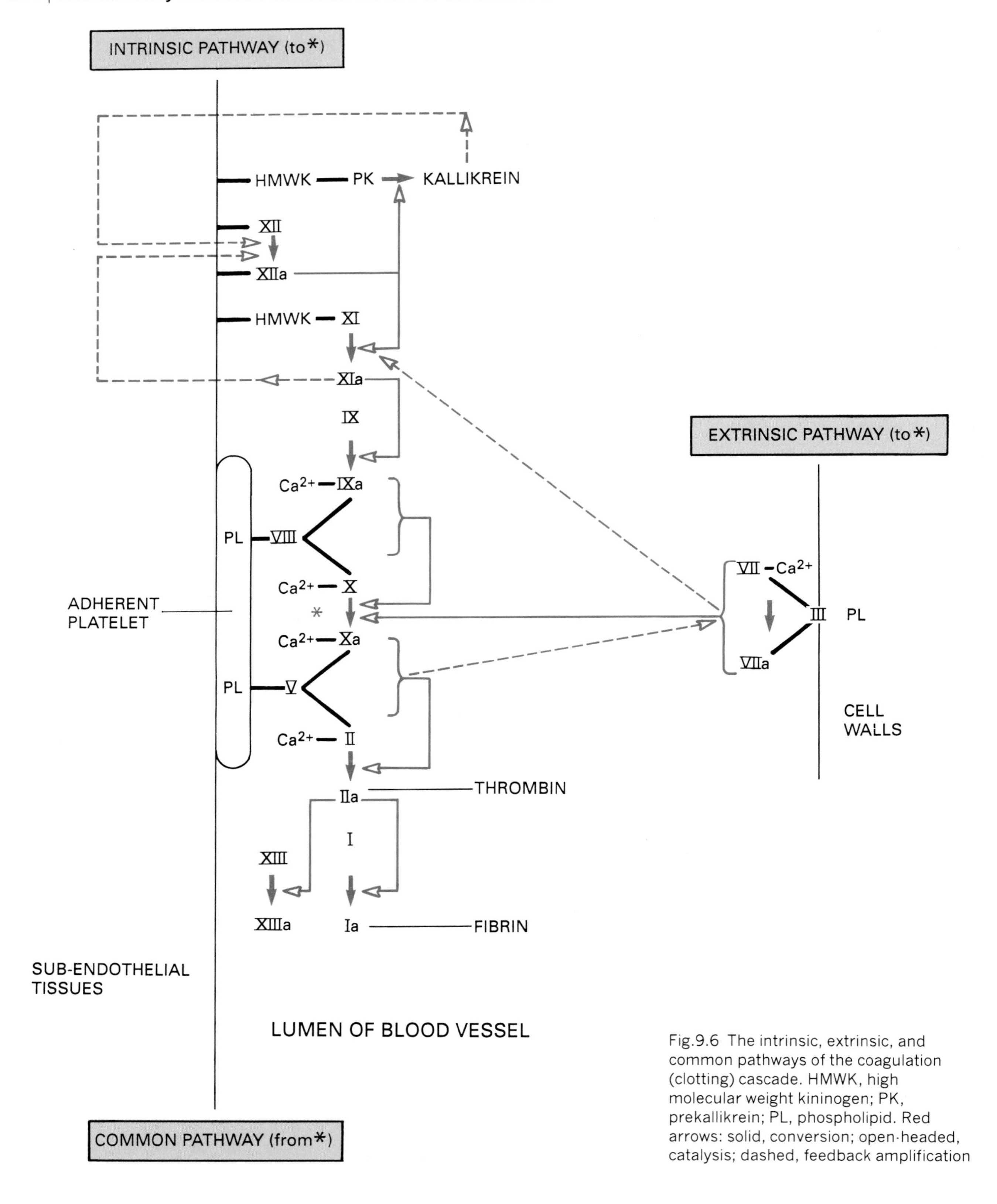

Fig.9.6 The intrinsic, extrinsic, and common pathways of the coagulation (clotting) cascade. HMWK, high molecular weight kininogen; PK, prekallikrein; PL, phospholipid. Red arrows: solid, conversion; open-headed, catalysis; dashed, feedback amplification

containing activated factor IX, factor VIII, calcium, and phospholipid which is provided by the *platelet* surface, where this reaction usually takes place. The platelet also provides a surface for the assembly of the complex and a protected environment away from the action of the plasma inhibitors that inhibit the clotting pathway. The precise role of factor VIII in this reaction is not clearly understood. Its presence in the complex is obviously essential, as evidenced by the serious consequences of factor VIII deficiency experienced by haemophiliacs. It may simply act as a stabilizing protein, as it has no intrinsic enzyme activity. Factor VIII is modified by thrombin (Fig.9.7), a reaction that results in greatly enhanced factor VIII activity, promoting the activation of factor X.

The extrinsic pathway

The extrinsic pathway is an alternative route for the activation of the clotting cascade. It provides a very rapid response to tissue injury, generating activated factor X almost instantaneously, compared to the seconds or even minutes required for the intrinsic pathway to activate factor X. The main function of the extrinsic pathway is to *augment* the activity of the intrinsic pathway. For an outline of the pathway see Fig.9.6.

There are two components unique to the extrinsic pathway, tissue factor or factor III, and factor VII. Tissue factor is present in most human cells bound to the cell membrane. The activation process for tissue factor is not entirely clear, and there is debate as to whether it is released from cells by mechanical means following tissue damge, or whether it is activated by proteases released during the trauma. Once activated, tissue factor binds rapidly to factor VII which is then activated to form a complex of tissue factor, activated factor VII, calcium, and a phospholipid, and this complex then rapidly activates factor X. There is reciprocal activation between the two factors generating factor VIIa and Xa. The amounts of activated factor VII increase dramatically once the activation has been triggered, due to activation by factor Xa, factor V, and phospholipid complex, which act as a positive feedback pathway.

The common pathway

The intrinsic and extrinsic systems converge at factor X to a single pathway which is ultimately responsible for the production of thrombin (factor IIa).

Clot formation

The end result of the clotting pathway is the production of thrombin for the conversion of fibrinogen to fibrin (Fig.9.7). Fibrinogen is a dimer, each half being composed of three paired chains. Fibrinogen is soluble in plasma. Exposure of fibrinogen to thrombin results in rapid proteolysis of fibrinogen and the release of fibrinopeptide A, a small peptide from the end of one pair of the fibrinogen polypeptide chains, the α chains. The loss of peptide A alone is not sufficient to render the resulting fibrin molecule insoluble, a process that is required for clot formation, but it tends to form complexes with adjacent fibrin and fibrinogen molecules. A second peptide, fibrinopeptide B, is then cleaved from the second polypeptide, the β chain, by thrombin, and the fibrin monomers formed by this second proteolytic cleavage polymerize

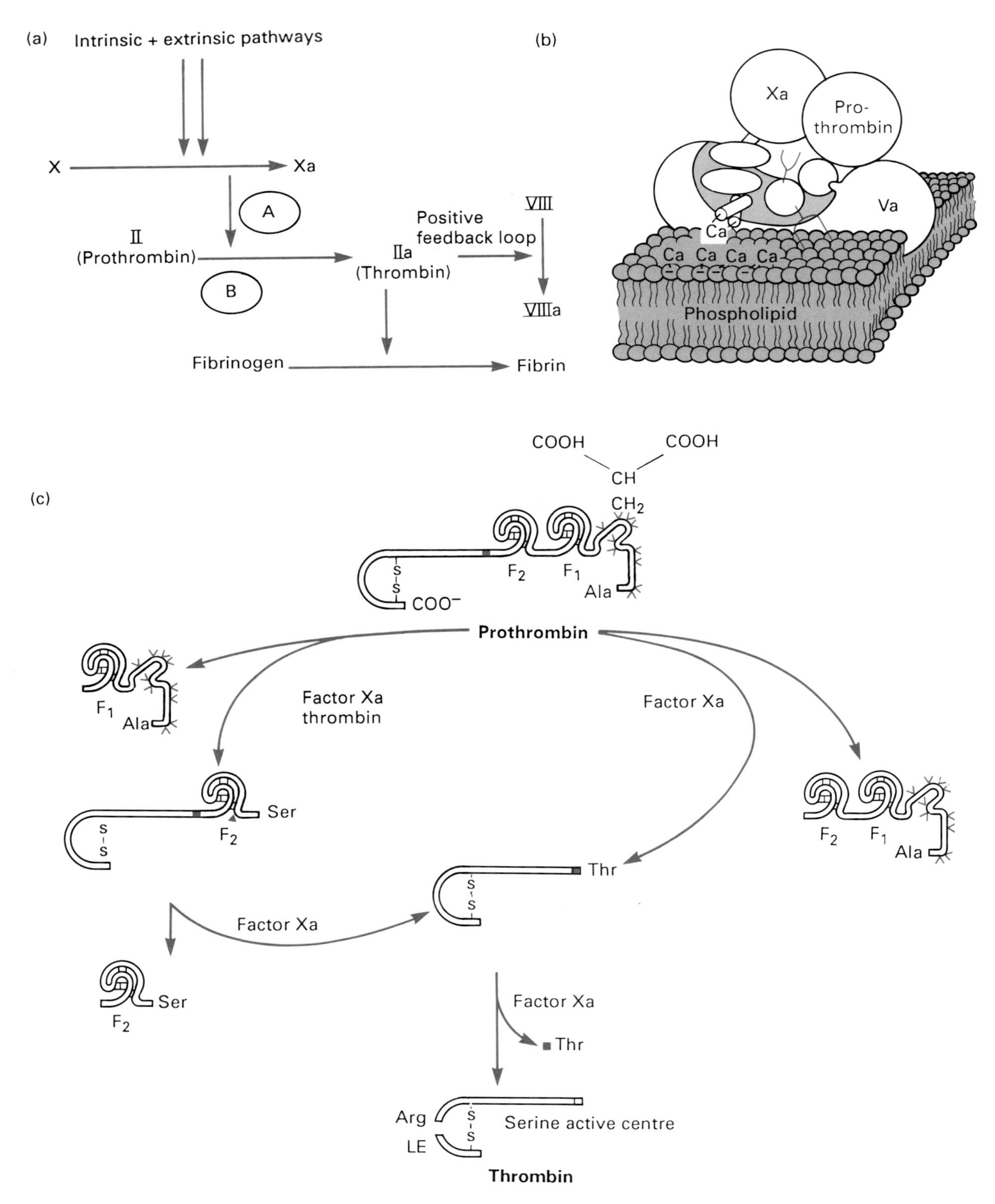

Fig.9.7 The common pathway. (a) An outline of the pathway. (b) A schematic drawing of the prothrombinase-converting complex, showing the possible interactions of prothrombin, factor Xa and factor Va. Cylinders represent the helices of a calcium-binding unit. (c) Events leading to the production of thrombin

spontaneously to form an insoluble gel. The polymerized fibrin, held together by noncovalent and electrostatic forces, is stabilized by the transamidating enzyme factor XIIIa, produced by the action of thrombin on factor XIII. These insoluble fibrin aggregates (clots), together with aggregated platelets (thrombi), block the damaged blood vessel and prevent further bleeding (see Section 9.2.1).

Interrelationships between the coagulation pathway and other plasma enzyme systems

Contact activation of the coagulation pathway, in addition to promoting blood clotting, results in the generation of plaminogen activator activity, which is involved in fibrinolysis or clot removal (see Section 9.3.2). Activated Hageman factor and its peptides can also initiate the formation of kallikrein from plasma kallikrein, and this triggers the release of bradykinin from kininogens in the plasma (Fig.9.1 and Section 9.3.3). Kinins are responsible for dilating small blood vessels, inducing a fall in blood pressure, triggering smooth muscle contraction, and increasing the permeability of vessel walls. In addition, activation of the coagulation pathway produces a vascular permeability enhancing factor, as well as chemotactic peptides.

It is obviously highly efficient to have the coagulation, fibrinolytic and kinin pathways, together with the initiation of inflammatory responses, linked through a single triggering event following tissue injury.

9.3.2 Fibrinolysis

Once haemostasis is restored and the tissue is repaired, the clot or thrombus must be removed from the injured tissue. This is achieved by the **fibrinolytic pathway**. The end product of this pathway is the enzyme plasmin, a potent proteolytic enzyme with a broad spectrum of activity. Plasmin is not only important in the fibrinolytic sequence, but it also forms a link with the other proteolytic cascade systems in plasma, for example it has been implicated in the activation of the alternative pathway of complement (see Section 9.4.5).

Plasmin is formed by activation of the proenzyme, plasminogen (Fig.9.8), by either plasma or tissue activators. The relative importance of these two activation mechanisms is not clear. Tissue plasminogen activators (Fig.9.9) are found in most tissues, except the liver and the placenta, where they are synthesized by endothelial cells and are found concentrated in the walls of blood vessels. The two best characterized are vascular activator (commonly known as tissue plasminogen activator—tPA) and urokinase. There is great interest in using tPA as a therapeutic agent for dissolving blood clots: the gene for tPA has now been cloned and the expressed gene product is available for clinical trials. Plasminogen activator is also a product of macrophages (Table 10.7). The level of tissue activator in the plasma is normally low, but can be increased by exercise and stress, and it is likely that most of the activator activity present in the plasma is released from vessel endothelium in response to fibrin.

Two forms of plasminogen are present in the plasma; one has a glutamic acid at the N-terminal of the polypeptide chain, and is called native or

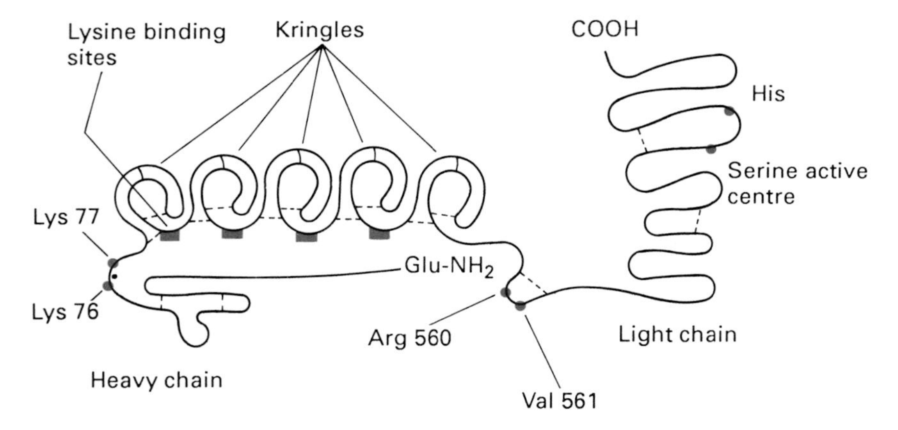

Fig.9.8 Diagrammatic representation of a plasminogen molecule (– – – represent disulfide bonds)

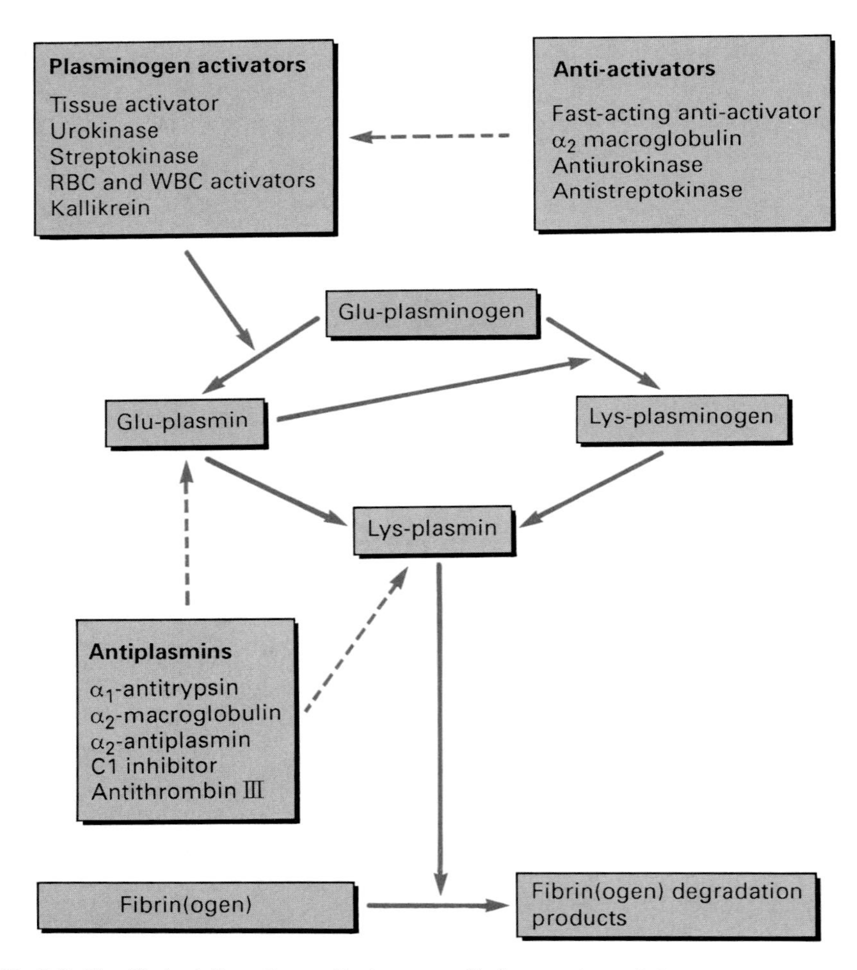

Fig.9.9 The fibrinolytic pathway. Facing page: fibrinogen degradation

glu-plasminogen, and the other a lysine. The latter form arises as a result of partial degradation of the parent molecule by autocleavage.

The structure of glu-plasminogen is shown in Fig.9.8. It contains five homologous looped structures known as kringles, and four of them have a lysine binding site which allows the molecule to interact with its substrate, for example fibrin and its inhibitors.

Triggering of fibrinolysis occurs when the plasminogen activator, plasminogen, and fibrin are all in close proximity. Both plasminogen and its activator bind avidly to fibrin as the clot forms. This close association prevents inhibition of plasmin activity by inhibitor, and allows proteolysis of the fibrin to proceed after the production of lys-plasminogen (Fig.9.9).

Conversion of plasminogen to plasmin can proceed via two routes. Most plasminogen activators cleave the Arg_{560}-Val_{561} bond (Fig.9.8) to form glu-plasmin. The two polypeptides produced by this cleavage event remain linked together by a disulfide bond. The C-terminal peptide contains the serine protease active site, but glu-plasmin is functionally ineffective in spite of this. To be functionally active it has to be converted to lys-plasmin by further proteolysis, predominantly between Lys_{76} and Lys_{77} (Fig.9.8). Both glu- and lys-plasmin attack the Lys_{76-77} bond in glu-plasminogen to form lys-plasminogen. Lys-plasminogen binds to fibrin before it develops protease activity.

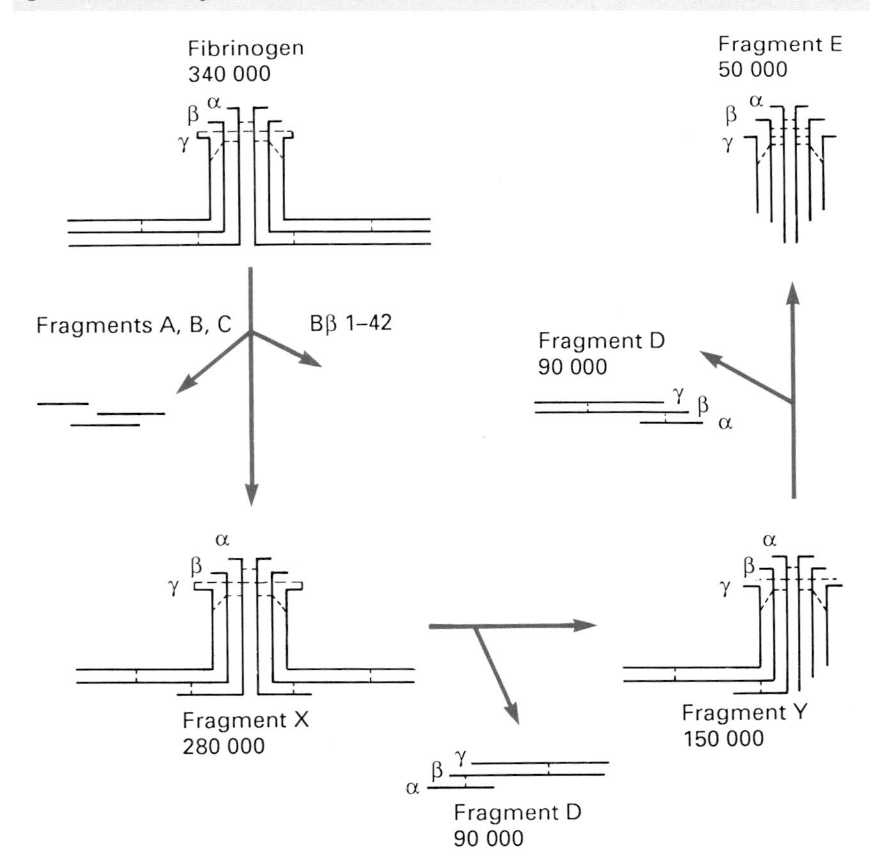

Plasmin attacks fibrin at a number of different sites, at least 50, reducing its size such that it no longer has haemostatic activity (e.g. Fig.9.9). Many fragments are formed during this process, and some retain the capacity to polymerize, thus some of the early degradation products can compete with fibrinogen for thrombin and act as inhibitors of clot formation. This may prevent the clot being removed before the tissue is repaired, but it is not clear whether this mechanism is important in the overall control of the clotting pathway.

9.3.3 Kinin formation

The kinins, bradykinin and lysylbradykinin, are important mediators of inflammatory responses. They are liberated from precursor molecules, kininogens, by the action of various proteases, collectively known as kininogenases. Three types of kininogen have been identified: high- and low-molecular weight kininogen (HMWK and LMWK respectively), and T-kininogen. These molecules are synthesized by hepatocytes and are released into the plasma, where in addition to releasing kinins, they function as (i) cofactors in the coagulation pathway (see Section 9.3.1); (ii) inhibitors of cysteine protease enzymes; and (iii) part of the acute phase response. The kinins are potent vasoactive basic peptides and their properties are wide ranging, including the ability to increase vascular permeability, cause vasodilation, pain, and the contraction of smooth muscle, and to stimulate arachidonic acid metabolism (see Section 9.2.3).

Three different pathways may lead to kinin formation during inflammation: (i) the generation of bradykinin as a result of activation of the Hageman factor (factor XII) and the production of plasma kallikrein; (ii) the production of lysylbradykinin by tissue kallikreins; and (iii) the action of cellular proteases in kinin formation. As this chapter is primarily concerned with the interrelationship of the plasma enzyme systems in inflammatory responses, this section will focus on the first of these three routes. For a review of the kinin system see Proud and Kaplan (1988).

The mechanism of bradykinin formation in plasma and in tissues is summarized in Fig.9.10. The generation of bradykinin in plasma is dependent on the interaction of the Hageman factor, prekallikrein, and HMWK with certain negatively charged surfaces. As mentioned above, these molecules are also the factors required for triggering the coagulation and fibrinolytic systems (Fig.9.6). In brief, HMWK and prekallikrein circulate in plasma as a 1:1 stoichiometric complex. This complex, together with the Hageman factor, binds to negatively charged surfaces or collagen. Once they are exposed by tissue damage, the Hageman factor is activated, prekallikrein is converted to kallikrein, and HMWK itself is digested to release bradykinin, a nine amino acid peptide. As is found for other vasoactive peptides, for example C3a, C4a, and C5a of the complement system, bradykinin contains a C-terminal arginine residue. (Kallikrein and other plasma enzymes, including plasmin, can also activate Hageman factor, see Fig.9.1, although autoactivation is potentially of greater significance than either of these other relatively inefficient processes. Cl inhibitor of the complement cascade, see Section 9.4.4, functions as a control protein for this event.)

As bradykinin is such a potent vasoactive peptide, its activity and its formation must be carefully controlled. Activation of the pathway is controlled internally, by the exposure of a suitable surface on which activation can proceed, and externally by the presence of inhibitors for each of the active components. C1 inhibitor controls the activity of the activated Hageman factor as mentioned above, while α_2-macroglobulin and C1 inhibitor act as kallikrein inhibitors. There are a variety of enzymes in plasma that control bradykinin activity, including carboxypeptidase N, which removes the C-terminal arginine residue, thus inactivating the peptide (Fig.9.10).

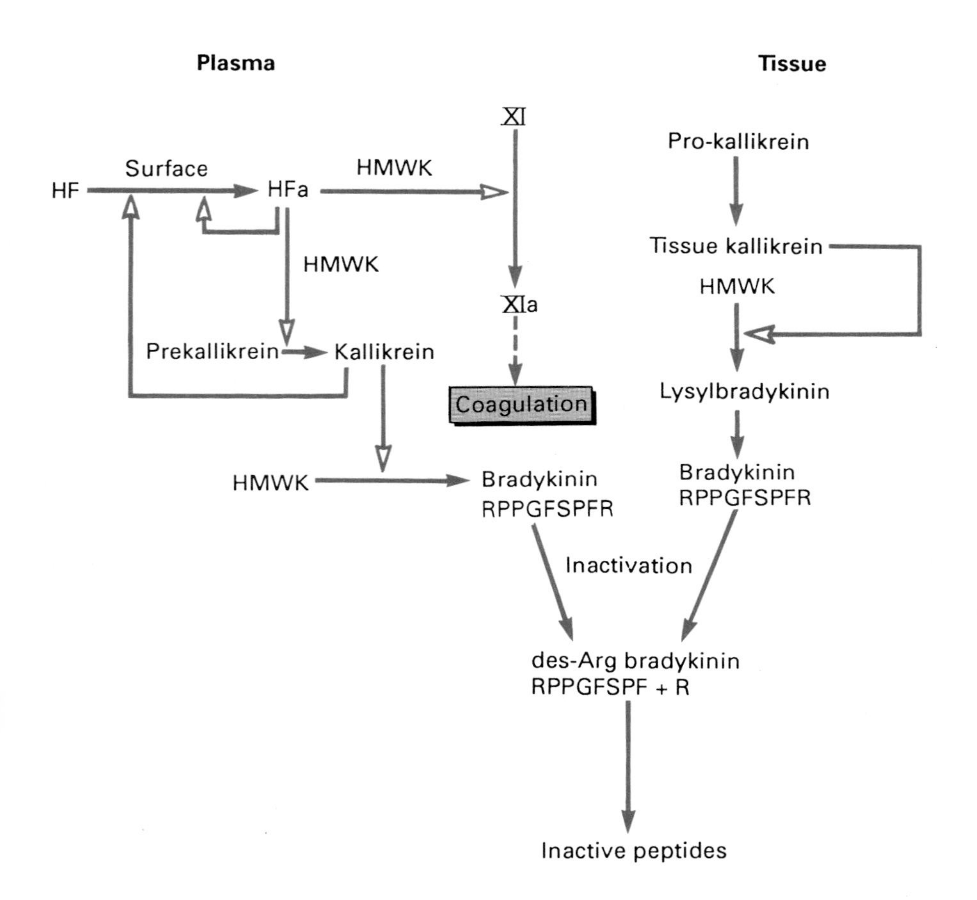

Fig.9.10 The kinin pathway. HF, Hageman factor; HMWK, high molecular weight kininogen

9.4 Complement

9.4.1 Overview

Complement components mediate several effector mechanisms of the immune response. Activation of the complement cascade facilitates the removal of foreign antigens from the body, and augments the activity of the humoral arm of the immune response.

Complement is a proteolytic cascade system present in the blood plasma of all vertebrates. It is composed of a series of proteins and glycoproteins which interact with one another at various stages during the cascade reaction. The majority of the complement components are present in the plasma in their inactive, precursor forms. Initiation of the complement cascade results in the sequential activation of each of the components in the pathway.

The main functions of the complement cascade are:

(1) **opsonization**—removal of foreign antigens coated with antibody and/or complement components by phagocytic cells, e.g. macrophages;

(2) **lysis** of invading micro-organisms and infected cells;

(3) **inflammation**—activation of the complement cascade results in the production of biologically active peptides with anaphylactic (anaphylatoxic) or chemotactic activity (Frank and Fries 1991). (**Anaphylaxis** is the term used to describe the inflammatory events which result in vasodilation and smooth muscle contraction; **chemotaxis** is the chemically directed migration of cells, see Section 9.2);

(4) **solubilization of immune complexes**.

Complement can be activated by two different routes, one termed the **classical pathway** and the other the **alternative pathway**. Activation of the complement system via either route can result in the same end-stage reaction, that is, the generation of a **membrane attack complex** which can mediate cell lysis. However, the initial events of each pathway are different (Fig.9.11). Activation of the classical pathway of complement always results in concomitant activation of the alternative pathway. However, the converse situation does not arise. Deficiencies in complement components can occur in some individuals, and investigation of these patients has helped to elucidate the events which occur during complement activation (Colten and Rosen 1992).

9.4.2 Nomenclature

The proteins known to be associated with the activation and regulation of the complement cascade are listed in Table 9.2. A general scheme for activation of complement components is in Fig.9.12, and the structure predicted for some of the molecules is shown in Fig.9.13 and in other figures throughout this section.

The components of the classical pathway are numbered 1 to 9. Each individual component is given a number, prefixed by the letter 'C', e.g. C1, C4, or C9, etc. (Table 9.2). The early components of the alternative pathway

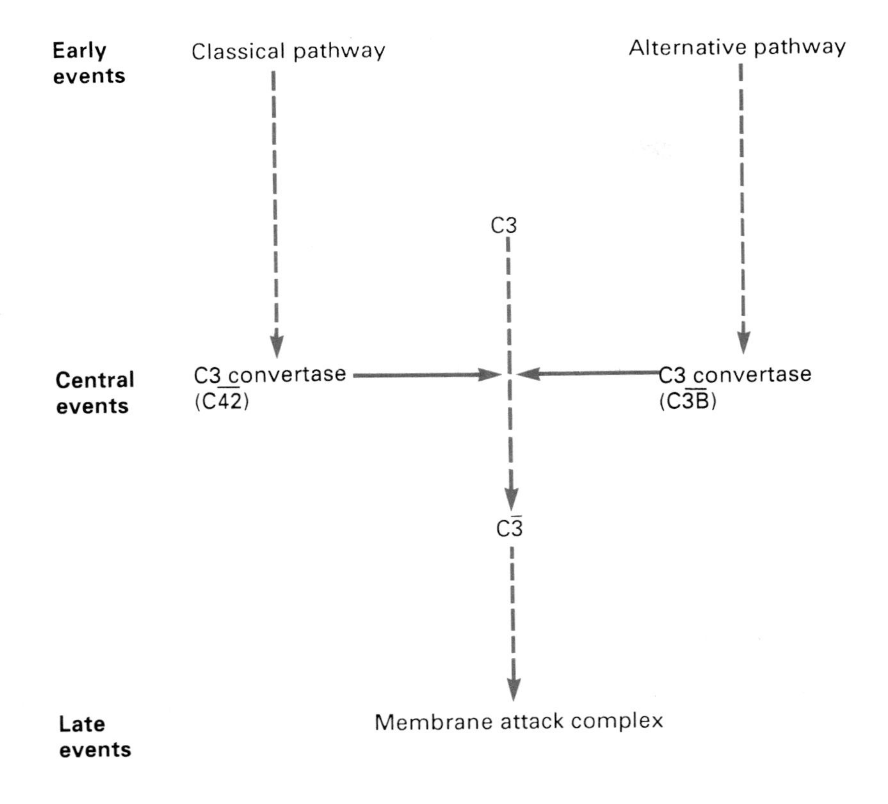

C$\overline{3}$: activated C3

Fig.9.11 The two pathways for complement activation

are known as factors, and each molecule is named by a letter, for example factor B. The alternative pathway uses identical components to the classical pathway in the later stages of activation, i.e. components C5–C9. C3 also participates in both pathways (Fig.9.11).

Activation of each of the components usually results from a proteolytic cleavage event which fragments the native molecule into more than one piece (Fig.9.12). The activated form of each component can be denoted in two ways, either for example as C$\overline{4}$, or as the fragments which are generated by the proteolytic cleavage step, for example C4a + C4b. In general, the 'active' fragment, i.e. the fragment which participates further in the complement cascade, generated by proteolytic cleave is designated the 'b' fragment and is usually the larger of the two peptides (Fig.9.12). The first nomenclature described above, i.e. C$\overline{4}$, also refers to the 'active' fragment generated by the proteolytic event (C4b).

The molecules of the complement system have been studied extensively, both at the gene and protein level. Information is available on the molecular structure of each of the components, their mechanism of activation, and the nature of the catalytic site. This information will not be discussed in detail in this chapter, as much of it has been reviewed recently in the literature (for a review see Reid 1989). The main characteristics of the components, control proteins and complement receptors are listed in Table 9.2.

Table 9.2a Components and factors of the complement system

		Mol. wt (kDa)	Number of chains in plasma form, prior to activation	Mol. wt of individual chains (kDa)	Chromosome location of gene	Plasma concentration µg/ml	Enzymatic site present in activated form (+) (natural substrate(s))
Classical pathway							
C1	C1q	462	18 (six A + six B + six C)	A: 26.5	1	80	–
				B: 26.5	1		
				C: 24	n.d.		
	C1r*	83	1		12	50	+ (C1r, C1s)
	C1s*	83	1		12	50	+ (C4, C2)
C3 convertase	C4	205	3 (α + β + γ)†	α: 97 β: 75 γ: 33	6	600	–
	C2	102	1		6	20	+ (C3, C5)
	C3	185	2 (α + β)†	α: 110 β: 75	19	1300	–
Alternative pathway							
	Factor D	24	1		n.d.	1	+ (B)
C3 convertase	Factor B	92	1		6	210	+ (C3, C5)
	C3	185	2 (α + β)†		19	1300	–
Terminal components							
MAC	C5	190	2 (α + β)†	α: 115 β: 75	n.d.	70	–
	C6	120	1		n.d.	64	–
	C7	110	1		n.d.	56	–
	C8	150	3 (α + β + γ)	α: 64 β: 64 γ: 22	1	55	–
	C9	71	1		n.d.	59	–

n.d. = not determined.
*C1r and C1s are present in plasma either as a tetrameric, $Clr_2.Cls_2$, or in a complex with C1q, $C1q{\cdot}C1r_2{\cdot}C1s_2$.
†These chains are initially synthesized as a precursor molecule, in a single polypeptide chain.

9.4.3 Activators of the complement cascade

Both the classical and the alternative pathway of complement can be activated by immune complexes. However, not all classes and subclasses of antibody can activate the classical and the alternative pathways if they are complexed with antigen. For example, if the immune complex is composed of IgM it can activate the classical pathway, but complexes containing IgA, IgD, and IgE do not. Data for the role of different classes and subclasses of human antibody in the activation of the complement are given in Table 9.3.

Neither the classical nor the alternative pathway can be activated by antibody alone; only antibody bound to antigen is an effective activator. Furthermore, experimental evidence suggests that the antibody must be

Table 9.2b Control proteins of the complement system

Protein	Mol. wt (kDa)	Approximate plasma concentration µg/ml	Specificity	Chromosome location of gene	Role
C1-Inhibitor	110	200	$C\overline{1}r$, $C\overline{1}s$	11	Forms 1:1 complex with both $C\overline{1}r$ and $C\overline{1}s$ in the $C\overline{1}$ complex and removes them
C4 binding protein (C4bp)*	500	250	$C\overline{4}$	1	Accelerates decay of $C\overline{42}$ and acts as a cofactor in the cleavage of C4 by factor I
Factor H	150	480	$C\overline{3}$	1	Accelerates decay of $C\overline{3B}$ and acts as cofactor in the cleavage of C3 by factor I
Factor I	88	35	$C\overline{4}$, $C\overline{3}$	4	Protease which inactivates C4 and C3 with the aid of cofactors C4bp, H, CR1, or MCP (Membrane cofactor protein)
Anaphylatoxin inactivator†	310	35	C3a, C4a, C5a	n.d.	Carboxypeptidase which inactivates the anaphylatoxins C3a, C4a, and C5a by removal of a C-terminal arginine residue
Properdin‡	220	20	$C\overline{3B}$	X	Positive regulator of the alternative pathway, and stabilizes the C3 and C5 convertases
S-protein (vitronectin)	83	505	$C\overline{5\text{-}9}$	n.d.	Up to three molecules of S-protein bind to $C\overline{5\text{-}7}$ preventing the complex from binding to cell surfaces

n.d. = not determined.
*C4 binding protein is a disulfide-bonded heptamer of identical subunits (~70 kDa).
†Anaphylatoxin inactivator contains three different polypeptides (83, 55, and 49 kDa). The stoichiometric composition of these polypeptides in the protein is not clear.
‡Properdin is a multimer of (56 kDa) subunits; the predominant form contains three or four subunits.

bound to the antigen by more than one of its binding sites before complement activation can be initiated. This requirement could represent an important control mechanism for the initiation of the complement cascade, as it ensures that large amounts of antibody of the appropriate specificity must be present in the plasma before activation occurs. Control of complement activation is essential for many reasons. For example, the end-stage reaction of complement activation via either pathway is the formation of membrane attack complex (MAC). This has no antigen specificity, and can result in membrane lysis and cell death. Because the membrane attack complex could potentially interact with any cell membrane, the events initiating the formation of the MAC must be carefully controlled to prevent lysis of normal host cells.

Activation of the classical pathway requires intact immunoglobulin molecules and involves interaction of complement components with the Fc region of the antibody. On the other hand, activation of the alternative pathway can be initiated *in vitro* by immune complexes containing $F(ab')_2$ fragments, which contain no Fc region (Fig.6.7). However, Fab–antigen complexes are ineffective, as the Fab fragment is monovalent and therefore crosslinking of the antigen cannot occur.

Table 9.2c Membrane-associated molecules which act as receptors or as regulators for activated complement components and their fragments*

Membrane molecule	Mol. wt† (kDa)	Binding specificity	Chromosome location of gene	Principal roles	Major human cell types positive
Complement receptor type 1 (CR1) (four structural allotypes)	type D 250 type B 220 type A 190 type C 160	C$\overline{3}$, C$\overline{4}$	1	Regulation of C$\overline{3}$ breakdown, binding of immune complexes to erythrocytes, phagocytosis and accelerates decay of the C3 and C5 convertases	E, B, G, M
Complement receptor type 2 (CR2)	145	C3d, C3dg, iC3b	1	Regulation of B cell function, Epstein–Barr virus receptor	
Membrane cofactor protein (MCP)	45–70	C$\overline{3}$, C$\overline{4}$	n.d.	Regulation of C$\overline{3}$ breakdown	B, T, N, M
Decay accelerating factor (DAF)	70	C$\overline{42}$, C$\overline{3B}$	1	Accelerates decay of the C3 and C5 convertases	E, L, P
Complement receptor type 3 (CR3)	165 (α) 95 (β)	iC3b	n.d.	Phagocytosis	G, M, φ
Glycoprotein p150, 95 (CR4)	150 (α) 95 (β)	iC3b	n.d. 21	Monocyte migration	G, M, φ
C3a/C4a receptor	n.d.	C3a, C4a	n.d.	Binding of anaphylatoxin C3a and C4a	G, A
C5a receptor	~45	C5a, C5a-des Arg	n.d.	Binding of anaphylatoxin C5a	G, A, M, φ, P
Homologous restriction factor	65	C8, C9	n.d.	Prevention of formation of MAC on homologous cells	E
C1q receptor	~65	C1q (collagen region)	n.d.	Mediates binding of immune complexes to phagocytic cells. Inhibition of IL-1 receptor expression by B lymphocytes	B, M, φ, P, D

n.d. = not determined.
Human cell types: E, erythrocytes; B, B lymphocytes; T, T lymphocytes; M, monocytes; φ, macrophages; G, granulocytes; N, neutrophils; L, leukocytes; P, platelets; A, mast cells; D, endothelial cells.
*Receptors for intact factor H and Ba have been found on monocytes, B lymphocytes and neutrophils.
†All appear to be single chain molecules except CR3 and p150,95 which have two non-covalently chains.

Note: Table 9.2 Adapted from Law, S.K.A. and Reid, K.B.M. (1988). *Complement*, In Focus Series, IRL Press, Oxford.

Non-immunoglobulin activators of both pathways of complement have also been described, although this route of activation is more commonly associated with the alternative pathway. An enormous variety of substances have been shown to initiate alternative pathway activation in the absence of antibody. The most relevant of these are the complex polysaccharide molecules of yeast cell walls, such as zymosan, and bacterial cell walls. Activation of the complement cascade by these substances represents a very important alternative to activation by immunoglobulins; as part of innate immunity it can occur immediately after infection without the lag period associated with the synthesis of antibody either in the primary or secondary

Precursor molecule
(zymogen)

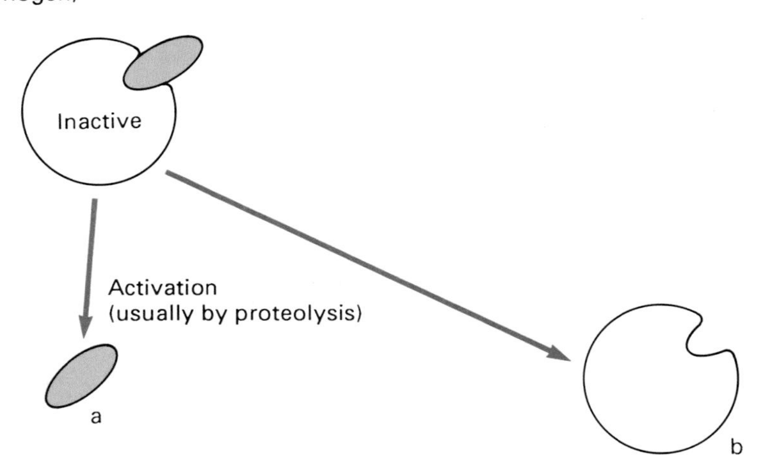

Smaller fragment (a)

Inactive with respect to the cascade reaction, but possesses other activities, e.g. can be chemotatic and/or anaphylatoxic

Larger fragment (b)

Participates further in complement cascade – active fragment

The activation process can be represented as, for example:

$$\text{either C4} \xrightarrow{\text{activation}} \text{C}\overline{4}$$

$$\text{or C4} \xrightarrow{\text{activation}} \text{C4b} + \text{C4a}$$

Fig.9.12 General scheme for the activation of complement components

Table 9.3 Activation of complement by immune complexes containing human immunoglobulins of different class and subclass

Class	Subclass	Classical pathway	Alternative pathway
IgM		++++	++
IgG	IgG1	++++	++ (IgG1–IgG4)
	IgG2	+	
	IgG3	+++	
	IgG4	–	
IgA		–	+++
IgE		–	++
IgD		–	++

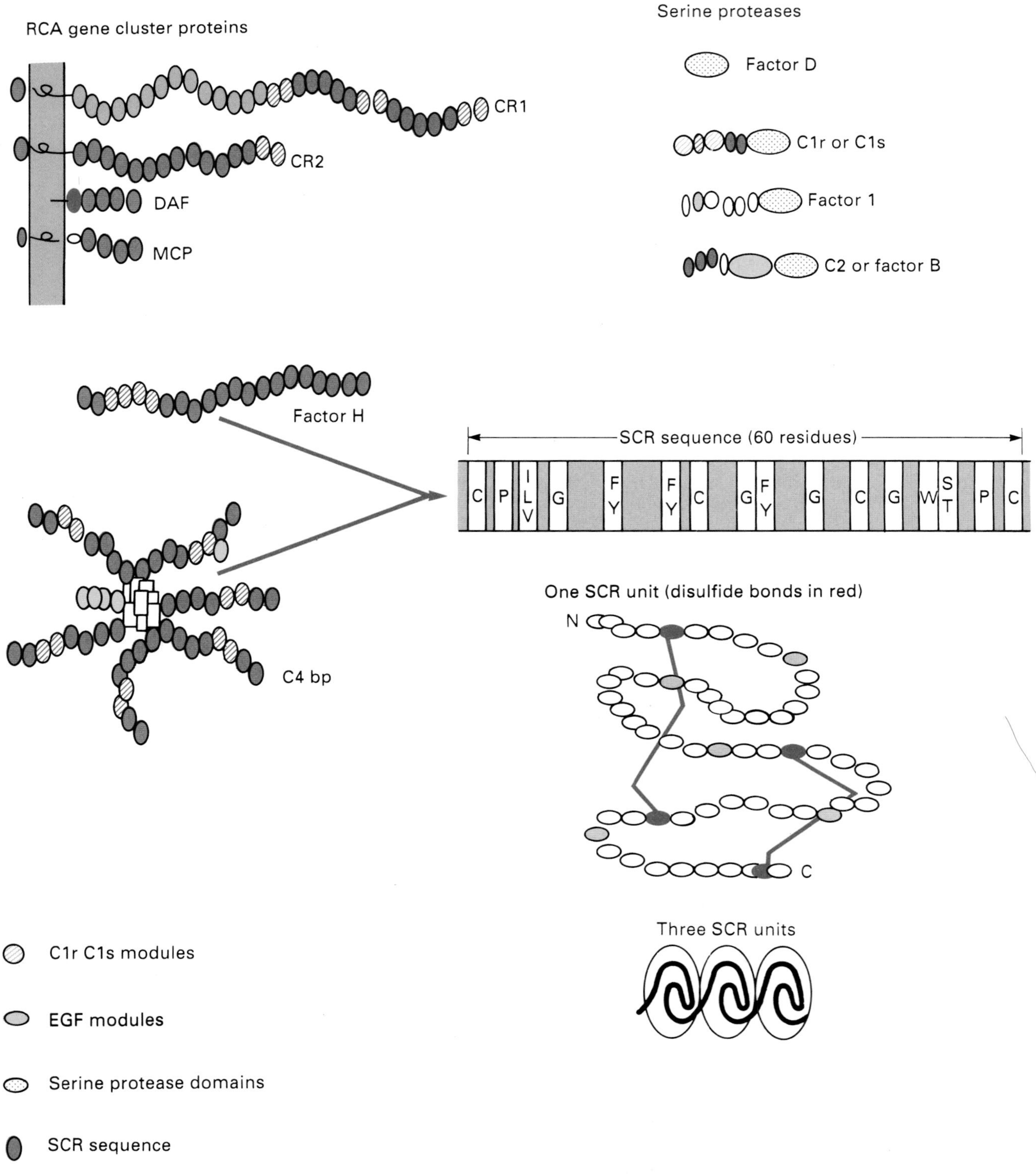

Fig.9.13 The structure of some complement components

adaptive immune responses. Pathogens have evolved various mechanisms to evade destruction by complement, for example by the development of structures which interfere with assembly of the C3 convertase (Cooper 1991).

9.4.4 The classical pathway—early events

The C1 complex

The classical pathway can be activated by immune complexes containing either IgM or certain subclasses of IgG (Table 9.3). Initiation of the activation process occurs by the binding of the first component, the C1 complex, to an immune complex (Fig.9.14(a)) (Sim and Reid 1991).

The C1 complex is made up of three proteins, C1q, C1r, and C1s (Table 9.2a). C1q binds directly to the Fc region of the immunoglobulin present in the immune complex. The $C_{\gamma}2$ domain of the Fc region is involved in this interaction for immune complexes containing IgG, and the $C_{\mu}3$ domain is involved in IgM complexes. The site of interaction between C1q and the antibody molecule was located using fragments of IgG. Experiments were carried out using complexes of antigen with the whole IgG molecule, $F(ab')_2$, Fab fragments (Fig.6.7), the whole antibody from which the C_H3 domain had been removed, or aggregated Fc portions of the molecule. In every case, except when $F(ab')_2$ or Fab fragments were used, C1q binding and classical pathway activation occurred. Although the C_H3 domain of the IgG molecule was found not to be essential for C1q binding and classical pathway activation, the reaction was more efficient when intact IgG was present in the immune complexes.

The number of antibody molecules which have to be present in the complex to initiate classical pathway activation has also been examined. In every case, at least two molecules of immunoglobulin in close proximity are required, indicating that C1q has to bind to two or more Fc regions to trigger the subsequent events. For IgG, between 10^2 and 10^3 antibody molecules may need to be present before two Fc regions are statistically close enough together (this obviously depends on the size of the antigen and the spacing of the antigenic determinants). The density of the antigenic determinants is also critically important for activation. If they are too far apart, the antibodies interacting with these determinants will not bind in sufficiently close proximity to one another to trigger activation.

This phenomenon can be demonstrated *in vitro* by using hapten-coupled sheep red blood cells (SRBC) and anti-hapten antibodies (Fig.9.15). When the density of the hapten on the sheep red blood cells is low, complement activation does not occur when antibody is present. Increasing the hapten density and therefore the number of antibody binding sites on the sheep red blood cell results in complement activation when antibody is present, due to the increased number of C1q binding sites on the antibody-coated SRBC (Borsos *et al.* 1981).

The structure of C1q has been analysed extensively and is shown in Fig.9.16. Its conformation has been visualized by electron microscopy, and has been described as resembling a bunch of tulips. It is nearly three times the size of IgG (462 kDa) (Table 9.2a). The molecule can be roughly divided into two parts: (i) the globular head regions which interact with the antibody molecules in the immune complex; and (ii) the collagenous tail which interacts with a tetrameric complex of two C1r plus two C1s molecules (Fig.9.17).

In the presence of calcium, C1q is associated with the other subunits of the C1 complex, components C1r and C1s. C1r and C1s exist together in the presence of Ca^{2+} as a tetrameric complex, $C1r_2$–$C1s_2$.

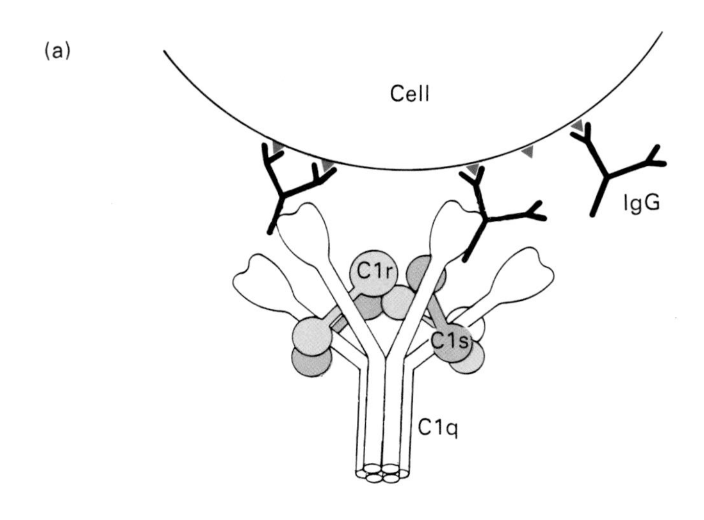

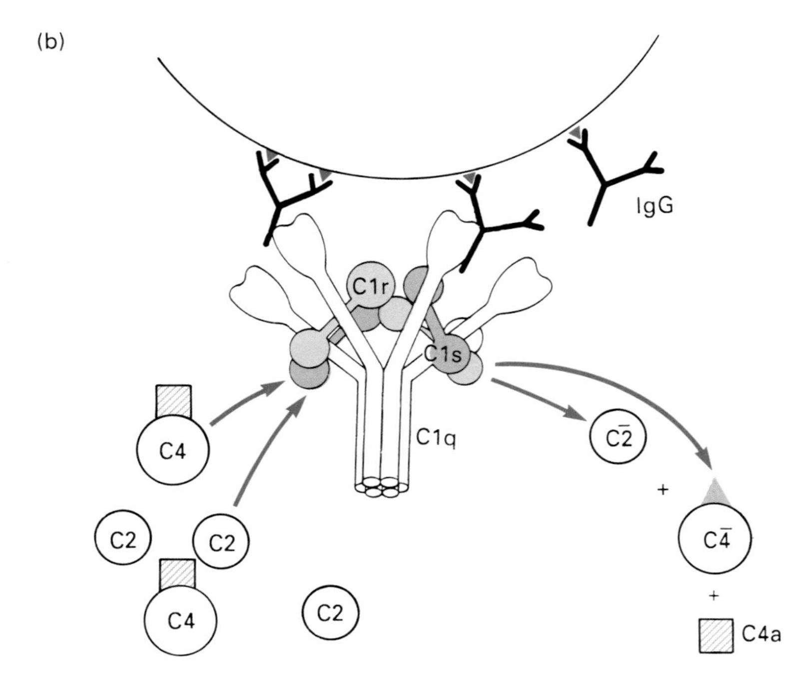

Fig.9.14 Activation of the classical pathway. (a) Schematic diagram to show the initiation of classical pathway activation. (b) Activation of C2 and C4. On facing page: (c) The events which lead to the generation of the classical pathway C3 convertase enzyme. (d) Generation of the C5 convertase enzyme. Adapted from Arlaud *et al.* (1987), *Immunology Today*, **8,** 106, with permission

C1r and C1s are very similar in structure; each has two globular domains, the larger of which is thought to contain the 'catalytic site' in activated $C\overline{1}r$ and $C\overline{1}s$. Each is a single polypeptide chain (83 kDa), which has serine protease activity once the component has been activated (Table 9.2a). C1r and C1s genes are found within 20 kb of each other in region p13 on human chromosome 12. The Clr_2–Cls_2 complex is rod-like in shape.

The interaction of the $C1r_2$–$C1s_2$ tetrameric complex with C1q is shown in Fig.9.17.

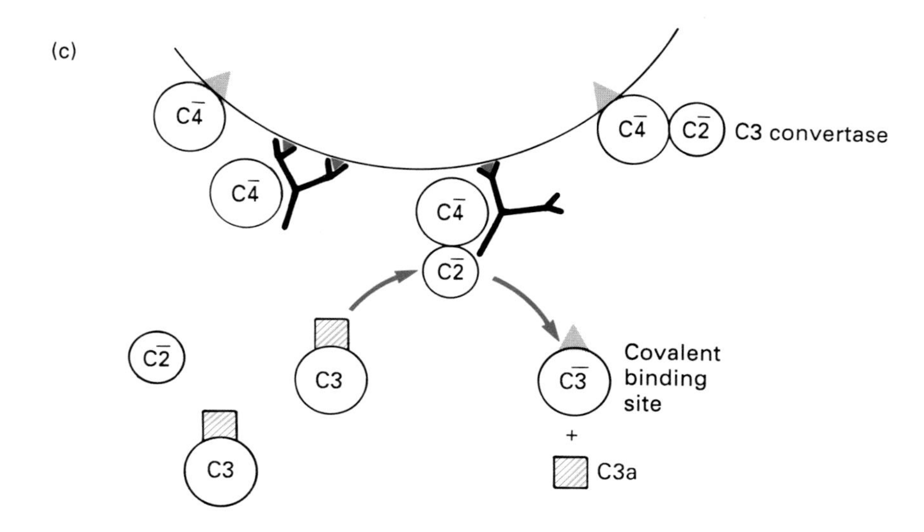

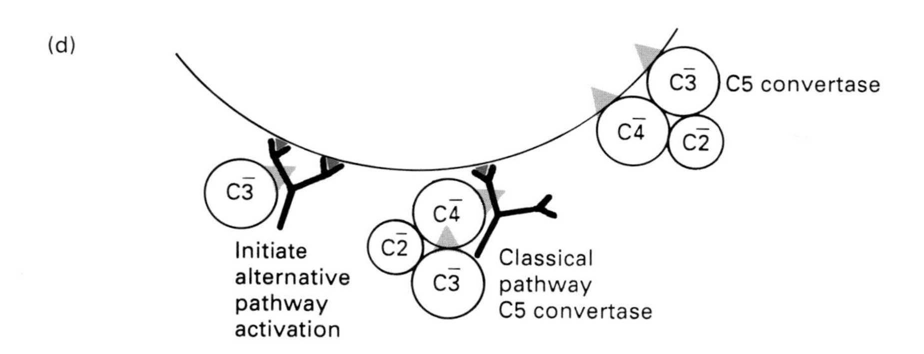

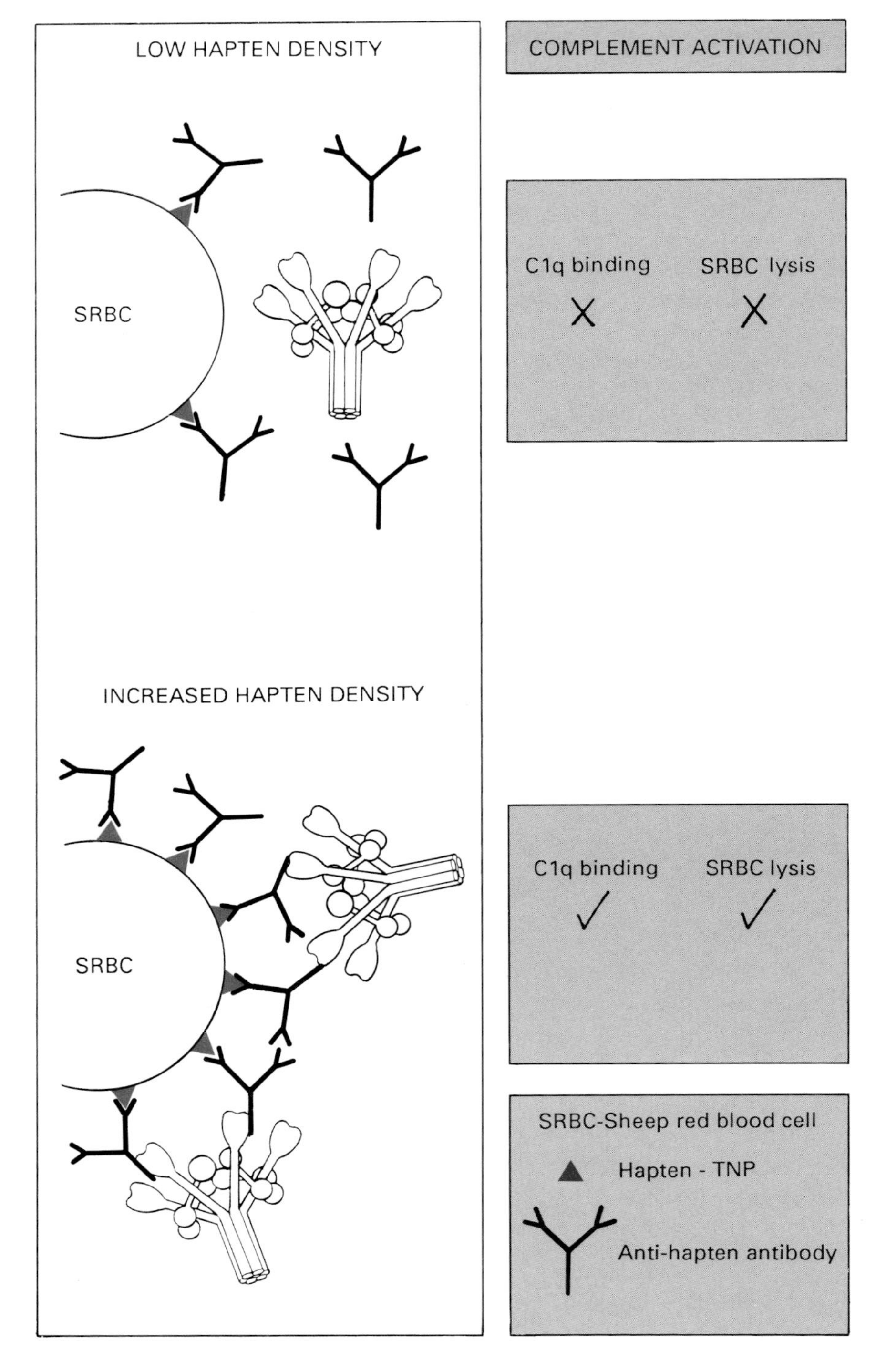

Fig.9.15 Antibody and antigen density is of critical importance for complement activation. Adapted from Arlaud *et al.* (1987), *Immunology Today*, **8,** 106, with permission

The activation of the C1 complex is under the control of the C1-inhibitor (C1-Inh) (Table 9.2b), which interacts with the unactivated complex, preventing the spontaneous activation of the proenzyme Clr. To trigger activation of the Cl complex, two or more heads of a C1q molecule have to bind to the antibody molecules in the immune complex (Fig.9.14(a)). This interaction probably induces a conformational change in C1q, releasing it from the control of C1-Inh and allowing auto-activation of C1r to take place.

Activation of C1r results in proteolytic cleavage of the polypeptide chain into two fragments, 55 kDa and 27 kDa, which remain disulfide bonded together. The smaller of the two peptides contains the active site, which is characteristic of the serine protease family and rapidly activates the proenzyme C1s present in the complex. C1s is responsible for the next phase of activation of the classical pathway, the activation of components C2 and C4.

After activation of C1 has taken place, C1-Inh rapidly forms a covalent, 1:1 complex with activated C$\overline{1}$r and C$\overline{1}$s, probably via their catalytic sites. The (C1-Inh)-C$\overline{1}$r-C$\overline{1}$s-(C1-Inh) complex is released, thus removing C$\overline{1}$r and C$\overline{1}$s from C1q, leaving the collagen regions of Clq free to interact with C1q receptors (Table 9.2c). The interaction of C1q with its cellular receptors results in a number of effector functions, including cell-mediated cytotoxicity, Fc receptor-mediated phagocytosis, stimulation of oxidative metabolism in polymorphonuclear leukocytes and inhibition of IL-1 production.

C1-Inh is critical for the control of the C$\overline{1}$ complex. Patients with hereditary angioedema have a deficiency in C1-Inh and as a result suffer from attacks of localized increased vascular permeability which may be due to over-activation of C2 and bradykinin (see Section 9.3.3). C1-Inh also inhibits a number of other plasma proteases, including kallikrein, plasmin, the Hageman factor, and factor XI (see Section 9.3).

Formation of the classical pathway C3 convertase

C4 and C2 are soluble serum glycoproteins, both of which are polymorphic and encoded within the major histocompatibility complex (MHC) (see Chapter 2), together with factor B of the alternative pathway (Carroll *et al.* 1984; Trowsdale *et al.* 1991). These complement components are known as MHC class III molecules (Table 9.4).

C4 is a three-chain molecule (Table 9.2a); the chains are held together by disulfide bonds. The messenger RNA for C4 is translated into a single polypeptide chain, which is then cleaved to form the three-chain structure by post-translational modification (Fig.9.18).

C4 is activated by C$\overline{1}$s of the C$\overline{1}$ complex. This proteolytic cleavage event releases a low molecular weight peptide (9 kDa) from the N-terminal end of the α chain of C4, called C4a (Fig.9.18). C4a is a weak anaphylatoxin, increasing vascular permeability. It diffuses from the site of complement activation and is also responsible for attraction of phagocytic cells (Table 9.5 and Section 9.4.7). The large fragment of C4, generated by C$\overline{1}$s, is called C4b or C$\overline{4}$; it participates further in the reactions of the complement cascade. The association of C4 and C1 may be via a set of repeating units of sequence located in the heavy chains (55 kDa) of Clr and Cls (Fig.9.13). These 60 amino acid short consensus repeats (SCR) are similar to those found in C3b

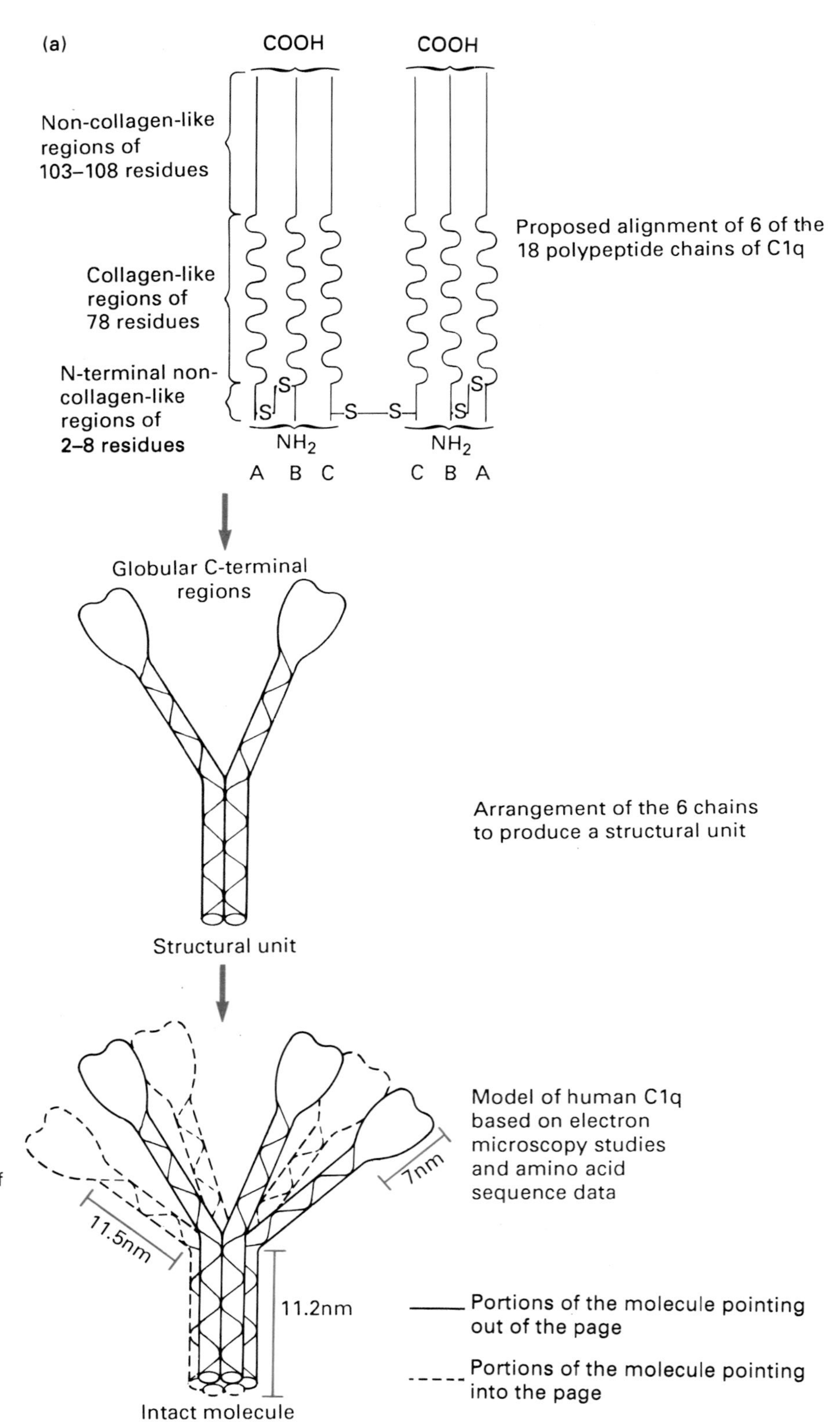

Fig.9.16 (a) Diagrammatic representation of the structure of C1q. (Adapted from Reid and Porter, *Biochem. J.*, **155**, 5–17.) (b) Limited proteolysis by pepsin produces a large collagen-like fragment which contains binding sites for the $C1r_2$-$C1s_2$ complex. Limited proteolysis by collagenase produces intact globular regions which contain binding sites for the Fc portion of immunoglobulin. (Adapted from Porter and Reid (1979), *Adv. Protein Chem.*, **33**, 1–71.) (c) Electron micrograph of a molecule of human C1q showing the apparent separation of proposed structural units. This indicates that strong non-covalent bonds holding the units together may be located at the N-terminal ends of the chains. (From Knobel *et al.* (1975))

(b)

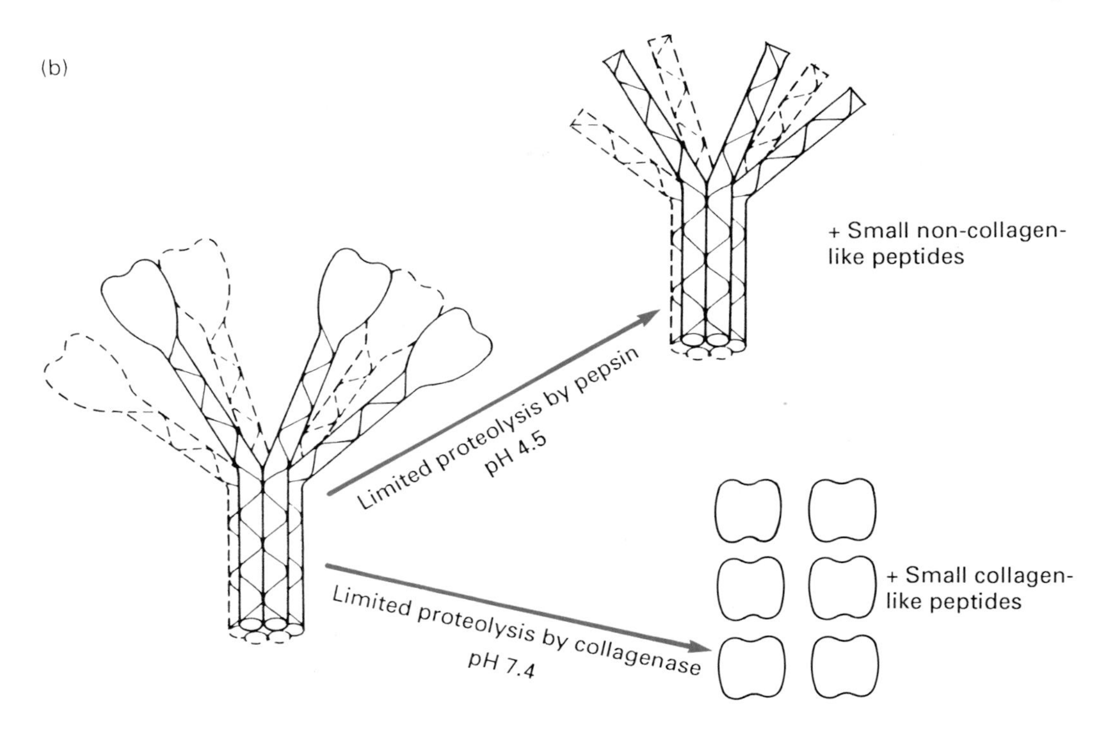
Limited proteolysis by pepsin
pH 4.5
+ Small non-collagen-
like peptides
Limited proteolysis by collagenase
pH 7.4
+ Small collagen-
like peptides

(c)

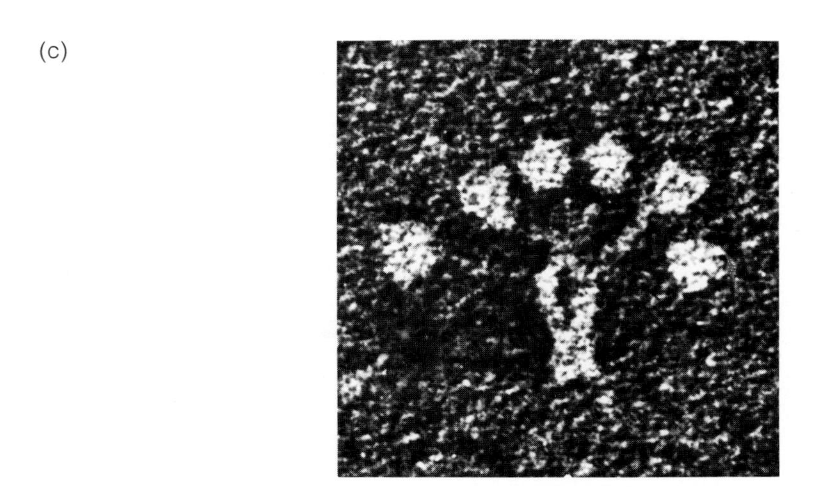

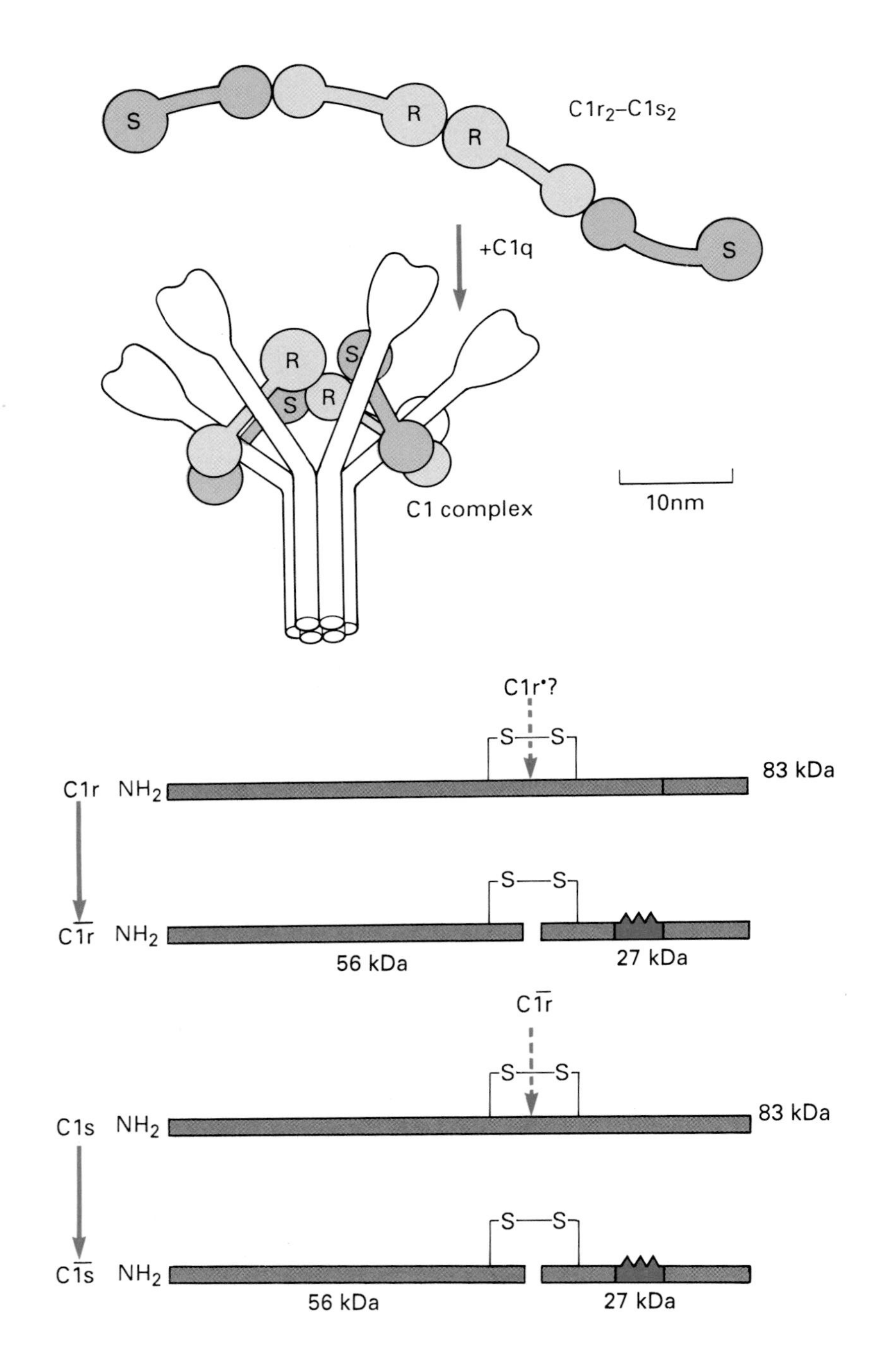

Fig.9.17 *Top*: Relationship between C1r and C1s, and their association with C1q to form the C1 complex. R and S denote the larger catalytic domains of C1r and C1s. The C1 complex may be formed by interaction between the rod-like $C1r_2{\cdot}C1s_2$ and the arms of C1q (Adapted from Arlaud *et al.* (1987). *Immunology Today*, **8**, 106). *Bottom*: Proteolytic cleavage events which occur during activation of C1r and C1s

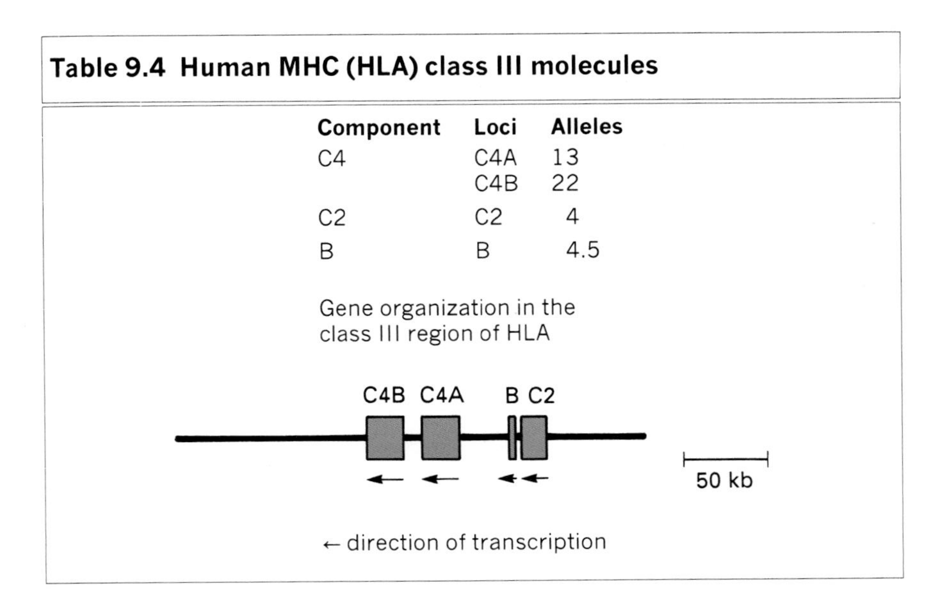

Table 9.4 Human MHC (HLA) class III molecules

Component	Loci	Alleles
C4	C4A	13
	C4B	22
C2	C2	4
B	B	4.5

Gene organization in the class III region of HLA

and C4b binding proteins (C3bp and C3bp, see below). Each $\overline{C1}$s activates many C4 molecules, resulting in amplification of the classical pathway.

$\overline{C4}$ has no proteolytic activity. However, immediately following activation by $\overline{C1}$s, it acquires the transient ability to bind covalently to proteins or polysaccharides including the Fab region of the antibody and the antigen molecules in an immune complex, or a site on the surface of the cell expressing the antigenic determinant (Fig.9.14(c)). After activation, $\overline{C4}$ molecules diffuse away from the site where the $\overline{C1}$ complex is bound and if within micro-seconds they encounter a suitable covalent binding site, they will bind, thus localizing the point or site of complement activation to an adjacent region of the immune complex.

This ability of $\overline{C4}$ form covalent bonds is shared by activated $\overline{C3}$ and is the property of a region of the α' chain of each molecule (Fig.9.19). Following activation, a reactive acyl group is generated in C4b or C3b which can then interact with either hydroxyl groups of carbohydrate residues to form ester bonds, or with amino groups of amino acids to form amide bonds (Fig.9.19). The covalent binding event is totally non-specific and the activated molecules can form covalent bonds with any suitable molecule. However, because the half-life of the active acyl group is so short, binding of C4b and C3b will only take place in the immediate vicinity of the C1 complex or the C3 convertase enzymes, respectively. If no binding sites are available, the reactive acyl group is hydrolysed by the aqueous environment and inactivated.

Covalently bound $\overline{C4}$ can interact with specific receptors, the complement receptor type 1, CR1 (CD35), on a variety of phagocytic cells (Table 9.2c). This interaction could be important for clearance of immune complexes.

Activated $\overline{C4}$ can also interact with the proenzyme C2. This interaction is Mg^{2+} dependent and if binding occurs close to $\overline{C1}$s, C2 is activated. This involves proteolytic cleavage of C2 into two fragments (C2b—30 kDa and

C2a—60 kDa, the catalytic subunit). Activation by $C\overline{1}s$ exposes the active site in the C2a fragment, which has serine protease activity when $C\overline{2}$ is associated with $C\overline{4}$. The bimolecular complex forms the C3 convertase enzyme $C\overline{42}$.

Control of classical pathway C3 convertase activity

General features. There are several levels of control of the classical pathway C3 convertase enzyme.

(1) The formation of the enzyme itself is controlled: a fully active C3 convertase enzyme is only generated when $C\overline{4}$, which is covalently bound

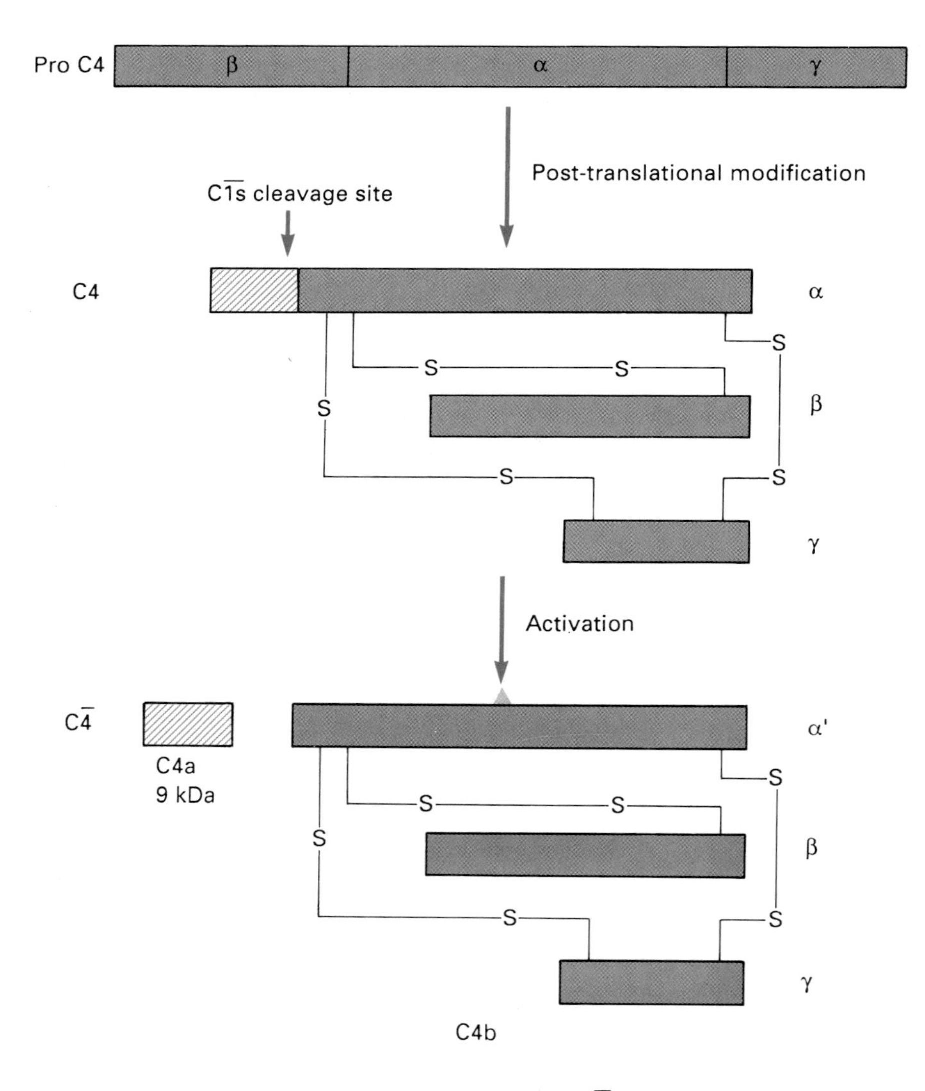

Fig.9.18 Biosynthesis and activation of C4

Table 9.5 Activities of the low molecular weight peptides produced during complement activation

Peptide	Biological activity
C3a	anaphylatoxic
C4a	anaphylatoxic (100x less potent than C3a) Chemotactic
C5a	anaphylatoxic (20x more potent than C3a) Chemotactic
$\bar{B}$	B cell stimulation

Activity C5a > C3a > C4a. Activity is controlled by anaphylatoxin inactivator, serum carboxypeptidase N, which removes the C-terminal arginine residues which are essential for their activity. However, C5a still has considerable chemotactic activity even in the absence of its C-terminal arginine.

to the activator, interacts with C$\overline{2}$. This requires that the two molecules are both present in their activated forms at precisely the same time and furthermore that they are in the correct location where they can interact with one another.

(2) The association of C$\overline{4}$ and C$\overline{2}$ is unstable. C$\overline{2}$ can be lost from the covalently bound C$\overline{4}$, thus destroying the C3 convertase activity. However, the bound C$\overline{4}$ can interact with a second C$\overline{2}$ molecule to form another C3 convertase enzyme.

(3) C$\overline{2}$ is labile and can rapidly lose its catalytic activity.

(4) The control proteins factor I and C4b binding protein (C4bp) (see below) can rapidly inactivate C$\overline{4}$ by proteolytic cleavage (Fig.9.18). Once inactivated, C$\overline{4}$ can no longer form the C3 convertase enzyme. Inactivation of C$\overline{4}$ by factor I and C4b binding protein (Table 9.2b) results in further cleavage of the α' chain of C$\overline{4}$, which fragments the molecule and releases the majority of it from the cell surface or the immune complex. Only the C4d fragment remains bound.

Factor I. This enzyme, formerly called C3b inactivator, a highly specific serine protease, can inactivate C$\overline{3}$ and C$\overline{4}$ in the presence of the appropriate cofactors, factor H or C4bp, respectively. It is a glycoprotein (88 kDa), composed of two polypeptide chains held together by disulfide bonds. It is responsible for cleaving the α' chains of C$\overline{3}$ and C$\overline{4}$ at two points (Figs. 9.20 and 9.21) resulting in the rapid loss of their biological activity, including their role in the C3 and C5 (see below) convertase enzymes. Factor I will not digest the native C3 and C4 molecules.

C4b binding protein. C4b binding protein is a cofactor for factor I, and together they regulate the activity of C$\overline{4}$ by degrading the molecule (Fig.9.20). C4bp is a multi-subunit protein (500 kDa) composed of seven identical peptide chains, and appears to have a 'spider-like' shape when analysed by electron microscopy (Fig.9.13). Each molecule has multiple binding sites for C4b located at the end of each of the spider-like arms. C4bp can inhibit the formation and accelerate the decay of the C3 convertase enzyme by displacing C2a. Once bound, C4bp acts as a cofactor for factor I.

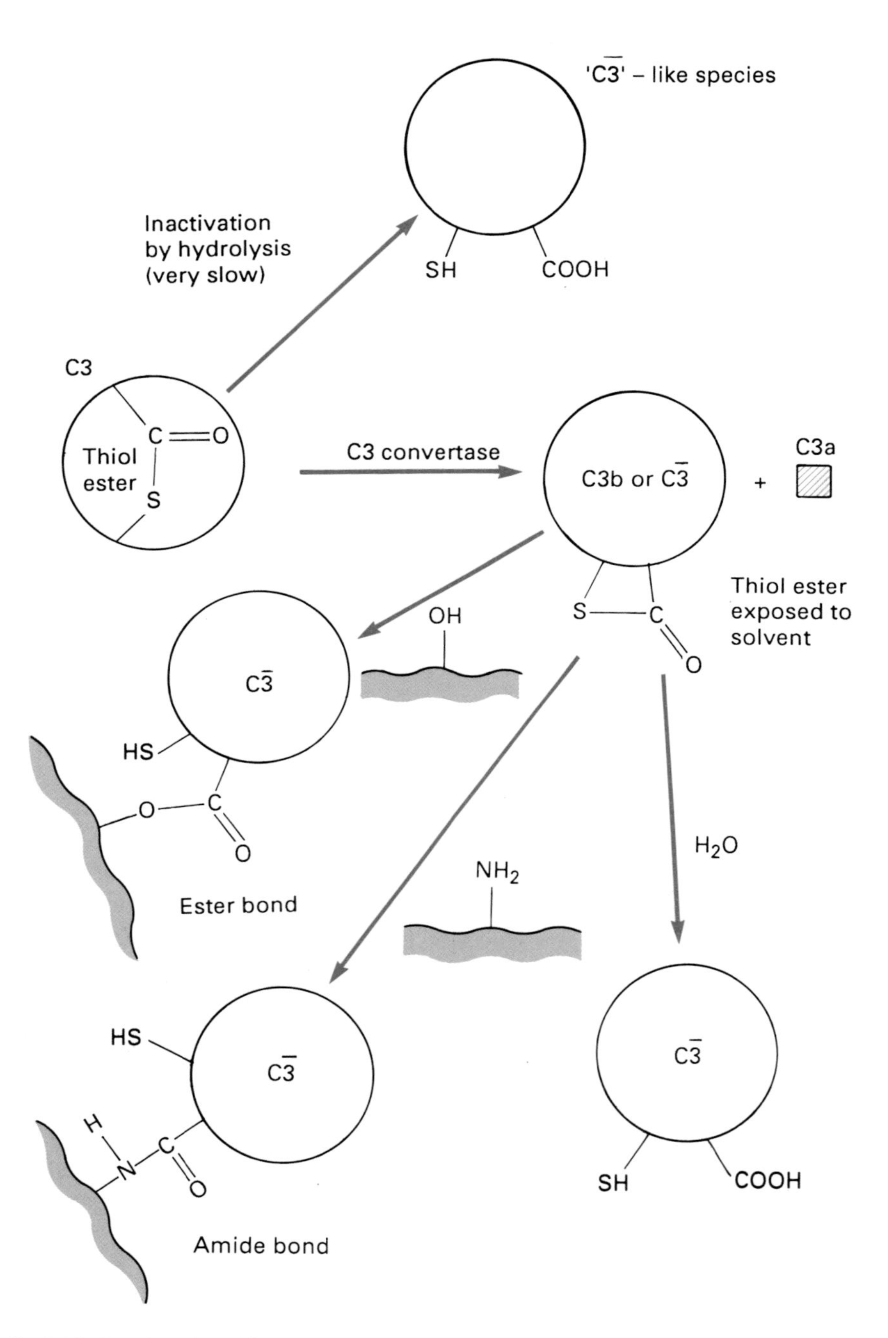

Fig.9.19 Covalent bond formation by activated C3 and C4 molecules or inactivation by hydrolysis. C3 is used in this illustration

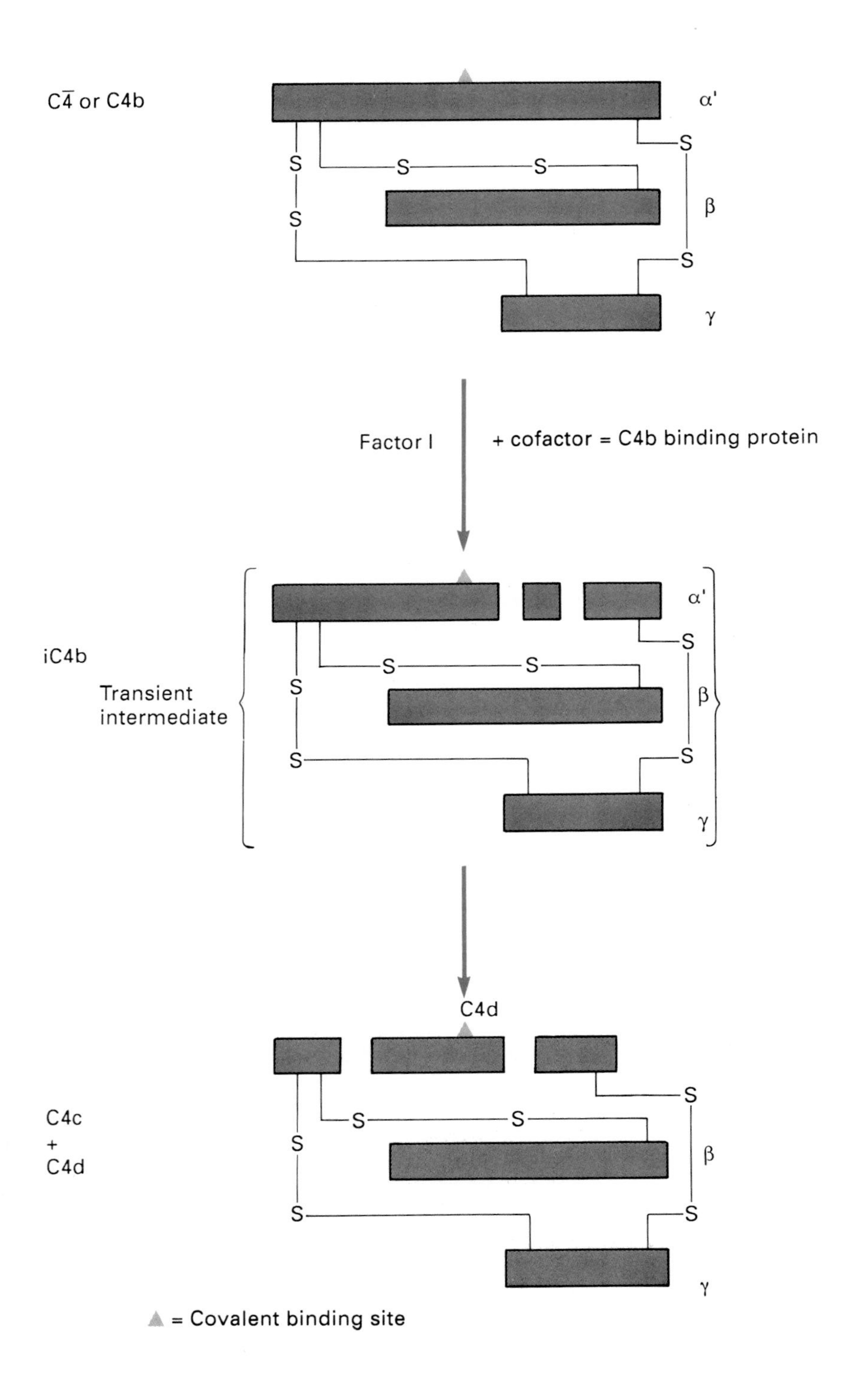

Fig.9.20 Control of classical pathway C3 convertase activity by factor I and C4b binding protein

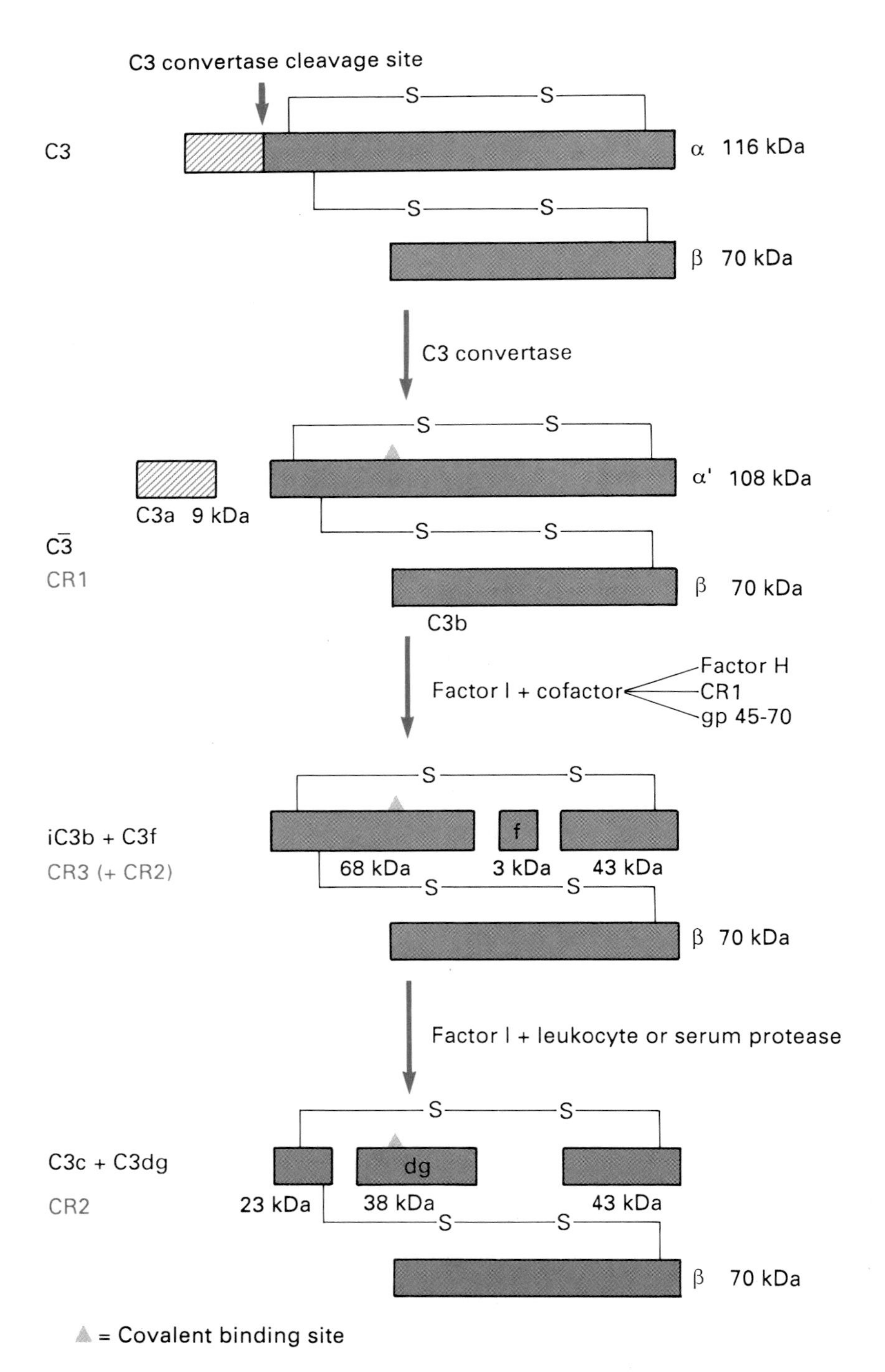

Fig.9.21 Fragments of C3 generated during activation and inactivation of the molecule. Correlation of these fragments with CR1, CR2, and CR3 receptor specificity

Control of C3 convertase activity is also mediated by complement receptor type 1 (CR1) and decay accelerating factor (DAF) (Table 9.2c). Both of these are membrane molecules. CR1 is a true membrane protein, anchored in the membrane by the transmembrane region of its polypeptide chain, whereas DAF is anchored by a phosphatidylinositol tail (see Fig.1.5). A third molecule, membrane cofactor protein (MCP) also regulates the activity of C4b (Table 9.2c). C4bp, CR1 (CD35), DAF, and factor H (see below) are all encoded together on chromosome 1 in both mice and humans. This gene cluster is referred to as the RCA or regulators of complement activation.

The central component of the complement system—C3

The classical pathway C3 convertase, $C\overline{42}$, is responsible for activating the central component of the complement system, C3. Human C3 is a two-chain protein, composed of an α and a β chain (Table 9.2a). It is activated by the C3 convertase enzymes of either the classical or alternative pathways, respectively $C\overline{42}$ and $C\overline{3B}$ (see Section 9.4.5). Activation by either enzyme results in cleavage of the α chain and the release of the low molecular weight peptide C3a. C3a is an active peptide possessing anaphylatoxic activity (Table 9.5 and Section 9.4.7). The large fragment, known as $C\overline{3}$ or C3b, is the 'active' fragment. C3 plays an important role in modulating the induction phase of immune responses (Erdei *et al.* 1991).

$C\overline{3}$, like $C\overline{4}$, can bind covalently to the complement activator. The mechanism of binding is the same for both $C\overline{3}$ and $C\overline{4}$ (Fig.9.19). As for $C\overline{4}$, if $C\overline{3}$ does not bind covalently to another molecule, either protein or carbohydrate, then the active acyl group generated by the activation event is hydrolysed in the aqueous environment. If $C\overline{3}$ fails to bind and remains in the fluid phase, it is rapidly inactivated by its control proteins, factor H (see below) and factor I.

The binding of $C\overline{3}$ to immune complexes results in disruption of the complex into smaller fragments. These can be removed from the circulation by interaction with CR1 receptors of phagocytic cells. Receptors for $C\overline{3}$, C3b, and its breakdown products, iC3b, C3d, and C3dg (Fig.9.21) are CR1 (CD35), which also binds $C\overline{4}$ (see above), CR2 (CD21), CR3 (CD11b/CD18), CR4 (CD11c/CD18) (Table 9.2c), and CR5 (Erdei *et al.* 1991). These molecules are present on various cell types and can mediate phagocytosis of the immune complexes (opsonization) and other effector responses of phagocytic cells (Section 10.4.5); CR2, which is apparently confined to B cells, delivers activation signals to the lymphocyte (see Section 8.4.2).

Control of C3 activity. As C3 participates in both the classical and alternative pathways its activity must be carefully controlled. The covalent binding activity of $C\overline{3}$ provides one means of control. Following activation, the ability of $C\overline{3}$ to bind covalently to either the alternative or classical pathway activator is only transient (Fig. 9.19). If $C\overline{3}$ is present in the plasma, it is rapidly inactivated by the control proteins factor I and factor H (Table 9.2b). Inactivation by the control proteins results in the proteolytic cleavage of the α' chain of $C\overline{3}$ at two positions to yield iC3b. Further proteolysis of iC3b by factor I and a trypsin-like enzyme produces the fragments C3c and C3dg (Fig.9.21). The factor I molecule is identical to that used to control $C\overline{4}$

activity, and it contains the active site of the control protein complex. However, it cannot inactivate $C\overline{3}$ in the absence of factor H or another cofactor with similar activity. The properties of factor I were given above.

Factor H, formerly known as β1H, acts as a cofactor for factor I in the degradation of $C\overline{3}$. It is the most abundant cofactor for $C\overline{3}$ present in the plasma. Factor H is a single polypeptide chain (150 kDa). It can also down-regulate $C\overline{3}$ cleavage by the alternative pathway C3 convertase by dissociating factor B from bound $C\overline{3}$. This parallels the activity of C4b binding protein in the degradation of C4. It is also likely that factor H regulates the activity of the C5 convertase by competing with C5 for binding to C3b. Factor H acts as a cofactor for the inactivation of $C\overline{3}$ by factor I.

Proteins interacting with C3b and C4b

All of the proteins that interact with C3b and C4b (factor B, factor H, C2, C4bp, CR1, CR2, DAF, and MCP) contain between two and thirty repeating units which have a similar, although not identical, structure of approximately sixty amino acids (Holers *et al.* 1992). These units conform to an SCR, or sequence consensus repeat, which has a framework of conserved amino acid residues: 1 Trp, 2 Pro and 4 Cys and Gly, and several hydrophobic residues are conserved in certain positions in the SCR (Fig.9.13). The significance of these SCRs in the binding to C3b and/or C4b is unclear. It has been suggested that each SCR forms a C3b or C4b binding unit; thus the more SCRs in a molecule, the higher its binding affinity for C3b or C4b should be, but unfortunately this hypothesis is not supported by the experimental data available.

Formation of the classical pathway C5 convertase enzyme

Formation of the classical pathway C5 convertase, $C\overline{423}$, is the final event unique to the classical pathway. C3 activated by the C3 convertase can interact with the already formed C3 convertase, $C\overline{42}$, to generate the C5 cleaving enzyme (Fig.9.14(d)). Interaction of $C\overline{3}$ with $C\overline{42}$ is essential to alter the substrate specificity of the active site in $C\overline{2}$, to enable the enzyme to activate C5. C5 and C3 are very similar in terms of their structure and mechanism of activation, and therefore only a minimal change in the active site of $C\overline{2}$ could result in this change in specificity. C5 does not have the capacity for covalent bond formation.

9.4.5 The alternative pathway

The alternative pathway was identified as a separate route for activation of complement in 1954 (Pillemer *et al.* 1954), although much of the characterization was not carried out until the 1970s, when guinea pigs that were deficient of the classical pathway component C4 were discovered. The alternative pathway provides an important route of complement activation. It is not reliant on the presence of pre-existing antibody and can be activated directly by, for example, yeast or bacterial cell wall polysaccharides, so the alternative pathway provides an important first line of defence against pathogens before the adaptive response is initiated. The molecules which participate in alternative pathway activation are listed in Table 9.2a and an outline of the events involved is shown in Fig.9.22.

Activation

The mechanism of activation of the alternative pathway remains open to discussion although, with the observation that activated C3 can bind covalently to cell surface molecules, it has become much clearer.

Activation of the alternative pathway only occurs in the presence of an activator when C3, present in the fluid phase, binds covalently via its reactive acyl group to the activator. This event generates a molecule of bound C$\overline{3}$ which can initiate alternative pathway activation (Fig.9.23(a)).

The initial molecule of bound C$\overline{3}$ can be generated by activation of the classical pathway (see Section 9.4.4). In this case, C3 is activated by the classical pathway C3 convertase, C$\overline{42}$ and the alternative pathway acts as an amplification mechanism in this situation. However, in the absence of classical pathway activation, the events that activate C3 and thus initiate the alternative pathway are less clear. Several hypotheses have been put forward, and two will be outlined below. Both of these propose 'low grade' degradation of C3 in the plasma and are known as the 'tick-over' hypotheses:

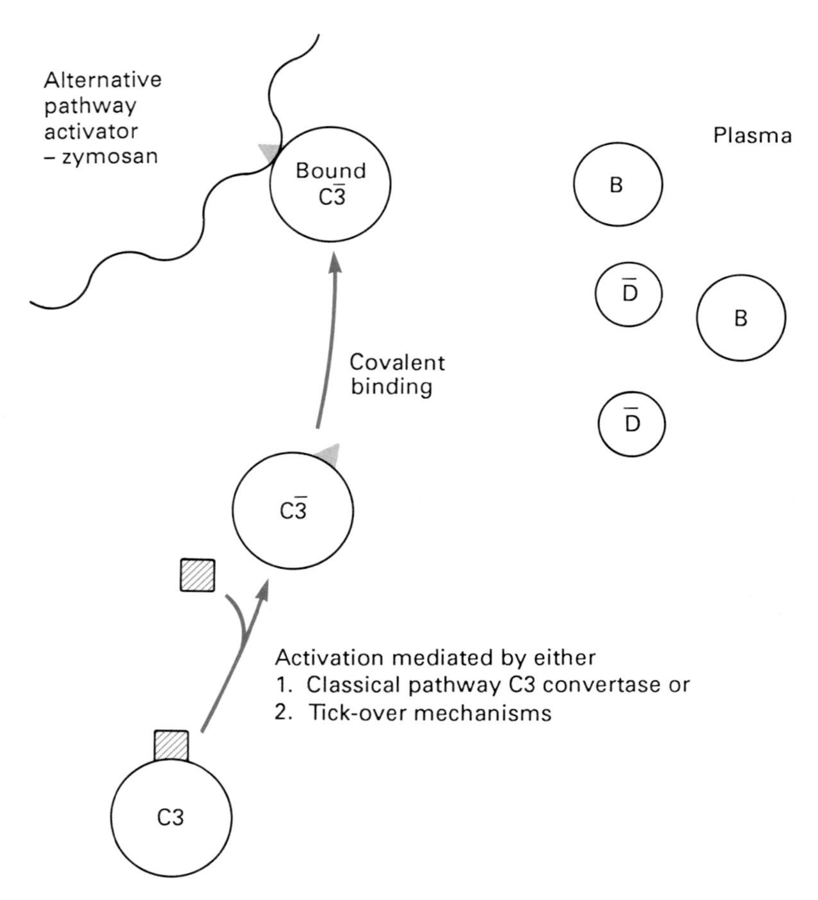

Fig.9.22 Schematic representation of the events which occur during activation of the alternative pathway. Covalent binding of C3 to an alternative pathway activator, zymosan, a yeast cell wall component (compare with Fig.9.23)

1. C3 can be cleaved by enzymes present in plasma other than the C3 convertases, such as plasmin. It has therefore been suggested that a low level of C3 cleavage is always ongoing in plasma in the absence of complement activation. In the absence of an alternative pathway activator, $C\overline{3}$ generated in this way is rapidly inactivated in the fluid phase by the control proteins factor I and factor H, and therefore cannot initiate alternative pathway activation. If it does bind to one of the host's own cells, then the membrane proteins CR1, DAF, and MCP (see Section 9.4.4) prevent the formation of the C3 convertase. However, in the presence of an activator, $C\overline{3}$ can bind covalently to the activator and is protected from inactivation by the control proteins. Thus it will initiate formation of the C3 convertase enzyme. In the presence of Mg^{2+}, $C\overline{3}$ bound to the activator can then bind factor B and activation of the pathway occurs (Fig.9.22). The precise mechanism of this activation is not clear. $C\overline{3}$ bound in a protective site could either result in an increase in the affinity of C3 for factor B, or prevent the binding of the control protein, factor H, to C3. There is some evidence to support this latter view.

2. The second proposal to explain the mechanism of alternative pathway activation suggests that the thioester bond in C3 can be broken without proteolysis (Fig.9.19). The bond is disrupted either by slow hydrolysis, small nucleophiles, or simply by perturbation of the structure to render the C3 inactive. In this state the inactivated C3 molecule, C3i, can combine with factor B and behave like 'C3'. It also behaves like 'C3' in other ways, for example it can bind to factor H and CR1 receptors. Therefore, spontaneous hydrolysis of C3 may result in the generation of the fluid phase C3 convertase enzyme ($\overline{C3B}$) which can generate more $C\overline{3}$. This will bind covalently in the presence of an alternative pathway activator and activate the pathway.

It is important to remember that all of these processes operate at a very slow rate in the fluid phase, and in the absence of an alternative pathway activator, C3b and C3i are rapidly destroyed.

Generation of the alternative pathway C3 convertase enzyme

The first molecule of covalently bound $C\overline{3}$ provides the site of initiation for alternative pathway activation (Fig.9.23(a)). Factor B, the alternative pathway counterpart of C2, then binds to the bound C3.

Factor B is a zymogen which is activated by another enzyme, factor $\overline{D}$ (Table 9.2a). Factor B can only be activated by factor $\overline{D}$ when it is bound to $C\overline{3}$, and this interaction requires the presence of Mg^{2+} (Fig.9.23(b)). Factor $\overline{D}$ is similar to C1s and is a serine protease. However, unlike C1s, factor $\overline{D}$ is always present in the plasma in its active form, but only in very low amounts, and it is highly substrate specific. Its only known substrate is factor B. Factor $\overline{D}$ activates factor B by cleaving it into two fragments Ba, 30 kDa and Bb, 60 kDa. The larger fragment, Bb, contains the active site, and when it is associated with $C\overline{3}$ the C3 convertase enzyme of the alternative pathway is generated, $\overline{C3B}$ or C3bBb.

Factor B and C2 are very similar in many respects, such as size, mechanism of activation, and substrate specificity. In addition, they are both class III major histocompatibility molecules (see Chapter 2; and Table 9.4) and are polymorphic. It has been suggested that they may have arisen by a gene duplication event.

The alternative pathway C3 convertase is responsible for the activation of C3, and in this respect is analogous to the classical pathway enzyme (Fig.9.21). $C\overline{3}$ generated by the alternative pathway enzyme can either:

(1) be used to form the C5 convertase enzyme $(C\overline{3)_n B}$; or

(2) more importantly, it can be used to generate more alternative pathway C3 convertase, by covalently binding to another site on the activator where it can associate with another molecule of factor B etc. (Fig.9.22).

The capacity of $C\overline{3}$ to form more of the same enzyme provides further amplification of the response and maximizes the production of C3b which has many important functions, including its involvement in phagocytosis and opsonization of foreign antigens.

Control of alternative pathway activity

The activity of the alternative pathway is controlled, in part, by the control proteins factors I and H, which act as regulators of C3 activity. The properties of these molecules are considered in Table 9.2b and were discussed in Section 9.4.4.

The molecule properdin also regulates the activity of the alternative pathway C3 convertase enzyme. However, in marked contrast to factors I and H it *stabilizes* the convertase activity (Table 9.2b). Properdin prevents the loss of $\overline{B}$ from the bound $C\overline{3}$ molecule. It combines with $C\overline{3}$ in either the fluid phase or when it is bound to a cell surface, and it promotes the assembly of the $C\overline{3B}$ enzyme, significantly increasing the half-life of the C3 and C5 convertases.

9.4.6 Terminal components and formation of the membrane attack complex

The activation of C5 is the last proteolytic event in the complement system. C5 is a two-chain molecule, 200 kDa, composed of an α chain and a β chain (Table 9.2a). It is activated in an identical way to both C3 and C4, and these three proteins form a family. Activation of C5 by either of the C5 convertase enzymes results in proteolytic cleavage of the α chain of C5 to release a low molecular weight peptide from the N-terminus, C5a. C5a has both anaphylatoxic and chemotactic activity. Although it does not participate further in the complement pathway, it is important for attracting other cells to the site of complement activation and causing a local inflammatory response (Table 9.5 and Section 9.4.7). C5a is probably the most important anaphylatoxin produced during complement activation. It is the most active of the three factors, C5a > C3a > C4a, and remains active even when the C-terminal arginine residue, which is critical for the activity of C3a and C4a, is removed.

C5b, the larger fragment generated on activation of C5, unlike C3b and C4b *cannot* bind covalently to the complement activator. It does, however, have an affinity for cell surfaces, through non-covalent interactions, particularly in the presence of C6 and C7, two components of the membrane attack complex (Fig.9.23(c)). In the absence of the late components of the complement cascade, C6–C9, C$\overline{5}$ rapidly loses its ability to interact with the cell surface. However, once activated, C$\overline{5}$ can form a stable complex with C6 in the plasma. This allows C$\overline{5}$ to retain its ability to interact with the cell membrane over a longer period of time. The C$\overline{56}$ complex has a molecular weight 330 kDa and contains a single molecule of each component. C$\overline{56}$ expresses a new antigenic determinant which is not present on either of the native molecules.

Once the C$\overline{56}$ complex has formed, attachment of C7 occurs rapidly to generate the C$\overline{567}$ complex (Fig.9.23(c)). In solution, the half-life of this

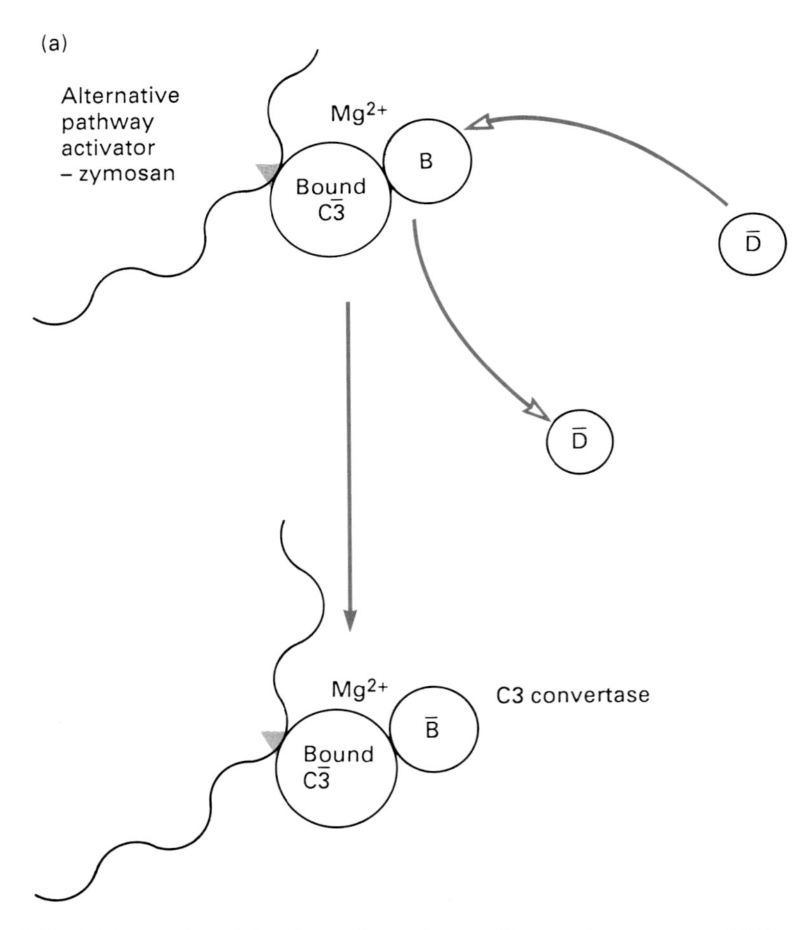

Fig.9.23 (a) Generation of the alternative pathway C3 convertase enzyme. (b) The amplification feedback mechanism of the alternative pathway and generation of the C5 convertase enzyme. (c) Diagrammatic representation of the formation of the membrane attack complex (MAC)

complex is short, but *in vivo* the $C\overline{567}$ complex rapidly binds to cell membranes. The interaction of $C\overline{567}$ with the target cell membranes can be prevented by a number of inhibitors. One class includes a series of polyanionic molecules, including dextran sulfate and DNA. Presumably these inhibitors function by blocking ionic interactions between $C\overline{567}$ and the membrane. Interestingly, serum lipoproteins can also act as inhibitors of the $C\overline{567}$ membrane interaction. This suggests that hydrophobic bonding may also be involved.

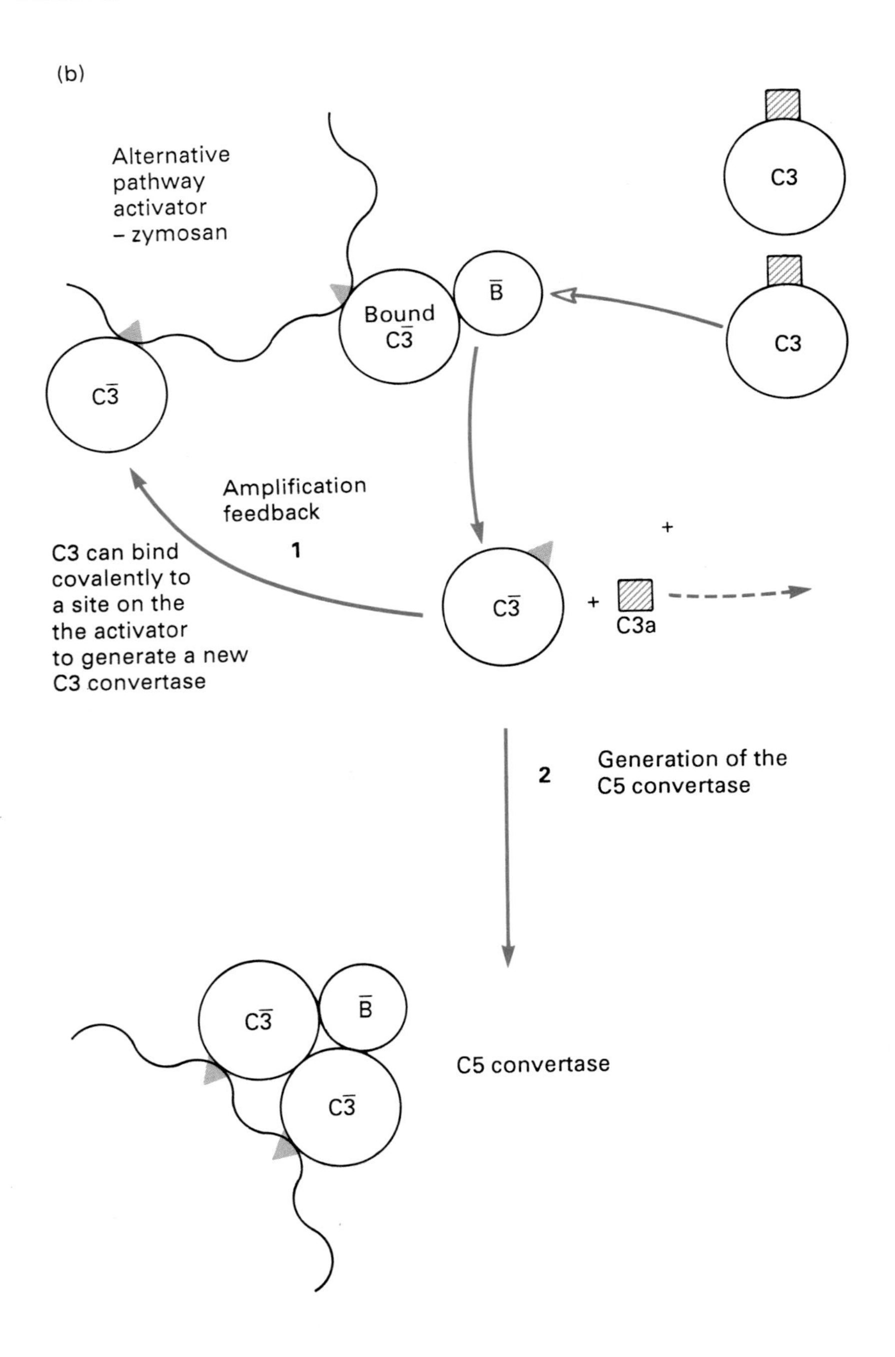

The binding of C8 to the C$\overline{567}$ complex takes place through a specific recognition site for C$\overline{56}$ present on the C8 molecule. Components C5, C6, C7, and C8 associate in the membrane attack complex on a uni-molecular basis and C$\overline{5678}$ is capable of mediating slow lysis of erythrocytes in the absence of the last component of complement, C9. However, the binding of C9 leads to the rapid lysis of the cells. The molar composition of the C$\overline{5\text{-}9}$ complex has been reported to be 1C5b:1C6:1C7:1C8:3-6C9 molecules, although the number of C9 molecules may be between 1 and 18 (Fig.9.23).

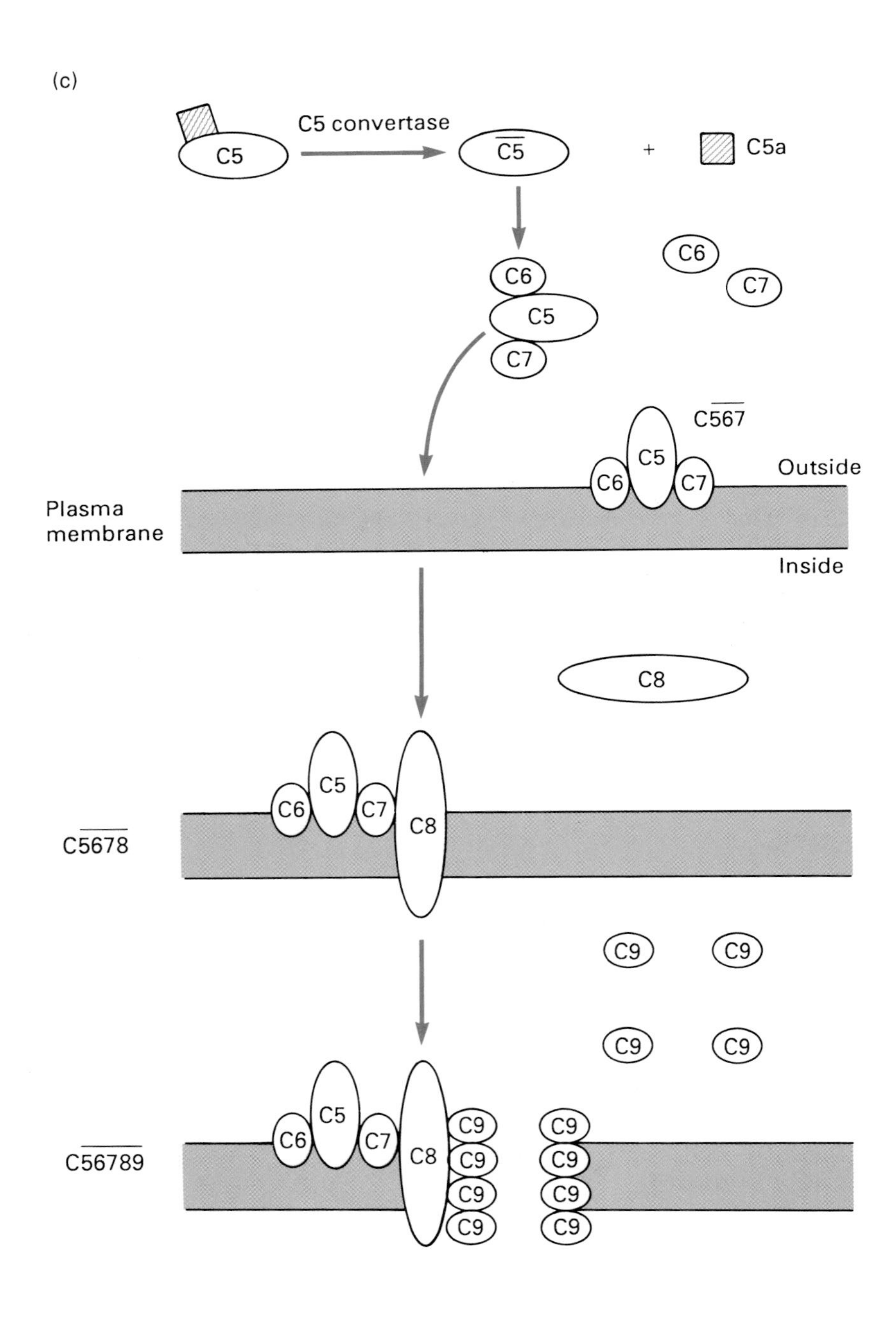

Electron microscopic studies have allowed the appearance of the membrane attack complex to be investigated. Current data suggest that the complex has a cylindrical stalk, at least part of which is embedded into the membrane. The remainder of the membrane attack complex projects above the membrane and forms a circular structure (see Fig.10.9). Immunochemical labelling of the membrane attack complex has shown that C9 is associated with the external ring-like component. These lesions are similar to those produced by the polyperforins secreted by natural killer cells and some cytotoxic T cells (see Section 10.3 and Fig.10.9) (Bhakdi and Tranum-Jensen 1991; Esser 1991).

Control of the membrane attack complex

To avoid bystander (i.e. non-specific), lysis of host cells, the activity of the $\overline{C567}$ complex is controlled by soluble inhibitors: lipoproteins, antithrombin III, and S-protein (or vitronectin) which are all present in the plasma; and membrane-bound inhibitors, e.g. CD59 and homologous restriction factor (HRF) which bind to C8 and C9 (Lachmann 1991).

S-protein (Table 9.2b) is the most efficient of the plasma inhibitors and probably is the most important for controlling the lytic potential of the membrane attack complex. It is a single polypeptide chain, 80 kDa, and is identical to vitronectin. Up to 3 molecules of S-protein can bind to the $\overline{C567}$ complex, preventing it from binding to cell surfaces, and these S-protein—$\overline{C567}$ complexes can bind C8 and up to 3 molecules of C9. They are cleared by vitronectin receptors that are expressed in a variety of cell types.

The membrane bound inhibitors, referred to collectively as homologous restriction factors, are anchored in the plasma membrane by glycolipid tails. They are present in the membranes of host cells to prevent damage by the membrane attack complex (MAC). CD59 binds to C8 and C9 in the MAC as it assembles, and is thought to prevent the unfolding of the first molecule of C9 that is bound to the MAC and to inhibit C9 polymerization so that lysis is prevented.

9.4.7 Summary of the functions and control of the complement system

1. Functions. Complement activation can result in:

(1) opsonization of particles of foreign antigen coated with antibody and/or complement proteins by phagocytic cells. (During complement activation immune complexes and other activators become coated with C3b and C4b. CR1 and/or CR3 receptors on the surface of phagocytic cells can interact with either of these complement components and promote phagocytosis of the particles, i.e. opsonization);

(2) inhibition of the formation and precipitation of immune complexes (see Section 6.2.4). This is an important function of the classical pathway during the early phases of antibody–antigen interaction. Once immune complexes have been formed, the alternative pathway plays an important role in their solubilisation. This involves both the antibody and

antigen becoming coated in C3b. A significant proportion of the bound C3b is in the Fab region which may promote disruption of the antibody–antigen lattice;

(3) lysis and cell death of the invading micro-organisms and bystander cells;

(4) the generation of biologically active molecules which contribute to an inflammatory response. These include the anaphylatoxins C3a, C4a, and C5a (see Sections 9.4.4 and 9.4.6).

All three anaphylatoxins are similar in structure and function. They are all single polypeptides of either 77 amino acids, C3a and C4a, or 74 amino acids, C5a, with molecular weights of between 9 and 11 kDa (Table 9.5). Each polypeptide is crosslinked by disulfide bonds, 3 in C3a and C4a, and 4 in C5a.

All three anaphylotoxins are biologically active and can induce smooth muscle contaction, increase vascular permeability and cause histamine release from mast cells and basophils. C5a is also chemotatic for neutrophils and monocytes, can augment cell adherence, enhance arachidonic acid metabolism (see Section 9.2.3), stimulate the production of toxic oxygen metabolites (Section 10.4.4), augment the expression of complement receptors, and increase IL-1 production by monocytes (see Section 9.5.1). C5a is up to 20 times more potent than C3a, which is in the order of 3000 times more potent than C4a.

The activity of the anaphylatoxins is regulated by removal of the C-terminal arginine residues by the enzyme carboxypeptidase N to produce the des-Arg derivatives which are less active. In addition, C5a receptors can bind, and thereby inactivate, C5a.

If these events were to occur in the wrong circumstances, they would have an adverse effect on the host organism. Thus control of the complement system is essential to avoid the wide-ranging consequences of unnecessary complement activity. The various levels of control are summarized below:

2. Activators. Specific activators are required to trigger activation of both pathways (see Section 9.4.3).

3. The half-life of the activated components. At each stage of complement activation, conversion of the next protein from its precursor to its active form must occur. The half-life of the activated component is short, and unless it interacts appropriately to continue the cascade, it is rapidly inactivated and can no longer participate in the pathway (see Section 9.4.4 and 9.4.5).

4. Control proteins. Apart from the intrinsic lability of the components of the system, a number of regulatory proteins also exist, each with the function of controlling a particular part of the complement cascade. The properties of each of the control proteins has been discussed at appropriate points in the text (Sections 9.4.4, 9.4.5, and 9.4.6).

9.4.8 Genetics and biosynthesis of complement proteins

Complement proteins are acute phase molecules which are synthesized rapidly following injury or infection. Most of the components normally present in the plasma are synthesized by the liver and the rate of synthesis can rapidly be increased in response to these events. The membrane-bound control proteins are synthesized by the cells on which they are expressed (Table 9.2b). It has also been demonstrated that peripheral blood monocytes and tissue macrophages are capable of synthesizing a wide variety of complement components and their regulatory proteins (Table 10.7). This may be important at extravascular sites, where the level of complement proteins is usually low.

Very little is known about the biosynthesis and secretion of the majority of components. A few studies have examined the synthesis of some of the multi-chain components, e.g. C3 and C4. C4 is synthesized as a single polypeptide chain (pro-C4) which is then cleaved into the three-chain structure before it is secreted from the cell (Fig.9.18). Messenger RNA for C4 when translated *in vitro* produces the pro-C4 molecule. Although C4 is a glycoprotein, the glycosylation events do not appear to be linked to the conversion of pro-C4 to C4, although the addition of agents which block glycosylation to cells in tissue culture does result in blocking of the secretion of the C4 molecule.

The amount of complement components present in the plasma can be modulated by some hormones, cytokines, and endotoxin. These modulators are generally thought to be of greatest significance for the biosynthesis of complement components by phagocytic cells.

The chromosomal location of some of the genes for complement components have been mapped. For example the *human* C3 genes are present on chromosome 19, the RCA gene cluster is on chromosome 1, and the genes C2, C4, and factor B are encoded by the short arm of chromosome 6 within the HLA complex. The latter assignments are interesting, as these components are encoded within the major histocompatibility complex and they form part of the class III family of major histocompatibility molecules (Table 9.4) (Carroll *et al.* 1984). In the *mouse*, the C4 genes have also been assigned to the major histocompatibility complex which, in this case, is located on chromosome 17.

All three complement components encoded by the HLA complex in man are polymorphic, as are other MHC molecules (see Chapter 2). The polymorphism of C4 molecules can be detected using serological techniques. The antigens involved were originally known as the Chido and Rodgers antigens, which were located in the C4d portion of the C4 molecule. The different forms of C4 are encoded by two different loci in man and the two forms are known as C4A and C4B. Their amino acid sequences have been determined, and there appear to be very few amino acid differences between the two molecules; only 13 changes out of the total of 1722 amino acid residues examined have been found. Variants can be typed according to their electrophoretic mobility and 13 alleles of C4A and 22 alleles of C4B have been identified so far. This parameter has also been used to define a system which can be used to classify the different forms of factor B and C2. The functional significance of the polymorphic forms of the complement components is not entirely clear, although it has been shown *in vitro* that they may form enzymes with different haemolytic efficiencies, e.g. C4A has been shown to be less effective in the lysis of erythrocytes than C4B. This may be one of the factors which defines susceptibility of an individual to disease.

9.5 Cytokines mediating inflammatory and effector functions

The properties of the cytokines, interleukin-1 (IL-1), interleukin-6 (IL-6), the tumour necrosis factors (TNFα and TNFβ), and the interferons (IFNα, IFNβ, and IFNγ) will be considered in this section as these are all soluble effector molecules that mediate powerful pleiotropic or multi-functional effects (see Section 4.4.1) during the effector phase of immune responses or inflammation. They are produced as part of the acute phase response, which can be triggered by infection, injury, tumour growth, and a variety of other stimuli (see Section 9.2.2). Each of the cytokines will be dealt with separately for convenience, but it is important to remember that they form part of the cytokine network and their biological activities and functions overlap. The redundancy between these cytokines may result from the need to back up systems such as these, that are critical for the elimination of pathogens and which play a key role in the body's response to tissue injury, autoimmunity, and inflammation (see Section 9.2). More information is required to clarify our understanding of the precise function of these cytokines *in vivo*, as the conditions used to assay their activity *in vitro* may not reflect the environment in which they act *in vivo*. For example, sudden changes in the concentration of a *single* cytokine would be unlikely to occur *in vivo*; much more likely would be changes in the concentration of several cytokines with related functions, e.g. IL-1, IL-6, and TNF. Furthermore, the amplification of immune and inflammatory responses as a result of the induction of cytokine gene expression by different cytokines operating in the regulatory network, may be a very important mechanism for the rapid activation of defence mechanisms when the host is confronted by a potentially lethal pathogen. However, as yet the interactions between cytokines operating in the network *in vivo* is not clearly understood.

9.5.1 Interleukin-1 and its receptor

IL-1 was originally named lymphocyte activating factor (LAF), because it was found to contribute to the stimulation of lymphocytes when suboptimal doses of mitogen were used. The technique used to define IL-1 activity was the so-called thymocyte proliferation assay. Thymocytes proliferate *in vitro* in the presence of a relatively high dose of a mitogen, such as conconavalin A (con A) or phytohaemagglutinin (PHA) (see Section 8.3.3). If the concentration of the mitogen present in the cultures is reduced below a certain threshold, then the thymocytes will no longer respond, unless a supernatant from a culture of activated macrophages is added. The agent produced by the activated macrophages which restores the proliferative capacity of the thymocytes has been characterized as IL-1.

Two forms of IL-1 have been identified, IL-1α and IL-1β. While these two molecules have distinct biochemical properties and only limited sequence homology, their biological activities are similar (see below). Why two forms of IL-1 are required is unknown, but one hypothesis suggests that each form may contact cell populations in different microenvironments; IL-1β acting as soluble mediator because it is secreted from the cells which produce it, and IL-1α being active during cell–cell contact, because it remains cell associated (di Giovine and Duff 1990).

IL-1 is part of the acute phase response that is triggered by infection, injury, tumours, and a variety of other immunological stimuli. Both forms of IL-1 are synthesized primarily by cells of the monocyte-macrophage lineage (see Section 10.4; Figs.1.11 and 1.31) in response to various inducers including endotoxin and TNFα. In addition IL-1 has been reported to be produced by other leukocytes including helper T cells, B cells, NK cells and neutrophils; certain epithelial cells, including thymic epithelium; endothelial cells; smooth muscle; astrocytes, microglia and glioma cells in the brain; keratinocytes of the skin; mesangial cells of the kidney, as well as by fibroblasts and chondrocytes of bone in response to various stimuli.

IL-1 genes and molecules

The genes for human IL-1α and IL-1β are encoded by separate, but closely-linked loci, located on chromosome 2. In the mouse, the genes are also located on chromosome 2, and are linked to the gene for β_2-microglobulin (see Section 2.8).

Each gene comprises seven exons; the gene encoding IL-1α covers 12 kb of DNA whereas that encoding IL-1β covers only 9.7 kb (Fig.9.24). Both genes are transcribed rapidly after stimulation (see below), but the mRNA for IL-1β accumulates at a faster rate (10–500 times) than that for IL-1α. As for other cytokines, the mRNAs for both forms of Il-1 have a short half-life, in the order of minutes. IL-1 protein is detectable within 30 minutes of stimulation and steady-state levels are reached within 18–20 hours. In contrast to the rate of transcription of the IL-1α and IL-1β genes, at the protein level IL-1α is more abundant than IL-1β. 60–70% of the IL-β produced by a cell is secreted into the extracellular environment, while the majority of IL-1α produced remains associated with the surface of the cell that produced it.

Both forms of IL-1 are synthesized as propeptides, 271 amino acids long, with a molecular weight of 31 kDa. IL-1 is mainly translated in the cytosol, as the immature polypeptide does not possess a signal sequence and therefore cannot insert into the endoplasmic reticulum. The mature forms of IL-1 are produced by post-translational modification from the C-terminal portion of the immature peptides (Table 9.6).

IL-1α contains 159 amino acids (17.5 kDa) and has an isoelectric point (pI) of 5, while IL-1β is slightly smaller comprising 153 amino acids (17.3 kDa) and has a pI of 7. Interestingly the propeptides for the two forms of IL-1 exhibit quite low amino acid homology, 26%; at the nucleotide level they show 46% homology. IL-1β is a globular protein containing 6 pairs of anti-parallel β strands.

IL-1 receptor

IL-1 mediates its action through a specific receptor, the IL-1 receptor (IL-1R) which is a transmembrane glycoprotein (80 kDa), expressed by many different cell types in the body (Fig.9.25). The IL-1R is a member of the immuno-

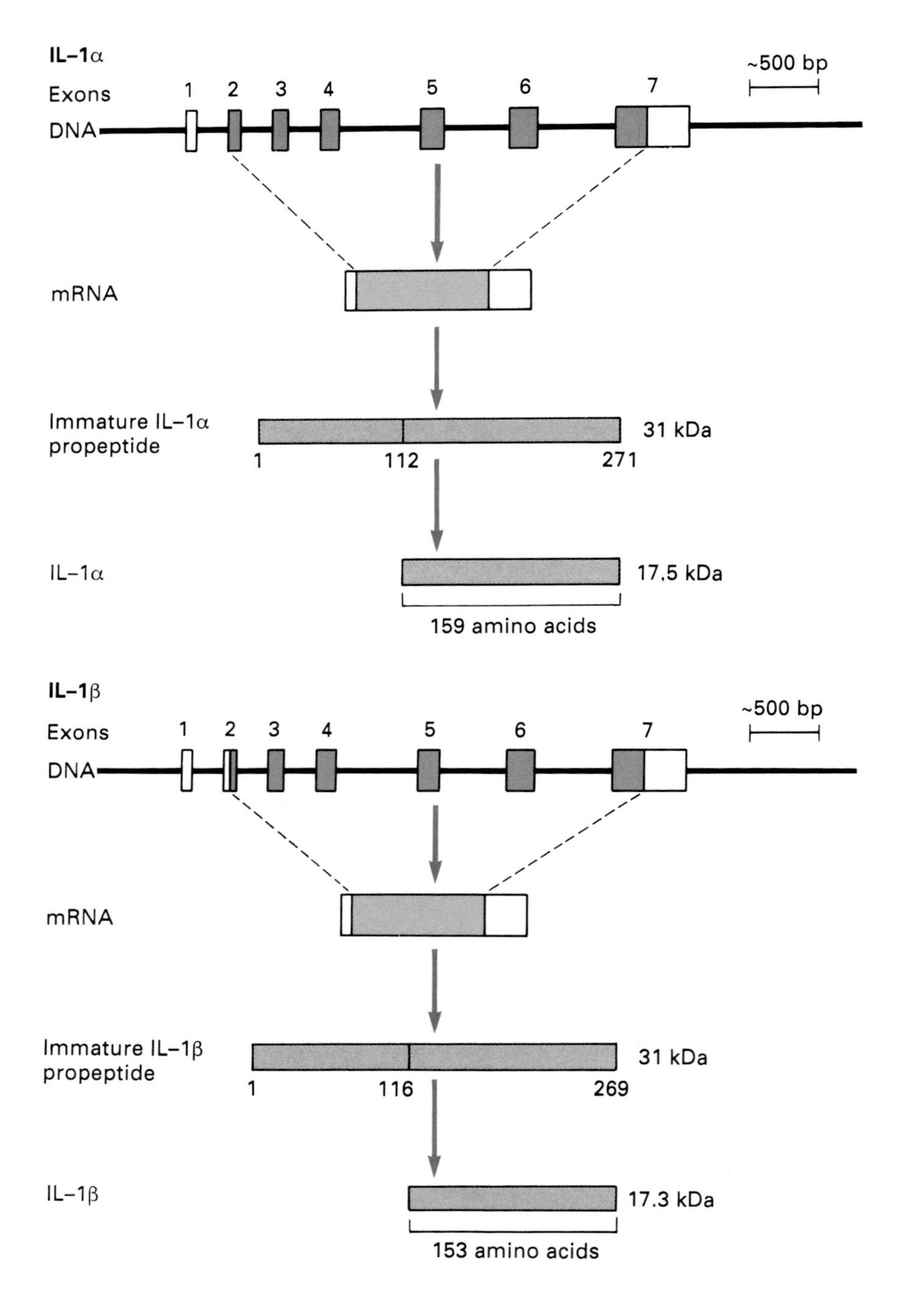

Fig.9.24 Biosynthesis IL-1α and IL-1β

Table 9.6 Human interleukin-1

Form	IL-1α	IL-1β
Original synonym	Lymphocyte activity factor (LAF)	
Main inducer	Infection and tissue injury	
Number of genes	1	1
Chromosomal location	2	2
Number of exons	7	7
Product: Immature	31 kDa	31 kDa
Mature	17.5	17.3
pI	5	7

globulin superfamily (see Section 7.4) as the extracellular portion of the polypeptide is organized into 3 domains which resemble those of immunoglobulins (Sims *et al.* 1988; Dinarello *et al.* 1989).

The IL-1R has a polypeptide backbone of 62 kDa with 7 potential N-linked glycosylation sites (see Fig. 1.4) in the extracellular portion of the polypeptide chain. The cytoplasmic tail of the molecule contains 217 amino acids, but the role of this portion in signal transduction is unknown (see below).

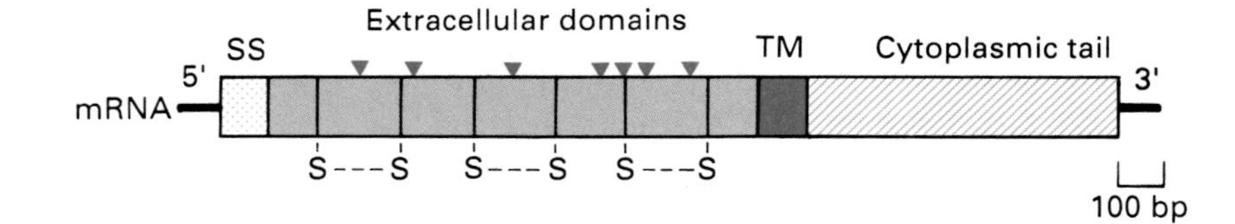

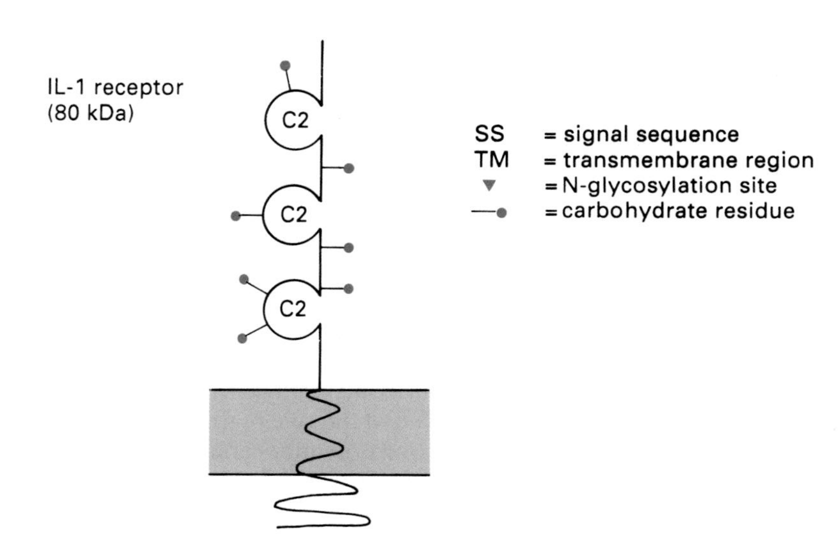

Fig. 9.25 IL-1 receptor: the mRNA encoding different regions of the polypeptide, and the membrane-bound glycoprotein molecule

A second form of the IL-1R (60 kDa) has been identified on B cells and fibroblasts. IL-1Rs of high ($4\text{–}15 \times 10^{-12}$ M), intermediate, and low (5×10^{-10} M) affinity have also been described, but the molecular basis for this is not understood. Whether more than one polypeptide chain is involved in forming the high affinity receptor, as is found for the IL-2R (see Section 4.4.3), is not known.

On many types of cells the receptor for IL-1α and IL-1β is identical, although the specific interactions that occur between the two forms of IL-1 and the receptor may be different.

For example, single point mutations or the addition of multiple amino acids to the amino terminus of IL-1β, as well as mutations in the middle section of the molecule, have a marked effect on both its biological activity and its interaction with the IL-1R. In contrast similar mutations in IL-1α have little effect on either its activity or its ability to bind to IL-1R.

Once the IL-1R has interacted with its ligand, the complex is internalized. The mechanism of signal transduction following binding of IL-1 to its receptor is unresolved at present and may depend on the cell type involved.

Cyclic AMP, protein kinase A, and NF-κB have been implicated as important in some studies using a pre-B cell and NK-like cell lines (Mizel 1990), whereas in other studies using fibroblasts, a pertussis-toxin-sensitive G protein and a serine kinase distinct from protein kinases A or C have been identified (O'Neill *et al.* 1990); Ca^{2+} influx and phosphatidylinositol breakdown (see Section 4.2.3) have also been described. A central feature of the mechanism of action of IL-1 is the induction of transcription of specific genes, the gene induced depending on the cell type involved. For example in B cells activation of the κ light chain genes has been reported, whereas in other leukocytes, induction of the CD25, IL-2R α chain gene, c-fos, c-myc, or c-jun have all been reported.

Regulation of IL-1 production and activity

Several natural feedback loops exist to control IL-1 activity and down-regulate its production. One of these loops involves the productoin of adrenocorticotropic and glucocorticoid hormones by IL-1 stimulation of the pituitary-adrenal axis and the subsequent inhibition of IL-1 production by the hormones. In addition, specific inhibitors of IL-1 bioactivity have been cloned and characterized, including (i) a protein (22 kDa) present in supernatants from cultured adherent mononuclear cells stimulated with immobolized immune complexes (Eisenberg *et al.* 1990; Hannum *et al.* 1990), and (ii) an IL-1R antagonist protein (25 kDa) which is produced by monocytes stimulated with phorbol myristate acetate (PMA) (Carter *et al.* 1990). Inhibitors which block the binding of IL-1 to its receptor have also been described in human urine (18–25 kDa) and in supernatants from cultured human monocytes after stimulation with granulocyte-macrophage colony-stimulating factor (GM-CSF). Further characterization of the role of these IL-1R antagonists *in vivo*, particularly in relation to inflammation and disease, is required although they have already been shown to reduce the lethal effects of endotoxin-induced shock (Ohlsson *et al.* 1990).

Stimuli for IL-1 production

IL-1 production can be stimulated in an antigen-dependent or independent fashion. *Antigen-dependent* stimulation can occur either directly, during antigen presentation when a T cell is in contact with an antigen presenting cell (see Section 3.2) or indirectly, as a result of the release of cytokines, such as IL-2 or IFNγ, from an activated T cell. *Antigen-independent* induction of IL-1 can occur when pathogens such as viruses or bacteria, or certain irritants such as asbestos or silica, come into contact with monocytes or macrophages. In addition other cytokines, GM-CSF, TFGβ, TNFα, IFNα and β, as well as IL-1 itself, can trigger IL-1 release from monocytes or macrophages.

Biological activities of IL-1

IL-1 has potent pleiotropic effects which can be either localized to the site of IL-1 release, so-called local effects, or systemic (Table 9.7). These local and systemic activities are vital to a large number of immunological, endocrinological, haemological, neurological, and metabolic responses that are triggered by certain pathogens or cancers (Fig.9.26). For example IL-1 : (i) is required by activated helper and cytotoxic T cells for progression from early to late G_1 in the cell cycle and for the expression of high affinity IL-2R, though not apparently during activation by dendritic cells (see Section 3.4.5); (ii) induces neutrophils to leave the bone marrow and enter the circulation (see Figs.1.34 and 1.35), and subsequently enter the tissues (Fig.1.36); (iii) acts as a chemoattractant for neutrophils, and results in their activation; (iv) enhances the production of collagen type IV by epidermal cells; (v) affects bone metabolism by inducing osteoblasts to proliferate and osteoclasts to resorb into bone; (vi) stimulates hepatocytes to produce other acute phase proteins, e.g. complement components (see Section 9.4); and (vii) induces endothelial cells to proliferate and increases their adherence to leukocytes. This is by no means a comprehensive list of the consequences of IL-1 production, but gives a feeling for the multifunctional nature of this cytokine.

High levels of IL-1, often in concert with other cytokines such as TNF and IL-6 (Table 9.8), can lead to tissue destruction and chronic inflammation (see Section 9.2). IL-1 participates in the acute phase response, in combination with TNF and IL-6, by triggering the synthesis of the acute phase proteins, which include some of the complement components, α_1-antitryppsin and haptoglobin in the liver. IL-1 has also been reported to be of major significance in the pathogenesis of some autoimmune disease, e.g. it inhibits insulin release from, and is toxic for, insulin-producing β cells in the pancreas; it has also been implicated in glomerulonephritis.

The effects of IL-1 can be inhibited by certain hormones; other cytokines, such as IL-4, IFNγ, and IL-1 itself; prostaglandin E_2 and histamine (see Section 9.2.3). The effects of IL-1 are also inhibited during starvation.

9.5.2 Interleukin-6 and its receptor

IL-6 is another typical example of a multi-functional cytokine. It was originally referred to by a number of different names each reflecting a characteristic of the numerous biological activities of the molecule. These included B cell differentiation factor (BCDF) and B cell stimulatory factor 2 (BSF-2) because of its role in B cell responses (see Section 8.4.3). It was also called IFNβ2 because it was originally thought to possess anti-viral activity, but this was later disproved. When the gene encoding the molecule responsible for each of these activities was cloned, it was clear that only one cytokine was responsible, IL-6 (Table 9.9).

Table 9.7 Effects of recombinant IL-1

Local effects of recombinant interleukin-1*

Non-immunological
- Chemoattractant (*in vivo*)
- Basophil histamine release
- Eosinophil degranulation
- Increased collagenase production
- Chrondrocyte protease release
- Bone resorption
- Induction of fibroblast and endothelial CSF activity
- Production of PGE_2 in dermal and synovial fibroblasts
- Increased neutrophil and monocyte thromboxane synthesis
- Cytotoxic for human melanoma cells
- Cytotoxic for human β islet cells
- Cytotoxic for thyrocytes
- Keratinocyte proliferation
- Proliferation of dermal fibroblasts
- Increased collagen synthesis
- Mesangial cell proliferation
- Gliosis

Immunological
- T cell activation
- IL-2 production
- Increased IL-2 receptors
- B cell activation
- Synergism with IL-4
- Induction and synergism with IL-6
- Activation of natural killer cells
- Synergism with IL-2, interferons, on NK cells
- Increased lymphokine production (IL-3, IL-6, IFN)
- Macrophage cytotoxicity
- Growth factor for B cells
- Increased IL-1 production

Systemic effects of recombinant IL-1†

Central nervous system	**Metabolic**
Fever	Hypozincemia, hypoferremia
Brain PGE_2 synthesis	Decreased cytochrome P_{450} enzyme
Increased ACTH	Increased acute-phase proteins
Decreased REM sleep	Decreased albumin synthesis
Increased slow-wave sleep	Increased insulin production
Decreased appetite	Inhibition of lipoprotein lipase
	Increased sodium excretion
	Increased corticosteroid synthesis

Haemologic	**Vascular**
Neutrophilia	Hypotension
Nonspecific resistance	Increased leukocyte adherence
Increased GM-CSF	Increased PGE synthesis
Radioprotection	Decreased systemic vascular resistance
Bone marrow stimulation	Decreased central venous pressure
Tumour necrosis	Increased cardiac output
	Increased heart rate
	Decreased blood pH
	Lactic acidosis
	Chemoattractant

*These effects are suggested by *in vitro* studies.
†These effects have been demonstrated by administration of IL-1 *in vivo*.

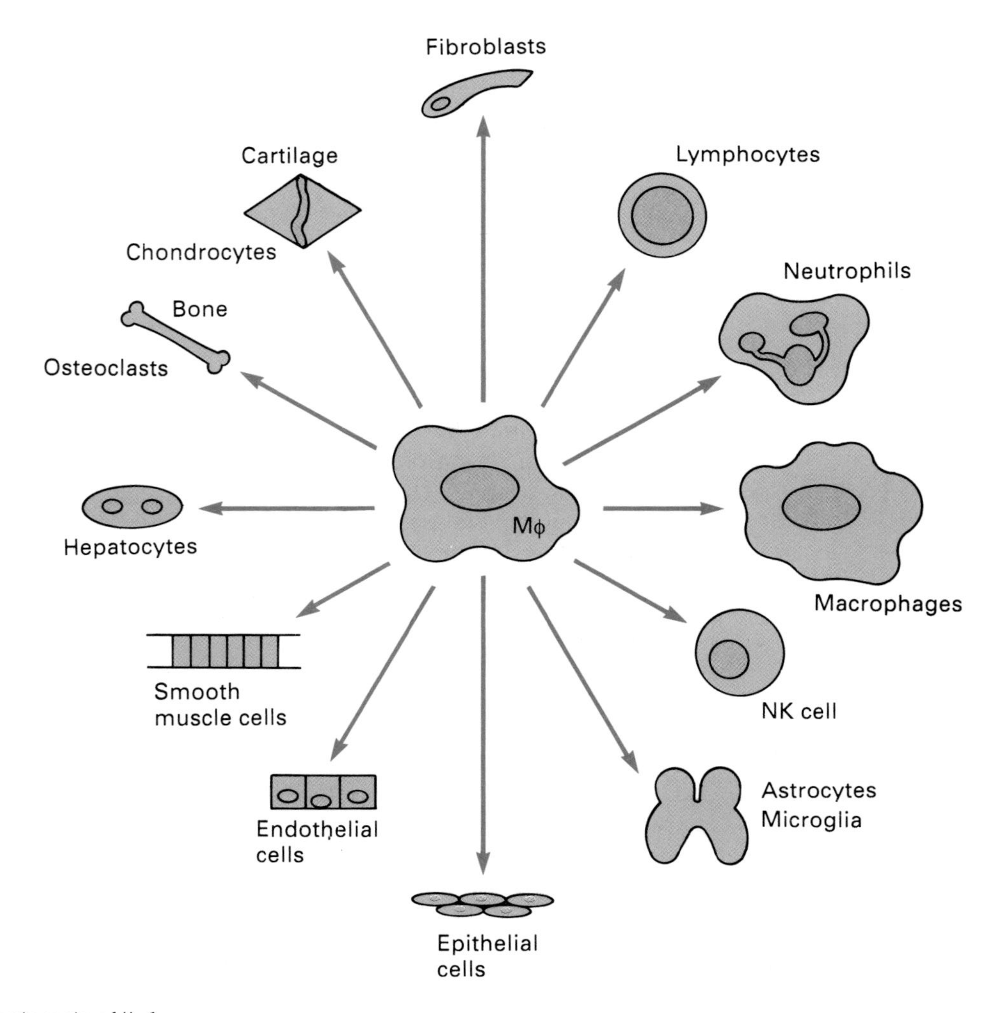

Fig.9.26 Cellular targets of IL-1

Table 9.8 Comparison of the biological activities of IL-1, TNF, and IL-6

Biological property	IL-1	TNF	IL-6
Endogenous pyrogen fever	+	+	+
Hepatic acute-phase proteins	+	+	+
T cell activation	+	±	+
B cell activation	+	±	+
B cell Ig synthesis	±	−	+
Fibroblast proliferation	+	+	−
Stem cell activation (haemopoietin-1)	+	−	+
Nonspecific resistance to infection	+	+	+
Radioprotection	+	+	±
Synovial cell activation	+	+	−
Endothelial cell activation	+	+	−
Induction of IL-1 and TNF from monocytes	+	+	−
Induction of IL-6	+	+	−
Haemodynamic shock	+	+	ND
Decreased albumin synthesis	+	+	ND
Activation of endothelium	+	+	ND
Decreased lipoprotein lipase	+	+	ND
Increased fibroblast proliferation	+	+	ND
Increased synovial cell collagenase and PGE_2	+	+	ND
Haemopoietin-1 activity	−	+	ND

IL-6 gene and the molecule

The gene for human IL-6 has been mapped to chromosome 7p21 in the human, and to a region of chromosome 5 in the mouse that is related to human chromosome 7. The IL-6 gene shows a high degree of homology, 65%, between the two species and comprises five exons (Fig.9.27). The human gene has three initiation sites for transcription and consensus sequences for cAMP and activator protein 1 (AP1) binding; in addition, glucocorticoid-responsive and enhancer elements have been identified.

Table 9.9 Human interleukin-6

Original synonyms include	BSF-2, IFNβ_2, BCDF
Main inducer	Infection and tissue injury
Number of genes	1
Chromosomal location	7p21
Number of exons	5
Product: Mature	21–28 kDa

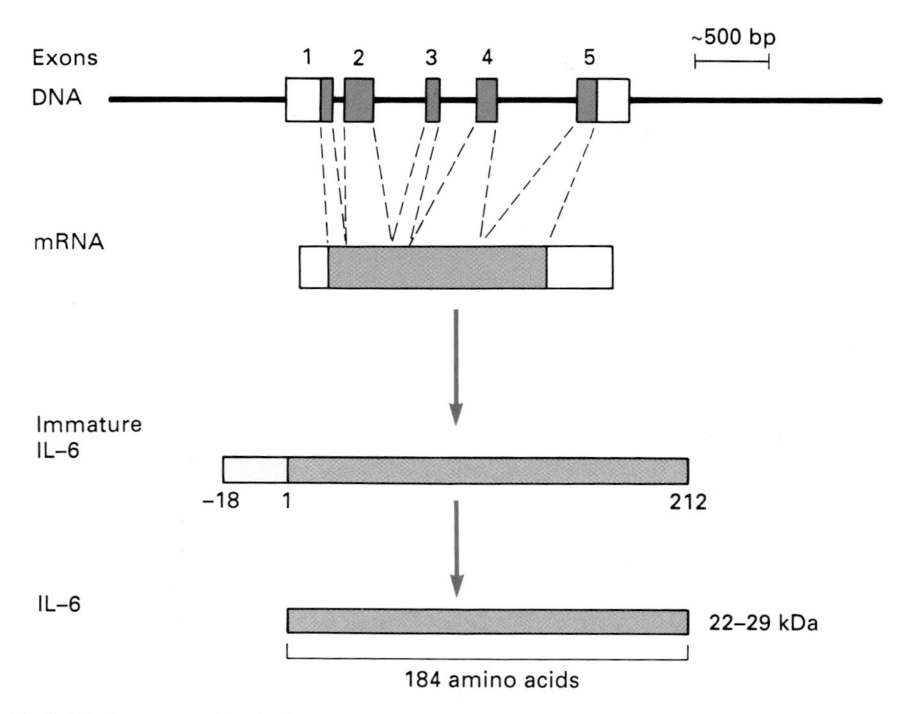

Fig.9.27 Gene encoding IL-6

Both the mouse and human proteins differ in molecular weight depending on the cellular source from which they are isolated, 22–29 kDa and 21–28 kDa respectively. The number of amino acids in each of the polypeptide chains is similar, 212 in human IL-6 and 211 in mouse. Heterogeneity in the molecular weight of IL-6 is thought to result from different modifications, including O-linked glycosylation in mouse and O- and N-linked glycosylation in human, as well as covalent modification by phosphorylation of serine residues at certain points in the polypeptide chain. A signal sequence is present in the immature polypeptide in both species (unlike IL-1, see p. 555).

The homology between human and mouse proteins is 42%, with the central region of the molecule being most highly conserved. Each molecule contains four cysteine residues in identical positions relative to one another in the primary stucture; this structural motif has also been found in other growth factors, including G-CSF. The N-terminal region of the polypeptide chain does not seem to be essential for the biological activity of IL-6, as deletions of up to 28 amino acids in this region have no effect. Similarly, glycosylation of the molecule does not appear to be essential for bioactivity as recombinant IL-6 produced by prokaryotes is active.

IL-6 Receptor

Both high (10^{-12} M) and low (10^{-9} M) affinity IL-6 binding sites have been identified on a variety of cells, including B and T lymphocytes; epithelial cells; fibroblasts; and cells of the nervous system. The IL-6 receptor (IL-6R) has been cloned (Yamasaki *et al.* 1988). It is a member of the immunoglobulin superfamily (see Section 7.4) and shares homology with similar domains in other receptors, such as the IL-1R (see Section 9.5.1); platelet-derived growth factor receptor (PDGF), and colony-stimulating factor-1 (CSF-1).

The IL-6R is a typical transmembrane protein of 468 amino acids (Fig.9.28). The polypeptide chain can be divided into the extracellular portion which contains 340 amino acids, a transmembrane region of 28 amino acids, and a cytoplasmic tail of 82 amino acids.

The cytoplasmic tail of the molecule does not appear to play a direct role in signal transduction following binding of IL-6 to the IL-6R. Instead the IL-6R polypeptide associates with another membrane glycoprotein, gp130 (130 kDa), which has been shown to transduce the signal (Hibi *et al.* 1990). The gp130 molecule does not bind to IL-6 itself, but is thought to be involved in the formation of high affinity binding sites for IL-6 when it associates with the IL-6R in the plasma membrane (compare IL-2R, Section 4.3.3).

The gp130 molecule acts as a transducer protein and consists of 918 amino acids with a single transmembrane domain. The extracellular portion of gp130 comprises six units of a fibronectin type III motif and is similar to that of some other cytokine receptors (see Section 4.3.3, p. 248).

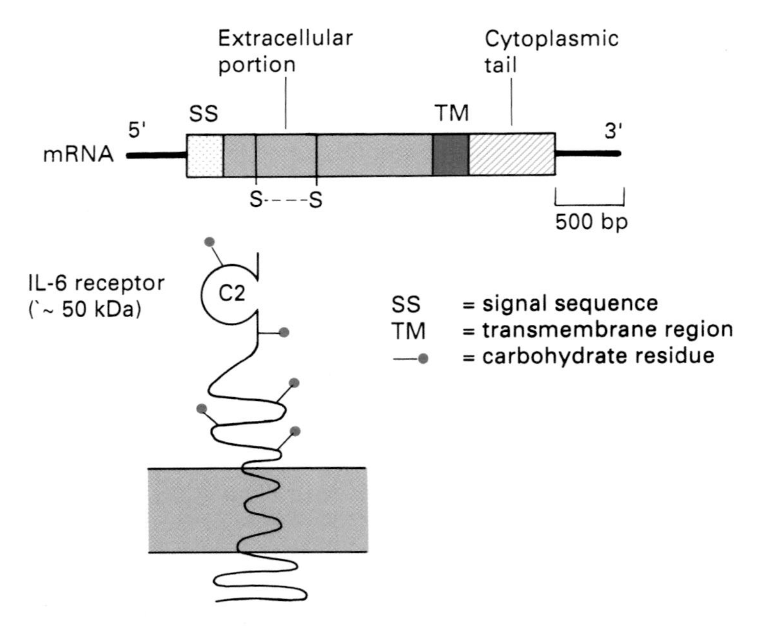

Fig.9.28 IL-6 receptor: the mRNA encoding different regions of the polypeptide, and the membrane-bound glycoprotein molecule

The signal transduction pathways that are triggered when IL-6 binds to the IL-6R-gp130 complex are not clearly understood, particularly in relation to the pleiotropic effects of this cytokine.

It has been proposed that IL-6-induced signal transduction in B cell hybridomas may proceed through a novel kinase pathway, which involves activation of the response genes TIS11 and the transcription factors AP-1 and junB. The events that occur downstream of the gp130 transducer may vary in different cell types, but these remain to be elucidated.

Stimuli for IL-6 production

IL-6 is not produced constitutively by normal cells. Production of this cytokine is readily stimulated by viral and bacteria infections, upon exposure to lipopolysaccharide (LPS) and in response to tissue damage or injury. In addition a variety of other cytokines can induce IL-6 biosynthesis, including IL-1 (see Section 9.5.1), TNFα (see Section 9.5.3), either alone or in combination with IFNγ (see Section 9.5.4), PDGF, IL-3, and GM-CSF, but the effect of a particular cytokine on IL-6 production depends on the cell type involved.

Biological activities of IL-6

IL-6 is pleiotropic, having many diverse biological activities *in vivo* and *in vitro* (Table 9.10) and it often acts in concert with other cytokines in a complex regulatory network (see Table 9.8; Fig.9.29). Many of the effects of IL-6 have been determined *in vitro* and the precise effects of IL-6 in some situations *in vivo* awaits clarification.

Acute phase response: IL-6 acts as an inducer molecule in the acute phase response, by controlling the transcription levels of acute phase proteins. IL-6 levels rapidly increase after infection and it contributes to the body's defence by inducing fever and stimulating the release of adrenocorticotropic hormone.

B cells: IL-6 is involved in B cell development and contributes to the generation of antigen-specific humoral responses (see Section 8.4.4). IL-6 has been shown to augment secondary, but not primary antibody responses and it can stimulate antibody production by Epstein–Barr Virus (EBV) transformed B cells *in vitro*, without triggering proliferation of the B cells. It also enhances the growth of B cell hybridomas and plasmacytomas *in vitro*, which is important as it could be used to facilitate the production of monoclonal antibodies (see Appendix). IL-6 may be involved in enhancing the growth of myelomas

Table 9.10 Effects of IL-6

Effects of IL-6
Enhanced growth of transformed B cells
B cell development
T activation
Haemopoeisis
Acute phase response

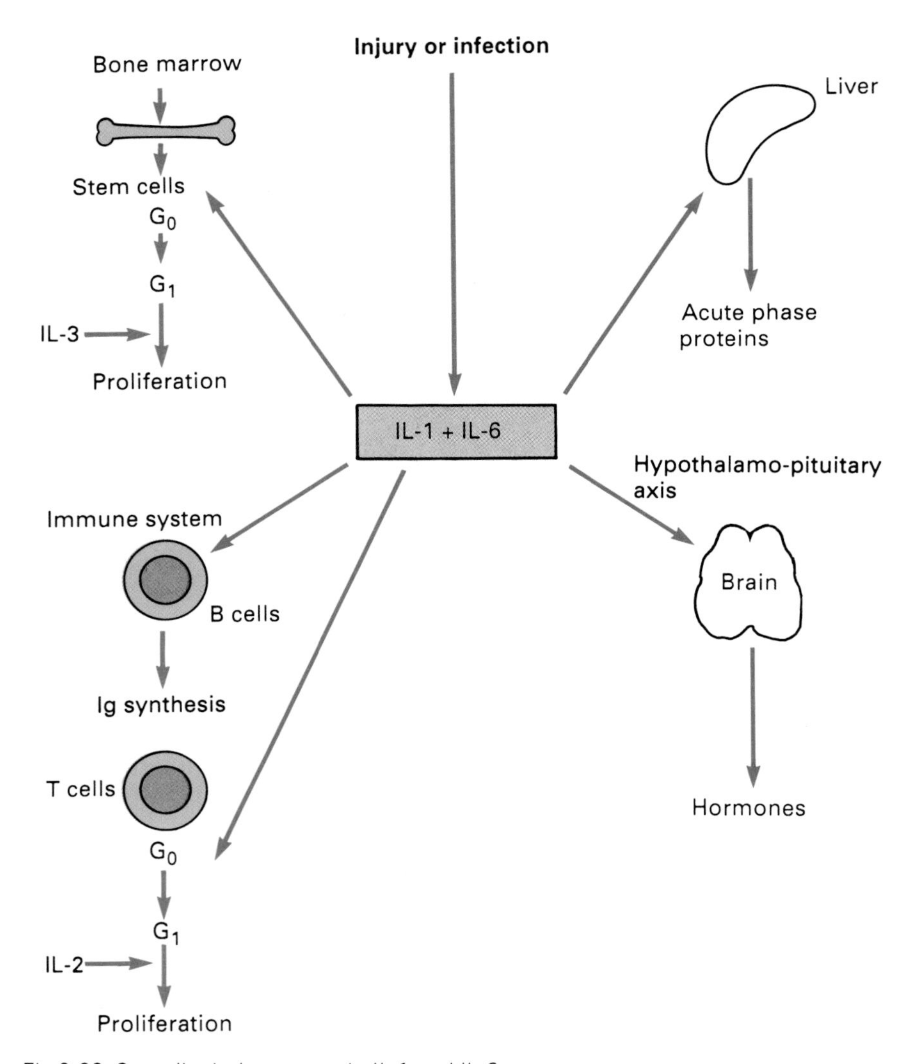

Fig.9.29 Co-ordinated response to IL-1 and IL-6

in vivo, suggesting a role for this cytokine in the development or progression of B cell tumours, although this remains to be confirmed.

IL-6 may act in synergy with other cytokines in some situations. For example in the mouse, a 50-fold increase in the secretion of IgM from B cells has been reported in the presence of a combination of IL-6 and IL-1, and synergy between these two cytokines has also been reported for human B cells stimulated with dexamethasone and IL-2.

T cells: IL-6 is involved in T cell activation and synergizes with IL-1 to control the initial stages of the activation process. One model proposed to explain the role of IL-6 and IL-1 in antigen-induced activation of resting T cells suggests that IL-6 causes the cells to move from G_0 to an early stage in

G_1 of the cell cycle (see Fig.1.14), where they express functional IL-2Rs. In the presence of IL-1 the T cells then move to the second stage of G_1 where the production of IL-2 is initiated (Van Snick 1990). Thus, IL-6 acts predominantly by enhancing the T cell's response to IL-2, and IL-1 predominantly by triggering IL-2 production. IL-1 and IL-6 have also been reported to act synergistically in the induction of CTL responses and IL-6 may play a role in stimulating the proliferation of thymocytes in association with IL-1.

Haemopoiesis: IL-6 may be one of the factors capable of inducing stem cells to enter the cell cycle, by acting synergistically with other factors, including IL-3, GM-CSF, and M-CSF.

9.5.3 Tumour necrosis factors and their receptor

The tumour necrosis factors (TNFs) are multifactorial cytokines that possess numerous biological activities involving the acute phase response to infection and injury, as well as regulation of cell growth and differentiation (Beutler and Cerami 1989; Vassalli 1992). This activity was originally ascribed to a molecule(s) named TNF because it caused necrosis when it was added to transformed cells both *in vitro* and *in vivo*. Two related TNFs exist, TNFα or cachectin and TNFβ or lymphotoxin (Paul and Ruddle 1988). The name cachectin was first used to describe a molecule that was found to cause wasting or cachexia in animals, particularly during chronic infections and malignancies. This was later found to be identical to TNFα which had been defined using tumour necrosis assays. TNFβ was originally thought to be the mediator of the 24 hour inflammatory response known as delayed type hypersensitivity (see Section 10.5.5). It was noted that a factor produced by activated T cells in the presence of antigen was also capable of killing other cells, by a process described as bystander killing. These biological activities were later found to be mediated TNFβ.

TNFα is produced predominantly by activated macrophages and lymphocytes, while TNFβ is produced by activated lymphocytes, but not macrophages. Resting cells from either of these lineages do not produce either of the TNFs (Table 9.11).

Table 9.11 Human TNFα and TNFβ

	TNFα	**TNFβ**
Synonyms	Cachectin	Lymphotoxin
Main inducer	Infection and tissue	Injury
	Activated macrophage	Activated lymphocytes
Number of genes	1	1
Chromosomal location	6	6
Number of exons	4	4
Mature protein	17.4 kDa	18.6 kDa
Immature protein	~25 kDa	~25 kDa

TNF genes and molecules

TNFα: The gene for TNFα is located on chromosome 6 in human and on chromosome 17 in mouse, in the class III region of the MHC (Trowsdale *et al.* 1991). (Figs.2.22 and 2.23).

The gene consists of four exons and spans 2.8 kb (see Fig.9.30). The mature form of *human* TNFα can exist either as a soluble, non-glycosylated molecule of 152 amino acids (17.4 kDa), or in a membrane-bound form on the surface of monocytes and some other cell types with cytotoxic activity (26 kDa). It is derived from a propeptide, that contains an additional 76 amino acids. In the *mouse*, TNFα is composed of 156 amino acids and the mature molecule may be glycosylated. It is also derived from a propeptide which contains an extra 79 amino acids. In both species the polypeptide contains a single disulfide bond and seven, regularly-spaced hydrophobic regions. The mature molecule can form multimers containing between 2 and 5 subunits, but it usually exists as a trimer and it is assumed that in this configuration TNFα is biologically active. The polypeptide chain of each subunit is folded into eight anti-parallel β strands, in a so-called 'jelly-roll' structural motif that is characteristic of viral coat proteins (Jones *et al.* 1989). TNFα is highly conserved between species, with 79% homology between human and mouse.

TNFβ: The gene for human TNFβ is also located on chromosome 6 in the class III region of the MHC (see Figs.2.22 and 2.23) and has 4 exons (see Fig.9.30).

Human TNFβ has a signal peptide of 34 amino acids and the mature protein comprises 171 amino acids (18.6 kDa). The molecule is glycosylated at amino acid residue 62 and the polypeptide chain does not contain a disulfide bond. In the *mouse* a signal peptide of 33 amino acids is present and the mature protein is 2 amino acids shorter than in human, comprising 169 amino acids. There is one potential site for glycosylation and, in contrast to human TNFβ, the mouse polypeptide contains a single cysteine residue. TNFβ can also form trimers and, as for TNFα, this is thought to the form that is biologically active.

Relationship between TNFα and TNFβ : TNFα and TNFβ are homologous molecules that appear to have evolved together, probably from a common ancestor. The genes for both molecules are closely linked on the same chromosome and the two cytokines seem to have similar functions acting through the same receptor (see below).

There is 28% homology between TNFα and TNFβ at the amino acid level and 46% at the nucleotide level. There are two conserved regions shared by the two molecules, at amino acid positions 35–66 and 110–133, and these are believed to be critical for the cytotoxic activity of the TNFs (see Section 10.3.4). In other respects the two molecules are dissimilar; for example, TNFβ does not contain the disulfide bond present in TNFα, and there is little similarity between the two molecules in this region, which may account for differences in their biological activity. In the 3′ untranslated region of both genes there are conserved sequences of tandem and overlapping repeats of an octamer sequence, TTATTTAT, which is also found in the same region of other cytokine genes, including IL-1, IFN, and GM-CSF. Such sequences may be important in the control of expression of these cytokines.

TNF receptors

The biological responses of TNFα and TNFβ are initiated by binding to specific cell surface receptors (TNFRs). Structural analysis of these receptors has

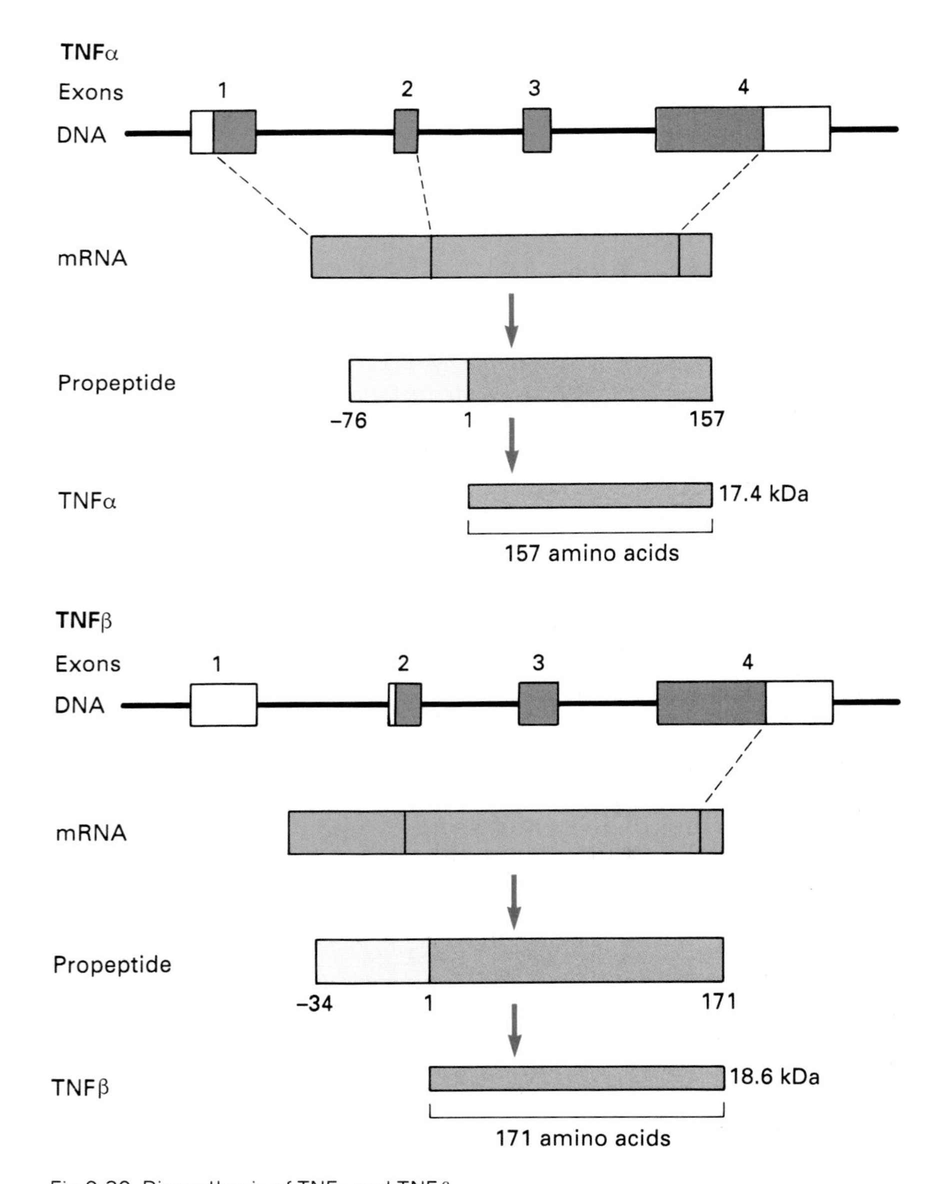

Fig.9.30 Biosynthesis of TNFα and TNFβ

shown that they are heterogeneous, with molecular weights ranging from 55–60 kDa, 70 kDa, and 80 kDa, and affinities of between 10^{-9} and 10^{-12} M. TNFRs are expressed by a variety of cell types and the heterogeneity observed in these molecules may reflect both cell specific and tissue specific expression.

Both TNFα and TNFβ bind to a 55 kDa receptor which has been cloned (Loetscher *et al.* 1990; Schall *et al.* 1990). The 55 kDa TNFR is a 415 amino acid polypeptide with a single transmembrane domain. The extracellular portion is rich in cysteine residues and is homologous to that of the nerve growth factor receptor. Signal transduction through the TNFRs after they have bound either TNFα or TNFβ and been internalized is not clearly understood.

Soluble TNF-binding or inhibitory proteins have been identified in human plasma and urine and these may be shed forms of TNFRs. Soluble TNFRs are present at a surprisingly high concentration in the plasma of healthy individuals, as well as in patients with autoimmune diseases. The function of this large pool of soluble TNFRs has been suggested to be either to absorb and inactivate TNF present in inappropriate sites or, probably more likely, to act as a reservoir of TNF that could be released slowly to maintain a low level of biologically active TNF.

Stimuli for the production of TNFα and TNFβ

TNFα is produced predominantly by activated macrophages (Table 9.11 and Section 10.4.3), but it can also be produced by activated lymphocytes, NK cells (see Section 10.3.4), B cells, tumour cell lines, smooth muscle cells, activated keratinocytes and cells in the brain (Table 9.12). TNFα is produced in response to a wide range of infections and to inflammatory stimuli, for example, viral, fungal, and parasitic infections, endotoxin, enterotoxin, and C5a (see Section 9.4). Other cytokines such as IL-1 and IL-2 can also stimulate its production.

TNFβ is produced predominantly by activated CD4 and CD8 single-positive T lymphocytes, usually as a result of MHC-restricted antigen presentation. Normal leukocytes such as monocytes, macrophages, or B cells do not sythesise TNFβ, although some transformed B cells, NK cells, and certain tumours may be able to produce this cytokine (Table 9.12).

Biological activities of TNFα and TNFβ

In many respects the biological activities of the TNFs are similar to those of IL-1 (Table 9.8). They have a wide variety of effects on a number of different cell types, including leukocytes, endothelial cells, fibroblasts, and tumours *in vitro* and some of these will be outlined below. Increased levels of TNFs have also been reported in certain disease states (see below). The cytotoxic effects of TNFα and TNFβ are summarized below (see also Section 10.3.4).

TNFα: Historically TNFα was thought to have direct cytostatic effects on tumour cells, inducing tumour necrosis as its name suggests. However, now it seems more likely that these effects are mediated *indirectly* by TNFα which triggers the synthesis and release of other mediators, such as prostaglandins,

Table 9.12 Cells which can synthesize TNFα and TNFβ

Source	TNFα	TNFβ
Activated macrophages	+++	—
Activated lymphocytes	+	+++
NK cells	+	+
B cells at certain stages	+	+
Tumour cell lines	+	+
Smooth muscle cells	+	—
Activated keratinocytes	+	?
Astrocytes and microglia	+	?

proteases, free radicals and lysosomal enzymes. The biological activities of TNFα include the following.

1. In general, TNFα augments the effector activity of leukocytes. This includes: (i) their ability to adhere to other cells; (ii) the capacity of macrophages to release inflammatory mediators such as IL-1 and PGE_2, to phagocytose particles and to produce reactive oxygen intermediates; (iii) increased expression of MHC class II antigens, high affinity IL-2R and TNFR by *human* T cells; (iv) increased cytotoxicity of NK cells and CTL; and (v) enhanced proliferation of activated B cells and increased antibody secretion in the presence of IL-2.

2. TNFα has multiple effects on endothelial cells (Pober and Cotran 1990). TNFα: (i) increases adhesion of lymphocytes and neutrophils to endothelial cells, which accounts for the profound neutropenia that occurs during septic shock (see below), as well as adhesion of eosinophils and basophils; (ii) induces the release of IL-1, IL-6, IL-8, and GM-CSF from endothelial cells; (iii) leads to a loss of fibronectin from basement membranes and reorganization of cytoplasmic actin, thereby increasing the permeability of endothelial cells; (iv) induces reorganization of monolayers of vascular endothelium and increases the size of the cells; and (v) together with TNFβ, modulates the expression of antigens by human endothelial cells, such as increasing the expression of MHC class I, class II, and certain endothelial cell antigens, such as MECA-325.

3. TNFα also has multiple effects on fibroblasts causing: (i) increased expression of MHC class I molecules; (ii) synthesis of other cytokines such as IL-1, IL-6, GM-CSF, M-CSF, and PGE_2; and (iii) synthesis of fibronectin, collagenase, and glycosaminoglycans, particularly hyaluronic acid. Depending on the dose of TNFα used it can either inhibit (low doses) or promote (high doses) the synthesis of collagen.

Some properties of TNFα *in vivo* are discussed in Section 10.4.4. Increased serum levels of TNFα have been observed in recipients of organ transplants, and it is possible that TNF may be an important mediator of tissue destruction during graft rejection. High levels of TNFα have also been found in patients with autoimmune diseases, where it may act to augment the inflammatory response, and in patients infected with HIV. Increased plasma levels of TNFα have been reported in patients with cancer, but factors that block TNFα activity have also been described in some of these patients.

TNFβ: The biological activities of TNFβ are outlined below.

1. TNFβ is cytotoxic. In general, infected or transformed cells are more susceptible to killing by lymphotoxin than normal cells. In some situations, TNFβ can synergize with IFNγ, which is thought to increase the sensitivity of some cells to TNFβ by inducing increased expression of TNFR.
2. Both TNFα and TNFβ have potent anti-viral effects, which in part may be related to their ability to induce IFNγ.
3. TNFβ can also induce necrosis of tumour cells and increase the expression of MHC antigens and adhesion molecules, such as ICAM-1, on endothelial cells, although in this respect TNFβ may be less potent than TNFα; both TNFs may synergize with IFNγ in this activity.
4. TNFβ has been shown to enhance the activity of polymorphonuclear leukocytes (PMN), including ADCC and phagocytosis.
5. TNFβ, TNFα has been reported to stimulate the resorption of bone by ostoclasts.

The precise role of TNFβ *in vivo* remains controversial (Paul and Ruddle 1988). It may play an important role in immunoregulation, as it can kill B cells that present antigen, as well as T cells producing TNFβ itself. It probably plays an important role in the body's defence against viral, bacterial, and parasitic infections. It contributes to tumour immunity and has been implicated as one of the key mediators of tissue destruction during graft rejection. It may also participate in inducing tissue damage during disease pathogenesis, although in these respects it is often difficult to dissect the role of individual cytokines operating in the complex regulatory network involving IL-1, IL-6, TNFs, and IFNs.

9.5.4 Interferons

Interferons (IFNs) are responsible for a phenomenon that was originally called '*viral interference*'. It was found that when an animal was infected with one virus it was often protected against infection by a different virus. IFNs act by inducing an anti-viral state in the target cells rather than through an interaction with the virus itself. They also have a variety of other effects, for example they inhibit the growth and proliferation of some tumours and normal cells and they modulate immune responses.

Types of IFN

There are three types of IFN: α, β, and γ (Table 9.13). IFNα and IFNβ were together originally known as type I interferon, and IFNγ was called type II interferon. A fundamental difference between these types was thought to be

Table 9.13 Human interferons

Type	α	β	γ
Original synonyms	leukocyte type I	fibroblast type I	immune, T type II
Main inducer	virus infection	virus infection	immune stimuli
Number of genes	>13	probably 1	probably 1
Chromosomal location	9	9	12
Introns in gene	none	none	3
Product:			
Signal peptide	23 AA	21 AA	20 AA
Mature peptide	166 AA (20 and 25 kDa)	166 AA (20 kDa)	146 AA (20 and 25 kDa)
Glycoprotein	some	yes	yes

that α and β forms were induced by viral infection of leukocytes and fibroblasts, respectively, whereas γ interferon was produced by T cells during immune responses. Because of this they were also formerly known as leukocyte IFN, fibroblast IFN and immune, or T, IFN respectively. However, these are misnomers because the production of these IFNs is not just restricted to these cell types. For example, some leukocytes, e.g. macrophages, null cells, and some lymphoblastoid cell lines, can be induced to make a mixture of IFNα and IFNβ, as can fibroblasts, while virally-infected T cells also produce IFNα; some NK cells produce IFNγ. The three types of IFN can be distinguished biochemically: IFNγ is sensitive to exposure to acid conditions (pH 2) whereas IFNα and IFNβ are resistant. However, IFNβ binds con A and is more hydrophobic than IFN.

IFN genes and molecules

The genes for human IFNs have been cloned (Fig.9.31). IFNα is a multigene family composed of at least 24 genes, some of which are alleles, while others are pseudogenes. In contrast, only single genes have been found for IFNβ and IFNγ. (We should note that IFNβ was sometimes called IFNβ1; another molecule, called IFNβ2, is in fact BSF-2 or IL-6, see Section 9.5.2).

The genes encoding IFNα and IFNβ are located on chromosome 9 in humans and chromosome 4 in mice. At least 24 loci for IFNα have been identified, of which 15 are functional genes and the remainder pseudo genes. The IFNβ gene is linked to the IFNα genes. An interesting feature of these genes is their lack of introns. This may be important for their expression during viral infection, when the normal cellular splicing mechanisms could be impaired. IFNα and IFNβ genes encode polypeptides of 166 amino acids that are respectively preceded by 23 and 21 residue signal peptides.

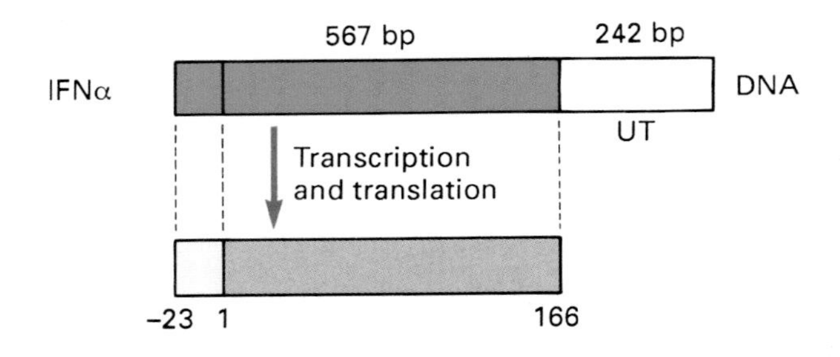

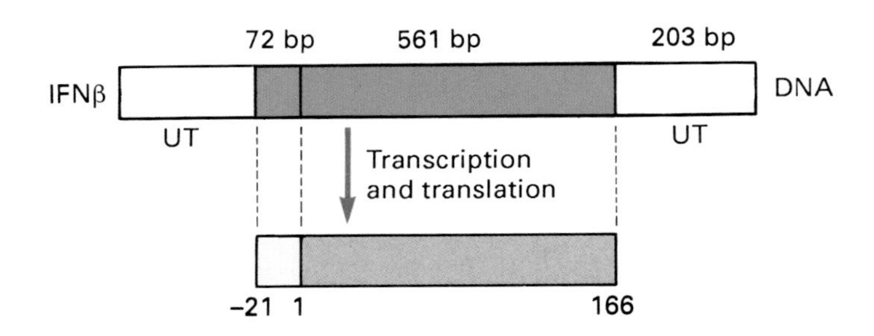

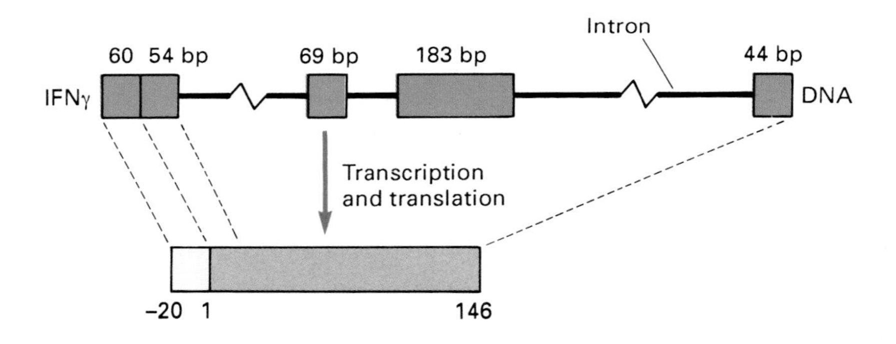

Introns reduced in scale by factor of 10

Fig.9.31 Genes for human interferons

In contrast, the IFNγ gene is located on chromosome 12 in humans and has three introns and encodes a mature peptide of 146 amino acids plus a 20-residue signal peptide. The predicted molecular weights of the IFNγ peptide is about 17 kDa but this is post-translationally modified to produce 20 kDa and 25 kDa products. IFNγ is a glycoprotein, as is IFNβ and some, but not all, IFNα, but the carbohydrate does not seem to be essential for antiviral activity. Other reports estimate IFNγ to have a molecular weight of about 35–70 kDa, suggesting that it might exist as a multimer, or coupled to serum proteins.

The sequences of IFNα are quite similar and there is about 80% homology between them. Between these and IFNβ there is about 30% homology at the amino acid level and 45% at the nucleotide level, suggesting they may have diverged from a common ancestor, and they are all located on human chromosome 9. However, there is very limited homology between these and the IFNγ gene which shows limited polymorphism between some individuals.

The induction of IFNα and IFNβ by viral infection, or in response to double-stranded RNA, requires gene transcription. In other words, when a leukocyte or fibroblast, for example, is infected with a virus, the IFN genes are switched on and transcribed into mRNA, which is then translated into the IFN product. Transcription of these genes is controlled by sequences that are 5′ (upstream) to the promoter for the structural gene. Here a purine-rich sequence of 42 base pairs is highly conserved in human IFN, while the corresponding sequence in the β gene is also found flanking genes that are responsive to steroid hormones, suggesting similar regulation.

IFN Receptors

IFNα and IFNβ bind to the same receptor, while IFNγ binds to another receptor. IFN action can be blocked by an antibody to a product of human chromosome 21, which is where the IFNα/β receptor is encoded. The gene encoding the IFNγ receptor has been mapped to chromosome 6 in human.

The IFNα receptor is a glycoprotein of 557 amino acids (63.5 kDa). The polypeptide chain contains a single hydrophobic transmembrane region and a cytoplasmic tail of approximately 100 amino acids that is rich in glycine residues (Uze *et al.* 1990). The receptor for IFNγ has also been characterized; the extracellular domain of 245 amino acids carries the ligand binding site and the large intracellular portion of the polypeptide (222 amino acids) may be involved in signal transduction (Auget *et al.* 1988).

Once IFN has bound to the cell, it can be removed without inhibiting development of the anti-viral state. Development of this state in the cell that actually produces the IFN (rather than in as-yet-uninfected targets) may require that the IFN is secreted and binds to the same cell since, for example, no anti-viral activity developed in fibroblasts that were stimulated to produce IFN in the presence of an anti-IFN antibody.

Stimuli for IFN production

Leukocytes mainly make IFN when they are infected with viruses. However, a number of natural and synthetic agents induce IFNγ and IFNβ in animals and/or cell cultures. These include: living and killed RNA and DNA viruses; double stranded RNA from fungi; synthetic compounds such as polyinosinic polycytidylic acid (poly I:C); bacteria that grow intracellularly; microbial products like LPS; and organic polymers such as pyran copolymers. Double-stranded RNA is an intermediate in some forms of viral replication, and seems to be an unusually effective stimulus for IFN production. As yet, we do not know how such a divergent group of agents can produce a similar cellular response. In contrast, T cells produce IFNγ when they are stimulated by specific antigen and MHC, or with mitogens (Fig.9.32).

Biological activities of interferons

IFN which is released from virally-infected cells (α, β) or from activated T cells (γ), enables other cells to become resistant to virus infection; viral replication is inhibited and the host cell is protected from lysis. IFNs first bind to cellular receptors on the target cells. This triggers gene transcription and translation of this new mRNA into proteins that are involved in producing the anti-viral state. Two enzymes, protein kinase and 2′,5′-oligoadenylate synthetase have been identified (see below), but others are probably also involved. An unusual aspect of IFN-induced viral resistance is that the target cell only becomes completely resistant when a virus subsequently enters the cell (Fig.9.33).

All three IFNs have a similar broad range of biochemical and biological activities, although there are important differences between them. Their effects require mRNA and protein synthesis by their target cells. By gel electrophoresis, no differences were detected between polypeptides induced in fibroblasts treated with IFNα and IFNβ, but additional products were induced by IFNγ. The overall effects of these IFN-induced products is to inhibit viral replication and this occurs at two or more levels by induction of:

(1) a protein kinase that phosphorylates a number of intracellular proteins, including some molecules that are required for protein synthesis, which is thereby inhibited. This protein kinase is activated when it binds double-stranded RNA, which is contained in the genome of certain classes of virus. Inhibition of host cell protein synthesis prevents synthesis of viral proteins.

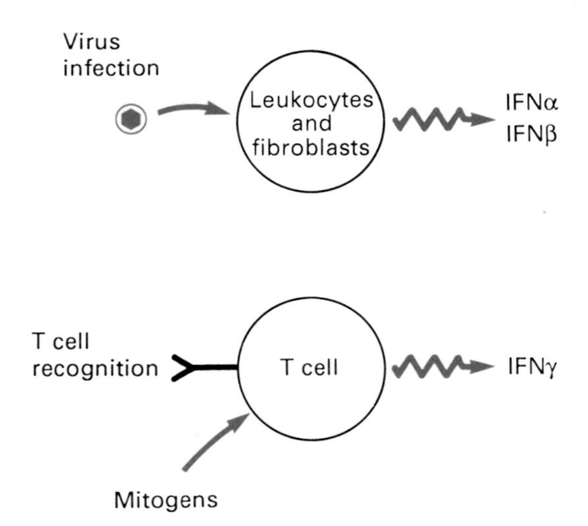

Fig.9.32 The principal stimuli of interferon production

(2) an enzyme called 2′5′ adenylate synthetase, which is also activated by double-stranded RNA. The active enzyme synthesizes poly-adenylates, from ATP, and these activate an endogenous endonuclease that degrades mRNA. Presumably this also destroys viral RNA, either in the genome (RNA viruses) or produced as an intermediate (DNA viruses).

Whether the above and/or other pathways are involved in the inhibition of cellular growth by IFNs is not firmly established. Different cell types can vary greatly in their sensitivity to the effects of IFNs, although tumour cells may generally be more sensitive than normal cells (see also effects of IFN on NK cells, Section 10.2.6). The growth inhibitory activity of IFNα and IFNβ, as well as the frequent deletion of these genes in acute lymphoblastoid leukaemia, suggests that IFNs may be potential tumour suppressors.

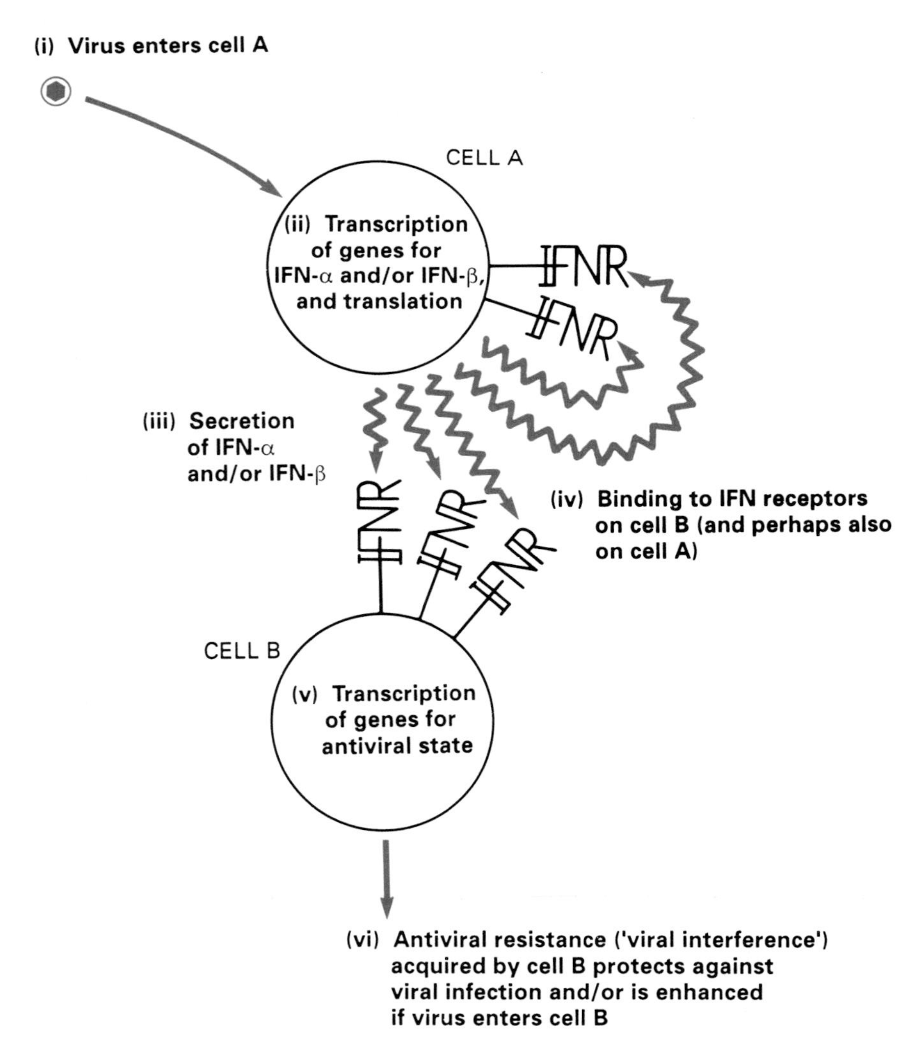
(i) Virus enters cell A
CELL A
(ii) Transcription of genes for IFN-α and/or IFN-β, and translation
IFNR
IFNR
(iii) Secretion of IFN-α and/or IFN-β
IFNR
IFNR
IFNR
(iv) Binding to IFN receptors on cell B (and perhaps also on cell A)
CELL B
(v) Transcription of genes for antiviral state
(vi) Antiviral resistance ('viral interference') acquired by cell B protects against viral infection and/or is enhanced if virus enters cell B

Fig.9.33 Induction of the antiviral state by interferons

Many studies of the effects of IFN have used preparations that may contain a mixture of IFNs. Now that cloned material is available, the different effects of various IFNs can be dissected. It is possible that different IFNs may act preferentially on different cell types, or give protection against certain viruses, or they may have different combinations of effects. It is becoming clear that IFNγ in particular, is an immunoregulatory molecule. It is perhaps the principal factor that activates the microbicidal activity of macrophages and it causes an increase in the mRNA for MHC class II molecules in macrophages and a number of other cell types, thus potentiating the immune response. There are reports that some of these effects are antagonized by IFNα and IFNβ. However, other investigators have proposed that only IFNγ is able to 'prime' macrophages, but that any IFN can then trigger these effects. IFNγ is probably involved in activation of NK cells (see Section 10.2) and it brings about quantitative and qualitative changes in the class of antibody made by B cells (see Section 8.4.4).

IFNs are also involved in the cytokine network and increase the production of both IL-1 and TNF (see Sections 9.5.1 and 9.5.3). IFNγ and TNFα have additive effects in the induction of expression of MHC molecules and IFNγ synergizes with TNFβ in inhibiting cellular growth. In contrast, IFNα and IFNβ apparently antagonize the induction of MHC class II gene expression by IFNγ. IFNγ regulates the production of IL-4 and IL-10 by T_H2 cells (see Section 11.3).

Further reading

Inflammation

Beer, D.J. and Rocklin, R.E. (1987). Histamine modulation of lymphocyte biology: Membrane receptors signal transmission and functions. *CRC Critical Reviews in Immunology*, **7**, 55–91.

Braquet, P. and Rola-Pleszczyinski, M. (1987). Platelet activating factor and cellular immune response. *Immunology Today*, **8**, 345–52.

Hanahan, D.J. (1986). Platelet activating factor: a biologically active phosphoglyceride. *Annual Review of Biochemistry*, **55**, 483–509.

Larsen, G.L. and Henson, P.M. (1983). Mediators of inflammation. *Annual Review of Immunology*, **1**, 335–59.

Phipps, R.P., Stein, S.H., and Roper, R.L. (1991). A new view of prostaglandin E regulation of the immune response. *Immunology Today*, **12**, 349–52.

Proud, D. and Kaplan, A.P. (1988). Kinin formation: mechanisms and role in inflammatory disorders. *Annual Review of Immunology*, **6**, 49–83.

Pruzanski, W. and Vadas, P. (1991). Phosphlipase A2—a mediator between proximal and distal effectors in inflammation. *Immunology Today*, **12**, 143–6.

Rola-Pleszczynski, M. (1985). Immunoregulation by leukotrienes and other lipoxygenase metabolites. *Immunology Today*, **6**, 302–7.

Complement

Immunology Today, **12(9)**, The biology of complement. (1992).

Campbell, R.D., Law, S.K.A., Reid, K.B.M, and Sim, R.B. (1988). Structure, organization and regulation of the complement genes. *Annual Review of Immunology*, **6** 161–96.

Colten, H.R. and Rosen, F.S. (1992). Complement deficiencies. *Annual Review of Immunology*, **10**, 809–34.

Cooper, N.R. (1991). Complement evasion strategies of microorganisms, *Immunology Today*, **12**, 327–31.

Davis, A.E. (1988). C1 inhibitor and hereditary angioneurotic eodema. *Annual Review of Immunology*, **6**, 595–62.9.

Erdei, A., Fust, G., and Gergely, J. (1991). The role of C3 in immune responses. *Immunology Today*, **12**, 332-7.

Frank, M.M. and Fries, L.F. (1991). The role of complement in inflammation and phagocytosis. *Immunology Today*, **12**, 322-6.

Holers, V.M., Kinoshita, T., and Molina, H. (1992). The evolution of mouse and human complement C3-binding proteins: divergence of form but conservation of function. *Immunology Today*, **13**, 231-6.

Lachmann, P.J. (1991). The control of homologous lysis. *Immunology Today*, **12**, 312-5.

Law, S.K.A. and Reid, K.B.M. (1988). Decay accelerating factor: biochemistry, molecular biology and function. *Annual Review of Immunology*, **7**, 35–58.

Reid, K.B.M. (1989). The complement system. In *Molecular immunology* (ed. Hames and Glover). pp. 189-241. IRL, Oxford.

Sim, R.B. and Reid, K.B.M. (1991). C1: Molecular interactions with activating systems. *Immunology Today*, **12**, 307-11.

Trowsdale, J., Ragoussis, J., and Campbell, R.D. (1991). Map of the human Major histocompatibility complex. *Immunology Today*, **12**, 443-6.

IL-1

di Giovine, F.S. and Duff, G.W. (1990). Interleukin 1: the first interleukin. *Immunology Today*, **11**, 13–20.

Dinarello, C.A., Clark, B.D., Puren, A.J., Savage, N., and Rosoff, P.M. (1989). The interleukin 1 receptor. *Immunology Today*, **10**, 49–51.

Mizel, S.B. (1990). Cyclic AMP and interleukin 1 signal transduction. *Immunology Today*, **11**, 390–1.

O'Neill, L.A.J., Bird, T.A., and Saklatvala, J. (1990). Interleukin 1 signal transduction. *Immunology Today*, **11**, 392–4.

IL-6

Van Snick, J. (1990). Interleukin-6: an overview. *Annual Review of Immunology*, **8**, 253–78.

TNF

Beutler and Cerami (1989). The biology of cachectin/TNF—a primary mediator of the host response. *Annual Review of Immunology*, **7**, 625–55.

Paul, N.L. and Ruddle, N.H. (1988). Lymphotoxin. *Annual Review of Immunology*, **6** 407–38.

Pober, S.P. and Cotran, R.S. (1990). Cytokines and endothelial cell biology. *Physiological Reviews*, **70**, 427–51.

Vassalli, P. (1992). The pathophysiology of tumour necrosis factor. *Annual Review of Immunology*, **10**, 411–52.

Literature cited

Auget, M., Dembic, I, and Merlin, G. (1988). Molecular cloning and expression of the human IFNγ receptor. *Cell*, **55**, 273–80

Beer, D.J. and Rocklin, R.E. (1987). Histamine modulation of lymphocyte biology: membrane receptors, signal transmission and functions. *CRC Critical Reviews in Immunology*, **7**, 55-91.

Beutler and Cerami (1989). The biology of cachectin/FNF — a primary mediator of the host response. *Annual Review of Immunology*, **7**, 623-55.

Bhakdi, S. and Tranum-Jensen, J. (1991). Complement lysis: a hole is a hole. *Immunology Today*, **12**, 318-20.

Borsos, T., Chapius, R.M., and Langone, J.J. (1981). Distinction between fixation of C1 and the activation of complement by natural IgM anti-hapten antibody: effects of cell surface hapten density. *Molecular Immunology*, **18**, 863-8.

Braquet, P. and Rola-Pleszczynski, M. (1987). Platelet activating factor and cellular immune response. *Immunology Today*, **8**, 345-52.

Carroll, M.C., Campbell, R.D., Bentley, D.R., and Porter, R.R. (1984). A molecular map of the human major histocompatibility complex class III region linking complement genes C4, C2 and factor B. *Nature*, **307**, 237-41.

Carter, D.B., Deibel, M.R., *et al.* (1990). Purification, cloning, expression, and biological characterization of an interleukin-1 receptor antagonist protein. *Nature*, **344**, 633–8

Colten, H.R. and Rosen, F.S. (1992). Complement deficiencies. *Annual Review of Immunology*, **10**, 809–34.

Cooper, N.R. (1991). Complement evasion strategies of microorganisms. *Immunology Today*, **12**, 332-7.

di Giovine, F.S. and Duff, G.W. (1990). Interleukin-1: the first interleukin. *Immunology Today*, **11**, 13-20.

Dinarello, C.A., Clark, B.D., Puren, A.J., Savage, N., and Rosoff, P.M. (1989). The interleukin-1 receptor. *Immunology Today*, **11**, 49-51.

Eisenberg, S.P., Evans, R.J., Arend, W.P., Verderker, E., Brewer, M.T., Hannum, C.H., and Thompson, R.C. (1990). Primary structure and functional expression from complementary DNA of a human interleukin-1 receptor antagonist. *Nature*, **343**, 341–6.

Erdei, A., Fust, G., and Gergely, J. (1991). The role of C3 in immune responses. *Immunology Today*, **12**, 332–7.

Esser, A.F. (1991). Big MAC attack: complement causes leaky patches. *Immunology Today*, **12**, 316–18.

Frank, M.M. and Fries, L.F. (1991). The role of complement in inflammation and phagocytosis. *Immunology Today*, **12**, 322–6.

Hanahan, D.J. (1986). Platelet activating factor: a biologically active phosphoglyceride. *Annual Review of Biochemistry*, **55**, 483–509.

Hannum, C.H., Wilcox, C.J., Arend, W.P., Joslin, F.G., Dripps, D.J., Heimdal, P.L., Armes, L.G., Sommer, A., Eisenber, S.P., and Thompson, R.C. (1990). Interleukin-1 receptor antagonist activity of a human interleukin-1 inhibitor. *Nature*, **343**, 336–40.

Hibi, M., Murakami, M., Saito, M., Hirano, T., Taga, T., and Kishimoto, T. (1990). Molecular cloning and expression of an IL-6 signal transducer, gp130. *Cell*, **63**, 1149–57.

Holers, V.M., Kinoshita, T., and Molina, H. (1992). The evolution of mouse and human complement C3 binding proteins: divergence of form but conservation of function. *Immunology Today*, **13**, 231–6.

Jones, E.Y., Stuart, D.I., and Walker, N.P.C. (1989). Structure of tumour necrosis factor. *Nature*, **338**, 225–8.

Knobel, H.R., Villinger, W., and Isliki, H. (1975). Chemical analysis and electron microscopy studies of human Clq prepared by different methods. *European Journal of Immunology*, **5**, 78.

Lachmann, P.J. (1991). The control of homologous lysis. *Immunology Today*, **12**, 312–15.

Larsen, G.L. and Henson, P.M. (1983). Mediators of inflammation. *Annual Review of Immunology*, **1**, 335–59.

Loetscher, H., Pan, Y.-C.E., Lanm, A.-W., Gentz, R., Brockhaus, M., Tabuchi, H., and Lesslauer, W. (1990). Molecular cloning and expression of the human 55 kd tumour necrosis factor receptor. *Cell*, **61**, 351–9

Mizel, S.B. (1990). Cyclic AMP and interleukin-1 signal transduction. *Immunology Today*, **11**, 390–1.

Ohlsson, K.I., Bjork, P., Bergenfeldt, M., Hageman, R., and Thompson, R.C. (1990). Interleukin-1 receptor antagonist reduces mortality from endotoxin shock. *Nature*, **348**, 550–2.

O'Neill, L.A.J., Bird, T.A., and Saklatvala, J. (1990). Interleukin-1 signal transduction. *Immunology Today*, **11**, 392–4. *Today*, **11**, 392–4.

Paul, N.L. and Ruddle, N.H. (1988). Lymphotoxin. *Annual Review of Immunology*, **7**, 625–55.

Phipps, R.P., Stein, S.H., and Roper, R.L. (1991). A new view of prostaglandin E regulation of the immune response. *Immunology Today*, **12**, 349–52.

Pillemer, L., Blum, L., Lepow, I.H., Ross, O.A., Todd, E.W., and Wardlaw, A.C. (1954). The properdin system and immunology. 1. Demonstration and isolation of a new serum protein properdin and its role in immune phenomena., *Science*, **120**, 279–85.

Pober, S.P. and Cotran, R.S. (1990). Cytokines and endothelial cell biology. *Physiological Reviews*, **70**, 427–61.

Proud, D. and Kaplan, A.P. (1988). Kininformation: mechanisms and role in inflammatory disorders. *Annual Review of Immunology*, **6**, 49–83.

Pruzanski, W. and Vardas, P. (1991). Phosphlipase A2 — a mediator between proximal and distal effectors in inflammation. *Immunology Today*, **12**, 143–6.

Reid, K.B.M. (1989). The complement system. In *Molecular immunology*, (ed. B.D. Hames and D.M. Glover), pp. 189–241. IRL Press, Oxford.

Schall, T.J., Lewis, M., Koller, K.J., Lee, A., Rice, G.C., Wong, G.H.W., *et al.* (1990). Molecular cloning and expression of a receptor for human tumour necrosis factor. *Cell*, **61**, 361–70

Sim, R.B. and Reid, K.B.H. (1991). C1: molecular interactions with activating systems. *Immunology Today*, **12**, 307–11.

Sims, J.E., March, C.J., Cosman, D., Widmer, M.B., MacDonald, H.R., McMahar, C.J., *et al.* (1988). cDNA expression cloning of the IL-1 receptor, a membrane of the immunoglobulin superfamily. *Science*, **241**, 585–9.

Trowsdale, J., Ragoussis, J., and Campbell, R.D. (1991). Map of the human major histocompatibility complex. *Immunology Today*, **12**, 443–6.

Uze, G., Lutfalla, G., and Gresser, I. (1990). Genetic transfer of a functional human IFNα receptor into mouse cells: cloning and expression of its cDNA. *Cell*, **60**, 225–34.

Van Snick, J. (1990). Interleukin-6: an overview. *Annual Review of Immunology*, **8**, 253–78.

Vassalli, P. (1992). The pathophysiology of tumour necrosis-factor. *Annual Review of Immunology*, **10**, 411–52.

Yamasaki, K., Taga, T., Hirata, Y., Yawata, H., Kaivanishi, Y., Seed, B., Taniguchi, T., Hirano, T., and Kishimoto, T. (1988). Cloning and expression of the human interleukin-6 receptor. *Science*, **241**, 825–8.

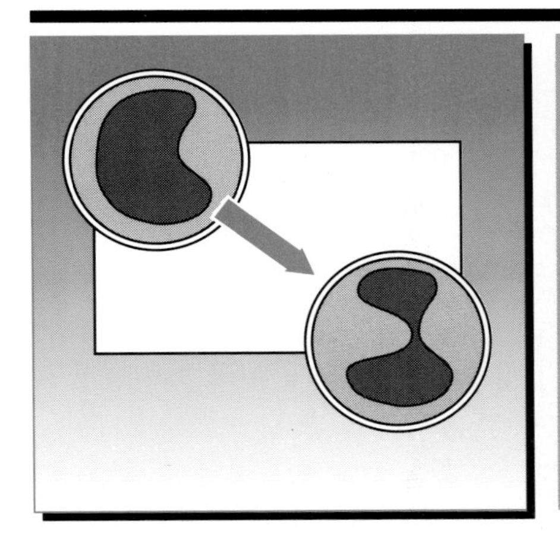

CELLULAR CYTOTOXICITY

Scheme 10. Cellular cytotoxicty

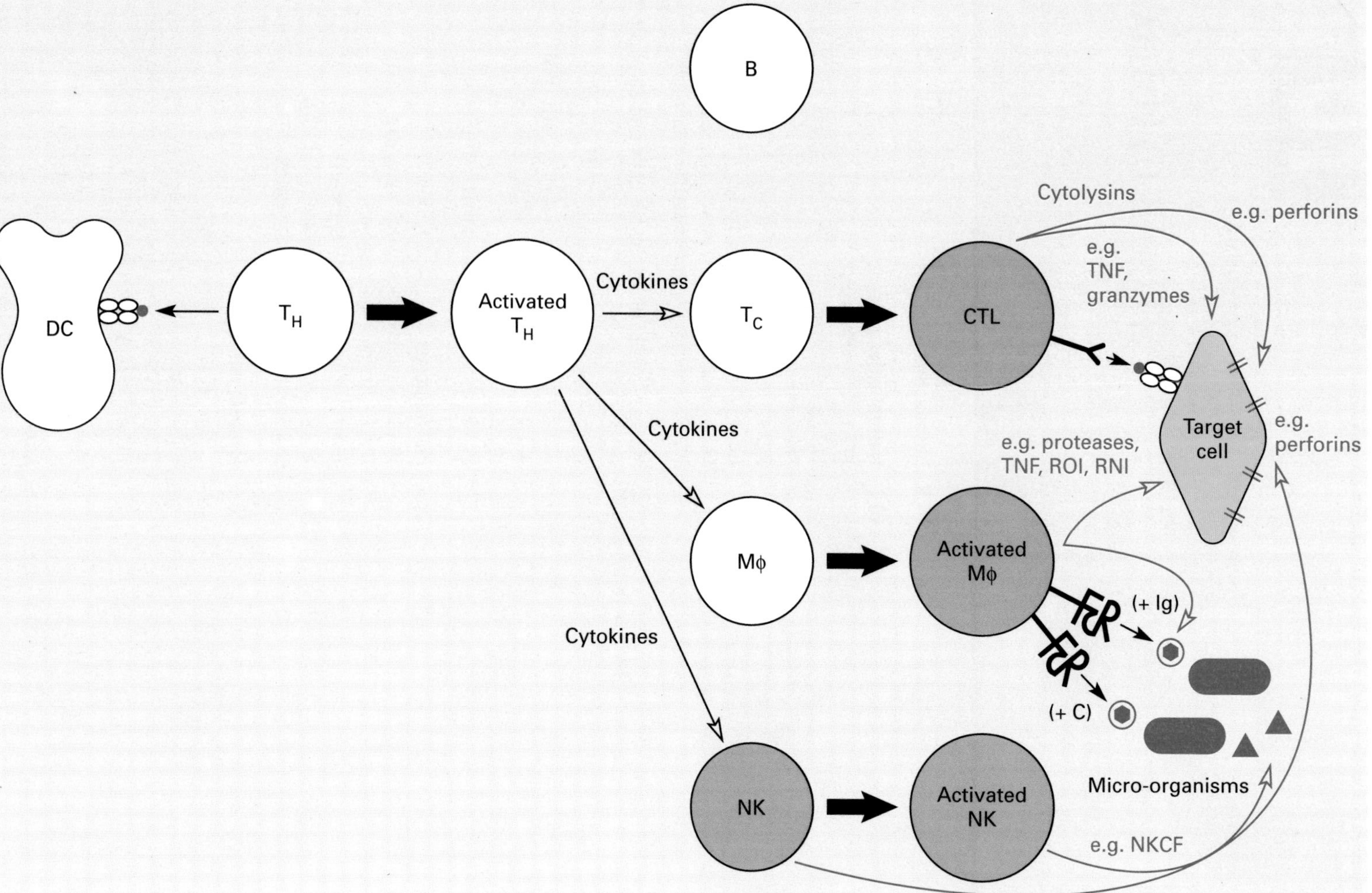

10.1 Introduction

Effector mechanisms bring about or 'effect' the removal of antigens from the body. These mechanisms can be mediated both by soluble molecules and by cells (e.g. Figs.1.24 and 1.28). Two soluble effector systems are mediated by antibodies and complement, which are discussed in Chapters 6 and 9, respectively. In this chapter, cellular effector mechanisms will be considered. The main focus will be on the mechanisms of killing by cytoxic T lymphocytes (CTL), natural killer (NK) cells, and macrophages. CTL belong to the adaptive arm of the immune system whereas NK cells and macrophages belong to the innate arm (Section 1.1.2). Hypersensitivity is briefly considered at the end of this chapter as a manifestation of 'inappropriate' effector mechanisms.

10.2 Natural killer cells

10.2.1 Characteristics

NK cells were initially identified in cell suspensions prepared from lymphoid tissues of mice and rats, particularly from the spleen, and were subsequently enriched from human peripheral blood as a population of **large granular lymphocytes**. NK cells can also be isolated from mucosal epithelia, such as the lamina propria of gut, but not from most other tissues.

Although human NK cells are morphologically defined as 'large granular lymphocytes' (LGL), only about three-quarters of blood cells with this morphology have NK activity (Timonen *et al.* 1981). These cells are non-adherent, non-phagocytic, and lymphoid in appearance. Their cytoplasmic granules are similar to those of monocytes in some respects, for example they contain β-glucuronidase, but they are different in others, for example they lack peroxidase. Some granules in human NK cells also have 'parallel tubular arrays' of unknown function.

One of the main characteristics of cells with 'natural killer' activity is their ability to kill a number of different cell targets in an *MHC-unrestricted* manner. Thus, cells with NK activity can kill targets that do not express 'classical' MHC molecules (e.g. MHC class I K and D molecules, as opposed to 'non-classical' MHC molecules such as Qa, Tla in the mouse; Fig.2.21). In contrast, killing by CTL is MHC-restricted and generally follows recognition of peptide–MHC complexes on the target cells (Table 10.1).Another characteristic of NK cells is their '*spontaneous*' cytotoxic activity, which can be demonstrated immediately after they are isolated from the animal, whereas the cytotoxic potential of CTL only develops after an immune response has been initiated (Table 10.2). Typically NK activity peaks at 3 days after infection by a pathogen, whereas CTL activity reaches a peak 7–9 days later. Originally the term 'NK activity' was applied to freshly-isolated cells with the capacity to kill defined target cells in an MHC-unrestricted manner, but this term is now also applied to cultured cells that have this ability.

Table 10.1 Cells with MHC-restricted and MHC-non-restricted NK-like cytotoxic activity

Cell type	MHC-restricted activity	MHC-non-restricted activity
CTL (especially those isolated early during immune responses *in vivo*)	Yes	No
CTL (especially those cultured in IL-2)	Yes	Yes*
NK cells (by definition)	No	Yes
Promonocytes	No	Yes*

* Such activity can be referred to as 'NK-like activity' to distinguish it from 'true' NK activity due to bona fide NK cells.

A major problem in trying to define NK cells is their heterogeneity. It seems likely that some cells in the body having NK activity do comprise a distinct lineage, although the relationship of this lineage to others is unknown. Certainly many NK cells can be distinguished from classical CTL, mononuclear phagocytes (monocytes and macrophages), and polymorphonuclear leukocytes (PMN) (Fig.1.11). However, there can be no doubt that cells of some other lineages may acquire NK-like activity at one stage or another of their development or activation (Table 10.1). In particular, MHC-restricted CTL can also acquire the ability to kill cells that are more generally targets for NK cells in an MHC-unrestricted manner, particularly after they have been cultured in IL-2 for some time (Acha-Orbea *et al.* 1983). Some precursors of mononuclear phagocytes may also be able to kill in this way. We shall therefore take the view that *at least some freshly-isolated cells belong to a distinct lineage of 'NK cells', but that other cells can develop 'NK-like activity', especially after culture.*

The definition of an NK cell is now taken to be a cell with the morphology of a large granular lymphocyte that has cytolytic activity even in the absence of target MHC class I and class II molecules, and with the following phenotype (Section 10.2.4). NK cells do not express CD3 or T cell receptors (neither the $\alpha\beta$ or $\gamma\delta$ forms) but they do express CD16 and NKH-1 (CD56: Leu-19) in the human, or NK-1.1/NK-2.1 in the mouse. A set of membrane molecules present only on NK cells, and which may be involved in recognition functions of these cells, has also been identified (Section 10.2.4).

10.2.2 Functions

It is clear that NK cells belong to the innate arm of the immune response because they have *spontaneous* cytotoxic activity against a variety of target cells, unlike T and B lymphocytes which only become activated after the

Table 10.2 Comparison of NK cells and freshly-isolated (*in vivo*-primed) CTL

Feature	NK cells	CTL
Arm of immune system	Innate	Adaptive
Mode of killing	Non-MHC-restricted killing of various target cells, also some extracellular bacteria *in vitro*	MHC-restricted killing of target cells
Activity	'Spontaneous' in fresh cells	Develops only after exposure to antigen
Antigen receptor	Not well defined	T cell receptor
Kinetics of response	Peaks 3 days after infection	Peaks 7–9 days after infection
Memory responses	No	Yes
Cytology	'Large granular lymphocytes'	Lymphocytes often lacking granules
Phenotype		
TcR-CD3	No	Yes
CD16	Yes	No
NKH-1 (Leu19)	Yes (human)	No
NK-1.1, 1.2	Yes (mouse)	No

antigen has been encountered (Table 10.2). Moreover, NK cells do *not* show secondary or memory responses. After production in the bone marrow, NK cells enter the circulation and become localized in certain tissues (Section 10.2.3). At this stage of development, their cytotoxic activity can be triggered within a few minutes of encountering an appropriate target cell. However, NK cells can be activated by cytokines to a different state in which they have greater cytotoxic activity, and typically 1–3 days after an infection *in vivo* they can develop into proliferating blast cells.

NK cells can kill a number of tumour cell lines, even from different species, in culture. Traditionally, tumour cell killing has been considered as a potentially important function of NK cells *in vivo*. However, they can also kill other targets, including certain virally-infected cells and some normal but immature cells. It is therefore possible that, *in vivo*, NK cells are responsible for (i) destroying virus-infected cells before CTL are activated, perhaps even before virus replication has begun, (ii) eliminating embryonic cells that might otherwise persist into adulthood, and (iii) they may also be involved in surveillance against certain types of tumour cells. The ability of NK cells to lyse some tumour cell lines has been used as an assay for NK function.

Cytotoxicity assays can be carried out with NK-sensitive targets such as the human K562 and mouse YAC-1 cell lines. These cells can be labelled with chromium-51 and incubated with a source of NK effector cells for a few hours at 37°C, after which the extent of cell lysis can be assessed by measuring the amount of chromium-51 released (see Appendix). Alternatively, one can allow the effector and target cells to form conjugates with each other, after which they can be distributed into a viscous medium such as agarose to prevent further interactions, and cytotoxicity can be assessed by staining dead cells with the vital dye trypan blue.

As noted earlier (Section 10.2.1), NK cells are not MHC restricted and killing does not require expression of 'classical' MHC molecules by the target cell. For example, the NK-sensitive target, K562, does not express classical MHC molecules. NK cells can also kill targets expressing xenogeneic MHC molecules, and there have been reports that they may be able to kill some extracellular bacteria directly. It has been suggested that NK cells may use 'non-classical' MHC molecules like the mouse Qa and Tla molecules (Section 2.8.1; p. 99) as restricting elements.

It was thought by some investigators that at least some NK cells lack *clonally-distributed* antigen receptors comparable to the T cell receptors or membrane Ig on B cells; if so, this would be an additional difference between NK cells and CTL.

One approach to examine this is to allow freshly-isolated NK cells to adhere to monolayers of one particular NK-susceptible cell line, and to determine the activity of the remaining non-adherent cells against other target cells. If the NK antigen receptors were clonally distributed, one would predict that this process would deplete only the specific NK cells, leaving NK cells with different receptors, and hence different specificities, within the non-adherent population. This result is indeed obtained if the experiment is done with a population of CTL, but the results have been equivocal when NK cells have been tested. Another approach is to compare the specificity of clones of NK cells (Section 10.2.5) with that of the parent cell from which they were derived: using CTL one would expect the clones to have the same specificity as the parent cell, but with NK cells contradictory results have been obtained. It is possible that the results of these experiments have been complicated by the presence of cells of other lineages (e.g. CTL) in the NK population.

In fact, NK cells may express receptors, distinct from T cell receptors, that could be clonally distributed to some extent (Section 10.2.4).

10.2.3 Origin

NK cells are leukocytes (Fig.1.31). If bone marrow chimeras (Section 5.3.2) are constructed using different strains of mice with genetically high or low levels of NK activity, the phenotype of the chimera correlates with that of the marrow donor. For example, if bone marrow from a strain with high NK activity is transplanted into a recipient that normally has low activity, the resultant chimera has high levels of NK activity. There are also mutant strains of mice, such as the **beige mouse** (Roder and Duwe 1979), that do not express NK activity; this phenotype can be reversed if they receive a transplant of bone marrow from a normal mouse. It is interesting to note that the beige mouse is rather prone to certain types of virally-induced tumours, but it is not clear if this shows that NK cells are more relevant to tumour destruction or to elimination of virus-infected cells. Because NK cells can be isolated from nude mice it is clear they do not require a thymus for their development; indeed such mice (and rats) generally have even greater NK activity than their euthymic counterparts.

10.2.4 Membrane molecules

Freshly-isolated human NK cells express antigens that are also present on myelomonocytic cells and granulocytes, such as CD16 and CD11b/CD18 (Fig.10.1). The CD16 antigen is an Fc receptor that is found on neutrophils (PMN) but *not* on freshly-isolated monocytes. There is, however, evidence that this molecule, the Fc_{γ}RIII receptor (Section 10.4.5), is expressed in different membrane-anchored forms on PMN and NK cells. CD16 on human NK cells, but not PMN, can mediate antibody-dependent cellular cytotoxicity and cytokine secretion (Section 10.2.7) (Anegon *et al.* 1988). CD11b/CD18 is the type 3 complement receptor (CR3), but this molecule does not mediate phagocytosis by NK cells, unlike its function on PMN and macrophages.

The majority of NK cells also express some molecules found on T cells, and their co-expression with markers of myelomonocytic cells strongly suggests that many NK cells belong to *neither* of these lineages. These molecules include CD2 (Section 4.2.7), and about 30–80% of *human* NK cells also express CD8 (Section 4.2.4) although *mouse* NK cells are reported to lack this antigen. Activated NK cells express CD25, one chain of the IL-2 receptor (Section 4.4.3), and sometimes both chains, as well as HLA-DR (MHC class II) molecules in the human. Mouse NK cells express Qa-5 (Section 2.8.1), which is also found on some T lymphocytes.

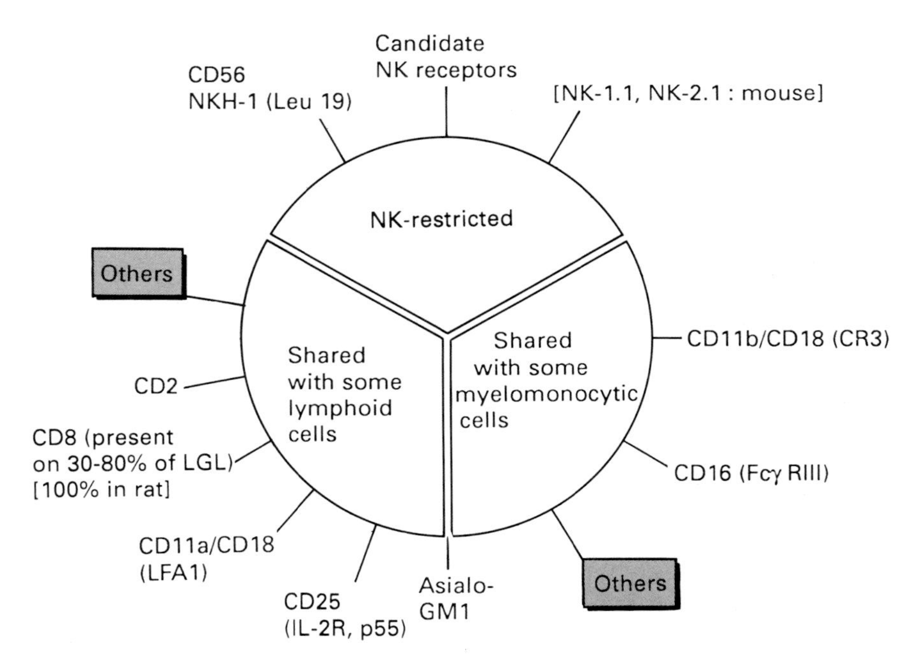

Fig.10.1 Some membrane molecules of NK cells in the human [and other species]

By definition, NK cells do not express CD3 or the $\alpha\beta$ or $\gamma\delta$ T cell receptors; in addition, human and mouse NK cells do not express CD4 or CD5 (Section 4.2.4) molecules. The absence of CD3 has been taken as evidence that while NK cells can express other T cell markers, they are distinct from the T cell lineage. It seems likely that at least some cells with NK activity that express CD3 may be 'contaminating' T cells bearing $\gamma\delta$ receptors, while others may be T cells with $\alpha\beta$ receptors that have acquired this activity in culture.

An additional problem in trying to define NK cells has been the paucity of NK-specific markers (Fig.10.1). Perhaps the most restricted marker in the *human* is NKH-1 (Leu19), CD56, which is a molecule of apparent molecular weight 200–220 kDa (Lanier *et al.* 1986). Depending on the particular blood donor, some 30–80% of human NK cells in peripheral blood may express a different antigen confusingly designated HNK-1 (Leu7; Abo *et al.* 1982). This was formerly thought to be a specific NK marker, but is now known to be expressed by $CD3^+$, $CD8^+$, DR^+ cells with little or no spontaneous cytotoxic activity, resembling normal activated human T cells; it is also present on the myelin-associated glycoprotein of nerve cells. In *mouse* an alloantigenic system of NK cells has been defined by monoclonal antibodies against a molecule called NK-1.1 and its alleles NK-2.1 and NK-3.1. NK cells express the asialo-GM1 glycolipid which is commonly used as a NK marker, although this molecule is also expressed by prothymocytes, monocytes, and granulocytes (Solomon and Higgins 1987).

Candidate structures for NK 'antigen' receptors have been identified on human NK cells, which are clonally-distributed, functional surface molecules. Two monoclonal antibodies, designated GL183 and EB6, defined four stable subsets of (resting and activated) NK cells depending on whether both, only one, or neither antigen was expressed (Moretta *et al.* 1990).

Treatment of $GL183^+$ NK cells with the corresponding antibody either enhanced or inhibited their cytotoxic activity against certain target cells; led to an increase in cytoplasmic Ca^{2+} ion concentration, indicating the molecule is involved in signalling events; and regulated responses to other stimuli such as anti-CD2 or anti-CD16 antibodies, and mitogens. The EB6 antibody could likewise increase or inhibit cytolysis of certain target cells by the corresponding subsets of NK cells. Moreover, there was a correlation between one of the subsets defined by these antibodies and the ability to recognize specifically a particular alloantigen.

The GL183 and EB6 antibodies immunoprecipitated either single 58 kDa, or non-covalently-associated 55 and 58 kDa, chains from the respective subsets of NK cells, and pre-clearing experiments (immunoprecipitation with one antibody, followed by the other) demonstrated that the antigens were expressed by distinct molecules. Peptide mapping revealed that the GL183 and EB6 molecules were quite similar in structure, although some unique peptides were present, suggesting they are members of the same family. It has been speculated that the molecules expressed by $GL183^+$ $EB6^-$ and $GL183^-$ $EB6^+$ NK cells may be associated with other chains, perhaps in a manner analogous to the $\alpha\beta$ and $\gamma\delta$ T cell receptor chains. A possibly related 54 kDa molecule has also been identified on a subset of mouse NK cells.

Although 'classical' NK cytotoxicity is not MHC-restricted, there are clear examples of recognition of allogeneic MHC molecules by cytolytic clones with the phenotype of NK cells (as in the study cited above, see box, and Suzuki *et al.* 1990, for example). There has been some progress in genetic mapping of the gene(s) encoding putative NK target structures in this mode of recognition (Ciccone *et al.* 1990), and one study to date has additionally identified the CD56 molecule (Leu 19, multiple isoforms of which are expressed on NK cells) as a potential target structure (Suzuki *et al.* 1991).

10.2.5 NK cell lines and clones

Freshly-isolated spleen cells from *nude mice* have NK activity, but this is lost when they are cultured. However, if a source of growth factors is added, such as supernatants from concanavalin A-stimulated T cells, it is possible to produce cell lines with NK activity in addition to MHC-restricted cytotoxicity. One of the active factors in the supernatants is most likely IL-2, because supernatants from which IL-2 is removed (using anti-IL-2 antibodies) do not support growth.

Human NK cells can also be grown in culture in the presence of IL-2, but under these conditions they can lose their granules, together with CD11b/CD18 and HNK-1 (Section 10.2.4), and they variably express CD2, CD3 and CD8. They also acquire the ability to kill targets that are not normally susceptible to lysis by fresh NK cells. Such cultured cells have been called **NK-like cells** to distinguish them from freshly-isolated NK cells (Table 10.3). However, the situation with NK lines may be complicated by the presence

Table 10.3 Terminology

Term	Definition
NK cell	Originally used to define freshly-isolated cells with morphology of LGL that spontaneously killed target cells in an MHC-non-restricted manner, particularly tumour cell lines (described as 'NK targets' or 'NK-sensitive cells'). Now restricted to $CD3^-$ TcR^- $CD16^+$ NKH-1^+ (or NK1.1, 1.2^+) cells with these features
Cells with NK-like activity	Cells in culture that have NK activity, but which may not be bona fide NK cells (e.g. CTL cultured in IL-2)
Anomalous killer (AK) cells	T cells and T cell clones cultured in IL-2 that acquire NK-like activity
Lymphokine-activated (LAK) cells	Freshly-isolated leukocytes (e.g. from peripheral blood) cultured in IL-2 that acquire NK-like activity
Cells exhibiting antibody-dependent cellular cytotoxicity (ADCC)	Cells with Fc receptors to which antibodies can bind and direct their cytotoxic activity towards a specific target
Killer (K) cell	A mononuclear cell with the capacity for ADCC

of contaminating cell types. For example, some cytotoxic T cell precursors from nude mice can also respond to IL-2, and cultured LGL from human blood may be overgrown by contaminating T cells.

Bona fide T cell clones that have been grown in IL-2 and have acquired NK activity have been termed **anomalous killer (AK) cells**. This should be contrasted to the term **lymphokine-activated killer (LAK) cells**, which refers to *primary* blood or spleen leukocytes (rather than T cell clones or lines) that have been cultured in IL-2 and have acquired the ability to kill certain tumour cells *in vitro* (Phillips and Lanier 1986). There has been much interest in LAK cells because of early reports they can cause regression of some solid tumours when administered to experimental animals and cancer patients, and suppress metastasis formation.

10.2.6 Regulation of NK activity by cytokines

Effects of cytokines on NK cells

The cytotoxic activity of NK cells can be enhanced by interferons (IFNs) and of these IFNγ may be the most effective (Section 9.5.4). Agents that induce IFN production, such as viruses, can enhance the activity of NK cells, although this can also occur by IFN-independent routes. IFNγ may increase the number of NK cells binding to their targets and the rate of lysis by these cells, as well as the ability of NK cells to recycle and destroy more cells. IL-2 can also increase the cytotoxic activity of NK cells, and it synergizes with IFNγ. These cytokines, alone or together, may cause maturation of NK cells in a manner analogous to development of CTL from precursors, and induce NK cell proliferation and/or activation.

Effects of cytokines on target cells

The effects of IFN are not limited to the NK effector cells (Fig.10.2). While IFNγ-treated NK cells have increased cytotoxic activity towards normal cells (as well as towards 'typical' NK targets), normal cells become resistant to this increased activity after exposure to IFNγ. Thus it has been shown that NK cells can bind equally well to untreated and IFN-treated fibroblasts, but only the untreated cells are lysed. Virally-infected cells (and tumour cells) are not protected by IFNγ in this manner perhaps because viral infection shuts off host-cell protein synthesis, which may be required to confer the resistance to NK activity (Trinchieri and Santoli 1978). Since this is an early event in viral infection, NK cells might be able to lyse these cells even before infectious particles have been assembled. While IFNγ makes normal cells resistant to the increased NK activity, it does *not* protect them against killing by CTL, or antibody plus complement treatment, and this is circumstantial evidence for a different recognition mechanism in NK cells and CTL (Section 10.3.2).

Other effects

NK activity can be inhibited by T cells and macrophages. Suppression by macrophages can be mediated by prostaglandins, and IFNγ *decreases* the susceptibility of NK cells to this inhibition. Whether NK cells can *secrete* IL-2 and IFNγ is not certain, but at least a subpopulation of LGL has been reported to produce IFNα, IFNγ, and IL-2, as well as IL-1, NK cytotoxicity factor (NKCF), lymphotoxin (TNFβ; Section 9.5.4), a colony stimulating factor (CSF), and a B cell growth factor (BCGF; Section 8.4.3).

10.2.7 Antibody-dependent cellular cytotoxicity

During antibody-dependent cellular cytotoxicity (ADCC), specific antibodies bind to Fc receptors (FcR) on effector cells and the cytotoxic activity of these cells is then directed towards the target for which the antibody is specific

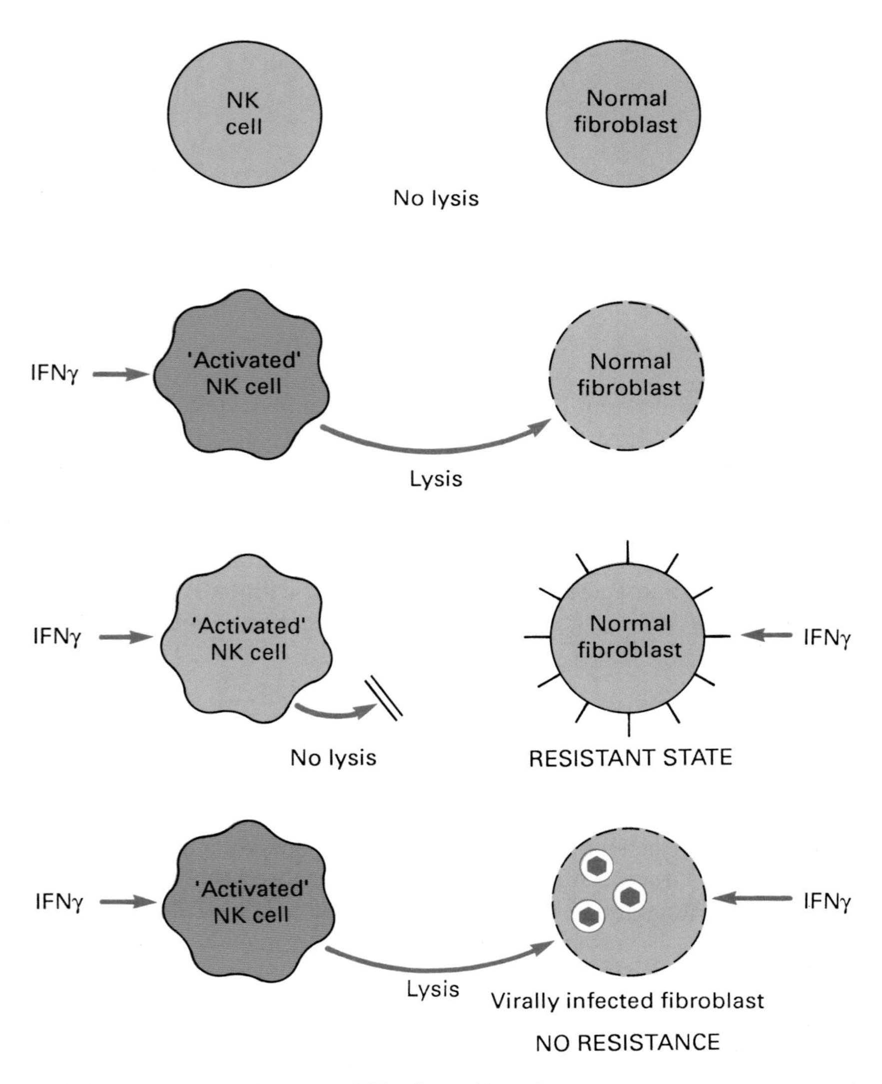

Fig.10.2 Effects of interferon-γ on NK cells and targets

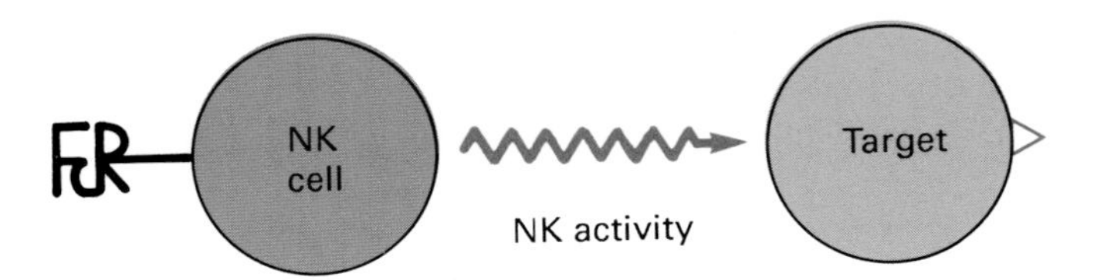

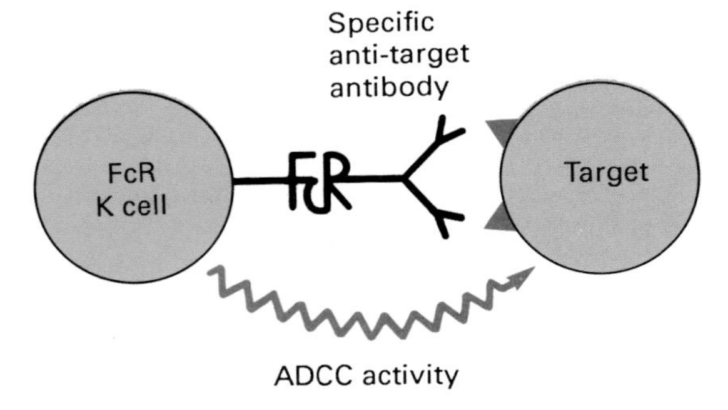

Fig.10.3 Antibody-dependent cellular cytotoxicity and NK activity

(Fig.10.3). Many different cell types, including monocytes, macrophages, neutrophils, and at least some cells with NK activity, can carry out ADCC against antibody-coated tumour cells (Ozer *et al.* 1979). All these effector cells have FcR, but not all cells with FcR can mediate ADCC.

The recognition phases for ADCC and NK activities are almost certainly distinct. ADCC, but not NK activity, can be inhibited by blocking Fc receptors with monoclonal anti-FcR antibodies, by treatment with PMA (which causes a loss of FcR from NK cells), and by blocking the function of bound antibodies with anti-Ig.

K cells

The term *Killer (K) cell* has been used to describe a mononuclear cell that carries out ADCC, that is, K cells are defined functionally (Table 10.3). Their relationship to NK cells and monocytes and macrophages is obscure, but some investigators believe that in the human the population of K cells in LGL overlaps, or is identical with, NK cells. It is not inconceivable that NK and K cell activity are functions of the same cell in different activated states, rather than of different cell types. In other words, a single population of cells may exhibit MHC-unrestricted NK activity (independent of Fc receptors) at one stage of development, but acquire ADCC activity at another. There is interest in using hybrid antibodies, to cross-link cytotoxic cells to defined target cells (e.g. tumour cells), which may then be killed by ADCC, NK activity, or other mechanisms discussed below (Titus *et al.* 1987; Colsky *et al.* 1988).

10.3 Mechanisms of killing by NK cells and cytotoxic T cells

How do NK cells and CTL kill their target cells? One of the most well-known consequences of complement activation is the production of 'holes' or pores in target cell membranes (Section 9.4.6; Fig. 9.23(c)). As discussed later in this section, similar pores were also seen in the plasma membrane of cells that were attacked by NK cells and by some CTL. These observations inevitably led to the idea that the mechanism of killing by complement, NK cells, and CTL might be similar. However, closer examination of the events that occurred within the target cells revealed that there were important differences between the consequences of complement activation on one hand and attack by NK cells and CTL on the other. In particular, rapid nuclear DNA degradation was often observed in cell targets of cellular cytotoxicity, but not during complement lysis.

Additional differences between killing by NK cells and CTL then became apparent. While the majority of NK cells have cytoplasmic granules containing molecules that can polymerize into pore structures within target membranes, freshly-isolated CTL frequently *lack* such granules but *are* fully competent killer cells. It now seems that what are termed granule-dependent, pore-forming mechanisms are important for killing by NK cells, but that granule-independent non-pore-forming mechanisms can operate in CTL. CTL can, however, acquire pore-forming mechanisms for killing *particularly after they have been cultured.* This section outlines the events that occur during cell-mediated cytotoxicity by these two types of effector cell.

10.3.1 Phases of killing

Killing by CTL and NK cells can be divided into several phases (Hiserodt *et al.* 1982) (Fig.10.4).

1. **Target cell recognition**. The effector cell binds to the target cell, and specific recognition occurs via the T cell receptor in the case of CTL, or via another structure for NK cells.

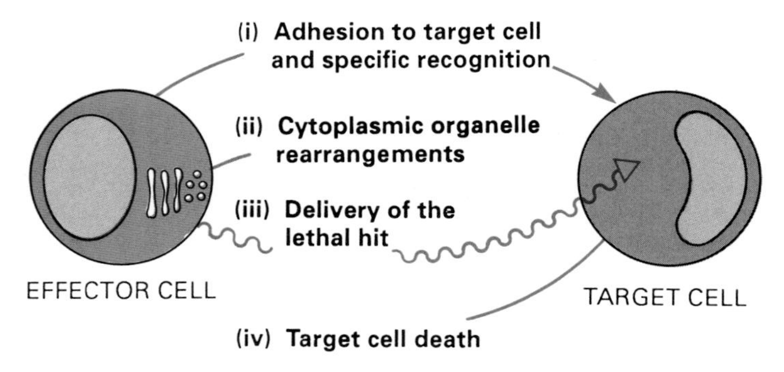

Fig.10.4 Phases of killing during cellular cytoxicity

2. **Cytoplasmic rearrangements** (i.e. reorganization of cytoplasmic organelles). As a consequence of target cell recognition, certain organelles become redistributed within the cytoplasm of the effector cell.
3. **Delivery of the 'lethal hit'** by the effector cell (also termed, by frustrated computer buffs, 'programming for lysis' of the target cells). During this phase a number of molecules that are toxic for the target cell are secreted by the effector cells. These toxic molecules are generally referred to as **cytolysins**, or leukolysins since they are *leuko*cyte-associated *ly*tic molecules. They can be classified as 'pore-forming' and 'non-pore-forming' molecules depending on whether or not they lead to the appearance of characteristic holes in target cell membranes.
4. **Target cell death.** Once the lethal hit has been delivered, destruction of the target cells is inevitable and death will occur even in the absence of the effector cell. Although for many years killing by cytotoxic cells was thought of as a 'murder' carried out by the killer cell, a more recent view is that the killer cell actually induces the target cell to 'commit suicide'.

10.3.2 Target cell recognition

Cytotoxic T lymphocytes

Molecules involved in T cell responses are discussed in Section 4.2. Here only a few aspects that are most pertinent to the effector function of CTL will be reviewed.

It seems probable that the CTL first adheres to the target cell in a relatively non-specific manner, and recognition via the T-cell receptor then occurs if its specific peptide–MHC complex is present (Fig.10.5).

In support of this idea, anti-CD3 antibodies do not inhibit effector-target cell binding but block cytolysis at a subsequent step, presumably during antigen recognition.

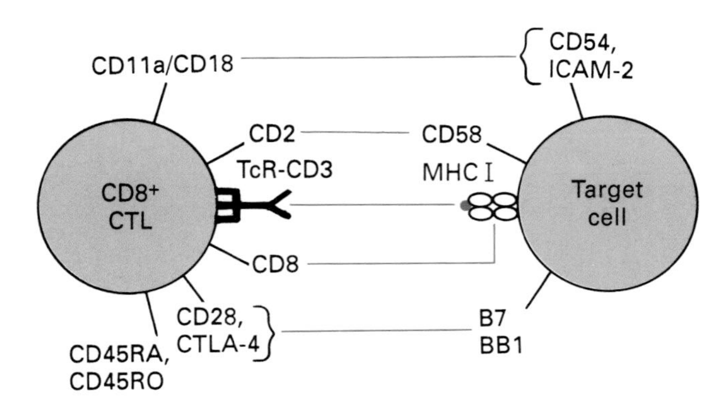

Fig.10.5 Some molecules mediating effector cell–target cell adhesion and antigen recognition by CD8⁺ CTL (see Section 4.2)

It seems most unlikely that the T cell receptor–CD3 complex itself would be able to mediate adhesion of a CTL to its target, and this function is almost certainly carried out by other specialized membrane molecules. Of these, the interaction between CD11a/CD18 (LFA-1) on the T cell and ICAM-1 or ICAM-2 on the target cell, as well as CD2-LFA3 interactions, may be most important.

This idea is partly supported by the observation that antibodies specific for CD11a/CD18 can inhibit cell conjugate formation. This adhesion step is inhibited in the presence of Mg^{2+} ions and at low temperatures (e.g. 4°C), a characteristic of this molecular interaction, as well as by metabolic inhibitors such as sodium azide.

Specific recognition of antigen-MHC is of course mediated by the T cell receptor–CD3 complex and, for many cytotoxic T cells, an interaction between CD8 on the T cell and MHC class I on the target cell is often required. Certainly anti-CD8 antibodies do not inhibit conjugate formation but, like anti-CD3 antibodies, block at a *later* stage.

There appears to be another step in the killing process after the Mg^{2+}-dependent and temperature-dependent adhesion step (above), that does not require cations (e.g. Mg^{2+} or Ca^{2+}) and which can be inhibited by cytochalasin, which interferes with assembly of actin filaments and thus disrupts microfilaments (Fig.1.6). This may be the stage in which T cell receptors are ligated and 'cross-linked' within the plane of the membrane.

Recognition via the T cell receptor and cell adhesion via other molecules can be bypassed by using lectins such as concanavalin A or phytohemagglutinin to bring the effector and target cells into contact. Under these conditions CTL can kill a variety of target cells non-specifically, in a process known as **lectin-dependent cell mediated cytotoxicity (LDCC)**. Normally, however, the T cell receptor imparts directionality to the interaction and 'focuses' the cytolytic activity of the CTL on to its target (Fig.10.6).

For example, Kuppers and Henney (1977) prepared CTL of different specificities: a-anti-d, and d-anti-b, where a, b, and d are different MHC genotypes (Fig.10.6). When these cells were cocultured, the a-anti-d effectors were able to recognize the d-anti-b effectors, but not vice versa, and killing only occurred in this direction.

Ligation of the T cell receptor–CD3 complex to its specific peptide–MHC complex, can be sufficient to prime cytotoxic T cells which then proliferate in the presence of cytokines.

This was shown by cross-linking the T cell receptor with a monoclonal antibody specific for a common determinant on T cell receptors expressed by about 25% of peripheral T cells in the mouse (Crispe *et al.* 1985). Proliferation of $CD8^+$ T cells was induced when they were incubated with the antibody coupled to Sepharose beads as an immobile support, in the presence of preformed cytokines. Of interest, $CD4^+$ T cells did not proliferate under these conditions, probably because they have a more stringent activation requirement, perhaps requiring dendritic cells (Section 3.4.3).

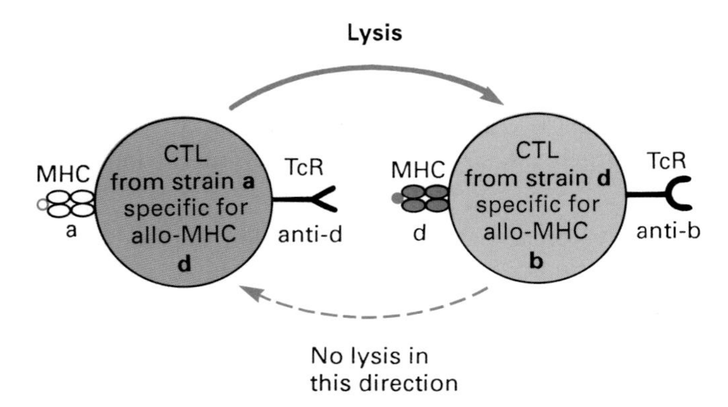

Fig.10.6 The T cell receptor imparts 'directionality' to killing by CTL (Kuppers and Henney 1976)

A number of cytokines may act on precursor cytotoxic T cells to trigger their development into mature effector cells, stimulate proliferation (clonal expansion), and perhaps to induce full cytotoxic activity (Fig.10.7)). Foremost amongst these are IL-2 and IL-4, which act through cytokine receptors that are expressed when the cytotoxic T cell is primed, perhaps simply following antigen recognition. Certainly IL-2 can drive proliferation of primed CTL, while IL-4 can induce cytotoxic activity (Table 4.18). Both cytokines are produced by activated helper T cells, but cytotoxic T cells may also be activated by a route that does *not* appear to require helper T cells.

For example, CTL can develop in mice treated with CD4 antibodies to deplete class II-restricted helper T cells, and dendritic cells can activate purified $CD8^+$ class I-restricted T cells in culture to proliferate and develop into cytotoxic T cells (Young and Steinman 1990); however, it is not clear whether a small subset of $CD8^+$ (class I-restricted) helper T cells is involved in the latter response.

NK cells

The precise nature of the receptor molecule(s) that mediates target cell recognition by NK cells, and the corresponding structure(s) on target cells that can be recognized, is as yet unknown, although candidate molecules for NK receptor(s) have been reported (Section 10.2.4). T cell receptor $\alpha\beta$ or $\gamma\delta$ molecules are not expressed on the surface of NK cells (by definition), and there are no rearrangements of the α, β, γ, or δ genes in these cells. Curiously, however, a 1.0 kb transcript from the β chain gene, with no V region, has been detected in some NK cells, together with T cell receptor δ chain transcripts of different sizes containing J and C regions but likewise with no V region. The significance of these truncated transcripts is unknown. It seems most likely that cells containing rearranged T cell receptor genes with NK-like activity are T cells that have acquired this function. There is evidence that cytotoxic T cells with NK-like activity use different receptors for MHC-restricted and MHC-unrestricted recognition because, for example, antibodies against the T cell receptor could block the former but not the latter function of a particular T cell clone (Binz *et al.* 1983).

What seems to be clear is that the mechanism of binding of an NK cell to its target is different to that of CTL binding (see above). NK binding requires Mg^{2+} ions but it can occur at 4°C as well as 37°C, and is not inhibited by

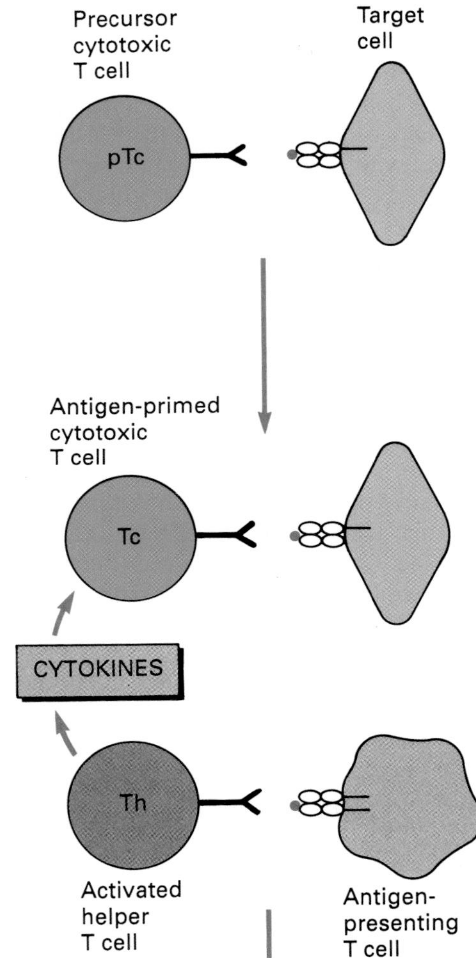
(i) Recognition of specific peptide – MHC complexes on target cell; expression of cytokine receptors
Precursor cytotoxic T cell
Target cell
pTc
(ii) Binding of T helper cell-derived cytokines: e.g. IL2 for proliferation; IL4 for development of full cytotoxic potential
Antigen-primed cytotoxic T cell
Tc
CYTOKINES
Th
Activated helper T cell
Antigen-presenting T cell
(iii) Target cell killing
CTL
Lysis
Effector cytotoxic T cell

Fig.10.7 A scheme for helper-dependent development of precursor cytotoxic T cells into mature CTL

azide. However, there are some similarities in that, for example, NK-target cell conjugation can be inhibited by antibodies against LFA-1, and killing can be inhibited by antibodies against CD45, which may be involved in NK and CTL function *after* target cell binding (Targan and Newman 1983).

10.3.3 Reorganization of cytoplasmic organelles

Profound changes occur in the cytoplasm of both CTL and NK cells after they have bound to their targets. A number of organelles become localized directly between the nucleus of the effector cell and the region of membrane in contact with the target cell. These include (Fig.1.6) cytoplasmic granules, the Golgi apparatus, and the microtubule-organizing centre (MTOC) to which cytoplasmic tubules, composed of tubulin and actin, are anchored, and which includes the centrioles. It is thought that organelle reorientation allows cytoplasmic granules and/or secretory vesicles to be directed towards the target cell, their toxic contents being discharged by exocytosis (Section 3.2.2) in a Ca^{2+}-dependent process. In fact reorientation of the MTOC and reorganization of cytoplasmic and cytoskeletal elements appears to be a general phenomenon during antigen recognition by T cells, since this also occurs during T-B interactions.

A single CTL may bind several target cells, and these are then killed sequentially, one at a time. This is accompanied by reorientation of the cytoplasmic contents of the effector cell to face each cell to lysed in turn (Kupfer *et al.* 1986). It is believed that the T cell receptors of CTL and the presumptive NK receptors are linked to intracellular components (e.g. cytoskeletal components like tubulin and actin), and that ligation of these receptors brings about the cytoplasmic rearrangements. Killing via Fc receptors, during ADCC (Section 10.2.7), is likely to be similarly coupled to intracellular components.

10.3.4 Delivery of the 'lethal hit'

The expression 'delivery of the lethal hit' could be a misnomer, because an effector cell might actually deliver not one but a series of 'hits' to the target cell before it is killed (Fig.10.8). It seems likely that this process occurs soon after the cytoplasmic organelles become reorganized (Section 10.3.3). If the membranes of target cells attacked by CTL or NK cells are examined, they are frequently found to contain holes or pores; similar structures are present after ADCC. These pores resemble those generated in target cells by the membrane attack complex of complement (Section 9.4.5). In the case of complement, these membrane lesions allow the entry of water into the cell, as well as loss of ions and polyelectrolytes, and hence destruction of the cell by osmotic or colloidal forces. It was originally thought that cytotoxic cells kill their targets by a similar mechanism, but it is now known that cytotoxic cells can also kill by mechanisms that do not lead to pore formulation in target membranes. A variety of enzymes and other molecules have been implicated in this mode of killing, although their precise functions and their relative importance compared to pore-forming systems have not yet been clearly defined. Many of these molecules are present in cytoplasmic granules

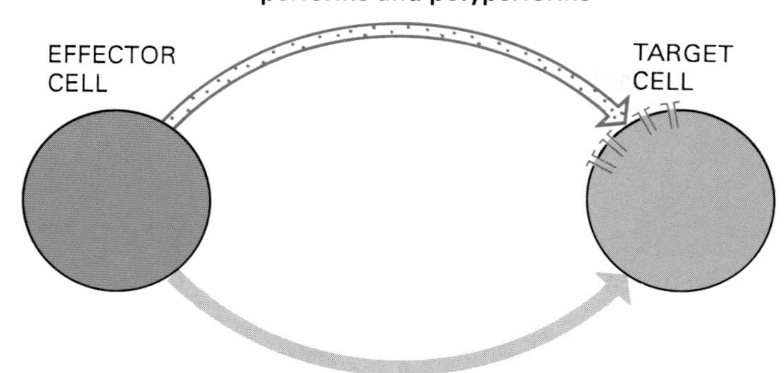

Fig.10.8 Mechanisms of cytotoxicity

of NK cells and cultured CTL, and a summary is given in Table 10.4. Some of these molecules are considered in more detail below.

Pore-forming mechanisms

Perforins and polyperforins. The membrane lesions in cells attacked by CTL or by NK cells have the shape of a torus and were found to have a hydrophobic domain at both ends (Fig.10.9). These pore-like structures were

Table 10.4 Molecules found in cytoplasmic granules of mouse CTL and NK cells*

Molecule	Distinctive features	Function
Perforin	Pore-forming protein	Lytic in the presence of calcium
Granzymes	Proteases	Proteolysis DNA damage?
Proteoglycans	Chondroitin sulfate A	Carrier molecule for granule proteins?
Leukolexin	TNF-related molecule M_r, 70 kDa	Induces calcium-independent slow lysis of some cells; DNA damage?
Chemoattractant	$M_r < 10$ kDa	Macrophage-activating activity, chemotaxis
Carboxypeptidase A	Exopeptidase	
Cathepsin D	Lysosomal enzyme	
Arylsulfatase	Lysosomal enzyme	
β-Glucuronidase	Lysosomal enzyme	
β-Hexosamidase	Lysosomal enzyme	

*Based on Tschopp and Nabholz (1990) and Podack *et al.* (1991) — see Further reading.

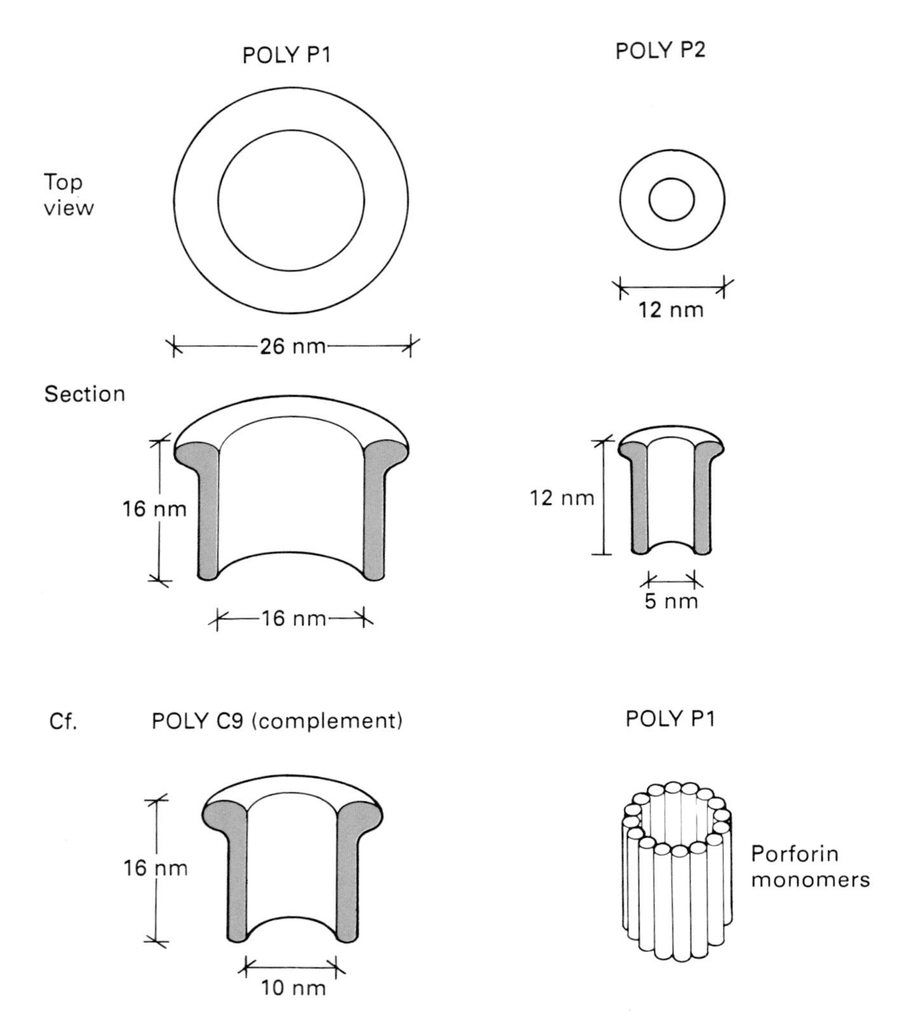

Fig.10.9 Polyperforins

called **polyperforins** because they are formed by polymerization of monomers termed **perforins**. Two main sizes of polyperforin lesions have been described, and called polyperforin 1 and 2 (polyP1 and polyP2) respectively (Podack and Dennert 1983); these are composed of different numbers of perforin monomers. The perforin molecule has a molecular weight of about 66–70 kDa, and the purified molecules polymerize in the presence of Ca^{2+} ions to produce a wide range of different polymers, some in excess of 1 million daltons. It is thought that at least 3–4 monomers are required to make a functional pore in a target cell membrane, and that they aggregate progressively up to about 10–20 monomers; polyP1 may be composed of about 18–20 perforin monomers. Perforins are stored in monomeric form within cytoplasmic granules of the effector cell and it seems likely that the monomers are released by exocytosis, become inserted into the target cell membrane, and then polymerize into polyperforins.

The perforin gene has been cloned, and the protein it encodes has two domains, one of which may bind Ca^{2+}. Molecular modelling of the other domain shows that it can be folded into an amphipathic α helix (Fig.3.34) which may be exposed by a conformational change that occurs in the molecule after it has associated with phospholipids in the target cell membrane. This region, which contains a lipid-binding domain, enables the perforin molecules to insert into the lipid bilayer of the target plasma membrane and to polymerize to form transmembrane pores.

Experiments have been carried out with purified cytoplasmic granules from a rat LGL tumour. While live LGL cells only killed certain tumour cell targets, the isolated granules were toxic to both NK-susceptible and NK-resistant targets, suggesting that their release and target cell specificity is controlled by effector cell recognition of its target. Target cell lysis occurred rapidly, and correlated with the appearance of pores, particularly polyP1. Their formation required Ca^{2+} ions, but exposure of the granules to calcium *before* they interacted with the target resulted in inactivation. Similar studies have also been performed on granules isolated from CTL (Podack and Konigsberg 1984).

Pores have also been found in the membranes of target cells damaged by ADCC. Using a variety of molecular weight markers, the diameter of pores formed after attack by human monocytes in this way is estimated to be about 10–12 nm. In contrast, antibody and complement treatment of cells results in the formation of pores, due to the membrane attack complex of complement, of about 50 nm diameter. Like polyperforin, the C5–C9 membrane attack complex contains multiple C9 molecules, thought to number 10–20.

Relationship between perforin and complement component C9. Cloning of the perforin gene has revealed homology to C9, including the presence of two cysteine-rich domains. The perforin gene also has homology with the complement components C6, C7, and C8 (Fig.10.10). Previously it was found that antibodies generated against cell membranes damaged by cytotoxic cells cross-reacted with the membrane attack complex of complement, and that antibodies produced against purified mouse perforin reacted with complement components C5 through C9 (Young *et al.* 1986). The determinants recognized were unmasked by reduction of disulfide bonds, suggesting they normally become exposed when they are activated in the membrane attack complex. Conversely polyclonal antisera against C9 can be used to purify perforins from LGL and NK cells. Nevertheless, the perforin gene consists of 3 exons (the coding regions are all in the second and third domains) whereas the C9 gene has 11 exons, suggesting that they have not evolved from a common ancestral gene (Lichtenfeld and Podack 1989).

Membrane damage by polyperforins. Polyperforins permeabilize cell membranes. Treatment of cells with perforins causes a depolarization of the membrane potential, indicative of a disturbance of the ionic equilibrium across the membrane, an influx of water, and the loss of polyelectrolytes and other molecules. Purified perforin and granules isolated from cytotoxic cells can also permeabilize liposomes, and when isolated perforins were incorporated into artificial membranes the latter were able to conduct up to 10^9 ions per second. However, isolated perforin is probably *not* able to mediate DNA fragmentation (Duke *et al.* 1989)

The correlation between the appearance of pores in target cell membranes and cell killing suggested this was likely to be an important cytolytic mechanism for CTL, NK cells, and for cells carrying out ADCC. Moreover, electrophysiological evidence has been obtained that one of the major granule proteins of human eosinophils, eosinophil cationic protein, can also form pores. Because a toxin isolated from *Entamoeba histolytica* can form similar

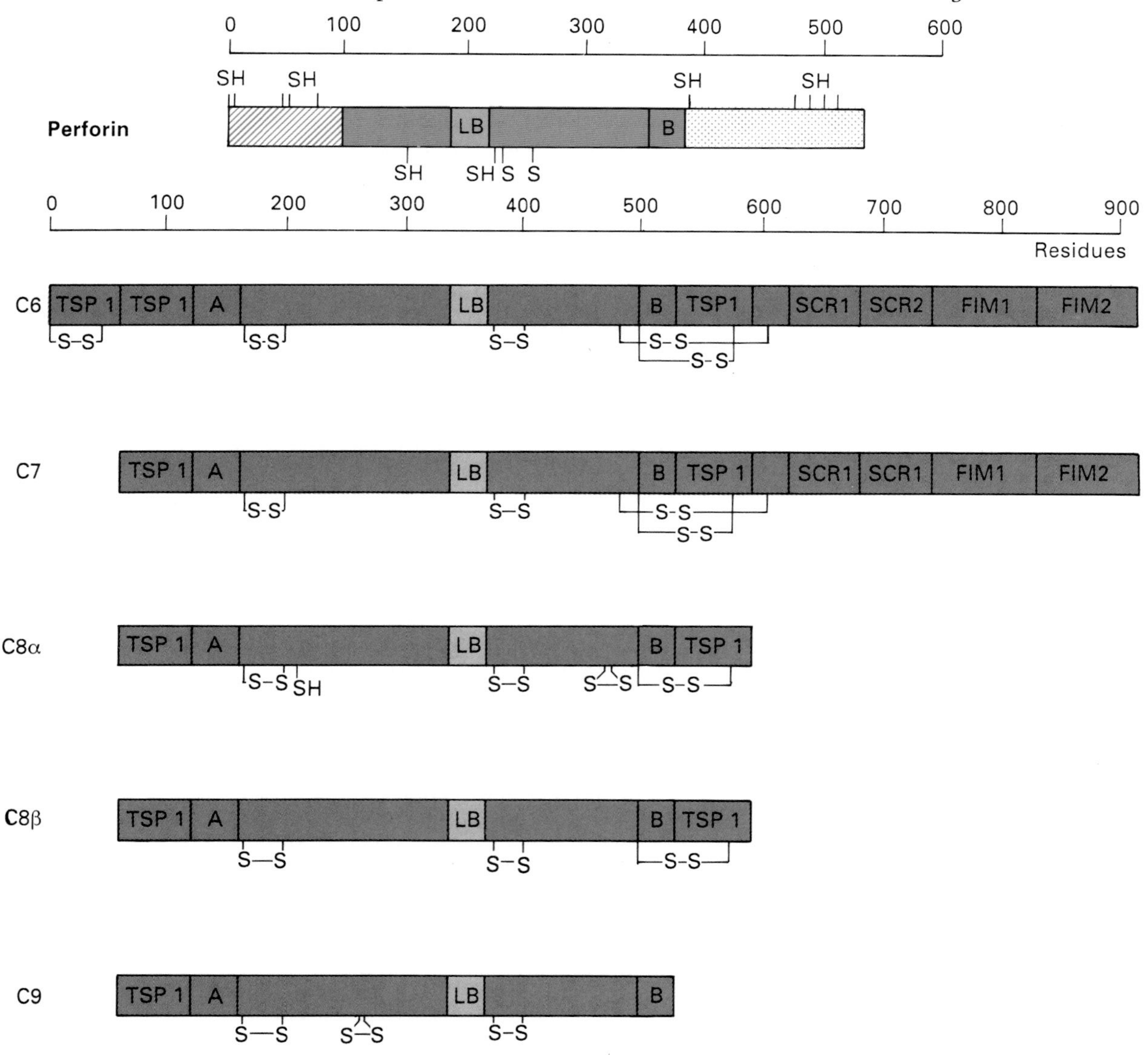

Fig.10.10 Structural organization of human perforin and the late complement components. Regions of perforin with homology to complement proteins are shown in solid red tint. LB, candidate for the lipid-binding domain; A, B, LDL receptor class A and B modules; TSP 1, type I thrombospondin module; SCR, short consensus repeat; FIM, factor I module; S–S, disulfide bonds and SH, cysteine (only some are indicated). Based on Tschopp and Nabholz (1990), with permission (© 1990 by Annual Reviews Inc.); see Further reading

lesions, this may be an evolutionarily ancient mechanism of cytotoxicity. However, pore formation is just one aspect of cellular cytotoxicity because a variety of non-pore-forming mechanisms have also been implicated (see below).

Non-pore-forming mechanisms

Evidence for perforin-independent and granule-independent cytotoxic mechanisms. A number of observations were not compatible with the idea that pore formation was *always* required for target cell lysis. These can be summarized as follows.

1. Using polyclonal antisera against perforin it was not possible to detect this molecule in lymphocytes obtained from the peritoneal cavity after an inflammatory response, which were nevertheless potent killer cells.
2. Several CTL lines that had cytoplasmic granules *lacked* pore-forming material, even though the cells were lytic, and even though subcellular fractions of these cells were toxic to target cells.
3. Some CTL generated in primary MLRs (see Appendix) did not possess cytoplasmic granules, yet they could kill target cells.
4. Some CTL are cytolytic in the *absence* of Ca^{2+} ions, and subcellular fractions can be toxic under the same conditions; cytolysis by polyperforins is strictly Ca^{2+}-dependent.
5. Antibodies specific for components of LGL granules could block NK activity as well as ADCC, but not CTL activity.
6. Perforin only appeared in some T cells when they were cultured in IL2 and this correlated with the acquisition of NK-like activity.
7. CTL can kill in the apparent absence of granule exocytosis.

It is possible to rationalize these findings in the following manner (Fig.10.11). NK cells, which generally contain granules and perforins, commonly use pore-forming mechanisms of cytotoxicity. In contrast, many *freshly-isolated* CTL do not contain granules or perforins, but can use non-pore-forming mechanisms of cytotoxicity (discussed below). *Cultured* CTL can, however, acquire granule-dependent and perforin-dependent mechanisms after continued exposure to IL-2 as in the case of many CTL lines and clones (Smyth *et al.* 1990).

Serine esterases. Eight serine esterases, named *granzymes* A–H, have so far been identified within granules of cytotoxic cells, which also contain perforins (Peters *et al.* 1991) (Table 10.5). The genes for these molecules have been cloned, and shown to be highly homologous to each other (Masson and Tschopp 1987). These enzymes are of interest in view of the similarity between perforin and certain complement components (see above), since the complement cascade is activated after cleavage of complement components by serine esterase-like enzymes. However, the purified granzymes A–H have no cytolytic activity and their function in delivery of the lethal hit has not yet been defined.

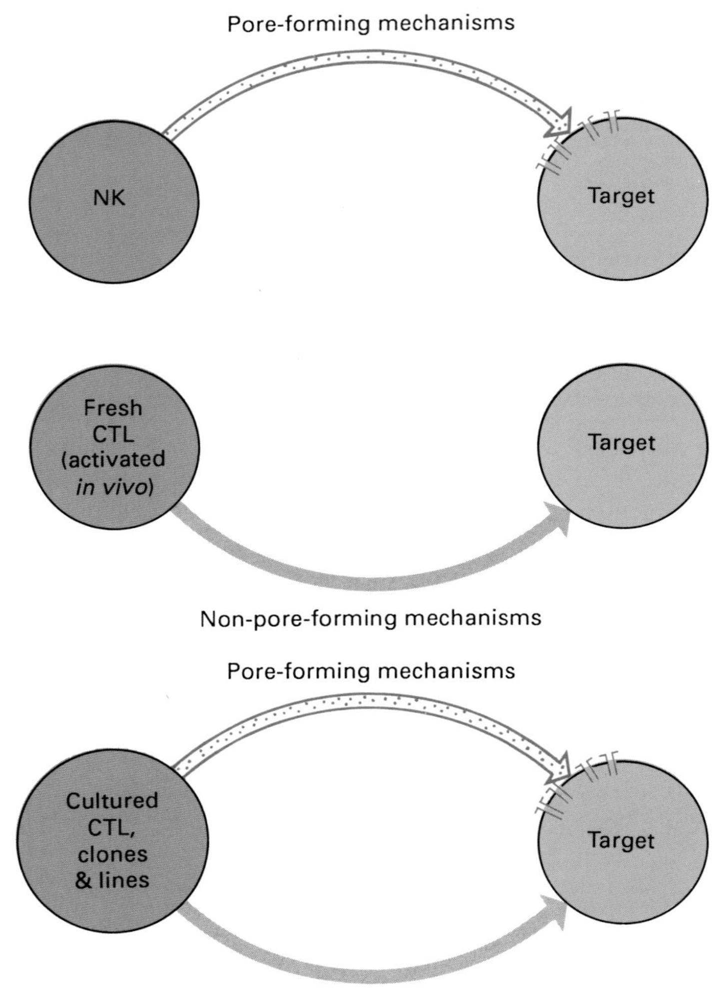

Fig.10.11 Possible differences in predominant cytotoxic mechanisms in different effector cells

The first two serine esterases to be defined at the cDNA level were termed CTLA-1 or CPP1, and HF (corresponding to granzymes B, G or H, and granzyme A, respectively). These molecules show between 40% and 35% homology to another serine protease found in rat mast cells (serine protease II). Of interest, the latter is 36% homologous to human factor D, the first component of the alternative pathway of complement (Section 9.4.5). The predicted structure for CTLA-1 (CCP1) is that of a characteristic serine protease, starting with the amino acid sequence Ile-Ile-Gly-Gly, and having a His-Asp-Ser sequence at the catalytic site. There is an Ala residue in the region of the molecule that most likely determines its substrate specificity, as in the rat mast cell serine protease II, and there are other similarities between these two molecules. However, HF has an Asp in its substrate specificity determining position, like factor D, which is likely to make it specific for cleavage between lysine and arginine residues (i.e. it may have trypsin-like activity).

Table 10.5 The serine esterase family in granules of CTL*

Serine sterase	Molecular mass kDa†	Characteristics
Granzyme A	$35^{r}/60^{nr}$	Trypsin-like specificity; disulfide-linked dimer
Granzyme B	29	Protease cleavage after Met
Granzyme C	27	
Granzyme D	35–50	Highly glycosylated
Granzyme E	35–45	Highly glycosylated
Granzyme F	35–40	Highly glycosylated
Granzyme G	33	
Granzyme H	31	

*Based on Tschopp and Nabholz (1990) — see Further reading.
†As determined by SDS-PAGE. Only the apparent mass of granzyme A changed when electrophoresis was carried out in reducing (r) versus non-reducing (nr) conditions.

Tumour necrosis factors. Tumour necrosis factors (TNFs), discussed in detail in Section 9.5.3, may be important in cellular cytotoxicity.

TNF was so-called because it caused necrosis when it was added to transformed cells in culture and *in vivo*. In addition to its cytotoxic effects, TNF also has cytostatic (growth inhibitory) effects on such targets. Two related TNFs are known: TNFα or cachectin and TNFβ or lymphotoxin. For some time TNFα was thought of primarily as a macrophage product, and TNFβ or *lympho*toxin as a *lympho*cyte product. Although macrophages are not thought to produce TNFβ, it is now known that activated lymphocytes can produce both molecules, and can in fact secrete more TNFα than TNFβ.

The precise role of TNFα and TNFβ in cellular cytotoxicity is uncertain. They may be present within granules of the effector cells or as free cytoplasmic molecules; membrane-bound forms of TNF have also been identified (Schmidt *et al.* 1987). TNFs and related molecules, including **leukalexins** (Liu *et al.* 1987), are released by effector cells during killing of target cells, and may cause DNA degradation. However, the isolated molecules act only slowly (over several hours), and other molecules may be required for rapid DNA breakdown and cell killing. Another point to note is that drugs affecting cytolysis by CTL have no effect on lymphotoxin production, and antibodies to lymphotoxin do not interfere with CTL lysis. Conceivably, however, this could simply reflect the coexistence of different cytolytic mechanisms in these cells.

An activity called **NK cytotoxic factor** (NKCF) may be different from TNF (Wright and Bonavida 1987). This soluble factor was found in cell-free supernatants of NK cells when they were cultured with NK-sensitive target cells. NKCF lysed NK-sensitive targets but not resistant cells, suggesting that the former may have a 'receptor' for NKCF, and this may be associated with pore formation. (IFNγ-treated susceptible targets were also lysed by pre-formed NCKF, but they did not induce its release from the NK cell).

Proteoglycans. A particular proteoglycan, chondroitin sulphate A, has been found in the granules of cytolytic cells and this is exocytosed from the cells when they bind to susceptible targets (Schmidt *et al.* 1985).

Proteoglycans often act as storage and/or carrier molecules for basically-charged proteases and other proteins. A good example of this is heparin, which is bound to histamine within mast cell granules (Section 10.4.5). One possibility is that chondroitin sulphate A is bound to perforin within cytolytic cell granules to protect the effector cell from self-lysis. After exocytosis the perforin is secreted, while the proteoglycan could conceivably remain associated with the cell membrane of the effector cell and protect it from lysis by perforin in the external environment.

10.3.5 Target cell death

The changes in cell structure that precede or accompany cell death in general can be characterized as either **necrosis** or **apoptosis** (Fig.10.12). Necrosis generally occurs in unfavourable conditions that can result, for example, from ischaemia, starvation, and various types of trauma. Apoptosis, on the other hand, is a form of 'programmed cell death' that occurs under normal physiological conditions such as during embryonic development when tissues are remodelled. It is also seen in the large number of thymocytes that die naturally within the thymus or after glucocorticoid treatment, and it can be

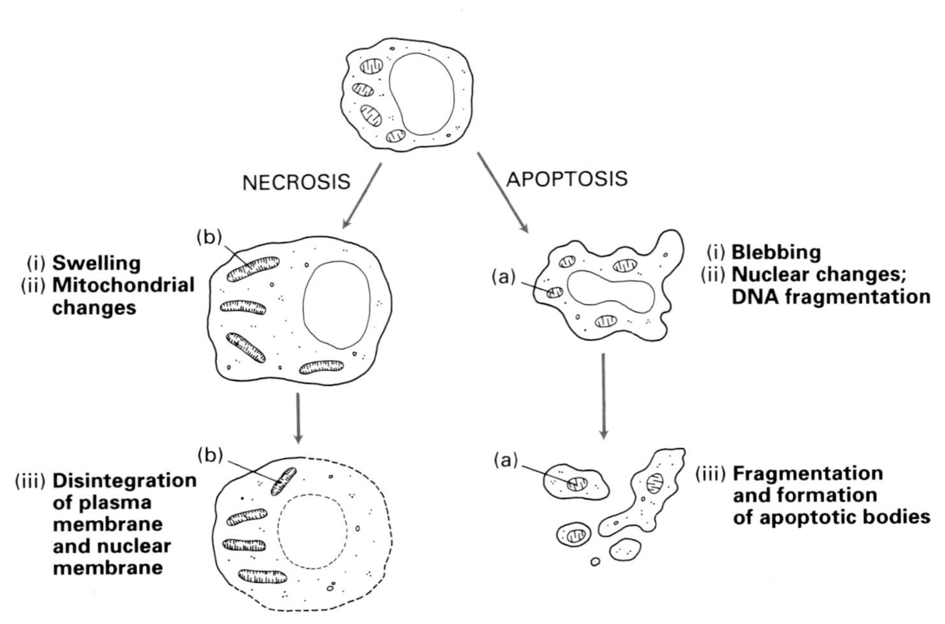

Fig.10.12 Two mechanisms of cell death: necrosis and apoptosis; normal (a) and abnormal (b) structures of mitochondria are indicated

induced in target cells treated with TNF (Section 10.3.4). While necrosis often results in loss of the normal tissue architecture, which is compensated to some degree by subsequent scarring, apoptosis leads to removal of cells in a controlled and organized manner without loss of overall structure and without scarring.

The membrane attack complex of complement causes cells to become necrotic. During complement-mediated cytotoxicity the earliest event following the insertion of transmembrane pores seems to be extensive damage to the plasma membrane and the subsequent swelling of the cell and its disintegration; changes in the nucleus of the cell only occur *later*. In contrast, *apoptosis* is often induced after attack by cytotoxic effector cells, and the earliest events include changes in the nuclear membrane and progressive disintegration of DNA *before* there is any obvious effect on the plasma membrane.

The characteristic microscopic appearance of a cell undergoing apopotosis is that of violent blebbing (zeiosis) of the nucleus, and fragmentation of the nucleus and cytoplasm into a number of 'apoptotic bodies' which subsequently disintegrate. If the DNA of such a cell is examined soon after attack by a cytotoxic effector cell, it is found to be rapidly degraded into fragments that are about 200 base pairs in length. The speed and efficiency of this process (which is dependent on the presence of Ca^{2+} ions) has led to the proposal that endogenous endonuclease(s) may be activated within the target cell (Duke *et al.* 1983). This auto-destructive process implies that the target cell has been induced to 'commit suicide' by the effector cell. It has been suggested that, in the case of a DNA virus that has infected a cell, the endonuclease(s) that are activated during apoptosis of the host cell might lead to nicking of the viral DNA and inhibit viral replication.

10.3.6 Protection from self-lysis

Cytotoxic cells have to protect themselves from lysis by the toxic molecules that are stored within them, or that are released during target cell lysis (Verret *et al.* 1987). In part, protection from toxic molecules within the cell comes about by storing some of these molecules within membrane-bound cytoplasmic granules. Both the family of serine esterases and perforins are stored in this way, possibly associated with proteoglycans. The potential role that proteoglycans might play in protecting the effector cell from its secreted cytolysins was noted earlier (Section 10.3.4). In addition, perforin needs to undergo a Ca^{2+}-dependent conformational change before it can insert into membranes, and the granules provide a relatively Ca^{2+}-free environment which may also facilitate safe storage. Other molecules may be present as inactive propeptides within the cell, that are processed to an active form during secretion. Conceivably the vectorial nature of cytolysin release on the target cell may mean that the effector cell only has to protect the small area of its plasma membrane that is directly in contact with the target cell.

Whatever the mechanism, it is clear that cytotoxic effector cells are not usually killed when they attack other cells. In this respect they are similar to macrophages (Section 10.4), but unlike neutrophils, for example, which are rapidly killed in inflammatory sites and have to be replaced by cells that are newly produced by the bone marrow during the course of the response.

10.4 Macrophages

The tissue macrophage belongs to the **mononuclear phagocyte system** (MPS; Table 10.6). This system also includes the circulating blood precursor to the macrophage, the monocyte, together with earlier forms in the bone marrow, the monoblast and promonocyte (Figs.1.31 and 1.32). Originally, blood monocytes and tissue macrophages ('histiocytes') were grouped together with reticulin-producing fibroblasts and with endothelial cells as a multi-organ system called the reticuloendothelial system (RES). This was before the distinct origins and functions of these different cell types were understood, for example neither fibroblasts nor endothelial cells have a *haemopoietic* origin, and the term RES is now rarely used although it is found in the older literature.

Differentiation of *haemopoetic* stem cells into the common granulocyte-macrophage precursor cell (CFU-GM; Fig.1.31) *in vitro* is induced by the combined actions of two colony stimulating factors (CSFs), IL-3, also called multi-CSF, and granulocyte-macrophage-CSF, GM-CSF (these and other cytokines, and their receptors, are reviewed in Arai *et al.* 1990). Subsequent development of CFU-GM along the macrophage lineage (i.e. monoblasts, promonocytes, monocytes, and mature macrophages), is stimulated by another cytokine, M-CSF or CSF-1.

The *receptors* for IL-3 and GM-CSF are heterodimers with distinct α chains of 80 kDa but a common β chain of 120 kDa; the receptor for CSF-1 is the 165 kDa product of the c-fms gene. The genes for the *cytokines* IL-3 and GM-CSF are closely linked to those for IL-4 and IL-5 on the same chromosome (Van Leeuwen *et al.* 1989).

Table 10.6 The mononuclear phagocyte system

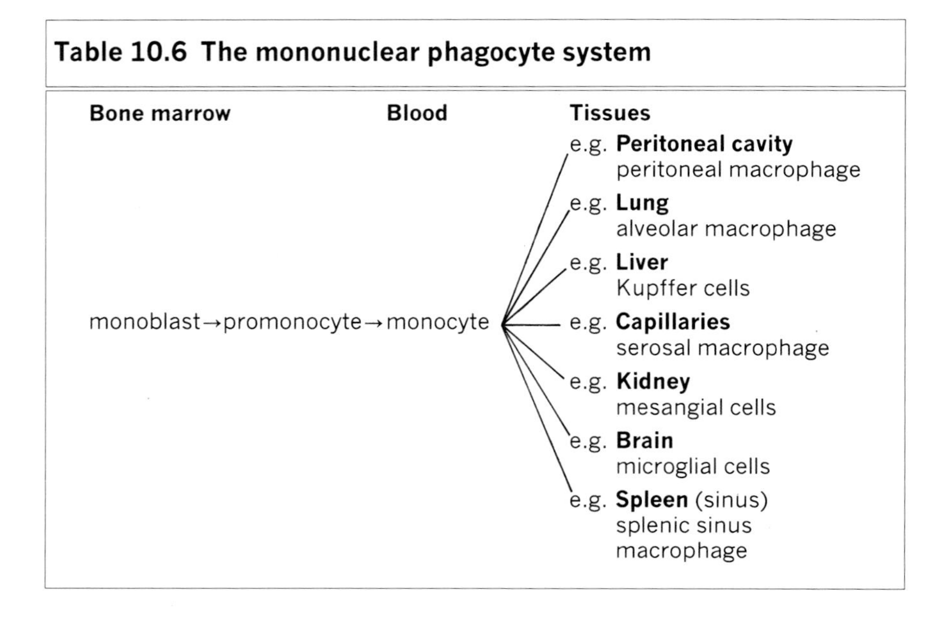

Monocytes migrate out of the blood into the tissues constitutively (that is, spontaneously and without a defined stimulus) and they can also be recruited into sites of inflammation and infection. Within a tissue, the monocyte differentiates into a **macrophage** and acquires a phenotype and specialized functions that are controlled by the environment in which it finds itself. Thus, macrophages in normal tissues can have different properties depending on where they are localized: peritoneal macrophages lining the peritoneal cavity, alveolar macrophages of the lung, Kupffer cells of the liver, microglial cells of the brain, and so on, can be distinguished from each other by a variety of criteria.

Macrophages can also change their phenotype and acquire or lose certain functions depending on the local stimulus, in a process termed macrophage activation (Fig.10.13). During this process the relatively quiescent *resident macrophage* can be stimulated by inflammatory responses to become an *inflammatory macrophage* that secretes a wide variety of products. These cells can also be triggered during immune responses to become *activated macrophages*, now with a greatly enhanced ability to kill microbes and tumours.

In the following subsections (10.4.1–10.4.4) we consider some properties of resident, inflammatory, and activated macrophages and review some of the mechanisms by which these cells kill a variety of different targets. In Section 10.4.5 we consider the function of some membrane molecules of macrophages, particularly Fc receptors and complement receptors. Endocytosis and antigen processing and presentation by macrophages are discussed in Section 3.2.2.

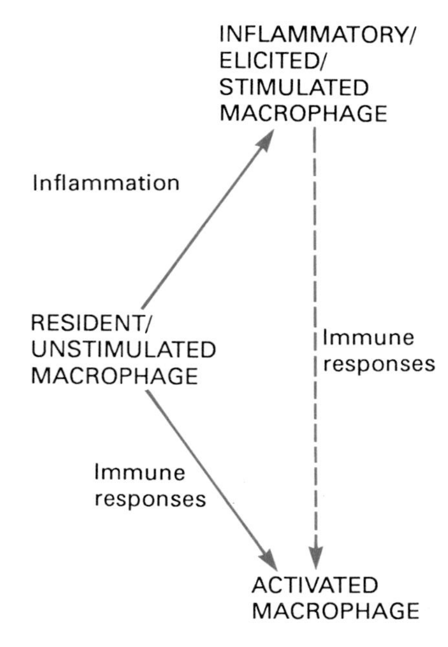

Fig.10.13 Macrophage activation

10.4.1 Characteristics of macrophages

Adhesion

Macrophages can be defined by a variety of criteria. Perhaps the most obvious property of macrophages in culture is their ability to adhere tenaciously to plastic and glass surfaces, although how this is achieved is still not clear. This allows macrophages to be separated from non-adherent cells like lymphocytes, and enriched macrophage populations can thus be obtained. Other cell types also adhere to tissue culture surfaces, especially fibroblasts, and to a lesser extent dendritic cells and some B lymphocytes (Fig.3.4). Adherent macrophages can be recognized by light microscopy because of their characteristic 'fried-egg' appearance soon after they have spread on the surface, by the presence of 'membrane ruffles' around the edge of the cell, and by their nuclear morphology, which is typically horseshoe or kidney-shaped (Fig.1.11). *In vivo*, macrophages employ a variety of different mechanisms to attach to other cells and components of the extracellular matrix (Section 4.2.6).

Adhesion molecules of monocytes include several discussed or noted in Sections 4.2.5 and 4.2.6, such as LFA-1 (CD11a/CD18), CR3 (CD11b/CD18), CD44 which binds hyaluran, a group of VLA proteins (VLA-4, VLA-5, and VLA-6; CD49/CD29), and LFA-3 (CD58).

Endocytosis

Another important property of macrophages is their possession of a large number of vesicles and other cytoplasmic inclusions. This is related to their very active endocytic capacity (Section 3.2.2). Macrophages avidly bind and ingest a wide variety of particles in culture, for example latex microspheres, although the molecules responsible for this particular interaction are not known. *In vivo*, macrophages are generally located so they can clear particles (e.g. bacteria) and debris from the blood and other tissue fluids by phagocytosis.

Fibroblasts are also phagocytic cells (this property being responsible for their inclusion in the RES; see above), but macrophages can be distinguished from these cells by their ability to phagocytose via Fc receptors which are not present on fibroblasts. For example, if macrophages are cultured with antibody-coated sheep erythrocytes, the erythrocytes become bound to the cell via the Fc portion of the attached antibody, and they are subsequently internalized and digested by the macrophage; as noted in Section 3.2.2, receptor-mediated phagocytosis occurs by a 'membrane-zippering' mechanism. The macrophage is also actively pinocytic, and the necessity for membrane recycling to compensate for the amount of membrane internalized during endocytosis is discussed in Section 3.2.2.

Secretion

Macrophages can be highly secretory cells (Table 10.7). Well in excess of 50 different biologically active substances have been defined that are secreted by macrophages at one stage or another (Nathan 1987—see Further reading). These molecules include enzymes and enzyme inhibitors, plasma proteins and mediators of inflammation (Sections 9.2 and 9.3), as well as toxic reactive oxygen and nitrogen intermediates (Section 10.4.4). Some of these products can be detected cytochemically in granules or in the cytoplasm of macrophages, and used to identify the cell.

For example, non-specific esterase and alkaline phosphatase are useful markers of macrophages, although they are not completely specific and their expression can vary depending on the stage of differentiation and activation of the cell. Assays for enzymes such as 5′-nucleotidase, which is expressed at the surface of the cell, can be used similarly, although this particular marker is lost when the macrophage becomes activated.

Macrophage markers

The advent of monoclonal antibodies (see Appendix) led to a revolution in the way we examine cells. One of the goals of this technique has been to identify antigens that can be used as specific markers and classify a cell as belonging to one lineage or another unequivocally. In practice, however, monoclonal antibodies against absolutely specific markers have been very difficult to produce. For example, we have seen in Section 4.2 that monoclonal antibodies against some T cell membrane molecules can be used to define markers of T cells, but none of these is entirely specific, perhaps with the exception of the T cell receptor (although even here there may be cross-

Table 10.7 Examples of secreted products of macrophages*

Bioactive lipids
6-ketoprostaglandin F_1; 12-hydroxyeicosatetranoic acid; leukotrienes B_4, C, D, E; platelet activating factors (PAF); prostacycline; prostaglandins E_2, F_2, I_2; thromboxane B_2
Coagulation components
Factors V, VII, IX, X; prothrombin; prothrombinase; plasminogen activator inhibitors; tissue thromboplastin
Complement components
C1, C2, C3, C4, C5; factors B, D, H (β1H), I (C3b inactivator); properdin
Cytokines and polypeptide hormones
IL-1, IL-1β, IL-6, TNFγ, IFNγ, CSF-1, G-CSF, GM-CSF, monocyte-derived neutrophil chemotactic factor
adrenocorticotrophic hormone (ACTH); angiogenesis factor; β-endorphin; erythropoietin; thymosin β4; transforming growth factor β (TGFβ)
Enzymes
Acid lysosomal hydrolases:
aryl sulfatase; cathepsins B and D; cholesteryl esterase; deoxyribonucleases; glycosidases; phosphatases; proteinases; ribonucleases; triglyceride lipase
Neutral proteinases:
angiotensin convertase; collagenases; elastase; plasminogen activator; stromelysin
Others:
arginase; lipoprotein lipase; lysozyme; phospolipase A_2; transglutaminase
Other plasma proteins
Acidic isoferritins; apolipoprotein; α_1-antiprotease; α_2-macroglobulin; chondroitin sulfate proteoglycans; fibronectin; gelatin-binding protein; lipomodulin; thrombospondin; transferrin; transcobalamin II
Reactive oxygen and nitrogen metabolites
O_2^-, H_2O_2, $OH^\cdot$, hypohalous acid, $NO^\cdot$, NO_2^-,
Steroid hormones
1α, 25-dihydroxyvitamin D3
Nucleotide metabolites
Cyclic AMP; deoxycytidine, thymidine, uracil, uric acid

*Adapted from Nathan (1987) — see Further reading

reactions with antibodies; see Section 11.1.3). A similar situation pertains to the definition of macrophages.

Some antibodies against antigens of monocytes and macrophages define determinants on molecules of known function. Examples of these are antibodies against some Fc receptors and the CD11b/CD18 type 3 complement receptor; these antibodies inhibit the function of these molecules when they bind to the cell. However, these molecules are also expressed by other cell types, such as B cells or granulocytes. A good marker for human macrophages (although it can be induced on neutrophils by cytokines) is CD14. This is a membrane receptor that recognizes a complex of endotoxin, LPS (Section 8.3.3), bound to an LPS-binding protein which is normally present in the plasma but at greatly increased levels during the acute phase response (Section 9.2.2) (Wright *et al.* 1990).

Other monoclonal antibodies define determinants on molecules that are of as yet unknown function. At the moment, the most specific for mouse macrophages is probably the F4/80 antigen, but even though this has proven to be an excellent marker for many cells of the macrophage lineage, not all macrophages express F4/80 and this molecule is also present on Langerhans cells, which are members of the dendritic cell family (Section 3.4.2). Taken together, the expression of a particular combination of markers can give a good idea of whether or not a particular cell is a macrophage, but none alone should be taken as positive proof. This of course also applies to the identification of other cell types.

10.4.2 Resident macrophages

In the absence of a stimulus, macrophages may remain relatively quiescent in a particular site for some time. For this reason these cells are known as an unstimulated or resident macrophages. They line the body cavities and blood capillaries, and have low secretory, cytocidal, and microbicidal activities. However, they do possess a variety of surface receptors, which allow them to respond to certain stimuli very rapidly (Fig.10.14).

1. Immunoglobulin Fc receptors (FcR) and complement receptors (CR) enable macrophages to bind particles that have been opsonized by antibody and/or complement. *Opsonization* is the process by which phagocytosis of molecules or particles can be facilitated after binding other molecules called **opsonins**. Antibodies and complement components can act as opsonins, as can the C-reactive protein which is produced during the acute phase response of inflammation (Section 9.2.2). Binding of opsonized molecules and particulates by the macrophage often results in their phagocytosis and destruction within the cell. Alternatively, the macrophage can secrete molecules into its environment that act directly on the opsonized substance, or which signal to other cells and facilitate its removal by different routes. FcR and CR receptors are discussed in detail in Section 10.4.5.

2. The mannosyl–fucosyl receptor (MFR) on mature macrophages can bind glycoproteins containing branched fucose-mannose residues and particulates like zymosan, a component of yeast cell walls; the gene for this receptor has been cloned (Ezekowitz *et al.* 1990). The events that follow binding may be similar to those that result from ligation of FcR or CR (above), but each receptor as a general rule mediates its own particular spectrum of effects.

3. Receptors for chemotactic peptides such as f-Met-leu-phe trigger movement of the cell towards the stimulus, and thereby stimulate macrophages to enter sites of inflammation (Snyderman and Pike 1984).

4. Fibronectin receptors mediate attachment of the macrophage to fibronectin deposited at sites of tissue injury (Bevilacqua *et al.* 1982). Macrophages have a number of other molecules, many of which are poorly characterized, that allow them to adhere to different surfaces.

5. Other recognition systems allow macrophages to bind and remove cells, such as effete erythrocytes in the liver and spleen. Resident macrophages are ideally placed to clear and digest cells and other debris and to remove particulate antigens when they enter the body or lymphoid tissues.

Some macrophages constitutively produce products such as lysozyme and lipoprotein lipase. Lysozyme kills some bacteria by destroying their cell wall peptidoglycans, while lipoprotein lipase is involved in the metabolism of serum lipids. Resident macrophages secrete a relatively limited number of other substances. They are, however, particularly good producers of arachidonic acid metabolites which are very important inflammatory mediators (Section 9.2.3). These molecules can be synthesized and secreted by macrophages when the FcR is ligated, or when both the MFR and CR are engaged, but not apparently by the latter alone.

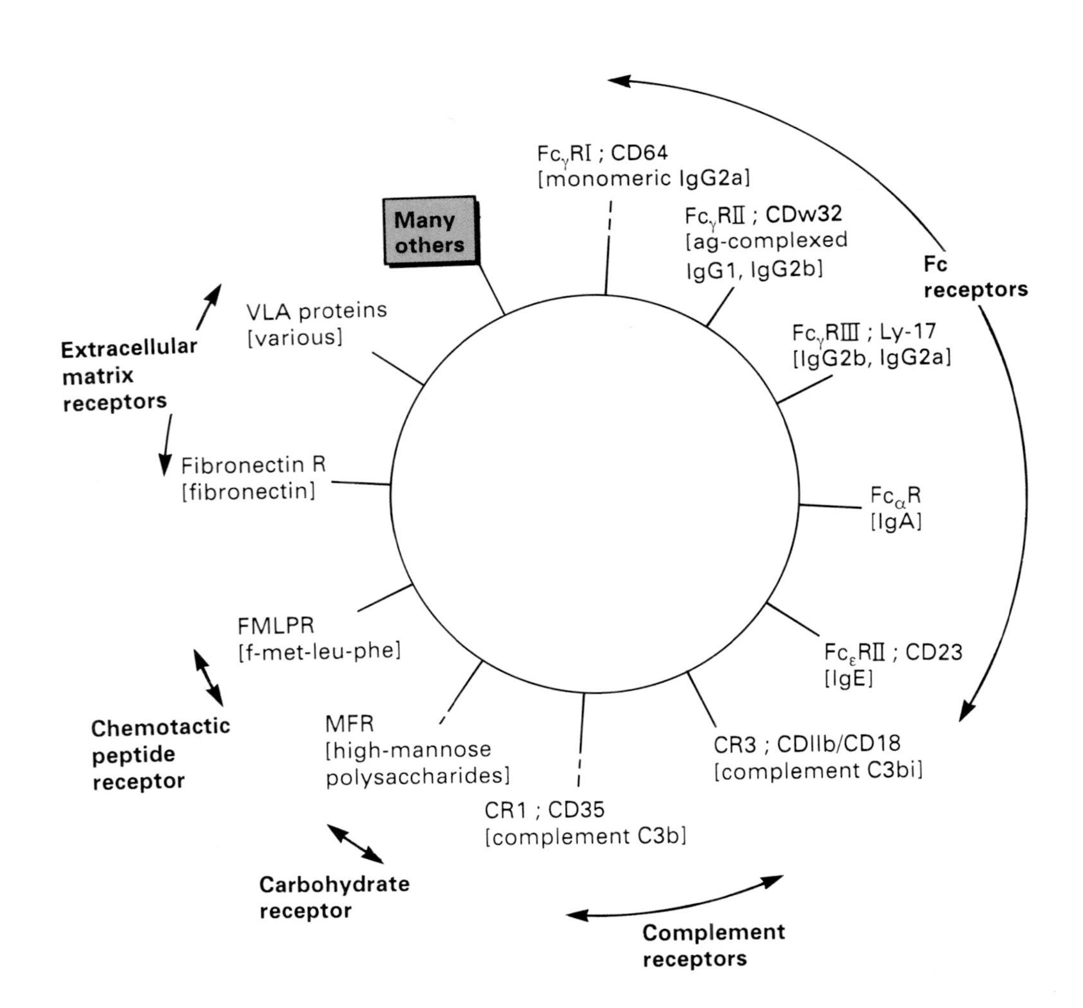

Fig.10.14 Some mouse macrophage receptors [and their ligands]

10.4.3 Inflammatory macrophages

A variety of inflammatory agents stimulate the macrophage to develop into inflammatory macrophages which can secrete high levels of a large number of different molecules (Table 10.8). However, these cells do not have significantly enhanced cytocidal activities compared to resident cells. Inflammatory macrophages can be obtained after injection of agents such as thioglycollate into the peritoneal cavity of mice, although this is a poorly-defined stimulus that seems to act nonspecifically. This causes an initial influx of polymorphonuclear leukocytes (neutrophils; Fig.1.11) over the first day or so, and then macrophages, derived from recruited blood monocytes, enter the peritoneal cavity over the next 2 or 3 days. Inflammatory macrophages are also termed 'stimulated' or 'elicited' macrophages, and their precise properties (e.g. their secretory profiles) depend on the actual stimuli that were used to elicit them.

Inflammatory macrophages have a marked increase in expression of a number of cell surface receptors, such as FcR and MFR. Furthermore, ligation of these receptors on inflammatory macrophages leads to a different set of responses compared to when the same receptors are engaged on resident cells, and yet another set of responses occurs in activated macrophages. Ligation of these receptors results in the release of a large number of soluble molecules by the inflammatory macrophage. These cells secrete a wide variety of **neutral proteases**, so-called because they are most active at neutral pH and therefore function in the extracellular milieu. These enzymes, such as elastase (which is also produced at low levels by resident cells) and collagenase, are probably involved in degradation of the extracellular matrix during wound healing, as well as during normal growth and tissue remodelling. The products that are formed can be further degraded, perhaps intracellularly, to produce various other mediators of inflammation (Section 9.2.3). As another example, the enzyme plasminogen activator is produced in large quantities by inflammatory macrophages, and converts plasminogen into plasmin. Plasmin in turn stimulates the complement and kinin systems, as

Table 10.8 Comparison of resident, inflammatory, and activated macrophages

	Resident Mφ	Inflammatory Mφ	Activated Mφ
Expression of membrane receptors	+	+++ (e.g. FcRI↑ MFR↑)	+ (MHC II↑)
Secretory activity	+	+++ (e.g. protease secretion ↑)	+
Cytocidal/cytotoxic activity	–	–	+++

– none; + some; +++ much

well as coagulation and fibrinolysis (Section 9.2.1). Macrophages also produce an inhibitor of plaminogen activator, indicating that they can control the activity of some of the products they secrete.

Inflammatory macrophages secrete a number of plasma proteins. These include all the early complement components (C1 through C5) together with those of the alternative pathway (Sections 9.4.4 and 9.4.5). Although the principal source of these proteins is normally the liver, the macrophage could be an important source of these molecules at extravascular sites during inflammation and especially before vascular changes have occurred that allow plasma proteins to enter the tissues. Macrophages also produce a number of coagulation factors (Section 9.3.1). Many by-products that are formed when the complement and coagulation systems are activated act on other cell types and presumably amplify and further regulate the inflammatory response. In addition, macrophages secrete molecules that can act directly on other cells, and contribute to tissue repair by fibroblasts and endothelial cell proliferation, for example. Macrophage mediators may act at distant sites and some, such as IL-1 (Section 9.5.1), regulate the production of acute phase proteins by liver hepatocytes.

It is notable that despite the increased secretion of an enormous number of products by inflammatory macrophages, they produce a lower level of prostaglandins (which tend to suppress immune responses; Section 9.2.3) than resident cells, and this may facilitate the subsequent induction of immune responses when the macrophage becomes activated.

10.4.4 Activated macrophages

Resident and/or inflammatory macrophages become activated when they are cultured with supernatants from stimulated T cells or recombinant IFNγ, a T cell product (Murray *et al.* 1985). Undoubtedly IFNγ is a major stimulus for macrophage activation, but GM-CSF and IL-4 have also been implicated as 'macrophage activating factors' (MAF) particularly for the induction of tumoricidal activity (Grabstein *et al.* 1986; Crawford *et al.* 1987). Since macrophages can be activated during infections in nude mice, it is presumed there must also be T cell-independent pathways for macrophage activation. Some of the effects of IFNγ on the macrophage are antagonized by IFNα and IFNβ (Section 9.5.4).

When mouse macrophages are activated, the expression of the MFR and one of the Fc receptors ($Fc_{\gamma}RII$) is decreased, while levels of another Fc receptor ($Fc_{\gamma}RI$) are increased (see below). Activated macrophages express asialo-GM1 which is found on resident alveolar macrophages as well as NK cells (Section 10.2.4), together with new antigens that have been defined by monoclonal antibodies that recognize determinants on activated but not resident cells. One notable consequence of macrophage activation is the greatly increased expression of MHC class II molecules by the cell (Ezekowitz *et al.* 1981). This presumably potentiates their interaction with activated helper T cells during the adaptive immune response (Section 1.2.3).

Activated macrophages generally have a lower secretory activity than inflammatory cells, and they produce a different spectrum of products. For example, they no longer produce elastase but they can secrete proteins such as apolipoprotein E, which is involved in transport of cholesterol to the liver, and transcobalamin which transports vitamin B_{12}; arachidomic acid metabolism is also altered (Scott *et al.* 1982).

Perhaps the main feature of activated macrophages is, by definition, a greatly increased capacity to kill a number of micro-organisms and some tumour cells. This can occur both by *oxygen- or nitrogen-dependent mechanisms*, in which **toxic reactive oxygen or nitrogen intermediates (ROI and RNI)** are produced, and by *ROI and RNI independent systems*, although their relative importance in different situations is unclear (Fig.10.15).

Reactive oxygen intermediates

Phagocytosis in activated macrophages, and to a lesser extent in inflammatory macrophages, is associated with the so-called **respiratory burst**. In this response, the consumption of molecular oxygen (O_2) by the cell increases, the hexose monophosphate shunt by which glucose-6-phosphate is used to produce NADPH and other reducing intermediates is activated, and various oxygen products are produced (Fig.10.16). Central to this pathway is the activity of a multi-component enzyme called NADPH oxidase, defects in which lead to 'chronic granulomatous disease' in humans. It is likely that the respiratory burst can be dissociated from the actual phagocytic event itself, because it can also be induced by a variety of other 'surface-active' stimuli, including some receptor–ligand interactions in the absence of phagocytosis (Yamamoto and Johnston 1984).

The oxygen species produced by macrophages include hydrogen peroxide (H_2O_2), the superoxide anion ($O_2^{\cdot -}$), the hydroxyl radical ($OH^{\cdot}$), and singlet oxygen ($^1O_2$). Superoxide is produced by inflammatory and activated macrophages, but it can be further converted into H_2O_2 by the action of the enzyme **superoxide dismutase** (Clark and Szot 1981) which is induced in activated macrophages. The cell seems to protect itself from attack by these molecules by utilizing biochemical pathways that centre around glutathione.

Another toxic system also exists in the monocyte as well as neutrophils which contain **myeloperoxidase**. In the presence of halide ions, particularly Cl^- and I^-, hypohalites and a number of other toxic products are formed, and together with H_2O_2 they form a very toxic anti-microbicidal system. It is possible that similar products can be produced by other routes in mature macrophages which do not contain myeloperoxidase. The cytocidal activity of activated macrophages against certain intracellular and extracellular pathogens, including parasites, yeasts, and bacteria, is correlated with their production of ROI (for example, Murray *et al.* 1985). Certain parasites have evolved mechanisms for evading this system, in one case, for example by producing catalase to destroy the H_2O_2 produced by the macrophage.

Production of ROI is also critical for the cytocidal activity of neutrophils, although the pathways by which these intermediates are generated may differ from those in the macrophage. It is notable that ROI are probably *not* important for killing by CTL and NK cells (Nathan *et al.* 1982).

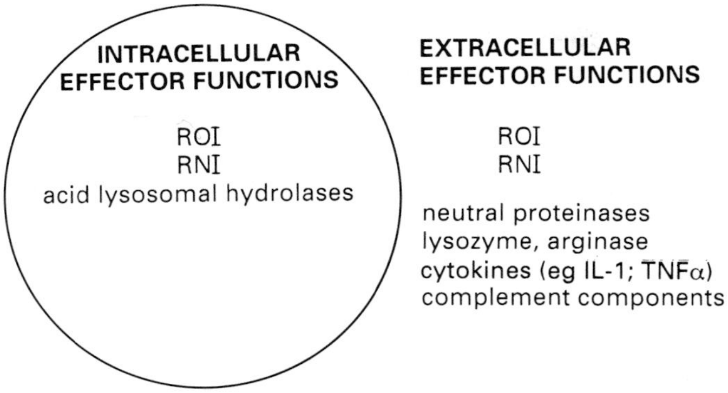

INTRACELLULAR EFFECTOR FUNCTIONS

ROI
RNI
acid lysosomal hydrolases

EXTRACELLULAR EFFECTOR FUNCTIONS

ROI
RNI

neutral proteinases
lysozyme, arginase
cytokines (eg IL-1; TNFα)
complement components

Fig.10.15 Some cytotoxic mechanisms of activated macrophages

Reactive nitrogen intermediates

One of the RNI produced by activated macrophages, nitric oxide (NO), can also be produced by other cell types and is important in a number of other systems in the body. For example, it is involved in neurotransmission in the central nervous system and in regulating muscle tone ('endothelial-derived relaxation factor'). Production of NO for these purposes occurs via a **constitutive, low-output route**, resulting in the production of femtomoles to picomoles (10^{-15}–10^{-12} mol) of NO per milligram of protein per hour. However, the NO used in cytotoxic responses of activated macrophages is generated by a **cytokine-inducible, high-output route**, with production being dramatically increased to nanomoles (10^{-9} mol) per milligram of protein per hour. It is an important effector molecule leading to the destruction of certain micro-organisms and neoplastic cells by activated macrophages, at least in the mouse (Liew *et al.* 1990). It should be noted that it is difficult to demonstrate production of NO by this route in human macrophages.

1 The hexose monophosphate shunt (pentose phosphate pathway) utilizes glucose-6-phosphate to produce NADPH:

$$6\ \text{G-6-P} + 12\ \text{NADP} \longrightarrow 5\text{G-6-P} + 12\ \mathbf{NADPH} + 6CO_2$$

2 The enzyme NADPH oxidase, located in the plasma membrane, catalyzes the conversion of molecular oxygen to the superoxide anion:

$$\text{NADPH} + H^+ + 2O_2 \longrightarrow \text{NADP}^+ + 2H^+ + \mathbf{2O_2^{\bullet -}}$$

3 Two molecules of superoxide can dismutate to hydrogen peroxide, spontaneously, or catalysed by the enzyme superoxide dismutase:

$$2O_2^- + 2H^+ \longrightarrow \mathbf{H_2O_2} + O_2$$

4 Reaction of hydrogen peroxide with superoxide can generate the very reactive hydroxyl radical, probably via an Fe^{2+}-catalysed Haber–Weiss reaction:

$$H_2O_2 + O_2^{\bullet -} \longrightarrow \mathbf{OH^\bullet} + OH^- + O_2$$

Fig.10.16 Outline of some reactions producing reactive oxygen intermediates in macrophages

Production of RNI by activated macrophages starts with the conversion of L-arginine and molecular O_2 to L-citrulline and NO (Fig.10.17). This is catalysed by the enzyme **nitric oxide synthase**, which is induced in activated macrophages, for example after treatment with IFNγ. Conversion of L-arginine to L-citrulline by this route is an unusual biochemical pathway, not inolving the more usual urea cycle via an ornithine intermediate with the production of urea.

NO is a highly reactive, short-lived molecule. It can react further with molecular O_2 to form nitrogen dioxide, NO_2, which subsequently reacts with water to produce nitrite (NO_2^-), as well as nitrate (NO_3^-) and hydrogen (H^+) ions; the NO_2^- can then, directly or indirectly, be converted to another toxic RNI, nitrogen dioxide (NO_2). Alternatively, NO can form a complex with Fe^{2+}-containing enzymes and this, for example, shuts off oxidative metabolism (Drapier and Hibbs 1988). It is worth noting that in this way activated macrophages can *limit* the availability of Fe^{2+} to intracellular pathogens such as Legionella (the bacterium responsible for Legionnaire's disease) and cause their destruction. This is accomplished in part by down-regulating surface expression of the transferrin receptor, which is responsible for the uptake of Fe^{2+}-complexed transferrin.

The NO molecule can also react with ROI. For example it can react with superoxide (O_2^-), resulting in its neutralization with the formation of NO_3^-, but in an acid environment, such as within lysosomes, peroxynitrite is formed which in turn produces the toxic hydroxyl radical (OH^-). The high-output route for NO generation can also be triggered in other cells such as hepatocytes, fibroblasts and endothelial cells when nitric oxide synthase is mobilized

1. The cytosolic enzyme nitric oxide synthase catalyzes the conversion of L-arginine and molecular oxygen to L-citrulline and nitric oxide (NO)

   $$\text{L-arginine} + O_2 \longrightarrow \text{L-citrulline} + \mathbf{NO^\bullet}$$

2. NO can react with Fe-S groups, resulting in the inactivation and degradation of prosthetic groups of certain enzymes in the mitochondrial electron transport chain, thus inhibiting oxidative metabolism

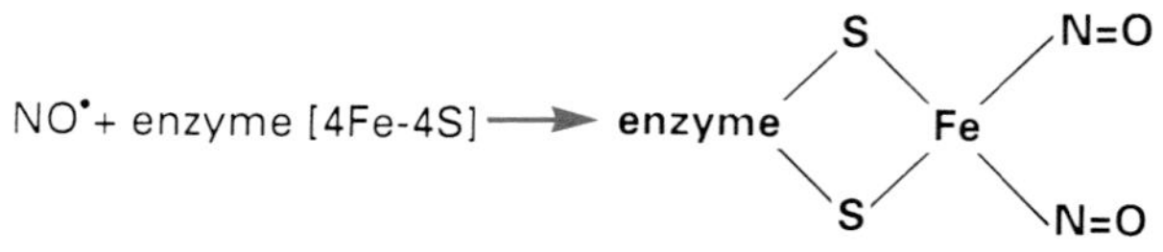

   $$NO^\bullet + \text{enzyme [4Fe-4S]} \longrightarrow \textbf{enzyme}\langle S_2 \rangle Fe(N{=}O)_2$$

3. Alternatively, NO can react with the superoxide anion radical to form the peroxynitrite anion – this decays rapidly after protonation to form the reactive hydroxyl radical $HO^\bullet$ and the nitric oxide free radical $NO_2^\bullet$

   $$O_2^{\bullet -} + NO^\bullet \longrightarrow \mathbf{ONOO^-}$$

   $$ONOO^- + H^+ \rightleftharpoons ONOOH \rightleftharpoons \mathbf{HO^\bullet} + NO_2^\bullet$$

Fig.10.17 Outline of some reactions producing reactive nitrogen intermediates in macrophages

in response to infection by certain micro-organisms. This indicates the importance of RNI for the cytotoxic activity of other cell types in addition to macrophages.

Oxygen- and nitrogen-independent cytotoxicity

Phagocytes can kill by mechanisms that do not require oxygen or nitrogen metabolites. Some of these systems may be more important for killing bacteria and parasites that are actually *within* the cell, than for extracellular organisms and tumour cells, although other mechanisms may be shared. Certain organisms, particularly Gram-negative bacteria, are killed by lysozyme which is secreted by macrophages. Other organisms can be killed intracellularly by phagocytes under acidic or alkaline conditions. Acid conditions are required for lysosomal enzymes to function in macrophages (acid proteases and cathepsins), while cationic proteins that are generally present in neutrophils but not in macrophages require alkaline environments.

At least ten anti-microbial proteins have been isolated from the azurophilic granules of human neutrophils (Fig.1.11). These include lysozyme, seprocidins which have homology to granzymes of cytotoxic T cells (Section 10.3.4), and a group of small (29–34 amino acid) peptides called defensins that have been shown to kill a variety of different micro-organisms, but which are *not* present in most macrophage populations.

While macrophages can use oxygen- and nitrogen-dependent mechanisms to kill some tumour cells extracellularly, for example by ADCC (Section 10.2.7), they can also kill under strictly anaerobic conditions although the precise mechanism for this is not yet known. Secretion of arginase by inflammatory and activated macrophages is toxic for certain pathogens (Olds *et al.* 1980), and depletes extracellular arginine, and this can be toxic for tumour cells *in vitro*. Macrophages also secrete molecules that are directly tumoricidal, and those identified so far include a cytolytic neutral protease and cachectin or TNFα.

Cachectin, TNFα. Cachectin, TNFα, and its relationship to lymphotoxin or TNFβ, was noted earlier (Section 10.3.4) and further details of these molecules are in Section 9.5.3. Here we outline a few of its properties *in vivo*.

Many infections or neoplastic invasive diseases result in abnormalities in the metabolism of glucose, proteins, and lipids. This can be manifested as cachexia or wasting, together with shock and sometimes death. Some years ago it was noted that rabbits infected with trypanosomes underwent similar physiological changes, and that in the final stages of infection their plasma became full of lipids, particularly triglycerides. This correlated with a dramatic reduction on the surface of adipocytes of the enzyme lipoprotein lipase, which normally breaks down triglycerides into fatty acids which are then absorbed and utilized by these cells. Using lipoprotein lipase levels as an assay, the molecule that was responsible for this effect was purified and named cachectin. This molecule turned out to be a major product of macrophages, although it can also be produced by activated lymphocytes. Its production was induced by stimuli such as LPS, or endotoxin, a bacterial cell wall product that in animals can induce shock and death. Independently, a molecule called TNF(α) had been purified from macrophages that was toxic for some tumour cell lines. When the sequences of cachectin and TNFα were compared, they were found to be the same.

It has been found that antisera against TFNα can protect animals from LPS mortality. However, if TNFα is administered to animals it can cause shock and death. TNFα is in fact about a million times more toxic than cyanide. Nevertheless, low levels of TNFα in tissues are associated with tissue remodelling. At higher levels, inflammatory and cytotoxic effects are seen. Increasing levels result in cachexia, organ failure, and ultimately irreversible shock and death of the animal. It is quite remarkable that the body produces a molecule that on one hand may kill tumour cells (although its precise role in this respect *in vivo* is still controversial) but which on the other hand may kill the very host that produces it. Possibly this has evolved for the benefit of the species as a whole, since there is less chance of an infection spreading through the population if badly infected individuals are killed in this way.

TNFα has multiple effects in the body and in many respects its actions are similar to those of IL-1 (compare Sections 9.5.1 and 9.5.3). Thus amongst its many reported effects, TNFα induces synthesis of acute phase proteins by the liver and stimulates collagenase production by synovial cells and the resorption of cartilage and bone. However, the situation is complicated by the fact that TFNα stimulates the release of IL-1 by macrophages, and both of these molecules (together with IFNγ) seem to be involved in a number of complicated circuits that regulate and control the production and activity of each other.

10.4.5 Some macrophage receptors

Cellular Fc receptors and complement receptors are examples of molecules that mediate between the innate and adaptive arms of the immune system and they are important molecules for the effector functions of macrophages. (Fig.10.18). Obviously, macrophages and complement belong to the innate arm of the immune system, while antibodies are produced by B cells of the adaptive immune system. However, macrophages can produce many complement components, and some activated complement products can in turn regulate B cell responses. Conversely, the complement cascade can be activated by certain classes of antibodies and macrophages have receptors for both antibodies and complement components, both of which are discussed in this section (see below). These are examples of the multitude of interactions that occur between cells and molecules of the immune system, and examples of interactions between these and other systems are noted in Section 9.2.1.

Fc receptors for IgG

Specificity of IgG receptors. Fc receptors can bind the Fc portion of particular antibody subclasses (Section 6.2.5). For example, three separate FcR for IgG have been identified on mouse macrophages which are specific for different subclasses of mouse IgG: FcRI binds monomeric IgG2a and is trypsin sensitive; FcRII binds antigen-complexed or aggregated IgG1 and IgG2b (although the purified molecule also binds IgG2a) and is trypsin resistant; and another FcR may bind IgG3. These receptors are also designated $Fc_{\gamma}RI$,

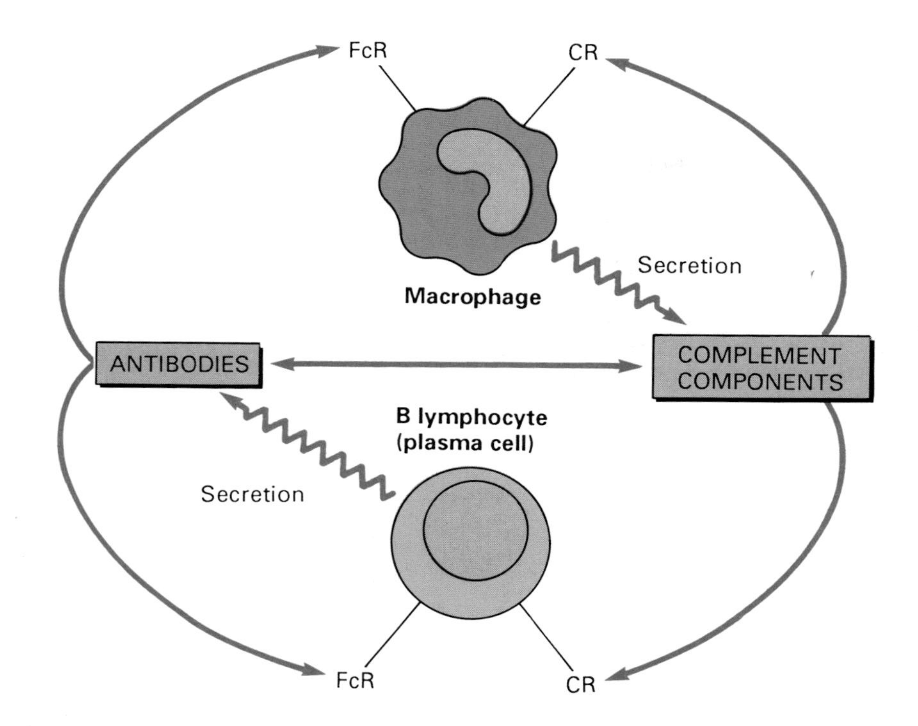

Fig.10.18 Examples of some links between innate and adaptive immunity

$Fc_{\gamma}RII$, and $Fc_{\gamma}RIII$, to distinguish them from Fc receptors for other classes of antibody. The subscripted gamma is sometimes omitted if the context is clear.

FcRI and FcRII can be independently modulated when macrophages are attached to surfaces coated with the appropriate subclass of antibody (Fig.10.19). When macrophages are plated on IgG2a-coated surfaces, they lose the ability to bind this subclass but they can still bind particles coated with IgG1 and IgG2b; when plated on IgG1-coated surfaces they lose the ability to bind particles coated with this subclass or with IgG2b, but they still bind IgG2a. This proves that the receptors are distinct molecules, and this approach can be used to examine other receptors, such as complement receptors (Griffith *et al.* 1975) (below).

Monoclonal antibodies specific for Fc receptors. The first monoclonal antibody to be produced against mouse FcRII (Ly17) was designated 2.4G2. This antibody was observed to inhibit the binding of IgG-opsonized erythrocytes to macrophages, and was subsequently used to purify the receptor.

On gel electrophoresis the molecule migrates as two bands with apparent molecular weights of about 60 kDa and 47 kDa, when obtained from macrophages, and as a single broad band between 45-70 kDa when prepared from other sources, such as B cells. The purified FcRII molecule has been reconstituted in artificial lipid layers and shown to act as a 'ligand-dependent ion channel', since binding of IgG causes a flux of ions across the membrane. It has been suggested that in macrophages this property may initiate the cellular response that follows ligation of the receptor although the molecular basis for this is not clear.

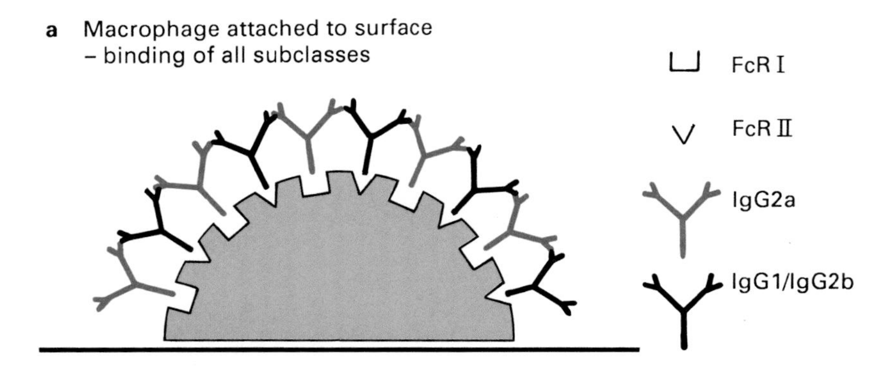

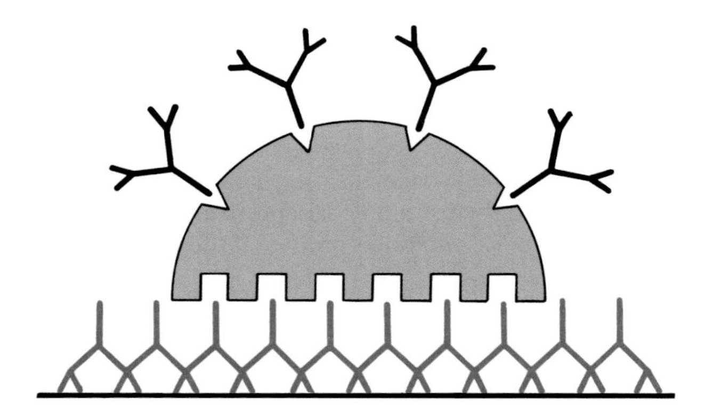

b Macrophage attached to surface coated with IgG2a
– modulation of FcR I; unable to bind IgG2a but binding
of IgG1/IgG2b unaffected

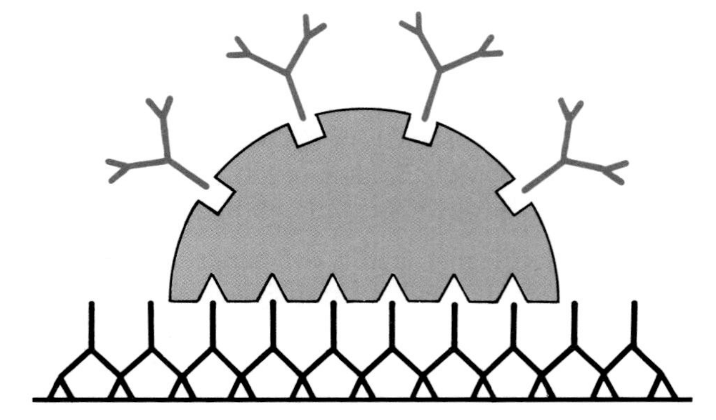

c Macrophage attached to surface coated with IgG1 or IgG2b
– modulation of FcRII; converse result to **b**

Fig.10.19 Independent modulation of FcRI and FcRII receptors

Antibodies against three different Fc receptors for IgG (particularly human IgG1 and IgG3, the most cytophilic subclasses) on human cells have also been produced. One of these, $Fc_{\gamma}RI$ (CD64), has a high affinity for monomeric IgG; it has a molecular weight of about 70 kDa, and is expressed on monocytes and macrophages. Another, called $Fc_{\gamma}RII$ (CDw32) binds aggregated IgG; it has a molecular weight of 42 kDa and is expressed on several different cell types including monocytes, neutrophils, eosinophils, and B cells, but not NK cells. A third IgG receptor on human cells is called $Fc_{\gamma}RIII$ (CD16) or $Fc_{\gamma}R_{lo}$. It also binds aggregated IgG and is expressed by some macrophages (but not monocytes), neutrophils, eosinophils, NK cells, and a small population of thymic T cells. This molecule is anchored to polymorphonuclear leukocytes via a phosphatidylinositol anchor (Fig.1.5), and is released by neutrophils when they are stimulated by the f-Met-leu-phe chemotactic peptide.

Cloning of the genes for IgG Fc receptors. A great deal has been learnt about the structure and function of Fc receptors from gene cloning and expression systems. A summary of the genes, different RNA transcripts, predicted structures, and cell-specific expression of these molecules, in mouse and human, is in Fig.10.20. For completeness $Fc_{\varepsilon}R$ receptors are included, and are discussed later.

Here, we will consider each type of $Fc_{\gamma}R$ receptor in turn. But first, we should note that all the α chains of these molecules are members of the immunoglobulin gene superfamily: $Fc_{\gamma}RI$ has three extracellular domains, whereas the others have two (which are homologous to the two membrane-distal domains of $Fc_{\gamma}RI$).

1. **$Fc_{\gamma}RI$**. In both the mouse and human one gene has been identified. This gene is expressed in monocytes and macrophages, and is up-regulated after IFNγ treatment, which also induces its expression on neutrophils. It is a single-chain structure on the cell surface.

2. **$Fc_{\gamma}RII$**. In the mouse, there is one gene for this molecule, but three are present in the human, and designated A (which is polymorphic), B, and C. It is also a single-chain structure. A notable feature of this receptor is that it is expressed by all cells bearing $Fc_{\gamma}R$ receptors *except* NK cells. In the human, $Fc_{\gamma}RIIA$ and C are expressed by monocytes and macrophages but not lymphocytes, whereas $Fc_{\gamma}RIIB$ is expressed by all three cell types. These genes are very homologous to each other: IIC (for short) and IIA differ only in their signal sequences, while IIB has some differences in its cytoplasmic tail compared to IIA.

Two different RNA transcripts for mouse $Fc_{\gamma}RII$, b1, and b2, can be produced by alternative splicing, and three for human $Fc_{\gamma}RIIB$, b1, b2, and b3, together with a fourth generated by alternative splicing of the signal sequence. (Two RNA transcripts are also produced from the human Fc RIIA gene by utilizing different polyadenylation signals).

Much has been learnt about the function and cell-specific expression of the mouse b1 and b2 transcripts. The b1 transcript contains an additional sequence in its cytoplasmic portion, compared to b2, and is predominantly expressed by B cells, whereas the b2 transcript is predominantly expressed by macrophages. These different forms mediate different functions in B cells and macrophages. Binding of immune complexes to the B cell Fc receptor mediates the delivery of inhibitory signals to the cell, but not internalization of ligands (Section 8.4.2), whereas binding of immune complexes or IgG-coated particles to the macrophage Fc receptor results in endocytosis via coated pits (Section 3.2.2), delivery to lysosomes, and the secretion of inflammatory cytokines and reactive oxygen intermediates by the cell.

When fibroblasts (COS cells) were transfected with the b1 (B cell) form of the $Fc_{\gamma}RII$ receptor, immune complexes were bound by the cell but there was no internalization. When the b2 (macrophage) form was transfected, the immune complexes were bound, internalized and delivered to lysosomes. But what happens if B cells, which do not normally phagocytose via the Fc receptor, express the b2 transcript? When this experiment was performed using a B cell line that lacked its own Fc receptors, it was found that the transfected b2 receptor mediated internalization of bound ligands, concomitant with a two orders of magnitude increase in the efficiency of antigen presentation to T cells (Section 3.2); in contrast the transfected b1 form mediated binding of ligands but not their internalization. Clearly, the additional sequence in the b1 transcript *prevents* internalization of ligands by the B cell. Other experiments indicate that the b1 form colocalizes with actin filaments which are apparently involved in tethering the receptor to the membrane.

Further experiments using different b1 and b2 constructs have localized sequences of the cytoplasmic tail involved in specific functions. For example, there is a homologous sequence in this region in the b2 forms of human and mouse $Fc_{\gamma}RII$ receptors. If this region is deleted, both localization of the receptor at coated pits and internalization are prevented. However, if the additional sequences in the b1 form are deleted, there is a random association of the receptors in and out of coated pits. These and other experiments have led to the idea that the extra sequence in the b1 form (i) is required to prevent localization at coated pits; (ii) is required for capping of Fc receptors; but (iii) is *not* involved in negative signalling functions to the B cell (Section 8.4.2).

3. **$Fc_{\gamma}RIII$.** In the mouse there is one gene for this molecule, which is expressed in macrophages, NK cells, and neutrophils. However, there are two genes in the human, designated Aα and B; the former encodes the transmembrane (TM) form of the receptor expressed by macrophages and NK cells, while the latter encodes the glycosyl-phosphatidylinositol (GPI) anchored form expressed by neutrophils.

The $Fc_{\gamma}RIII$ molecule of macrophages, like $Fc_{\gamma}RI$ and $Fc_{\gamma}RII$, mediates internalization of antibody-coated particles. However, it is the only Fc receptor for IgG to be expressed by NK cells where it mediates ADCC (Section 10.2.7). Multivalent ligation of the GPI-linked $Fc_{\gamma}RIII$ receptor on neutrophils signals an increase in cytoplasmic Ca^{2+} ion concentration in the cell (not seen when the $Fc_{\gamma}RII$ receptor is cross-linked), and promotes internalization of certain opsonized particles and the extracellular lysis of others (although it appears to be less associated with phagocytosis than the TM-linked form and the other $Fc_{\gamma}R$ receptors).

The human $Fc_\gamma RIIIB$ receptor is a single chain molecule at the cell surface. However, the mouse $Fc_\gamma RIII\alpha$ and human $Fc_\gamma RIIIA\alpha$ chains are associated with at least two other transmembrane molecules, designated $Fc_\gamma RIIIA$, $Fc_\gamma RIIIA\zeta$, and probably also $Fc_\gamma RIII\beta$, to form a functional receptor. Different combinations of these molecules can be associated with the α chains; for example, $\alpha\gamma_2$, $\alpha\zeta_2$ and $\alpha\gamma\zeta$ have all been identified in macrophages and NK cells.

The γ chain of the $Fc_\gamma RIII$ receptor, and the β chain if present, also associate with a different (but related) α chain to form the $Fc_\varepsilon RI$ receptor, for example on mast cells where the molecules associate in an $\alpha\beta\gamma_2$ combination. *The ζ chain of the $Fc_\gamma RIII$ receptor turns out to be the same molecule as the ζ chain associated with the TCR-CD3 complex on T cells* (Section 4.2.3). Undoubtedly much will be learnt of the signalling functions of these molecules in the near future.

Regulation and function. The expression of FcRI and FcRII on mouse cells is regulated according to the state of macrophage activation (Sections 10.4.2–10.4.4). Thus, expression of *both* receptors is increased on inflammatory cells, whereas there is a selective increase in FcRI but decreased expression of FcRII on activated macrophages.

Binding of immune complexes or antibody-coated particles to FcR has a number of consequences which include the following:

(1) phagocytosis by macrophages in all states of activation;

(2) release of arachidonic acid metabolites from resident and inflammatory macrophages;

(3) of hydrolytic enzymes particularly by inflammatory macrophages;

(4) triggering of the respiratory burst and the release of toxic ROI by activated macrophages;

(5) antibody-dependent cellular cytotoxicity (ADCC; Section 10.2.7). There is little doubt that tumour cells can be killed by macrophages carrying antibodies directed against them. At least in part, ADCC by macrophages involves the production of ROI, although there are likely to be other mechanisms for tumoricidal activity (e.g. TNF production).

Role of different IgG receptors on the same or different cells. The presence of *different* receptors for various subclasses of antibody on the *same* cell allows antigen to be directed along specific cellular routes (Fig.10.21). Depending on which subclass of antibody is produced in the course of an immune response, different cellular receptors are engaged (e.g. see Section 8.4.5). This can determine, for example, whether prostaglandins or ROI are secreted by the cell (e.g. via $Fc_\gamma RI$ versus $Fc_\gamma RII$ on mouse macrophages).

Expression of the *same* FcR by *different* cell types, such as $Fc_\gamma RII$ of mouse macrophages and neutrophils, facilitates a coordinated but diverse response to be made against a given antigen. The same receptor on different cells may mediate similar or different effects, and sometimes the same receptor can exist in different forms on different cells, such as the CD16 $Fc_\gamma RIII$ molecule on human macrophages (TM form) and NK cells (GPI form), which mediate different types of response (Section 10.2.4).

In addition, *different* IgG receptors can be expressed by *different cell types* and even on different stages of the same cell lineage. For example, freshly-isolated human monocytes do not express the $Fc_{\gamma}RIII$ receptor but it is present on Kupffer cells and macrophages in the red pulp of spleen, where it may be responsible primarily for the clearance of immune complexes *in vivo*; it is also expressed on circulating neutrophils.

IgE and IgA receptors

Fc_{ε}RII of macrophages (Fig.10.22). Macrophages have IgE receptors designated CD23, or $Fc_{\varepsilon}RII$ (Fig.10.20) (Spiegelberg *et al.* 1983). This molecule, which is also expressed by B cells and eosinophils, mediates cytotoxicity against parasites and synergizes with CR3 (CD11b/CD18). It may promote phagocytosis of IgE-coated particles, and mediates the release of lysosomal enzymes and inflammatory mediators including leukotrienes, prostaglandins,

RECEPTOR	STRUCTURE	GENES [RNA transcripts] (and associated molecules) Mouse	Human	CELL EXPRESSION
$Fc_{\gamma}RI$ 'high affinity IgG receptor'	a	$Fc_{\gamma}RI$	$Fc_{\gamma}RI$ CD64	– monocytes/macrophages – inducible on neutrophils
$Fc_{\gamma}RII$ 'low affinity IgG receptor'	b	$Fc_{\gamma}RII$ [b1, b2] Ly-17	$Fc_{\gamma}RIIA$ $Fc_{\gamma}RIIB$ $Fc_{\gamma}RIIC$ [b1, b2, b3] CD32	– all FcR$^+$ cells *except* NK cells; – mouse b1 on B cells, b2 on macrophages; – human $Fc_{\gamma}RIIA$ on monocytes; $Fc_{\gamma}RIIB$ on monocytes, macrophages and lymphocytes; $Fc_{\gamma}RIIC$ on monocytes and neutrophils
$Fc_{\gamma}RIII$ 'low affinity IgG receptor'	c TM, d GPI	$Fc_{\gamma}RIII\alpha$ Ly-17	$Fc_{\gamma}RIIIA\alpha$ $Fc_{\gamma}RIIIB$ CD16	– in mouse, on neutrophils, NK cells, mast cells, inducible on macrophages; – in human, $Fc_{\gamma}RIIIA\alpha$ on macrophages NK cells, T subset, inducible on monocytes; human $Fc_{\gamma}RIIIB$ on neutrophils
		(IIIAγ; IIIAζ, IIIAβ?)		
$Fc_{\varepsilon}RI$ 'high affinity IgE receptor'	e	$Fc_{\varepsilon}RI\alpha$ $Fc_{\varepsilon}RI\beta$ $Fc_{\varepsilon}RI\gamma$	$Fc_{\varepsilon}RI\alpha$ $Fc_{\varepsilon}RI\beta$ $Fc_{\varepsilon}RI\gamma$	– $Fc_{\varepsilon}RI\alpha$ and $Fc_{\varepsilon}RI\beta$ on mast cells and basophils
$Fc_{\varepsilon}RII$ 'low affinity IgE receptor'	A member of the C-type lectin family: related to MEL-14, GMP-140, ELAM-1 (Section 4.2.6)	$Fc_{\varepsilon}RII$ CD23	$Fc_{\varepsilon}RII$ CD23	– macrophages

Fig.10.20 Fc receptors: (above) schematic structures, genes and RNA transcripts, and cellular expression; (facing page) predicted protein structures in more detail (Redrawn from Ravetch and Kinet (1991), with permission (© 1991 by Annual Reviews Inc.); see Further reading)

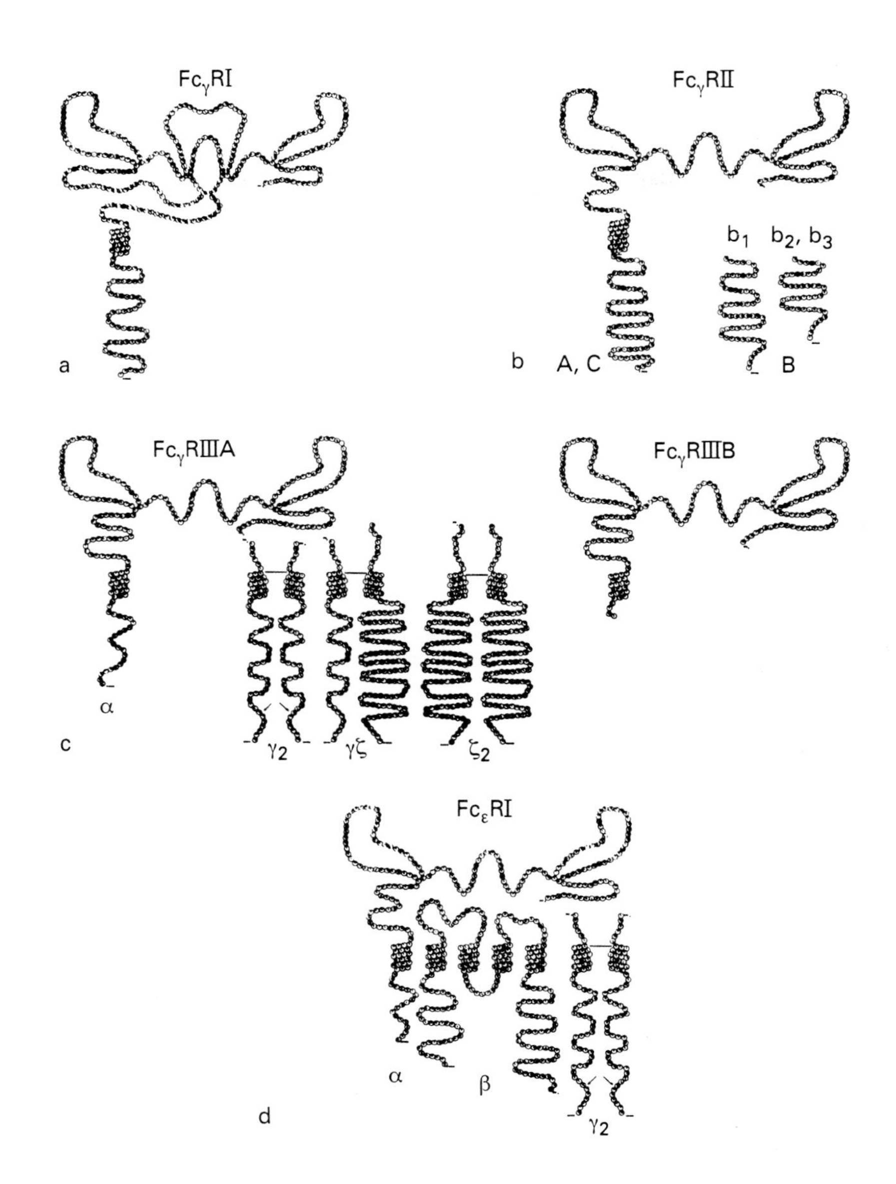

and perhaps IL-1. CD23 is distinct from the more well-known IgE receptors of mast cells and basophils, $Fc_\varepsilon RI$, in that it has a much lower affinity for IgE (10^7 M^{-1}) and a molecular weight of 45–50 kDa (possibly associated with a β chain of 25–33 kDa, although this may be derived by processing of the molecule).

The $Fc_\varepsilon RI$ receptor of mast cells and basophils. In contrast to the macrophage $Fc_\varepsilon RII$ receptors, $Fc_\varepsilon RI$ has a high affinity for IgE and is a tetramer of an α chain of 45 kDa plus a β chain of 33 kDa associated with two γ chains of 9 kDa (Fig.10.20) (Blank *et al.* 1989). This receptor on mast cells and basophils mediates allergic responses, or type I hypersensitivity reactions (Section 10.5.1). A classic example of this is hay fever, where IgE specific for pollen antigens becomes bound to IgE receptors on mast cells. Subsequent binding of the antigen causes cross-linking of the receptors, an influx of calcium ions into the cell, and explosive degranulation of the cell with the release of preformed mediators. At the same time, newly-synthesized arachidonic acid metabolites and cytokines are released (Plaut *et al.* 1989). A variety of other agents, including the anaphylatoxins C3a and C5a (Section 9.4.7), can trigger mast cell degranulation and the release of mediators independently of the Fc receptor.

One of the principal mediators released from mast cells is **histamine**. This preformed molecule is the major vasoactive amine in man, and it causes smooth muscle contraction and an increase in mucous production and vascular permeability. Histamine is also chemokinetic and chemotactic for neutrophils. It inhibits T cell functions, perhaps via suppressor T cells (Section 11.2.2), and inhibits B cell differentiation. Histamine is complexed in mast cell granules to the proteoglycan heparin, which is best known for its anticoagulant activity. These granules also contain a variety of other chemotactic and cell-activating factors and enzymes. There are species differences in that mast cells of rodents but not humans contain **serotonin**, which has similar effects to histamine, and mice are quite resistant to the effects of histamine itself.

It is notable that IgE is the predominant class of antibody produced during parasitic infections. Eosinophils, which possess both IgG and IgE receptors, seem particularly important in defence against parasites. Eosinophils produce large amounts of cationic proteins when they bind *IgG*, including the so-called major basic protein, which is stored in granules in a crystalloid core. However, in response to *IgE*, the main substance released is eosinophil peroxidase. Eosinophils produce even more ROI than neutrophils under certain circumstances, but whether the FcR is involved in this is not fully established.

$Fc_\alpha R$ of macrophages. An Fc receptor for IgA has been characterized on human monocyte/macrophages and granulocytes (Maliszewski *et al.* 1990). This 60 kDa glycoprotein molecule can bind monomeric or polymeric IgA1 and IgA2 molecules. Cloning of the gene for this molecule has revealed that it is homologous to the $Fc_\gamma R$ and $Fc_\varepsilon RI\alpha$ receptors (Fig.10.20), with two extracellular Ig-like domains.

Complement receptors

Activation of complement on the surface of micro-organisms or cells generates both surface-bound and freely-diffusible cleavage products (Sections 9.4.4, 9.4.5). The cell-associated products can result in formation of the C5–C9 membrane attack complex with subsequent cell lysis (Section 9.4.6). Other surface-bound products, particularly those derived from C3, opsonize particles and their promote attachment and/or phagocytosis by macrophages and neutrophils, and they can also activate B cells. The diffusible products act as mediators of inflammation and they bind to specific complement receptors on a wide variety of cell types (Tables 9.2c and 10.9). Three different receptors for C3 products have been described, which are designated CR1, CR2, and CR3. Monocytes and macrophages express CR1 and CR3, while

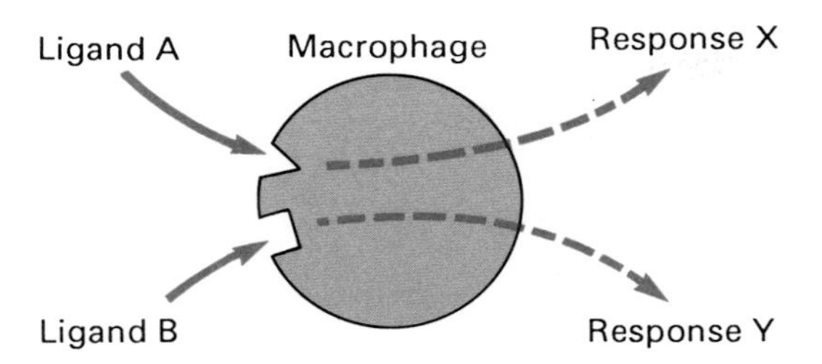

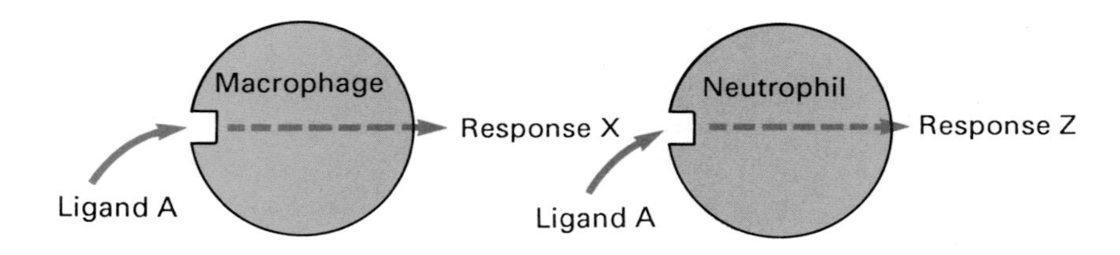

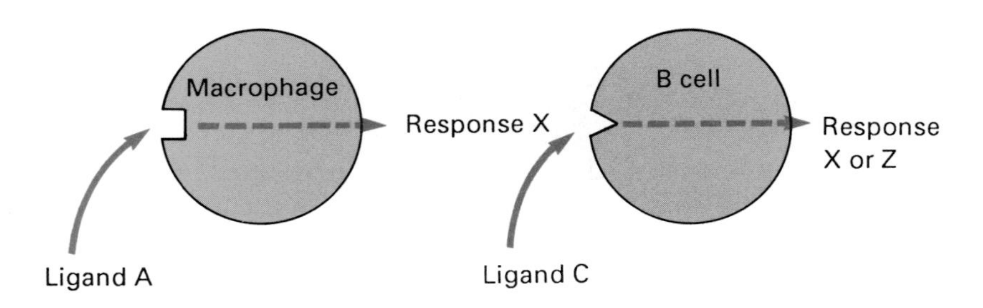

Fig.10.21 Consequences of different receptor distributions

Table 10.9 Some complement receptors

Receptor	Specificity	Cell distribution	M_r
CR1 (CD35)	C3b (iC3b, C4b)	Mφ†, neutrophils, B cells, etc.	Polymorphic: four forms from 190–280 kDa
CR2 (CD21)	C3d (iC3b)	B cells	145 kDa
CR3 (CD11b/18)*	iC3b	Mφ, neutrophils, NK cells, DC‡	165 kDa, 95 kDa

*See Section 4.2.6. † Mφ = macrophage ‡ DC = dendritic cell

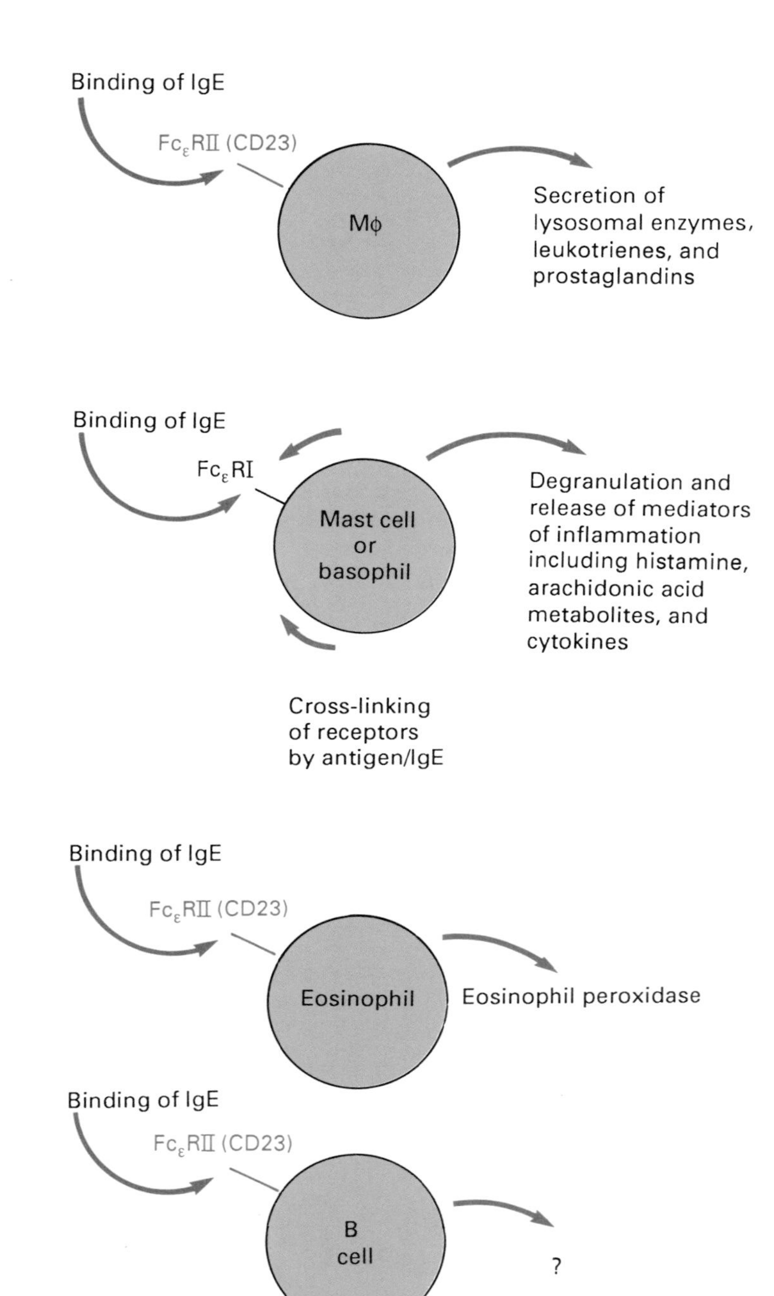

Fig.10.22 Fc_{ε} receptors for IgE

CR2 is expressed on B cells in association with other molecules (Section 8.4.2).

C3 receptors of macrophages. CR1 (designated CD35) is mainly specific for C3b, but it also binds iC3b and C4b (Klickstein 1988). This receptor is polymorphic and the various allotypic variants have molecular weights between 160 and 260 kDa. Macrophage CR1 mediates attachment of complement-coated particles as well as phagocytosis in certain circumstances. It is also a cofactor for cleavage of Cb3 to iC3b and further to C3dg (Fig.9.21). The receptor is upregulated on phagocytic cells by C5a and other chemotactic peptides. CR1 is widely distributed on different cell types, including macrophages, neutrophils, B cells, T cells, mast cells, fibroblasts, endothelial cells, and erythrocytes.

CR3 designated CD11b/CD18 is primarily specific for iC3b, the cleavage product of C3b, although it also binds C3dg, which may be the actual ligand for the receptor within iC3b (Wright *et al.* 1987). A number of monoclonal antibodies have been produced against this molecule, one of which is called Mac 1 in the mouse, and they define a heterodimer with a β chain that is shared with other membrane components LFA-1 and p150,95 (see Section 4.2.6). The CR3 receptor is also present on neutrophils, B cells, T cells, mast cells, and erythrocytes, and it is weakly expressed by dendritic cells. Other functions of this molecule are discussed in Section 4.2.6.

Phagocytosis via complement receptors. CR1 and CR3 can bind suitably coated particles, but the actual circumstances in which they promote phagocytosis are somewhat controversial (Fig.10.23). It appears that phagocytosis only occurs in the presence of a 'second signal', one of which is ligation of fibronectin receptors on the same cell. **Fibronectin** is a plasma protein that is deposited at sites of inflammation and wound healing, and its receptor on human monocytes and macrophages has a molecular weight of about 110 kDa. Using cultured human monocytes it was found that phagocytosis of complement-coated particles only occurred when *both* complement and fibronectin receptors were simultaneously occupied, although interaction with other proteins, particularly serum amyloid P and laminin (Section 4.2.6), was also effective (Wright *et al.* 1983). This suggests that phagocytosis by complement receptors is triggered when the cell binds inflammatory mediators or attaches to some components of the extracellular matrix. Both CR1 and CR3 showed similar behaviour, and this was unrelated to the number of receptors expressed at the cell surface. The receptors were also activated for phagocytosis by treatment of the cells with phorbol myristate acetate, PMA, which stimulates phosphorylation of CR1 in phagocytic cells, but not phosphorylation of CR3 or FcR.

The ability to phagocytose via complement receptors in the presence of a 'second signal' is related to the developmental or activation stage of the cell. Freshly-isolated human monocytes and resident mouse macrophages are normally unable to phagocytose via complement receptors, but inflammatory and activated cells can. However, there is little change in the numbers of receptors that are expressed during macrophage activation. Neutrophils also phagocytose via complement receptors, but this requires both the presence of fibronectin and additional exposure to chemotactic peptides.

It is possible that macrophage complement receptors and FcR may cooperate to promote phagocytosis, and this could be important to assist the removal of immune complexes. However, the presence of complement in immune complexes does not always enhance their degradation. For example, if the number of FcR is high, a large amount of C3b in the complex actually *inhibits* degradation; these properties could determine the site at which complexes are eliminated *in vivo*.

Zymosan (a yeast cell wall component) can be phagocytosed via MFR by freshly-isolated monocytes, but it can also be taken up via CR3 in the absence of exogenous complement. As noted in Section 10.4.3, macrophages produce many complement components and it may be that these are activated on the surface of zymosan by the alternative pathway (Section 9.4.5) (Ezekowitz *et al.* 1984). This could be important for clearance of particles at extravascular

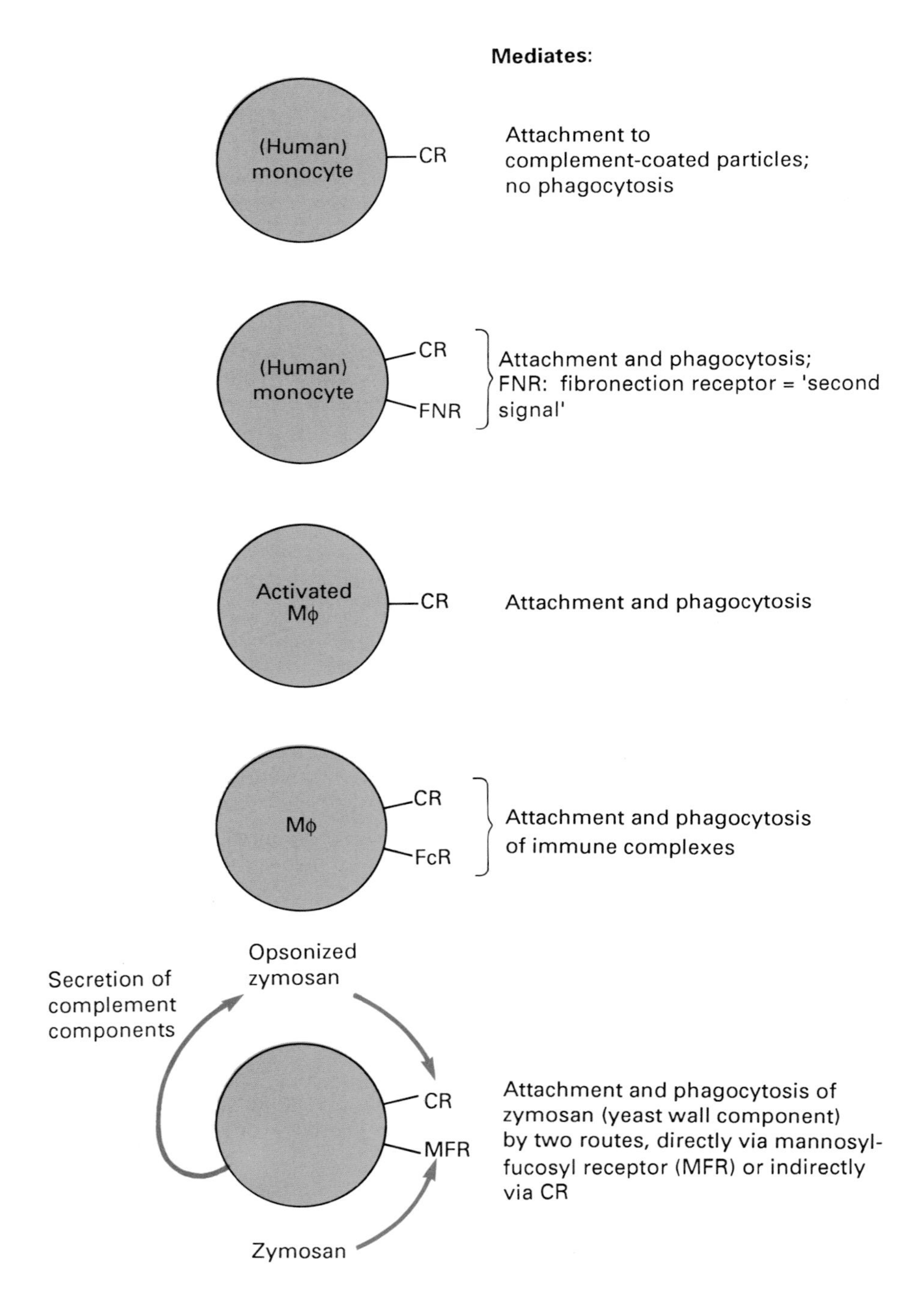

Fig.10.23 Function of complement receptors on monocyte/macrophages and synergy with other receptors

sites where complement is limiting, but a similar system may allow Leishmania parasites to penetrate macrophages.

Ligation of complement receptors on human monocytes and mouse macrophages does *not* trigger the release of H_2O_2 or arachidonic acid metabolites by the cell, unlike ligation of FcR. The inability of these cells to produce ROI under these circumstances might be important in preventing tissue damage at certain sites, for example during clearance of immune complexes in the liver. Nevertheless, liver Kupffer cells are unable to produce reactive oxygen intermediates, even after activation by IFNγ (Lepay *et al.* 1985). These are examples of how ligation of different receptors on the same cell can result in different responses (Fig.10.21), and of how cells of the same lineage can acquire or lose certain functions depending on their local microenvironment.

Other complement receptors of macrophages

The macrophage has a variety of other complement receptors in addition to CR1 and CR3.

1. C1q binds to the macrophage after C1r and C1s have been cleaved from C1 by C1 inhibitor (Section 9.4.4); C1 itself does not bind. The C1q receptor (Table 9.2c) and the FcR act synergistically to promote phagocytosis of immune complexes, and the C1q receptor may also be present on neutrophils and B cells.
2. Macrophages have C5a receptors (Table 9.2c), since this component is chemotactic for monocytes, as well as neutrophils, and ligation of this receptor is reported to stimulate IL-1 release by macrophages. The gene for the C5a receptor has been cloned and found to be similar in structure to the receptors for the f-Met-leu-phe peptide and IL-8, both of which are chemotactic for neutrophils. [IL-8, also called NAP-1, can be secreted by a variety of different cell types including macrophages, and is a member of the **intercrine** family, a group of homologous cytokines with proinflammatory and restorative functions (Oppenheim *et al.* 1991)].
3. Factor Bb of the alternative pathway inhibits the migration of mononuclear phagocytes, and is thus termed a 'migration inhibition factor' or MIF, and it causes spreading and adhesion of mouse macrophages and human monocytes. These effects are presumably mediated by a specific receptor.
4. The macrophage may also express a receptor for factor H.

10.5 Hypersensitivity (Fig.10.24)

Hypersensitivity is the result of an immune response that occurs inappropriately, or to an exaggerated degree. In this type of reaction, the adaptive immune response triggers effector mechanisms that cause inflammation (Section 9.2.1) and/or tissue damage. If these mechanisms are directed towards the body's own components this is manifested as **autoimmunity**,

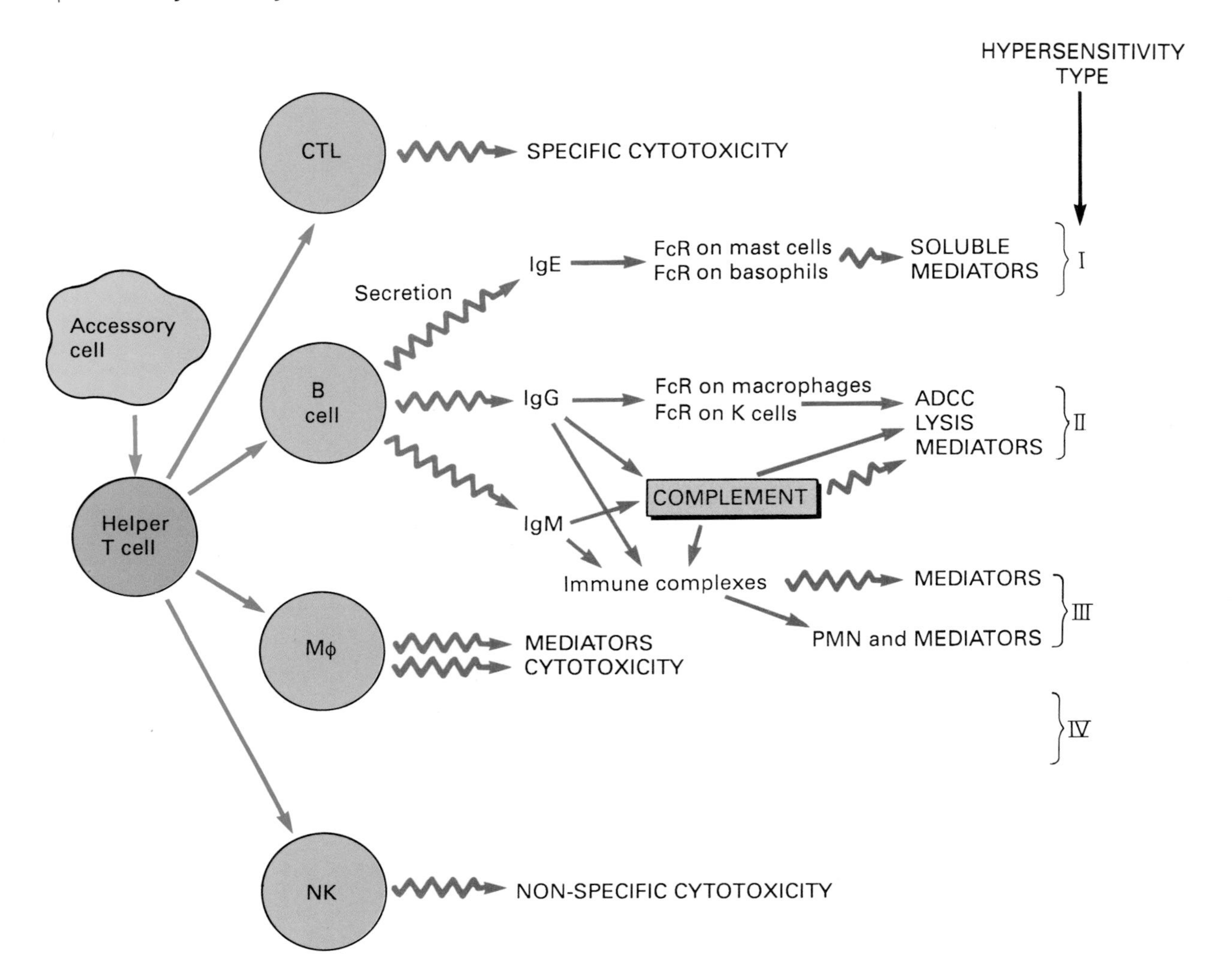

Fig.10.24 Effector mechanisms in hypersensitivity

and they can also be directed against grafted organs in **transplantation reactions**.

Four main types of hypersensitivity have been defined, but these represent extremes and each probably rarely occurs in isolation. Three forms are antibody-mediated, while the fourth is mediated by T cells and macrophages (Fig.10.25). Each will be described very briefly, so that some of the mechanisms discussed in this and other chapters (e.g. Chapter 9) can be put in a different context.

10.5.1 Type I hypersensitivity

Type I, or immediate, hypersensitivity is now known as **allergy**, although this was not the original meaning of the term. Perhaps one of the best known examples is IgE sensitization to environmental antigens such as pollen which causes hay fever. For unknown reasons, exposure to pollen may cause the production of IgE antibodies which bind to Fc receptors on mast cells. As noted in Section 10.4.5 (see IgE receptors), subsequent encounter with the antigen causes degranulation of the mast cells and the release of preformed mediators of inflammation. The clinical features are known as atopy, and this type of hypersensitivity sometimes causes fatal **anaphylactic reactions**. These responses can occur within minutes of exposure to the antigen, for which reason they are also referred to as **immediate hypersensitivity**.

10.5.2 Type II hypersensitivity

This type of hypersensitivity is mediated by antibodies which bind to Fc receptors on killer (K) cells and macrophages. They lead to antibody-dependent cellular cytotoxicity (ADCC; Section 10.2.7) or phagocytosis of the target. Alternatively, these antibodies fix complement and lyse the target cell and/or generate activated C3 components which recruit other effector cells.

10.5.3 Type III hypersensitivity

This is caused by immune complexes, which can be formed between antigens and antibodies. These complexes, instead of being removed by the reticulo-endothelial system, are deposited in some organs during persistent or repeated infections and in some autoimmune conditions. Inflammation occurs, partly as a result of the interaction between the antigen–antibody complexes and complement, which generates the anaphylatoxins C3a and C5a, and partly by molecules secreted by recruited and activated neutrophils.

10.5.4 Type IV hypersensitivity

Type IV, or **delayed-type hypersensitivity**, results from the action of cytokines that are produced by activated T cells. (These cells were formerly thought of as a distinct subset called T_{DTH}, but are now thought simply to be helper T cells.) These cytokines include IFNγ which activates macrophages (Section 10.4.4) and others that recruit different cell types. Type IV hypersensitivity as a class includes several types of reaction that are roughly defined according to the time of onset and the cells involved:

1. The **Jones–Mote reaction** is a response that typically occurs within 24 hours and is characterized by basophil infiltration in the subepidermis of skin.
2. **Contact sensitivity** and **tuberculin reactions** appear from 48–72 hours after antigen has been administered to a sensitized individual.
3. **Granulomatous responses** occur weeks after exposure to antigen, for example in response to a persistent agent with the macrophage.

Type IV hypersensitivity is thought to occur during transplantation reactions, but there is controversy as to its importance in damaging tissues during graft rejection compared to the effects of cytotoxic T lymphocytes.

Further reading

Natural killer cells

Burton, R.C., Koo, G.C., Smart, Y.C., Clark, D.A., and Winn, H.J. (1988). Surface antigens of murine natural killer cells. *International Review of Cytology*, **111**, 185–210.

Ortaldo, J.R. and Herberman, R.B. (1984). Heterogeneity of natural killer cells. *Annual Review of Immunology*, **2**, 359–94.

Trinchieri, G. (1989). Biology of natural killer cells. *Advances in Immunology*, **47**, 187–376.

Young, L.H., Liu, C.C., Joag, S., Rafii, S., and Young, J.D. (1990). How lymphocytes kill. *Annual Review of Medicine*, **41**, 45–54.

Mechanisms of cytotoxicity by NK and CTL

Bonavida, B. and Collier, R.J. (ed.) (1987). *Membrane-mediated cytotoxicity*, Vol.45. Alan R. Liss, New York.

Cohen, J.J., Duke, R.C., Fadok, V.A., and Sellins, K.S. (1992). Apoptosis and programmed cell death in immunity. *Annual Review of Immunology*, **10**, 267–93.

Di Virgilio, F., Pizzo, P., Zanovello, P., Bronte, V., and Collavo, D. (1990). Extracellular ATP as a possible mediator of cell-mediated cytotoxicity. *Immunology Today*, **11**, 274–7.

Grimm, E.A. and Owen-Schaub, L. (1991). The IL-2 mediated amplification of cellular cytotoxicity. *Journal of Cell Biochemistry*, **45**, 335–9.

Herberman, R.N., Reynolds, C.W., and Ortaldo, J.R. (1986). Mechanism of cytotoxicity by natural killer cells. *Annual Review of Immunology*, **7**, 309–37.

Kupffer, A. and Singer, S.J. (1989). Cell biology of cytotoxic and helper T-cell functions: immunofluorescence microscopic studies of single cells and cell couples. *Annual Review of Immunology*, **7**, 309–37.

Martz, E. and Howell, D.M. (1989). CTL: virus control cells first and cytolytic cells second? DNA fragmentation, apoptosis and the prelytic halt hypothesis. *Immunology Today*, **10**, 79–86.

Podack, E.R. (ed.) (1988). *Cytolytic lymphocytes and complement: effectors of the immune system*. CRC Press, Boca Raton, Florida.

Podack, E.R., Hengartner, H., and Lichtenheld, M.G. (1991). A central role of perforin in cytolysis? *Annual Review of Immunology*, **9**, 729–57.

Storkus, W.J. and Dawson, J.R. (1991). Target structures involved in natural killing (NK): characteristics, distribution, and candidate molecules. *Critical Reviews in Immunology*, **10**, 393–416.

Tschopp, J. and Nabholz, M. (1990). Perforin-mediated target cell lysis by cytolytic T lymphocytes. *Annual Review of Immunology*, **8**, 279–302.

Macrophages

General

Adams, D.O. and Hamilton, T.A. (1984). The cell biology of macrophage activation. *Annual Review of Immunology*, **2**, 283.

Asherson, G.L. and Zembala, M. (ed.) (1989). *Human monocytes*. Academic Press, New York.

Evered, D., Nugent, J., and O'Connor, M. (ed.) (1986). Biochemistry of macrophages. Ciba Foundation Symposium, Vol.118.

Cytokines and mediators (see Chapters 4, 8, and 9)

Hibbs, J.B. *et al* (1990). In: *Nitric oxide from* L*-arginine: a bioregulatory system* (ed. S. Moncada and E.A. Higgs), pp.189–223. Excerpta Medica, Amsterdam.

Nathan, C.F. (1987). Secretory products of macrophages. *Journal of Clinical Investigation*, **79**, 319–26.

Parker, C.W. (1987). Lipid mediators produced through the lipoxygenase pathway. *Annual Review of Immunology*, **5**, 65–84.

Stenson, W.F. and Parker, C.W. (1980). Prostaglandins, macrophages, and immunity. *Journal of Immunology*, **125**, 1–5.

Receptors

Ahearn, J.M. and Fearon, D.T. (1989). Structure and function of the complement receptors, CD1 (CD35) and CD2 (CD21). *Advances in Immunology*, **46**, 183–219.

Conrad, D.H. (1990). $Fc_{\varepsilon}RII$/CD23: the low affinity receptor for IgE. *Annual Review of Immunology*, **8**, 623–46.

Fanger, M.W., Shen, L., Graziano, R.F., and Guyre, P.M. (1989). Cytotoxicity mediated by human Fc receptors for IgG. *Immunology Today*, **10**, 92–9.

Fearon, D.T. and Wong, W.W. (1983). Complement ligand-receptor interactions that mediate biological responses. *Annual Review of Immunology*, **1**, 243–71.

Keegan, A.D. and Paul, W.E. (1992). Multichain immune recognition receptors: similarities in structure and signaling pathways. *Immunology Today*, **13**, 63–8.

Kinet, J.B. (1989). Antibody-cell interactions: Fc receptors. *Cell*, **57**, 351–4.

Ravetch, J.V. and Kinet, J.-P. (1991). Fc receptors. *Annual Review of Immunology*, **9**, 457–92.

Ross, G.D. and Medoff, M.E. (1985). Membrane complement receptors specific for bound fragments of C3. *Advances in Immunology* **37**, 217–67.

Unkeless, J.C., Scigliano, E., and Freedman, V.H. (1988). Structure and function of human and murine receptors for IgG. *Annual Review of Immunology*, **6**, 251–81.

Other

Gleich, G.J. and Loegering, D.A. (1984). Immunobiology of eosinophils. *Annual Review of Immunology*, **2**, 429–59.

Kitamura, Y. (1989). Heterogeneity of mast cells and phenotypic change between subpopulations. *Annual Review of Immunology*, **7**, 59–76.

Stevens, R.L. and Austen, K.F. (1989). Recent advances in the cellular and molecular biology of mast cells. *Immunology Today*, **10**, 381–6.

Literature cited

Abo, T., Cooper, M.D., and Balch, C.M. (1982). Characterization of HNK-1+ (Leu-7) human lymphocytes. I. Two distinct phenotypes of human NK cells with different cytotoxic capability. *Journal of Immunology*, **129**, 1752–7.

Acha-Orbea, H., Groscurth, P., Lang, R., Stitz, L., and Hengartner, H. (1983). Characterization of cloned cytotoxic lymphocytes with NK-like activity. *Journal of Immunology*, **130**, 2952–9.

Anegon, I., Cuturi, M., Trinchieri, G., and Perussia, B. (1988). Interaction of Fc receptor (CD16) ligands induces transcription of interleukin 2 receptor (CD25) and lymphokine genes and expression of their products in human NK cells. *Journal of Experimental Medicine*, **167**, 452–72.

Bevilacqua, M.P., Amrani, D., Mosesson, M.W., and Bianco, C. (1981). Receptors for cold-insoluble globulin (plasma fibronectin) on human monocytes. *Journal of Experimental Medicine*, **153**, 42–60.

Binz, H., Fenner, M., Frei, D., and Wigzell, H. (1983). Two independent receptors allow selective target lysis by T cell clones. *Journal of Experimental Medicine*, **157**, 1252–60.

Blank, U., Ra, C., Miller, L., White, K., Metzger, H., Kinet, J.P. (1989). Complete structure and expression in transfected cells of high affinity IgE receptor. *Nature*, **337**, 187–9.

Ciccone, E., Pende, D., Viale, O., Tambussi, G., Ferrini, S., Biassoni, R., Longo, A., Guardiola, J., Moretta, A., and Moretta, L. (1990). Specific recognition of human $CD3^-$ $CD16^+$ natural killer cells requires the expression of an autosomic recessive gene on target cells. *Journal of Experimental Medicine*, **172**, 47–52.

Clark, R.A. and Szot, S. (1981). The myeloperoxidase-hydrogen peroxide-halide system as effector of neutrophil-mediated tumor cell cytotoxicity. *Journal of Immunology*, **126**, 1295–301.

Colsky, A.S., Stein-Streilein, J., and Peacock, J.S. (1988). Surrogate receptor-mediated cellular cytotoxicity. A method for 'custom designing' killer cells of desired target specificity. *Journal of Immunology*, **140**, 2515–19.

Crawford, R.M., Finbloom, D.S., Ohara, J., Paul, W.E., and Meltzer, M.S. (1987). B cell stimulatory factor-1 (interleukin-4) activates macrophages for increased tumoricidal activity and expression of Ia antigens. *Journal of Immunology*, **139**, 135–41.

Crispe, I.N., Bevan, M.J., and Staerz, U.D. (1985). Selective activation of Lyt 2+ precursor cells by ligation of the antigen receptor. *Nature*, **317**, 627–9.

Drapier, J.C. and Hibbs, J.B. (1988). Differentiation of murine macrophages to express nonspecific cytotoxicity for tumor cells results in L-arginine-dependent inhibition of mitochondrial iron–sulfur enzymes in the macrophage effector cells. *Journal of Immunology*, **140**, 2829–38.

Duke, R.C., Chervenak, R., and Cohen, J.J. (1983). Endogenous endonuclease-induced DNA fragmentation: an early event in cell-mediated cytolysis. *Proceedings of the National Academy of Sciences USA*, **80**, 6361–5.

Duke, R.C., Persechini, P.M., Chang, S., Liu, C.C., Cohen, J.J., and Young, J.D. (1989). Purified perforin induces target cell lysis but not DNA fragmentation. *Journal of Experimental Medicine*, **170**, 1451–6.

Ezekowitz, R.A.B., Austyn, J., Stahl, P.D., and Gordon, S. (1981). Surface properties of bacillus Calmette–Guérin-activated mouse macrophages. Reduced expression of mannose-specific endocytosis, Fc receptors and antigen F4/80 accompanies induction of Ia. *Journal of Experimental Medicine*, **154**, 60–76.

Ezekowitz, R.A.B., Sim, R., Hill, M., and Gordon, S. (1984). Local opsonization by secreted macrophage complement components. Role of receptors for complement in uptake of zymosan. *Journal of Experimental Medicine*, **159**, 244–60.

Ezekowitz, R.A., Sastry, K., Bailly, P., and Warner, A. (1990). Molecular characterization of the human macrophage mannose receptor: demonstration of multiple carbohydrate recognition-like domains and phagocytosis of yeasts in Cos-1 cells. *Journal of Experimental Medicine*, **172**, 1785–94.

Grabstein, K.H., Urdal, D.L., Tushinski, R.J., Mochizuki, D.Y.,

Price, V.L., Cantrell, M.A., Gillis, S., and Conlon, P.J. (1986). Induction of macrophage tumoricidal activity by granulocyte-macrophage colony-stimulating factor. *Science*, **232**, 506–8.

Griffin, F.M. Jr, Bianco, C., and Silverstein, S.C. (1975). Characterization of the macrophage receptor for complement and demonstration of its functional independence from the receptor for the Fc portion of immunoglobulin G. *Journal of Experimental Medicine*, **141**, 1269–77.

Hiserodt, J.C., Britvan, L.C., and Targan, S.R. (1982). Characterization of the cytolytic reaction mechanism of the human natural killer (NK) lymphocyte: resolution into binding, programming and killer cell-independent steps. *Journal of Immunology*, **129**, 1782–7.

Klickstein, L.B., Bartow, T.J., Miletic, M., Rabson, L.D., Smith, J.A., and Fearon, D.T. (1988). Identification of distinct C3b and C4b recognition sites in the human C3b/C4b receptor (CR1, CD35) by deletion mutagenesis. *Journal of Experimental Medicine*, **168**, 1699–717.

Kupfer, A., Singer, S.J., and Dennert, G. (1986). On the mechanism of unidirectional killing in mixtures of two cytotoxic T lymphocytes. Unidirectional polarization of cytoplasmic organelles and the membrane-associated cytoskeleton in the effector cell. *Journal of Experimental Medicine*, **163**, 489–98.

Kuppers, R.C. and Henney, C.S. (1977). Studies on the mechanism of lymphocyte-mediated cytolysis. IX. Relationships between antigen recognition and lytic expression in killer T cells. *Journal of Immunology*, **118**, 71–6.

Lanier, L.L., Le, A.M., Civin, C.I., Loken, M.R., and Phillips, J.H. (1986). The relationship of CD16 (Leu-11) and Leu-19 (NKH-1) antigen expression on human peripheral blood NK cells and cytotoxic T lymphocytes. *Journal of Immunology*, **136**, 4480–6.

Lepay, D.A., Nathan, C.F., Steinman, R.M., Murray, H.W., and Cohn, Z.A. (1985). Murine Kupffer cells. Mononuclear phagocytes deficient in the generation of reactive oxygen intermediates. *Journal of Experimental Medicine*, **161**, 1079–96.

Lichtenheld, M.G. and Podack, E.R. (1989). Structure of the human perforin gene. A simple gene organization with interesting potential regulatory sequences. *Journal of Immunology*, **143**, 4267–74.

Liew, F.Y., Millott, S., Parkinson, C., Palmer, R.M., and Moncada, S. (1990). Macrophage killing of Leishmania parasite in vivo is mediated by nitric oxide from L-arginine. *Journal of Immunology*, **144**, 4794–7.

Liu, C.C., Steffen, M., King, F., and Young, J.D. (1987). Identification, isolation, and characterization of a novel cytotoxin in murine cytolytic lymphocytes. *Cell*, **51**, 393–403.

Maliszewski, C.R., March, C.J., Schoenborn, M.A., Gimpel, S., and Shen, L. (1990). Expression cloning of a human receptor for IgA. *Journal of Experimental Medicine*, **172**, 1665–72.

Masson, D. and Tschopp, J. (1987). A family of serine esterases in lytic granules of cytolytic lymphocytes. *Cell*, **49**, 679–85.

Moretta, A., Bottino, C., Pende, D., Tripodi, G., Tambussi, G., Viale, O., Orengo, A., Barbaresi, M., Merli, A., Ciccone, E., and Moretta, L. (1990). Identification of four subsets of human $CD3^-$ $CD16^+$ natural killer (NK) cells by the expression of clonally-distributed functional surface molecules: correlation between subset assignment of NK clones and ability to mediate specific alloantigen recognition. *Journal of Experimental Medicine*, **172**, 1589–98.

Murray, H.W., Spitalny, G.L., and Nathan, C.F. (1985). Activation of mouse peritoneal macrophages *in vitro* and *in vivo* by interferon-γ. *Journal of Immunology*, **134**, 1619–22.

Nathan, C.F., Mercer-Smith, J.A., Desantis, N.M., and Palladino, M.A. (1982). Role of oxygen in T cell-mediated cytolysis. *Journal of Immunology*, **129**, 2164–71.

Olds, G.R., Ellner, J.J., Kearse, L.A., Kazura, J.W., and Mahmoud, A.A.F. (1980). Role of arginase in killing of schistosomula of *Schistosoma mansoni*. *Journal of Experimental Medicine*, **151**, 1557–62.

Oppenheim, J.J., Zachariae, O.C., Mukaida, N., and Matsushima, K. (1991). Properties of the novel proinflammatory supergene 'intercrine' cytokine family. *Annual Review of Immunology*, **9**, 617–48.

Ozer, H., Strelkauskas, A.J., Callery, R.T., and Schlossman, S.F. (1979). The functional dissection of human peripheral null cells with respect to antibody-dependent cellular cytotoxicity and natural killing. *European Journal of Immunology*, **9**, 112–18.

Peters, P.J., Borst, J., Oorschot, V., Fukuda, M., Krahenbuhl, O., Tschopp, J., Slot, J.W., Geuze, H.J. (1991). Cytotoxic T lymphocyte granules are secretory lysosomes, containing both perforin and granzymes. *Journal of Experimental Medicine*, **173**, 1099–109.

Phillips, J.H. and Lanier, L.L. (1986). Dissection of the lymphokine-activated killer phenomenon. Relative contribution of peripheral blood natural killer cells and T lymphocytes to cytolysis. *Journal of Experimental Medicine*, **164**, 814–25.

Plaut, M., Pierce, J.H., Watson, C.J., Hanley-Hyde, J., Nordan, R.P., Paul, W.E. (1989). Mast cell lines produce lymphokines in response to cross-linkage of $Fc_{\varepsilon}RI$ or to calcium ionophores. *Nature*, **339**, 64–7.

Podack, E.R. and Dennert, G. (1983). Assembly of two types of tubules with putative cytolytic function by cloned natural killer cells. *Nature*, **302**, 442–5.

Podack, E.R. and Konigsberg, P.J. (1984). Cytolytic T-cell granules. Isolation, structural, biochemical, and functional characterization. *Journal of Experimental Medicine*, **160**, 695–710.

Roder, J.C. and Duwe, A. (1979). The beige mutation in the mouse selectively impairs natural killer cell function. *Nature*, **278**, 451–3.

Schmidt, D.S., Hornung, R., McGrath, K.M., Paul, R., and

Ruddle, N.H. (1987). Target cell DNA fragmentation is mediated by lymphotoxin and tumor necrosis factor. *Lymphokine Research*, **6**, 195–202.

Schmidt, R.E., McDermott, R.P., Bartley, G.T., Bertovich, M., Amato, D.A., Austen, K.F., Schlossman, S.F., Stevens, R.M., and Ritz, J. (1985). Specific release of proteoglycans from human natural killer cells during target lysis. *Nature*, **318**, 289–91.

Scott, W.A., Pawlowski, N.A., Murray, H.W., Andreach, M., Zrike, J., and Cohn, Z.A. (1982). Regulation of arachidonic acid metabolism by macrophage activation. *Journal of Experimental Medicine*, **155**, 1148–60.

Smyth, M.J., Ortaldo, J.R., Shinakai, Y., Yagita, H., Nakata, M., Okumura, K., and Young, H.A. (1990). Interleukin 2 induction of pore-forming protein gene expression in human peripheral blood $CD8^+$ T cells. *Journal of Experimental Medicine*, **171**, 1269–81.

Snyderman, R. and Pike, M.C. (1984). Chemoattractant receptors on phagocytic cells. *Annual Review of Immunology*, **2**, 257–81.

Solomon, F.R. and Higgins, T.J. (1987). A monoclonal antibody with reactivity to asialo-GM_1 and murine natural killer cells. *Molecular Immunology*, **24**, 57–65.

Spiegelberg, H.L., Boltz-Nitulescu, G., Plummer, J.M., and Melewicz, F.M. (1983). Characterization of the IgE Fc receptors on monocytes and macrophages. *Federation Proceedings*, **42**, 124–8.

Suzuki, N., Bianchi, E., Bass, H., Suzuki, T., Bender, J., Pardi, R. Brenner, C.A., Larrick, J.W., and Engelman, E.G. (1990). Natural killer lines and clones with apparent antigen specificity. *Journal of Experimental Medicine*, **172**, 457–62.

Suzuki, N., Suzuki, T., and Engelman, E.G. (1991). Evidence for the involvement of CD56 molecules in alloantigen-specific recognition by human natural killer cells. *Journal of Experimental Medicine*, **173**, 1451–61.

Targan, S.R. and Newman, W. (1983). Definition of a 'trigger' stage in the NK cytolytic reaction sequence by a monoclonal antibody to the glycoprotein T-200. *Journal of Immunology*, **131**, 1149–53.

Timonen, T., Ortaldo, J.R., and Herberman, R.B. (1981). Characteristics of human large granular lymphocytes and relationship to natural killer and K cells. *Journal of Experimental Medicine*, **153**, 569–82.

Titus, J.A., Perez, P., Kaubisch, A., Garrido, M.A., and Segal, D.M. (1987). Human K/NK cells targeted with heterocrosslinked antibodies specifically lyse tumor cells in vitro and prevent tumor growth *in vivo*. *Journal of Immunology*, **139**, 3153–8.

Trinchieri, G. and Santoli, D. (1978). Anti-viral activity induced by culturing lymphocytes with tumor-derived or virus-transformed cells. Enhancement of human natural killer cell activity by interferon and antagonistic inhibition of susceptibility of target cells to lysis. *Journal of Experimental Medicine*, **147**, 1314–33.

van Furth, R. (ed.) (1992). *Mononuclear phagocytes. Biology of monocytes and macrophages.* Kluwer Academic, Dordrecht.

Van Leeuwen, B.H., Martinson, M.E., Webb, G.C., and Young, I.G. (1989). Molecular organization of the cytokine gene cluster, involving the human IL-3, IL-4, IL-5 and GM-CSF genes, on human chromosome 5. *Blood*, **73**, 1142–8.

Verret, C.R., Firmenich, A.A., Kranz, D.M., and Eisen, H.N. (1987). Resistance of cytotoxic T lymphocytes to the lytic effects of their toxic granules. *Journal of Experimental Medicine*, **166**, 1536–47.

Wright, S.C. and Bonavida, B. (1987). Studies on the mechanism of natural killer cell-mediated cytotoxicity. VII. Functional comparison of human natural killer cytotoxic factors with recombinant lymphotoxin and tumor necrosis factor. *Journal of Immunology*, **138**, 1791–8.

Wright, S.D., Craigmyle, L.S., and Silverstein, S.C. (1983). Fibronectin and serum amyloid P component stimulate C3b- and C3bi-mediated phagocytosis in cultured human monocytes. *Journal of Experimental Medicine*, **158**, 1338–43.

Wright, S.D., Reddy, P.A., Jong, M.T., and Erickson, B.W. (1987). C3bi receptor (complement receptor type 3) recognizes a region of complement protein C3 containing the sequence Arg-Gly-Asp. *Proceedings of the National Academy of Sciences USA*, **84**, 1965–8.

Wright, S.D., Ramos, R.A., Tobias, P.S., Ulevitch, R.J., and Mathison, J.C. (1990). CD14, a receptor for complexes of lipopolysaccharide (LPS) and LPS binding protein. *Science*, **249**, 1431–3.

Yamamoto, K. and Johnston, R.B. (1984). Dissociation of phagocytosis from stimulation of the oxidative metabolic burst in macrophages. *Journal of Experimental Medicine*, **159**, 405–16.

Young, J.D., Liu, C.C., Leong, L.G., and Cohn, Z.A. (1986). The pore-forming protein (perforin) of cytolytic L lymphocytes is immunologically related to the components of membrane attack complex of complement through cysteine-rich domains. *Journal of Experimental Medicine* **164**, 2077–82.

Young, J.W. and Steinman, R.M. (1990). Dendritic cells stimulate primary human cytolytic lymphocyte responses in the absence of CD4+ helper T cells. *Journal of Experimental Medicine*, **171**, 1315–30.

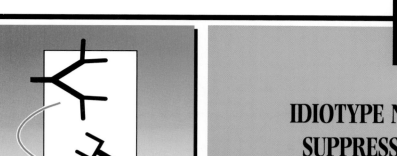

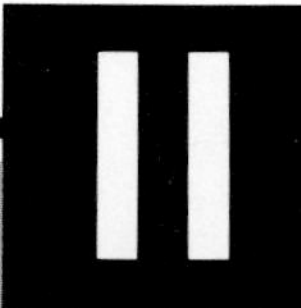

11

IDIOTYPE NETWORKS AND SUPPRESSOR PATHWAYS

Scheme 11. Idiotypes and suppressor pathways

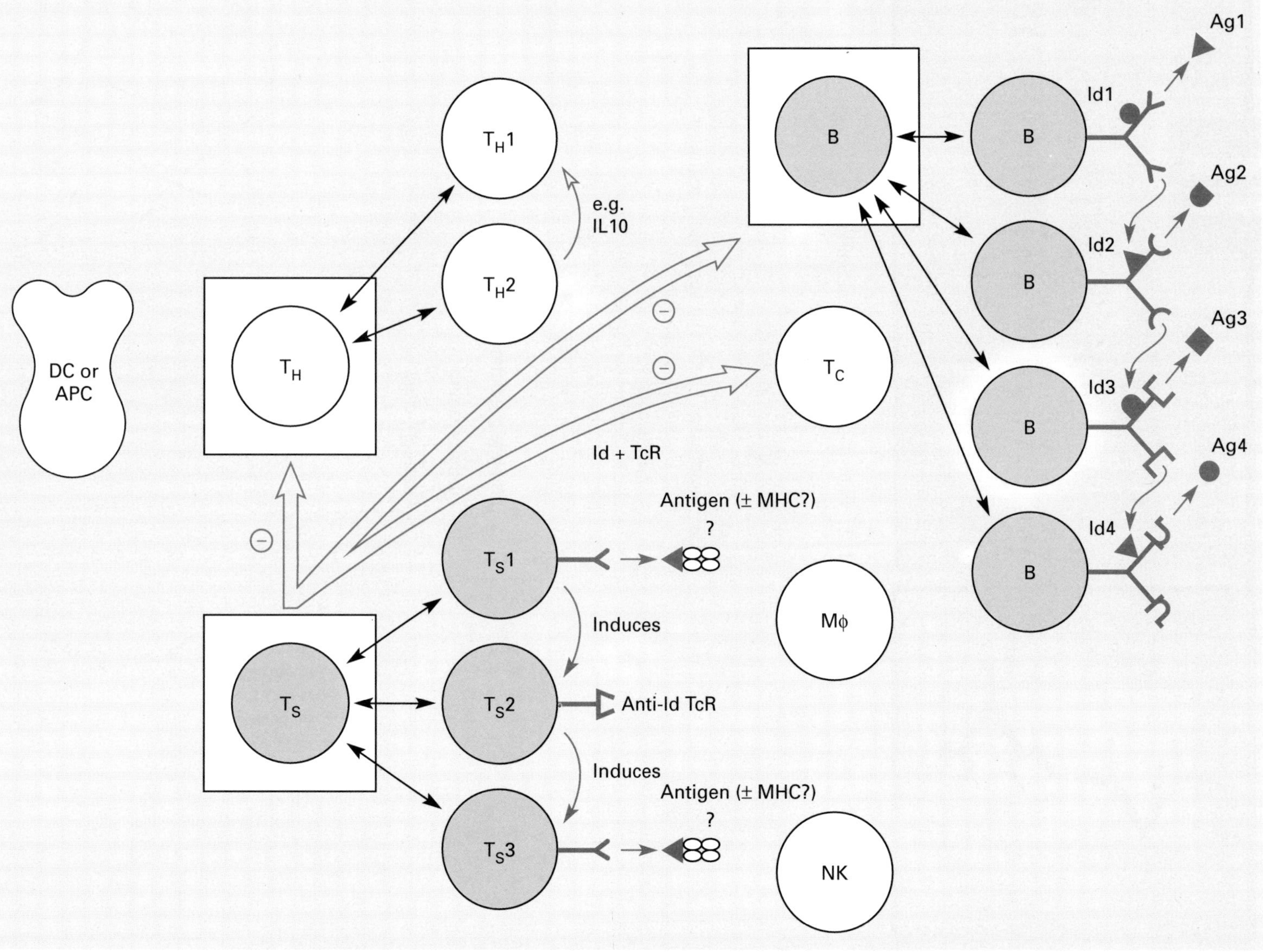

DC or APC

T_H

T_H1

T_H2

e.g. IL10

T_S

T_S1

T_S2

T_S3

Induces

Anti-Id TcR

Induces

Antigen (± MHC?)

?

Antigen (± MHC?)

?

Id + TcR

B

T_C

Mϕ

NK

Id1

Id2

Id3

Id4

Ag1

Ag2

Ag3

Ag4

11.1 Idiotypes and idiotypic networks

In our analysis of immune responses to foreign antigens so far, we have seen that different antigens can induce B cell and T cell responses, and that antigen recognition is mediated by specific antigen receptors on lymphocytes. By now it should be obvious that much of our understanding of the immune system derives from studies using components of the immune system itself as experimental tools. For example, monoclonal antibodies (see Appendix) have been raised against other antibodies, and T cell receptors, (Chapter 6) and used to study their structure and function, and some monoclonal antibodies can inhibit or stimulate cellular responses that would normally be induced by foreign antigens (e.g. Sections 8.4.2, 4.2.1). It is clear, then, that *antibodies and T cell receptors can also be thought of as antigens.*

In principle, any molecule that has diverged in evolution can be immunogenic in a genetically different individual of the same or a different species; thus immune responses can be produced against polymorphisms or species-specific differences in a molecule (see Appendix). In general, however, most cellular components are not immunogenic within a given individual (e.g. Sections 8.6, 5.2.5). In this section we discuss why antibodies and T cell receptors might be different in this respect from other self molecules. While genetically different individuals of the same or another species can make immune responses against these molecules, *certain portions of lymphocyte antigen receptors may be immunogenic even within the same individual that produces them.*

11.1.1 Idiotypes of antibodies

Antibodies as antigens: idiotopes and idiotypes

When an animal is immunized against an antigen, an immune response is made to its antigenic determinants, or **epitopes** (Fig.1.18). In the same way, an animal can be immunized against a purified antibody. The resulting immune response can be specific for polymorphic determinants, such as allotypes, and for species-specific differences such as isotypes (Section 6.2.2; Fig.6.10). Part of the immune response against the antibody in question can also be directed against **idiotopes**.

Idiotopes are antigenic determinants on the variable regions of antibodies, which are formed by the association of the variable domains of Ig heavy and light chains (Section 6.2) (Fig.11.1) (Kunkel *et al.* 1963). An idiotope is thus one particular epitope of an antibody that could, for example, be detected by the binding of a monoclonal antibody to it. Frequently, several different idiotopes can be expressed by an antibody and these can be detected in a normal oligoclonal (Section 1.1.3) response. The sum of all the idiotopes expressed by a particular antibody is called its **idiotype** (Fig.11.2). This can be thought of as the antigenic characteristic of the antibody's variable region, and especially of its antigen-binding site (although constant region domains can influence the structure of variable domains and hence idiotypes; Morahan *et al.* 1983). A response against the idiotype of an antibody is called an

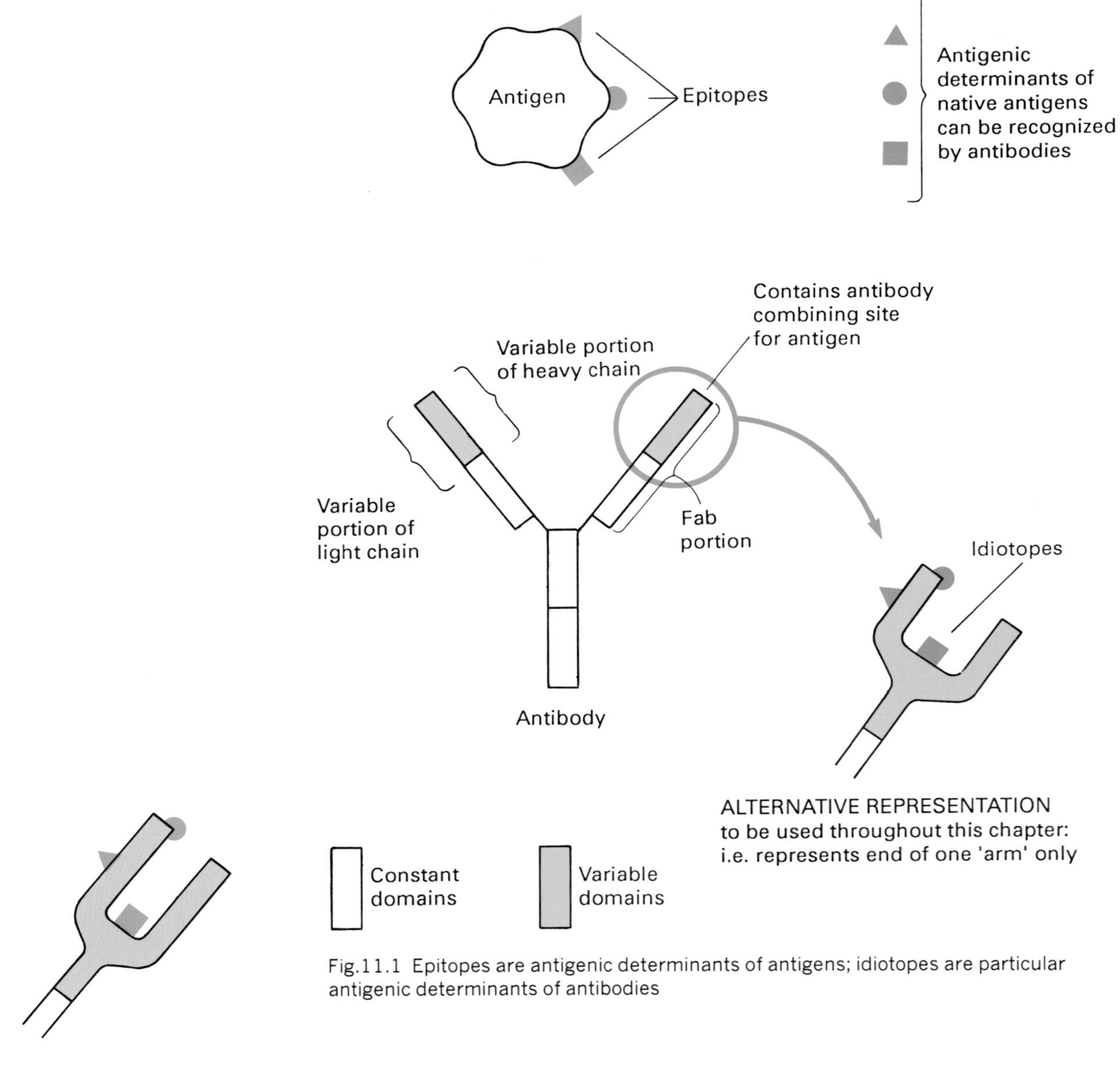

Fig.11.1 Epitopes are antigenic determinants of antigens; idiotopes are particular antigenic determinants of antibodies

Antigenic determinants of the variable region of the antibody **= idiotopes**

Fig.11.2 The idiotype of an antibody is the sum of its idiotopes

anti-idiotypic response, and the antibodies generated are termed **anti-idiotypic antibodies** (Fig.11.3).

While individuals do not normally make immune responses against their own molecules, idiotopes may be different in that at least some of them may be immunogenic within the same individual (Rodkey 1974). Much of an antibody molecule is of course not immunogenic within that individual (Fig.11.4). For instance, every individual will be immunologically tolerant to immunoglobin constant regions, just as for any other 'self' molecule, because these are present in all antibodies. In contrast, the variable region of a particular antibody may represent a unique structure that has appeared for the first time within a certain individual who, consequently, may not be

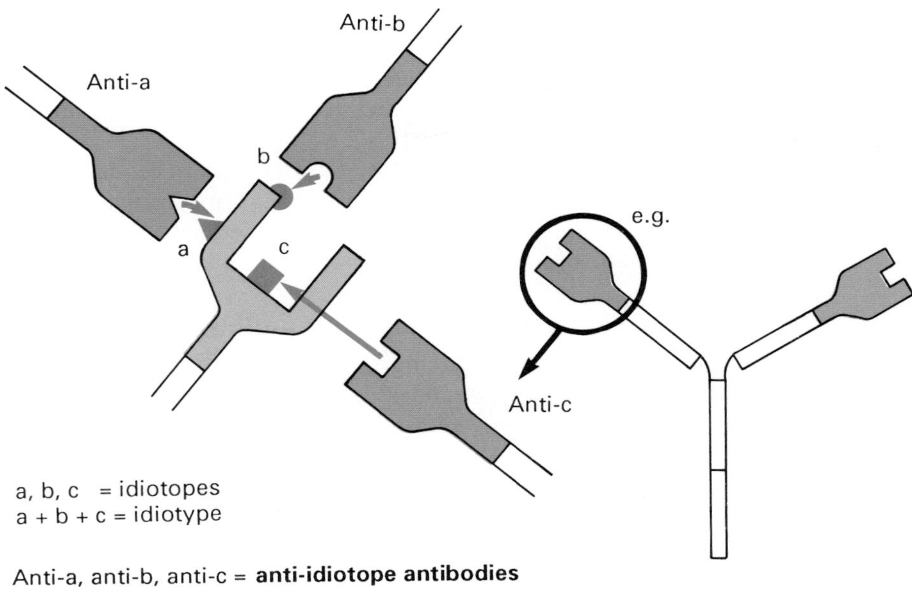

Fig.11.3 Anti-idiotype antibodies

tolerant to it (although this idea does not take into account mechanisms for tolerance such as clonal energy in the periphery — Section 5.2.5). For example, a 'new' V gene might be expressed in a newly-formed B cell, a 'new' VDJ combination might have been constructed, or somatic mutation might have occurred (Section 7.2). It has been argued that since the individual will already be tolerant to the rest of the molecule, the immune response to a newly-formed antibody is limited to the variable region determinants alone, the idiotopes. In this sense an antibody may be viewed by the immune system simply as another antigen (although it is important to remember that, unlike other antigens, antibodies have Fc portions with recognition and regulatory functions within the immune system; see Section 8.4.2).

An immune response that is produced by an individual against idiotypes of their *own* antibody molecules is called an **auto-anti-idiotypic response**. The antibodies that are elicited are termed **auto-anti-idiotypic antibodies**. Similar anti-idiotypic responses may also be made by genetically-different individuals of the same or a different species, although in this case the majority of lymphocytes responding will be specific for other antigenic determinants (i.e. allotypes or isotypes respectively).

The particular set of idiotopes recognized in an anti-idiotypic response, which defines the idiotype of any given antibody, depends on the immune response in question, and thus the definition of an idiotype is not an 'absolute'.

For example, in one particular study it was found that the spectrum of anti-idiotypic antibodies prepared against a certain myeloma protein (an antibody produced by a clonal B cell tumour) differed depending on whether they were raised in mice of the same (isogeneic) strain as the myeloma or a different (allogeneic) strain: in the isogeneic situation, the anti-idiotypic antibodies did *not* cross-react with a different myeloma protein, whereas they did in the allogeneic situation.

The spectrum of anti-idiotypic antibodies raised in a different species is also likely to be different, but nevertheless an identical or very similar region of the antibody can be recognized, sometimes including the very same determinants (or idiotopes) that are associated with the antigen-binding site.

Production of anti-idiotypic antibodies

One approach to making an anti-idiotypic antibody is to immunize an animal against an antigen, purify the antigen-specific antibodies that are produced, and use these to immunize another animal (Fig.11.5). The second animal may then produce anti-idiotypic antibodies which can in turn be purified.

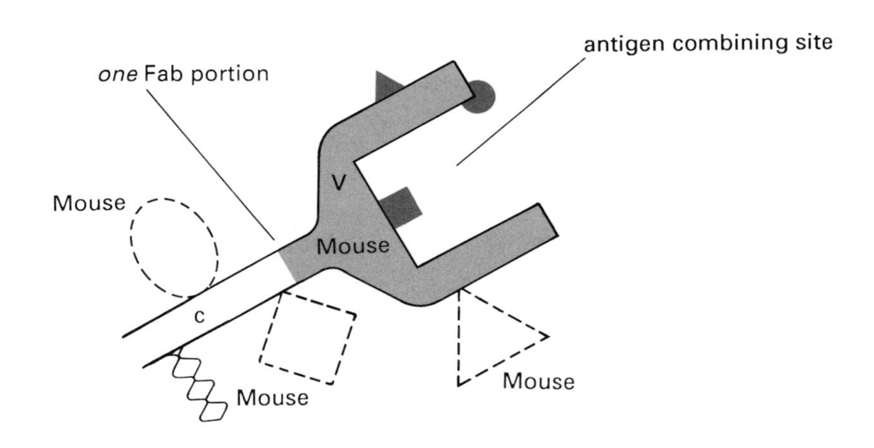

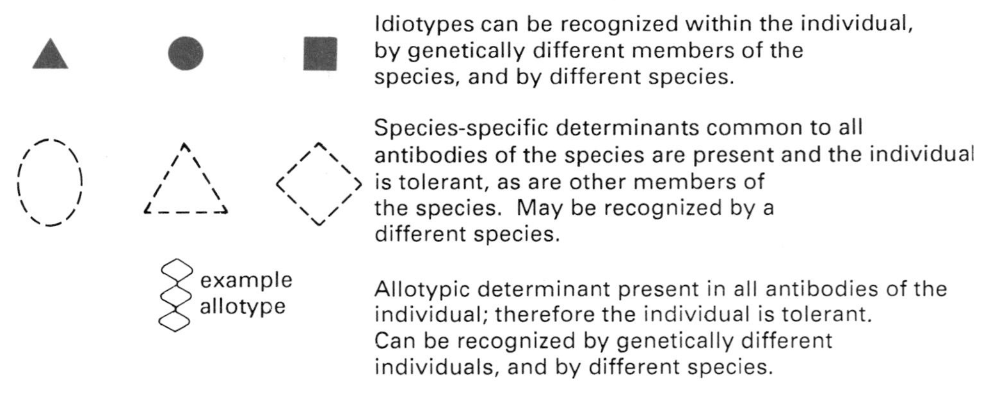

Idiotypes can be recognized within the individual, by genetically different members of the species, and by different species.
Species-specific determinants common to all antibodies of the species are present and the individual is tolerant, as are other members of the species. May be recognized by a different species.
example allotype
Allotypic determinant present in all antibodies of the individual; therefore the individual is tolerant. Can be recognized by genetically different individuals, and by different species.

Fig.11.4 Idiotypes are not the only antigenic determinants on an antibody

Normally, the serum of an animal immunized against a complex antigen contains a heterogeneous group of antibodies. However, if antigens with a relatively simple structure are used, only a limited number of B cell clones will respond and a limited number of different antibodies are present in the serum. One example is when mice of the A/J strain are immunized against the bacterial component, streptococcal A carbohydrate. Antibodies specific for streptococcal A carbohydrate can be purified from the immune serum by affinity chromatography on columns containing the antigen (see Appendix), and used to immunize another animal to produce an anti-idiotypic antiserum. To generate a good anti-idiotypic response it is often necessary to couple the first antibody to a 'carrier', such as KLH, or to use other immunization strategies. (The necessity for these procedures has, in fact, led some investigators to question the physiological relevence of anti-idiotypic antibodies in normal animals.) The anti-idiotypic antibodies can in turn be purified by affinity chromatography on columns containing antibodies that express the idiotype (e.g. Fig.11.6). Alternatively, monoclonal anti-idiotypic antibodies can be produced by making hybridomas from B cells of the immunized animal (see Appendix).

Another way to produce anti-idiotypic antibodies is to immunize against a homogeneous myeloma protein or a monoclonal antibody expressing the idiotype. Sometimes the antigen specificity of the myeloma protein is known. For example, a particular mouse myeloma protein called **A5A** recognizes the streptococcal A carbohydrate mentioned above. Anti-idiotypic antibodies against this myeloma protein were found to recognize the majority of antibodies produced in the normal response of A/J mice against this antigen. For this reason A/J mice are said to produce a predominantly **A5A idiotypic response** to streptococcal A carbohydrate. Many similar examples are known (Table 11.1, and see below).

Anti-idiotypic antibodies produced within the same, or a different, species from which the primary antibody was derived are known as **homologous** and **heterologous** anti-idiotypic antibodies respectively.

If a rabbit, for example, is immunized against a mouse antibody like the A5A myeloma protein, much of the response will be directed to species-specific determinants present on mouse antibodies in general, as well as to the idiotype. However, it is possible to absorb the rabbit antiserum with normal mouse antibodies (or other myeloma proteins) to remove reactivity to these common determinants, but to retain the anti-idiotypic activity specific for the myeloma protein used for immunization (Fig.11.6). If the immunization is between different strains of the same species, genetic differences such as allotypes may be detected, and to overcome this, homologous anti-idiotypic antibodies are generally produced in allotype-matched animals.

Nomenclature. Antigens induce antibodies expressing particular idiotypes. These in turn can stimulate the production of anti-idiotypic antibodies. In their own turn, the latter may induce anti-(anti-idiotypic) antibodies, and so forth. To avoid this cumbersome terminology, an alternative nomenclature can be used (Fig.11.7). The antigen is said to induce **ab1 antibodies** (bearing the idiotype), which in turn can elicit **ab2 antibodies** (the anti-idiotype), followed by **ab3 antibodies** (the anti-anti-idiotype), and so forth. This procedure, in which antibodies are purified and used in turn to raise other antibodies, is sometimes referred to as an **immunization cascade**. As will become clear below, a variety of terms are used to describe different types of idiotype. Although this is sometimes necessary, immunologists do have a

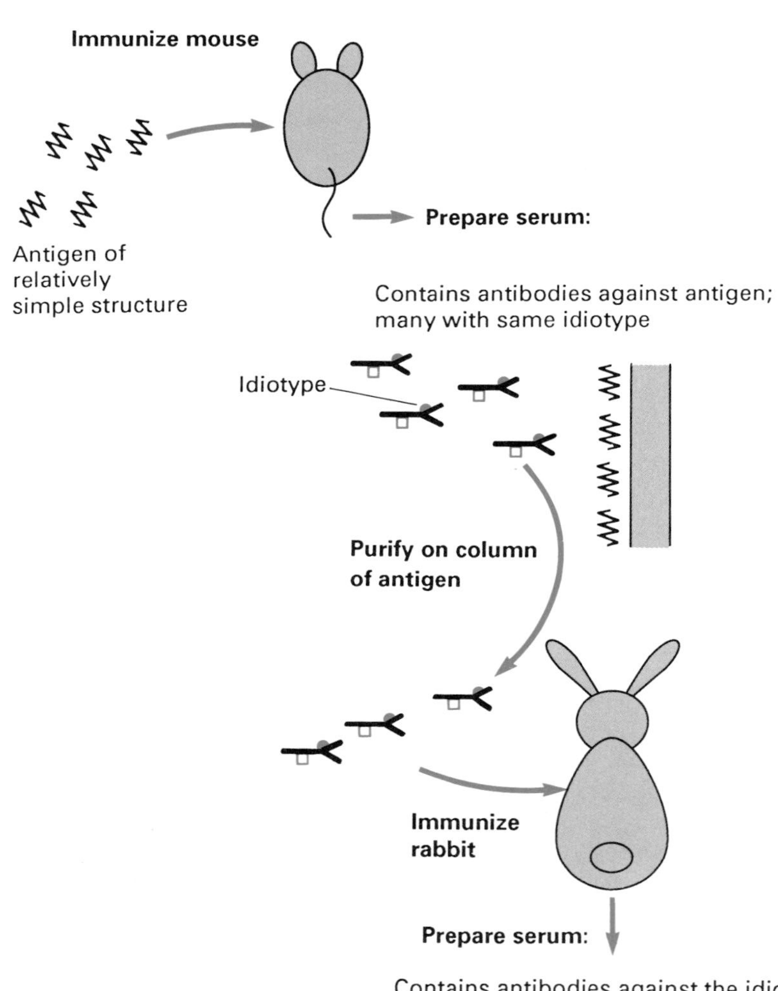

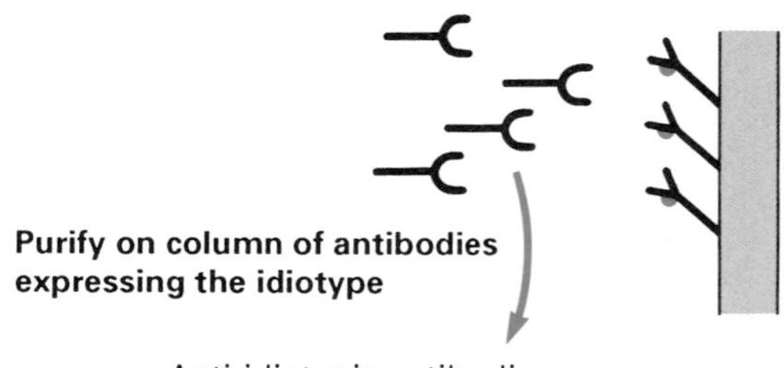

Fig.11.5 The principle of production of an anti-idiotypic antibody

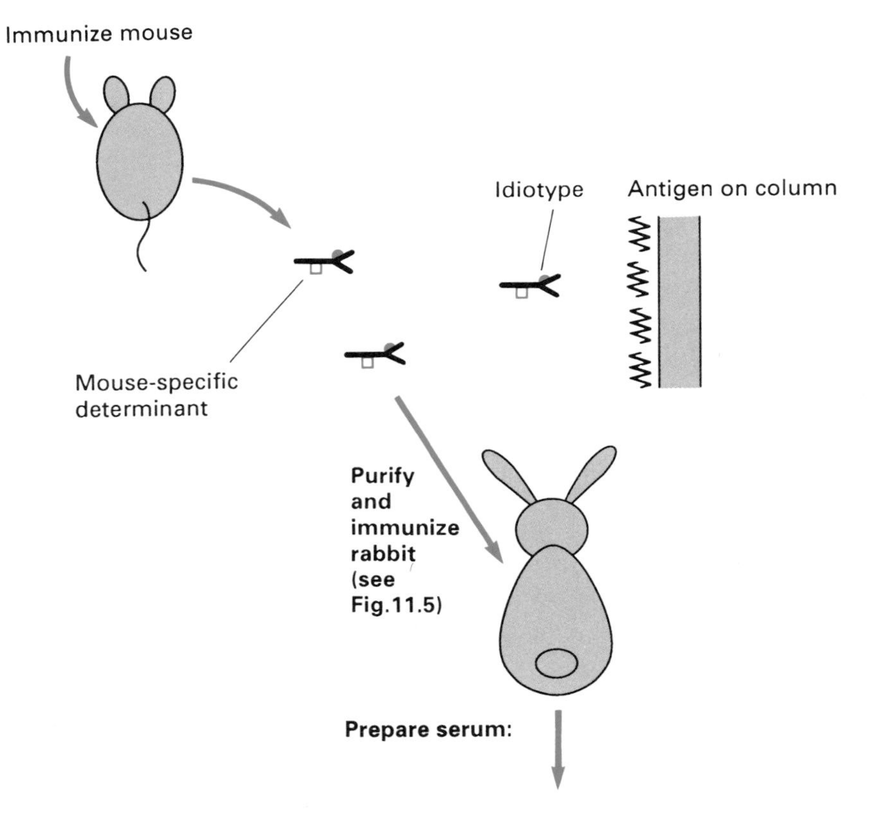
Immunize mouse
Idiotype
Antigen on column
Mouse-specific determinant
Purify and immunize rabbit (see Fig.11.5)
Prepare serum:

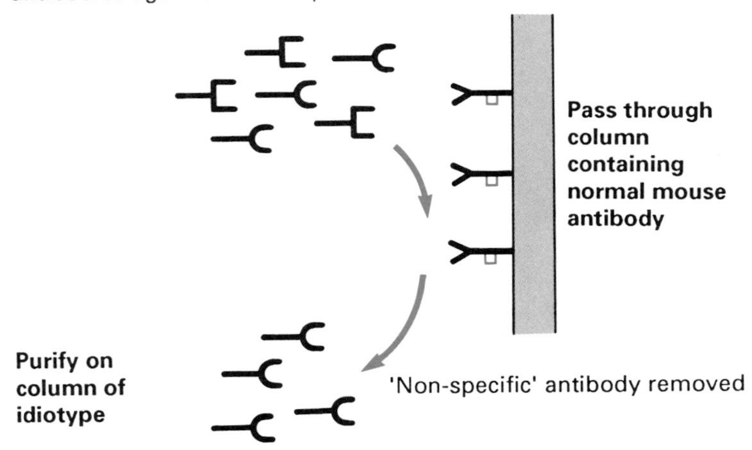
Contains antibodies against idiotype, and also antibodies against mouse-specific determinants
Pass through column containing normal mouse antibody
Purify on column of idiotype
'Non-specific' antibody removed

Fig.11.6 Removal of antibodies not directed against idiotypes

peculiar penchant for inventing new jargon. However, readers who are not already acquainted with these terms are urged to persevere, because the underlying concepts are often quite straightforward.

Combining site associated and non-site associated idiotypes

The antibody combining site for antigen is called the **paratope**. Thus the *paratope* of an antibody binds to an *epitope* on the antigen. Idiotopes are

Table 11.1 Examples of idiotypes associated with particular antibody responses of normal mice

Antigen	Idiotype	Comment
Carbohydrates		
α-(1→3) dextran	IdI(M104) IdI(J558)	Designated according to unique Ids of two particular myeloma proteins (M104 and J558): these and many anti-dextran abs produced, e.g., in BALB/c also share a cross-reactive id: IdX
β-(2→1) fructosan	E109IdX	
levan	A48Id	Myeloma protein = ABPC 48
Strep A CHO	A5AId	The predominant id produced in e.g. A/J mice in response to this antigen
Haptens:		
p-azophenyl-arsonate (ARS)	ArsIdX	A cross-reactive idiotype (also termed CRI_{ARS}) expressed by some anti-ARS abs, e.g. in normal A/J but not BALB/c mice
hydroxy-3-nitro-iodophenylacetyl (NIP)	Ac38, Ac146	Examples, respectively, of Ids located outside or within the antigen-binding sites of certain anti-NIP abs
phosphocholine (PC)	T151dX	The predominant id found, e.g., in BALB/c anti-PC abs, but other ids such as M511 and M603 are also expressed
trinitrophenyl (TNP)	460Id	Many abs and T cells produced in BALB/c anti-TNP responses express this dominant id

situated both within the paratope and outside of it, and anti-idiotopic antibodies (i.e. ab2; see above) can have different properties depending on precisely where the specific idiotope (on ab1) is located. The main difference between ab2 antibodies is whether or not they bind close to, or within, the paratope and inhibit binding of the antigen to the antibody (ab1). (Fig.11.8). In the case of monoclonal ab2 antibodies prepared against the idiotype of an antibody, ab1, which in turn recognizes a particular epitope like a hapten, there are three main possibilities.

1. Some ab2 antibodies recognize idiotopes that are not associated with the combining site, and which do not inhibit binding of the hapten to ab1. These have been termed ab2α antibodies.
2. Other ab2 antibodies recognize idiotopes close to the combining site, and therefore inhibit binding of the hapten, for example by steric hindrance. These have been termed ab2γ antibodies.
3. A third type of ab2 antibody recognizes idiotopes that actually *form* parts of the antigen combining site of ab1. Such ab2 antibodies would obviously compete for and often inhibit binding of the hapten, and they have been termed ab2β antibodies. The remarkable thing about ab2β antibodies is that because their binding site is at least in part complementary to the ab1 combining site, *they may resemble the actual antigenic determinant*

recognized by ab1, and therefore mimic the antigen. Ab2β is said to represent (or carry) the **internal image** of the antigen for which ab1 is specific. Ab2β antibodies have also been termed homobodies.

In summary, idiotopes can be classified according to whether they are recognized by ab2α, β, or γ antibodies. This classification depends on whether ab2 inhibits binding of the antigen to ab1 and/or resembles the so-called internal image of the antigen. Other categories of idiotypes have been proposed according to their *functional* properties (Ertl and Bona 1988), and alternate classifications should become possible as their *structural* correlates are defined.

Recurrent and private idiotypes

An alternative classification leads to the definition of recurrent and private idiotypes, but this is based simply on the *frequency* with which a given idiotype is detected in antibody responses (Fig.11.9).

Some idiotypes are regularly found (i.e. they 'recur') in certain immune responses. An example of a *recurrent* idiotype is the A5A idiotype expressed by antibodies produced during immune responses to streptococcal A carbohydrate in A/J mice (see above). The same recurrent idiotype can even be expressed by antibodies of a different species. Recurrent idiotypes have also been termed public, major, or cross-reactive idiotypes (from which we get the 'CRI' or 'IdX' terminology; Table 11.1). In contrast to recurrent idiotypes, *private* or minor idiotypes are expressed irregularly in some individuals but not in others, and may or may not be expressed in genetically identical individuals. The total repertoire of idiotypes in the immune system, according to this classification, is the sum of the recurrent and the private idiotypes.

Recurrent idiotypes that are produced by most individuals of a particular strain of a given species can be encoded by a single VJ or VDJ combination of germline gene segments. In other cases small groups of related V gene segments are used, and these can include some of their somatically-mutated progeny. For example, the response of mice carrying the Igh^b antibody locus to the hapten NP is dominated by a recurrent idiotype called the **NP^b** idiotype,

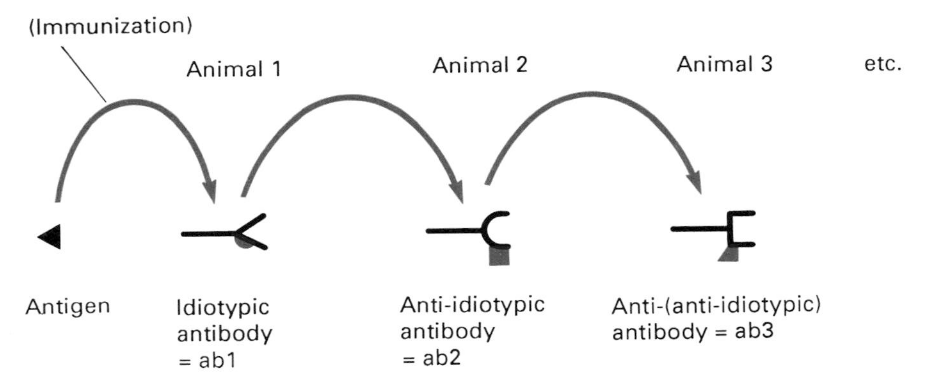

Fig.11.7 Immunization cascade: nomenclature

due to the expression of a few closely related VDJ gene segments. In contrast, some private idiotypes are generated by somatic mutation of Ig genes of a certain individual, or they can be formed by unique rearrangements of VDJ segments.

The genetic basis for expression of several other idiotypes has also been examined. For example, the **T15** idiotype is dominantly expressed by BALB/c mice during responses to phosphocholine (PC). This idiotypic response is

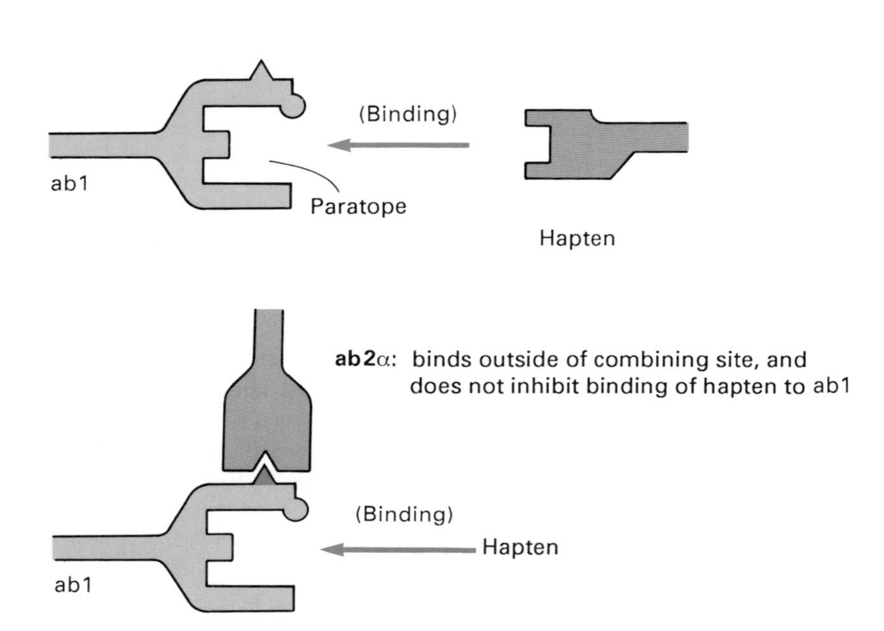

ab2γ: binds close to combining site, and inhibits binding of hapten to ab1

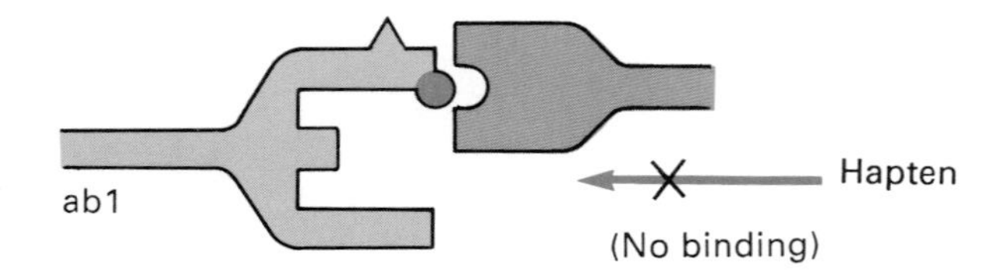

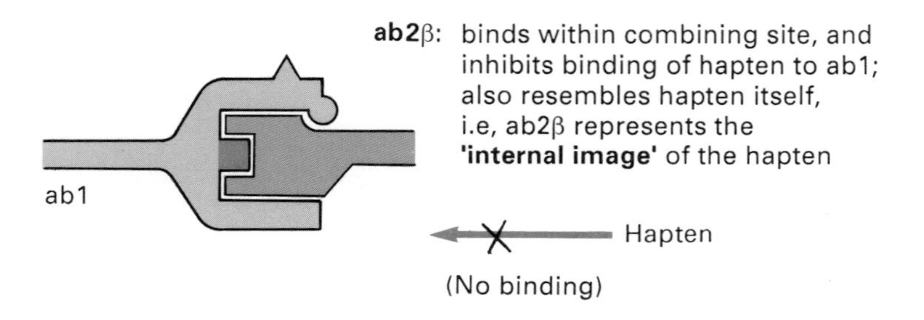

Fig.11.8 Classification of antibodies against idiotypes: combining site associated and non-site associated idiotypes

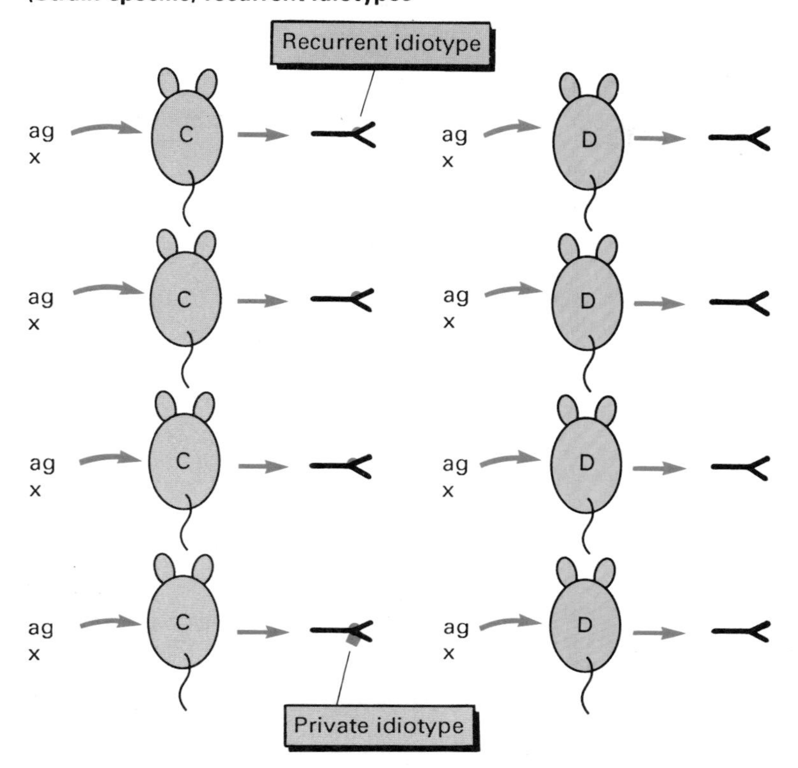

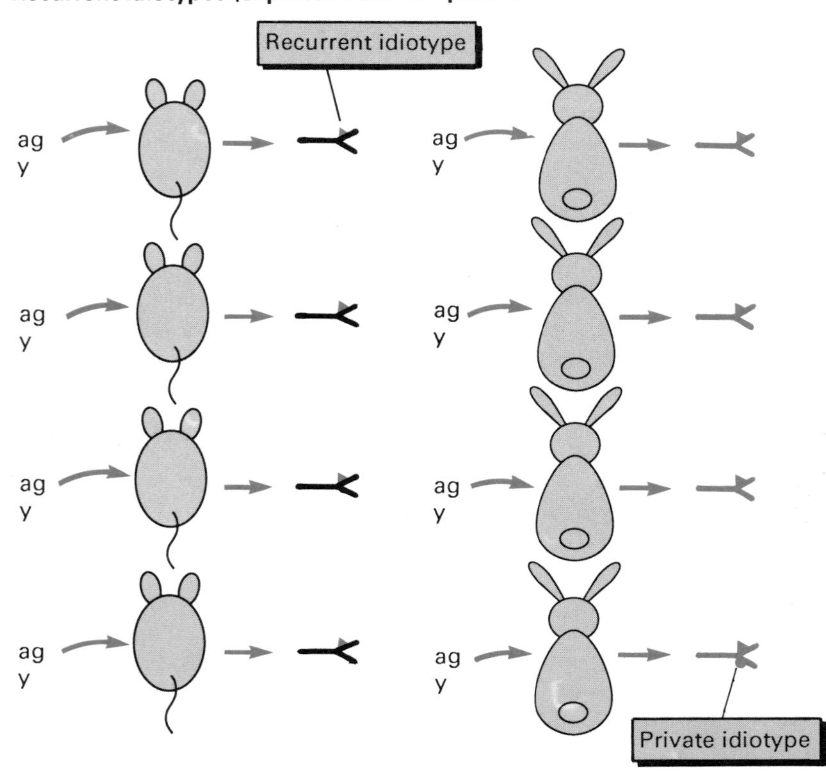

Fig.11.9 Recurrent and private idiotypes

due to the useage of a single germline V_H segment and a limited number of J_H segments, together with pairing of a heavy chain to one of only three $V\kappa$ light chains. The presence of somatic mutations has been demonstrated in both chains, and N-junctional diversity (as well as gene conversion) has also been implicated in generation of the repertoire of $T15^+$ antibodies against PC (Perlmutter *et al.* 1984).

Some recurrent idiotypes might be expressed because the conformations of different antigen-combining sites are similar, even though they are constructed from different VJ or VDJ segments. This is particularly so where recurrent idiotypes are expressed in different species, since these could be more similar in *conformation* than in primary sequence (although sometimes V_H segments are conserved between species). A special case of this is presumably where the anti-idiotype represents the internal image of the antigen, as for example in the case of certain anti-viral antibodies, which can be idiotypically similar in different species.

Parallel sets

Some idiotypes are associated with a particular antigenic specificity (for example many antibodies with the A5A idiotype are specific for streptococcal A carbohydrate) but this is not always so, because idiotypic determinants can be shared by antibodies with different or even unknown specificities (Oudin and Cazenave 1971). These are called parallel sets or non-specific immunoglobulins, and they seem to be generated spontaneously in many immune responses and in autoimmune situations (Migliorini *et al.* 1987) (Fig.11.10). Parallel sets may arise because a particular subgroup of V genes is expressed, the members of which are structurally related but give rise to different specificities. An example is the T15 group of V_H genes, only *one* of which encodes the **T15 idiotype** expressed by many antibodies specific for phosphocholine (see above), while other T15 V genes can be expressed by antibodies specific for unrelated antigens. The same is true for the J558 V_H family, the **J558** idiotype being expressed by some anti-dextran antibodies (Victor-Kobrin *et al.* 1985*a*) (Table 11.1).

The other side of the coin is that one can almost always find antibodies that are specific for a particular antigen but which do *not* express the same idiotype. Some of these may be derived by somatic mutation. It has been shown that an idiotopic determinant can be lost by a substitution of as little as one amino acid residue, although this alteration may have no detectable effect on the antigen specificity (Radbruch *et al.* 1985). Alternatively, these antibodies may be similar in conformation rather than in amino acid sequence (see above).

Same specificity; different idiotypes

Parallel sets:
Different specificity; same idiotypes

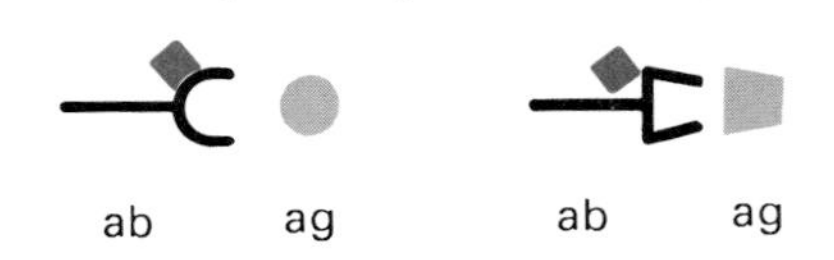

Fig.11.10 Parallel sets

11.1.2 Idiotypic networks

Idiotypes as the internal images of epitopes

According to the **network theory**, originally proposed by Jerne (1974), the immune system can be thought of as a 'network of interacting idiotypes' (Fig.11.11). Within this network, each antibody possesses an antigen-combining site (paratope) and expresses a particular set of idiotypic determinants (idiotopes); these enable the antibody *both* to interact with an antigen and to be an antigen itself, respectively. It is further envisaged that for every antibody in the immune system there is a complementary antibody, or set of antibodies, with their own paratope(s) and idiotopes. By analogy, for every ab1 there is a set of ab2; for each ab2 there is a set of ab3; and so forth. This leads to the idea that antibodies and B cells form a dynamic entity, or network, that contains mutually-interacting components which in principle could regulate and control the activities of each other.

To this general concept can be added the idea of internal images (Section 10.1.1), in that an anti-idiotope is specific for another receptor within the immune system (i.e. the idiotope) but it could potentially recognize some other determinant in the external world (e.g. an epitope of a foreign antigen). *The essential point is that the entire repertoire of idiotopes could represent the internal images of most, if not all, naturally-occurring epitopes.* The remarkable corollary to this hypothesis is that because each idiotope elicits a complementary anti-idiotope, a receptor for an epitope on a foreign antigen could be formed *before* the antigen is actually encountered. And we know that the immune system works by selection from a vast pool of 'ready-made' receptors, in accordance with the idea of clonal selection (as well as by selection of newly-formed antibodies during affinity maturation) (Section 1.1.3).

To generalize this, consider an antibody (ab1) and a complementary anti-idiotypic antibody (ab2) within a given individual (Fig.11.11). Ab1 is potentially capable of binding a particular epitope (ep1). Ab2, the (auto)-anti-idiotypic antibody which binds to the idiotope of ab1 (id1), can also use its combining site to recognize another epitope (ep2) in the external world. Of course, in this sequence, ab1 and ab2 are experimentally defined simply according to the order of immunization. However, if ab2 had been used for immunization it might equally have induced ab1, as well perhaps as ab3 directed against ab2. It is important to emphasize that the idiotype and anti-idiotype are operationally defined and that the immune system may not discriminate between them (but see regulatory idiotypes; below).

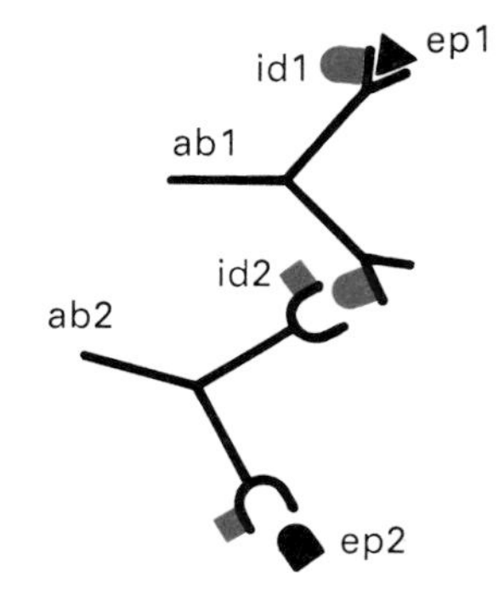

ab = antibody; id = idiotope
ep = epitope of antigen in external world

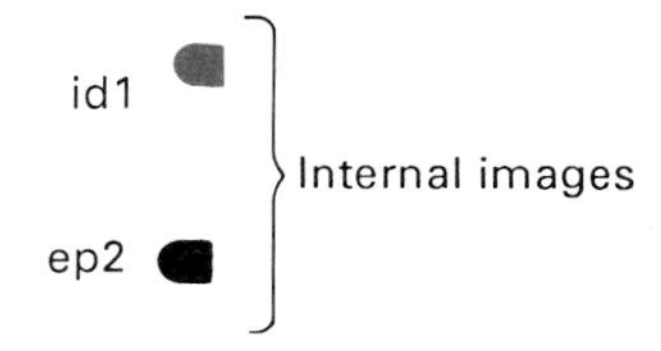

Fig.11.11 Idiotypic networks (see text)

Evidence for idiotopes as internal images

Some of the most remarkable evidence that idiotopes can actually represent the internal images of antigenic determinants in the external world, has been obtained using anti-idiotypic antibodies (ab2) prepared against hormone-specific antibodies (ab1) (Fig.11.12). Several groups of investigators have found that subsets of ab2 antibodies can mediate hormone-specific effects in the absence of the hormone itself.

For example, it was found that a rabbit ab2 antibody against a rat anti-insulin antibody could bind to the insulin receptor and stimulate glycolysis in the cells expressing the receptor, while another ab2 antibody against an anti-alprenolol antibody bound to the beta-adrenergic receptor and caused an increase in cellular adenylate cyclase activity. Some ab2 antibodies can also inhibit binding of ligands, such as the f-Met-leu-phe peptide, to their receptors.

In some cases these antibodies may represent the internal image of the antigen. For example, in a different system, an ab2 antibody specific for the haemagglutinin molecule of a retrovirus, inhibited binding of the virus to its cellular receptor; the V region of this antibody was found to contain a sequence that closely resembles a region within the haemagglutinin molecule (Bruck *et al.* 1986).

Anti-idiotypic antibodies have been used therapeutically. By virtue of their ability to bind the cellular receptors for viruses, or to mimic them, some of these antibodies can neutralize viruses and reduce or prevent infection (Gaulton and Greene 1989). They have also been used to induce immunity to other pathogens such as bacteria (McNamara *et al.* 1984) and parasites (Grzych *et al.* 1985), and their capacity to act as vaccines for certain types of tumours is being actively investigated (Raychaudhuri *et al.* 1986). This is one of the most exciting prospects for these reagents, since in principle it allows an animal to be immunized against an antigen without exposure to the antigen itself. The advantages of this approach for vaccination against pathogenic organisms is obvious. It is clearly advantageous to be able to induce immunity to viruses and other pathogens by administering anti-idiotypic antibodies prepared in a different species.

For example it has been shown that an ab2 antiserum specific for antibodies against the tobacco mosaic virus (TMV) can induce neutralizing ab3 anti-TMV antibodies, resembling ab1, in a different species. In another case, human ab1 antibodies specific for hepatitis B have been used to raise ab2 antibodies in rabbits, which in turn were used successfully to vaccinate chimpanzees against this virus. This suggests that the reciprocal route might be achievable for human vaccination.

Open and closed networks

The concept of idiotypic networks owes much to the results obtained from the experimental immunization cascade (Fig.11.7). Such an idiotypic cascade

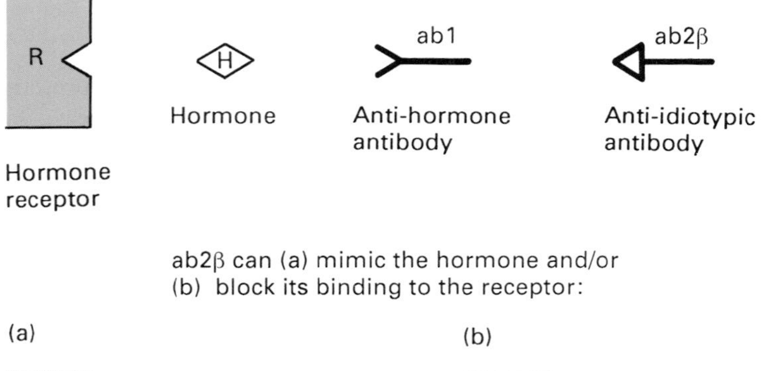

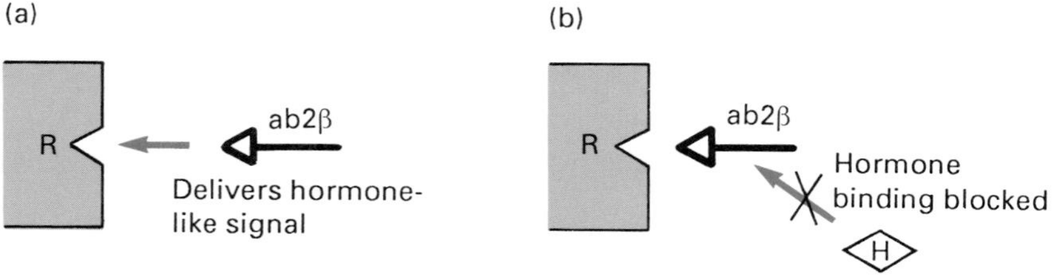

Fig.11.12 Idiotypes as internal images of epitopes as revealed by binding of ab2 to hormone receptors

Fig.11.13 An open-ended view of idiotypic networks

has been demonstrated in mice of the same strain or in allotype-matched rabbits. However, we should stress that these are experimental approaches that are frequently induced by non-physiological means (e.g. modification of antibodies and the use of adjuvants during immunization), and they are not necessarily representative of the (auto-) response that may occur within a given individual. Nevertheless, the spontaneous formation of ab2 antibodies after immunization with antigen, or of ab3 after immunization with ab1, has been demonstrated, suggesting that this might be part of a physiological network (for example, Schechter *et al.* 1982). But can one go on generating successive antibodies indefinitely (ab1, ab2, ab3, . . .), each with their own particular idiotypes?

One extreme view of the network is that it is 'open-ended' and that every antibody induces another set of antibodies each with their own different idiotopes (Fig.11.13). The network as originally proposed was also *vectorial* in nature, since it proceeded in one direction but not in the other. It was hypothesized that ab1 *stimulated* ab2 production, but that ab2 *suppressed* ab1; in other words, an interaction via the paratope was stimulatory, while an interaction via the idiotopes was suppressive (Fig.11.14). (Note that the regulatory function of antibody Fc receptors could provide a precedent for such a scheme; see Section 8.4.2). These interactions were proposed to regulate the concentration of antibodies expressing particular sets of idiotypes that were produced during an immune response. In this model, the role of antigen during immunization was to *perturb* the pre-existing equilibrium within the network.

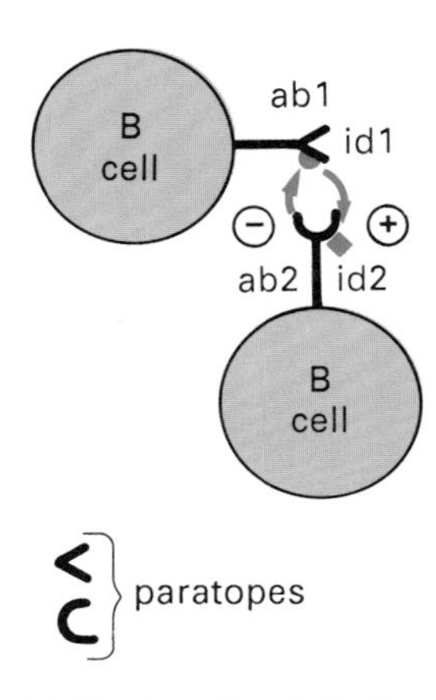

Fig.11.14 The hypothesis that interaction between receptors via paratopes is stimulatory, but via idiotopes is suppressive; an equilibrium is thus established, which is perturbed in the presence of antigen

There are problems with both the vectorial and open-ended network hypotheses. In particular, there is evidence that diversity does *not* increase along an immunization pathway, at least not to the degree suggested. This implies that idiotypic networks may be more *circular* in nature (Fig.11.15). Consider the extreme and idealized example of an internal image. An epitope on an antigen might elicit an antibody ab1 which in turn could induce ab2β, the internal image of the epitope. In this case, the relevant part of ab2β might look just like the antigen itself to the immune system (and we have already

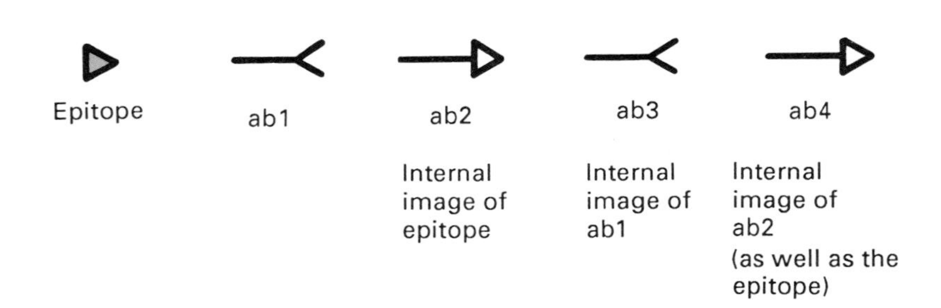

Fig.11.15 A more circular or 'reflective' view of idiotypic networks

discussed some of the instances in which ab2 can behave like antigen; see above). However, the next antibody in the chain, ab3, could represent the internal image of ab1, and in the same way ab4 could resemble ab2, and so forth: hence the circularity. We have used internal images simply as an example, but this circularity may not be confined just to this situation. Such three- and four-membered networks of idiotypes have indeed been demonstrated.

For example, one particular study examined the antibodies formed in an immunization cascade starting with an ab1 antibody, a mouse myeloma protein, ABPC 48, specific for a bacterial levan (Bona *et al.* 1981). Ab4 was found to resemble ab2 in that it could bind to ab1; they did not, however, appear to be identical, since their affinities were different. In addition, ab1 seemed to be similar to ab3 since it could inhibit the binding of ab3 to ab4. A further study used an ab1 antibody against insulin, and it was found that not only could ab2 stimulate signalling via the insulin receptor (see above), but that ab3 resembled ab1 in its ability to bind to the hormone itself. Even so, there are reports that parallel sets (Section 10.1.1) can be generated in similar immunization cascades, in that ab3 can be idiotypically similar to ab1 but it may *not* bind to the original antigen.

In summary, there is evidence that the idiotypic network is not completely open-ended but that it may be more circular or reflective. One problem is that, very often, only a small proportion of antibodies within ab2, ab3, and so on, behaves as we have just discussed; much of the antibody produced can be idiotypically unrelated.

We should mention that there is a report of perhaps the most extreme form of circularity: an antibody with a paratope that binds the hapten for which it is specific (phosphocholine) and a complementary idiotope within the same antibody molecule (Kang *et al.* 1988). This means the antibody, which was termed an autobody, can bind to itself.

Regulatory idiotopes

Paul and Bona (1982) proposed the concept of 'regulatory idiotopes' largely to explain the finding that in some immunization cascades (Fig.11.7), ab3 can resemble ab1, ab4 can resemble ab2, and so forth (Fig.11.16). What they suggested was that antibodies to conventional epitopes may have several idiotopes (detected experimentally), but that only some of these idiotopes can actually elicit anti-idiotypic responses within the individual or in syngeneic animals. These were thus termed regulatory idiotopes.

According to the idea of regulatory idiotopes, the ab2 antibodies described above would primarily be directed against regulatory idiotopes on ab1. However, these ab2 antibodies themselves are assumed *not* to possess regulatory idiotopes, and when ab2 antibodies are used for immunization it was proposed that they elicit ab3 with regulatory idiotopes similar to ab1 (rather than being directed against idiotopes on ab2); ab4 in turn would be similar to ab2, lacking regulatory idiotopes. In effect, this means that immunization with an antigen induces ab1 antibodies that carry regulatory idiotopes and

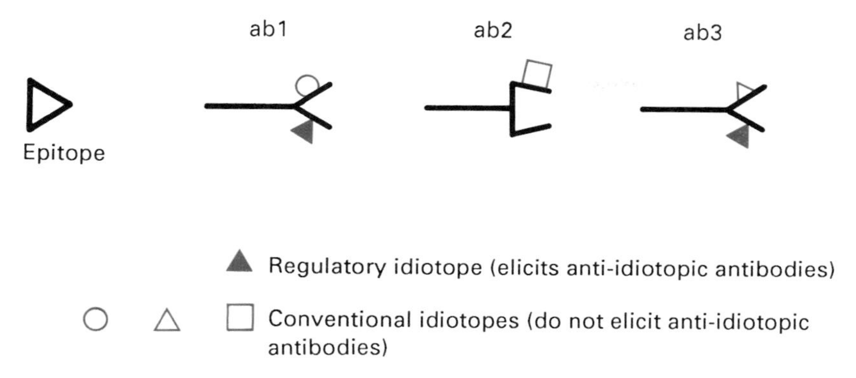

Fig.11.16 Regulatory idiotypes

these stimulate ab2 antibodies against these idiotopes. Alternatively, one could use ab2 to induce antigen-specific antibodies that carried the idiotope (or, if the same idiotope was also expressed by unrelated antibodies, one would induce a parellel set; see for example, Victor-Kobrin *et al.* 1985*b*). Possession of a regulatory idiotope by a large proportion of antigen-specific antibodies would, it has been argued, allow these antibodies to be regulated together.

One explanation offered for this hypothesis is that the germline encodes regulatory idiotopes which are 'designed' to be targets for regulatory mechanisms. Other idiotopes might arise somatically, either as a result of imprecise joining of V, (D), J gene segments or somatic mutation (Section 7.2.3), and could be outside of this control. The concept of regulatory idiotopes is also consistent with another proposal for 'germline circles', in which the germline is thought to encode 'families of idiotypically-interacting antibodies'. In essence, the suggestion is that antibody genes may have evolved so that they *already* contain the information to generate most paratopes and complementary idiotopes, within the germline of any individual. As a crude example, for which we must stress there is no direct evidence, a heavy chain V gene might encode a binding site (paratope) for an idiotope that is encoded by a D gene segment in the same genome. If this were so then the basic information for components of the idiotypic network would be genetically predetermined in any individual.

11.1.3 T cell idiotopes

Idiotypic connectivity

So far, (Sections 11.1.1, 11.1.2) we have considered idiotypes on antibodies, but recurrent idiotypes can also be expressed by T cell receptors (Fig.11.17). This was shown by early work that tried to define the nature of the T cell receptor. Some anti-idiotypic antibodies prepared against idiotypes on B cells apparently reacted with idiotypes on T cells, and a limited amount of biochemical data supported this idea (Martinez *et al.* 1986). At the time, this was taken by some investigators to imply that the variable regions of the T cell receptor were encoded by Ig genes. Since we now know that antibodies

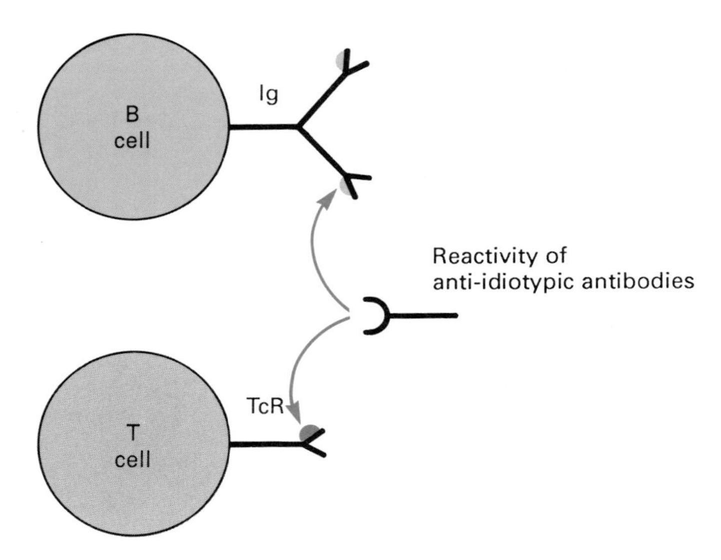

Fig.11.17 T cell receptors can express recurrent antibody idiotypes

and T cell receptors are encoded by completely different genes, it seems likely that the cross-reactive recognition of B cells and T cells by anti-idiotypic antibodies was due to conformational similarity between these molecules. That is, parts of membrane-bound Ig molecules on B cells and T cell receptors may have a similar 'shape' even though they are formed from different amino acid sequences. This should not be too surprising, since we have already given examples of anti-idiotypic antibodies that recognize hormone receptors (Section 11.1.2). In one sense the hormone receptor can be thought of as expressing an idiotope, although it is conventionally referred to as an epitope because it is not present on an antibody. Clearly, hormone receptors and antibodies are encoded by entirely different genes, yet there may be 'cross-reactive' recognition of both by the very same antibodies.

Because antibodies and T cell receptors may express similar idiotypes, it is possible that T cells could be linked to the B cell idiotypic network. This phenomenon, as well as idiotypic interactions between B cells, is referred to as **connectivity** in the network. For example, anti-clonotypic antibodies against T cell antigen receptors can inhibit or stimulate T cell responses (Section 4.2.1) and such antibodies, were they to occur naturally, might modulate T cell responses. Such connectivity has been shown to be present in unimmunized mice (Couthino *et al.* 1987, and see below). Ab2 (anti-idiotypic) antibodies can also 'prime' T cells that provide help for B cell antibody responses (see below). Ignoring for a moment MHC restriction, T cell receptors might be able to form idiotypic-anti-idiotypic networks, exemplified by the interactions that are thought to occur between suppressor T cell subsets, which are discussed in Section 11.2.2. Idiotype-specific helper T cells are another example, because they seem to recognize idiotypes on antigen-specific helper cells in some B cell responses (see below).

Idiotype-specific helper T cells

'Conventional' antigen-specific helper T cells recognize peptides bound to MHC molecules (Section 3.2.1) and are required for many antigen-specific B cell responses (Section 8.3.1). There also seem to be T cells that may be required to help B cells produce a particular isotype, allotype, or idiotype on a specific antibody, and these helper cells may not be MHC-restricted in the classical sense. Precisely what idiotype-specific helper cells recognize is not clear. There are reports they can bind directly idiotype-coated tissue culture plates and that they are depleted from populations enriched for MHC-restricted helper cells. Some investigators have suggested these T cells might be composed of two or three interacting subsets, in which case there is a parallel with suppressor T cell subsets; Section 11.2.2.

In several systems, antibody production by B cells seems to require co-operation between a conventional antigen-specific, MHC-restricted T helper cell and an idiotype-specific helper T cell (Fig.11.18).

Thus, B cells make antibodies against phosphocholine that express the T15 idiotype (Section 10.1.1) apparently only in the presence of *both* types of helper cell, although whether these T cells are required at the *same* time or function *sequentially* is not fully established. Strains of mice that are genetically incapable of making antibodies expressing this idiotype do not have T15 idiotype-specific helper T cells (Bottomly and Mosier 1979). Another example is the response to lysozyme (HEL) in which a particular (i.e. dominant) idiotype is expressed in the secondary but not the primary antibody response (Sercarz and Metzger 1980). Here it seems that conventional helper T cells help the B cells make anti-HEL antibodies, but the appearance of the dominant idiotype is controlled by idiotype-specific helper cells which act later.

However, in other situations, conventional carrier-specific, MHC-restricted helper T cells can be sufficient, alone, to support idiotype-dominant responses (Hathcock *et al.* 1986).

B cell idiotypes and the T cell repertoire

Some investigators have proposed that the *generation* of idiotype-specific helper T cells, and perhaps isotype and allotype-specific helper T cells as well, is *dependent* on the presence of the idiotype-bearing antibody *during* T cell development. This brings us to the controversial area of whether, during ontogeny, T cells can be selected to recognize B cell idiotypes, as well as MHC molecules (Section 5.3.1).

In one study it was found that if mice that normally produce the T15 anti-phosphocholine antibody response are treated with anti-μ antibodies from the time of birth to eliminate their B cells, they no longer have T15 idiotype-specific helper T cells and they cannot make this response, even though they have normal levels of 'conventional' helper cells and can make other antibodies against phosphocholine (Bottomly *et al.* 1980). These results have been interpreted as showing that the idiotype (T15) must be present on B cells *before* T15-specific helper cells can develop.

In another study, the idiotypes shared between T cell receptors and antibodies were examined after anti-μ treatment of mice (Martinez *et al.* 1985). B cells and antibodies produced in Balb/c mice in response to the hapten TNP express an idiotope which is also present on the TNP-specific myeloma protein MOPC 460 (Fig.11.19). Up to two-thirds of the helper T cells that participate in the TNP response also express this idiotope (note that this is an example of a strain-specific, recurrent idiotope). Mice treated with anti-μ early in life no longer expressed this idiotope on their T cells, although it was expressed if the treatment was started after 3 or more weeks of neonatal development. These results again suggest that B cells are required for T cells to express a particular idiotype, especially early in life. Whether this is a general phenomenon is not known, although we should note that the evidence that anti-idiotypic antibodies against idiotypes on B cell Ig actually react with T cell receptors, rather than cross-react with some other T cell component, is often circumstantial.

The observations above (see box) and other findings have led some investigators to suggest that the development of idiotype-specific helper T cells, and the expression of certain idiotypes on T cells, depends on an interaction between T cells and the idiotypes of B cell Ig molecules. If this occurs only early in life then it explains the paradoxical results of other investigators who, for example, constructed bone marrow radiation chimeras (Section 5.3.2) from mouse strains expressing different idiotypes, and found that

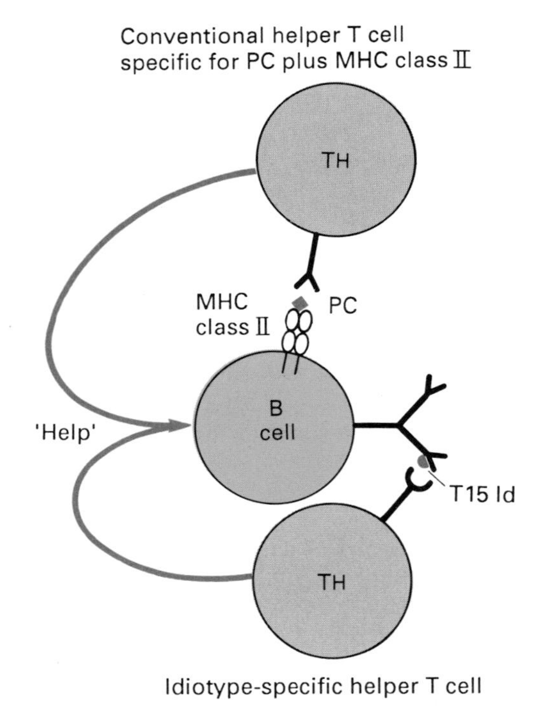

Fig.11.18 Proposed co-operation between two types of helper T cells for a T15-idiotype antibody response against phosphorylcholine

expression of T cell idiotypes seems to be antibody-*independent*. This area is becoming still more intriguing in light of the existence of $CD5^+$ B cells in the thymus of mice (Section 8.5) and the demonstration that these cells can influence the selection of developing T cells (e.g. tolerance induction — Sections 5.2.5 and 5.3.5). Certainly, connectivity between idiotypically-interacting B cells can also occur early in life (Vakil and Kearney 1986), but sometimes is established only later (Bona *et al.* 1979).

11.1.4 Idiotypic regulation *in vivo*

Effects of administering idiotypic antibodies

Antibodies can profoundly enhance or suppress the expression of other antibodies when they are administered to animals. Many studies have examined the effect of administering idiotypic or anti-idiotypic antibodies on the expression of the corresponding idiotypes and anti-idiotypes. The results obtained have been complex and often appear contradictory. It has been pointed out that part of this problem is semantic: the immune system does not necessarily 'know' whether an antibody used for immunization is an idiotype or an anti-idiotype since these are, after all, merely operational terms (with the possible exception of regulatory idiotopes, if they exist; Section 11.1.2). In many cases the results were obtained under non-physiological conditions, with unusually high doses of antibody, although it does seem that administering very small doses of these antibodies can also have profound effects.

A priori, the administration of an idiotype or an anti-idiotype is likely to have complex effects on the immune system, and the antibody could deliver either stimulatory or inhibitory signals to the lymphocytes that bear corresponding receptors. For instance, there is evidence that anti-idiotypes can mimic antigens and activate B cells, helper T cells, suppressor T cells, and cytotoxic T cells, and/or induce the appearance of B or T cells with complementary idiotypes. It also seems likely that immunization against an idiotype can induce the anti-idiotype, or vice versa, to be produced and it can be difficult to decide whether the observed effects are due to one or the other of them.

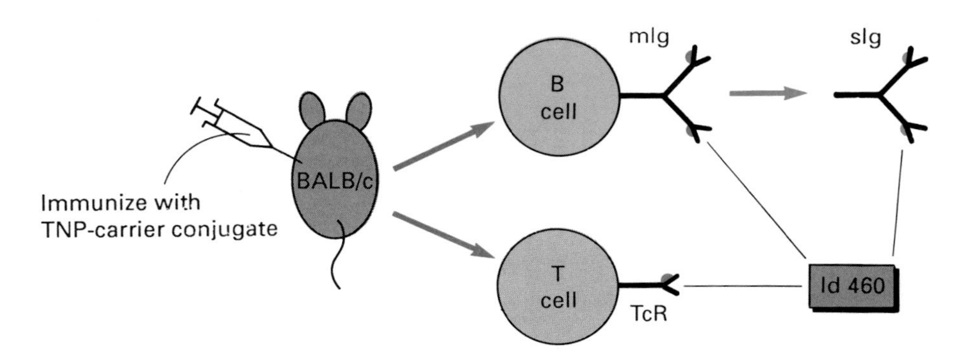

Fig.11.19 Many B cells and T cells participating in the immune response to the trinitrophenyl (TNP) hapten in BALB/c mice express the same idiotype, Id460, of a myeloma protein of BALB/c origin, called MOPC 460, which is specific for TNP (not shown)

There are further complications in that lymphocyte subsets interact with each other, such as in the case of idiotype-specific T helper cells in B cell responses (Section 11.1.3), and suppressor T cell subsets which may also interact through idiotype-anti-idiotype recognition (Section 11.2.2). Clearly, it becomes quite tortuous to decide whether a particular effect is due to activation or inhibition of a stimulatory or suppressive cell that carries either an idiotype or an anti-idiotype!

Oscillatory effects have also been seen. In the case of antibodies, ab1 may elicit an auto-ab2, which in turn stimulates auto-ab3, some of which resembles the original ab1, and so forth. Such effects are also seen after immunization with an antigen (Kelsoe and Cerny 1979). Therefore interpretation of the data from such experiments may depend on precisely how the antibodies were administered, the temporal sequence of events, and the time of assay. Despite these difficulties there have been some useful attempts to rationalize the findings, and some phenomena do seem reproducible, such as idiotype suppression (see below).

Effects of high-dose and low-dose antibodies. In an attempt to unify some of the experimental observations, Rajewsky and Takemori (1983) proposed the following (Table 11.2).

1. At 'low' doses (nanograms), antibodies enhance the expression of *complementary* binding sites. Thus, treatment with a low dose of an idiotype causes increased production of the *corresponding anti-idiotype*, and vice versa. This could be due to activation of T and B cells expressing the complementary receptors.
2. In contrast, administration of higher doses (micrograms) of antibodies stabilizes the expression of those receptors, so that high dose idiotype increases production of the *same idiotype*. (This could be because suppressor T cells are induced that inhibit the activation of cells with complementary receptors, and thus a high dose of an idiotype may actually *suppress* production of the corresponding anti-idiotype, or vice versa.)

As examples of the consequences of administering anti-idiotype or idiotype *in vivo*, we consider **idiotype suppression** and the **activation of silent (idiotype) clones** (see below). Our previous comments concerning the complexity and kinetics of such responses should, however, be borne in mind.

Table 11.2 Effects of administering idiotypic and anti-idiotypic antibodies (hypothesis of Rajewsky and Takemori 1983)

Dose given	Antibody given	Consequence	
'Low' (ng)	idiotype	anti-idiotype	increased
	anti-idiotype	idiotype	
'high' (μg)	idiotype	idiotype	stabilized
	anti-idiotype	anti-idiotype	

Idiotype suppression (Fig.11.20). If an animal is injected with an anti-idiotype, this can result in long-lasting *suppression* of clones of lymphocytes with the *corresponding* idiotypes. This idiotype suppression can be induced by exposure to an anti-idiotypic antibody at birth (**neonatal suppression**) or *in utero* (**maternal suppression**) (see, for example, Augustin and Cosenza 1976, and Weiler *et al.* 1977). It is more difficult to induce idiotype suppression in adults.

The T15 idiotype is dominantly expressed on antibodies against phosphocholine (Section 11.1.1). This idiotypic response is not seen in adults that had been treated, as neonates, with anti-T15 antibodies. In other words, the T15 response is suppressed. Nevertheless, an anti-phosphocholine response can still be generated that allows the emergence of T15-negative clones which would not otherwise be produced. Idiotype suppression of the T15 response appears to be due both to a direct effect of the anti-idiotype on precursor T15-positive B cells and perhaps the generation of suppressor cells. Idiotype suppression cannot be induced with $F(ab')_2$ fragments, suggesting that the antibody Fc portion may also play a role, perhaps by binding to Fc receptors on B cells (Section 8.4.2).

Activation of silent clones. A silent idiotype is one that is not normally produced after a particular immunization regimen. However, silent idiotypes can be induced if an idiotype is injected into an animal at birth, and this idiotype will dominate the response later in life (Fig.11.21).

For example, some strains of mice do not produce antibodies with the A48 idiotype when they are immunized with a bacterial levan, but this idiotype response can be produced if they are treated as neonates with antibody bearing this idiotype (Rubinstein *et al.* 1984). (Speculatively, neonatal treatment with A48 makes the animal tolerant and unable to produce anti-A48, and this might suppress the A48 response of adults; compare with suppression of the T15 response, above.)

Silent idiotypes can also be activated after treatment with anti-idiotypes (Hiernaux *et al.* 1981), which can even induce the expression of silent idiotypes in a different species (Fig.11.22).

For example, antibodies made in mice and rabbits against tobacco-mosaic virus (TMV) normally express different idiotypes. In one particular study, rabbit ab2 antibodies were prepared against an idiotype of a rabbit ab1 antibody specific for TMV (Francotte and Urbain 1984). These rabbit ab2 antibodies were then used to induce ab3 antibodies in *mice*. These ab3 antibodies were found to be directed against TMV, even though the mice had never seen the virus. Furthermore, these *mouse* antibodies now expressed idiotypes that were normally associated with *rabbit* ab1 antibodies, despite the fact that these idiotypes are not found in normal mice immunized with TMV. Even more remarkable is the fact that the ab2 antibodies were directed to a *private* idiotype on ab1, rather than to a recurrent idiotype (see Fig.11.9).

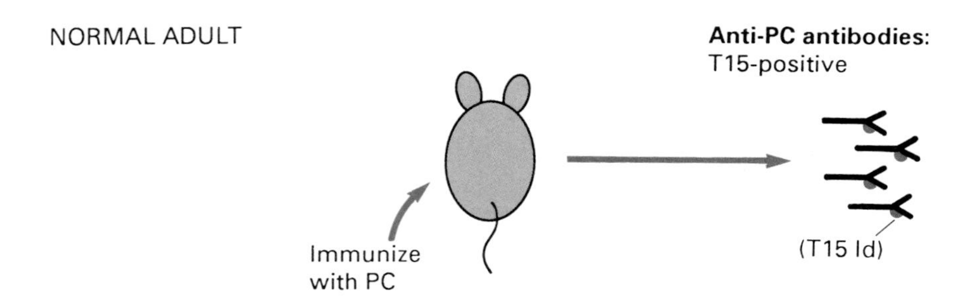

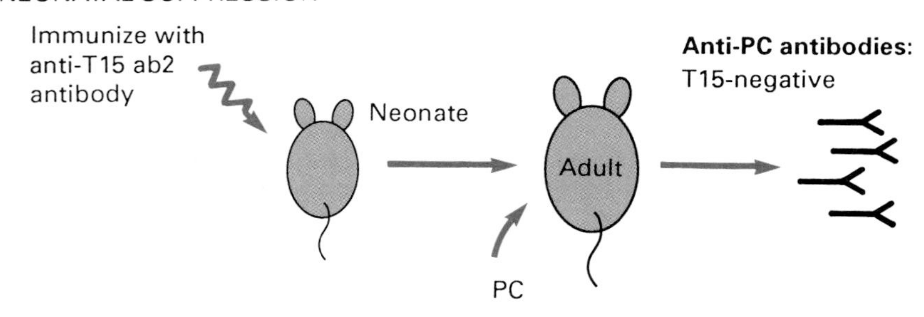

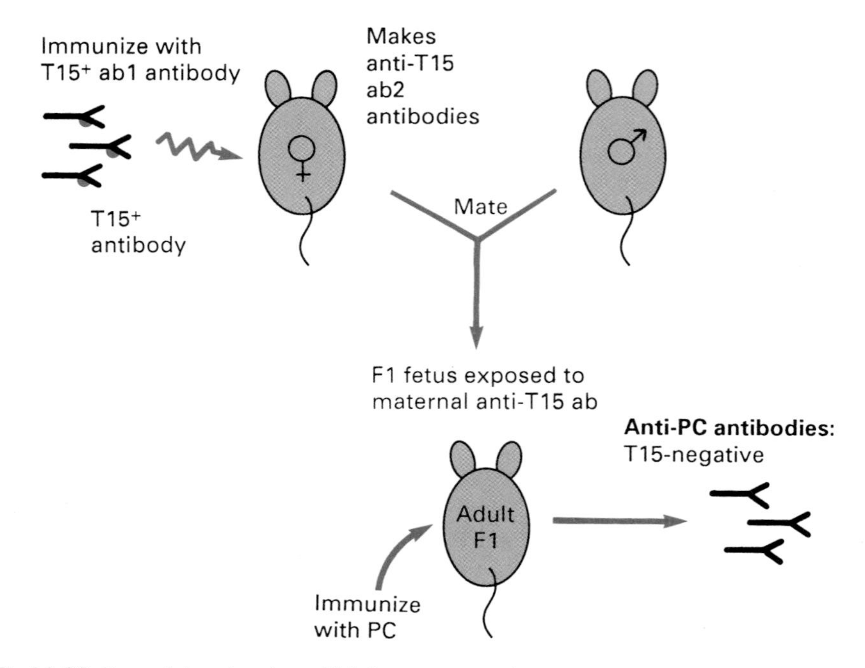

Fig.11.20 Neonatal and maternal idiotype suppression

Immunization with T cell receptor peptides can prevent autoimmune disease

Multiple sclerosis (MS) is an autoimmune disease of humans mediated by T cells specific for a protein of the myelin sheaths of nerves. It results in chronic inflammation within the central nervous system, with extensive macrophage and T cell infiltration and demyelination of nerves. A similar disease can be induced in rats and mice by immunization with guinea pig myelin basic protein (MBP) and certain peptides of MBP. This disease is called **experimental autoimmune encephalomyelitis (EAE)** and is considered to be a model for human MS. The term 'encephalitogenic' is used for T cells and peptides that can elicit EAE.

EAE can be induced in naïve recipients by adoptive transfer of $CD4^+$ T cells from MBP-immunized animals, specific for peptides of MBP bound to MHC class II molecules. It has been found that the cells that transfer disease utilize a very limited repertoire of T cell receptor genes, in particular the $V_\beta 8.2$ and to a lesser extent the $V_\alpha 2$ gene segments in mouse, and their homologues in rat (Section 7.3.1). A limited repertoire of MBP-reactive T cells is also activated *in vivo* in MS patients, which preferentially express the human $V_\beta 17$ and/or $V_\alpha 12$ segments depending on the peptide specificity, and a limited repertoire is also demonstrable in plaque tissue from the brains of these patients. However, the reasons for this limited repertoire usage are not yet clear.

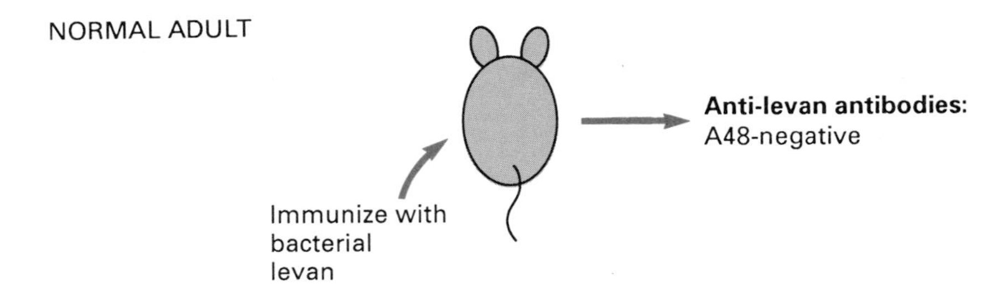

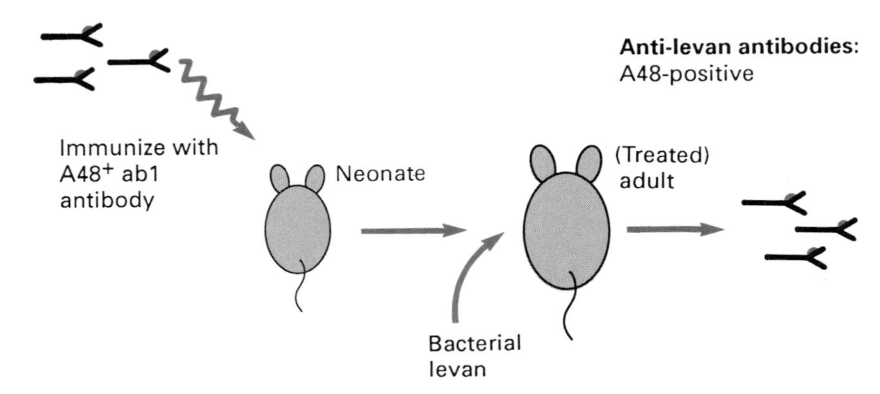

Fig.11.21 Activation of silent clones within a species

One strategy for preventing EAE is to immunize rodents with analogues of encephalitogenic peptides, the aim being to prevent binding of the disease-inducing peptides to the MHC class II molecules. Some of these analogues do indeed bind more strongly to class II molecules and prevent EAE when the animals are co-immunized with MBP. In some cases they prevent recognition by encephalitogenic T cell clones. However, there is as yet no evidence that this is the mechanism for prevention of disease; indeed some analogues *can* be recognized by encephalitogenic T cells but nevertheless do not induce EAE. Instead, it now seems likely that protection is mediated by regulatory interactions of T cells, which redirect the response towards non-encephalitogenic determinants of MBP.

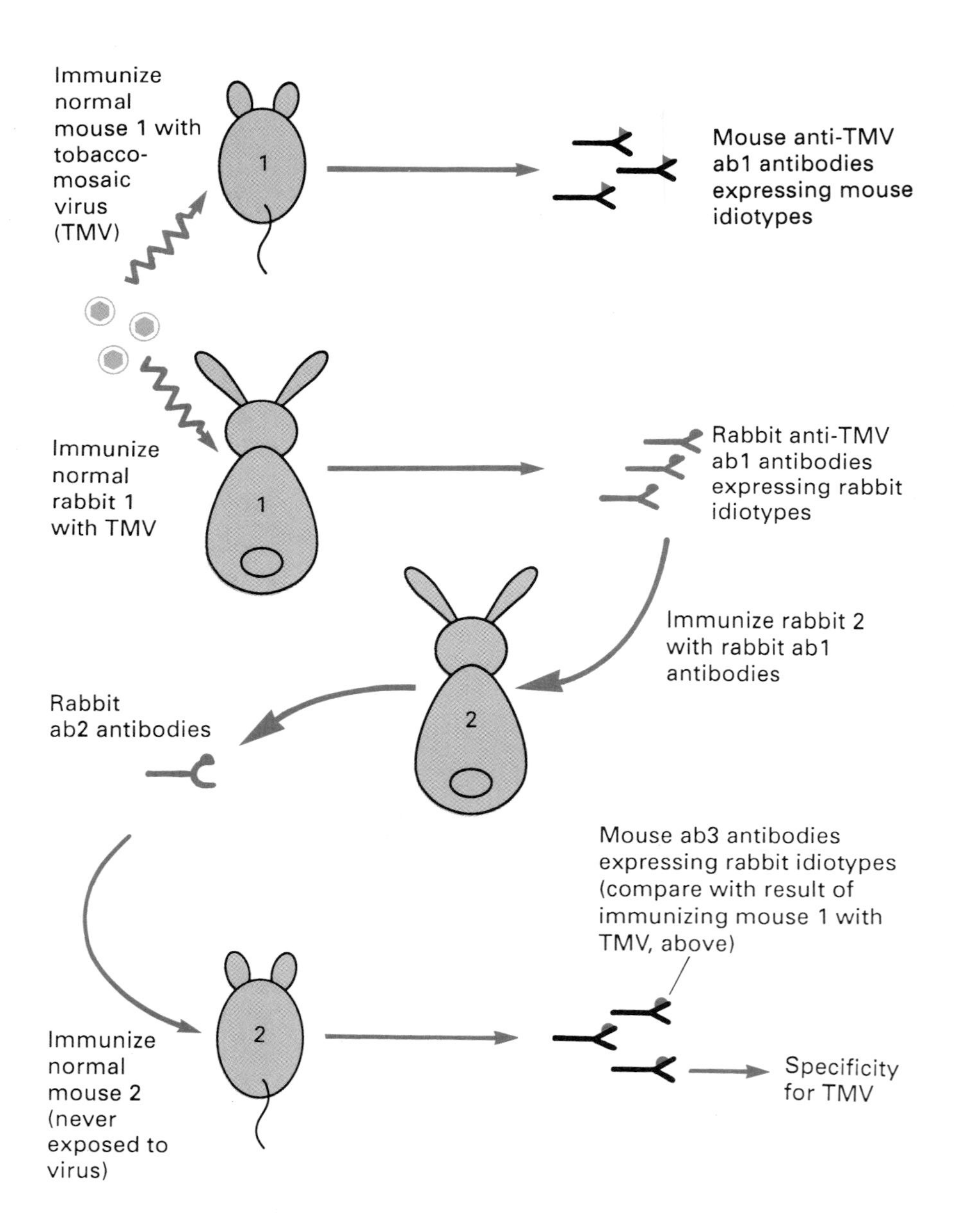

Fig.11.22 Activation of silent clones across a species

That the T cell receptor of encephalitogenic clones could be a target for regulatory cells was first suggested by the observation that immunization with attenuated clones of these cells could prevent EAE. Subsequently it was found that antibodies specific for $V_\beta 8^+$ T cell receptors (Section 5.2.5; Fig.7.19) can prevent EAE in mice. It has also been shown that disease can be prevented in rats by immunization with peptides corresponding to defined regions of the TcR $V_\beta 8$ chain. For example, immunization with a synthetic peptide corresponding to residues 39–59 of $V_\beta 8$, which includes the hypervariable CDR2 region (Section 6.3.2; Fig.6.28), prevents induction of EAE by the major encepalitogenic peptide of MBP (residues 72–89), and can reverse disease in rats that are showing clinical signs of EAE (Offner *et al.* 1991).

Immunization with the $V_\beta 8$ peptide resulted in the production of antibodies specific for this peptide, as well as peptide-specific, class I-restricted regulatory $CD8^+$ T cells. These cells, which were not cytotoxic, responded specifically to $V_\beta 8^+$ T cells *in vitro*, in the absence of accessory cells. Since it is unlikely that the regulatory T cells can recognize the intact $V_\beta 8^+$ T cell receptor, one possibility is that they are specific for an endogenous processed peptide of the *receptor* that is presented in association with MHC class I molecules at the cell surface. If so, recognition of MBP-reactive, $V_\beta 8^+$ T cells could interfere with their activation.

The regulatory T cells, but not the antibodies, were found to confer protection from EAE when they were adoptively transferred to naïve recipients. These T cells altered the pattern of T cell responses in the recipients, such that they responded well to the specific V_β peptide (even though they had not been immunized with the peptide itself) but poorly to MBP and its encephalitogenic 72–89 peptide; the converse pattern was seen with T cells from control rats that had not received regulatory cells. Therefore, administration of these cells appears to shift the recipient's response away from MBP.

The CDR2 region of the T cell receptor is thought to comprise part of the peptide–MHC binding site of T cell receptors and can be thought of as idiotypic in nature. Therefore immunization with a peptide containing this region, as in these experiments, could perturb a pre-existing regulatory T cell network, perhaps through idiotypic interactions. The regulatory T cells induced by the peptide act as suppressor cells in this system, and could be involved in putative suppressor cell circuits (Section 11.2) or in the regulation of cytotoxic T cells directed at the encephalitogenic T cells mediating EAE. Immunization with peptides corresponding to defined TcR regions might also prove useful in other autoimmune diseases, because a limited repertoire of receptors has also been demonstrated in experimental models of diabetes and in rheumatoid arthritis in humans.

11.2 Suppressor T cells and suppressor pathways

11.2.1 Immune suppression

The suppression of an immune response represents the converse of its induction. An antigen may stimulate an immune response, but it is logical that once the antigen has been destroyed the response should either wane of its own accord or be switched off. It seems likely that as antigen is successfully removed from the immune system the response should die down automatically (although antigen may be retained for some time after the response has finished, for example on the surface of follicular dendritic cells; Sections 3.4.2, 8.3.1), but mechanisms for active suppression of the immune response also seem to be required. Immune suppression can be brought about by a variety of mechanisms which have been mentioned already, and some examples follow (Fig.11.23).

1. Dendritic cells which are required to activate helper T cells (Section 3.4.3) may be regulated and/or killed by NK cells and perhaps by cytotoxic T cells specific for the peptide–MHC complexes they express (Section 10.3).
2. Activated T cells utilize and remove essential growth factors such as IL-2 (Section 4.4.3). For example, IL-2 is internalized via high-affinity receptors, so reducing its local concentration. In addition, helper T cells become refractory to the effects of IL-2 upon continued stimulation by this molecule.
3. Cytotoxic T cells (and NK cells) may kill the *target* cells that stimulated their development (Sections 10.2, 10.3), and human cytotoxic T cells, which express MHC class II molecules, can be killed by $CD4^+$ cytotoxic T cells (Ottenhoff and Mutis 1990).
4. Antibodies can inhibit B cell responses by binding to Fc receptors on B cells (Section 8.4.2) and may be regulated by idiotype-anti-idiotype interactions as discussed in Section 11.1.
5. Macrophages can produce prostaglandins, which non-specifically inhibit lymphocyte responses (Section 9.2.3). Before and after activation by molecules such as IFNγ, macrophages secrete a variety of toxic oxygen and nitrogen metabolites and enzymes, some of which can deplete essential nutrients for lymphocyte growth (Section 10.4). Macrophages and other cells can also kill specific targets by ADCC when they have bound a specific antibody (Fig.10.2.7). It is also worth noting that tumour cells can suppress the activation of macrophages by secreting a molecule called macrophage deactivating factor (Ding *et al.* 1990)
6. It should be remembered that immune responses are tightly linked to the events of inflammation and that the adaptive immune system may be regulated according to the strength and extent of the particular inflammatory response and the type of mediators produced (Section 9.2.1).
7. Suppression of immune responses by cytokines is discussed in Section 11.3.

11.2.2 Suppressor T cells

Suppressor T cells (Ts) are thought to be a distinct class of regulatory cells comprising different subsets of T cells whose interactions are mediated by soluble suppressor factors (TsF). Unlike conventional helper and cytotoxic T cells, at least some Ts have been thought to bind free antigen without the requirement for MHC-associative recognition.

The phenomenon of immune suppression, sometimes mediated by T cells, does appear to be real. But at the outset we should stress that, despite many hypotheses and much speculation, it is at present poorly understood.

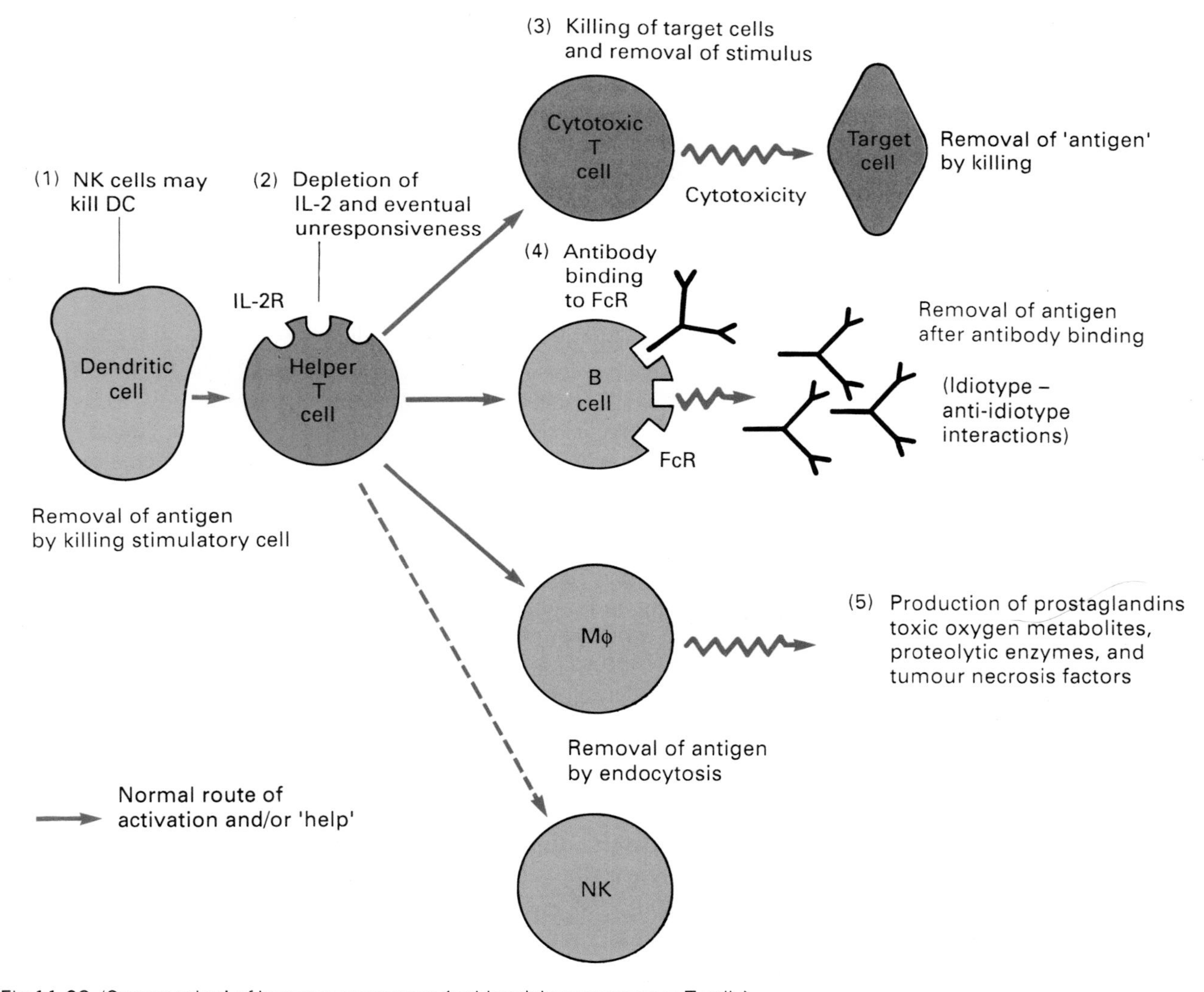

Fig.11.23 'Suppression' of immune responses (not involving suppressor T cells)

Markers and assays

Markers. There are no clearly-defined specific markers for Ts, and the definition of a lymphocyte as a suppressor cell depends on functional assays, as is also the case for helper and cytotoxic T cells. There has been a tendency to refer to the CD4 and CD8 molecules as markers of the helper and cytotoxic/suppressor subsets, respectively. However, as discussed in Section 4.2.4, the expression of CD4 and CD8 by T cells correlates with their restriction specificity (recognition of peptides bound to MHC class II and class I molecules, respectively) rather than function. Even so, the CD4/CD8 phenotype has been used to delineate various subsets of Ts in humans, rats, and mice. In mice, the expression of CD5 (Ly1; Section 4.2.4) and Thy1 (Section 4.2.8) by suppressor cells has also been examined; expression of Thy1 was often taken as the primary evidence that the suppressor cells were in fact T cells.

Perhaps the most controversial marker of all is mouse IJ. This was thought to be expressed by many mouse Ts, and some investigators argued that it was also present on accessory cells required for the generation of suppressor cells. However, as discussed in Section 2.8.2, although IJ was genetically mapped into the MHC class II region, no gene that specifically encoded IJ was found when this region was investigated at the molecular level. Another MHC (class I) molecule, Qa-1 (Fig.2.24), was also reportedly expressed by some Ts, but this is not a specific T cell marker. The results of phenotypic analysis using these markers are discussed later in this section (11.2.2).

Assays. To demonstrate Ts *in vitro*, presumptive Ts are added to an 'indicator' system (a measurable immune response) and the percent suppression is assessed.

One could, for example, measure the inhibition of proliferation in a mixed leukocyte reaction, the number of plaque-forming cells in an antibody secretion assay, or chromium release in a cytotoxicity assay (see Appendix). This approach is, however, subject to artefacts due to inhibition of indicator cell-cell contact, depletion of essential nutrients from the culture medium, and alterations of the kinetics of the response (i.e. the cells may simply grow more slowly).

Alternatively, Ts can be detected by *in vivo* adoptive transfer assays. In this type of assay, putative suppressor cells are injected into an animal that is making an immune response. The cell inoculum may be tested for its effects on a number of *in vivo* immune responses such as delayed-type hypersensitivity responses or allograft rejection in the recipient. Before suppression is demonstrable in these systems, it is frequently necessary to manipulate the recipient, typically by sublethal irradiation or with anti-lymphocyte serum therapy, for reasons that are not understood.

Systems in which suppressor cells have been detected

The first report of T suppressor cells was by Gershon and Kondo (1982). Mice were treated with high doses of sheep erythrocytes to induce suppressor cells (see below), and T cells from these mice were transferred to irradiated recipients together with normal spleen cells. The response of the latter to sheep erythrocytes was found to be suppressed in the presence of the former T cells.

Since that time, specific Ts have been reported in a wide variety of systems, using a combination of the above and other assays. We shall outline just a few (see also suppression of autoimmune disease by regulatory T cells, Section 11.1.4).

Antibody production. Ts may control antibody production *in vivo*, classic examples being idiotype suppression and allotype suppression (Section 10.1.4.).

As already discussed, idiotype suppression can be induced by administering anti-idiotypic antibodies *in vivo* or by using the idiotypic antibody coupled to syngeneic cells. While the anti-idiotype does in some cases appear to activate lymphocytes and elicit Ts, it may also inhibit idiotype-bearing pre-B cells directly.

Another instance in which Ts have been implicated is in the phenomenon of **allotype suppression** (Herzenberg and Herzenberg 1974). Allotype suppression is induced when an F1 fetus is exposed to maternal antibodies directed against the paternal antibody allotype (Fig.11.24). When mature, the F1 animal does not produce antibodies that express the paternal allotype because Ts have been generated. This can be shown by adoptive transfer of these cells to secondary recipients, when production of antibodies with the relevant allotype is found to be suppressed.

T cell helper and proliferative responses. Ts have been found in some responses controlled by Ir genes (see Section 5.4 for a discussion of Ir responses).

For example, (Fig.11.25) some strains of mice can make immune responses to the synthetic antigen GAT (i.e. they are responders), and this is controlled by H-2 IA genes. Other strains do not respond when they are immunized against GAT on its own, even though they can respond if they are immunized with GAT coupled to an immunogenic carrier. In this case, GAT-specific helper T cells can be detected in such non-responder strains. However, if the animal is pre-immunized with GAT alone, it no longer responds to GAT plus carrier because suppressor cells have been induced that inhibit the latter response (Kapp *et al.* 1974). GAT unresponsiveness in these strains, then, is due to the development of active suppressor cells after immunization, and this also appears to be controlled by the MHC class II region. (In the human, so-called DY determinants, which may be associated with novel MHC class II molecules not encoded by the HLA-DR, DP, or DQ loci, can stimulate autoreactive T cells with suppressive activity; Pawelec *et al.* 1988). Some fragments of protein antigens such as lysozyme and β-galactosidase also preferentially activate Ts in non-responder animals, and prevent an immune response when the animals are challenged with the whole molecule. These determinants have been termed **suppressor epitopes** (Fig.11.26) (Sercarz *et al.* 1978).

Tolerance. Ts have been implicated in tolerance, and this form of 'immunological unresponsiveness' can be defined as the '*specific depression of an immune response induced by previous exposure to the antigen*'. Operationally, it is the opposite of a secondary response. Note that this definition excludes unresponsiveness due to the genetically-controlled *inability* to respond (e.g. an Ir defect; Section 5.4) or in the case of inherent non-immunogenicity of the antigen (e.g. a hapten alone; Section 8.3.1).

As an example, a large dose of sheep erythrocytes can induce unresponsiveness in adult mice (as in the original experiments by Gershon and Kondo, see above) and this **'high-zone' tolerance** is a consequence of Ts being induced. Very low doses of antigens can also, perhaps surprisingly, lead to

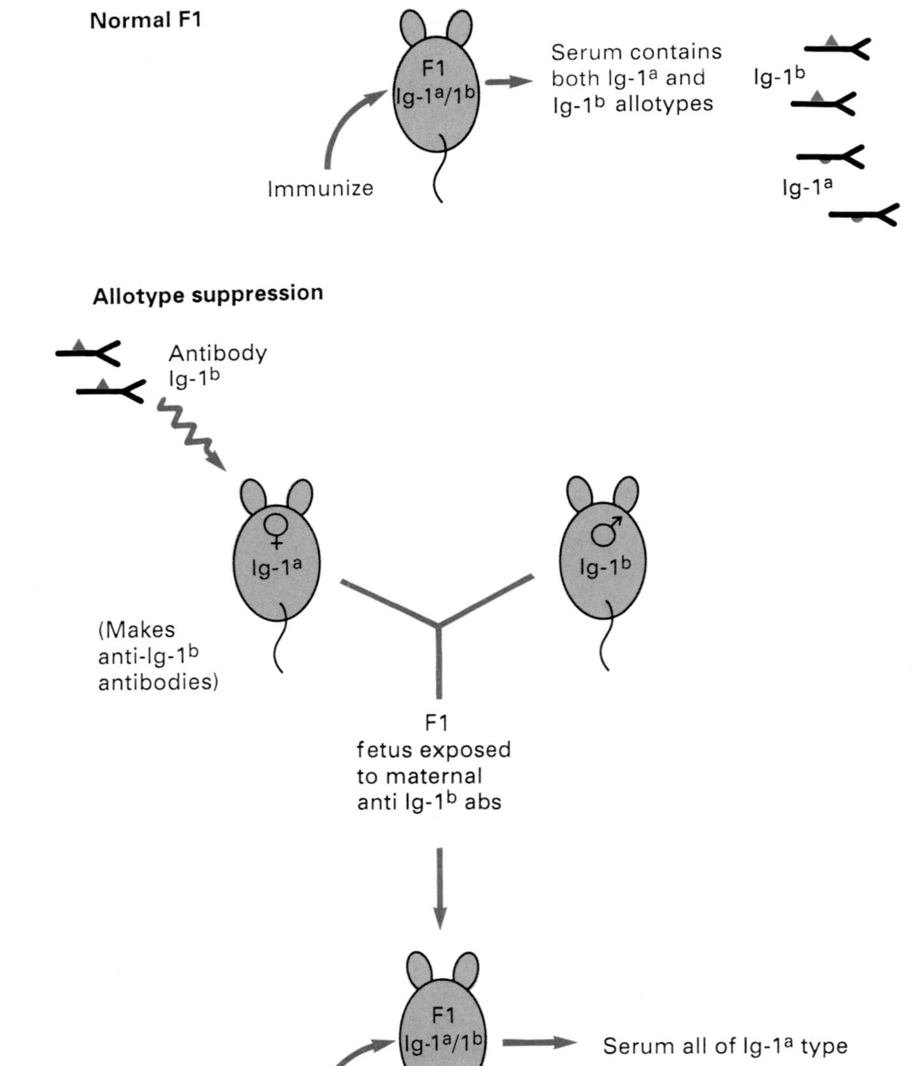

Fig.11.24 Allotype suppression

antigen unresponsiveness, and Ts have been implicated in some models of this phenomenon of **'low-zone' tolerance** (Stumpf *et al.* 1977).

Ts have been found in a wide variety of other instances of immunological unresponsiveness, such as allograft tolerance and in tumour immunity. The generation of Ts is, of course, only one mechanism for tolerance induction. Others include clonal deletion and clonal anergy, discussed in detail in Section 5.2.5. It is often difficult to decide which of these mechanisms is most important in various types of tolerance to self molecules, although Ts have been implicated in some cases.

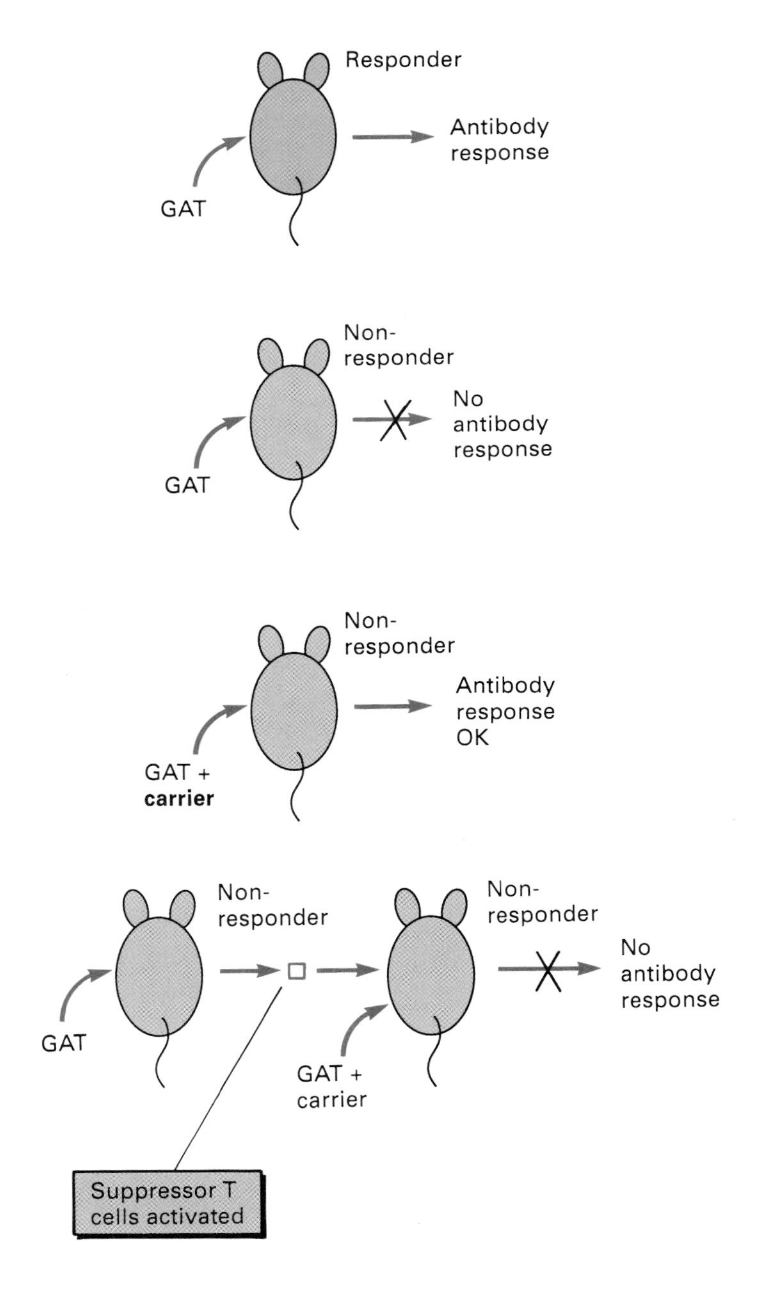

Fig.11.25 Suppressor T cells induced to antigens under Ig gene control

Neonatal animals. Non-specific suppressor cells can be found in neonatal animals and can also be generated in culture when spleen cells are incubated with concanavalin A. They also appear after certain procedures such as total lymphoid irradiation when the lymphoid tissues of adult animals are subjected to repeated doses of relatively small amounts of radiation to a high cumulative dose. The relationship of these so-called **natural suppressor cells** and other mechanisms to the normal state of natural self-tolerance is obscure.

Interacting subsets

It has been proposed that two, three, or even four distinct Ts subsets may interact to bring about suppression. Attempts were made to define these subsets on the basis of the markers mentioned earlier (CD4, CD8, CD5, IJ, Qa-1); the expression of idiotypic or anti-idiotypic determinants; and their sensitivity to cyclophosphamide treatment or irradiation *in vivo*. In addition, macrophage-like accessory cells were thought to be involved at different stages and were even considered to be the final effector (suppressive) cells according to some schemes.

The 'first' T cell in the pathway to suppression was sometimes called the **suppressor inducer (Tsi) cell**. It can be thought of as a helper cell for induction of suppressor cells, as opposed to a helper cell for development of cytotoxic T cells or B cell responses. The T cell that actually causes suppression was often called the **suppressor effector (Tse) cell**. Other cell types were implicated between Tsi and Tse. We shall briefly consider three pathways that have been proposed by different investigators, and note that these were not necessarily mutually exclusive (Fig. 11.27).

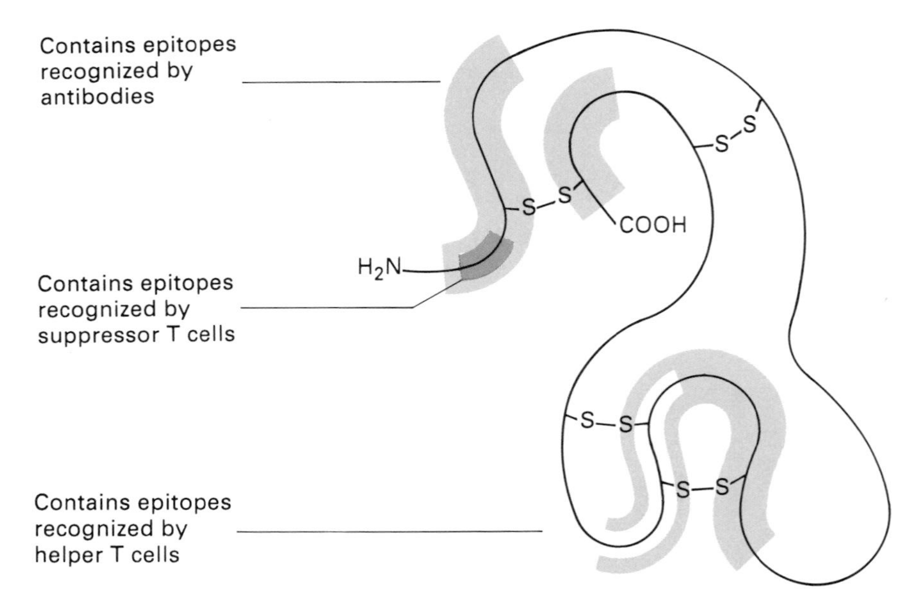

Fig. 11.26 Epitopes of hen-egg lysozyme (the polypeptide is represented as a continuous thin black line)

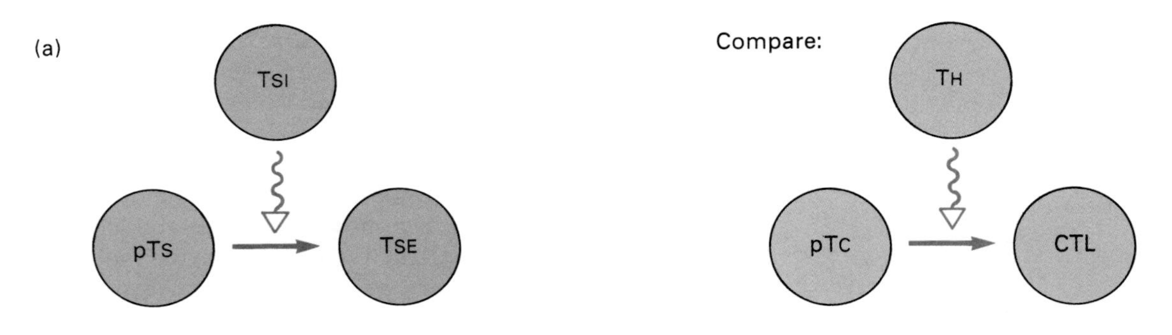

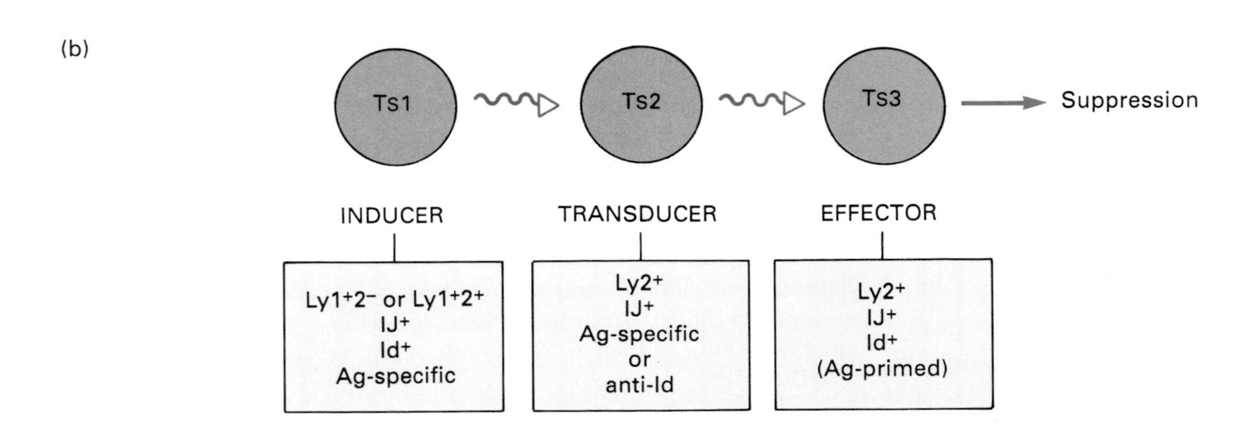

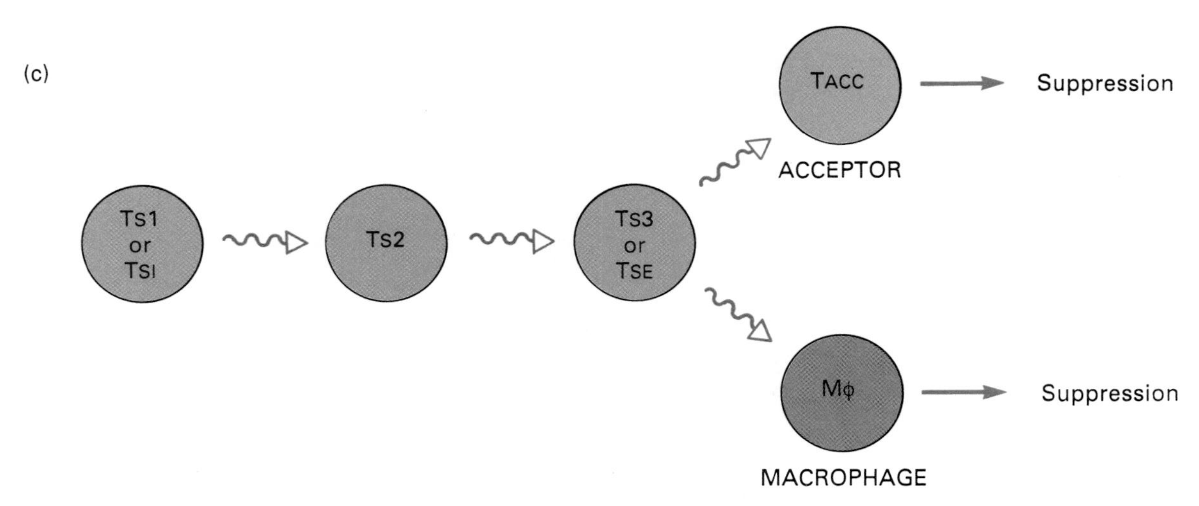

Fig.11.27 Models for interactions of putative suppressor T cell subsets and, in addition, (a) comparison with helper-dependent activation of CTL; (b) phenotypes and antigen (Ag) or idiotypic (Id) specificities; and (c) additional cellular interactions leading to suppression

1. In one pathway, Tsi caused a so-called 'suppressor precursor' cell (pTs) to differentiate into Tse. This scheme is reminiscent of the role of helper T cells in the development of mature cytotoxic T cells from their precursors.

2. Another pathway that was proposed consisted of three distinct subsets, termed Ts1, Ts2, and Ts3. In this **suppressor cell cascade** the subsets were thought to be linked to each other through idiotype-anti-idiotype interactions. This means that the first cell, Ts1 (which may be the same as Tsi above), is antigen-specific and expresses a particular idiotype. Ts2 is not specific for an antigen *per se*, but recognizes the receptor on Ts1, and is loosely referred to as an 'anti-idiotypic' cell. Ts2, in its turn, elicits the Ts3 subset (possibly the same as Tse above) which now carries the same or a related idiotype to Ts1. Ts2 was termed a 'transducer' cell because it receives a signal from Ts1 and transmits, or transduces, it to Ts3, which actually suppresses the immune response in question. Any of these cells could also develop from precursor cells.

In this scheme, the receptor of Ts2 can roughly be thought of as the internal image of the antigen that is recognized by Ts1, although this is not proven. This scheme should be compared to the antibody series ab1, ab2 and ab3, and to the concept of regulatory idiotypes (Section 10.1.2). If a parallel is drawn between Ts subsets and idiotopes on antibodies, one could argue there may again be two levels of interaction: antigen induces Ts1 that expresses an idiotope; this in turn induces Ts2 specific for the idiotope; and so on. According to some investigators, Ts1, Ts2, and Ts3 could be discriminated according to their surface phenotypes (Germain and Benacerraf 1981) (Fig.11.27(b)).

It has been reported that treatment of populations of mouse suppressor cells with anti-CD8 (anti-Ly2) antibodies kills the suppressor T cells. However, treatment with anti-CD5 (anti-Ly1) antibodies prevents the *development* of these effector cells, although it should again be noted that CD5 is actually expressed by all T cells to a greater or lesser degree (Section 4.2.4). Thus it was suggested that the suppressor inducer cell, Ts1, had the phenotype $Ly1^+ Ly2^-$ (like many helper cells) but that the final effector cell, Ts3, was of the $Ly1^- Ly2^+$ phenotype. Similar studies also suggested that Ts2 may be an $Ly1^+ Ly2^+$ cell. Some studies in the rat have come up with similar schemes, but with CD4 instead of CD5 as a marker. The relevance of CD4 and CD8 on these presumed subsets (i.e. their MHC restriction) was not adequately explained.

3. In the two pathways described above, the final effector cell was Ts3. Other schemes suggested that a different T cell, the T acceptor cell, was 'armed' by a product of Ts3 and actually caused suppression. Alternatively, this might have been brought about by a similarly 'armed' (or perhaps activated?) macrophage.

The target of the 'final' suppressor cell in these pathways could in principle be any specific lymphocyte, such as the helper T cell responsible for providing help for antibody production by B cells or the generation of cytotoxic T cells. The target may also be the effector cell itself, such as a B cell, a cytotoxic

T cell, or a T cell responsible for delayed-type hypersensitivity. The suppressor effector cell was thought to be able to inhibit the functions of its own inducer cell and for this reason the pathway was termed **feedback suppression**. The actual mechanism(s) of suppression was not understood, and while it was apparently antigen-dependent it was said to involve 'non-specific mediators'.

11.2.3 Suppressor factors

It was proposed that the interaction between different Ts subsets was mediated by suppressor factors: different Ts subsets produced different factors, and these were termed TsF1, TsF2, and TsF3, by analogy with the cells producing them (see Section 10.2.2) (Fig.11.28). According to this concept, TsF1 could be used to replace the function of the original cell. For instance, either Ts1 cells or TsF1 factors were active in the induction phase of an immune response, and either the cells or their factors could induce Ts2 cells when they were administered to normal syngeneic recipients or added to cultured spleen cells. Again, Ts2 cells or TsF2 factors could suppress the immune response, but they acted in the effector phase.

In many cases, crude cell extracts were used as a source of suppressor factors. For example, sonicates of spleen cells or thymocytes from animals making IgE responses to the hapten TNP, were found to inhibit this immune response when they were injected into animals. Even though these extracts were reportedly antigen-specific, and worked as well as intact T cells from

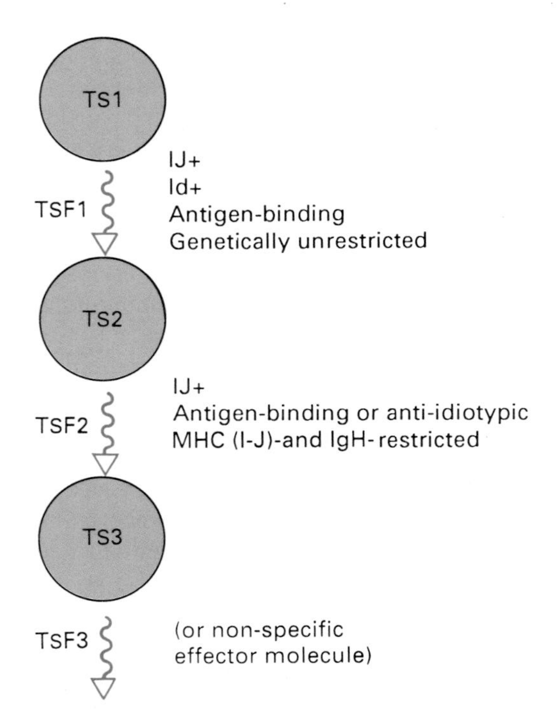

Fig.11.28 Putative T suppressor factors (TsF) and their characteristics

the animals, their precise nature was not defined. Other presumptive antigen-specific suppressor factors were derived from T cell lines or hybridomas, which should have facilitated their purification and characterization, but no consensus was ever reached as to their structure (or indeed existence). They were obtained from mouse Ts specific for heterologous cells, synthetic copolymers, globular proteins and haptens, respectively including sheep erythrocytes, GAT, KLH and NIP.

Where examined, most TsF appeared to be idiotype or anti-idiotype positive in that they were reported to react with antisera against idiotypes on antibodies. In particular, these factors seemed to share determinants with immunoglobulin heavy chain variable regions, but not their constant regions. Many also had IJ determinants, as shown by binding of antisera and monoclonal anti-IJ antibodies. The problem with these markers is that they were defined by antisera raised against antigens on different cell types (such as Ig on B cells) or that were not well defined (such as IJ). Nevertheless, there were reports that TsF1 and TsF3 bound to specific haptens, and that TsF2 bound to idiotypes expressed by soluble antibody molecules that were coupled to tissue culture plates or to columns.

Suppressor factors were reported to be one- or two-chain molecules, and their markers and apparent molecular weights varied enormously (Fig.11.29) (Webb *et al.* 1983, see Further reading).

We shall mention a few of the many factors that were described, although it could be argued that these are now only of historical interest.

A suppressor factor from an $Ly1^-$ $Ly2^+$ Ts clone, specific for glycophorin on sheep erythrocytes, was reported to be a single chain with a molecular weight of about 70 kDa (Fig.11.29, top). This was split by the enzyme papain into two subunits with different functions: one of 25 kDa retained the capacity to bind antigen while the other of 45 kDa was suppressive. This seemed analogous to the effects of papain on antibodies, where antigen-binding Fab fragments of 22 kDa are produced together with Fc portions of 50 kDa which mediate the activity of different Ig subclasses (Section 6.2.1). In contrast, another TsF from hybridomas that were specific for erythrocytes had a molecular weight of around 200 kDa, giving some idea of the great variability in size that has been reported.

Some suppressor factors also possessed determinants characteristic of the cells supposedly producing them. For example, a KLH-specific TsF was reportedly composed of two chains. One bound the antigen and had Ig-related determinants, while the other was IJ-positive, but both chains were needed for suppression. These polypeptides were separately translated from hybridoma mRNA in a frog oocyte system, apparently to produce the functional factor. A similar approach was also used to translate mRNA for a GAT-specific TsF. The latter, however, turned out to be a one-chain molecule with a molecular weight of about 24 kDa that possessed antigen-binding sites and was both idiotype- and IJ-positive.

The reasons for the apparent diversity of suppressor factors, assuming that they exist, are not clear. It could, for example, reflect the existence of fundamentally different types of molecule. Alternatively, it was proposed that different cells could produce different chains which then associated to form

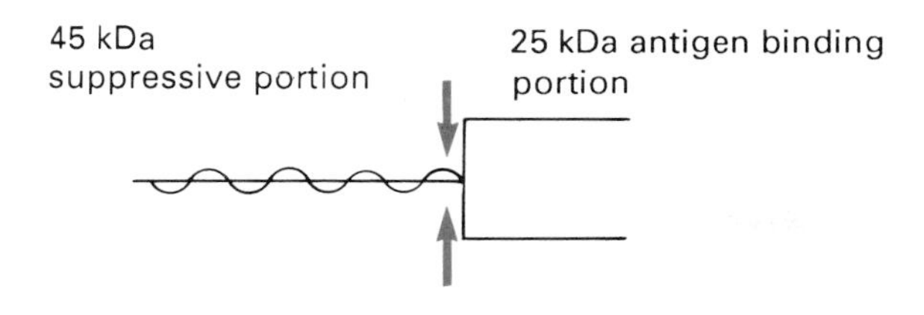

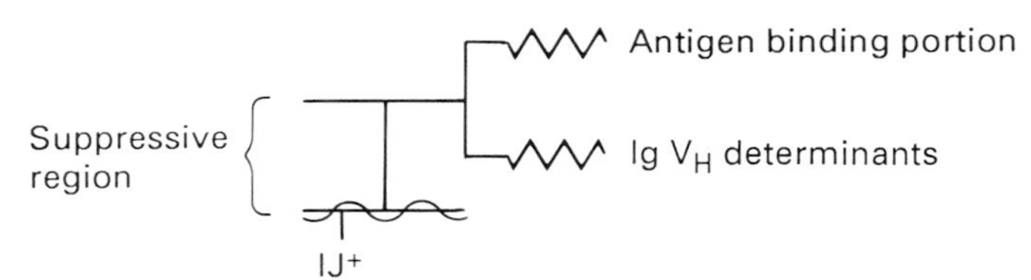

Fig.11.29 Examples of hypothetical suppressor factors

the 'active' TsF. For example, it was suggested that an IJ^+ chain from the inducer cell might become complexed with an antigen-specific chain from the effector cell, and this combination might specifically target the suppressive signal.

The antigen specificity or idiotype specificity of suppressor factors might have argued for a secreted or otherwise soluble form of the antigen-specific Ts receptor. In this case sharing of idiotype and Ig heavy chain determinants would be compatible with the presumed conformational cross-reactivity of these antisera with the T cell receptor. It could also explain some of the reported genetic restrictions between the interacting subsets and their factors, some of which appear to require compatibility at Ig heavy chain and/or MHC genes (see, for example, Eardley *et al.* 1979). However, we shall, for the reader's benefit, not pursue this further.

Significantly less is currently known about the T cell receptors expressed by suppressor cells than on helper and cytotoxic cells. From some studies it seems that the TcR β chain constant region genes in suppressor cell hybridomas may actually be *deleted*, rather than productively rearranged as in cytotoxic and helper cells although this is by no means an absolute rule. Nevertheless, according to one report to date, CD3-associated heterodimers are expressed by suppressor hybridomas. Therefore it is possible that either a different 'isotype' of T cell receptor is used in Ts, perhaps permitting a secretory form to be produced, or that different gene pools are utilized. This could also facilitate the recognition of antigen in the absence of MHC, which seems to be a feature of many Ts that can, for example, be bound to plastic tissue culture dishes coated with antigen. On the other hand, other suppressor cells apparently express an $\alpha\beta$ T cell receptor (Modlin *et al.* 1987).

11.2.4 Accessory cells in suppressor pathways

The idea of a suppressor cell cascade was further complicated by the suggestion that antigen-presenting or accessory cells were required to activate Ts1 and Ts3 cells, for example (Fig.11.30). The cell responsible for Ts1 induction was thought to express MHC class II molecules, but it appeared to resemble a macrophage rather than a dendritic cell, and was presumed to express

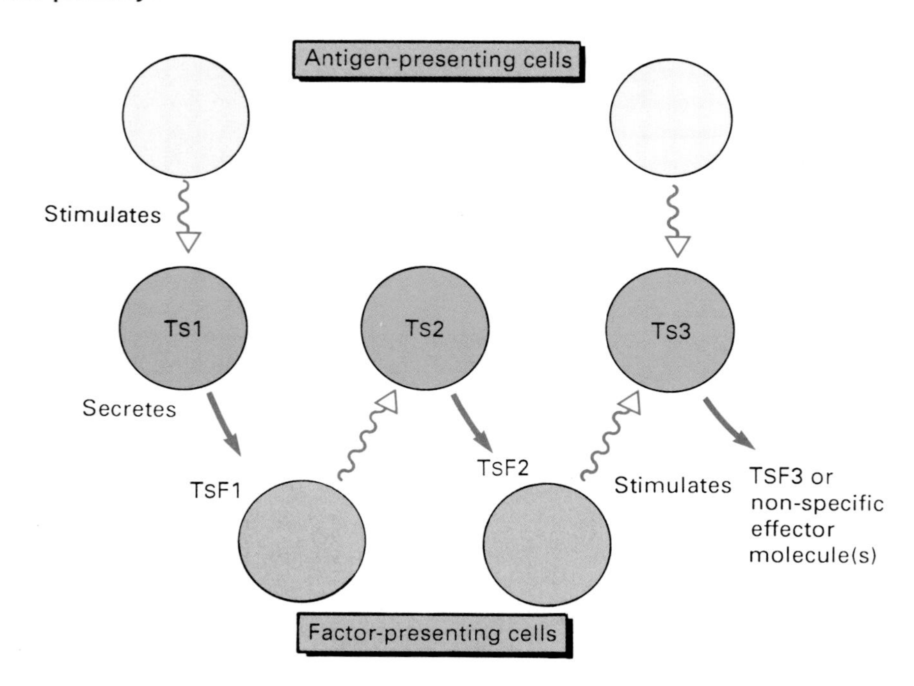

Fig.11.30 Accessory cells in suppressor pathways: a speculative scheme

IJ determinants. Some investigators even suggested that the TsF itself may need to be 'presented' to the next T cell in the cascade by what was termed a 'factor-presenting cell' that also expressed IJ. The existence of such cells still awaits confirmation.

11.2.5 Contrasuppression

We have considered regulatory circuits that can result in suppression of immune responses (Sections 11.2.2–11.2.4). Another circuit was suggested to result in what was termed **contrasuppression** (Gershon *et al.* 1981). This was defined as an activity that *interfered with suppression* to allow immunity to develop, the final contrasuppressor cell apparently making the helper T cell *refractory* to the suppression caused by a suppressor effector cell.

By analogy with the feedback suppressor circuits discussed above (Section 10.2.2), contrasuppression was thought to involve inducer, transducer, and effector cells (Fig.11.31). These were all reportedly distinguishable on

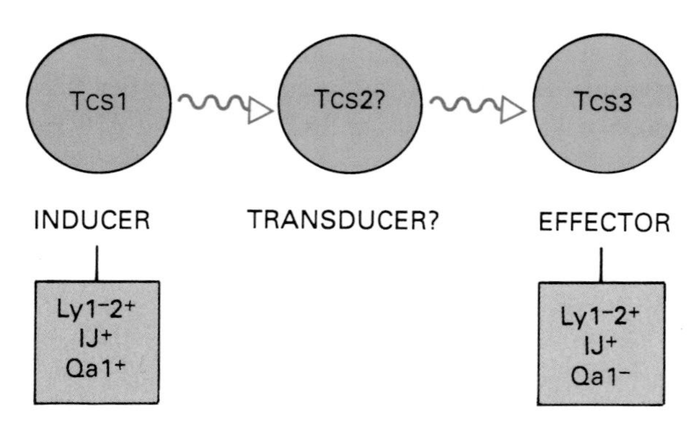

Fig.11.31 Subsets in contrasuppression

the basis of the expression of CD4, CD5 and CD8, their IJ phenotype, and the absence of Qa1 on some cells in the contrasuppressor but not suppressor circuits, and the effector cell of contrasuppression adhered to a particular lectin called *vicia villosa*. There is, however, more recent evidence that $\gamma\delta$ T cells can function as contrasuppressor cells (Fugihashi *et al.* 1992).

11.2.6 Veto cells

Another cell with suppressive activity for immune responses is the so-called **veto cell** (Fig.11.32). The main feature of a veto cell is that when it is recognized by a T cell, the latter becomes inactivated. *Note that the cell that is inactivated is a T cell that actually recognizes the veto cell.* It is not yet clear whether this inactivation is a result of suppression or killing. The relationship between veto cells and other suppressor cells also remains to be elucidated. It is notable, though, that the suppressor effector cell (Tse) and anti-idiotypic T cells could have veto activity against specific helper T cells that recognize peptide–MHC complexes and/or idiotypes on the former cells.

Veto cells seem to belong to the T cell lineage. They can be obtained from bone marrow, thymus and fetal liver of normal mice, and also from spleens of nude mice, suggesting that at least some of these cells may be immature T cells or earlier precursors. However, they are not detectable in freshly-isolated preparations of spleen or lymph node cells from normal mice, although they can develop in cultures of spleen cells.

Veto cells have been described in a number of systems. For example, cytotoxic T cells can be susceptible to inactivation by veto cells. Cytotoxic cells seem to be sensitive at their precursor stage, after activation but before they have completely matured into an active cytotoxic effector cell (Miller 1980). It appears that the T cell that is inactivated recognizes peptide–MHC complexes that are expressed by the veto cell. Non-cytotoxic suppressor cells with veto activity for CTL have also been cloned from human bone marrow cultures (Takahashi *et al.* 1990). They were found to contain full-length transcripts for T cell receptor γ genes, although surprisingly no DNA rearrangement of these, nor of α, β, and δ, genes could be detected. These cells, which are specific for *self* MHC molecules, may play a role in the extrathymic induction of self-tolerance (Section 5.2.5.) (Hiruma *et al.* 1992).

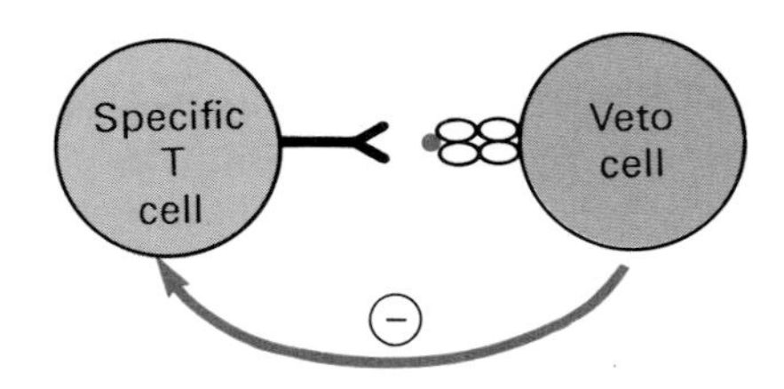

Fig.11.32 Veto cells

11.3 Immune regulation by cytokines

Cytokines (Section 4.4.1) acting alone or in combination can both stimulate and inhibit immune responses. The production of cytokines with suppressive activities is thus one mechanism by which T cells can suppress immune responses (Section 11.2). Of course, the situation is complicated because the same cytokine can have different effects when it binds to its receptor on different cell types, or to the same cell at different stages of development or activation. Moreover, the cellular response can be altered completely in the presence of other cytokines; thus the spectrum of cytokines produced during the course of an immune response determines its nature. Nevertheless, the pathways by which cytokines regulate immune responses, and their contribution to host defence against infection by various organisms, are now beginning to be unravelled, and some are discussed in this section.

Despite the focus of this section on regulation by cytokines through binding to specific cell surface receptors, it is important to bear in mind the importance of *intermolecular* interactions between molecules expressed by different cell types. This is well-illustrated by the effects of monoclonal antibodies specific for molecules expressed by lymphocytes (Chapters 4 and 8). Such interactions determine whether, for instance, $CD8^+$ T cells develop into CTL or into cells with the capacity to suppress antibody production by B cells.

At least in part this correlates with whether or not they express the CD28 molecule (Section 4.2.8), and whether they interact with a $CD4^+$ T cell expressing CD45RA or CD45RO (Section 4.2.5). Even so, the outcome is likely to be controlled by cytokines. Those produced by subsets of $CD4^+$ T cells regulate the expression of genes for cytokines or cytolysins by $CD8^+$ T cells, and in turn some cytokines secreted by subsets of $CD8^+$ T cells may have inhibitory effects on B cell responses.

It is also important to note that the function of cytokines can be regulated by other mechanisms, in addition to regulation of gene expression for cytokines and their receptors. At the level of the *cytokine* itself, these include the production of soluble cytokine receptors and other cytokine-binding proteins, and at the *receptor* level, by the production of receptor agonists and down-regulation of cytokine receptors by extracellular proteases or altered recycling and intracellular degradation of the molecule; examples of some of these are in Section 9.5.

Cytokines with suppressive functions

The patterns of cytokines produced by different 'subsets' of $CD4^+$ T cells and stimulation of different cellular responses are noted in Sections 4.4.2 and 8.4.5. Here we focus on three particular cytokines, IFNγ produced by T_H1 cells, and IL-4 and IL-10 produced by T_H2 cells, as examples of some inhibitory or suppressive functions associated with these molecules and cells.

Counter-regulation by IFNγ and IL-4. IFNγ is an important macrophage activating factor (Section 10.4.4). Amongst its effects on macrophages are the induction of potent microbicidal mechanisms, increased expression of the

($Fc_{\gamma}RI$) Fc receptor for mouse IgG2a, the principle class of antibody produced by B cells in response to IFNγ (thus directing certain extracellular organisms to macrophages), and increased expression of the IL-2 receptor. The latter is required for the synergistic effects of IFNγ with IL-2, also produced by T_H1 cells, in the induction of tumouricidal activity in macrophages. Many of the macrophage activating effects of IFNγ are *inhibited* by IL-4, produced by T_H2 cells.

IL-4, together with other cytokines produced by T_H2 cells, induces the production of IgE by B cells which binds to specific Fc receptors on a variety of cell types, including eosinophils and mast cells, whose production and activation is controlled particularly by these cytokines (see below) and can be inhibited by others. An additional effect of IL-4 is to reverse the FcR receptor-mediated inhibition of B cell responses (Section 8.4.2). This function can in turn be inhibited by IFNγ, which also inhibits the proliferation of T_H2 cells in response to cytokines such as IL-4. It is clear, therefore, that cytokines can have antagonistic effects on each other, thereby altering the course of an immune response.

Interleukin-10. IL-10 was first defined as a 'cytokine synthesis inhibitory factor' (CSIF) (Fiorentino *et al.* 1989). This cytokine is produced by T_H2 cells and inhibits the synthesis of most if not all cytokines by T_H1, but not T_H2, cells in response to antigen-presenting cells. However, IL-10 does not inhibit cytokine production in response to immobilized anti-CD3 antibodies or the mitogen concanavalin A, leading to the suggestion that this cytokine may act at the level of the accessory cell, perhaps by inhibiting the delivery of costimulatory signals (e.g. Section 4.2.8). Despite its effects on cytokine synthesis, IL-10 does not inhibit antigen-specific proliferation of T_H2 cells, provided that IL-2 is supplied exogenously, another example of counter-regulation in cytokine responses. Similar effects of IL-10 have also been observed with CTL clones exhibiting the T_H1 pattern of cytokine secretion: inhibition of cytokine production but not proliferation of the cells.

IL-10 is secreted by T_H2 cells as a 35–40 kDa homodimer, and is also produced by $CD5^+$ B cells (Section 8.5) and some mast cell lines, but not by T_H1 cells or CTL. The IL-10 gene has been cloned in mouse, and human (Viera *et al.* 1991). The mouse IL-10 gene encodes a 178 amino acid polypeptide (including an 18 amino acid signal sequence) which contains two N-glycosylation sites. Differential glycosylation accounts for the different monomeric molecular weights observed for IL-10, of 16 kDa and 20 kDa.

Monoclonal antibodies specific for IL-10 have been produced, some of which block CSIF activity almost completely, and it is likely that IL-10 is exclusively responsible for this function of T_H2 cells (Mosmann *et al.* 1990).

Because of the suppressive effects of IL-10 on cytokine synthesis, T_H2 cells suppress responses of the T_H1 subset as well as macrophages (note also the effects of IL-4 on macrophage activation; above).

At the same time, IL-10 has stimulatory effects on other cell types. For example, it acts in synergy with IL-3 and IL-4 to promote the optimal growth of mast cells (although IL-10 alone does not support growth of these cells). Therefore, as well as T_H1 cells, T_H2 can act as suppressor cells for different cell types or each other, but their additional stimulatory functions divert the course of an immune response in a different direction.

Cytokines and host defence

By virtue of their ability to promote the differentiation, development and activation of different cell types, cytokines are obviously central to host defence against infection. However, some pathogenic organisms have subverted the functions of different cytokines for their own use, for example, by stealing the genes for these molecules and incorporating them into their own genomes, or by inducing a particular pattern of cytokine synthesis in the host that directs the immune response away from them. Of course, during evolution, the host's response is to evolve mechanisms for counteracting the various strategies used by pathogens. Analysis of who evolved which mechanism for what purpose is therefore complicated, and it is sometimes difficult to determine whether a particular response acts for the benefit of the host (resulting in immunity) or the pathogen (resulting in immunopathology); sometimes however this is absolutely clear, with elimination of the pathogen or death of the host!

One example of a pathogen that has apparently hijacked a cytokine gene is the Epstein–Barr virus (EBV), which causes infectious mononucleosis and Burkitt's lymphoma in humans. This virus contains an open reading frame, called BCRF1, which has extensive homology to the human IL-10 gene. BCRF1 is homologous only to the region of the IL-10 gene encoding the mature protein, but not the signal sequence or the 5′ and 3′ untranslated regions. The BCRF1 product, after expression in a suitable system, inhibits IFNγ synthesis by T cells and perhaps also by NK cells (Hsu *et al.* 1991). It has therefore been suggested that this virus has captured a processed IL-10 gene, and that selective pressures maintain its similarity to the cellular gene. The BCRF1 product may protect the virus from cytocidal activities that would otherwise be induced by IFNγ in the infected host cell. Some poxviruses may have adopted a similar strategy, but by acquiring a *cytokine receptor* from the host, because their genomes contain a region homologous to the TNF receptor.

Different subsets of T cells are preferentially induced during particular infections. For example, infection with certain parasitic helminths (worms) such as *Heligmosomoides polygyrus*, *Nippostrongylus brasiliensis*, and *Schistosoma mansoni*, results in the increased production of eosinophils and mast cells, and greatly increased serum levels of IgE, responses associated with T_H2 cytokines (reviewed in Finkelman *et al.* 1991). In case of the former organism, these cytokines seem to be essential for host protection, whereas in the latter two they either are not required or are deleterious respectively. In some cases, then, these responses may be an immunopathological consequence of infection, particularly, for example, because most of the IgE antibodies produced are not actually specific for the organism.

On the other hand, infection of mice with the protozoan *Leishmania major* results in either T_H1 or T_H2 cells depending on the strain, and this will be considered in more detail here (reviewed in Locksley and Scott 1991). Some strains of mice (e.g. C57BL/6 and C3H/HeN) are resistant to Leishmania and successfully overcome infection, whereas others (e.g. Balb/c) are susceptible to progressive and fatal disease. These outcomes are respectively associated with expansion of T_H1 and T_H2 cells in the different strains. For instance, if T_H2 cells are ablated from susceptible mice, by depletion of CD4 T cells or sublethal irradiation, they become capable of overcoming the infection.

It has been shown that immunization of mice with different components of Leishmania selectively induces different subsets of $CD4^+$ T cells. Immunization of resistant strains of mice with a peptide from part of the major surface glycoprotein of this organism, gp63, results in DTH responses (Section 9.5.4) and protective immunity associated with expansion of T_H1 cells. On the other hand, immunization of susceptible strains with a peptide that is expressed by Leishmania in tandem repeats, enhances disease progression associated with expansion of T_H2 cells.

At present it is not clear why these peptides selectively induce different subsets of T cells, but this could be due to differences in their T cell receptor repertoires, or the mode of antigen presentation. With respect to the latter, responses of T_H1 cells can be induced by 'non-B' accessory cells, whereas those of T_H2 cells can be induced by these cells as well as B cells; possibly these antigens are processed and presented differently by different accessory cells. These subsets also have different costimulatory requirements, T_H2 cells responding simply to T cell receptor occupancy and costimulation by IL-1, whereas T_H1 cells may have more stringent requirements for activation (e.g. immunostimulation; Section 3.4.3).

The immunity associated with the induction of T_H1 cells seems likely to result from the particular spectrum of cytokines produced by these cells. Leishmania is an intracellular pathogen of macrophages, and activation of macrophages by IFNγ leads to the elimination of the organism. However, production of IL-4 by T_H2 cells inhibits the macrophage activating effects of IFNγ, and cytokine synthesis by T_H1 cells is also inhibited by IL-10 (see above). IL-4 is likely to be a major mediator of progressive disease because susceptible strains can overcome infection if they are treated with anti-IL-4 antibodies. These observations demonstrate the importance of cytokines in host defence, and some of their suppressive effects are also clearly implicated in disease.

Further reading

Idiotypes and idiotypic networks

Bona, C. (ed.) (1988). *Elicitation and use of anti-idiotypic antibodies and their biological applications*. CRC Press, Boca Raton, Florida.

Davie, J.M., Seiden, M.V., Greenspan, N.S., Lutz, C.T., Bartholow, T.L., and Clevenger, B.L. (1986). Structural correlates of idiotopes. *Annual Review of Immunology*, **4**, 147–65.

Finberg, R.W. and Ertl, H.C. (1986). Use of T cell specific anti-idiotypes to immunize against viral infections. *Immunological Reviews*, **90**, 129–55.

Gaulton, G.N. and Greene, M.I. (1986). Idiotypic mimicry of biological receptors. *Annual Review of Immunology*, **4**, 253–80.

Janeway, C. (1990). Immunotherapy by peptides? (News). *Nature*, **341**, 482–3.

Jerne, N.K. (1984). Idiotype networks and other preconceived ideas. *Immunological Reviews*, **79**, 5–24.

Kohler, H., Urbain, J., and Cazenave, P.A. (ed.) (1984). *Idiotypy in biology and medicine*. Academic Press, New York.

Marcos, M.A.R., de la Hera, A., Gaspar, M.L., Marquez, C., Bellas, C., Mampaso, F., Toribio, M.L., and Martinez, C. (1986). Modification of emerging repertoires by immunosuppression in immunodeficient mice results in autoimmunity. *Immunological Reviews*, **94**, 51–74.

Martinez, A.C., Pereira, P., Toribio, M.L., Marcos, M.A.R., Bandeira, A., de la Hera, A., Marquez, C., Cazenave, P.-A., and Coutinho, A. (1988). The participation of B cells and antibodies in the selection and maintenance of the T cell repertoire. *Immunological Reviews*, **101**, 191–215.

Moller, G. (1987). Anti-idiotype antibodies as immunogens. *Immunological Reviews*, **90**, 5–155.

Periera, P., Bandeira, A., Coutinho, A., Marcos, M.A., Toribio, M., and Martinez, C. (1989). V-region connectivity in T-cell repertoires. *Annual Review of Immunology*, **7**, 209–49.

Rakewsky, K. and Takemori, T. (1983). Genetics, expression, and function of idiotypes. *Annual Review of Immunology*, **1**, 569–607.

Sim, G.K., MacNeil, I.A., and Augustin, A.A. (1986). T helper cell receptors: idiotypes and repertoire. *Immunological Reviews*, **90**, 49–72.

Varela, F.J. and Coutinho, A. (1991). Second generation immune networks. *Immunology Today*, **12**, 159–66.

Zanetti, M. and Katz, D.H. (1985). Self-recognition, autoimmunity, and internal images. *Current Topics in Microbiology and Immunology*, **119**, 111–26.

Suppressor T cells and suppressor pathways

Asherson, G.L., Colizzi, V., and Zembala, M. (1986). An overview of suppressor cell circuits. *Annual Review of Immunology*, **4**, 37–68.

Bloom, B.R., Salgame, P., and Diamond, B. (1992). Revisiting and revising suppressor T cells. *Immunology Today*, **13**, 131–6.

Cooper, J., Eichmann, K., Melchers, J., Simon, M.M., and Weltzein, H.U. (1984). Network regulation among T cells: qualitative and quantitative studies on suppression in the non-immune state. *Immunological Reviews*, **79**, 63–86.

Dorf, M.E. and Benacerraf, B. (1984). Suppressor cells and immunoregulation. *Annual Review of Immunology*, **2**, 127–58.

Fink, P.J., Shimonkevitz, R.P., and Bevan, M.J. (1988). Veto cells. *Annual Review of Immunology*, **6**, 115–37.

Flood, P.M. (1985). The role of suppressor cells in maintaining tolerance to self molecules—a commentary. *Journal of Molecular and Cellular Immunology*, **2**, 140.

Lanzavecchia, A. (1989). Is suppression a function of class II-restricted cytotoxic T cells? *Immunology Today*, **10**, 157–69.

Murphy, D. (1987). The I-J puzzle. *Annual Review of Immunology*, **5**, 405–27.

Rammensee, H.G. (1989). Veto function *in vitro* and *in vivo*. *International Review of Immunology*, **4**, 175–91.

Sanders, M.E., Makgoba, M.W., and Shaw, S. (1988). Human naïve and memory T cells: reinterpretation of helper-inducer and suppressor-inducer subsets. *Immunology Today*, **9**, 195–9.

Sercarz, E. and Krzych, U. (1991). The distinctive specificity of antigen-specific suppressor T cells. *Immunology Today*, **12**, 111–18.

Strober, S. (1984). Natural suppressor (NS) cells, neonatal tolerance, and total lymphoid irradiation. Exploring obscure relationships. *Annual Review of Immunology*, **2**, 219–37.

Webb, D.R., Kapp, J.A., and Pierce, C.W. (1983). The biochemistry of antigen-specific T-cell factors. *Annual Review of Immunology*, **1**, 423–38.

Cytokines with suppressive functions (see also Chapters 4, 8, 9, and 10)

Mosmann, T.R. and Moore, K.W. (1991). The role of IL-10 in crossregulation of T_H1 and T_H2 responses. In *Immunoparasitology today* (ed. C. Ash and R.B. Gallagher), pp. A49–A53. Elsevier Trends Journals, Cambridge.

Sher, A. and Coffman, R.L. (1992). Regulation of immunity to parasites by T cells and T cell-derived cytokines. *Annual Review of Immunology*, **10**, 385–409.

Literature cited

Augustin, A. and Cosenza, A. (1976). Expression of new idiotypes following neonatal idiotypic suppression of a dominant clone. *European Journal of Immunology*, **6**, 497–501.

Bona, C., Mond, J.J., Stein, K.E., House, S., Lieberman, R., and Paul, W.E. (1979). Immune response to levan. III. The capacity to produce anti-insulin antibodies and cross-reactive idiotypes appears late in ontogeny. *Journal of Immunology*, **123**, 1484–90.

Bona, C.A., Heber-Katz, E., and Paul, W.E. (1981). Idiotype-anti-idiotype regulation. I. Immunization with a levan-binding myeloma protein leads to the appearance of auto-anti-(anti-idiotype) antibodies and to the activation of silent clones. *Journal of Experimental Medicine*, **153**, 951–67.

Bottomly, K. and Mosier, D.E. (1979). Mice whose B cells cannot produce the T15 idiotype also lack an antigen-specific helper T cell required for T15 expression. *Journal of Experimental Medicine*, **150**, 1399–409.

Bottomly, K., Janeway, C.A. Jr., Mathieson, B.J., and Mosier, D.E. (1980). Absence of an antigen-specific helper T cell required for the expression of the T15 idiotype in mice treated with anti-μ antibody. *European Journal of Immunology*, **10**, 159–63.

Bruck, C., Co, M.S., Slaoui, M., Gaulton, G.N., Smith, T., Fields, B.N., Mullins, J.I., and Greene, M.I. (1986). Nucleic acid sequence of an internal image-bearing monoclonal anti-idiotype and its comparison to the sequence of the external antigen. *Proceedings of the National Academy of Sciences USA*, **83**, 6578–82.

Coutinho, A., Marquez, C., Araujo, P.M.F., Pereira, P., Toribio, M.L., Marcos, M.A., Martinez, C. (1987). A functional idiotypic network of T helper cells and antibodies limited to the compartment of 'naturally' activated lymphocytes in normal mice. *European Journal of Immunology*, **17**, 821–5.

Ding, A., Nathan, C.F., Graycar, J., Derynck, R., Stuehr, D.J., and Srimal, S. (1990). Macrophage deactivating factor and transforming growth factors-β1 -β2 and -β3 inhibit induction of macrophage nitrogen oxide synthesis by IFN-gamma. *Journal of Immunology*, **145**, 940–4.

Eardley, D.D., Shen, F.W., Cantor, H., and Gershon, R.K. (1979). Genetic control of immunoregulatory circuits. Genes linked to the Ig locus govern communication between regulatory T-cell sets. *Journal of Experimental Medicine*, **150**, 44–50.

Ertl, H.C. and Bona, C.A. (1988). Criteria to define anti-idiotypic antibodies carrying the internal image of an antigen. *Vaccine*, **6**, 80–4.

Finkelman, F.D., Pearce, E.J., Urban, J.F., and Sher, A. (1991). Regulation and biological function of helminth-induced cytokine responses. In *Immunoparasitol today* (ed. C. Ash and R.B. Gallagher), pp. A62–A66. Elsevier Trends Journals, Cambridge.

Fiorentino, D.F., Bond, M.W., and Mosmann, T.R. (1989). Two types of mouse helper T cell. IV. Th2 clones secrete a factor that inhibits cytokine production by Th1 clones. *Journal of Experimental Medicine*, **170**, 2081–95.

Francotte, M. and Urbain, J. (1984). Induction of anti-tobacco mosaic virus antibodies in mice by rabbit antiidiotypic antibodies. *Journal of Experimental Medicine*, **160**, 1485–94.

Fujihashi, K., Taguchi, T., Aicher, W.K., McGhee, J.R., Bluestone, J.A., Eldridge, J.H., and Kiyono, H. (1992). Immunoregulatory functions for murine intraepithelial lymphocytes: γ/δ T cell receptor-positive (TCR$^+$) T cells abrogate oral tolerance, while α/β T cells provide B cell help. *Journal of Experimental Medicine*, **863**, 695–707.

Gaulton, G.N. and Greene, M.I. (1989). Inhibition of cellular DNA synthesis by reovirus occurs through a receptor-linked signalling pathway that is mimicked by anti-idiotypic, antireceptor antibody. *Journal of Experimental Medicine*, **169**, 197–211.

Germain, R.N. and Benacerraf, B. (1981). A single major pathway of T-lymphocyte interactions in antigen-specific immune suppression. *Scandinavian Journal of Immunology*, **13**, 1–10.

Gershon, R.K. and Kondo, K. (1971). Infectious immunological tolerance. *Immunology*, **21**, 903–14.

Gershon, R.K., Eardley, D.D., Durum, D.D., Green, D.R., Shen, F.W., Yamauchi, K., Cantor, H., and Murphy, D.B. (1981). Contrasuppression. A novel immunoregulatory circuit. *Journal of Experimental Medicine*, **153**, 1533–46.

Grzych, J.M., Capron, M., Lambert, P.H., Dissous, C., Torres, S., and Capron, A. (1985). An anti-idiotype vaccine against experimental schistosomiasis. *Nature*, **316**, 74–6.

Hathcock, K.S., Gurish, M.F., Nisonoff, A., Conger, J.D., Hodes, R.J. (1986). Influence of helper T cells on the expression of a murine intrastrain cross reactive idiotype. *Proceedings of the National Academy of Sciences USA*, **83**, 155–9.

Herzenberg, L.A. and Herzenberg, L.A. (1974). Short-term and chronic allotype suppression in mice. *Contemporary Topics in Immunobiology*, **3**, 41–75.

Hiernaux, J., Bona, C., and Baker, P.J. (1981). Neonatal treatment with low doses of anti-idiotypic antibodies leads to the expression of a silent clone. *Journal of Experimental Medicine*, **153**, 1004–8.

Hiruma, K., Nakamura, H., Henkart, P.A., and Gress, R.E.

(1992). Clonal deletion of postthymic T cells: veto cells kill precursor cytotoxic T lymphocytes. *Journal of Experimental Medicine*, **863**, 863–8.

Hsu, D., Malefyt, R., Fiorentino, D., Dang, M.-N., Vieira, P., deVries, J., Spits, H., Mosmann, T., and Moore, K. (1991). Expression of interleukin 10 activity by Epstein–Barr virus protein BCRF1. *Science*, **248**, 830–2.

Jerne, N.K. (1974). Towards a network theory of the immune system. *Annals of Immunology (Paris)*, **125C**, 373–89.

Kang, C.Y., Brunck, T.K., Kieber-Emmons, T., Blalock, J.E., and Kohler, H. (1988). Inhibition of self-binding antibodies (autobodies) by a V_H derived peptide. *Science*, **240**, 1034–41.

Kapp, J.A., Pierce, C.W., Schlossman, S., and Benacerraf, B. (1974). Genetic control of immune responses in mice. V. Stimulation of suppressor T cells in nonresponder mice by the terpolymer L-glutamic acid 60-L-alanine 30-L-tyrosine 10 (GAT). *Journal of Experimental Medicine*, **140**, 648–59.

Kelsoe, G. and Cerny, J. (1979). Reciprocal expansions of idiotypic and anti-idiotypic clones following antigen stimulation. *Nature*, **279**, 333–4.

Kunkel, H.G., Mannik, M., and Williams, R.C. (1963). Individual antigenic specificities of isolated antibodies. *Science* **140**, 1218–19.

Locksley, R. and Scott, P. (1991). Helper T-cell subsets in mouse leishmaniasis: induction, expansion and effector function. In *Immunoparasitol today* (ed. C. Ash and R.B. Gallagher), pp. A58–A61. Elsevier Trends Journals, Cambridge.

McNamara, M.K., Ward, R.E., and Kohler, H. (1984). Monoclonal idiotope vaccine against *Streptococcus pneumoniae* infection. *Science*, **220**, 1325–6.

Martinez, C., Bragado, R., de la Hera, A., Toribio, M.L., Marcos, M.A., Bandeira, A., Pereira, P., and Coutinho, A. (1986). Functional and biochemical evidence for the recognition of T cell receptors by monoclonal antibodies to an immunoglobulin idiotype. *Journal of Molecular and Cellullar Immunology*, **2**, 307–13.

Migliorini, P., Ardman, B., Kaburaki, J., and Schwartz, R.S. (1987). Parallel sets of auto-antibodies in MRL-lpr/lpr mice. An anti-DNA, anti-SmRNP, anti-gp70 network. *Journal of Experimental Medicine*, **165**, 483–99.

Migliorini, P., Ardman, B., Kaburaki, J., and Schwartz, R.S. (1987). Parallel sets of auto-antibodies in MRL-lpr/lpr mice. An anti-DNA, anti-SmRNP, anti-gp70 network. *Journal of Experimental Medicine*, **165**, 483–99.

Miller, R.G. (1980). An immunological suppressor cell inactivating cytotoxic T-lymphocyte precursor cells recognizing it. *Nature*, **287**, 544–6.

Modlin, R.L., Brenner, M.B., Krangel, M.S., Duby, A.D., and Bloom, B.R. (1987). T-cell receptors of human suppressor cells. *Nature*, **329**, 541–5.

Morahan, G., Berek, C., and Miller, J.F. (1983). An idiotypic determinant formed by both immunoglobulin constant and variable rgions. *Nature*, **301**, 720–2.

Mosmann, T.R., Schumacher, J.H., Fiorentino, D.F., Leverah, J., Moore, K.W., and Bond, M.W. (1990). Isolation of monoclonal antibodies specific for IL-4. IL-5 and IL-6, and a new Th2-specific cytokine, Cytokine Synthesis Inhibition Factor, by using a solid-phase radioimmunoadsorbant assay. *Journal of Immunology*, **145**, 2938–45.

Offner, H., Hashim, G.A., and Vandenbark, A.A. (1991). T cell receptor peptide therapy triggers autoregulation of experimental encephalomyelitis. *Science*, **251**, 430–2.

Ottenhoff, T.H., and Mutis, T. (1990). Specific killing of cytotoxic T cells and antigen-presenting cells by CD4$^+$ cytotoxic T cell clones. A novel potentially immunoregulatory T–T interaction in man. *Journal of Experimental Medicine*, **171**, 2011–24.

Oudin, J. and Cazenave, P.A. (1971). Similar idiotype specificities in immunoglobulin fractions with different antibody functions or even without detectable antibody function. *Proceedings of the National Academy of Sciences USA*, **68**, 2616–20.

Paul, W.E. and Bona, C. (1982). Regulatory idiotopes and immune networks: a hypothesis. *Immunology Today*, **3**, 230–4.

Pawelec, G., Fernandez, N., Brocker, T., Schneider, E.M., Festenstein, H., and Wernet, P. (1988). DY determinants, possibly associated with novel class II molecules, stimulate autoreactive CD4$^+$ T cells with suppressive activity. *Journal of Experimental Medicine*, **167**, 243–61.

Perlmutter, R.M., Crews, S.T., Douglas, R., Sorensen, G., Johnson, N., Nivera, N., Gearhart, P.J., and Hood, L. (1984). The generation of diversity in phosphorylcholine-binding antibodies. *Advances in Immunology*, **35**, 1–37.

Radbruch, A., Zaiss, S., Kappen, C., Bruggeman, M., Beyreuther, K., and Rajewsky, K. (1985). Drastic change in idiotypic but not antigen-binding specificity of an antibody by a single amino-acid substitution. *Nature*, **315**, 506–8.

Raychaudhuri, S., Sacki, Y., Fuji, H., and Kohler, H. (1986). Tumour-specific idiotype vaccines. Generation and characterization of internal image tumor antigen. *Journal of Immunology*, **137**, 1743–9.

Rodkey, L.S. (1974). Studies of idiotypic antibodies. Production and characterization of autoantiidiotypic sera. *Journal of Experimental Medicine*, **139**, 712–20.

Rubinstein, L., Ming, Y., and Bona, C. (1984). Idiotype-anti-idiotype regulation. II. Activation of A48Id silent clone by administration at birth of idiotopes is related to activation of A48Id specific helper T cells. *Journal of Experimental Medicine*, **156**, 506–21.

Sercarz, E.E. and Metzger, D.W. (1980). Epitope-specific and idiotype-specific cellular interactions in a model protein antigen system. *Springer Seminars in Immunopathology*, **3**, 145–70.

Sercarz, E.E., Yowell, R.L., Turkin, D., Miller, A., Araneo, B.A., and Adorini, L. (1978). Different functional specificity repertoires for suppressor and helper T cells. *Immunological Reviews*, **39**, 108–36.

Schechter, Y., Maron, R., Elias, D., and Cohen, I.R. (1982). Autoantibodies to insulin receptor spontaneously develop as anti-idiotypes in mice immunized with insulin. *Science*, **216**, 542–5.

Stumpf, R., Heuer, J., and Kolsch, E. (1977). Suppressor T cells in low zone tolerance. I. Mode of action of suppressor cells. *European Journal of Immunology*, **7**, 74–85.

Takahashi, T., Mafune, K., and Maki, T. (1990). Cloning of self-major histocompatibility complex antigen-specific suppressor cells from adult bone marrow. *Journal of Experimental Medicine*, **172**, 901–10.

Vakil, M. and Kearney, J.F. (1986). Functional characterization of monoconal auto-anti-idiotype antibodies isolated from the early B cell repertoire of BALB/c mice. *European Journal of Immunology*, **16**, 1151–8.

Victor-Kobrin, C., Manser, T., Moran, T.M., Imanishi-Kari, T., Gefter, M., and Bona, C.A. (1985*a*). Shared idiotopes among antibodies encoded by heavy chain variable region (V_H) gene members of the J558 V_H family as a basis for specificity. *Proceedings of the National Academy of Sciences USA*, **82**, 7696–700.

Victor-Kobrin, C., Bonilla, F.A., Bellon, B., and Bona, C.A. (1985*b*). Immunochemical and molecular characterization of regulatory idiotopes expressed by monoclonal antibodies exhibiting or lacking β 2-6 fructosan binding activity. *Journal of Experimental Medicine*, **162**, 647–62.

Vieira, P., de-Waal-Malefyt, R., Dang, M.N., Johnson, K.E., Kastelein, R., Fiorentino, D.F., deVries, J.E., Roncarolo, M.G., Mosmann, T.R., and Moore, K.W. (1991). Isolation and expression of human cytokine synthesis inhibitory factor cDNA clones: homology to Epstein–Barr virus open reading frame BCRFI. *Proceedings of the National Academy of Sciences USA*, **88**, 1172–6.

Weiler, I.J., Weiler, E., Sprenger, R., and Cosenza, H. (1977). Idiotype suppression by maternal influence. *European Journal of Immunology*, **7**, 591–7.

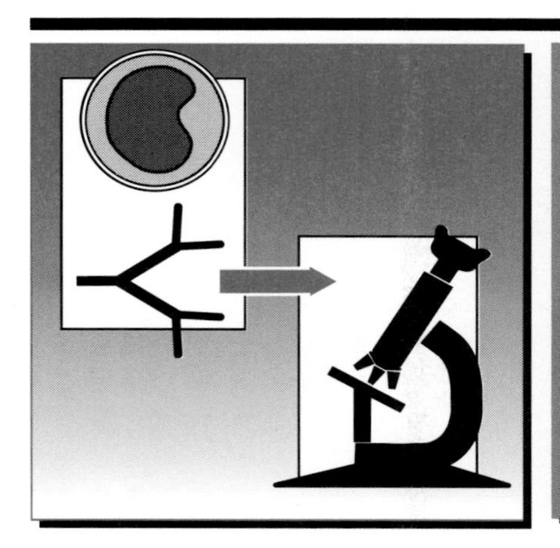

APPENDIX: CELLULAR AND MOLECULAR TECHNIQUES

A1 Production and use of antibodies

A1.1 Production of antisera

(a) **Xenoantisera** are used to detect differences between species.

Xenoantisera can be prepared to soluble or cellular antigens, e.g. hen egg lysozyme, human lymphocytes, mouse immunoglobulins.

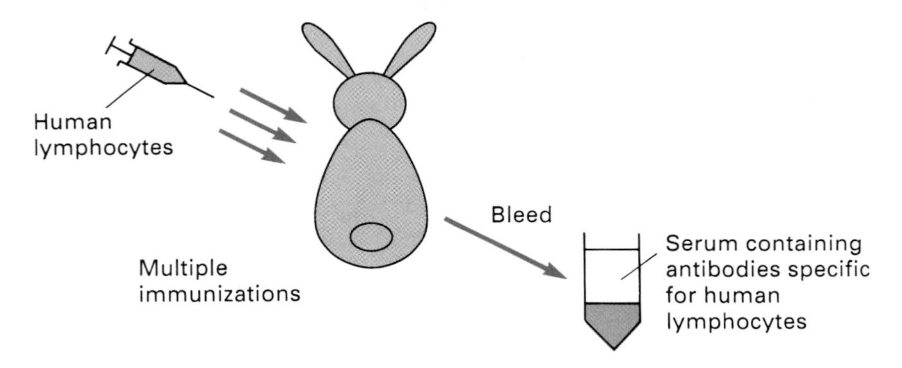

Rabbit anti-human lymphocyte serum

(b) **Alloantisera** are used to detect allelic differences between different members of the same species.

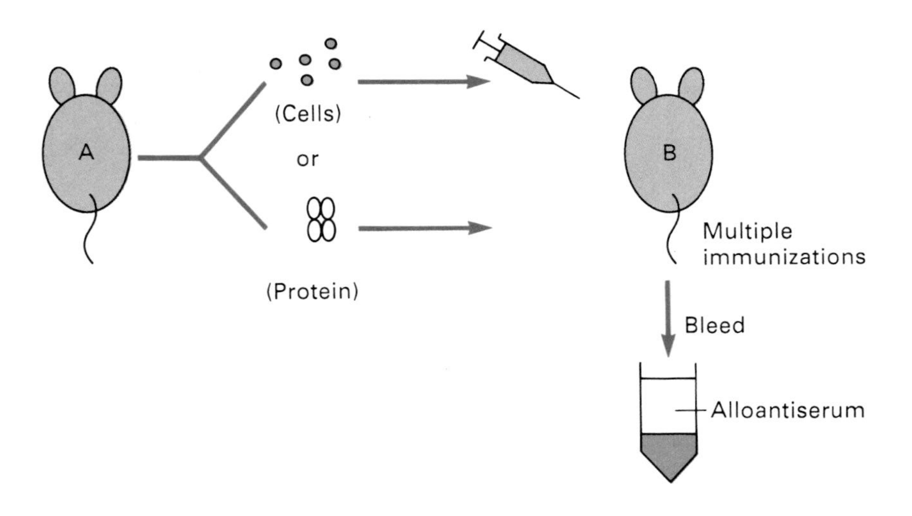

Xeno- and alloantisera contain a mixture of antibodies, each specific for a different epitope present in the immunizing antigen.

A1.2 Production of monoclonal antibodies

This technique allows large of amounts of a single antibody, specific for one epitope present on the immunizing antigen, to be produced.

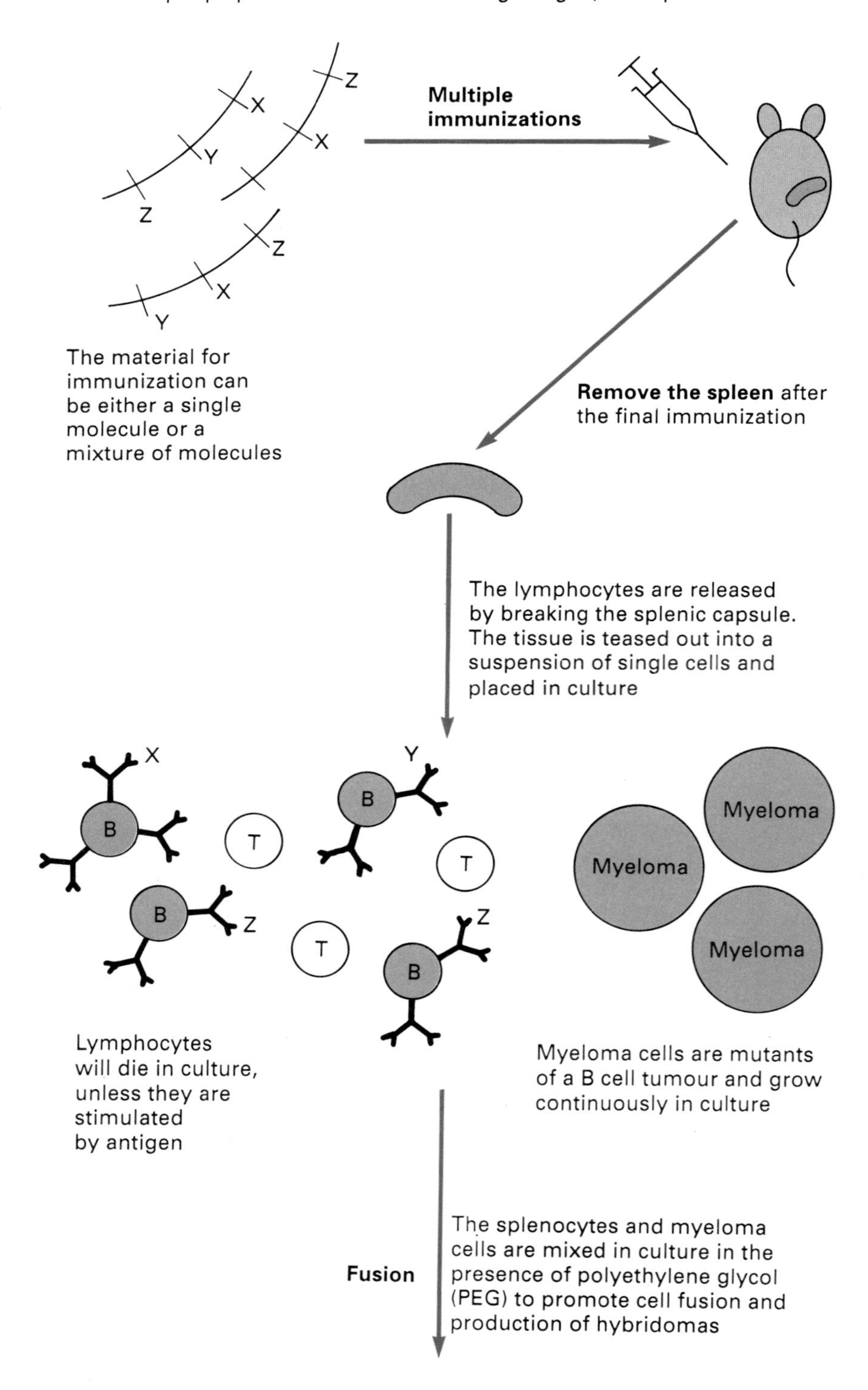

X
Y
Z
Z
X
Z
X
Y
Multiple immunizations
The material for immunization can be either a single molecule or a mixture of molecules
Remove the spleen after the final immunization
The lymphocytes are released by breaking the splenic capsule. The tissue is teased out into a suspension of single cells and placed in culture
X
B
T
Y
B
T
B
Z
T
Z
B
Myeloma
Myeloma
Myeloma
Lymphocytes will die in culture, unless they are stimulated by antigen
Myeloma cells are mutants of a B cell tumour and grow continuously in culture
Fusion
The splenocytes and myeloma cells are mixed in culture in the presence of polyethylene glycol (PEG) to promote cell fusion and production of hybridomas

T

B

Hybridoma

Hybridoma

X

Z

X

T

Z

Myeloma

In vitro selection

Selection media:
hypoxanthine, aminopterin,
thymidine (HAT)

1. Unfused B and T cells will die due to lack of antigen stimulation
2. Unfused myeloma cells will die – they are unable to grow in the presence of HAT
3. Hybridomas proliferate

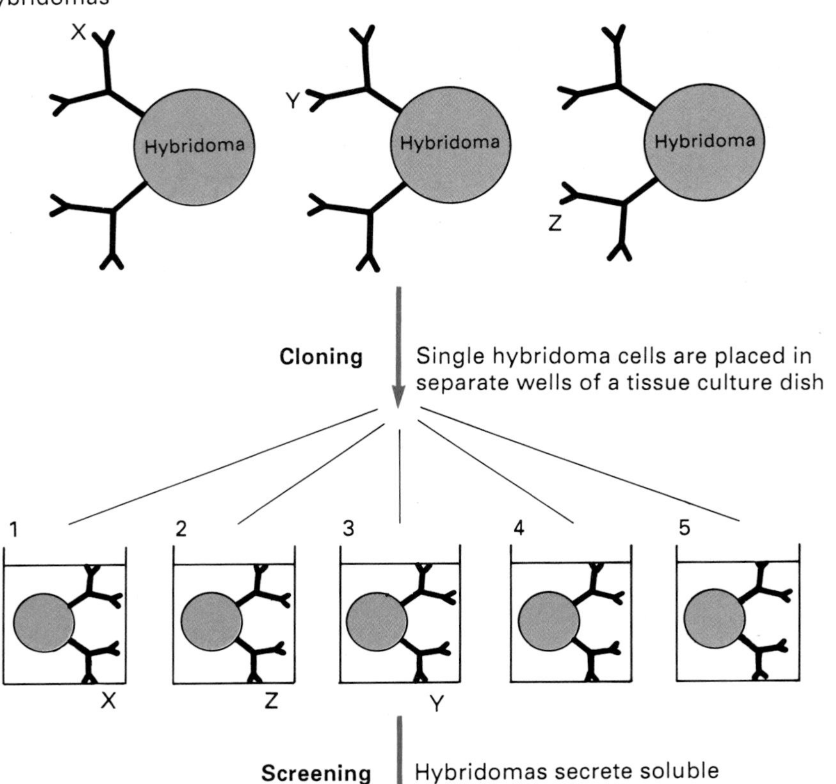

Screening

Hybridomas secrete soluble antibody into the tissue culture medium

Screen or assay tissue culture medium for antibody binding to the immunizing antigen. If the cloning is successful the supernatant from each cloning well will contain antibody produced by a single B cell hybridoma (monoclonal)

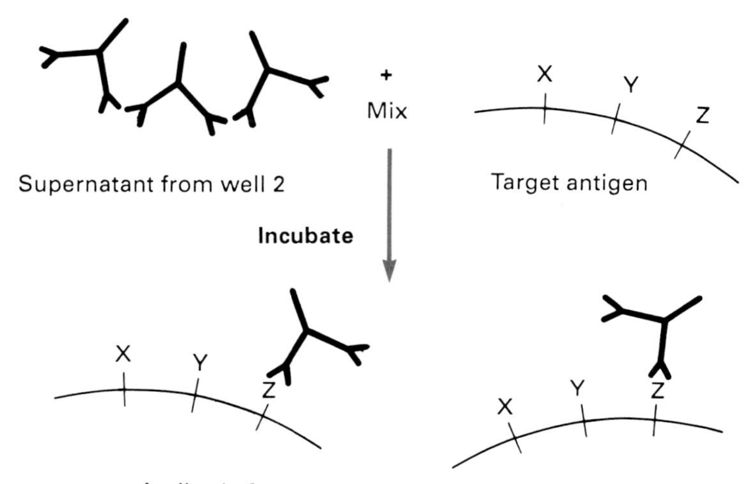

Antibody from well 2 will bind to the antigen Z

Antiserum specific for mouse immunoglobulins is used to detect binding (e.g. rabbit-anti-mouse Ig: RAM)

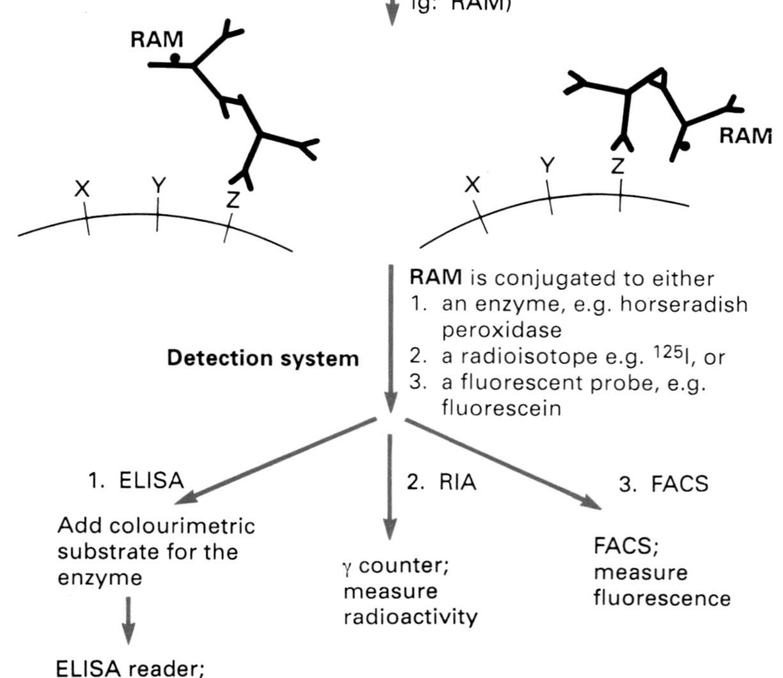

ELISA = enzyme-linked immunosorbent assay
RIA = radioimmunoassay
FACS = fluorescence-activated cell sorter

A1.3 Immunoprecipitation

Antisera and monoclonal antibodies can be used to characterize the molecule or antigen with which they interact. The technique of immunoprecipitation can be used to isolate the antigen from a mixture of molecules before further biochemical analysis is performed, e.g. using either gel electrophoresis in one or two dimensions or isoelectric focusing.

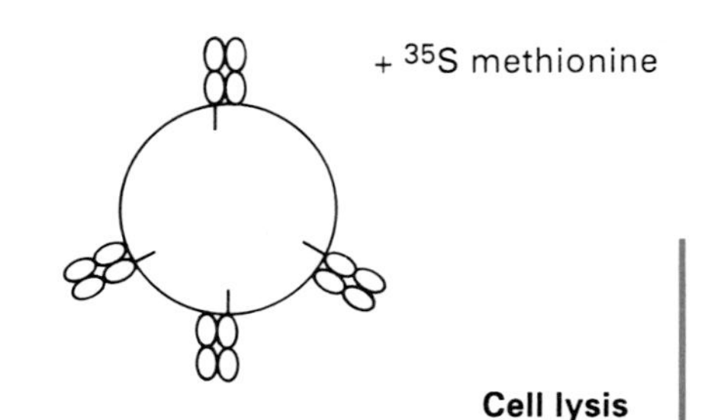

+ ^{35}S methionine

Cells expressing the antigen, for example, are incubated with a radioactive amino acid to label newly-synthesized protein, here, MHC class I molecules

Cell lysis

For intracellular proteins – cells are disrupted by physical techniques

For membrane proteins – cell membranes are solubilized with detergent

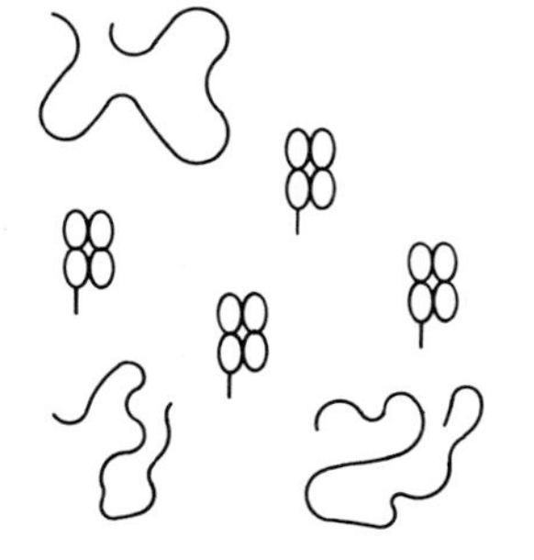

+

Incubate the antibody and lysate together

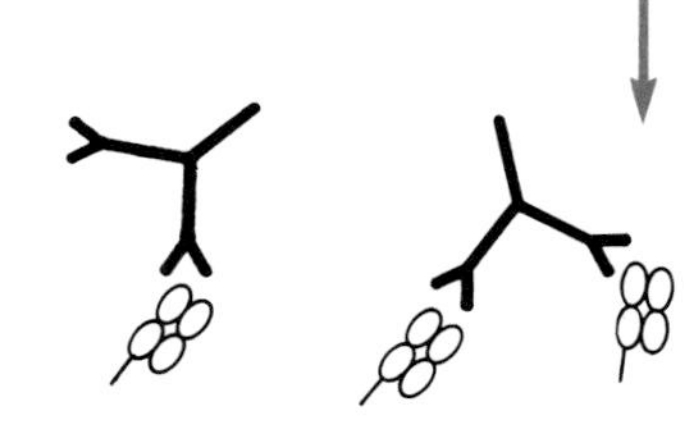

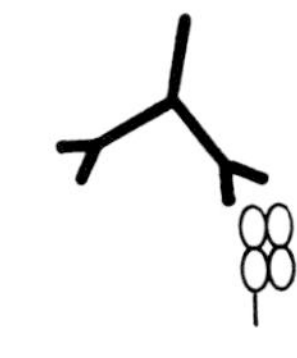

Add RAM coupled to Sepharose or magnetic beads to cross-link the immune complexes

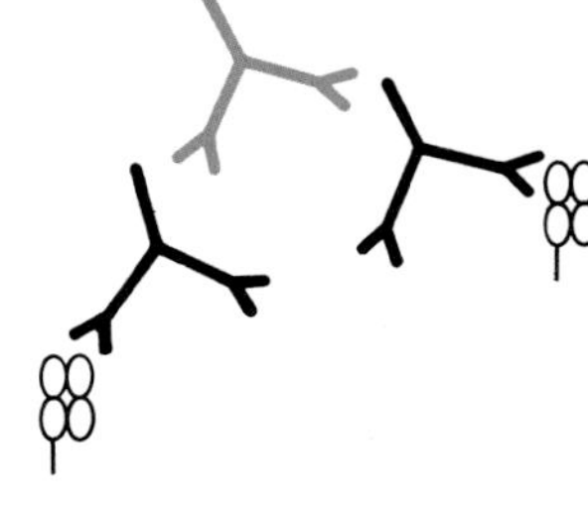

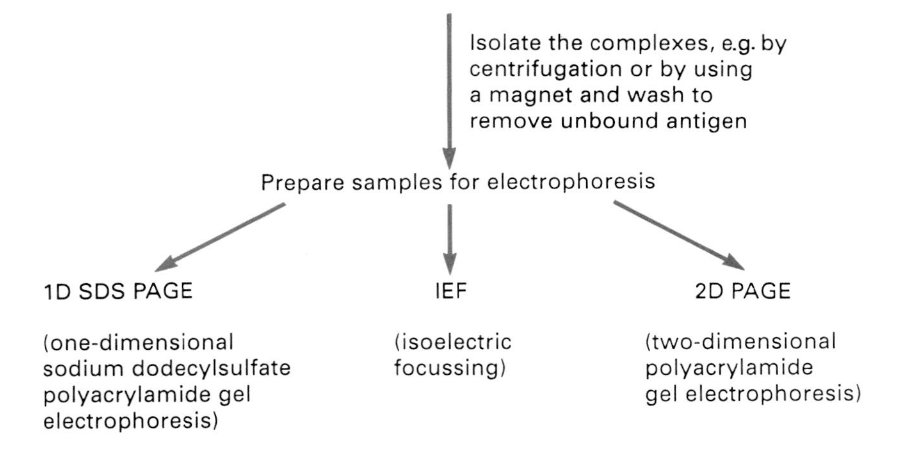

(a) 1D SDS polyacrylamide gel electrophoresis

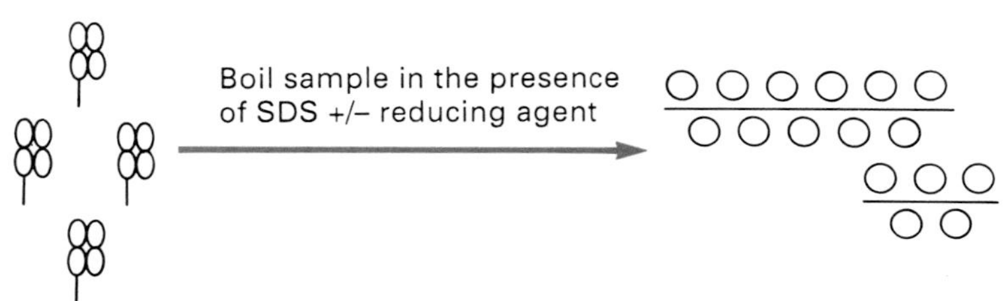

Non-covalent interactions between polypeptide chains are disrupted and the proteins are denatured. In the presence of reducing agents disulfide bonds are broken. The proteins bind SDS and become negatively charged

SDS polyacrylamide gel stained with Coomassie blue to visualize the proteins

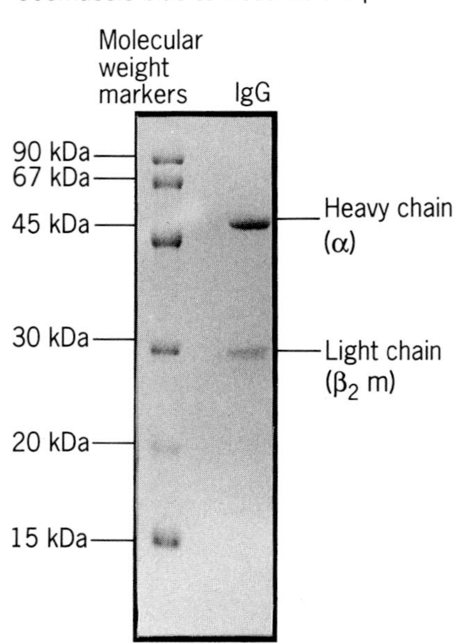

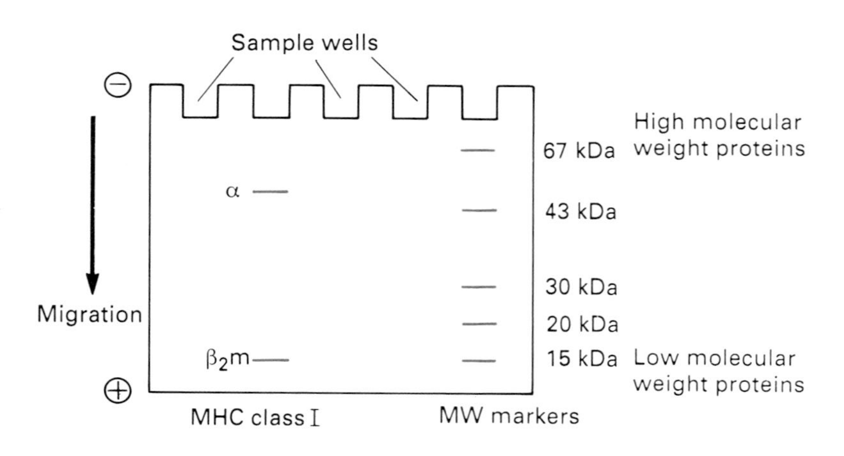

Proteins are separated in the gel according to molecular weight. The level of cross-linking in the gel can be manipulated to allow separation of proteins in different MW ranges. Gels are stained with Coomassie blue to visualize the protein or, if samples are radioactively labelled, exposed to X-ray film.

(b) Isoelectric focusing

Equilibrium IEF – proteins migrate to their isoelectric point.

Ampholines and acrylamide are mixed when the gel is cast.

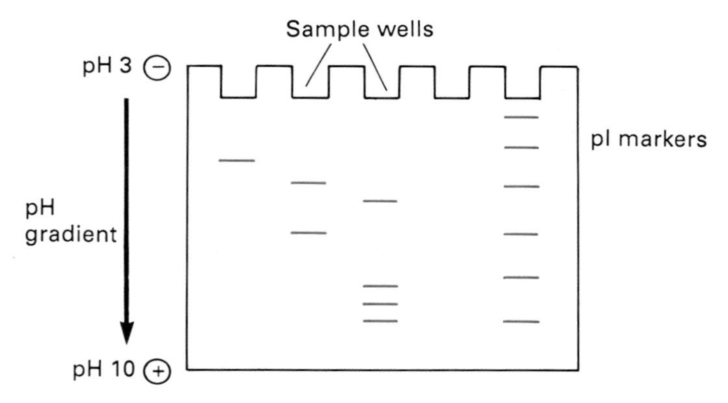

Autoradiograph of an isoelectric focusing gel

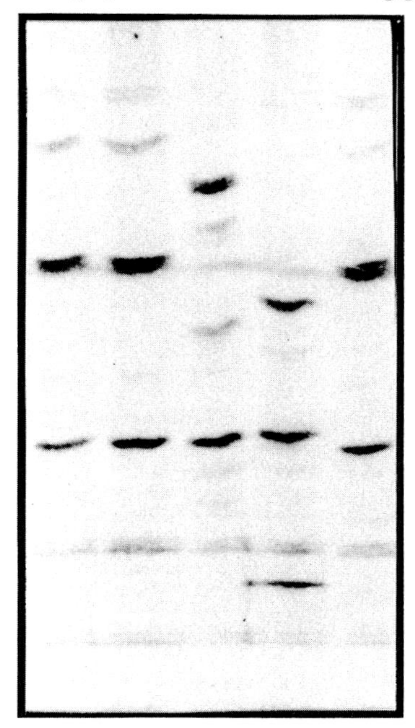

IEF of different HLA class I alleles

The authors wish to thank Damian Counsell, Institute of Molecular Medicine, Oxford for providing this gel

(c) 2D gel electrophoresis

Complex mixtures of proteins can be resolved using this technique.

1st dimension

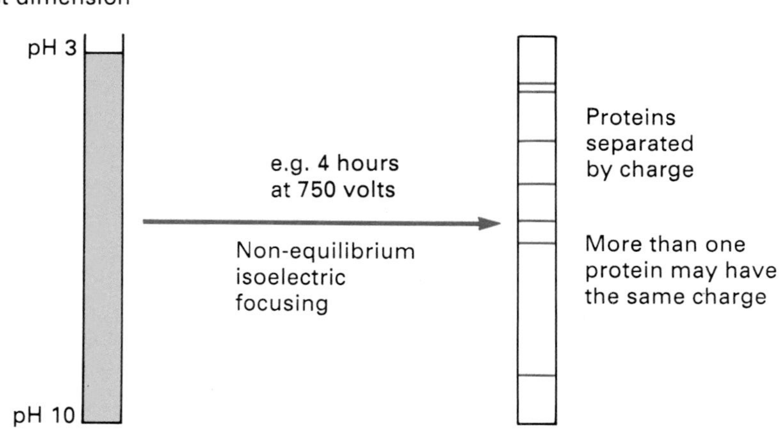

2nd dimension

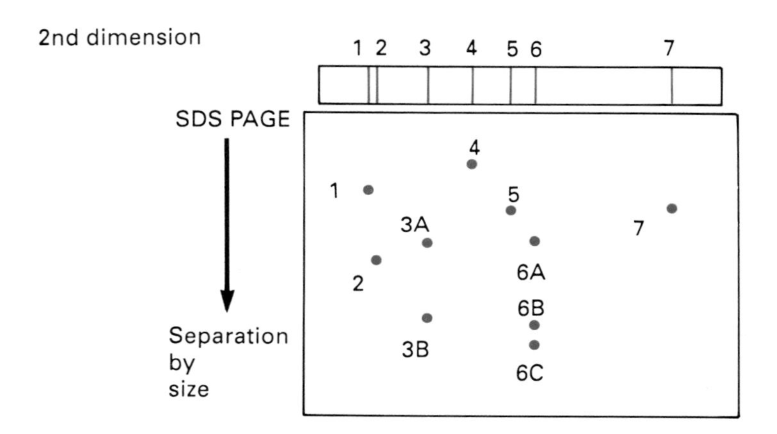

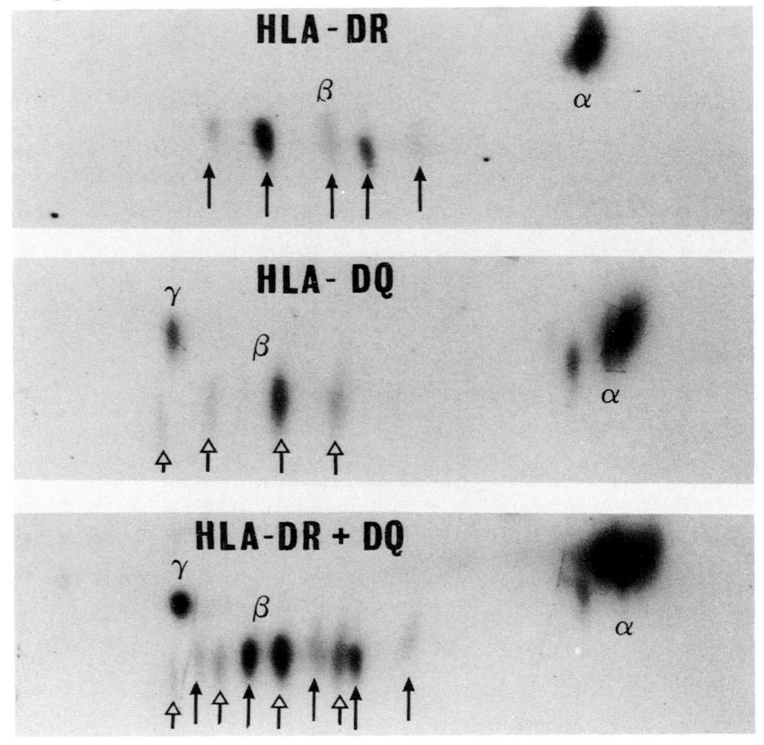

A1.4 Immunocytochemistry

Indirect peroxidase technique

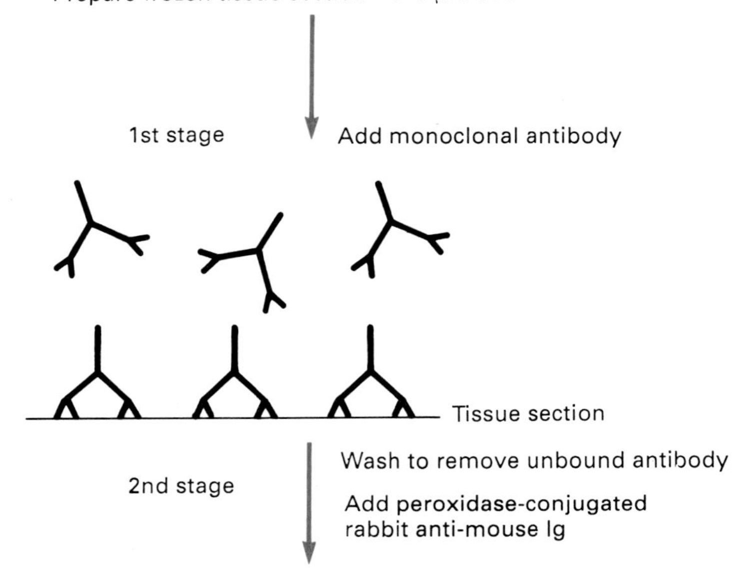

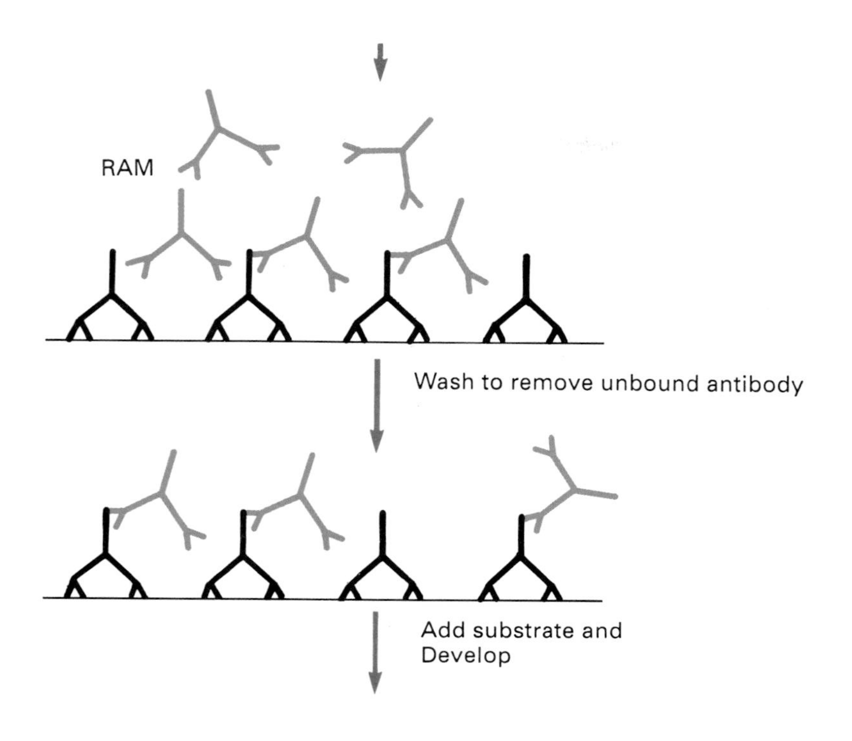

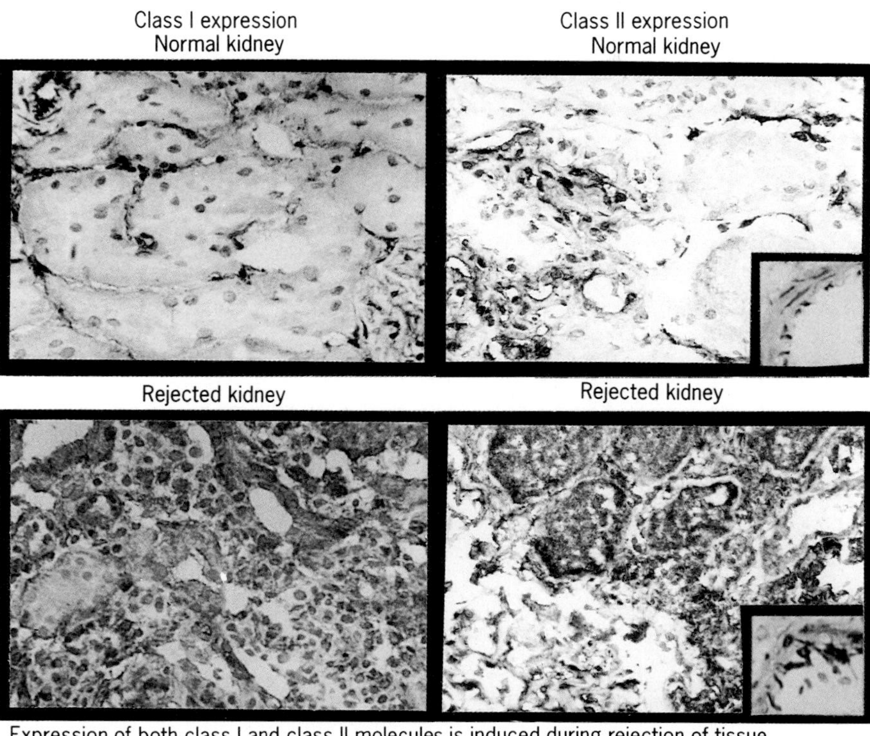

Expression of both class I and class II molecules is induced during rejection of tissue transplanted into an individual who is not identical with the kidney donor for major and minor histocompatibility antigens

A1.5 Affinity chromatography

Antibodies can be coupled covalently to a solid support and used to purify the molecule with which they interact.

Example Purification of a class I molecule HLA-A2

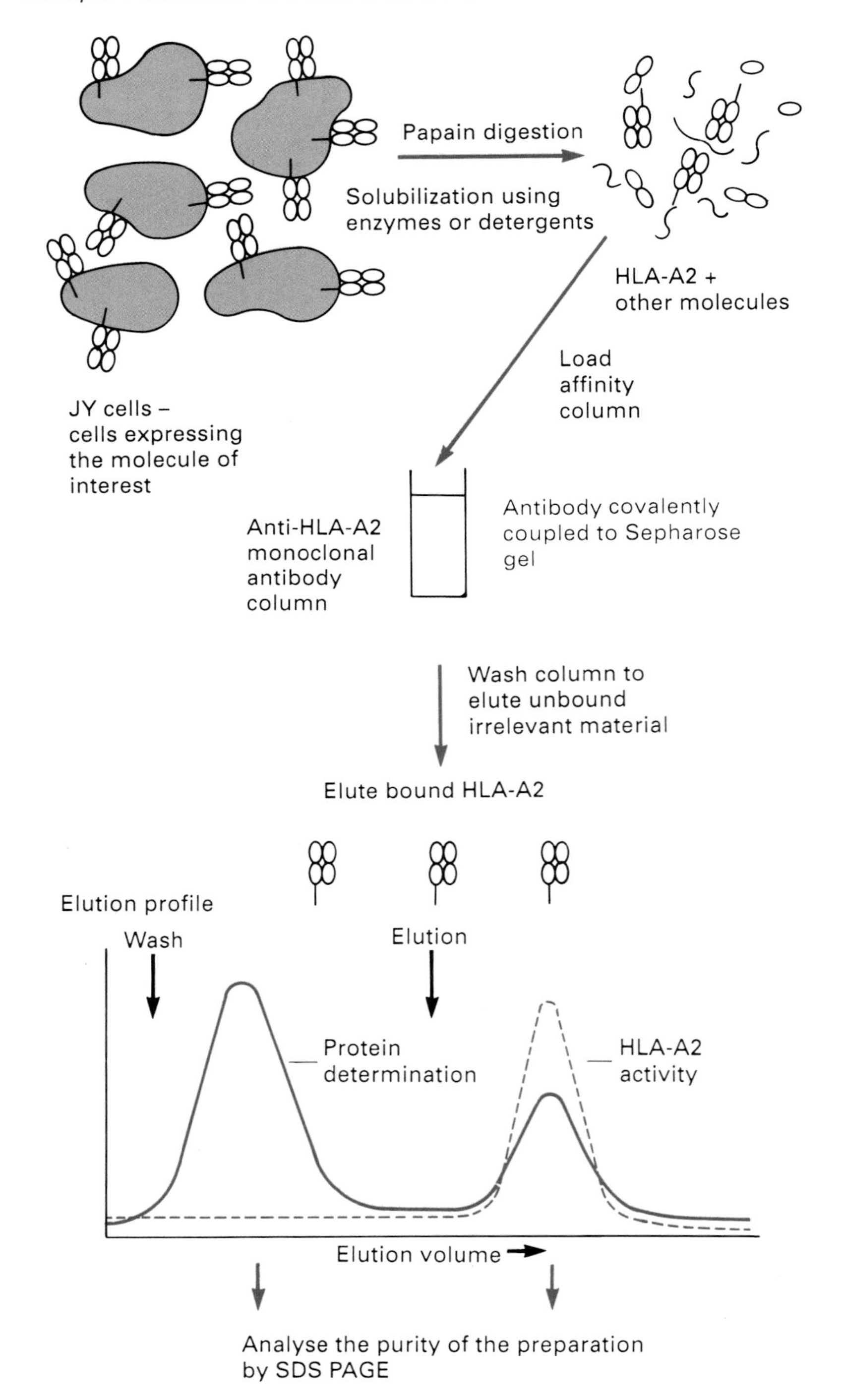

A1.6 Western blotting or immunoblotting

This technique can be used to analyse the composition of a mixture of proteins.

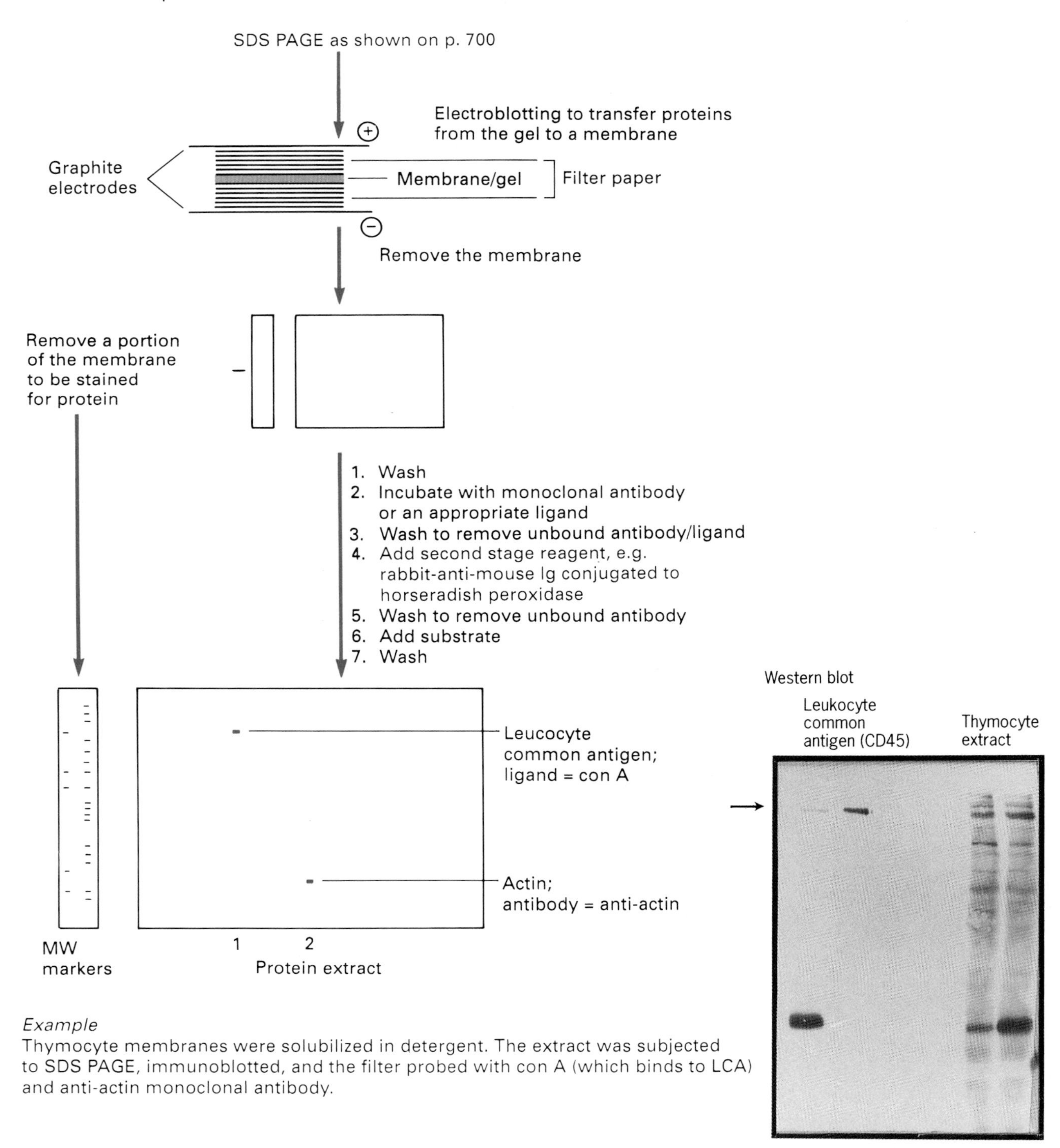
SDS PAGE as shown on p. 700
Electroblotting to transfer proteins from the gel to a membrane
Graphite electrodes
Membrane/gel
Filter paper
Remove the membrane
Remove a portion of the membrane to be stained for protein
1. Wash
2. Incubate with monoclonal antibody or an appropriate ligand
3. Wash to remove unbound antibody/ligand
4. Add second stage reagent, e.g. rabbit-anti-mouse Ig conjugated to horseradish peroxidase
5. Wash to remove unbound antibody
6. Add substrate
7. Wash
Leucocyte common antigen; ligand = con A
Actin; antibody = anti-actin
MW markers
1
2
Protein extract
Western blot
Leukocyte common antigen (CD45)
Thymocyte extract

Example
Thymocyte membranes were solubilized in detergent. The extract was subjected to SDS PAGE, immunoblotted, and the filter probed with con A (which binds to LCA) and anti-actin monoclonal antibody.

A1.7 Antibody-mediated complement-dependent cytotoxicity assay

This technique is most commonly used in tissue typing.

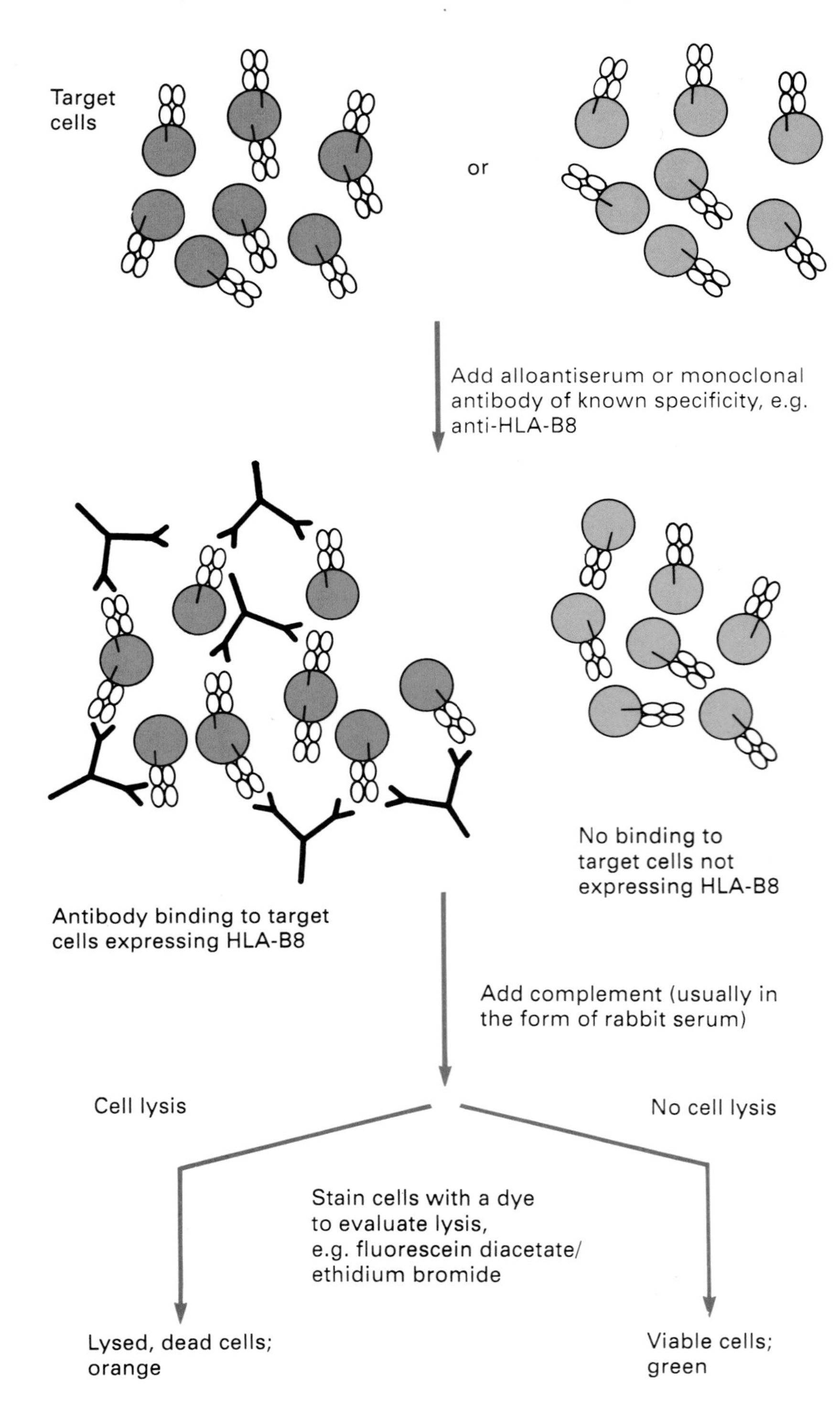

A2 Cellular techniques

A2.1 Cell proliferation assays

(a) Proliferative responses to soluble antigens

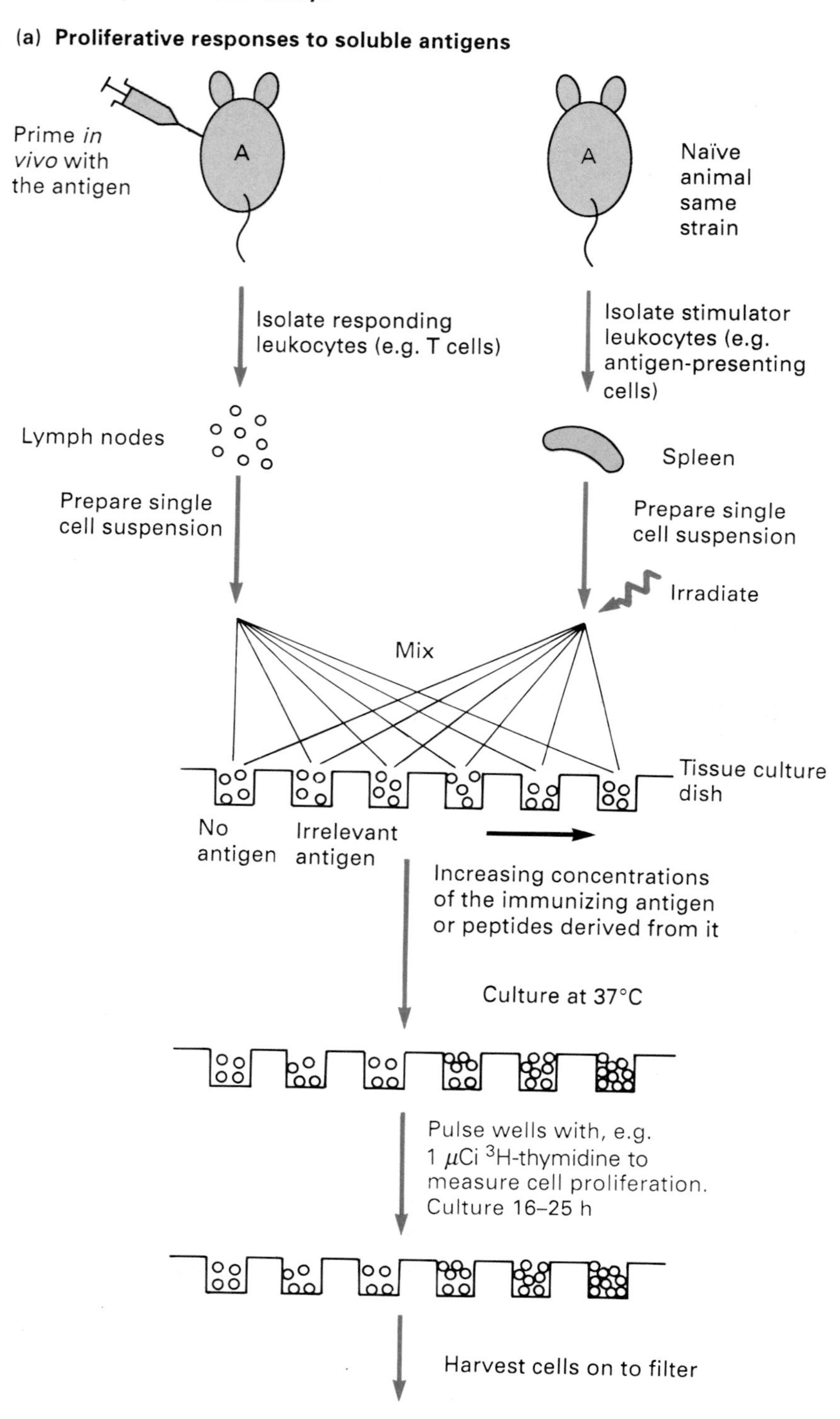

Prime *in vivo* with the antigen
A
A
Naïve animal same strain
Isolate responding leukocytes (e.g. T cells)
Isolate stimulator leukocytes (e.g. antigen-presenting cells)
Lymph nodes
Spleen
Prepare single cell suspension
Prepare single cell suspension
Irradiate
Mix
Tissue culture dish
No antigen
Irrelevant antigen
Increasing concentrations of the immunizing antigen or peptides derived from it
Culture at 37°C
Pulse wells with, e.g. 1 μCi ^{3}H-thymidine to measure cell proliferation. Culture 16–25 h
Harvest cells on to filter

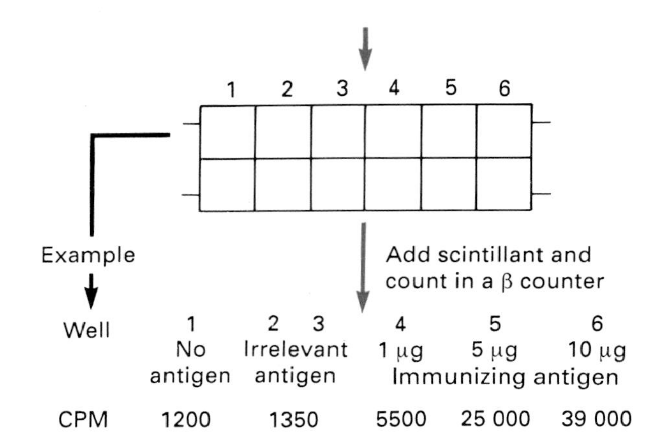

(b) **Mixed leukocyte reaction (MLR)**

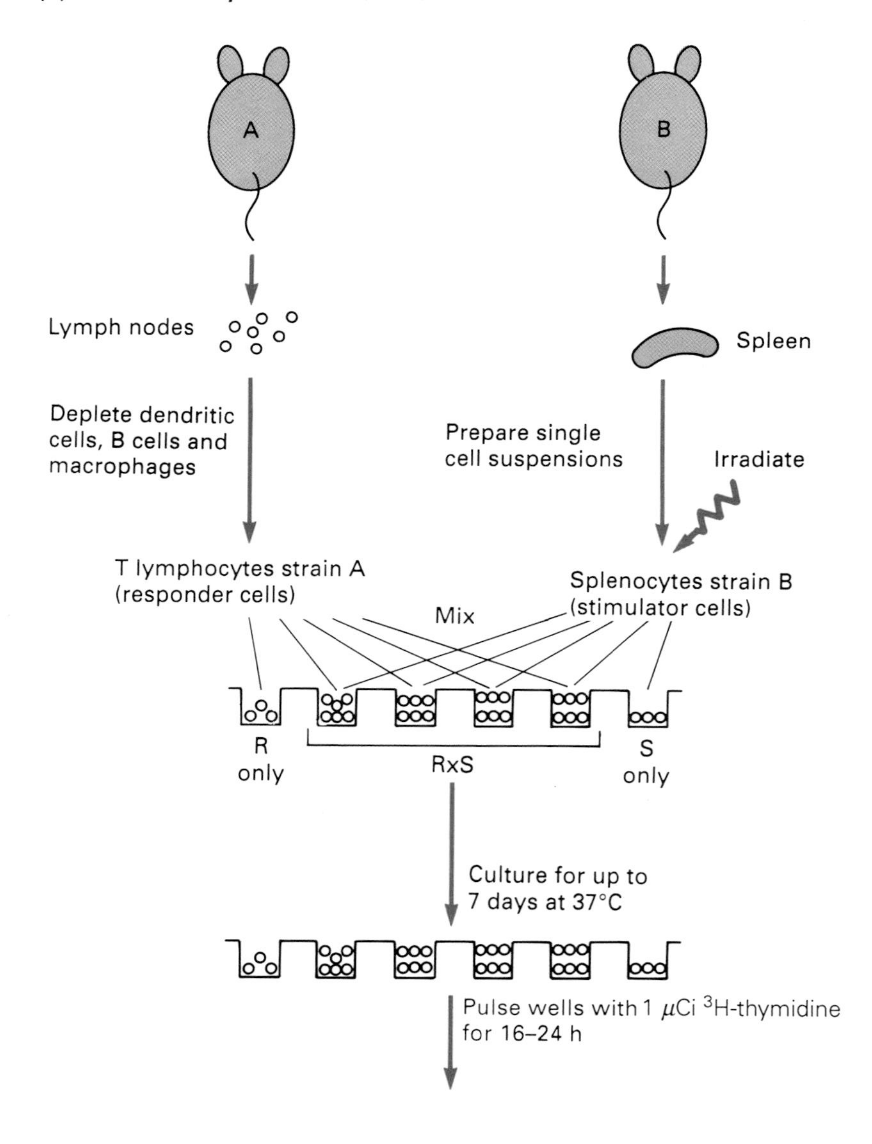

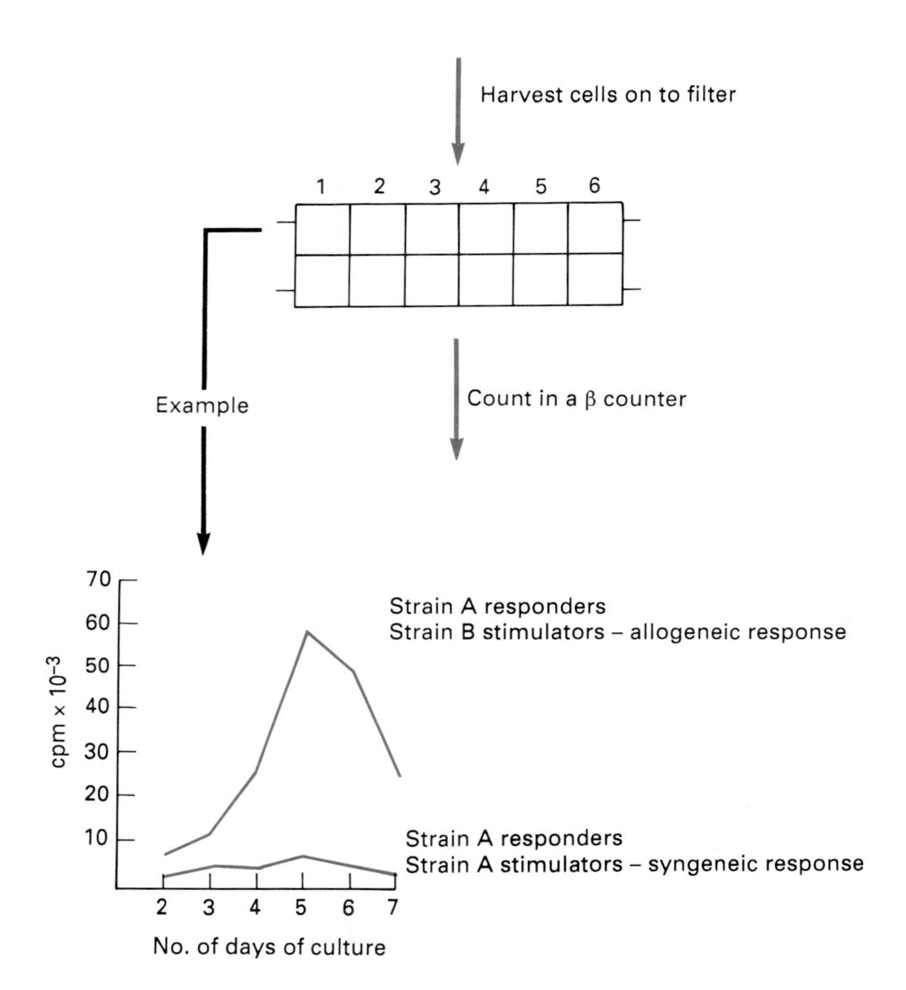

A2.2 Cytotoxic T lymphocyte assay (CTL assay or ^{51}Cr release assay)

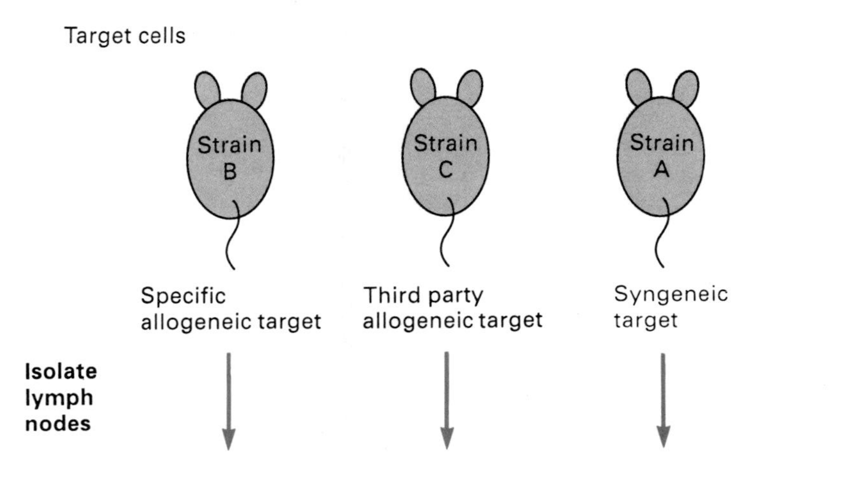

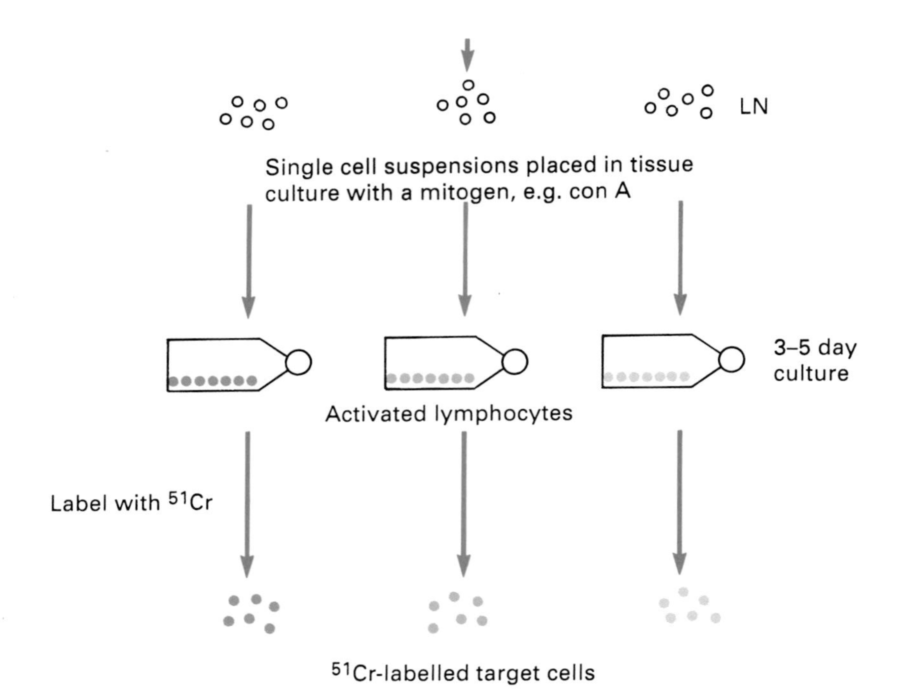

LN
Single cell suspensions placed in tissue culture with a mitogen, e.g. con A
3–5 day culture
Activated lymphocytes
Label with 51Cr
51Cr-labelled target cells

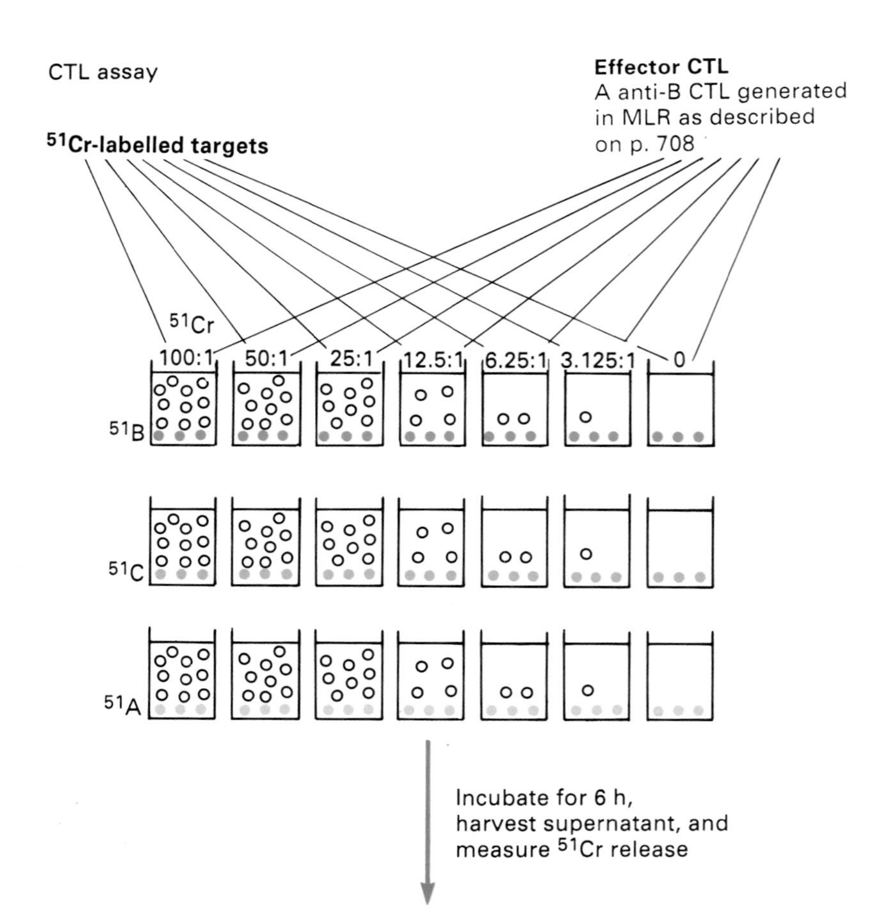

CTL assay
51Cr-labelled targets
Effector CTL
A anti-B CTL generated in MLR as described on p. 708
51Cr
100:1
50:1
25:1
12.5:1
6.25:1
3.125:1
0
51B
51C
51A
Incubate for 6 h, harvest supernatant, and measure 51Cr release

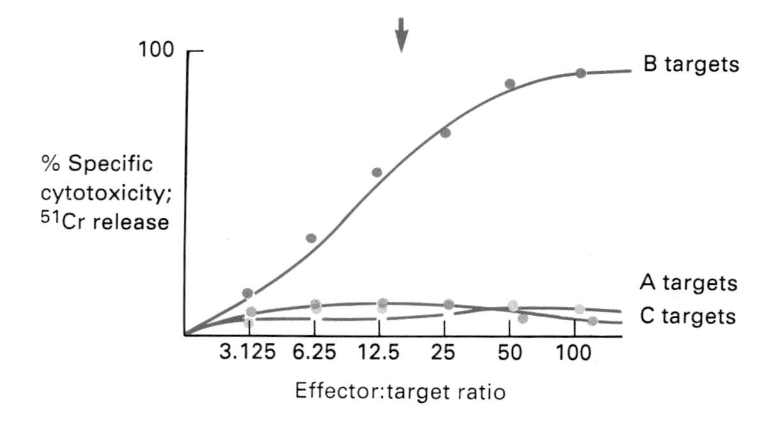

A2.3 Plaque-forming cell (PFC) assay

This assay is used to measure antibody production by B lymphocytes (plasma cells). [Note: sheep red blood cells can be coupled to other antigens for use in this assay.]

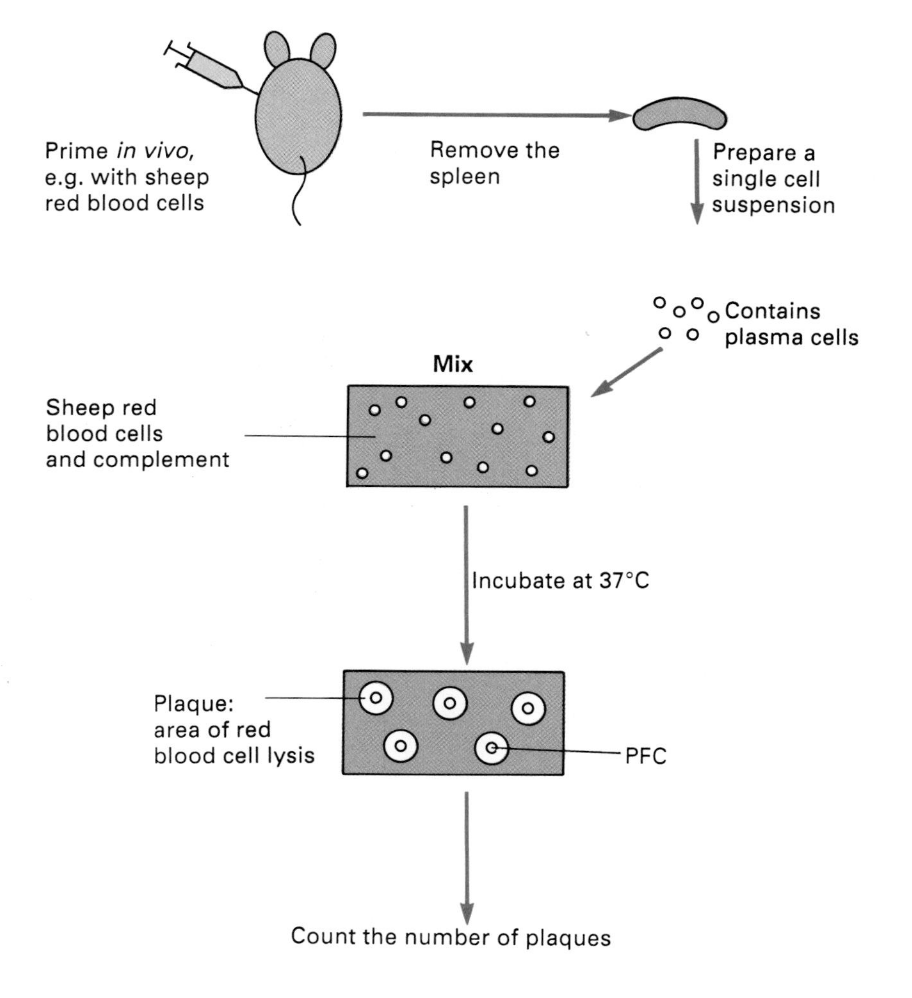

A3 Molecular techniques

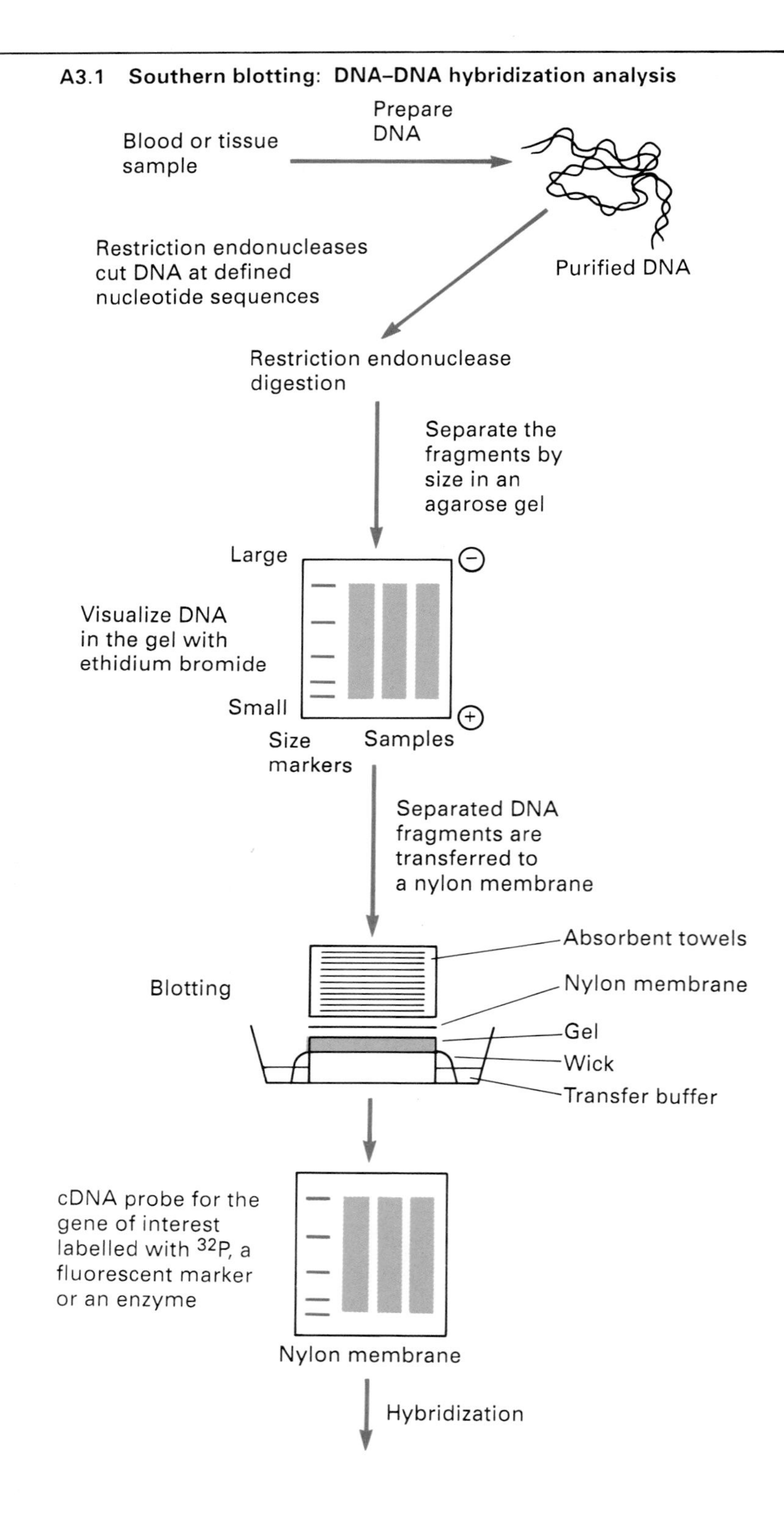
A3.1 Southern blotting: DNA–DNA hybridization analysis
Blood or tissue sample
Prepare DNA
Purified DNA
Restriction endonucleases cut DNA at defined nucleotide sequences
Restriction endonuclease digestion
Separate the fragments by size in an agarose gel
Large
Small
Visualize DNA in the gel with ethidium bromide
Size markers
Samples
Separated DNA fragments are transferred to a nylon membrane
Blotting
Absorbent towels
Nylon membrane
Gel
Wick
Transfer buffer
cDNA probe for the gene of interest labelled with ³²P, a fluorescent marker or an enzyme
Nylon membrane
Hybridization

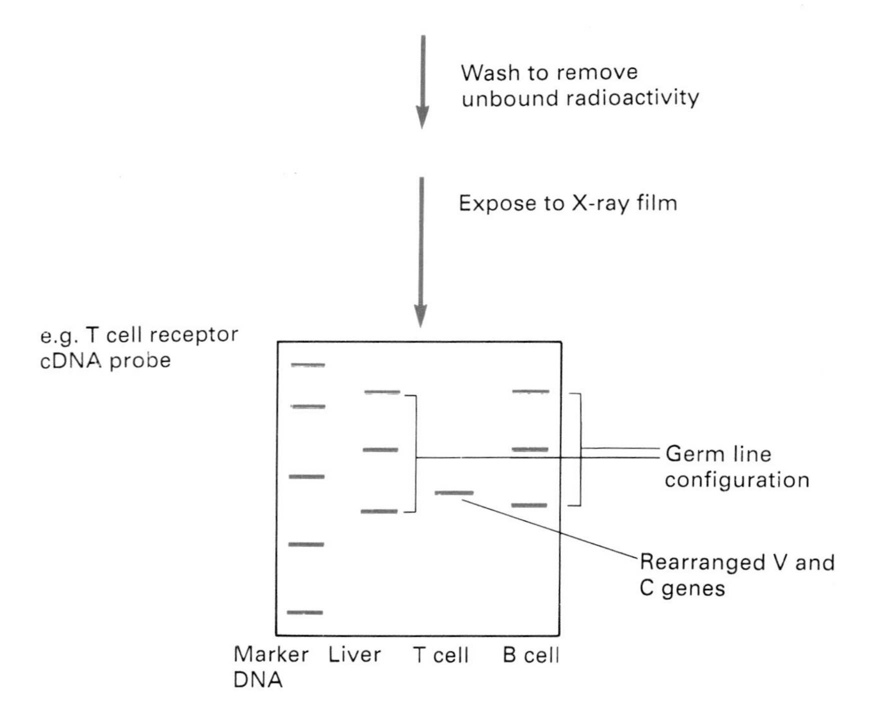

Southern blot—restriction fragment polymorphisms HLA–DR

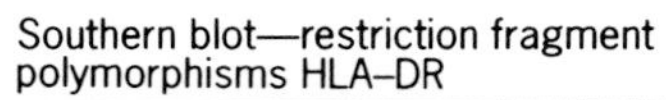

This technique can be used to analyse DNA differences (restriction fragment length polymorphisms; RFLPs) due to

1. gene rearrangements, i.e. for T cell receptor and immunoglobulin genes;
2. gene polymorphisms, such as those that exist in MHC genes.

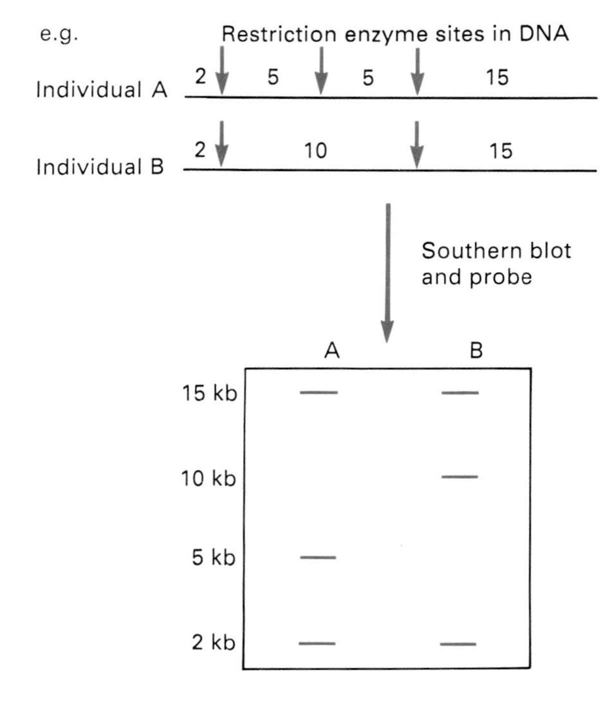

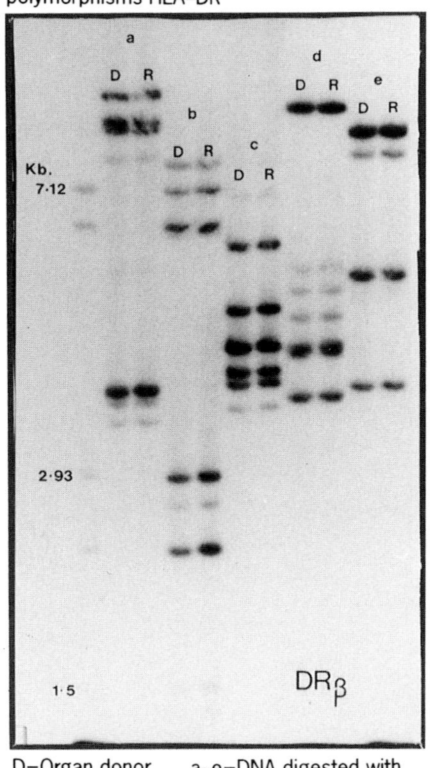

D=Organ donor
R=Recipient

a–e=DNA digested with different restriction endonuclease enzymes

A3.2 Northern blotting: RNA–DNA hybridization

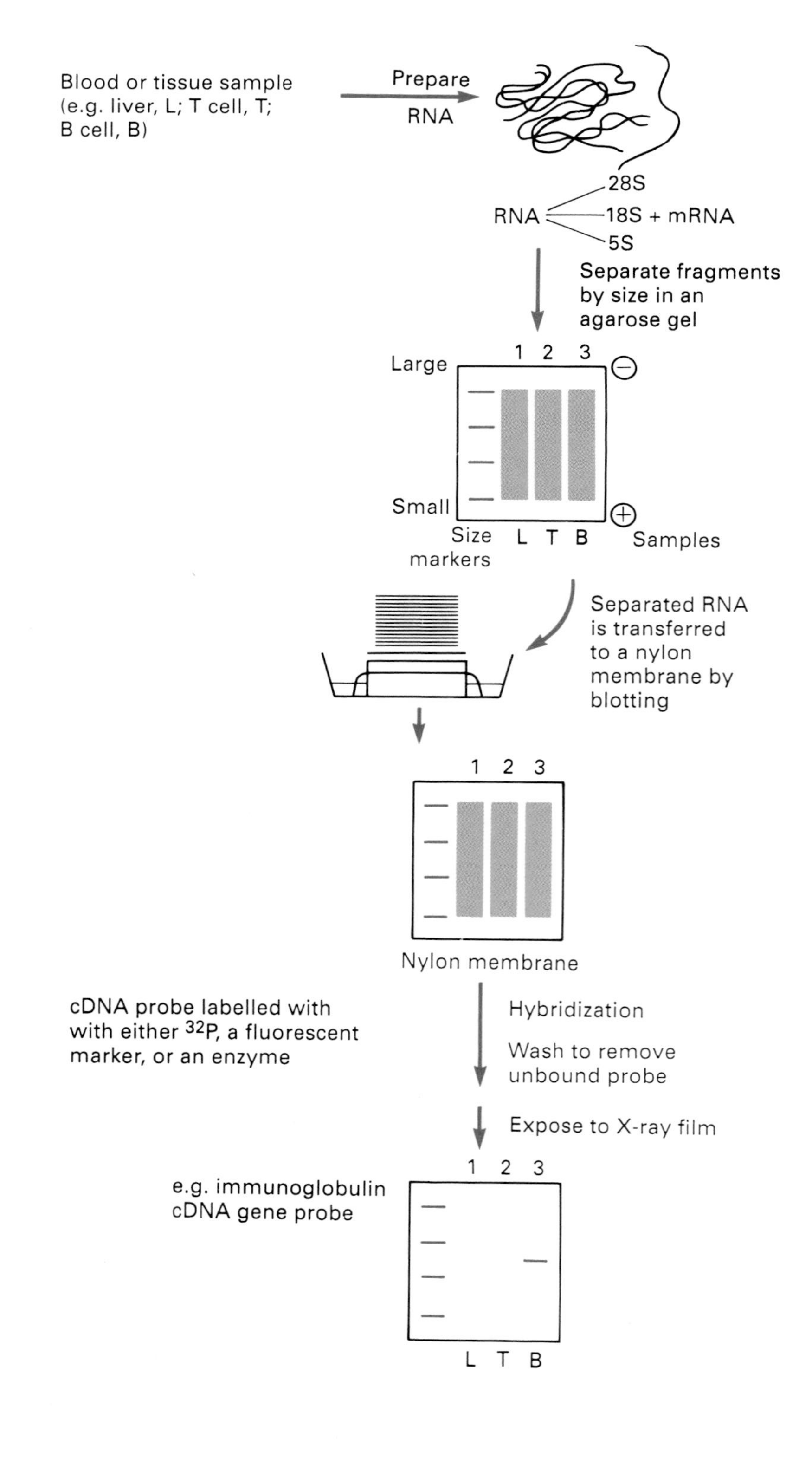

A3.3 Polymerase chain reaction (PCR)

This technique is a fast and efficient method for amplifying known sequences of DNA.

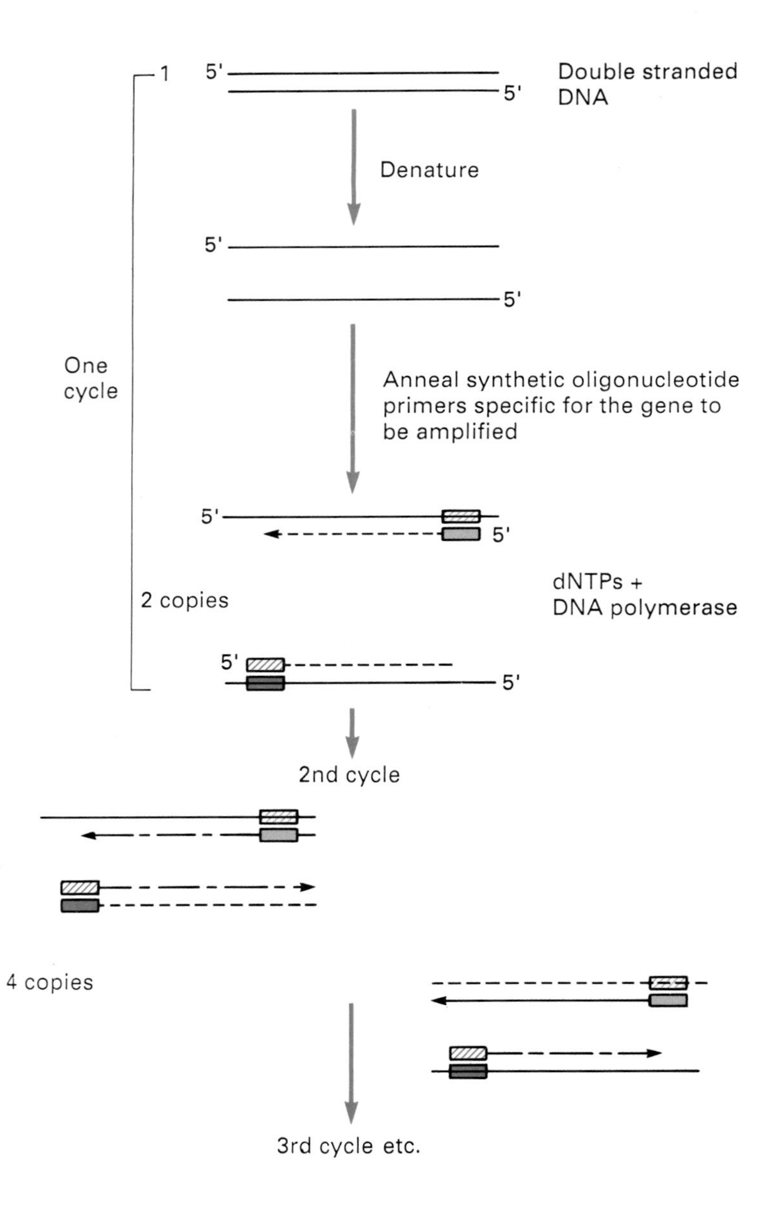

A3.4 DNA-mediated gene transfer (DMGT): gene transfection

This technique allows a new gene to be introduced into a cell that does not normally express it.

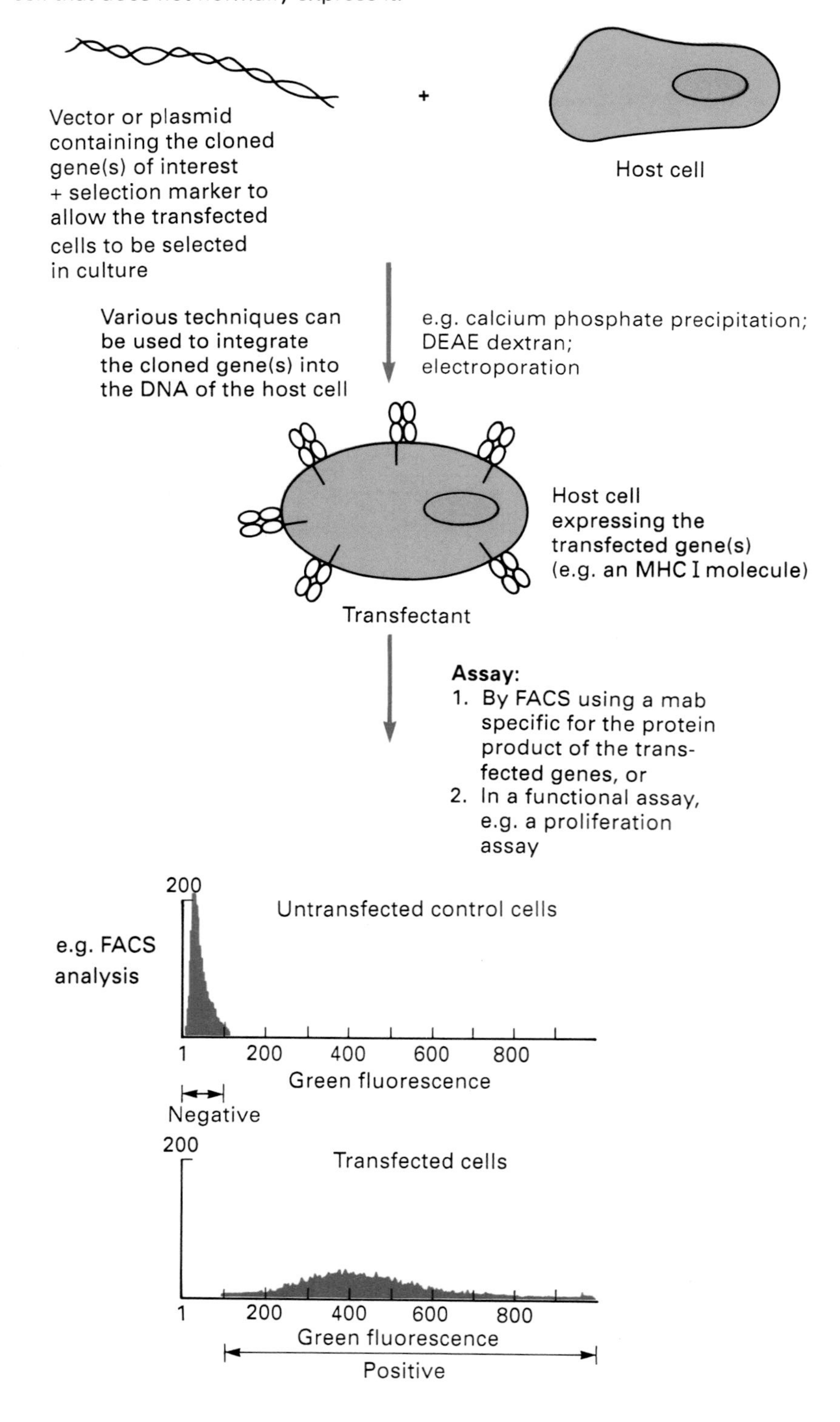
Vector or plasmid containing the cloned gene(s) of interest + selection marker to allow the transfected cells to be selected in culture
+
Host cell
Various techniques can be used to integrate the cloned gene(s) into the DNA of the host cell
e.g. calcium phosphate precipitation; DEAE dextran; electroporation
Host cell expressing the transfected gene(s) (e.g. an MHC I molecule)
Transfectant
Assay:
1. By FACS using a mab specific for the protein product of the transfected genes, or
2. In a functional assay, e.g. a proliferation assay
e.g. FACS analysis
200
Untransfected control cells
1
200
400
600
800
Green fluorescence
Negative
200
Transfected cells
1
200
400
600
800
Green fluorescence
Positive

A3.5 Production of transgenic mice

This technique allows a cloned gene to be introduced into the germline DNA of an animal, e.g. a mouse. The gene can be transmitted to the progeny.

Three techniques are described here for the production of transgenic mice:
(a) microinjection of pronucleus;
(b) retrovirus-mediated gene transfer;
(c) embryonal stem (ES) cell-mediated gene transfer.

(a) Microinjection of the pronucleus

1. Isolation of fertilized egg
2. Microinjection of cloned DNA into the pronucleus
3. Reimplantation into pseudopregnant female
4. Analysis of offspring using DNA–DNA hybridization (Southern blotting) to establish that the transgene has been permanently incorporated in the genome

This technique was first reported by Gordon *et al.* (1980) and the term 'transgenic mouse' was applied by Gordon and Ruddle a year later (1981). Expression of the transgene can be directed to certain tissues in the body by incorporating tissue-specific promoters into the DNA construct injected into the pronucleus. For example, by using the insulin promoter, expression of the transgene can be confined to the β cells of the pancreas.

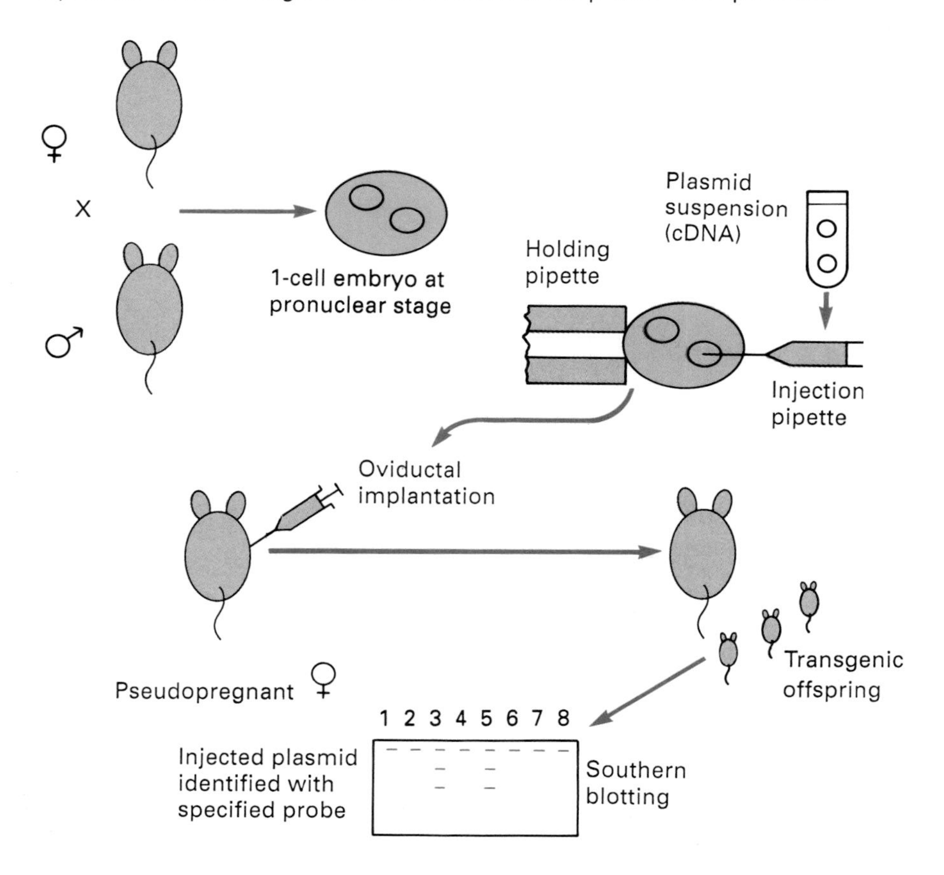

(b) Retrovirus-mediated gene transfer

1. Mouse embryos are incubated with the recombinant retrovirus Moloney murine leukemia virus (MoMuLV) containing the gene of interest.
2. After insertion of the retroviral DNA into the genome, the embryos are reimplanted into a pseudopregnant female and the progeny allowed to develop.

Recombinant infectious retrovirus vectors which contain cloned genes can be be engineered and used to produce transgenic mice by retrovirus-mediated gene transfer (e.g. Janner *et al.* 1985).

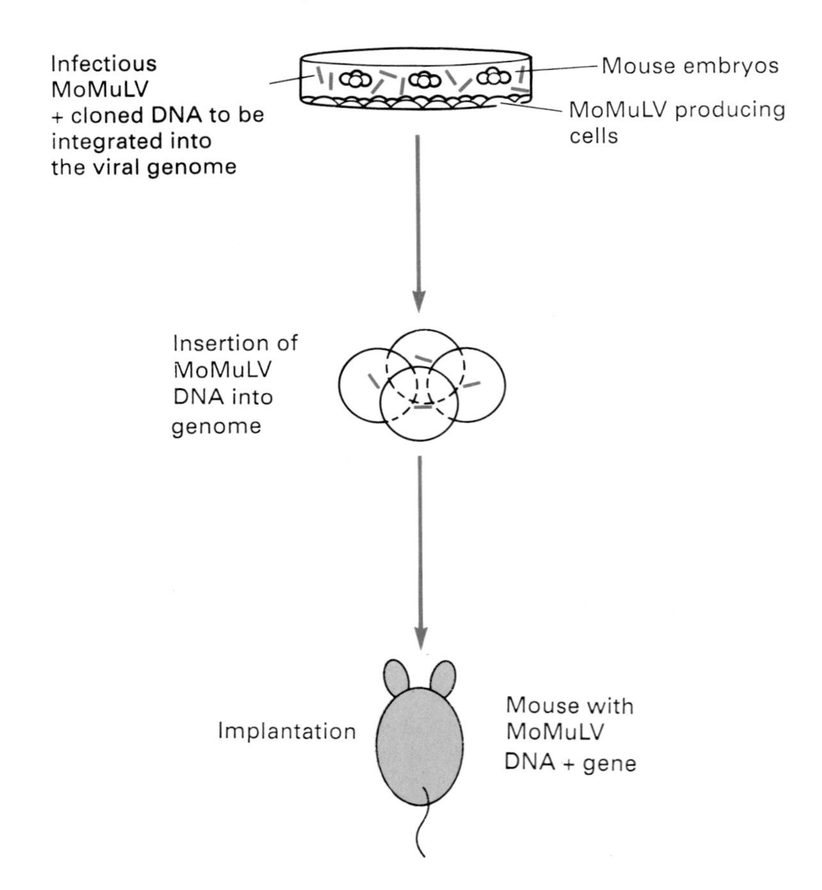

(c) Embryonic stem (ES) cell-mediated gene transfer

1. Embryonic stem (ES) cells are permanently established cell lines derived from early mouse embryos which can be maintained permanently in culture. Cloned genes can be incorporated into the ES cells by DNA-mediated gene transfer (DMGT).
2. Transfected mouse ES cells are injected into a normal mouse blastocyst and transferred to a pseudopregnant foster female.
3. ES cells contribute to the germ line of the mosaic progeny and therefore breeding using sperm from the founder mouse can lead to transmission of the transfected genes through the germ line to establish transgenic mice.

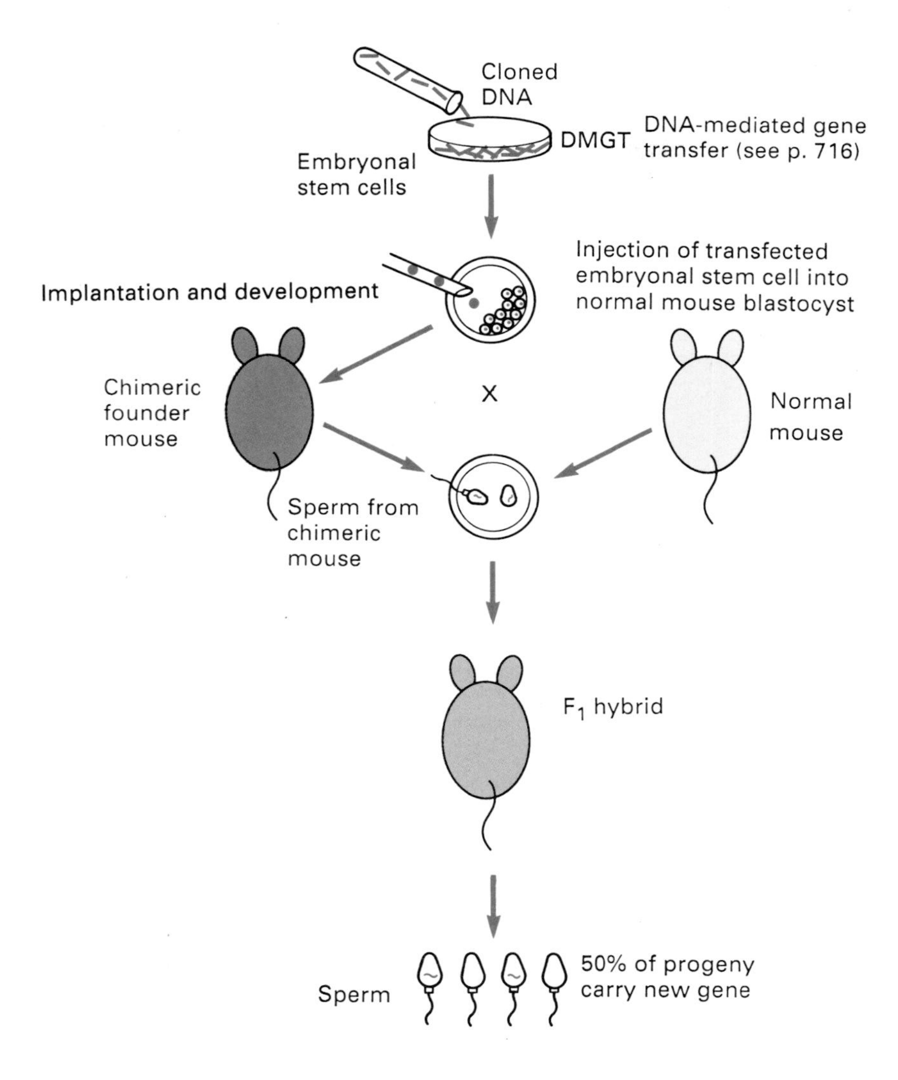

Table A.1 Advantages and disadvantages of the three gene-transfer methods described (pp. 717–19)

Method	Advantages	Disadvantages
Microinjection	Technically easy; high transgene expression. Combination of technical simplicity and expression make it the most widely used technique	Significant host DNA rearrangements; integration of large concatemers
Retrovirus infection	Technically easiest; single proviral DNA makes it well suited to cloning insertional mutations	Low expression; limit to the size of the foreign gene insert
ES cells	Can select for expression or cellular mutation prior to production of the transgenic animal	Technically most difficult; germline transmission may not occur

Further reading

Gordon, J.W. and Ruddle, F.H. (1981). Integration and stable germ line transmission of genes injected into mouse pronuclei. *Science*, **214**, 1244.
Gordon, J.W., Scangos, G.A., Plotkin, D.J, Barbosa, J.A., and Ruddle, F.H. (1980). Genetic transformation of mouse embryos by microinjection of purified DNA. *Proceedings of the National Academy of Sciences USA*, **77**, 7380.
Koller, B.H. and Smithies, O. (1992). Altering genes in animals by gene targeting. *Annual Review of Immunology*, **10**, 705.
Janner, D., Haase, K., Mullican, R., and Jaenisch, R. (1985). Insertion of the bacterial gpt gene into the germ line of mice by retroviral infection. *Proceedings of the National Academy of Sciences USA*, **82**, 6927.

A3.6 Homologous recombination

This technique is used to target and replace genes in the germline, either to investigate the function of the gene by introducing a defective gene during the recombination event (e.g. gene 'knock-out' experiments) or, in the future, to replace a defective gene with a normal gene (i.e. 'gene therapy').

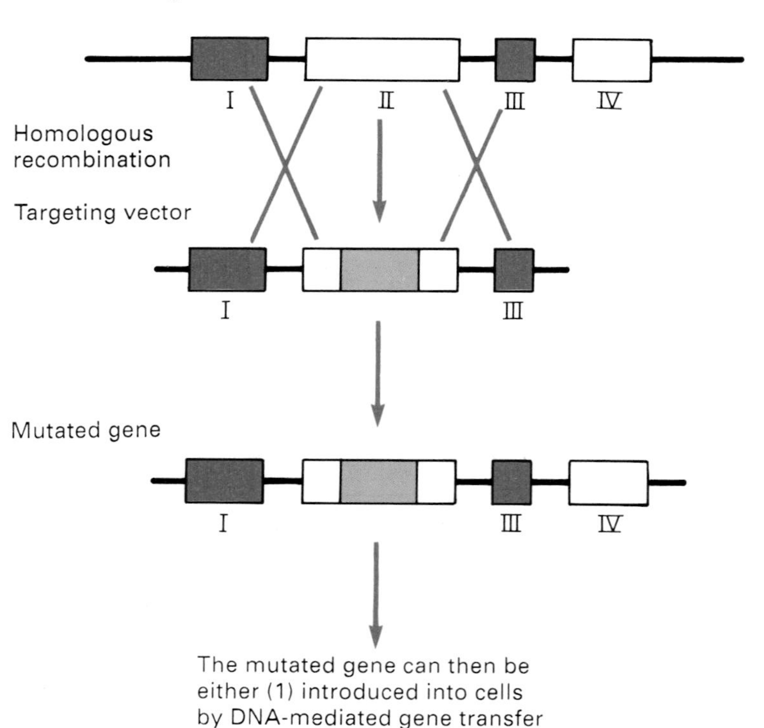

INDEX